COLD REGIONS ENGINEERING

Cold Regions Impacts on Transportation and Infrastructure

PROCEEDINGS OF THE ELEVENTH INTERNATIONAL CONFERENCE

May 20–22, 2002
Anchorage, Alaska

EDITED BY
Kelly S. Merrill

ASCE American Society of Civil Engineers
1801 ALEXANDER BELL DRIVE
RESTON, VIRGINIA 20191–4400

Abstract: The 11th International Conference on Cold Regions Engineering was held in Anchorage, Alaska May 20–22, 2002 and was organized by the Technical Council on Cold Regions Engineering (TCCRE) of the American Society of Civil Engineers. The proceedings papers address all aspects of cold regions engineering, including transportation design, infrastructure construction, frozen ground issues for all types of facilities, and water resource and water treatment challenges. The twenty-fifth anniversary of the Trans-Alaska Pipeline System was recognized by the inclusion of 11 papers that discuss lessons learned from this great civil engineering project.

Library of Congress Cataloging-in-Publication Data

International Conference on Cold Regions Engineering (11th : 2002 Anchorage, Alaska)
Cold regions engineering : cold regions impacts on transportation and infrastructure : proceedings of the eleventh international conference, May 20-22, 2002, Anchorage, Alaska / edited by Kelly S. Merrill.
p. cm.
Includes bibliographical references and index.
ISBN 0-7844-0621-9
1. Transportation engineering—Cold weather conditions—Congresses. 2. Civil engineering—Cold weather conditions—Congresses. 3. Cold regions—Congresses. I. Merrill, Kelly S. II. Title.

TA1005 .I525 2002
620'.411--dc21 2002018539

Library of Congress Catalog Card No: 2002018539
ISBN 0-7844-0621-9
Manufactured in the United States of America.

Preface

The 11th International Conference on Cold Regions Engineering was organized by the Technical Council on Cold Regions Engineering (TCCRE) of the American Society of Civil Engineers (ASCE). The conference was held in Anchorage, Alaska, from May 20 to 22, 2002. 2002 marks the 32nd anniversary of the 1st International Conference on Cold Regions Engineering, held in Fairbanks, Alaska, in 1970. Alaska is extremely pleased to host the 11th edition of the series in 2002. The historical look back at the origin of this conference is particularly appropriate given that 2002 is the 150th Anniversary of ASCE.

2002 is also the 25th Anniversary of the Trans-Alaska Pipeline System, the winner of the 1978 ASCE Outstanding Engineering Achievement (OCEA) Award. The pipeline anniversary is recognized by the inclusion of 11 papers in these Proceedings dealing with lessons learned from one of the great civil engineering projects of all time. Several papers are included concerning the Whittier Access Project, the winner of the 2001 OCEA Award. The conference presented an excellent opportunity to recognize the history and heritage of civil engineering in cold regions, while also looking ahead to begin solving the problems of tomorrow.

The Conference theme "Cold Regions Impacts on Transportation and Infrastructure" was selected to provide a forum for a wide range of cold regions issues in both the lower 48 states of the United States as well as cold regions areas across the world. The Conference addresses all aspects of cold regions engineering, including transportation design, infrastructure construction, frozen ground issues for all types of facilities, and water resource and water treatment challenges. Papers were developed by academics, researchers, and practitioners to provide a well-balanced cross-section of the state of the art in cold regions engineering. Papers representing authors from the United States, Canada, Russia, Japan, China, and Finland are included.

All papers presented in these Proceedings have undergone at least two peer reviews by qualified individuals working in related fields. The time and energy expended by the authors in writing the papers, and by the reviewers in developing substantive comments, is greatly appreciated. Literally thousands of volunteer hours have been expended by the authors and reviewers to produce a quality Proceedings. All papers in the Proceedings are eligible for discussion in the Journal of Cold Regions Engineering as well as being eligible for ASCE awards.

This preface would not be complete without acknowledging the tireless efforts of the Local Organizing Committee in preparing for this Conference. Many thanks to the organizing committee members and others who have helped make the Conference and the Proceedings successful. A special thanks to Linda Oles for her diligent work in handling the administrative tasks associated with compilation of these Proceedings.

Kelly S. Merrill, P.E.
Proceedings Editor

Contents

Trans-Alaska Pipeline System (TAPS)

Other Pipelines

Northwest Alaska Resource Transportation Analysis

Whittier Access Project

Antartica

Highway Design

Airstrip/Roadway Foundations

Pavements

Street and Roadway Management Systems

Bridges

Railroad Operations

Structural Design

Thermal Performance of TAPS Heat Pipes with Non Condensable Gas Blockage

Steve Sorensen, P.E.[1], Jason Smith[2], John Zarling, P.E. Ph.D.[3]

Abstract

Over 124,300 heat pipes were installed in vertical support members on the Trans-Alaska Pipeline System. During infrared, IR, surveys in the early 1980's it was observed that cold topping was occurring on some of these heat pipes. The cold topping was due to an accumulation of non-condensable gases, a by-product of corrosion or chemical dissociation of anhydrous ammonia, in the problem heat pipes. Alyeska Pipeline Service Company under took an experimental program, beginning in the fall of 2000, to measure the heat transfer performance of NCG blocked heat pipes.

Results of this test program are presented and discussed showing the reduction in heat transfer performance of the heat pipes as a function of hydrogen blockage levels. Infrared images are also presented showing the increase in cold topping as the amount of hydrogen in the heat pipe is increased. These IR images are correlated with the degradation in heat transfer performance. A thermodynamic analysis is also presented showing the relationship between the mass of hydrogen within the heat pipe and apparent blockage as a function of evaporator and condenser temperatures.

Introduction

The 1,300 km Trans-Alaska Pipeline System, TAPS, transverses Alaska from Prudhoe Bay on the North Slope to its southern terminus, the Port of Valdez. Pipeline construction began in 1974 and was completed in 1977 with first oil flow occurring that summer. Along this route, the pipeline crosses permanently and seasonally frozen soils. Because warm oil flows through the pipeline, about 680 km of the pipeline was built above ground to avoid burying a warm pipeline in areas of thaw unstable permafrost. An above ground pipeline support assembly was designed to prevent thaw settlement of this ice-rich permafrost. The above ground, 1.22 m diameter pipeline, is attached to a shoe assembly that rests on crossbeam supported by two vertical support members, VSM's. Each VSM and crossbeam structure is called a bent. Bents are spaced approximately 18 m apart along the elevated portion of the pipeline.

Passive heat transfer devices, referred to as heat pipes, were included in the VSM design in those regions where warm permafrost was observed. The cooling effect of passive heat transfer devices on the soils surrounding the VSM's prevent thaw degradation of the permafrost, as well as increase the adfreeze strength of the

----

[1]Alyeska Pipeline Service Company, Fairbanks, Alaska;
SorensenSP@alyeska-pipeline.com
[2]VECO Alaska, Inc., Fairbanks, Alaska; SmithJK@alyeska-pipeline.com
[3]Zarling Aero and Engineering, Fairbanks, Alaska; zae@gci.net

pile-soil bond. North of the Brooks Mountain Range it was not generally necessary to install heat pipes in the VSM's because the permafrost is colder. However, south of the Brooks Range, where discontinuous permafrost occurs and the permafrost can be closer to its melting temperature, two heat pipes were installed in most VSM's (VSM's with heat pipes installed are referred to as thermal piles or thermal VSM's.). The total number of heat pipes installed is approximately 124,300.

Heat pipes, as used on TAPS, consist of sealed steel pipes ranging in lengths from 8.5 m to 20.1 m. The lower portions of the heat pipes, below the top of the VSM's, have an outside diameter of 51 mm with a wall thickness of 6.3 mm. The upper portions of the heat pipes, above the top of the VSM's, have an outside diameter of 76.2 mm with a 19.0 mm wall thickness. These dimensions result in a uniform inside diameter of 38.1 mm. The heat pipes are charged with anhydrous ammonia, NH_3, as the working fluid. About 2% - 3% of the heat pipe's internal volume contains liquid ammonia with the remaining volume filled with its vapor. An extruded aluminum fin assembly, of either 1.22 m or 1.83 m length, is pressed onto the top end of each heat pipe. Heat pipes 11.3 m in length or less were fitted with the 1.22 m finned section and heat pipes of greater length received a 1.83 m finned section.

Heat pipes work on the principal of continuous evaporation and condensation of a working fluid. During the wintertime, when the air is colder than the ground, heat from the ground causes the liquid in the lower portion of the heat pipe to vaporize. The vapor flows upward from the evaporator section (below ground) to the condenser section (above ground) where the vapor condenses back to the liquid phase because of cooling by ambient air. The condensate flows downward, by gravity on the internal wall surface of the heat pipe, where it absorbs heat from the ground and is re-evaporated to continue the process. The upward vapor flow is the result of a pressure gradient between the evaporator and condenser sections of the heat pipe. The vapor pressure of the ammonia increases in the evaporator section due to the addition of ground heat, and decreases in the condenser section as a result of the heat rejected through the finned section. During the warmer periods of the year, when the temperature of the air is warmer than the saturation temperature of the liquid ammonia, the heat pipe is dormant.

As part of the TAPS's thermal pile maintenance program, surveillance of the heat pipes using Infrared (IR) viewing equipment has been used to determine if heat pipes were continuing to transfer heat from the ground. During periods when the heat pipes are actively transferring heat, the finned condenser sections or "radiators" will appear "hot" or "glow" in the infrared portion of the radiation spectrum. Johnson (1983) reported that during the 1980 surveillance effort, many of the heat pipes appeared to have "cold tops" and that probably caused by non-condensable gasses, NCG. This condition was recognized as a potential problem during design, although the extent and magnitude were not predictable. A typical infrared image of a "cold top" or blocked heat pipe next to a fully functional heat pipe is shown in Figure 1. The NCG blocks the upper portion of the finned section and cause a reduction in performance of the heat pipe. IR surveys conducted in subsequent years showed that the level of blockage and number of heat pipes blocked appeared to increase over time. Following the 1990/1991 survey, Williams (1991) reported that the number of heat pipes with detectable blockage was 46% based on a grading process estimated to have an error of ±10%. At that time about 28% of those heat pipes were reported to have blockage equal to or greater than 20%. A blockage level of 30% or more was set to identify heat pipes that were candidates for repair. In 1991 the total number of heat pipes with over 30% blockage was reported at about 10% of the total population heat pipes.

Alyeska Pipeline Service Company, APSC, has conducted extensive investigations to establish cause(s) for the occurrence of the NCG's and potential solutions to the problem. Edelstein (1989) verified that the non-condensing gas was predominantly hydrogen with lesser amounts of nitrogen, naphthenate grease, water, and detergent also identified. He noted that only those heat pipes that were installed on the pipeline have experienced NCG blockage, whereas NCGs have not been found in stored surplus heat pipes.

In 1983 a "getter" repair method was initiated, which utilized a hydrogen absorbing metal halide, zirconium-dimanganese. The getter material was placed in a cavity in a tapered steel pin. These pins had an opening on their tapered end allowing the hydrogen to be absorbed by the getter material. The larger pins, with a getter capacity of 165 grams, are about 254 mm long and 32 mm in diameter with a tapered end having a nominal 16 mm diameter. These, or similar but smaller, pins were installed through the top sidewall of those heat pipes deemed in need of repair. To install the getter pins, a tapered hole was drilled almost through the thick, upper end of the heat pipe wall extending above the fin assembly. The getter pins were then pressed through the remaining heat pipe wall, using a hydraulic press assembly. The repair required accurate drilling and close tolerance fit-up for a satisfactory friction fit between the heat pipe wall and the getter pin wall. By 1990, 2,777 heat pipes were repaired with "getter" pins of varying metal halide capacity.

The getter pin method was successful in relieving hydrogen blockage in most cases. However, in-service observations of getter pin repaired heat pipes have identified a notable percentage of heat pipe failures, probably due in part to the installation method. Although the getter pins initially appeared to "absorb" hydrogen accumulation as designed, about 15% of the repaired heat pipes experienced leakage of the anhydrous ammonia to the atmosphere shortly after installation and resulting in "dead" units. The getter pin program was discontinued in 1993 with the intention of revisiting the friction fit design, and identifying a more effective repair strategy.

In this context, it was widely held that a general solution to NCG blockage of heat pipes lies in predictive repair by relieving molecular hydrogen from the system, or by simply replacing the entire ammonia volume, and not in preventive measures. Because of the large population of heat pipes and increasing number of heat pipes with blockage, it was important to identify those actually needing repair. In order to find the population of heat pipes in need of repair, it was first necessary to define performance of the heat pipes as a function of blockage level. Thus was born the subject of this paper.

Problem Statement

Heat pipe performance became a concern when IR surveys showed an increasing trend in both the number of NCG affected heat pipes and in the amount of NCG accumulation in the TAPS heat pipes. Further analysis of partially blocked heat pipe performance determined that NCG affected heat pipes still retain some capacity to transfer thermal energy from surrounding soils. A portion of the heat transfer capacity of the blocked heat pipe was thought to still occur in the finned condenser section, with additional heat rejection occurring in the section between the bottom of the finned section and the top of the pad, along the above ground portion of the VSM. Conduction along the aluminum fins, as well as some level of convection and condensation within the blocked section, are postulated as the reasons for continued heat transfer. The safety factor associated with the original design criteria coupled with carefully analyzed IR imagery provided a valid argument that blocked heat pipes were still thermally effective during partially blocked conditions.

In an effort to fully understand the performance of a blocked heat pipe with respect to the minimum requirements established in the original APSC design basis, a series of performance tests were conducted during the winters of 1999-2000 and 2000-2001. These tests on TAPS heat pipes were conducted using a full-scale apparatus. The primary initiative of the testing was to establish overall heat pipe performance with respect to varying levels of hydrogen blockage. In the process of modeling and testing heat pipes with NCG's, a series of observations were made concerning the effectiveness of IR imagery, as well as the general behavior of the ammonia-hydrogen mixture. This test program on TAPS heat pipes provided an essential component for development of an enhanced repair methodology.

Test Program

Four heat pipes were tested during the 2000-2001 testing season, each placed in its own calorimeter bath to simulate ground heat flow and temperatures for the embedded portions of the units. The four identical calorimeter baths, fabricated by Flowline Alaska, were 5.8 m long, 200 mm diameter steel pipes. An oversize end cap was welded on the bottom and a raised face weld neck flange was welded on the top of each calorimeter pipe. Urethane rigid foam plastic insulation of 76 mm thickness and an outer sheet metal shield housed the inner pipe. Each bath pipe was filled with proproylene glycol and water with a freeze point of $-32^{o}C$ after the installation of a submersible circulating pump and heat pipe. Figure 2 shows a detailed section view of the calorimeters and Figure 3 shows the four calorimeters with heat pipes installed.

Dryden Instrumentation of Anchorage, Alaska installed the temperature sensing instrumentation on the four heat pipes consisting of three thermistor strings: String 1--nine thermistors installed along the root of adjacent aluminum fins on the condenser section, String 2-- five thermistors installed along the shielded section below the finned section, and String 3--three thermistors installed along the evaporator section to be submerged in the bath. A 9.5 mm thick soft foam plastic pad was placed over the thermistors so they would sense temperatures approximating condenser and evaporator surface temperatures. Along the root of adjacent fins, the foam pad consisted of a strip cut with a trapezoidal cross-section to fit at the bottom of the V. All other pads were about 76 mm by 76 mm in size. A daub of thermal paste was placed at each thermistor to further reduce contact resistance. The instrumentation was designed and installed so it could be removed and reinstalled if additional tests were required.

A "hot tap" was completed above the finned section on each of the four test heat pipes. A 19 mm Thread-O-Let was welded on the heat pipe and then the hot tapping process was performed through a ball valve attached to the heat pipe. This hot tap was then plumbed to a manifold that allowed for hydrogen injection and pressure measurement of the vapor in the heat pipe.

Thermistors to sense bath temperature were installed at the top, mid-point and bottom of each bath. A submersible circulating pump and 600 and 900-watt immersion heaters were connected to a watt transducer to measure power input to each bath. A temperature controller was used to maintain the temperature of each bath within 0.05 C° of a manually set bath temperature. An anemometer with a 0.9 m/s threshold monitored winds speed adjacent to the finned condenser sections of the heat pipes. Ambient outdoor temperature was measured with a radiation shielded themistor in a louvered enclosure. A data logger recorded the data sensed by the thermistors, wind speed transducer and power transducers.

Heat Pipe Preparation

APSC has a large quantity of "new" heat pipes stored at the Nordale Yard east of Fairbanks, Alaska. This inventory of heat pipes is maintained to serve as replacements where pipeline repairs and/or maintenance are required. Nine 9.5 m long "new" heat pipes were secured for the test program

The liquid levels of the ammonia in each of the nine heat pipes were measured. This was done by elevating the condenser end of the heat pipe to achieve an angle of about 30° with respect to the ground. An ultrasonic detector was then moved along the lower end of the heat pipe to locate the liquid-vapor interface.

The total quantity of ammonia in the heat pipes can be calculated based on the measured liquid level interface, temperature of the heat pipe and the known internal dimensions of the heat pipe. The liquid density, ρ_l, and vapor specific volume, v_v (reciprocal of density), are published for ammonia in the two-phase region as a function of temperature, ASHRAE (1993). The mass of ammonia, m, can be determined with these data in conjunction with the volumes of liquid, V_l, vapor, V_v, and total volume, V, or:

$$m = m_l + m_v = V_l \rho_l + \frac{V_v}{v_v} = V_l \rho_l + \frac{V - V_l}{v_v}$$

Using the above equations in conjunction with the measured ambient temperature and liquid level, and heat pipe dimensional data, the average total mass of ammonia was calculated for the test heat pipes at 0.199 kg. A 1.22 m long aluminum finned section was then hydraulically pressed on the condenser sections of each of the heat pipes.

The four heat pipes were installed in their respective calorimeters baths and data were gathered and IR images taken to confirm all heat pipes were performing similarly with no blockage observed. One heat pipe remained the control throughout the testing program.

Hydrogen Injection

Hydrogen injection was performed through the manifold attached to the hot tap on the top of the heat pipe. A two-stage hydrogen regulator was fitted to a 22.7 kg bottle of compressed hydrogen with the regulator set to a pressure of 550 kPa. A 737 mm long section of 12.7 mm diameter steel pipe (internal volume - 93.4×10^3 mm^3) attached to the manifold was purged with hydrogen and then filled to a pressure of 550 kPa. The fill valve to the hydrogen bottle was closed and the needle valve connecting to the heat pipe was opened until the hydrogen pressure bled down to 410 kPa. The mass of hydrogen injected through this known volume was estimated using the Ideal Gas Law. Typically, about 14×10^{-6} kg of hydrogen was injected into three test heat pipes during each charge cycle. Following injection, data were collected and IR pictures taken over a 1-2 week period before another injection cycle was performed.

Data Recording

Bath temperatures during the 2000–01 test program was maintained at -0.5°C yielding an evaporator surface temperature of about –1°C. Power to each of the baths and wind speed were averaged over a one hour time period and stored in the data logger. Ambient, fin, bath and evaporator temperatures were stored at the end of each hour.

Data were downloaded from the data logger to an Excel spreadsheet. Heat pipe conductance, $C=Q/\Delta T$, was calculated by subtracting the calorimeter wall heat loss from the energy input. Heat loss through the calorimeter wall was estimated based on the thermal conductivities of the steel pipe and urethane foam insulation and temperature difference between the glycol and outdoor ambient air.

Some of the data recorded for the period of November 28 - December 5, 2000 are presented in Figures 4 and 5. Because the heat removal rate by the heat pipes depends on evaporator – ambient air temperature difference and wind speed, care in the interpretation of these data is required. Figure 4 shows the energy removed from the calorimeter bath by the heat pipe and the ambient air temperature variation. During the cold spell that occurred on December 2, the heat pipe energy removal rate increased as expected because of the increased temperature difference between the evaporator and ambient air. Figure 5 shows the energy removed by the heat pipe and average wind speed. During periods of increased wind speeds, the energy removal rate by the heat pipe also increased as expected.

Ammonia and Hydrogen Mixture

The total pressure within the heat pipe is the sum of the partial pressures of ammonia and hydrogen. The partial pressure of the ammonia is dependent on the evaporator temperature, i.e. the liquid pool and liquid film in the lower portion of the heat pipe. Data from ASHRAE (1993) can be used to determine the partial pressure of the ammonia based on temperature of the liquid phase. The partial pressure of the hydrogen can be calculated based on the mass of hydrogen contained in the heat pipe using the equation of state for an ideal gas. However, when the evaporator section is heated and the condenser section is cooled, the resulting temperature difference initiates vaporization and condensation, which causes a pressure difference in the gas mixture (evaporator pressure higher than condenser). The pressure difference results in the flow of the gas mixture upward to the finned condenser. Ammonia in the mixture begins to condense resulting in a mixture enriched in hydrogen. As ammonia continues to vaporize and sweep the hydrogen upward, the mixture in the evaporator section becomes hydrogen poor. The partial pressure of the hydrogen in the evaporator section decreases and increases in the condenser section. As equilibrium is established and as long as the vapor velocity exceed the diffusion velocity of the hydrogen, most of the hydrogen is swept to the condenser section.

The partial pressures of the hydrogen and ammonia in the heat pipe can be estimated. Assuming pure ammonia in the evaporator section, the partial and total pressures are equal and can be determined from ASHRAE (1993) based on the evaporator temperature. In the condenser section, the partial pressure of the ammonia can be estimated based on the condenser wall temperature, ammonia dew-point temperature. The partial pressure of the hydrogen is equal to the total pressure minus the partial pressure of the ammonia in the condenser section. Ammonia condenses to its saturation pressure at the equilibrium wall temperature in the blocked section. Because of conduction along the condenser section, convective mixing, and diffusion of hydrogen and ammonia at the interface, there is no distinct line of separation between the “pure” ammonia and hydrogen-ammonia mixture, but a zone of transition.

The amount of hydrogen blockage can be estimated based on the mass of hydrogen and temperatures of the evaporator and condenser sections. The following equation expresses the percent volume of the condenser section blocked by the mixture of hydrogen and ammonia.

$$\frac{V_H}{V_{COND}} = \frac{m_H R_H T_{COND}}{(P_{TOTAL} - P_{AM})} \frac{1}{V_{COND}} 100\%$$

In the above equation T_{COND} is surface temperature of blocked section of the condenser, P_{TOTAL} is total pressure in heat pipe based on evaporator temperature, P_{AM} is partial (vapor) pressure of ammonia in blocked condenser section based on T_{COND}, V_H is volume of hydrogen-ammonia mixture, V_{COND} is internal volume of condenser section, m_H is mass of hydrogen and R_H is the gas constant for hydrogen.

The above equation neglects hydrostatic and dynamic pressure drops occurring between the evaporator and condenser sections. This formulation was used to develop a normalized blockage chart. Because the performance testing conducted during the winter of 2000-2001 was done at an evaporator temperature of $-1C^o$, that condition was chosen as a normalizing value. Normalized blockage as a function of condenser temperature and evaporator temperature are shown in Figure 6. Heat pipe testing at several evaporator temperatures is being conducted to validate these predicted trends.

Figure 6 shows blockage as a function of evaporator temperature (pressure) and condenser temperature. As the condenser temperature is decreased at a fixed evaporator temperature, the partial and vapor pressure of the ammonia decreases resulting in an increase in the partial pressure of the hydrogen in the hydrogen-ammonia mixture and a subsequent decrease in its volume and blockage. Alternatively, as the evaporator temperature is decreased at a fixed condenser temperature, the total pressure is reduced and the volume of hydrogen-ammonia mixture in the condenser section expands resulting in an increase in blockage.

Performance as a Function of Hydrogen Blockage

As a conservative approach, the original APSC design basis for thermal VSM's used zero wind speed to establish thermal VSM performance. Therefore, the recorded data from these tests were grouped as a function of wind speed and amount of hydrogen injected. Zero wind speed data were defined as one-hour averaged anemometer readings less than 0.2 m/s.

Figure 7 shows TAPS heat pipe conductance versus blockage for both 1.22 m and 1.83 m finned sections. The curve for the 1.22 m finned section is based on data collected during the test program. Data used to generate this curve was statistically evaluated to represent the 90% confidence interval of the overall collected data. The curve developed for the 1.83 m finned section was calculated from the results of testing the 1.22 m finned section using geometric and heat transfer scaling.

Figure 8 presents heat pipe conductance versus wind speed from 0 to 1.8 m/s at various levels of NCG blockage. The heat pipe conductance increases with increasing wind speed as expected. The zero blockage conductances are greater than those reported by Haynes and Zarling (1988) for a similar heat pipe with the evaporator tilted 0 to 9 degrees from the horizontal. The higher conductances are likely due to the vertical orientation of the evaporator, larger fin radiation loss due to a lower surrounding temperature, larger evaporator-condenser temperature differences, and a 2.4 m bare pipe within a metal collar between the calorimeter bath and finned section.

Visual evaluation of blockage levels was determined using infrared cameras. Two composite IR photos and measured fin surface temperature variations along the finned condenser sections are shown in Figure 9.

Conclusions

The 2000-2001 test program obtained the data necessary to determine TAPS heat pipe thermal degradation as a function of NCG blockage. During heat pipe-testing, measured quantities of hydrogen were introduced into three of the test heat pipes. The thermal performance at various levels of NCG blockage was determined by measurements of power supplied to the evaporator section and evaporator and ambient air temperatures. These data were used to develop a relationship for heat pipe conductance as a function of observed and measured NCG blockage.

A relationship has also been presented that suggests that the observed NCG blockage level is a function of ambient air and evaporator temperatures. Further testing is being conducted to substantiate this hypothesis.

Acknowledgements

The authors express their thanks for the support and assistance provided for this project by Rod Crew, Senior Project Manager, Alyeska Pipeline Service Co.; Everett Hightower, Project Manager, Chugach North Technical Services; Christopher Sexton,

Project Superintendent and Kyl Walker, General Foreman, Houston/Nana Construction Services, Aaron Kasgnoc, Mechanical Engineer, VECO Alaska Inc.; and James Dryden, Dryden Instrumentation Inc.

References

ASHRAE Handbook of Fundamentals. (1994). Refrigerant Tables and Charts, Refrigerant 717 (Ammonia), Properties of Saturated Liquid and Saturated Vapor.

Edelstein, F. (1989). "Evaluation of Gas Blockage – Trans-Alaska Pipeline System Heat Pipes, Phase 2 Test Program", Grumman Space System, Bethpage, New York.

Haynes, F.D. and J.P. Zarling. (1988). "Thermosyphons and Foundation Design in Cold Regions", *Cold Regions Science and Technology*, Vol. 15, 251-259.

Jahns, H. O., T. W. Miller, L. D. Power, W. P. Rickey, T. P. Taylor and J. A. Wheeler. (1973). "Permafrost Protection for Pipelines", *Proceedings of Second International Permafrost Conference*, Yakutsk, USSR, National Academy of Sciences, 673-684.

Johnson, E. R. (1983) "Performance of the Trans-Alaska Oil Pipeline", *Proceedings of the Fourth International Permafrost Conference*, Fairbanks, Alaska, National Academy of Sciences, 109-111.

Williams, R. J. (1991). "1990/1991 Heat Pipe Analysis", Personal Communication, Alyeska Pipeline Service Company.

Figure 1. Unblocked and Partially Blocked Heat Pipe

Figure 3. Four Calorimeter Set-up

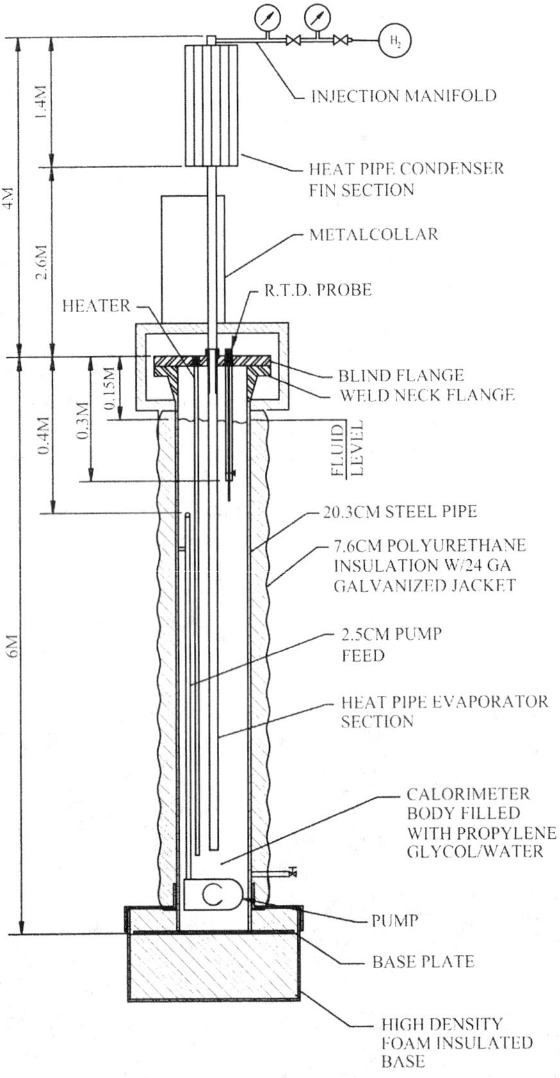

Figure 2. Blockage Test Calorimeter

Figure 4. Heat pipe unit performance and ambient temperature as a function of time.

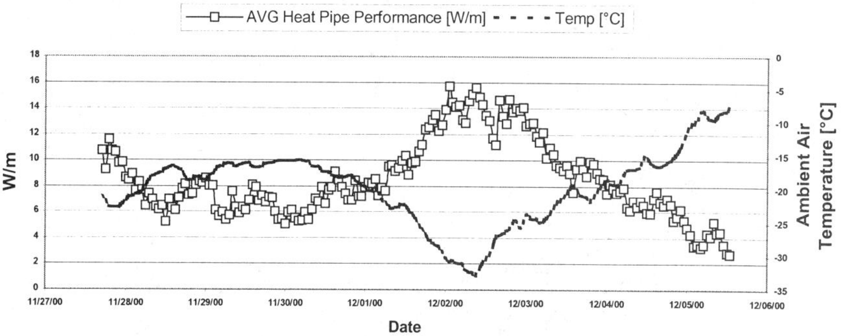

Figure 5. Heat pipe perfromance and wind speed variation

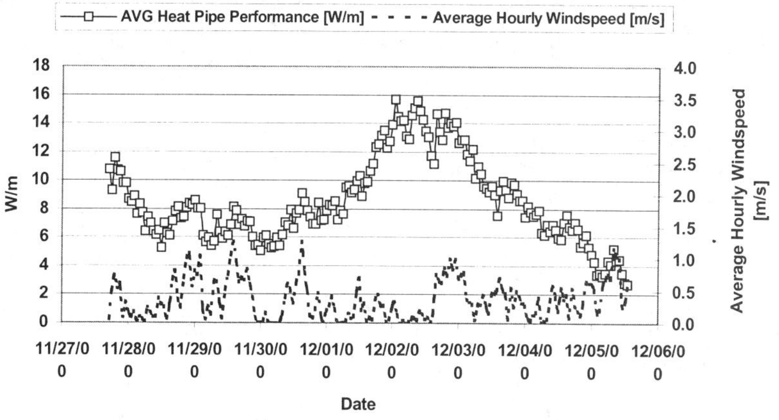

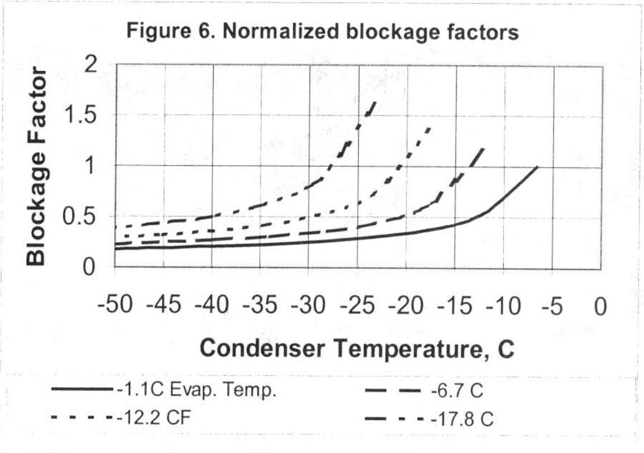

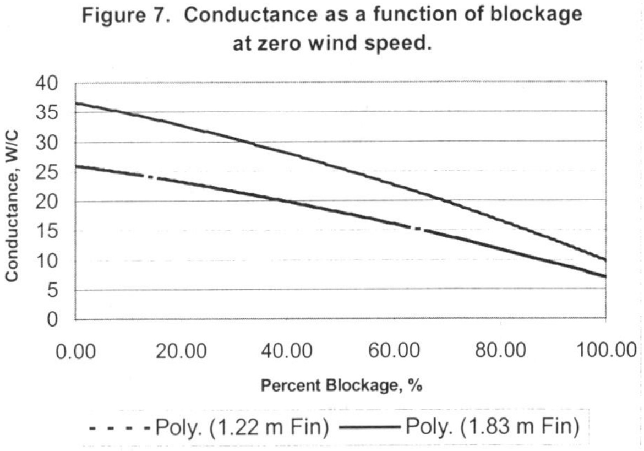

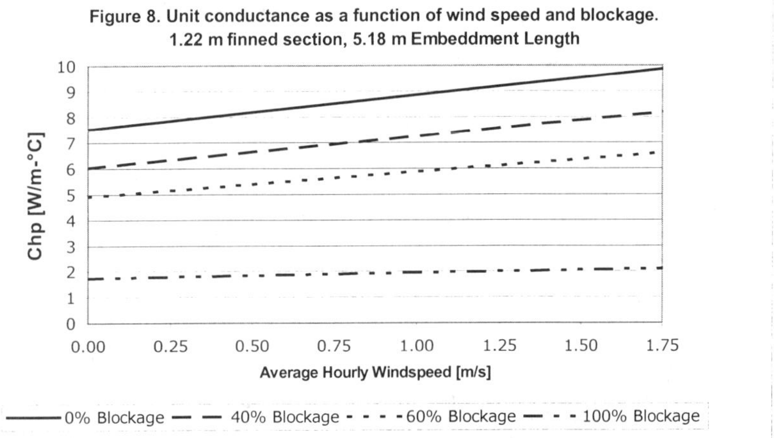

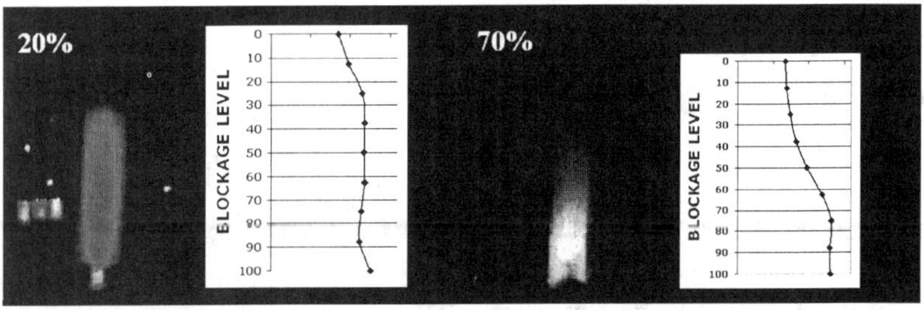

Figure 9. IR and fin temperature profiles for blocked heat pipes

Lessons Learned for River Crossing Designs From Four Major Floods Experienced Along the Trans Alaska Pipeline

Wim M. Veldman, M.ASCE, P.Eng.[1]
John Ferrell[2]

Abstract

In its 800 mile (1280 km) route from Prudhoe Bay to Valdez, the 48-inch (1200 mm) Trans Alaska oil pipeline traverses varied hydrologic zones ranging from the barren Arctic North Slope to the rugged Brooks and Alaska ranges to the glacier fed streams in the Valdez area.

The pipeline involves 80 major buried or elevated river crossings and more than 50 miles (80 km) of instream and floodplain alignments either deep buried or protected by river training structures. The design faced unique and challenging conditions on a project wide scale such as massive channel icings, sudden and dramatic flood releases from glacier-damned lakes and sediment deposition and channel changes in alluvial fans and steep braided rivers.

The hydrologic and river engineering data and knowledge base available at the time of TAPS' design and construction in 1973 – 1977 was limited particularly in the uninhabited north half of the project. The performance of river training structures – key to the integrity of the line in many locations – in steep braided arctic rivers had been largely untested.

This paper addresses the general performance of the river crossings and river training structures, now in place for 25 years, via lessons learned from four major floods. Several floods were well in excess of the Pipeline Design Flood. A record high outburst flood from a glacier- damned lake was experienced at a major bridge crossing. Recently, severe icings and flooding was experienced at two locations where previously no significant icing problems occurred.

The river training structures have performed well. Primarily, the post-flood works required have involved new structures where the pipeline is parallel to the major river systems. The lessons learned are valuable for the continued safe operations of TAPS as well as for the planning and design of future pipelines, especially those located in Arctic regions.

----

[1]President, Hydroconsult, Calgary, Canada and consultant to Alyeska Pipeline Service Company since 1973
[2]Senior Civil Engineer, Alyeska Pipeline Service Company, Fairbanks, Alaska

Reasons for Extensive River Crossings and Instream Alignments

The moding of the TAPS hot oil line, buried or elevated, was dependent generally on the percentage of silt or finer material. The buried mode was employed in those areas having less than 6% silt or finer material content. The elevated mode was used where the silt/finer material content was in excess of 6%.

As the elevated mode was considerably more expensive than the buried mode, the focus of the alignment and design process in 1973 was to maximize the length of buried line. River and floodplain areas are generally comprised of granular material and thus were prime candidates for buried pipeline routing. The narrow river valleys in the mountain ranges often dictated, for topographic reasons, an alignment close to or even wholly instream. Overall, the modes are split nearly 50/50 between buried and elevated.

Alyeska developed detailed hydrologic, hydraulic and river engineering criteria and design methodologies for the river crossings and floodplain segments (Alyeska, 1974).

Reasons for the Use of Extensive River Training Structures

Extensive revetments (bank armoring), guidebanks (to guide flow into bridges) or spurs (dikes perpendicular out from the banks to deflect the main channel flow) were used to:

- enhance the design of and shorten the length of bridges,
- protect floodplain transitions from the buried to the elevated mode,
- protect the shallow buried portions of the line from main channel flows and velocities,
- maintain overbank or floodplain flow conditions, and,
- expedite construction.

The last reason was often the key to the usage of structures especially spurs. Extensive deep burial of a 48-inch (1200 mm) concrete coated pipe in a wide wet ditch during extreme temperatures posed severe scheduling and equipment capability and availability concerns.

Overview of the Flood Case Histories

The locations and dates of the four flood cases evaluated, ranging in locations from the north slope to near the south end of the line, are illustrated on Figure 1. They are:

- the August 1992 flood on the lower Sagavanirktok River on the North Slope. The estimated flow was more than two times the Pipeline Design Flood. Significant bank erosion at Milepost (MP) 47 necessitated emergency bank protection works followed by the addition of three new spurs each about 800 feet (244 m) long. Several spurs were also overtopped and washed out in the MP 20 area necessitating the reconstruction, raising and addition of spurs as well as the construction of a flood relief channel in the floodplain. Figures 2 –

6 illustrate the hydrologic and hydraulic data, and comparative air and site photos.

- the June 2001 breakup flows on the Sagavanirktok River in the MP 63-85 reach. Severe icings in the main channel diverted the total breakup flow into spurs protecting the elevated and buried line. Several spur sections were washed out. This area had experienced no significant icing problems for 25 years. Figures 7 – 10 present comparative air and site photos.
- the August 1994 flood on the Middle Fork Koyukuk River between MP 229 and 243, about a 1:100-year event (U.S. Department of the Interior, personal communication and Jones and Fahl 1994), caused bank erosion and channel changes which necessitated the construction of bank revetments and spurs. The existing spurs performed very well. Since the 1994 flood, a number of smaller flows have resulted in additional bank erosion which has necessitated the construction of several other revetments. Figures 11 – 14 illustrate comparative air and site photos and hydrologic data.
- in August 1997, outburst floods occurred simultaneously from several glacier damned lakes located about 40 miles (64 km) upstream on the Tazlina River at MP 687. The flood was well in excess of previous floods and the Pipeline Design Flood. Significant bank erosion towards the north pier of the 650 foot (198 m) suspension bridge necessitated construction of a rock revetment. Although not an immediate bridge integrity concern, the potential impact of the next major flood warranted the armoring. Figures 15 – 16 illustrate the hydrologic and bank erosion data and air and site photos.

Approach Used and Components Considered in the "Lesson Learned" Case Histories.

This paper is not intended to be a detailed hydrologic, hydraulic or river engineering analysis of the conditions and impact of the four flood case histories – the hydrologic analysis alone would suffice for an individual paper - but rather the general lessons learned with respect to:

- bank erosion, as applied to pipeline sagbend setback or selecting an alignment parallel to a river,
- icings, as applied to the height of river training structures or bridges
- performance of river training structures, as applied to their feasibility, practicability and limitations in protecting the pipeline in a riverine environment,

Much of the information presented herein is derived from Alyeska's internal reports and the chief author's analysis and reports prepared during operation. The figures illustrate current site, historic and hydrologic conditions and should be reviewed in their entirety prior to proceeding to the next sections.

Bank Erosion

Lessons Learned : Common practice is to determine potential bank erosion from evidence from comparative airphotos or surveys (U.S. Department of the Interior 1972,1975,1976,1977). Some researchers have proposed mathematical models which compute bank shear forces and resultant bank erosion rates for specific locations and specific hydraulic conditions.

The TAPS experience suggests the following lessons learned:

- the vast majority of bank erosion occurs during high flow events. At a sharp bend on the Middle Fork Koyukuk River at MP 231.5, very similar to that shown on Figure 11, the 85 feet (25.9 m) of erosion experienced during the 100-year flood in 1994 exceeded the total erosion in the previous 17 years. For well-defined bends (Figures 11 – 13), the general zone of maximum potential bank erosion is well established although even in these cases, the maximum point of attack and thus erosion potential may vary depending on upstream approach flow conditions. (Figures 12 and 13).
- the magnitude of erosion is dependent on whether the bank is thawed or frozen at the time of the flood as well as the height of the bank. Similar bends at MP 229 and 230 (Figures 11 – 13) have experienced different erosion rates. Due to the much greater volumes of frozen material to be eroded, the higher bank at MP 230 (Figure 13) erodes at a slower rate. In the buried mode, the thaw bulb of TAPS' buried line can envelop much of the bank area subject to erosion (Figure 12) thus increasing the potential for erosion, whereas in the elevated mode (Figure 13), the pipeline has little or no impact on the thermal regime of the bank thus reducing its rate of erosion. At MP 85, high flows during breakup in June 2001 induced little or no erosion of a frozen bank even at a well-developed sharp bend (Figures 9 and 10C).
- even in almost straight incised river reaches, small changes in flow alignment upstream, such as the development of a side bar, can result in significant bank erosion during a flood (Figures 15 – 16). Extrapolating bank erosion rates prior to 1997 (Figure 15) would not have predicted the 75 feet (23 m) of erosion experienced as a result of the record 1997 flows. If the bank erosion had not been arrested by constructing a rock revetment, the extrapolation of historical erosion, including that which occurred in 1997, would underestimate future erosion potential as the 1997 flood-induced bank indentation (Figure 16D) would have increased future erosion potential.
- on wide braided rivers, accurately predicting potential bank erosion is extremely difficult and speculative for a specific location since it requires a prediction of gravel bar formations which influence flow patterns (Figure 3). Whereas the downstream progression or migration of a well-developed channel can be determined reasonably well for normal to moderate floods (Figure 5), record summer floods such as occurred in 1992 (Figure 2), cause dramatic channel changes as a result of the high sediment load conveyed. At one location, the change could induce significant bank erosion, such as the 220-feet (67 m) of erosion which occurred at MP 47 (Figure 4), whereas at

other locations, the bank erosion could be minimal to nil depending on the angle of attack and location of the main channel within the wide braided river.

Design Guidelines for Bank Erosion: Based on TAPS' lessons learned from the last 25 years, the following design guidelines are suggested:

- comparative airphotos particularly immediately prior to and after a major flood are most useful as the majority of bank erosion occurs during high flows.
- use site-specific erosion rates from comparative air photos of incised rivers provided that significant floods have occurred during the time interval of photos. Significant for this purpose is defined as a late summer flood event (when the banks are thawed) at a magnitude at least equal to 50% of the design flood magnitude.
- for a sharp bend, assume the design flood and thus maximum bank erosion will occur during the life of the project.
- for wide-braided gravel bed rivers, assess historic bank erosion evidence for the length of similar river conditions. If a significant flood has occurred during the time interval of the photos, use the maximum evident erosion along the total length of similar river conditions as the potential erosion that could occur at a specific location.
- for this braided river situation, analyze whether to design for the maximum historic erosion experienced along the river or the average bank erosion depending on:
  - the nature of the works being protected,
  - whether emergency works, if and when necessary, are feasible to construct,
  - the nature and heights of the banks,
  - the life of the project. For a 25+ year project life, designing for the maximum evident erosion would be appropriate whereas for a 5 – 10 year project life, the average bank erosion in the study area could be appropriate.

Icings

Lessons Learned: Thick ground-fast ice develops in Arctic rivers where:

- frozen or impermeable bed sections force the winter flow to the surface to freeze in a series of thin overflows,
- spring-fed tributaries overflow wide braided rivers.

Commonly accepted factors that promote significant icings are low snowfall and low temperatures particularly during the late fall freeze-up period and a high base flow prior to freeze-up (Carey 1973, Kane 1981). As the integrity of the spurs and thus the pipeline was dependent on the design icing level and resultant spring breakup water levels – initial breakup flows are over the ice – understanding the behavior and maximum potential of icings was key to TAPS' floodplain design. After concerted field surveys, research, and discussions with State personnel and researchers at the University of Alaska, Fairbanks, in 1973 – 1976, it was concluded that:

- temperatures and snowfall in the fall are not reliable predictors of the potential severity of icings,
- site specific conditions such as culverts and bridges have a dramatic impact on icings,
- physical floodplain features such as low wide terraces limit the maximum possible overflow levels.
- in areas where tributary overflows occur into wide rivers from one side, the icing level across the width of the river can vary by several feet and therefore,

a site specific "design icing level" for a crossing or structure is not possible to predict.

These lessons learned were confirmed during construction in 1975 –1977 and in the operations phase particularly during the 2001 breakup on the Sagavanirktok River as follows:

- the 1975 breakup on the Dietrich River, following a low snowfall and temperatures in the preceding fall – classic conditions for severe icings - caused significant icings, overflows and washout of the pipeline workpad several miles north of the Dietrich construction camp. A long dike was consequently constructed to protect the workpad and elevated line.
- the following winter was characterized by high snowfall and warmer temperatures, conducive to low icing development according to generally accepted principles. Little icing formed in the area of the newly constructed dike. The warmer temperatures however enabled the tributary inflow to overflow the Dietrich River for several miles downstream prior to freezing and consequently caused significant flooding of the Dietrich construction camp.
- the 2000 –2001 winter on the North Slope was uncharacteristically warm. The 2001 spring breakup was late but then occurred suddenly in early June. At MP 63 and 85, severe icings across the entire width of the Sagavanirktok River caused the breakup flow to spill into the ice-free floodplains where several spurs were breached (Figures 8 and 10).
- in assessing overall breakup conditions along the river in the MP 63 – 85 reach, it was concluded that the unusual locations for the severe icings – no significant icings had been reported in these locations since commencement of operations – were due to a greater downstream progression of the overflow due to the warmer temperatures, similar to the Dietrich camp flooding situation described earlier.

Design Guidelines: The following design guidelines are suggested:

- recognize obvious ice-inducing structures such as bridges or culverts. Use historical evidence from existing structures to determine maximum icing levels and from this, breakup water levels,

- determine the main winter inflow sources, the prime trigger for icings on wide, shallow braided rivers. Use historical evidence to establish upstream and downstream limits of major icings,
- within these broad limits, assume that the maximum-recorded icings can occur anywhere at some time during the life of the project,
- if the total tributary inflow is from one side of a wide river valley, as in the case of the Dietrich River, recognize that initial breakup flows will be concentrated along the opposite or "lower" side of the river's ice. If structures are located along this lower side, they will annually bear the brunt of the initial spring breakup flow.

Performance of Spurs

Lessons Learned: Extensive spur fields exist along the Sagavanirktok River on the non-vegetated north slope (Figures 3 – 10) as well as in the vegetated floodplains of the Middle Fork Koyukuk River south of the Brooks Range (Figure 14). Some of the lessons learned in the last 25 years are:

- a spur spacing of two (2) to four (4) times their effective length as measured from the spur nose to the buried or elevated pipe, has been generally effective in deflecting the main flow in non-vegetated areas such as the Sagavanirktok River (Figures 3 and 6). Where the spacing used was greater, well developed channels have and can migrate towards the pipeline.
- in vegetated floodplains, smaller sub-channels can form in irregular patterns between the spurs, depending on the influence of vegetation or sediment deposition, to the point that additional spurs need to be constructed (Figure 14). Site specific channel changes in vegetated floodplains are difficult to predict although on a global scale, areas of potential channel switching or channel enlargement can be predicted.
- the stability of riprap on spur noses or revetments is dependent on the angle of flow attack into the nose and breakup / ice jam induced velocities. Sharp attack angles, common in braided rivers (Figure 5) require increased riprap sizes or flatter slopes than similar hydraulic conditions in relatively straight incised streams (Figure 15). At MP 47, stranding of ice during breakup (Figure 5) produced head losses and velocities well in excess of those computed from open channel flow models such as the U.S. Army Corps of Engineers' HEC-2.
- spurs are an economic alternative in wide braided river compared to continuous bank revetment as only the noses of the structures need armoring. As the amount of armoring needed on a spur nose is virtually the same for long or short spurs, long spurs on wide braided rivers are most economic. Where the flow is generally perpendicular to the spurs (Figure 9), riprapping only the spur nose area is adequate. Where a sharp angle of attack occurs into the spur field (see Spur 6, Figure 5A and Figure 14) additional riprap may be required on the shank to the point that spurs are less economic than a revetment.

- where spurs are constructed across a main flow channel (Figure 4), the flow needs to be deflected to form new channels beyond the noses of the spurs. If the spurs are constructed in the winter, little or no scouring occurs during the frozen bed conditions during the first breakup, the result being increased heads on the spurs. At MP 47, Spur 7 was breached in the first breakup in 1993 (Figure 3B) in part due to this reason.
- in vegetated floodplains with sub-channels, (Figure 14), additional spurs constructed following the 1994 flood were shorter in length compared to the original existing adjacent spurs in order to prevent blockage of sub-channels that would lead to high heads and velocities until a new channel was scoured out through the vegetated floodplain. On the Sagavanirktok River at MP 20, a bypass channel was excavated across a bar in conjunction with a new spur constructed in 1993 to provide a flow relief to the partial flow blockage caused by the new spur (Figure 6).
- small sub-channels in a spur field in a steep river can lead to spur overtopping, even at relatively low flows, if the blocked sub-channel flow is unable to "escape", back out to the main channel. This occurred at several locations immediately upon start-up of the line in 1977. The solution was to field-check all spur heights relative to the local terrain and add fill, where necessary, to ensure a minimum spur height of four feet relative to the highest local grade. With this fix, similar problems have not re-occurred.
- a minimum freeboard of three feet, above the design flood level, was used to establish the spur crest elevation in the initial design phase. On a wide braided river such as the Sagavanirktok on the North Slope, this provided for a flow more than 2 times the design flood before the spurs were overtopped (Figure 5). As a result of ponding created by long spurs particularly in steep vegetated floodplains interspersed by subchannels, the water level at the bank can be several feet higher that that computed from hydraulic models at the spur nose.

Design Guidelines for Spurs: The following design guidelines are suggested:

- a spur spacing not exceeding 2 – 3 times their effective length (Figure 6),
- armoring of the spur noses as per common guidelines. If and when dramatic angles of flow attack into the spur field occur in the future, add more riprap on the spur shank when necessary,
- if spurs block sub-channels, ensure the flow has an escape route to the main channel and/or that the spur crest level is at least 4-feet (1.2 m) above the highest adjacent terrain that restricts the escape of the sub-channel flow,
- carefully consider the usage of spurs in vegetated floodplains interspersed with subchannels subject to change. As channel changes and thus flow patterns in the spur field are difficult to predict in these circumstances, consider continuous revetments as an option.
- for long spurs on steep rivers, where the differential head across a spur can be significant, evaluate the need for:
  - a sloping spur profile to increase freeboard at the bank,

  - increasing the design velocity for the riprap on the spur nose, as compared to that computed by the main channel model, by at least 25% to account for local acceleration around the spur nose.
- establish the freeboard and spur crest elevation by considering:
  - the water level versus flow relationships - for a "flat" rating curve, common in wide braided rivers, less freeboard is necessary.
  - the available hydrologic data base,
  - uncertainties with respect to computing open water flood levels or over-ice breakup water levels.

Acknowledgements

The permission of Alyeska Pipeline Service Company to utilize data, air and site photos is greatly appreciated. They were key to illustrating the lessons learned. The lessons learned and general design guidelines outlined herein were developed by the chief author of this paper and do not necessarily represent Alyeska's design criteria or approach at this time.

The 1974 river design criteria for TAPS was primarily developed by the late Dr. T. Blench, P.Eng. and Dr. R. Cooper, P.Eng. and Dr. V. Galay, P.Eng. of Northwest Hydraulic Consultants, Canada. Their knowledge and foresight was a significant factor in the overall sound performance of the river sections of TAPS. The chief author of this paper was NHC's senior staff member assigned full time to the construction phase of TAPS in 1973 – 1976 and has provided continuous follow-up consulting to TAPS in the operations phase.

References

Alyeska Pipeline Service Company (1974). *"Summary Report, River and Floodplain Design Criteria for the Trans Alaska Pipeline System,* Houston, Texas 1974.

Carey, K.L. *"Icings Developed from Surface Water and Groundwater".* Monograph 111-D3, Cold Regions Science and Engineering Laboratory, May 1973.

Jones, S.H. and Fahl, C.B. (1994). *"Magnitude and Frequency of Floods in Alaska and Conterminous Basins of Canada"* U.S. Geological Survey, Water-Resources Investigations, Report 93-4179 prepared in cooperation with the State of Alaska and Federal Highway Administration, Anchorage, Alaska.

Kane, D.L. *"Physical Mechanics of Aufeis Growth."* Institute of Water Resources, University of Alaska, Fairbanks, February 1981.

United States Department of the Interior, Geological Survey. (Various years). *"Water Resources Data for Alaska".* (Data as per hard copy reports, CD-ROM database or from the Internet and interim data via telephone).

United States Department of the Interior, Geological Survey (1972). *"Channel Erosion Surveys along Proposed TAPS Route,* July 1971", Anchorage, Alaska.

United States Department of the Interior, Geological Survey (1975). *"Channel Erosion Surveys Along Proposed TAPS Route*, 1975", Anchorage, Alaska.

United States Department of the Interior, Geological Survey (1976). *"Channel Erosion Surveys Along Proposed TAPS Route,* 1976", Anchorage, Alaska.

United States Department of the Interior, Geological Survey (1977). "*Channel Erosion Surveys Along Proposed TAPS Route, 1977", Anchorage, Alaska.*

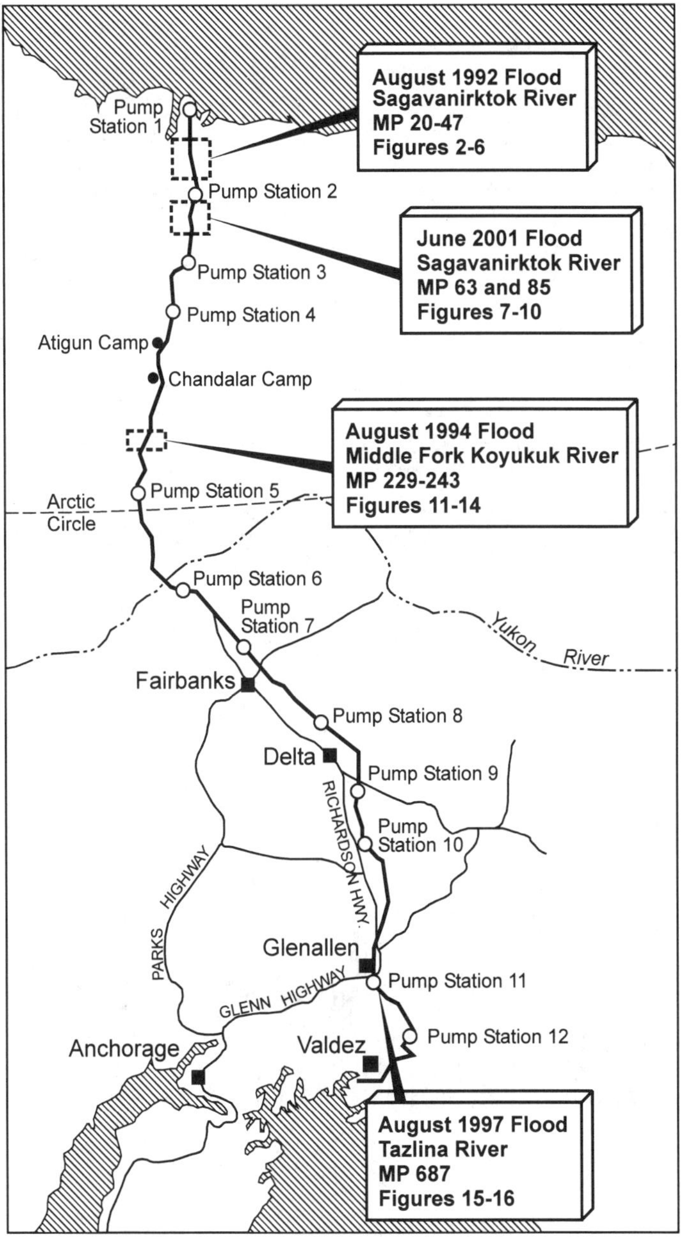

Figure 1. Location Map of Trans Alaska Pipeline Illustrating Flood Areas Evaluated.

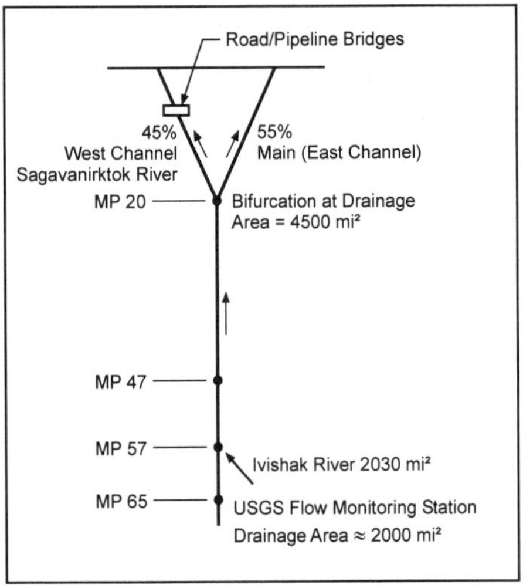

2A. Schematic of Sagavanirktok River Delta

NOTES: RE: COMPUTING 1992 FLOOD PEAK

1. The USGS gage is located upstream from the Ivishak River watershed, the main flood-producing area during the 1992 flood.
2. The peak flow in the MP 20-47 section in August 1992 was estimated by two methods as follows:
   - from the peak water level measured at the West Channel bridges and from a Stage-Discharge relationship at these bridges developed from multiple flow measurements by ARCO.
   - from qualitative observations of the bifurcation during numerous overflights, the West Channel flow is about 45% of the total flow in the Sagavanirktok River.
   - from 1992 high water mark surveys at MP 20, river surveys and backwater calculations for the initial design of the pipeline in 1973-1974, the peak 1992 flow upstream of the bifurication was estimated.
3. The two methods resulted in about the same peak flow of 300,000 cfs upstream from the bifurcation.
4. For comparison purposes, the Pipeline Design Flood at MP 20, shown on the graph below, is pro-rated to the flow in the West Channel (45% of the total flow). Thus the 1992 flood was in excess of 2 times the Pipeline Design Flood.

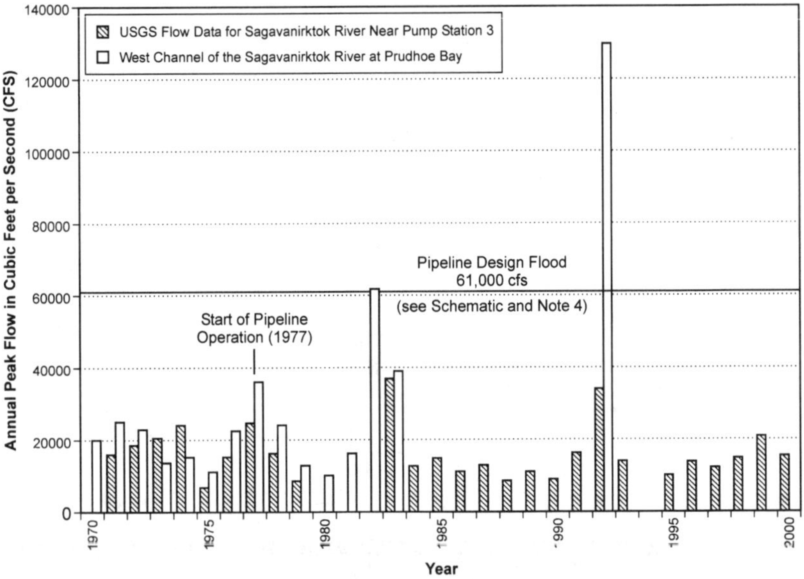

2B. Maximum Annual Flow

Figure 2. Analysis of August 1992 Flood Sagavanirktok River, Milepost 20-47 Annual Peak Flows

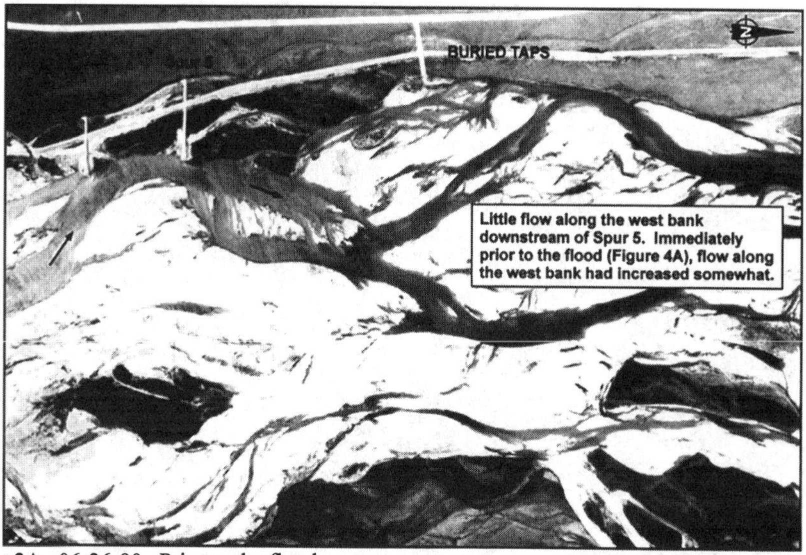

3A. 06-26-90. Prior to the flood.

3B. 07-14-93. Spurs 6, 7 & 8 constructed April 1993 following the 1992 flood.

Figure 3. Analysis of August 1992 Flood - Sagavanirktok River, Milepost 47 Overall Airphoto Comparison Illustrating Global River Changes Due to the August 1992 Flood

4A. 06-29-92. Shortly prior to the flood.

4B. 06-20-96. Post flood spurs (Nos. 6, 7 and 8) were constructed in April 1993.

Figure 4. Analysis of August 1992 Flood - Sagavanirktok River, Milepost 47 Close-up Airphoto Comparison Illustrating Local Flow Changes and Bank Erosion Due to the 1992 Flood and the Construction of the Spurs

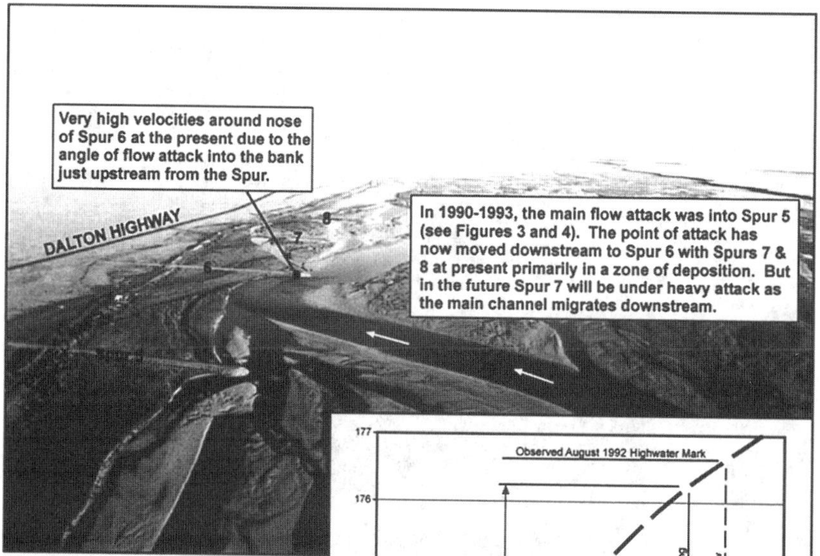

5A. Sept. 2001 flow conditions.

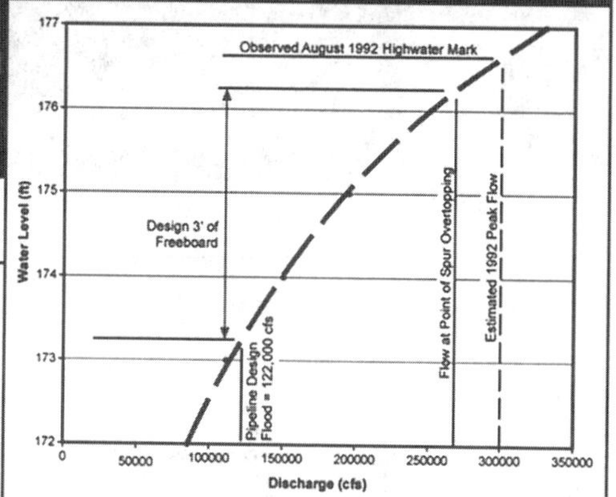

5C. Water level versus flow at MP 20 Spurs

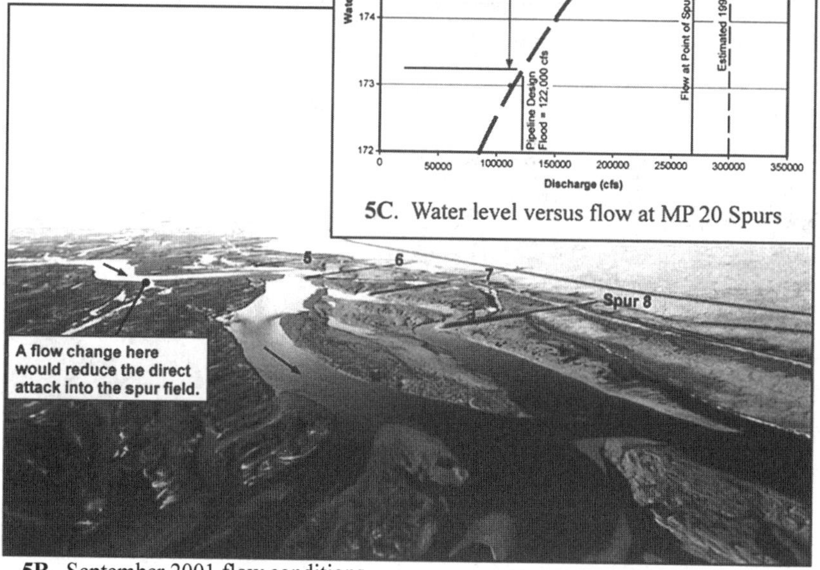

5B. September 2001 flow conditions.

Figure 5. Analysis of August 1992 Flood
Sagavanirktok River, Milepost 47 and 20
2001 Site Photos of the Spur Field and Flow Conditions

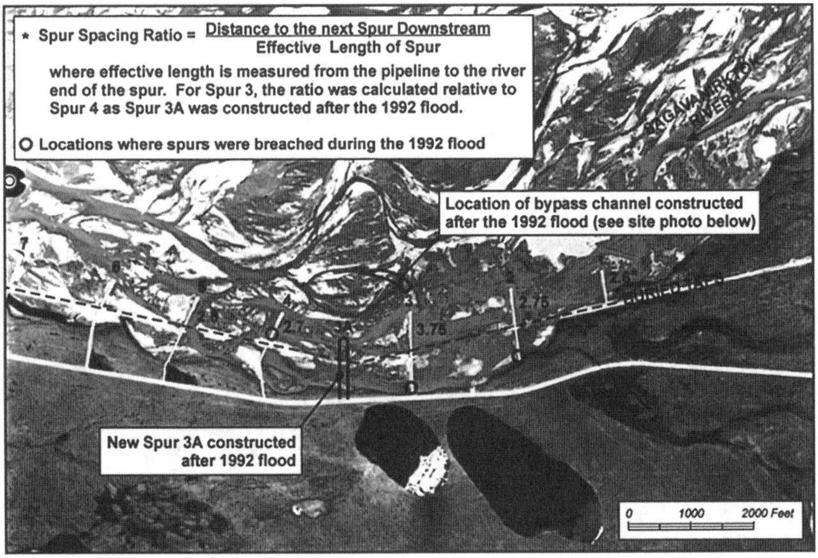

6A. June 25, 1983 Airphoto.

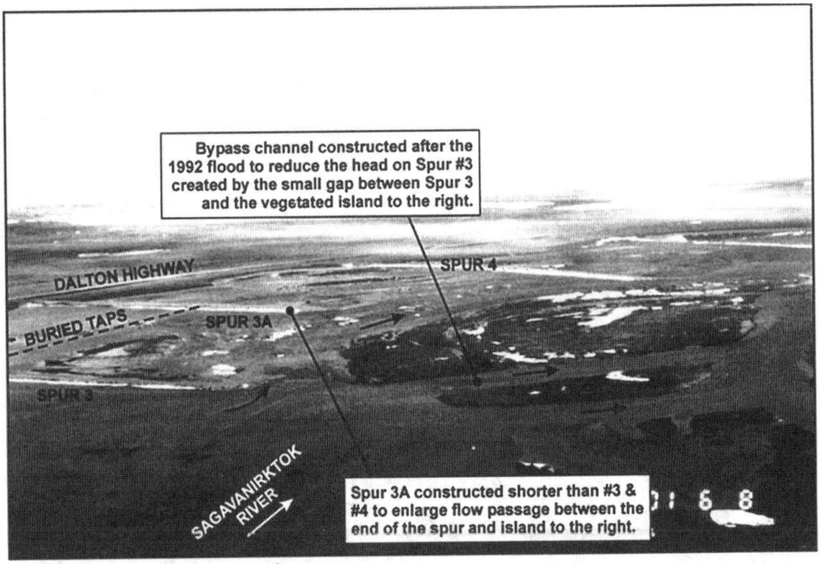

6B. June 8, 2001 Breakup flow.

Figure 6. Analysis of August 1992 Flood
Sagavanirktok River, Milepost 20
Impact of 1992 Flood and Remedial Measures

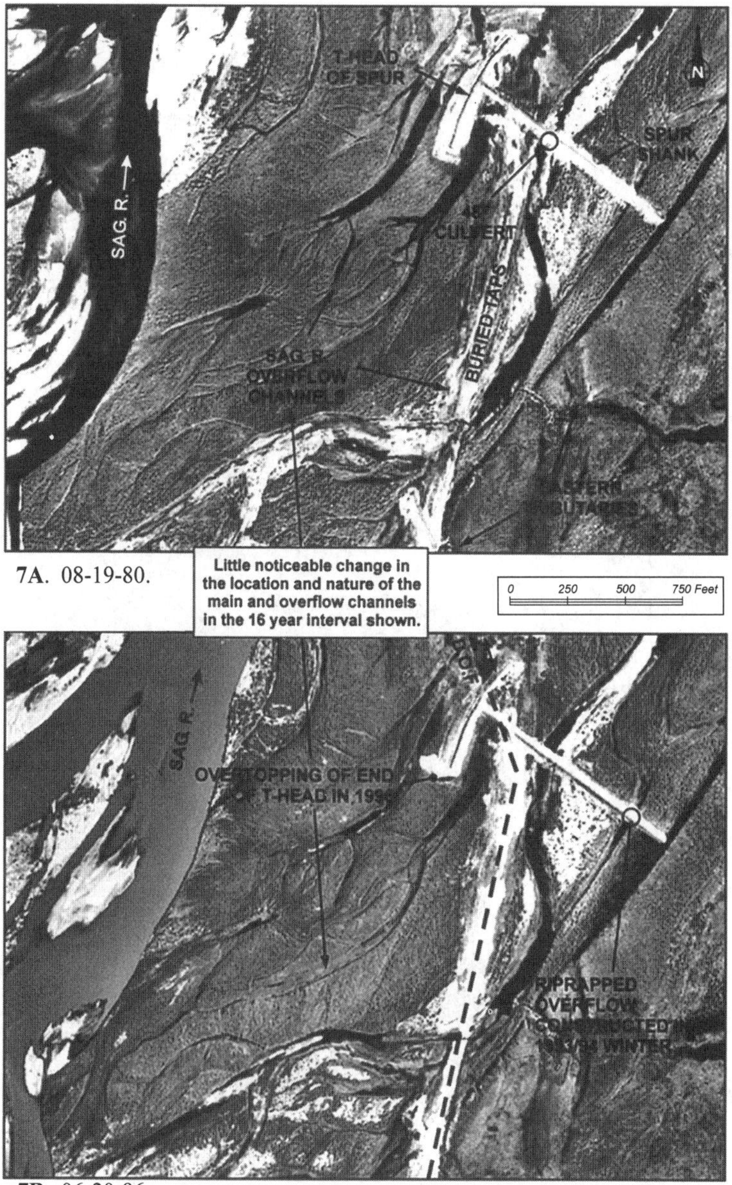

7A. 08-19-80.

7B. 06-20-96.

Figure 7. June 2001 Breakup Flow Sagavanirktok River, Milepost 63 Comparative Airphotos

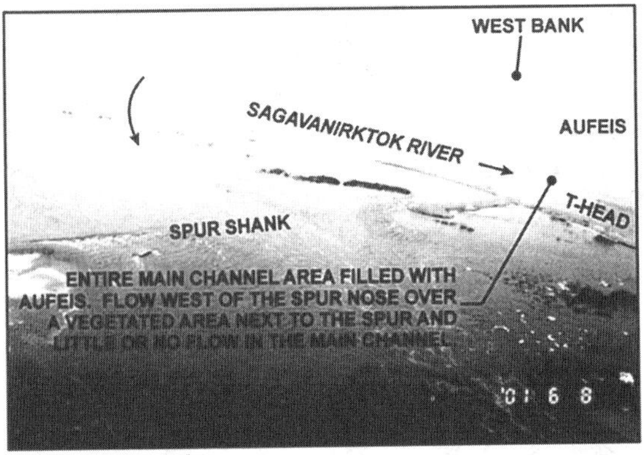

8A. June 8, 2001.

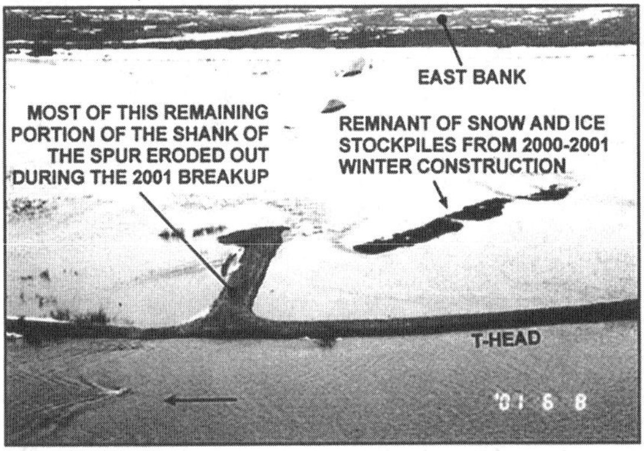

8B. June 8, 2001.

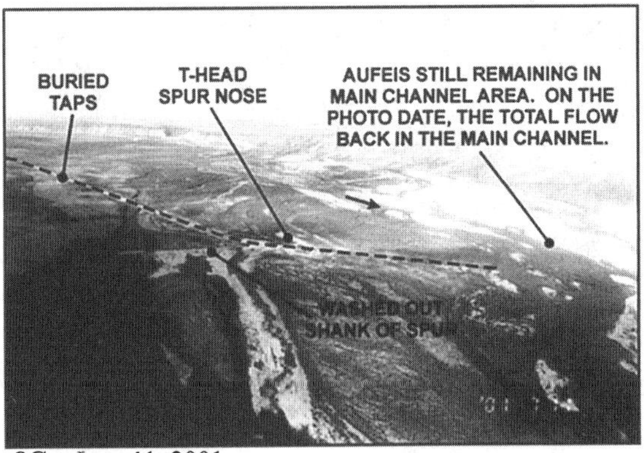

8C. June 11, 2001

Figure 8. June 2001 Breakup Flow Sagavanirktok River, Milepost 63 Site Photos June-July 2001

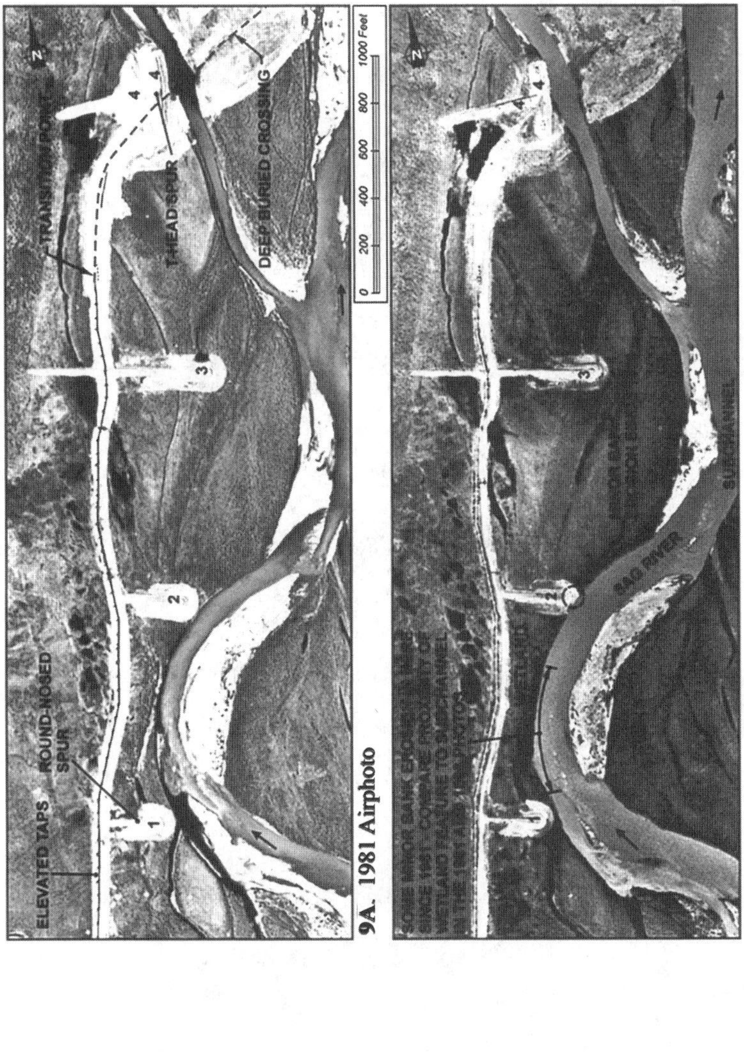

9A. 1981 Airphoto

9B. 1996 Airphoto

Figure 9. June 2001 Breakup Flow Sagavanirktok River, Milepost 85 Comparative Airphotos

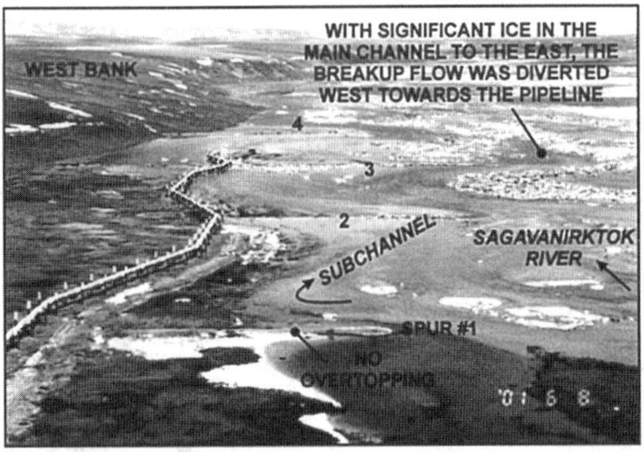

10A. June 8, 2001

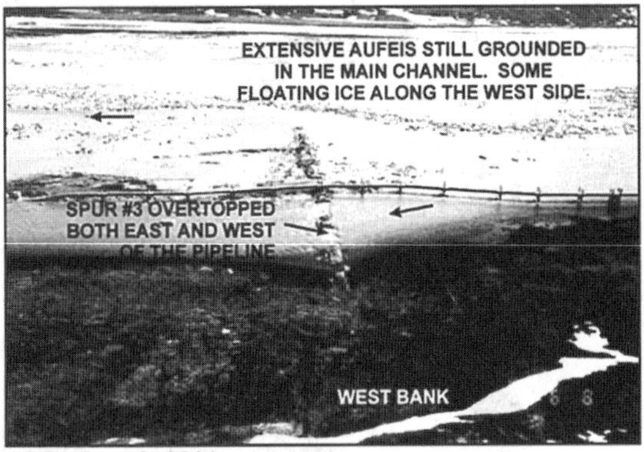

10B. June 8, 2001

10C. July 10, 2001

Figure 10. June 2001 Breakup Flow Sagavanirktok River, Milepost 85 Site Photos June-July 2001

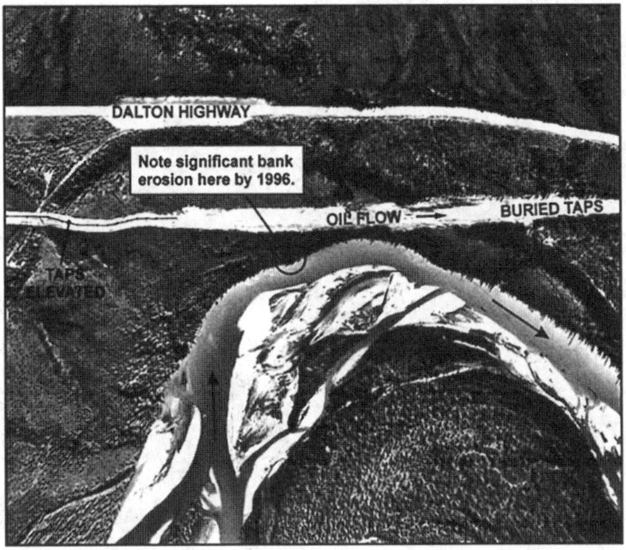

11A. 07-03-82

0 250 500 750 1000 Feet

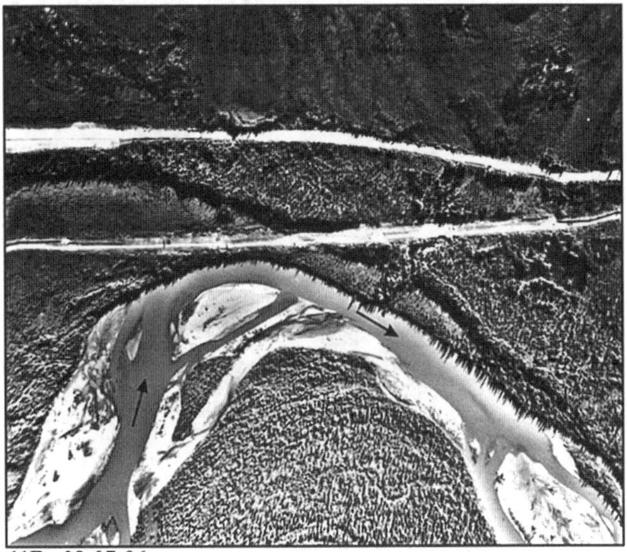

11B. 08-07-96

Figure 11. August 1994 Flood
Middle Fork Koyukuk River, Milepost 229
Comparative Airphotos

12A. July 1998

12B. September 2001. Riprap revetment constructed in April 2001.

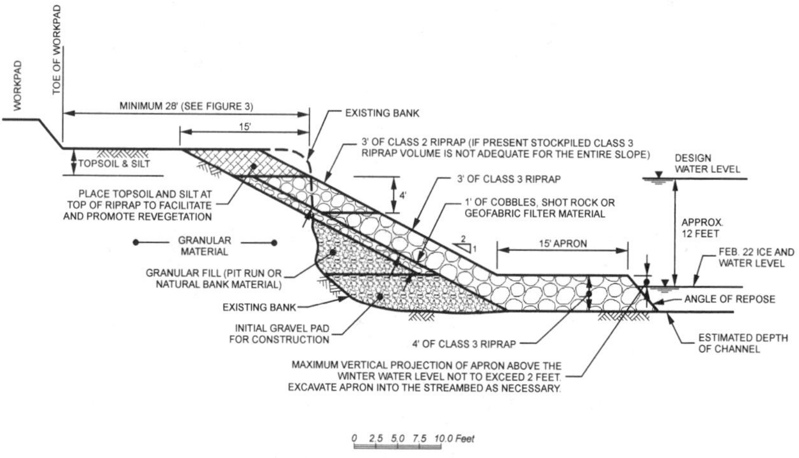

12C. Revetment Design

Figure 12. August 1994 Flood
Middle Fork Koyukuk River, Milepost 229
1998-2001 Site Photos and Revetment Design

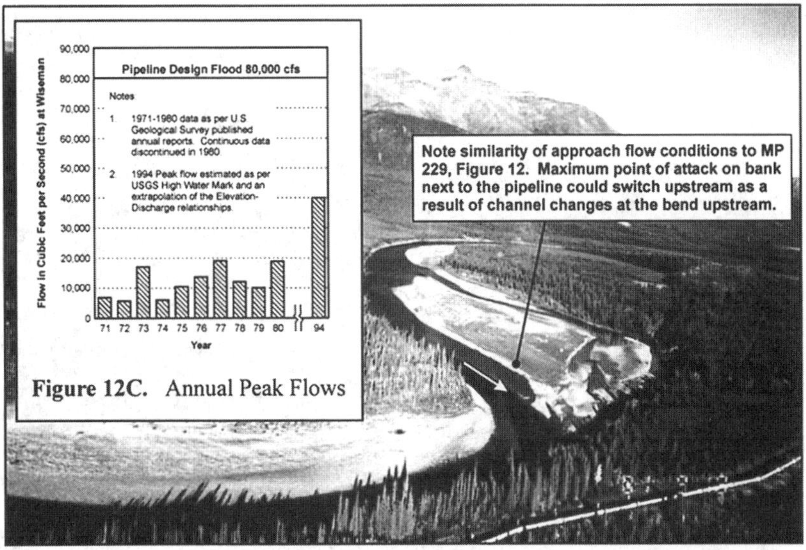

Figure 12C. Annual Peak Flows

13A.

13B.

Figure 13. August 1994 Flood
Middle Fork Koyukuk River, Milepost 230
2001 Site Photos

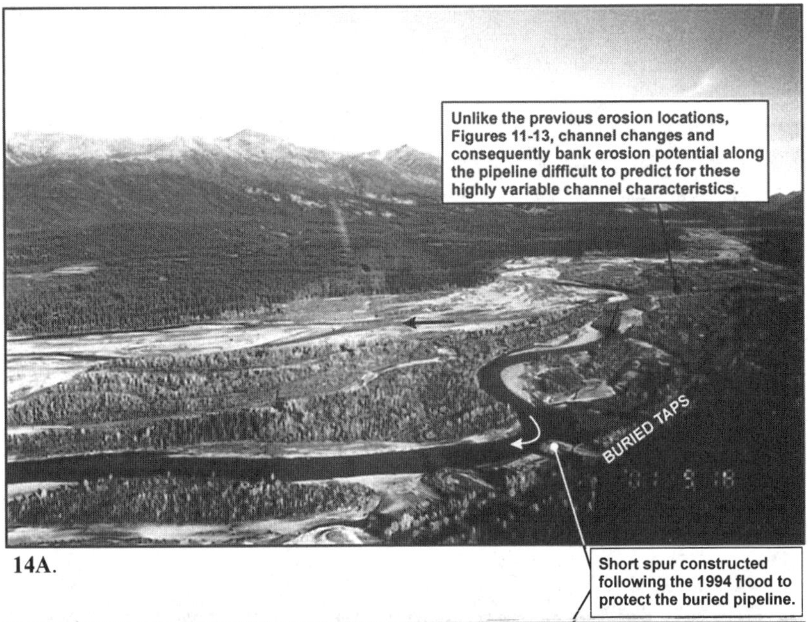

14A.

14B.

Figure 14. August 1994 Flood
Middle Fork Koyukuk River, Milepost 243
2001 Site Photos of Spur Field

NOTES:

1. Flows from 1949 to 1972 as per USGS data and as per "Summary Report, River and Floodplain Design Criteria", Alyeska Pipeline Service Company, March 1, 1974 Revision.

2. The peak flow in 1997 - the USGS gage was terminated in 1972 - was estimated from the high water mark and computed Stage-Discharge Relationship for the Richardson Highway Bridge located less than 5 miles downstream of the Pipeline Bridge.

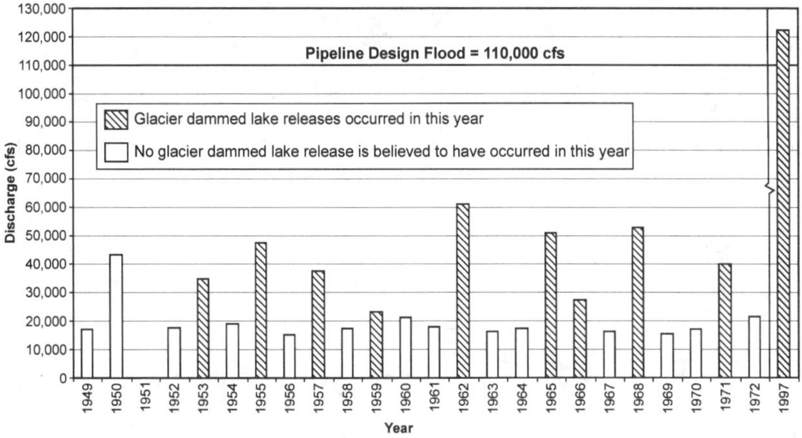

15A. Maximum Annual Flow - Tazlina River near Glennallen

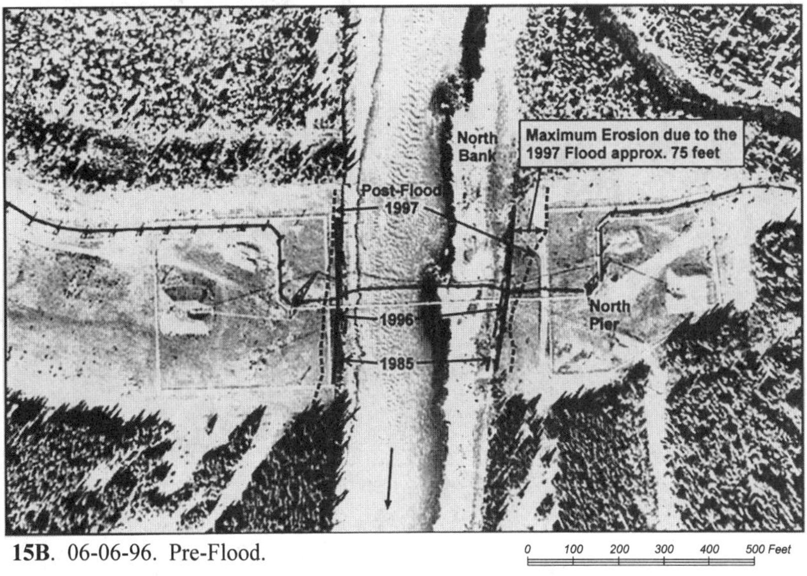

15B. 06-06-96. Pre-Flood.

Figure 15. Tazlina River Bridge, Milepost 687
Annual Peak Flows and Measurements of Bank Erosion

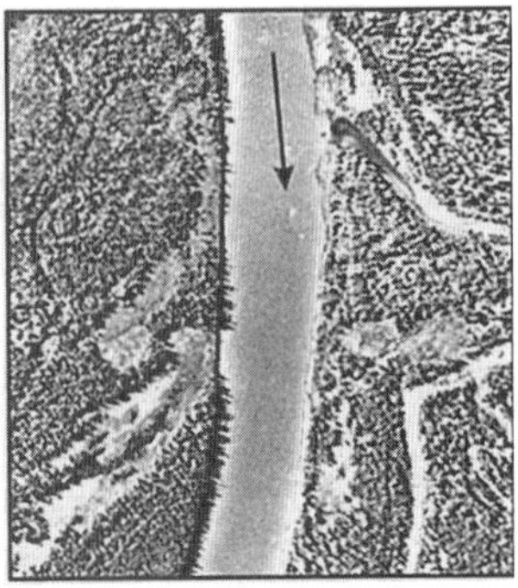

16A. August 1974

16D. October 1997

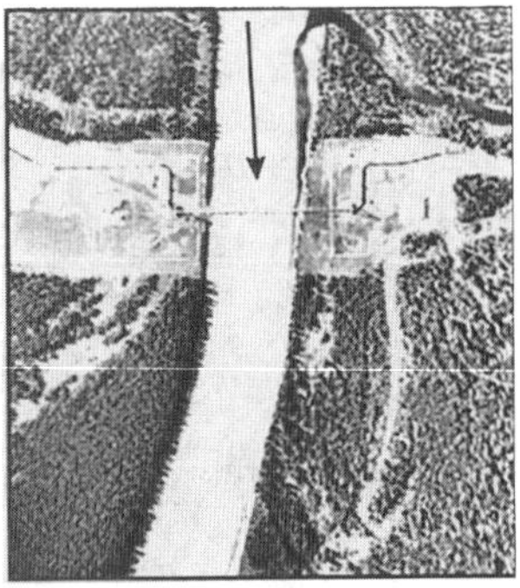

16B. August 1985

16E. October 1997

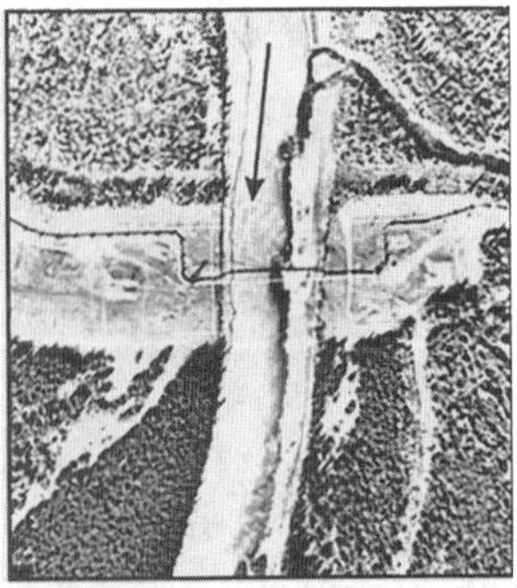

16C. August 1996

16F. October 1997

Figure 16. Analysis of August 1997 Flood Tazlina River Bridge, Milepost 687
Pre-Flood Airphotos and Post-Flood Site Photos

Performance Of The Squirrel Creek Slopes Steep Slopes On Discontinuous Permafrost

Rupert G. Tart, Jr.[1] and John E. Ferrell[2]

Abstract

The two slopes approaching Squirrel Creek were modified during construction because natural slopes in the area of similar steepness were displaying landslides. The original design plans called for the pipe to be buried on these slopes. When drilling found discontinuous, warm permafrost along these slopes, the pipe design mode was changed to aboveground with heat pipes to maintain frozen ground. Since construction of the Trans Alaska Pipeline System (TAPS), Alyeska Pipeline Service Company (ASPC) has been monitoring instrumentation on these slopes and performing periodic maintenance. This paper will describe the history of the performance of these slopes and the maintenance activities that have been employed to address some of the issues that have developed along the slopes since construction. Documented solifluction-like movements, changes in ground thermal state, frost heave, and thaw settlement have all been monitoring and addressed with maintenance.

Introduction

The Trans Alaska Pipeline extends from the Valdez Marine Terminal on Valdez Sound at the south end to Pump Station 1 near Prudhoe Bay. Progressing north from Valdez the pipeline crosses the Chugach Mountains then enters the Copper River Basin paralleling the Richardson Highway.

The Copper River Basin is the southernmost limit of permafrost along the pipeline route. Most of the basin is underlain by discontinuous permafrost. Much of the route through basin is underlain by overconsolidated glacial clay and silt formations.

In the Copper River Basin, the pipe crosses several east-west trending drainages that direct water from the Chugach Mountains to the Copper River. At pipeline milepost 717, about 83 miles north along the pipeline from Valdez, the pipeline crosses Squirrel Creek as shown in Figure 1. Both the north and south approaches to Squirrel Creek were steep and presented concerns to the pipeline designers. Along Squirrel Creek and other similar drainages in the Copper River Basin, natural landslides were observed on some of the slopes approaching these creeks and rivers as shown in Figure 2.

Before detailed field studies began, the preliminary design for the slopes at Squirrel Creek included three different modes. A sketch depicted the three modes is given in Figure 3. At the creek crossing the pipe was to be conventionally buried. The north slope was to be moded aboveground and the south slope was to be moded special burial.

---

1 Principal, Golder Associates Inc., Anchorage, Alaska
2 Civil Engineer, Alyeska Pipeline Service Company, Fairbanks, Alaska

During the first geological reconnaissance of these slopes, Alyeska engineers and geologist began to focus on mode changes that would be more appropriate for these steep slopes.

Special burial requires artificial refrigeration to maintain frozen ground to support the pipeline. Further analyses determined that the aboveground mode with thermal piles would be more appropriate in this area.

Crossing Squirrel Creek borings indicated both frozen and thawed soil profiles depending on the distance from the creek. The underlying sands and gravel were thought to be stable but there was some question as to the stability of the intermittent zones that were frozen, especially during an earthquake. For this reason this zone which was originally moded below ground was remoded above ground, again with thermal piles to decrease pile length and promote refreezing of thawed zones. The aboveground mode was originally selected for the north approach to Squirrel Creek and was not changed.

The other design issue was the steepness of the slopes approaching the creek. It was decided to flatten the slopes to no steeper than 27% in the Copper River basin formations. Thus, most of the south approach needed to be cut and a short section of the north slope had to be cut to meet this criteria. The photograph in Figure 4 shows the cut benches on the south approach to Squirrel Creek looking from the short cut section on the north side of the creek.

Because the two slopes approaching Squirrel Creek were a concern to the designers, it was recommended that these two slopes be monitored after construction to confirm that the slopes were performing as planned. Until about 1989, no significant deviations from the performance expected by the designer had been noted.

Deep Thaw

During a slope inspection in 1989, it was noted that one bent had a hanging shoe; the crossbeam had moved down enough that the shoe was no longer being supported as showed in Figure 5. This caused the initiation of a significant research effort by Alyeska. Survey data was studied and modified approaches to the acquisition and reporting of this data were established. A geotechnical drilling program was initiated to determine why the VSM's were apparently settling. The records from infrared surveys of the heat pipes were examined to determine if there had been heat pipe failures.

The research revealed the following:

- The heat pipes were functioning properly. Even in locations where the annual heat pipe surveys have indicated the heat pipes were performing inefficiently, the heat pipes were still removing heat. No VSM settlement as has been related to a heat pipe failure.
- The surveys indicated that many of the VSMs in this area were moving during the year.
  - Some were heaving
  - Some were settling
  - Some were leaning
- The borings and thermistors revealed that there was an ice lens below the depth of the VSM bottoms and that the thaw front had reached the upper part of this ice lens.

- Borings close to the VSM's indicated that there was still a frost bulb around the VSMs.

The bent settled more than one foot before, it was stabilized by adding deep, free-standing heat pipes. The settlement of this bent is inferred in the photograph in Figure 5, which shows the separation of about 0.3 m between the shoe and the crossbeam. At this location, a geotechnical investigation found an ice lense below the bottoms of the VSMs. The ground above the ice lens was thawed although there were significant freeze bulbs still in existence around the VSMs. The continued thawing of this ice lens was halted by installing free-standing heat pipes which extended through the ice zone. It was concluded that the wide zone of vegetation clearing in this area coupled with the fact that this was a south-facing slope, resulted in the development of a deep thaw bulb around the VSMs. The impact of this thawing would not have been as significant if the ice lens had not been under the VSMs. Figure 6 depicts how this thaw bulb likely progressed to the ice lens.

Heaving Thermal VSMs

Another concern developed early at Squirrel Creek. Some shoes were no longer centered on the cross beams and some VSMs were leaning. In addition, wet zones appeared during some periods of the year on the pad and mounding and voids formed around the VSMs. The initial response was to evaluate the impact of these unexpected observations on the integrity of the pipeline system. In some cases the contact between the shoe and beam was not uniform or only one skid of the shoe was touching the beam. In Figure 7, the heave process is depicted as VSM heave is beginning to initiate in mid-winter. There is a frozen bulb around the VSM and the frozen active layer has connected with the frozen bulb. The freeze front has not reached the bottom of the permeable workpad. Water is perched on the low permeability silts and clays under the workpad. It continues to flow downslope being trapped by the freeze fronts from the active layer and the frozen bulb around the VSM. The greater availability of water upslope of the VSM results in downslope, lateral heaving as the water freezes and there is more expansion at the upslope freeze fronts than at the downslope freeze fronts around the VSM. The VSM heaves and leans downslope as shown in Figure 8.

Structural analyses were initiated to determine the impact of the lost of support of a bent on the integrity of the pipe. These analyses showed that the loss of 2 or possibly more adjacent bents would not cause overstressing of the pipe. Therefore it was concluded that the complete loss of two adjacent bents, or possibly more bents, would not endanger pipeline integrity.

Since pipeline integrity was not an issue, the appearance of the VSMs and the shoe contacts were a concern from the standpoint of a deviation from the expectations of the designers. A monitoring program was established that tracked the long term movements of the VSMs and shoe movements.

Initially the VSM movements were attributed to movements of the slope since this slope had been a concern during design and construction and since this slope was designated during design as a slope to monitor throughout the life of the pipeline. Therefore, as soon as VSM movements were noted, geotechnical studies were initiated during which long term instrumentation was installed on the slope to monitor thermal changes, groundwater conditions, and slope movements. Thermistor strings, piezometers, slope indicator casings, sondex settlement casings, and shoe movement instruments were installed. For more than 12 years this instrumentation has been continuously maintained and monitored.

The thermistor strings have shown some slight warming, but no significant areas of thermal state change. Figure 9 shows typical results from a thawed area on the south approach. Since their installation, areas that were thawed have remained thawed and areas that were frozen have remained frozen. The major changes have been in the active layer, as one would expect. The active layer freezes each fall trapping near surface groundwater flowing over the slope as shown in Figure 7. As the active layer freezes, it begins to heave the VSMs, as shown in Figure 7. The trapped water flows to the freeze front resulting in more heave as the front progresses downward. Because the heave is occurring on a slope, the VSMs lean downslope as they heave. Also, as the trapped groundwater freezes in the gravelly workpad capping the slope, it loosens this fill.

When thaw begins at the ground surface the loosen gravel fill becomes soft and wet until it has time to settle and drain. While it is wet and loose it sometimes can have near surface downslope movements similar to the solifluction movements seen in permafrost terrain such as the area shown in Figure 10. Mounds are created around the VSMs as shown in Figures 11 and 12. In some places the thawing results in springs and pipes as the frozen layer becomes thin enough to release the trapped groundwater. See Figure 13.

The inclinometers confirm that the only movement on the south side of Squirrel Creek is the solifluction of the active layer. An example of this movement as depicted by plotting the slope indicator data is presented in Figure 14. At other bents on these slopes the inclinometers have shown limited zones of downslope movements resulting from deep thawing and movements away from growing freeze bulbs.

Analyses concluded that the clearing of this south facing slope had resulted in the development of a deep thaw bulb around and eventually under the VSMs. The solution was the installation of deep heat pipes near the VSMs to prevent further thaw below the VSMs. After 10 years this solution appears to be working.

Liquefaction

The liquefaction criteria used by the designers concluded that saturated granular formations that had been recently frozen became liquefiable upon thawing. Thus, some of the area immediately adjacent to Squirrel Creek would be considered liquefiable when these areas thawed. Bent 12 located about 200 m north of Squirrel Creek falls into this classification.

During design one of the primary methods of mitigating liquefaction was to use thermal VSMs to support the pipe. The final mode selected for crossing Squirrel Creek was the use of thermal VSMs to support the aboveground pipe. The VSM logs through this area showed some locations to be frozen and others to be thawed, warm discontinuous permafrost. A boring in 1992 found loose, saturated granular soils adjacent to Bent 12, which had settled. None of the other borings in this area have identified loose granular deposits. The granular materials overly lean stiff clays. These sands and gravel deposits are thicker at the creek and feather out up the slopes. The deepest zones of these deposits is thought to be 15 to 20 m in thickness.

To stop the settlement of Bent 12, the VSMs were pushed into the ground by statically loading the bent. Steel split rings are welded to each VSM to provide support point for the crossbeam. The split rings were raised until the load of the pipe and shoe were being supported on the crossbeam. As the VSMs settled more, the split rings were adjusted again until the pipe was being supported. When the VSMs had settled about 4 to 6 ft they began to have sufficient resistance to provide support to the pipe as shown in Figure 15. Alyeska continues to monitor this bent and will adjust the split rings to maintain load on the VSMs in the future, if necessary.

Prior to these adjustments, adjacent bents did not settle significantly, but maintained the pipe support taking the increased loads which resulted from the loss of support at Bent 12. Considering this, and gravelly nature of the formation, it has been concluded that there was likely an ice pocket near Bent 12 which probably thawed and resulted in the movement at that bent. As shown in Figure 16, the other adjacent bents not moved nearly as much as Bent 12. Areas of large movement usually indicate areas of settlement or heave. Small movement zones are indicative of stable VSMs.

Although studies continue in this area, the current consensus is that there is no significant liquefaction hazard at this site because:

- The boring was performed at the time active thawing was occurring away from the thermal VSM. Since the initial thawing which was occurring at the time the boring was made, it is believed there has been some consolidation of the coarse granular soils and some increase in strength.
- It is believed that there is a year-around freeze bulb around each thermal VSM that will significantly increase its load capacity throughout the winter months, if not for the entire year.
- The settlement that has occurred is believed to be limited to a relatively small area where deep ice lenses were under the VSM bottoms.
- Squirrel Creek is a high energy stream which has not likely deposited large volumes of the fine sands and silty sands associated with liquefaction.
- Because of the free draining properties of the materials adjacent to this stream, it is expected that any adjustment during a seismic event will be not be large enough to move VSMs significantly.

Summary

The two slopes approaching Squirrel Creek were an engineering challenge during the design and construction of the Trans Alaska Pipeline. There were many concerns during design because of the limited amount of available geotechnical data and the lack of reported experience from others on the performance of steep slopes in a discontinuous permafrost environment. Since construction, much has been learned through observations and monitoring. There have been no massive slides as some had thought may occur. The granular workpad, when it becomes loose and saturated in some areas can move in the spring and summer. These movements make maintenance access more difficult for limited periods but do not appear to have more substantial impacts.

There has been some movement of VSMs from both settlement and heave. In both cases the amount of movement does not appear to have adversely impacted the integrity of the aboveground pipeline system. The slopes continue to be monitored and maintained and much more will be learned about the performance of steep slopes in permafrost in the future by Alyeska engineers.

Acknowledgement

There authors appreciate the support and assistance provided by Alyeska Pipeline Service Company in the preparation of this paper. All of the work reported in this paper was developed as part of Alyeska's ongoing work on these slopes. Special thanks to Mr. Elden Johnson and Mr. David Norton who were instrumental in making the information presented available .

Figure 1. Squirrel Creek Crossing

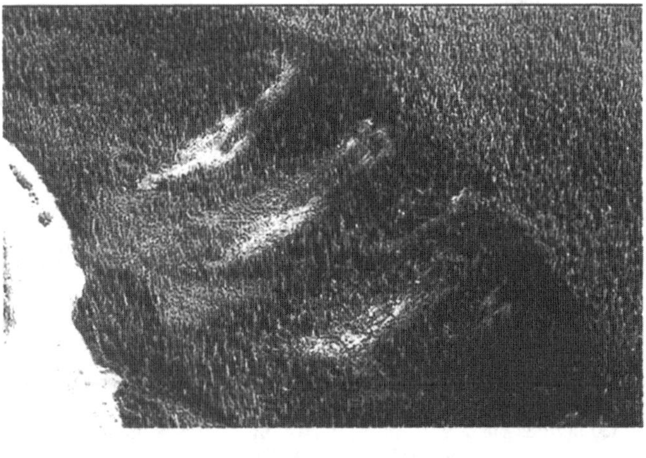

Figure 2. Natural Slides in Squirrel Creek Area

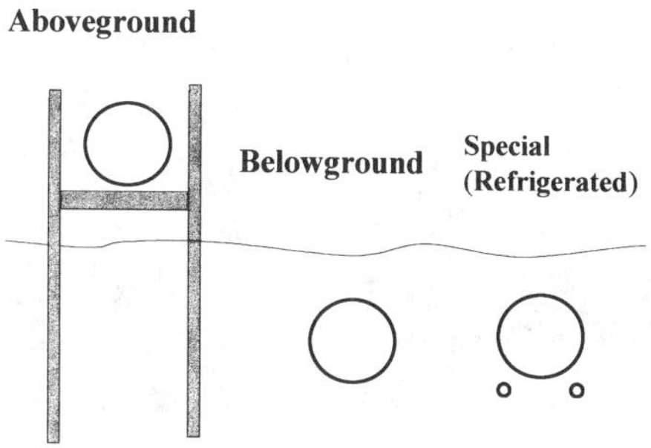

Figure 3. Initial Pipeline Modes at Squirrel Creek

Figure 4. Benches on South Slope

Figure 5. Hanging Shoe

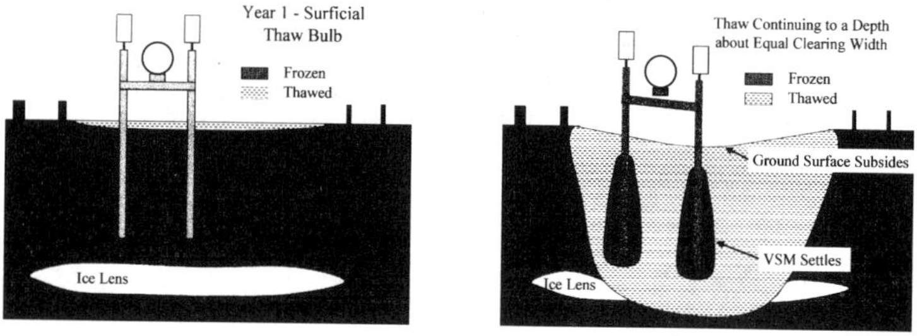

Figure 6. Thawing Around VSMs

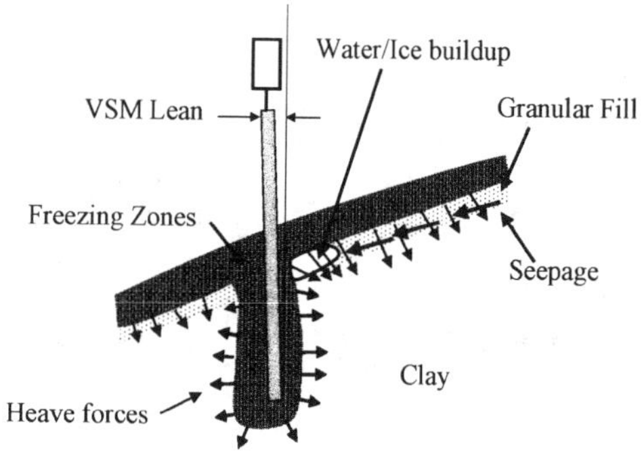

Figure 7. Heaving on a Slope

Figure 8. Leaning VSM

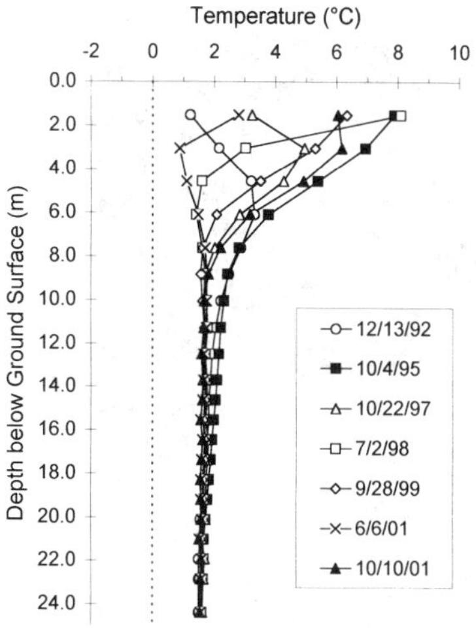

Figure 9. Long Term Temperature Results

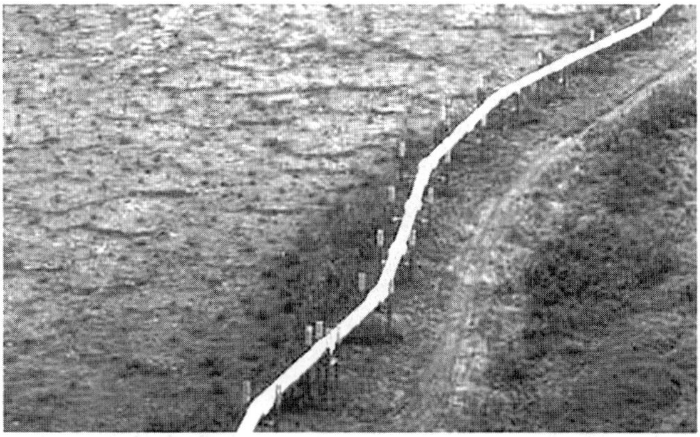

Figure 10. Solifluction Lobes Along TAPS

Figure 11. Mounding Around VSM

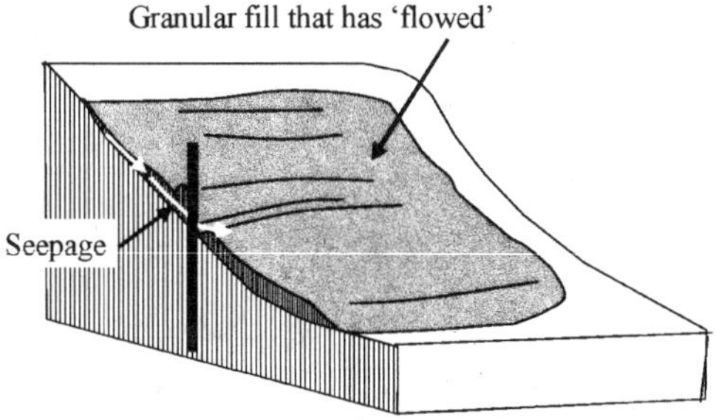

Figure 12. Skin Flow Around VSM

Figure 13. Piping Through Workpad

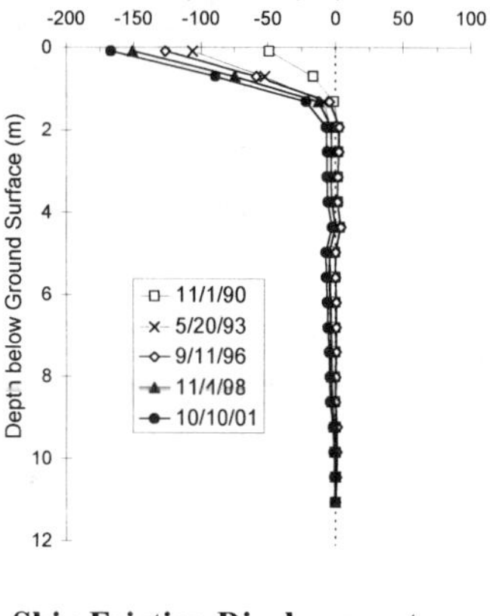

Figure 14. Skin Friction Displacements

Figure 15. VSMs That Have Settled

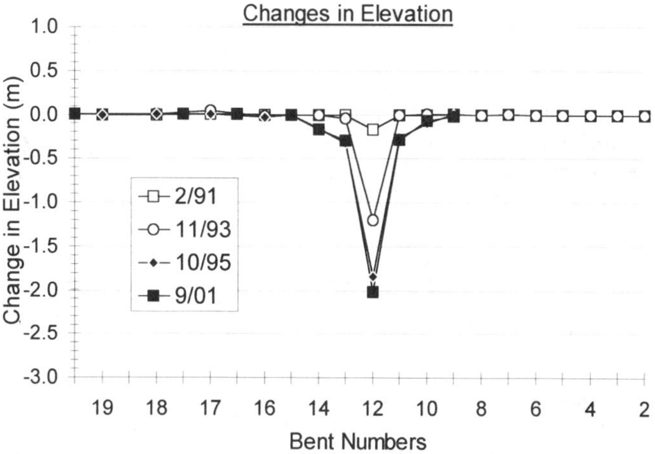

Figure 16. VSM Elevation Changes

Analytic Model for the Evaluation of Inclined Piling in Permafrost

Steven Sorensen P.E. [1]
Keith Meyer Ph.D., P.E. [2]
Paul Carson P.E. [3]

Abstract

There are approximately 78,000 Vertical Support Members (VSM) on the Trans-Alaska Pipeline System (TAPS). A small fraction of VSM have tilted over time, primarily as a result of frost heave. Corrective maintenance to address this situation, which can include VSM replacement, is very expensive. As an alternative, a model was developed to determine if, and when, VSM integrity would be compromised beyond acceptable factors of safety to the point where the pipeline foundation would be jeopardized. The model was finite element based, using non-linear soil elements to simulate the subsurface resistance. This paper reports on the model and a summary of the results.

Introduction

The 1290-km Trans-Alaska Pipeline System (TAPS) traverses Alaska from Prudhoe Bay on the North Slope to its southern terminus, the Port of Valdez. Approximately 680 kilometers of the pipeline was built aboveground to avoid burying a warm oil pipeline in areas of thaw unstable permafrost. A typical pipeline support bent is comprised of two 46-cm diameter pipe pilings, referred to as Vertical Support Members (VSM), that support a cross beam and shoe assembly which in turn support the pipeline, see Figure 1. The shoe assembly can slide laterally and longitudinally on the cross beam with a static coefficient of friction of 0.1. Bents are spaced approximately 18 meters apart along the elevated portion of the pipeline.

[1] Structural Engineer, Alyeska Pipeline Service Company, 701 Bidwell Street, Fairbanks, AK 99701; phone 907-787-5582; sorensensp@alyeska-pipeline.com
[2] Vice President, Alaska Operations, Michael Baker Jr., Inc., 4601 Business Park Boulevard, Suite 42, Anchorage, AK 99503; phone 907-273-1649; kmeyer@mbakercorp.com
[3] Structural Engineering Lead, Alaska Operations, Michael Baker Jr., Inc., 4601 Business Park Boulevard, Suite 42, Anchorage, AK 99503; phone 907-273-1616; pcarson@mbakercorp.com

The construction specifications for TAPS allowed a one-percent deviation from vertical for installation of the VSM. Three-percent tilt is allowed as an operational constraint prior to mandatory reporting. Over the 22 years of operation, a small fraction, 250 out of 78,000 VSM have tilted from vertical between three and eighteen percent. The majority of tilting VSM are associated with areas of fine grained soils, high moisture content active layer, and thawing of the deeper regions of marginally frozen or warm permafrost. The general mechanism of tilting begins with warming of the marginal permafrost soils and loss of adfreeze bond between the VSM and surrounding soils. Frost heave during late winter pushes the VSM upward. Because of eccentricities in the VSM to cross beam connection, or longitudinal and lateral forces from pipeline gravitational forces on slopes, the VSM tends to tilt slightly as it is heaved upwards. As freeze thaw cycles continue from year to year, tilting becomes more pronounced.

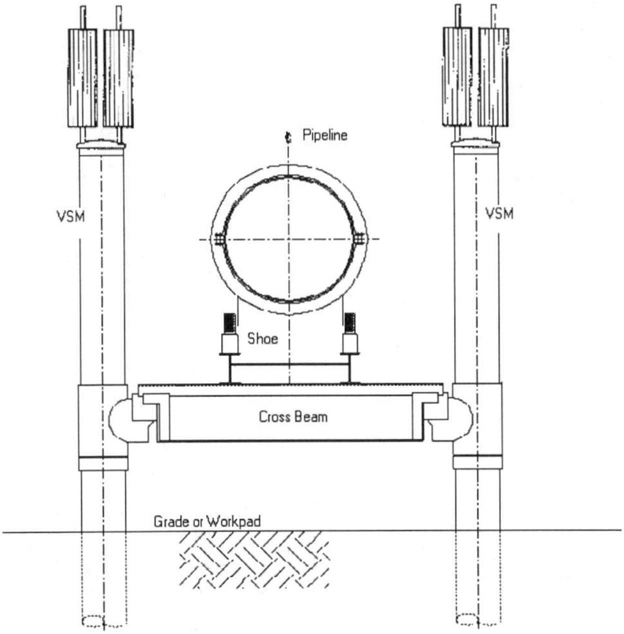

Figure 1. An Aboveground VSM Support Assembly

Route surveillance has recorded several VSM that have continued to tilt over time. Some VSM have been replaced because of appearance, however no VSM ever failed from tilting. The cost of replacing VSM is very high because of the remote nature of the pipeline and because so few VSM are affected over a wide area, reducing construction efficiencies. As an alternative, an analytic tool has been developed to determine if, and when, VSM integrity would be compromised beyond acceptable factors of safety to the point where pipeline foundation would be jeopardized. The model was finite element based, using non-linear soil elements to simulate the subsurface resistance.

Problem

VSM performance became a concern at several locations along the pipeline when surveillance over many years showed an increasing trend in both the number and in the degree of tilt greater than three percent.

As a function of the original design, loss of two consecutive supports could be tolerated without significant risk to pipeline integrity (ref. 1). Additional analysis of the pipeline under loss of support scenarios effectively demonstrated that under normal operational conditions, loss of several supports could be tolerated in the short term (ref. 3). A review of the lateral load criteria for VSM design, using conservative allowable soil strengths, demonstrated that the original support system design was conservative and robust (ref. 2). However, the ability of tilting VSM to support loads was not well defined. Some VSM were replaced because of appearance and without clear understanding of their load carrying capacity.

In an effort to define operational performance of tilting VSM with respect to the load requirements established in the original Alyeska Pipeline Service Company (APSC) design basis, a more definitive model of VSM behavior in four typical soil types was needed. "Pile-PC®" is a digital analysis program utilizing large-deflection concepts and non-linear soil resistance functions for the analysis of single piles. It utilizes Microsoft® Excel® as the user interface for both the input and output. The user inputs the required information into the Microsoft Excel input sheet, and then the analysis is initiated based on these values. The analytical program is written in Visual Basic for Applications (VBA) and the code itself is appended to the Microsoft Excel workbook – i.e., no separate code application installation is required.

Analytical Approach

There are two aspects of this problem that require advanced structural considerations. The first is the effect of the compressive load "P" on the pile acting through the lateral displacement "Δ". This "P-Δ" effect causes an additional moment on the pile in addition to the lateral load acting through the pile stick-up. For small displacements, this can often be ignored, but for this problem which examines the structural state to relatively large tilt angles, it is required. The second effect is the non-linear resistance of the soil, especially the lateral soil resistance. To examine these effects for the large number of cases involved, an easy-to-use analytical program that incorporates both these effects was required. The finite element approach was selected as the analytical basis. It was decided to implement this approach fully in Microsoft Excel to avoid the development of a new user interface and to facilitate its use by engineers.

In the Finite Element approach to structural analysis, the structure is idealized into an assembly of structural elements. The elements are attached to adjacent elements at node points, which may be actual joints or fictitious points obtained by subdividing a structural member into a number of smaller members. To determine the structural characteristics of the entire assemblage of elements, first it is required to ascertain the stiffness of each individual element, i.e. the response of the unassembled elements to the various types of loadings the individual element will or could resist.

There are a variety of well-known techniques for deriving the relationships between the force applied to an individual, unassembled member and the deflection of the member resulting from the application of the force. Of these methods, the simplest is the unit-displacement method wherein a displacement equal to unity (units are implied by the type of deformation) is applied to the individual element and the resulting force response is found. This procedure leads directly to the required matrix equations relating element forces to their corresponding displacements (ref. 4).

Two-Dimensional Elastic Small-Deflection Beam Formulation

The relationship between the forces and displacements on a two-dimensional (planar) beam can be fairly easily derived from elementary beam mechanics, based on small deflection theory. In this formulation, there is a linear relationship between the applied forces **P** and the displacements **U**, i.e., $\mathbf{P}=\mathbf{K_E U}$, where $\mathbf{K_E}$ is a matrix of constants relating the forces to the displacements under the assumptions of elasticity and small deflection theory.

The final structural matrix formulation for the two-dimensional beam is given below, while Figure 2 shows all the implied structural actions.

$$\begin{Bmatrix} F_1 \\ F_2 \\ F_3 \\ F_4 \\ F_5 \\ F_6 \end{Bmatrix} = \begin{pmatrix} AE/L & & & & & \\ 0 & 12EI/L^3 & & \text{(Symmetrical)} & & \\ 0 & 6EI/L^2 & 4EI/L & & & \\ -AE/L & 0 & 0 & AE/L & & \\ 0 & -12EI/L^3 & -6EI/L^2 & 0 & 12EI/L^3 & \\ 0 & 6EI/L^2 & 2EI/L & 0 & -6EI/L^2 & 4EI/L \end{pmatrix} \begin{Bmatrix} u_1 \\ u_2 \\ u_3 \\ u_4 \\ u_5 \\ u_6 \end{Bmatrix}$$

$$\mathbf{P} = \mathbf{K_E} \quad \mathbf{U}$$

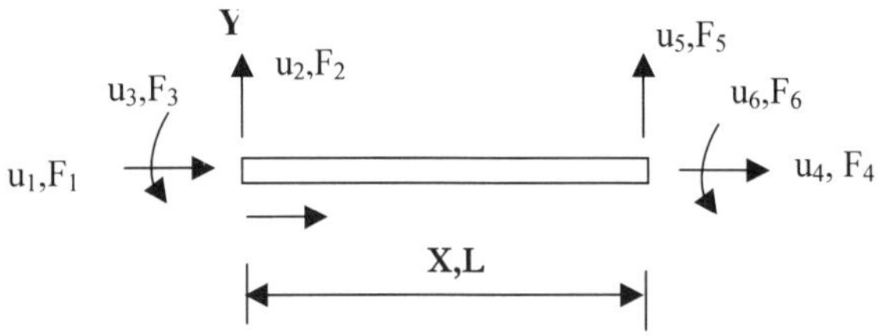

Figure 2. General Beam Force-Displacement Relations

Extension of Beam Formulation for Large Deflections

When large deflections are considered, the linear relationship of $P=K_EU$ discussed in the previous section can no longer be used. To treat the effects of large displacements, the solutions for U are obtained in steps, with the relationship between the forces and displacements updated in preparation for further steps. In addition to the K_E matrix that relates P and U, we add the effects of large deflections with the matrix K_G so:

$$\mathbf{P= (K_E + K_G)U}$$

K_G is known as the "geometric stiffness" matrix. K_G is initially set to the null matrix.

The values for the geometric stiffness are derived by considering higher-order terms in the relationship between the strains induced in the member and the displacements of the member. The small-deflection strain-displacement relation for the axial stress is:

$$\varepsilon_{xx} = \partial u_x/\partial x$$

When the next term in the expansion for this strain is added the relation is:

$$\varepsilon_{xx} = \partial u_x/\partial x + ½(\partial u_y/\partial x)^2$$

Note that, as expected, when the displacement u_y is small with respect to the length of the structure under consideration, the second term does not contribute a significant amount. For example, if the displacements are on the order of 1% of the length, then we can approximate the effect of the second term as on the order of $(0.01)^2$ / 0.01, or about one-hundredth, the effect of the first term. This relatively small influence can be, and is, often neglected in structural evaluations. Using this strain-displacement relation in the development of the beam force-displacement relations results in the following extra terms:

$$\mathbf{K_G} = \begin{pmatrix} 0 & 0 & 0 & 0 & 0 & 0 \\ 0 & T/L & 0 & 0 & -T/L & 0 \\ 0 & 0 & 0 & 0 & 0 & 0 \\ 0 & 0 & 0 & 0 & 0 & 0 \\ 0 & -T/L & 0 & 0 & T/L & 0 \\ 0 & 0 & 0 & 0 & 0 & 0 \end{pmatrix}$$

where T is the tension in the beam element. This formulation for $\mathbf{K_G}$ is known as the "string stiffness." Even higher-order formulations are possible, although this addition often suffices for most structural calculations.

If the beam is in tension, the term has the effect of "stiffening" the beam in response to lateral forces. Thus, this formulation is required for the analysis of cable systems, which develop resistance to lateral loads by tensioning the cable members. Conversely, if the tension in the beam is negative, the term has the effect of weakening the system. The consideration of this term is required for a P-Δ analysis for beams, which considers the effect of the axial load acting through the lateral beam displacement, causing increased moment.

To use this formulation, note that the value of T is not initially known. Therefore, the analysis first completes the initial structural evaluation assuming that $\mathbf{K_G=0}$. The first evaluation thus uses only the elastic, small deflection, formulation. Once complete, an initial estimate of the value of T is known, and can be used to form the first estimate of $\mathbf{K_G}$. This new $\mathbf{K_G}$ is added to the elastic matrix (which itself is constant), and the analysis re-run to form an updated estimate of T and thus an updated estimate of $\mathbf{K_G}$. This process is continued until convergence, i.e., until the value of T used to form $\mathbf{K_G}$ is equal to, or within some specified tolerance of, the value of T that results from the analysis. Alternatively, small steps can be taken in the loading so that $\mathbf{K_G}$ can be evaluated at the end of each load step and used as the starting point for the next load step. Another solution strategy is to combine both techniques and execute small load steps and iterate until convergence at the end of each of these incremental steps. This latter strategy is the solution method used by Pile-PC.

Soil Springs Formulation

The soil spring formulation is easily derived from elementary mechanics. For the soil spring elements in Pile-PC, it is assumed that one end of the spring is fixed and the other is attached to a node on the beam. As the beam deflects, the soil reacts through a spring constant "k" as shown in Figure 3.

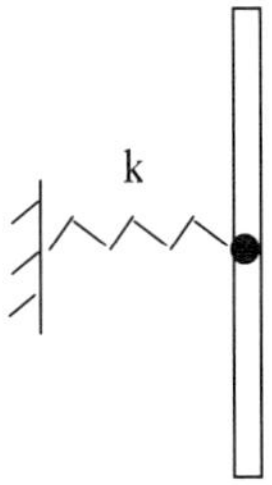

Figure 3. Soil-Springs

Thus, the matrix form is identical to the simple one-dimensional equation:

$F_s = ku$

Where

F_s = Soil Resistance to the beam deflection

k = Effective stiffness of the soil

u = Beam deflection

Note that the figure depicts lateral soil springs, but that the same formulation holds for axial soil springs. Both lateral and axial soil springs can be modeled in Pile-PC.

The value of "k" is not required to be constant throughout the analysis, i.e., the soil stiffness may be assumed non-linear. In particular, a value of "yield" is input denoting the end of the initial linear soil response of $F_s = k_1 u$. Thereafter, the soil is assumed to follow another different linear path $F_s = k_2 u$, which simulates the drop-off in soil strength following yield as shown in Figure 4.

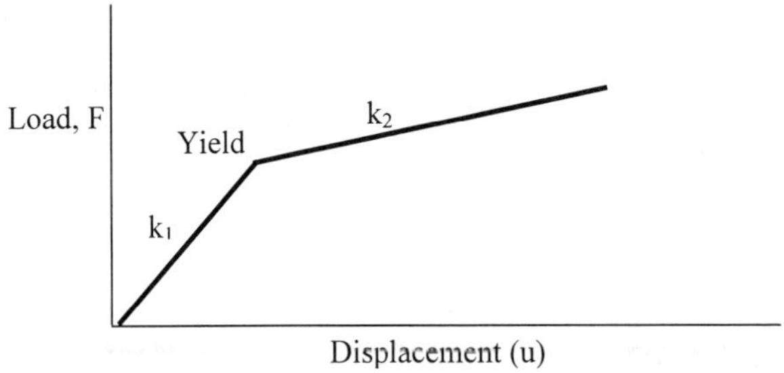

Figure 4. Soil-Spring Load Deflection Relation

In the Pile-PC formulation, the load-deflection equations for the soil springs are elastic even though they are bilinear. That is, the soil is assumed to unload along the same path that it loads. This formulation is appropriate for increasing loading situations, although should be used with care where the load history is assumed to reverse.

Global Formulation

Once the stiffness properties of the elements are formulated, the individual element matrices must be assembled to represent the overall structural matrix. This is done by uniquely assigning node numbers to all joints or divisions of the elements that make up the structure and then using proper matrix addition to add the effects of the individual elements to the overall or "global" matrix. The solution is given by $\mathbf{U} = \mathbf{K}^{-1}\ \mathbf{F}$, where $\mathbf{K}^{-1}$ is the inverse matrix of $\mathbf{K}$. In Finite Element practice, the inverse of a matrix is rarely found since the computations are extensive. Alternatively, the system of equations are solved simultaneously by first decomposing the N by N structural matrix into an equivalent system composed of an upper and lower matrix product, which can then be easily solved.

The matrix evaluation results in the initial estimate for all of the displacements of the structure. Using these displacements, the individual elements are reassessed using the new information. For the beams, the initial estimate of the axial force, if any, can be found and used to formulate each beam element geometric stiffness which, when added to the beam elastic stiffness matrix, forms the estimate of the element combined stiffness for the next iteration. Similarly, the displacements in the soil springs are evaluated to see if the soil stiffness is beyond the yield value and, if so, the new value of the soil stiffness is substituted for the next load step. Using these new individual element matrices, the global stiffness matrix is reassembled and the analysis continued until convergence to a specified tolerance is achieved.

Implementation

The VBA program reads all the data from the input sheet, and automatically divides the structure into a number of smaller elements. Once the data is further checked, the more formal finite element part of the program activates to formulate the individual member stiffnesses, assemble the global equations, and steps through the loading. Once complete, a new output worksheet is made in the same Microsoft Excel workbook, and all displacements, member forces, stresses, and soil resistances are output. A plot of the deformed pile shape is automatically inserted on the output worksheet, before the program terminates and control is transferred back to the user. At this time, the user can perform any further Microsoft Excel functions to investigate this case, or use the results from a number of output worksheets to plot trends.

As an example of the utility of this program, a further Microsoft Excel worksheet was developed which stepped through a series of initial tilt angles, each case automatically filling out the input sheet and producing a separate output worksheet until all cases were completed. The trend of increasing the tilt angle was then automatically plotted.

Although many analytical programs accept Microsoft Excel worksheets as input for their programs, the idea of developing an advanced analytical tool for finite element analysis working completely within Microsoft Excel is novel. It avoids the need for specialized user-interfaces and installation techniques. More importantly, it places an advanced tool within a widely distributed and used program, and makes the powerful Microsoft Excel functions available to more easily analyze and work with results. A further step in the implementation will be the development of a plasticity module to allow full inelastic analysis.

Results

A series of structural analyses were conducted using the finite-element program. The analyses were initially conducted using plumb VSM with a six-meter embedment in each of the four soil types described in the APSC Design Basis. The analyses consisted of placing a vertical and lateral load on a VSM as shown in Figure 5 with an initial Top of Beam to Top of Pad (TOB-TOP) height. The loading is ramped until either the soil around the VSM failed or the stress in the steel exceeded AISC allowable stresses. This was completed for TOB-TOP heights of 0.6 to seven meters in increments of 0.9 meters and the deflections calculated at the completion of load or at "failure" for each combination. These deflection values were converted to percent tilt and used to create a series of graphs defining the maximum anticipated VSM tilt at various TOB-TOP heights based solely on a "load-driven" mechanism. After completing the initial analyses, two things became apparent:

1. The three- percent value being used to flag tilting VSM is justified. This is approximately the lower limit of tilt calculated for all combinations of load and TOB-TOP height that resulted in stresses less than the AISC allowable level (i.e., tilt at the maximum AISC allowable stress were as low as three percent for some combinations analyzed).

2. The analyses show that the magnitude of actual field observations of tilting VSM cannot be attained for reasonable combination of loads (i.e., the analyses show that the loads required to produce higher percent tilts are of such a magnitude that other aboveground hardware components would have already probably failed). This led to the conclusion that the majority of the field observations must be a result of geotechnical movement rather than bending deflection.

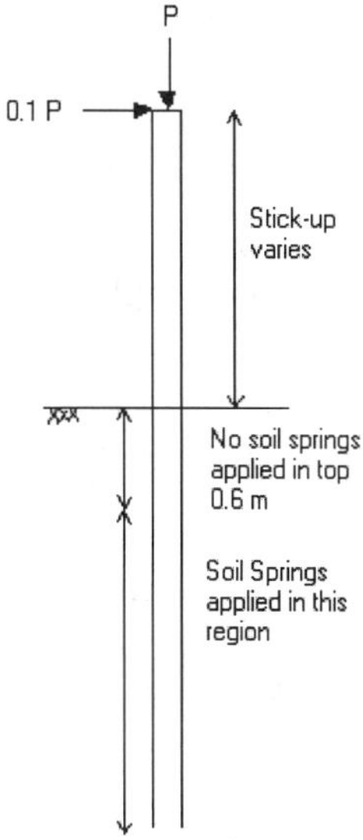

Figure 5. Pile Geometry Analyzed

The second series of analysis was completed assuming an initial tilt angle. This is more representative of actual field conditions. The final product was a series of curves, illustrated in Figure 6, for each soil type and embedment depth combination analyzed. These curves define the percent tilt and TOB-TOP height combinations at which "failure" is reached for a set VSM load. "Failure" is defined as reaching the soil resistance limits or an overstress condition in the steel based on AISC allowables.

Typically, the operational criteria for maintenance at three percent tilt is conservative to ensure against VSM overstress or soil failure. To determine if corrective maintenance on VSM exhibiting tilts greater than three percent is necessary, an evaluation is required. VSM may be evaluated initially using the graphs produced from the second series of analyses described above. If VSM do not pass this initial evaluation, more intensive assessment is required and site-specific information must be obtained that includes current soils information and temperature data. The VSM can then be re-evaluated using the finite element analysis model with the site-specific information. Based on the results of this re-evaluation further maintenance and/or monitoring action can be initiated.

The weak thawed soil charts for the nearest corresponding embedment length were used for initial evaluations on 92 tilting VSM. By plotting the intersection of the tilt and TOB-TOP on the appropriate chart, a comparison of the recorded load to the allowable load was made. This initial evaluation determined 83 of these VSM to be adequate based on the information gathered. Additional information is being gathered for the remaining tilting VSM as of printing of this paper.

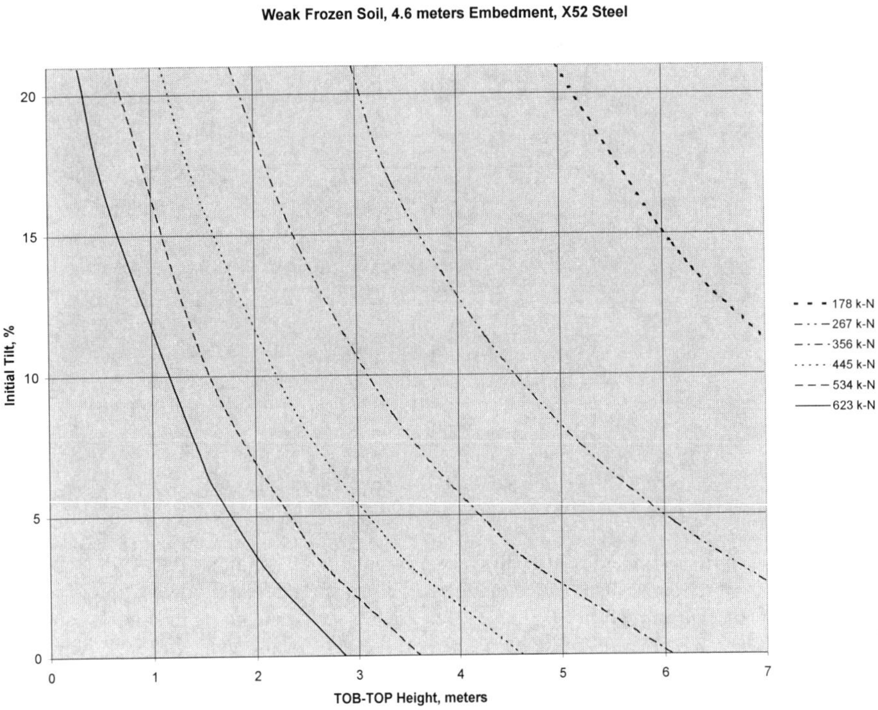

Figure 6. Example Results plot

Conclusions

The finite element model has allowed APSC to perform an initial assessment of tilting VSM based on information from as-built data and the tilt angle. If the VSM fails this initial screening evaluation, an additional assessment can be initiated using the finite element model and current geotechnical information gathered from the field. Since the majority of tilting VSM are found in weak frozen or weak thawed areas, the failure mode is overturning and not overstress of the pile. Of the 78,000 VSM installed, 250 exhibit tilt in excess of three percent. Of the tilting VSM studied to date, approximately ten need further evaluation using current site-specific data.

References

Alyeska Pipeline Service Company (1973), "Summary Report, Design Criteria and Stress Analysis of Vertical Support Members," September 14, 1973

Alyeska Pipeline Service Company (1976), "Lateral Load Criteria for Aboveground Supports"

Alyeska Pipeline Service Company (1977), "Aboveground Pipeline Response to Settlement of Vertical Support Members"

Przemieniecki, J.S. (1968), *Theory of Matrix Structural Analysis*, McGraw-Hill Company, New York, NY.

Performance of the PVR Rock Slope At Valdez Marine Terminal

Graeme Major[1], Rupert G. Tart, Jr[2]., and Deborah J. Krontz[3]

Five areas within Alyeska Pipeline Service Company's (Alyeska's) Valdez Marine Terminal (VMT) have steeply cut rock slopes that have been reinforced with rock bolts and monitored since the VMT was constructed. Monitoring has included regular reading of pneumatic piezometers that were installed in the slopes during construction. Flag levels were established for each piezometer on these slopes based on stability calculations or historical water levels.Annual reports are prepared to compare groundwater levels at each piezometer during that year to the established flag levels. If flag levels are exceeded, engineering reviews are made to assess the potential impacts of these excursions. Some piezometers have required maintenance and repair, and some rockbolts have required maintenance to maintain them to the standards assumed by the slope designers. Scaling and localized shotcreting have be required in some areas to control the potential for rockfall. This paper addresses the history of one of the VMT slopes, presents some of the long-term monitoring data, discusses the meaning of the data, and describes the maintenance activities that have taken place on the slope.

Introduction

The VMT was constructed from 1975 through 1977 on a series of rock cut excavations and benches, and rockfill benches. The larger facilities, including oil storage tanks, are generally located on bedrock foundations on the excavated benches. The Power and Vapor Recovery (PVR) Slope is the excavated rock slope immediately south of the Power and Vapor Recovery plant. It is the largest rock cut at the terminal and has been continuously monitored since it was completed. It has required periodic maintenance, and has been the subject of several engineering evaluations. Although other slopes at the VMT require monitoring and maintenance, this paper focuses on the monitoring and continuing evaluations of the PVR Slope.

A large failure developed during the initial excavation of the slope in 1975, as shown in Figure 1. The failure occurred largely along foliation planes of the phyllitic bedrock, although it was apparently initiated following exposure of some high-angle, cross-cutting structures near the toe of the excavation. A program of engineering evaluation and design was implemented to mitigate the risks of possible future slope failures. The final slope design included reinforcement of the phyllite slope with 30-foot long, 1-inch, resin anchored, fully grouted, tensioned high strength Dywidag threadbars. Lock-off loads of 60 kips were designed to produce an adequate safety factor against failure for the slope

---

1 Principal, Golder Associates Inc., Reno, NV
2 Principal, Golder Associates Inc., Anchorage, AK
3 Civil Engineer, Alyeska Pipeline Service Company

under static conditions for the documented geological and groundwater conditions. The bolt capacity between the lock-off load and the full design capacity of 100 kips is available to accommodate earthquake loads. The final slope design included:

- Excavating the slope to parallel the foliation planes,
- Rockbolting the slope to increase stability under static conditions and earthquake loading,
- Constructing a buttress at the toe of the slope to enhance toe stability and for rockfall protection, and
- Installing a groundwater monitoring system of pneumatic piezometers to enable confirmation that piezometric levels do not exceed those assumed for the slope design.

Alyeska has monitored the slope since construction, and has performed periodic maintenance. Maintenance has included regular scaling and cleanup of loose rock as necessary. Two major rock bolt replacement and re-tensioning programs were performed in 1992 and 2000. Rockbolts were replaced or re-tensioned in areas where weathering and degradation of the slope surface had removed the rock support from below the rockbolt bearing plates. This typically occurred in the vicinity of fault zones, where the sheared and altered rock is more susceptible to the effects of weathering. A coating of fiber-reinforced shotcrete was applied to some weathered zones to stabilize the slope by protecting the rock against further degradation.

Slope monitoring has included continual visual reconnaissance of the slope by terminal personnel, and regular detailed inspections of the slope by experienced rock engineering professionals. The monitoring has also included the regular measurement of groundwater levels behind the slope. The remainder of this paper focuses on this groundwater monitoring program and its results.

Pneumatic piezometers

A total of 15 Glötzl pneumatic piezometers were installed behind the PVR slope during construction to monitor groundwater pressures. Each piezometer was installed in a specially designed boring with a sand backfill around the piezometer in the sensing zone. Continuing reviews of the piezometer monitoring system have indicated that 14 of the original piezometers are still functioning reliably after some 25 years of service. Locations of the PVR slope piezometers are shown on Figure 2. The intake and return lines for piezometers at the PVR slope have been terminated in manifold boxes that are protected from snow and rockfalls, and enable readings to be taken from convenient locations.

Flag Criteria

Alyeska assigned "Flag" piezometric levels for each piezometer in the PVR cut based on stability calculations representing site conditions during a contingency earthquake. Flag levels are defined as a total head above the site datum. Total head at each piezometer is the sum of the elevation head (the height of the piezometer above the site datum), and the pressure head that is measured by the piezometer.

Stability analyses used to develop flag levels are based on average piezometric levels through a slope section, so individual high piezometric readings do not necessarily indicate a stability or safety problem. Also, the volume of water retained in the joints (which are generally developed along the foliation planes) is limited by the size of the joints. This low fracture porosity resulting from discontinuous foliation joints, as illustrated in Figure 3, is suspected of preventing the effective drainage of the slope after the installation of the piezometers. No significant water flow or reduction in the piezometric pressure was noted when the slope designers first tried to reduce piezometric pressures with drainage. It is probable that any movement of rock blocks bounded by foliation or other joints would immediately reduce any locally high groundwater pressures. It is also probable that some of the joints are not hydraulically connected to other joints in the rock mass resulting in local, short-term high water levels which are not representative of the rock mass.

If the average piezometric levels in an area remain higher than the flag levels for a significant time, or continue to rise, a review of site conditions would be warranted before taking any remedial action such as drainage enhancements. To date this has not occurred on the PVR slope.

Piezometer Monitoring Results

Piezometers have been monitored regularly since they were installed. Flag levels and recent (September, 2000) piezometric elevations for the piezometers installed in the PVR slope are shown Table 1. The column "Flag minus Head" represents the amount by which the current reading is below (positive) or above (negative) the flag level.

Plots of individual piezometer measurements and annual averages of piezometer monitoring data are maintained for each piezometer from the time of installation until the present., which also show the flag levels. Plots of typical results for Piezometer 40 are presented in Figures 4 and 5.

Readings taken from 1975 through 1983 appear to be generally more consistent and are likely more reliable than those taken from about 1983 through 1994. Pronounced variability in measured piezometric levels for a number of piezometers during the period approximately 1983 through 1994 is interpreted to result largely from operator procedures and/or poor instrument performance. Substantially decreased variability in monitoring results is evident subsequent to 1994, when improved monitoring procedures and the exclusive use of a new Glötzl readout unit were implemented.

An evaluation of piezometer readings was undertaken to evaluate trends in changing groundwater conditions, and to identify problems with the pore pressure cells or measuring systems. The evaluation was undertaken by area because if significant variations in groundwater levels occur, they would influence and be detected by more than one piezometer within an area. This approach helps to identify individual anomalies that may result from instrument errors, or from local conditions that are not representative of a larger area that may be significant for slope stability. Important items addressed in this evaluation included:

- Performance of individual piezometers
- Historic trends and possible temporal trends
- Significance of monitoring results relative to flag levels

These items are discussed in the following sections.

Performance of Individual Piezometers

Field performance checks of pneumatic piezometers included checking the Inlet and Return lines for leaks. In general, both the field checks of the instrumentation, and the consistency of the recorded monitoring data, indicate that the piezometer monitoring system has performed reliably since it was installed more than 20 years ago. However, leaks have developed in some lines of the pneumatic system, or at the pneumatic piezometer diaphragms, and have affected some piezometer measurements.

Where the leak is only in the Return line, this should not generally affect the measurement taken using the Inlet line. Where the leak is only in the Inlet line, it is generally possible to obtain a valid reading by pressurizing the Return line. When leaks are indicated for both lines, which could also be caused by a leaking diaphragm, the accuracy of any estimate of piezometric head is unknown, but is probably reliable if the leakage rate is small. On the PVR slope the Return line was leaking on P-47 which had no impact on the readings except that the Return discharge could not be observed. Inlet line leaks on P-45 resulted in erroneous readings; subsequent readings made using the Return line for input produced more reasonable results. In addition, leaks that required repairs were noted for a number of lines for the piezometers monitored from the manifold boxes.

Historic Trends

Since 1994, readings have been mostly consistent and generally similar to the pre-1984 readings. In a limited number of cases, there appears to be a change in piezometric level since construction, although such changes are generally small. Table 1 summarizes piezometric levels for piezometers that have been functioning reliably since construction., based on the "Average Annual Level" plots for each PVR piezometer. It is evident that most changes between the pre-1984 and current readings are small.

Possible Temporal Trends

Groundwater levels are often affected by short-term and long-term precipitation patterns. Annual precipitation at Valdez from 1973 through 2000 has varied between approximately 45 inches and 90 inches, averaging approximately 68 inches during this period. Annual rainfall at Valdez is shown in Figure 6. There appears to be a generally rising trend from the time of construction through about 1993. However, since that time, precipitation has been near or below average.

Total precipitation of 62.4 inches in 2000, the most recent monitoring year, was a little below average; precipitation was a little above average at 72.4 inches in the preceding year. Rainfall has generally been close to average over the past four years, so measure groundwater conditions over that period are likely a reflection of average groundwater level variations at the site.

As noted previously, identifiable changes in groundwater levels are generally small, seldom more than a few feet over a period of more than 20 years.

Significance of Excursions Above Flag Levels

Flag levels for the Power and Vapor Recovery Cut are based on stability analyses that include a contingency earthquake. This implies that slope stability may be compromised if piezometric levels exceed flag levels during a design-basis earthquake. Out of the 14 functioning piezometers in the PVR cut, the September 2000 monitoring identified the following piezometric levels at or above their flag levels.

The piezometric level at Piezometer 47 was always close to the flag level from the time of installation, but appears to have risen by approximately eight to ten feet in the more than 20 years since construction. The piezometric level at Piezometers 40, 49, and 50 were also always close to the flag level, and have risen approximately two to six feet since construction. These excursions are only local phenomena of limited magnitudes; the average piezometric levels throughout the slope are well below the flag and design levels as shown in Table 1. The slope is therefore considered to be in general compliance with the design-basis groundwater assumptions.

General Conclusions on Groundwater Conditions

The piezometer monitoring system has mostly performed well over the more than twenty years that it has been installed. Piezometer readings taken using suitable standard procedures with properly calibrated instruments indicate that most piezometer installations show little change in groundwater levels through the years. The minimal relative changes can be seen in the piezometer levels shown at the two cross section locations shown on Figure 2. These cross-sections with piezometric levels and flag levels shown are presented in Figures 7 and 8. These show that recent groundwater levels are generally close to the lowest levels measured at the time of construction; there has generally been little change in overall groundwater levels at the site since that time.

Most piezometers show little or no systematic seasonal variation in groundwater levels. Where seasonal variations may exist, they are generally small and of little practical significance, since they do not lead to excursions above flag levels. Apparent long-term changes in groundwater levels are indicated by some piezometers. However, the magnitudes of such changes are generally small, mostly only a few feet over a period of more than 20 years. There appears to be a trend toward slow increases in groundwater levels near the base of the PVR Cut.

Our reviews of monitoring data have concluded that piezometer measurements taken using proper procedures demonstrated that most piezometer installations showed little change in groundwater levels since the time the facilities were constructed, and that most piezometers show little or no systematic seasonal variations in groundwater levels. Four of the 14 functioning piezometers in the Power and Vapor Recovery cut indicate groundwater levels above flag levels. Three of these are between 0.3 and 3.3 feet above flag level, while the fourth is 9.1 feet above flag level. However, the average piezometric elevations in the PVR cut have remained below flag levels. The slope is therefore considered to be in compliance with the assumptions used for stability analyses on which the design is based.

Acknowledgements

There authors appreciate the support and assistance provided by Alyeska Pipeline Service Company in the preparation of this paper. The work reported in this paper has been developed as part of Alyeska's continuing maintenance and monitoring of this slope. Special thanks to Mr. Pat McDevitt and Mr. David Norton who were instrumental in making the information presented available. The photograph in Figure 1 was provided by Mr. Thomas Krzewinski.

Table 1

September 2000 Piezometer Summary

Piezometer Number	Elevation Head (ft)	Pressure Head (ft)	Total Head (ft)	Flag Level (ft)	Flag Level minus Head(ft)	Change since Pre-1984 Readings (ft)
36	467.8	28.3	496.1	504	7.9	1
37	441.9	77.6	519.5	522	2.5	1
40	390.3	60.0	450.3	450	(0.3)	2
41	384.5	82.2	466.7	470	3.3	1
42	377.7	75.0	452.7	465	12.3	(8.0)
43	381.4	42.6	424.0	430	6.0	0
44	360.9	35.4	396.3	400	3.7	7
45	358.2	3.1	361.3	410	48.7	1
46	361.0	28.2	389.2	390	0.8	6
47	338.7	130.4	469.1	460	(9.1)	11
48	371.6	11.7	383.3	390	6.7	0
49	337.9	43.7	381.6	380	(1.6)	6
50	328.5	54.8	383.3	380	(3.3)	5

Figure 1. Failure of the PVR Slope in 1976

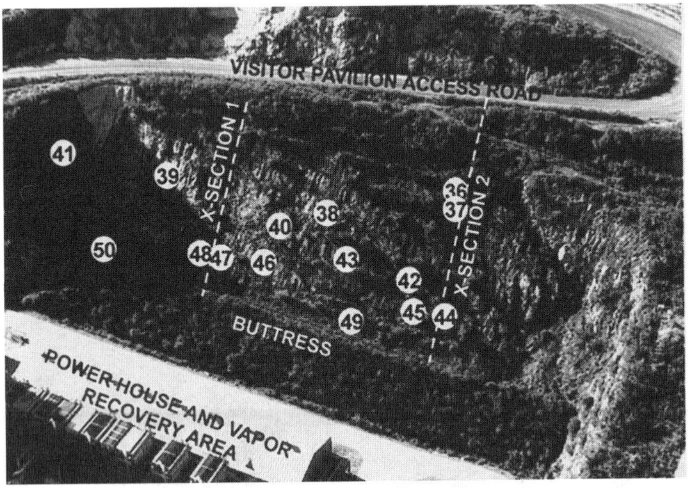

Figure 2. Locations of Piezometers on PVR Slope

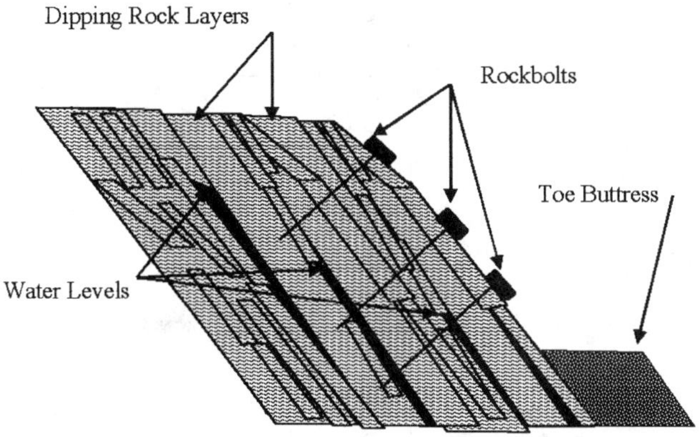

Figure 3. Conceptual Water Levels in Steeply Dipping Rock Strata

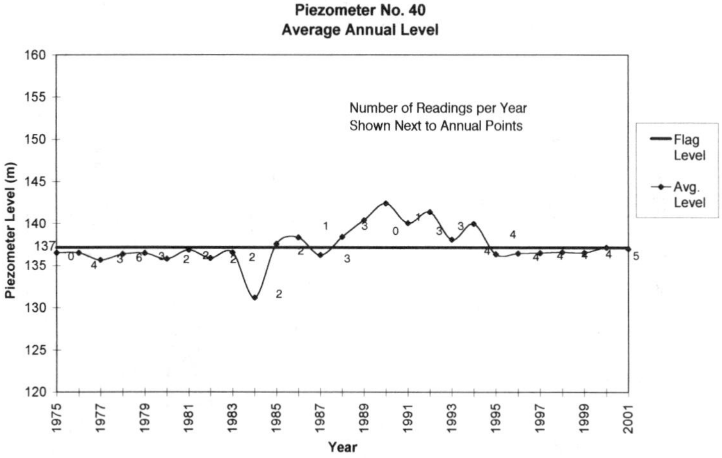

Figure 4. Average Annual Water Level at Piezometer 40

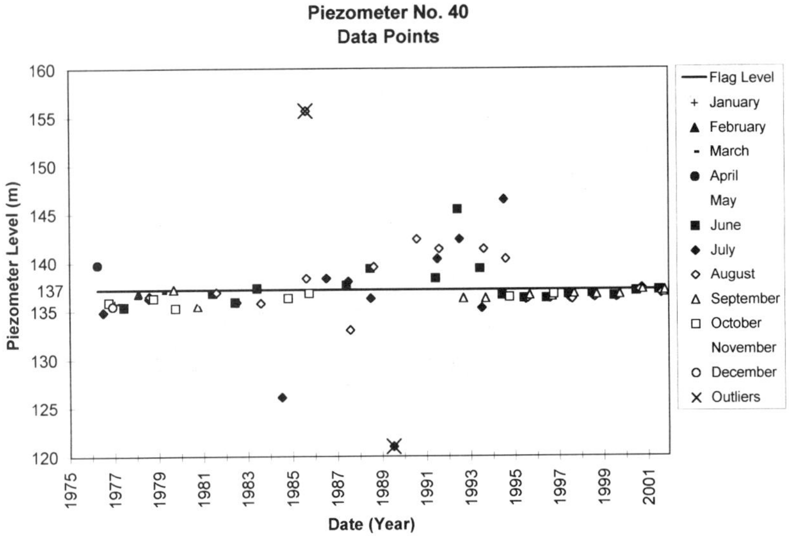

Figure 5. Monthly Water Levels at Piezometer 40

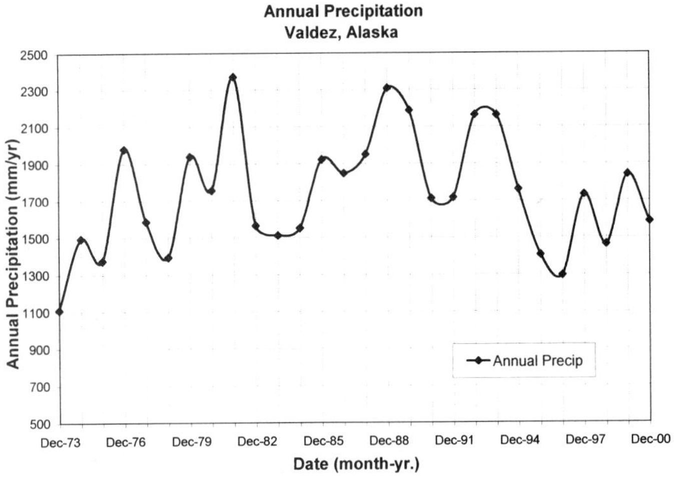

Figure 6. Average Precipitation at Valdez

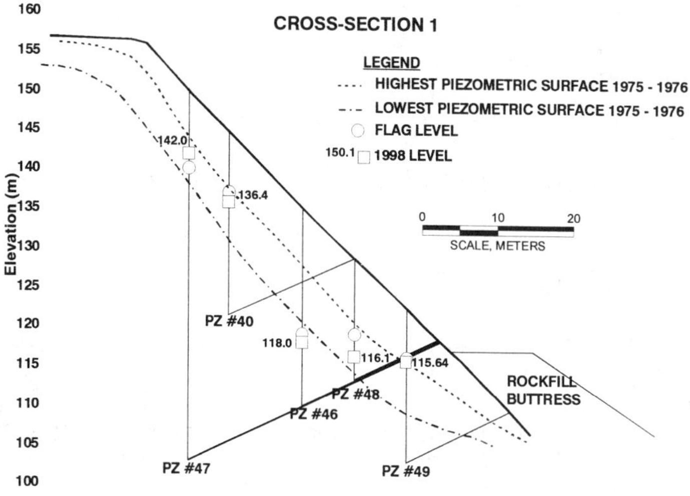

Figure 7. Comparison of 1999 Water Levels to Flag Levels

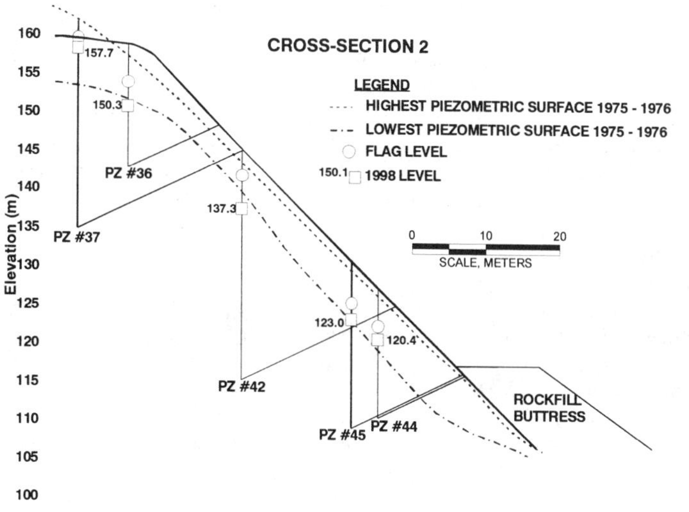

Figure 8. Comparison of 1999 Water Levels to Flag Levels

Evaluation of Buried Secondary Containment Liners

Steven L. Anderson, P.E.[1], Alexandre Lai, P.E.[2], and Rupert "Bucky" Tart, Jr., P.E.[3]

Abstract

Buried secondary containment liners at several pump station tank farms along the Trans Alaska Pipeline System (TAPS) were evaluated in 2000 to determine if they complied with the Alaska Department of Environmental Conservation oil pollution prevention regulations under the Alaska Administrative Code, Title 18, Chapter 75 (18 AAC 75). These secondary containment systems (SCSs), which were constructed in the mid to late 1970's, are diked areas composed of a flexible membrane liner covered by an average of about 0.61 m of soil that surround the pump station tank farms.

Golder Associates Inc. performed a comprehensive assessment of the tank farm liner systems at five pump stations that included a qualitative examination of project records, the development of a visual inspection checklist, and select test pits at each pump station tank farm to confirm that the liners met regulatory requirements. The field investigation also included using geophysical techniques, such as ground penetrating radar and electric leak detection surveys, to identify potential liner damage locations. Based on the liner damage identified at each pump station tank farm SCS, leakage rates were calculated and compared to leakage rates through a soil liner with a permeability of 1×10^{-6} cm/sec.

Introduction

Under the Alaska Administrative Code, Title 18, Chapter 75 (18 AAC 75), the Alaska Department of Environmental Conservation (ADEC) requires that secondary containment systems (SCSs) at existing facilities be "sufficiently impermeable...to protect groundwater from contamination and to contain a discharge until it can be detected and cleaned up."

During meetings with the Alyeska Pipeline Service Company (Alyeska) in 1999, the ADEC requested a comprehensive assessment of the tank farm liner systems at Pump Stations 3, 5, 7, 9, and 12 to approve Alyeska's "Oil Discharge

[1] Senior Geotechnical Engineer, Golder Associates, Inc.
[2] Risk Management Coordinator/Senior Engineer, Alyeska Pipeline Service Company
[3] Principal, Golder Associates, Inc.

Prevention and Contingency Plan." These SCSs, which were constructed in the mid to late 1970's, are diked areas composed of a flexible membrane liner, typically reinforced chlorinated polyethylene (CPER), covered by an average of about 0.61 m of soil that surround the pump station tank farms. The comprehensive assessment was to include a qualitative examination of project records and select test pits to confirm the tank farm liners condition.

Qualitative Examination of Records

The qualitative examination of records included interviewing people familiar with the historical pump station tank farm operations and an investigation of construction archives, engineering reports, and spill records. Based on the interviews and qualitative historical document review, the liner installation had been monitored for quality control and the installation met the intent of the plans and specifications.

Golder Associates Inc. (Golder) generated a report that summarized the findings of the historical document review and recommended test pit locations in areas where the liner integrity may be suspect. In addition, a visual inspection checklist was developed to inspect the surface condition of the cover soils for gross problems that increase the likelihood or indicate the presence of leaks.

Field Investigation

The field investigation at Pump Stations 3, 5, 7, 9, and 12 was performed to directly observe and evaluate the tank farm secondary containment liner at these facilities. The field investigation included a visual inspection of the SCSs, selection of test pit locations based on the visual inspection and qualitative record review, excavation of test pits, and documentation of the exposed liners condition. In addition, work at Pump Station 5 included a geophysical survey using ground penetrating radar (GPR) and electric leak detection (ELD) methods, selection of three 15 m by 15 m trench areas to perform the ELD survey, and exposing the liner at up to 16 significant anomaly locations identified during the ELD survey. These tasks are summarized in the following sections.

Visual Inspection. Visual inspection of the protective cover soils was performed to identify gross problems that increase the likelihood or indicate the presence of leaks. A visual inspection checklist developed during the qualitative examination of records was used. Some of the suspect areas included deep ruts, animal burrows, surficial cracks and depressions, and areas where evidence of structural movement was apparent.

Test Pits and Trenches. The location selection and excavation methods used for the test pits and trenches are described below.

Location Selection. With the exception of two pump stations, the results of the visual inspection and the qualitative record review were generally used to identify test pit locations at the tank farms. When potential leak locations could not be identified by visual inspection, test pits were located randomly in an area where there was no record of previous test pits.

At Pump Station 5, the majority of test pits were located based on results from the ELD survey, except for one test pit, which was positioned at a penetration boot that had no record of being repaired. Two of the three trenched areas at Pump Station 5 were located based on anomalies identified during the GPR survey performed as part of this investigation and a 1992 GPR survey. The third trench was selected based on the visual inspection of the cover soils.

Additional test pits were also excavated along the east slope of the east cell at Pump Station 7 to further define the extent of large liner damage that was observed earlier in the investigation. Cover soils along the east slope in this area were observed to have grading marks from what appeared to be a front loader bucket. Test pits were initially located about every 3 m along the east slope. If large liner damage was uncovered in the test pits, then the test pits were expanded about 1.5 m on either side or to the next adjacent test pit where no liner damage had been found.

Excavation. Excavation of the protective cover soils to expose the liner was done using air knives, a vacuum truck, and metal and plastic shovels according to prepared instructions. A small backhoe was also used at Pump Station 5 to remove the first foot of cover soils for the test pits. The protective cover soils were typically excavated with extreme care so as not to damage the liner. However, some liner damage did occur occasionally as a result of the excavation.

Geophysical Survey. A geophysical survey that included GPR and ELD was performed by Golder at Pump Station 5. This geophysical survey was performed in addition to the visual inspection of the cover soils to identify potential leak locations.

Ground Penetrating Radar. GPR is a geophysical technique that transmits electromagnetic waves into the ground that are reflected at subsurface interfaces and subsequently detected by a receiving antenna. A subsurface profile is acquired by moving the transmitter and receiver antennas along a traverse and continuously recording a series of soundings. The resulting reflection data are displayed on a graphic recorder as a continuous profile depicting the subsurface strata.

Reflections arise due to contrasts in the electrical properties (conductivity and dielectric properties) of subsurface materials. These properties are a function of water content, grain size, mineralogy and the presence of petroleum products. The best subsurface penetration is obtained through hard material such as rock or concrete and in medium to coarse-grained sediment. The poorest penetration is in fine-grained sediment (silt and clay) or conductive material such as salt water.

The GPR survey was conducted with a GSSI System 8 with a 400-MHz antenna. The 400-MHz antenna produces a high-resolution image of the subsurface with a maximum penetration depth on the order of 6 m to 9 m. Field procedures included laying out a grid system on the ground surface, calibrating the instrument, profiling along the alignment, interpreting the data, and marking the location of GPR anomalies on the ground with spray paint. The results of the GPR survey are discussed in more detail below.

Electric Leak Detection Survey. The ELD system is an electrical technique that is closely related to the direct current electrical resistivity method. To locate the leaks, an electrical current is applied to the ground between electrodes. One electrode is located inside of the cell above the liner and the other is located outside of the cell in the soils surrounding the liner. A second set of moving electrodes is then used to measure the electrical potential at closely spaced points throughout the area of concern. If the liner is intact, very little electrical current will flow through the highly resistive plastic liner. However, if a hole exists in the liner there will be an increase in the current flow through the point of leakage that will create an electrical potential value. Careful mapping of these electrical potentials indicates the location of perforations in the liner. These perforations can range in size from less than a millimeter to very large tears in the liner.

Results of Geophysical Survey. The GPR anomalies were categorized as either an undulation of the liner or a low amplitude reflection of the liner. The undulating anomaly of the liner is related to the subsurface elevation change of the underlying material. The low amplitude signal, in some areas, may be related to a change in the material overlying the liner.

An example of the contour plan of current potential data developed during the ELD survey and the approximate location of the test pits is shown in Figure 1. The anomalies were identified as those with readings larger than 950 mV at Trench 1, larger than 150 mV at Trench 2, and larger than 80 mV and lower than –400 mV at Trench 3. The majority of anomalies were confirmed with test pits.

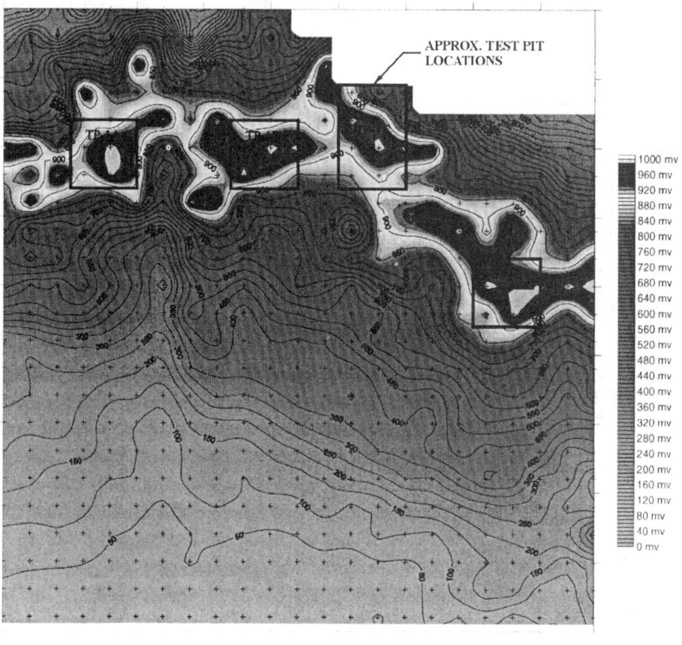

Figure 1. ELD Survey Results for Trench 1 at Pump Station 5

Based on the results of the excavation program, the ELD was precise and effective at locating leaks in the liner. However, results from the GPR survey were ambiguous and undefined.

Liner Inspection. The liner was visually inspected for damage after the cover soils had been removed and the liner had been swept clean and washed off. The liner inspection was done according to developed procedures, which included a geomembrane liner inspection checklist. The geomembrane liner inspection checklist included a description of the protective cover soils, the location and size of the exposed liner, the condition of the liner, and photographs and sketches of the test pits. The geomembrane liner inspection checklists for the test pits at Pump Station 5 are summarized in Table 1.

The liner was typically inspected using visual methods only. However, pinholes and other small liner damage that could not be visually verified to perforate the liner were tested using a vacuum box. The condition of the liner seams was observed for fishmouths and/or unbonded lengths with less than 5.1 cm overlap. Unbonded seams with greater than 5.1 cm overlap and at least 61 cm of protective cover were considered to be passable based on previous work by Golder.

Observations made during the visual inspection of the trenches and test pits are summarized below.

Protective Cover Soils. The protective cover soils were inspected for thickness, aggregate particle size, and aggregate angularity. The protective cover soils were typically composed of a sandy gravel surface course overlying a sand cushion layer that averaged about 7.6 cm thick. At Pump Station 5, the sand cushion layer was not discernable from the surface course materials. However, sandy soil was typically observed above the geomembrane liner.

Table 1. Summary of Liner Evaluation Checklist for Pump Station 5

TEST PIT OR TRENCH NUMBER	LOCATION (SLOPE OR FLOOR)	PROTECTIVE COVER THICKNESS (MIN. - MAX., cm)	SAND CUSHION THICKNESS (MIN. - MAX., cm)	LINER COLOR	LINER DAMAGE (YES OR NO)	SEAM BONDING AND CONDITION		
						HEAT FUSION	ADHESION	UNBONDED
TRENCH 1	FLOOR	---	---	GRAY	YES	OKAY	OKAY	FISHMOUTH
TP-1A	FLOOR	81 - 86	0	GRAY	YES	OKAY	FISHMOUTH	---
TP-1B	FLOOR	76 - 81	0	GRAY	YES	---	OKAY	---
TP-1C	FLOOR	76 - 79	0	GRAY	YES	OKAY	---	OKAY
TP-1D	FLOOR	74 - 76	0	GRAY	YES	OKAY	---	---
TRENCH 2	FLOOR	---	---	GRAY	YES	OKAY	OKAY	FISHMOUTH
TP-2A	FLOOR	61 - 81	0	GRAY	YES	OKAY	OKAY	---
TRENCH 3	BOTH	---	---	GRAY	YES	OKAY	---	---
TP-3A	BOTH	46 - 76	0	GRAY	YES	OKAY	---	---
TP-3B	FLOOR	76	0	GRAY	YES	OKAY	---	---
TP-4	FLOOR	124	0	GRAY	YES	---	---	< 5.1 cm OVERLAP

NOTES: 1. ALL LINER EXPOSED WAS REINFORCED.
2. LINER AND SEAMS WERE INSPECTED VISUALLY. VACUUM BOX USED OCCASIONALLY TO DETERMINE WHETHER DAMAGE PENETRATED LINER.
3. SEAM BONDING AND CONDITION INDICATES WHAT TYPE OF SEAM(S) WAS EXPOSED AND ITS CONDITION. UNBONDED SEAM GENERALLY CONSIDERED TO BE OKAY WITH > 5.1 CM OVERLAP AND AT LEAST 61 CM OF PROTECTIVE COVER SOILS.
4. - - - = NOT APPLICABLE

Liner Damage. The following types of liner damage, and their definitions, were observed at the pump station tank farms:

- Perforation (holes and tears): damage that perforates or penetrates the liner
- Abrasion: damage that marks or scores the layer of geomembrane liner above the reinforcing scrim, but does not perforate the entire liner
- Delamination: damage when a laminated geomembrane layer separates from the reinforcing scrim, but does not necessarily perforate the entire liner
- Fishmouth: irregular contact surface between the top liner and the bottom liner in a seam
- Unbonded seam with less than 5.1 cm overlap

The majority of damage observed was caused by perforations that ranged in size from a pinhole (defined in this study as a perforation having a diameter less than or equal to 0.32 cm) to tears up to 2.1 m wide. Seam damage in the form of fishmouths and unbonded seams with less than 5.1 cm overlap were also observed. Sketches of the two types of fishmouths observed are shown in Figure 2.

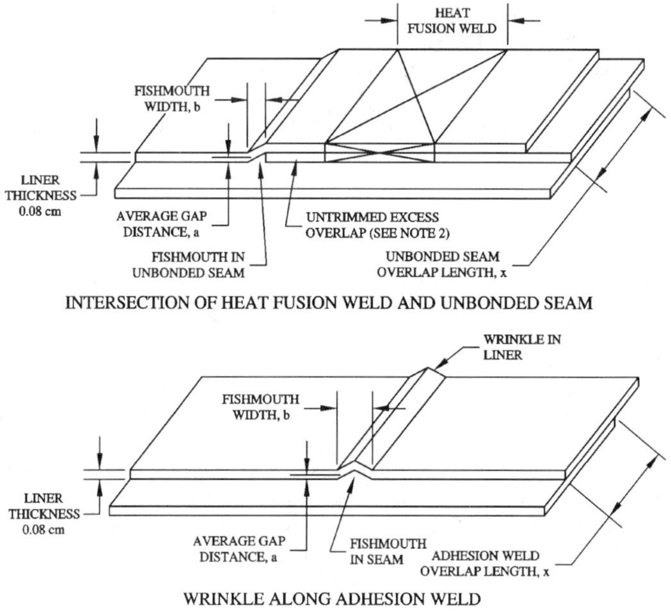

NOTES: 1. DIMENSIONS USED TO CALCULATE FLOW RATE THROUGH FISHMOUTH AS FLOW BETWEEN TWO FIXED PLATES.

2. NO FLOW ASSUMED THROUGH UNTRIMMED EXCESS OVERLAP WITH AT LEAST 61 CM OF SOIL COVER.

Figure 2. Examples of Liner Seam Fishmouths

Liner Evaluation

The secondary containment systems were evaluated based on the condition of the liner observed in the test pits and trenches excavated at the pump station tank farms. The liner evaluation included a review of literature and past reports, estimation of leakage rates through observed liner damage, and comparison of an overall leakage rate to the flow rate through a clay liner with a permeability of $1x10^{-6}$ cm/sec.

Literature and Report Review. Literature and past reports were reviewed to determine what has been done previously to estimate the leakage rate through geomembrane liner defects (Giroud, et al., 1997; Texas Agricultural Experiment Station, 1987). Of all the literature reviewed, tests performed by Golder in 1999 were the only work done that simulated conditions such as exist at the pump station tank farms secondary containment systems, i.e. thickness and permeability of cover soils, head of liquid on top of geomembrane, and type of geomembrane. Therefore, information from this 1999 report was used.

Leakage Rate Estimation. Leakage rates were estimated for liner perforations and for fishmouths in the liner seams. A hydraulic head of 1.2 m was typically used, which was based on 95% of the largest tank volume over the lined area of the smallest tank farm cell. The methods used to estimate these leakage rates are described below.

Leakage Rate Through a Perforation. The leakage rate through a perforation, either a hole or tear, was estimated using the equation for flow through a submerged weir:

$$q = Ca\sqrt{2gh} \qquad \text{(Daugherty, et. al., 1985)}$$

where: q is the leakage rate through the perforation (m^3/sec converted to lpd); C is a coefficient that is a function of the inlet geometry and is between 0 and 1 (dimensionless); a is the area of the perforation (m^2); g is the acceleration due to gravity (m/sec^2); and, h is the total hydraulic head (m).

A coefficient of 0.6, which is the coefficient for free flow through an orifice (Giroud, et al., 1997), was used in the 1999 Golder study to estimate flow through a geomembrane liner perforation overlain by 61 cm of cover soils that had a permeability of $1x10^{-3}$ cm/sec. However, calculations using this coefficient overestimated the measured flow rate through a soil cover overlying a geomembrane perforation by four orders of magnitude. Therefore, data from the 1999 Golder study were used to back-calculate a new coefficient value of $8x10^{-6}$. This coefficient was then used to calculate the leakage rates for a variety of perforations as shown in Figure 3.

Leakage Rate Through a Fishmouth. The leakage rate through a fishmouth was estimated using the equation for flow between two fixed plates:

$$Q = \frac{ba^3}{12\mu}\Delta P \qquad \text{(White, 1979)}$$

where: Q is the leakage rate through the fishmouth (m^3/sec converted to lpd); b is the width of the fishmouth (m); a is the average gap distance between the geomembranes (m); μ is absolute viscosity of the fluid (10^{-3} N-sec/m^2 at room temperature for water); ΔP is the pressure differential $[-d(P+\rho gz)/dx]$ or $(P_1-P_2)/x$; P_1 is the pressure head at the top of the seam (water head (m) converted to N/m^2); P_2 is the pressure head at the bottom of the seam (assumed to be zero); x is the distance along the flow path between the seam (seam overlap length) (m); ρ is the mass density of the fluid; and, z is the elevation between the top and bottom of the seam (assumed negligible).

Figure 2 shows two types of seam fishmouths and the dimensions used to estimate the leakage rate. The average gap distance was estimated from field observations to be equal to the thickness of the liner at a maximum, about 0.08 cm. The fishmouth widths were measured in the field or calculated assuming a 2 horizontal to 1 vertical slope. The seam overlap lengths were measured in the field.

The calculated leakage rates for the seam fishmouths observed at Pump Stations 3 and 5 are shown in Figure 4. However, since these calculated leakage rates do not consider the reduction of flow due to the cover soils, they are likely to be overly conservative. In addition, based on the previous tests performed by Golder in 1999, the calculated fishmouth leakage rates are also likely to be overly conservative by one to several orders of magnitude.

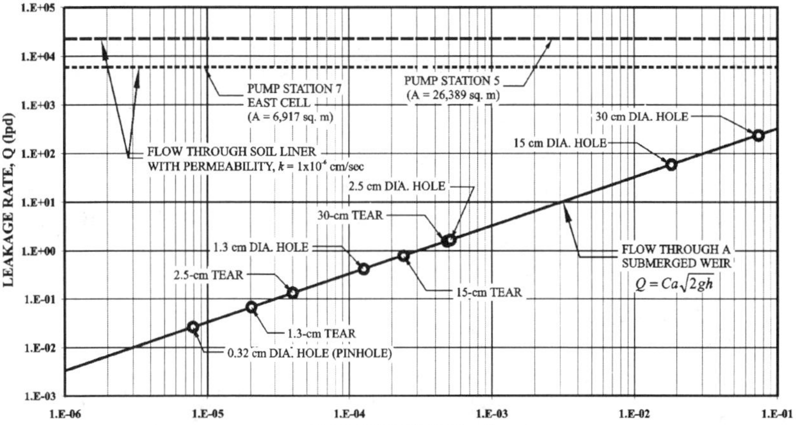

NOTES: 1. FLOW THROUGH A SUBMERGED WEIR CALCULATED ASSUMING A COEFFICIENT (C) OF 8X10-6 AND A HYDRAULIC HEAD (h) OF 1.2 m.

2. FLOW THROUGH A SOIL LINER BASED ON DARCY'S LAW, $q = kiA$, ASSUMING A HYDRAULIC GRADIENT (i) OF UNITY.

3. AREA OF TEAR IN LINER ASSUMES WIDTH OF 0.16 cm.

Figure 3. Comparison of Liner Perforation Leakage Rates

Acceptable Leakage Rates. Under the current ADEC 18 AAC 75.990(124) regulation for "sufficiently impermeable," compiled October 1999, a soil liner with a permeability of $1x10^{-6}$ cm/sec would be acceptable for use as newly installed secondary containment. Darcy's Law was used to determine the flow rate through a soil liner and is defined as:

$$q = kiA \qquad \text{(Sowers, 1979)}$$

where: q is the flow rate (m^3/sec converted to lpd); k is the coefficient of permeability ($1x10^{-6}$ cm/sec as defined by ADEC regulation 18 AAC 75.990(124) converted to m/sec); i is the hydraulic gradient or head loss over the flow path, $\Delta h/l$, assumed to be equal to 1; Δh is the head loss (m); l is the flow length (m); and, A is the lined area of the pump station tank farm (m^2).

The hydraulic gradient, or head loss over the flow path, of unity assumes that the depth of saturated soil liner thickness is equal to the hydraulic head that is forcing water downward into the subbase soil below the soil liner. Low permeability soil liners are generally 0.3 m to 0.9 m thick. Therefore, since we are typically assuming a hydraulic head of 1.2 m for calculating the leakage rates through a damaged liner, using a hydraulic gradient of unity to calculate the flow rate through a low permeability soil liner is rather conservative. In other words, using a larger hydraulic gradient would yield higher "acceptable" flow rates through the soil liner.

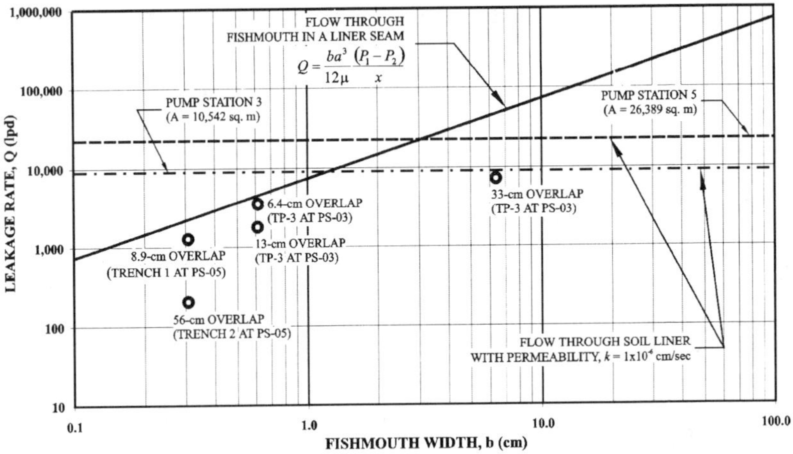

NOTES: 1. FLOW THROUGH FISHMOUTH IN A LINER SEAM ANALYZED AS FLOW BETWEEN TWO FIXED PLATES, ASSUMING A 0.08-cm GAP (a), AN ABSOLUTE VISCOSITY (μ) OF 0.001 N-SEC/M^2, A PRESSURE HEAD AT THE TOP OF THE FISHMOUTH (P_1) OF 1.2 cm, AND A SEAM OVERLAP (x) OF 5.1 cm. CALCULATION DOES NOT CONSIDER REDUCTION OF FLOW DUE TO COVER SOILS.

2. FLOW THROUGH A SOIL LINER BASED ON DARCY'S LAW, $q = kiA$, ASSUMING A HYDRAULIC GRADIENT (i) OF UNITY.

Figure 4. Comparison of Liner Seam Fishmouth Leakage Rates

The approximate lined area for each pump station tank farm was estimated from scale drawings of each facility. The soil liner leakage rates for select pump stations are shown for comparison to estimated leakage rates for perforations and fishmouths in Figures 3 and 4, respectively.

The number and type of observed liner defects were used to predict the number of defects that could exist at each facility by extrapolating the liner area exposed in the test pits and trenches over the total lined area. Defects of concern were identified as holes, ranging in size from 0.32 cm diameter (pinholes) to 2.5 cm diameter, and 15-cm tears because of their frequency. Other damage such as fishmouths, defects larger than 15 cm, and unbonded seams with less than 5.1-cm overlap were discounted for the following reasons:

Fishmouths. Fishmouths were discounted because they were observed to be uncommon and have an estimated leakage rate that is very low.

Large Defects. The only large defects that were larger than 15 cm were observed along the east slope of the east cell at one pump station, which was considered an isolated event.

Unbonded Seams With Less Than 5.1 cm Overlap. This type of defect was only observed at one pump station where the older liner boot had been cut. The liner boot has since been repaired.

Calculation of Total Leakage Rates. The total leakage rate at each facility was determined by summing the product of the number of predicted defects and their estimated leakage rate. The total leakage rate at each pump station tank farm was much less than the acceptable leakage rate as calculated under the ADEC regulation.

Conclusions

Golder preformed a comprehensive assessment of the tank farm secondary containment liners at Pump Stations 3, 5, 7, 9, and 12 that included a qualitative examination of project records and a field investigation to confirm the tank farm liners condition. The following conclusions were developed as part of the comprehensive assessment:

- The historic inspection records indicated that the liner installation was monitored for quality control and that the installation met the intent of the plans and specifications. There was no record of collateral damage.
- The SCSs at the pump station tank farms appeared to be "sufficiently impermeable" because the leakage rate through the predicted number of defects at each pump station was many times lower than the flow rate through a system with a $1x10^{-6}$ cm/sec soil liner.
- Liner damage that perforated the liner was only found at three pump stations.
- The majority of liner damage at each pump station tank farm was due to pinholes with a diameter that was less than or equal to 0.32 cm. With one exception, no liner damage was observed that was larger than 15 cm.
- The large liner damage observed along the east slope of the east cell at one pump station was considered an isolated event.

References

Alaska Department of Environmental Conservation (1992); "Oil and Other Hazardous Substances Pollution Control," ADEC, 18 AAC 75, May 1992.

Alaska Department of Environmental Conservation (1996); White Paper on Secondary Containment, Titled "Sufficiently Impermeable Secondary Containment Systems," ADEC, Division of Spill Prevention and Response, Industry Preparedness and Pipeline Program, March 11, 1996.

Alaska Department of Environmental Conservation (1999); "Oil and Other Hazardous Substances Pollution Control," ADEC, 18 AAC 75, October 1999.

Daugherty, R.L., Franzini, J.B., Finnemore, E.J. (1985); Fluid Mechanics with Engineering Applications, 8th ed.; McGraw-Hill, Inc., 1985.

Giroud, J.P., Khire, M.V., and Soderman, K.L. (1997); "Liquid Migration Through Defects in a Geomembrane Overlain and Underlain by Permeable Media," Special Issue on Liquid Migration Control Using Geosynthetic Liner Systems; Edited by T.S. Ingold, Volume 4, No. 3 and 4, Industrial Fabrics Association International, 1997.

Sowers, George F. (1979); Introductory Soil Mechanics and Foundations: Geotechnical Engineering, 4th ed.; Macmillan Publishing Co. Inc., New York.

Texas Agricultural Experiment Station (1987); "Quantification of Leak Rates Through Holes in Landfill Liners," Reproduced by the National Technical Information Service, Springfield, Virginia, Publication No. PB87227666, August 1987.

White, Frank M. (1979); Fluid Mechanics; McGraw-Hill, Inc., 1979.

Assessment, Design, and Some Lessons Learned From Application of Natural Channel Design Principles in Conversion of a Multiple Thread to Single Thread Channel on an Alluvial Fan

Christopher H. Roach P.E.[1], MacNamara C. Shoulders[2], and Robert F. Carlson Ph.D.[3]

Abstract

McCallum Creek crosses the Trans Alaska Pipeline in the Central Alaska Range near Paxson, Alaska. It was converted from a braided channel on an alluvial fan to a stable single thread channel at the pipeline crossing. Objectives included controlling the position of the channel at the pipeline crossing and minimizing aufeis accumulations through the converted reach. A natural channel design methodology was used, which involves determining the dimension, pattern, and profile of a stable reference reach for the associated valley type and stream type and emulating those characteristics in the design reach. A failure mode analysis was conducted. Mitigative measures were incorporated into the design for each potential failure mode. The channel was lowered to reduce surface energy loss and associated winter aufeis accumulations. The site has been subjected to at least one bankfull flow event since construction was completed in May 2000. Performance has been good during the 2000 and 2001 summer seasons. Aufeis accumulations have not recurred since the channel conversion. Lessons learned are presented from this and several similar projects constructed along TAPS in 2000 and 2001.

Introduction

This paper describes the assessment, design, and some lessons learned from converting a braided channel on an alluvial fan to a stable single thread channel. The work was a unique application of natural channel design, which has found increasing favor in recent years for stream restoration.

Natural channel design (NCD) involves determining the dimension, pattern, and profile of a stable reference reach for the associated stream type, valley type and hydro-physiographic region and emulating those characteristics in the design reach. *Stability* has been defined as the ability of a stream, over time, to transport the flows and sediments of the watershed while neither aggrading nor degrading and while maintaining a consistent dimension, pattern, and profile (Rosgen, 1996). The active channel is designed for the bankfull discharge while flood flows are accommodated in the floodplain. Sediment transport calculations are performed to verify that the hydraulic conditions of the design channel are adequate to maintain sediment throughput.

Channel controls include a combination of structures and re-vegetation for bank protection and grade control. Bank protection is provided by vane structures that can be constructed from rocks, logs, or a combination thereof. Grade control structures typically include cross-vane weirs for small to moderate size streams and w-weirs for

1 Senior Civil Engineer, M. ASCE, Alyeska Pipeline Service Company, Fairbanks, Alaska 99706
2 Environmental Coordinator, Alyeska Pipeline Service Company, Fairbanks, Alaska 99706
3 Professor, Department of Civil and Environmental Engineering, University of Alaska, Fairbanks, Fairbanks, Alaska 99775

larger channel widths. Vanes project upstream, are angled at 20 to 30 degrees away from the bank, slope down at 2% to 7% from the bankfull stage to the invert at the bed, and typically project out into the channel up to one-third the bankfull width, Figure 1. Vane structures, including cross vane weirs and w-weirs, provide bank protection by altering the hydraulic characteristics of the flow near the banks. Velocity head is reduced while elevation head is increased, resulting in a reduction of velocity, shear stress, and stream power adjacent to the bank and a re-direction of near-bank flows out toward the center of the channel. The characteristics, history, design, and placement of these structures are discussed by Johnson *et al.* (2001) and Rosgen (2001). Re-vegetation of the banks and floodplain decreases the time necessary for re-colonization by riparian plant species, minimizing bank erosion and floodplain degradation should a flood occur shortly following construction.

The Rosgen stream classification system was used in the data collection and assessment process (Rosgen, 1996). This system classifies morphologically similar streams into eight major *stream types*. Each of the eight stream types is further divided by up to six sub-categories depending on the 50th percentile bed pavement size, from bedrock to silt/clay. An alphanumeric designation is used to describe each resulting stream type. Distinguishing morphological criterion includes *entrenchment ratio*, *width to depth ratio*, *sinuosity*, *slope*, and channel bed *pavement* size (D_{50}). The classification system is based on the assumption that valley characteristics will directly influence the resulting stream morphology. *Valley types* are distinguished by down-valley gradient in percent slope and a qualitative description of vertical containment (entrenchment), side slope, and predominant morphological character.

Site Assessment

McCallum Creek flows westward across the Richardson Highway and Trans Alaska Pipeline (TAPS) about 30 km (18 mi) north of Paxson in the Central Alaska Range. The Richardson Highway Bridge is approximately 220 m (725 ft) upstream from the pipeline crossing. McCallum Creek joins Phelan Creek about 0.8 km (0.5 mi) downstream from the pipeline crossing. The reach between the Highway Bridge and Phelan Creek crosses a broad alluvial fan with a down valley gradient of approximately 1.5 percent. This reach has historically been a braided channel relatively free of riparian vegetation, Figure 2.

Upstream from the Richardson Highway Bridge, McCallum Creek flows through a relatively narrow valley with moderately sloping sides and valley floor gradients of about 3 percent for the first several kilometers. Further upstream, the valley width reduces and the gradient increases to about 8 percent as the stream approaches its headwaters. An access road runs up the McCallum Creek valley, crossing the creek by a single lane bridge about 2.4 km (1.5 mi) upstream from the Richardson Highway.

Phelan Creek valley in the vicinity of McCallum Creek has a broad valley floor with moderately sloping sides and extensive glacial outwash deposits. The down valley gradient ranges between 0.6 percent and 1.9 percent, with an average of about 1.2 percent. The predominant channel characteristic is a wide braided pattern with minimal vegetative influence and the active channel limits corresponding to the floodplain boundaries. Two reaches of Phelan Creek have a stable single channel morphology: one through the Highway Bridge at Richardson Highway Mile Post (RHMP) 201.4 where the stream flows through glacial terrace deposits, and another adjacent to an alluvial terrace deposit near RHMP 206, Figure 3.

Extensive annual aufeis accumulations have recurred in the Phelan Creek valley, including McCallum Creek in the vicinity of the pipeline and Highway Bridge where it flows across the alluvial fan. Aufeis accumulations appear to be related to channel morphology in this region. Throughout the braided channels, aufeis accumulations are extensive and re-occur annually. Aufeis has been non-existent at both the stable single channel reaches of Phelan Creek.

Phelan Creek provides year round habitat for several species of resident fish including Burbot, Sculpin, Dolly Varden and Round Whitefish. McCallum Creek provides seasonal habitat for grayling and round whitefish. Grayling spawn from May through June, and round whitefish spawn in the fall. Both species utilize the clear water tributaries of larger systems for spawning habitat. The fry usually remain close to the spawning area during their first summer. The juveniles and sub-adults prefer backwaters and sloughs for summer feeding. After spawning, the adults migrate to summer feeding areas and generally take up residence in pool habitats. Grayling and whitefish feed almost exclusively on aquatic or terrestrial insects. Prior to the stream conversion the reach in the vicinity of the pipeline crossing had no pool habitat and no riparian habitat.

Glacial melt, snowmelt, and rainfall runoff influence the flow regime at McCallum Creek. Spring snowmelt runoff and summer glacial melt provide sustained but low to moderate flows. Large flood events are associated with rainfall runoff or a combination of rainfall runoff and snowmelt. The Richardson Highway Bridge at McCallum Creek was the site of a US Geological Survey (USGS) crest-stage gage from 1967 to 1991. An initial approximation of the bankfull discharge was estimated from the 1.5-year flood (Log Pearson III), at 11.3 cubic meters per second (cms) (400 cubic feet per second (cfs)), Figure 4. The maximum known flood was on August 13, 1967 at a discharge of 28.6 cms (1010 cfs) (Jones and Fahl, 1994). The 1967 flood was a regional event that caused major flooding throughout the Tanana River basin. The USGS maintains a long-term daily gage near the terminus of the Gulkana glacier on Phelan Creek, about 9 km (5 mi) from the study site.

Prior to the channel conversion, riprap protected dikes ran approximately 228 m (750 ft) downstream from the Highway Bridge along each side of the stream to the pipeline crossing. The dikes had been placed to limit the lateral migration of the stream at the pipeline crossing and prevent erosion of the adjacent work pad. The site had historically required regular maintenance by APSC, as floods would cause streambed aggradation leading to overtopping of the dikes. Once the dikes were overtopped, the stream would down-cut through the dikes and form a gully as erosion proceeded rapidly during the flood event, leading to wash out of the dikes and work pad in the affected area. This was not a threat to pipeline integrity due to deep burial depth, but it created the need for frequent maintenance at the site. Following dike and work pad repairs in 1997 and again in 1999, APSC maintenance crews expressed a desire for a long-term solution.

Reference Reach

A stable reference reach was located at the access road bridge crossing, Figure 5. A topographic survey was taken of the channel and a portion of the floodplain over a 191 m (627 ft) reach, 116 m (380 ft) upstream and 89 m (292 ft) downstream of the bridge site. Nine cross-sections were developed from the topographic survey. Follow up surveys were conducted to collect more detailed cross section and longitudinal profile data and to characterize the bed surface pavement and sub-pavement. The reference reach was initially classified as a B3 stream type, although this was later changed to a C4b stream type after more detailed surveys were completed, Table 1. Calculated water surface profile elevations for the estimated bankfull discharge corresponded very closely with the field-identified bankfull stage indicators, so the initial bankfull discharge estimate was judged to be acceptable.

Channel Conversion Design

Objectives included controlling the position of the creek at the pipeline crossing; protecting fish passage and fish habitat at the site; and minimizing, to the extent possible, aufeis accumulations at the reach. The design concept involved converting the braided channel to a stable single thread channel between the highway-bridge and pipeline crossing, Figure 6.

Four factors have been attributed to the causes of braiding in gravel bed rivers: large and sudden variations in discharge, steep channel slopes, a high sediment supply, and highly erodible banks (Drage, 1977). Additionally, a relationship between channel braiding and aufeis is evident considering the distribution of aufeis throughout the braided channels of Phelan and McCallum Creeks.

The concept for a channel conversion was developed after observing braided channel to single channel transitions in Phelan Creek for several years. In reviewing the four factors commonly identified as the cause of braiding, it was evident that each must be addressed in order to effect a change in channel morphology. Discharge, slope, sediment supply, and bank composition were considered in relation to the morphology of the channel transitions observed in each of the single channel reaches in Phelan Creek. The same discharge was being transmitted through both the braided and single channel reaches, although the flood stage may be attenuated slightly in the braided channel reach due to a relatively larger flood plain area. The sediment throughput appeared to be equal based on no significant aggradation or degradation trends at the transitions. The channel slopes were relatively consistent through the transitions. Thus, it appeared that the dominant factor in braiding at these reaches was the bank composition. At each of the single channel reaches the stream had cut through deep deposits that resulted in a bank composition that was relatively more resistant to erosion than in the braided reaches.

It was therefore postulated that if a change in bank composition could be made, then converting the braided channel to a single channel would be feasible, provided that the channel could be designed to transmit the sediment load. As a design condition, bank composition may be addressed through bank protection, in this case rock vanes. The sediment supply and channel slope becomes a design condition: the channel dimension is balanced with the pattern and profile to develop a combination of depth and slope, respectively, that will induce adequate shear stress to transmit the maximum bedload sizes expected to be routed through the reach.

Two routing options were considered: the first running straight along the top of the alluvial fan following the most direct path to Phelan Creek; the second following a mild curve upon exit from the Highway Bridge opening, down the side of the fan, and then along the right (northern) boundary of the fan to Phelan Creek. The first route had the flattest gradient (negatively influencing sediment transport capacity) but it had the advantage that the stream could still shift laterally across the alluvial fan where the single channel reach terminated and would therefore minimize the potential risk of head-ward aggradation into the design channel. The second route provided the benefit of a greater slope through the design reach but there was a concern that potential deposition near the slope break at the lower limits of the reach could lead to failure of the channel through head-ward aggradation. Therefore, the selected route was along the top of the fan, recognizing that the drawback of a flatter slope was balanced by the minimization of risk from head-ward aggradation.

The morphology and hydraulic characteristics of the reference reach were emulated as closely as possible at the design reach, Table 1. The bankfull channel was designed for the cross-sectional area of the reference reach while the floodplain was designed for flows exceeding bankfull. In order to minimize aufeis buildup in the new channel, the bed elevation was lowered approximately 1 meter (3 feet) from its original elevation along most of the reach to lower the water table and allow the adjacent bank fills to provide more surface insulation.

Sediment transport calculations included critical shear stress, critical dimensionless shear stress, and unit stream power. Relationships for critical shear stress vs. maximum particle size transported (Leopold, 1994) were utilized in estimating critical shear stress. Critical dimensionless shear stress may be calculated from the ratio between channel bed pavement and sub-pavement 50th percentile sizes (Andrews, 1983). The minimum mean depth required to transport a given bedload size can then be calculated from Shields entrainment function once the critical dimensionless shear stress has been determined. Published data by Rosgen (1996) and Annable (1995) of critical dimensionless shear stress by stream type was used for estimating critical dimensionless shear stress for the converted reach. This was not calculated directly due to anticipated bed coarsening and the resultant changes in the relationship between the channel bed pavement and sub-pavement sizes. Unit stream power is the time rate of work per unit width of channel. Throughput rates for both suspended sediment and bed material load have been correlated to unit stream power (Yang, 1996). Unit stream power was used as a means of evaluating hydraulic similitude between the reference and design reach.

Maintaining sediment transport capacity required the width to depth ratio to be minimized in the design reach. A straight pattern (plan-view) was selected to maximize slope for sediment transport. Even with minimizing the width to depth ratio and maximizing the slope, the design shear stress and unit stream power was not as great as at the reference reach, Table 1. However, sediment transport capacity was judged to be adequate given that the vanes produce strong secondary circulation cells and increase velocity gradients, which increases localized bed shear stresses and corresponding sediment transport capacity.

A water surface profile analysis was performed to verify the stage of both the bankfull and flood flows (USACE, 1996). The calculated stage of the bankfull discharge was within 0.05 m (0.18 ft) of the design bankfull channel, which was considered acceptable. The stage of the maximum flood of record was found to be approximately 0.6 m (2 ft) below the top of the dikes. The recurrence interval for the overtopping discharge exceeded 1000 years, which indicated that the elevation of the dikes was adequate for extreme flows.

Four rock vanes and three cross vane weir structures were selected to provide bank protection and grade control for the re-constructed reach. The rock vanes were placed on alternating sides of the channel at the estimated intersect position of the maximum velocity core at each opposing bank. Scour potential over the pipeline crossing was addressed through appropriate placement of the cross-vane weir structures. An initial re-vegetation effort was undertaken immediately following construction to provide bank stability and hydraulic roughness through the first summer. In the fall of 2000 an additional row of willows was placed as live siltation (Muhlberg *et al.*, 1998) along each bank slightly above the first row, Figure 7.

Due to the unique nature of the work, a failure mode analysis was conducted. Mitigative measures were determined for each potential failure mode, Table 2. Potential project risks were discussed with personnel responsible for the area. They chose to support the effort as a means of evaluating the feasibility and long-term effectiveness of the design concept.

Lessons Learned

Several lessons learned came as a result of this and several similar projects implemented along TAPS during the 2000 and 2001 seasons; the most relevant of which are described below:

- Determination of the stable channel form is the single most important criterion for natural channel design. A definition of *stability* has been presented earlier, and is stated in such a way that changes in dimension, pattern, profile, and bed material particle size distribution may be quantified. Recent publications on stream restoration and enhancement have placed considerable emphasis on the use of alternative bank protection measures, including bioengineering techniques and various flow control structures. However, bank protection and grade control measures are a secondary consideration. If *stability* is not addressed in the design then these measures alone may not be adequate to meet the restoration or enhancement objectives. The root cause of any observed channel instability must be addressed directly in the design if stream restoration or enhancement works are to be successful.
- Results from two flow seasons indicate that the design concept is viable. The lack of any aggradation or degradation in the channel indicates that, at the flow rates experienced, the sediment balance is being maintained.
- Changing the stream morphology to a single thread channel appears to have eliminated aufeis development in the converted reach. This indicates promise as a means of controlling aufeis in braided channels. Further evaluation will be necessary to quantify the influence of the channel conversion on aufeis growth.

- If cross sections and longitudinal profiles will be cut from a topographic survey, the survey should be conducted in sufficient detail that all relevant morphological features are captured. This includes surveying the full width of the flood prone area, the presence of terraces, bankfull stage, and sufficient shots in the active channel to adequately define the dimension, pattern, and profile. The reference reach was initially assessed through a topographic survey plotted in 0.6-m (2-ft) contour intervals. The cross-sections and longitudinal profile were then cut from the topographic survey. This resulted in some loss of accuracy and lead to suspicions by the design team that the morphology of the reference reach had not been captured as accurately as would have been desired. After construction, follow-up cross section and longitudinal profile surveys did indicate some discrepancy with the original topographic data, although it was not considered great enough to require modifications.
- Channels must be shaped as closely as possible to a naturally stable state to minimize post-construction adjustment. The design cross-section was a trapezoid proportioned for the required cross sectional area, mean depth, and maximum depth. The over-bank region was sloped directly from the bankfull stage to produce the desired entrenchment ratio. When this shape was built, rapid and immediate bank erosion resulted following water-in. Since the over-bank slope started directly from the bankfull stage, this resulted in over-steepened, over-heightened banks. Additional work was required, by trial and error, to excavate a channel and floodplain shape that was stable. Subsequently, attention was given to the shape of the channel and floodplain at several reference reach sites. The conclusion was that naturally stable stream channels are not shaped like a trapezoid. In several following restoration projects, reference reach cross-sections were evaluated to determine the correct shape. Cross sections were normalized to relate channel depth at a given percentage of the bankfull width to the bankfull maximum depth. This allowed design and construction of channels that were shaped as closely as possible to a naturally stable state. Floodplain benches at the bankfull stage were used so channel adjustments did not result in an over-steepened, over-heightened bank. This approach has been successful in preventing the erosion initially observed during the McCallum Creek project.
- The design called for material excavated from the channel to be side-cast and used to build dikes to define a floodplain and contain flood flows. The capacity of the floodplain dikes was evaluated and was found to be adequate to contain flows in excess of a 1000-year recurrence interval. However, this will lead to a water-surface elevation that is higher than the landform on the backsides of the dikes because the alluvial fan is concave downward in cross section. It would have been preferable to lower the channel more so that dikes projecting above the adjacent landform would not be necessary. This would eliminate the perched water surface elevation for flood flows and the associated transverse hydraulic grade through the dikes, thus reducing risk; and it would be more visually attractive by eliminating the look of a diked system. This approach was used on a similar alluvial fan channel conversion project constructed at Darling Creek in 2001. Further monitoring and comparison of these two sites will be necessary to

quantify the influence of adjacent bank fills on surface energy loss and aufeis growth.

- Use symmetric structures in a high-energy straight reach. Alternating vanes were used for bank protection and cross-vane weirs were used for grade control, recognizing that cross vanes also provide bank protection. The alternating vanes were placed so that the flow deflected off one bank would intersect the opposing bank some distance downstream, and the next opposing vane was located near where the flow intersection occurred. This is working well at McCallum Creek; the vanes appear to be properly located, as no bank erosion has occurred. However, it has lead to some additional risks. The B3/B4 stream type is a relatively high gradient, high-energy stream with considerable erosive capacity. With a straight reach of this stream type, use of symmetric structures such as cross vane weirs throughout would be more appropriate, rather than alternating vanes. This would protect both banks, provide a greater degree of grade control, and keep the flow-lines oriented downstream and off the banks.
- Careful attention must be given to structure geometry and quality of construction, some important considerations follow: (1) Maintaining bank angles between 20 and 30 degrees, with angles less than 25 degrees preferred. Angles less than 20 degrees still provide bank protection but result in long structures requiring more rock. Angles greater than 30 degrees may cause bank erosion by eddy rotation, protect less bank, and place greater demand on the structure due to sharp bank and dip angles – resulting in an increased risk of structure damage and scour of the footers. (2) Maintaining a consistent slope (dip) into the bed without gaps or low-spots in the vane. Gaps and low spots create flow convergence, reduce the effectiveness of the structure, and create the potential for bank erosion – particularly if they occur high in the structure near the tie-in with the bank. (3) Correct placement of the footers to fully support the top rocks; if the top rocks are supported by gravel rather than bearing on the footers then flows may scour the underlying gravel away causing the top rocks to settle and the structure to become ineffective. This is particularly true if gaps are present that can increase the scour potential. (4) Design specifications typically include bank angle, dip angle, and intersect with the bed at approximately 1/3 the bankfull width. In some cases these criterion cannot all be satisfied simultaneously and a compromise must be made. Carefully evaluating the structure geometry relative to the channel will help determine the best fit. (5) Sills must be placed so that they do not protrude above the floodplain. If sills project above the floodplain they can cause a hydraulic jump and erosion downstream of the sill during floods. Place sills flush with the floodplain or slightly below grade and cover with a veneer of fill to blend in smooth.
- Use of bank and floodplain re-vegetation in combination with structures has proven effective in minimizing post construction erosion. Immediately following construction the floodplain and banks can be highly disturbed, which places a greater emphasis on re-vegetation.
- Effective communication between the equipment operator and the on-site construction supervision is critical to achieve a good outcome. Building the structures involves careful placement of rocks with varying sizes and shapes that

must be fit together to form the required geometry. This is a trial-and-error process and may seem slower than many operators are accustomed to. The operator and the on-site supervision must be willing to dismantle and rebuild a structure if the geometry is not correct and should understand this is likely to occur. In-process verification of structure geometry is helpful to avoid backtracking. Production has come in at a maximum of about three rock vanes or two cross-vanes per 12-hour shift once the hoe operator becomes familiar with building the structures. Experience on TAPS has shown that the incidence of re-work has increased when production rates were pushed beyond this. This applies to streams with a bankfull width of about 15-m (50-ft).

- Adequately sized rock is a critical component of the structures. Boulders with a minimum equivalent spherical diameter of 1-m (3-ft) were used, figured as the average of the dimension of the three primary axes. This has proved to be a reasonable minimum size for handling and placement. The maximum size range is dependent on the lifting capacity of the equipment. Work along TAPS utilized a Hitachi 450 excavator equipped with a hydraulic thumb and a GVW of approximately 50,000-kg (110,000 lb.). This provided lifting capacity for boulders up to about 2-m (6.5-ft) equivalent spherical diameter. In-process inspection of the rock has proved invaluable to ensure that the specified rock gradation is loaded for transport from the source. This project used approximately 42 cubic meters (500 cubic yards) of rock, or 0.16 cubic meters per lineal meter (0.67 cubic yards per lineal foot).

Summary

This project involved converting a multiple thread channel on an alluvial fan to a stable single thread channel using a natural channel design (NCD) methodology. Design objectives included controlling the channel location at a belowground crude-oil pipeline crossing; minimizing, to the extent possible, aufeis accumulations through the converted reach; and maintaining fish passage and fish habitat. The site has been subjected to at least one bankfull flow event since construction was completed in the spring of 2000. No significant bank erosion, aggradation, or degradation was apparent following the event. Post-flood maintenance of one cross vane weir was required during the summer of 2000 to change the structure geometry due to some shifting of the rocks. Observations during the summer 2001 season indicate that the structures all performed well. Aufeis accumulations have not recurred since the channel conversion project. Lessons learned from 14 NCD projects constructed along TAPS in 2000 and 2001 were presented. The cost for this work was approximately 25% of a comparable project with riprap-protected guide-banks based on historical APSC data.

The site is included in a two-year monitoring study by the University of Alaska, Fairbanks. APSC and the Alaska Science and Technology Foundation (ASTF) provided funding. Objectives include evaluating post-construction geometric response, hydraulic and sediment characteristics, and calibrating hydraulic models to field data. Conclusion of the study is scheduled for fall 2002.

Acknowledgements

Alyeska Pipeline Service Company is gratefully acknowledged for the use of data and photographs. Special recognition is extended to Wayne Wilson, Ross Oliver, Ed Carns, Chuck Heller, and Elden Johnson for their individual contributions to the success of this work. Greg Kinney and Paula Williams are thanked for their review of this paper. Appreciation is expressed to Dave Rosgen and Robert Delk for sharing lessons learned from many years in the field of stream classification, assessment, and restoration. Opinions expressed herein are solely those of the authors. A preliminary version of this work was published previously (Roach *et al.*, 2001).

References

Andrews, E.D. (1983). "Entrainment of gravel from natural sorted riverbed material." *Geological Society of America Bulletin* Vol. 94.

Annable, W.K. (1995). *Morphological relationships of rural water courses in Southwestern Ontario for use in natural channel designs*, Master of Science Thesis, The University of Guelph, School of Engineering, Guelph Ontario, Canada.

Drage, B.T. (1977). Hydraulic engineering investigation of braided rivers flowing from the eastern Brooks Range, Alaska. Master of Science Thesis, University of Alaska, Fairbanks, College of Engineering, Fairbanks, AK.

Johnson, P.A., R.D. Hey, M. Tessier, and D.L. Rosgen (2001). "Use of vanes for control of scour at vertical wall abutments." *J. Hydr. Engrg.*, ASCE, Reston, VA.

Jones, S.H., and C.B. Fahl (1994). "Magnitude and frequency of floods in Alaska and conterminous basins of Canada." *Water-Resources Investigations Report 93-4179*, U.S. Geological Survey, Anchorage, AK.

Leopold, L.B. (1994). *A view of the river*, Harvard University Press, Cambridge, MA.

Muhlberg, G.A., N.J. Moore, and F. Inoue (1998). "Streambank Revegetation and Protection." *Technical Report No. 98-3*, Ak. Dept. of Fish and Game, Anchorage, AK.

Roach, C.H., M.C. Shoulders, and R.F. Carlson (2001). "Conversion of a multiple thread to single thread channel on an alluvial fan." In Hayes, D.F. (ed.) *Designing Successful Stream and Wetland Restoration Projects*, ASCE, Reston, VA.

Rosgen, D.L. (1996). *Applied river morphology*, Wildland Hydrology, Pagosa Springs, CO.

Rosgen, D.L. (2001). "The cross-vane, W-weir and J-hook vane structures…their description, design, and application for stream stabilization and river restoration." In Hayes, D.F. (ed.) *Designing Successful Stream and Wetland Restoration Projects*, ASCE, Reston, VA.

US Army Corps of Engineers (1996). *HEC-RAS river analysis system version 1.2*, Hydrologic Engineering Center, Davis, CA.

Yang, Y.T. (1996). *Sediment transport theory and practice*, McGraw-Hill, New York, NY.

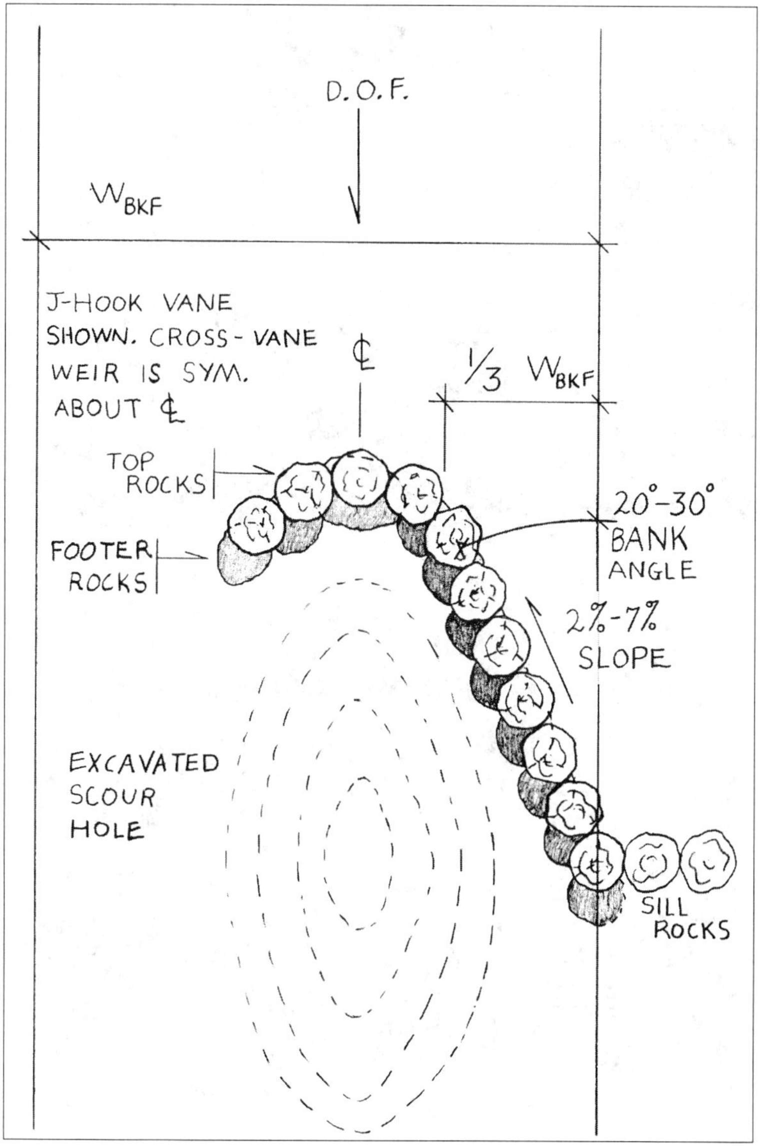

Figure 1. Typical Vane Structure (Plan View)

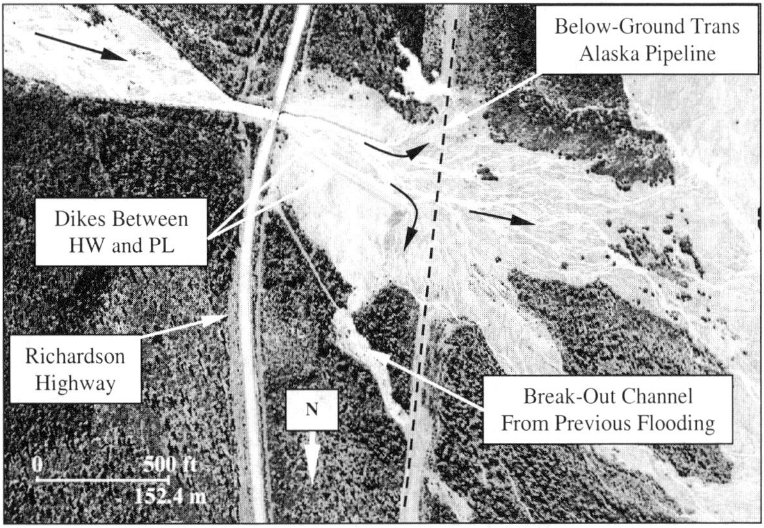

2 (a).

2 (b).

Figure 2. McCallum Creek Prior to Channel Conversion

Figure 3. Braided Channel to Single Channel Transition, Phelan Creek

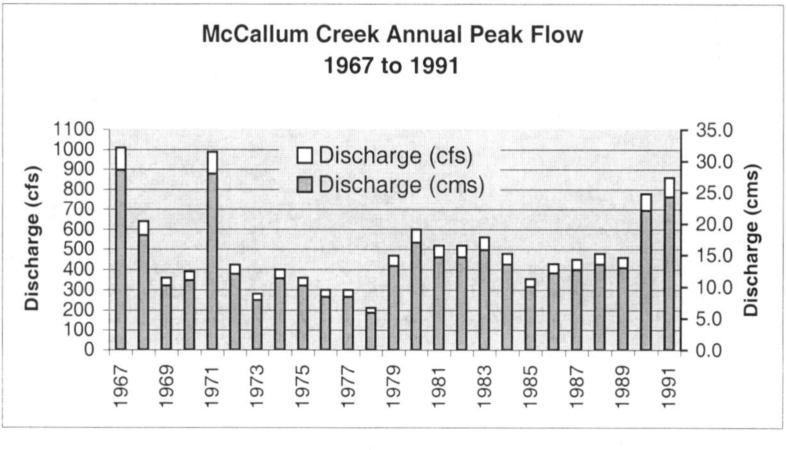

Figure 4. Peak Flow Histogram, USGS McCallum Creek Gage

Variable [SI] (Eng.)	Reference Reach		Pre-Conversion		Design		As-Built	
	[SI]	(Eng.)	[SI]	(Eng.)	[SI]	(Eng.)	[SI]	(Eng.)
Valley Type	II	II	III	III	III	III	III	III
Stream Type	$C4_b$	$C4_b$	D3	D3	B3	B3	B3	B3
Morphology	Rapids	Rapids	Braided	Braided	Rapids	Rapids	Rapids	Rapids
Drainage Area [km^2] (mi^2)	39.4	15.2	40.1	15.5	40.1	15.5	40.1	15.5
Glacial Influence (%)	19	19	19	19	19	19	19	19
Mean Annual Precipitation [cm] (in)	152	60	152	60	152	60	152	60
Mean Basin Elevation [m] (ft)	1487	4880	1487	4880	1487	4880	1487	4880
Bankfull Discharge [cms] (cfs)	11.3	400	11.3	400	11.3	400	11.3	400
Cross-Section Area [sm] (sf)	6.4	69	9.4	101	6.3	68	4.6	50
Bankfull Width [m] (ft)	12.2	40	31.7	104	7.3	24	9.4	31
Bankfull Mean Depth [m] (ft)	0.5	1.7	0.3	0.97	0.9	2.8	0.5	1.6
Bankfull Max Depth [m] (ft)	0.9	2.9	0.5	1.5	1.1	3.5	1.3	4.2
Ratio Max to Mean Depth	1.7	1.7	1.5	1.5	1.3	1.3	2.6	2.6
Width Depth Ratio	23	23	107	107	9	9	19	19
Flood Prone Area Width [m] (ft)	30.2	99	N/A (1)	N/A (1)	14.6	48	19.8	65
Entrenchment Ratio	2.5	2.5	N/A (1)	N/A (1)	2.0	2.0	2.1	2.1
Bankfull Mean Slope (%)	2.8	2.8	1.3	1.3	1.3	1.3	1.4	1.4
Bankfull Mean Velocity [m/s] (ft/s)	1.8	5.8	1.2	4.0	1.8	5.9	2.4	8
Shear Stress [kg/sm] (lb/sf)	13.5	2.8	3.8	0.8	9.1	1.9	6.4	1.3
Critical Dim. Shear Stress	0.096	0.096	0.020	0.020	N/A (2)	N/A (2)	N/A (3)	N/A (3)
Unit Stream Power [kg/m^2s] (lb/ft^2s)	79	16	15	3	54	11	51	11
Note 1: Width of Flood-Prone Area at 2 x d_{max} is above alluvial fan landform (infinite), so E.R. not calculated								
Note 2: Considered tabulated values from Rosgen (1996) and Annable (1995). Not calculated directly in design due to anticipated shift in D_{50} (Pavement/Sub-Pavement)								
Note 3: Not calculated. Bed pavement and sub-pavement samples not obtained due to flow conditions								

1 (a). Geometric and Hydraulic Data

% Finer	Reference Reach		Project Reach	
	Pavement (mm)	Sub-Pavement (mm)	Pavement (mm)	Sub-Pavement (mm)
D_{100}	350	180	256	180
D_{84}	180	120	110	100
D_{50}	55	65	73	14
D_{16}	16	4	35	3
Sample Location	Approx. 30 m (100') above access road bridge.	Approx. 90 m (300') above access road bridge	At XS D.S. of Highway Bridge	U.S. of Hwy bridge - from lower 1/3 of mid-channel bar

1 (b). Sediment Data

Table 1. Morphological Data for McCallum Creek

Figure 5. Reference Reach

Potential Failure Mode	Mitigative Measure
General aggradation	Design channel with high shear stress and unit stream power
Headward aggradation from deposition in downstream braided reach	Route channel along top of alluvial fan. No slope break at termination of single channel reach. Allows braided reach downstream to migrate laterally across full width of fan
General degradation	Use robust grade control structures at midpoint and end of single channel reach to prevent down-cutting (Cross-Vane Weirs)
Bank erosion & collapse from lateral migration	Protect bank using rock vanes, cross-vane weirs, and re-vegetation
Stream bank breach: piping failure of dikes bounding floodplain.	Keep channel bed elevation as low as possible to minimize transverse hydraulic grade through dikes, maximize through-thickness of adjacent dikes
Stream bank breach: overtopping during floods	Provide adequate capacity for both bankfull and extreme flood events within adjacent dikes
Stream bank breach: overtopping during spring flows over aufeis	Install robust floodplain sills to prevent downcutting. Could install spring overflow channel at upstream limits of single channel reach and/or cut a pilot channel in the ice directly over the main channel if warranted in future. Non-issue if conversion eliminates aufeis.

Table 2. Failure Mode Analysis Summary

Figure 6. McCallum Creek Following Channel Conversion

Figure 7. Re-Vegetation Using Live Siltation, Fall 2000

Analysis of Aboveground Pipeline Insulation in Contact with Vertical Support Members

Keith Meyer, P.E., Ph.D[1]
Paul Carson, P.E.[2]
James Maple, P.E., Ph.D.[3]

Abstract

The aboveground support system of the Trans Alaska Pipeline System was extensively analyzed during the initial design phase of the pipeline to ensure integrity during operational and contingency loading conditions. These analyses assumed the position of the pipeline on the beam supported by the Vertical Support Members (VSM) were within defined operational bounds. The original design analyses did not include an analysis of the pipe when the pipe position on the support beam during operational conditions placed the insulation of the pipe in contact with a VSM. During an audit of the pipeline, the pipeline insulation was identified to be in contact with a VSM at a number of locations, and identified as a potential risk to the pipeline integrity during a seismic event. This paper describes the series of analyses performed to address this operational situation, as well as the relatively small, remedial construction required as a result of these analyses and the recommended continued monitoring program.

Introduction

In 1993, the US Department of the Interior, Bureau of Land Management (BLM) conducted an audit of the Trans Alaska Pipeline System (TAPS) operated by Alyeska Pipeline Service Company (APSC). Concerns were raised with regard to pipeline integrity during a seismic event in view of the identified contacts between the pipeline insulation modules and the Vertical Support Members (VSM). A series of structural and dynamic analyses were conducted to evaluate the pipeline system under this condition. This paper discusses these analyses and presents the findings.

[1] Vice President, Alaska Operations, Michael Baker Jr., Inc., 4601 Business Park Boulevard, Suite 42, Anchorage, AK 99503; phone 907-273-1649; kmeyer@mbakercorp.com

[2] Structural Engineering Lead, Alaska Operations, Michael Baker Jr., Inc., 4601 Business Park Boulevard, Suite 42, Anchorage, AK 99503; phone 907-273-1616; pcarson@mbakercorp.com

[3] Principal, J. A. Maple & Associates, (Deceased, August 2001)

Pipeline Description

TAPS is a 1.22-meter (48-inch) diameter crude oil pipeline that has transported over 13 billion barrels of oil from the oil fields in and around Prudhoe Bay, Alaska to the Valdez Marine Terminal, where the oil is loaded on ocean-going tankers for delivery to refineries. The pipeline was constructed in the mid-1970's with the first oil reaching Valdez, Alaska in July 1977. The pipeline is approximately 1,280 kilometers (800 miles) in length, crosses three major mountain ranges, several major rivers (including the Yukon River), and three fault zones. The subsurface conditions range from continuous permafrost for the first 448 kilometers (280 miles) to discontinuous permafrost for the next 736 kilometers (460 miles) across the interior of Alaska to sporadic permafrost through the Chugach Mountain Range as the pipeline traverses the last 96 kilometers (60 miles) to its terminus. The pipeline was generally buried wherever thawed or thaw-stable soils were encountered. Where this criterion could not be met, the pipeline was constructed on specially designed aboveground supports (APSC 1997).

This aboveground mode is built in a zigzag configuration to allow for expansion or contraction of the pipe due to temperature changes by converting changes in pipeline length into lateral (sideways) movements. To permit lateral motion, the pipe is mounted on a structure that can slide on crossbeams installed between VSM (piles). This design also allows for motions caused by an earthquake. Slightly more than half of the entire pipeline is built aboveground.

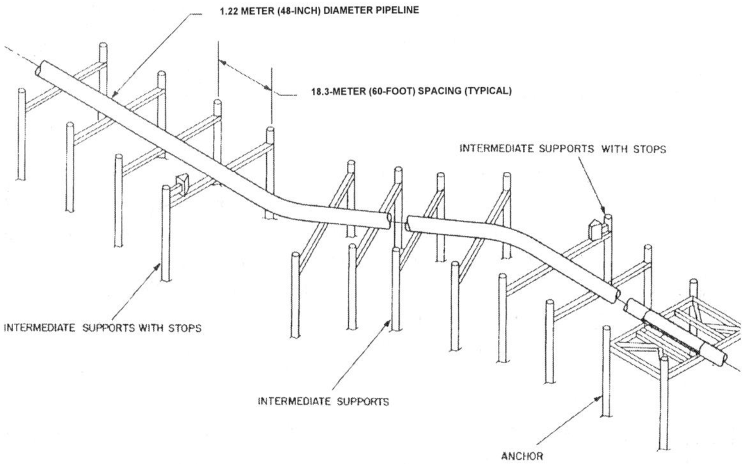

(Courtesy of Alyeska Pipeline Service Company)

Figure 1. Schematic of Standard Zee Configuration

The aboveground system is broken into short lengths (244 to 549 meters or 800 to 1,800 feet) which are traversed by one of five types of "configurations"

(APSC 1997). The standard zee configuration (Figure 1) is the most common and the analyses presented in this paper are for this type configuration. At each end of a typical configuration, the pipe is restrained by the installation of an anchor support. The anchor supports consist of four VSM, a structural steel platform and a friction slide plate assembly. The friction slide assembly is design to resist an initial differential force in the longitudinal direction (along the pipe) before sliding and damping dynamic forces.

At approximately 18.3-meter (60-foot) spacing in between the anchors, the pipe rests on intermediate supports (Figure 2). Intermediate supports typically consist of two VSM and a structural steel crossbeam. In a relatively few cases, a third VSM was installed to increase resistance to lateral loads.

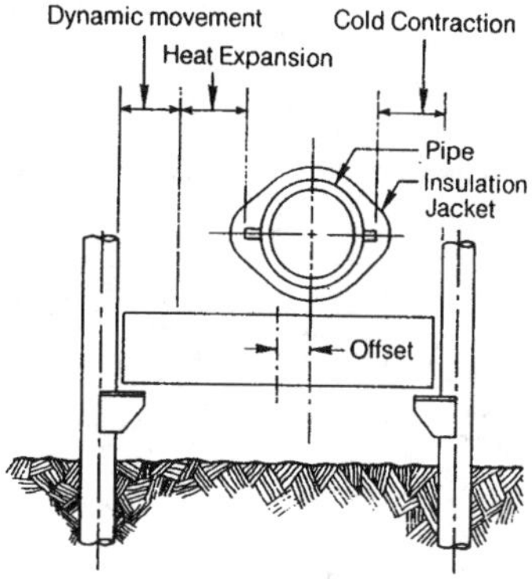

(Courtesy of Alyeska Pipeline Service Company

Figure 2. Cross Section of Intermediate Support VSM

Original Design Analysis

In the original design, all aboveground configurations for TAPS were analyzed under static conditions (gravity, design pressure and design temperature differential; Figure 3). In addition, dynamic analyses were conducted to determine the movement of the pipe on the supports during a seismic event. The results of these analyses were used to size the support hardware. Crossbeams of varying length were installed to accommodate the predicted lateral movements of the pipeline. The beams were positioned off center with respect to the pipe centerline to allow for thermal expansion as the pipe heated up to the design temperature. The side to which the pipe was expected to move when heated was called the "hot" side, while the opposite side was the "cold" side. Seismic bumpers were designed for selected locations to control seismic movement of the pipeline (Figure 4).

The original design was based on a bare pipe scenario. Clearances were designed without consideration of the insulation modules, which were assumed to be sacrificial during a contingency seismic event. There are 40.6 centimeters (16 inches) from the outside surface of the shoe insulation module to the pipe surface. Similarly, there are 16.5 centimeters (9.5 inches) from the outside edge of the insulation to the face of the pipe clamp flange.

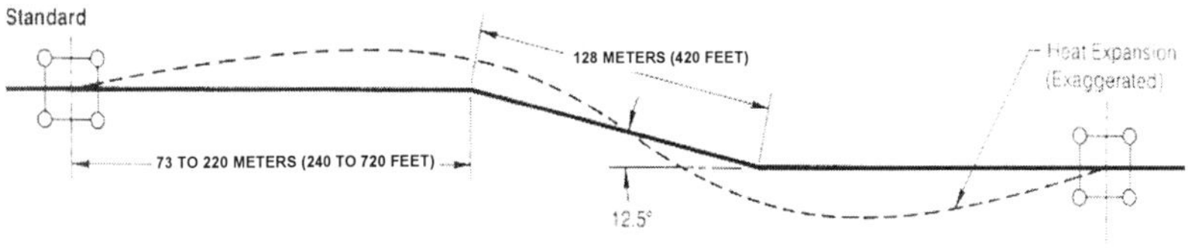

(Courtesy of Alyeska Pipeline Service Company)

Figure 3. Movement of Pipeline in a Standard Zee Configuration

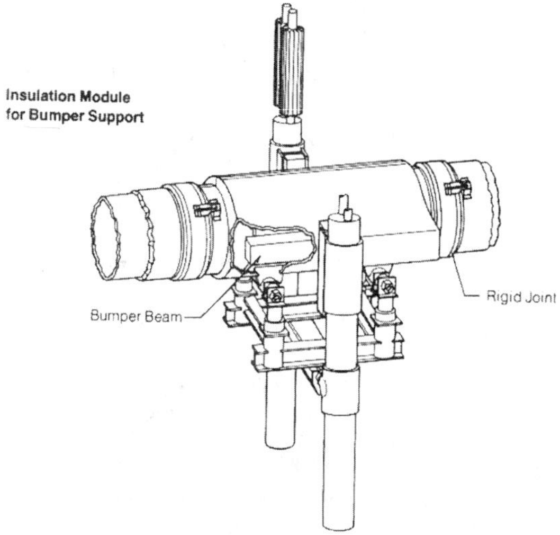

(Courtesy of Alyeska Pipeline Service Company)

Figure 4. Intermediate Support with Bumper Beam

Statement of Concern

During an audit conducted by the BLM in 1993, the pipeline location within several aboveground configurations was observed to differ from the original design. In particular, the pipeline insulation at several intermediate bents was seen to be very close, and in some cases in contact, with the VSM. This raised concern as to whether earthquake induced motion and forces might exceed those motions and forces computed by the original design analyses. A study was initiated to determine the integrity of the pipeline under the observed conditions.

Analytical Approach

The analysis focused on developing a computer modeling procedure that would reasonably replicate the conditions for the majority of contacts. The best apparent method is to model the pipe in the least energy position by using the deflected shape, from changes in pressure and temperature as modeled by assuming zero friction, between the shoes and the crossbeam, as a basis for analysis. This differed from the original design analysis, which analyzed the movement of the pipeline under static conditions (e.g. dead weight, pressure, and temperature differential) for a 10% friction interface. The original design assumption of 5% coefficient of friction provides a conservative upper bound for the evaluation of displacements for short-term loads when modeling a 10% friction interface. (The value of 10% coefficient of friction was also analyzed to produce the bounding force values for the static load cases). The deflected shape for the case of a zero coefficient of friction was demonstrated to be significantly different from the 5% friction case used in the original design. The main differences encountered are the switching of the "hot" and "cold" sides for several supports and significant amplification of deflections in the vicinity of some seismic bumpers. The first difference becomes especially significant in conjunction with the fact that the centerlines of the beams were originally offset toward the design "hot" side determined from the 5% friction assumption. The seismic bumpers generally negate effects of the second difference (amplified deflections). Using the zero friction model as basis for analysis produces good results when compared with actual field conditions. This is consistent with "shakedown" effects for structural systems where elements involving frictional constraints tend to migrate over time to a lower energy state.

Once these deflected shapes of the models were calculated and calibrated to observed pipe locations, the dynamic effect is then imposed on these deflected shapes to find the additional movement due to seismic excitation. The dynamic analyses themselves retained the assumption of 10% friction coefficient, consistent with the original design submittal, since the friction coefficient is considered fully applicable for short duration loadings.

Seismic Analysis

The pipeline alignment is divided into five distinct seismic zones. The northern part of the line is designated as a relatively low, 5.5 Richter Magnitude (RM) zone, while the two highest (8.0 and 8.5 RM zones) are located in the southern end of the line and near the fault crossing zones close to the Alaska Mountain Range, respectively. These two RM zones were assigned the same design ground accelerations and displacements and are, therefore, normally analyzed together.

The design spectrum is the basic descriptive technique for dynamic excitation for structures. It is much more descriptive of excitation than a single "g" level since it is well known that response acceleration levels vary with frequency. There are two basic spectra for the TAPS project, developed during the original design, which correspond to the "Operational" and "Contingency" loading. Operational loading defines the excitation level which the project structure must withstand and remain fully operating, while the contingency loading is the excitation level which the structure must withstand without catastrophic failure. At the contingency level, it is recognized that some damage may occur, and that a safe shutdown of the structure may be required for inspection and repair. Often, the operating level is used to compare structure excitation so that members remain within stress allowables consistent with wind loading and other occasional loads, while contingency allowables are increased to the yield value.

There are several specific techniques to approach the analysis of structures subjected to dynamic loading, but they can all be loosely categorized into three main approaches: Equivalent Static techniques, Response Spectrum/Modal techniques, and Time History techniques. The Equivalent Static Procedure and Response Spectrum Procedure, which are the easiest to apply and also the most commonly used procedure, have implicit assumptions which make it inapplicable to the TAPS pipeline. First, these techniques assume the structure is to have a linear response, or at least linear about a defined equilibrium position. The presence of friction and nonlinear stops and gaps violates this assumption. Also, these common techniques assume that the excitation at the base of the structure is uniformly applied in time, i.e. all supports are excited in phase by the same time loading function. However, ground excitation is best characterized by a traveling wave, which imparts motion to the surface as it travels at a seismic wave velocity. This wave velocity (typically 1000 m/sec or 3283 ft/sec) imparts a lag to excitation for structures which have a considerable length relative to the wave velocity. For these reasons, only the Time History Techniques are suitable for the analysis of TAPS aboveground configurations.

Unfortunately, the Time History Technique is also the most demanding dynamic analysis procedure since it imparts the excitation as a prescribed function of excitation versus time to the structure, and the structure is monitored over the duration of excitation for response. This requires a discretization of the excitation input with time, an analysis of all structural response during each discrete time interval, and then evaluation of the new state of the structure which forms the initial basis for the next excitation discrete interval. This technique has the advantage that nonlinear events can be incorporated, and commercial computer programs are available for this. Because of the small axis of most structures, however, it is unusual to find an analysis program that will readily analyze the lagged excitation for long axes structures, such as the aboveground pipeline configurations of TAPS.

The analytical technique requires a time history of the motion and subsequent integration of the response. The design spectra for the project form the basis for the aboveground seismic loading. Earthquake time history motions were artificially generated using random generation routines governed by statistical procedures so as to replicate the overall frequency-response characteristics of the basis spectra. In order to ensure that the response is bounded by the analysis, it is standard practice to generate an ensemble of these histories each of which have slightly different frequency characteristics. Five such earthquake time histories were generated. Each record is an acceleration time history with accelerations recorded at a constant time interval of 0.03 seconds. There are a total of 1,664 accelerations for each record, so each record simulates an earthquake history with duration of 49.89 seconds. The seismic motion input required is the same used by the original design, generated by Dames and Moore at that time.

The analytical technique is to either run all five earthquake records for each direction of motion (while running another one of the five histories in the orthogonal direction) or to identify which combination produces the response envelope for the structure class under consideration. The original design studies concluded that a combination of the earthquake record #3 applied parallel to the seismic wave travel direction and the earthquake record #2 applied normal to the seismic wave travel direction (the “3x2 combination”) produced the maximum response. This study also found that the “3x5 combination” could produce the maximum response for some configurations for the new set of initial condition, so all analyses were subjected to both earthquake combinations.

In addition, it is recognized that the "attack angle" of the earthquake (the angle between the seismic wave front and the structure axes) should be specified so as to produce the maximum response. This procedure was evaluated during the design phase where it was determined that an attack along the longitudinal axes of the configuration was most conservative.

Analysis Tools

During the original design, the TAPS project recognized that time history analysis was applicable to the analysis and design of the TAPS pipeline. Lacking an applicable computer analytical tool, the unique program called DrainPIPE was developed specifically for the TAPS project by Graham Powell, Ph.D., former Professor at the University of California, Berkeley. This program allows friction, gaps, and stops at supports for the pipeline, as well as proper consideration of the traveling wave excitation. The DrainPIPE program was originally written in FORTRAN to run on a CDC 6600 mainframe computer. The original card deck of the program source code was translated to DOS and recompiled using the Lahey FORTRAN compiler for use on the personal computer.

Output DrainPIPE files from the original design and new input decks for the PC version were prepared to ensure that the results for the analyses would be comparable. The DrainPIPE program is a complex numerical simulation program, and every analysis involves hundreds of time integration steps. Thus, only a small number of comparable results that use all options and element types of the program is required to instill confidence that the DOS program version is operational. The DrainPIPE PC version of the program was validated by comparing analytical results on the PC program with available original design output files.

The same input earthquake records were used for all RM zones, and for both the Operational and Contingency loadings, but scaled so as to recognize the different excitation levels while still preserving the frequency content of the records.

The following scale factors are applied to acceleration.

8.0/8.5 RM, Contingency	Unscaled (i.e. scaling factor = 1.0)
7.5 RM, Contingency	Scale factor = 0.67
7.0 RM, Contingency	Scale factor = 0.44
5.5 RM, Contingency	Scale factor = 0.30

The Operational loading is found by taking 50% of the corresponding contingency scale factors for acceleration.

Viscous damping values for typical structural analyses use estimates of 2% to 5% for analysis. However, the DrainPIPE analyses explicitly models energy dissipation, using frictional elements and stop elements with slip after yield. Therefore, only a nominal value of 0.5% damping is input to DrainPIPE so as to control the numerical integration procedure.

Structural elements were modeled to incorporate not only their initial elastic resistance, but also their yield points and resistance after yield. Successive yield points were input so as to simulate successive failures of the component systems – for example insulation crushing then followed by bending failure of the VSM and then ovallization of the pipe.

Results

The 7-4-7 Zee configuration is the most prevalent used in the design of the pipeline and therefore was chosen as the "Base Case" for the analyses. This configuration consists of three straight segments joined by two 12.5-degree 36.6-meter (120-foot) radius field bends (See Figure 3). The first segment from the anchor is twelve 18.3-meter (60-foot) nominal spans with a horizontal 12.5-degree field bend at the end of the segment. Next is a 128-meter (420-foot) diagonal segment consisting of seven 18.3-meter (60-foot) nominal spans and another 12.5-degree bend in the opposite direction from the first bend. The final 219.5-meter (720-foot) segment is parallel to the beginning segment, offset 27.7 meters (91 feet), and ends at another anchor. Adjacent to this "target" configuration, a further configuration was added beyond each anchor to replicate the influence of these adjacent "boundary" configurations on the target configuration.

The base case scenario includes the following model parameters:

1. Initial gaps between the insulation module and the VSMs at the pipe centerline level at all supports are based on static analysis results for a frictionless case.

2. No friction preload on intermediate supports. DrainPIPE has a provision to preload the frictional supports, using the results of previous static analyses. This preload was not used, which is consistent with the results of the frictionless static analysis case.

3. Anchor preload forces of 333.6 kN (75 kips) directed inward toward the target configuration were imposed on the anchors at the ends of the target configuration, with no preload forces imposed on the anchors at the ends of the boundary configurations. This is consistent with the original design analysis.

The DrainPIPE analysis results are examined for nodal and elemental responses on the target configuration. Responses on the boundary configurations, though reported by DrainPIPE, are ignored since the boundary configurations are only modeled to provide boundary conditions to the target configuration, i.e. their response is incidental to the focus of the analytical investigation.

The analysis results for this example configuration is shown in Table 1. The different results correspond to global (X and Y) displacements. The long legs of the 7-4-7 configurations were modeled as parallel to the X global direction, and the Y global direction is perpendicular to these legs. The maximum X displacement for the 8.5 RM case is shown as 17.0 centimeters (6.7 inches) which is a combination of elastic deformation and slip. This occurred at the first anchor of the configuration. The minimum X displacement is given as –10.2 centimeters (-4.0 inches), but this occurred at the last bend of the configuration. The maximum Y displacement was 48.5 centimeters (19.1 inches) and occurred at the last support before the first of the configuration, while the minimum Y displacement of –49.0 centimeters (-19.3 inches) occurred at the last bend of the configuration. For all analytical results, the DrainPIPE analyses were scanned to find the minimum and maximum values regardless of their location in the target configuration.

The anchor slip at the first anchor in these example results of the target configuration is noted as 16.3 centimeters (6.4 inches). This value indicates that the anchor has overcome its initial frictional resistance and has slipped. At the other end of the configuration, the anchor shows a displacement of –7.2 centimeters (-2.8 inches), with a slip of –6.3 centimeters (-2.5 inches). The same modeling was used, so it can be concluded that this anchor has also overcome its initial frictional resistance.

The analyses indicate that all four bumpers will be impacted with the maximum bumper force approximately 97.9 kN (22 kips) occurring at the bends nearest the transverse leg. Note that for the bumper and intermediate reactions, the frictional force is not added to the stop reactions. Contact at other points is also noted at the center of the 219.5-meter (720-foot) leg, and at the support next to the bumper, for a total of four intermediate contacts. The maximum reaction at these points is approximately 89.0 kN (20 kips) located at the support next to the bumper.

This type of table was completed for the hundreds of DrainPIPE analyses concluded. To expedite the analyses and ensure consistent quality, additional digital procedures were developed to extract summary tables from the DrainPIPE results for comparison and evaluation. In addition to the example Zee configuration, the analysis considered other types of standard configurations such as transitions from belowground to aboveground moding. Finally, special analyses were conducted for configurations that required site specific evaluation such as the TAPS configurations containing bridges and mainline valves.

A review of all configurations analyzed indicates that the pipe stress does not exceed the design criteria as presented in DB-180 (APSC 1997). The highest intermediate contact forces are reported from the analysis of the Zee at just less than 89 kN (20 kips) near the center of the longitudinal leg where a zero initial gap was specified from static analysis. The 720 Vee configuration (two straight segments connected by 12.5-degree bends to effect a change in alignment direction) produces the highest bumper force of 360.3 kN (81 kips). This is in good agreement with the original analysis results for the design of TAPS, which also indicated that the 720 Vee case controlled the bumper forces.

Table 1 – Analysis Summary

Case	Nodal Displacements (cm)				Anchor Slip (cm)		Intermediate Bents		Bumper Bents	
	Max X	Min X	Max Y	Min Y	Pos	Neg	Contacts	Max Force (kN)	Contacts	Max Force (kN)
8.5/8-C	17.0	-10.2	48.5	-49.0	16.3	-6.4	4	88.1	4	95.6
8.5/8-O	9.9	-8.1	35.6	-31.8	9.1	-5.1	3	28.9	2	59.2
7.5-C	7.6	-6.6	29.5	-25.4	9.4	-3.8	2	24.5	2	30.7
7.5-O	4.1	-4.1	14.7	-13.7	2.5	-1.5	2	17.8	2	28.0
7.0-C	3.0	-3.3	10.4	-7.4	1.3	-1.0	2	17.3	2	25.8
7.0-O	1.8	-1.8	4.8	-5.6	0.3	-0.3	2	15.1	2	16.5
5.5-C	1.5	-1.5	4.1	-5.3	0.0	0.0	2	13.3	2	14.7
5.5-O	0.8	-1.0	2.0	-3.8	0.0	0.0	2	6.7	2	9.3

Future Evaluation Procedure

Based on the results of the analyses, a decision tree was developed to guide further assessment of VSM contact. The decision tree is considered a conservative procedure for screening contacts so as to determine the need for additional analyses and/or remedial measures. The decision tree calls for site specific analysis of the contacts, and appropriate remedial action whenever the insulation module of an intermediate bent is indented more than 3.8 centimeters (1.5 Inches). Intermediate bent contacts with up to 3.8 centimeters (1.5 inches) of indentation go through several screening steps before they can be considered acceptable. Contacts in the 5.5 or 7.0 RM zone with up to 3.8 centimeters (1.5 inches) of indentation are considered acceptable, based on the results of dynamic analyses for those areas. Dynamic analyses show that maximum expected lateral movements, due to seismic excitation, including the restraining effect of the insulation module, are approximately 5.1 centimeters (2 inches) and 12.7 centimeters (5 inches) in the 5.5 and 7.0 RM zones, respectively. These values would be expected to increase if the restraining effect of the insulation was removed. These values are the maximum anticipated movements anywhere in the configuration and actual movements at contact locations could be much less. The maximum values are well below the actual clearance of 20.3 centimeters (8 inches). This is calculated from 24.1 centimeters (9.5 inches) between the module and the pipe clamp flange minus the maximum 3.8 centimeters (1.5 inches) of indentation.

In the 7.5, 8.0 and 8.5 RM zones, the maximum expected movements without the restraint afforded by the insulation module could be greater than 20.3 centimeters (8 inches), and so cannot be immediately eliminated as acceptable under contingency seismic loading. For these locations, the expected displacement at the contact location must be identified by including the effects of the restraint afforded by undamaged insulation modules in further bounding dynamic analyses.

Conclusion

In conclusion, the contact situation at transitions can be forecast reliably with assuming zero friction in static analyses with operating loads corresponding to the current operating condition. Starting with these initial conditions, time history dynamic analysis is then used to predict expected motion and forces during the design earthquake events. For this analysis, the specially formulated DrainPIPE program, which incorporates the needed elements of friction, successive yielding of structural components, and time-lagged excitation corresponding to the traveling wave velocity, is the best available tool for use in the calculations. A project procedure for utilizing the tool was developed to expedite the hundreds of analyses performed, as well as to ensure consistent quality of the analysis.

The results confirmed that the TAPS aboveground configurations are generally able to resist expected seismic forces. In those few cases where TAPS criteria could have been exceeded during seismic events, field retrofits of additional seismic bumpers were completed. In addition, a monitoring tool, based on a flow chart developed in concert with the analysis was developed for continued monitoring of the configuration.

References

Alyeska Pipeline Service Company (1997), "Design Basis Update, DB-180, Third Edition, Revision 0," Anchorage, Alaska.

Geometry Monitoring of the Trans-Alaska Pipeline

James D. Hart[1], Graham H. Powell[2], David Hackney[3], Nasir Zulfiqar[4]

Abstract

An important aspect of the ongoing operation of the buried sections of the Trans-Alaska Pipeline System (TAPS) is the curvature imposed on the pipeline by differential settlement of the surrounding soil as a result of thawing of ice-rich subsurface soils. High pipe curvature is a source of concern since it can potentially lead to wrinkling of the pipe wall and/or tensile fracture of the pipe at girth welds. Alyeska Pipeline Service Company has monitored pipeline settlement using different techniques since start up of the line in 1977.

This paper provides a brief historical overview of how the techniques used to monitor pipe settlement have evolved over the life of TAPS. The paper also provides a discussion of the how the pipe geometry is currently monitored using the BJ Pipeline Inspection Services (BJ) Geopig® "smart" pig. The discussion includes a brief description of the Geopig measurements which are numerically processed to obtain the three-dimensional pipeline displacement and curvature (bending strain) profiles. Several illustrations of this technology are presented. The paper concludes with a summary of how Alyeska's state-of-the art pipeline geometry monitoring program is used to help make decisions related to maintaining the structural integrity of TAPS.

Introduction

It was recognized during the TAPS design that a hot oil pipeline to transport North Slope crude oil to Valdez would have to cross permanently frozen ground. In areas of known thaw unstable ground, the pipe was elevated to keep this ground from thawing. In areas were the pipe was buried, it was recognized that all the thaw unstable ground may not be identified, and differential settlement could potentially occur. Differential settlement of the soil in which a pipeline is buried can impose

[1] President, SSD, Inc., Reno, NV, USA
[2] President, Graham H. Powell, Inc., Danville, CA, USA
[3] Engineering Coordinator, Alyeska Pipeline Service Company, Fairbanks, AK, USA
[4] Project Engineer, SSD, Inc., Reno, NV, USA

significant curvatures and strains on the pipe. If left unchecked, excessive pipe curvature and strain can eventually lead to damage of the pipe, for example due to localized wrinkling of the pipe wall and/or tensile fracture of the pipe at girth welds.

Based on this concern, a monitoring program was required by the Stipulations to the Grant of Right of Way. Alyeska Pipeline Service Company (Alyeska) has monitored potential settlement areas of TAPS using different techniques since start up. This paper provides an overview of how the techniques used to monitor pipe settlement have evolved over the life of TAPS. The pipeline geometry is currently monitored using the BJ Geopig "smart" pig. The Geopig measurements provide a remarkably accurate means of computing the precise geometry of the pipeline including locations with high curvature, ovality or denting of the pipe wall, and locations of girth welds and other features.

History of Alyeska's Pipeline Geometry Monitoring Program

Since start-up of TAPS in June of 1977, a variety of techniques have been used to monitor pipe settlement. Initially, the pipeline was monitored using an inline inspection device called the "Super Pig" (Anderson, 1979). The data from this pig was difficult to interpret and in 1979, it became stuck at a check valve and was destroyed during recovery.

In 1980, Alyeska began development of a technique to estimate pipeline curvature based on the use of polynomial curve fits of the pipe profile based on measurements at a series of discrete monitoring points (Simmons, 1988). The idea was to measure top-of-pipe elevations at discrete points across a settlement zone and compare these measurements with the as-built top-of-pipe elevations. The discrete measurement points consist of aluminum "monitoring rods" that are permanently cad-welded to the top of the buried pipe. The tops of the monitoring rods are exposed at the ground surface to facilitate measurement of elevation changes using conventional surveying techniques. A series of monitoring rods across a settlement zone provided a basis for estimating the pipe profile through the settlement zone and allowed for observations on the rate of settlement. Alyeska has over 2300 of these monitoring rods in place throughout 612 km (380 miles) of buried line. The curvature estimation method uses a divided difference approach to establish an interpolating polynomial for the pipe elevation profile. The resulting profile polynomial is then differentiated twice to obtain estimates of curvature. The accuracy of this procedure is influenced by the degree of the polynomial, the spacing of the monitoring rods and the accuracy of the pipe elevation measurements. Based on experience, it was determined that a 5th degree polynomial used with a monitoring rod spacing of between 3 and 15 meters (10 and 50 feet) and an elevation accuracy of ± 0.6 cm (± 0.02 feet) resulted in acceptable estimates of the pipeline curvature.

In 1981 Alyeska began using a Deformation Pig developed by Tubescope, formerly Vetco Pipeline Service Company (Tubescope, 1981). This "caliper pig" proved very accurate in measuring and locating dents, ovalities and pipeline wrinkles. The main disadvantage of the Deformation Pig is that it would not provide an indication that the pipe was settling and bending until the pipe wall had experienced plastic deformation on the compression side and formed a wrinkle. Alyeska has also used an electronic device to monitor the elevation of below ground pipe. This device, which was dubbed the TSI for Technical Service Instrument (TSI, 1981) used a radio signal induced in the pipe and a directional antenna used by a field crew to measure the depth of the mainline pipe below the ground surface. Unfortunately, the TSI is only accurate to within 0.3 m (1 foot) in flat ground. This survey method, while useful for locating the pipe was not adequate to measure progressively increasing bending deformations.

Each of these monitoring methods had deficiencies and it became increasingly clear that a more advanced method was required. The declassification of cruise missile technology by the U.S. Government in the late 1980s allowed for the development of an internal inspection device based on a sophisticated Inertial Navigation System (INS) called the Geopig. In 1992, the first run of the Geopig through TAPS was successfully completed. The Geopig data provided not only the pipe elevation and vertical curvature profiles but also the horizontal profile and horizontal curvature of the pipe as well as accurate locations of girth welds, pipe ovality and field (sag, over, side and compound) bends. The development of the Geopig tool advanced Alyeska's pipeline geometry monitoring program to the cutting edge of this technology.

Description of the Geopig and Geopig Data

The 1219 mm (48-inch) diameter Geopig tool is illustrated in Figure 1. The pig is supported on four flexible urethane support cups; a pair near the front and a pair near the back of the pig body. The distance between the centers of the front and back cups is about 1.67 meters (5.5 feet). The Geopig is equipped with a counterweight in order to maintain a constant roll angle in the pipe. Due to its self weight, the pig is in more intimate contact with the 6:00 position than with the 12:00 position of the pipe cross section. This is evident from the substantial amount of cup wear (up to 0.25 cm or 0.1 inches) on the bottom (6:00) side of the cups. The "steady-state" elevation of the pig body at the front supports is lower than the elevation of the pig body of the back supports, indicating that the pig travels through the pipeline in a slightly "nose heavy" position. The pig travels through the line at the same speed as the oil flow, which for recent TAPS flow rates is about 2.5 m/sec (100 inches/sec).

The Geopig collects data from the following independent types of instruments; gyroscopes, accelerometers, calipers, and odometers. The pitch, roll, and yaw rotations of the pig are recorded using three gyroscopes at a sampling frequency of 50 Hz. The three-dimensional accelerations of the pig are recorded using three accelerometers at a sampling frequency of 50 Hz. Four odometers measure the position of the pig at a sampling frequency of 4 Hz. In its original configuration, the Geopig had a total of 96 sonar transducers (80 on the front sonar ring, and 16 on the back sonar ring) which measured the shape of the cross section of the pipe at a sampling frequency of 32 Hz. Recently, the Geopig sonar calipers have been replaced with one ring of 64 contact caliper "fingers" around the circumference. There is also a separate analog girth weld detector system. More details regarding the Geopig data and its accuracy can be found in (Czyz, 2000).

The most important data for use in curvature screening is the gyroscope and odometer data. The resolution of the gyroscope data is 2 arc sec (10 μrad). The odometer data is interpolated to the same spacing as the gyroscope data, which at the 50 Hz sampling rate, results in data spacing of approximately 5 cm (2 inches). Note that the BJ GeoDisplay® software (BJ, 2001) scales and rotates the odometer data in order to provide a match with GPS position control points located every 5 to 10 kilometers along the TAPS alignment. This procedure results in an odometer accuracy of 1 in 2000 (i.e., $\Delta L/L=1/2000$).

The second most important data from the Geopig is the caliper data, which can be useful in investigating the effects of dents and/or asymmetric ovality in conjunction with the gyroscope and odometer data. The calipers can measure the inside radius of the pipe to within an accuracy of ±1.27 mm (±0.05 inches). The calipers are sampled at a frequency of 128 Hz. BJ indicates that the caliper data is able to successfully locate incipient wrinkles with amplitudes less than or equal to 6.4 mm (0.25 inches) over lengths ranging from 30 to 71 cm (12 to 28 inches). The caliper data is the key to an accurate centerline correction that can be applied when the path of the pig is significantly different from the path of the pipe centerline (such as when it travels over an asymmetrical ovality or a dent). In this situation, the dent/ovality feature can erroneously manifest itself in the gyroscope data as a pipe curvature feature. Centerline correction is discussed in a later section of this paper.

Geopig Data Processing Applications

A detailed discussion of the numerical methods used to process the Geopig data to obtain pipe curvature and three-dimensional coordinates is beyond the scope of this paper. Some of the procedures are discussed in other references (Czyz, 2000). In essence, the gyroscope and odometer data is numerically differentiated to obtain the curvature of the pig path and numerically integrated to obtain three-dimensional coordinates of the pig path. This section provides some illustrations of results from Geopig data processing.

Illustration of Noise in Geopig Gyroscope Data

Figure 2 presents an illustration of the pitch data from the 1994 Geopig run across the Salcha River. Figure 2(a) presents the relative pitch angle vs. odometer data and Figure 2(b) presents the corresponding power spectral density spectrum of the pitch data. The location of the high curvature zones are coincident with the steep sections of the pitch profile. In order to focus on the frequency range where most of the energy is concentrated, the spectral plots are shown only over a zoomed frequency range from 0 to 10 Hz. As previously noted, for a given frequency, the equivalent wavelength is estimated by dividing the average pig speed by the frequency. The pitch spectrum plots indicates the presence of a component of "pitch energy" centered at a frequency of about 8 Hz which corresponds to a feature length of about 34 cm (13.5 inches). Subsequent investigation of the noise, including examination of similar noise features in pipe joints from three different pipe manufacturers indicates that the noise features are actually different for different pipe manufacturers. The most likely cause of the noise is that the pig is responding in a pitching motion as it traverses small, periodic diameter changes (expander marks) in the pipe joints that result from the (UOE) pipe manufacturing process. None of the available evidence indicates that the noise is in any way related to real pipe curvature demand. In order to illustrate the influence of this noise on the computed curvature, the Salcha River pitch data was processed to compute curvature using a range of gage lengths. Figure 3 presents vertical bending strain profiles computed from the pitch and odometer data across the Salcha River using different gage lengths (from about 3 meters to 0.3 meters). These plots show that high-frequency oscillations are present in the computed curvature and that the amplitude of the noise increases with decreasing gage lengths (especially for gage lengths shorter than the length of the pig). Again, the dominant wavelength of this "noise" is about 34 cm (13.5 inches). It is apparent that the "noise" component has a significant effect on the computed bending strain (i.e., curvature).

Effect of Filtering

Figure 4 presents the same information as shown in Figure 3 except in this case the data has been low pass filtered using a filter cut-off frequency of 1.5 Hz. This cut-off frequency was selected to match the length of the Geopig and because it will remove both the low amplitude noise in the signal near 4 Hz and the more significant noise near 8 Hz. The effect of the filtering is very pronounced since it virtually eliminates the bothersome noise and provides a much clearer picture of the pipe curvature.

Discussion of Gage Length

There are two lengths involved in the curvature calculation, namely the pig length and the gage length. The pig length is approximately 1.67 m (5.5 feet), and is fixed. The gage length can be varied, and has a minimum value of approximately 10 cm (4 inches). The pitch (or azimuth) of the pig is the difference in elevation over the pig length, divided by the pig length. It is important to note that this is not the exact pitch at the center of the pig, but an average value over the pig length. The calculated curvature is the difference in this average pitch divided by the gage length. No matter how short the gage length, the pitch will always be an average value, and hence the exact maximum curvature can never be calculated (except in the rare case where the curvature is uniform over a substantial length). The calculated curvature will, however, be progressively more accurate as the gage length is reduced. The Geopig has remarkable accuracy, with a pitch resolution of 2 arc seconds, or 0.00001 radians. For a 10 cm (0.33 foot) gage length, an error of 0.00001 radians in pitch corresponds to an error of 0.00001 radians / 10 cm = 0.000001 cm^{-1} = 0.00003 ft^{-1}. This is only about 1.5% of the wrinkling curvature capacity of TAPS pipe (the curvature capacity varies from roughly 0.00005 cm^{-1} to roughly 0.00008 cm^{-1}, depending on the pressure, wall thickness and other parameters). Hence, the pig is sufficiently accurate to allow a gage length of 10 cm (4 inches) to be used for detailed curvature comparisons (based on a pig speed of 2.54 m/sec (100 inches/sec) and a sampling rate of 50 Hz). However, in order to use this short of a gage length, low-pass filtering of the data is typically required.

Centerline Correction Processing

At one of the Salcha River crossing high curvature locations, the pipeline curvature is of overbend type (i.e., with the longitudinal compression strain located at the bottom of the pipe). The caliper data from the Geopig shows that there is what can be characterized as an asymmetrical ovality of the pipe at this location. This ovality extends over a length of roughly 6 to 7.6 m (20 to 25 feet) and has a maximum depth at the bottom of the pipe of about 2.54 cm (1 inch). As the Geopig traverses this location, it must ride up and over this ovality, thus causing the pitch angle recorded by the pig to deviate from that of the centerline. As a result, the calculated curvature has two components, one associated with the overbend curvature and one associated with the ovality. At the point of maximum curvature, the two components have the same sign, and hence the calculated curvature is significantly larger than the overbend curvature alone. Because the gyroscope measures the pitch (slope) of the pig, rather than that of the pipe, the pitch measurements must be corrected to obtain the pitch of the pipe. This "pig-to-pipe attitude correction" or centerline correction can be performed using the caliper data. A very simple approach is described below.

Figure 5 shows contours of the pipe ovality along the bottom half of the pipe at this location. The figure also shows a dashed line associated with the 6:00 o'clock position of the pipe. The maximum height of the idealized ovality is just under 2.54 cm (1 inch) and it extends over a length of about 6 m (20 feet), i.e., from circular to maximum asymmetrical ovaling and back to circular over about 6 m. It is assumed that an ovality of 1 unit on one side of the pipe corresponds to a change in the location of the effective pipe axis of 0.5 units (i.e., a maximum axis shift equal to one half of the depth for a one-sided ovality). The assumed centerline profile is shown in Figure 6(a). By running a "mathematical pig" with a pig length of 1.67 m (5.5 feet) over this profile, it is possible to develop a corresponding profile of pig pitch angle based on the elevations of the front and back pig support cups. The resulting pitch profile is shown in Figure 6(b). Based on this pitch profile, the vertical bending strains computed using different gage lengths are shown in Figure 6(c). The curvature (or bending strain) of the pipe centerline can now be estimated by subtracting the curvature due to the ovality from the total vertical curvature. Application of a centerline correction resulted in a 40% to 50% reduction in the curvature demand at this location.

It is also possible to remove the effect of the unsymmetrical ovality using low pass filtering, provided that the length of the ovality feature is less than the length of the true curvature feature. Note that in order to be consistent, removal of dent/ovality features from the curvature demand requires that the effect of the curvature and ovality features, on the curvature capacity, are properly considered.

Conclusions

Since the Geopig first successfully traversed TAPS in 1992, the Geopig data has been looked at in a variety of new and different ways and a keen understanding of how to best use the data from this remarkably sensitive and accurate tool has been developed.

This paper has provided an overview of various techniques used to process the data from the Geopig. Alyeska uses these techniques for routine screening as well as detailed evaluation of data from a Geopig run. Advanced processing techniques can be used to examine high curvature areas, to make run-to-run profile comparisons or to perform centerline corrections at locations where high curvature is coincident with un-symmetric ovalities or dents. Ultimately, decisions regarding the pipeline structural integrity are based on a demand-capacity comparison. The computed curvature demand is compared to estimates of the curvature capacity which depend on the internal pressure, the wall thickness, the proximity to pipe girth welds and other factors. Special care and engineering judgment must be exercised to ensure that the calculated results are consistent and meaningful especially in relation to filtering of the data and selection of the gage length for calculating curvature demand for use in screening the data vs. the gage length for performing detailed demand-capacity comparisons.

Acknowledgements

The work described in this paper is based largely on projects undertaken by SSD, Inc. for Alyeska Pipeline Service Company. Alyeska's support for this work and permission to publish this paper is most gratefully acknowledged. The authors would also like to thank Dr. Jaroslaw A. Czyz and Mr. Joel Falk of BJ Pipeline Inspection Services for their efforts in support of these projects.

References

Anderson, D.R. (1979). "System Monitors Trans-Alaska Pipeline Curvature", Oil & Gas Journal, June 25th.

BJ Pipeline Inspection Services (2001). "GeoDent® and GeoDisplay® Software", Calgary, Alberta, Canada.

Czyz, J.A., and Falk, J. (2000). "Use of Geopig for Prevention of Pipeline Failures in Environmentally Sensitive Areas." The Pipeline Pigging, Integrity Assessment and Repair Conference, Houston, February.

Simmons, G.G., and Alto, J.V. (1988). "Pipeline Curvature by Polynomial Approximation." Proceedings of the Conference on Pipeline Infrastructure, ASCE Pipeline Division, Boston, Massachusetts, June 6th & 7th.

Technical Service Instrument: Model "FerroPhon" by Herman-Sewerin GmbH, Gutersloh, Germany, 1981.

Tubescope/Vetco Pipeline Service Company, 48-inch Diameter Deformation Pig for Alyeska Pipeline Service Company, Houston Texas, 1981.

Figure 1: Illustration of 48-inch Diameter Geopig
(Image Courtesy of BJ Pipeline Inspection Services)

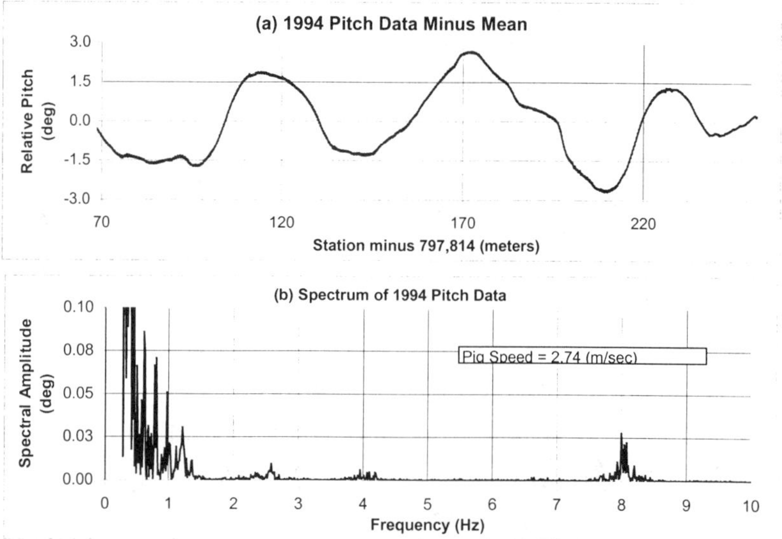

Figure 2: Illustration of Pitch Data from Salcha River Crossing

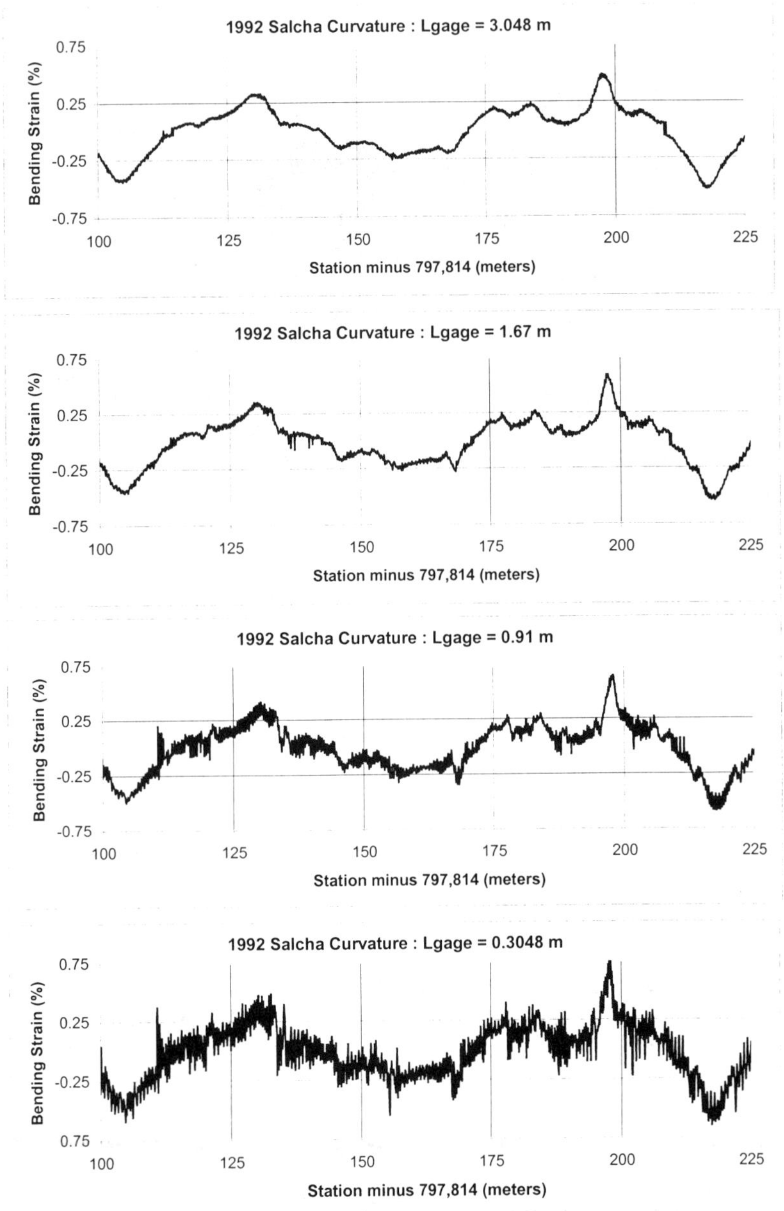

Figure 3: Illustration of Unfiltered Curvature Data from Salcha River Crossing

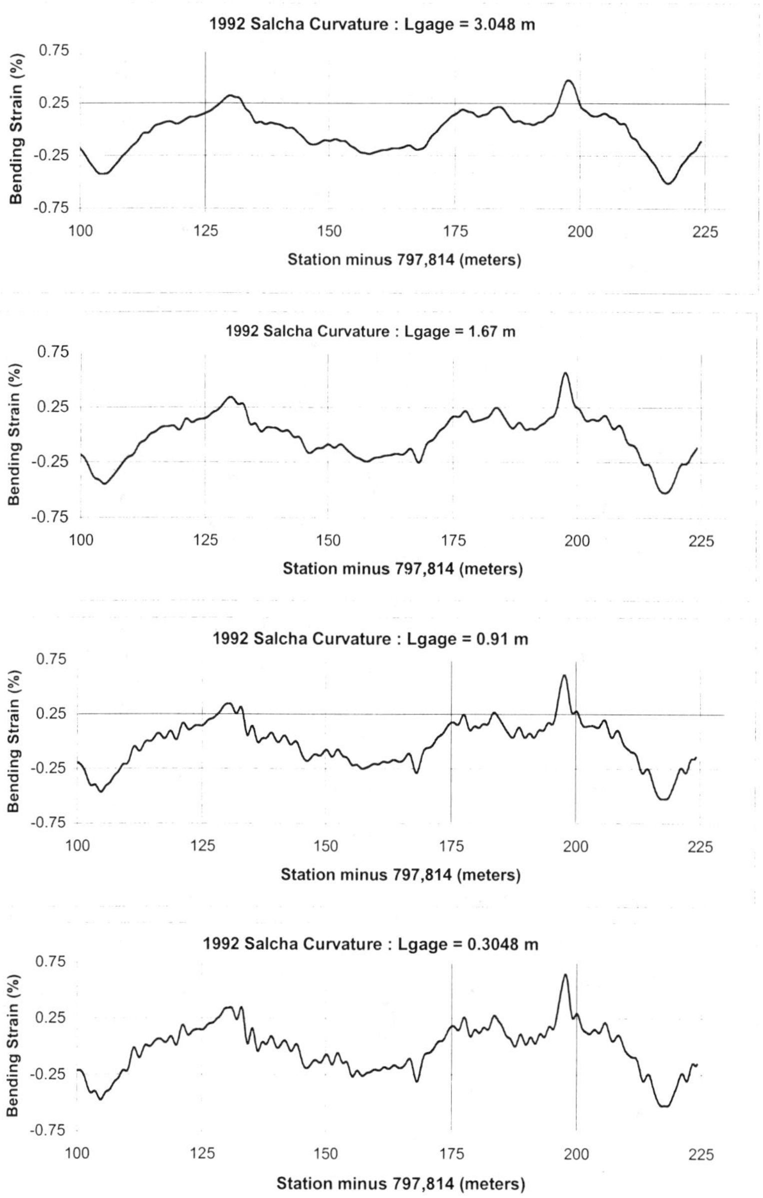

Figure 4: Illustration of Filtered Curvature Data from Salcha River Crossing

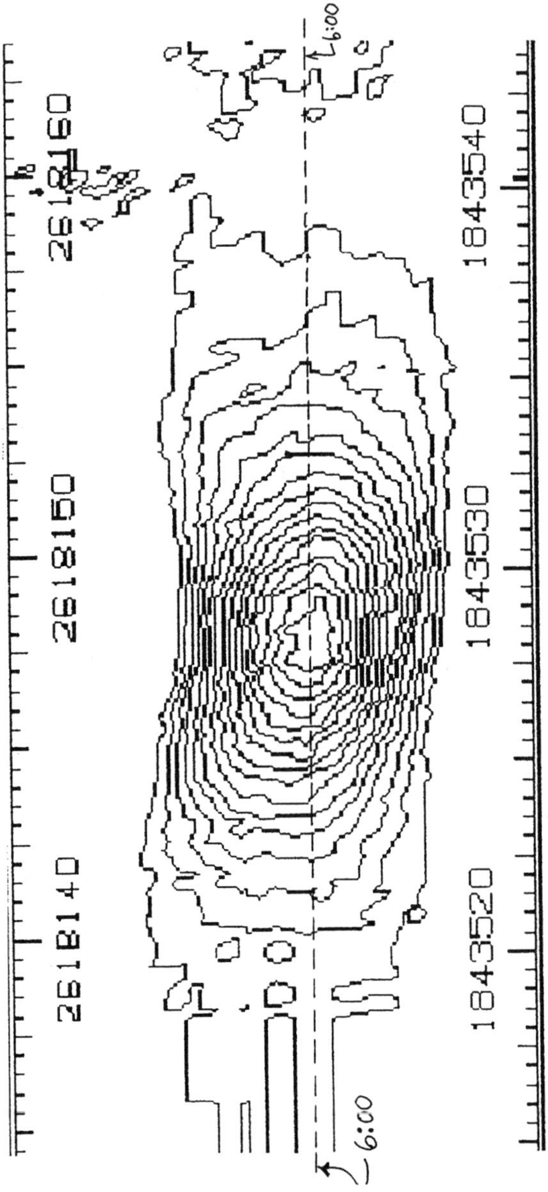

Figure 5: Contours of Ovality at Salcha River High Curvature Zone

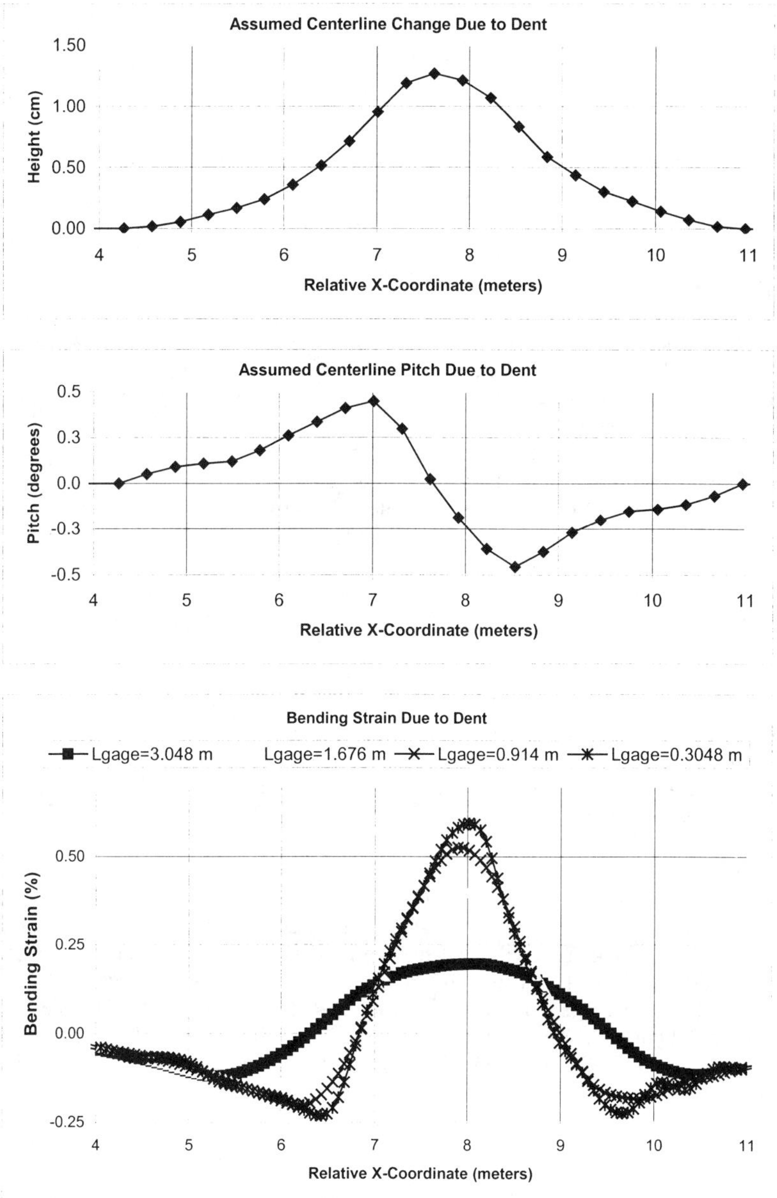

Figure 6: Illustration of Profile Due to Dent at Salcha River High Curvature Zone

Useful Life of the Trans-Alaska Pipeline

J. David Norton, P.E.[1], and Janna Endell Miller, P.E.[2]

Abstract

Useful life is defined as a combination of *design life, physical life,* and *economic life.* The useful life of TAPS can be described as the period during which the pipeline provides a safe, environmentally sound, economically viable transportation link to get Alaska North Slope crude oil to market. Useful life is also one of the three primary elements for establishing the duration of the federal and state rights-of-way for TAPS, the others being serving a public purpose or benefit, and the ongoing facility costs.

This paper addresses the following issues with special emphasis on facility aging and life-cycle issues associated with operating a warm crude oil pipeline in cold regions:

- TAPS design life is based on the incorporation of robust components coupled with system monitoring programs designed to detect and counteract aging and facility use factors.
- TAPS physical life can be virtually unlimited given the execution of appropriate maintenance, repair and replacement programs.
- TAPS economic life is governed by the extent of recoverable North Slope crude oil reserves. Predictions show these reserves being produced in quantities sufficient to support the continued operation of the pipeline well into the 21st century.

Introduction

The Trans Alaska Pipeline System (TAPS) has transported over 13 billion barrels of Alaska North Slope crude oil since start-up in 1977. The pipeline currently carries nearly 20 percent of all crude oil produced in the U.S., and it can continue for decades to provide a critical link in the supply of a significant share of the nation's crude oil.

In its 800-mile (1288 km) length from the North Slope oil fields to the port of Valdez, the pipeline crosses hundreds of miles of state and federal land. Before constructing and operating the pipeline on those lands, the TAPS Owner companies were required to obtain long-term, renewable rights-of-way from the government land

---

[1] Consultant, J.D. Norton and Associates; dnorton@alaska.net

[2] Reliability Centered Maintenance (RCM) Implementation Lead, Alyeska Pipeline Service Company; millerjk@alyeska-pipeline.com

management agencies. These rights-of-way, issued in 1974 for 30-year terms, expire in 2004. The TAPS Owners' application for renewal of these federal and state rights-of-way seeks the maximum terms for the renewals.

In considering the length of a renewal term, federal and state right-of-way laws invoke several criteria — the useful life of the pipeline, its public purpose, and its cost. The focus of this paper is the useful life of the pipeline, which comprises both physical and economic utility. The physical life span of a pipeline is determined primarily by both the quality of its original design and its upkeep. The economic life of a crude oil pipeline is determined by how long it can provide its owners with a reasonable economic return and attract shippers.

Useful Life

While the term "useful life" is not defined in the applicable laws or regulations, the useful life of TAPS seems clearly meant to describe the remaining life of the pipeline, which is a function both of its physical life span and its economic utility. The physical life of the pipeline is determined principally by the nature of its original design and subsequent maintenance, repair, and replacement activities. The economic life of the pipeline is determined by how long it provides reasonable economic return, considering original investment and operating costs.

For purposes of this document, *useful life* is defined as a combination of *design life, physical life,* and *economic life.* The useful life of TAPS can be described as the period during which the pipeline provides a safe, environmentally sound, economically viable transportation link for Alaska North Slope crude oil.

TAPS useful life can continue well beyond the maximum allowable 30-year right-of-way renewal because:

- TAPS design life is based on the incorporation of robust components coupled with system monitoring programs designed to detect and counteract aging and facility use factors.
- TAPS physical life is considered virtually unlimited given the execution of appropriate maintenance, repair, and replacement programs.
- TAPS economic life is governed by the extent of recoverable North Slope crude oil reserves. Predictions show these reserves being produced in quantities sufficient to support the continued operation of the pipeline, with tariff rates that attract shippers, well into the 21st century.

The succeeding sections discuss, in turn, design life, economic life and physical life in more detail.

Design Life

Engineers developed design criteria for TAPS based on assumptions that protection of the Alaska environment was paramount and that, as the only oil transportation link to Alaska's North Slope, the pipeline had to function reliably and safely, with sufficient structural integrity to resist arctic conditions over an indeterminate period. These significant technical challenges resulted in a design that incorporates many redundancies and safety factors to account for known and unpredictable future conditions. The pipeline design was intentionally robust, and 25 years of operation have provided the opportunity for a critical evaluation of the design assumptions and features (TAPS Owners, 2001b). That evaluation has confirmed that the design decisions were correct.

Key TAPS design features addressed technical challenges such as support of a warm-oil pipeline in permafrost and seismic risks to pipeline integrity.

Design elements included assumptions and features that anticipated the effects of aging, such as:

- Estimates of thaw settlement allowances for buried pipeline segments,
- Pipe movement allowances in the above-ground design to provide for crude-oil temperature changes over time, and
- Analysis of soil creep or frost jacking at aboveground pipe supports.

TAPS performance is evidence that the design tolerances and protective features have been more than adequate to meet the challenges. Pipeline integrity has been stable, and age-dependent issues have been manageable over time. Buttressing the design features are surveillance and monitoring programs that continually assess the viability and functioning of the system and gauge the status of the pipeline system against the original design standards. Maintenance and repair programs keep TAPS in a safe, reliable state that protects the surrounding environment from adverse impacts from TAPS operations.

The design life of TAPS is a concept used by engineers to provide a basis for time-dependent economic analysis of alternative materials and techniques for the original pipeline design. For continued operations, the performance of design features that mitigate the effect of cold region operation or seismic risk is continually evaluated and upgrades and modifications are made whenever appropriate.

While some early statements regarding the intended service life of the pipeline estimated 30 years, in fact the pipeline design was not based on retirement or cessation of operations 30 years from the start of operations in 1977. These statements are derived from original estimates of the life of the proven Prudhoe Bay field reserves (Norman, 1971), which were used to justify the field owners' decision to proceed with TAPS construction. However, given the size of the Prudhoe Bay field, the possibility of further North Slope discoveries, and the consequences of a design failure, the pipeline was designed and built so that it could be physically operated indefinitely while meeting all safety and environmental criteria. Of course, Prudhoe Bay has produced far more reserves and has a far longer life span than originally predicted. Moreover, other significant North Slope fields have been, and still are, being developed. The original design goal — to incorporate features which would facilitate a virtually unlimited useful life of the system — was well selected.

In fact, the initial duration of the Federal Grant and State Lease was set at 30 years because that was the maximum period allowed by law, not because of any concern that the pipeline system would not last longer than 30 years.

Economic Life

The economic life of TAPS is essentially determined by whether there is sufficient North Slope crude oil economically available to justify continued operation of TAPS.

U.S. government and State of Alaska predictions of North Slope oil production show that, although production from current fields will continue to decline, by 2020 production will level off in the range of 500,000 barrels per day until the end of the period in 2034. These conservative estimates are based principally on well-established decline curves for existing fields and on a small increment in future production to be supplied by increased production from existing fields and by production from very modest discoveries. The estimates do not include the assumption that any oil will be produced from new major discoveries or from areas that are currently closed to exploration and production. More than sufficient economically recoverable oil is available to support the operation of the pipeline.

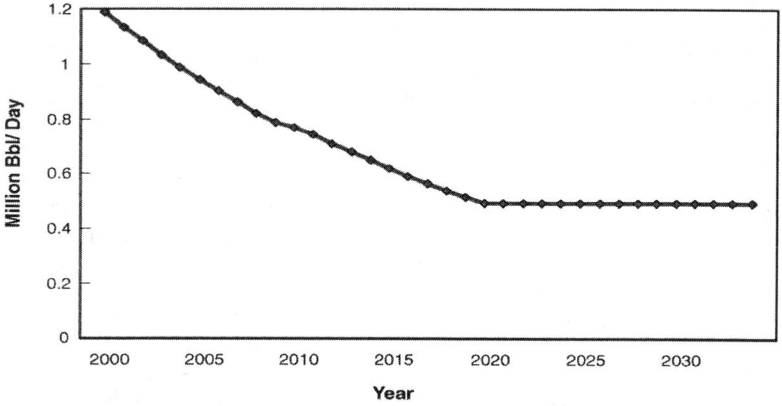

Figure 1 - Projected Throughput to 2034 (Source: TAPS Owners 2001a)

Currently, TAPS throughput is about 1 million barrels per day (bbl/day). Throughput projections indicate that throughput will continue to decline and then stabilize at about 500,000 bbl/day (TAPS Owners, 2001a) (Figure 1). At pipeline startup in 1977, throughput was about 300,000 bbl/day, and at peak operation in 1988, throughput was over 2 million bbl/day (Figure 2). Two conclusions can be drawn from this operating experience:

1) TAPS has already operated in the throughput ranges expected to occur in the next 30 years, and
2) Since throughput has decreased by over one million bbl/day in 12 years, dealing with an additional decline of 500,000 bbl/day in the next 30 years is well within technical and operating abilities.

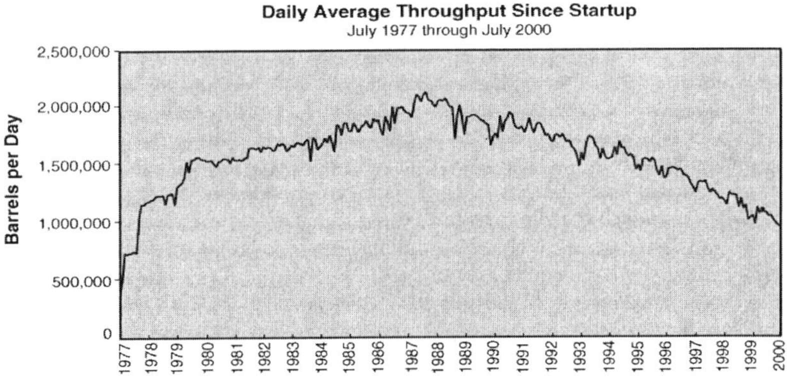

Figure 2 - Daily Throughput

Physical Life

TAPS physical life will last as long as the integrity of the pipeline and facilities is maintained adequately to allow continued safe and environmentally sound transport of crude oil. Alyeska possesses one of the industry's most rigorous maintenance programs, key aspects of which are continual monitoring and replacement of TAPS components, where advisable, either to ensure system integrity or to take advantage of technological improvements and efficiencies. Just one indication of the massive nature of that program is the approximate $25 million to $50 million spent each year to detect and control corrosion.

TAPS is proven to be reliable — from July 28, 1977, through December 31, 2000, the pipeline operated for over 204,000 hours and was shut down for a total of only 852 hours, giving it a reliability rate of 99.6 percent. The pipeline also has an excellent performance record with respect to leaks. Since startup, over 13 billion barrels of crude have been transported, with only 6 major leaks (i.e., greater than 1,000 barrels) on the mainline pipeline totaling approximately 31,600 barrels, one-half of which occurred through a sabotage incident at Steele Creek. The expectation is that these results will be improved on in the future, given diligent upkeep and further developments in such areas as in-line inspection and leak detection technology which Alyeska has pioneered in past years.

Since startup of TAPS in 1977, Alyeska has continued to improve and expand its initially comprehensive programs to detect and repair potential problems that might threaten the integrity of the pipeline system. Alyeska has developed innovative, state-of-the-art methods to monitor the condition of the pipeline and associated facilities. Where corrosion, settling, or other problems have been detected, prompt repairs — including the replacement of several sections of mainline pipe — have been made. These programs verify that the initial design specifications and construction methods were robust and lasting even in harsh arctic conditions.

Pipeline Longevity/Performance Studies

In addressing TAPS longevity, it is useful to compare the TAPS operating period with that of other pipelines. Some pipelines have been in good operating condition for more than 50 years (Muhlbauer, 1996).

A study of Cook Inlet, Alaska, oil pipeline performance performed by Alaska Department of Environmental Conservation (ADEC) noted:

> "The fact that the pipelines have reached their original design life does not imply that the lines have become inadequate or unsafe. The integrity of an older pipeline is a function of how well the line has been maintained, the type of throughput, and how the current operating conditions compare with the original design conditions. With proper maintenance the remaining life of a pipeline can be several multiples of the original design life." (Visser et al., 1993)

Several studies have examined the effect of aging on pipelines. In one recent study, a European pipeline consortium collected data over a 25-year period on the performance of cross-country oil pipelines in Western Europe (Lyons, 1998). The data were analyzed to record the pipeline system development over time, quantify environmental performances, and reveal trends in causes of spills. The following summarizes the findings of the study:

- In 1971, 70 percent of the pipelines inventoried were 10 years old or less, but by 1995 only 8 percent were 10 years old or less and 30 percent were over 35 years old.
- Pipeline spills averaged fewer than 14 per year and most were very small. Less than 5 percent of the spills were responsible for 50 percent of the gross volume spilled.
- Over the 25 years, the frequency of spills improved from 1.2 spills per 1,000 kilometers (620 miles) of pipeline to 0.4 spills per 1,000 kilometers.
- The two most important causes of spills are third-party accidents and mechanical failure, with corrosion in third place, and operational and natural hazards making minor contributions.

The study concluded that there is no evidence that the aging of a pipeline system increases risk. The development and implementation of new techniques, such as internal inspection using smart pigs, hold out the prospect that pipelines can continue reliable operations for the foreseeable future.

In assessing TAPS longevity versus performance, it is useful to review oil spill statistics over time. If aging of TAPS increased risk, an upward trend in oil spills would be noted. Such an analysis was done for the draft *Environmental Report for Trans-Alaska Pipeline System Right-of-Way Renewal* (TAPS Owners, 2001) (Figure 3) presents volumetric spill rates by year for the pipeline. There is substantial variability, but also evidence of a downward trend in volumetric spill rates in later years. Except for a sabotage incident near Livengood, all large pipeline spills occurred during the first five years of operation of TAPS.

A linear regression line (the dashed line in Figure 3) has a negative slope, indicating decreasing volumes spilled. Nonetheless, the predictive power of the linear trend model is not high, indicating that year-to-year variability is large relative to any time trend. For this reason, it is conservatively assumed that the volumetric spill rate is constant over time. Consequently, oil spill statistics do not indicate a pipeline nearing the end of its useful life.

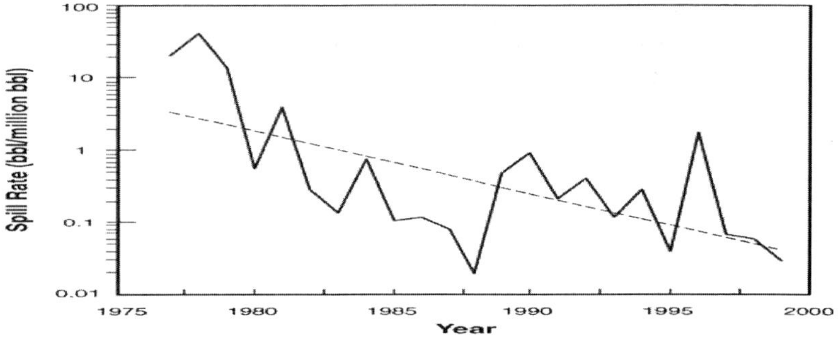

Figure 3 - Volumetric Spill Rates by Year

Age-Related Effects on Pipelines

Pipeline age itself has no metallurgical effect on the microcrystalline structure of steel that would cause the strength and ductility of the pipe to degrade over time. However, there are several ways that the age of a pipeline can influence the potential for failures, including corrosion, fatigue, and manufacturing and construction methods (Muhlbauer, 1996; Crocket and Maguire, 1999). An indirect effect of age is the increased time when a pipeline is exposed to the threat of "outside-force damage" (damage caused by accidental impact from an external force such as a vehicle or heavy equipment).

The following discussion highlights these effects on pipelines in general. Following this discussion is a section describing techniques and programs used specifically on TAPS to mitigate the threat of these age-related effects.

- **Corrosion** is related to the environmental conditions surrounding the pipe. It is reasonable to assume that with the passage of time, the opportunity for undetected (and hence, uncontrolled) corrosion and/or fatigue effects increases. Pipe-coating systems are susceptible to deterioration over time from mechanical abrasion and chemical reactions from absorbing gases and liquids in the surrounding environment. Pipeline operators use corrosion-control programs to counter the threat from corrosion.
- **Fatigue stresses** that result in cracks are a potential effect of age in metal pipelines. Fatigue cracks, if unchecked over time, can lead to pipe failure. Pressure fluctuations during pipeline operation over long periods can lead to fatigue. Common practices used to detect fatigue are fatigue monitoring, hydrostatic testing, and in-line inspection with crack detection tools.
- **Manufacturing and construction methods** can affect physical life. Poor field welds, incomplete fusion of longitudinal pipe seams, and defects in the steel manufacturing process could contribute to pipe failure over time. Although most defects of these types are detected before initial pipeline operation, some defects could manifest themselves over time as pressure cycles, fatigue stresses, and external impacts affect pipeline performance. Radiographic examinations of

welds, construction and manufacturing inspections, hydrostatic testing, and in-line inspections are methods by which this threat to pipeline integrity is detected.

- **Outside-force damage** is an indirect age-related effect on physical life simply by the increased time of exposure to potential incidents. Damage caused by outside forces is usually localized and is minimized by information dissemination (e.g., posted notices, public awareness campaigns), surveillance, and monitoring.

Mitigating the Effects of Age on TAPS

Potential effects of age discussed above are countered on TAPS through surveillance and maintenance programs to identify flaws in coatings, provide adequate cathodic protection and monitor pipe condition through in-line inspection. Pumps, turbines and other components of the pipeline are repaired and replaced as required. If the potential age-related effects are properly controlled, the physical life of the steel pipe is considered essentially unlimited. On TAPS, potential age-related threats are mitigated as described below.

Corrosion. Monitoring of corrosion protection is accomplished in several ways. Cathodic protection monitoring of mainline pipeline takes place annually. Data are gathered from test stations, buried corrosion coupons, cased road crossings, and the fuel gas pipeline. Cathodic protection data also are gathered at buried propane tanks, pump stations, and the Valdez Marine Terminal. Rectifiers are checked six times a year.

Inhibitors are used to control corrosion in isolated and low-flow or seldom-flow piping in pump stations and in road-crossing casings. Internal coupons, which verify the effectiveness of the inhibitors, are removed and analyzed twice yearly. In-line inspection tools ("smart pigs") are used to monitor corrosion and curvature on the mainline pipeline. Data are collected, stored, evaluated, and trended.

Fatigue. Cracks from fatigue stresses can affect the physical life of metal pipelines. On TAPS, two potential fatigue-stress scenarios exist: structural resonance of the piping and pipeline pressure-cycling.

Structural resonance of piping occurs in the piping manifolds of mainline pumps when the pump impeller spins at a rate that can excite the piping or its appurtenances at their natural structural frequency, resulting in high vibration and stress levels. Structural resonance manifests itself only in piping and appurtenances adjacent to the mainline pumps. To mitigate the effects, operators routinely check for fatigue damage to piping near the mainline pumps and implement corrective measures as required to maintain system reliability.

Pressure-cycling is a concern only in areas on the mainline pipe where dents, sleeves, or similar anomalies can result in localized pipe-wall bending stresses as the pipe goes through changes of pressure. The degree of potential fatigue damage depends on the number of cycles and the stress magnitude for each cycle. For dents, sleeves, and other pipeline anomalies, the potential fatigue stresses can be high during shutdowns and restarts, but the number of cycles is low. In areas where the cycles may be high, such as at the base of slackline areas, the pressure deviations and resultant stresses are low. These slackline and dent areas, such as Thompson Pass, have been studied, and either fatigue life has been determined to be unlimited or corrective actions have been implemented. (Baskurt et al., 1998; Hart et al, 1998; Norton et al., 1998; Stevick et al, 1998; Tart and Hughes, 1998; Tonkins et al., 1998)

Manufacturing and Construction Methods. Manufacturing defects or poor construction methods could have a deleterious effect on the longevity of a pipeline. However, TAPS was built under the most stringent criteria available and was closely inspected, tested, and monitored. The following examples of pipe materials, welding, and valves indicate the level of mitigating measures employed.

Mainline Pipe. TAPS pipe is carbon steel (API 5LX and 5LS) with five different pressure capabilities. There are two wall thickness 0.462-inch (11.7 mm) and 0.562-inch (14.3 mm) and three specified minimum yield strengths 60,000; 65,000; and 70,000 psi (413,400; 447,850; and 482,300 kPa).

The design basis criteria for allowable curvature were based on pre-construction mainline pipe testing at the University of California at Berkeley (Bouwkamp and Stephen, 1974). In order to study the potential behavior of the pipeline prior to construction, test specimens were subjected to a number of different load conditions. The basic parameters in these studies were the internal pressure and the temperature differential between tie-in or installation temperature and operating temperature. To evaluate these phenomena under increasing lateral loads, a total of seven specimens under different pressure and simulated temperature conditions were investigated. These tests continued until the pipe wall buckled. For five specimens, tests continued until the pipe wall ruptured. Furthermore, one test specimen was used to study the effect of a pressure drop on pipe-wall stability.

Recent advances in analytical techniques have led Alyeska to develop tools to evaluate the below-ground pipe at specific locations on the basis of the demand on the pipe and the pipe's particular capacity to resist bending. These studies and tools allow Alyeska to confirm that, in light of current technical knowledge, TAPS pipe continues to retain original throughput performance capabilities.

Welding. Welding on the pipeline, whether during construction or repair, must:

- Be performed by qualified welders in accordance with approved procedures;
- Be protected from weather conditions such as precipitation;
- Be performed in a manner to prevent, repair, or remove defects; and
- Undergo nondestructive testing (radiography and hydrotesting).

All mainline pipe was hydrotested and all welds inspected by radiography before TAPS was commissioned. In the event that pipeline repairs require relocation or replacement, all replacement components undergo hydrostatic testing, and all mainline welds are inspected to ensure the integrity of the relocated or replacement pipe. Alyeska does not use any replacement pipe or component that has not been hydrostatically tested in conformance with Department of Transportation standards or that fails to meet hydrostatic testing standards.

Valves. Alyeska specifications for the design of mainline valves require conformance to American Petroleum Institute standard API 6D, as well as several other requirements. Department of Transportation regulations under 49 CFR 195 also require that new valves meet the test requirements of API 6D, which covers valves of 2-inch nominal pipe diameter and larger.

The Department of Transportation regulations also require pipeline operators to maintain those valves "required for safe operation" in good working order. Alyeska maintains its valves in good working order and demonstrates the functionality of the valves through partial closure of the valves twice a year. In addition to the regulatory requirements, Alyeska implemented the TAPS Valve Program in 1997 to validate the condition of these valves and to perform testing to determine if the valves performed in accordance with sealing criteria developed by Alyeska. Valves that do not meet performance criteria are repaired or replaced. (Aus et al., 2000; Jackson and White, 2000; Pomeroy and Norton, 2000; Weber and Malvick, 2000)

Outside-Force Damage. Approximately half of the 800-mile (1288 km) pipeline is above ground and, therefore, potentially subject to damage caused by accidental impact from an external force such as a vehicle or heavy equipment. Access to most of the right-of-way is determined by the federal (Bureau of Land Management) and state (Alaska Department of Natural Resources) landowners. In addition, Alyeska limits access to the pipeline through signs and with locked gates on access roads to the right-of-way. For locations where access roads pass under the pipeline, "headache" bars have been installed to ensure that vehicles have enough clearance under the pipe.

For below-ground pipe, all excavation within the right-of-way must be authorized by Alyeska. Warning signs along the pipeline contain a 24-hour telephone number to contact the Controllers at the Operations Control Center (OCC). Callers with excavation requests are connected with the appropriate personnel to coordinate excavation requirements. In addition, Alyeska conducts an educational program to help the public, government organizations, and people engaged in excavation-related activities to recognize a crude-oil pipeline emergency and to report it to Alyeska and/or other emergency response organizations.

Alyeska's pigging program monitors the pipeline for external damage using state of the art ultrasonic technology to detect pipe-wall thinning from gouges and scrapes. Other curvature pigs detect dents and ovalities. Leak detection systems monitor for leak loss from any source, including outside-force damage.

Conclusion

The useful life of TAPS is the period during which an economic benefit is derived from continuing operations and during which the pipeline can be operated safely and without harm to the environment. The useful life of TAPS will continue well beyond 30 years for at least three key reasons.

1) The physical life of TAPS is virtually unlimited assuming continued appropriate maintenance and surveillance. In-line inspection tools (smart pigs) for pipeline corrosion, deformation and settlement are run on rolling three-year

cycles. If corrosion or other damage is found that would reduce the capacity or pressure capability of TAPS, repairs are made. Pumps, turbines and other components of the pipeline are repaired and replaced as required.

2) The economic life of TAPS will continue for the foreseeable future based on pipeline throughput estimates from a variety of public sources.
3) The design life of TAPS is a concept used by engineers to provide a basis for time-dependent economic analysis of alternative materials and techniques for the original pipeline design. For continued operations, the performance of design features that mitigate the effect of cold region operation or seismic risk is continually evaluated and upgrades and modifications are made whenever appropriate.

Acknowledgements

The authors wish to acknowledge Alyeska Pipeline Service Company for providing support for this paper. The authors especially wish to acknowledge James F. Johnson, James Lukin, Bill Christian, Dan Maxim, and Rebecca McAllister.

References

Aus, W.T., Pomeroy, G.W., and Norton, J.D. (2000). "Mainline Valve Operations and Maintenance Program on the Trans-Alaska Pipeline." *In:* Proceedings of the ETCE/OMAE Conference, New Orleans, Louisiana.

Baskurt, U.J., Bradshaw, P, and Hart, J.D. (1998). "Slackline Testing of the Trans-Alaska Pipeline System at Thompson Pass." *In:* Proceedings of the International Pipeline Conference, Calgary, Alberta, Canada. June, 1998.

Bouwkamp, J.G., and Stephen, R.M. (1974). "Full-Scale Studies on the Structural Behavior of Large Diameter Pipe Under Combined Loading." Report No. UC-SESM 74-1. University of California, Berkeley. January, 1974.

Crocket, P.A., and Maguire, R. (1999). "Pipeline Failure Management." *In: Ageing Pipelines, Optimizing the Management and Operation: Low Pressure-High Pressure.* Professional Engineering Publishing Limited for the Institution of Mechanical Engineers, Bury St. Edmunds and London, United Kingdom.

Hart, J.D., Powell, G.H., Maple, J.A., Stevick, G.S., and Norton, J.D. (1998). "Fatigue Damage Calculations for Dented and Ovalled Section of the Trans-Alaska Pipeline System at Thompson Pass." *In:* Proceedings of the International Pipeline Conference, Calgary, Alberta, Canada. June, 1998.

Jackson, M.D., and White, A.M. (2000). "Mainline Valve Testing on the Trans-Alaska Pipeline System." *In:* Proceedings of the ETCE/OMAE Conference, New Orleans, Louisiana.

Lyons, D. (1998). "Western European Cross Country Oil Pipelines 25 Year Performance Statistics, Report #2/98." Prepared for Conservation of Clean Air and Water in Europe, oil pipelines management group. Brussels, Belgium. June, 1998.

Muhlbauer, W.K. (1996). "Pipeline Risk Management Manual." Gulf Publishing Co., Houston, TX, pg. 92.

Norton, J.D., Tonkins, W., and Flanders, B. (1998). "Alyeska Remedies TAPS' Pipeline Vibrations at Thompson Pass." *In: Oil and Gas Journal*. June 1, 1998.

Norman, C. A. (1971). "Economic analysis of Prudhoe Bay oil field," in Adelman, M.A., Editor, *Alaskan Oil, Costs and Supply,* Praeger Publishers, New York, NY.

Pomeroy, G.W., and Norton, J.D. (2000). "Use of Isolation Stopples for Valve Replacement on Trans-Alaska Pipeline." *In:* Proceedings of the International Pipeline Conference, Calgary, Alberta, Canada.

Stevick, G.S., Hart, J.D., and Flanders, B. (1998). "Fatigue Curves for Damage Calculations for a Dented and Ovalled Section of the Trans-Alaska Pipeline System." *In:* Proceedings of the International Pipeline Conference, Calgary, Alberta, Canada. June, 1998.

TAPS Owners (2001a). "Draft Environmental Report for Trans Alaska Pipeline System Right-of-Way Renewal." Trans Alaska Pipeline System Owners, May 2, 2001.

TAPS Owners (2001b). "Duration of Right-of-Way Renewal for the Trans Alaska Pipeline System." Trans Alaska Pipeline System Owners, March 23, 2001.

Tart, R. Jr., and Hughes, J. (1998). "In-situ Evaluation of Pipeline Bedding and Padding Spring Constants." *In:* Proceedings of the International Pipeline Conference, Calgary, Alberta, Canada. June, 1998.

Tonkins, W.G., Baskurt, U.J., and Hart, J.D. (1998). "Trans-Alaska Pipeline System Linewide Slackline Investigations for Potential Pipe Vibrations." *In:* Proceedings of the International Pipeline Conference, Calgary, Alberta, Canada. June, 1998.

Visser, R.C., Duthweiler, F.C., and Carlile, H.B. (1993). "Oil Pipeline Risk Assessment - Cook Inlet, Alaska." Prepared by Belmar Management Services for Alaska Department of Environmental Conservation. November, 1993.

Weber, B., and Malvick, M. (2000). "Trans-Alaska Pipeline System Mainline Block Valve Risk Assessment." *In:* Proceedings of the ETCE/OMAE Conference, New Orleans, Louisiana.

Environmental Impact of 25 Years of Trans-Alaska Pipeline Operation

J. David Norton, P.E.[1], Brett Jokela, P.E.[2], Erik Haas[3], Robert G. Dugan[4]

Abstract

In 2001, the Trans Alaska Pipeline System (TAPS) owners produced a comprehensive analysis of the impacts associated with the renewal of the TAPS rights-of-way. The resulting report, *Environmental Report for the Trans Alaska Pipeline System Right-of-Way Renewal*, (TAPS Owners, 2001) described the physical, biological, and socio-economic impacts of right-of-way renewal — and of non-renewal. The Environmental Report followed the guidelines of the Council on Environmental Quality for preparing environmental impact statements (EISs). Since projections of future impact of TAPS operation closely follow post-construction impacts since 1977, an assessment of the impact of 25 years of operation of a warm oil pipeline in cold regions can be described. This paper reviews the results of the report with special emphasis on physical impacts.

Because the pipeline system has been in continuous operation since 1977 and is under stringent regulatory controls, its effects on the physical environment — terrestrial, air quality, and water quality — are known to be relatively minor and can reasonably be expected to continue without significant change. Justification for predictions of low impact can be summarized in the following conclusions:

- The TAPS pipeline and related facilities already exist with known, observable impacts;
- Major changes to the pipeline system or to the affected physical environment are not expected;
- New surface-disturbance areas associated with TAPS will be small and isolated;
- There would be no unavoidable adverse effect on the physical environment that would not be mitigated to the extent technically feasible;
- Declining throughput could result in lower emissions and discharges affecting air and water quality.

[1] Consultant, J.D. Norton and Associates; dnorton@alaska.net
[2] Principal Engineer, MWH; brett.jokela@mw.com
[3] Principal, HMH Technical Environmental Consulting; hmh-ehaas@alaska.com
[4] Senior Engineering Geologist, Golder Associates, Inc.; bob_dugan@golder.com

Introduction

In 2001, the Trans Alaska Pipeline System (TAPS) owners produced a comprehensive analysis of the impacts associated with the renewal of the TAPS rights-of-way (ROW). The resulting report, *Environmental Report for the Trans Alaska Pipeline System Right-of-Way Renewal*, describes the physical, biological, and socio-economic impacts of right-of-way renewal — and of non-renewal.

The proposed action (right-of-way renewal), considered in the Environmental Report is unusual in that the action does not involve construction of a project but rather its continued existence and operation. As a result, the report describes the affected environment as it now exists following construction and 25 years of operation of TAPS. That report is based on a review of TAPS operational impacts to date and forms the basis for this paper.

Background

The trans-Alaska pipeline is an 800-mile (1288 km), 48-inch (1.2 m) diameter crude oil pipeline that is elevated above ground for 420 miles (676 km) of its length and buried for the other 380 miles (612 km). Eleven pump stations were built to move the oil through the pipeline, and four of these are now on operational standby. The marine terminal at Valdez has storage facilities for over 9 million barrels of oil and four tanker-loading berths. The peak average daily throughput of 2.03 millions barrels per day was reached in 1988. The current rate is approximately 1 million barrels per day.

TAPS is unique among pipelines both for its engineering achievements and for the public controversy generated at the time it was originally proposed. Debates surrounding land claims by Alaska Natives and preparation of an Environmental Impact Statement (EIS) under the newly enacted National Environmental Policy Act (NEPA) delayed construction from 1969 until 1974.

The Department of the Interior released a final EIS for the project in March 1972, but the project was authorized only when President Richard Nixon signed the Trans Alaska Pipeline Authorization Act on November 16, 1973. These federal and state rights-of-way, issued in 1974 for 30-year terms, expire in 2004. The TAPS Owners' application for renewal of these federal and state rights-of-way seeks the maximum duration for the renewals, 30 years.

Summary of Environmental Consequences of TAPS ROW Renewal

Most environmental assessments and EISs deal largely or exclusively with future projects. The principal focus of the analysis in the Environmental Report is the continuation of a system that has been in operation since 1977. Years of operations data and experience with TAPS allowed for the development of accurate estimates of future effects. The following sections summarize the effects predicted for renewal and non-renewal of the TAPS ROW.

Effects of Continuing TAPS Operations until 2034

The renewal of TAPS ROW agreements involves continued operation of TAPS for an additional 30 years. Following is a summary of the physical, biological, and social effects of the proposed action.

Physical Characteristics

Because the pipeline system has been in continuous operation since 1977 and is under stringent regulatory controls, its effects on the physical environment — terrestrial, air quality, and water quality — are known to be relatively minor and can reasonably be expected to continue without significant change under the proposed action. Justification for predictions of low impacts can be summarized in the following conclusions:

- The TAPS pipeline and related facilities already exist with known, observable impacts;
- Major changes to the pipeline system or to the affected physical environment are not expected during the right-of-way renewal period;
- New surface-disturbance areas associated with TAPS will be small and isolated;
- There would be no unavoidable adverse effect on the physical environment that would not be mitigated to the fullest extent technically feasible;
- Declining throughput will result in lower emissions and discharges affecting air and water quality.

Biological Resources

Biological resources potentially affected by continued TAPS operation include vegetation and wetlands, fish, birds, and terrestrial and marine mammals. The ecosystems affected by the operation of TAPS and associated activity for almost 25 years are healthy. With the exception of very limited local impacts, the vegetation, fish, and wildlife along TAPS have not been impacted at the resource population level. TAPS, as it exists today, is simply another feature on the landscape, to which the flora and fauna have habituated. Even Alaska's North Slope, with extensive oil fields, has a healthy community of flora and fauna. Populations of large and small mammals, birds, and fish are healthy despite development of the oil fields. In Prince William Sound, some populations of seabirds and sea otters were reduced substantially by mortality from the *Exxon Valdez* oil spill. However, despite concerns over toxicity from residual oil, populations are generally recovered or recovering. In all three areas — the TAPS right-of-way, the Alaska North Slope, and Prince William Sound — there are local impacts, but overall, vegetation communities and fish and wildlife populations have fared well and will continue to thrive in the future.

Social Systems

The renewal of the TAPS right-of-way will provide the opportunity to produce an additional estimated 7 billion barrels of oil from the existing North Slope oil fields and will increase the likelihood of commercializing some 30 trillion cubic feet of currently "stranded" natural gas. It will also result in great economic benefits for the

U.S., the state of Alaska, local governments, and residents of Alaska. Because TAPS transports a significant portion of the oil produced in the U.S., TAPS' continued operation will reduce the U.S. balance-of-trade deficit by an estimated $112 billion dollars during the renewal period. (This projection does not include potential gas commercialization and is based on a low oil price of $16 per barrel.) Employment opportunities will be enhanced. Social change will continue with both positive and potentially negative effects. Subsistence resources could be affected if there is a large oil spill, but the TAPS Ship Escort/Response Vessel System and the phase-in of double-hull tankers greatly reduce the risk of such an event.

Effects of Closing TAPS after 2004

The no-action alternative involves the termination of TAPS operation and the dismantling, removal, and restoration (DR&R) of TAPS facilities. DR&R of TAPS will also result in the end of Alaska North Slope oil production. Following is a summary of the physical, biological, and social effects of the no-action alternative.

Physical Characteristics

In the no-action alternative, it is assumed that the above-ground facilities related to TAPS will be removed during a three-year period of DR&R. During that time, major activities will involve the physical removal of equipment and subsequent transportation to disposal or recycling sites. For a relatively short time, these activities will result in disruption to the terrestrial environment. These short-term impacts along the TAPS right-of-way include the potential for spills, increased use of heavy vehicles and traffic with attendant increase in equipment emissions and dust. After DR&R, it is likely that some of the TAPS workpad, access roads, and the Dalton Highway will remain in place.

Biological Resources

As long as there are mitigation and prevention measures, the direct impacts of DR&R on biological resources will be limited and manageable. After DR&R, there will be no direct impacts of TAPS. The ecosystems along TAPS, on the North Slope, and in Prince William Sound have fared well during the last 25 years of oil-industry operations. With the exception of some disturbance during the DR&R period, the environment will essentially return to its pre-oil-industry state through a combination of active restoration and natural ecosystem succession under the no-action alternative. The use of natural resources, primarily fish and wildlife, may increase following the closing of TAPS as employment and the state economy dramatically decline. However, it is also possible that the human population (and fish and wildlife harvests) will decrease in response to the economic decline.

Social Systems

Non-renewal of the TAPS right-of-way would have devastating effects on the economy of Alaska and would significantly impact the U.S. balance of trade. The opportunity to produce additional oil from the North Slope oil fields will be eliminated, and the U.S. balance-of-trade deficit would increase by an estimated $112 billion dollars between 2004 and 2034. The likelihood of commercializing North Slope natural gas will be significantly reduced without the oil-production infrastructure. Lost revenues to the State of Alaska ($14.2 billion), local governments ($6.5 billion), and residents of Alaska will cause a severe drop in employment, loss of social services, and economic hardships.

Physical Effects of Continued TAPS Operation

The remainder of this paper provides additional summary emphasis on physical effects of TAPS operation on the terrestrial, water, and air environments. In this way the impact of 25 years of operation can be assessed.

Soils and Permafrost. Continued operation of TAPS will impact some parts of the terrestrial environment because of maintenance activities, corrosion digs, construction projects for pipeline-related facilities, and the continued presence of a warm-oil pipeline in permafrost terrain. Maintenance since startup has caused localized temporary land disturbance but has generally stabilized the ground on and adjacent to the ROW. Since nearly all maintenance activities occur on or along existing stabilized embankments, new major long-term changes to the terrestrial environment are not expected.

With respect to soils and permafrost, the impacts of continued TAPS operation are minimal. Most of the thermal impacts have already occurred or are significantly slowing. Continued monitoring and maintenance will identify and repair any areas where settlement or heave may exceed operational standards.

The pipeline was designed and built to maximize pipeline stability and minimize environmental impact. It was typically buried conventionally where the soils were unfrozen or thaw-stable and was elevated in zones of thaw-unstable permafrost. Since construction, the soils in some permafrost areas have been affected by thawing. The thaw bulb that has developed around the buried pipe has reached equilibrium and is unlikely to continue to grow because the pipeline temperature is decreasing with decreased throughput. The thaw bulb could eventually shrink in the cold continuous permafrost north of the Brooks Range. Shrinkage of the thaw bulb could contribute to frost heave in some areas. However, the impact of continued pipeline operation in areas of permafrost is likely to be minimal and well within operational limits.
Workpad and access-road embankments built over the relatively warm permafrost south of the Brooks Range have compressed or disturbed the vegetative cover that formerly protected the permafrost. This has caused the ground to absorb more radiant heat and has resulted in the thawing of near-surface permafrost under the embankments. This is especially evident in areas of above ground pipeline mode (Photo 1).

Lowering of the permafrost table has caused settlement where the near-surface soils contained excess ice before thawing. The settlement has caused localized disruption of drainage and formation of ponds in the depressions. The ponds rarely extend more than a few yards from the edge of the embankments. As the top of the permafrost continues downward, the rate of thaw generally decreases. Continued operation of the pipeline is not likely to further affect this process significantly because the effect of the above-ground pipeline is minimal compared to the effect of the workpad that is already in place. In addition, the thaw depths under the workpad are already approaching the anticipated maximum in some locations, and the VSMs have settled in only a few areas.

In addition to the effects on drainage, the continued downward migration of the permafrost table may locally affect VSM stability. Where permafrost thaws below the VSMs, settling or seasonal frost jacking of VSMs may occur. In the future, localized repairs will likely be necessary. VSM movements have generally been mitigated by adjusting and providing uniform support to the pipe, adding free-standing heat pipes, and using other techniques such as applying wood chips to insulate the workpad.

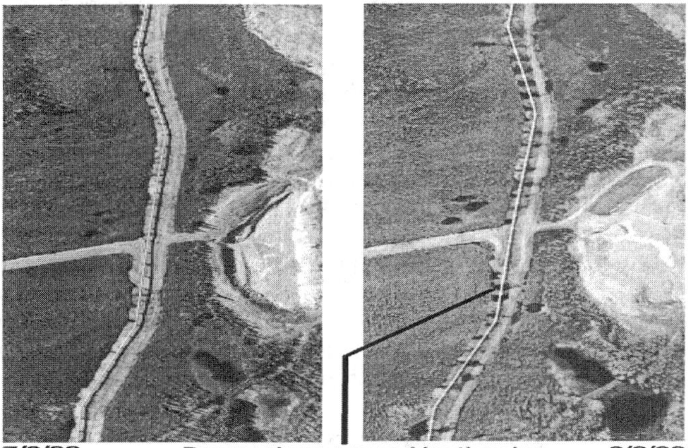

7/3/82 *Photo 1* - Depressions caused by thawing permafrost between bends 6/8/92

Monitoring programs report that about 200 of approximately 78,000 VSMs installed on TAPS show signs of movement (up, down, leaning) that exceeds maintenance standards and may require intervention. None of these affect the structural integrity of the associated aboveground pipeline support system. Many of these VSMs are on slopes in marginal permafrost areas and may have to be replaced with longer VSMs installed to a greater depth. All of these impacts have been, and will continue to be, mitigated through Alyeska's routine monitoring and maintenance program.

Loss of pipeline support due to thawing permafrost has occurred since construction in a few areas. Notable examples include:

- Pipe deformation resulting from thaw settlement that resulted in a leak on the north side of Atigun Pass at milepost (MP) 166 in 1979. The permafrost at this location was not identified during TAPS construction.
- Several feet of vertical settlement for a segment of buried pipeline due to thawing of ice-rich soils in the Dietrich River floodplain at MP 200 in 1985. The permafrost at this location was not discovered during TAPS construction.
- Minor movement of VSMs on slopes at Pump 11 Hill (MP 687) and Squirrel Creek (MP 717) due to thaw settlement and frost jacking of VSMs.
- Pipe settlement and deformation at MP 734.

The leak in Atigun Pass reached the Atigun River and caused some impact. The Dietrich River settlement resulted in pipeline realignment, causing temporary disturbance to the ground surface over a limited area. Maintenance and repairs associated with the movement of VSMs on the slopes have not affected areas off the workpad.

Rivers and Floodplains

The impact of the pipeline on the behavior of rivers and floodplains depends on whether river training structures are used, the type of structure, and whether the pipeline is located in the active or main channel area or in vegetated floodplain fringe areas.

Buried river crossings with no river training structures have little or no impact on the behavior of rivers. As part of construction, riverbeds were restored to the original grade. Typically, some settlement occurred in backfilled trenches during the first year due to the voids created by thawing of frozen backfill. With the first moderate to high flow during spring runoff or rainfall, the natural bedload filled in the settled areas. The banks were typically restored to their natural conditions or, for erosion control purposes, cobbles or rock placed on them. After a single year of operation, there was little or no evidence of construction except for loss of vegetation at the banks.

Most of the bridged river crossings have some type of river training structure to guide flow. The structures are necessary to reduce scour and bank erosion and thus protect the integrity of the bridge piers and abutments. The Yukon River bridge and the Atigun River bridge north of Pump Station 4 are examples of crossings with no river training structures. These bridged crossings have little or no impact, except for the local impact of the piers on flow. The piers result in a local deepening of the river immediately in front of and alongside the piers.

Water Resources

Pipeline operations require fresh water for potable water for manned facilities, equipment washing, dust abatement on roadways and pads, and hydrostatic testing. The Alaska Department of Natural Resources regulates use of Alaska's water resources and issues permits for temporary or long-term water appropriations. Alyeska has certificates of appropriation for water use at permanent facilities, including each pump station except Pump Stations 1 and 6. Water used at Pump Station 1 is purchased from the North Slope Borough's Service Area 10 water utility. A well at 5-Mile Camp is used as a water source for Pump Station 6. Each active pump station typically consumes between 4,500 and 7,500 gallons per day, mostly for domestic uses.

The Valdez Marine Terminal (VMT), which uses significantly more water than other facilities, has a certificate of appropriation for withdrawals from Allison Creek. Since October 1995, average water withdrawals under this appropriation have amounted to over 111,000 gallons per day. Industrial water uses at VMT include power-plant boiler-water, stack-scrubber systems, steam cleaning of equipment, and other washdown processes.

Alyeska maintains additional temporary water-use permits for facilities such as mobile construction contingency facilities (MCCFs) and for special projects.

Volumes of water for temporary use vary significantly. The largest single project for which temporary-water-use permitting was necessary occurred in 1997, when 7.4 million gallons were withdrawn from East Lake, near PS 1, for tank cleaning and testing at Pump Station 1.

Continued operation of TAPS will require continued use of water resources to support operations and maintenance activities. Wastewaters will continue to be treated, discharged, and assimilated by upland and freshwater receiving environments along the pipeline. Marine waters of Port Valdez will continue to be used to assimilate treated discharges from the VMT, including sanitary wastewater and ballast water.

Water Use and Sanitary Discharge at Pump Stations. As throughput of oil declines, ramping down of additional pump stations will mean reductions in staff at pump stations that are placed on standby. Furthermore, automation of certain operations will allow reduction of field crews at other sites. Reductions in staffing at the pump stations and camp facilities will result in a parallel drop in domestic water use for drinking water and sanitation at each facility.

At Pump Stations 1, 3, and 4 wastewater is disposed by injecting it into the exhaust stacks of the pump turbines. This means of disposal requires sufficient stack temperatures to ensure vaporization, volatilization, and disinfection of the wastewater plant effluents. Reduced throughput may affect the temperatures of pump engine exhaust, necessitating use of alternative means for wastewater disposal, such as hauling wastewater or by on-site package treatment plants.

Secondary biological sewage treatment and effluent disposal to wetlands are expected to continue for Pump Stations 5 and 6 (when the permanent living quarters are active) in accordance with discharge limitations imposed by state and federal permits. Permitted discharges are expected to be assimilated by local water resources with no significant effect on productivity or viability of aquatic ecosystems.

The septic system leachfields at Pump Stations 7, 9, and 12 will be nearing their typical useful life in the next decade and will be replaced if necessary.

Industrial Water Use and Discharges. In addition to the pump station and camp domestic-water needs, water will continue to be used for a variety of industrial activities. Industrial water needs are usually associated with intermittent and temporary activities, and are likely to have a wide variance from the average projected use. Discharges result from dewatering of excavations, which could be required for:

- Follow-up investigations from corrosion survey data analysis,
- Mainline valve inspections,
- Mainline cathodic-protection monitoring and remediation,
- Maintenance of existing river-training structures or the addition of new structures, or
- Maintenance and repair of the fuel gas line.

Effects of stormwater runoff into waters along the pipeline route are not significant, because such runoff carries no introduced pollutants and results only in transient increases in sediment load.

Discharges to Port Valdez. Reduced throughput of oil will reduce the number of tanker visits, and segregation of ballast water in double-hulled tankers will reduce the average volume of ballast water treated on a per tanker basis. Average annual BWTF discharge is expected to eventually stabilize between one-quarter and two-thirds of the current long-term historical average flow from the BWTF, currently about 14 million gallons per day (53,000 m^3/day). Additionally, the total loading of pollutants in Port Valdez has significantly decreased due to improvements in BWTF processes that have taken place over the years, such as the installation of the biological treatment tanks, air strippers, and monitoring instrumentation. The pollutant loading is expected to continue to decrease with reduced TAPS throughput.

Other oily wastewaters derived from VMT operations (Non-Ballast water) will continue to be routed through the ballast water treatment facility (BWTF). These include blowdowns, steamrack discharge, and stormwater. Discharges from these VMT operations currently account for approximately 1.36 million gallons per day (5,300 m^3/day) of the wastewater treated and discharged by the BWTF.

Reduced hydraulic loadings will affect the dynamics of treatment at the BWTF by increasing the hydraulic retention time in the process units during normal operations. Longer detention times could lead to slightly higher levels of contaminant removal prior to the biological treatment processes. The reduced contaminant load going to the biological tanks, coupled with longer detention times, may require adjustments to keep the biological process viable. Alyeska sponsors research on the response of BWTF processes on variable flows, in conjunction with the engineering faculty of the University of Alaska Anchorage (Woolard and Luetters, 1997).

Air Quality Impacts

In order to evaluate the potential impacts of the next 30 years of TAPS operation, several aspects need to be examined. First, the extent of existing impacts from the operation of TAPS during the last 25 years should be considered. Impacts from the operation and, to a minor extent, from construction need to be evaluated since the continued operation of TAPS may entail some construction activities. Secondly, projected operational configurations subject to long-range operations plans must be evaluated. These plans may increase or decrease future emissions and consequently can impact the environment. A third aspect is to search for and to evaluate any dynamic effects that may cause future impacts that significantly differ from past effects.

As TAPS throughput has steadily declined since 1989, there has been a corresponding TAPS-wide net decrease in emissions because of the reduced demand on pumping capacity (and the reduced burning of fuel). Consequently, the ambient air quality impacts associated with the TAPS pump stations are no higher now and are likely lower than they were during the peak throughput years of 1988-89. Major improvements were made to the control of hydrocarbon emissions at the VMT, and no emission increases are expected beyond the 1998 post-startup levels of the tanker vapor control system.

Ambient air-pollutant levels at North Slope production areas have not shown any measurable changes since the start of the air-quality monitoring program in 1986. It can thus be concluded that no additional degradation of the ambient air quality has occurred beyond air pollution levels prior to 1986 due to exploration and new production activities. An extrapolation into future operating phases may be appropriate. However, careful consideration must be given to any known potentially adverse effects. One such known and potentially adverse effect is the steady increase in the H_2S content of the fuel gas consumed at the North Slope facilities and Pump Stations 1 to 4. The fuel-gas H_2S content has increased from levels in the single digits typical in the early years of operation to approximately the mid-twenties today. On a rough scale it can be concluded that the H_2S content has doubled over the last 20 years — with a resultant doubling of the SO_2 emission rates for all stationary fuel-gas-burning equipment (SO_2 results from the combustion of H_2S). However, before the conclusion is made that emission rates have doubled, the actual fuel-gas consumption rates would have to be assessed for any specific facility.

Also, it is incorrect to assume that actual increase in SO_2 emissions would automatically cause a proportionate increase in measurable ambient impacts. As discussed earlier, the Prudhoe Bay monitoring sites have not measured any increases in ambient SO_2 levels in recent years despite an increase in fuel-gas H_2S levels. Part of the reason is that actual emissions of SO_2 are quite low (about 1 percent) compared with other pollutants. Thus, any changes in the emission rates of SO_2 will likely have little effect on measured levels or airborne pollution and practically no effect on any climatic factors. Existing global climatic trends like atmospheric warming will likely have a significantly stronger effect on local climate regimes. Currently, all pump stations can operate all mainline turbine units and other stationary equipment up to their maximum permitted levels. Generic modeling has been conducted based on the worst-case normal operating conditions. Consequently, it is unlikely that the potential emission increases from the remaining stations in future throughput cycles will create any additional impacts over existing or past levels, which were found to be in compliance with the ambient standards evaluated in the modeling study.

Global Climate Change

There are numerous ongoing debates on the rate and duration of the current apparent warming trend in Alaska. Global-warming forecasts indicate that the high latitudes of the earth may warm by several degrees by the middle of this century as a result of an effective doubling of carbon dioxide (CO_2), methane (CH_4) and other greenhouse gases (Esch and Osterkamp, 1991; Nixon, 1991, Vyalov et al., 1993, 1998). One study in interior Alaska indicated that an air temperature shift to a warmer regime began with the winter of 1976. After 1976, the mean annual air temperature in Fairbanks increased 1.5°C for the next 20 years as shown in (Osterkamp and Romanovsky, 1999). These studies suggest that a warming trend is occurring, although the cause of the temperature increase is not well understood and may be related to the cyclic nature of the climate.

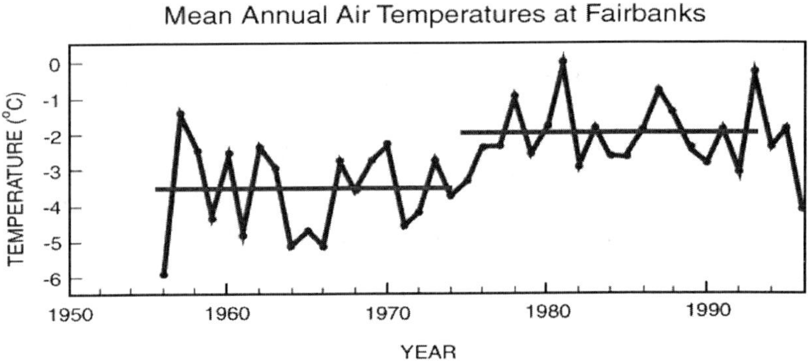

Figure 1 - Air Temp

The likely range of mean annual air temperature increases over the ROW renewal period is 2°C to 5°C if the warming trend observed in the Fairbanks area continues. Over time, warmer air temperatures would increase ground temperatures. This would result in warming of the relatively cold permafrost on the North Slope and Brooks Range and lowering of the permafrost table in southern portions of the pipeline where permafrost temperatures approach 0°C.

A thaw bulb has already developed in some permafrost areas along the ROW as a result of the construction of the workpad and other facilities. Pipeline construction unavoidably compressed or disturbed the vegetative cover that formerly protected the permafrost. A gradual increase in the size of the thaw bulb can be expected as a result of continued climatic warming.

Since the beginning of TAPS operations in 1977, the consequence of any climate change-induced permafrost thawing to pipeline operations has been negligible (Cole et. al., 1999). Continued warming of the air temperature will most likely also have negligible impact on TAPS operations or pipeline integrity for the following reasons:

- Where above-ground pipe is located in areas of relatively warm permafrost, heat pipes are used to help maintain frozen conditions.
- Heat pipes can be added, if necessary, to VSMs that do not currently require them.
- In areas that are ice-rich (i.e., areas of high moisture content), the rate of ground temperature change will be slow, especially at depth.
- Continued monitoring and maintenance will identify and repair any areas where settlement or heave may exceed operational standards.

Conclusion

Continued operation of TAPS is expected to have only minor impact on the physical environment. Justification for prediction of low impact is based on the fact that pipeline facilities have been in place for 25 years with known effects and observable impacts. Major changes to the pipeline system are not expected, new surface-disturbance areas will be small and isolated, and unavoidable adverse effects would be mitigated to the extent technically feasible. Moreover, declining throughput could result in lower emissions and discharges affecting air and water quality.

Acknowledgements

The authors wish to acknowledge Alyeska Pipeline Service Company for support in preparing this paper. The authors wish to specifically thank Jim Lukin, Ray Jakubczak, Wim Veldman, Dan Maxim, Matt Cronin, Rebecca McAllister and Anne Brown.

References

Cole, H., Colonell, V., and Esch, D., (1999). "The Economic Impact and Consequences of Global Climate Change on Alaska's Infrastructure." University of Alaska Fairbanks Center for Global Change and Arctic System Research, Proceeding of a Workshop - Assessing the Consequences of Climate Change for Alaska and the Bering Sea Region.

Esch, D.C., and Osterkamp, T.E. (1991). "Cold Region Engineering: Climate Warming Concerns for Alaska." Journal of Cold Region Engineering, 4(1):6-14.

Nixon, J.F. (1991). "Effect of Climate Warming on Pipe Creep in Permafrost." Journal of Cold Region Engineering, 4(1):67-73.

Osterkamp, T.E., and Romanovsky, V.E. (1999). "Evidence for Warming and Thawing of Discontinuous Permafrost in Alaska." Permafrost and Periglacial Processes, 10:17-37.

TAPS Owners (2001). "Draft Environmental Report for Trans Alaska Pipeline System Right-of-Way Renewal." Trans Alaska Pipeline System Owners, May 2, 2001.

Vyalov, S.S., Gerasimov, A.S., Zolotar, A.I., and Fotiev, S.M. (1993). "Ensuring Structural Stability in Permafrost Ground Areas at Global Warming of the Climate." Pages 955-960 in: Proceedings of the Sixth International Conference on Permafrost, Beijing, China.

Vyalov, S.S., Gerasimov, A.S., and Fotiev, S.M. (1998). "Influence of Global Warming on the State and Geotechnical Properties of Permafrost." Pages 1097-1102 in: Proceedings of the Seventh International Conference on Permafrost, Yellowknife, Northwest Territory.

Woolard, C., and Luetters, S.T. (1997). "Starvation Response in Biological Treatment Systems." *in:* Proceedings of the Seventieth Annual Water Environment Federation Conference and Exposition. Chicago, Illinois.

A Method For Design of Flexible Buried Structures in Permafrost Areas

By David C. Cowherd[1]

Introduction

Permafrost areas present unique problems to the design and performance of buried flexible structures. All buried flexible structures depend on soil-structure interaction for support. Generally the structure will not support itself under the imposed loads, and the backfill around the structure must be designed to provide support. An analysis of the soil-structure interaction must be made. The natural soil outside the select backfill envelope must also provide sufficient support. The natural soil outside and beneath the structure is, therefore, very important to the stability of the structure. In addition to normal design considerations a buried flexible structure on permafrost must take into account the potential for thawing of the permafrost, ice lenses, excavation of permafrost, and other considerations unique to arctic areas.

This paper presents the methodology for evaluation of a buried flexible structure consisting of a circular flexible metal pipe, 22 feet (6.7m) in diameter in a permafrost area. The hypothetical structure is to be placed beneath a stockpile consisting of 80 feet (24.4m) of zinc ore weighing 185 pcf (29 kN/m^3). In addition to the normal problems associated with high loadings on flexible buried structures, this structure is to be founded in an ice rich permafrost area.

Theory of Flexible Buried Structures

Two factors enter into the evaluation of the stability of flexible buried structures; yielding of the steel, and deflection stability. Yielding is a relatively straight forward analysis calculated by determining the load in the structure walls divided by the yield stress. The method for determining the structural stability is a simple ring compression analysis. Ring compression stress depends only on the external loads applied to the structure. The method for calculating ring compression and the associated factor of safety is shown schematically in Figure 1.

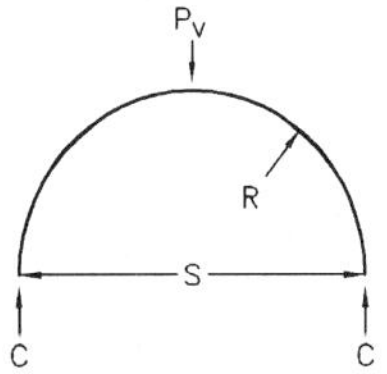

$$C = P_V \times \frac{S}{2}$$

C = RING COMPRESSION, lb/ft (kN/m)
P_V = DESIGN PRESSURE, lb/ft^2 (kN/m^2)
S = SPAN, in ft (m)

Figure 1 Ring Compression

[1]CEO and Chief Engineer, CBC Engineers and Associates, Ltd., 125 Westpark Road, Centerville, Ohio 45459; phone 937-428-6150; cbceng@erinet.com

The vertical load (P_v) is calculated from the dead load on the top of the structure plus the live load. The full dead load (DL) is equal to the unit weight of the fill (γ_m) times the depth of cover (H):

$$DL = \gamma_m \times H$$

Where:

DL = dead load pressure in 1bs./ft.^{2} (kN/m^2)
γ_m = unit weight of overlying fill in 1bs./ft.^{3} (kN/m^3)
H = total height of cover over pipe in feet (m)

Knowing the ring compression and the cross-sectional area of the pipe wall, the actual ring compressive stress f_c can be calculated as:

$$f_c = \frac{C}{A}$$

Where:

f_c = actual compressive stress
C = ring compression
A = cross-sectional area of plate

The factor of safety for wall area is $S.F._{area} = fyield \div f_c$

It is not sufficient to merely check the yield factor of safety of the structure, particularly in soft or yielding ground. It is possible that a structure could have a relatively high factor of safety against yield as calculated by ring compression and yet fail due to excessive deflection if the top arc of the structure becomes too flat. A technique for evaluating the degree of flatness by measuring the mid-ordinates of the top arc of the structure has been developed (Cowherd, D.C., and Perlea, 1991). Table 1 shows the relationship between shortening of the top mid-ordinate of the structure to the safety of the structure .

Table 1
% Top Mid-Ordinate Change And Remedial Action

Mid-Ordinate Percent Change	Depth of Cover (ft.) (m)	Recommended Action
<15	Any	No action required.
15 - 20	Over 6.0 (1.8m)	No action required.
15 - 20	Under 6.0 (1.8m)	Monitor on 6-month interval
20 - 25	Over 6.0 (1.8m)	Reduce live load allowed to 90% of H-20 and monitor on 6-month intervals
20 - 25	Under 6.0 (1.8m)	Reduce live load allowed to 75% of H-20 and monitor on 6-month intervals.
25 - 30	Over 6.0 (1.8m)	Reduce live load allowed to 75% of H-20 and monitor on 6-month intervals.
25 - 30	3.0 (0.9m) - 6.0 (1.8m)	Reduce live load allowed to 50% of H-20 and monitor on 6-month intervals.
25 - 30	Under 3.0 (0.9m)	Reduce live load allowed to 50% of H-20 and do detailed analysis
>30		Close structure until detailed analysis is done.

Note: Detailed analysis to include soil borings to determine expected additional movement.

As a load is applied to the structure by the fill and live load above it, the load is transferred into the backfill and soil around the structure (Figure 2).

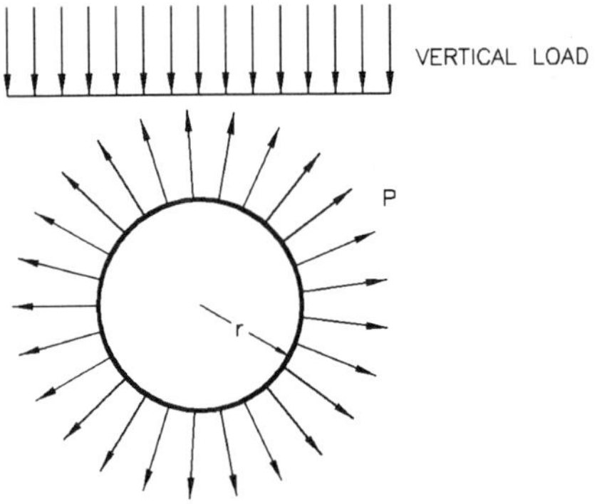

Figure 2 Stress Delivered To Soil

As the load is applied to the backfill and original soil around the structure, the backfill and original soil consolidate, allowing the sides of the structure to move out and the top to move down. The amount the backfill consolidates is dependent on the compaction of the backfill. The amount the natural soil consolidates is dependent on the compressibility characteristics of the soil. If the top of the structure reaches a sufficient degree of flatness, the structure will collapse even though the factor of safety in yielding might be quite high. It is, therefore, imperative that the backfill around the structure, and the natural soil outside the backfill envelope, be of a sufficient degree of compaction that they will not allow the sides of the structure to move far enough into the backfill to allow detrimental flattening.

Additional Problems in Permafrost Regions

This problem is greatly exacerbated in ice rich permafrost regions. Depending on the nature of the natural soil; whether it is frost susceptible or not; whether the permafrost is ice rich or not, and the temperature inside the structure; the permafrost could be melted to significant depths creating considerable loss of support for the base beneath the structure, or the soil beside the structure. Thawing of the permafrost beneath the backfill could lead to considerable settlement of the backfill beside the structure, leading to downdrag on the structure as the heavier fill tries to move down past the "hollow" structure. This situation adds considerable load to the top of the structure creating flattening of the top arc. Thawing of ice-rich permafrost in the walls of the excavation outside a select backfill zone could also create a loss of support for the sides of the structure allowing movement of the sidewalls of the structure outward, permitting the top to move down. This possibility is shown in Figure 3.

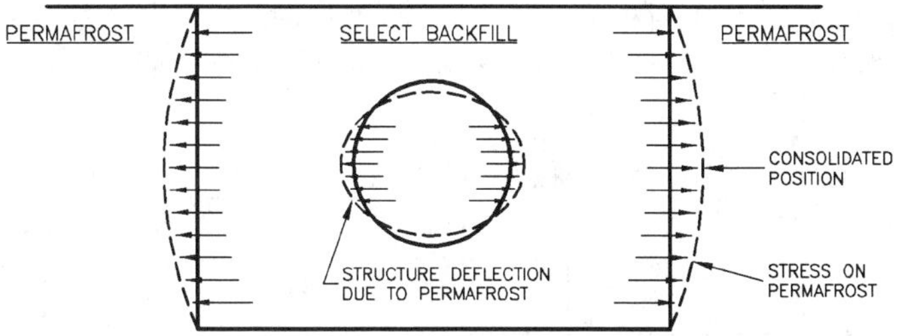

Figure 3 Structure Deflection Due To Permafrost

Another problem in permafrost areas is the potential formation of ice-lenses in frost susceptible soils. In frost susceptible soils freezing of the backfill could create large pressures on the side of the structure tending to cave in the side walls to the inside of the structure, if the backfill is a frost susceptible material. Because of these conditions it is necessary that select backfill around a buried flexible structure be non-frost susceptible in permafrost regions.

In the design of a flexible buried structure in non-arctic areas, it is generally necessary to consider:

1) the foundation soil upon which the structure and backfill will bear,
2) the nature of the select backfill material around the structure,
3) the width of the select backfill material; and
4) the loads on the structure.

The additional problems to be considered in arctic areas are:

1) the depth and nature of the active zone,
2) the potential effect of frost in the backfill and natural material,
3) the effect on the permafrost of potential heating within the structure; and
4) methods of preventing (or minimizing) thawing of the permafrost.

The design of buried flexible structures in the arctic must take into consideration all of these factors in addition to the conventional design considerations.

Hypothetical Evaluation of a Flexible Buried Structure in the Arctic

Structural Evaluation

The structure in question is designed as a smooth wall steel structure, ¾ inches thick, with reinforcing ribs eight (8) inches long by one (1) inch thick placed at four (4) feet on center. The total ring compression load (C) in each wall of the tunnel is $C = P_v \times S/2$ where P_v is the vertical load and S is the structure span:

$C = 185$ pcf x 80 ft. x 22 ÷ 2 = 162,800 lb/ft (2,376 kN/m)

The area per foot of wall is ¾ in. x 12 in./ft. = 9 in.^{2}/ft. The stress in the wall is 162,780 lb/ft ÷ 9 in.^{2}/ft. = 18,087 p.s.i. (124.7 MPa) or well below the yield stress. The structural factors of safety are calculated using standard

AASHTO Specifications (AASHTO, 1996), and CANDE-89 computer program (US-DOT, 1989). The factors of safety calculated using this technique are:

Table 2
Factors of Safety - Fabricated Steel Plate with Structural Flanges

Alternative		Factor of Safety			
	Thrust Stress	Displacement	Elastic Buckling	Performance factors	
				Max. Stress	Buckling
Structural Flanges with Fabricated Steel Plate Pipe 3/4" (19mm) Thick Plate	4.20	15.86	4.00	1.64	1.94

Foundation Evaluation

Having determined the pipe structure is safe in a conventional thrust and buckling analysis, the effects of constructing such a structure on, and in ice-rich permafrost must be considered. This evaluation must include a consideration of the foundation beneath the structure, design of the backfill around the structure, the natural soils in the walls of the excavation outside the backfill; and the design of measures to minimize thawing of the permafrost.

The soils at the site are assumed to be relatively consistent, consisting of a silty material, classified as a ML according to the Unified Soil Classification System, having a dry unit weight (γ_d) of 80 pcf (12.6 kN/m^3), and a moisture content of about 42%. Ice lenses are assumed to exist in the soil mass. The permafrost is estimated to extend to a depth of some 600 feet (183m) at the site. The bearing capacity of the frozen, ice-rich, soil can be calculated from the equation $q_r = Nf_c\sigma \div F.S.$ where Nf_c is taken as 5.14 since the structure foundation, consisting of the concrete floor, can be taken as continuous. The long range shear strength of the soil (σ) is equal to $\sigma_f = 5 + 2.25(32 - Ts)$ in psi (McFadden, T.T. and Bennett, F.L., 1991). The average temperature of the soil is 18°F (-7.8°C). The frozen soil ultimate bearing capacity is therefore $q_u = 5.14(5 + 2.25[32-18]) = 187.6$ psi or 27,104 psf (1,298 kPa). This bearing capacity is more than sufficient as long as the permafrost stays frozen. The total thrust in both walls of the structure is transferred into the concrete bearing slab which is 12.5 feet (3.8m) wide. The bearing stress applied to the soil is thus 16,145 psf (773 kPa) which is well below the bearing capacity. The stress generated by the structure onto the backfill is Pv • Rt ÷ Rs where Rt = top radius and Rs = side radius, or for a circular structure the stress = 185 x 80 x 11/11 = 14,800 psf (708 kPa). This is the stress applied to the select backfill at the wall of the structure. With the select backfill 22 feet (6.7m) wide (for reasons later discussed) the stress at the wall of the excavation is 0.4 x 14,800 = 5920 psf (283 kPa), which is also well below the expected bearing capacity. All of the factors of safety are sufficient as long as the permafrost remains frozen, therefore, steps must be taken to ensure minimal thawing of the permafrost.

Design To Minimize Thawing Of Permafrost

The mean annual temperature in the site area is taken to be 20° Fahrenheit (-6.7° Celsius). After an evaluation of the nature of the permafrost material, an evaluation is made of the distress which could occur to the structure if the permafrost were to melt to various depths. The structure is to be placed in a trench some 29 feet (8.8m) deep. The design of the width of the trench to provide sufficient select backfill was evaluated by determining what movement could occur if the permafrost outside the select backfill were to thaw. Two aspects of thawing of the original permafrost were explored, namely thawing of the permafrost below the structure and thawing of the permafrost beside the structure.

Based on the nature of the natural soil, (silty ice-rich permafrost) if the soil were to thaw it would be quite compressible allowing considerable settlement of the backfill in the trench around the structure. The structure itself places a smaller load on the original soil than does the backfill material around the structure with the load of the stockpile above the backfill. Based on the nature of the silty natural soil, it was estimated that the backfill would try to move down some 4 feet (1.2m) due to a permafrost melt to a depth of 10 feet (3m) below the bottom of the structure, while the structure itself would move down much less than the backfill. This would create large downdrag forces on the structure and would lead to ultimate collapse of the structure. Due to the great depth of the permafrost, it is not practical to excavate the permafrost. It is therefore necessary to take two precautions to help alleviate the problem of consolidation due to melting permafrost. These are:

1) Over-excavate beneath and beside the structure and provide a non-frost-susceptible backfill.
2) Provide thermo-siphons (convection tubes) to remove potential heat from the soil.

Design Of Select Backfill

For this example the accumulated freezing index is 7300°F (4038°C) days, while the accumulated thawing index is 2000°F (1093°C) days. The number of freezing days per year is 230 while the number of thawing days is 135. The mean annual temperature is 20°F (-6.7°C). This information is used to calculate the depth of frost and depth of thaw, to help ascertain the potential thickness of the active layer. The thickness of the active layer is important because if the ice-rich permafrost thaws, excessive movement of the flexible buried structure could be experienced. A gravel pad beneath, and to the sides, of the structure to mitigate the potential thaw depth is provided. This procedure will both remove soil that could thaw, and provide for deeper freezing of the area to keep the permafrost high. The depth of freeze or thaw is calculated from the following equation.

$$X_{f,t} = \lambda \left| \frac{(2)\,(24)\,k_{f,t}\,n\,\theta_A}{m.c.\,(\gamma_d)\,L} \right|^{1/2}$$

Where:

$X_{f,t}$	=	depth of frost (f) or depth of thaw (t)
$k_{f,t}$	=	the thermal conductivity of the soil [use k_f (frozen) for depth of freeze and k_t (thawed) for depth of thaw] (BTU/hr ft °F or W/m °C)
θ_A	=	the accumulated air freezing (or thawing index (°F-days or °C-days)
γ_d	=	the dry density of the soil (lb_m/ft^3 or kN/m^3)
$m.c.$	=	the moisture content of the soil by weight (% ÷100)
L	=	the latent heat of fusion of the water (143 BTU/lb_m or 333 kJ/kg)
n	=	the ratio of the surface freezing index to the air freezing index, or surface thawing index to air thawing index for depth of thaw. See Tables 4.2.1 through 4.2.4
λ	=	thermal coefficient

(McFadden, T.T. and Bennett, F.L., 1991)

The depth of freeze and thaw are calculated for both the silty soil and the proposed sand and gravel select backfill. The following parameters were used to made these calculations:

Table 3
Parameters for Calculation of Depth of Freeze and Thaw

Freezing index θ_A = 7300°F (4038°C) dry	$\gamma_{d\ silt}$ = 80 pcf (12.6 kN/m^3)
Thaw index θ_A = 2000°F (1093°C) dry	$M.C._{silt}$ = 36%
# Freezing days/yr. = 230	$n_{freezing\ silt}$ = 0.3
# Thawing days/yr. = 135	$n_{freezing\ gravel}$ 2.9
Mean annual temp = 20°F (-6.7°C)	$n_{thaw\ gravel}$ = 1.0
$\gamma_{d\ gravel}$ = 135 pcf (21.2 kN/m^3)	$n_{thaw\ silt}$ = 2.0
$M.C._{gravel}$ = 6%	

Utilizing these parameters, the depth of freeze and thaw for the fine grained silty permafrost and the sand and gravel select backfill are calculated. The results are tabulated in Table 4.

Table 4
Depth of Freeze/Thaw (ft.)

Condition	(α) Thermal Ratio	μ Fusion Parameter	X_i	λ	K	n	(X) Depth of Freeze/Thaw Ft. (m)
Silt $_{freeze}$	1.26	0.032	0.46	0.85	14.0	0.3	16.0 (4.9)
Silt $_{thaw}$	0.81	0.077	0.46	0.83	7.5	1.0	10.9 (3.3)
Gravel $_{freeze}$	0.42	0.31	0.13	0.75	20.0	0.9	50.3 (15.3)
Gravel $_{thaw}$	0.41	0.381	0.13	0.95	20.0	2.0	37.6 (11.5)

The active zone in the silty permafrost according to the above calculations is 10.9 feet (3.3m). It is, therefore, decided to extend the excavation for the structure to a total depth of 37 feet (11.3m) to provide a 5 foot (1.5m) layer of sand and gravel over the pipe and then to mound sand and gravel to a height of 6 feet (1.8m) over the top of the gravel surface in an attempt to keep the permafrost frozen.

Since the depth of thaw for the gravel is 37.6 feet (11.5m), it is desirable to place non-frost-susceptible gravel to a thickness of 22 feet (6.7m) (one structure diameter) to the sides of the structure. The final design is for 11 feet (3.35m) of non-frost-susceptible gravel beneath the structure, 22 feet (6.7m) (one structure diameter) beside the structure, and 5 feet (1.5m) over the structure. Figure 4 shows the backfill section.

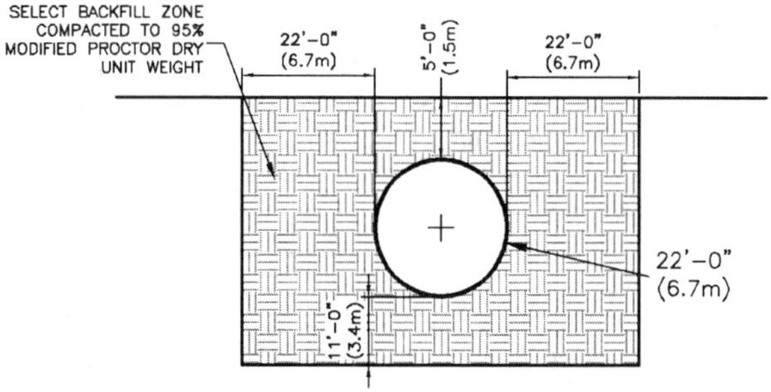

Figure 4 Select Backfill Around 22'-0" (6.7m) Dia. Pipe Structure

The design requires over-excavating the silty permafrost to a depth of 11 feet (3.35m) below the base of the structure and to a distance of one-structure span (22 feet) (6.7m) outside the walls of the structure, and refilling with a non-frost susceptible sand and gravel. The non-frost susceptible sand and gravel is designed to meet the gradation requirements of AASHTO A-1a with the further requirement that the percentage of fines be less than 3%. Table 5 shows the specified gradation.

Table 5
AASHTO Classification

Group Classification	A-1-a
Sieve	Percent Passing
No. 10 (2.00mm)	50 max.
No. 40 (0.425mm)	30 max.
No. 200 (0.075mm)	2 max.

The specified backfill is to contain less than 2 percent passing the number 200 sieve (0.075mm), and be relatively well-graded to meet the requirements of non-frost susceptible material. The select backfill must be a non-frost susceptible material as the build-up of any ice lenses in the backfill adjacent to the structure could be disastrous to the structure. Non-frost susceptible materials are defined as "*Soils which because of their coarse grained structure, capillary action between the grains is either too small or nonexistent so that no 'wicking action' is present in these soils. They do not, therefore, develop segregated ice*" (McFadden, T.T. and Bennett, F.L., 1991).

A distance of one times the span of the structure is determined to be sufficient. Since the depth of freeze for the backfill is greater than the depth of thaw it is felt the permafrost will enter the gravel backfill. The foundation preparation therefore consists of over-excavating only 11 feet (3.35m) of permafrost vertically and to a distance of some 22 feet (6.7m) to either side of the structure measured from the springline of the structure (Figure 4). This excavation should be refilled with non-frost susceptible sand and gravel around and under the structure. The sand and gravel is specified to be compacted to 95% of the maximum modified Proctor (ASTM D1557) dry unit weight, in order to achieve a relatively rigid mat, and minimize differential settlement if the permafrost were to thaw. The sand and gravel backfill achieves and provides stability from the following points of view.

1) It provides a compacted mat of material that will minimize uneven settlement which could create distress to the structure.
2) It provides an insulating layer over the permafrost to help keep the permafrost high. It is estimated the permafrost will rise into the gravel blanket.
3) It increases the bearing capacity of the soil.

Even though it is believed the granular backfill will keep the permafrost frozen, a parametric study to assess the consequences of thawing the soil outside the limits of the select backfill using the computer programs SOILEVAL (Cowherd, D.C., and Perlea, V.G., 1990) was performed. The model for this study consisted of using the soil parameters for the select backfill, and then varying the distance to which the permafrost was assumed to thaw beyond the select backfill. This study allowed a calculation of the distance to which the sides of the structure could move out (and the top down) due to the thaw-strain and consolidation of the melted permafrost. These results were analyzed using the program MULT-SPAN (Cowherd, D.C., and Perlea, 1991) to determine if the expected flattening would be detrimental to the structure. The thickness of thawed material was analyzed in 10 foot (3.0m) increments. This was done by assuming all the permafrost soil outside the select envelope thawed, with 10 feet (3.0m) adjacent to the excavation remaining frozen, then 20 feet (6.0m) and so on. This procedure yielded a distance from the edge of the structure that the permafrost could thaw without detrimental consequences to the structure. The results of this study are shown in Table 6.

TABLE 6
RESULTS OF PARAMETRIC OF STUDY

PERMAFROST CONDITION	FLATTENING OF TOP RADIUS OF STRUCTURE (%)
All Permafrost Thawed	23.4
First 10' (3m) of Permafrost Frozen	12.5
First 20' (6m) of Permafrost Frozen	5.5

The parametric study showed that if all the permafrost outside the select fill were to thaw, the top of the structure would move down 2.57 feet (0.8m). This would be detrimental to the structure. If 20 feet (6m) of wall of the excavation remained frozen, the top movement would be 7.3 inches (185mm). This is not detrimental to the structure. It was decided that added precautions would be taken to keep the first 20 feet (6m) of the wall of the excavation outside the select backfill frozen.

Even though the calculations showed the permafrost should remain frozen below the select backfill, precautions were also taken in this area. It was decided to add thermo-siphons (natural convection devices) to extract heat from the soil both outside the walls of the excavation, and below the select backfill within the excavation.

Design of Thermo-Siphons (Convection Devices)

Convection tubes are designed beneath and beside the structure to remove heat and to help keep the permafrost frozen in case sufficient heat leaks through the sand and gravel pad to create a melting of the permafrost beneath the structure. The convection tubes beneath the structure consist of 12 inch (305mm) diameter pipes placed on either end. The convection devices are placed on 11 foot (3.35m) centers horizontally in the sand and gravel backfill and vented on both ends to allow cold air in and warm air out. A positive means of venting the heat is designed. This is accomplished by connecting one end of the convection devices to a header pipe, and extending the pipe outside the backfill area to vent into the atmosphere. In order to ensure adequate venting, a fan is attached to the header pipe to vent the warm air and pull cold air through the system. Since the convection devices are buried and it will not be possible to get to the devices to replenish fluids, a convection siphon is designed instead of using a system with fluids. In order to ensure that there is an exchange of air, a cold air sump which opens into the convection devices is constructed outside the structure. Figure 5 shows the convection devices under the structure.

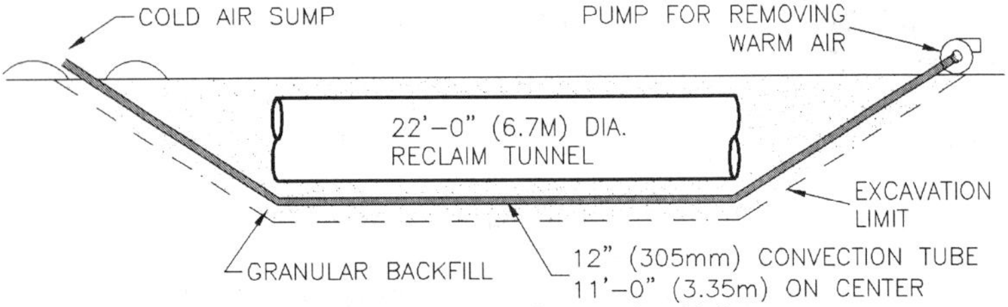

Figure 5: Horizontal Convection Tubes

The thermo-siphons for the side walls are designed to be vertical siphons with interior baffles to facilitate a flow of air. It is probable that the stockpile could extend over the top of the thermo-siphons, or that they could be damaged by material falling from the stockpile or equipment. Consequently, it is decided that a passive system, as designed for the pad beneath the structure, should be utilized. These convection tubes are designed to be 12 inches (305mm) in diameter and 10 feet (3m) apart in both directions with the first row being 5 feet (1.5m) from the edge of the excavation and the second 15 feet (4.6m) from the edge. In this manner the first 20 feet (6.0m) of the walls of the excavation should remain frozen. Again, a header pipe is connected to the siphons with a cold air sump on one end and a suction fan attached to the other end. Figure 6 shows the design of the vertical thermo-siphons.

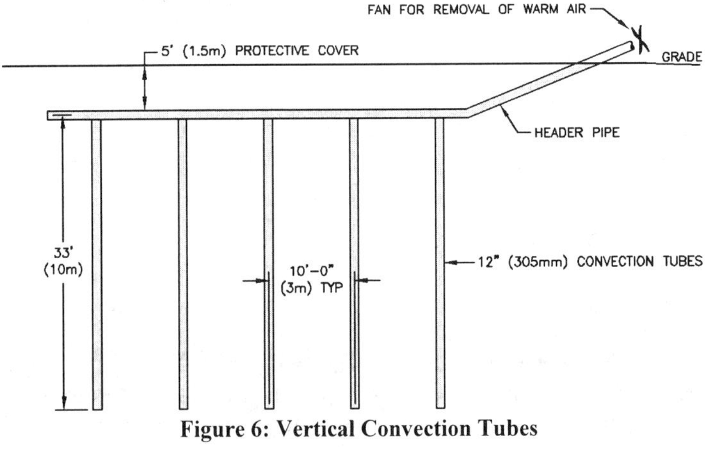

Figure 6: Vertical Convection Tubes

Conclusions

The design of flexible buried structures in permafrost areas presents unique problems. These problems can be overcome by designing provisions to ensure that the permafrost remains frozen, or does not thaw to an extent that would create a problem to the flexible buried structure. The techniques that are required for such structures include the following:

1. Design of appropriate undercuts and blanket of non-frost susceptible material beneath the structure.
2. Design of select backfill consisting of non-frost susceptible material to a sufficient distance beside structure.
3. Provisions to minimize thawing of permafrost both beneath and beside the structure.
4. Special techniques for thawing the permafrost for excavation for the structure.
5. Provisions for monitoring the degree of flattening of the structure.
6. Provisions for re-thawing permafrost if thaw occurs.

These structures can be utilized economically as opposed to other type of structures in arctic areas, however, the special precautions and evaluation due to arctic temperatures and permafrost, must be taken into account and provisions for overcoming these obstacles designed into the plan.

REFERENCES

Cowherd, D.C., and Perlea, V.G., (1991) "Evaluation of Flexible Culvert Behavior" Transportation Research Record No. 1315, National Research Council, Washington, D.C.

AASHTO, (1996) "Standard Specification for Highway Bridges - Section 12", Sixteenth Edition.

US-DOT, (1989) "CANDE-89 - Culvert Analysis and Design Computer Program", Publication No. FHWA-RD-89-169, June.

McFadden, T.T. and Bennett, F.L., (1991) "Construction in Cold Regions" - A Guide for Planners, Engineers, Contractors, and Managers, John Wiley & Sons.

Cowherd, D.C., and Perlea, V.G., (1990) "A Method of Evaluating and Projecting Future Movement of In-Situ Pipes" in Proceedings of the First National Conference on Flexible Pipes "Structural Performance of Flexible Pipes", Sargand, Mitchell & Hurd (eds.), Published by Balkema, Rotterdam.

Ground Temperature Variation Induced by A Buried Chilled Gas Pipeline

Scott Huang[1], Satoshi Akagawa[2], Toshiya Tanaka[3], Takashi Ono[4], Yuhki Nasu[5], Kazuhiko O'Hashi[6], and Masami Fukuda[7]

Abstract

In order to understand the complex behavior of gas pipelines in arctic regions, a full-scale experimental pipeline was constructed near Fairbanks, Alaska in December 1999. The pipeline was laid 1.8 m below the ground surface and covered with 0.9 m thick silty-sandy soil. The first 30 m of the line is in permafrost and the remaining section is in unfrozen ground. During the first phase of monitoring, chilled air with temperature between –4°C and –14°C was circulated through this 105 m long pipeline.

The ground temperature measurements indicated that a frost bulb had developed in the non-permafrost ground. The mean monthly average temperature measured by the thermal fences indicated that the ground 2.0 m deep and 1 m away from the pipe center was reduced from –1.22°C in December 1999 to –4.6°C in December 2000. During the same period, the mean ground temperature at the 4 m depth gradually diminished from 0.18°C to –0.72°C and from –0.1°C to –0.18°C at the 8 m depth.

Vertical upward movement of the pipeline caused by a frost front migrating out from the pipeline was noted. The maximum movements of the pipeline were 99.0 mm in the unfrozen area, 90.1 mm near the permafrost-non permafrost boundary, and 69.1 mm in the permafrost area.

An empirical equation was developed to describe the ground temperature change. The statistical analysis indicated that air temperature had a less effect on the immediate ground (i.e. < 2 m from the pipeline center) than the pipeline temperature.

----

[1]Professor, Department of Mining and Geological Engineering, University of Alaska Fairbanks, Fairbanks, AK 99775, USA; phone: 907-474-6880; ffslh@uaf.edu
[2]Professor, Graduate School of Engineering, Hokkaido University, Sapporo, Japan
[3]Senior Engineer, Energy Industries Engineering Division, NKK Corporation, Yokohama, Japan
[4]Professor, Department of Civil Engineering, Hokkai-Gakuen University, Sapporo, Japan
[5]Manager, Energy Facilities Engineering Division, Nippon Steel Corporation, Kanagawa, Japan
[6]General Manager, Energy Facilities Engineering Division, Nippon Steel Corporation, Tokyo, Japan
[7]Professor, Institute of Low Temperature Sciences, Hokkaido University, Sapporo, Japan

Introduction

A successful means to deliver natural gas from gas reservoirs in the far north regions to consumers requires a careful execution of a multi-billion dollars task which involves many aspects of economic, engineering, and environmental studies. One of the potential engineering-related problems encountered while constructing a gas pipeline in arctic and sub-arctic regions is the compromise of structural integrity of the pipes. The pipelines can be impaired by a number of factors such as changes of temperature in the immediate ground surrounding the buried pipes and the induced bending and torsion in the pipe segments laid in the boundaries between permafrost and non-permafrost terrains. Leakage of methane to the atmosphere from broken pipes also can have a harmful green-house effect.

The loss of pipeline structural integrity due to repeated frost heave and thaw settlement of the ground in which the pipes are laid is one of the important environmental and engineering issues that should be addressed and analyzed thoroughly. Release of methane from leaky pipelines can be a significant factor contributing to the global warming effect. Explosions of natural gas in the transmission lines because of fractured pipes have been well documented. A high demand on natural gas by the industrial countries has compounded the problem. Development of the gas fields in high arctic regions such as East Siberia, Russia and North Slope, Alaska will certainly pose an even greater danger to the delicate environment than otherwise.

In order to understand the complex behavior of a production gas transmission line in cold regions, a full-scale experimental gas pipeline near Fairbanks, Alaska was constructed. This test facility was built in December 1999 to study the changes of ground thermal regime, vertical movement of the pipeline, and the induced bending stress in the pipes. The facility is located 3.8 km on the Chena Hot Spring Road, Fairbanks, Alaska.

Test Site

Six 0.9 m diameter pipe segments (Grade API X-65 with a wall thickness of 8.5 mm) ranging from 16 m to 17.5 m in length were welded together to form a pipeline of approximately 105 meters. During the first phase of monitoring, pressurized chilled air with temperatures between –4°C and –15°C (Figure 1) was supplied by a 206-m air line (Figure 2). The pipeline was laid 1.8 m below the ground surface and covered with 0.9 m thick silty-sandy soil. The first 30 meters of the line are in permafrost and the remaining section is in unfrozen ground. The pipes in the permafrost section were covered with 1 m thick dry sand and backfilled with crushed in-situ soil and a thin layer of uncrushed in-situ soil (Figure 3). For the pipes in the unfrozen ground, 0.5 m thick dry sand was used instead.

This pipeline was instrumented with a series of thermistors, strain gauges, heave rods, deformation gauges, and several other monitoring devices (Huang, 2001). The instrument setup is utilized to collect the movement and strain data which are the indicators of the pipeline responses caused by changing ground temperature and soil conditions.

Site Geology. Before installation of the experimental pipeline at site, field drilling was performed in June 1999 and again in December 1999 to delineate the distribution of permafrost in the area (AGRA Earth & Environmental, 2000). From the drilling data, the depth to the permafrost table varied from 0.76 m to a depth exceeding 6.38 m. Along the alignment of the pipeline, the permafrost table was about 2.28 m deep on the east side and 4.56 m on the west side. Seasonal frost extended to 2.0 m during drilling in December 1999. Silt with low plasticity is the dominant soil in the top 1.5 to 2 m. A fibrous, brown peat layer of about 0.5 m overlies a second silt layer of 5 m thick. At some localities, this second silt layer is frozen and contains a trace amount to occasional small ice crystals. An example cross-sections showing the subsurface geology was constructed from December 1999 drill data (Figure 4).

Instrumentation. The instrumentation plan included installation of thermistors, strain gauges, heave rods, deformation gauges, heave plates, and surface settlement plates. One weather station, 3 water level monitoring wells, and 3 boreholes for downhole video camera were installed at site as well. Temperatures of the immediate ground and the pipeline, strains in the pipes, and heaving of the top 1 m foundation soil below the pipes are recorded by 3 HP data acquisition systems with a frequency of twice-a-day. Movements of the pipeline and overburden soils and groundwater levels are measured once every two weeks manually. A rotating laser is used to measure the changes of pipeline level.

Nine thermistors were attached to the upper surface of the pipeline to measure the temperature change of circulating air. Three thermal fences were constructed with two (TFA and TFB) in the unfrozen section and one (TFC) in permafrost. Thermal fence A has 6 strings with one on the north side of the pipeline and five on the south side. The thermistor spacing varies from 0.25 m to 1.0 m. The maximum depth is 8.25 m. The locations of thermistor strings measured from the pipeline center are 1 m, 2 m, 3 m, 4 m, and 6 m on the south side and 1 m on the north side. Thermal fence B has 3 strings with one on the north side and 2 on the south side. The thermistor spacing varies from 0.25 m to 1.0 m with a maximum depth of 8.0 m. The locations of thermal strings measured from the pipeline center are 1 m and 3 m on the south side and 1 m on the north side. Thermal fence C has 4 strings with 1 on the north side and 3 on the south side. The thermistor spacing varies from 0.5 m to 1.0 m. The maximum depth is 7.0 m. The locations of thermistor strings measured from the pipeline center are 1 m, 3 m, and 6 m on the south side and 1 m on the north side.

Five deformation gauges were installed in the trench before the pipes were placed. The deformation gauges measure the ground movement due to frost heaving and thaw settlement. The gauges locate at 1 m below the bottom of the trench. The positions of the gauges are 27.85 m, 30.96 m, 32.33 m, 37.04 m, and 68.85 m measured from the inlet riser of the pipeline.

Twenty-eight heave rods were welded to the upper surface of the pipes. Vertical movement of the pipeline is surveyed once every two weeks. In addition to the heaving of the pipeline, vertical movement of the covering soil above the pipes are measured using a set of 11 surface settlement plates. Forty-one strain gauges measure the changes in pipe curvature and length, which allow calculations of bending stresses on pipes and provide data for fatigue analysis.

Field Measurements and Analysis

Ground Temperature Measurement. Temperatures of the ground surrounding the pipes were measured by three thermal fences. Thermal fence A (TFA) locates 88 m away from the inlet riser. Six example temperature profiles measured by the TFA strings are shown in Figures 5 and 6. The measurement recorded on December 12, 1999 was considered as the reference temperature. Figure 5 shows three temperature curves of the ground 1 m from the pipe. The initial ground temperature were -0.52°C at the 2-m depth and -0.16°C at the 8.25-m depth. After six months of cooling, the ground temperatures were decreased by 1.6°C at the 2-m depth and –0.03°C at the 8.25-m depth. Measurements of these two depth locations one year later (i.e. December 31, 2000) did not show any significant changes as compared with the measurements taken on June 11, 2000. Similar observations were noted for the ground temperature profiles 6 m away from the pipeline center (Figure 6). The initial ground temperatures were 0.71°C at the 2-m depth and 0.11°C at the 7-m depth. They respectively decreased to -0.13°C and –0.08°C on June 11, 2000. A year later, the ground temperatures 2 m and 7 m below the surface were 0.2°C and –0.08°C, respectively.

The mean monthly average temperature measured by three thermal fences indicated that the ground 2.0 m deep and 1 m away from the pipe center was reduced from –1.22°C in December 1999 to –4.6°C in December 2000. Reduction of the mean ground temperature gradually diminished from 0.18°C to –0.72°C at the 4 m depth and from –0.1°C to –0.18°C at the 8 m depth.

Monthly temperature profiles of the ground 1 m away from the pipeline center are shown in Figure 7. The trends of monthly average temperature were similar to the daily temperature trends. The cooling effect had shifted the ground temperature toward freezing side. The monthly average temperatures of the ground 2 m deep were reduced from –0.47°C in December 1999 to –2.53°C in December 2000. However, the monthly average air temperature was –25.5°C in December 1999 and –16.4°C in December 2000. During the same period, the ground temperature at the 8.25-m depth decreased slightly from –0.16°C to –0.21°C.

Pipeline Movement. Vertical movement of the foundation soil 1 m below the pipes was measured by five deformation gauges. The vertical upward movements of the foundation soil reached a maximum in mid May 2000. It appeared that the freezing front caused by the chilled pipeline migrated downward to the 1-m depth in about 6 months. Beyond that, the deformation gauges registered a near constant value. The maximum ground heave was about 60 mm at 30.96 m from the riser. The deformation gauge 4, 37.04 m from the riser, experienced a sharp jump in mid April 2000. Monthly heaving of the foundation soil is plotted against the gauge locations in Figure 8. There was a higher vertical movement in the permafrost and unfrozen ground boundary. It was also true for the pipeline to experience a higher movement rate in the boundary area.

Vertical movement of the pipeline was manually surveyed once every two weeks. Figure 9 shows the daily survey measurements of 28 heave rods. The graph displays a relatively large vertical movement at location about 33 m from the riser. This observation was in agreement with the measurements of the deformation gauges. From September 2000 to December 2000, the pipeline moved upward abruptly. The survey data also indicated that the pipes in the permafrost section had experienced some upward movement. The maximum movements of the pipeline were 99.0 mm in the unfrozen area, 90.1 mm near the permafrost-non permafrost boundary, and 69.1 mm in the permafrost area.

Temperature Data Analysis. Freezing and thawing of ground during seasonal temperature changes are responsible for variations in soil properties and behavior of the foundation materials (Andersland and Ladanyi, 1994). Heaving of the foundation soil underneath pipeline is further affected by the pipeline temperature. Prediction of the depth and lateral extent to which the ground surrounding the pipeline is affected requires knowledge of the frost migration.

Because of the direct influence on ground heaving and pipeline movement, ground temperature distribution and its corresponding variation were analyzed statistically. The monthly average ground temperatures of the three thermal fences were calculated and compared with the average ground temperature in the initial month (i.e. December, 1999). The temperature difference between each subsequent month and the reference month was formulated to determine the combined effect of pipeline and ambient air temperatures. Parameters considered in the analysis included depth from the ground surface, distance from the pipe outer surface, air temperature, pipe temperature and the elapsed time starting mid December 1999. Other parameters which might have some degree of influence on ground temperature variation but were not included in the current analysis were the soil properties, initial ground temperature distribution, and ground water condition. From the statistical analysis, an empirical equation with a high degree of correlation was established to describe the ground temperature changes.

$$\Delta T = i + a/l + b/z + c/(l*z)$$

where ΔT: ground temperature change, ^{o}C

l: the distance measured from the outer surface of pipe , m

z: the depth from the ground surface, m

i, a, b, c: intercept and coefficients depending on the elapsed time in days, monthly ambient air temperature, and average pipe temperature (Results of the statistical analysis on TFA-N, TFB-S1, and TFC-S1 are listed in Table 1).

The general observation from the analysis indicates that the air temperature had a less effect on the immediate ground (i.e. < 2 m from the pipeline center) than the pipeline temperature during the monitoring period. The dominant effect of pipe temperature might have been caused by a frost front migrating beyond the first 2 m distance. Although less significant than the pipe temperature, the air temperature effect slightly increased in winter months. This observation was similar for both unfrozen and frozen ground. The cooling effect caused by the chilled pipeline on ground 3 m away was not as apparent as that on the immediate ground. Initially, the cold air had a relatively significant effect on the ground temperature variation. However, its influence decreased from July 2000 to November 2000 and regained the magnitude thereafter. During the same period of time, the pipeline temperature had an opposite effect. At the present time, the equation developed is applicable only for the immediate ground. As additional information such as soil properties and geology becomes available, the equation will be expanded and refined to include ground beyond the 2 m distance.

Conclusions

Based on the field data collected from December 1999 to December 2000, the following observations and analysis are summarized:

- The ground temperature measurements indicated that a frost bulb had developed in the non-permafrost ground. The effect of cooling had shifted the ground temperature toward freezing side. As an example, changes of the ground temperature recorded by TFA-N1 thermal string indicated that the monthly average temperatures of the ground 1 m from the pipeline center and 2.0 m deep was reduced from –0.52°C in December 1999 to –2.12°C in December 2000. Change of the ground temperature 8.25 m deep was –0.03°C during the same period.
- From the measurement of deformation gauges, it appeared that the freezing front caused by the chilled pipeline migrated downward to 1 m deep in about 6 months. The maximum ground heave was about 60 mm at 30.96 m from the riser. Deformation gauge 4, 37.04 m from the riser, had experienced a sharp jump in mid April 2000.
- The maximum movements of the pipeline were 99.0 mm in the unfrozen area, 90.1 mm near the permafrost-non permafrost boundary, and 69.1 mm in the permafrost area.
- The statistical analysis indicated that air temperature had a less effect on the immediate ground (i.e. < 2 m from the pipeline center) than the pipeline temperature. However, the variation of ground temperature is complicated and likely affected by other factors such as in-situ soil moisture content, groundwater, snow cover in winters, etc.

References

AGRA Earth & Environmental (2000). *Report on the design and construction of the Fairbanks frost heave facility*, AGRA Earth & Environmental Limited, Calgary, Alberta, Canada.

Andersland, O.B. and Ladanyi, B. (1994). *An introduction to frozen ground engineering*, Chapman & Hall, New York

Huang, S.L. (2001). "Gas pipeline project in Fairbanks, Alaska", *2001 Annual research report to Japan Science and Technology Corporation*, University of Alaska Fairbanks, Fairbanks, Alaska.

Acknowledgement

The authors with to thank the Japan Science and Technology Corporation (JST) for the financial support of the study.

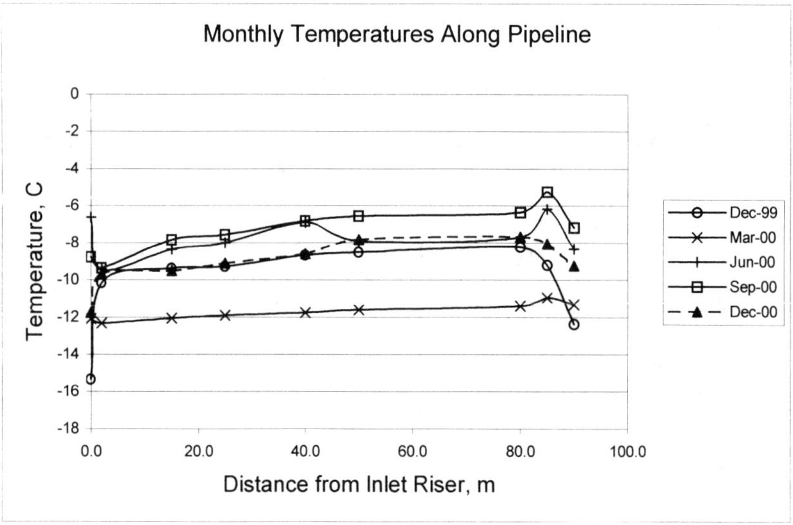

Figure 1. Monthly average air temperature in the pipeline.

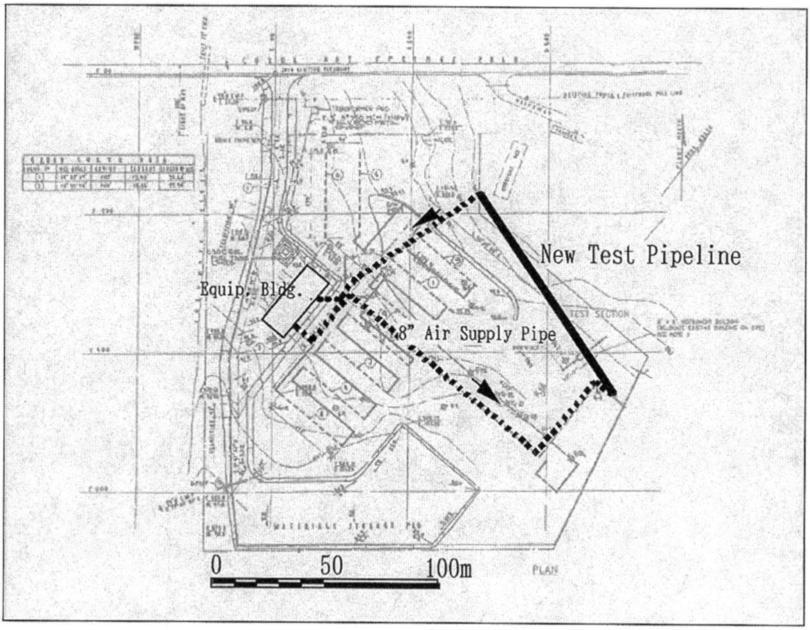

Figure 2. Map showing the pipeline and air supply line at the test site.

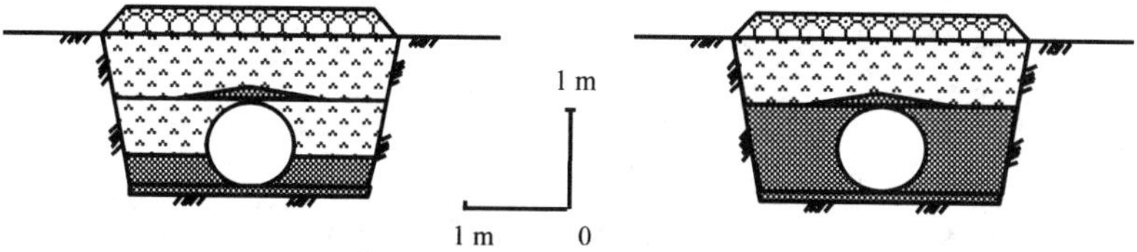

Figure 3. Cross-sections of pipes in non-permafrost section (left) and permafrost section (right) with dry sand as bedding material and crushed native soil (middle layer) and non-crushed native soils (top layer) as covering materials.

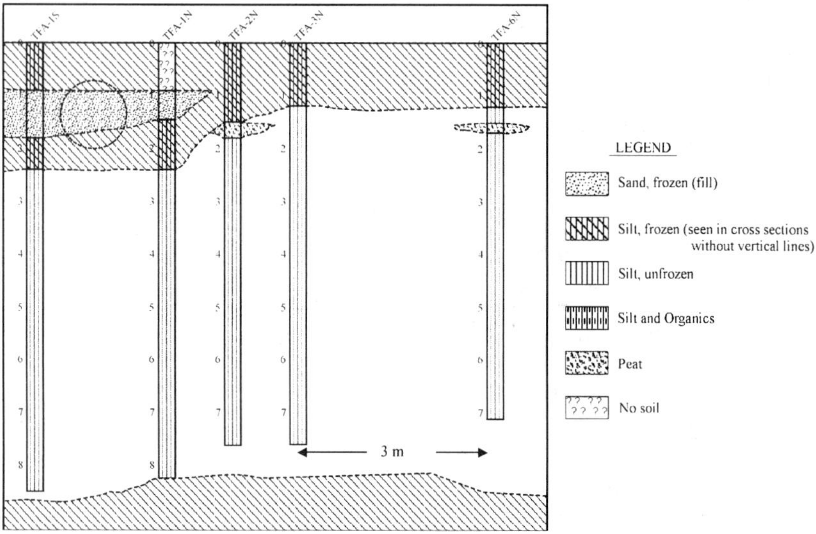

Figure 4. Subsurface geology near thermal fence A.

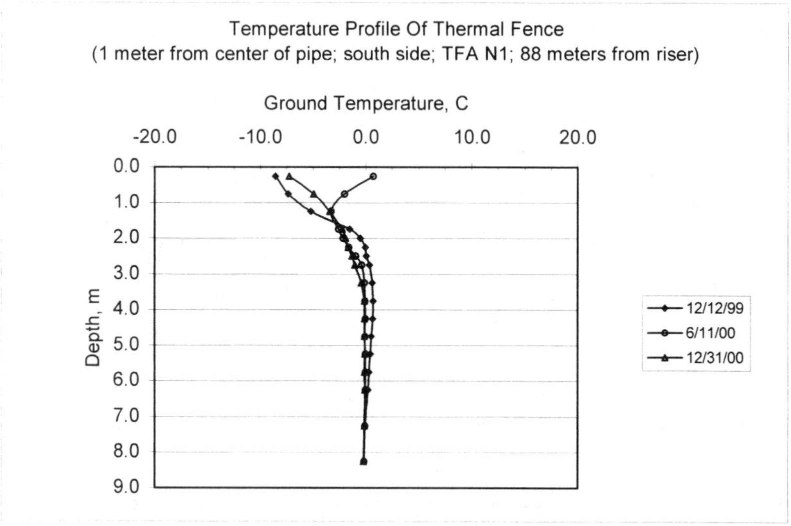

Figure 5. Ground temperature profiles of TFA located one meter away from the pipeline center.

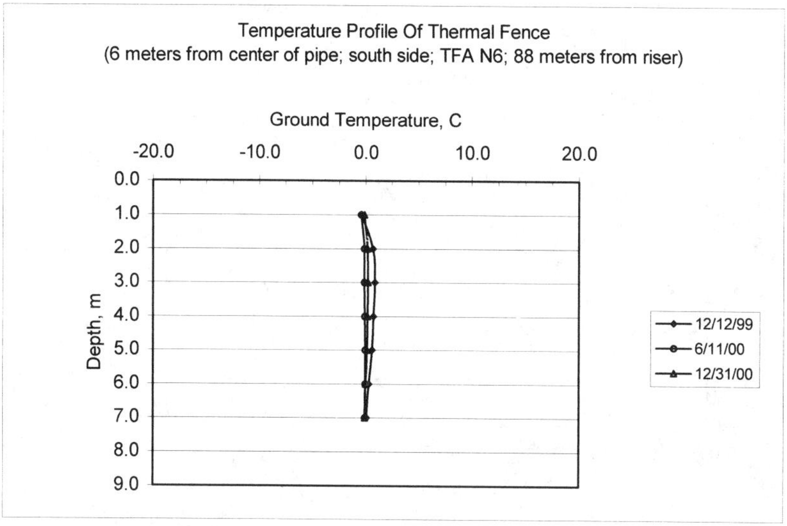

Figure 6. Ground temperature profiles of TFA located six meters away from the pipeline center.

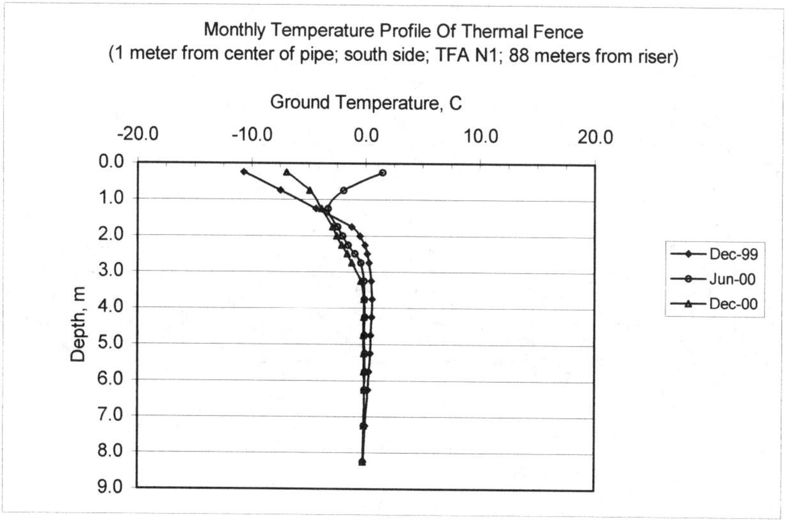

Figure 7. Monthly ground temperature profiles of TFA located one meter away from the pipeline center.

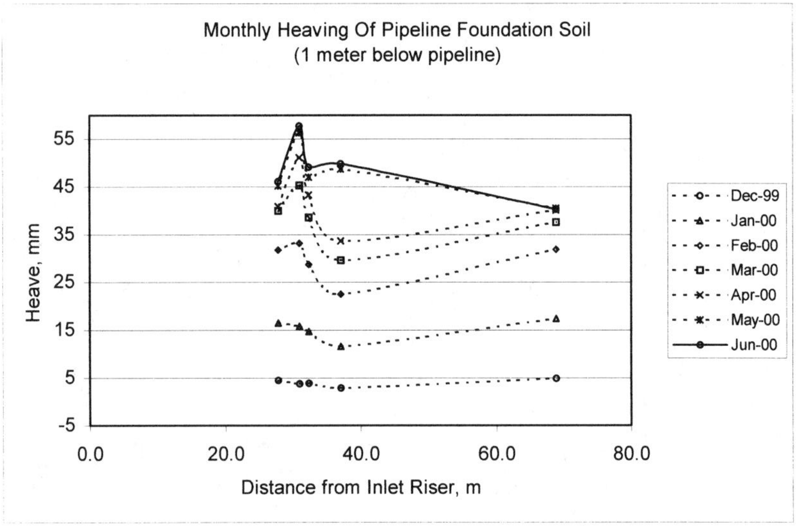

Figure 8. Profiles of daily vertical movement of foundation soil at 1 meter below the pipeline bottom surface.

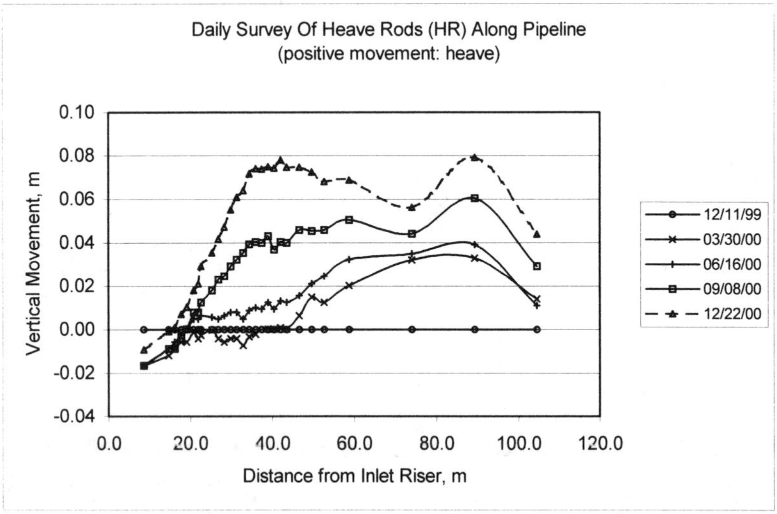

Figure 9. Profiles of vertical movement along the pipeline.

Table 1. Equations of intercept, coefficients a, b, and c for three thermister strings 1 m away from the pipeline center.

TFA-N1	
intercept	*i = -0.0733 - 0.1695 *a - 0.1354*b*
coef. a	*a = -1.6646 - 0.0098 time - 0.070 monthly air temp*
coef. b	*b = -5.4293 - 0.0118 time - 0.5107 avg pipe temp*
coef. c	*c = 8.9838 + 0.0152 time + 0.0824 monthly air temp + 0.6546 avg pipe temp*
TFB-S1	
intercept	*i = 0.0478 - 0.1857 *a - 0.1517*b*
coef. a	*a = -2.9341 - 0.0171 time - 0.1257 monthly air temp*
coef. b	*b = -1.8279 - 0.0235 time - 0.5146 avg pipe temp + 0.1174 monthly air temp*
coef. c	*c = 8.9054 + 0.0278 time + 0.1589 monthly air temp + 0.5482 avg pipe temp*
TFC-S1	
intercept	*i = 0.179 - 0.2023 *a - 0.056*b*
coef. a	*a = -3.9722 - 0.0168 time - 0.1151 monthly air temp*
coef. b	*b = -1.41 - 0.0256 time - 0.1308 avg pipe temp*
coef. c	*c = 9.5757 + 0.027 time + 0.1655 monthly air temp + 0.3549 avg pipe temp*

Wind-Induced Vibration Monitoring Program For Alpine Pipelines

James D. Hart[1], Dean Carson[2], Nasir Zulfiqar[3], Scott D. Lust[4]

Abstract

As part of the Alpine development project, Phillips Alaska, Inc. has installed a 305 mm (12-inch) diameter utility line and a 356 mm (14-inch) diameter produced oil line between the Alpine and Kuparuk oil fields on the Alaskan North Slope. In sections of the pipeline alignment that are oriented nominally perpendicular to the prevailing wind directions, a system of pipeline vibration dampers (PVDs) has been installed to mitigate potential fatigue damage due to wind-induced vibration (WIV). As part of the right-of-way agreement for the pipelines, Phillips has implemented a pipeline WIV monitoring program with the objectives of insuring that the PVD systems are performing as designed and to confirm that the spans without PVDs are not experiencing frequent and significant WIV. The Alpine WIV monitoring systems represent the most extensive, long-term WIV monitoring program ever undertaken on the North Slope. This paper provides an overview of the Alpine pipelines WIV monitoring program including; a description of the layout of the instrumentation, the data acquisition and the power supply systems, and a description of commissioning these systems. A summary of the WIV monitoring approach and the pertinent wind observation and vibration data gathered is also provided. The paper concludes with a summary of the overall findings and conclusions from this program.

Introduction

The Alpine pipelines run approximately 48 km (30 miles) between Alpine and Kuparuk. The Alpine pipelines were designed by Michael Baker Jr., Inc. As part of the overall design, wind-induced vibration (WIV) evaluation and pipeline vibration damper (PVD) tuning studies were performed by SSD, Inc. as a sub-contractor to Baker. The utility line has a wall thickness of 8.4 mm (0.330 inches) and the produced oil line has a wall thickness of 7.9 mm (0.312 inches). Both lines are fabricated from API 5LX-65 grade pipe material. The lines are constructed in a typical arctic cross-country configuration, which consists of long straight run sections between thermal anchors and expansion loops.

[1]President, SSD, Inc., Reno, Nevada, jim.hart@ssdinc.com
[2]Senior Engineer, Phillips Alaska Pipelines., Anchorage, Alaska, dcarson@ppco.com
[3]Project Engineer, SSD, Inc, Reno, Nevada, nasir.zulfiqar@ssdinc.com
[4]Mechanical Engineer, Michael Baker Jr., Inc., Anchorage, Alaska, slust@mbakercorp.com

Within the long straight pipe runs, the pipes span 19.8 m (65 feet) between vertical support members (VSMs). With this span length, these pipelines are the longest spanning 305 mm (12-inch) and 356 mm (14-inch) diameter lines on the North Slope. These lines have primary vertical mode vibration frequencies in the range from about 1.5 to 5 Hz. Based on WIV analyses of the pipelines, it was concluded that they could be susceptible to significant WIV in pipeline sections oriented nominally perpendicular to the prevailing wind directions. Based on this assessment, a system of PVDs was installed on both pipelines in segments with orientations between N45°W and N30°E. Phillips Alaska, Inc. implemented a pipeline WIV monitoring program to insure that the PVD systems are performing as designed and to confirm that the spans without PVDs are not experiencing frequent or significant WIV.

Description of Test Stations

The Alpine pipelines are being monitored at three stations along the alignment. The stations are monitoring the wind speed, wind direction, air temperature and the mid-span vertical acceleration of pipelines using digital data logger systems. Within each test station, the motion of each pipeline is monitored within three different spans (referred to as the "U" or "upstream" span, the "K" or "key" span, and the "D" or "downstream" span). Each test station is also being monitored with a "scratch gage" (as shown in Figure 1) which measures the maximum displacement of the pipe in the "U" span.

Test Station 1 is a 17 span section located to the south-southwest of Alpine. It has an azimuth of 170.7° and hence is approximately perpendicular to the prevailing wind directions. Test Station 2 is a 19 span section located West of the Kuparuk Drill Site 2H road crossing. The azimuth of this section is 82.5°, which is nearly parallel to the prevailing wind directions and outside of the design wind fan. Test Station 3 is a 14 span section located near Kuparuk CPF2. As shown in Figure 2, Stations 1 and 3 are located within the design wind fan and hence have PVDs installed, while Station 2 is outside the design wind fan and hence no PVDs are installed.

Data Acquisition System

The data acquisition system selected for the field monitoring program is the Campbell CR5000 system by Campbell Scientific, Inc. (Campbell, 2000). This system is capable of measuring input channels at measurement rates of up to 5000 samples/sec. The logger is programmed to collect data in tables which are written to a data card. This logger unit was selected to provide a balance between processor speed and relatively low power demands.

Data Logger Power Supply and Housing

The test stations are located at least 1 km from the nearest existing power source. A local power source for each test station was determined to be the most appropriate power design solution. This power source consists of a Teledyne Energy Systems

Model 2T thermo-electric generator or "TEG" (Telen, 2000), battery storage system, associated fuel system, and an overall housing. The TEG unit uses propane as the source fuel. Both 12 volts of DC electrical current and heated air are generated as outputs. The fuel system for each TEG consists of three 18 kg (40 lbs) propane tanks, a common manifold, and pressure hose. The automobile sized 100 amp-hour sealed gel cell storage battery supplies TEG ignition power and short-term system voltage. The enclosure housing, heated by the rejected hot air provides a suitable environment for these systems and the Campbell CR5000 unit.

The enclosure is composed of foam filled plywood panels with a full sized hinged lid, hinged end panel access door, and removable side ports. Small thermostatically controlled ventilation fans were incorporated for the maintenance of a nominal temperature range between –12.2°C and 26.7°C (10°F and 80°F). A fuel system solenoid valve is included to shut the TEG down if the enclosure temperature gets too warm (43°C or 110°F). The TEG exhaust is routed from the dual exhaust stacks to a common header and then out the side of the enclosure. For reliability, the control system is completely independent of the logger control system.

Accelerometers

The primary measurements of interest are the pipe mid-span accelerations. The accelerometers selected for this project are Crossbow single-axis units (Crossbow, 2000) with an onboard voltage regulator. These devices are high stability, capacitive beam type accelerometers with a bandwidth from DC to 125 Hz and a span of ±2g. These accelerometers are attached at the mid-spans of the U, K and D spans of the pipes oriented to measure the vertical acceleration of the pipe.

Weather Stations

The additional measurements of interest are the wind speed, wind direction and the air temperature. The wind speed and direction are measured using a R. M. Young Wind Monitor (Campbell, 1996). The anemometers are located a minimum of 9 meters (30 feet) away from the pipelines with the elevation of the propellers nominally equal to the elevation of the pipelines. The anemometers are oriented to be aligned with the Alaska State Plane coordinate grid. The air temperature is measured using a R. M. Young RTD Temperature Probe (Campbell, 1997).

Data Logger Program

A site-specific data logger program was developed to control the operation of the CR5000 systems at each of the three WIV monitoring stations. The pipe motion trigger condition is that either the standard deviation of the pipe acceleration or the standard deviation of the corresponding integrated pipe velocity is above a triggering threshold value. The largest motions from the monitored mid-spans of the utility line and the monitored mid-spans of the produced oil line are compared to the acceleration and velocity trigger threshold values. If a motion trigger occurs, the logger is programmed to record the data to a 60 second "motion event" file on a data card. For wind, the trigger condition is that the mean value of the perpendicular component of the wind speed is within a certain range and that the corresponding turbulence intensity (defined as the ratio of the standard deviation of the wind speed to the mean wind speed) is below a certain threshold. If a wind trigger occurs, the logger is programmed to record a "wind event" file, which contains a synopsis of the data conditions over the duration of the wind event.

Once a pipe motion or wind triggered event is complete, the data logger will return to monitoring mode and will trigger again as soon as another trigger condition is met or until the data card is full. The number and type of the triggers is computed and stored every 15 minutes along with the other pertinent logger status information.

Monitoring Procedure

Once the WIV monitoring stations were installed and commissioned, the monitoring plan was to let the systems gather data for a time period of approximately 3 weeks. At the end of each monitoring period, weather permitting, a data retrieval team was sent to the field to collect the data cards containing the data, replace them with new cards, re-supply the TEGs with propane, and to document the scratch post information and reset the scratch posts. In addition, the overall system was checked out and any observed problems were addressed and/or repaired.

Free Vibration Tests

SSD has a significant amount of experience in performing free-vibration ("pluck") tests on operating pipelines to investigate the dynamic characteristics. A pluck test involves statically loading a span into a deflected shape then quickly releasing the load to initiate free-vibration. The pipe was jacked into a deflected configuration in the K span using the loading set ups shown in Figure 3. A wooden pallet loaded with approximately 20 sand bags provided a reaction weight of over 454 kg (1000 lbs). The top and bottom rigging straps were jacked towards each other using a come-along until the pallet began to lift off of the ground. With the pipe in the deflected configuration, the data logger was manually triggered on and then after a delay of about 5 seconds, piece of all thread holding the load was cut using bolt cutters initiating free-vibration.

An example of the acceleration Fourier Amplitude Spectrum obtained from a typical pluck test is shown in Figure 4 with a comparison of the experimental and analytical vibration frequencies (obtained using AutoPIPE (Rebis, 2001) modal analysis). The pipe frequencies and the frequency range of the vertical primary modes (as well as the secondary modes) were identified in the instrumented spans and confirmed to provide a close match with the corresponding analytical values. The damping ratios in the pipes with PVDs (1.7% to 2.2% of critical) and without PVDs (0.4% to 0.8% of critical) were also estimated. Additional pluck tests on the installed PVDs confirmed that the damper frequencies correspond very well to the "target" pipeline frequencies. More details on PVD design can be found in Reference (Norris, 2000).

Summary of Wind Direction Measurements

Figure 5 presents a comparison of the histograms of raw wind direction (degrees, clockwise from North) at Station 3 over monitoring periods between November 2000 and September 2001. The dashed vertical lines shown on this figure correspond to wind directions that are perpendicular to pipeline orientations within the design wind fan (N45°W and N30°E). It is observed that the wind histograms for all sites have similar distributions (e.g., prevailing wind directions from essentially East-North-East with significant occurrences of winds from West-South-West). The solid lines oriented perpendicular to the boundaries of the Kuparuk design fan shown quite reasonably encompass or span the prevailing wind direction bins.

Summary of Event Counters

Figure 6 presents a comparison of the number of wind and motion triggered events recorded at Stations 1, 2 and 3 over the course of the monitoring to date. It is observed that the number of wind-triggered events is nominally proportional to how perpendicular the test station is to the prevailing wind directions. The number of motion triggered events shown in this figure is composed mainly of transient pipe response events or spurious triggers caused by accelerometer drift which have nothing to do with pipeline WIV. Only a handful of these motion triggers actually correspond to WIV. The absence of any significant WIV events in the presence of wind conditions favorable to WIV represents a very significant finding for the WIV monitoring program.

Description of Pipe Motion Events

Once a pipe motion event is stored, the data over the entire event duration is post-processed to provide various measures of the pipeline response and the wind conditions. Some events were subject to a more refined level of screening including processing of the acceleration signals using a Fast Fourier Transform to establish the frequency content of the motion.

Once "transient" and "spurious velocity" triggers are excluded from the motion event counts, a handful of "real" motion events remained which are associated with low level WIV. An examination of the frequency content of the pipe vibration indicates that the 305 mm diameter and 356 mm diameter pipe motions were both broad-banded extending over the entire frequency range of the primary, secondary, and tertiary vibration modes of both pipes. The estimated maximum pipe stress ranges in both lines was less than 21 N/mm^2 (3 ksi), i.e., the pipe stress ranges are benign. The design of the Alpine lines assumed that the perpendicular component of wind with any crossing angle would produce WIV. The data show that the crossing angle of the wind must be within approximately 10° of being perpendicular to the pipe to induce WIV, confirming that the design assumption was conservative.

Summary of Wind Triggered Events

As shown in Figure 6, a total of nearly 145,000 wind-triggered events have been recorded by the loggers at Stations 1, 2 and 3 to date. This means that the perpendicular component of the wind speed was within the 12.9 km/hr to 25.8 km/hr (8 mph to 16 mph) range where there is the potential for vortex shedding at and near the primary mode frequencies of the two pipelines. The number of events is also consistent with the azimuth of the test stations in relation to the Kuparuk wind fan as shown in Figure 2.

Data from the wind-triggered events was processed to provide various measures of the pipeline response and the wind conditions. Figure 7 provides an example of the key wind and motion parameters during each of the events recorded at one monitoring period. During many of these events, low level "background" pipe wind-induced vibrations were occurring. However, even during the strongest of these events, the pipe motions were on the order of 10% of the level required to cause a motion trigger. There were no narrow-banded, resonant WIV events recorded during the data collection period at any test station. The estimated maximum stress range during these conditions was in the range from 7 to 21 N/mm^2 (1 to 3 ksi). With respect to the potential for causing fatigue damage at pipeline girth welds, we consider stress ranges below 40 N/mm^2 (5.8 ksi) to be entirely benign based on project specific fatigue criteria. This supports that the PVDs (where required) are performing as designed and that PVDs are not required in locations where they are not installed. Spans without PVDs (e.g., in Station 2) have not experienced WIV significant enough or with enough frequency to cause weld fatigue damage or failure.

Summary of Scratch Gage Data

When the data retrieval team visited the test stations, the scratch gage plates were either photographed and the pointers reset or the plates were retrieved and replaced with new plates. For many of the periods, the plates exhibited the presence of longitudinal scratches, which are associated with longitudinal movement of the pipes. Close examination of the plates indicates a distributed sequence of numerous "dwell spots" along many of the longitudinal scratches where the scratch gage pointer has worn into the scratch plate. These spots appear to be associated with the time periods between incremental longitudinal shifts of the pipes. The vertical dimension of these dwell spots provides an indication of the level of vertical pipe vibration that occurred since the gages were set. In general, the vertical dimension of the dwell spots was approximately 3 mm (1/8-inch), peak-to-peak, which correspond to "Level Green" screening conditions (i.e., benign WIV stress ranges). We believe that the vertical pipe motions that caused the dwell spots are associated with essentially ambient "background" level WIV response of the pipelines similar to what was observed in the digital data.

Conclusion

Phillips Alaska, Inc. implemented a pipeline WIV monitoring program with the objectives of ensuring that the PVD systems are performing as designed and to determine if the segments within the design wind fan are the only locations that require PVDs. In addition, the program was intended to confirm that the spans without PVDs are not experiencing frequent and significant WIV. These objectives have been accomplished. In terms of the number of instruments deployed and the number of monitored days, the Alpine WIV monitoring systems represent the most extensive, long-term WIV monitoring program ever undertaken on the North Slope. At Stations 1, 2 and 3, both the digital data and the scratch gage observations consistently indicate that the WIV response of both pipes has been benign for the entire monitoring period to date. Only broad-banded, background level WIV has been observed. This provides an excellent indication that the PVDs are performing as expected and that spans without PVDs are not experiencing frequent or significant WIV. Perhaps the most useful finding to date is the absence of any significant WIV in the presence of wind conditions that could lead to WIV. The locally measured wind direction histograms are consistent with the expected ENE and WSW wind directions and the data indicates that the Kuparuk design wind fan sufficiently encompasses or spans the prevailing wind direction bins. The design wind fan is adequate to identify the sections requiring WIV mitigation and does not need to be broadened for the Alpine pipeline system. The maximum estimated WIV stress ranges are less than 21 N/mm^2 (3 ksi) compared to a project specific 40 N/mm^2 (5.8 ksi) endurance limit for pipeline girth welds. This indicates that zero fatigue damage has been accumulated due to WIV stress ranges.

Acknowledgments

The work described in this paper is based on a project undertaken by Michael Baker Jr. Inc. and SSD, Inc. for Phillips Alaska, Inc. Phillips' support for this work and permission to publish this paper is most gratefully acknowledged. Special thanks is due to Mr. Dan Rice of the Joint Pipeline Office - Alaska State Pipeline Coordinator's Office for the valuable review, input and guidance that he provided over the course of this project. The authors would also like to acknowledge the diligent efforts of Mr. Marcus K. Olson of Michael Baker Jr. Inc. (Anchorage) in development of the data logger power supply and housing and participation in the commissioning tests of these systems. The authors are also grateful to Mr. Michael Adams of Campbell Scientific, Inc. in Logan, Utah for his support.

References

AutoPIPE Pipe Stress Analysis and Design Program, User Reference, Rebis, Walnut Creek, CA., 2001.

Campbell Scientific, Inc., "05103 and 05305 R. M. Young Wind Monitor – Instruction Manual", September 1996.

Campbell Scientific, Inc., "43347 RTD Temperature Probe and 43408 Aspirated Radiation Shield – Instruction Manual", January 1997.

Campbell Scientific, Inc., "CR5000 Measurement and Control System", Operators Manual, June 2000.

Crossbow High Stability, LF Series Accelerometers, Models CXL02LF1Z and CXL02LF1, 1999-2000 Crossbow Product Catalog.

Norris, M. A., Ptak, K. R., Zamora, B. A., and Hart, J.D., *"Implementation of Tuned Vibration Absorbers for Above Ground Pipeline Vibration Control",* International Pipeline Conference, Calgary, Alberta, Canada, October 2000.

Telan Operation and Maintenance Manual, 2T and 2DT Series, Teledyne Brown Engineering, Energy Systems, Hunt Valley, MD, February 2000.

Figure 1. Typical Installation of Scratch Gages for Alpine Pipelines

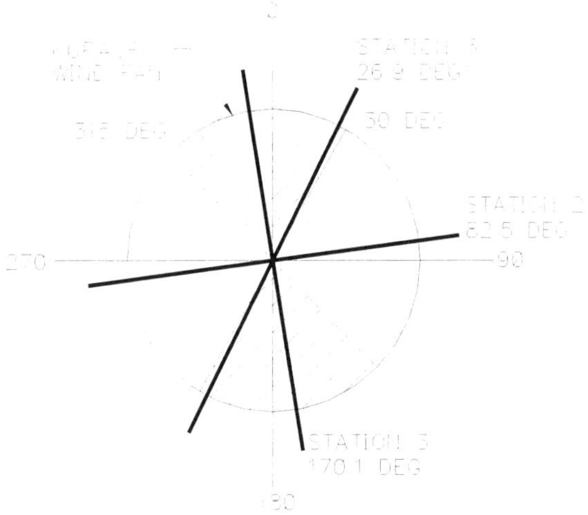

Figure 2. Kuparuk Wind Fan

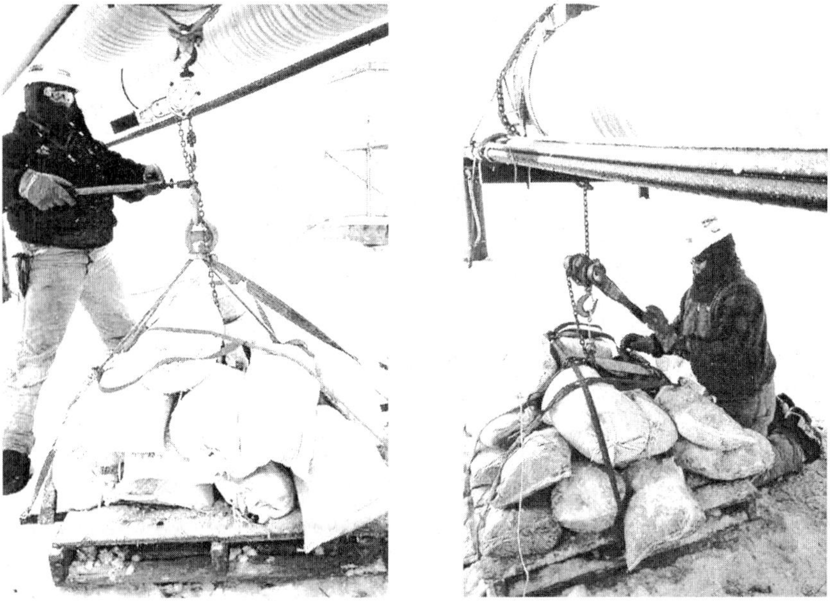

Figure 3. Jacking Reaction Weight and Releasing to Initiate Free-Vibration

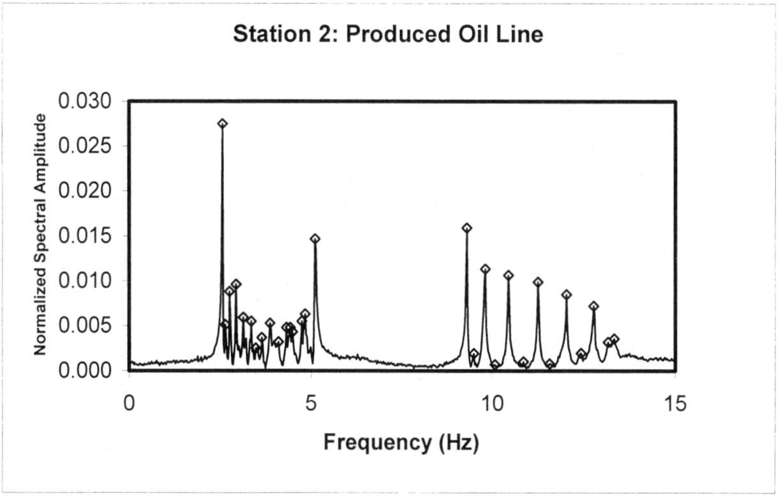

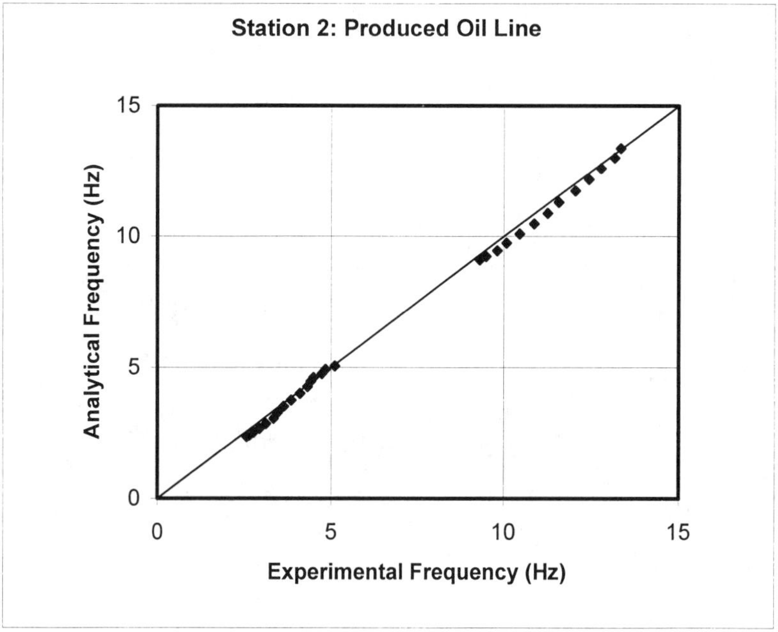

Figure 4. Example Primary and Secondary Mode Frequency Response from Free-Vibration Test and Comparison with Analytical Frequencies

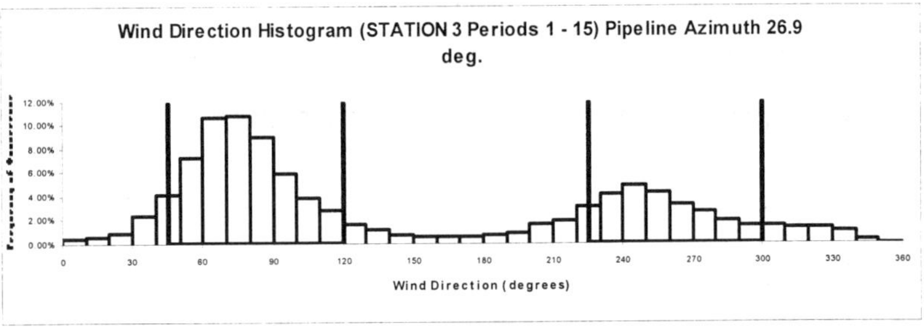

Figure 5. Typical Wind Direction Histogram

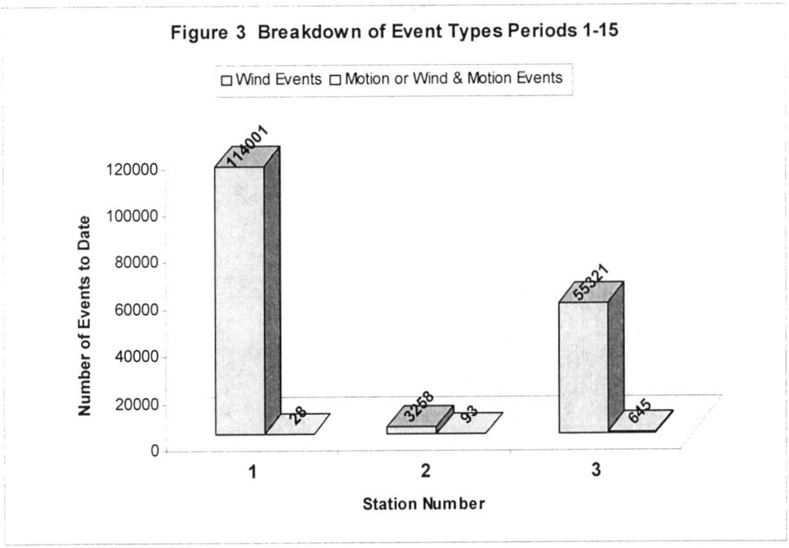

Figure 6. Event Counts Block Diagram Periods 1-13

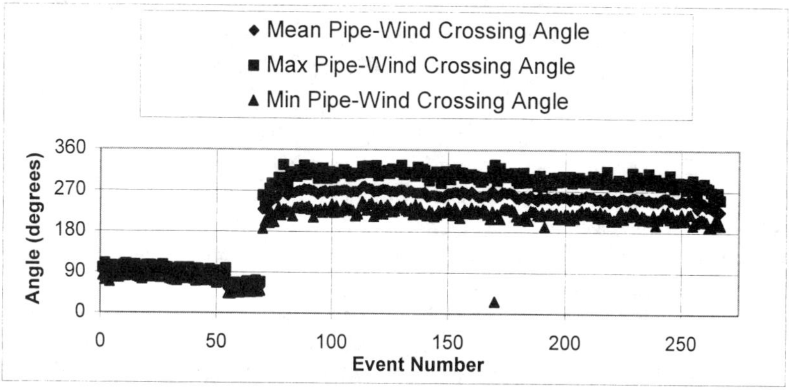

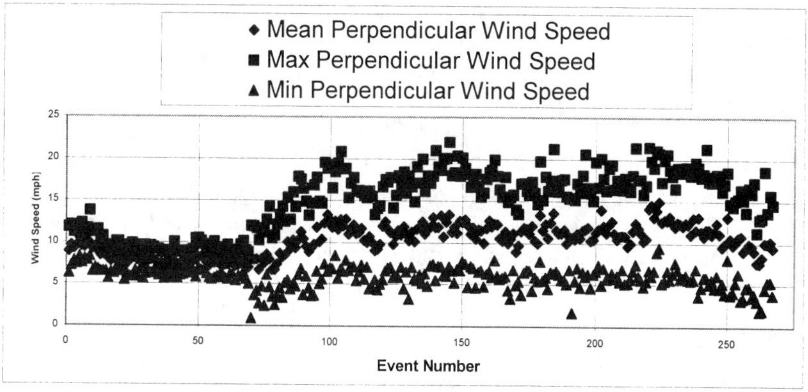

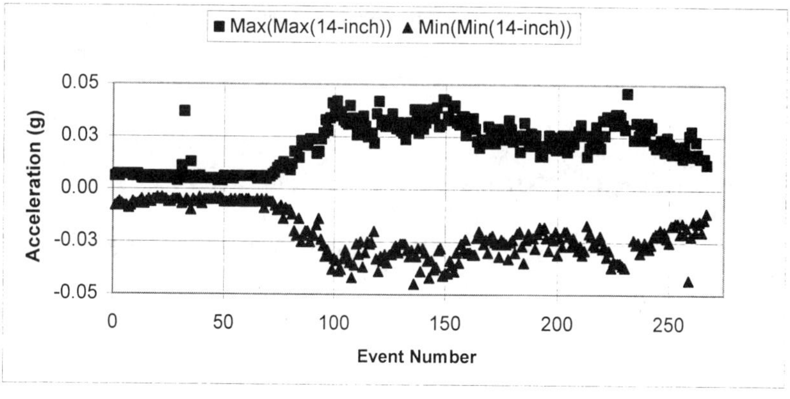

Figure 7. Example Results for a Monitoring Period

An Examination of Resource Transportation Alternatives in Northwest Alaska

John L. Aho, Ph.D.[1], Jeff Ottesen[2], Don Galligan[3],
Dan Sterley, P.E.[4], Patrick Burden[5], and Bill Wong, P.E.[6]

Abstract

This paper briefly describes the information gathered and conclusions reached during Phase 1, Potential Corridors Identification, of the Resource Transportation Analysis (RTA). The RTA is a component of the Northwest Alaska Transportation Plan, which will provide a strategy to guide transportation infrastructure development in Northwest Alaska.

The study area covers approximately 564,300 square kilometers of northern Alaska. It encompasses the entire North Slope Borough (NSB) west of the Dalton Highway, the Northwest Arctic Borough (NWAB), the Yukon River Basin west of the Dalton Highway and north of Nenana, and the Seward Peninsula and Norton Sound region (Figure 1).

Resource districts in the study area are the DeLong Mountains District, Ambler Mining District, Seward Peninsula District, Ruby District, Wiseman and Chandalar districts, and North Slope District, which includes the Western Arctic District. The paper summarizes the known reserves of mineral resources, including oil and gas, lead, zinc, copper; coal, gold, platinum, and other hard minerals.

The environmental and land status, transportation options, and financial issues associated with each potential mining district are summarized in this paper. Cold regions impacts on the study include climate, sea ice, and permafrost.

Introduction

Northwest Alaska has areas of high mineral potential. Many areas already have been mined or are recognized to contain significant mineral deposits. Representative resources are noted below.

---

[1]Principal Project Manager, CH2M HILL, 301 W. Northern Lights Blvd., Suite 601, Anchorage, Alaska 99503

[2]Statewide Planning Chief, ADOT&PF, 3132 Channel Dr., Room 200, Juneau, Alaska 99801-7898

[3]Environmental Specialist, HDR, Inc., 2525 C Street, Suite 305, Anchorage, Alaska 99503

[4]Senior Project Manager, CH2M HILL, 301 W. Northern Lights Blvd., Suite 601, Anchorage, Alaska 99503

[5]Economist, Northern Economics, Inc., 880 H Street, Suite 210, Anchorage, Alaska 99501

[6]Transportation Specialist, Sandwell Engineering, Inc., 1045 Howell Street, Suite 700, Vancouver, BC, Canada V6Z2A9

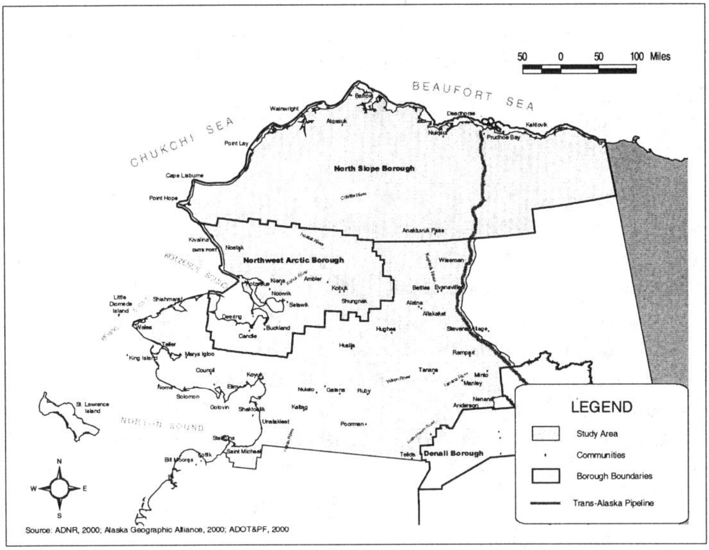

Introduction

Northwest Alaska has areas of high mineral potential. Many areas already have been mined or are recognized to contain significant mineral deposits. Representative resources are noted below.

- Several major mineral deposits in the Ambler District are precluded from development because of the lack of transportation infrastructure.
- Cominco Limited operates the Red Dog Mine near Kotzebue and has additional proven high-grade ore bodies nearby that could be developed in the next 5 to 10 years, adding to or sustaining the current volumes handled by the DeLong Mountains Transportation System (DMTS).
- The Seward Peninsula, Ruby, Wiseman, and Chandalar districts contain known reserves of gold.
- The National Petroleum Reserve-Alaska (NPR-A) has recognized mineral potential, but development is hindered by because of land use restrictions.
- As many as four major coal mines could be developed in Northwest Alaska, with annual production on the order of 10 million tons each. NPR-A is estimated to contain possibly a quarter of all coal in the United States–coal that is high grade and low sulfur.

- The Western Arctic region west of NPR-A and adjacent to the Chukchi Sea may contain reserves of as much as 1 trillion tons of coal. At a minimum, 125 million tons of these resources are strippable by modern mechanized mining. Further exploration would likely delineate other strippable reserves.
- Gravel and crushed rock are available from quarries that lie along or near potential railroad routes.

Many issues and concerns must be addressed during development of transportation and resource projects in Northwest Alaska. Some of these issues and concerns are typical of transportation and mineral development projects in general; others are unique to the region. Because of the concerns, issues, and potential impacts identified during the planning phases for resource transportation projects, as well as issues such as the lack of transportation infrastructure in the Northwest region and the presence of vast tracks of federal conservation units (national park and wildlife refuges), a wide scope of public and agency comments is anticipated.

Native corporations in the study area have indicated that they are interested in the development of resources in Northwest Alaska. However, they are concerned about the effects of development on subsistence resources and culturally sensitive areas. In addition, although the regional Native corporations may support resource and transportation development in the region, many villages are concerned about local impacts; in particular, the effects of linking rural areas to the population centers of the state (Fairbanks and Anchorage). The interests of resource development and transportation infrastructure need to be weighed against encroachments into conservation units and impacts to the rural lifestyles of the people in the region.

Environmental and land status

Environmental. Environmental permitting and land status concerns will be key issues for resource development and transportation projects in Northwest Alaska.

Permitting requirements discussed in this paper are preliminary and speculative. Given the vast size of the study area and the wide range of development alternatives, identifying and discussing details of the environmental regulatory requirements for potential projects is not possible at this time. The initial phase of the work only attempts to address those issues and concerns that seem most obvious in this early stage of the RTA study.

Recognizing the amount of federal land located within the study area and the apparent length associated with most of the transportation corridors identified, it is expected that the National Environmental Policy Act (NEPA)will serve as the platform on which local government and federal and state agencies will base many of their perceptions, comments, and permitting requirements. The resulting environmental document likely will be an environmental impact statement. In preparation for future discussion, planners should identify the full scope of environmental permitting requirements as development alternatives are refined.

Land status

Adding complexity for resource development and transportation projects in Northwest Alaska is the diverse nature and makeup of land ownership and land status types. Federally managed conservation units pose unique challenges to transportation projects in the region.

Further, transportation projects in Northwest Alaska could facilitate patent issuance for state-selected, Native-selected, and top-filed lands, which are currently under the management of the Bureau of Land Management. Under the terms of the Alaska Native Claims Settlement Act (ANCSA) and Alaska National Interest Lands Conservation Act (ANILCA), Native corporations have priority over the state in patenting top-filed lands. The process of conveying land will likely involve a relatively large number of stakeholders (state, federal, Native, and private) and could add significant time to the overall planning and development process for resource transportation and extraction projects in the region.

Transportation

Existing and Potential Tidewater Port Sites

Several studies have identified existing ports or potential port sites that may be suitable for handling Western Arctic coal. However, the findings indicated that there are no existing or potential port sites in the Western Arctic area that would be suitable for loading coal on a continuous basis due to heavy winter sea ice. These studies also indicated that Nome and Kotzebue have dock facilities that may be used to load coal on a seasonal basis because of winter ice conditions.

Other studies that focused on hard minerals have identified additional potential port sites along the Northwest Alaska coast. These sites include the DMTS port, Port Clarence, Cape Nome, Cape Darby, and ports and port sites in Southcentral Alaska.

Cominco Alaska is currently working closely with the Alaska Industrial Development and Export Authority and NANA Regional Corporation on a detailed feasibility study to construct a direct loading terminal at the DMTS port. To date, extensive environmental and technical studies have been completed to arrive at the most feasible concept. The proposed concept would be to construct a trestle extending as far as 800 meters offshore to a new berth at an existing water depth of 9 meters. A navigation channel and ship-turning basin would be dredged to a depth of about 16 meters to accommodate fully loaded Panamax vessels. The new port could handle both concentrates and coal.

One key constraint that must be considered in selecting a port site is the available water depth (at least 16 meters as noted above). Another key constraint is the number of days available for shipping during the ice-free season. The average shipping days range from 60 days at Omalik Lagoon, to 100 days at the DMTS Port, to 140 days at Cape Nome, and to 365 days at the port of Seward in Southcentral Alaska.

The length of the shipping season can be extended by the use of ice-strengthened or icebreaking vessels. Analyses in related transportation studies indicated that the season at each of the Northwest Alaska port sites during which oceangoing vessels could be loaded directly from the shore-based terminal could be increased by about 45 days with ice-strengthened vessels and by another 120 days with icebreakers.

Potential Transportation Options

The potential transportation systems examined include the common methods already in use in the state, such as railroad, roads, winter trails and ice roads, river barges, aviation, and petroleum pipelines. Less conventional systems such as pneumatic capsule systems, slurry pipelines, large air cushion vehicles (ACVs), overland conveyors, and transmission of coal energy in the form of electrical power were also considered. Although transmission of electrical power usually is not thought of as a transportation system per se, the possible cost of delivering coal energy from mine to consumer by wire was compared with that of truck hauling by road and rail delivery to provide a more complete understanding of options.

The only transportation modes that have been used and proven for long-distance hauling (more than 160 kilometers) of high-volume commodities on a year-round basis and in an arctic climate are railroad and road alternatives.

The remaining transportation modes are better suited for short-distance hauling, seasonal operations, or both. In some cases, the modes have not been proven to be reliable for commercial applications.

Existing Facilities and Infrastructure

Lynden Transport is currently hauling by truck about 14 million tons per year of lead-zinc concentrates from Red Dog Mine 84 kilometers to the seaport. The company has concluded that the most feasible mode for transporting Northwest Alaska coal to the Red Dog seaport is also by truck. The DMTS road is deemed an excellent "northern" road (Haugen, 2000a).

The main constraints affecting the capital and operating costs of truck hauling are road surface, road grades, seasonality, allowed hauling weights, labor availability, fuel cost and availability, and parts and equipment access.

The Alaska Railroad Corporation operates the only freight railroad in Alaska. The mainline runs from Milepost 1.5 at the port of Seward to Milepost 470 at the yard complex at the City of Fairbanks. The railroad interchanges freight in the ports of Seward, Anchorage, and Whittier. Major commodities hauled are petroleum products, coal, gravel, and intermodal containers.

The DMTS road and Dalton Highway are the existing major roadways within the study area, excluding any inter-village roads.

Other road systems within the region include roads connecting Nome to the communities of Council and Teller, Elliott Highway beginning in Fox and ending in Manley on the Tanana River, Ruby-Poorman Road, and the Ophir-Sterling Landing Road.

River barges are used extensively on the Kuskokwim and Yukon rivers to deliver bulk fuels and general cargo to communities along the rivers during the ice-free season. The water depths on the rivers dictate the size of barges used. Most of the barges used vary in draft from 1 to 3 meters and carry from 100 to 1,000 tons.

The Kobuk River is navigable during the summer months by barge traffic to service larger communities such as Noorvik, Kiana, Ambler, Shungnak, and Kobuk. This river flows through the heart of the Ambler District.

Potential Corridors

Eleven potential corridors that could link resource districts to tidewater, either directly or through an existing transportation system, were identified. Figure 2 shows the corridors and the mineral districts. The corridors are briefly described below.

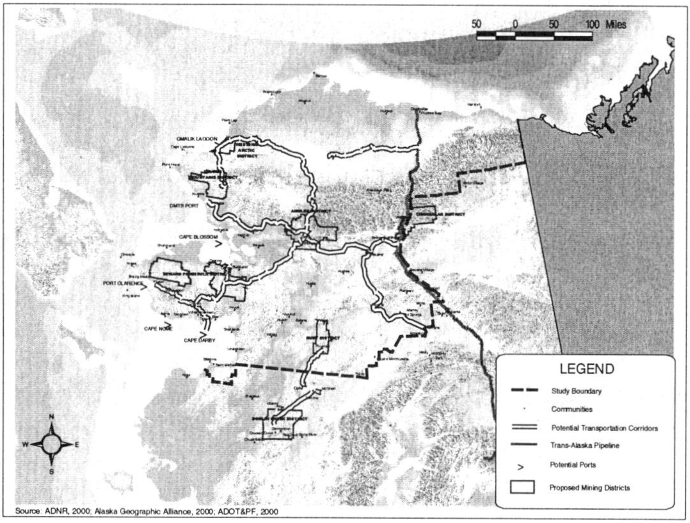

Western Arctic to DMTS corridor. This 150 kilometer corridor extends south from the Western Arctic District to the DMTS District, where the existing DMTS facilities provide tidewater access.

Ambler to DMTS corridor. This 390 kilometer corridor connects the Ambler District with the DMTS Districts to the west.

Ambler to Seward Peninsula corridor. This 450 kilometer corridor travels southwest from the Ambler District to the Seward Peninsula District where access to tidewater could be obtained at Cape Nome, Cape Darby, or Port Clarence.

Ambler to Dalton Highway corridor. Traveling 340 kilometers, this corridor extends east from the Ambler District to the Dalton Highway.

Ambler to Nenana corridor. This 640 kilometer link connects the Ambler District to Nenana, which has rail and highway access permitting travel to Southcentral Alaska ports.

Western Arctic to Seward Peninsula corridor, southern route. This 1025 kilometer link joins the Western Arctic District to a potential Seward Peninsula port site. The corridor combines the Western Arctic to DMTS, Ambler to DMTS, and Ambler to Seward Peninsula corridors described above.

Western Arctic to Seward Peninsula corridor, northern route. This 990 kilometer link also joins the Western Arctic District to a potential Seward Peninsula port site, but follows a northern alignment. It runs north along the Brooks range before turning south to the Ambler District, where the route is the same as for the southern option.

Western Arctic to Nenana corridor, southern route. This 1180 kilometer route combines the Western Arctic to DMTS, Ambler to DMTS, and Ambler to Nenana corridors described above.

Western Arctic to Nenana corridor, northern route. This 990 kilometer route uses the northern alignment along the north side of the Brooks range before turning south to the Ambler District, where the route is the same as for the southern option.

Northern Brooks Range to Dalton Highway corridor. Covering 300 kilometers, this corridor extends east from the Northern Brooks Range to the Dalton Highway.

Donlin Creek to Manley corridor. This 325 kilometer corridor uses two links: (1) from the Donlin Creek area northeast to Takotna and along the existing Sterling Landing to Ophir Road to Ophir, 190 kilometers; and (2) from Ophir to Poorman, then along the existing Ruby-Poorman Road to Ruby on the Yukon River, and finally to Manley by way of the Tanana River, 135 kilometers. In Manley, highway access to port sites is available.

Marine Transportation

Mineral concentrates are generally transported in vessel sizes ranging from Handy to Panamax, and coal is generally transported in vessel sizes ranging from Panamax to Cape class. Panamax vessels are considered the smallest size for transporting coal; most of the world coal trade moves in the larger Cape size or larger vessels. To export a large volume of coal on a consistent basis, a port needs to be capable of loading at least Panamax vessels. Bulk carrier specifications are given in Table 1.

Development Schedules

Construction of a new overland transportation system spanning Northwest Alaska will be a major undertaking and a lengthy process in terms of a development schedule. Although construction of a new port terminal will also be a major undertaking, it will be focused at one location. The last major transportation and terminal project completed in Northwest Alaska is the DMTS road (1988) and port (1990).

Table 1. Bulk Carrier Specifications

Ship Class	Deadweight Tons	Overall Length (feet)	Beam (meters)	Light Draft (meters)	Loaded Draft (meters)
Handy Size	10,000-30,000	180	25	3.2	10
HandyMax	30,000-50,000	220	30	3.6	11
Panamax	50,000-100,000	240	32	4.3	12
Cape Size	100,000+	280	42	4.9	15

The components that will have the greatest uncertainty in terms of timing are the environmental documents, permitting, and obtaining right-of-ways. First and foremost of challenges to construction in Northwest Alaska is permitting. Addressing regulatory agency concerns about environmental impacts is an expensive and time-consuming task.

Financial Issues and Constraints

Northwest Alaska is a vast storehouse of natural resources, many of which have not been explored to determine their economic feasibility. The discussion presented here is based on known resources, and new discoveries could alter these recommendations. What is not likely to change is the need for discovering large, rich deposits to overcome the high cost of developing resources in Northwest Alaska and to better utilize the capacity of any proposed transportation system to move the necessary quantity of resources to market at a reasonable cost.

The cost of road and rail construction has been severely affected by policy changes in material procurement procedures. It is difficult to acquire permits to use alluvial sources from riverbeds.

Use of upland sources results in higher costs and risks because of moisture content, permafrost, and poor quality of material.

DeLong Mountains District

The DeLong Mountains District is in the northwest part of the RTA study area and in the northwest portion of the NWAB. This district is already a world-class mining province, with in-place infrastructure, substantial reserves of high-quality mineral deposits, and potential for additional discoveries. Many of the threshold constraints that hinder development in other remote areas were resolved as part of Red Dog Mine development. Therefore, future development of the district resources can be seen as incremental expansions and additions to existing infrastructure in an established mining district.

Development of new deposits may require spur roads off and, perhaps, upgrade of the DMTS road, or other new roads. Future increases in zinc export volumes from Red Dog or other nearby mineral prospects will require further port-related improvements to handle greater throughput.

Because of the location and its existing infrastructure, the DeLong Mountains District may play a key role in constructing transportation systems for development and transport of mineral resources in the Ambler and Western Arctic districts.

Ambler Mining District

The Ambler District is in the Baird Mountains on the south flank of the Brooks Range. Through the years, Kennecott Exploration Company has conducted several feasibility studies and evaluated several transportation options to support mine development. To date, none of these transportation options has been pursued for two main reasons. First, the cost to build needed transportation infrastructure made the project financially unattractive. Second, after passage of ANILCA, the district became virtually surrounded by national park lands (Gates of the Arctic National Park and Preserve, Kobuk Valley National Park) and wildlife refuge lands (Selawik National Wildlife Refuge). It is difficult to secure an overland transportation corridor across these park lands.

Kennecott and NANA Development Corporation officials and the Division of Geological and Geophysical Survey, Alaska Department of Natural Resources, cite the lack of an approved surface transportation corridor as a hindrance to further exploration investment in anticipation of eventual development. The upshot is that the mineral resources of the Ambler District remain promising, but incompletely explored and undeveloped.

Information provided by Kennecott, NANA, and others indicate that several transportation-related strategies might enhance the economic feasibility for mining development in the Ambler District: approval of an overland access corridor to a coastal port site or to the Interior rail or road systems; cost sharing with other major projects for development of multi-user transportation infrastructure; public finance of transportation infrastructure to defray the front-end capital burden; and access to Western Arctic coal energy to displace fuel imports and for onsite ore processing that would permit export of refined metals rather than ore concentrates.

Several of the mineral deposits in the Ambler Mining District can be characterized as world-class, but development costs, and costs to transport the ore to market have precluded development to date. Implementation of one or more of the transportation-related strategies identified above could result in development of one or more deposits.

Seward Peninsula District

The Seward Peninsula is in the westernmost part of the RTA study area. As an established mining district with a long history of gold exploration and mining, the Seward Peninsula has an existing internal highway and mining road system that already provides access to or near most mineral resources of commercial interest. Local road improvements would be needed to support development of some prospects, but the proximity of access to local prospects reduces the importance of road improvements and increases the importance of the resource characteristics of the prospect. Provision of improved roads is unlikely to be a major determinant in assessing the viability of most prospects in the Seward Peninsula District.

The Seward Peninsula also has long-term potential to play a transportation role in mining developments farther north. The Seward Peninsula is a potential location for road or railroad routes and port facilities to support future development of mineral resources in the Ambler, Western Arctic, and DeLong Mountains districts.

The transportation facilities that would comprise a Norton Sound port plus the overland railroad or road to the Ambler and Western Arctic districts represent a long-term option for development to advance export of mineral resources from those districts. Access to a Norton Sound port is one of several transportation options that might be evaluated to support Ambler District mineral development. It is very unlikely that a single mining development could pay for a lengthy road or railroad corridor to Norton Sound. Until a number of prospects or deposits in the region are developed, which could jointly provide sufficient volumes and financially support the transport system, it is unlikely that a transportation link to a port on Norton Sound would prove beneficial to resource projects within the study area.

Ruby District

The Ruby District is inside the southcentral boundary of the RTA study area. The limitations and deficiencies of the existing transportation system impair commercial development of the known gold placer resources in the Ruby District. They also foreclose access for development of the more significant mineral resources of the Iditarod and McGrath districts.

Long-term potential for significant resource development in the larger region, including the Iditarod and McGrath districts, could be enhanced by development of improved surface access to and through the Ruby District. It is uncertain at this time whether the improved surface access would reduce costs of transporting equipment, materials, and supplies to the extent needed to achieve feasibility for the known resources in the Iditarod and McGrath districts. The Tanana and Yukon Rivers would likely provide access and egress to the Ruby District.

Wiseman and Chandalar Districts

The Wiseman District straddles the Dalton Highway north of Fairbanks, and the Chandalar District lies directly east of the Wiseman District on the south flank of the Brooks Range. A recommendation to further evaluate the need for surface transportation access to the Chandalar copper belt assumes that ongoing exploration by Doyon, Limited, will confirm mineral reserves that can be commercially developed with road access. The envisioned access corridor is relatively free of several issues that surface transportation corridors often pose. For example, there do not appear to be significant land status, subsistence, or environmental issues associated with the corridor.

North Slope District

The North Slope District is north of the Brooks Range and the Colville River. It includes the Western Arctic District, which is recognized for its coal resources south of Point Lay along the Chukchi Sea coast. A great deal of research has been done to examine whether Western Arctic coal is an economically feasible resource. One report found that "in some cases, the quality of the coal was able to overcome the transportation disadvantages noted" (Northern Economics, 1997a).

Given the potential size of the coal resources in the North Slope District and the ability of these resources to provide energy for in-state use, both industrial and utility-based, several corridors to develop these coal resources and the metalliferous lode deposits in the district are recommended. One corridor would link the western portion of the North Slope District (the Western Arctic District) with the Red Dog Mine, and a second corridor would link the eastern portion of the district with the Dalton Highway corridor. The second corridor might also provide access to potential oil and gas resources in the northern foothills of the Brooks Range.

Transportation costs and seasonal shipping issues make it difficult for the Western Arctic coals to achieve feasibility with the traditional business model of exporting or shipping coal to market. A mine-mouth power project that provided electric energy to the Red Dog mine and other potential users, with a transmission line along the identified transportation corridor, is considered a more viable business model than shipping the coal. After the mine is operating and fixed costs are covered by the mine-mouth power project, the coal could be competitive in domestic or international markets because it would only have to cover the variable costs of mining and transportation.

Oil and Gas

Oil and gas are clearly the most valuable resources in the study area. Many studies have indicated that petroleum reserves are abundant in northern Alaska.

Because of the potential reserves that exist on the coastal portions of the North Slope, it is recommended that corridors to support oil and gas be studied in the next phase of the RTA.

Industry currently supplies its own transportation systems (including ice roads and airstrips) along the arctic coast, and faces several issues concerning future expansion of the existing transportation network.

One possible approach to providing road access to oil and gas resources and minimizing impacts on local residents might be to develop the corridor along the north foothills of the Brooks Range, and then develop north-south corridors to known reserves. This network could reduce the cost to develop smaller discoveries, increasing the amount of economically recoverable reserves in the study region.

Markets for Hard Minerals and Coal

Information is presented on the prices and markets for many minerals found in the study area. More research is needed in the next phase to determine whether it is economically viable to develop the known mineral resources for export at world commodity prices. Coal and a few other minerals may have markets in the state that could support their development.

Funding Conclusions

The transportation alternatives being analyzed in the RTA will require considerable funding resources. A key element in assessing the viability of the transportation corridors being studied is determining initial capital and the operation and maintenance costs and whether the funding is available to finance them. The funding requirements for transportation projects needed in Northwest Alaska far exceed the amount of funds available. The identification of funding mechanisms is particularly important because the Alaska Department of Transportation and Public Facilities has stated that standard funding programs will not be the sole source or major source of funds for industrial development of roads or railroads. Given the limited funds available for RTA projects, project sponsors must employ more innovative financing mechanisms. The Alaska Industrial Development and Export Authority may be one means to finance these projects. The potential resource development projects identified in the Phase 1 report will need to contribute to the financing required for development of beneficial transportation projects.

References

Alaska Department of Transportation and Public Facilities, *Resource Transportation Analysis Phase 1-Potential Corridors Identification.* August 2001

Haugen, David, Lynden Inc., *Telephone Conversation with Bill Wong.* Sandwell Engineering, Inc, 2000a

Northern Economics, Inc., *Domestic Coal Handling Study.* 1997a, October

Whittier Access Project: Conversion of Existing 4 km (2.5 Mile) Long Railroad Tunnel to a Dual-Use Single-Lane Highway/Railroad Facility

Paul Witt PE[1]
Thomas Moses PE[2]
Frank P. Frandina PE[3]

Abstract

The Whittier Access Project is the 8.4 km (5.2 mile) extension of the Portage Glacier Highway to Whittier Alaska. This project provides the first highway access to the port community of Whittier, Alaska—a vital cargo port, recreation area, and tourist destination located on magnificent Prince William Sound. This paper gives an overview of the history of the project, alternatives considered, and some of the challenges successfully conquered during the construction. The harsh environmental conditions at this location required an innovative design to provide a facility that would meet the needs of the public.

Project History – A Very Hard Placed To Access

The planning, design, and construction of the Whittier Access Project was one of the most unique projects ever conceived in a cold region. The uniqueness of the project is exemplified by its having earned eight national awards in engineering and environmental excellence, including the American Society of Civil Engineers' Outstanding Civil Engineering Achievement for 2001, Washington, D.C., April 28, 2001.

The severe arctic conditions represented in this microclimatic area affected every aspect of the job. The high wind, subzero temperatures, and heavy precipitation required unique solutions through all phases of the project. While some new technological ground was broken, the majority of these unique solutions were the result of combining existing technologies in new ways.

---

[1] HDR Alaska, Inc., 2525 "C" Street, Suite 305, Anchorage, AK 99503, (907) 274-2000, pwitt@hdrinc.com
[2] Alaska Department of Transportation and Public Facilities, 4111 Aviation Drive, Anchorage AK 99502, (907) 269-0432, tom_moses@dot.state.ak.us
[3] Hatch Mott MacDonald, Hatch Mott MacDonald, 438 Main Street, Suite 700, Buffalo, NY 14202, (716) 854-1181, ffrandina@hatchmott.com

Figure 1. Locomotive at tunnel entrance.

Figure 2. Pickup truck exiting tunnel.

The project consisted of extending the Alaska Highway System from the Portage Glacier Highway through a new 122 m (400-ft) tunnel, 8.4 km (5.2 mi) of roadway, and converting the 4-km (2.5-mi) Anton Anderson Memorial Railroad Tunnel to a dual-use one-way rail/highway tunnel (see Figures 1 and 2).

Whittier's History

In order to best understand the Whittier Access Project, a bit of Whittier's history is helpful. Whittier is a small community located on pristine Prince William Sound in Alaska (see Figure 3). The city was originally chosen as an ice-free port by the U.S. military in the early days of WWII because of its potential as a far north, natural deep-water port, its proximity to central Alaska, and its severe weather conditions and geography which minimized the chance of a successful air or sea attack.

Figure 3. Map showing location of project.

Whittier is located at the end of a fjord and has steep mountainsides that drop vertically into the water on three sides. The U.S. Army determined that Whittier averages over 5 m (196 in) of precipitation a year (much of it snow), with low clouds dangerous to aircraft. Portage Pass, just east of Whittier, produces winds in excess of 193 km/h (120 mph) at ground elevation, with higher velocities near the mountain pass.

During 1941 and 1942, the U.S. Army Corps of Engineers directed the blasting of a 4.2-km (2.6-mi) hard rock tunnel through the mountains surrounding Whittier. The tunnel (named the Anton Anderson Memorial Tunnel) and a shorter 1.6-km (1-mi) tunnel were designed to connect Whittier by rail to the mainline of the railroad (now owned by the Alaska Railroad Corporation (ARRC)), located 12.9 km (8 mi) away.

Present Day

Whittier's importance as a port increased until the 1960s, when its military significance began to decline. This turn of events opened the opportunity for private ownership and further development that exists today.

Surrounded by the Chugach Mountains on three sides and the Gulf of Alaska on the other, Whittier has always been a port for the Alaska Marine Highway System (a system of ferry ships connecting communities in Southeast and Southcentral Alaska). However, Whittier was never connected directly to the state's highway system. Instead, the ARRC linked the two "highways" on either side of Whittier by shuttling automobiles and pickup trucks on flatcars through the tunnels (see Figure 4).

Whittier's geographic location makes it a gateway to the heart of Alaska for freight ships and cruise boats, as well as a window to Prince William Sound for people working in the fishing industry, for recreational boaters, and for tourists coming from Anchorage and other southcentral Alaska communities. For local residents, the town is tantalizingly close to Anchorage, the commercial center of Alaska, and the state's highway system—but not accessible enough by the rail shuttle because of the high cost of freighting a car or truck by railroad to the Seward Highway, the infrequent railroad schedules, and round-trip travel time. In the winter, the train ran only three days per week. The tunnel is now open seven days per week for vehicle and train traffic. A cost comparison for travel to Whittier is shown in Table 1.

Table 1. Travel cost comparison.

Vehicle Class	Railroad Fare	Highway Tunnel Toll 2001
A (passenger car)	$50-112	$15
B (passenger car w/trailer)	$78-148	$40
C (Bus)	$400-1000	$125
D & E (Truck)	$190-764	$125-250

Figure 4. Loading flatcars for Whittier.

Alternatives Considered

The many studies that have analyzed access to Whittier since WWII have made it clear that the demand for convenient and affordable access exceeded what was supplied by the railroad shuttle operation. Growing demand for access to Whittier spurred the Alaska Department of Transportation and Public Facilities (ADOT&PF) to initiate an additional study in 1993. The stated purpose of the study was to provide: (1) greater capacity to meet the demand in the tunnel corridor; (2) access that was more frequent, convenient, affordable, and safe; and (3) a cost-effective contribution to Alaska's economy.

Alternatives considered in the study were:

- Increasing conventional rail service, i.e., continue to load highway traffic onto flatcars.
- Utilizing a high-speed electric rail service.
- Constructing various highway routes over the mountains.
- Constructing a highway with tunnels through the mountains.
- Constructing a highway to Maynard Mountain and enlarging the Anton Anderson Tunnel to carry highway traffic along the existing rail bed in a dual street-railroad concept.

Each of the five alternatives and various modifications was extensively studied. A short description of each alternative and the conclusion reached follows.

Increasing conventional rail service called for adding passenger coaches to the existing trains and having more frequent service. This alternative was eliminated due to the users' desires to have more frequent and convenient service than could be provided with this concept. This form of access would not meet the estimated traffic demand of transporting 600 vehicles per hour for the design year. Another drawback was the known problem of driving highway vehicles onto open flatcars (see Figure 4) that were often snow-covered and needed considerable manpower to keep cleared of the worst of the ice.

Utilizing high-speed electric rail service would have increased passenger service to Whittier, but was rejected because it would not increase automobile access to the marine highway system.

Constructing various highway routes over the mountains was rejected as an alternative partially because the grades needed for the new roads could not be easily met. More significantly, the location of the new roadway would require that a large portion of Portage Lake be filled in for embankment. Further study determined that a road traversing Portage Pass through the mountains would not be safe nor could it be maintained year-round. It was determined that the road would have to be closed for likely weeks at a time because the velocity of the wind, 161-km/h (100 mph) and higher, would not allow the roadway crews to keep up with the drifting snow. The routes considered would be open to the full force of the wind without being able to use snow fences or wind deflectors effectively.

Constructing a highway and tunnels through the mountains was rejected primarily because of cost. A new tunnel would likely have to be fully lined and be at least 4.2 km (2.6 miles) in length. This alternative would also require additional encroachment onto U.S. Forest Service lands.

Constructing a highway to Maynard Mountain and enlarging the Anton Anderson Tunnel to carry highway traffic along the existing rail bed in a dual street-railroad concept (see Figure 5) became the preferred alternative. While it was not the most conventional solution, it met all the criteria for the project, and it could be built with the funds available. The volume of rail and highway traffic was deemed to be compatible for the project corridor, and its safety issues and operating concerns were worked out in conceptual designs. The ARRC supported this alternative as part of the railroad's charter to promote economic development of the state.

The team also looked at widening the tunnel to accommodate rail and highway traffic simultaneously, but this alternative was rejected for both engineering and operating reasons. It was determined that by properly locating and designing the support buildings, tunnel portals, and staging areas, drifting snow and avalanche debris could be controlled and this option could be operated all winter long.

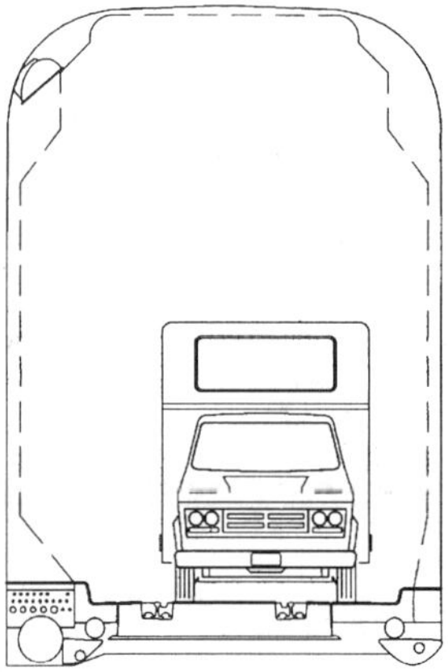

Figure 5. Concept showing road and tracks.

Further analysis of this alternative found that the combined rail-highway tunnel offered the citizens of Alaska the best combination of cost, serviceability, and ability to maintain operations in the arctic environment. The landmark effort boasts several "firsts," such as:

- The longest highway tunnel in North America.
- The longest combined rail-highway use tunnel in the United States.
- The first U.S. tunnel to use a combination of portal and jet ventilation fans. Cost effective 55.9 kW [75-hp] jet fans are used for normal ventilation. However, during periods of high wind 223.7-kW [300 hp] portal fans are used to overcome outside winds and pressure differentials.
- A unique, computerized traffic-control system that regulates both rail and highway traffic.
- A tunnel facility designed to operate in temperatures down to –40°C (–40°F) and in winds up to 240 km/h (150 mph), with the Whittier portal building designed to withstand a major avalanche.
- The first highway or rail project in Alaska to utilize the design-build approach

After a conceptual design was agreed upon, prescriptive and performance criteria were written in the form of a request for proposals for a design-build contract. Kiewit Construction Company was awarded the contract with Hatch Mott MacDonald designated as their lead designer. HDR was retained by ADOT&PF to assist in managing the design and construction of the project on site.

The highway approach to the Anton Anderson Memorial Tunnel, the two bridges, and the shorter 122-m (400-foot) tunnel, were constructed as a conventional design-bid-build with CH2M Hill as the designer and Herndon and Thompson, Inc. as the contractor. The final paving of the road, tunnel lining, IC control system, and lighting of all tunnels were part of Kiewit's contract.

Designed to Operate in Severe Weather

Determining the design parameters of the project was somewhat problematic (see Figure 6). As part of the conceptual design, anemometers were set up in the tunnel to gage wind speed over the course of a year. Gathering data outside the tunnel required a more creative approach since no historic data were available other than that the Begich–Boggs Visitor Center at Portage Lake normally experienced winds in excess of over 161 km/h (100 mph). The lack of data was because the visitor center's anemometer becomes inoperable at higher wind speeds. It was a standing order on the railroad that high-profile rail cars should not be left on the siding on the west side of the tunnel (in Bear Valley) because there was a history of these cars being blown off the tracks by the winds.

Figure 6. Winter snow levels in Portage Valley.

In the recent past, ARRC made two attempts to string communication wire in Bear Valley. The first attempt resulted in some of the telephone poles being snapped off in a manner of days. The second attempt utilized special poles approximately 0.6 m (24 inches) in diameter and 1.8 m (6 feet) high. These poles lasted several years before meeting the same fate. Forensic engineering techniques determined that it took an approximate 193-km/h (120-mph) gust to break the pole. Setting a safety factor of 1.25, the design wind speed for the project was set at 241 km/h (150 mph).

Records show that Bear Valley can receive as much as 13.7 m (45 ft) of snow a year and that the project site normally experiences temperatures in the range of +26 to –40 °C (+79 to –40° F). This range was utilized as a design parameter. Snow drifting analysis was used to check the building/structure locations and entrances to minimize snow removal costs.

There was also the problem of water in the tunnel. During the winter in Alaska, this water normally freezes to ice. The construction procedures had to deal with up to 0.91 m (3 ft) of ice, chipping it back with jackhammers at each shift and warming the tunnel with 4.184×10^6 J/g-°C (1 million BTU) portable diesel heaters. The tunnel doors were kept closed as much as possible to hold in the heat. However, the air had to be monitored for CO and particulate matter. At times, the doors were kept open and the portable jet fan device was fashioned onto a rail car and brought into the tunnel. The fan device often had to be warmed with propane heaters before it could function.

In areas where water was seeping into the crown of the tunnel, structural sections consisting of insulated, corrugated metal plate with a hydronic heating system between the tunnel lining and the rock are used. Water infiltration is generally limited to the shear and broken zones generally near the tunnel portals. Total infiltration has been estimated to peak at 946 l/min (250 gpm) in the spring, and varies from 151 l/min (40 gpm) to 265 l/min (70 gpm) the remainder of the year.

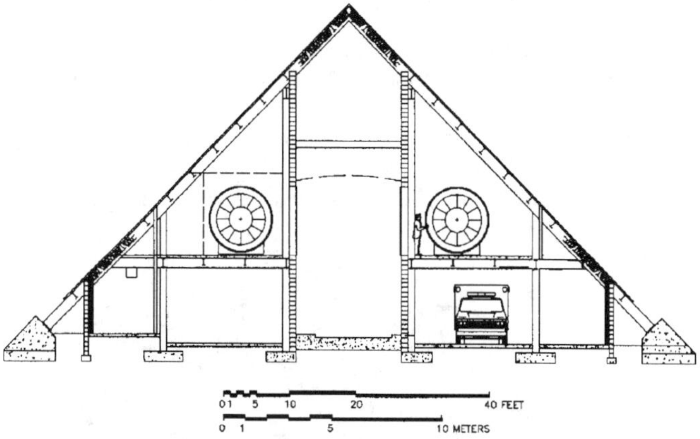

Figure 7. Sketch of tunnel portal.

The portal buildings (see Figures 7 and 8) have structural steel frames and reinforced masonry walls. The buildings are designed to safely absorb the shock of any potential avalanche, and their A-frame shape will split any snow slides and reduce the snow loads. Engineers designed the portal roof on the Whittier side, where there is the greatest avalanche potential, with 355-mm (14-in) thick prestressed concrete on a structural steel frame to withstand forces of 4882 kg/m^2 (1,000 lb/ft^2). On the Portage Valley side, the insulated metal roof is designed to withstand forces of 1074 kg/m^2 (220 lb/ft^2). As shown in Figure 8, the portal buildings successfully withstood several avalanches during the first year.

Figure 8. Avalanche debris at tunnel entrance.

Construction Challenges—Cold, Wind, Rain, Snow, and Avalanches

The existing rail link to Whittier was the only link to the outside world for local residents and carries 30% of the ARRC freight traffic. Due to the heavy tourism traffic in the summer that ran trains all day, 7 days per week, in the summer, the construction had to take place in the winter (see Figure 9).

Figure 9. Construction equipment in Bear Valley.

In addition to being limited to working during winter, all blasting that was needed for increased rail clearance and to create areas for safehouses and jet fans, and all installation of video cameras, drainage, surveillance tracking radar, communication, and lighting had to take place between the daily trains. Work continued around the clock in temperatures that reached –40 °C (–40° F), in winds as high as 193 km/h (120 mph), and in snows totaling 40 feet in the winter of 1998.

Only nine construction outages totaling 44 days were needed over the 2-year construction period. During these limited shutdowns, crews worked around the clock to install the new roadway system (see Figure 10). Specialized construction methods and equipment were designed to optimize construction activities and ensure clean air in the tunnel, especially during cold conditions. The work plan also accommodated extreme weather and emergency medical evacuations.

Figure 10. Needle-beam crane inside the tunnel.

The sequence used during the shutdowns started with removing the existing track rails and "rolling out" the old ties. Bulldozers then removed the ballast and waste material, always moving the waste material from the west end of the tunnel towards the Whittier side of the tunnel. The material was loaded into customized open-ended railroad dump cars and pulled to Whittier by a diesel-powered car mover.

When dozing was completed, crews set up pumps to "de-water" the construction area. As many as 20 centripetal pumps were used at a time. The rock floor was excavated as needed for clearance, with generally no more than 152 mm (6 in) of rock excavated. The majority of the rock excavation for clearance was drilled and shot from the tunnel roof. Roof excavation ranged from 0 to 2.13 m (0 to 7 ft) above the track area.

After excavation, a geotextile fabric (non-woven Armco 4553) was installed over the tunnel floor, along with at least 152 mm (6 in) of new stone base material. The stone base consisted of a non-frost-susceptible, well-graded angular stone. Material was classified non-frost susceptible (NFS) per Municipality of Anchorage Soil Frost Classification Group (see Table 2).

Table 2. NFS Aggregate Base

Sieve Designation	Percent Passing by Weight
25 mm (1")	100
19 mm (3/4")	70-100
9.5 mm (3/8")	45-79
4.75 mm (#4)	25-58
2.36 mm (#8)	15-47
600 μm (#30)	7-26
300 μm (#50)	5-19
75 μm (#200)	0-6

The stone was crushed on-site from waste material left over from rock cuts made on the roadway portion of the job. The material was brought in on open-ended railroad dump cars from the west, stockpiled, and then distributed with front-end loaders. The base material was then compacted with vibratory steel drum rollers, and a leveling sand layer was installed on top of the stone base to within a 3.18-mm (0.125-in) tolerance.

The design called for the highway surface within the tunnel to be constructed of longwearing concrete. The concrete used in the Startrack® panels is 41.63 Mpa (6000 lb/in^2) with a minimum air entrainment of 6%. The mix was specifically designed for this project and the expected severe freeze-thaw conditions. Each Startrack® panel (manufactured by Oldcastle Corporation) measures 2.29m by 2.44 m by 0.28 m (7.5 ft by 8 ft by 11 in) and has a textured concrete driving surface. Rails are embedded in the panel grooves and fastened down with elastic Pandrol® clips. To help move water from the roadway surface, the panels were crowned to drain towards the curbs

The concrete road panels (see Figure 11) were set in place with heavy equipment specifically designed for this job. Kiewit developed a specialized one-of-a-kind needle beam crane (see Figure 10) that could build track/roadway ahead of itself. The needle beam was converted by Kiewit's underground division from a crane used to lay large-diameter pipe on a mine project.

Figure 11. Laying trackbed inside tunnel.

The panels were brought from the west on a flatcar. The back end of the needle beam picked up the panels from the flat car and stored them on its center section. Its cantilevered front end then distributed the panels ahead of the needle beam on the prepared base. The needle beam was rail-mounted, so small 15-foot pieces of rail were installed in the panels on a temporary basis. The needle beam was basically building track and road ahead of itself. Finally, new permanent rails were installed into the concrete panels. As soon as each shutdown period ended, trains began using portions of the new combined road-and-track. Workers installed more than 1,800 of the pre-formed concrete highway/railroad panels necessary to "pave" the 4 km (2.5-mi) tunnel.

Summary and Conclusion

The project opened to highway traffic on June 7, 2000, on time and on budget. Successfully utilizing the design/build concept, the project saved at least a year over conventional design bid build.

Balancing the needs for functionality in a severe climate with cost effectiveness resulted in a number of firsts. Firsts in ventilation and intelligent traffic control systems made the project possible. The success of this facility demonstrates that rail and highway traffic can safely and effectively use the same tunnel.

The Whittier Access project was truly a one of a kind project that had many unique design requirements. Through a combination of ingenuity, partnering and hard work, often in brutal weather conditions, the project objectives were met.

While the combination of cold, wind, and precipitation created significant challenges for the project team, they were able to work together to meet those challenges. Engineers, from the owners, consultants, and contractors pooled their experience and knowledge to find the best solution.

References

Alaska Department of Transportation and Public Facilities. 1993 *Whittier Access: Preliminary Alternatives Report* Anchorage AK.

Alaska Department of Transportation and Public Facilities. 1993 Public Facilities. 1993 *Working copy of RFP and Memorandum of Understanding, Project No. ANCH-0496 (5)/52371*, Anchorage, AK

Alaska Department of Transportation and Public Facilities. 1995 *Whittier Access Project Revised Draft Environmental Impact Statement Technical Reports* Anchorage AK.

Alaska Department of Transportation and Public Facilities. 2000 *Public Hearing, Toll Information Sheet*, Anchorage, Alaska.

Alaska Office of History and Archaeology. 1993. *The Portage-Whittier Transportation Corridor: A Discussion of Historic Significance.* Anchorage Alaska

Moses, Tom, Witt, Paul, and Frandina, Frank 2000 "Two-In-One Tunnel" *Civil Engineering* April, pp 48-53

Western Region Climate Center. 2000 "Alaska Climate Summaries –Whittier" http//wrcc.sage.dri.edu/summary/climsmak.html

Additional Information

Alaska Department of Transportation and Public Facilities. 2002 "Whittier Access Website" www.dot.state.ak.us/creg/whittiertunnel/index.html

Black, Monica. 1999. "Alaska Tunnel Gets Millennial Makeover." *Railway Track and Structures*. May pp 28-30.

Burke, Jack. 2000. "Building the Road to Whittier in Alaska." *World Tunneling*. June

Sorensen, Eric. 2000. "Tunnel Vision Opens New Vistas for Alaska Outpost." *The Seattle Times*, April 23, A12.

Wallace, Janey. November 2001. "Credit Where Much is Due." *T & T North America*. November, pp. 20.

Design of the Reconstruction of the Anton Anderson Memorial Tunnel

Frank P. Frandina PE[1]
Paul Witt PE[2]

Abstract

The Anton Anderson Memorial Tunnel was recently reconstructed to become the longest highway tunnel in North America and the only tunnel shared by both highway and rail traffic. The project is located about 50 miles southeast of Anchorage in an area with little historical data that could be used to develop design criteria. This paper illustrates the development of the design criteria for the project, identifying the roles of the owner, owner's consultants and the design-build team's designers. Several examples are presented including the development of design criteria for egress/refuge, wind loading, snow loading including avalanche loading, tunnel drainage, ventilation, ice control structures and the unique traffic control system that allows joint highway and rail use of the tunnel.

Project History

Shortly after completion of the Alaska Railroad between Seward and Anchorage in 1923 (confirm), it was realized that winter maintenance of the track through the steep mountainous, avalanche prone slopes was often impossible. Surveys had been conducted of a spur to the natural deep-water port in Whittier that would eliminate the most severe sections, but it was not until the advent of World War II that the spur to Whittier was built. During 1941 and 1942, the U.S. Army Corps of Engineers directed the blasting of the 2.6-mile (4.1 km) hard rock tunnel through the mountains surrounding Whittier. The tunnel (named the Anton Anderson Memorial Tunnel) and a shorter 1-mile tunnel were constructed to connect Whittier by rail to the mainline of the Alaska Railroad Corporation (ARRC), located 8 miles away.

---

[1] Chief Engineer and Project Design Manager, Hatch Mott MacDonald, 438 Main Street, Suite 700, Buffalo, NY 14202, (716) 854-1181, ffrandina@hatchmott.com

[2] Project Manager (Owner's Engineer), HDR Alaska, Inc., 2525 "C" Street, Suite 305, Anchorage, AK 99503, (907) 274-2000, pwitt@hdrinc.com

Whittier, a small fishing village soon became a major military supply base. During the Korean War, the base at Whittier expanded and so did rail traffic through the tunnel. As the only land route in and out of Whittier, the tunnel had to transport both freight and passengers. The demilitarization of Whittier has a corresponding increase in commercial and later recreational traffic. Still the railroad was the only mode of transportation and a vehicle shuttle service was instituted in the mid 1960's whereby vehicles would be driven onto railroad flatcars and transported through the tunnel.

Soon the capacity of the shuttle was reached during the summer tourist season and the Alaska Department of Transportation and Public Facilities (ADOT&PF) conducted a series of studies that eventually resulted in the current project.

The completion of the Whittier Access Project in June 2000 has provided the first direct road access to the Prince William Sound community of Whittier. This 5.2-mile (8.3 km) long project consists of three new bridges, a new 445-foot (136 m) long tunnel and the conversion of the existing 2.6-mile (4.1 km) long Anton Anderson Memorial railroad tunnel into a dual use highway/railroad facility. Wide enough for a single lane of one-way traffic, this unique project is not only the longest highway tunnel in North America but also the only combined public highway-rail use tunnel in the world.

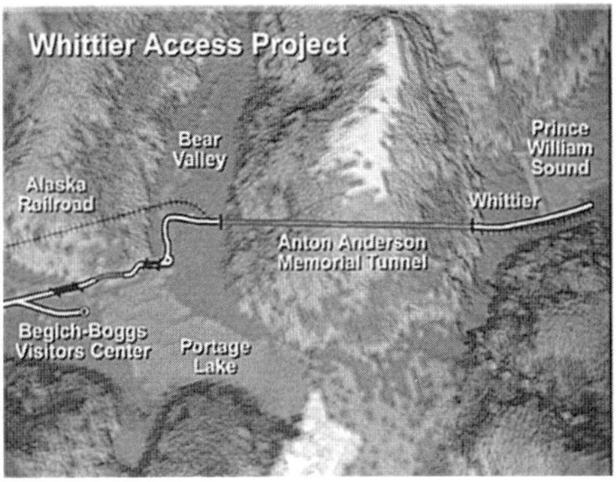

Figure 1. Location Map

Figure 2. Completed Tunnel

Project Requirements

With the advice of their design consultant, HDR Alaska, the ADOT&PF decided on Design-Build-Operate Contract. A first for the ADOT&PF, this unique form of construction contract reduced the time needed to design and construct the tunnel segment by 16 months. The two-year operation portion of the contract is to ensure that the tunnel is operating efficiently and safely prior to turning it over to Alaska DOT/PF. This form of project delivery also shared the responsibility for development of the design criteria among the owner (ADOT&PF), the owner's design consultant (HDR), and the design-build contractor's (Kiewit Construction) designer, (Hatch Mott MacDonald (HMM).)

The request for proposals (RFP) established the key project requirements including:

- When in the highway mode, the tunnel shall be capable of carrying 800 cars per hour with traffic reversals twice per hour.
- The tunnel control system must complete the transfer from highway mode to railroad mode within 15 to 30 minutes of a request for tunnel occupancy from the railroad.
- The tunnel facility shall be designed to operate in temperatures down to minus 40 degrees F (-40 C) and in winds up to 150 mph (240 kph).
- When outside of avalanche zones, the buildings must be designed to withstand snow loads of 220 pounds per square foot (10.5 kpa).
- Water leaking into the tunnel must be prevented from freezing within the tunnel and diverted from the roadway.
- Provisions must be made for either a safe exit route from the tunnel or providing places of safe refuge for use during a fire.

Hatch Mott MacDonald (HMM) had the responsibility for the detailed design, which often included development of design standards for this one of a kind facility. HMM was assisted by 5 design subconsultants: Rim Architects, Peratrovich, Nottingham & Drage (PN&D), RSA Engineering, Dryden and LaRue, and AGRA E&E.

Emergency Egress/Refuge

The preferred method for accommodating emergency egress/refuge requirements is usually via a separate protected exit route. In the 1960's, the Army constructed an 8 foot (2.4 m) diameter tunnel about 150 feet (45 m) north of the main tunnel for a pipeline to transport petroleum, oil and lubricants, or POL from the Whittier port to an air force base near Anchorage. Initial consideration was given to the construction of 8 crosspassages from the AMM tunnel to the POL tunnel.

Unfortunately the POL tunnel was at a higher elevation and the exit route would not be able to meet ADA requirements. The POL tunnel would also need significant rehabilitation to provide lighting and ventilation and to eliminate tripping hazards.

Therefore, a series of 8 refuge areas, or "safehouses" were constructed through the tunnel at spacings no greater than 1600 feet (490 m). Each safehouse is separated from the tunnel by a fire rated masonry wall and had an independent fresh air supply, toilet facilities, blankets, water and communications with the tunnel control room. The design capacity of each safehouse was established at 55 persons, which is based on the capacity of the largest tour busses on the road.

To prevent potential overcrowding of the safehouses during an incident, the traffic control system was designed to meter the traffic so that the 55-person design capacity would not be exceeded. Busses are metered at a 45 second spacing, which will result in a 1650 feet (500 m) separation at the 25 mph (40 kph) speed limit. Similarly, cars are metered at a 2.5 second spacing which will result in less than 18 vehicles near any safehouse at any time. To prevent overcrowding of the safehouse closest to an incident, traffic signals in the tunnel will stop traffic throughout the tunnel immediately upon confirmation of an incident.

Snow Load

Without any National Weather Service monitoring stations nearby, meteorological data was obtained from local sources, often anecdotal. On the Whittier side of the mountain, military records had recorded basic meteorological data that could be used to establish a design ground snow. However on the Bear Valley side, the only records were the daily plowing logs kept by the owner of a lodge adjacent to the USFS Visitor Center well west of the tunnel's ancillary structures and snowfall records from the Alyeska ski area about 15 miles (24 km) further west.

Snowfall totals can vary widely over short distances. The building code for Anchorage (50 miles (80 km) away) requires a snow load of 40 pounds per square foot, but the limited data suggested that Anchorage requirements would be grossly inadequate near the tunnel. After evaluating the available data, it was decided to design the structures for a ground snow load of 220 psf (10.5 kpa).

Figure 3. Tunnel Portal, Winter '98 - '99

When the construction site was mobilized in fall of '98, the designers moved to the site from Anchorage at the same time. The first winter storm in early November lasted 8 days and dropped 11 feet (3.4 m) of snow on the Bear Valley work site. The snow caused the roofs of the construction trailers to sag requiring daily removal of snow from the trailer roofs. Preliminary design of the exterior structures had just begun and the design team quickly adopted the philosophy of steeply sloped roofs to minimize snow accumulation on the roofs.

Avalanche Loading

The ADOT&PF furnished a series of earlier studies of the avalanche potential within the project limits. Fortunately, the existing western tunnel entrance in Bear Valley was outside of the avalanche zone and the staging area and other facilities were sited to avoid the avalanche and run-out zones. The designers from the lower 48 quickly learned that slopes with trees might be safe from avalanches, but slopes littered with broken trees should be avoided.

Figure 4. Original Portal & Snowshed

Figure 5. Portal Building, Spring '00

Unfortunately, the eastern portal in Whittier was in an avalanche zone. The existing heavy gage corrugated metal plate (CMP) snowshed covering the track between the end of the tunnel and the old portal building was severely deformed from numerous snow slides over the years. The project architects at Rim Consultants worked closely with the structural engineers from PN&D and developed a portal structure with a peaked roof that would split and divert any sliding or rolling snow away from the roadway and track.

An avalanche consultant was retained to estimate the loadings from a "100-year" avalanche on the proposed structure. Loads due to direct impact and the static deposition of snow were evaluated over time from the start of the avalanche. The study concluded that the main roof structure at the Whittier portal should be designed for 830 psf (40 kpa) plus impact.

The roof design is 14 inch precast concrete planks supported on a heavy structural steel frame. PN&D estimated the impact factor for the building to be 1.57 resulting in an equivalent static load on the roof of 1300 psf (62 kpa). Between the portal building and the tunnel, HMM designed a reinforced concrete shell around the existing CMP snowshed, using the existing CMP as a form for the new concrete. This permitted construction of the snowshed reinforcement to proceed without disrupting train traffic.

The design was tested not once, but three times during the winter of '99-'00 when record snowfall totals resulted in three significant avalanches before construction was completed.

Tunnel Drainage

The RFP required the collection and disposal of all water in the tunnel. However, the only recorded estimate of inflow was from a single observation in January 1986. The project schedule required that the drainage system be designed before an annual cycle of inflow observations could be made. Nevertheless, periodic estimates were made of the observed flows during winter '98-'99 to determine a design flow prior to the installation of the drainage system in spring '99. The water quantities in the tunnel would increase during the spring as snow on the mountain melted, but the design could not wait.

The existing tunnel drainage system was open ditches on either side of the track that discharged into closed pipes or a sump at the Whittier portal. The railroad had later installed sumps and pumps at a few locations near both portals. It is apparent that much of the water in the tunnel ditches was lost in the invert during the transport process, since test holes in the tunnel invert were often dry. Because the project improvements would include capturing water entering through the tunnel roof and walls and channel the inflows into a closed drainage system, the total discharge from the tunnel would be higher than what was observed. An allowance had to be made to account for the invert losses to insure that the closed drainage system would be capable of conveying the total captured flows.

Many methods were used to estimate the inflows through the length of the tunnel. Since the railroad had installed sumps near both portals, estimates were made of the area of contribution and sump discharge. During the track reconstruction, the dewatering operation within each invert replacement zone typically involved the capture of water upstream of the work zone and pumping through pipes to the other end of the work zone. Using a five-gallon pail and a watch, discharge estimates were made. A visual estimate was also made of the percent of water in the invert that was actually captured.

It was determined that most of the water enters from the tunnel walls and roof near the Bear Valley portal, with another concentration near the Whittier portal and a lesser amount distributed in a few areas in the tunnel interior. 35% of the inflow is in the first 1100 feet (340 m) from the Bear Valley portal, 15% in the next 1940 feet (590 m), 15% in the last 520 feet (160 m) at the Whittier portal, with the remaining 35% spread throughout the tunnel interior. There may also be some inflow up from the tunnel floor, but these would have a relatively minor contribution and are probably limited to a few shear zones.

Based on the limited three months of observations of flows, HMM estimated that the steady state flow would be about 40 (150 l/m) to 70 gpm (260 l/m). But interpretation of anecdotal data from railroad workers estimated that a total inflow of 300 gpm (1130 l/m) was probable in the spring. Due to the uncertainty of designing the drainage system with less than a full year of monitoring, HMM designed a system capable of collecting and transporting 690 gpm (2600 l/m). Additional observations were made in the following months as installation was progressing and a peak flow was observed in mid May '99 of 250 gpm (950 l/m) confirming the validity of the original estimate.

The first component of this system is a pair of 8 inch (200 mm) corrugated steel pipe (CSP) storm drains to remove surface water runoff from the road through slotted curbside inlets. Water drains into these pipes and is carried by gravity through the length of the tunnel to Whittier. The second component is a perforated CSP pipe system to capture water that seeps through the tunnel's rock walls into the sub-base. This pipe drains the water to a sump, located every 1600 feet (500 m), where it is then pumped into the main gravity-feed storm drain for removal from the tunnel.

Ice Control Structures

The RFP required that any water seeping into the tunnel be diverted from the pavement / track and that ice not form on any surface in the tunnel. Ice control had been a continual problem for the ARRC, who was budgeting about $200,000 per year for ice control. Otherwise, ice on the track could cause derailments and ice on the walls and tunnel roof would rapidly reduce the tunnel clearances and create additional hazards as the ice fell during thaw periods.

Two basic approaches were investigated, active and passive systems. Passive systems provide an insulated lining that retain enough of the latent heat in the rock to keep any seepage water in liquid form until it reached the drainage system. A few thousand feet (1000 m) inward from the portals, the rock in the tunnel was a constant 47 degrees F (-8 C) year round. Active systems rely on external heat sources to keep the water liquid.

Initial designs were based on passive systems and thermodynamic models were created to confirm the design. Unfortunately, the lack of site specific data required many assumptions into the model. Uncertainties related to the volume of winter traffic created uncertainties as to the volume of subzero dilution air that the ventilation system would have to pump through the tunnel sucking away the latent heat from the tunnel. While the design / build team felt that a passive system should have worked, the cost to install an active system was felt to be prudent considering the cost and schedule risk of retrofit should the passive system fail to prevent icing.

The ice control system consists of heavy gage galvanized CMP panels with 2 inches (50 mm) of insulation covering a hydronic heating system. The hydronic heating system consists of hot water heated by a gas-fired boiler, circulating through a network of hoses between the rock and the insulated panels. Seepage water is collected in a gutter at the base on the liner and conveyed to the drainage system.

Figure 6. Ice Forming on Track Under Water Seep

Figure 7. Completed Ice Control Structures

The Design/Build team briefly considered grouting the rock at the high inflow areas near the portals to reduce the quantity of inflow and possibly also reduce the ice control lengths, but the effectiveness of a grouting program could not be performed without a trial grouting program that the schedule would not permit.

Wind Loading

While drafting the RFP, HDR recognized the unusually high wind speeds occurring outside the tunnel on a fairly regular basis. The only historic data available was from the USFS Portage Glacier Visitor Center, which routinely recorded winds in excess of over 100 miles per hour (160 kph), before its anemometer became inoperable, sometimes blowing away. It was a standing order on the railroad that high profile rail cars should not be left on the siding on the west side of the tunnel (Bear Valley siding) since there was a history of these having been blown off the tracks by the winds.

In the recent past ARRC had made two attempts to string communication wire in the valley. The first attempt resulted in some of the telephone poles being snapped within a few days. The second attempt utilized special poles approximately 24 inches (600 mm) in diameter and 6 feet (1.8 m) high. These poles lasted several years before meeting the same fate. The ARRC then dug a trench and placed the communication line underground, away from the harsh winds. By utilizing forensic engineering techniques, HDR determined that it took an approximate 120 mph (190 kph) gust to break the pole. Adding 25% to the "observed" wind speed, the design wind speed for the project was set at 150 mph (240 kph).

HMM then designed all buildings, light standards, traffic signals and signage to the 150 mph (240 kph) wind speed. As a result, ADOT&PF standard details had to be reanalyzed and often reinforced to accommodate the unusually high wind. Any skeptics that might have thought the design was unnecessarily robust dropped their objections after a storage trailer anchored with tornado tie downs was blown over. All major structures have fared well through the first year, but a few weeks before the tunnel opened, a wind gust snapped a railroad crossing gate.

Ventilation

The ventilation system had a number of design requirements:

- Dilution of vehicle exhaust fumes to safe levels
- Purging of train exhaust fume prior to vehicle entry
- Control of smoke to permit evacuation or rescue in the event of a fire

The ventilation system had to be able to positively control air movements in the tunnel despite the often brutal conditions outside. The RFP also specified that the ventilation system should be able to routinely function against a tunnel headwind of 5 feet/sec (1.5 m/s) and to also be able to accommodate the most extreme winds that could be anticipated.

To provide basic design data, an anemometer was installed in the tunnel, and one year worth of natural air velocities in the tunnel were recorded prior to issuance of the RFP. HMM's designers conducted a number of ventilation simulations and determined that the power requirements for the ventilation system was highly sensitive to the friction of air moving along the unlined tunnel. Because of the paucity of published data on friction in unlined rock tunnels, a full scale test to measure the friction factor in the tunnel was then initiated. The solution was to initiate a flow of air through the 2.6-mile (4 km) tunnel with the piston action of regularly scheduled trains, so that the airflow decay rate could be measured. With measurements of the steady state airflows before the train entered the tunnel and periodic measurements after the train left the tunnel, the friction factor was able to be determined with credibility.

HMM ultimately selected a design that incorporates both energy-efficient reversible jet fans attached to the tunnel ceiling and the more traditional portal fans at the tunnel entrances. This is the first time in the United States that jet fans, which look like engines on jet aircraft, were used in conjunction with portal fans, which sit at the tunnel portals. There are six 75-horsepower (56 kw) jet fans available for normal operations and four 300-horsepower (225 kw) portal fans for emergencies and to purge the tunnel. In case of emergency, the powerful portal fans will blow smoke and fumes away from any vehicles stopped in the tunnel. During normal operations, the more efficient, reversible jet fans keep the tunnel air healthy at a lower operating cost.

Figure 8. Jet Fans Hung from Roof

Traffic Control System

The agreement between the ARRC and ADOT&PF dictates that the ARRC will have priority usage of the Whittier tunnel over highway usage. "Priority usage" is defined to mean that highway usage is suspended and ARRC trains can use the tunnel within a minimum period of time following notification to the tunnel operator. The period of time is to be the minimal amount of time required to clear traffic from the tunnel, and between fifteen to thirty (15-30) minutes.

Two computer systems are dedicated to help the tunnel operations staff operate the tunnel in a safe and efficient manner while complying with the ARRC/ADOT&PF agreement. Together, the Tunnel Control System (TCS) and Train Signal System (TSS) make it possible for cars and trains to safely take turns traveling through the tunnel.

The TCS is responsible for all vehicle movement within the tunnel. Assembled from intelligent traffic system (ITS) components, the TCS tracks each vehicle and meters traffic through the tunnel. One of its main functions is to monitor the direction of vehicle movement and, through its control of the traffic signals and gates, allow vehicles to travel only in one direction at a time. The TCS also monitors tunnel operations via live video covering the length of the tunnel. If a car stops for any reason, the vehicle detection equipment will alert the tunnel control operator in the Tunnel Control Center, direct the video cameras to automatically display that area of the tunnel, and shut appropriate gates. The TCS also continually monitors and controls the lighting and ventilation systems.

In a similar fashion, the TSS is responsible for train movement through the tunnel. This system controls train switches and signals and ensures that trains only move through the tunnel in one direction at a time and only when there are no vehicles in the tunnel.

Because these systems are linked, the TCS and TSS together ensure that cars and trains never meet in the tunnel. When the tunnel is in the railroad mode, the TSS controls the trains and locks out the operation of the TCS system. All highway gates are closed and the highway signals are set to red. When the tunnel is in the highway mode, the TCS locks out the TSS until highway vehicles are cleared from the tunnel.

When in the highway mode, if the tunnel is desired by the railroad, the transition from highway to rail is handled much like any standard railroad-highway grade crossing, except that this crossing is 2.6 miles (4 km) wide. Track circuits in the TSS detecting an approaching train signals the TCS to immediately close approach gates to prevent any additional cars from entering the tunnel. The TCS monitors cars leaving the tunnel and prompts the tunnel operator to perform a visual confirmation (via CCTV) that the tunnel is empty before the tunnel transfers to the TSS and rail mode is completed.

Figure 9. Vehicle Queuing in Whittier Staging

Summary

The Whittier Access project was truly a one of a kind project that had many unique design requirements. Through a combination of American ingenuity, partnering and hard work, often in brutal weather conditions, the project objectives were met.

References

Alaska Department of Transportation and Public Facilities (ADOT&PF). 1998 *Working copy of RFP and Memorandum of Understanding, Project No. ANCH-0496 (5)/52371*, Anchorage, AK.

Water And Ice Control In The Anton Anderson Memorial Tunnel, Whittier, Alaska

Christopher Tattersall[1]
Richard Mitchells[2]
Eric Gallion[3]
Robert Weller[4]

Abstract

Recently converted to a dual use highway/railroad tunnel, the 4.0 km Anton Anderson Memorial Tunnel in south central Alaska provided more than its share of water- and ice-related problems to the design/build contractor. Previously, water inflows and associated ice build-up had been significant maintenance tasks for the Alaska Railroad in winter months. There had also been on-going degradation of the track structure due to frost heaving and the effects of water and ice. On June 7, 2000, the converted tunnel was opened to the public with a new roadway structure and an active ice control system. This paper details the approach and solution to the problem of ice build-up.

Introduction

It has been said that most of the problems civil engineers tackle are caused by, aggravated by, or simply have to do with, water. As ice, water exhibits some unique properties. Ice is less dense than water, implying that water expands when it freezes. This provides engineers in cold climates with a wide range of problems to solve, including protecting building foundations from freezing (or in permafrost zones from thawing) and providing drainage for soil structures which do freeze, such as embankments and road subgrades. Another property of water in its frozen state is its low coefficient of friction. The accumulation of ice on roadways is dangerous and northern highway maintenance departments expend a lot of effort addressing this.

The 4000 m Anton Anderson Memorial Tunnel, located in south central Alaska's Chugach Mountains, has been plagued by water- and ice-related problems since its construction in the 1940's. The tunnel runs through Maynard Mountain, a 1200 m high peak separating the ice-free port of Whittier from the main highway network of central Alaska. The upper portion of the mountain is a permanent ice

[1] Consultant, Hatch Mott MacDonald, Mississauga, ON.

[2] Consulting Engineer, Restoration Science & Engineering., Anchorage, AK.

[3] Supervising Mechanical Engineer, Hatch Mott MacDonald, Buffalo, NY.

[4] Supervising Civil Engineer, Hatch Mott MacDonald, Buffalo, NY.

field. On average, the region receives more than 1300 mm of rain in the summer and 10 m of snow in the winter. Much of this water finds its way into the unlined rock tunnel though joints and shear zones in the rock mass. The rock is a poorly metamorphosed greywacke with a high compressive strength, but it weathers rather rapidly, especially in the presence of running water. Water inflow historically has created operational and maintenance problems for the Alaska Railroad.

In 1998 the State of Alaska awarded a design-build contract to convert the railroad tunnel into a dual use highway/railroad tunnel. A key part of this conversion was a requirement to address the following ice and water concerns:

- Frost heave of the highway/rail invert.
- Hazardous driving conditions and increased potential for train derailment due to ice build-up on the highway/rail surface.
- Reduction of the tunnel section and train/vehicle clearances due to build-up of ice on the tunnel roof and walls.
- Increased risk of falling objects during thaw periods.

This paper explains how these concerns were addressed and the solutions that were provided.

Existing Conditions

The Whittier Access road parallels the old Portage Pass, a traditional overland route from the waters of Prince William Sound to Turnagain Arm and Cook Inlet. This narrow neck of land with a single mountain pass through the Chugach Mountains is all that keeps the Kenai Peninsula from being an island. This mountain range is also the traditional battleground between low pressure weather systems winding in from the Gulf of Alaska and arctic high pressure systems often stationed over central Alaska. This results in extremely variable weather conditions in the area, depending on which system is predominating. In winter months, the arctic high has the effect of lowering air temperatures to the -30°C range and setting up a steep pressure gradient between the two tunnel portals. With cold air flowing in the tunnel towards Whittier at up to 20 km/h, the east end of the tunnel would experience icing problems for almost a thousand metres.

To help combat the freezing problems, the Alaska Railroad installed doors on each portal, and would keep them closed until just prior to the passage of a train. The doors were manually operated by crews stationed in Whittier and Portage. Even with doors, ice removal was a major maintenance issue. Crews were required to remove ice growths from the sidewalls and crown to maintain train clearance. The track would heave, forcing train speeds to be reduced. The rails would get buried in the ice, requiring crews to painstakingly chip railheads and flangeways clear. Figure 1 is a photograph from railroad archives showing typical ice removal work.

The competing weather systems are also responsible for very high snowfalls, well over 10 m. Although not directly related to the ice control problem, avalanches

Figure 1. Ice removal undertaken by Alaska Railroad crews

did take their toll on portal structures and equipment in the vicinity of the portals, especially at the Whittier end.

Design Options

The first design decision was whether to use a passive or active ice control system. A passive system would provide an insulated barrier between the tunnel air and the rock surface of the tunnel, using the latent heat in the rock mass and water to prevent freezing. This barrier would also deflect the incoming water away from the road surface and into the drainage system. Active systems would combine a similar barrier with an external heat source to prevent the freezing of incoming water.

The advantage of an active system is its ability to reliably prevent water inflows from freezing under extreme temperature and wind conditions. However the construction and operating costs of an external heat source can be significant. Also a critical design issue is space constraints in the tunnel for heating equipment. The main advantage of a passive system would be the elimination of these cost and space requirements.

Thermodynamics

When water freezes, distinct changes occur at the molecular level. As the water molecules arrange themselves into crystals, the average separation between molecules increases, leading to a reduction in density of about 9 percent. In the process, they release a substantial amount of energy, more than seven times the energy released in cooling from 20°C to 0° (see Figure 2).

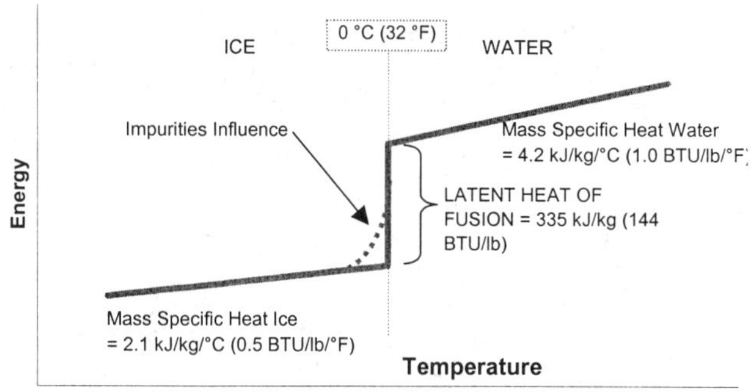

Figure 2 Properties of water in the vicinity of 0°C

The principle upon which the ice control design was based was to prevent the transfer of this latent heat of fusion in order to keep the water its liquid state. The rock temperature in the interior of Maynard Mountain has been measured to be a constant 3°C. The temperature of the air in the tunnel, however, changes with the time of year, setting up heat flow. Conservation of energy requires that this heat flux is the amount of heat transferred to the air in the tunnel, which is in turn a function of the air temperature (also density and humidity) and velocity. Theoretically, for a given air temperature and velocity, the temperature of the surface of the rock, and the water flowing through it, can be determined by a two-dimensional steady-state heat transfer analysis.

In practice such an analysis does not account for two aspects of the ice control needs in the Anton Anderson Memorial Tunnel. Firstly, there is a significant time lag between a change in temperature of the air and the corresponding change in temperature of the rock, so that steady state conditions are rarely seen. This means that although the air temperature may plunge to -40°C for several days, the system does not reach a steady state before the temperature begins to rise again. An economical design, therefore, could be based on higher air and rock surface temperatures than recorded minimums. Secondly, the temperature of the air varies along the length of the tunnel. This meant there was a three-dimensional aspect to the problem addressing how far from the portals ice control should be provided.

This second problem became the key design issue. It was concluded that a numerical model could not provide a definitive answer for two reasons. It was possible to model the effects of cold air moving through a tunnel by selecting a design wind speed, temperature, duration and heat transfer coefficients. However, a slight change in any variable could change the length of the "freeze zone" by hundreds of metres. Secondly, although we did have good empirical information on the extent of freezing in the railroad tunnel from existing maintenance records, the operating plan for the converted dual-use facility, which included powerful new ventilation fans, was heavily dependent on the level of traffic demand. There was therefore no way to predict accurately what the new air flow characteristics would be from the historical record.

In the end, this 3-dimensional aspect was eliminated by the contractual constraints of budget and schedule. As a baseline for bids during the tendering of the contract, the extent of ice control zones had been set at a total length of 146 m, a length which had come out of a 1980s tunnel condition survey. After the award, there was no additional budget available to increase this length. Instead of an analytical approach, therefore, ice control locations totaling 146 m were prioritized based on the railroad's maintenance records and a tunnel leakage survey carried out by the authors, verified/adjusted several times during construction.

Two-dimensional modeling was used to establish that 50 mm of polyethylene insulation with an air gap behind would reduce the heat transfer rate enough to prevent freezing of the water on the surface of the rock. This assumed an average range of winter conditions based upon records from the two closest recording sites: Alyeska ski resort and the Town of Whittier. It was conceded that there might very well be conditions under which a passive system would not prevent freezing, but it was decided to proceed with the fabrication and installation of a passive ice control unit as a trial section and to monitor rock and air temperature during the first winter of construction.

Ice Control Structures

Several designs for the ice control structures were compared early in the design process. One design, similar to what is currently in use on Norwegian road tunnels in rock, involved panels of expanded polyethylene foam (typically Ethafoam® by Dow) with a reinforced shotcrete protective cover. The polyethylene panels are held to the rock by short rock bolts with sealing collars. Reinforcing mesh is fastened to the ends of the bolts with butterfly plates and nuts. Difficulties with installation logistics experienced by the contractor as well as reports of some problems with existing installations in Norway, led to this design being dropped.

Other ice control structures were investigated, including a thin-gauge metal lining with pumped cellular concrete behind. This scheme was estimated to be the cheapest in terms of installed cost, but there were serious concerns as to the durability of the insulating concrete when subjected to the anticipated freeze-thaw cycles.

Prefabrication options were most desirable since the construction logistics for this project were the single biggest challenge facing the contractor. Every train passage required the single lane rail tunnel to be cleared. The tunnel was too narrow for construction vehicles to pass, and at the same time diesel fumes accumulated, leading to delays for ventilating. The contractor was able to perform several activities in the tunnel at a time, but all materials and equipment had to be pre-staged. The elimination of one in-tunnel activity, such as the construction of a shotcrete and Ethafoam® ice control lining, represented a significant savings in cost and time.

The final design selected (Figure 3) used a heavy-gauge corrugated liner plate in the form of an arch with a layer of insulation fastened to it. Although more

Figure 3. Multi-plate arch ice control structure

expensive, this system had a big advantage over the Norwegian design in that it could be prefabricated. The structural arch was designed to withstand the effects of wind blast and suction from the passage of a train as well as the weight of potential ice build-up on the inside. The 3 mm corrugated galvanized sheet steel was braced back to the rock face with grouted all-thread bars. Individual sheets were fabricated in standard sizes to facilitate erection as well as fabrication. Two radii were required to construct the arch.

Insulation consisted of sheets of 50 mm Ethafoam® expanded polyethylene foam panels. Water was kept on the "warm side" by covering the insulation with a membrane of the type normally used as ice and water shielding for metal roofing installations. Insulated and waterproofed multi-plate modules were prefabricated on site. The contractor used one lane of the nearby Portage Lake Tunnel as a workshop, as it provided a well-lit and protected area to assemble the components into modules. The completed modules were assembled on an arch-shaped jig mounted to a rail car. This provided the work platform for the installation of the multi-plate arch in the tunnel and could be raised into position using hydraulics.

The multi-plate arches were set on cast-in-place concrete strip footings at walkway level. A drainage channel formed into the back of the footing, drained into 50 mm pipes connected to the roadway drains. The channels and drain feeder had to be heated, as analysis showed that the footings were locations of relatively high heat loss.

The construction contract included provision for additional measures to keep water off the roadway surface. Water panning fabricated from zinc-plated sheet steel was provided in wet areas away from the portal "freeze zones". In these locations, the intent was to direct the water to the tunnel sidewalls away from the roadway, not pipe water to the roadway drain pipes directly.

Heat Supply

The difficulty and diminishing returns in providing more than 50 mm of insulation led the designers to make provision to change to an active ice control system. Providing an external heat source would enable extreme cold air temperatures to be handled and provided a fall-back in case the trial section did not perform as expected. A hot air system that would blow warm air between the insulation and the rock was first investigated. Although very effective from a heat transfer perspective, there were some practical difficulties providing and locating the heat source as well as a serious cost and schedule concern related to additional rock excavation necessary to provide clearance for ductwork. This approach was abandoned.

The intention with the trial ice control section had been to make provision for future heat supply. The easiest way of doing this was to install an array of aluminum conduits behind the lining into which self-regulating heat tape could be installed later if results from the trial installation indicated the need for supplied heat. Calculations from the 2-D model indicated that under the worst conceivable conditions these should be rated at 6.5 W/m and spaced at 1.2 m intervals. This amount of heat would keep the water from freezing and it was felt that this provided a 50 percent margin of redundancy in case of failure of some of the heat tapes.

When it became clear that the construction schedule would not allow a trial ice control section to be installed in the first winter, the impetus for going to an active system increased. Rather than risk having to remobilize to repair a number of failed ice control structures in an operating tunnel, the contractor decided to proceed with providing an active heat source for all ice control areas. Although easy to install, electric heat tapes are expensive and have a short mean operating life. Also, providing sufficient power to numerous locations in the tunnel would have been very expensive and would have required upsizing the main power transformer. With the high cost of electricity in the region, operating costs were also a concern.

A hydronic system was then investigated, using a gas-fired boiler with circulating glycol/water heat distribution piping (Figure 4). This technology has been used in several industrial ice control applications in Alaska including the front driveway slab of a hotel in nearby Girdwood and in the patented Groundheater® portable heating system, which was used by the contractor in the tunnel to control ice build-up during construction. Because both applications had proved to be reliable and effective in the local environment and were relatively inexpensive to operate, this type of heating system was selected for the tunnel.

The key design issue for the hydronic system was the effective transfer of heat to the zone behind the insulated lining. Design calculations called for 12 mm pipe at 1.2 m spacing. HDPE plastic pipe was selected for this application, as it met the requirements for flexibility, ease of welding, durability and cost. In the ice control areas, chain link mesh was installed around the arches to protect the hydronic lines

Figure 4. 220 kW (750,000 BTU/h) boiler and circulation pumps for hydronic system

and the waterproofing sheet from spalled rock pieces: this mesh provided a convenient place for attaching the piping.

The hot water/glycol mixture was circulated to each ice control area through 50 mm insulated HDPE piping from high efficiency boilers, one located in each portal building. A valve box at each ice control area provided a location for controlling the flow to individual heat circuits and allowed for adjusting the heat balance and bleeding air from the system (see Figure 5).

One of the more challenging exercises was locating the boiler exhaust flues. The portal buildings were tucked up against the mountain and subject to very high snow accumulation. The boilers had restricted flue lengths, a function of their high efficiency design, and routing these to an area where they would not be buried in snow or damaged required the boiler locations in the portals to be carefully chosen.

The materials used were selected to meet the rigors of the tunnel environment with minimal maintenance. For protection the supply and return piping was buried behind the roadway curbs and all joints were welded and pressure tested. Although not yet subjected to worst case design conditions, the design of the active heating system for the ice control areas is considered to be energy efficient, cost effective and robust.

Figure 5. Valve box adjacent to each ice control zone

Drainage

Two aspects of water removal from the tunnel set the requirements for the design of tunnel drainage: capture and removal of water from the road subgrade and capture and removal of water from the roadway surface. The fact that the tunnel has a uniform gradient of around 0.5 percent, falling towards Whittier, meant much of the drainage would be by gravity.

The subgrade down to rock invert was to be replaced over the whole length of the tunnel with new crushed material. Providing a drainage system for this structure was very important given the rail loading and the difficulty in carrying out future subgrade repairs. Early in the project it was obvious that a central drain cut into the rock invert running the length of the tunnel would be ideal, however the requirement to maintain rail traffic meant that access to the middle of the invert was severely restricted. Instead, subgrade drainage was concentrated on both sides of the invert.

A program of invert probing pointed to a problem that had been suspected initially: the rock elevation below the subgrade varied considerably and high spots were going to be natural barriers to gravity drainage. In order to minimize the requirement for rock cutting or pumping over these high spots, a compromise between depth of perforated collector drain and number of lift stations was reached. After mapping the complete invert rock profile, a scheme requiring eight lift stations for each side of the roadway was designed. By fortunate coincidence, these pumps were close to the locations of the eight safehouses, with their available power supply points.

Unlike the subgrade drainage piping, which followed the profile of the rock invert, the roadway drains were set a constant depth relative to road surface. This meant gravity drainage was possible for the whole length of the tunnel, with all tunnel water flowing to the Whittier portal. Because of the very congested space (see Figure 6), drain size was limited to 200 mm corrugated steel pipe (CSP). The capacity of these was well in excess of observed flows from previous inspection reports, as well as the authors' own observations. It is recognized that the pipe diameter is small by highway standards.

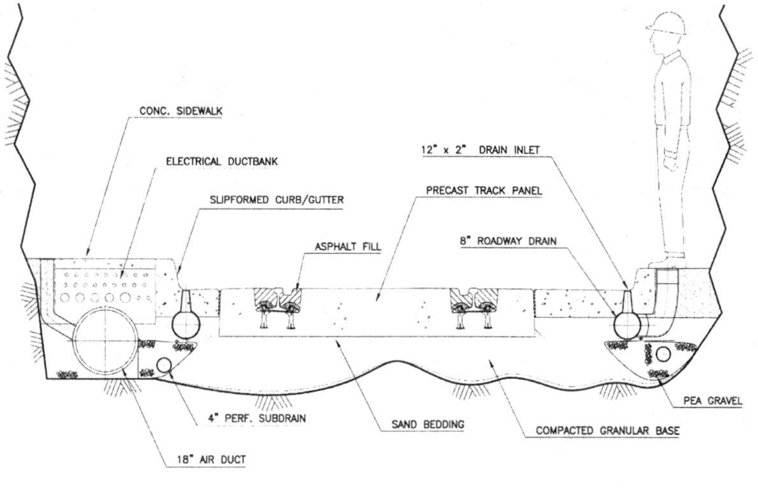

Figure 6. Tunnel invert cross section

Freezing culverts is a common winter problem in Alaska and in the design of the roadway drains this was a serious concern. A wide range of options was considered and the final design included several lines of defense. Instead of typical catch basins, water on the road surface is channeled to the drain pipes through 300 mm long by 50 mm wide slot drains spaced at 30 m centers. The size of the openings was deliberately kept small to minimize heat loss from the flowing water and separate capped clean-outs were provided to permit rodding and flushing of the pipes. As a final precaution against freezing, the Bear Valley ice control system was configured to supply a small amount of hot water to the upstream end of both drain pipes. In the event that drains do freeze, 300 m of 20 mm galvanized thaw piping was installed at the Bear Valley end (30 m at the Whittier end), which, when fed from a portable steam generator, will allow the pipes to be thawed out.

Summary

Dealing with water and ice-related issues was a major component of the design of this tunnel. Broadly speaking, the approach was as follows:

1. Collect water flow where it enters the tunnel and channel into the drainage piping.
2. Keep the water in its liquid state by insulating it from the air stream and adding supplemental heat when conditions require.
3. Protect the subgrade from the problems of the past by (a) using clean granular material and (b) draining the subgrade as much as practical.

The winter of 2000/2001 was very mild compared with the two previous winters. The ice control system performed well in it's first season of operation, but has not been tested in the temperature ranges for which it was designed. By the time this paper goes to press its effectiveness will have been tested by a second winter.

Sustainable Design Strategies for the Modernization of the Amundsen-Scott South Pole Station

Joseph J. Ferraro, AIA[1]
William D. Brooks, AIA[2]

Abstract

Located on the polar plateau at an elevation of 2,835 meters (9,301 feet) above sea level, the South Pole is arguably the most pristine and extreme environment on earth. To the worldwide community of scientists, it offers a unique platform for scientific research including aeronomy, astrophysics, glaciology, climatology, and seismology.

The Amundsen-Scott South Pole Station (United States) is situated on the plateau within one hundred meters of the geographic South Pole. Operational since 1956, Amundsen-Scott is undergoing its second significant reconstruction and modernization. To extend the station's useful life, minimize the operational energy it consumes, and provide a healthy habitat for its occupants, the new station incorporates a number of sustainable design strategies. Many of these strategies are uniquely suited to the climatic conditions of the polar plateau.

This paper will briefly introduce the philosophy of sustainable design, describe the primary sustainable characteristics of the Amundsen-Scott South Pole Station, and discuss how the station's innovative design could set the standard for future development of the polar plateau.

Sustainable Design

Often referred to as "Green" design, sustainable design is a relatively new term that stems from a growing awareness of the negative environmental impacts caused by conventional design and development. In recognition of finite global resources, sustainable designs revolve around a philosophy of developing sites and facilities in ways that minimize or eliminate environmental impacts, thus

[1] Partner, Ferraro Choi And Associates, Honolulu, Hawaii
[2] Director of Architecture, Ferraro Choi And Associates, Honolulu, Hawaii

sustaining resources indefinitely. Sustainable designs typically incorporate strategies such as energy efficiency, improving indoor environmental quality, the use of renewable materials and sources of energy, recycling, and conservation of water and natural habitats. Of all the strategies that are employed, energy efficiency can have the most significant effect on reducing environmental impact, and is therefore often given the highest priority.

A universal characteristic of projects that are energy efficient is that no two designs are alike. The reason for this diversity is that buildings must incorporate a passively designed architectural envelope that responds to the climatic conditions specific to a site to be truly energy efficient.

Sustainable development in cold regions is no exception. A building design that is ideal for an Alaskan winter may not be ideal for conditions in Antarctica. Similarly, a building designed to respond to climatic conditions on the coast of Antarctica will have little in common with one designed for the polar plateau.

A second priority of sustainable design is to provide a superior quality indoor environment for building occupants. This is accomplished by providing appropriate ventilation and controls, selecting materials and finishes that minimize or eliminate the emission of volatile organic compounds (VOC), and the introduction of daylight and views to the outdoors. All of these strategies diminish the probability of a "sick building" and are common in well-designed sustainable facilities. The relative importance of indoor environmental quality increases with the amount of time occupants must spend inside.

In Antarctica, where winter-over personnel must spend seven months indoors during the Austral winter, providing a high quality indoor environment is paramount. Because the winter environment precludes outdoor activity, it is probably more analogous to think of Amundsen-Scott in terms of a large space station or ship at sea, than to consider it a conventional building.

Construction History at Amundsen-Scott Station

Currently in the second of five construction phases, the modernization of Amundsen-Scott is scheduled to be complete and operational in 2006. The revitalized station combines proven strategies for extreme cold regions with new, sustainable design solutions that have evolved as specific responses to the climatic conditions of the polar plateau.

First operational in 1956, the original Amundsen-Scott Station buildings were constructed of wood and canvas. The relentless annual snow deposition of 0.2 meters, bitter winter winds, soon led to improvements that consisted of corrugated steel arches that were placed over the living and working facilities. The arches provided needed protection from the elements, and were designed to withstand being buried beneath the ever rising snow plain and annual snowdrifts (See Figure 1).

The need to clear increasingly more snow each summer to regain access to the station, coupled with evidence of increasing structural fatigue resulting from supporting the weight of 18 seasons of accumulated snow, led to the original station being replaced in 1975. The replacement station's design perpetuated the idea of placing buildings within steel arches that could withstand being buried, and also placed facilities within a dramatic 51 meter diameter geodesic dome. The landmark dome housed laboratories, offices, dining, and berthing space for a winter-over population of 28, while the adjacent arches enclosed a garage/shop, fuel storage, the power plant, and other support areas. The dome resisted being buried by snowdrifts better than the arches did, but over time, the unequal snow loading imposed on the dome's base ring began to compromise its structural integrity. Compounding the problem, the station was struggling to accommodate a summer season population that had grown to 150, five times its design size. By the end of the 1980's, it was evident that the station was again physically and programmatically near the end of its useful life, and would need to be replaced.

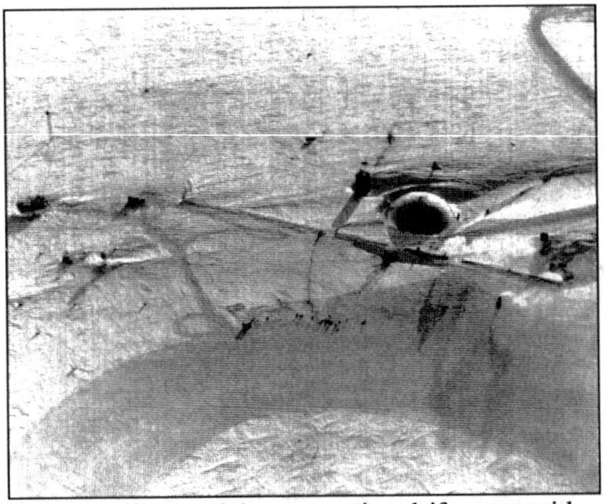

Figure 1. Convential construction drifts over with snow and becomes buried

In early 1991 the National Science Foundation (NSF) orchestrated a high-level design retreat in Enfield, New Hampshire to develop concepts for the new station. In June 1991 NSF published the results in a document entitled the Enfield Concept Design. Notable in the concept design was the idea that the habitat portion of the new station could be above-grade in elevated, jackable structures.

The concept that emerged from the Enfield retreat established a sustainable design direction that was a significant departure from the history of the conventional and inefficient below-surface facilities at Amundsen-Scott. It envisioned a state-of-the-art station that would have a minimum design life of 25 years, breaking the historical pattern of abandoning and rebuilding. It also envisioned a larger, modern, more energy efficient station that would accommodate an expanding science program and improve station living conditions.

In total, the program called for over 6,000 square meters of floor area to provide for berthing, dining, laboratories, a medical clinic, recreation, administration, storage, and services support spaces. The above-surface portion of the station would be designed to serve as a complete habitat for scientists and station operations personnel, accommodating a winter-over population of 50, and a summer population of up to 150. The elevated habitat would be interconnected with tunneling passageways to below-surface support spaces including a one-megawatt power plant, bulk storage facilities, fuel storage facilities, and a garage/shop (Figure 2).

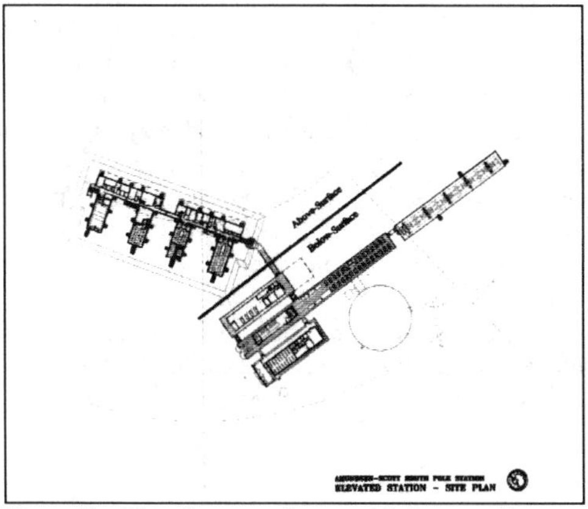

Figure 2. The above-surface and below-surface facilities at the South Pole Station

Sustainable Design Goals

The unique sustainable aspects of Amundsen-Scott were developed in response to four critical design objectives:

- To minimize impact to the South Pole environment
- To ensure a station useful life of 25 years or more
- To be highly energy efficient
- To provide a high quality indoor environment

Minimizing Impacts to the South Pole Research Environment

Ongoing and future scientific research is the primary justification for the Amundsen-Scott Station. Research at the station includes aeronomy, astrophysics, glaciology, climatology, and seismology. To ensure that station development would not compromise the station's scientific mission, it was necessary to perform a detailed site analysis to optimize the location of the new elevated station and related support facilities.

The analysis consisted of a review of the existing and proposed station components in relation to climate, the airstrip and typical flight patterns, the effects of structure generated snowdrifting, the annual migration of the polar ice cap (10 meters per year), and the existing science sectors surrounding the site. The results of the analysis indicated that the station activities could be defined as four primary zones, labeled as Quadrants A - Operations Sector, B - Clean Air Sector, C – Quiet Sector, and D – Downwind Sector (See Figure 3).

Quadrant A was determined to be the best site for the new station. Development in this quadrant would avoid wind borne contaminants that could affect atmospheric research in the Clean Air Sector, was sufficiently far from seismic research in the Quiet Sector, did not pose height problems for balloon experiments in the Downwind Sector, facilitated easy access to the airstrip, and localized snowdrifting to areas which were indentified to remain undeveloped. The complete site analysis can be found in the Amundsen-Scott South Pole Station *Conceptual Design and Programming Study*, prepared for the National Science Foundation in March, 1994 (See List of References).

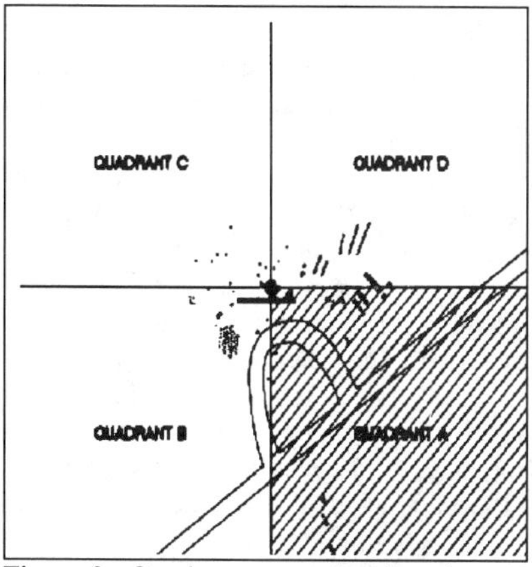

Figure 3. Quadrants were developed to determine the best site for the station

Other efforts undertaken to protect the environment in the immediate vicinity of the station included:

- Replacement of the existing fuel storage system of on-grade synthetic rubber bladder style tanks without secondary containment or leak detection to cradle slung single wall rigid fuel tanks on secondary containment sleds with an electronic fuel leak detection system.
- Replacement of widespread outdoor cargo storage with consolidated indoor cargo storage facilities.
- Renovate and reuse the existing below-grade steel arch structures.
- Retrograde (removing from the site) of the dome, miscellaneous unused structures, and all (except sewage) waste.
- Separating and recycling solid waste products.

Ensuring a Station Useful Life of 25 Years or More

Historically, the greatest challenge to achieving sustainable facilities at Amundsen-Scott has been contending with annual snow deposition and drifting. All conventional structures built on-grade at the South Pole are covered by winter snowdrifts after only one or two seasons. The drifted structures are further buried by the typical annual snowfall of 0.2 - 0.3 meters. As there is no thaw cycle at the Pole, they eventually become too deeply buried to practically excavate and access. Eventually, the buried structures deform under the ever-increasing weight of the snow, and must be abandoned.

The new design for the habitat portion of Amundsen-Scott has overcome this historical dilemma with an elevated, linear complex of two interconnected C-shaped buildings that are configured and oriented to control snow drifting.

The snow drifting design evolved from a series of predictive studies at the Canadian based research facility of Rowan Williams Davies & Irwin (RWDI), including water flume and wind tunnel testing, Computational Fluid Dynamics, and Finite Area Element computer modeling techniques. The studies verified that the two buildings, interconnected and elevated approximately 3 meters above the surface with long axis of the complex oriented perpendicular to the prevailing winter winds, would perform well. The windward face of each building and the interconnecting link is chamfered to smoothly channel the wind beneath the complex. Forced to accelerate, the wind carries the snow well past the buildings where it is deposited in long leeward drifts. A windward drift also forms just in front of the station as a result of momentary turbulence where the wind encounters the building face. Over time, the leeward and windward deposits will tend to accumulate around the ends of the station and connect into a rough crater shape. Eventually, the surrounding snow field will build to a height where it will prevent the wind from channeling beneath the station (See Figure 4).

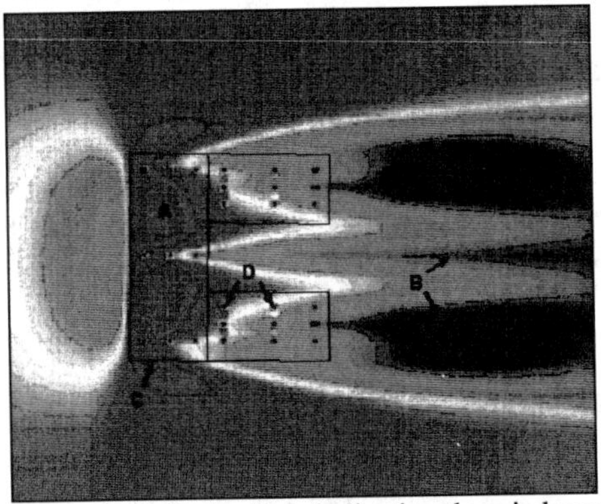

Figure 4. CFD Analysis indicating the windward and leeward snow drift patterns

To overcome the potential drifting at this point, the station is designed to be jackable, so that the entire complex can be raised approximately 4 meters (one floor level), and the cycle of effective drifting mitigation will resume. The duration of the initial cycle has been effectively increased by initially erecting the station on a compacted snow berm that is itself 2 meters higher than grade. Subsequent studies indicate that when all factors are considered, the station's ability to control drifting could continue until the windward drift approaches the height of the building's mid section. This prediction indicates that it may not be necessary to raise the building for well beyond 25 years.

Energy Efficiency

Energy efficiency is a key design strategy for the modernization of Amundsen-Scott. Virtually all energy consumed by the station is produced by diesel-fired generators. The generators and all mobile equipment are fueled with AN-8 diesel fuel because of its low freeze point. In addition to power produced by the generators, a modest amount of electrical energy is harvested during the short Austral summer from a 30 kilowatt array of photovoltaic panels. Wind-generated power has been problematic to date due to the difficulty in lubricating and maintaining moving parts in the extreme cold temperatures present at the site. A pilot wind generator project has been performing well at the site. If successful, a modest wind farm will be erected adjacent to the station. No other practical sources of energy are available.

The station's fuel storage resupply capacity (currently 1.5M Liters) and cost at approximately $5 per liter, constrains the amount of power that can be produced over the course of the 7-8 month Austral winter (during which there is no way to travel to and resupply the station). With this well defined limitation on the fuel, energy efficiency became the key strategy in determining the physical size and power consuming attributes of the station (See Figure 5).

Figure 5. Below-surface fuel storage facility

Energy efficiency strategies include:

- A passive envelope design that is configured and oriented to control snow drifting, thus minimizing fuel needed to operate snow clearing equipment to regain access to the station each season.
- An extremely thermally efficient and vapor tight architectural building envelope, incorporating insulation values of R=70 for the roof and soffit, R=50 for walls, a tight and self-healing vapor barrier (Vapor resistance = .007 perms), insulated freezer style doors (R=50), and high performance triple insulated windows (R=6.66).
- A forced water/glycol heat transfer system circulates recovered waste heat from the power generation plant heat recovery system to provide space heating for all occupied areas of the elevated station, and to melt snow for the domestic water system. This system is also employed as a snow melting medium for the station's deep water well.
- Use of CO2 sensors to determine both occupancy and air quality, and enable the modulation of indoor ventilation on an on demand basis.
- Use of heat recovery from exhaust systems to pre-heat outside air in ventilation systems.
- Modulating outside air based upon exhaust fan operation to reduce overuse of outside air when not needed and eliminate uncontrolled drafts when exhaust fans are energized.
- Use of variable-air-volume system to provide only the air flow needed to meet current space demands, and use of variable speed drives to minimize electrical energy use at the fan when volume requirements are reduced.
- Localized boilers to produce on-demand domestic hot water requirements.
- "Cold-soaking" the summer berthing areas during the winter (approximately 20 percent of the elevated station floor area, to a temperature of 8 degrees C.)
- Use of light interior finishes with medium to high reflectance values, task lighting, and low ambient light levels to minimize power required for electric lighting.
- Use of daylighting during the summer, to minimize the need for electrical lighting.
- Proper detailing of penetrations and openings to minimize the requirement for heat tracing.
- Use of certified EnergyStar office and computer equipment.
- Full commissioning to ensure systems achieve design performance targets.

Indoor Environmental Quality

A primary objective of any sustainable design is to provide a high quality indoor environment, to ensure occupant health, well being, and productivity. At Amundsen-Scott, the need for a high quality indoor environment is accentuated. The winter-over population must spend the sunless, seven month Austral winter entirely indoors. Outside temperatures range from an average of minus 49.4 degrees C. to minus 82.8 degrees C. To maintain a healthy living environment, and to enhance occupant productivity, the indoor environment at Amundsen-Scott has been carefully designed to incorporate the following strategies:

- Smoking is restricted to a single, small smoking room with an independent ventilation system to effectively eliminate environmental tobacco smoke.
- CO2 monitoring is integrated into the ventilation system to ensure CO2 levels do not exceed code prescribed maximums.
- Increased ventilation effectiveness is provided by low-velocity air distribution, and careful placement of supply and exhaust outlets to minimize draft sensation.
- Low or no-emitting materials and interior finishes are utilized to stay under volatile organic compound emission limits. Examples of finishes include a loose-laid, interlocking modular athletic flooring system and free-laid carpet tile.
- Indoor pollutant source control is accommodated by providing walk-off mats at and coat rooms adjacent to entries with separate ventilation systems to store fuel-soiled work attire.
- Extensive use of individual temperature and lighting control is provided, with the most controllability occurring in private spaces such as berthing suites.
- Use of radiation heating equipment where possible to provide even, comfortable heating.
- Use of hydronic heating in general, resulting in more gentle temperature changes, and reducing the need for ventilation airflow.
- High performance windows allow daylighting and views in public and berthing spaces are extensively used during the short Austral summer. Translucent blinds protect against glare, and vapor tight, removable window panels prevent heat loss and icing during the winter and also mitigate light emission to astronomy experiments.
- Appropriate use of color and variation of finishes helps to prevent interior sameness and monotony.
- Proper acoustic design maintains speech and/or sound privacy for most occupied spaces.

Conclusion

This new generation of sustainable design for polar buildings provides both environmental and energy efficient aspects. This building type also provides a higher margin of safety and quality of life to the station's occupants. The design has met its sustainable design goals to protect and sustain the pristine polar plateau research environment and redirect operational expenses to science projects (See Figure 6).

At the date that this project design was initiated the U.S. Green Building Council's Leadership in Energy & Environmental Design (LEED) rating system had not been established. If a review were to take place at the time of this writing it is conceivable that the project could be considered for a Gold certification when it is completed in 2006. Due to its unique location and other project constraints, it is also conceivable that this project might be the basis for a new division of LEED criteria and a standard for new Antarctic building design for all Antarctic Treaty nations.

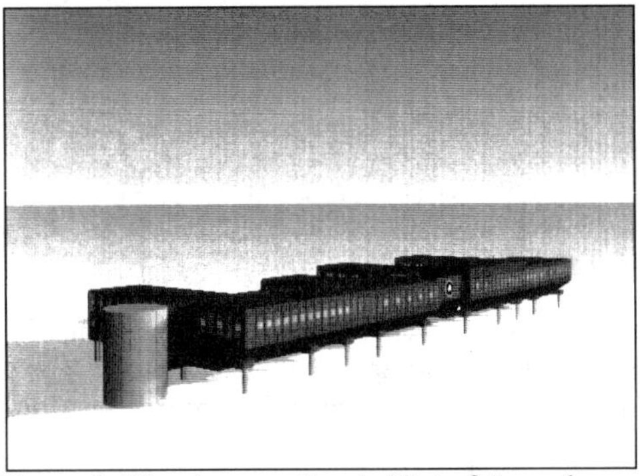

Figure 6. The above surface habitat of Amundsen-Scott South Pole Station.

References:

LEED: A Look at the Rating System That's Changing the Way America Builds, Article from Environmental Building News, June 2000

Brooks, William D., AIA, Elevated Station Design for the South Pole Redevelopment Project at Amundsen-Scott South Pole Station, Technical Paper published in "Cold Regions Engineering – Putting Research Into Practice", Proceedings of the 10th International Conference on Cold Regions Engineering, 1999, New Hampshire

Ferraro, Joseph J., AIA, Master Plan for a New South Pole Redevelopment Project, Technical Paper in "Cold Regions Engineering – Putting Research Into Practice", Proceedings of the 10th International Conference on Cold Regions Engineering, 1999, New Hampshire.

Basis of Design, South Pole Station Modernization, Amundsen-Scott South Pole Station, Antarctica, by Ferraro Choi And Associates Ltd, prepared for the National Science Foundation, April 9, 1999

Public Technology and U.S. Green Building Council, Sustainable Building Technical Manual: Green Building Design, Construction, and Operations, Washington, DC, Public Technology, Inc., 1996

Conceptual Design and Programming Study, Amundsen-Scott South Pole Station, South Pole Replacement Facilities, by Ferraro Choi And Associates Ltd, prepared for the National Science Foundation, March 1994

Performance-Based Analysis of ARFF Requirements for Airfields at McMurdo and South Pole Stations, Antarctica

W. Mark Cummings, P.E.[1] and Douglas Carpenter[2]

Abstract

A study was conducted to determine the adequacy of the aircraft rescue and fire fighting (ARFF) capabilities at the U.S. Antarctic Program's (USAP) airfields in Antarctica. Realizing that many of the civilian and military standards and regulations that normally govern the ARFF requirements for airports would not be applicable at the USAP airfields, a performance-based approach was used to analyze the risks and potential fire hazards associated with airport operations in the Antarctic environment. Typically, existing ARFF regulations require that significant quantities of a foam/water agent be available to support aircraft fire fighting and rescue operations. Given the environmental conditions in Antarctica and the effect they have on potential fire hazards, the analyses indicate that adequate protection can be provided using primarily a dry chemical agent, thus obviating the need to provide special measures to prevent the foam/water agent from freezing. However, very little data are available to allow a quantification of the amount of dry chemical agent needed and the methods/strategies for proper application in combating the various postulated fires scenarios.

Introduction

A performance-based analysis of the aircraft rescue and fire fighting (ARFF) requirements was performed for the airfields at both McMurdo and South Pole Stations in Antarctica. This analysis was sponsored by the National Science Foundation (NSF)[3] as part of an effort to assess the level of ARFF capabilities needed to adequately protect the aircraft and personnel operating at these airfields. This study was precipitated by concerns raised to the NSF regarding the adequacy of the existing aircraft rescue and fire fighting (ARFF) capabilities at the U.S. Antarctic Program (USAP) airfields. The premise behind these concerns was the fact that existing ARFF capabilities were significantly less at the USAP airfields as compared to what is normally required by civilian and Department of Defense (DoD) regulations and guidelines governing airport operations. Rather than

[1] Principal/Sr. Engineer, Fire Risk Management, Inc., 721 Bath Road, Wiscasset, ME 04578; phone 207-882-5471; wmark@fireriskmgt.com

[2] Vice President, Combustion Science & Engineering, Inc., 8940 Old Annapolis Road, Columbia, MD 21045; phone 410-884-3266; dcarpenter@csefire.com

[3] Though the NSF sponsored this analysis, all opinions and recommendations contained within this paper are strictly those of the authors and should not be construed as representing the position of the NSF.

simply performing a code compliance review and then developing and justifying a list of variances, a performance-based approach was used to identify and evaluate the potential fire hazards that actually exist at the USAP airfields and to subsequently recommend protection strategies to effectively mitigate those hazards. Three key features to this approach include:

1. Recognizing that environmental conditions can have a significant impact on reducing (or exacerbating) potential fire and/or life safety hazards,
2. Recognizing that most prescriptive codes and regulations that exist are designed to mitigate hazards associated with "normal" operational conditions, and
3. Applying an engineering analysis that allowed the evaluation of the actual hazards present to dictate the needed fire and life safety protection requirements.

This paper outlines the methodology used in conducting the performance-based analysis and briefly presents both the results and the subsequent recommendations for providing adequate ARFF capabilities at the USAP airfields.

Background

The first scientific research base in the Antarctic region was established as part the International Geophysical Year (IGY) between July 1957 and December 1958. At the end of IGY, the NSF was made responsible for U.S. Antarctic research and in 1970, the NSF was made responsible for the overall U.S. Antarctic Program research and operational support. Besides funding research, the NSF also provides scientists with logistical, operational, and laboratory support in Antarctica.

The NSF operates essentially three (3) airfields within Antarctica at any given time during the summer flying period. These consist of Williams Field and either the Sea/Ice Runway or Pegasus airfield at McMurdo Station and an airfield at South Pole Station. Both the airfield at the South Pole and Williams field are packed-snow runways and can only support ski-equipped aircraft. The Sea/Ice Runway and Pegasus airfield consist of ice runways and can accommodate wheeled aircraft.

The air logistical support for the USAP is contracted as part of a continuing U.S. Military program that supports Antarctic research, Operation Deep Freeze. The New York Air National Guard (NYANG) operates up to six ski-equipped Lockheed LC-130 cargo aircraft that are used primarily to fly personnel and equipment between McMurdo and South Pole Stations. In addition, the United States Air Force Air Mobility Command provides a number of supplemental C-141, C-17, and C-5 aircraft that provide roundtrip flights between Christchurch, NZ and McMurdo Station for the transport of passengers and cargo.

Prior to 1998, the U.S. Navy operated the LC-130 aircraft and also provided the requisite ARFF capabilities. Subsequent to the Navy relinquishing their logistical support to the USAP, the ARFF capabilities have been provided by contractor personnel. Concern was raised by the Military over the fact that existing ARFF capabilities do not comply with any of the existing codes or regulations, civilian or military. The rationale for this situation appears to be twofold; 1) issues associated with operating in the harsh Antarctic climate, and 2) the fact that the airfields, with the possible exception of South Pole Station, are meant to be "temporary" and all airport structures and equipment must be mobile to facilitate removal at the end of each summer flying season.

To ascertain what ARFF capabilities are actually needed to provide adequate, cost-effective protection at the various USAP airfields, the NSF elected to conduct a performance-based hazard analysis. The goal of this analysis was to identify and evaluate potential fire hazards, review existing ARFF standards and regulations to assess their basis and applicability to the USAP airfields, and to evaluate the needed ARFF capabilities that are appropriate for the unique Antarctic environment, while providing an adequate level of protection.

Hazard Assessment

Initial plans for this study included the use of a more risk-based analysis. However, detailed data were not available to perform the more traditional risk assessment. As such, an alternate method was used that involved developing a more global, qualitative estimate of the risk(s) associated with Antarctic flight operations. This approach provided the latitude needed to select appropriate protection strategies, while not compromising, and in many cases improving, safety. The following paragraphs briefly outline the steps taken in performing this assessment.

Review of existing ARFF requirements. A review was performed of all pertinent requirements contained in existing, applicable military and civilian ARFF regulations, standards, and/or recommendations. This included, the National Fire Protection Association (NFPA) standards, the Federal Aviation Administration's (FAA) Federal Aviation Regulations (FAR), the International Civil Aviation Organization (ICAO) standards, and a number of DoD and U.S. Air Force instructions and guidelines. It is relevant to note that for civilian airports operating in the United States with scheduled commercial service, only the FAA regulations (FAR Part 139) are legally required, though it is known that some airports choose to apply some of the NFPA standards. The ICAO's standards apply only to airports in those countries, excluding the United States, which are contracting members of the ICAO. The DoD instructions apply only to military airfields with "permanently assigned" aircraft. For this reason other U.S. Air Force instructions were also included, since they provide additional guidance regarding air operations that more closely match those being performed at the USAP facilities.

This review focused primarily on three (3) subject areas that were considered most pertinent to this analysis; 1) the number of ARFF vehicles and manning, 2) ARFF response times, and 3) extinguishing agent types and quantities.

<u>*ARFF Vehicles and Manning*</u>: Each of the different governing documents assigns an index or category to an airport. These are generally based on the size of the largest aircraft that would be expected to operate at that airport, although both the FAA and ICAO do provide some latitude based on the actual volume of aircraft operations. Williams Field and South Pole Station operate only LC-130 aircraft. The Pegasus and Sea-Ice Runways are designed to accommodate LC-130, C-141, C-17, and C-5 aircraft. Therefore, the categories or indices assigned Pegasus and the Sea-Ice Runways must reflect C-5 operations, since this is the largest aircraft. Table 1 lists the airport categories or indices that would be assigned to each of the USAP airfields.

Table 1 – USAP Airfield vs. Assigned Airport Category/Index

Airfield	FAA (Index)	NFPA (Category)	ICAO (Category)	DoD (Aircraft Length Threshold)
Pegasus/ Ice Runway	D	9	8	> 53 m
Williams Field	B	6	5	< 53 m
South Pole Station	B	6	5	< 53 m.

The number of required ARFF vehicles varies slightly from document to document, but is a function of the category or index assigned to the airport. Table 2 lists both the number of vehicles that are (currently) normally deployed at each of the USAP airfields and the number that would be required by each of the different civilian and military governing documents.

Table 2 – USAP Airfield vs. Required Number of ARFF Vehicles

Airfield	Vehicles Currently Available	Minimum Required Number of Vehicles				
		FAA	NFPA	ICAO	DoD	Air Force (Pamphlet 32-2004)
Pegasus & Sea-Ice Runways	2	3	4	3	4	4
Williams Field	3	1	2	1	3	4
South Pole Station	1	1	2	1	3	3

Each of the governing documents addresses the manning levels that are to be available to operate the vehicles and to perform fire fighting and rescue operations. The NFPA, FAA, and ICAO do not specify exact numbers of personnel, but simply state that "sufficient" personnel will be available during air operations to perform the requisite ARFF functions. In general, the DoD

Instruction and the Air Force Pamphlet recommend that, as a minimum, three (3) personnel be assigned to each of the required ARFF vehicles.

ARFF Response Time: "Response time" for ARFF is defined by NFPA 403 (1998 edition) as; "The total period of time measured from the time of an alarm until the first ARFF vehicle arrives at the scene of an aircraft accident and is in position to apply agent." Though each of the governing documents has a slightly different metric with regards to how they measure response time performance, each requires that responding vehicles reach the aircraft within 3 minutes. Only the NFPA requires a response within 2 minutes.

Extinguishing Agents: All documents require that significant quantities of a foam/water agent be provided. The commercial documents also include a requirement for a "complementary" agent, typically a dry chemical. The amount of agent required to be carried on ARFF vehicles and the rate at which the agent must be applied is dependent on the category or index of the airfield. The capacity of agents listed in Table 3 is the aggregate for all ARFF vehicles in service at the airfield.

Table 3 – Available Fire Suppression Agent vs. Requirements

Airfield	Agent	Currently Available	Governing Document Agent Requirements				
			FAA	NFPA	ICAO*	DoD	USAF**
South Pole Station	Foam/Water	0 l	5678 l.	14,173 l	5402 l	34,069 l	10,735 l
	Dry Chemical	113 kg	227 kg	204 kg	180 kg	NS	NS
Williams Field	Foam	4542 l	5678 l.	14,173 l	5402 l	34,069 l	10,735 l
	Dry Chemical	1293 kg	227 kg	204 kg	180 kg	NS	NS
Pegasus / Ice Runways	Foam	4542 l	15,142 l	36,226 l	18,200 l	45,425 l	22,591 l
	Dry Chemical	1225 kg	227 kg	408 kg	450 kg	NS	NS

* - For ease of comparison, Metric values have been replaced with equivalent values using English units

** - values are based on "Greater Risk/Loss" category of Table A2.1, Attachment 2 of Pamphlet 32-2004

NS - not specified

Characterize ARFF requirements. To develop mitigation strategies that are consistent with the intent of the ARFF codes and standards, it is necessary to understand the mission objectives and rationale for the ARFF services requirements that are outlined in the governing documents. Generally stated, each of the governing documents indicates that the mission and/or objective of providing ARFF services is to save lives by providing a safe path through burning aircraft fuel by the exterior application of an extinguishing agent. Only the new DoD regulations and existing NFPA standards contain discussion regarding interior aircraft fire suppression and rescuing (extracting) victims who cannot escape on their own.

It is recognized that the facilities and environmental conditions that exist at the USAP airfields are significantly different from those normally encountered at established commercial or military airports. Existing ARFF codes and standards were developed to address conditions and hazards that significantly differ from those that can exist in the Antarctic. For example, while the ARFF codes and standards assume that ice and snow will be present on runways and adjacent areas in cold climates, they do not specifically address runways that are constructed of ice and snow.

The fire fighting capabilities required by each of the different organizations appears to have been derived from a common basis. Research was conducted in the 1940's by the FAA Technical Center in Atlantic City, NJ that evaluated the potential size and spread of fuel fires resulting from an aircraft crash [Geyer, 1969]. This information was used by an early ICAO panel, the Rescue and Fire Fighting Panel (RFFP), to develop a parameter known as the "Theoretical Critical Fire Area." The RFFP's definition for the critical area is "the area adjacent to an aircraft in which fire must be controlled." The critical area represents the area around the aircraft's fuselage that needs to be protected to prevent the skin of the aircraft from melting due to exposure to a severe fire.

The common theme that is evident in all of the code requirements is that the primary fire hazard to be mitigated is a large, post-crash (jet fuel) pool fire adjacent to the aircraft's fuselage. All existing codes are designed to ensure that airports have the capability to rapidly deploy and apply a foam/water suppression agent, which is normally the most appropriate agent to control and/or extinguish a pool fire.

The requisite response times for ARFF services are based on the anticipated failure of the aircraft's aluminum skin when exposed to a severe fire. The intent is to ensure that the firefighting activities are implemented in sufficient time to prevent "burn-through" of the aircraft's skin and to support the successful evacuation or rescue of personnel (passengers & crew). The requirements that stipulate the number of ARFF vehicles that are to be available are simply a function of the two previous requirements; type and amount of suppression agent needed and response time. This requirement recognizes that a single vehicle may not be capable of supporting all fire fighting and rescue requirements and that larger vehicles having greater capabilities may not be able to meet the response time requirements.

Assess the relative risk of Antarctic flight operations. It would seem intuitively obvious that the inherent risks associated with flight operations in severe environmental conditions would be greater than those in more temperate climates. However, the increased level of pilot training and their familiarity with both the USAP airfields and the climatic conditions might also be a factor in reducing the overall risks associated with Antarctic flight operations. The initial goal for this assessment was to support an estimate of the ability to mitigate risk by employing ARFF intervention strategies. Unfortunately, adequate data to support such a detailed analysis were not available. As such, a more global approach to establishing the relative risk of Antarctic flight operations was taken, using information obtained during a site visit to the airfields in Antarctica and by reviewing documents and data provided by the NSF and other pertinent resources [U.S. Navy (1990), Meller (1993)].

Risk is often defined as the consequence of an event in dollars multiplied by the probability of occurrence for that event over a given time interval [Modarres, 1993]. However, describing risk as a cost was determined not to be the most appropriate method for this analysis, due to the wide range of cargo transported on-board the aircraft and the resistance to place a specific value on human life. For engineering applications, risk can be defined as the probability of a hazard resulting in an accident [Crowl, 1990]. This definition fit well with the general parameters established for this part of the analysis, which was to assess the Antarctic flight operational risk. Therefore, risk was defined as the probability of any given flight, during landing, take-off, or ground operations experiencing an accident.

Unfortunately, no cohesive set of data that specifically detail Antarctic flight incidents at the McMurdo or South Pole airfields was identified during the information-gathering portion of this study. Incident data on other Arctic rim aircraft operations were reviewed. Various locations, such as airfields in Iceland, were contacted in an attempt to gather information. However, the records that could be located were inconsistent, limited in detail, and for the most part did not include information on ARFF operations. As a result, a broader, more general set of data detailing aircraft incidents was pursued. Several organizations were contacted, including the Air Force, Air National Guard, Navy, Coast Guard, FAA, and the National Transportation Safety Board (NTSB). With the exception of the Naval Safety Center, the armed forces were generally hesitant to provide any accident data and to pursue this route would have resulted in a significant delay in completing the overall analysis. The FAA deferred to the NTSB as having the best source of data on flight incidents.

Accident rates for domestic operations were calculated from the NTSB data and compared to the results from the sparse Antarctica data to establish a baseline and a point of comparison for an accepted risk (flying in the US) and the uncharacterized risk (flying in Antarctica). The NTSB data for commercial flights revealed a downward trend in accident frequency over the past decade, with the expected accident rate for the last reported year (2000) being approximately 2×10^{-6} accidents per flight. Reviewing the available data for flights in Antarctica revealed an accident rate on the order of 2×10^{-4} per flight, thus indicating a greater risk of an accident as compared to domestic flight operations. However, these data do not include information regarding either the severity of the accident or whether or not any fire was involved.

An important point to note is that the existence of ARFF operations has no impact on the frequency of aircraft incidents that can propagate into accidents requiring a response from the ARFF operations. ARFF operations only provide a mechanism for mitigating the consequences of accident involving flight operations after one occurs. The purpose for this comparison was simply to establish the likelihood that ARFF services would be needed in Antarctica, based on the accepted risk of traditional commercial aviation. Since there are no data available that address whether or not ARFF services have ever had an impact on the outcome of an accident, it was not possible to quantify any risk reduction that might be associated with having ARFF services available.

Credible Fire Scenarios. It is realized that certain fire hazards are inherent with the operation of any airport and that some of these will exist at the USAP facilities. It is also recognized that there are certain fire scenarios that may not be possible or relevant at the NSF airfields due to the specific environmental or operational conditions. An initial list of potential fire hazard scenarios associated with airport operations was developed and then evaluated to ascertain which scenarios might not be credible, given the specific conditions that exist at the NSF airfields. The purpose for determining credible fire hazard scenarios was to facilitate an assessment of the ARFF capabilities needed to mitigate the expected fire hazards at the NSF airfields. As such, each fire scenario was evaluated in the context that it is the initial fire to which the ARFF responders would be expected to mitigate. The following paragraphs briefly address the fire scenarios initially identified and whether each was, or was not, considered credible.

Engine and APU Fires: Development of a leak of flammable or combustible fluid (e.g. fuel, hydraulic or lubricating oil) in the vicinity of an ignition source within the engine of both an aircraft and the auxiliary power units (APUs), used to supply power to aircraft prior to starting their engines, is considered to be a credible event. These fires tend to be fairly localized near the source of the fuel. Even when it is not possible to remotely isolate the source of the fuel, the likelihood of such fires spreading beyond the area (engine) of origin is remote.

Fires Resulting from failures of Jet Assisted Take-Off (JATO) Rockets: JATO systems are mostly used for take-offs from remote field camps, but occasionally have been used at the established airfields, such as Williams Field. A review of the NSF incident data revealed that this scenario has occurred, primarily due to a failure of the external mounting brackets that allowed the rocket to impact the aircraft. Also, if only one set of JATO rockets fires, this creates an asymmetric power condition that could cause an aircraft to depart controlled flight (including take-off rollout) and possibly crash. JATO rockets use a propellant that includes an oxidizer mixed with the solid fuel that provides a flame hot enough to ignite the aluminum skin of an aircraft, as well as any flammable/combustible liquid. This is a credible scenario.

Aircraft Fueling/Defueling Fires: Fuel spills do occur with the fueling and de-fueling of aircraft in Antarctica. However, during these evolutions, personnel are standing by with portable extinguishers to quickly identify a spill, immediately shut down pumping operations, and extinguish any fire that might have been initiated. Even if a fuel spill does occur during one of these evolutions, it is still very unlikely that a fire will result, for a number of reasons. Administrative procedures are in place to prevent the introduction of ignition sources near the fueling operations. Also, at both Williams Field and South Pole Station airfields, it is impossible to develop a fuel pool. Any fuel that is spilled will immediately permeate down through the snow and away from any ignition source. Additionally, a fuel spill on the ice runways is also considered an unlikely source of a fire. Due to the temperature of the ambient air and the fact that any spill will be on an ice or snow surface, ensures that the fuel will be well below its flashpoint. Therefore, even if the fuel can develop into a pool, a significant level of sustained energy will be required to both ignite the fuel and allow it to grow to the point of self-sustained burning. Though unlikely to occur, this scenario was maintained as a credible fire scenario.

Hot Brakes and Wheel Fires: The airfield surfaces in Antarctica are either packed snow or ice. Only ski-equipped aircraft land on the packed snow runways and thus, no brakes are used. Although aircraft utilizing the ice runways at low speeds may apply brakes, they are not very effective at landing speeds and the majority of braking is done by reversing engine thrust. Therefore, it is not considered likely that the brakes would ever become hot enough to be a credible ignition source for either the tires or hydraulic oil that might be present.

Combustible Metal Fires: For this scenario to occur, a combustible metal must be brought to its ignition temperature. The ignition temperature of titanium is 1600°C, while the diffusion flame temperature associated with flammable and combustible spray fires is nominally 600°C. The relatively high ignition temperature can usually only be obtained as a result of friction. Thus, the potential for combustible metals to be the first item ignited is unlikely and would typically be the result of another initiating fire scenario.

Aircraft Interior Fire. All cargo that is to be transported on military aircraft must meet the requirements of Air Force Joint Manual 24-204, *Preparing Hazardous Materials for Military Air Shipments*. However, several incidents of improperly prepared cargo have resulted in the potential for a fire or explosion to occur. Typically, these have been in-flight incidents and mainly involved fumes/vapors accumulating within the aircraft. However, the possibility does exist that an internal fire could occur while an aircraft is on the ground and that ARFF services would be called upon.

Post-Crash Fuel Tank Failure: A number of fire scenarios may be postulated as a result of an aircraft's fuel tank(s) being ruptured, subsequent to the aircraft's impact with the ground, another aircraft, or some airfield structure. Aircraft fuel tanks are mainly located in the wings, which can often be separated from the fuselage in a crash, causing a breach in the fuel tanks. An example of such an incident occurred in 1972 at McMurdo's Ice Runway when a taxiing LC-130 struck a 1.5 meter (5 feet) high snow berm, causing a wing to come in contact with the ground and resulting in a fire. Since details are limited, it is presumed that the damage to the wing caused fuel to be released and impinge on the hot surfaces of an engine.

Mitigation Strategies and ARFF Technical Requirements. Given that fire hazards do exist at the NSF airfields, measures must be implemented to mitigate their potential impact on USAP operations and personnel. An underlying assumption must be that a fire scenario has progressed to the point that ARFF services are required. This means that somewhere in the progression of events, standard operating and emergency procedures already in place and designed to initially prevent these hazards from occurring, have failed. Equally, this assumption must include the fact that these fire protection strategies only address scenarios that occur within an airfield's "ARFF zone of response."

Though both detrimental to human performance and limiting in the selection of viable fire fighting agents and delivery systems, the extreme cold climates and terrain features within the Antarctic region can have a very beneficial impact on restricting the fire hazards normally associated with aircraft operations. With the possible exception of fires initiating within the interior of an aircraft, the analyses of the credible fire scenarios indicate that none of the fire hazards is likely to be the primary threat to survivors of a crashed aircraft. Given that a fire does initiate within the interior of an aircraft, it would still be unreasonable to expect that an ARFF response could occur soon enough to have an impact on the successful evacuation of onboard personnel. Fire tests on commercial passenger and military cargo aircraft demonstrate that untenable conditions within the cabin can occur within minutes of fire initiation [(Gassmann, 1966), (Dussa, 1989)].

The climatic conditions and terrain at the various USAP airfields are such that any of the credible fire scenarios are most likely going to be confined to a fairly localized area near the fuel source. Both dry chemical and halon extinguishing agents have a proven history of successfully combating localized fires. The existing requirements for the specified types and amounts of fire suppression agent are driven by the need to fight large jet fuel pool fires, hence the requirements for large quantities of a foam/water agent. Based on the results of this analysis, it will be either impossible or very unlikely for a pool fire of any significant size to develop, depending on which of the USAP airfields is being considered. As such, the need to provide large quantities of a foam/water agent is unnecessary at the USAP airfields.

The primary advantage of a foam/water agent is that it both "smothers" a fuel pool fire and prevents re-ignition by isolating the liquid fuel from an ignition source and preventing the formation of flammable vapors above the fuel's surface. Chemical agents are primarily designed to suppress a fire by either "knocking" the flame away from the fuel source or by chemically interrupting the combustion process. Even if a pool fire was able to develop on an ice runway, the combination of a chemical agent and the cold Antarctic environment will provide much of the same advantages as those provided by a foam/water agent. The dry chemical agent will knock down the flame and by applying a solid blanket of agent on the fuel's surface, any flames can be pushed away from areas where the fire has already been extinguished. This process, coupled with the fact that the very low ambient temperatures will essentially preclude any fuel vapor formation, provides similar firefighting capabilities to those of a foam/water.

No information was identified that provided sufficient data to support a definitive quantification of the amounts of dry chemical agent that would be needed to combat the limited pool fires described above. Although the use of twin agent systems (foam and dry chemical used in tandem) is a common strategy in combating pool fires, with each agent having beneficial firefighting properties, there apparently has never been any significant motivation to develop tactics for the use of only a dry chemical. This is an area where additional data are needed to better support recommendations for the amounts of dry chemical that should be carried on the USAP ARFF vehicles. However, due to the anticipated localized nature of the potential fires, it would be expected that very large quantities would not be needed. Equally, the number of personnel needed to apply the agent to a localized fire should also be limited. It is postulated that no more than one or two dry chemical hose lines (one or two persons per line) would be needed to combat aircraft expected fires at the USAP airfields.

Removal of personnel from the scene of a crashed aircraft needs to be performed in a very expeditious fashion. This sense of urgency stems not only from hazards associated with a post-crash fire, but from the effects of the harsh environment to which crash survivors are likely to be exposed. Since it has been evaluated that most post-crash fires will be localized in nature, the risk to crash survivors from fire may be minimal as compared to that associated with injured personnel being exposed to the extreme cold. Given the potential for injuries that may be sustained in a crash, it is reasonable to assume that survivors may not be mobile or able to properly protect themselves from the cold temperatures. It is imperative that ARFF capabilities include the ability to rapidly reach a crash to assist survivors and quickly remove them from the outside environment.

All the ARFF vehicles supporting the McMurdo Station airfields are tracked vehicles. The top speed of the fastest vehicle is on the order of 24 kph (15 mph). Due to their weight, their operation is confined to the packed surfaces, which limits where they can be staged during flight operations. It is anticipated that the first responding vehicle will require more than 4½ to 5 minutes to reach the far ends of the runways. Given the fact that the ARFF truck at South Pole Station is a wheeled vehicle, it is likely that it could reach the far ends of the runway within the requisite 2 or 3 minutes, if it was manned at the time of an event. However, current operating procedures are such that during flight operations the truck is positioned near the runway (adjacent the fueling pits), but is not manned. It was stated by SPS personnel that it takes about 4 minutes, at best, to respond to a crash alarm and reach the ARFF truck.

Since the firefighting capabilities of the ARFF vehicles should primarily include moderate quantities of chemical agents and not large quantities of a foam/water agent, it should be possible to provide a vehicle with a configuration that is both adequate for the terrain and that can achieve response times of 2 or 3 minutes.

Conclusions and Recommendations

The use of a performance-based approach to evaluate the ARFF requirements at airfields in Antarctica indicates that trying to meet standard ARFF codes and regulations is unwarranted. Many of the hazards that can exist at airports in more temperate climates, and for which the codes and regulations were developed, simply do not exist in Antarctica. The anticipated pool fires that are the basis for the requirements for large quantities of a foam/water agent are either impossible or highly unlikely to develop at the USAP airfields. The evaluation of the fire hazards that can likely occur at the USAP airfields indicates that they will likely be highly localized and that use of a chemical agent to mitigate these hazards is a viable option. These facts, coupled with the complications associated with the use of water-based agents in an extremely cold environment, lead to the conclusion that chemical agents are the most appropriate choice as the primary fire suppression agent at the USAP airfields. Furthermore, it has been evaluated that there should be little or no impact on the ability to provide effective aircraft fire protection by not having a foam/water agent available. However, it was also recommended that actual fire testing of pool fires on ice and frozen surfaces be conducted to validate this theoretical evaluation and to provide data regarding the use (tactics) and amounts of chemical agent that are needed to provide effective fire control and suppression. The possibility exists that such data could also be very useful in providing alternative, cold weather fire fighting tactics at any airport subject to very cold temperatures.

The evaluations associated with this study indicate that the quantity of dry chemical fire fighting agent provided by the existing ARFF equipment at McMurdo should be adequate. This assessment is based solely on the premise that aircraft fires at the USAP airfields will likely be very localized in nature. Again, this can only be substantiated through the conduct of tests designed to provide the necessary data. A recommendation was made to increase the quantities of dry chemical agent available on the ARFF truck at the South Pole Station. Thought additional agent is contained in the three (3) portable units near the refueling pits, this could not be used to readily support firefighting elsewhere at the airfield.

The two to three minute response time promulgated in the various governing documents is technically based on the time for a pool fire to burn through an aircraft's fuselage. Since significant pool fires are unlikely to occur, it would appear that the basis for specifying a response time should be more a function of effecting a timely rescue of injured personnel. The technical basis for the response time of USAP ARFF units should address issues of providing prompt medical aid for injuries and trauma associated with a crash and minimizing exposure to the harsh Antarctic climate.

With respect to the number of vehicles and level of staffing, it was recommended that one rapid response fire-fighting vehicle be provided for all airfields supporting only LC-130 flight operations. In addition, a second rescue vehicle be provided that can support medical aid and the rapid removal of survivors from the crash scene. Only one (1) ambulance is currently available at the McMurdo airfields. This would not be able to meet the recommended requirement when more than one airfield is in operation simultaneously. Currently, no vehicle is assigned to South Pole Station to perform this function. For the airfields supporting C-141, C-17, and C-5 flight operations, it was recommended that two rapid response fire-fighting vehicles be provided, along with the rescue vehicle. During periods when both Williams Field and one of the ice runways is in operation, an alternate rescue vehicle could be staged at one of the airfields that would not be staffed, such as a bus equipped with supplies to support a triage area at the crash site. Given the evaluated fire scenarios, the recommended staffing levels for each ARFF vehicle was two.

References

Crowl, C.A., and Louvar J.F., (1990). *Chemical Process Safety: Fundamentals with Applications*, Simon and Schuster Company, Englewood Cliffs, New Jersey.

Dussa, R., Fiala, R., Wagner, R., and Zenses, B., (1989). *Full Scale Study of a Cabin Fire in an A300 Fuselage*, AGARD, Aircraft Fire Safety, October 1st, 16 p.

Gassmann, J. J., (1966). *Characteristics of Fire in Large Cargo Aircraft*, Technical Report No. ADS-73, National Aviation Facilities Experimental Center.

Geyer, G. B., (1969). *Effect of Ground Crash Fire on Aircraft Fuselage Integrity*, Report No. NA-69-37, Federal Aviation Administration, National Aviation Facilities Experimental Center, Atlantic City, NJ.

Modarres M., Kaminskiy M., and Krivtsov V., (1999). *Reliability Engineering and Risk Analysis*, Marcel Dekker, Inc., New York.

Mellor, M., (1993). *Notes on Antarctic Aviation*, CRREL Report 93-14, Cold Regions Research & Engineering Laboratory, U.S. Army Corps of Engineers, Hanover, NH, August.

U. S. Navy, (1990). *Report of Operation Deep Freeze 1989 – 1990*, Commander of Navy Support For Antarctica, United States Navy.

Tunneling at the South Pole

Michael R. Walsh, PE[1]

Abstract

Tunneling in cold regions presents many unique engineering challenges. When that region is the South Pole, these challenges become extreme. In 1996, the U.S. Army Cold Regions Research and Engineering Laboratory first deployed the South Pole Tunneling System. This system was designed and built specifically for machining tunnels in the firn (highly compacted snow) of the Antarctic ice cap. The tunneling machine creates 2- x 3-m tunnels through the material at depths of up to 16 m. A pneumatic chip disposal system transfers the debris to the surface, where a centrifugal fan powers the disposal system. This system has been used to create two tunnels for water and power transport in support of the new South Pole Station currently under construction. This paper will discuss some of the engineering challenges faced by both the designers and operators of the equipment over the course of four deployments.

Introduction

Tunneling operations present many unique challenges to engineers of every discipline. No two tunneling jobs are alike, even when driving parallel tunnels. For every challenge, there is an opportunity, and at the South Pole, both challenges and opportunities come wrapped in ice. With a mean annual temperature of almost –50°C, the challenges are obvious at Pole. Even in the austral summer, the mean temperature for the warmest month hovers around –28°C. With wind speeds as high as 24 m/s, special equipment, procedures, and crew are all required. As the station is located at the Pole, the physiological elevation can reach 4,000 m, affecting both the equipment and the hardy souls operating it (NSF, 1994).

In 1991, engineers from the U.S. Army Cold Regions Research and Engineering Laboratory (CRREL) in Hanover, New Hampshire, accepted the challenge of the National Science Foundation (NSF), the U.S. Government entity with

[1] Mechanical Engineer, U.S. Army Cold Regions Research and Engineering Laboratory, 72 Lyme Road, Hanover, NH 03755-1290 mwalsh@crrel.usace.army.mil

responsibility for the U.S. Antarctic Program, to devise a self-contained system for the construction of unlined tunnels at the South Pole Station. These tunnels are to provide for safe and secure transport of personnel and utilities between the various buildings and structures that have grown up around the central dome of the station. The tunnels are unlined to reduce the amount of material transported to the Pole. With the need for a new, expanded station being advocated by NSF, the construction and use of the tunnels will allow the old station to be connected to the new facility without the problems generated by surface structures and utilities.

In 1991, CRREL began the process of designing the South Pole Tunneling system, a tunneling system based on a custom-built horizontal road-header-type tunneling machine. After production and testing of the system in New Hampshire, the system was deployed in Antarctica to drive two tunnels at the South Pole. This paper describes the system developed at CRREL, some of the problems encountered and overcome, and the results of the tunneling operations at the Pole.

The South Pole Tunneling System

The CRREL South Pole Tunneling System is composed primarily of five major subsystems. These are the horizontal roadheader tunnel excavation machine (the tunneler), the chip disposal system, the surface drill rig, the generator set (genset) module, and the workshop. The tunneler and chip disposal system were technically the most challenging parts of the system. The design of these two components has been evolutionary over the course of the seven seasons the equipment has been in Antarctica. The current state of the equipment is likely the final iteration of the design for this system. The description that follows is that of the system used during the 2001–2002 season (Walsh 1999).

Before going into detail on the system components, this paper will present a brief system overview that will help orient the reader. Fig. 1 shows how the system is integrated. As can be seen, the fan and drill are used on the surface, while the tunneler and chip conveyance ducting are below the surface. A vertical section of fixed-length ducting segments connects the two levels. The tunneler machines the face of the tunnel, depositing the chips at the base of the face. A snowblower, extensibly attached to the front of the tunneler, is fed into the chip pile, centering and feeding the chips into the ejector pipe going over the tunneler. Behind the tunneler, a series of telescoping duct assemblies directs the chips to the transition sled, which is also attached via a short section of adjustable ducting to the vertical pipes. At the surface, the vertical pipes are attached via a flexible hose and a section of telescoping ducting to the centrifugal fan (blower), which powers the chip conveyance system and blows the chips clear of the area. The drill rig drills 30-cm diameter holes used for the vertical chip conveyance tubing and the power cord that supplies power from the sled-mounted generator module on the surface to the tunneler. The generator also provides power to the centrifugal fan and workshop. The workshop is used as a base of operations and to make minor repairs to the equipment.

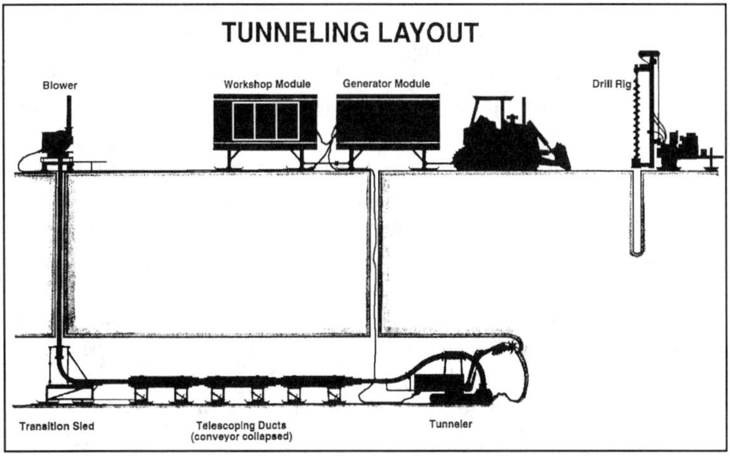

Figure 1. The CRREL South Pole Tunneling system.

The Tunneler

The tunneler concept is based on a small tracked excavator with an industrial snowblower mounted to the front for chip removal (Fig. 2). The design is similar to that of a horizontal-drum roadheader-type tunnel excavating machine. An electrohydraulic power pack is mounted on an extended frame behind the existing diesel motor on the excavator for operating in the confined space of the tunnel.

A hydraulically powered rotating shaft with a series of adjustable cutter arms was designed and installed for machining the tunnel. The tracks on the excavator were extended 60 cm to the rear of the machine to lower the ground pressure and balance the machine. A pivoting, extensible frame was fabricated for the attachment of the snowblower to the front of the machine. This allows the snowblower to be lifted

Figure 2. The tunneling machine.

and lowered as well as extended and retracted to aid in accessing the debris pile at the base of the tunnel face during tunneling operations. An extensible chip ejector system was designed to bring the milled debris from the snowblower to an attachment point at the rear of the machine. Tilt indicators provide the operator with information on the pitch and roll of the tunneler, as well as the angular position of the boom. A vacuum gauge in the cab connected to the ducting indicates the status of the disposal system.

The tunneler employs a transverse rotational machining system for milling the face of the tunnel (Mellor, 1975, 1977). This system, which is called the "drum," is actually made up of a series of diametrically opposed cutters on arms (Fig. 3). These arms are mounted to hubs along the axis of rotation such that 360° rotation in 15° increments can be made to the angular position of each of the cutters. The cutters are toothed, with an 11° clearance angle and a 24° top rake. The cutter material is flame-hardened AISI 4150 steel. Cutter geometry is based on work done by Ueda and Kalafut (1989). Power calculations were based on data on snow properties from Mellor (1964) and Gow and Ramseier (1964). Because of the brittle nature of the material to be machined, no chip breaker was machined into the cutter. The cutters are tack welded to their seats.

A total of 26 cutter arms are used to span the 1.84-m-long drum. The drum is 0.6 m in diameter with a maximum infeed depth of 15 cm. Infeed is limited by the flange diameter of the hubs used for mounting the arms. The number of arms can be adjusted using spacers and various width cutters. Three cutter widths are currently used: 5, 6.4, and 7.6 cm, with the 6.4-cm cutters the most common. The cutters are toothed, with 1.27-cm-wide teeth opposing non-toothed sections on the opposing cutter arms. This results in larger disaggregated chips during machining and less work for a given amount of material removed.

Figure 3. View of cutter drum.

Hydraulic power for the tunneler is provided by two sets of tandem pumps. The three pre-existing excavator control circuits normally powered by the machine's diesel engine are fed by two electrically driven pumps, a 328-cm^3/s gear pump and a 574-cm^3/s gear pump operating through a 50:50 splitter. A 1,150-cm^3/s gear pump powers the drum, while a 574-cm^3/s gear pump powers the snowblower. Maximum demand for the hydraulics is approximately 43 kW. The closest-sized electric motor commercially available is a 37.3-kW model, less than the maximum demand load but sufficient for the anticipated peak operational spikes. A high-efficiency 460-V motor is used for this application. The motor and pumps are protected from high startup torque and inrush amperage with a soft-starter, rated to 75 kW. The soft starter limits inrush current and allows longer ramp-up time for the motor.

Chip Disposal System

A pneumatic chip conveyance system was designed to move the tunneling debris from the face of the tunnel to the surface. The system is driven by a large centrifugal blower located on the surface (Fig. 4). The blower is attached to a sled-mounted vibration isolation platform designed to be towed by a snow cat or moved with a fork-equipped tractor. A tapered inlet has a built-in shutoff (blast gate) to assist in startup, and the outlet has a directional chute for guiding the debris. Power for the fan is supplied by a 37-kW electric motor.

In the tunnel, the chips are conveyed longitudinally through a series of four sets of expanding tubes on adjustable trucks. Directly behind the tunneler is a two-tube unit, called a "pup truck," that allows 1.5 m of travel. This permits the operator to move the tunneler back and forth short distances without stressing the larger units. The three large units consist of three concentric tubes, two sliding, and allow 6 m of travel before full extension each. Collapsed total length is 16 m; fully extended it is

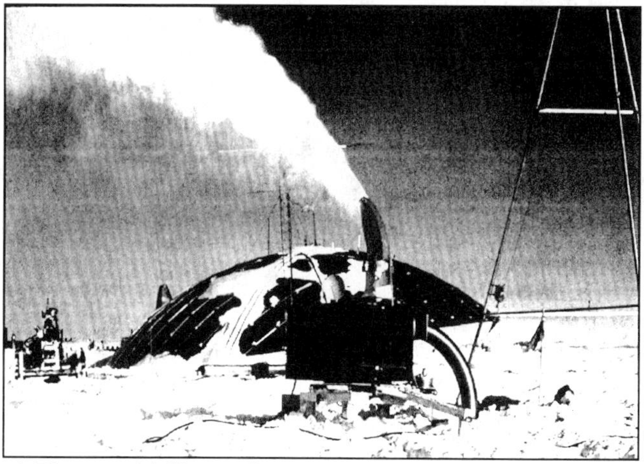

Figure 4. Blower and drill on surface. The South Pole Dome is in the background.

35.5 m. Each sliding tube joint has a set of metal fingers and a silicone annular gasket that allows sliding motion but little suction loss. A 30-cm section of flexible pipe is attached between assemblies to compensate for misalignment. The ends of the assemblies can be vertically adjusted using built-in screw jacks. The ducts are tied tothe trucks and each other through a series of cables. This transmits the force of extension and travel through the cables and trucks rather than through the ducts during forward progress. Mechanical stops contain the sliding duct tubes when collapsed.

To redirect the chips from horizontal to vertical, a transition sled was fabricated. This is primarily a long flexible hose supported on a lightweight, ski-mounted frame. The end of the hose that connects to the expanding tubes has limited vertical freedom but a 1-m horizontal range. A series of vertical tubes, lowered from the surface using a tripod, guides the chips from the tunnel to the surface. These tubes, ranging in length from 60 cm to 3 m, are attached using low-profile clamp rings. The lowest vertical pipe segment has a nested flexible hose for vertical adjustment and alignment to the transition sled adapter. Total vertical adjustment is 1 m. At the surface, a flange on the uppermost tube anchors the assembly. A flexible hose and sliding tube assembly connects the blower assembly to the vertical pipes.

Drill Rig

The drill rig is a Simco 2400 SK-1 trailer-mounted drill with special CRREL-designed skis to allow easy transport on snow. The controls have been slightly modified to allow speed control when operating the up-feed as well as the down-feed of the drill head. Access holes for power and chip removal are drilled using 30-cm-diam. single-flight augers. The auger is in 1-m lengths, and three assemblies have been made up using three segments each, for a total length of 9 m per assembly. One segment is left for attachment to the augerhead when starting the hole. Two augerheads are available: a short double-bit head and a longer single-bit head. The double-bit head is generally used because of its superior cutting characteristics in the hard firn of the South Pole.

A 1-m-diam. backboring bit, designed and fabricated at CRREL, is used for forming emergency egress shafts from the tunnel to the surface. A 30-cm access hole must first be drilled at the location of the egress shaft. Drill string is then lowered through the hole and the bit attached. The rotating bit is then raised to form the hole. A 30-cm plug on the upper end of the bit keeps the bit centered in the access hole.

Generator Module

The generator module is a self-contained unit consisting of a Caterpillar 205 kVA (derated to 180-kVA), 460-VAC, 3-phase genset in an air-transportable aluminum-framed wooden module. A 1,360-L fuel tank is located within the module, with the module's floor constructed to serve as a secondary containment structure in case of a fuel or oil spill. Power is fed into an eight-breaker panel through a 300-A circuit breaker. Power connections are made at a panel outside the module opposite the breaker panel. Type-W no. 6 4-conductor power cord fitted with locking end connectors distributes power to the blower, the tunneler, a warm-up shelter, and the workshop. The cord sets are 23 m long each and can be connected in series.

Workshop

The workshop is enclosed in a module of similar construction to the generator module. Sections of the two sides have Trombe walls built in to utilize passive solar heating. The workshop contains tools, equipment, and hardware necessary to support the operation, as well as a base station for radio communications between the tunnel and the surface. A safety cabinet within the workshop is used for storing hazardous materials associated with the project. Although cramped, the workshop is a welcome relief from the cold and wind on the surface.

Deployments

In late 1995, NSF approved four members of the CRREL design team for deployment to Antarctica to test the machine at the South Pole Station. A proof-of-concept tunnel, originally planned for a length of 30 m at a depth of 11 m, was expanded to 125 m at a maximum depth of 16 m. It would also be a "working tunnel," a utilidor, used for the station's wastewater system. The tunneling system was assembled in January 1996 on the surface at the South Pole Station and tested against a mound of snow formed while trenching for the tunnel entry point. Problems surfaced because of the low temperatures experienced but were either worked through or fixed. With system adjustments made, the tunneler and fan were lowered into the starting trench and testing of the tunneling system commenced. Shortly after starting, the fan motor burned out, likely because of damage during prior tests in the United States. Concurrently, problems developed with the flexible stainless steel hose used to direct chips from the snowblower to the ejector tube at the rear of the tunneler. Overheating of the hydraulic oil was caused by the warmer-than-expected temperatures in the tunnel (-40° vs. -54°C). Failure of a pump shaft brought the tests to a close after only 8 m of progress.

Over the course of the summer following the initial deployment, the tunneling system was reexamined and the hydraulics simplified to improve reliability and reduce heat generation. A rigid ejector chute for the front end of the tunneler was designed and manufactured to replace the flexible tube that had been problematic in January. The pup truck was fabricated at this time, and a choke for the snowblower designed. A new pressure relief valve and pump assembly were purchased to replace damaged equipment.

In November 1996, CRREL returned to Pole to modify the tunneler and complete the initial tunnel. Tunneling commenced on 22 November and was completed on 3 December. This was the first complete test of the system, including shifting surface equipment to new access holes and running the system with the fan on the surface. Additional problems and some old ones persisted over the course of completing the tunnel. Hydraulic oil temperature was still running high; some components, especially the snowblower lift/extension frame, needed to be more robust; and the debris transfer system needed optimizing.

The success of the tunneling system in 1996 led to its being scheduled for a much longer tunnel complex. These tunnel, a main tunnel over 600 m in length with five 60-m branches, one 45-m branch, and one 19-m branch, would require a simpler, more reliable, system than that employed for the first tunnel. The system was reexamined and some engineering and design work done to make the system more user-friendly. In January 1998, a CRREL engineer and technician traveled to McMurdo Station in Antarctica to modify the equipment. Missing equipment and the late arrival of cargo resulted in the need to return the following January to complete the modifications and test the system at Pole. At this time, we trained a contract miner, John Wright, in the operation of the equipment.

In November 2000, the modified tunneling system was deployed for the new tunnel complex. Many of the problems encountered in the 1996 deployments had been resolved, but hydraulic issues persisted. The failure of the drum motor, which prematurely ended the season, was the major impediment to both production and progress. Originally attributed to faulty pressure relief valves, it was discovered near the end of the following deployment (2000–2001 season) that the failures were due to an oversized pump overdriving the motor. This component was targeted for replacement after the 1998–99 deployment but had been missed. Some mechanical problems persisted with the heavily used snowblower extension frame, but the overall reliability of the system was better than most commercial systems.

The 2001 – 2002 season, the last anticipated for the tunneling system, is underway at the time of the writing of this paper. The branch tunnels are being driven both by hand (19-m branch) and by machine. The correct drum pump has been installed and the equipment is operating much more reliably. The tunneling crew has been augmented with an engineer for this season. The tunnel complex should be completed by the end of the season in February.

Performance

Performance data for the four tunneling deployments are incomplete. Data for the late 1996 season can be found in Table 1. The overall production rate for the machine while operating was around 1 m/h. The target production rate is 3 m/hr. Reliability data were not recorded at that time (Walsh et al., 2001).

Data obtained from Mr. Wright over the course of the three seasons under his lead indicate a similar production rate. Although tests conducted in 1999 indicated a higher production rate is possible (2 m/hr), tunnel straightness and slope, monitored using a laser level, were a greater concern than performance for this longer tunnel. Detailed reliability data were recorded for this tunnel (Wright 2000). Overall downtime was 51%. This compares quite favorably with industry downtimes of 60-70%. This downtime is exacerbated by the incredibly difficult logistics support at the South Pole as well as the extremely harsh working environment.

Table 1. Deployment analysis – November/December 1996

Date	Time On Site (hrs)	Tunneling Time (hrs)	Progress (m)	Production Rate (m/hr)	Tunnel length* (m)	Notes
11/22	12	3	4.6	1.5	12.2	Start in P.M.
11/23	11.5	8	7.6	1	19.8	
11/24	9	4.5	1.5	0.3	21.3	Training day
11/25	12	10	9.1	1	30.4	
11/26	19.5	4.5	5.5	1	35.9	Drill holes/Redeploy
11/27	20.5	11	9.8	1	45.7	Fix pump/blown hoses
11/28	24	16	16.8	1	62.5	Drill holes/Redeploy
11/29	20	15	15.2	1	77.7	Fix blown hoses
11/30	12	7	7.6	1	85.3	Genset problems
12/1	19	3.5	3.1	1	88.4	Drill holes/Redeploy
12/2	20.5	17	21.3	1.3	109.7	
12/3	12	4	6.1	1.5	115.8	Finish in A.M.

*Cumulative. Includes progress from November 1996 deployment.

Engineering Issues

The unique operating conditions at Pole drove the engineering and design of the tunneling system. At the beginning of the project, working parameters were set and the engineering followed. The two driving parameters were the ambient temperature and the snow strength. The values given for these parameters were both incorrect.

The air temperature greatly affected the hydraulics temperature. The higher ambient temperature in the tunnel (-40°C) resulted in borderline hydraulic operating temperatures. The complexity of the original hydraulic system also contributed to the generation of waste heat. Simplification of the hydraulics over the course of the deployments, removal of flow control valves, and removal of heat tracing, insulation, and guarding during operations have alleviated this problem. Oil temperatures are now in the 26°-54°C range during machining operations.

The original design depth for the tunnel was 11 m. In January of 1996, the tunnel invert was increased three times until a depth of 16 m was reached. Only the limited supply of vertical tubing prevented the tunnel floor from being lowered further. The chip disposal system was only marginally affected by the increase in depth. Problems occur, however, with the increase in snow strength with depth.

At a floor depth of 11 m, the unconfined compressive strength of the snow in the tunnel will be approximately 7 kg/cm^2 at the roof and 9.3 kg/cm^2 at the invert. At an invert depth of 16 m, the strengths become 10.6 and 13.9 kg/cm^2 (Gow and Ramseier 1964). Snow density was predicted to be 0.5 g/cm^3. Actual density measurements from the mid-height of the tunnel indicate a density of 0.55 g/cm^3. This is likely higher than predicted because this area was highly trafficked during the construction of the present station in early 1970s. Using the density rather than the depth relationship, snow strength can approach 21 kg/cm^2, three times that used in the design. This greatly affects the ability of the tunneler to achieve the target production rate and also will affect the wear on the machine, although cutter wear was not an issue. Unfortunately, the current system cannot be upgraded to the power levels required to meet the original production rate.

The pneumatic chip conveyance system worked quite well in its final iteration. This had been the impediment on the tunneling system's predecessors. Weakness of the snowblower frame and poor impeller location limited the potential of the system. A better feed path of the debris into the airstream will greatly enhance the efficiency of the system. As this is projected to be the last season of use for this system, it is unlikely that these issues will be resolved.

Results

The current configuration of the tunneling system is capable of operating at an average production rate of just over 1 m/hr at a 16-m depth, although tests indicate a production rate of 2 m/hr is possible. The maximum sustained (>4 hrs) production rate from previous deployments was 1.5 m/hr (1996). The maximum operating depth was approximately 16 m from surface to floor, 5 m greater than the design depth. Maximum total linear run for snow removal is 59 m, including an 18-m lift from the tunnel floor to the eye of the centrifugal blower fan. The maximum length tunneled during one shift was 13 m, and the maximum one-day progress was 21.3 m. Approximately 47% of the shift time on average was devoted to productive tunneling operations. The remainder was absorbed in preparations for moving the equipment, heading measurements, preparing vertical holes, equipment repair, downtime due to meals and logistics, and training.

Several engineering issues remain to be resolved. These include the change in power requirements caused by the increased operating depths, modifying the debris disposal system to enable more efficient infeed of snow into the airstream, and addressing several mechanical weak points. Some of these modifications are beyond the capabilities of the current configuration (Walsh 2001).

Conclusions

The CRREL South Pole Tunneling System is a viable option for creating unlined tunnels at the South Pole. System modification and simplification since the initial deployment have resulted in increased reliability and productivity. A complete rebuild of the tunneler, transforming it from a prototype to a "preproduction" system, is recommended but not planned at this time. An economic analysis of the system is premature at this time as the system still needs to be optimized. The original analysis called for a production rate of approximately 3 m/hr to be more cost-effective than a cut-and-cover system. However, that analysis was based on a crew of eight, whereas operations over the four tunneling seasons normally occurred with a crew of two to four, making the system cost-competitive with the alternatives (manual excavation, cut-and-cover), especially at the depths now encountered. In any case, the use of tunnels for passage and utilities is clearly preferable over surface structures from the standpoint of safety and maintenance of equipment, especially during the extreme cold and darkness of winter.

The unlined tunnel driven in 1996 is currently being utilized at the South Pole Station for the wastewater outfall line. The new tunnel system, to be completed this season (February 2002), will go on line at the end of the season to transport fresh water to the new South Pole station (Fig. 5). Both tunnels will serve as utilidors for power to remote locations.

Figure 5. Installation of utilities in the new tunnel.

References

Gow, A. J., and R.O. Ramseier (1964). *Age hardening of snow at the South Pole.* CRREL Research Report 112, USA CRREL, Hanover, NH..

Mellor, M. (1964). *Properties of snow.* CRREL Monograph III-A1, USA CRREL, Hanover, NH..

Mellor, M. (1975). *Mechanics of cutting and boring, Part 1: Kinematics of transverse rotation machines.* CRREL Special Report 22-6, USA CRREL, Hanover, NH.

Mellor, M. (1977). *Mechanics of cutting and boring, Part IV: Dynamics and energetics of transverse rotation machines.* CRREL Report 77-19, USA CRREL, Hanover, NH.

National Science Foundation (1994) "Facts about the United States Antarctic Program." Publication NSF 94-4, NSF Office of Polar Programs, Arlington, VA..

Ueda, H. T., and John Kalafut (1989). *Experiments on the cutting process in ice.* CRREL Report 89-5, USA CRREL, Hanover, NH.

Walsh, M.R. (1999) *The CRREL South Pole tunneling system.* CRREL Report 99-1, USA CRREL, Hanover, NH.

Walsh, M.R., Wright, J.H., and Jensen, E. I. (2001) "A tunnel boring machine for the South Pole." In *Proc.2001 Rapid Excavation and Tunneling Conf.*, Soc. For Mining, Metallurgy, and Exploration, Inc., Littleton, CO, 687-698.

Walsh, M.R. (2001) "Tunneling System for the South Pole." *J. Cold Regions Engineering*, ASCE, Vol. 15(4), p.231-236.

Wright, J.H. (2000) "South Pole tunnel project, 1999–2000: After action report." Prepared for the National Science Foundation and Antarctic Support Associates (Unpublished).

Selecting a Guardrail End Terminal for High Snowfall Regions

Stephen P. Mattingly[1] and Zhongguo (John) Ma[2]

Abstract

This paper investigates claims of high guardrail damage rates for guardrail end terminals in high snowfall areas and recommends a selection of terminals for use in high snowfall areas based on an evaluation of end terminal specifications and a survey of eight northern tier state Departments of Transportation (DOT) that use these terminals. In areas of significant snowfall levels and cold temperatures, agencies need an end terminal that can withstand these conditions, resist damage during snow removal activities, and be readily repaired without having to excavate or drive posts into frozen ground. Furthermore, any viable end terminal must meet the National Cooperative Highway Research Program (NCHRP) Report 350 standards. This paper recommends eliminating all terminals that do not use sleeved steel posts. Wooden posts tend to shear-away at the top of the soil tube whereas steel posts can be replaced year-round.

Part of minimizing the danger to the end terminals during snow removal activities relies on more effective marking strategies and improved awareness and caution among winter maintenance crews. While this study ranks all of the alternative end terminals, those without steel posts and a limited percentage of posts sleeved cannot be recommended. Two terminal types, the sequential kinking terminal (SKT) and the ET-2000, emerge from the evaluation as the preferred designs for high snowfall regions because they use steel posts and sleeve all of their posts. While many of the other alternatives may prove adequate for many sites, only the ET-2000 and SKT are recommended for use in high snowfall regions at this time. More research is required to identify a single terminal as the best alternative because the user data exhibits limited variability.

[1] Assistant Professor, Department of Civil and Environmental Engineering, University of Alaska-Fairbanks, P.O. Box 755900, Fairbanks, AK 99775-5900; phone 907-474-7977; ffspm@uaf.edu
[2] Assistant Professor, Department of Civil and Environmental Engineering, University of Alaska-Fairbanks, P.O. Box 755900, Fairbanks, AK 99775-5900; phone 907-474-6733; ffzm@uaf.edu

Introduction

Background. The W-beam guardrail has traditionally been the first choice for use in protecting the motoring public from serious roadside hazards. This wide usage results from its favorable safety record, ease of construction and repair, and low cost. One trouble spot for this system has been the difficulty in safely treating the end of the barrier. As a result, since the 1960s, research in guardrails has been focused on guardrail end treatment. Tests indicated that the standard terminal section (37 ½ ft. long) was extremely dangerous (Cichowski *et al.*, 1961). A collision with an untreated guardrail will have severe results for the vehicle and its occupants. Safety improvements in terminal design have continued from these initial studies. The most recent advancement in terminal design focuses on new Federal Highway Administration (FHWA) requirements.

An important milestone in the development of the terminals is the FHWA rule that requires all safety devices installed on federal-aid highways after September 1998 must meet the new NCHRP Report 350 standards (Ross *et al.*, 1993). The development of a new generation of GETs to meet the new NCHRP 350 standards relies on numerous research efforts that result in new terminal designs. This new generation includes the slotted rail terminal (SRT) (Mak *et al.*, 1997), the ET-2000 (Sicking *et al.*, 1989), the beam-eating steel terminal (BEST) (Pfeiffer and Sicking, 1994), the sequential kinking terminal (SKT) (Sicking *et al.*, 1998a), and the flared W-beam guardrail terminal (FLEAT) (Sicking *et al.*, 1998b). These terminals are a subset of all terminals that have met the new NCHRP 350 standards. Denman and Welch (1999) describe the development and testing required to meet NCHRP Report 350 (Test Level 3) standards. They test the redirective and gating end terminal (REGENT) system attached to both wood and steel downstream guardrail posts. The report by Snyder (1995) describes a demonstration program of two end terminal treatment systems, the C-A-T and the BRAKEMASTER, conducted by the Pennsylvania Department of Transportation. His work appears quite interesting because, over a thirty-two month period, it examines installation procedures, accident-related data, costs, and service and maintenance information for both systems. Unfortunately, that study has no direct impact on this research because it examines systems designed prior to NCHRP Report 350. This study's investigation did not find any more recent demonstration programs that look at the new treatment systems.

Problem Statement. This paper examines claims that the new GETs do not always withstand winter snow removal activities effectively. This may be true because the development of all new terminals is based on ultimate crash-loading tests. Previous GET research fails to address normal service loading, such as loads caused by actual snow plowing and blowing operations. As a result, the SRT-350s have not held up well with many reportedly damaged as snow was pushed against them by snowplows.

Research Objectives. Since all GETs will likely remain susceptible to damage during snow removal activities, this paper seeks to identify the safest, most reliable, end-terminal which can be repaired above ground during winter months. Above ground repair in cold regions is critical so that any damaged end sections can be repaired in a timely manner without excavating frozen ground. The ideal GET for use in high snowfall regions must resist damage due to snow removal activities. The variety of new terminals that meet NCHRP 350 standards, another critical criteria for GET acceptance, makes the identification of appropriate terminals for use in the high snowfall areas difficult.

This study has two phases. The first investigates damaged end terminals and seeks to identify possible causes and solutions. The study uses a field investigation and survey to acquire this information. The second phase attempts to identify the best end terminals available for use. This evaluation process incorporates GET vendor specifications and a small survey of other northern-tier states.

End Terminal Damage Investigation

Field Investigation. The research team confirmed that the new slotted rail terminals (SRT-350s) are susceptible to damage through field visits. They believe that this damage may have been caused by direct contact from the snowplow equipment or by snow-only contact, i.e. too much snow had been pushed against the slotted W-beam through snow removal operations parallel to the rail. The research team cannot identify the exact cause of this damage; however, by comparing the slotted portion of the W-beam with the unslotted portion, the research team observed that the slotted portion has a lower bending resistance during snowplow operations. Maintenance crews indicate that wooden posts used for terminal treatment tend to shear away from their soil-tube bases. When this occurs, the post replacement effort is prohibitive because the base must be excavated in frozen ground. During the field investigation, the researchers identify two different terminal marking strategies, carsonite markers and flexi-flags on spring steel. The more effective strategy marks both the terminal and guardrail ends with six-foot flexi-flags on spring steel. Additionally, the spring steel seems resistant to damage from winter maintenance activities. The other, less effective, approach marks only the terminal end with an eight-foot carsonite marker, which is susceptible to damage by snow removal activities.

Snow Removal Strategy. Typically, winter maintenance techniques follow a basic pattern. First, high-speed, front-mounted snowplows are used during the snowstorm and immediately following the storm. This occurs because they move faster than the rotary plows and graders. In general, high-speed, front-mounted snowplow operations seem to pose little danger to the terminals. Between storm events, the maintenance crews attempt to thoroughly clean the road surfaces using graders and slopers. This initial grader action seems to be the action that may precipitate a terminal failure. According to maintenance operators, the graders will force large quantities of snow directly into the guardrail end terminal. When a grader gets too close to the terminal and pushes snow against the GET, the now forces may cause the terminal to fail. While maintenance operations may place the terminal in peril, but all damage is likely to result from equipment contact with the terminal.

Experience shows that maintenance crews may need to adjust their snow removal strategies. In one area, maintenance crews relied on their typical plowing strategy and cleared the area next to the terminal during the first winter after the installation of new terminals. After the snow season, half of the new terminals were damaged sufficiently to require replacement. The following winter the snow clearance strategy was adjusted to avoid the area around the terminal. Only one of the twelve terminals was damaged sufficiently to require replacement. When maintenance crews fail to take care around new terminals, they will become severely damaged.

Preliminary Survey of Northern-Tier States. The researchers surveyed other states that experience large snowfalls to determine their experiences with damage to the new end terminals during snow removal activities. The investigation received responses from the following states: Colorado, Delaware, Idaho, Illinois, Maine, Michigan, Minnesota, Montana, New Hampshire, North Dakota, Oregon, South Dakota, Utah, Vermont, Washington, Wisconsin and Wyoming. None of these states identify terminal failures due to snow-only contact except in a single heavy snow area within the State of Washington, where snowfall averages anywhere from a season average of 350 inches to over 700 inches and a single event can be three to four feet. The States of Minnesota and Maine identify significant damage due to equipment contact with the terminals. Many of these states express great interest in the danger of snow-only terminal failure; however, the slow proliferation of the new terminals and the recent mild winters in many of these states make any definitive statement on the existence of snow-only failures throughout the United States impossible.

Effective marking seems to be an effective technique for reducing equipment collisions with the terminals. The strategies for marking the terminals vary somewhat from state to state; however most states use plow markers or fiberglass poles with reflectivity. Another popular marker is simply an extended steel delineator post. Almost all of the agencies only mark the terminal ends. The State of Wyoming emphasizes the use of raised delineators to divide traffic as well as mark the edge of pavement. The State of Minnesota would like to educate their operators in an effort to reduce terminal damage. They would like to use a video that makes operators aware of the damage that they can cause and the costs associated with its repair.

End Terminal Rating and Evaluation

Research Approach. The results from the second phase are based on a vendor survey and a user (DOT) survey to gather data for specific evaluation criteria. These criteria are:

- NCHRP 350 compliance – Test Level 3
- Percentage of terminal posts sleeved
- Materials cost
- Availability of steel post
- Overall terminal rating
- Resistance to snowload
- Probability of above ground repair

Two data gathering approaches are utilized because the vendors are most likely to provide detailed product information and specifications; however, the users seem more likely to provide accurate, unbiased data regarding product performance criteria. The rest of this section describes the potential evaluation criteria and the process for obtaining the data.

The vendor survey focuses on obtaining a detailed product description. The most important criterion for this study is that all alternatives must at least be compliant with the NCHRP 350 Test Level – 3. The evaluation criteria focus on repairing and replacing GETs during winter conditions and cost. Sleeved posts tend to be easier to replace during the winter as opposed to unsleeved posts that must be driven into frozen soil. This study identifies the number of sleeves for each of the products. As previously discussed, maintenance workers prefer steel posts to wood posts, especially when unsleeved, because they can be driven into frozen ground more readily and they do not shear-away at the top of the sleeve. For each of the products, the vendors identify if a steel post can be specified. All of the data for these criteria must be obtained from product vendors and can be collected by using the vendor websites and conducting brief telephone interviews with sales representatives.

In order to identify a large selection of alternatives, this study uses an Internet search to identify seven major end-terminal vendors that distribute to State DOT's. In the cases where companies did not have adequate web facilities, the study acquires the vendor contact information from other agencies. After examining a vendor's website and marketing materials, a researcher contacts the vendor for a phone interview, which may last between fifteen and thirty minutes. As much information as possible is gleaned from the Internet and marking materials review, but all of this data is confirmed during the interview and any gaps filled. By using Internet research and DOT contacts, this study examines seven vendors: Bryson Products, Trinity Industries, Universal Industrial Sales, Gregory Highway Products, R.G Steel Corp., Road Systems, and Highway Safety Corp. This study examines the wide variety of GETs that the vendors produce including, FLEAT, SKT, REGION, BREAKMASTER, SRT 350, ET- 2000, CAT, and BEAT. The state of Wyoming has developed an end-terminal called the WYBET, which Trinity Industry distributes.

The user survey asks for an unbiased assessment of the alternative GETs over the criteria identified by the research team. While fifteen state DOTs were considered and contacted as part of this survey, the study identified personnel in eight of the targeted states that had enough knowledge of the GETs to participate in the survey. Wyoming, Montana, Minnesota, Michigan, Utah, Oregon, Washington, and Idaho provided data for this study. Once again, winter repair and replacement of the GETs figures prominently in the evaluation. The agencies provide their assessment of the viability of above ground replacement for each of the GETs that they use as a percentage of overall repairs. Beyond the previous criterion that closely mirrors similar criterion from the vendor survey, the DOTs assess the GETs' snow load resistance on a one to seven point scale. These study criteria are rated on a one to seven point scale. Finally, the agencies give each terminal an overall rating from one to seven. This data is collected using a phone survey.

Vendor Survey Results. Table 1 presents the results from the vendor survey. The vendors generally lack knowledge regarding the installation costs of their products, and even if they possess some knowledge of these figures they may be loathe to provide a potential installation price. The material prices for the end terminals vary widely between $800 and $2,800; however, when the CAT and BREAKMASTER terminals are excluded, the price only varies between $800 and $1,400. Many terminal types have steel posts available while a few must be either steel or wood. While the vendors fail to provide installation costs, they provide a wealth of information regarding their product's specifications.

Table 1. Vendor Survey Results

Manufacturer	Terminal Name	Number of Sleeves and Posts	Actual Cost Materials	Steel Post?
Bryson Products, Mississippi	FLEAT	2, 7 posts	$1,200	Wood/Steel
	SKT	2/4/8, 8 posts	$1,200	Wood/Steel
	REGIONT	2, 11 posts	$1,200	Wood
	BREAKMASTER	Concrete Slab	$2,500	Steel
Trinity Industries	SRT 350	2, 6 posts	$900	Wood/Steel
	ET 2000	8, 8 posts	$1,200	Wood/Steel
	CAT	6, 6 posts	$2,500	Wood
Universal Industrial Sales, Utah	BEAT	No Sleeve	N/A	Wood
	FLEAT	2, 7 posts	$805	Wood/Steel
	SKT	2/4/8, 8 posts	$1,015	Wood/Steel
	REGION	2, 11 posts	$800	Wood
	BREAKMASTER	Cemented Anchors	$2,800	Wood/Steel
Gregory Highway Products, Ohio	SKT	No Sleeve, 8 posts	$1,400	Wood
	FLEAT	2, 7 posts	$1,000	Wood/Steel
R.G. Steel Corp., Pennsylvania	FLEAT 350	2, 7 posts	$1,100	Steel
	REGION	2, 11 posts	$800	Wood
	SKT 350	4, 8 posts	$1,400	Steel
Road Systems	FLEAT	2, 7 posts	$1,100	Wood/Steel
	SKT	2/4/8, 8 posts	$1,400	Steel
Highway Safety Corp., Ohio	FLEAT 350	2, 7 posts	$1,100	Wood/Steel
	SKT 350	4, 8 posts	$1,400	Steel

In all cases, the vendors have no reference with which to evaluate the resistance of their products to snow removal activities. Above ground repair seems possible with all of the terminals as long as the installation utilizes soil tubes. Similarly, all vendors state that steel posts are necessary for any wintertime removal or replacement. The wood posts on some end-terminals tend to break apart when driven into frozen ground. One vendor provides an example of this wintertime repair. The North Dakota DOT uses guardrail terminals with steel posts and soil tubes, and they can successfully execute winter repairs and replacements. North Dakota experiences a high snowfall and represents a good example of the needs faced by other high snowfall regions.

Northern Tier State DOT Survey Results. Although the researchers attempted to contact numerous northern tier states during this study, only eight state DOTs elected to participate. While these sample sizes appeared to be small (even smaller with respect to individual terminal types), they actually represented a significant percentage of the population of northern tier states, where at most twenty states could be classified as northern tier states. In this context, the study received responses from forty percent of the population with response rates for individual terminal types possibly being higher or lower than this value because the actual population of terminal usage remained unknown.

Table 2 provides summary data from the northern tier state DOT survey. The sample sizes for each terminal type vary widely with all eight states using the FLEAT terminal and only one using the SKT and BREAKMASTER terminals. With the exception of the probability of above ground repair, the study rates the following criteria on a seven-point scale with seven being the best. The overall ratings for the terminals range between 5.6 for the CAT and 6.67 for the WYBET terminals. Resistance to snow load ranges between four and a half and six. The probability of above ground repair varies between sixty and seventy percent. Fifty percent of the agencies indicate that winter replacement of the terminals is possible for all terminal types as long as the posts are sleeved. In the opinion of the DOTs, there seem to be little difference between the different terminal types over many of the criteria. These findings appear discouraging because the guardrail end-terminal vendors and state DOTs possess limited knowledge of the winter replacement of GETs.

Table 2. Northern Tier State DOT Survey Results

Terminal Type	Sample Size	Overall Rating (1-7)	Resistance to Snow Load (1-7)	Prob. of Above Ground Repair (0-100%)
WYBET	3	6.67	5.00	60.00
SRT	7	6.14	5.00	61.43
FLEET	8	6.13	5.25	62.50
CAT	5	5.60	5.60	64.00
REGION	3	5.67	5.33	63.33
SKT	1	6.00	6.00	60.00
BREAKMASTER	1	6.00	6.00	70.00

Evaluation. To perform a quantitative evaluation, each attribute must be rescaled on a zero to one scale because the attributes require a common scale before they can be added together. After summing all of the attributes together, a final score can be calculated and used to rank-order the alternatives. In this evaluation, all of the attributes receive equal weighting (but this does not have to be the case), and Table 3 presents the evaluation results for all of the considered terminals. To simplify the presentation, the user score combines the following attributes, overall rating, resistance to snow load, and probability of above ground repair, and the vendor score represents these attributes, steel post available, cost and percentage of terminal posts sleeved. For this analysis, the user score for the ET-2000 is given by an average of all user values because no external agencies using it responded to this study. The overall ordinal ranking is adjusted to range between zero and one with one representing the greatest preference. The final adjusted values are sorted from the highest to lowest values.

Table 3. Guardrail End Terminal Evaluation

Vendor	Terminal Type	Vendor Score	User Score	Overall Score	Adjusted Rank
Universal Industrial Sales, Utah	SKT	2.89	2.27	5.16	1.00
Bryson Products, Mississippi	SKT	2.80	2.27	5.07	0.97
Road Systems	SKT	2.70	2.27	4.97	0.93
Trinity Industries	ET 2000	2.80	2.14[1]	4.94	0.92
Universal Industrial Sales, Utah	FLEAT	2.28	2.19	4.47	0.75
R.G. Steel Corp., Pennsylvania	SKT 350	2.20	2.27	4.47	0.75
Trinity Industries	SRT 350	2.28	2.14	4.42	0.73
Gregory Highway Products, Ohio	FLEAT	2.19	2.19	4.37	0.72
R.G. Steel Corp., Pennsylvania	FLEAT 350	2.14	2.19	4.32	0.70
Road Systems	FLEAT	2.14	2.19	4.32	0.70
Bryson Products, Mississippi	FLEAT	2.09	2.19	4.27	0.68
Highway Safety Corp., Ohio	FLEAT 350	1.85	2.19	4.04	0.60
Highway Safety Corp., Ohio	SKT 350	1.70	2.27	3.97	0.57
Bryson Products, Mississippi	BREAK...	1.15	2.37	3.52	0.41
Trinity Industries	CAT	1.15	2.17	3.32	0.34
Universal Industrial Sales, Utah	REGION	1.18	2.13	3.32	0.34
R.G. Steel Corp., Pennsylvania	REGION	1.18	2.13	3.32	0.34
Bryson Products, Mississippi	REGIONT	0.98	2.13	3.12	0.27
Gregory Highway Products, Ohio	SKT	0.70	2.27	2.97	0.21
Universal Industrial Sales, Utah	BEAT	0.50	1.94	2.44	0.03
Universal Industrial Sales, Utah	BREAK...	0.00	2.37	2.37	0.00

1 –This value is given by an average of all user values because no external agencies using it responded to this study.

The SKT emerges as the highest rated terminal followed by the ET-2000. These terminals differentiate themselves from the rest of the alternatives because they have the option of having every post in the terminal sleeved, which improves winter reparability. Both of these terminals suffer from a lack definitive results from the user survey; however, the limited variability between terminal types in the user survey minimizes any concerns that may result. Selecting between these two terminal types poses a more significant difficulty because both of their specifications seem to meet an ideal design for high snowfall regions where steel posts are used with soil tubes and the aforementioned lack of variability among the user survey results limits its effectiveness in differentiating amongst alternatives. Most of the other alternative end terminals can use steel posts and some of their posts are sleeved, which makes them suitable for use in high snowfall regions; however, their ranking does not match the top two alternatives. The REGION, CAT and BEAT terminals should not be considered for use in high snowfall regions because they only use wooden posts, which can not be driven into frozen ground. The BREAKMASTER also does not appear appropriate for high snowfall regions because it utilizes cemented anchors. Both the SKT and ET-2000 represent the highest ranked candidate terminals for use in high snowfall areas.

Critical Analysis. This study's limited sample size with respect to the user survey may degrade the final results; however, within each DOT, the number of respondents that seem qualified to address the larger picture appears small. In this respect, the data set may be adequate; however, the large variation that exists between some agencies reduces the confidence in the final results. If a significantly larger number of qualified respondents can be identified, a larger more in depth survey may yield more reliable results.

For many of the attributes, most of the alternatives receive similar scores. This lack of differentiation amongst alternatives implies either a lack of knowledge or similar performance. Throughout both surveys, this project measures many attributes over a seven-point scale. The use of the seven-point scale is widely accepted; however, it still carries the inherent difficulty that the scale is not standardized across different respondents. For example, one respondent may believe that the overall maintainability of a given terminal is a five while another respondent from the same agency may believe that its maintainability is only a three. The small sample sizes coupled with these inherent difficulties and lack of variability amongst the user responses degrades the effectiveness of the user survey in differentiating amongst the alternatives.

Conclusions

Study Results. All end terminals remain susceptible to damage due to snow removal activities. This paper recommends some changes in winter maintenance strategies and identifies two candidate terminals for use in high snowfall regions. Effective marking and training seem to be the best strategies for reducing damage to GETs, which results from snow removal activities. The snowplow operators must learn to respect the new terminals more and provide them with a wide berth. Marking both the end of the terminal as well as the beginning of the flare of the terminal seems critical for helping the operators avoid contact with the terminal. With both locations marked, an operator can carefully avoid the entire length of the terminal. An education program, such as a video that emphasizes the damage and its cost, may also assist in reducing the number of terminals damaged due to snow removal equipment contact.

When a GET is inevitably damaged during the winter, the maintenance crew needs to be able to repair or replace the terminal above ground (i.e. no excavation of frozen soil). The findings from this study support the concept of the end terminals being repaired or replaced during winter months. The vendors unanimously advocate the use of steel posts with soil tubes for facilitating winter replacements. The agencies do not seem to require the use of steel posts, just soil tubes; however, the vendors' support for steel posts is difficult to ignore. This study believes that all terminals that do not have a steel post option should not be used in high snowfall regions. As a result, the REGION, CAT and BEAT terminals should not be considered for use in high snowfall regions. The BREAKMASTER does not appear appropriate because it utilizes cemented anchors. Additionally, all of these terminals fell at the bottom of the rank ordering of alternatives. Based primarily on their use of steel posts and soil tubes for every terminal post, the SKT and ET-2000 represent the highest ranked candidate terminals for use in high snowfall areas. The evaluation process shows little differentiation exists between these terminals and both should be considered as the recommended terminals.

Future Research. Verifying the existence on snow-only failure of guardrail end terminals represents a significant area for future research. If the new terminals are failing due to snow-only contact, the FHWA may need to reevaluate their design criteria. With the limited differentiation between the SKT and ET-2000, a demonstration program may prove the most effective approach for identifying a single terminal for preferred usage. The demonstration program can supplement or supercede the data that this study identifies. This new evaluation needs to investigate installation procedures, costs, accident-related data, snow load survivability and maintenance for each selected system over at least two winters. While a more aggressive and detailed survey effort may prove beneficial, the researchers believe that further data collection may be unlikely to provide further insight because detailed knowledge about end terminals seems lacking. Furthermore, knowledge seems lacking specifically about winter performance and above ground repairs. A thorough investigation into above ground repairs may also prove beneficial because this represents one of the most critical attributes under investigation and the results from the user survey show limited differentiation amongst alternatives.

Acknowledgement

The research underlying this paper has been sponsored by the Alaska Department of Transportation and Public Facilities (ADOT&PF). Support of Billy G. Connor, Research Manager, Clint Adler, Research Engineer, ADOT&PF is gratefully acknowledged. The contents of this report reflect the views of the authors who are responsible for the facts and the accuracy of the data presented herein.

References

Cichowski, W. G., *et al.* (1961). "Appraisal of Guardrail Installations by Car Impact and Laboratory Tests." Proceedings of the fortieth annual meeting, Highway Research Board, 40, 137 – 178.

Denman, O. S. and Welch, J. B. (1999). "Development of a Flared End Terminal That Meets Criteria of NCHRP Report 350: the REGENT System." *Transportation Research Record*, TRB, 1690, 17-30.

Mak, K. K., Ross, H. E., Bligh, R. P., and Menges, W. L.. (1997). "NCHRP Report 350 Testing of W-Beam Slotted-Rail Terminal." *Transportation Research Record*, TRB, 1599, 22 – 31.

Pfeiffer, B. G., and Sicking, D. L. (1994). "Development of a Metal Cutting Guardrail Terminal." Final Report to Interstate Steel Co., Report TRP-03-43-94. Midwest Roadside Safety Facility, University of Nebraska-Lincoln.

Ross, H. E., jr., Sicking, D.L., Zimmer, R. A. and Michie, J. D. (1993). *Recommended Procedures for the Safety Performance Evaluation Highway Features*. National Cooperative Highway Research Program, Report 350. Transportation Research Board, National Research Council, National Academy Press, Washington, D.C.

Sicking, D. L., A. B. Qureshy, and H. E. Ross. (1989). "Development of Guardrail Extruder Terminal." *Transportation Research Record*, TRB, 1233, 34 – 42.

Sicking, D. L., Reid, J. D., and Rohde, J. R. (1998a). "Development of Sequential Kinking Terminal for W-Beam Guardrails." *Transportation Research Record*, TRB, 1647, 89 – 96.

Sicking, D. L., Reid, J. D., and Rohde, J. R. (1998b). "Development of Flared Energy-Absorbing Terminal for W-Beam Guardrails." *Transportation Research Record*, TRB, 1690, 8 – 16.

Snyder, R. R. (1995). *In-Service Evaluation of the BRAKEMASTER AND C-A-T Crash Cushions*. Final Report prepared for U.S. Department of Transportation, Federal Highway Administration, Washington, D.C. Pennsylvania Department of Transportation, Harrisburg, Pennsylvania.

The Effect of Insulation on Road Surface icing

Henry Gustavsson[1]
Olli Ravaska[2]
Åke Hermansson[3]

Abstract

A presently ongoing Nordic project "Exclay Internordic Geoproject" aims at research into an overall utilization of expanded clay (exclay) in road embankments. One part of the project is concentrated on skid-resistance of exclay insulated and lightened roads, as insulation close to the road surface may prevent heat transfer from the underlying embankment and cause icing on the road. For this purpose two test sites were constructed with exclay insulation at different depths. This paper discusses temperature, dewpoint, skid-resistance measurements and temperature modelling at the Tuupakka test site and the applicability of the results to the purposes of the project.

Introduction

One way to prevent frost damage to road pavements is by means of the embankment insulation. The optimal location of the insulation for frost protection purposes is at the top of the embankment just below the AC layer. For reparation and reconstruction works, this location is also the most cost effective, due to the reduced need for excavation and filling. It has been found, however, that such a location of the insulation has a harmful disadvantage: under certain circumstances, icing can unexpectedly be formed on the road, causing severe problems for traffic.

In Sweden polystyrene insulations in road construction were examined in 70's and 80's (Gandahl, 1981). It was observed that temperatures on an insulated road surface were lower and the surface was more exposed to icing than having a conventional, mineral road embankment.

----

[1]Research Eng., Helsinki University of Technology, Dept. of Civ. and Env. Eng., P.O. Box 2100, FIN-02015 HUT, Finland; henry.gustavsson@hut.fi
[2]Professor, Helsinki University of Technology, Dept. of Civ. and Env. Eng., P.O. Box 2100, FIN-02015 HUT, Finland; olli.ravaska@hut.fi
[3]Researcher, Swedish National Road and Transport Research Institute, VTI, Highway Engineering, SE-581 95 Linköping, Sweden; ake.hermansson@vti.se

A construction having a 500 mm covering layer above the insulation was exposed to icing for 4.5 % longer time than a non-insulated surface. The corresponding time for a structure with an insulation directly below the AC-layer was 30% longer. Based on these results it was decided that at least 500 mm covering should be used above the insulation in Sweden. In Finland the road authorities ended up to a required thickness of 700 mm for the covering; at the same time in Norway only 300 mm was required. In these studies it was found out that also other material properties affect the surface temperatures: a non-insulated road structure with very coarse material (like crushed stone) below the pavement surface, can be even more susceptible to icing than a road with polystyrene insulation located at a depth of 350 mm. If the thickness of the covering could be reduced without increasing the risk of icing, the use of various insulating materials could be increased.

Air temperature and road surface temperatures may differ greatly from each other. The temperature on the road surface in Scandinavia may be many degrees lower than that in the air in the winter (Kilpeläinen and Ravaska, 1998), with the reverse being true in the summer. In the winter, the relative humidity of the air in Southern Finland is near 90 % on average, and the dewpoint is often 0 to 4 °C lower than the air temperature. When the surface temperature under these circumstances drops below the dewpoint and below the freezing point, vapour in the air condensates and hoar frost is formed on the road surface. This is particularly harmful during the first cold nights of the autumn. The ground is still warm, but the insulation at the top of the road embankment prevents the heat to flow towards the pavement surface and thus the potential of icing exists. Together with the use of summer tires and inexperience with ice on the road after the summer season, this makes driving dangerous. The areas where an insulated zone begins immediately after a non-insulated one may be extremely dangerous.

A few years ago, the Exclay Internordic Geoproject called "LWA-geolight" was initiated by Optiroc Group aiming at research into an overall utilization of expanded clay (termed as exclay in the following) in road embankments. Two test sites have been constructed: one in Helsinki, Finland (Forsman, Valtonen, Gustavsson, 2000) and one in Trondheim, Norway. The Finnish part of the project is concentrated on skidding caused by exclay used as insulation and lightening. The Helsinki University of Technology is involved in the project, performing skid-resistance measurements on the test roads as well as heat transfer analyses and temperature measurements in order to gain a better understanding of the physical behaviour of the whole road structure. This section of the project started at the end of 1999 by means of friction measurements and computer modelling.

Mechanical and thermal properties of exclay

Expanded clay (exclay) is manufactured by burning of granulised clay at high temperatures in an oven while adding air during the process. Exclay can be used either for frost insulation purpose in frost susceptible soils or for lightening of road embankments on soft soils to prevent settlement. The material is often called

LWA (Light Weight Aggregate) or LECA (Lightweight Expanded Clay Aggregate) characterizing its low unit weight. Heat conductivity of the material is normally 0.15 to 0.20 W/mK and dry unit weight after compaction 3 to 4 kN/m^3. The angle of friction varies from 35° to 40° and the resilient stiffness 100 to 200 MPa, depending on the stress level and the grain size distribution. It can be used as natural granular materials (gravel) in road construction, if the stress level is kept low enough to prevent crushing of the grains. It is provisionally recommended to keep the vertical stress level below 100 kPa. This can usually be ensured in a paved road by using a covering of at least 300mm above the material. More detailed information of the properties can be found in a SINTEF report (Furuberg et al. 2000).

Test site and structures

The four test structures discussed here are located in a bus ramp of the Tuupakka junction of the Helsinki Ring Road III (Valtonen, Forsman, 2001). During the summer of the year 2000, test structures were constructed near Helsinki at locations in which expanded clay was used as lightening in transfer zones together with a piled slab, soil replacement and deep mixing. The subgrade at the site was soft clay with a depth of 5 - 10 m. Four of the test structures were equipped with temperature sensors from which the data has been transferred to researcher´s computer by means of a wireless GSM-data transfer system. The normative grain size of the exclay was 10 to 20 mm and layer thickness of 0.7 to 1.45 m were used. The base course materials were crushed rock and reclaimed concrete. The asphalt concrete layer was 100 mm thick. The test structures were opened for normal traffic in the beginning of October 2000.

Temperature measurements and climatic data

Air and road surface temperature measurements have been performed since October 2000 at the Tuupakka test site. The relative humidity and solar radiation measurements were taken at the Helsinki airport meteorological stations. The dewpoint can be calculated from relative humidity and air temperature. At the Helsinki airport there are also different kinds of sensors that indicate the condition of the road surface (snow, ice, hoar frost).

Table 1 shows the monthly averages for air and surface temperatures in Tuupakka for a reference structure without insulation and for exclay (denoted as EC in the following) insulated structures with a covering layer of 700 mm, 500 mm and 350 mm respectively. As the autumn 2000 was exceptionally warm, results here start on December. Temperatures did not drop substantially below zero before late December. As expected, it can be seen from Table 1 that the lowest average surface temperature in the winter 2000-2001 was measured in the exclay structure with a covering of 350 mm. It was not expected, however, that the temperature differences would be that small. Also surprisingly, the average surface temperatures of the other exclay structures (EC+500 mm and EC+700 mm) were higher than those of the reference structure.

Table 1. Monthly average temperatures at the Tuupakka test site.

Month	Surface temperature		(°C)		Air temp, (°C)
	Ref.	EC+700	EC+500	EC+350	
December 00	1.68	1.81	1.75	1.47	1.47
January 01	-1.63	-1.42	-1.50	-1.72	-1.46
February 01	-6.44	-6.28	-6.32	-6.71	-7.10
March 01	-1.20	-0.98	-1.03	-1.36	-2.68
Average	**-1.90**	**-1.72**	**-1.77**	**-2.08**	**-2.44**

Daily variation of the temperatures as well as the dewpoint is displayed in Fig. 1 in which hourly values of five days in January are plotted. The figure clearly presents the same situation as discussed above.

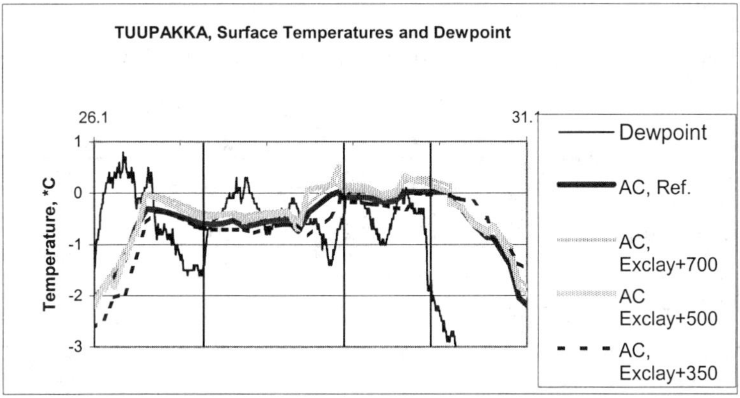

Figure 1. Temperature and dewpoint variation during 5 days in January. The vertical lines show the time when friction measurements were done.

Because the Helsinki airport is located only about 5 km from the Tuupakka test site, the airport meteorological station observation data of the relative humidity and dewpoint measurements could be used to estimate the risk of icing in Tuupakka (see Table 2). The average air temperature seems to be a little lower at the airport (0.80 °C) and the AC-surface temperature at the airport is also lower than at the reference structure at Tuupakka (0.66 °C). The difference between the air temperature and the dewpoint was 1.88 °C at the airport. Table 2 also presents the calculated dewpoint difference (2.68 °C) at Tuupakka, using the Tuupakka air temperature and the airport dewpoint temperature.

Table 2. Average monthly temperatures, relative humidities, dewpoints and dewpoint differences in Helsinki Airport and Tuupakka.

Helsinki Airport						Tuupakka
Month	**Air temp °C**	**AC-surf °C**	**Rh %**	**DewPoint °C**	**DDPHki °C**	**DDPTuup °C**
Dec 00	0.99	0.78	91.57	-0.11	1.10	1.58
Jan 01	-2.15	-2.37	92.54	-3.19	1.05	1.74
Feb 01	-7.99	-7.33	85.19	-10.07	2.08	2.97
Mar 01	-3.80	-1.33	79.79	-7.11	3.31	4.42
Average	**-3.23**	**-2.56**	**87.27**	**-5.12**	**1.88**	**2.68**

The number of hours during which the surface temperature is below zero, is presented in Table 3.

Table 3. Number of hours for surface temperature below 0 °C at the Tuupakka test site.

	Surface temperature < 0 °C (h)			
Month	Ref.	EC+700	EC+500	EC+350
December 00	309	305	312	328
January 01	637	589	609	661
February 01	635	602	617	663
March 01	459	440	445	495
Sum (hours)	**2040**	**1936**	**1983**	**2147**

Table 4 presents the number of hours when the surface temperature is below zero and the dewpoint temperature is higher than surface temperature. Under these circumstances a risk of hoar frost and icing is assumed to exist.

Table 4. Number of hours for surface temperature < 0 °C and dewpoint temperature higher than surface temperature in Tuupakka.

	Surface temperature < 0 °C and Tsurf < DP (h)			
Month	Ref.	EC+700	EC+500	EC+350
December 00	31	28	40	49
January 01	134	112	125	175
February 01	168	146	142	173
March 01	54	58	66	90
Sum (h)	**387**	**344**	**373**	**487**

Table 4 shows that the surface temperature of EC+350 mm was exposed to icing/hoar frost for the longest time (surface temperature below zero and the dewpoint temperature higher than the surface temperature). The corresponding time for the reference structure was 100 hours less and for the other exclay structures (+500 mm and +700 mm) it was even less than for the reference structure. Table 3 also shows the same order: the surface temperature of EC+350 was below zero for the longest time, the reference structure was the next, EC+500 the third and the surface of EC+700 was below zero for the shortest time.

Temperature modelling

The surface temperatures of the test structures were calculated using a simulation model developed at the National Road and Transport Research Institute (VTI) in Sweden (Hermansson, 2000). The simulation model uses climatic data in the form of hourly values for incident short wave radiation, air temperature and wind velocity. The reflected portion of the incident short wave radiation is given by the albedo of the pavement surface. The remaining portion is absorbed by the pavement surface, which causes the surface temperature to rise. Long wave radiation and convection losses are also taken into account.

For the calculation of heat transfer, the ground is divided into cells, thinner near the surface and thicker at a deeper level, where temperature variation is smaller. Each cell is assigned values for temperature, porosity and degree of water saturation, of which only the temperature varies during the simulation. Thermal conductivities and thermal capacities are calculated according to Sundberg (1988), using information of water content, porosity, soil material and bulk density.

Each cell is given a temperature at the start of the calculation. The model then calculates a new temperature for each cell - several times for each simulated hour. This is done in accordance with a generally accepted heat transfer theory, i.e. heat flows from warmer to colder cells in relation to temperature difference and thermal conductivity. The temperature change depends in its turn on the amount of energy received and the thermal capacity of the material in the cell. During the calculations, the exchange of energy of the uppermost cell through the pavement surface in the form of radiation and convection is also handled.

The cell division is made down to a depth of five meters below the surface where the temperature is supposed to be constant, at a level of the annual mean air temperature.

Input data to the simulation

For the simulation, usually hourly values for solar radiation, air temperature and wind speed are used. For this study, the air temperature was monitored every 3 hour right at the test section. During 3 a.m. to 6 a. m. the air temperature was measured every 1.5 hour. Horizontal solar radiation was measured hourly at the Helsinki airport located 5 km north east of the test section. Wind speed is very essential for pavement surface temperature during sunny summer days when the surface temperature differs much from the air temperature. This is not the case during winter and as no wind speed measurements were available, the simulation was performed assuming the weather to be calm.

Comparison between measured and calculated temperatures

The calculated temperatures are presented in Fig. 2. For comparison, measured temperatures at a depth of 20 mm also appear in the same figure. As one can see, the agreement between the measured and the calculated temperatures is quite good.

The number of hours when the road surface temperature is below 0 °C was calculated from the simulation results of a period from October 6th to March 18th. The results of the simulation and observations are given in Table 5. The simulation gives a little shorter time for the road surface to be below 0°C than the observations. The difference between the calculated and observed time is about 5% for the reference structure and 8% for the insulated road.

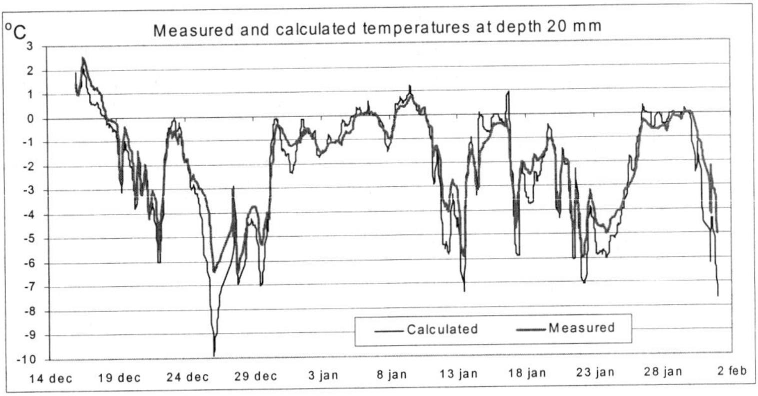

Figure 2. Calculated and measured pavement temperatures for exclay+350mm structure in December and January at a depth of 20 mm below the road surface.

Table 5. Simulated and observed number of hours for surface temperature < 0 °C from 6th October to 18th March.

Structure:	Reference		Exclay+350 mm	
	Simulated	Observed	Simulated	Observed
Time for $T_{surf} \leq$ 0 °C (h)	1793	1887	1842	2009

Figure 3 shows the results of the simulation at the beginning of January. During the first ten days the road surface temperature is lower than that in the air. When the air temperature falls, the surface temperatures do not reach as low values immediately. The surface temperature of the insulated road seems to be somewhat lower than the temperature at the reference structure as it was expected from the observations.

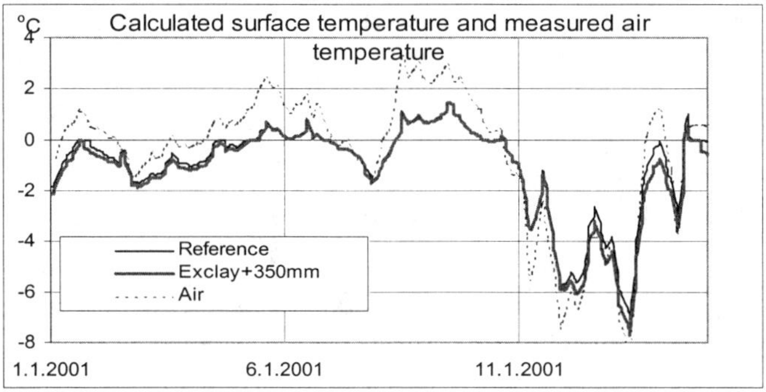

Figure 3. Air temperatures and simulated road surface temperatures for reference structure and exclay +350 mm.

Skid-resistance measurements

The research programme also included skid-resistance measurements on the test structures, in order to verify the results in practice. These measurements were made using a portable friction tester (PFT) owned by The Swedish National Road and Transport Research Institute and a C-Trip friction meter installed in a passenger car (Levomäki,Valtonen, 2001).

The PTF equipment is a hand operated friction meter incorporating three wheels, one of which measures surface friction. The weight of the device is 38 kg and the operating speed is 0.5 m/s. The device provides an average friction based on eight measurements respective to every 104 mm. The C-Trip measurement unit is comprised of an electronic device installed in a car, which measures the friction caused by deceleration during braking. Correct measurement provides that the breaks are not locked, and an unlocked ABS break system in a car facilitates the measurements. In addition, the condition of the tires plays an important role in the accuracy of the measurements. A comparison study of five friction meters, including the PTF and C-Trip friction meter, was carried out in Finnish Lapland in December 2000 (Malmivuo, 2001).

The average skid-resistance measurement results obtained from the left and right tracks are presented in Figs. 4 and 5 (Levomäki, Valtonen, 2001). The results can be summarized as follows:

- PFT offers higher friction values than C-Trip, at high friction levels in particular

- highest and lowest friction values in both types of measurement are obtained at about the same time
- C-Trip results as measured at the same time vary more than PFT results for different structures

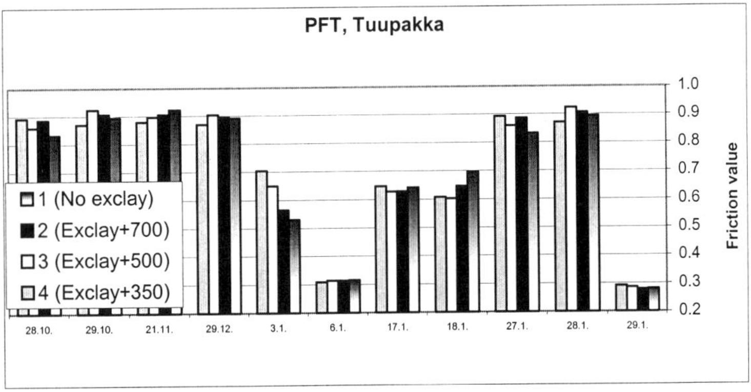

Figure 4. Average PFT results from Tuupakka (Levomäki, Valtonen, 2001).

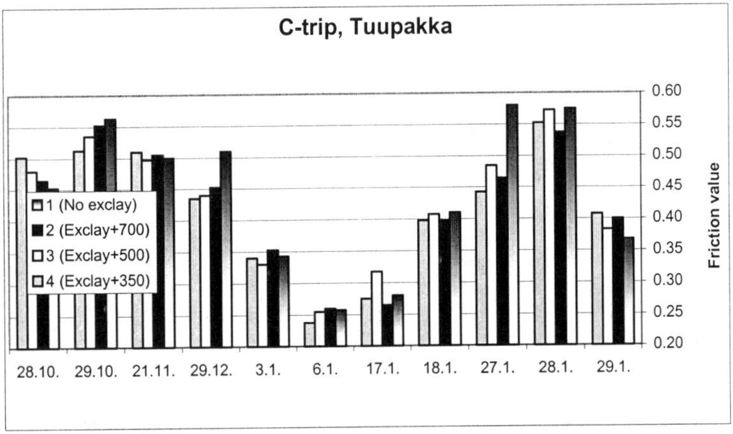

Figure 5. Average C-Trip results from Tuupakka (Levomäki, Valtonen, 2001).

Differences in the friction values, e.g. small values on the 6th and 17th of January, can be explained by the weather and ice conditions at the site during the measurements. On 6th Jan it was snowing and on 17th Jan the road surface was covered with frost and ice.

As a whole, no certain conclusions can be drawn in regard to the skid-resistance measurements and different road test structures. One explanation for this is road-salting. The test structures are located in places where salt is used during icy and snowy conditions.

Fig. 6 presents temperature and dew point measurements at the Helsinki airport during January 2001, also including the warnings and alarms for snow, icing and hoar frost (measured and calculated) with respect to salting. The figure also presents the Tuupakka test road temperature measurement results in addition to the exact skid-resistance measurement times, shown by vertical lines through the figure.

The intention was to measure the skid-resistance in accordance with the alarm system described above. Fig. 6 demonstrates, however, that no measurements were performed at the times when icing or hoar frost would have been expected. The dewpoint during each measurement occasion was below the road surface temperature. It is also worthy of note that only one measurement was performed during the period most important for the purposes of this study, i.e. December 2000, during which the frost period started and the most dangerous icings on the roads were supposed to be formed.

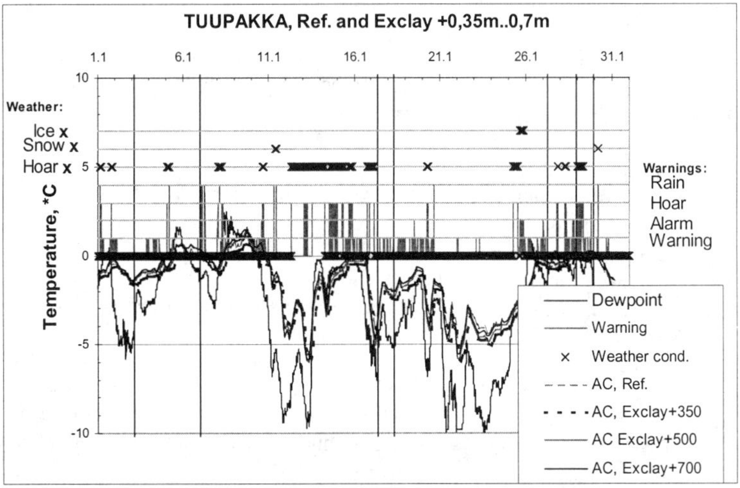

Figure 6. Temperatures, dewpoints, warnings and alarms in Tuupakka test road and Helsinki airport in January 2001. Vertical lines through the figure denote the friction measurement times and short (0 to +5) vertical lines the weather station alarms and warnings.

Conclusions

The aim of the present study was to clarify how the location of the exclay insulation in a road embankment affects icing and skid-resistance on a road surface. Temperature and skid-resistance measurements were performed on a test road having insulation at the depths of 350 mm, 500 mm and 700 mm respectively. The results of the temperature measurements can be summarized as follows:

- comparing the different test structures, differences in the surface temperatures were relatively small - also including the reference structure with no insulation
- lowest average surface temperature in the winter 2000-2001 was measured in the exclay structure with a covering of 350 mm, as it was expected
- on the test structures with covering thicknesses of 500 mm and 700 mm even higher surface temperatures than on the reference structure were measured, which was not expected

The skid-resistance measurements did not substantiate this study mainly because:

- salting was used on the test roads
- they were not concentrated on the autumn because of an exceptionally warm weather
- timing for measuring icing and hoar frost was not successful

On the basis of this study the layer thickness above an exclay insulation could be 500 mm without any substantial change in the thermal regime of the embankment. The results of this preliminary study are not, however, very convincing and therefore no final conclusions may be drawn unless a substantial amount of additional research is completed. Lessons have been learned, however, from this study and the future measurements can be planned and implemented in a more successful way.

The research will continue during 2002 financed mainly by Optiroc Group, Nordic Industrial Fund and in Finland also the Finnish Road Administration and Kaitos Ltd. In the winter 2001-2002 temperature and friction measurements will be made in Finland (Tuupakka) in co-operation with SCC-Viatek Ltd. and Helsinki University of Technology. In Norway (Trondheim) temperature and friction measurement using PFT will be made by SINTEF (The Foundation for Scientific and Industrial Research at the Norwegian Institute of Technology). Also the modelling of surface temperatures will be continued at VTI (Swedish National Road and Transport Research Institute).

References

Forsman, J. Valtonen, J., Gustavsson H. (2000). *Exclay Internordic Geoproject. Finnish Activities 1999-2000.* Measuring of Friction, Temperature Modelling, Test Structures, Report II. SCC-Viatek Ltd., Espoo. 62 p.

Furuberg, T., Hoff, I., Solheim, O.M. (2000). Environmental friendly insulation products for the construction and building business. Summary report DP1: Leca frost protection in roads, railways ditches. *SINTEF report STF22 F00601.*

Gandahl, R. (1981). Plastic Foam Insulation of Roads. *Frost resistance capacity, partial insulation and frost heaving, special transitions, icing and economy.* No. 214A. National Road and Transport Research Institute (VTI), Linköping, Sweden.

Hermansson, Å. (2000). Simulation Model for Calculating Pavement Temperatures Including Maximum Temperature. *Transportation Research Record 1699*, paper No. 00-131, TRB, National Research Council, Washington D.C. 2000, pp 134-141.

Kilpeläinen, M., Ravaska, O. (1998). Modelling of Road Surface Temperatures in Winter. *Proc. of IX Int. Conf. on Cold Regions Engineering, Duluth , Minnesota, 1998, Cold Regions Impact on Civil Works*, ASCE. pp. 304-316.

Levomäki, M., Valtonen, J. (2001). *Exclay Internordic Project, Finnish Activities 2000-2001*, Report III: Measuring of Friction. Helsinki University of Technology, Laboratoty of Highway Engineering.

Malmivuo, M. (2001). Kitkamittauslaitteiden vertailututkimus 2000. *Tiehallinnon selvityksiä 6/2001 (In Finnish)*, Finnish Road Administration, FINNRA, Helsinki, ISBN 951-726-723-1.

Sundberg, J. (1988). Thermal Properties of Soils and Rocks. *SGI Report No. 35*, SGI, Linköping

Valtonen, J., Forsman, J. (2001). Exclay Internordic Geoproject (LWAgeolight). "The depth of an insulation layer". *The VII International Conference on Durable and Safe Road Pavements. Kielce 8.-9.5.2001.*

Traction Performance of Transit and Paratransit Vehicles on Ice and Snow Surfaces

Lutfi Raad[1], John Lu[2], and Leroy Hulsey[3]

Abstract

In this study, the traction performance of transit and paratransit vehicles was evaluated on snow and ice surfaces. Field tests were conducted in Fairbanks, Alaska, using three types of vehicles: 1) a 41-passenger transit bus; 2) a 32-passenger transit bus, and 3) a 9-passenger paratransit vehicle. Each vehicle was tested for different combinations of tire types that included highway tires, snow tires, studded-siped tires, highway 3-rib tires, all-season tires, and snow siped tires. Winter traction performance was evaluated using a number of tests that included stopping distance, starting traction, hill-climbing, cornering, and controllability. Results indicate that winter traction performance is significantly influenced by vehicle type, tire combination used, and the road surface condition (compacted snow or ice). Research findings and recommendations for appropriate tire combinations that are most suitable for winter traction are presented.

Introduction

Winter traction performance of transit and paratransit vehicles has a direct impact on their safe operations during winter. In cold regions, transit and paratransit vehicle drivers suffer difficulties when driving on snow or ice surfaces, especially when starting, stopping, cornering, and hill climbing. The main reason for such difficulties is the reduced coefficient of friction between tire and road surface. The direct consequences are increased stopping distances, reduced acceleration ability (i.e. reduced starting traction), reduced maximum cornering speed (to avoid loss of control during cornering), decreased hill climbing ability, and loss of control with immediate stopping.

[1]Professor, Department of Civil and Environmental Engineering, Transportation Research Center, University of Alaska Fairbanks, Fairbanks, Alaska 99772
[2]Assistant Professor, Department of Civil Engineering, University of South Florida, Tampa, Florida 33620
[3]Associate Professor, Department of Civil and Environmental Engineering, Transportation Research Center, University of Alaska Fairbanks, Fairbanks, Alaska 99770.

While the driver's behavior and the roadway surface condition can be treated as given, transit and paratransit vehicle characteristics and tire characteristics are the major factors contributing to traction performance on snow and ice surfaces. In past years, research studies have been conducted to evaluate vehicle tire traction performance on snow and ice surfaces. Studded tires have been used during winter to increase traction for safety reasons. However, recent studies conducted by Lu et al. (1994) and Lu (1995) show that studded tires may not enhance traction performance as generally perceived. There is also plenty of evidence that they cause roadways to wear out faster, they increase airborne dust on and around roads, and they increase environmental noise. These results have been presented by researchers in the United States (Christman and Sime 1978, Lu et al. 1994, Lu 1995), Canada (Smith and Schonfeld 1970), Japan (Konagai et al. 1993), Sweden (Nordstrom and Samuelsson 1991), Germany (Hiersche et al. 1974), France (Requirand 1976) and other countries. Most of these studies focussed on comparing different tires. The test vehicles used in these studies were passenger cars, or small -sized vehicles, such as pick-up. No past study has evaluated the winter traction performance of transit and paratransit vehicles (mostly large-sized and middle-sized vehicles). There are not clear indicators for public transportation system managers to use in selecting the right tires and vehicles for winter operations. Although many managers prefer studded tires for winter operations, increased roadway maintenance cost associated with them has limited the use of studded tires, especially for large-sized vehicles. In addition, many non-studded winter tires have been developed in recent years.

Field Tests

Field tests were conducted in Fairbanks where the yearly average temperature is -0.9°C (30.4°F) and the yearly average snowfall is 2.88 m (113.4 inches). The average winter season lasts about 8 months. All tests were performed during the months of February and March and each test consisted of at least 3 to 6 trials the average of which was reported. In addition, a G-Analyst force meter was used to measure the maximum acceleration (G) incurred during each test. Testing on compacted snow and icy surface was done in the parking lot of the Fairbanks North Star Borough Department of Transportation and on the surface of a gravel pit off Davis Road in Fairbanks, respectively. The snow on the surface of the gravel pit was ploughed prior to each test, when necessary, in order to expose the underlying icy surface. In this regard, the icy surface was not glared but was relatively rough.

Three types of vehicles were used in these tests: 1) a 41-passenger transit bus (Gillig Phantom model 1981, 10.7 m long, rear wheel drive, all-wheel air brakes, 15,436 kg weight); 2) a 32-passenger transit bus (Gillig Phantom model 1994, 9.1 m long, rear wheel drive, all-wheel air brakes, and 11,386 kg weight), and 3) a 9-passenger paratransit vehicle (Ford E-350 model 1994, 6.1 m long, rear wheel drive, all-wheel hydraulic brakes, 4,639 kg weight). Each vehicle was tested for different combinations of tire types that included highway tires, snow tires, studded-siped tires, highway 3-rib tires, all-season tires, and snow-siped tires. All tire types and combinations were selected by the Fairbanks North Star Borough Department of Transportation. The characteristics of these tires were as follows:

Van Tires

Type	Size	Load Range	Brand
Highway	245/75R16	E	Goodyear Radial
Snow	245/75R16	E	Goodyear Workhorse
Studded-Siped Snow	245/75R16	E	Goodyear Workhorse (Extra Grip)

Bus Tires

Type	Size	Load Range	Brand
Highway	11-22.5	H	Goodyear (Unisteel II)
Highway 3-rib	11-22.5	G	Goodyear G-286
All-Season	11-22.5	H	Goodyear G-159
Snow-Siped	11-22.5	G	Goodyear G-362

The tire combinations used for each vehicle include:

Van

Combination Type	Steer Axle	Drive Axle
(a)	Highway	Highway
(b)	Highway	Snow
(c)	Studded-Siped-Snow	Studded-Siped-Snow

9.1 m Bus

Combination Type	Steer Axle	Drive Axle
(a)	Highway 3-Rib	Highway
(b)	Highway	All-Season
(c)	3-rib-Siped	Snow-Siped

10.7 m Bus

Combination Type	Steer Axle	Drive Axle
(a)	Highway	Highway
(b)	Highway	Highway 3 rib
(c)	3-rib-Siped	Snow-Siped

The type of tests and corresponding parameters used are summarized below:

1. Stopping distance test: stopping distance, and maximum G after the vehicle has reached 40.2 km/hr (25 mph) and the brakes were fully applied.
2. Starting traction test: time to reach a speed of 40.2 km/hr (25 mph) under full engine power starting from zero velocity, and the corresponding maximum G.
3. Cornering test: limiting speed before lateral skidding occurred, and maximum G,
4. Hill climbing test: maximum hill climbing grades.
5. Controllability test: maximum lateral displacement and rotation upon hard braking.

Details on the type of vehicles and tires used, and the relationships between maximum measured G, coefficient of friction, maximum hill climbing grade, and limiting cornering speeds are presented elsewhere (Lu et al. 1994, Lu 1995). The governing relationships used in this study are summarized below.

For stopping distance test, and also for the starting traction test, the maximum measured G (i.e. Gmax) and the average friction coefficient μ between the tires and the pavement surface are related by:

$$\mu = G_{max} \qquad (1)$$

For cornering test,

$$\text{Maximum Cornering Speed (km/hr)} = (\text{Radius in meters} \times G_{max} \times 127.4)^{0.5} \qquad (2)$$

For hill climbing test, the maximum hill climbing slope β_{max} is given by:

$$G_{max} = \tan(\beta_{max}) \qquad (3)$$

Results

Results are summarized in Tables 1 – 4. These results indicate the following:

1. It is evident that for similar tire combination and surface conditions, the tested transit and paratransit vehicles had different traction performance.

2. On snow surfaces, the van exhibited shorter stopping distances (17 m to 25 m) than either the 9.1 m bus (36 m to 44 m) or the 10.7 m bus (34 m to 36 m). The starting traction time of the van was also smaller (i.e. better traction) and ranged from 6 sec to 9 sec in comparison with 10 sec to 12 sec for the larger transit buses. For cornering tests, there was no significant difference in cornering speed; with the van showing a slight edge (29 km/hr to 34 km/hr). Similarly, the maximum G values exhibited by the van were essentially greater than those for the buses. However, the trend in variation of maximum G values for different tests was similar for the van and the transit buses. The van also had better hill climbing ability than the 9.1 m bus. The smallest hill climbing ability was for the 10.7 m-bus with maximum grade of about 15%.

3. On icy surfaces, the van had shorter stopping distances (17 m to 47 m) than the transit buses (57 m to 75 m). The van also exhibited less starting traction time of 5.9 sec (except for tire combination a) in comparison with the buses traction time (10 sec to 17 sec). The 10.7 m bus exhibited the worst traction time performance (13 sec to 17 sec). Differences in cornering speeds for the van and the buses were not statistically significant. The maximum G values were in general greater for the van (except for tire combination a) than the transit buses.

4. The use of winter tire conditions (studded-siped tire combination) for the van was most influential in improving stopping distance traction and starting traction on icy surfaces. Not so in case of larger transit buses, where the siped-tire combination exhibited less stopping distance and starting traction time compared to other tire combinations used. It is interesting to note in this case that the maximum climbing grade on icy surface for a van with studded-siped tires was 27% in comparison with 14%-17% range for transit buses with siped tires combination. Maximum highway grades, however, do not usually exceed 10%.

5. Highway tire combinations when used in the van had in general, better traction performance relative to other tire combinations than when used in transit buses. For example, on icy surfaces, the highway tires combination in the 9.1m -bus had a stopping distance of 60 m compared with 75 m for the siped tires combination. The differences in traction performance could be attributed to type of tire, vehicle weight, vehicle geometry, engine power, braking efficiency, among other variables.

Table1 Summary of Traction Tests Results for the Van

Tire Combination	Surface	Stopping Distance		Starting Traction		Cornering Test		Hill Climbing (max. grade, %)
		Distance (m)	G_{max}	Time (sec)	G_{max}	Speed (km/hr)	G_{max}	
a	snow	25.6	0.29	9.2	0.15	28.9	0.37	15
b	snow	21.9	0.32	6.0	0.22	33.8	0.45	22
c	snow	17.4	0.38	6.0	0.26	32.2	0.43	26
a	ice	46.6	0.21	17.7	0.12	27.3	0.32	12
b	ice	26.2	0.24	5.9	0.22	32.2	0.41	22
c	ice	17.4	0.38	5.9	0.27	30.6	0.39	27

Table 2 Summary of Traction Tests Results for the 9.1 m Bus

Tire Combination	Surface	Stopping Distance		Starting Traction		Cornering Test		Hill Climbing (max. grade, %)
		Distance (m)	G_{max}	Time (sec)	G_{max}	Speed (km/hr)	G_{max}	
a	snow	43.9	0.22	11.3	0.20	27.3	0.29	20
b	snow	54.9	0.20	10.2	0.20	29.0	0.37	20
c	snow	36.3	0.27	10.3	0.20	30.6	0.39	20
a	ice	60.4	0.21	10.3	0.20	27.3	0.30	20
b	ice	57.0	0.22	10.0	0.21	30.6	0.37	21
c	ice	74.7	0.16	14.0	0.14	29.0	0.34	14

Table 3 Summary of Traction Tests Results for the 10.7 m Bus

Tire Combination	Surface	Stopping Distance		Starting Traction		Cornering Test		Hill Climbing (max. grade, %)
		Distance (m)	G_{max}	Time (sec)	G_{max}	Speed (km/hr)	G_{max}	
a	snow	35.1	0.21	11.6	0.16	32.2	0.42	16
b	snow	34.1	0.19	12.3	0.15	29.0	0.33	15
c	snow	36.6	0.21	12.5	0.15	32.2	0.40	15
a	ice	58.2	0.18	17.3	0.16	32.2	0.41	16
b	ice	57.3	0.18	15.6	0.16	30.6	0.38	16
c	ice	60.0	0.17	12.8	0.17	29.0	0.37	17

Table 4 Summary of Controllability Tests Results for all Vehicles

Tire Combination	Surface	Van		9.1 m Bus		10.7 m Bus	
		Displacement (m)	Rotation (deg.)	Displacement (m)	Rotation (deg.)	Displacement (m)	Rotation (deg.)
a	snow	0.53	0.4	0.18	3.2	1.4	0.4
b	snow	N/A	N/A	0	1.0	0.25	2.7
c	snow	N/A	N/A	1.6	2.8	0.10	0.4
a	ice	0.36	0.3	0.64	3.0	1.4	1.5
b	ice	N/A	N/A	1.2	1.1	0.25	2.4
c	ice	N/A	N/A	0.84	6.3	1.2	0.4

6. Available controllability data do not seem to indicate significant effect of vehicle size on lateral displacement and rotation. The data in this study are not complete, and more work is needed in this regard.

7. Although the evaluation of traction performance requires a variety of tests to assess vehicle-surface friction and response, the stopping distance traction should be given more weight in the selection of tire combination for a given vehicle for winter conditions. This assumes that the selected tire combination has "acceptable" traction performance for starting traction, cornering, and hill climbing. Judging by the results of this study, the following tire combinations are recommended for winter traction of transit and paratransit vehicles:

Type of Vehicle	Steering Axle	Drive Axle
Van	Studded-Siped-Snow	Studded-Siped-Snow
9.1 m -Bus	Highway	All-Season
10.7 m -Bus	Highway	Highway 3rib

Summary

In this study, winter traction performance has been investigated for transit and paratransit vehicles on snow and ice surfaces. Three types of vehicles were used: a 9-passenger van, a 9.1m (32 passenger) bus, and a 10.7 m (42 passenger) bus. A number of tire combinations with different traction aids has been investigated. The traction tests involved stopping distance, starting traction, cornering, hill climbing, and controllability. Results show that the van had in general better winter traction on snow and ice surfaces in comparison with the transit buses. The type of tire combination used had also significant effect on traction performance. Maximum G values were lowest for starting traction tests, followed by stopping distance tests and cornering tests. Controllability test data did not show significant effect of vehicle size on lateral displacement or rotation.

Acknowledgment

This project was funded by TransNow (USDOT Regional Center). The authors also wish to acknowledge the excellent help and support received from Fairbanks North Star Borough Department of Transportation (FNSBDT) who provided both vehicles and manpower during the course of the research work. Specifically, the help and feedback from Max C. Lyon, Jr., Transportation Director of FNSBDT, is greatly appreciated.

References

Christman, R., and Sime, J. M. Summary Report: Performance and Effects of Studded Tires in Connecticut. Report No. 343-3-78-5, Connecticut State Department of Transportation, Wethersfield, Connecticut, 1978.

Hiersche, E. U., Schmitz, H., and Pressler, G. Effect of Studded Tires on the Flow of Traffic. Strassenbau und Strassenverkehrstechnik No. 170, 1974, pp. 147.

Konagai, N., Assano, M., and Horita, N. Influence of Regulation of Studded Tire Use in Hokkaido, Japan. Transportation Research Record No. 1387, 1993, pp. 165 – 169.

Lu, J., Junge, D., and Esch, D. Winter Tire Traction Evaluation. Final Report, Report No. INE/TRC 94.03, Transportation Research Center, University of Alaska Fairbanks, 1994.

Lu, J. Studded Tire Performance and Safety. Final Report, Report No. INE/TRC 94.06, Transportation Research Center, University of Alaska Fairbanks, 1995.

Nordstrom, O., and Smuelsson, E. Road Grip of Winter Tyres on Ice. VTI Report 354, Swedish Road and Traffic Research Institute (VTI), 581 01 Linkoping, Sweden, 1991, pp. 15 – 28.

Requirand, R. Studded Tyres: 1975 Experiment, Tests on Car Circular Test Track. Bulletin de Liaison des Ponts et Chaussees, No. 82, 1976, pp. 57 – 68.

Smith, P., and Schonfeld, R. Pavement Wear Due to Studded Tires and the Economic Consequences in Ontario. Highway Research Record No. 331, 1970, pp. 54 – 78.

Performance of base/subbase materials under frost action

Vincent C. Janoo[1]

Abstract

Several Federal Aviation Administration (FAA) specifications for base and subbase materials were evaluated with respect to frost heave and thaw weakening. The study included both unstabilized and stabilized materials. Most of the unstabilized materials allowed up to 15% fines passing the no. 200 sieve, with the exception of the crushed aggregate base, which allowed no more than 8% fines, and the sand–clay base, which allowed up to 25% fines. The Corps of Engineers (COE) frost-susceptibility criterion was used to delineate the frost susceptibility, while the time to drain a base or subbase layer was used to determine the thaw-weakening potential. The subbase (P-154) and the crushed aggregate base (P-209) met the COE frost-susceptibility criterion, but none of the materials met the COE drainage requirements. However, if the fines contents of the base or subbase were reduced to less then 3%, they did meet the criterion. The resilient modulus of cement-treated soils remained at the level found before freezing, when the cement content was greater than 5%.

Introduction

Pavements in seasonal frost areas freeze and thaw annually, often undergoing frost heave and thaw weakening in addition to load-induced distress. As frost penetrates into the pavement structure, moisture is drawn to the freezing front by capillary action. Ice lenses are formed and material is displaced, usually vertically. The displacement is translated to the surface and is called frost heave. Thaw weakening takes place when the segregated ice lenses melt. Depending on the permeability of the subsurface layers, the excess moisture from the melting ice can become trapped between the surface course and the frozen layer below. When a wheel load is applied to the surface, the water is unable to dissipate. This causes an undrained loading condition and positive pore pressures are developed in the saturated layer. The increase in pore pressures reduces the effective stress that the layer can tolerate, thus reducing the bearing capacity of the pavement structure.

To deal with frost heave and thaw weakening, the subgrade has been emphasized, and bases and subbases are usually thought to be resistant to both. However, the presence of fines (quantity and type) in the base or subbase layer can change its frost-heave susceptibility. In general, as the fine content increases, the susceptibility to frost heave increases. Granular unbound bases containing frost-susceptible materials may also weaken significantly during thaw-weakening periods, owing to increased saturation and reduced permeability.

With the above in mind, the frost susceptibility and thaw-weakening potential of several of the Federal Aviation Administration (FAA) base and subbases used in airport pavement construction was studied. This study was based on the allowable

[1] Research Civil Engineer, ERDC/CRREL, 72 Lyme Road, Hanover, NH 03755, phone 603 646-4702, Vincent.C.Janoo@erdc.usace.army.mil

gradation, range of plasticity index, and other pertinent information found in the specifications in FAA (1989) Advisory Circular AC150/5370-10A. No laboratory or field tests were conducted. This paper summarizes the study and additional details can be found in Janoo et al. (1997).

Base/subbase material specifications

Some of the base and subbase specifications evaluated are presented in Table 1. The materials were further divided into stabilized and unstabilized. The FAA designation is presented within parenthesis in Table 1 and a brief description of each is presented. Details can be found in Janoo et al. (1997). The allowable percentage of fines (passing the 0.074-mm sieve) in the unstabilized materials is presented in Table 2.

The allowable liquid limit and plasticity index (PI) for the subbase, aggregate base, crushed aggregate base, and the sand–clay base were 25% or less and no more than 6%, respectively. An additional specification was identified for use of subbase and crushed aggregate base in cold regions: the amount of material finer than 0.02 mm must be less than 3%.

The stabilized materials—lime-treated base, soil cement base, cement-treated base, and econocrete subbase—were evaluated in terms of durability and reduction of strength when subjected to freeze–thaw cycling. Details of the various stabilized materials can be found in Janoo et al. (1997).

Table 1. Division of FAA materials into stabilized and unstabilized categories.

Unstabilized	Stabilized
Subbase (P-154)	Lime-treated base (P-155)
Aggregate base (P-208)	Soil-cement base (P-301)
Crushed aggregate base (P-209)	Cement-treated base (P-304)
Sand-clay base (P-213)	Econocrete subbase (P-306)

Table 2. Allowable fines in and PI of the unstabilized materials.

Material	Allowable fines (%)	Plasticity Index (PI)
Subbase (P-154)	0–15	≤ 6
Aggregate base (P-208)	5–15	≤ 6
Crushed aggregate base (P-209)	3–10	≤ 4
Sand-clay base (P-213)	4–25	≤ 6

Frost susceptibility

Chamberlain (1981), after reviewing frost-susceptibility index tests used in the U.S. and abroad, concluded that the criterion developed by Casagrande in the early 1930s was best. Casagrande (1931) found that the frost susceptibility (in terms of rate of heaving) of a soil could be indexed to the amount of material finer than 0.02 mm it contained. Later, the Corps of Engineers (COE) adopted this criterion for classifying the frost susceptibility of base, subbase, and subgrade soils. In most cases for granular materials, if the amount of material finer than 0.02 mm is kept below 3%, it could be classified as non-frost-susceptible. As the frost susceptibility of a material increases, so does the rate and magnitude of frost heave. Only two of the FAA materials (P-155 and P-209) met the COE non-frost-susceptibility criterion. Since there were no data in the specification on material finer than 0.074-mm, the frost susceptibility of the remaining base course materials was determined on the basis of the amount passing the 0.074-mm sieve.

Many U.S. state Departments of Transportation and international agencies use the amount passing the 0.074-mm sieve for estimating the frost-susceptibility of the soil. The allowable amount passing that sieve varies among agencies, ranging between 5 and 15%. However, most states or provinces in the "traditional" cold regions limit the amount passing the 0.074-mm sieve to no more than 10%. Townsend and Csathy (1963) found that indexing frost susceptibility to the amount passing the 0.074-mm sieve was conservative, i.e., although the criterion was good for finding frost-susceptible soils, it sometimes classified non-frost-susceptible material as frost-susceptible. Chamberlain (1981) evaluated the reliability of the various criteria and concluded that if the amount of fines passing the 0.074-mm sieve was limited to 7% or less, there was an 80% chance of properly identifying a frost-susceptible soil.

Using data from laboratory frost heave tests of Kaplar (1974), Janoo et al. (1997) found that, for coarse grained materials, the critical amount finer than 0.02 mm was approximately half the amount passing the 0.074-mm sieve, as long it was less than 18%. The COE criterion was modified, allowing a coarse grained mixture having no more than 6% passing the 0.074-mm sieve to be identified as a non-frost-susceptible material. Another indicator of frost susceptibility is the Plasticity Index (PI). Croney and Jacobs (1967) reported that, as the PI increased, the magnitude of frost heave decreased. Their conclusion was based on PI values ranging between 10 and 80%. Using laboratory frost heave test results from Kaplar (1974) for coarse-grained gravelly base course materials (GW, GP, GW-GM, GP-GM, GM, GW-GC, GP-GC, GM-GC, GC), Janoo et al. (1997) found that as the PI increased, the average rate of heave also increased, as long as the PI was less than or equal to 7%. At PI values greater than 7%, the heave rate decreased, which agreed with the results from Croney and Jacobs (1967). Using the allowable PI values for the subbase and base classified the materials as being of medium frost susceptiblity (Janoo et al. 1997).

For evaluating the FAA's base course materials, three criteria were used. One was the Asphalt Institute criterion, which defines a frost-susceptible soil as one that contains more than 7% passing the 0.074-mm sieve, the second was the COE criterion that the amount finer than 0.02 mm should be one-half of the amount passing the no. 200 sieve, and the third was the PI. The frost-susceptibility ratings are shown in Table 3. Depending on the gradation used, the aggregate base and the sand-clay base are frost susceptible. A "Yes" in Table 3 means that it meets the respective criterion for non-frost-susceptible soils. The PI criterion classified the materials as having low to medium frost susceptibility. Using the single grain size criterion is at times contradictory (see P-154, Table 3) and may not be the best way to characterize the frost susceptibility of base course materials. Another factor not included in this analysis is the effect of the maximum aggregate size and the percentage of material larger than the 4.75-mm sieve size. The author believes that these have a significant effect on the frost susceptibility of the material. However, with existing criteria, the best way to make the base and subbase material non-frost-susceptible is to reduce the amount passing the 0.074-mm sieve to 6%. In addition, if these materials are to be used in cold regions, a second requirement should be added that the material finer than 0.02 mm in the gradation be less than 3%.

Table 3. Frost-susceptibility classification of FAA-specified base–subbase materials.

FAA designation	% passing no. 200 sieve	Allowable % finer than 0.02 mm	Asphalt Institute criterion	COE criterion	PI Criterion
Subbase (P-154)	< 15	< 3	No	Yes	Medium
Aggregate base (P-208)	5–15		Yes, ≤ 7%	Yes, ≤ 6%	Medium
			No, > 7%	No, > 6%	
Crushed aggregate base (P-209)	0–8	< 3	Yes, ≤ 7%	Yes	Low
			No, > 7%		
Sand–clay base (P-213)	4–25		Yes, ≤ 7%	Yes, ≤ 6%	Low to medium
			No, > 7%	No, > 6%	

Stabilization admixtures used for reducing frost heave include chemicals to prevent freezing of the soil water (such as calcium chloride or sodium chloride), chemicals to aggregate the soil particles into larger units, or chemicals that waterproof the soil particles and cements (lime, cement, and asphalt).

Data on the effect of stabilization admixtures on the frost susceptibility of base course material are limited. Croney and Jacobs (1967) reported that adding 2% cement to frost-susceptible gravels was sufficient to significantly reduce frost heave. Lambe et al. (1971) reported that soils stabilized with 3% asphalt emulsions reduced frost heave below any meaningful level. Work done with fine grained soils indicates that the addition of 3% lime or 3% portland cement may effectively reduce frost heave in stabilized soils. However, adding less than 3% Portland cement to base courses containing either ML-CL or CH fine materials may actually increase the heave rate.

Thaw weakening

Base and subgrade materials are thaw weakened when water from surface infiltration (melting snow) or melting ice lenses is unable to drain. This undrained condition will accelerate pavement damage when it is loaded. Examining COE field tests in the 1940s on base courses in the northern tier of states, Johnson (1974) reported that bases with 10% fines passing the no. 200 sieve showed serious thaw weakening and significantly reduced (up to 70%) bearing capacity (in terms of CBR). This reduced bearing capacity has to be related to the drainability of the base course, which in turn is affected by the amount of fines it contains. Therefore, the critical property of the base–subbase material during thaw-weakening is its permeability (vertical and horizontal). The horizontal permeability does not have to be the same as the vertical.

One way to assess the thaw-weakening potential of the base course is to estimate how long it will take to drain to at least a saturation level of 80%. Haynes and Yoder (1963) found that granular materials subjected to repeated loading became unstable when the degree of saturation was greater than 80%. Based on accelerated pavement testing, Thompson (1969) reported from in-situ moisture measurements that a pavement test section failed owing to the reduction in the bearing capacity of the base course. The moisture content in the base course varied between 86 and 90% saturation, whereas the subgrade moisture content did not change with the exception of the top 25 mm (1 in.).

The drainage of airport bases was analyzed using a model developed by Casagrande and Shannon (1951), which was based on field observations during the spring thaw of six airfields in Maine, Michigan, Wisconsin, and North and South Dakota by the COE. The model assumes symmetry along the centerline of the pavement and that the base is saturated; the equations represent drainage for one-half of the base course, ABCD (Fig. 1). The base drains in two parts. First, it is saturated and the free water surface changes from AD to AC (Fig. 2). This assumption is based on the premise that water is allowed to drain freely through the open face CD. Second, the water surface changes from AC to BC. Other assumptions are that the centerline and the bottom of the base course are impervious and that the phreatic surface is a straight line. This assumption of an impermeable subgrade is valid during spring thaw when the subgrade is still frozen.

An additional assumption made for this study is that the base is sloping. The calculations are done in two stages. Stage 1 is as follows, with $0 < U \leq 50\%$:

$$T = 2US - S^2 \ln \frac{S + 2U}{S}$$

Stage 2 is, with $50\% < U \leq 100\%$:

$$T = S + S \ln \frac{2S - 2US + 1}{(2 - 2U)(S + 1)} - S^2 \ln \frac{S + 1}{S}$$

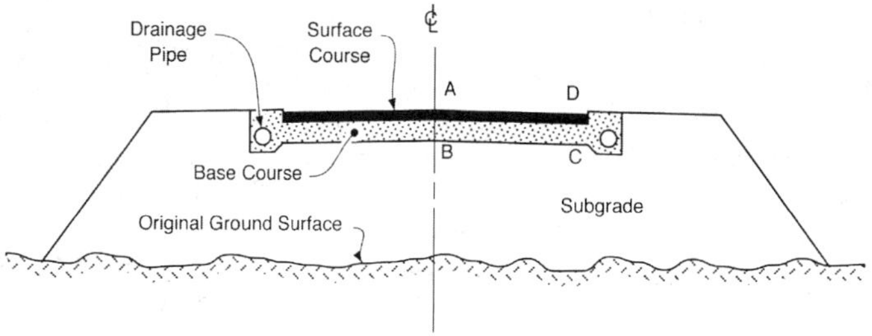

Figure 1. Cross section of pavement.

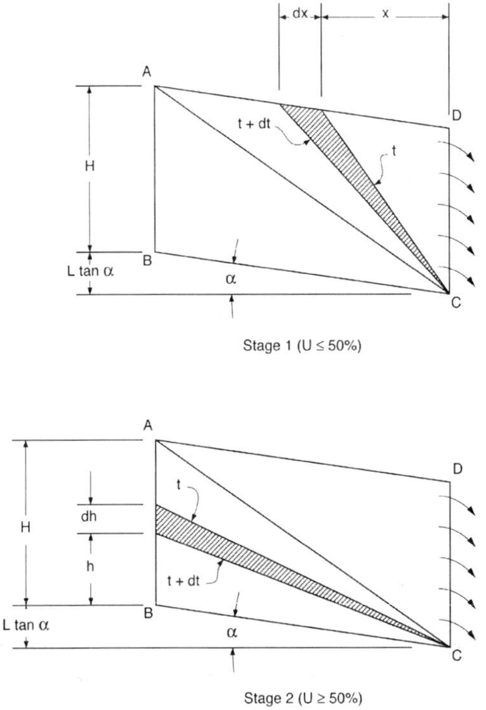

Figure 2. Casagrande and Shannon model for base course drainage.

where U = drained area/total area; S = slope factor = H/L tan α (see Fig. 2); T= time factor = $2tkH/cnL^2$; $c = 2.45 - 0.8/\sqrt[3]{S}$; L = one-half width of pavement; H = thickness of base course; α = angle of slope of base layer; t = time; k = permeability of base course; n = porosity of base course.

The drained area/total area or degree of drainage (U) was calculated for the aggregate base presented in this paper. U for other bases and subbase materials can be found in Janoo et al. (1997). How quickly a base course will drain essentially depends on its coefficient of permeability k, effective void ratio (or porosity), in-situ density, and soil structure. The permeability of soil is usually presented as a function of its gradation and fine contents. However, the *type* of fines also affects k. Barber and Sawyer (1952) reported k values for graded aggregate materials as a function of gradation, fine contents, and type of fines. The effect of the fine type can change the magnitude of the base course by a factor of 100 (Janoo et al. 1997). For the gradation with 3, 5 and 10% fines, the k values used were 3×10^{-2}, 3×10^{-3}, and 3×10^{-4} cm/s, respectively. The base thickness was 61 cm. The effective porosity of the base course was 0.15. The results are shown in Figure 3. The COE criterion for drainage of a base course of 50% within 10 days is only met by the aggregate gradation containing no more than 5% fines.

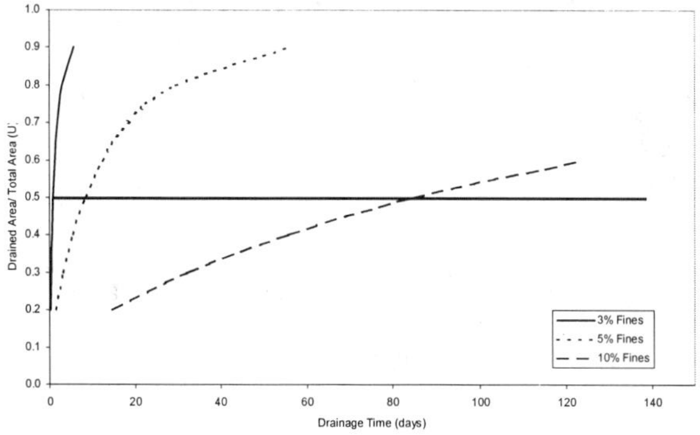

Figure 3. The effect of fine content on the drainability of the base course.

The degree of saturation (S) can be related to the degree of drainage (U) with the following relationship developed by Carpenter et al. (1981):

$$S = 1 - \alpha(U)$$

$$\alpha = (\%\text{gravel})(\text{gravel factor})+(\%\text{sand})(\text{sand factor})$$

where α is an index to represent the drainability of the base course and is a function of the percentage of gravel, sand, fines, and type of fines. The percentage of gravel, sand, and fines in the gradation, the gravel and sand factors, and the α values are presented in Table 4. The degree of saturation as a function of drainage days for various fines contents in the aggregate base course is presented in Figure 4. To reach 89% saturation, the 3, 5, and 10% fines base course needed 10 hours, 21 days, and about 150 days, respectively.

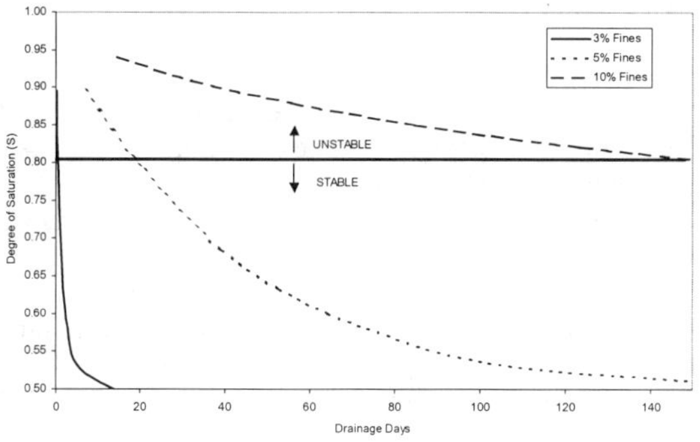

Figure 4. Change in the degree of saturation of the base course with time.

Table 4. Percentages of gravel, sand, and fine content, factors, and α.

Gravel	Sand	Fines	Gravel Factor	Sand Factor	α
50	47	3	70	57	62
50	45	5	60	50	53
50	40	10	40	25	30

Mechanical properties during freeze–thaw

Cole et al. (1987) reported that, as thaw progressed at a local airport, the resilient modulus of the subbase and base course materials changed by a factor of 100 from the frozen to thaw state. The subbase fell within the FAA P-154 specifications, while the base fell within the P-208 specifications. The percentage passing the no. 200 sieve for either material was approximately 12%.

Other indications of the reduction of the bearing capacity of base and subbase were obtained from field CBR tests by the COE in the mid-1940s. In-place CBR tests were conducted on the top of the base or on top of the subgrades at several airfields (flexible) during the fall and in the thaw-weakening period in the spring. Table 5 shows the change in average CBR during the normal and thaw-weakening periods on top of the base and subbase layers. There is a reduction of 13 to 62% in CBR during the thaw-weakening period.

Table 5. Change in CBR during thaw-weakening period.

Layer	USCS	Percent passing no. 200 sieve	CBR (%)		
			Normal	Thaw-weakening period	Percent reduction
Subbase	GM	10–20	64	33	48
Base	GM-SM	N/A*	37	27	27
Base	GC	6–12	37	14	62
Base	GW	N/A*	58	28	52
Base	GW	N/A*	24	21	13

N/A – Data not available.

Stabilized materials such as soil–cement base (P-301), cement-treated base (P-304), and econocrete subbase (P-306) are classified as cementatious materials. There are very limited data on the engineering properties of stabilized soils subjected to freeze–thaw cycling. These soils undergo large volume changes when subjected to changes in temperature and moisture, with moisture being the more critical factor. Volume changes can lead to cracking of stabilized layers. Yoder and Witczak (1975) suggest that freeze–thaw cycling may actually be destructive to lime-treated soils.

The thaw-weakening resistance of these stabilized materials are inferred from the freeze–thaw durability tests (D559 and D560, ASTM 1992, 1996). The results of these tests simply show whether or not the material would retain a specified percentage of weight at the end of 12 cycles of freeze–thaw cycling. If it does, then it is considered to be frost resistant. The results from the durability test cannot be used as engineering properties for mechanistic design.

Kettle and McCabe (1985) presented data on the freeze–thaw resilient modulus and tensile strength of cement-treated soils. The soils tested ranged from a dense graded stone to silty clays. The gradations of the various soils are presented in Figure 5. Tables 6 and 7 summarize the tensile strength and M_r. The before-freeze data in Table 6 refer to the average strength of the material prior to freezing. The residual refers to strength of the thawed specimens after the ASTM freeze–thaw test (D-560), i.e., after 12 freeze–thaw cycles. The results indicate that the strengths of most of the soils at 5% cement treatment are drastically reduced after 12 freeze–thaw cycles. At 15% cement level, most of the soils actually gained some strength. Brandl (1981) also reported results where low lime contents actually reduced the tensile strength of the mixture. He attributed this to poor cementing between the soil particles. The results after freeze–thaw cycling (≤10% cement) indicate either poor cementation or breaking of the soil–cement bonds.

The resilient modulus tests reported in Table 7 were done at 5% of the tensile strength load level. A Poisson's ratio of 0.25 was assumed. At 5% cement content, in most cases, the resilient modulus decreased with increasing freeze–thaw cycles. Increasing the cement content to 10% and above actually increased the stiffness of the material by about 10 to 30% after freeze–thaw cycling.

Summary and Conclusion

The FAA subsurface material specifications for unstabilized materials allowed up to 15% fines passing the 0.074-mm sieve, with the exception of the crushed aggregate base, which allowed no more than 8% fines, and the sand–clay base, which allowed up to 25% fines. With respect to frost action, the following conclusions were reached.

- With the exception of the FAA subbase and crushed aggregate materials, the remaining unstabilized soils did not meet the COE frost-susceptibility requirements of no more than 3% finer than 0.02 mm.

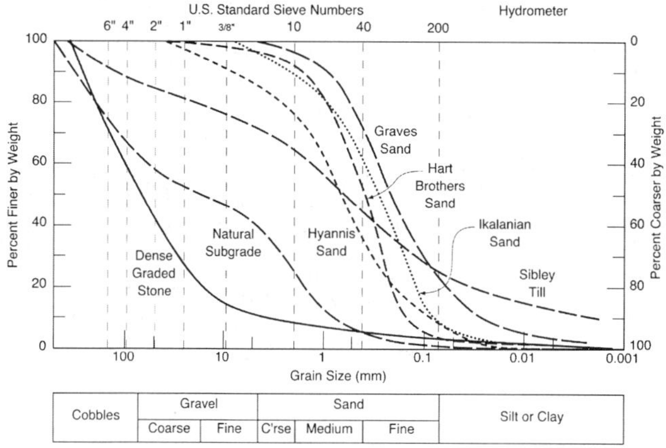

Figure 5. Grain size distribution of test soils and natural subgrade.

Table 6. Summary of tensile strength data.

Material	Percent fines passing 0.074-mm sieve	Cement content (%)	Tensile strength (kPa)	
			Before freeze	Residual
Hart (SM)	19	5	250	20
		15	540	720
Graves (SM)	33	5	76	35
		10	221	165
		15	460	475
Sibley (SC-SM)	42	5	610	550
		10	710	720
		15	1010	1180
DGS (SP-SM)	9	5	400	40
		10	790	420
Hyannis (SM)	23	5	69	21
		10	280	120
		15	410	490
Manchester silt (CL-ML)	93	5	7	0
		10	14	1
		15	25	24
Ikalanian (SM)	35	5	9	1
		10	27	20

Table 7. Summary of resilient modulus data.

Material	Cement content (%)	Pre-test modulus (MPa)	Thaw modular ratio (%)		
			1-cycle	3-cycle	12-cycle
Hart (SM)	5	1931	0.64	0.30	0.04
	10	17582	0.98	1.00	1.01
	15	22629	0.91	0.98	1.00
Graves (SM)	5	1820	—	0.23	0.10
	10	4599	—	1.17	1.32
	15	9494	—	1.17	1.18
Sibley (SC-SM)	5	11784	—	1.00	1.08
	10	10915	1.04	1.06	1.20
	15	14824	0.96	1.13	1.21
DGS (SP-SM)	5	21892	0.95	0.02	0.00
	10	28056	0.88	0.85	0.69
Hyannis (SM)	5	3234	0.88	0.35	0.25
	10	7315	0.91	0.45	0.41
	15	10977	0.99	1.02	1.04
Ikalanian (SM)	5	3592	—	0.50	0.35
	10	7460	—	1.06	1.15
Fairbanks (ML)	5	7522	0.96	1.16	1.17
	10	14976	1.00	1.11	1.15

- When the percentage passing the 0.074-mm sieve was used to determine frost-susceptibility, it found to be was conservative, i.e., materials that would pass the COE criterion failed that criterion. Use of the the PI caused the base and subbase material to fall into the low to medium frost-susceptibility area.
- For frost-susceptible gravel soils, 2% cement is required to change it to a non-frost-susceptible material.
- Thaw weakening of the base material was inferred from the time it took to drain an airport pavement by 20, 50, and 100%. The COE drainage requirement is that 50% of the base be drained within 10 days. The drainage model developed by Casagrande and Shannon showed that none of the materials passed the COE criterion. Some of the materials came close to it, suggesting that if the fines contents were further reduced to 2 or 3%, they may meet the criterion.
- Available data showed a hundred-fold reduction in frozen and thawed moduli of materials that meet the subbase and base specifications. Field test data showed a reduction in the bearing capacity of base and subbase ranging from 13 to 60%.
- For stabilized materials, the results of the ASTM (1992) D-560 durability test can be used to infer the frost-resistance (thaw-weakening) properties of the material. For cement-treated soils, the tensile strength of the material decreased with increasing freeze–thaw cycles. Also, when 15% cement was added to the soil, there was no reduction in the tensile strength after 12 freeze–thaw cycles. The resilient modulus of cement-treated soils remained at its before-freeze level when cement contents greater than 5% were added to the soil.

References

ASTM (1992). "Standard test method for freezing and thawing compacted soil–cement mixtures" (ASTM D 560-89), *Annual Book of ASTM Standards*, vol. 04.09, section 4, Philadelphia, Pennsylvania, American Society for Testing and Materials.

ASTM (1996). "Standard test method for wetting and drying compacted soil–cement mixtures" (ASTM D 599-89), *Annual Book of ASTM Standards*, vol. 04.01, section 4, Philadelphia, Pennsylvania, American Society for Testing and Materials.

Barber, E.S., and Sawyer, C.L. (1952). "Highway subdrainage," *Public Roads*, 26(12), 251–268.

Brandl, H. (1981). "Alteration of soil parameters by stabilization with lime," *Proceedings, 10th International Conference of Soil Mechanics and Foundation Engineering, Stockholm*, vol. 3, p. 587–594.

Carpenter, S.H., Darter, M.I., and Dempsey, B.J. (1981). "A pavement moisture accelerated distress (MAD) identification system," Vol. II, FHWA Report, FHWA-RD-81-080.

Casagrande, A. (1931). "Discussion of frost heaving," *Proceedings, Highway Research Board*, 11, 168–172.

Casagrande, A., and Shannon, W.L. (1951). "Base course drainage for airport pavements," *Proceedings, American Society of Civil Engineers*, 77, 1–23.

Chamberlain, E.J. (1981). *Frost susceptibility of soil: Review of index tests*, USA Cold Regions Research and Engineering Laboratory, Monograph 81-2.

Cole D.M., Bentley, D.L., Durell, G.D., and Johnson, T.C. (1987). *Resilient modulus of freeze–thaw affected granular soils for pavement design and evaluation, Part 3, Laboratory tests on soils from Albany County Airport*, USA Cold Regions Research and Engineering Laboratory, CRREL Report 87-2.

Croney, D., and Jacobs, J.C. (1967). *The frost susceptibility of soils and road materials*, Road Research Laboratory, Ministry of Transport, U.K., RRL Report LR 90.

FAA (1989). *Standards for specifying construction of airports*, Federal Aviation Administration Advisory Circular, DOT/FAA/AC 150/5370-10A.

Haynes, J.H., and Yoder, E.J. (1963). "Effects of repeated loading on gravel and crushed stone base material used in the AASHTO road test," *Highway Research Board Record,* 39, 82–96.

Janoo, V.C., Eaton, R., and Barna, L. (1997). "Evaluation of airport subsurface materials," USA Cold Regions Research and Engineering Laboratory, CRREL Special Report 97-13.

Johnson, T.C. (1974). *Is graded aggregate bases the solution in frost areas?* USA Cold Regions Research and Engineering Laboratory, Miscellaneous Paper 710.

Kaplar, C.W. (1974). *Freezing test for evaluating relative frost susceptibility of various soils*. USA Cold Regions Research and Engineering Laboratory, Technical Report 250.

Kettle, R.J., and McCabe, E.Y. (1985). "Mechanical stabilization for the control of frost heave," *Canadian Journal of Civil Engineering*, 12(4), 899–905

Lambe, T.W., Kaplar, C.W., and Lambe, T.J. (1971). *Additives for modifying the frost susceptibility of soils, Part II*, USA Cold Regions Research and Engineering Laboratory. Technical Report 123, Pt. 2.

Townsend, D.I., and Csathy, T.I. (1963). *Soil type in relation to frost action*, Ontario Department of Highways, Joint Highway Research Program Report No. 15, Civil Engineering Report No. 29.

Thompson, O.O. (1969). *Evaluation of flexible pavement behavior with emphasis on the behavior of granular layers*. Ph.D. dissertation, University of Illinois.

Yoder, E.J., and Witczak, M.W. (1975). *Principles of Pavement Design,* 2nd edition, New York, John Wiley and Sons, Inc.

Seasonal Deterioration of Unsurfaced Roads

S. Shoop[1], R. Haehnel[1], V. Janoo[1], and D. Harjes[1]

Abstract

Spring thaw is a critical time of year for road deterioration. The types of deterioration or distresses most common during spring on unpaved roads are rutting and roughness, which is often in the form of washboards (also called corrugations). These distresses were observed over several years and were studied using standard pavement deterioration models, such as MMOPP (Mathematical Model of Pavement Performance). These types of models are usually designed for paved roads, and are sometimes limited in their options for soil characterization and manipulation of the forces at the road–tire interface. Therefore, a finite element model of the tire–road interaction was also developed. Various finite element models were generated to simulate an unpaved secondary road subjected to local vehicle traffic. These models were implemented in a three-dimensional dynamic analysis with a wheel rolling over the soil. Simulations of the unpaved road were used to study the effects of vehicle speed, load, suspension system, wheel torque, wheel slip, etc., on rutting and washboard formation. Results are discussed with regard to field measurements and observations.

Introduction

The objective of this project was to simulate seasonal damage to secondary roads, particularly with respect to spring thaw. This was done initially through an evaluation of existing road performance models for use on unpaved, secondary roads. The prediction of road deterioration is important not only for maintenance scheduling, but also, for the military, for predicting the effect of road deterioration on vehicle speeds for operations, for rapid deployment, operational tempo, and especially for maintenance of supply routes. With the new emphasis of our armed forces on peacekeeping missions, skirmishes, and operations other than war, the use of secondary roads for rapid deployment will be critical, and the deterioration of these roads must be avoided. Rutting and washboard formation are primary distresses on secondary roads and can seriously slow vehicle throughput.

[1] US Army Corps of Engineers, Engineer Research and Development Center, Cold Regions Research and Engineering Laboratory, Hanover, NH, 03755-1290, sshoop@crrel.usace.army.mil

The approach to the study of the seasonal effects on secondary road deterioration consisted of three major efforts:

1. Use the traditional pavement performance models for simulating deterioration of unpaved roads.
2. Observations of road distress throughout the year, but particularly during the spring.
3. Finite element modeling of distress formation on unpaved roads (rutting and washboard).

Modeling Road Deterioration with MMOPP

Several models have been developed for the analysis of pavement performance. Of these, MMOPP (Mathematical Model of Pavement Performance) (Ullidtz, 1998), and HDM-IV (Highway Design and Maintenance IV) (World Bank, 1998) were evaluated for use on unpaved roads. The HDM-IV model uses empirically derived equations for predicting highway deterioration and maintenance. The MMOPP is mechanics-based, has more flexibility in the input of the pavement structure and vehicle characteristics, and, therefore, was chosen to evaluate performance of unpaved roads under a variety of conditions.

The input parameters for MMOPP consist of defining the pavement geometry and materials, climate, vehicle loading, and traffic. The road structure is considered as a series of layers of specified thickness and material properties (elastic, plastic, and fatigue parameters). The traffic is simulated by a number of passes per year, with an option for growth rate, for a given set of vehicles. The vehicles are defined based on how they load the pavement through the application of a 1/4-car vehicle dynamics model with specified tire pressure, wheel distance, wheel mass, suspended mass, spring constants, suspension damping, etc. The pavement and vehicle inputs are illustrated in Figure 1.

The climate is defined using seasons, or in this case, average monthly air temperatures for the local area, as well as factors indicating the seasonal or temperature impact on the different moduli of the pavement structure. Pavement material properties and their seasonal variations were estimated based on seasonal Falling Weight Deflectometer (FWD) testing of local pavement sections and pavement test sections in CRREL's Frost Effects Research Facility.

The output of the MMOPP is road deterioration as a function of time. Road deterioration is calculated based on accumulated plastic strain through the pavement section, which can then be related to Rut Depth (Fig. 2) or roughness. The roughness calculation in MMOPP is a function of the variation of road material properties and vehicle dynamic load along the road length. These parameters are statistically varied along the road length at 300-m increments, with the resulting variations in properties, loading, and plastic strain, defining the road roughness. Roughness is then reported as the International Roughness Index, IRI, which is defined as the accumulated vertical suspension travel divided by the length of road traveled, illustrated in Figure 2.

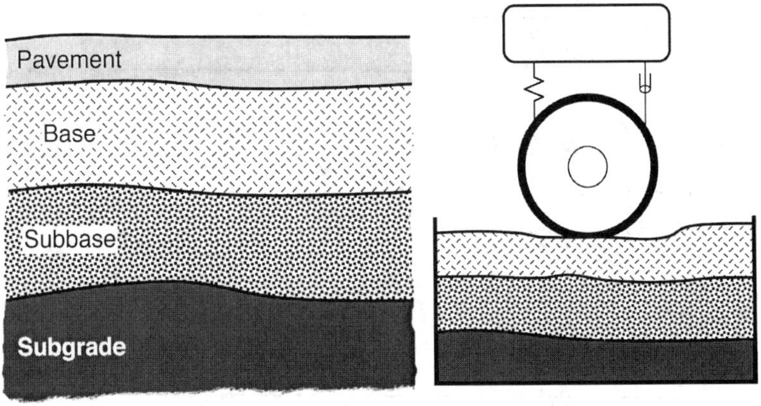

Figure 1. A typical four-layer road structure requiring input of thickness, elastic, plastic, and fatigue parameters (left); and the ¼-car model requiring input on tire and suspension characteristics and vehicle weight.

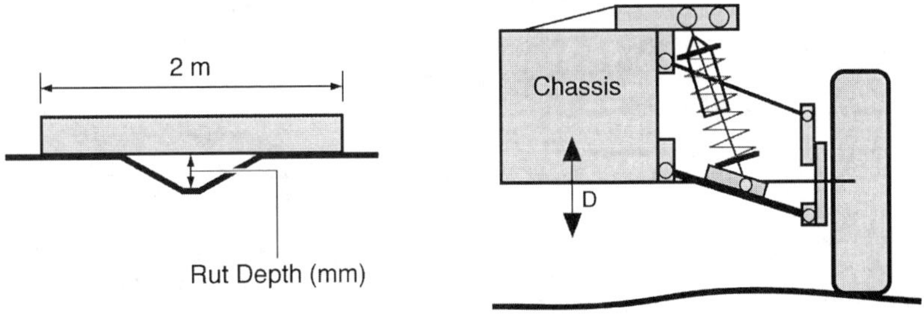

Figure 2. Rut depth is defined as the maximum depression under a 2-meter straightedge (left). IRI (International Roughness Index) is equal to the ratio of the accumulated vertical suspension travel, D, to the length of road traveled (right).

MMOPP Results

The results of the MMOPP analysis for the CRREL Instrumented Vehicle (CIV) on the local roads in Hanover, New Hampshire were used to assess the effects of types of traffic, paved vs. unpaved roads, and similarity of local climate effects to those in the Balkans. Selected results are graphically expressed as rut depth or IRI as a function of time in Figure 3. The springtime deterioration is apparent in each of the graphs as a jump in the rut depth or IRI during the spring of each year.

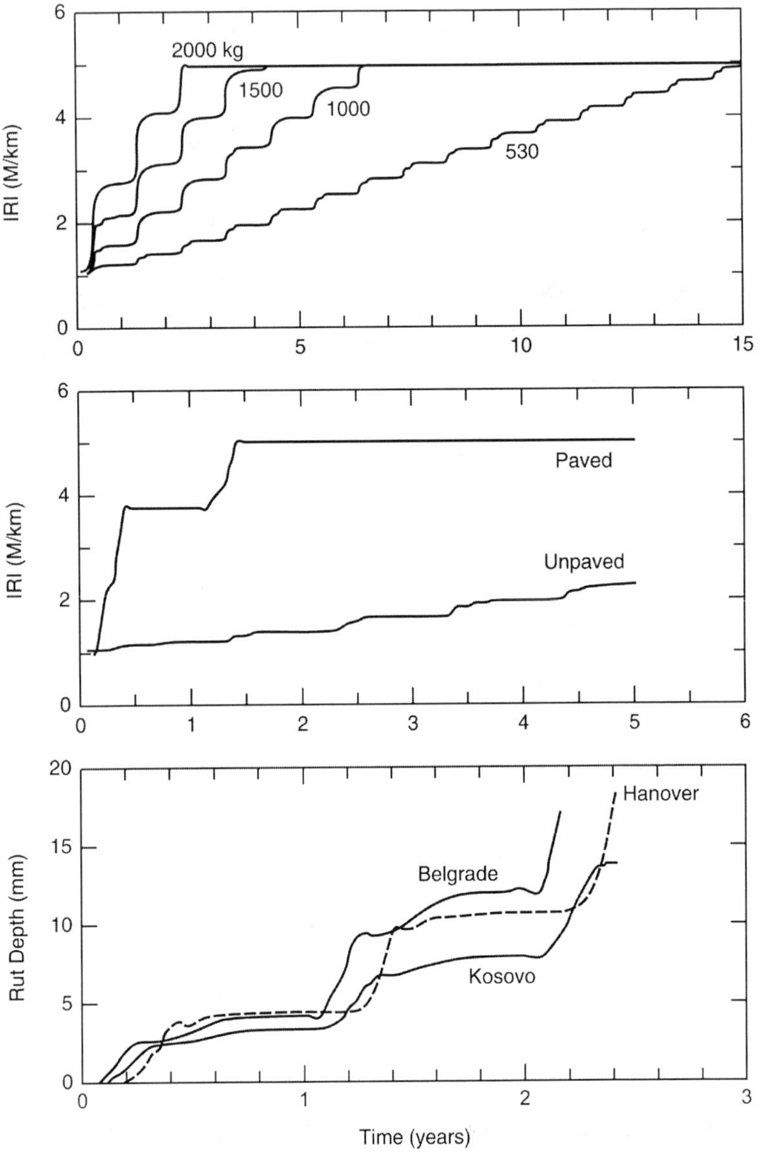

Figure 3. MMOPP predicted IRI and Rut Depth for unpaved secondary roads showing the effects of vehicle weight (1/4-car weight, top), paved and unpaved (center), and climate (bottom) on road deterioration.

The model predicts that either lower speeds or increased vehicle weights will increase the rate of pavement deterioration, as expected. Also, the deterioration of the unpaved section was predicted to be four times faster than the same construction with the addition of a 5-cm asphalt layer.

The last task was to compare the deterioration of local unpaved roads to similar roads in the Balkans, to determine if local roads could be used as experimental surrogates. The predicted deterioration of the local roads was similar to that predicted in the Balkans, based on the climate of selected cities, given similar unpaved road materials and construction technique.

Although results provide valuable insight on the impacts of different variables, the model was originally designed for modern pavements and thus may not address some of the mechanisms of deterioration of secondary, unpaved roads. This may be particularly true for the large deformation that can occur in unbound material of the surface layer during thaw weakening, the roughness produced by frost heaves on marginal roads, and the corrugations or washboards. For example, for modern pavements and highways, an IRI value of 6 is considered "Very Poor," requiring immediate attention. In comparison, the IRI for paved secondary roads easily reaches 10 to 15 M/km, and short washboard sections on unpaved roads can have an IRI of 83 M/km!

Field Observations of Seasonal Distresses on Unpaved Roads

To evaluate the nature of the seasonal distresses to secondary, unpaved roads, a field observation schedule was developed. The observations were aimed at evaluating both the nature and the cause of the distress. Rutting was a primary distress during the thaw weakening of springtime (Fig. 4), and the formation mechanism was straightforward. However, the cause of the formation of washboards was more elusive. Initially, the common thinking among roadway personnel was that washboards formed only during the dry summer months, and were probably caused by movement of the dry granular surface material. However, springtime washboards, formed when the road is wet and soft and deforms in a plastic manner, are also documented (Fig. 4).

The springtime washboards appear to be slightly different in geometry (wavelengths of 84 cm vs. 25 cm in the summer, and peak-to-peak amplitude of 3.8 cm vs. 1.3 cm during the summer), and occur primarily only within the wheelpaths, although this may be because the spring thaw necessitates more frequent road repair. During all seasons, the washboards commonly occurred during transitional areas on the road, such as around corners, up or down slopes, or moving from paved to unpaved sections.

A literature review and field study of washboard formation is documented in Monaghan et al. (in progress). Her review of the published literature indicates that washboards were believed to form primarily during dry periods and their formation was the result of a disturbance of the road surface (such as a pothole or other change in elevation) initiating a periodic response in the vehicle suspension, thus a periodicity in the pavement loading. However, the primary occurrence within the transitional zones indicates a formation mechanism aside from that provided by a periodic vertical load resulting from an impulse load or vertical step. In fact, a pilot

Figure 4. Rutting on unpaved secondary roads during spring thaw (top). Springtime washboard formation on curves and slopes (center) and measurement of washboard geometry (bottom).

study by Liston (personal communication), in which a vehicle was repeatedly driven over a test area with a 4-cm vertical step, indicated that a vertical step alone was not sufficient for washboard formation and that lateral forces and slip may contribute. The common factor in each of the transitional locations is a change in velocity or direction of the vehicle, which would result in additional horizontal forces on the surface and/or slip of the tire on the road surface. Monaghan indicates that very few analyses of washboard formation exist, and no models account for horizontal forces or slip of the tire. Pavement performance models like MMOPP and HDM IV also do not consider the horizontal forces at the tire–road interface, even though those forces are considerable during acceleration, deceleration, and cornering. One way to evaluate these as mechanisms of road deterioration and washboard formation is by using a dynamic finite element simulation of wheel and suspension interacting with an unpaved road.

Finite Element Simulations

A high-resolution model of the formation of unpaved, secondary road distresses was completed using finite element simulations. Separate models were used to evaluate the formation of rutting during spring thaw and for washboard formation.

Rutting Model. The rutting model consisted of a wheel modeled as a rigid analytical surface with the dimensions of the CRREL Instrumented Vehicle tire. An unsprung mass is placed at the hub to simulate the weight of the vehicle and tire along with the rotational inertia. The thawing road was modeled as a thawed sandy soil overlaying a rigid base (the frozen layer). The soil was modeled with eight-node continuum elements near the wheel and four-node infinite elements at the edges of the mesh. The soil was modeled using a modified cap Drucker–Prager constitutive law as formulated in the ABAQUS finite element code (HKS, 1998). The frozen layer was modeled by applying a zero vertical displacement boundary condition at the bottom of the soil layer. The cap Drucker–Prager material model, with linear elasticity, was used for the continuum elements, and a linear elastic model was used for the infinite elements. Coulomb friction was applied at the tire–soil interface (μ = 0.65). The model has 11,520 continuum elements and 1,418 infinite elements. For these simulations, 10 cm of thaw was simulated (Fig. 5). The material model of the thawing soil was generated from extensive laboratory testing on a local road material commonly used on secondary roads. The soil is a frost susceptible silty sand, tested under thawing conditions (loose and wet). The development of the material model has been presented elsewhere (Shoop et al., 2001) and is documented in Shoop et. al. (in progress).

The model simulates a wheel settling onto the thawing soil under gravity forces, accelerating to 1 m/s, and then maintaining that velocity as the wheel rolls to the end of the finite element domain. The tire was rolled over the thawing surface under varying tire slip configurations, where slip, s, is defined as

$$s = 100\% \left(\frac{\omega r - u}{u} \right) \qquad (1)$$

and ω is the angular velocity of the wheel, r is the wheel radius, and u is the axle or hub velocity. The convention is that a negative slip indicates braking slip.

Because of the strong effect of the wheel slip and torque on rutting, the following cases were modeled:

1. The wheel was towed through the soil at a fixed axle speed (1 m/s) and the wheel was allowed to freely roll over the surface (freewheel model).
2. The wheel was driven over the surface at a fixed axle speed of 1 m/s and a fixed angular velocity to maintain a slip of 0, 20, 50, and 80% (accelerating slip).
3. The wheel was towed over the surface with a fixed slip of –20, –50, –80 and –100% (braking slip).

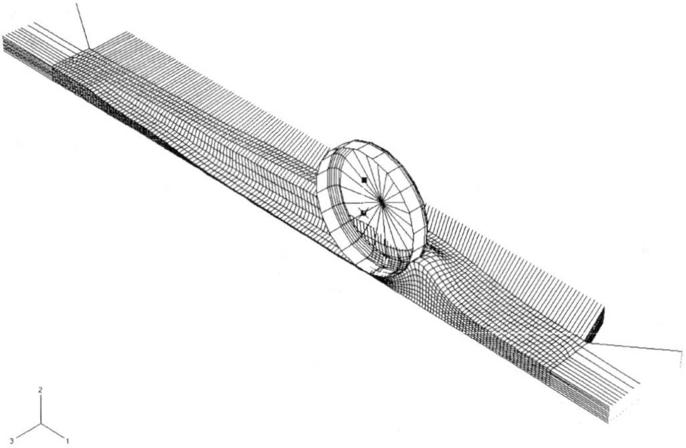

Figure 5. Model of a free-rolling rigid wheel on thawing soil (half model with longitudinal symmetry plane).

Washboard Model. The simulation used to study washboard formation was quite different due to the dynamic nature of the problem. The model of the vehicle was a 1/4-car model, which included the suspension simulated using a spring and dashpot, as well as the sprung mass of the vehicle chassis. These values were taken from measurements on the CRREL Instrumented Vehicle (Lacombe, personal communication). Also, because of the need to evaluate the effect of speed, and because higher speed would require a very long mesh of the road, the road model was changed to an Adaptive Lagrangian–Eularian (ALE) formulation where the mesh remains stationary while the soil moves through the mesh beneath the rolling wheel at the speed of the vehicle. Monitoring the surface of the mesh at the edge of the simulation, after it is exposed to the vehicle loading, captures the deformed profile of the road. Monitoring at this location, instead of directly under the wheel, allows for any elastic rebound, and records only the plastic or permanent road deformation. Several cases were modeled to study the effects of speed, wheel torque, slip, suspension, and impulse load. They were

1. Towed free-rolling wheel, effect of speed (2.2, 5, 15, 25 mph).
2. Drive vs. towed wheel (at 2.2 mph).

3. Towed wheel, effect of slip (0, –10%, –50%, free rolling).
4. Effect of suspension (suspension vs no suspension).
5. Effect of impulse load (2000 N for 0.1 second vs. gravitational inertia).

The soil used in the washboard model is similar to the soil in the rutting model, but in a compacted state.

FEM Results

Figure 6 is a snapshot taken from the simulation of the free-rolling wheel in thawing soil. Qualitatively this model agrees well with field observations. Figure 6 shows moderate plowing of the soil in advance of the wheel, and the wheel sinking through nearly the full depth of the thawed soil layer such that the wheel is essentially resting on the top of the frozen layer.

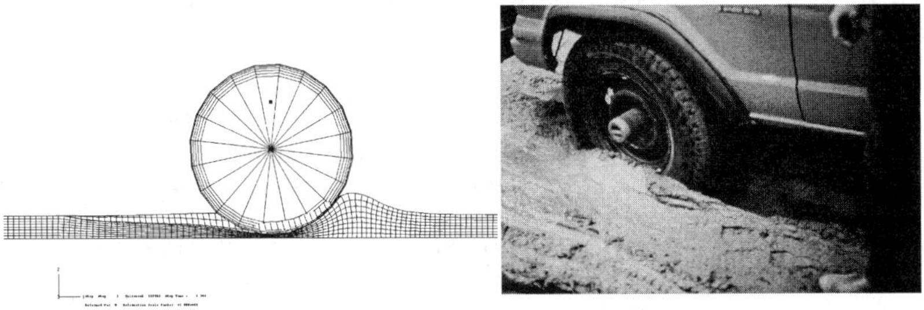

Figure 6. Free-rolling wheel simulation (left, simulation time is 2.34 s, slip is approximately -116%), and CRREL Instrumented Vehicle in thawing soil (right).

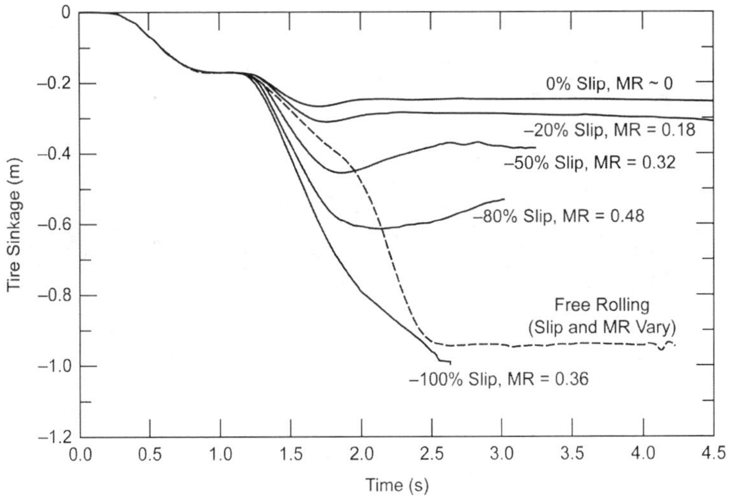

Figure 7. Sinkage of the tire in the thawing soil, with motion resistance coefficient (MR) labeled. Soil depth is 0.1 m.

Figure 7 shows a comparison of the depth to which the tire sinks in the soil for various wheel slip. At zero slip the wheel sinks only 2.2 cm into the thawed soil, yet as the absolute value of slip increases, the tire sinks further into the soil until the slip reaches 100% and the wheel sinks nearly through the full depth of the thawed soil.

The relationship between slip, sinkage, and motion resistance is readily apparent in Figure 7. In general, motion resistance (resisting horizontal force divided by vertical load on axle) steadily increases with increasing slip (and sinkage) with a maximum occurring between –80% and –100% slip. For the free-rolling tire, the slip and sinkage varies throughout the simulation. Interestingly, the results for the free-rolling wheel show that when the wheel is slipping around –100 to –116%, the average resistance coefficient is about 0.323, which is about the same resistance as the –100% slip model. However, once the free-rolling wheel maintains a steady slip of about –92%, the resistance increases by threefold; this is possibly due to the continuing buildup of soil in front of the wheel; in the field, this wheel would become immobilized.

For the washboard formation model, the vertical deformation of the soil as a function of distance along the road is shown in Figures 8 and 9. These simulations are for only one pass of the vehicle; therefore the soil deformation is small. Figure 8 shows the effects of vehicle speed on soil deformation. The vehicle speed and natural frequency of the suspension produce the longest wavelengths in the deformed soil, with higher speeds resulting in longer wavelengths. After these suspension effects die out, shorter wavelength soil deformation persists. The effect of the suspension is less at the lower speeds.

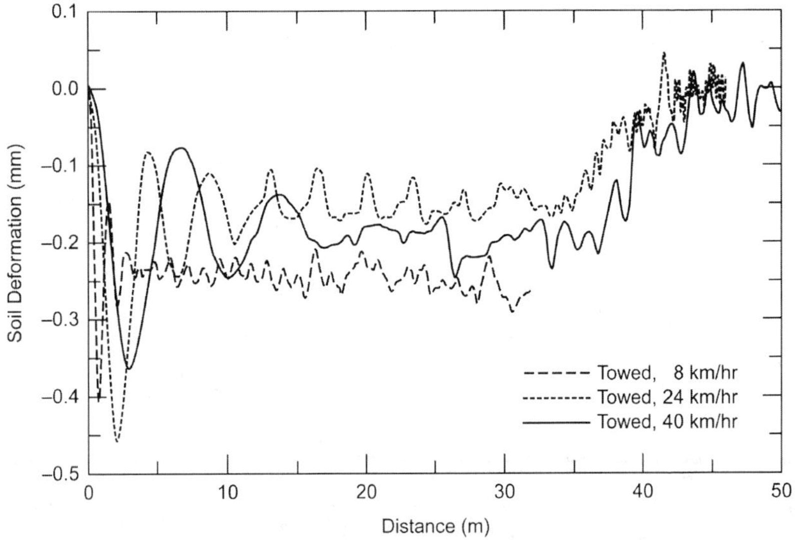

Figure 8. Effect of vehicle speed on washboard formation. Deformation decreases at 30 to 35 m because of damping in the suspension and soil.

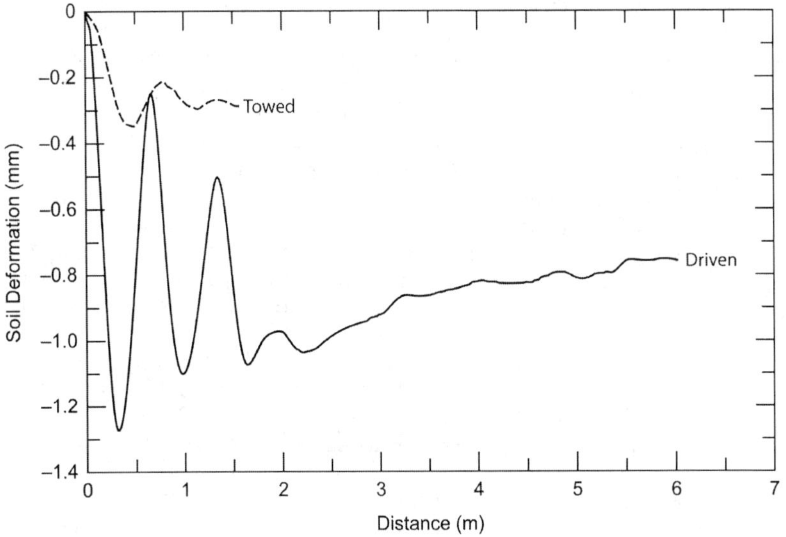

Figure 9. Soil deformation under a driven vs. towed wheel.

A spectral analysis of the soil deformation from the models identified several dominant frequencies; the lowest frequency identified ranged from 1.5 to 2 Hz, which is close to that of the natural frequency of the suspension (1.59 Hz). These frequencies correspond to the long wavelengths visible in the deformation record (Fig. 8). The other predominant frequencies identified correspond to wavelengths in the range of 55 to 65 cm, which fall within the range of wavelengths measured in the field (25 to 84 cm). When the vehicle speed is low, the effect of the suspension dies out more quickly and the deformation is of more random and shorter wavelength (like that of the short wavelengths in Figure 8).

The effect of the additional horizontal force provided by a driven wheel over a free-rolling towed wheel is shown in Figure 9. The amplitude of deformation caused by the driven wheel is up to three times greater than a free-rolling wheel because of the additional horizontal forces at the interface.

Summary and Conclusions

Spring thaw is a critical time for road deterioration. The most common distresses (rutting and increased roughness or washboarding) occur during the spring on unsurfaced roads. These distresses were observed over several years and were studied using standard pavement deterioration models, like MMOPP (Mathematical Model of Pavement Performance). These types of models are usually designed for paved roads, and are limited in their options for soil characterization and manipulation of the forces at the tire–road interface. Therefore, a three-dimensional dynamic finite element analysis of a wheel rolling over unpaved roads was used to study the effects of vehicle speed, load, suspension system, wheel torque, wheel slip, etc., on rutting and washboard formation.

Based on field observations, predictive models, and finite element simulations we found that

1. Pavement degradation models (e.g., MMOPP) can be used to simulate performance of unpaved roads; however, they are not designed to capture the very large deformations that occur on unpaved roads during spring thaw.
2. Field observations of damage to unpaved roads indicate that spring thaw is critical for rut formation, and that washboards (corrugations) can form during both wet and dry periods.
3. Existing washboard formation and pavement deterioration models do not account for horizontal forces or wheel slip, which is critical on unsurfaced roads. Spring and summer washboards form on steep slopes, curves, and road transitions, where the vehicle applies large horizontal loads on the surface through acceleration, deceleration, and steering.
4. Three-dimensional finite element models of a wheel rolling over an unpaved road effectively capture the physics of the problem and can be used to study rutting and washboard formation.
5. Finite element (FE) simulations show the strong influence of wheel slip on rut depth (known as slip-sinkage).
6. Field observations and FE simulations suggest that washboard wavelength is a function of the suspension natural frequency, the vehicle speed, and the soil properties.
7. FE simulations predict that a driven wheel causes more deformation in the road surface than a towed wheel because of the large horizontal load on the surface.

References

Hibbitt, Karlsson & Sorenson, Inc. (HKS) (1998) ABAQUS Theory Manual, Hibbitt, Karlsson & Sorenson, Inc., Pawtucket, RI.

Lacombe, J. (2001) Personal communication.

Liston, R. (2001) Personal communication.

Monaghan, S.K., Liston, R., and Young, B. (In progress). A data report on washboard formation, ERDC-CRREL Technical Report.

Shoop, S.A., R. Affleck, V. Janoo, R. Haehnel, E. Berney, D. Smith, J. Peters (2001) Modeling Soft Soil Behavior Under Vehicle Loads, 2001 Mechanics and Materials Summer Conference, ASCE-Engineering Mechanics Division Conference on Inelastic Behavior of Geomaterials, San Diego, CA, June 27–29, 2001, pg 106–107.

Shoop, S.A., Affleck, R., Janoo, V., Haehnel, R., Barrett, B. (In Progress). Constitutive Modeling of Lebanon Sand After Freeze–Thaw, ERDC-CRREL Report.

Ullidtz, P. (1998) Modelling Flexible Pavement Response and Performance, copyright by Polyteknisk Forlag, Narayana Press, Gylling, Denmark.

World Bank (1998) The Highway Design and Maintenance Standards Model, Technical Description of the HDM-IV Model, HMD-4, Vol. 2.

The Use Of Geocomposite Drainage Layers To Mitigate Frost Heave In Soils

By Mark D. Evans[1], Karen S. Henry[2], Scott A. Hayden[3] and Morgan Reese[4]

Abstract

Frost heave in fine-grained soils below roadway pavements causes extensive and costly damage. In order to mitigate frost heave and associated damage, a geocomposite drainage net was installed in highway test sections along a 3.0 km reconstructed portion of US Route 1A in Frankfort and Winterport, Maine, to function as a capillary barrier to reduce frost heave and a drainage layer to accelerate thaw recovery. This section of road had previously been plagued by pavement cracking, rutting, and potholes requiring frequent and costly maintenance to maintain a trafficable pavement surface. Thermocouple strings were embedded in the test sections at six-inch (152-mm) vertical intervals and temperatures were recorded with depth within the section. Ambient air temperatures and roadway elevations were also documented. This paper deals only with the influence of the geocomposite on frost heave. The freezing index, frost penetration and heave for the winter of 1997-1998, immediately after construction are reported. The results suggest some benefit when the geocomposite is placed within the subgrade. It appears that the primary benefit of the geocomposite is that it provides dewatering benefit during spring thaw—probably controlling the elevation of the water table. There may also be an effect of the cut/ fill depths that masked the effectiveness of the geocomposite drainage layers.

Introduction

Frost heave in fine-grained soils below roadway pavements causes extensive damage. In order to mitigate frost heave and its associated damage, a geocomposite drainage net was installed in highway test sections along a 3.0 km reconstructed portion of US Route 1A in Frankfort and Winterport, Maine, to function as a capillary barrier and a drainage layer. Pavement cracking, rutting and potholes, had previously plagued this section of road requiring frequent and costly maintenance to maintain a trafficable pavement surface (Figure. 1).

1 Associate Prof., Fellow, ASCE, US Military Academy, Civil and Mech. Eng., West Point, NY 10996
2 Research Civil Engineer, Member, ASCE, Civil and Geotechnical Eng. Research Branch, US Army Cold Regions Research and Eng. Laboratory, Hanover, NH 03755
3 Soils Research Scientist, ME Dept. of Trans., Quality and Tech., 219 Hogan Rd, Bangor, ME 04401
4 2LT, US Army, Germany

Geocomposite drainage nets were installed in three different configurations into sections of US Route 1A during its reconstruction in the summer of 1997. During the subsequent winter of 1997-1998, the test sections were monitored to see what effect, if any, the geocomposites were having on frost heave of the newly reconstructed roadway. The following sections briefly describe potential benefit from installing capillary barriers, the test sections, and frost penetration and heave recorded during the winter of 1997-98 on the test sections.

Mitigating Frost Heave In Soils

Three things are required for soil to frost heave: (1) freezing temperatures, (2) a water supply, and (3) frost-susceptible soil (Henry and Holtz, 2001). One means of limiting the water supply to frost-susceptible soil that is exposed to freezing temperatures is to place a capillary barrier layer between the alleged water supply (the water table) and the freezing front to stop the upward migration of soil water (Henry and Holtz, 2001). A capillary barrier is a layer of relatively large-pored material placed in finer-pored soil to divert downward infiltrating water and/or to cut off the capillary flow of water induced by evaporation or freezing. Taber (1929) conducted experiments that showed that a layer of coarse-pored sand placed between the water table and the depth of freezing could significantly reduce frost heave. Use of large-pored materials (sands and gravels) placed above the water table to prevent water from rising in soils to reduce frost heave was described by Casagrande (1938), Beskow (1946), Rengmark (1963), and Taivenen (1963). However, there is apparently no more recent literature referring to such practices to reduce frost heave in pavements. In this study, a geocomposite drainage net was used as a capillary barrier in three test sections to test the efficacy of these layers in mitigating frost heave.

Project Description

Hayden and Dunn (1996) performed a subsurface geotechnical investigation of the soils along the roadway test section. Soils encountered were mostly moist clays of the Presumpscot Formation with water contents greater than 20%. These soils classified as AASHTO soil group A6, and as CL in the USCS (Henry and Affleck, 1998).

Thirteen different test sections were installed and monitored in this project as described by Hayden et al. (1998). Three of the sections were installed to specifically investigate the influence of the geocomposite on frost heave. The three test sections with the geocomposite inclusions were designated as D1, D2, and D3 (see Figures 2 and 3). In D1, 460 mm of of subgrade was excavated then replaced and compacted on top of the geocomposite (CL, A-6 soil). The symbol in Figure 2: ● ● ● ● ● indicates the location of the drainage geocomposite layer(s). In D2 the geocomposite was placed at the subbase/ subgrade interface and in D3 it was placed at the top and bottom of the subbase. Once the frost penetrates below the geocomposite, no benefit can be provided by it. Therefore, the most benefit was expected in D1, provided that the water table remained below the geocomposite during the winter. This is because the subgrade is significantly more frost-susceptible than the subbase.

The geocomposite used is Tenax Tendrain 100-2, tri-planar geonet sandwiched between two needle-punched, nonwoven geotextiles. Construction of the

Figure 1. Photo of original road in poor condition

Figure 4. Photo of end drain.

geocomposite drainage layer is shown in Figure 4 and the material properties are shown in Table 1. The geocomposite drainage layer was connected to collection pipes on both sides of the road and drained into collection stations described by Hayden (1998).

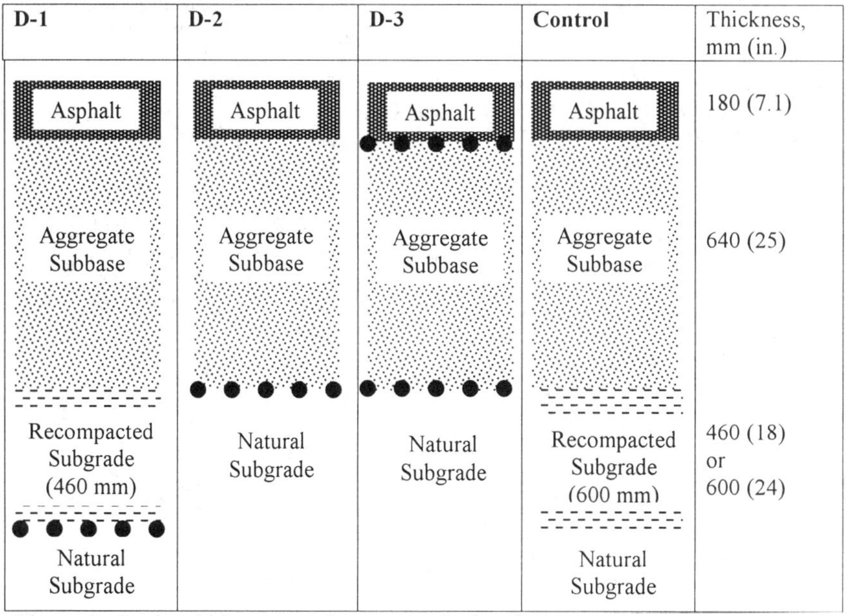

Figure 2. Section thicknesses and geocomposite location.

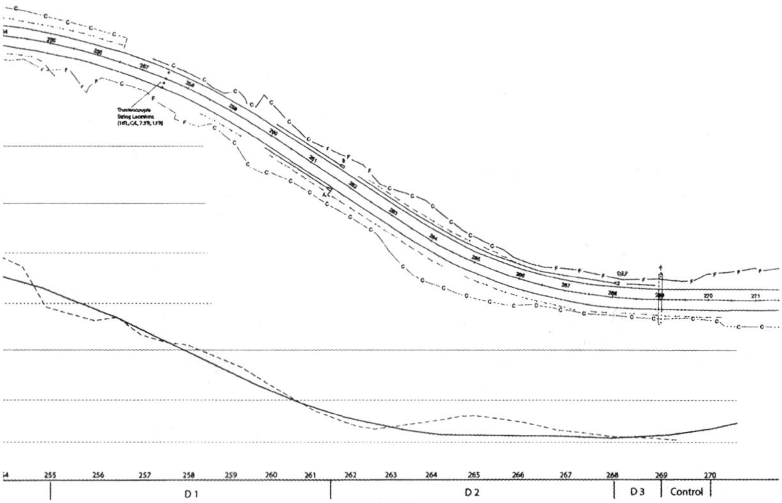

Figure 3. Plan view (top) and elevation (bottom) of test sections D1 through the Control Section. The slope from 255+00 to 261+16 is about 4.6%.

Table 1. Geocomposite properties.

Component/ Property	ASTM Test Method	Minimum Permissible Values
GeoNet Core		
Tensile Strength (Machine Direction)	D4595	14 kN/m
Compressive Behavior (% Retained Thickness)		
@2400 kPa (short term)	D1621	50%
@1200 kPa (5000 hrs)	D1621	60%
Resin Density	D1505	0.940 g/cm^3
Resin Melt Index	D1238	1.0 g/ 10 min.
Thickness	D5199	7.6 mm
GeoTextile		
Apparent Opening Size	D4751	0.125 mm
Permittivity	D4491	1.26 sec^{-1}
Permeability	D4491	0.3 cm/sec
Grab Tensile Strength	D4632	1000 N
Geocomposite		
Ply Adhesion	D413	14.6 N/m

Geocomposite Placement and Instrumentation

Test section D1 was designed to incorporate a capillary barrier in the subgrade, so that upward migrating soil water would be intercepted, thus preventing the upper layers of the subgrade from heaving. The geocomposite was placed directly onto the natural subgrade in D2. Test section D3 was just like D2 except that a second geocomposite was placed on top of the aggregate subbase and the asphalt was placed above it. Thus, in D3 water should be intercepted both above and below the aggregate subbase, minimizing the opportunity for it to freeze in the subbase layers.

During construction of the drainage test sections and the control test section, zones of very soft clay soils were encountered (Hayden, et al. 1998). They deflected with foot traffic and created problems with geocomposite placement—sometimes forming a wave ahead of construction vehicles. This indicates the very saturated nature of the soils at some locations in the subgrade.

Thermocouples were installed in the subbase and subgrade at station 257+50 (ft) -- 10 ft (3.1 m) left of the centerline, and 7.5 ft and 15 ft. (2.3 m and 4.6 m) to the right of the centerline. Vertical strings of thermocouples were installed at depths of 8 to 50 inches (203 to 1270 mm) and from 51 to 77 inches (1295 to 1956 mm) in 6-inch (152 mm) increments. Thermocouples above and below the geocomposite were placed on separate dowels and installed so that no holes were made in the geocomposite.

Six tilt-buckets were installed in the drainage sections to monitor water flow from the geocomposite. Three buckets were installed in section D-1, one in D2, and two in D-3. The capacity of each bucket was known and the number of times it dumped was recorded by a traffic counter (Hayden et al. 1999). Total flow from each of the drainage layers was thus collected and recorded as reported in Christopher, et al (1999).

Freezing Index And Frost Heave Measurements

Freezing Index data were compiled based on surface temperatures (Figure 5). The rate of accumulation of degree freezing days is relatively constant from December through mid-February, indicating that the rate of heat loss at the pavement surface was relatively constant during that time. The elevation of the pavement surface was surveyed periodically during the winter of 1997-1998 at 50ft (15m) intervals to measure frost heave.

Freezing temperatures penetrated to a depth of between 2.5 ft and 3.5 ft (820 to 1150 mm) between December 1997 and January 1998 (Figure 6). Station 257+50 ft, where these temperatures were recorded, is near the center of D1, and the geocomposite is located at a depth of 1.28 m (4.2 ft). Thus, freezing temperatures never reached the geocomposite during the months that the temperatures were recorded. Freezing temperatures penetrated quickly in mid-December but then slowed from mid-December to late January, indicating that there was a change in the thermal properties of the soil as the rate of heat loss remained constant during that time. The frost was in the subgrade layer on 5 January 1998. On 10 February, the date that maximum frost heave was recorded, the frost depth ranged from 1.03 to 0.74 m (3.38 to 2.41 ft) (Table 2). No data were available after 28 January 1998.

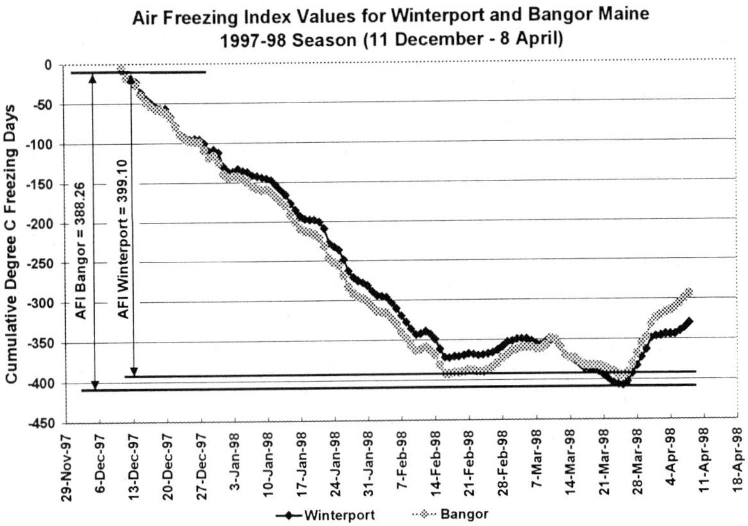

Figure 5. Air freezing index at Winterport and Bangor, ME, 1997-1998

Table 2. Frost depth on 10 February 1998 at 257+50, Test section D1.

Location	Depth of frost (m/ ft)
10 ft (3.04 m) Left of Centerline	1.03/ 3.38
Centerline	No data recorded
7.5 ft (2.29 m) Right of Centerline	0.81/ 2.67
15 ft (4.57 m) Right of Centerline	0.74/ 2.41

Frost Heave, Control Section

Frost heave was recorded at several centerline stations in the control section from mid-December 1997 to mid-April 1998 (Figure 7). These data were recorded to a precision of 0.005 ft. (1.5 mm). A maximum value of 0.5 in. (13 mm) of heave was recorded during early to mid-February for the stations monitored. Note that almost all of the heave occurred after 5 January, indicating that very little heave occurred in the subbase and most of it occurred in the subgrade. The survey data also reveal an accumulated settlement value of approximately 0.5 in. (13 mm) in the spring. The total recorded deflection of the ground surface, from December to April, is approximately 1 in. (25 mm) for this control layer.

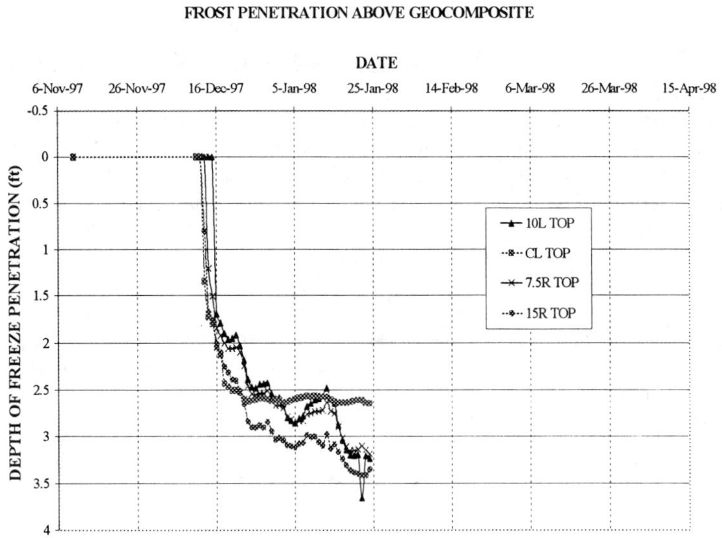

Figure 6. Frost penetration depth (feet) versus date at station 257+50 ft, near the center of D1 (1 ft. penetration = 305 mm).

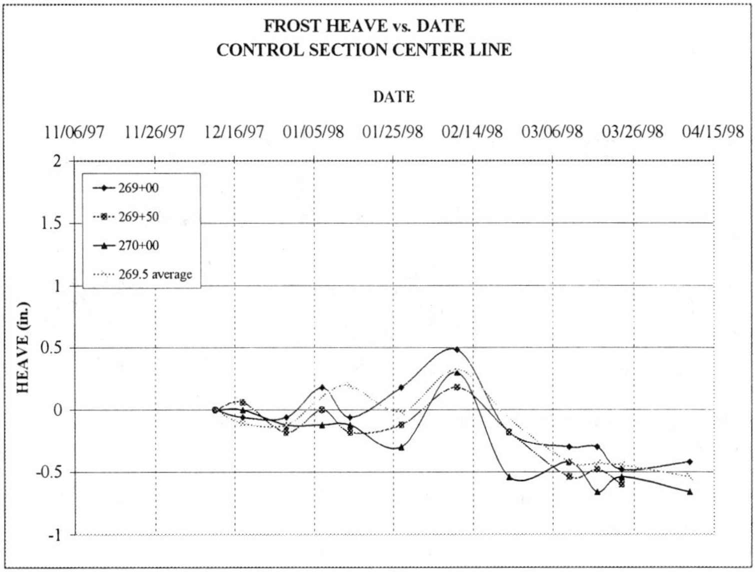

Figure 7. Frost heave (in.) versus date for the control section (1 in. heave = 25 mm)

Frost Heave, D1

Frost heave data recorded at several stations in section D1 from mid-December 1997 to mid-April 1998 are shown in Figure 8. While the reader may not be able to distinguish the individual data symbols, data for all 12 stations are plotted to show the full range of data recorded. The maximum heave measured ranged from 0.5 to 1 in. (13 to 24 mm). The maximum heave occurred in mid-February, coinciding with the maximum air freezing index (AFI), after which the data trends gradually back to zero heave by mid April. Significant heave occurred at some locations through mid December, meaning that the subbase layer experienced frost heave. This, in turn, suggests that there is a very high water table. The fact that heave occurred after 5 January (i.e., in the subgrade) quite possibly suggests that water was able to migrate to the freezing front above the geocomposite. Very little settlement below the initial elevation was recorded in this layer. The total recorded deflection of the ground surface, from December to April, is approximately 1 in. (25 mm) for this drainage layer.

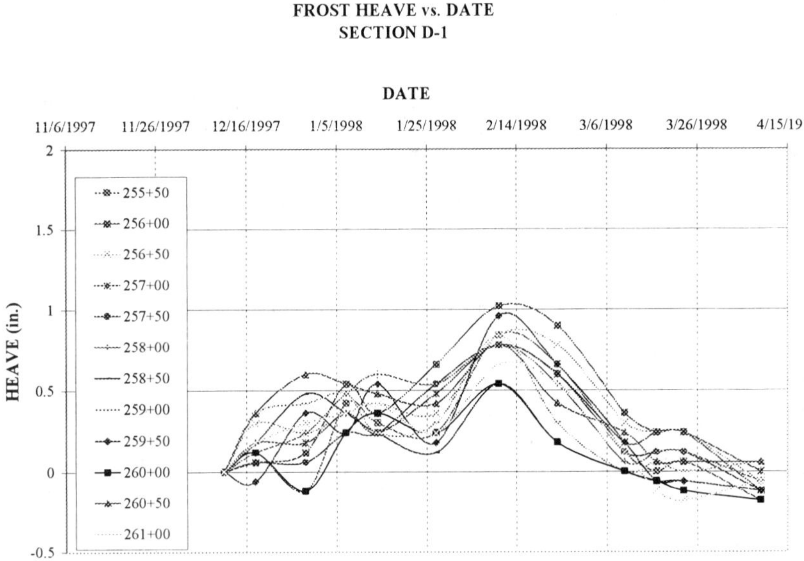

Figure 8. Frost heave (in.) versus date for section D-1 (1 in. heave = 25 mm).

Figure 9 shows the location of the water table with respect to the subbase/ subgrade interface from August 1998 through May 1999 for three locations near D1. (This information is not available for the monitoring period of 1997-98.) The water levels do not indicate the exact water level in the road structure; but they give a general idea of the location of the water level, and its variability, in the subgrade adjacent to the road. Near station 255+50, the water level remains above the level of the geocomposite during the entire time that measurements were made, with one possible exception (in October 1998). Thus, in this location, the geocomposite cannot function to break capillary flow of water and reduce frost heave. This compromises its ability to reduce frost heave in the rest of the test section as water that has migrated above the geocomposite in one section can then migrate in response to suction gradients to lower water content sections. For the water levels recorded at 256+75 and 261+60, the water level is far enough below the geocomposite to indicate good potential for frost heave reduction.

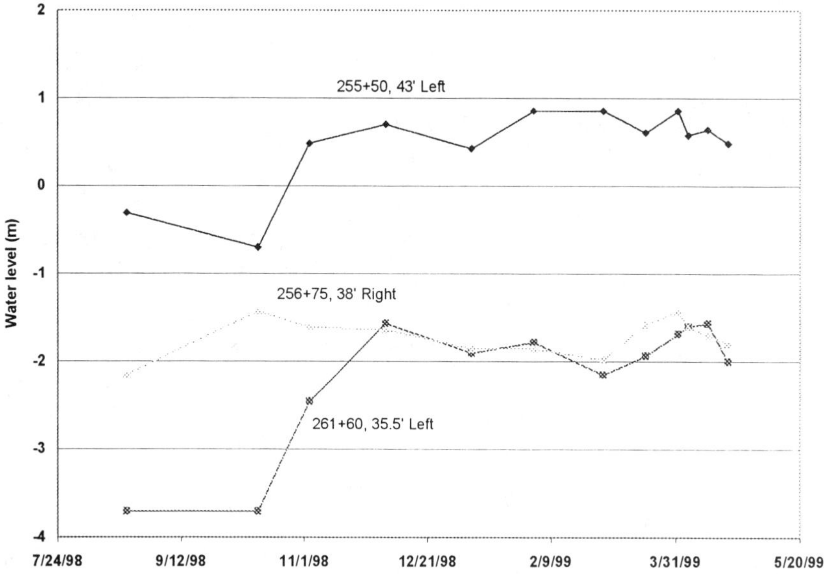

Figure 9. Elevation of the water table with respect to the top of the subgrade at three locations near D1.

Water levels monitored directly in the roadway in April of 2001 were 4.4-4.8 ft (1.34-1.46 m) below the pavement surface at station 260+00—this is approximately the level of the geocomposite. Indeed, tilt bucket measurements made in this test section in 1998 and 1999 indicate a total monthly discharge from test section D1 of approximately 10 times that discharged from any other section during spring thaw and at least ten times the input due to precipitation—indicating saturated (or, supersaturated) conditions at the geocomposite level. It appears that the geocomposite controlled the level of the water table in D1 during the spring thaw.

Frost Heave D2

Frost heave recorded at several centerline stations in D2 are shown in Figure 10.

There is a wide range of heave values in this section. While 9 of 12 stations indicate peak heave of less than 1 in., three stations indicate heave up to 2 in. during early to mid-February. Similar to D1, heave occurred early in the winter, indicating that the subbase layer is frost-susceptible and heaves. On average, the heave in D2 was greater than the heave in D1. The three stations that heaved the most had very large heave rates between 5 January and 14 February, when the subgrade was freezing. There was an average accumulated settlement of approximately 0.25 in. (7 mm) lower than the starting point in the fall. The average total recorded deflection of the ground surface, from December to April, is approximately 1 in. (25 mm).

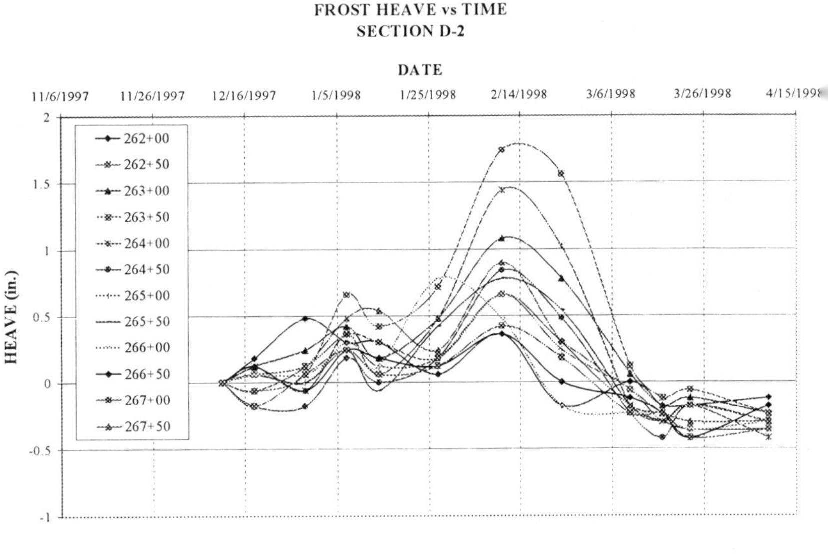

Figure 10. Frost heave (in.) versus date for section D2 (1 in. heave = 25 mm).

Water level readings taken in the roadway during April of 2001 indicate that the water level at station 263+50 was 1.2 ft (0.37 m) below the asphalt surface on 13 April and decreased to 2.9 ft (0.88 m) below the surface on 24 April. The geocomposite, located at the subbase/ subgrade interface is located at a depth of 2.68 ft (0.81m). Therefore, the water level was very high—in the subbase layer—early in the month. The water level might be so high in this test section that it is possible that the geocomposite will rarely function as a capillary barrier.

Frost Heave D3

Frost heave data recorded at two centerline stations in section D3 from mid-December 1997 to mid-April 1998 are shown in Figure 11. A maximum value of 0.5 in. of heave was recorded during early to mid-February for the stations monitored. There was an average settlement of about 0.5 in. (7 mm) in the spring. The total recorded ground surface deflection from December to April is approximately 1 in. (25 mm).

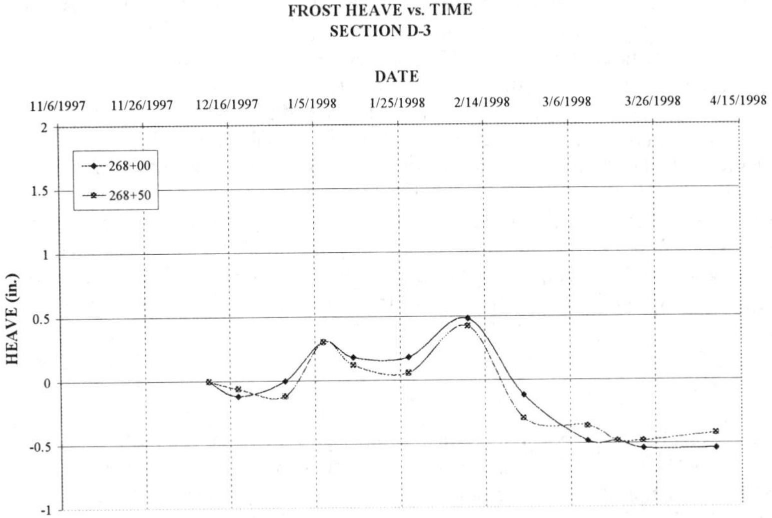

Figure 11. Frost heave (in.) versus date for section D-3 (1 in. heave = 25 mm).

Discussion and Conclusions

Figure 12 shows a combined plot of heave (on 10 Feb 1998) and cut/ fill depth versus station from station 254 to 270. The 10 February 1998 date was selected since it showed the greatest heave values recorded for the period monitored. In general, fill

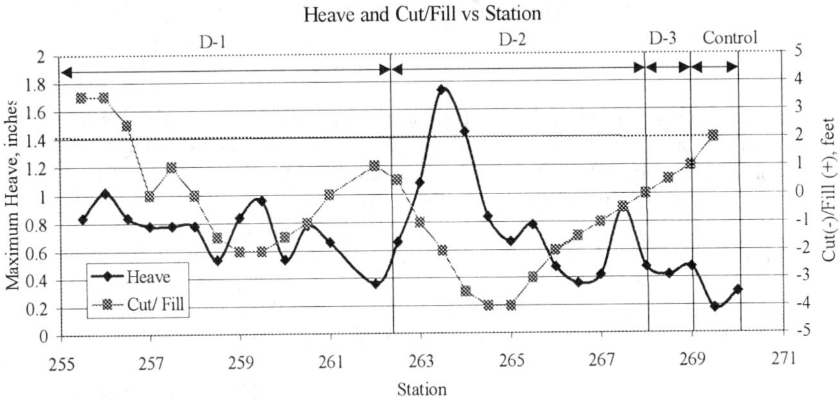

Figure 12. Heave and Cut/ Fill depth (feet) versus station (1 in. heave = 25 mm, 1 ft. cut/ fill = 305 mm)

areas or shallower cuts showed less heave while greater cut depths showed greater heave. It is possible that the deeper cuts brought the road subbase soils closer to the natural groundwater surface while greater fill depths separated the subbase layers from the natural, prevailing ground water surface. However, as discussed above, it is also possible that D1 heaved less than D2 because of the fact that the geocomposite reduced the amount of water that migrated upward during freezing.

The data collected and presented in this paper suggests that the geocomposite was somewhat effective in mitigating frost heave in the subgrade in D1. Observations of the water table elevation suggest that the water table in D1 is so high in some locations that it is at or above the level of the geocomposite. This would seriously degrade the geocomposite's capillary barrier action during freezing. It appears that the primary benefit of the geocomposite is that it is providing dewatering benefit during spring thaw—probably controlling the elevation of the water table. There may also be an effect of the cut/ fill depths that masked the effectiveness of the geocomposite layers.

References

Beskow, G. (1946). Supplement: Some results of Scandinavian soil frost research 1935-1946," in Translation of "Soil freezing and frost heaving with special applications to roads and railroads. The Swedish Geological Society, C, No. 375, Year Book no. 3 (Translated by Jorg O. Osterberg). Technological Institute, Northwestern University, in CRREL Special Report 91-23, U.S. Army Corps of Engineers Cold Regions Research and Engineering Laboratory, Hanover, NH, pp. 161-169.

Casagrande, L. (1938) Examination of the sub-soil of roads. In Proceedings, Eighth International Road Congress, The Hague, 27 p.

Christopher, B.R.; S.A. Hayden and A. Zhao (1999) Roadway base and subgrade geocomposite drainage layers. Testing and Performance of Geosynthetics in Subsurface Drainage, ASTM STP 1390, J.S. Baldwin and L.D. Suits, Eds., American Society of Testing and Materials, West Conshohocken, PA.

Hayden, S. and P. Dunn (1996) The use of geocomposites with reinforcement, separation and drainage applications for highway reconstruction along a 1.9 mile portion of Routs 1A in the towns of Froankfort and Winterport. Unpublished report, Maine Dept. of Transportation, Augusta, ME, 18 p.

Hayden, S. A., B.R. Christopher, D.N. Humphrey, C. Fetten and P.A. Dunn (1998) Instrumentation of reinforcement, separation and drainage geocomposite test sections used in the reconstruction of a highway in Maine, in Proceedings, Ninth International Conference on Cold Regions Engineering, Duluth, MN, pp. 420-433.

Hayden, S.A., D.N. Humphrey, B.R. Christopher, K.S. Henry and C. Fetten (1999) Effectiveness of geosynthetics for roadway construction in cold regions: Results of a multi-use test section, in Proc., Geosynthetics '99 Conference, Boston, MA, 16 p.

Henry, K.S. and R.D. Holtz (2001) "Geocomposite capillary barriers to reduce frost heave in soil," *Canadian Geotechnical Journal,* Vol. 38, pp 678-694.

Henry, K.S. and R.A Affleck (1998) "Freezing tests on lean clay with Tenax Tri-planar geocomposite capillary barrier," report to Tenax, Inc., in fulfillment of 1997 Cooperative Research and Development Agreement between the U.S. Army Cold Regions Research and Engineering Laboratory and Tenax, Inc., of Baltimore, Contract Report CRREL CON97, 18p.

Rengmark, F. (1963). Highway pavement design in frost areas in Sweden. Highway Research Record, No. 33, pp. 137-157

Taber, S. (1929). Frost heaving. Journal of Geology, 38, pp. 303-317, in CRREL Special Report 91-23, Historical perspectives in frost heave research, U.S. Army Cold Regions Research and Eng. Laboratory., Hanover, NH, pp. 29-35.

Taivenen, O. A. (1963). Preventive measure to reduce frost action on highways in Finland. Highway Research Record, No. 33, pp. 202-216.

Polymer-Modified Asphalt Emissions from Alaskan Hot Plants – A Questionnaire Study

Todd Bethard[1], Assoc. Member and Hannele Zubeck[2], Assoc. Member

Abstract

A questionnaire and a literature study were conducted to collect information from private contractors regarding the emissions produced when using polymer-modified asphalt (PMA) pavements in cold regions. The purpose of the study was to investigate if the emissions could be eliminated or reduced with existing experiences and information.

The emphasis of the study was to investigate factors such as equipment, method of mixing asphalt and aggregate, fuel used to heat aggregate, and the materials used that might be causing the emission problems.

On the basis of the research, it was concluded that the existing mixing plants are mostly drier drums older than 10 years, and the asphalt is mixed with aggregate through superheated air in the drum. The most commonly used fuel is diesel #2. The materials used included base asphalts AC-5 and AC-2.5, SBS and crumb rubber modifiers, and proprietary blends. Based on the small amount of materials used by the questionnaire responders, it was impossible to determine which materials create most emissions. Recommendations were given to reduce emissions by using batch plants or modifying existing dryer drums to include special compartments for the mixing of asphalt with the aggregate, and by not exceeding the ideal mixing temperatures.

Introduction

A questionnaire study was funded by the Alaska Department of Transportation and Public Facilities (AKDOT&PF) to collect information from private contractors

[1] Graduate Student, University of Alaska Anchorage School of Engineering, School of Engineering, 3211 Providence Drive, Anchorage, AK 99508; phone 907-274-2000; tbethard@hdrinc.com

[2] Assistant Professor, University of Alaska Anchorage School of Engineering, School of Engineering, 3211 Providence Drive, Anchorage, AK 99508; phone 907-786-1970; hannele.Zubeck@uaa.alaska.edu

regarding the emissions produced when using polymer-modified asphalt (PMA) pavements in cold regions. The use of PMAs has been recommended to control pavement rutting, cracking and premature aging. However, a concern has been raised that PMAs increase emissions from hot plants. At the same time, asphalt plant emissions are increasingly scrutinized by the Environmental Protection Agency (EPA) and Alaska Department of Environmental Conservation (ADEC) due to tighter regulations brought about by the Clean Air Act. Successful use of PMAs in cold regions depends on the ability of paving contractors to meet specifications, create the proper mix, install the pavement properly, and keep the production of asphalt-aggregate mix in accordance with the guidelines set forth by government agencies.

PMAs have been used in pavements for up to 15 years. Due to higher viscosity, the thePMA's are mixed with aggregates at higher temperatures than traditional asphalt cements (Zubeck et al 1999), which may increase emissions for certain material combinations. The other factor that may increase the mixing temperature is the effect of ambient or outside air temperature. It is more critical with PMA mixes than conventional mixes, which means that the ideal mixing temperatures for PMAs must be maintained high. This is a specific concern in Alaska, where, in general, air temperatures are much lower than other states. Most agencies specify the minimum air temperature at which the asphalt mix can be placed to reduce constructability problems and to increase the life of the pavement. Alaska prohibits placing HMA unless the roadway surface temperature is 5°C or warmer.

The purpose of the study was to investigate the emissions from Alaskan hot mix plants producing PMA mixtures and determine if the emissions could be eliminated or reduced through past experiences and existing information. This was done by collecting information from local and out of state contractors. The emphasis of the study was to investigate factors that might be causing the emission problems. These factors include:

- mix producing equipment (age and type),
- the materials used (type of base asphalt, polymers, other additives)
- method of mixing asphalt and aggregate, and
- fuel used to heat aggregate.

The investigation was done as a questionnaire and literature study and is summarized in this paper.

Questionnaire Study

A total of thirteen contractors in Alaska, Arizona, and California were contacted by either phone or mail. Seven contractors who responded that they are using PMA mixtures (or thought they had used them in the past) received a questionnaire given in Table 1. A summary of contacts, responses and anonymous responders is given in Table 2. The list of contractors-producing asphalt cements in Alaska was received from the AKDOT&PF. The others were received through personal contacts. After the questionnaire was sent out three times, three of the four contractors using PMAs responded.

Table 1. Questionnaire

1.	What is the make and model of the hot mix equipment used at the plant?
2.	Is it a batch-type hot plant or a drier drum mix plant?
3.	What is the age of the equipment?
4.	Have any modifications, improvements, or other alterations been made to the equipment in an attempt to reduce emissions? What and when?
5.	What type of emissions scrubber is currently used on the plant?
6.	Which method of mixing the asphalt and aggregate is used? • The asphalt is sprayed through the super-heated air in the drier drum. • The asphalt is added to the aggregate in a separate custom-built compartment. • The asphalt is added in the pug mill of a batch plant. • Other: ________________
7.	What are the typical mixing temperatures with: • Polymer-modified asphalts • AC-5 • AC-2.5 • Other grades (give grade):
8.	Is there a specific mixing temperature at which you notice more or less emissions?
9.	What fuel type is used to fuel the burners at the plant?
10.	Do you store the asphalt-aggregate mixture after mixing? Where and for how long?
11.	What types of materials are used? • Polymer type/grade: • Polymer concentration (if known): • Base asphalt (source and grade): • Proprietary products: • Non-polymer modifiers:
12.	Are air quality records available for the plant (records while using polymers and records when polymers were not added)?
13	Has the plant had an air emission source test (permit compliance test required by ADEC) done while producing the polymer mix? • If so, what were the derived particulate matter emission rates with polymers? • What were they without the polymers?
14.	Have any specific problems been encountered which would affect the quality of emissions from the plant?

Table 2. Contractors Contacted

Contractor	Phone Interview: Do you use PMAs?	Questionnaire Mailed:	Response Received:
Company #1, AK	YES	YES	YES
Company #2, AK	NO	NO	NO
Company #3, AK	NO	NO	NO
Company #4, AK	NO	NO	NO
Company #5, AK	NO	NO	NO
Company #6, AK	NO	NO	NO
Company #7, AK	YES	YES	YES
Company #8, AK	Maybe in the past	YES	NO
Company #9, AK	NO	NO	NO
Company #10, AZ	YES	YES	YES
Company #11, AZ	YES	YES	NO
Company #12, CA	No interview	YES	NO
Company #13, AZ	No interview	YES	NO

Summary of Responses

General: PMA use in Alaska is not routine practice, and only a few contractors have dealt with the emissions implications of this type of application. The responses of the three contractors that contributed to the study are given in Table 3 and summarized in the following sections.

Equipment: Questions about the equipment were asked 1) to develop a database, and 2) because of the authors' hypothesis that emissions from batch plants are less than that from drier drum plants, and that modifications such as special compartments to mix asphalt with aggregate reduce emissions.

All responders are using drier drum plants that are over 10 years old. There is no common brand of drum plant being used neither in Alaska nor in Arizona. All plants are outfitted with baghouse filter emission scrubbers. None of the responders reported modifications to their plants. However, answers to question 6 (about the mixing method) revealed that in two plants, one in Alaska and one in Arizona, the asphalt is sprayed in a "special compartment" or "special injection chamber" added to the pug mill. These are the kinds of modifications that the authors were expecting to find.

Asphalt-Aggregate Mix Production: Questions about mix production were asked to create a database and because of the authors' hypothesis that the following factors increase emissions:

- Asphalt is sprayed or injected onto the aggregate in the drum through superheated air that is actually hotter than the specified mixing temperature. The gases in the drum may be higher than 300°C, whereas the needed spraying temperature is much lower. (Figure 1, FHWA, 1999)
- Increased mixing temperature increases emissions in some cases.
- A specific fuel used to heat the aggregate.
- Long storage time of asphalt-aggregate mixture.

Two of the three responders reported using special "injection chambers or compartments" in which the asphalt is mixed with aggregate to reduce emissions and asphalt aging. The typical mixing temperatures for the PMAs varied from 155 to 168°C, for the AC-5's from 135 to 149°C and for the AC-2.5 a mixing temperature of 135°C was reported. One of the responders from Alaska reported that emissions increased with increased mixing temperature. The other responder from Alaska reported increased emissions above the indicated ideal mixing temperatures. The responder from Arizona did not experience more emissions with increased mixing temperature.

All responders used diesel #2 as a heating fuel for the plant. The Arizona contractor also uses reclaimed oil and natural gas. The storage time for the asphalt-aggregate mixture varies from zero to several hours. One of the responders in Alaska reported a maximum storage time of four hours.

Table 3. Questionnaire Response Questions 1-5

Location	**Company**	**1. Make/ Model**	**2. Batch or Drier Drum?**	**3. Age of Plant (years)**	**4. Modifi-cations?**	**5. Type Emissions Scrubber**
Anch. AK	Company #1	DM 65 Barber Green Drum Mix	Drier drum	20+	No	Bag house
Anch. AK	Company # 7	Boeing 400	Drier drum	12	No	Bag house
Phoenix, AZ	Company #10	9 Drum Plants of different configura tions	Drier drum	<15	No	Bag house

Table 3. Continued, Questionnaire Responses Questions 6-10

Location	**Company**	**6. Mixing Method**	**7. Typical Mixing Temperatures**	**8. Effect of Mixing Temperature?**	**9. Type of Fuel**	**10. Storage of mixture**
Anch. AK	Company #1	Asphalt pumped in at end of drier at metered quantity.	PMA: 155°C AC-5: 135°C AC-2.5: 135° C	As temperatures rise, so do emissions	Diesel #2	None
Anch. AK	Company #7	Special injection chamber in the drum to minimize air exposure	PMA: 168°C AC-5: 138-149°C	Increased emissions as temperatures exceed the indicated temperatures	Diesel #2	4 hour max storage time in silo
Phoenix, AZ	Company #10	Sprayed through super heated air in drum, added to aggregate in special compartment, added in the pug mill	PMA: 166°C AC-5: N/A AC-2.5: N/A Other: 157°C	No	Reclaimed Oil Diesel #2 Natural Gas	Several hour storage in silos

Table 3. Continued, Questionnaire Responses Questions 11-14

Location	**Company**	**11. What types of material are used?**	**12. Air Quality Records**	**13. Permit requirement?**	**14. Specific Problems?**
Anch. AK	Company #1	Polymer: Rubber Concentration: 3% Base Asphalt: AC-5, AC-2.5 Proprietary Products: None Non-Polymer Modifiers: None	DEC Required stack test completed	N/A	Jobs specifying warm ATB should include specifications for virgin rock and not RAP. The RAP produces blue smoke as the asphalt is burned off.
Anch. AK	Company #7	Polymer: SBS: PG58-28 Concentration: 3% Base Asphalt: Williams-Tesoro-AC-5 Proprietary Products: None Non-Polymer Modifiers: None	No	N/A	No
Phoenix, AZ	Company #10	Terminal blends Crumb Rubber	No	No	No

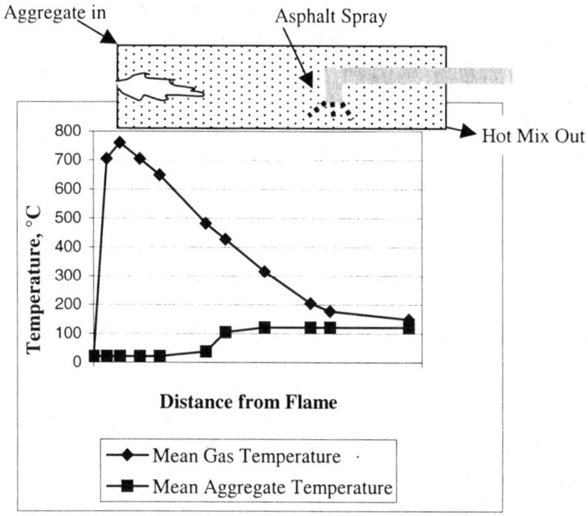

Figure 1. Temperature Profile in Drum Mixer (reproduced from FHWA, 1999)

Materials: Questions about the materials were asked to create a database and because the authors' hypothesis that the following factors may increase the amount of emissions:

- Mixing polymers with base asphalts that are not developed for polymer-modified asphalts,
- Certain polymer types,
- Increased polymer concentration, and
- Additional modifiers, e.g. other oil products that may produce emissions when heated.

Based on the study by Zubeck et al. (1999) pre-blended, often proprietary, blends do not produce excessive emissions. These products are the end result of a comprehensive research and product development program that considers mixing temperatures and emissions.

The Alaskan responders reported that they mix a polymer with local base asphalts, AC-5 or AC-2.5. They are not using pre-blended proprietary products or non-polymer modifiers. The reported polymers were "rubber," which may be a styrene-butadiene-rubber (SBR), or styrene-butadiene-styrene (SBS). The polymer concentration was 3%. The responder from Arizona uses "terminal blends" that the authors call proprietary blends. They also use crumb rubber from recycled vehicle tires.

Air Quality Records: In order to find out which factors affect the amount of emissions, contractors were asked if they have air quality measurements on file.

None of the responders had any records collected. One of the contractors from Alaska reported that they completed the stack test required by the ADEC. None of the responders had the air emission source test conducted (compliance test required by ADEC).

Specific Problems: Two of the responders had no specific problems. The third reported that recycled asphalt mixture (RAP) produces blue smoke when heated at high temperature.

Literature Study

In addition to the questionnaire study, information was researched from other sources. The following sections include summaries from emission publications by the EPA and the National Asphalt Pavement Association (NAPA). The AKDOT&PF's requirements are also summarized below.

The EPA (2000) has developed a means of estimating the amount of emissions from about 350 drum plants. In 1996, over 500 million tons of hot mix asphalt (HMA) were produced in the United States. Of this, an estimated 1,300 drum plants produced approximately 260 million tons. This yields approximately 200,000 tons a year for a typical drum plant. Approximately 85% of the plants being constructed today are drum mix plants.

HMAs produce two major categories of emissions: ducted sources (those vented to atmosphere via a stack, vent, or pipe), and fugitive sources (emitted directly from the source to open air). The primary emission source of a drum plant is the mixers. These mixers produce particulate matter (PM) along with a variety of other gaseous pollutants and are the focus of the EPA study. The PM emissions associated with these HMA production facilities include PM-10 and PM-2.5 (particulate matter less than 10 and 2.5 micro meters diameter respectively), hazardous air pollutants (HAP), and metals and associated organic compounds. Other sources of emissions include storage silos, truck load-out emissions, hot oil heaters, and emissions from the HMA in the truck beds.

The EPA has developed a set of factors to estimate the quantity of emissions released by HMA plants. The factors were derived by dividing the measured emission rates from the plant by the HMA production rate (the amount of asphalt produced). The factors can be used to estimate the quantity of emissions from a HMA plant for a given period of time. These estimates and factors are based on several assumptions, such as: mixers are fired with natural gas, mixer emissions are controlled with fabric filter, PM emissions from load out and silo filling are all PM-10, and the annual HMA production rate for a typical drum mix facility is 200,000 tons/year. Unfortunately these assumptions do not apply for any of the questionnaire responders. The EPA (2000) shows in table form the estimated annual emissions for a typical drum mix plant and presents a detailed list of specific emission types and quantities for drum mix plants. The data provided by the EPA is only in draft form, and revisions are pending. The NAPA disputes some of the data contained in the EPA report (NAPA 1993). However, these emission estimates provide insight into the composition of typical HMA drum plants across the United States.

NAPA has also compiled information regarding construction practices and mixing temperatures for HMA companies to reduce emissions. The NAPA emphasizes the importance of asphalt storage and HMA mixing temperatures. The recommended temperatures for PG grades are given in Table 4.

Agency requirements: The viability of PMA use in Alaska is dependent on the ability of the contractor to meet the regulations placed on them by the EPA / ADEC. The agencies have implemented a stack test to determine the amount of emissions produced by a hot mix plant. The requirements for hot mix asphalt plants found in the Code of Federal Regulations are lengthy. They can be found in 40 CFR Part 60 Subpart i, Standards of Performance for Hot Mix Asphalt Facilities and are summarized below:

- Defines Hot Mix Asphalt Facility as any facility used to manufacture hot mix asphalt by heating and drying aggregate and mixing with asphalt cements.
- No facility, after the completion of the initial performance test, may discharge gases that contain particulate matter in excess of 90mg/dscm or exhibit 20 percent opacity.
- Owner/operator may use only reference methods and procedures for testing and monitoring as specified in the CFR.
- Determination of compliance with the particulate matter standards is as follows:
    - According to the standard for particulate matter concentration, the sampling time and sample volume for each run shall be at least 60 minutes and 0.90 dscm.
    - The opacity needs to determined following a standardized method outlined in Section 60.11.

Stroup-Gardiner and Lange (2002) investigated emission potentials of ten binders from various sources around the country, two of them being modified binders. The goals of their study was to identify key odor-related VOCs and determine if odor reducers interact with them, to develop a simple laboratory test to monitor mass loss and opacity with time, and to identify critical temperatures above which problems can be expected. They identified seven key odor classes of VOCs associated with HMA production. Odor reducers significantly reduced the amount of odor compounds. Combination of two laboratory tests, mass loss rate and opacity measurements were found to have potential for identifying critical temperatures above which smoke and emissions can be expected. Also, the Penske Marten flash point had a fair correlation between visible emissions.

Table 4. Suggested Storage and Mixing Temperatures to Reduce Emissions (NAPA 2000)

Binder Grade[1]	HMA Asphalt Tank Storage Tank Temperature Range (°C)	HMA Plant Mixing Temperature Range (°C)
PG 46-28	127-143	116-146
PG 46-34	127-143	116-146
PG 46-40	127-143	116-146
PG 52-28	127-146	116-149
PG 52-34	127-146	116-149
PG 52-40	127-146	116-149
PG 52-46	127-146	116-149
PG 58-22	138-152	127-154
PG 58-28	138-152	127-154
PG 58-34	138-152	127-154
PG 64-22	141-157	129-160
PG 64-28	141-157	129-160
PG 64-34	141-157	129-160
PG 67-22	146-149	138-163
PG 70-22	149-163	135-166
PG 70-28	141-160	135-163
PG 76-22	157-166	141-168
PG 76-28	154-163	138-166
PG 82-22	157-168	143-171

[1] PG grades are not necessarily polymer modified.

Summary of Findings

Based on the Environmental Protection Agency's (EPA) studies, asphalt plants produce emissions and as a consequence EPA regulates them. According to the questionnaire study conducted, the responders did not have specific problems related to emissions. However, one of the Alaskan responders reported increased emissions with increased mixing temperatures. Another contractor using proprietary blends reported no increased emissions with temperatures. The responders used drier drum plants, and two of the three contractors have modified the plants to reduce emissions. There was not enough data to determine if any specific material or heating fuel causes excessive emissions.

Based on this study, the following recommendations to reduce emissions are given:

- Avoid spraying the asphalt onto the aggregate through the superheated air in the drum by modifying the plant or using a batch plant.
- Use pre-blended proprietary polymer-modified asphalts or conduct a product development program to reduce emissions.
- Do not exceed indicated ideal mixing temperatures.

Due to the low number of Alaskan contractors using polymer-modified asphalts and being able to respond to the questionnaire, future studies should be conducted to get more responses from other cold climate states. A laboratory and field study should be conducted to verify the suitability of simple laboratory tests, such as Penske-Marten flash point, for Alaskan conditions.

References

Environmental Protection Agency, EPA (2000) "Hot Mix Asphalt Plants Emission Assessment Report (DRAFT)." EPA-454/R-00-0XX.

Code of Federal Regulations, 40 CFR Part 60 Subpart i, Standards of Performance for Hot Mix Asphalt Facilities.

Federal Highway Administration, "Hot Mix Asphalt for the Undergraduate," (1999), http://www.fhwa.dot.gov/asphtech/junior/lectures/block4/b4lect1.htm.

National Asphalt Pavement Association, Asphalt Institute, State Asphalt Pavement Associations (2000) "Best Management Practices to Minimize Emissions During HMA Construction." EC 101.

National Asphalt Pavement Association Task Force, NAPA (1993) "Evaluation of Stack Emissions from HMA Facility Operations." Special Report 166.

Stroup-Gardiner, M. and Lange, C., (2002), "Hot Mix Asphalt Smoke and Emissions Potential," TRB 81st Annual Meeting, CD-ROM, January 13-17, 2002, Washington, D.C.

Zubeck, H. K., Raad, L., Ryer, J. (1999) "Constructability of Polymer-Modified-Asphalts and Asphalt-Aggregate Mixtures in Alaska.", FHWA-AK-RD-99-1, ftp://ftp.dot.state.ak.us/pub/nres/Research%20Reports/FHWA-AK-RD-99-1.pdf

Development of a new mechanistic index to predict pavement performance during spring thaw

Guy Doré, ing. PhD.[1]
Cedric Imbs[2]

Abstract

In northern climates, frost action is a major cause of pavement deterioration. It is well known that freezing temperatures in frost susceptible subgrade soils cause the pavement to heave. The resulting displacement imposed to the pavement can be substantial, reaching 100 to 200 mm in severe cases. When spring thaw occurs, segregated ice melts generating high pore pressures in frost susceptible subgrades. The bearing capacity loss can be substantial, leading to important structural damage to pavement exposed to severe and uncontrolled loading conditions.

Considerable research efforts have been devoted to the characterization of soils and material properties and their variation as a function of moisture and temperature variations. These research initiatives are typically conducted in the field using deflexion measurements or in the laboratory using cyclic triaxial testing. The use of these research results typically requires assessing seasonal fatigue damage using analytical techniques to compute stresses and strains at critical location in pavement structures and empirical damage models.

A new approach based on a mechanistic index is proposed. The index is taking into consideration the amount of water accumulated by the freezing process, the rate of thawing in pavement layers and in the subgrade soil and the rate of consolidation of the pavement structure. The paper will describe the theoretical development of the new index and will present correlation between the index and the characteristics of observed weakening (intensity and duration) during spring thaw.

[1] Professor, Civil engineering department, Laval University, Quebec City, Canada G1K 7P4 Tel: (418)656-2203, Fax: (418)656-2928, Guy.Dore@gci.ulaval.ca
[2] Graduate student, Civil engineering department, Laval University, Quebec City, Canada, G1K 7P4

Introduction

Bearing capacity loss of pavement structures during spring thaw has long been recognised as being a major cause of pavement deterioration in cold regions. Highway agencies managing pavement networks in seasonal frost conditions are dealing with the problem using special pavement design procedures and load restriction policies. In most cases, empirical procedures are used

Purpose of the paper

The purpose of the paper is to introduce the concept of a thaw weakening index which would allow for the characterization of soil and pavement material with respect to their behaviour in site specific thawing conditions. It is expected that the mechanistic based index will allow for the prediction of the intensity of the weakening and the duration of the recovery period for each pavement layer individually and for the whole pavement structure.

Theoretical background

During winter, frost penetrates in pavement materials and subgrade soils. While progressing in the pavement structure, frost cause interstitial water to expand and can also cause segregation ice to form in the unbound granular materials. Notwithstanding the fact that these phenomena are generally considered insignificant, they cause pavement materials to loosen. Heaving of pavement surfaces reaching 10-15 mm have been observed on experimental test sites in the province of Quebec before the frost front reached the subgrade soil. When the frost front reaches frost susceptible subgrade soils, water is sucked toward the frozen fringe where ice lenses are formed. Heave of the pavement surface resulting from these phenomena can reach and even exceed 150 mm for climatic conditions prevailing in the province of Quebec. During spring, thaw penetrates into the pavement structure and releases the water accumulated in the interstitial and segregation ice. High water contents combined with lower densities are essentially responsible for the weakening of pavement materials and subgrade soils. The strength is then progressively recovered as soils and materials consolidate (drain) over time.

Figure 1 illustrates that frost heave and thaw consolidation (as a function of time) can be divided in four distinct phases. In the second part of the figure, these phases are illustrated in a density-moisture space. Assuming initial density an water content near optimal conditions, frost penetration generates high suction at the segregation front. Available water is then forced to flow toward the freezing front increasing the water content of the freezing soils (phase A). During the ice segregation process, water is pumped through the frozen fringe causing an important volume increase (reduction of the soil dry density) near the segregation front (phase B). During these two phases, the soil structure is essentially supported by frozen water.

When thaw progresses through the soil layer, water is gradually released from interstitial and segregation ice. This phase, denoted C on the figure is considered as the critical period during the thawing process. At the scale of a very small soil element located at the thaw front, the effective soil strength tends toward zero due to high pore pressures prevailing at that location. If the amount of ice in the soil element and if the thaw rate tend to be low while the drainage capacity and, as a result, the consolidation rate are high, the effect of the strength loss of the soil element is going to be short and limited to a small area. Therefore, the weakening occurring at the thaw front is going to have a limited effect on the bearing capacity of the soil layer. Increasing ice content and/or thaw rate combined with decreasing drainage capacity will cause the thickness of the weakened area and the duration of the weakening to increase. The effect on the bearing capacity of the layer can then be considerable.

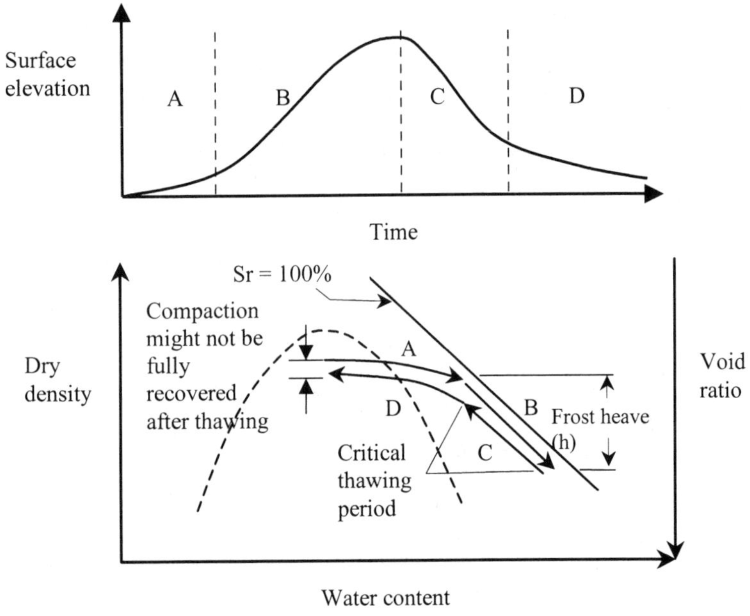

Figure 1. Volume change with time (top) and moisture density relationship (bottom) during the four phases of the freeze-thaw process.

After pore pressures are dissipated in the thawed soil, drainage of interstitial water will continue until equilibrium is reached between the flow gradient and soil suction. During that phase (D), additional settlement is caused by increasing matric suction.

Thaw weakening is thus a complex process which is essentially a function of three major factors:

1. The amount of frost heave occurring in the considered layer
2. The rate at which the layer is thawing
3. The rate at which the layer consolidates

Frost heave occurring in a layer will cause water to accumulate in pores and in ice lenses increasing thus the effective porosity of the material. Water accumulation as interstitial ice and more specifically as segregation ice can be seen as a weakening potential. Frost heave can be measured in laboratory freezing tests or in the field. It can also be estimated using the segregation potential theory proposed by Konrad and Morgernstern (1981) and according to which, frost heave in a given soil layer is equal to:

$$h = 1{,}09nD + \left(1{,}09SP \bullet GradT \bullet \Delta t\right) \qquad \text{(equation 1)}$$

Where: n is the porosity
SP is the segregation potential of the soil
GradT is the thermal gradient in the frozen fringe while the layer is freezing
Δt is the time required to freeze the layer having a thickness D

During thawing, the rate at which the thaw front progresses in the layer will control the rate at which water is released in the material. The rate at which the layer consolidates is in turn an indication of the capacity of the material to drain the released water and to reduce pore pressure. Based on these principles, Nixon and Morgernstern (1971) have proposed the concept of the Thaw-Consolidation Ratio to help predict consolidation behaviour of thawing permafrost. The ratio (R) is defines as follows:

$$R = \frac{\dot{x}}{\dot{h}} = \frac{\alpha}{2\sqrt{c_v}} \qquad \text{(equation 2)}$$

Where $\dot{x}$ is the progression rate of the thaw front
$\dot{h}$ is the consolidation rate
α is the coefficient of $\sqrt{t}$ in the modified Berggren equation
c_v is the coefficient of consolidation

Later, Sarelainen (1997) used the Thaw-Consolidation theory to better explain pore pressure build-up and assess effective stress in thawing soils.

The proposed index, referred to as the thaw weakening index (TWin), combines the weakening potential represented by the total heave normalized by the thickness of the considered layer with the thaw-consolidation ratio as follows:

$$TWin = \frac{h}{D} \times \frac{\dot{x}}{\dot{h}} = \frac{h}{D} \times \frac{\alpha}{2\sqrt{c_v}} \quad \text{(equation 3)}$$

The dimensionless index incorporates therefore most of the factors contributing to thaw weakening behaviour of a given material in a specific environment. Frost heave represents the weakening potential accumulated by frost action. The rate of thawing is in turn a function of the climatic conditions during spring (heat transmitted to the pavement system) and the resulting thermal response (heat absorbed) of the material. Finally, the rate of consolidation is the resulting mass transfer (drainage) and volume change (settlement) in the pavement system.

Figure 2 illustrates a typical variation of stiffness of soils and pavement materials during a freeze-thaw process. Stiffness typically increases at pavement freezes. At the beginning of the thawing period, stiffness drops and reaches a low level (S_s) near the end of thawing. Stiffness is then slowly recovered as pavement structure is drained and consolidates to reach a "normal" level (S_r) several weeks and sometimes several month later. For the present study, the intensity of thaw weakening will be represented by the modular ratio or the ratio of the low spring stiffness level and the normal stiffness level (S_s/S_r). The duration of the weakening period is defined by the time when the stiffness level is below the recovered summer level ($T_r - T_s$).

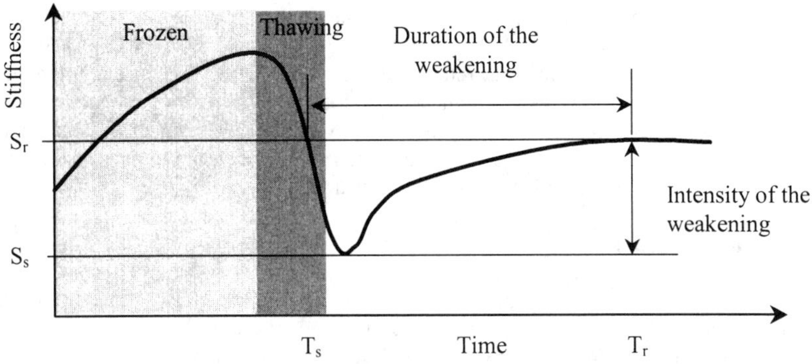

Figure 2: typical variation of stiffness of soils and pavement materials during a freeze-thaw process.

It is postulated that the Thaw Weakening index can adequately characterize soils and pavement materials with respect to their thaw weakening behaviour in any given conditions. It is thus expected that the index can be used to predict the intensity as well as the duration of spring thaw weakening of a pavement system as a function of site specific conditions.

Field evaluation of the thaw weakening index

Field data from a test site intensively monitored during spring thaw was used to assess the validity of the thaw weakening index. The test site is located on provincial highway 155, 25 km south of the city of Trois-Rivières. Highway 155 is a major transportation route carrying an average of 5000 vehicles per day. The test site includes four test sections. The reference section, reflecting normal flexible pavement design practice in the province of Quebec, was used for the validation. The normal freezing index for the area is 1000 °C*d resulting in an average frost penetration of 1,5 m.

The pavement structure for section 1 is composed of 180mm of asphalt concrete over a 300 mm thick crushed-rock granular base and a 450 mm thick granular subbase. The pavement structure is underlain by a 300 mm thick layer of silty sand over clay.

Pavement instrumentation

Section 1 of the St-Celestin test site is the most heavily instrumented section of the project. It includes 7 thermistor strings and 2 series of moisture sensors (time domain reflectometry (TDR) probes) installed along a cross section of the pavement. It also includes a frost tube, a frost heave reference and a piezometer. Pavement response sensors were also installed in section 1. They include two fiber-optic strain gages and a multi-depth deflectometer (MDD). The strain gages were glued at the bottom of laboratory-molded asphalt cores retrofitted in the asphalt concrete layer. The MDD system is composed of three rods anchored at the interface between each layer and a fourth (reference) rod anchored at a depth of 2.5 m. The measurement system is made of four fiber-optic displacement sensors mounted in a case attached to the asphalt concrete layer. The pavement response sensors are illustrated in figure 3.

Testing program

The St-Célestin test site has been intensively monitored since fall 1998. The pavement evaluation program includes pavement condition monitoring (temperature and moisture), pavement response tests under controlled loading (falling weight deflection, inclinometer and ovalization tests) and pavement performance evaluation (roughness, surface distress and rut depth). These pavement response and pavement conditions monitoring activities were performed every week during spring thaw (March and April) and every second week during the recovery period (May and June). Other tests were performed during summer, fall and winter to measure typical pavement condition and response under load during these periods.

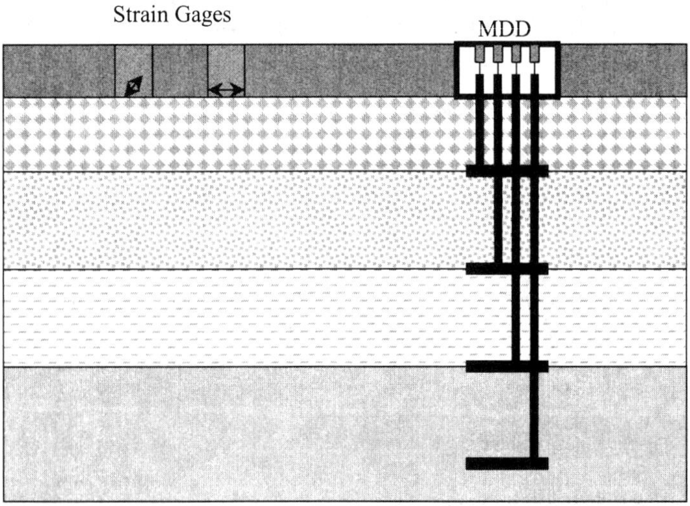

Figure 3. Layout of the pavement response sensors

During the spring of 2001, pavement response measurements with the strain gages and the MDD were added to the monitoring program. Tests were then performed twice a week during spring thaw, every week during the first month of the recovery period (May) and every second week afterwards. A standard "Benkelman beam" truck moving at 50 km/h was used as standard load for the tests. Each test consisted of five valid[3] truck passes that were used to compute the average deflections and strains. The deflection in each layer was obtained by subtracting the deflections measured on the rods anchored at the top and at the bottom of the layer. The strain for a specific layer was then readily obtained by dividing the deflection measured in the layer by the layer thickness.

Test results

The evolution of the freezing and the thawing fronts during winter and spring 2000-2001 are shown in the top part of figure 4. In the bottom part of the figure, the corresponding heave-consolidation curve is illustrated. These data were used to compute frost heave (h) for each individual layer. Frost heave accumulated in a specific layer while the frost front was progressing through the layer was used for that purpose. These data were also used to compute the thaw rate $\dot{x}$ and the consolidation rate $\dot{h}$ for each layer as shown in figure 4. Knowing the thickness of each layer, the thaw weakening index was readily calculated for each layer.

Figure 5 illustrates the evolution of the strains for the strains in the granular subbase during spring thaw

The computer program ELSYM 5 was used to compute representative stresses in each of the layer as a function of the moisture and temperature conditions for each test date. These stresses where then used to compute an equivalent modulus for each layer and for each test date. The ratio of the modulus calculated for the peak strain value (during spring thaw) and the modulus calculated for the summer strain value was used as the weakening measurement for each individual layer.

Table 1 is a summary of the test results for the St-Celestin test road. Thaw weakening indices computed from test data from the Numi-Pulsa test road obtained from Palolahti et al. [1993] and from the Vormsund test road obtained from Nordal and Refsdal [1994] were added to the table. Where available, information on soil and material gradation (passing 0.08mm) was also added to the table.

---

[3] A test was considered valid when the center of the dual tires (rear axle) was within 50 mm of the center of the instruments.

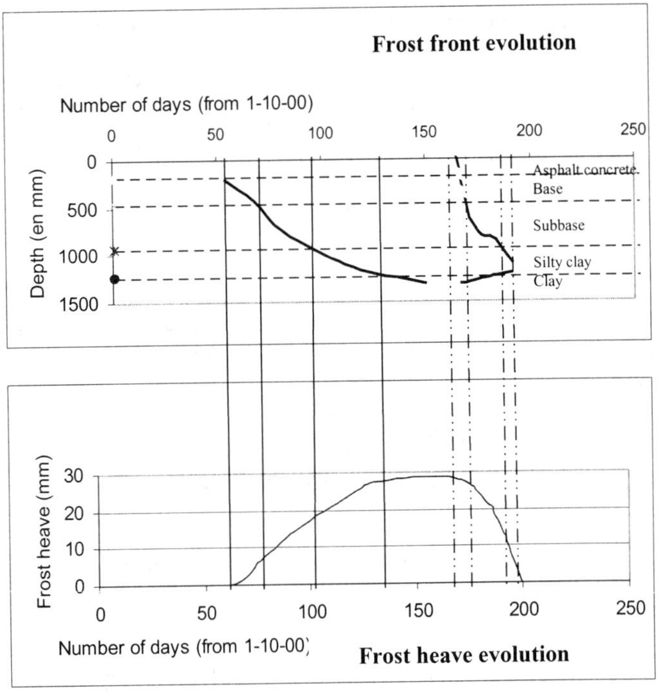

Figure 4 Frost heave and thaw consolidation data for the St-Celestin test road.

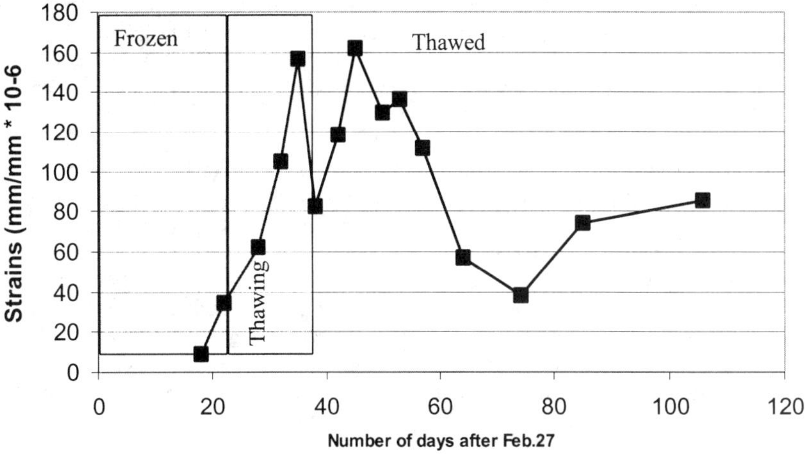

Figure 5. Evolution of the vertical strain in the granular subbase

	St Célestin			Nuumi- Pusula		Vormsund
	2000-01			1991-92		
Layer	Base	Subbase	Silty-sand subgrade	Gravelly sand (base)	Clay subgrade	Clay subgrade
Thickness (mm)	275	475	255	630	400	∞
Thaw rate (mm/d)	98.2	28.3	34.2	20.3	19	17.8
Consolidation rate (mm/d)	0.1	0.56	1.41	0.95	2.35	1.61
Frost heave (mm)	4.6	12.2	11.7	19.4	30.6	72
Thaw weakening index (Twin)	16.43	1.30	1.11	0.66	2.60	2.15
Passing 0,08 mm	8%	12%	25%			
Modular ratio	1.59	1.84	1.18		1.90	

Discussion

The thaw weakening index was difficult to measure for the base material at the St-Celestin test road due to the short duration of the thawing period for that specific layer. The consolidation ratio is not considered to be reliable and the thaw weakening index is thus not considered valid. However, the high value of the index is probably an indication of the high sensitivity of a dense graded, crushed gravel base material to thaw weakening as suggested by the relatively high modular ratio for that layer. Very high thaw rates combined with high moisture contents, relatively low permeabilities and very high load levels probably contribute to this sensitivity.

The thaw weakening index measured for all other layers are consistent with the expectations. Comparison between different types of subgrade soils show that fine-grained soils have higher thaw-weakening indices than coarser soils. Values ranging from 2.15 to 2.60 were obtained for clays while a value of 1.11 was obtained for the St-Celestin silty sand. In the pavement granular materials, the (silty) sand subbase shows a much higher thaw-weakening index (1.30) than the gravely sand at Nuumi-Pusula (0.66). The fact that the subbase at St-Celestin (12% passing 0.08 mm) has a higher thaw weakening index than the silty sand subgrade (25% passing 0.08mm) is probably related to a low rate of consolidation of the first layer which, in turn, is probably associated to poor drainage conditions caused by the frozen underlying soils.

Finally, excluding the base layer at the St-Celestin test road (see previous comment), the modular ratios measured at the two test sites are consistent with the thaw-weakening index, i.e.: higher ratios are obtained for higher indices.

More field data are obviously required to validate the concept of the thaw-weakening index. In the mean time, a laboratory test, incorporating a full freeze-thaw cycle is under development. The test will allow the characterization of the frost heave and thaw weakening behaviour any soil or pavement material under controlled laboratory conditions. It is expected that that laboratory-measured indices will be transposable to any field conditions by adjusting the thaw rate to the site condition.

Conclusion

A new mechanistic index has been developed to predict the mechanical behaviour of soils and pavement materials in thawing conditions. The index is essentially based on theoretical considerations. It has also been validated using limited field data from three test sites, which tend to support the validity of the proposed concept. The thaw-weakening index is a promising tool to characterize soils and pavement materials with respect to their spring-thaw behaviour in any given site specific conditions. The index can be predicted using the segregation potential theory, consolidation characteristics of the thawing soils. It can also be derived from field measurements of frost heave, thaw rate and consolidation rate. It is also expected that the thaw-weakening index can be derived from laboratory measurements using a special freezing cell, which is currently under development.

Aknowledgements

The authors are grateful to the Quebec Ministry of Transportation (MTQ) and the Quebec fund for research (FCAR) for their financial and technical support to the project.

References

Konrad J-M. and Morgenstern N.R. (1981). "The segregation potential of freezing soils", Canadian Geotechnical Journal, vol 18, pp.482-491

Nixon J. and Morgenstern N.R. (1971). "One dimensional consolidation of thawing soils", Canadian Geotechnical Journal, vol. 8, no.4, pp.558-565

Nordal R.S. and Refsdal G. (1994) "Frost protection in design and protection" VTT symposium 94, Frost in geotechnical engineering, Vol.1, pp.127-163

Palolahti A., Slunga S., and Saarelainen S. (1993). "Determination of elastic stiffness of thawing subgrade soils", Frost in geotechnical engineering, Phukan (ed.), Balkema, pp.65-68

Saarelainen S. (1997). "Field and laboratory methods for determining properties in thawing soils", Ground freezing 97, Knutsson (ed.) Balkema, pp.53-61

Pavement Evaluation in Cold Regions

Vincent Janoo and Edel Cortez[1]

Abstract

Pavement evaluation in cold regions is focused on the distress to pavements during the thaw-weakening period. To minimize the damage to the pavement structure, state DOTs apply load restrictions for a certain period of time. The length of the load restriction period has a significant impact on the logging and trucking industries. Over the years, state and federal agencies have developed empirical and semi-mechanistic approaches for predicting the onset of the critical period during thaw, load restriction durations, and the allowable loads on pavements. This paper reviews the current state of the art in pavement evaluation in cold regions.

Introduction

Evaluation of pavements in cold regions includes the effects from the combination of traffic loading and environmental conditions. In cold regions, pavements are subjected to annual freeze–thaw cycles. These freeze–thaw cycles can be beneficial and detrimental to the pavement structure. During the freezing process, the existing moisture in the subsurface layers freezes and bonds the soil particles together. Also, water from below is drawn to the freezing front and creates enriched ice lenses. The bearing capacity of the frozen soil is increased and the pavement structure is able to carry higher loads than it was designed for. During these times, overloading of the pavement structure is allowable with very little damage to it. However, if the subgrade is highly variable, surface cracking can occur as a result of differential frost heave.

During spring or during periods of winter thaw, melting of the ice lenses formed during the freezing period occurs. Depending on the drainage characteristics of the pavement structure and the permeability of the subsurface layer, an undrained loading condition can be created from traffic on the surface. During these times, the bearing capacity of the pavement structure is severely reduced. To minimize the damage to the pavements during spring thaw, many agencies impose load restrictions on their pavements. The length of the load restriction can have a severe impact on movement of goods and material and thus on the economy of the region. Therefore, a pavement evaluation method that has the ability to predict the onset of thaw weakening, the reduced load, and the length of the load restriction period is critical.

The purpose of this paper is to review existing methods for predicting critical and thaw weakening periods and to propose a mechanistic-based pavement evaluation process.

---

[1] Research Civil Engineer, ERDC/CRREL, 72 Lyme Road, Hanover, NH 03755-1290

Pavement evaluation

For detrimental effects of frost action to occur, such as frost heave during the winter or pavement rutting during the spring, the three conditions of freezing temperatures, frost-susceptible soils, and high water table must exist. If any of these three conditions is missing, frost heave or thaw weakening is minimal. Also, if the water table is located at depths of 5 m or deeper, the detrimental effect due to frost action is minimal. A typical stiffness of the base and subgrade in a pavement structure subjected to freeze–thaw is shown in Figure 1. The thaw weakening period is made up of the critical and recovery periods. The critical period is the time when the stiffness of the layer drops from its pre-frozen value to its minimum value. Another definition of the critical state often cited is from the beginning of thaw to its lowest value. The recovery period is the time between when the stiffness is at its lowest value to its pre-frozen state.

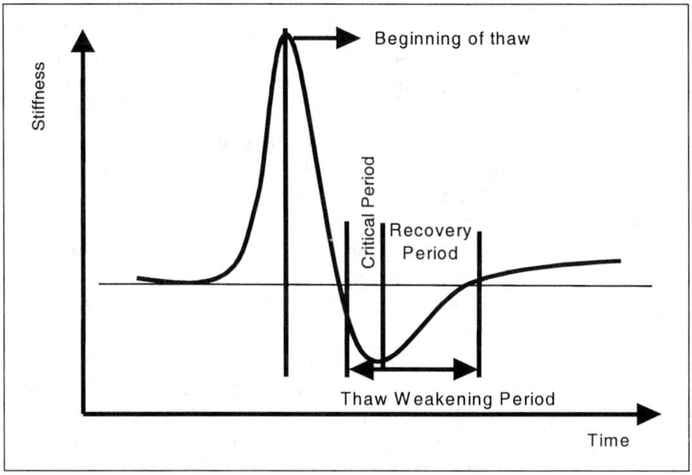

Figure 1. Illustration of terms used in seasonal frost areas.

Critical period

There are several methods for predicting the onset of the critical period. For example, based on theoretical studies and field observations, the Washington and Minnesota DOTs (Rutherford et al. 1985, VanDeusen et al. 1998) found that the onset of the critical period could be estimated from the air thawing Index (TI). They estimate the critical period to begin when the TI is greater than 15 °C days.

The daily thawing index (TI) is calculated as follows:

$$TI = \sum_{i=1}^{n} \left(T_{mean} - T_{ref} \right) \tag{1}$$

where

T_{mean} = average daily temperature (° C)
T_{ref} = reference temperature (° C)
n = number of days into thaw.

The reference temperature ranges between a constant of –1.7 °C (Washington) to a variable (Minnesota). The variable reference temperature ranges from –1.5 °C to 10 °C and is a function of the increase in sunshine during the spring. In Minnesota, once the TI exceeds the 15 °C days, load restriction is placed three days after.

The Corps of Engineers (COE) developed a model for the Federal Aviation Administration (FAA) for predicting thawing events during the winter. The objective was to forecast when heavier than design aircrafts could use existing airfields in the winter. The process initially involves determining the differential air-freezing index (AFI) to create an additional frost penetration equal to half of the maximum base course aggregate size. This differential AFI is assumed then to equal the number of thawing degree-days (TI) required for thaw penetration into the base course equal to half of the maximum base course aggregate size. The reason for half the maximum aggregate size thaw penetration is to limit the thaw penetration so that the base layer remains stable. Exceeding this constitutes the beginning of the critical period. This procedure has been computerized (THAWFORECAST) to connect to an Internet weather service, download the five-day temperature forecast, calculate the thawing degree-days, check against the critical TI value, and post appropriate warnings.

In COE evaluation procedure, the beginning of the thaw weakening is based on the air-freezing index. The air-freezing index (AFI) is the peak difference in the cumulative daily mean temperature. Locations with low AFI usually have low frost penetration and many freeze–thaw cycles, whereas areas with high AFI values usually have deep frost penetration and a limited number of freeze–thaw cycles.

In the thawing index method, the focus is on spring thaw and it is possible to miss the effect of the thaw cycles during the winter as the number of thaw cycles during the winter will affect the bearing capacity of pavement structures. In the COE method, the onset of the thaw-weakening period is tied to the AFI. The following criterion was developed for the beginning of the thaw weakening period: (1) If the AFI ≤ 550 C degree-days, the thaw period begins at the end of ***one half*** of the length of the freezing season. (2) If the AFI > 550 C degree-days, the thaw period is estimated to begin one month prior to the end of the freeze–thaw period. Based on historical data, a database has been established that provides information on the mean beginning and end dates for the freezing season for national and international sites. The database WORLDINDEX can be obtained from the COE PCASE Web site (http://www.pcase.com). The time for the beginning of the thaw period is considered conservative and the COE is looking into reducing the period as follows: (1) If the AFI ≤ 550 C degree-days, the thaw period begins one month prior to the end of the freezing season. (2) If the AFI > 550 C degree-days, the thaw period is estimated to begin at the end of the freezing season. The beginning of the critical period is set to three days after the beginning of thaw.

Thaw weakening and recovery periods

There are several empirical procedures available for determining the duration of load restriction. The length of the thaw-weakening period can be estimated from the air freezing index (AFI). For example, Washington DOT estimates the length of the thaw period (D) by the following equation:

$$D = 0.018AFI + 25 \quad (2)$$

where

D = length of the thaw period (days)
AFI = air freezing index (°C days).

Equation (1) is reported to be valid for AFI range of 200 to 1000 °C days.

Minnesota DOT extended the Washington DOT method based on results from Mn/Road. They developed an equation to predict the length of the thaw period, and the period of load restriction is the time between the end of thaw period determined from (D_{end}) and the onset of the critical period based on the date when the TI was above 15 °C days. In addition to the AFI, the frost depth (P) is used to determine the thaw duration:

$$D_{end} = 0.15 + 0.010\ AFI + 1.91\ P - (1290 * P/AFI) \quad (3)$$

where

D_{end} = length of thaw period, days
P = frost depth (m)
AFI = air freezing index (C degree-days).

The frost depth (P) is estimated by the following equation:

$$P = 0.328 + 0.0578\sqrt{AFI}. \quad (4)$$

The COE method takes into account the frost susceptibility of the soil and the number of freeze–thaw cycles. In the COE evaluation process, the length of the thaw-weakening period is based on the frost susceptibility of the subgrade soil. The frost susceptibility of a soil is classified into eight groups ranging from non-frost-susceptible (NFS) to highly frost-susceptible (F4). Table 1 provides the time for load restriction as a function of frost susceptibility of the soil. Table 1 is for highway pavement and should be used as a guideline; local experience should take precedence. This period begins at the end of the freezing season and includes the time for the pavement to recover to its non-frost condition.

Table 1. Length of load restriction at the end of complete thaw for highways.

Frost susceptibility soil classification	Load restriction period (days)
F1/F2	14
F3	30
F4	45

Strength loss during thaw

It is common to relate the reduction in the bearing capacity of the base/subbase and subgrade layers during the spring thaw to the change in the resilient modulus of the layers. There are several methods available today that predict the change in the resilient modulus and the associated damage to the pavement during the thaw-weakening period. One is the State of Wisconsin, UWFROST model; the other is the COE FROSTLEEP model. FROSTLEEP is under development and is not available to the public at this time. The makeup of UWFROST and FROSTLEEP are similar and the generic submodels making the system are shown in Figure 2. Both models predict the subsurface temperature and moisture content as a function of depth and time. Note that the FHWA Integrated Climatic Model can also be used for this purpose.

Data required by both models to predict subsurface temperatures, moisture, and frost and thaw depths include air temperatures, layer thickness, and layer material properties (thermal and hydraulic). The predicted subsurface temperatures and moistures are used to predict the resilient modulus of the subsurface layers as a function of time.

The difference in the two models is in the modulus prediction algorithms. UWFROST, being regionally based, has a modulus prediction model, based on field FWD tests in Wisconsin. From Falling Weight Deflectometer (FWD) data, Jong et al. (1998) developed two critical periods for a pavement structure during the spring thaw period. The first critical period starts when the thaw reaches the top of the base course. This is based on the discovery, from the FWD data, of a sharp decrease in the base and subgrade modulus during this period. This period tended to last about one month in Wisconsin. The second critical period in a pavement structure is the recovery period; this period lasted up to four months (Fig. 3). The normalized modulus (R_m) is defined as follows:

$$R_m = \frac{M_r(t)}{M_r(t_s)} \qquad (5)$$

where

$M_r(t)$ = resilient modulus at any time t,
$M_r(t_s)$ = resilient modulus in late summer.

The times were further normalized as follows (see Fig. 3):

$$T_t = \frac{t - t_t}{t_e - t_t}$$

$$T_r = \frac{t - t_e}{t_s - t_e} \qquad (6)$$

where

T_t = normalized time for weakening period
T_r = normalized time for recovery period
t_t = time at beginning of thaw
t_e = time at end of thaw
t_t = time when pavement is fully recovered.

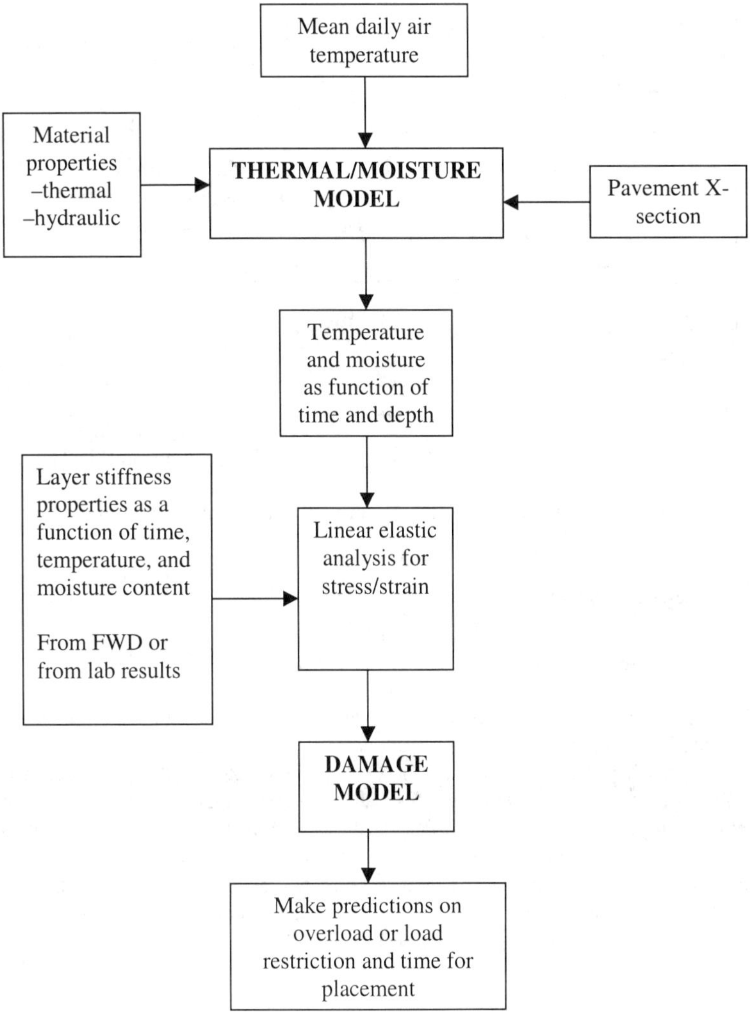

Figure 2. Mechanistic/empirical model for overload and load restrictions.

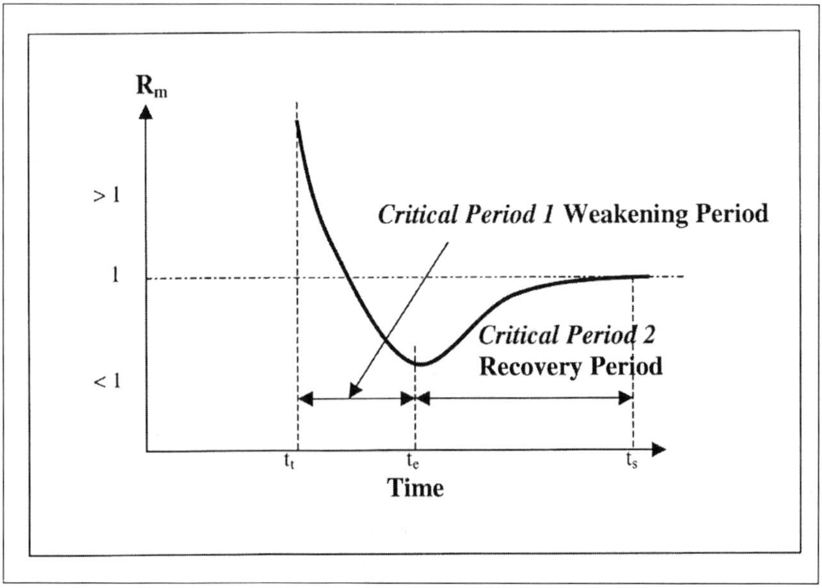

Figure 3. Normalization method during thaw period (Jong et al. 1998).

Based on regression analysis, the normalized modulus (R_m) can be predicted from the following equation:

$$R_m = \alpha + \beta T_i + \delta T_i^2 \quad (7)$$

where α, β, δ are regression coefficients and $T_i = T_t$ or T_r.

In terms of a mechanistic pavement evaluation process, the material properties as a function of temperature and moisture content can be determined as follows. (This is an approach being considered by the COE.) The resilient modulus of the layer is estimated from laboratory and field test data. Material models range from simple to complex, involving several variables, such as stress state, moisture tension, and others. For asphalt concrete (AC) layer, a combination of models is used to predict the asphalt concrete modulus. At temperatures greater or equal to 1 °C, the Ullidtz (1987) model based on back-calculated asphalt concrete modulus is used. At temperatures below 1 °C, the model developed by Schmidt (1975) is used.

Schmidt

$$E = 10^{\left(6.285 - 1.931E-2(T) - 3.280E-4\left(T^2\right) - 1.888E-5\left(T^3\right) + 1.175E-7\left(T^4\right) + 1.502E-8\left(T^5\right) - 2.022E-10\left(T^6\right)\right)}. \quad (8)$$

Ullidtz

$$E_t = 15{,}000 - 7900\log(T). \tag{9}$$

Based on back-calculation of asphalt concrete modulus during the thaw period, Janoo and Berg (1991) developed the following equation:

$$E_{AC} = 5994 - 242(T) \tag{10}$$

where

E_{AC} = asphalt concrete modulus (psi) (1 psi = 0.006894757 MPa)
E_t = asphalt concrete modulus (MPa)
T = pavement temperature (°C).

For the base and subbase layers, data based on laboratory results from Johnson et. al. (1986) are used. The frozen modulus of the base and subbase is a function of the unfrozen moisture content, which in turn is a function of the temperature. Based on the same study, Johnson et al. developed models for predicting the base and subbase modulus during thaw. They predicted the resilient modulus as a function of moisture content, stress state, and dry density.

Frozen

Base

$$M_r = 18.9\left(\frac{w_u}{0.075}\right)^{-4.82} \tag{11}$$

Subbase

$$M_r = 81.8\left(\frac{w_u}{0.055}\right)^{-4.02} \tag{12}$$

$$w_u = 0.03(-T)^{-0.25} \tag{13}$$

Thawed

Base

$$M_r = 2.56\times10^4\left(\frac{101.36-\psi}{\psi}\right)^{-1.99}(J_1)^{0.37}(\gamma_d)^{2.90} \tag{14}$$

Subbase

$$M_r = 8.00\times10^8\left(\frac{101.36-\psi}{\psi}\right)^{-2.99}(J_1)^{0.37}(\gamma_d)^{-5.55} \tag{15}$$

where

M_r = resilient modulus (MPa)

w_u = unfrozen water content
T = freezing temperature (°C)
ψ = moisture tension (kPa)
J_1 = first stress invariant (kPa)
γ_d = dry density (Mg/m^3).

Note that the frozen and thawed modulus is dependent on other factors, such as gradation and percent passing the 0.074-mm sieve.

For frozen and thawing sand subgrades:

Frozen

$$M_r = 2.59\left(\frac{w_u}{0.29}\right)^{-0.85}(J_1)^{0.93} \tag{16}$$

$$w_u = 0.0314(-T)^{-0.29} \tag{17}$$

Thawed

$$M_r = 1.35\times10^6(101.36-\psi)^{-2.36}(J_1)^{-3.25}(\gamma_d)^{-3.06} \tag{18}$$

where

M_r = resilient modulus (MPa)
w_u = unfrozen moisture content
ψ = moisture tension (kPa)
J_1 = first stress invariant (kPa)
γ_d = dry density (Mg/m^3).

The model is similar for the base and subbase with the exception that a stress term is added to the subgrade in the frozen condition. For frozen clays, Bigl and Berg (1996) presented the following model:

$$M_r = 846(w_{u-g})^{-2.161} \tag{19}$$

where $w_{u\text{-}g}$ = gravimetric unfrozen water content.

Simonsen et al. (1999) reviewed the results from resilient modulus testing of various subgrade soils in New Hampshire. For frozen soils, they recommended the following models as a function of soil type:

Glacial till (A-4) and silty fine sand (A-2-4)

$$M_r = k_1 T^{k_2} \sigma_d^{\,k_3} \tag{20}$$

where for the glacial till, k_1 = 3648, k_2 = 0.71, and k_3 = 0.34. For the silty fine sand, k_1 = 3069, k_2 = 0.60, and k_3 = 0.38.

Fine sand (A-1-a)

$$M_r = k_1 |T|^{k_2} (J_1)^{k_3} (\sigma_d)^{k_4} \tag{21}$$

where $k_1 = 16069$, $k_2 = 0.49$, $k_3 = -3.0$, and $k_4 = 2.6$.

Marine clay (A-7-5)

$$M_r = k_1 |T|^{k_2} \sigma_3^{\ k_3} \tag{22}$$

where $k_1 = 363$, $k_2 = 1.42$, $k_3 = 0.46$.

A simpler model relating the subgrade modulus to temperature and moisture content during freezing and thawing was developed using data from Montana field sites (Asphalt Institute 1982).

$$M_r = k_1 (T_{ref} - T)^{k_2} (w_v)^{k_3} \tag{23}$$

where

M_r = resilient modulus (MPa)
T = subgrade temperature (°C)
T_{ref} = reference temp (20 °C)
w_v = volumetric moisture content.

For an A-2-4 soil, $k_1 = 105$, $k_2 = 0.2$, and $k_3 = 0.4$. For an A-4 soil, $k_1 = 200$, $k_2 = 0.7$, and $k_3 = 0.7$. The resilient moduli are then used to calculate the strains at the bottom of the AC layer and at the top of the subgrade. The following failure criteria are available:

Asphalt concrete horizontal strain criteria

$$N_a = 10^{(2.68 - 5\log|\varepsilon_t| - 2.66\log E_a)} \quad \text{(U.S. Army 1988)} \tag{24}$$

$$N_a = a\varepsilon_t^{\ b} E_a^{\ c} \quad \text{(Coetzee and Connor 1990)} \tag{25}$$

where, when $E_a \geq 10342$ MPa, a = 3.364E6, b = –7.370 and c = –4.470; and when $E_a < 10342$ MPa, a = 6.565E6, b = –5.764, and c = –3.640,

N_a = number of load applications
E_a = modulus of asphalt concrete (psi) (1 psi = 0.006894757 MPa)
ε_t = tensile strain at bottom of AC layer
a,b,c = regression coefficients.

Equation (25) was developed from lab fatigue tests as a function of temperature (–26 ° C to 27 ° C) and is significant to pavements during the winter.

Subgrade

$$N_s = 10000\left(\frac{A}{\varepsilon_c}\right)^B \text{ (U.S. Army 1988)} \quad (26)$$

where

N_s = allowable number of load repetitions
ε_c = vertical compressive strain on top of subgrade
A = $0.000247 + 0.00245 \log E_s$
B = $0.0658E_s^{0.559}$
E_s = subgrade modulus (psi), (1 psi = 0.006894757 MPa).

The damage to the pavement structure during the freeze–thaw period is referenced to the damage expected from a standard load during the late summer. Based on the predicted strains, the pavement structure can be overloaded in the winter so long as the strains remain below the reference strain. In the thaw period, again with reference to the late summer strains, the appropriate load restriction (magnitude) can be applied. Forecast of either the overload or load restriction periods can be made based on current and five-day air temperature forecasts.

Summary and conclusions

- There are several empirical models that predict the onset of the critical period and length of the thawing period. Several of the models use the air temperature to determine the time when the critical thawing index is reached. The length of the load restriction period in some cases was set to a constant time period. The empirical models are very site-specific, with the exception of the COE method.
- The COE empirical method accounts for the freeze–thaw cycles seen in areas of low freezing index. It is recommended that, in these areas, the thaw weakening period start approximately one month prior to the end of freezing. The COE also accounts for the subgrade type and its effect on the critical and recovery period. Different periods as a function of soil type were presented and this period is added to the end of the freezing period.
- Semi-empirical mechanistic evaluation procedures for pavements in cold regions were reviewed. These methods allow overloading of the pavement structure, set load reduction, and load reduction periods.

References

Bigl, S.R., and R.L. Berg (1996) Modeling Mn/Road test sections with the CRREL mechanistic pavement design procedure. U.S. Army Cold Regions Research and Engineering Laboratory, Special Report 96-21.

Coetzee, N.F., and B.G. Connor (1990) Fatigue characteristics of Alaskan pavement mixes. Transportation Record No. 1269, Transportation Research Board, National Research Council, Washington, D.C., p. 168–175.

Janoo, V.C., and R.L. Berg (1991) Layer moduli determination during freeze–thaw periods. Nondestructive deflection testing and backcalculation for pavements. *Proceedings, Transportation Research Record No. 1377*, Transportation Research Board, p. 26–35.

Johnson, T.C., R.L. Berg, E.J. Chamberlain, and D.M. Cole (1986) Frost action predictive techniques for roads and airfields: A comprehensive survey of research findings. U.S. Army Cold Regions Research and Engineering Laboratory, CRREL Report 86-18.

Jong, Da-Tong, P.J. Boscher, and C.H. Benson (1998) Field assessment of changes in pavement moduli caused by freezing and thawing. Transportation Research Record No. 1615, Transportation Research Board, p. 41–48.

Rutherford, M.S., J.P. Mahoney, R.G. Hicks, and T. Rwebingira (1985) Guidelines for spring highway use restrictions. Report No. FHWA-RD-86-501, Washington State Department of Transportation.

Schmidt, R.J. (1975) Use of ASTM tests to predict low-temperature stiffness of asphalt mixes. Transportation Research Record No. 544, Transportation Research Board, p. 35–45.

Simonsen E., V. Janoo, and U. Isacsson (1999) Resilient properties of unbound road materials during seasonal frost condition. On thaw weakening of pavement structures, Report TRITA-IP FR 99-50, Royal Institute of Technology, Division of Highway Engineering.

Ullidtz, P. (1987) *Pavement Analysis*. Elsevier Science.

VanDeusen, D., C. Schrader, D. Bullock, and B. Worel (1998) Recent research on springtime thaw weakening and load restrictions in the State of Minnesota. Transportation Research Record No. 1615, Transportation Research Board, p. 41–48.

U.S. Army (1988) Pavement design for roads, streets, and open storage areas, elastic layered method. Department of the Army, Washington, D.C., Technical Manual TM 5-800-09.

City of Duluth, Minnesota
Long-Term Street Improvement Program

History, Past Performance, Current Practice and Future Considerations

John J. Hinzmann, Jr., P.E., Member ASCE [1]; Thomas G. Krzewinski, P.E., Fellow ASCE [2]; David J. Prusak, P.E., Member ASCE [3]

Abstract

This paper continues to document the long-term street reconstruction program in Duluth, Minnesota as previously described in a paper published in 1998 by some of the same authors (Prusak and Krzewinski, 1998).

The program started in 1993, construction started in 1994. At the completion of the 2001 construction season, 38.5 miles of road have been reconstructed. The 2002 program provides for the reconstruction of an additional 6 miles of road. This program is slated to continue until all 355 miles of city streets have been reconstructed.

This paper will discuss the performance of streets in place, expected life cycle costs, current roadway sections, and future modifications to the program.

Introduction

The harsh winters experienced in the City of Duluth can produce frost depths of 5 to 10 feet. The predominant soils in the area are highly frost-susceptible, and in most cases water is readily available to the frost front. Much of the infrastructure is within the zone of reconstruction, and many of the utilities have been in place for seventy years or more. Early in the project conception, a determination was made that it was not cost effective to reconstruct roadways to be frost-free. Instead, the focus became creating a frost-tolerant design.

---

[1] Project Engineer, Short Elliott Hendrickson Inc. ® (SEH), 831 East Second Street, Duluth, Minnesota 55805.

[2] Principal Geotechnical Engineer, American Engineering Testing, Inc., P.O. Box 16008, Duluth, Minnesota 55816.

[3] Chief Engineer of Utilities, City of Duluth, 211 City Hall, Duluth, Minnesota 55802.

Currently, depths of reconstruction vary from 2 to 3 feet in the roadway. The 2 foot section has become the minimum standard/typical, and justification must be supplied to the city for deeper sections when conditions warrant. Utility lines are below the depth of the roadway reconstruction, resting at depths varying from 3 to 10 feet, and in rare cases even deeper. Since frost-tolerant design relies on uniform frost heave, any repairs or modifications of these lines create additional problems for the design and construction of the roadbed.

The City of Duluth has also embarked on a maintenance program for its reconstructed streets. The program currently consists of crack repair and spot repairs only, with overlays anticipated in the future as necessary.

History

Funding is, of course, always of major concern for large ongoing projects such as these. The City of Duluth is both fortunate and farsighted to have funding available from a local casino as a windfall. The City receives annually an amount that will pay for between 75% and 80% of the roadway costs. The benefiting property owners pay the remaining 20% to 25% of the cost through an assessment. The individual utility authorities pay the additional cost for system updates. The assessment history since project inception is as follows:

Year	Assessed Cost Per Front Foot	Miles Reconstructed
1994	$28.50	4.5
1995	$30.00	6.0
1996	$30.00	5.8
1997	$30.00	6.1
1998	$34.00	6.0
1999	$35.00	5.4
2000	$38.00	4.9
2001	$40.00	4.2

The total dollars currently available for this program limit the amount of reconstruction to approximately 4 to 5 miles per year. Early in the program, funding was available for longer lengths of reconstruction. The City of Duluth has prioritized areas to be worked on first by a combination of road condition and utility needs. This prioritization is laid out as a five year program, and updated annually.

Geology and Climate Data

Roadways in the City of Duluth have historically performed poorly with frost boils, potholes and alligator cracking common, and significant patching and repairs a normal part of street maintenance. Curbs and drainage structures are commonly heaved and tilted. Utilities have experienced frost-related damage with uplifted culvert ends, broken pipes, and frequent service interruptions. The residents accepted this poor performance since they were unaware that solutions existed.

Duluth experiences a long, cold winter. The mean annual air temperature is 38.2 degrees F. The design freezing index for Duluth is 3,130 degrees F – days (Steurer 1989). Duluth is also marked with moderately sloping ground. Numerous creeks and drainage ways traverse the slopes. Water is readily available as the ground seasonally freezes.

The predominant soil type in the Duluth area is a glacial till that varies from a silty sand to a sandy silt. In portions of the City, the till is slightly cohesive and is classified as a clayey sand. This material is highly frost susceptible and heaves dramatically as the ground freezes. The second most predominant soil type is a lacustrine clay soil. This material overlies the till soils in the lower lying neighborhoods. This material is moderately to highly frost susceptible. The heave in this soil type is slightly less dramatic. Both soil types experience a significant reduction in strength associated with springtime thawing of the active layer.

Typical Street Before Reconstruction

Typical Street After Reconstruction

Typical Sections

Shown on the following pages are typical sections that have been used for these projects. Figure 1 shows the section in predominant use currently. Figure 2 illustrates a thicker section that has been used in situations where more adverse conditions exist such as steep slopes or very poor subsoils such as peat. Variations of these sections include modification of geotextile fabric type or total elimination of the fabric. All sections include perforated edge drain on both sides.

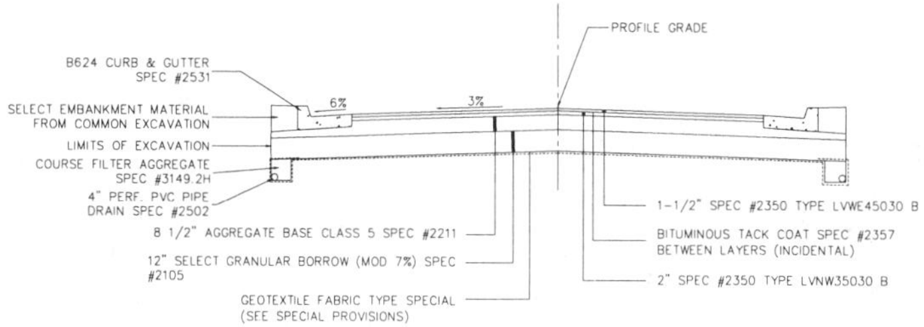

Figure 1. Typical Section - 24" Total

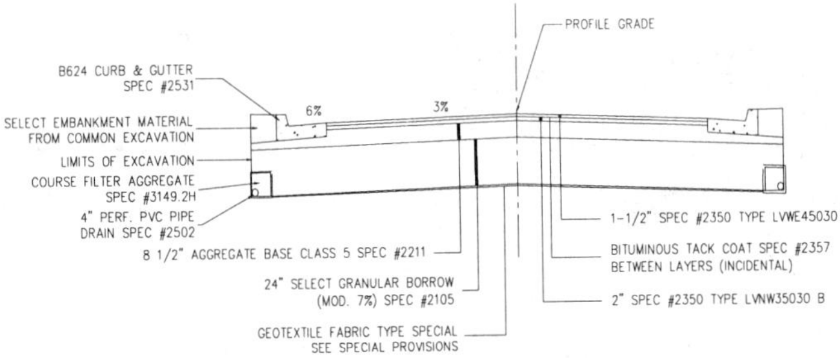

Figure 2. Typical Section - 36" Total

Figure 3 shows a typical utility transverse crossing, and Figure 4 shows a typical utility modification along the roadway. In cases where the native soil is not used to backfill deep trenches along the roadway, perforated pipe is sometimes used to relieve any water that may concentrate in these trenches. This can be a critical element of the design because, again, the design is frost-tolerant only, and the frost movement experienced on the roadway needs to be uniform.

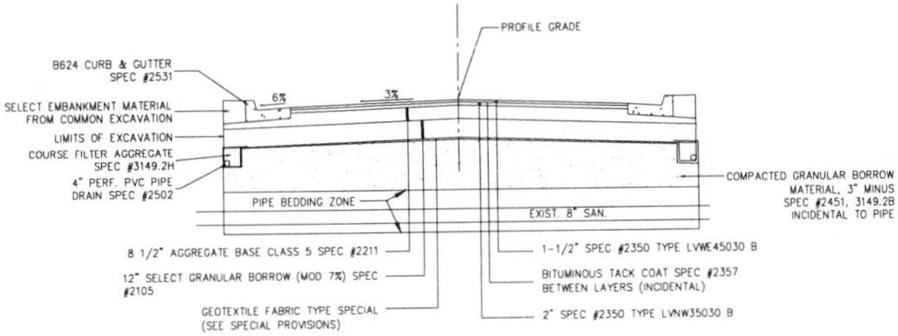

Figure 3. Typical Section - Transverse Utility

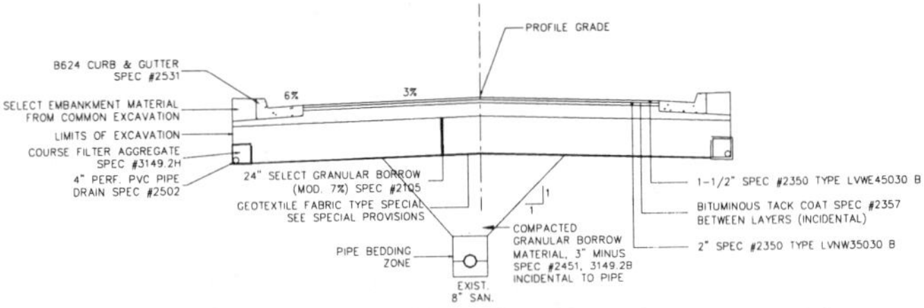

Figure 4. Typical Section - Longitudinal Utility

Geotextile Fabric

The current City specification for geotextile requires the use of a high strength non-woven fabric which provides separation between the subbase and subgrade, provides a capillary break to water moving towards the frost front, promotes drainage through the fabric to the subdrains and provides some degree of subgrade strengthening. It was determined that the non-woven fabrics could be used to minimize the migration of fines from the subgrade into the non-frost susceptible subbase. Maintaining the integrity of this subbase material was determined to be critical to the long-term performance of the roadway. This also made it more likely that the future reconstruction would be limited to overlays, or milling and repaving. In the past, several varieties of both woven and non-woven fabrics have been incorporated into street reconstruction.

One of the options considered was the cost of the fabric, and how it relates to the addition of free-draining sand in lieu of the fabric. The fabric bid prices have been about $.90 to $1.00 per square yard. Non-frost susceptible sand for these jobs has been about $9.00 to $10.00 per cubic yard. Simple math yields the cost of the fabric would equate about 3½" of sand.

Unsuitable Conditions

Often, the soil conditions are not suitable to support the roadway. The three most common conditions are when rock, peat, or deep uncontrolled fills are encountered. Rock is typically handled as shown in Figure 5 for roadway conditions. Rock trenches are generally filled with a non-frost-susceptible material.

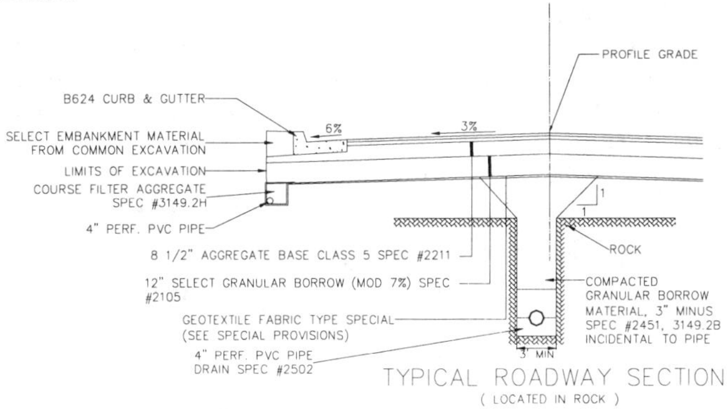

Figure 5.

When peat is encountered, however, there are a variety of solutions that have been attempted, ranging from removal of the entire peat area under the roadway, if small enough, to the addition of geogrid to span the peat area.

The unsuitable conditions have been left to individual designers, and the solutions have varied widely.

Deep miscellaneous and uncontrolled fills are common in the City of Duluth. The sloping terrain was marked by numerous deep erosional gullies, seasonal stream beds, and creek valleys before development occurred. The early developers used whatever material was available to level the playing field and allow movement across these gullies/valleys. Sometimes they placed culverts to allow water to pass, and at other times they chose to relocate the drainage and or connect nearby drainage ways. Attempting to trace the old drainage ways is difficult. Many are currently under significant structures and major roadways. For a frost-tolerant design to be effective, these deep uncontrolled fill areas must be located and considered during roadway design. The bottom of the fill units are sometimes easy to determine with standard drilling techniques since the topsoil layer is commonly left in place at the bottom of the fill unit. Roadway design has addressed these areas on an individual basis with solutions ranging from complete removal of the fill and replacement of the culvert, to continuing the subcut through the deep fill area with no special correction.

The sloping ground in Duluth also makes reconstruction difficult when road widening or straightening cuts into adjacent side hills. Retaining walls have been incorporated into several of the projects. The grade issues which occur when attempting to reconnect existing driveways have been a challenge.

Project Costs

As shown in the history portion of this paper, the initial costs have been slowly rising as would be expected from inflation alone. We would like to present a chart showing a further breakdown of the project construction costs and begin a discussion of maintenance costs. The following shows by year the costs of various elements of the projects, namely: Utility, Storm Sewer (>18" diameter), and Roadway (includes sanitary, water and gas).

Year	Miles Constructed	Utility	Storm Sewer	Roadway	Total
1994	4.5	N/A		N/A	$4,750,000
1995	6.0	$719,575		$5,333,021	$6,052,600
1996	5.8	$952,931		$5,029,539	$5,982,470
1997	6.1	$863,864	.	$5,803,079	$6,666,943
1998	6.0	$754,901		$6,495,508	$7,250,409
1999	5.4	$739,457	$276,096	$5,557,691	$6,573,244
2000	4.9	$807,910	$251,996	$4,699,704	$5,759,609
2001	4.2	$798,480	$383,970	$4,182,057	$5,364,507

These are project costs, and include both engineering design and construction administration.

As indicated earlier, the City of Duluth has embarked upon a program for maintenance of the reconstructed street. It is not completely clear yet as to whether monies for this program will be from casino monies or the general fund from the City of Duluth. To date, there is very little data available as to the costs for maintenance and what may be done in the future.

Typical Crack Repair
Street Improvement Program Street (3 years old)

Typical Crack Repair
Other Street (12 years old)

Performance

The performance of the completed roads has been good overall. Several minor problems have been observed on some of the reconstructed roads and they can be summarized as follows:

- Irregular, longitudinal cracks have been noted.
- A separation between the pavement and the curb has developed.
- Minor heaves or settlements have occurred.
- The location of utilities is obvious.
- Several roads have performed differentially over short segments.

The causes of the problem areas have not been determined but speculation includes poor control during construction, deep fills associated with utility construction adjacent to a typical 2 foot section, unusual water conditions, variable pavement quality, etc.

Conclusion

The work done to date has consisted of several different sections, and to date they have not exhibited significantly different maintenance costs. This, of course, may change, and we must therefore continue to monitor individual areas to help determine the future course of this program.

It is also important to treat the typical section that has evolved as a minimum standard. It is important to engineer for the sometimes subtle subsurface differences in the segments being reconstructed. A proper subsurface exploration program and the delineation of constraints to design is important for the ultimate success of the program.

The program appears to be working well. Approximately 1/8 of the City streets have been reconstructed over the past 9 years. Maintenance for this very valuable resource is underway. There are yet, of course, as in all programs of this nature, unanswered questions. What will the long range maintenance costs be? Could a better section be designed to enhance the life cost of roadways? Should different surfacing options be examined? The City of Duluth is planning to do a detailed investigation this summer in areas where the frost tolerant design has been distressed; evaluating pavement, gravel, subbase, insitu subgrade and water table. This will be compared to sections which have performed better in an effort to answer some of the questions raised. We would expect to report on these findings in a future paper.

References

Kinney, T.C., "Use of Geosynthetics in Road and Airfield Construction in Cold Regions," *TCCRE Monograph Roads and Airfields in Cold Regions*, Vinson, T.S., ed., ASCE, New York, NY, (1986) pp. 271-287.

Krzewinski, T.G. and Tart, R.G., "Thermal Design Considerations in Frozen Ground Engineering," *ASCE Technical Council on Cold Regions Engineering (TCCRE),* Monograph Series, (1985).

Prusak, D.J. and Krzewinski, T.G., "City of Duluth Long-Term Street Improvement Program-History, Current Practice, and Future Considerations," *Proceedings, 9th International Cold Regions Engineering Specialty Conference ASCE*, Duluth, MN, (1998), pp. 397-407.

Steurer, P.M., "Methods Used to Create an Estimate of 100 Year Return Period of the Air-Freezing Index," *National Oceanic and Atmospheric Admin (NOAA) National Climatic Data CTR*, U.S. Dept. of Commerce, (1989).

Tart, R.G., Musial, M.R., and Krueger, M.E., "Pavement Distress Caused by Deep Heave in Anchorage, Alaska." *Proceedings, 8th International Cold Regions Engineering Specialty Conference*, ASCE, Fairbanks, AK, (1996) pp 923-934.

Reduced Tire Pressure: Review of An Alternate Road Usage Technique for Reducing Springtime Damage on Low Volume Roads

Maureen A. Kestler[1], Richard L. Berg[2], Sae-Im Nam[3], and Charles E. Smith[4]

Abstract

Although major highways are designed to withstand high volumes of traffic year round, low volume roads are highly susceptible to damage from trafficking during spring thaw. To reduce pavement damage, states, counties, and Federal agencies often post load restriction signs. However, companies whose livelihood depends on trucking can suffer major economic losses while waiting for thawing roads to recover and load restriction signs to be removed. Additionally, restriction-exempt vehicles, e.g., snowplow, fuel, and garbage trucks, can cause damage on closed roads during spring thaw. Pavement damage can be particularly severe in regions with multiple freeze-thaw cycles. Similar damage occurs in non-frost climates during wet seasons.

The US Department of Agriculture Forest Service has conducted extensive studies showing significant benefits are derived from using trucks with reduced tire pressures on gravel- and non-surfaced roads: Aggregate design thickness can be reduced by up to 50%, reduced rutting results in reduced required maintenance/grading, erosion can be reduced by up to 84%, tire life is increased, and a smoother vehicle ride results in reduced driver injury and fewer vehicle repairs. The Forest Engineering Research Institute of Canada and other agencies have conducted further studies, all of which show appreciable reduction in road damage by using reduced tire pressure. In seasonal frost areas, road maintenance agencies can either shorten the length of the seasonal road closure by allowing lower tire pressure vehicles to resume hauling before full recovery is achieved (yet produce no more damage than on a fully recovered road with conventional tire pressures) or shorten the road closure season even more by allowing hauling with reduced tire pressure to yield an acceptable level of damage. This paper

---

[1] USDA Forest Service, 719 Main Street, Laconia, NH 03246, Tel: 603-528-8778, mkestler@fs.fed.us

[2] FROST Associates, 6 Floyd Ave, West Lebanon 03784, NH, Tel: 603-298-5698, bergnhfla@aol.com

[3] Research Scientist, Hanover, NH 03755, Tel: 603-643-9568, sae.im.nam@valley.net

[4] US Army Cold Regions Research and Engineering Laboratory, Lyme Road, Hanover, NH 03755, Tel: 603-646-4491, csmith@crrel.usace.army.mil

provides a non-technical discussion of several tire pressure studies, the results of which can be used to develop an alternate road usage technique to optimize the balance between reducing road damage and maximizing truck-dependent local economies, particularly in seasonal frost areas.

Introduction

Pavement damage is a function of pavement structure (materials and thickness), environmental factors, and applied traffic. Damage to low volume roads (LVRs) is magnified when the latter two occur in combination, as is the case for trafficking during spring thaw. This paper serves as a primer to spring thaw road damage and the effect of tire pressure on pavement damage, and provides qualitative recommendations on road usage techniques that utilize reduced tire pressure. The first part of the paper:

- Provides a review of pavement damage and failure mechanisms
- Outlines the environmental factors that affect strain, and reviews the mechanics of freezing and thawing of pavement structures
- Outlines trafficking factors that affect strain

In light of the above issues, the paper then:

- Discusses selected tire pressure studies for thaw and non-thaw conditions
    - USDA Forest Service (USFS) test trials at numerous locations (Steward 1994, Kestler and Nam 1999, and unpublished)
    - USFS and US Army Cold Regions Research and Engineering Laboratory (CRREL) analytical study, and field testing for a consortium of departments of transportation (DOTs), Federal agencies, and private companies (Kestler and Nam 1999, and unpublished)
    - Test trial on thawing gravel at Ethan Allen, Vermont (unpublished)
    - Canada test trials (Blair 2001, Davies 1997a, Davies 1997b, Bradley 1997, and unpublished)
- Relates results from tire pressure studies to possible road management techniques for reducing damage to LVRs during spring thaw

In all instances throughout this paper, reference to reduced tire pressure is rigorously reference to an optimal tire sidewall deflection of approximately 20%. Therefore, optimal tire pressures for loaded conditions will differ from those for unloaded conditions.

Failure Mechanisms

Structural deterioration and, ultimately, failure of a flexural pavement are defined by cracking of the asphalt concrete (AC) and rutting in the wheelpaths. It is generally accepted that cracking, or fatigue, is the result of excessive tensile strain at the bottom of the AC due to repeated flexural stress, and that rutting is the result of excessive compressive strain at the top of the subgrade [1] (Yoder and Witczak, 1975). These strains are typically limited during pavement design to prevent cracking and rutting.

Environmental Factors; Freezing And Thawing Of Pavements

LVRs subjected to seasonal freezing are highly susceptible to damage from traffic during mid-winter and spring thaws. As freezing temperatures penetrate moist, frost susceptible subgrades, water migrates upward from the water table or from any moisture beneath the freezing front, ice lenses are formed, and the pavement heaves. When ice lenses melt, excess water remains trapped above still frozen layers. Without adequate drainage, the pavement responds to unconsolidated, undrained conditions by exhibiting a temporary loss of support capacity, known as thaw weakening. Trafficking during thaw-weakened conditions results in heightened strains that exceed design levels, and deterioration occurs at an accelerated rate until heaved soil consolidates and increased pore pressures dissipate.

A disproportionately high percentage of pavement damage occurs during these short thaw-weakened periods. Timed properly, prohibiting hauling can eliminate road damage (no excessive strains, therefore no damage). Alternatively, larger footprints of lower pressure tires result in reduced strains at critical locations, therefore reduced damage. But can the larger tire print reduce the strain enough so that imposed damage is considered acceptable? If so, it is conceivable that reduced tire pressure can either be used instead of or in combination with load reductions, or the window of prohibited hauling can be shortened, thereby allowing economically-needed hauling to resume sooner.

Applied Traffic

- Extensive studies have been conducted on ***axle load***. It is fairly universally accepted that increased load increases strains at critical locations within the pavement structure and accelerates pavement deterioration.
- ***Axle configuration*** has received minimal attention in the literature. Sebaaly (1992) concluded that tandem axles produce lower tensile strains but higher compressive strains than single axles under the same load. Tandem axles produce less fatigue than two single axles with a similar load.
- ***Tire type*** has also received minimal attention in the literature, but Sebaaly (1992) showed tire types did affect load equivalency factor (LEF), which was higher for

---

[1] Researchers are not in total agreement, particularly about the location of the vertical compressive strain. However, these are generally accepted views, and are the beliefs used in the tire pressure studies discussed.

single wide base tires than for duals, for rutting and cracking. Tire type is receiving renewed attention by several States interested in the effect of heavy agricultural equipment on LVRs.

- ***Tire contact pressure*** is the actual pressure at the tire/road surface interface. This is complex and difficult to model because of the variety of tire types and tread patterns. It also changes significantly under different tire pressures.
- Pavement response models assume ***tire shape and pressure distribution*** to be circular and uniform, respectively. They also assume tire pressure equals contact pressure. Using the finite element program ABAQUS, the USFS and CRREL calculated 17% higher tensile strains at the bottom of the AC beneath the low pressure than those at comparable points beneath a round tire of uniform pressure distribution. Results indicate that layered linear elastic analysis methods, generally used for tire pressure studies, may underestimate low tire pressure damage. However, the magnitude of the error was insignificant compared to the difference in strain and damage between high and low pressure tires. (Kestler and Berg unpublished)
- ***Tire pressure*** was initially of interest to the military because of its affect on mobility. However, studies by the USFS and others since the 1980s have shown benefits to extend beyond mobility to include significant reductions in damage to LVRs. Several studies are summarized below. It is believed that relating tire pressure study results to road management techniques has the potential to significantly extend pavement life in areas of seasonal freezing.

Tire Pressure Studies

General

Optimal tire pressures can be achieved by simply inflating and deflating tires to target pressures. Lower tire pressures have been shown to reduce damage, but are recommended for low speeds only. Manually setting to a constant lower pressure for low speed hauls is commonly termed constantly reduced pressure (CRP). Although a time-consuming process, no capital investment is entailed in employing CRP. Alternately, trucks can be equipped with a central tire inflation system (CTIS) that enables the driver to adjust tire pressure from within the cab while the vehicle is in motion. Systems can be built into a truck or retrofitted. Capital investment, which can be as high as $15K or $20K, may at first appear significant, but the cost can be recouped quickly. At present, the military has 45,000 CTI equipped vehicles. Canada's count is on the rise at 325 (Bradley). The USFS has slightly over 100. Otherwise very few agencies or companies employ many CTI equipped vehicles.

History

Variable tire pressure (VTP) technology was used as early as World War I. Both the US and former Soviet Union had crude systems during World War I. The technology was dropped, then re-emerged when the US military needed such a system for increased mobility. In 1983 the USFS launched an extensive program to evaluate the feasibility of using CTI. The primary reason was to reduce road maintenance and reduce aggregate thickness. This reduction is due to the larger tire footprint resulting from lower tire pressure. Construction and maintenance cost savings were estimated at over $20 million per year.

USDA Forest Service Test Trials

The USFS initiated a large-scale test program in 1983. Much testing was done within the USFS; other test components were accomplished outside the USFS under USFS guidance, e.g., Nevada Automotive Test Center (NATC), the Waterways Experiment Station (WES) now part of the US Army Engineer Research and Development Center (ERDC), and a smaller analytical study by CRREL (now also part of ERDC).

- *NATC:* NATC tests indicated that lower tire pressure could reduce road maintenance costs by up to 80% (Ashmore and Steward 1994).
- *WES:* VTP tests conducted at WES, conducted in conjunction with Federal Highway Administration funded tests, showed up to 25% savings on surfacing materials for gravel roads (Smith 1993).
- *CRREL:* Analytical studies are discussed as part of a separate section below.
- *USFS:* Numerous demonstration tests, including mobility tests, were conducted from 1983 through the mid 1990s within the USFS. A few sites / tests include:
    - *Klamath National Forest (NF)*
    - *Moores Creek timber sale, Boise NF*
    - *Alabama loop test*
    - *Drawbar pull tests*
    - *Oklahoma field harvesting operations*
    - *Olympic NF*
    - *Carr Creek timber sale, Hiawatha NF*
    - *Prong flight timber sale, Siuslaw NF*
    - *Florence, OR*
    - *Goodyear tire study*
    - *Lowell, Willamette NF*
    - *Commensurate share test*

Benefits shown by the above USFS demonstrations (Steward 1994) include, but are not limited to:

- 50% reduction in required road thickness (larger tire footprint; lower stresses/strains)
- Reduction in road surface maintenance (kneading effect; less rutting)
- Up to 84% reduction in road-related sediment (less rutting; eliminates channelized water to transport sediment)
- Reduced driver injury (smoother ride)
- Reduced vehicle operation / maintenance costs (smoother ride; less jolting)
- Elimination of rocks getting stuck between the dual tires; fewer blowouts
- Increased tire life

USFS and CRREL Analytical study, and Field Testing for Consortium of DOTs, Federal agencies, and private companies

Because of the positive results of USFS testing on non-thawing gravel- and non-surfaced roads, CRREL and the USFS conducted a series of computer simulations on a matrix of tire pressures, AC pavement structures, and environmental conditions (Kestler and Berg unpublished). All included thawing. For the conditions modeled, results from the US Army Corps of Engineers mechanistic design procedure showed:

- Using reduced tire pressures could significantly extend AC pavement life.
- Peak strains coincided with thaw cycles.
- Peak strains induced by high tire pressure were approximately twice those induced by low tire pressure.
- 414 kPa (60 psi) served as a threshold tire pressure under which negative damage occurred, even during trafficking through spring thaw.
- The influence of reduced tire pressure on pavement life decreases with increasing AC thickness. This observation conforms with Mahoney (1994).

With the addition of several other agencies, a multi-phased effort of controlled testing was recently conducted in the Frost Effects Research Facility at CRREL, under both thawing and non-freeze-thaw conditions.

- *Non freeze-thaw, 5 cm (2 in.) AC atop 20 cm (8 in.) crushed aggregate atop 1.5 m (5 ft) silt subgrade:* Trafficking using a heavy vehicle simulator (HVS) with varying tire pressures resulted in increased pavement damage with increased tire pressure (Figure 1) (Kestler unpublished).
- *Thawing conditions, bituminous surface treatment (BST) atop 20 cm (8 in.) crushed aggregate atop a silt subgrade:* Failure, due to rutting, occurred more quickly for 689 kPa (100 psi) than for 414 kPa (60 psi). Additional detail is provided in Kestler and Nam (1999). However, the sections were so weak and saturated that, under real field conditions, a similar pavement trafficked with low pressure would have failed after approximately one week with just a few trucks daily. This agreed with Sweet (1994), who analytically showed reduced tire pressure reduces springtime road damage, however, he also concluded that, for very weak pavements, the benefits of low tire pressure may still not be economic.

Analysis for the final phase with varying pressures on a thawing AC pavement is currently underway. Preliminary analysis shows that for AC pavements undergoing thawing, heavy traffic can be resumed one to two weeks prior to full recovery provided low tire pressure is employed (unpublished). This yields nearly the same cumulative damage as allowing trafficking to resume at high pressure after full recovery.

Field Trafficking Using a Central Tire Inflation (CTI) Equipped Vehicle at Ethan Allen Training Center, Vermont

A small qualitative demonstration was conducted at Ethan Allen Training Center in Vermont during spring 2000. Equipment was provided by the Vermont National Guard. Ethan Allen site arrangements were made by Capt Bartleau, and testing was conducted by the authors. There are no publications discussing this qualitative trial.

Two adjacent gravel-surfaced test sections undergoing thawing were trafficked using a Heavy Expanded Mobile Tactical Truck Pallet Loading System (HEMTT-PLS) with a CTIS. Test Section I was trafficked using low tire pressure (241 kPa [35 psi]), Test Section II was trafficked using medium tire pressure (379 kPa [55 psi]). Grades and orientation were similar: densities and near surface moisture contents for Sections I and II were 135 +/- 5 PCF and 7 +/-1%, respectively. Because the moisture content of Test Section I was *slightly* higher than that of Test Section II, Test Section I was trafficked using the low tire pressure -- to guarantee results would not be biased.

Because this was not a funded project, no formal data were recorded. However, visual observation, photos, and video qualitatively showed significantly greater rutting occurred on the medium tire pressure section than on the low tire pressure test section. Figure 2 shows the condition of both test sections after 50 vehicle passes.

The vehicle continuously circled around on non-surfaced areas. The area subjected to medium pressure was rutted. In contrast, the area subjected to low pressure remained smooth, appearing as if it had undergone compaction with a kneading compactor (Figure 3). Additionally, the driver (same driver for both sections) reported a much smoother ride while trafficking using low tire pressure than medium tire pressure.

Results were in conformance with USFS CTI field testing, i.e., reduced tire pressure not only resulted in reduced damage to LVRs, but it also healed existing ruts. However, the significance of this small-scale trial at Ethan Allen was that it was conducted on thawing soils. Although analytically calculated for thawing gravel-surfaced and unsurfaced roads, field trials on thawing gravel-surfaced roads were limited to those, of which there were several, conducted in Canada.

Canada Test Trials

Although extensive evaluations have been conducted in the US, only localized use has been adopted. In contrast, Canada has done much testing, and is far more receptive to using VTP. Saskatchewan began implementing reduced tire pressure in the mid 1990s. Davies (1997a and 1997b) provides success stories. Several timber companies in other provinces have also successfully used VTP. Some of the Forest Engineering Research Institute of Canada's (FERIC's) many observations are briefly discussed below.

Many timber access roads in Canada are constructed to minimum standards because access is either short term or is only required intermittently. Although springtime load restrictions can reduce pavement damage, adverse economic ramifications can be significant. Timber companies can stockpile pulpwood or chips before spring thaw, but this requires accelerating operations prior to thaw. It increases inventory costs, and aged chips reduce paper quality. If stockpiling does not last through spring thaw, the mills may have to shut down, thereby requiring additional startup costs (Bradley).

Over the past several years, FERIC has conducted considerable VTP testing for reducing road damage, including thawing soils (Blair 2001, Bradley 1997, Bradley). Field trials using several pressures and reduced load (analyzed using ELSYM5) showed that reduced pressure reduces critical tensile strains more than does reduced load. However, reducing the load has a greater effect on the compressive strain than does reducing the tire pressure (although tire pressure reduction still reduces road damage) (Blair 2001). During tests in Alberta, reduced tire pressure reduced rut development by up to 90%, and healed rutting roads, particularly when trucks varied their wheelpaths (Bradley). Although subtleties differ, FERIC's many test trials all show pavement life can be extended by using reduced tire pressure.

Other Studies

Several others have conducted studies evaluating various aspects of VTP. Bradley (1997) provides a comprehensive overview. With minor variations, the consensus is reduced tire pressure results in a larger tire footprint, which in turn results in lower stresses and strains, and, ultimately, results in less damage for LVRs.

Qualitative Recommendations for Springtime Road Management Techniques

General guidelines are currently being developed for recommended uses of VTP to minimize springtime pavement damage for specific conditions (pavement structures, trafficking levels, and environmental conditions). Although this will not provide a universal optimal tire pressure and corresponding length of time for use, a methodology will also be established for agencies to determine optimal tire pressures for their particular situations. In the meantime, qualitative recommendations follow.
Since CTI is not standard in commercial trucks in the US, it would be necessary to use CRP on low speed roads during spring thaw for the following recommendations.

Restriction exempt vehicles: If a road is typically closed to traffic during spring thaw, the use of reduced tire pressure by restriction exempt vehicles, such as snow plow vehicles, fuel trucks, and garbage trucks will reduce springtime road damage, regardless of site specific conditions. The resulting extension of pavement life has the potential to be significant, because for some very weak pavements, just a few trucks a day during a one or two month spring thaw period can do far more damage than design-level traffic the entire remainder of the (non-thaw-weakened) year.

Replacing road closure with tire pressure reduction: Although reduced tire pressure has been shown to reduce road damage, a straight substitution of reduced tire pressure for the total duration of a former road closure is probably not economically wise for AC roads unless the amount of damage is considered acceptable. If intentions are to substitute tire pressure reduction for road closure of an AC road, one should first calculate stresses and strains for the pavement structure in question assuming no traffic during spring thaw. Then calculate strains using reduced tire pressures, and ask if the resulting cumulative damage is acceptable. Since gravel- and non-surfaced roads are easier to repair, a straight substitution of reduced pressure for road closure depends on whether subsequent road maintenance/blading is deemed worth-while.

***Combined pressure and load reduction**:* In some instances, a combination of reduced tire pressure and reduced load may produce a level of cumulative damage that is deemed acceptable for an AC road. Again, this is site specific for the particular pavement, environmental conditions, and trafficking level.

Resuming hauling with reduced tire pressure prior to full recovery: Other than the blanket recommendation that restriction-exempt vehicles use reduced tire pressure when forced to travel on closed roads, perhaps the most promising result of springtime tire pressure studies is that the window of road closure could be shortened with no additional damage over high pressure tires resuming trafficking after full recovery. Although preliminary analysis indicates one to two weeks for the conditions evaluated, further analysis is needed to determine how much time for a variety of conditions.

Enforcement: Enforcement of lower tire pressure use has been a topic of controversy.

The USFS formerly managed numerous timber sales in which the timber purchaser was often responsible for road maintenance. This provided incentive for the timber purchaser to utilize lower tire pressures on the logging trucks. However, the number of timber sales in the National Forests has dropped significantly in the past several years. While this eliminates incentive for logging trucks to use lower tire pressure on national forest lands, the incentive is similar for private timber/paper companies to implement reduced tire pressure on their own roads (as well as on other roads on route to the mill if hauls are relatively short or low speed). So, in the timber industry, enforcement should not be too major an issue.

A portable Global Positioning System (GPS) based gauge that records pressure over time was recently developed by FERIC. A device of this nature could be required by, or provided to trucks that apply for a permit to haul during spring thaw. Periodic checking of the device would provide a means of enforcement.

Interestingly, once truck drivers in Saskatchewan started using lower tire pressures, they are said to have loved it. Perhaps the smoother drive and added traction will provide the needed incentive for drivers themselves to want to use reduced tire pressure, which will also reduce road damage and vehicle maintenance.

Summary

During the past several years, considerable research has been conducted exploring the benefits of using reduced tire pressure. Benefits (for aggregate- and non-surfaced roads) observed by the USFS include reduced required aggregate thickness, reduced road maintenance, reduced road-related sediment, increased mobility, reduced driver injury, reduced vehicle operation and maintenance costs, and increased tire life. Additional studies by others showed reduced tire pressures likewise extended AC life.

Tire pressure studies have been extended to thawing roads, and it has been shown that damage can also be reduced in areas of seasonal freezing. Current practice in seasonal frost areas is to post load restrictions. However, companies whose livelihood depends on trucking can suffer major economic losses while they await removal of load restriction signs. In such areas, VTP may provide an optimal balance between reducing road damage and maintaining local economies. Canada, which relies heavily on its timber industry, has had much success in its test trials and implementation attempts.

Additional analysis is still underway. However, results from completed studies in combination with preliminary analysis on recent tests indicate a number of possible methods for lengthening the haul season. Possible methods include 1) replacing road closure with tire pressure reduction (not recommended for most AC pavements if not willing to accept some level of damage, however more acceptable on gravel surfaced roads when grading can readily repair damage), 2) replacing road closure with combined pressure and load reductions, and, 3) allowing trafficking to resume prior to full recovery using reduced tire pressures – possibly the most promising potential technique. For the conditions modeled, preliminary analysis shows almost no additional damage for allowing truck traffic to resume one to two weeks before full recovery than for allowing high tire pressure vehicles to resume hauling after complete recovery. Regardless of conditions, using reduced tire pressure on restriction exempt vehicles, e.g., snow-plow vehicles, fuel trucks, and garbage trucks, on roads closed to trucks during spring thaw will in all instances reduce damage.

In conclusion, using trucks with reduced tire pressure provides a promising technique for reducing road maintenance costs, extending pavement life, and improving local economies in seasonal frost areas.

Acknowledgments

The authors wish to thank the USFS, the Vermont National Guard, the Ethan Allen Training Center, and CRREL.

References

Ashmore, S. Colin and John E. Steward (1994). Effects of Vehicle Type and Tire Pressure on Road Maintenance, Nevada Automotive Test Center, Carson City, NV.

Blair, Colin W. (2001). Investigation into Improved Road-Friendliness Offered by Optimized Tire Pressures on Thawing Thin Pavement Roads, FERIC Report, Vancouver, B.C., Canada.

Bradley, Al - Personal Communication, Saskatchewan Highways and Transportation.

Bradley, Al - Presentation at a 1995 Central Tire Inflation Meeting in Louisville.

Bradley, Allan (1997). The Effect of Reduced Tire Inflation Pressure on Road Damage: A literature Review, FERIC Special Report SR-123, Forest Engineering Research of Canada, Vancouver, B.C., Canada.

Davies, Tom, Tami Wall, Al Carpentier (1997a). Saskatchewan Highways and Transportation Reduced Tire Pressure Initiative, presented at the 76th Annual Transportation Research Board Meeting, Washington, D.C.

Davies, Tom, Megan Brown, Al Carpentier, Doug Clark, Ralph Paragg, and Tami Wall (1997b). CRP Tire Heating Trials Federated Coop – Summer 1996, Saskatchewan Highways and Transportation – Internal Report.

Kestler, Maureen A. and Sae-Im Nam (1999). Proceedings of the ASCE Cold Regions Engineering Specialty Conference, Lincoln, NH.

Kestler, Maureen A. and Richard L. Berg (unpublished) report submitted to USFS.

Mahoney, Joe P., Ronald L. Copstead, Robert R. Keller, and Brian R. Sweet (1994). The Potential Use of Central Tire Inflation During Highway Load Restriction Periods, SP1061, Tire Inflation Systems, Society of Automotive Engineers, Warrendale, PA.

Sebaaly, P. and N. Tabatabaee (1992). Effect of Tire Parameters on Pavement Damage and Load Equivalency Factors, Journal of Transportation Engineering, Volume 118, No. 6, November/December.

Smith, D. M. (1993). Effects of Variable Tire Pressure on Road Surfacings, Volume 2: Analysis of Test Results, Technical Report GL-93-20, Vicksburg. MS.

Steward, John E. (1994). Tire Pressure Control Through Central Tire Inflation – The Lightest Foot on the Road, International Road Federation Conference, Calgary, CA.

Sweet, Brian (1994). Effects of Lower Tire Pressure on Frost Weakened Roads, Masters Thesis, University of Washington.

Yoder, E.J. and M. W. Witczak (1975). Principles of Pavement Design 2nd Edition.

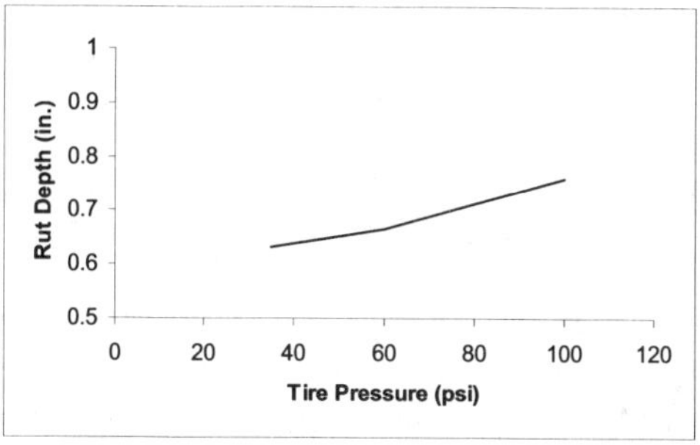

Figure 1. Rut depth profile after 20,000 passes, non freeze-thaw conditions.

a. After 50 passes low pressure. b. After 50 passes medium pressure.

Figure 2. Test sections after equivalent number of vehicle passes.

a. Field traversed by low tire pressure. b. Field traversed by medium tire pressure.

Figure 3. Non-surfaced field in which the HEMTT circled around to traffic sections.

Investigate – Analyse – Make Decisions
The Procedure for Rehabilitation of Roads

Taina Rantanen[1]

Abstract

The essential question in the rehabilitation procedure of the roads is how to find the right activities on a certain road section so that the quality of the service and the life cycle costs are on right level. Pavement management system has proved to be a valuable tool on network level. On project level we have good tools for investigating and analysing the condition state of the roads. The tools for choosing and dimensioning different rehabilitation methods are not yet ready. A lot of development work is going on and is still needed. This paper describes the investigation and repair methods that are already in use. Because the theories are still "open" on many areas so many of the notes are based on the experience the author has gained during the 14 years of working with pavement design.

Introduction

The demand to get more benefit for less money and to save the natural resources are the basis for developing the rehabilitation procedure of the roads in Finland. The aim is to do right activities at right time and right place. The procedure starts with Pavement Management System (PMS) on the network level and ends up with rehabilitation plans for a single road.

The development of the pavement management system started in Finland in late 1980's. Earlier there had been several research projects to define the prediction models for the change of the condition state. The equipment to measure the road condition was being developed at the same time. PMS has been in use for 14 years countrywide. It has been and is a useful tool for planning the rehabilitation actions on network level.

[1]MSc, Project Manager, A-Tie Oy (A-Tie Engineering Ltd),
Satakunnankatu 24 A, FIN-33210 Tampere, FINLAND, phone +358 3 2468 523; taina.rantanen@atie.ains.fi

Thanks to the system there are now lots of condition data for over a decade, which can be used as a starting point for different analysing and planning projects. The condition data serves as telling how the road looks like. Different investigations are needed to get the answer why the road has deteriorated. There are many good methods that are already in use: falling weight deflectometer, ground penetrating radar, different laboratory investigations concerning the properties of the materials.

The Use of Road Condition Measurements

The pavement management system contains two data banks: condition data and road data (general information of the road). The condition data consists of rut depth, roughness, defects and bearing capacity. Ruts, evenness and defects are each given as an average for each 100-meter section of the road, bearing capacity is measured in longer sections.

With PMS one is able to analyse this data to

- evaluate the present state of the road network
- predict the changes in condition state
- set goals for the future and plan maintenance policy
- make annual rehabilitation programs and
- evaluate the effects of the planned maintenance actions.

PMS shows the road sections needing rehabilitation actions. The next stage is to make plans in more detail. The condition data is a starting point to evaluate the structural condition and the type of rehabilitation action that is needed. Each measured condition parameter refers to certain problems on the road structure. The most commonly used indicators are demonstrated in table 1. The road can be divided into segments according to the problems and further investigations can be planned and focused on right locations.

Table 1. The information of different condition state indicators

Condition state indicator	**Possible cause**
Ruts (mm)	Lack in bearing capacity, material problems on bearing layer/asphalt
Defects - Sum of defects (m^2) (calculated from defects by using weighed factors /1/) - Longitudinal cracks (m) - Alligator cracking (m^2)	- Gives the most deteriorated sections - Uneven frost heaves, lack in bearing capacity - Lack in bearing capacity, material problems on bearing layer/asphalt
Evenness IRI (mm/m)	Uneven frost heaves; the structure may be mixed with the frost susceptible base during thawing

In the data bank there is no information of the exact location of individual defects. The cause of a defect is though different if the crack situates beside/in the rut or in the middle of the road. The first one refers to lack in bearing capacity, the latter to uneven frost heave. In-situ observations are therefore always needed to get this information. They should be made during the thawing period when all the defects are open and easy to observe and for example frost heaves are to be noticed. A video film is also useful but it doesn't replace the in-situ observations.

Investigation Methods

Every road and each section of the road doesn't need the same investigations. According to the site observations and condition data it is possible to determine the possible causes of the structural problems and to choose the right investigation procedure. One should bear in mind that the results of the investigations in many methods depend on the time of measurement.

The most commonly used investigation methods for road structures are falling weight deflectometer (FWD) and ground penetrating radar (GPR). Taking samples for laboratory tests and other "traditional" investigations are also needed.

Falling weight deflectometer

Different parameters can be calculated from the FWD-measurement data. The ones that have proved to be most useful in practice are listed below. The research concerning FWD-data hasn't so far given exact threshold values for describing the state of the condition, however, the order of magnitude is known. The values given here are based on experience.

- Bearing capacity E_2 is calculated from the deflection d_0 at the loading point. The value depends on both the road structure and the subsoil.
- Surface curvature index SCI (we use d_0-d_{300}) tells about the condition of the upper part of the pavement. The bigger the value is the weaker is the layer. Values about 200μm and more tell about a weak layer on the upper part of the structure.
- d_6 (the deflection at the distance of 1200 mm from the loading point) tells about the stiffness of the subsoil. Values near zero refer to rock and values over about 300 to peat.
- The modulus of elasticity of different layers can be back calculated from the deflection values. Different computer programs are available for the analysis.

When using the data one should always consider the environmental effects by the time of investigation. If for example the content of fines in the bearing layer is high the bearing capacity of the layer is weak during thawing, and as a consequence alligator cracking occurs. If the FWD-measurements are carried out in late summer the structure appears to be in good condition because the weak layer has become stiffer by reduced moisture content. If the problem is ignored and only repaving is carried out it is most likely to have alligator cracking quite soon in the new pavement.

It would be perfect to have the measurements during the thawing time. There is though always a risk to have still frost and the measurements are not reliable. It's therefore recommended carrying out the FWD-measurements in early summer or –especially on gravel roads- in autumn, when the moisture content of the structure and subsoil is high.

Ground penetrating radar (GPR)

With GPR we can get continuous information about

- the thickness of different structure layers
- the discontinuities in the layers (excavations etc. made earlier)
- the subsoil
- the moisture contents of different layers

To get the information out of GPR-measurements it is essential to make drillings and take soil samples. The FWD helps to interpret the measurements. When the measurements are carried out in winter the discontinuities in the frozen zone help to localise possible rock.

The GPR-technique is under constant development and has become one of the basic methods in evaluating the reasons for the deterioration of the road.

Laboratory investigations

Research work on materials is an essential part of developing the rehabilitation of the roads. If the defects refer to problems in materials then laboratory testing is needed. The moisture susceptibility is examined by tube suction test /2/ or by determining the water adsorption.

Laboratory testing is also needed when we are planning to improve the existing material by adding for example bitumen. A lot of research work is going on also in this field and we are waiting for new results to apply in designing/dimensioning the rehabilitation actions.

Analysis of the Results

The more there is information about the damages and the structures of the road the more confident we are to make the right conclusions and decisions. If all the information is referring to the same problems, it is easier to choose the right and most cost effective rehabilitation action.

With this analytical thinking we are able to divide the road to sections due to the problems and choose the right actions. Here is an example:

- Section 1: lack in bearing capacity due to problems in materials
  => STABILISATION (usually bitumen stabilisation, a common method to improve the existing materials)
- Section 2: longitudinal cracks caused by uneven frost heave
  => STEEL NET REINFORCEMENT /3/ (a commonly used method, usually installed in the non bound bearing layer)
- Section 3: damages caused by the weak subsoil during thawing
  => NEW STRUCTURES (RECONSTRUCTION)
- Section 4: no structural problems, defects caused by the ageing of the asphalt
  => REPAVING

Decision Making

One aim of the procedure is to find the locations where reconstruction is the only option and where other cheaper methods (stabilisation, steel nets etc) can be used. Especially when we are talking about low volume roads there isn't usually enough money for "safe" solutions. We must choose the second best and, what is most important, we must know the procedure that is out of question (waste of money). Usually we have to take risks and allow for example minor frost heaves and longitudinal cracks occur after repairing.

All the decisions should follow the general policy of the road maintenance. Everybody in the process (decision-makers, designers, constructors) from the first PMS-made rehabilitation programs to the implementation phase should be aware of and agree the target: to get the road serve as well as is needed. Every road and especially every section of a single road doesn't need the same maintenance.

Conclusions

The development of the pavement management system and the condition measurements has been a good starting point in getting the rehabilitation of the road network under control. The system itself is working well. Lots of efforts are still needed and are already going on to get the whole chain of actions to implement the same target: right actions on right places.

We have now good methods for investigating and analysing the condition state of the roads. We don't still have enough knowledge of how the rehabilitation actions have really affected the road condition. That is significant in choosing the right actions, in developing the PMS prognosis and in determining the models for lifecycle costs. Especially, the low volume roads have problems: the structures of the roads are so inhomogeneous that it's hard to find a model describing their behaviour under dynamic load.

References

/1/ Rantanen Taina (1993). Road condition measurements and pavement management in Finland, FinnRa; Helsinki, TIEL 3200177E

/2/ Saarenketo Timo, Kolisoja Pauli, Vuorimies Nuutti, Ylitapio Seppo (2001). Suction and deformation properties of base course aggregates. FinnRa, Helsinki

/3/ Using steel nets as reinforcement in road structures (1999,only in Finnish), Innogeo Oy, Tammet

Deck Wearing Surfaces for the Yukon River Bridge

J. Leroy Hulsey[1], Lutfi Raad[2] and Billy Connor[3]

Abstract

The Yukon River Bridge in interior Alaska is recognized as an outstanding civil engineering design. The structure is a six-span 884.4-m steel bridge on a 6 percent grade. Two 155.9-cm wide by 414.0-cm deep steel box girders support an orthotropic steel deck.

The structure carries the oil pipeline, vehicles, tourism busses and heavy trucks. In addition to highway traffic and the oil pipeline, there is potential for the structure to carry a future gas line. Weather can vary from harsh winter temperatures combined with snow and ice to mild summer temperatures.

This structure was designed in the early 70's with a temporary 127-mm two-layer timber deck-wearing surface. Since then, timber running planks have been replaced three times with similar timbers (1981, 1992, and 1999). In 1992, in addition to running planks, timber deck planks were replaced and the steel deck was cleaned and coated against corrosion.

A successful wearing surface for this bridge must be cost effective, lightweight, abrasion resistant and provide traction; especially during winter months. A wearing surface selection criterion is discussed and wearing surfaces on similar bridges are examined. Two examples illustrate a design procedure for selecting a wearing surface on this structure; design charts are used. The design charts account for truckloads and temperature change. No charts are available for traction or abrasion.

Introduction

This paper presents a discussion of wearing surfaces for orthotropic steel decks. Wearing surfaces for the Yukon River Bridge orthotropic steel deck are discussed and two example wearing surfaces are selected.

Background. Consider a wearing surface over the othotropic steel deck on the Yukon River Bridge. The driving surface is on a 6 percent grade and the traffic count is light compared to other bridges of its type.

The deck was designed for a wearing surface weight of 1,436 Pa and HS20-44 highway loading. A 127-mm thick temporary-wearing surface was installed in 1976. This was a two-layer untreated Grade 2 timber deck. Transverse timber planks were bolted to the steel deck. Timber running planks parallel to traffic were installed over the deck planks.

[1]Assoc. Prof. of Civil & Env. Eng., Univ. of Alaska Fairbanks; ffjlh@uaf.edu,
[2]Prof. of Civil & Environmental Engineering, University of Alaska Fairbanks,
[3]Statewide Research Manager, Alaska Dept. of Transportation & Public Facilities.

Average daily traffic (adt) in 1990 was 350 vehicles per day (vpd). Average daily traffic for year 2010 is estimated at 425 vpd. About 15 percent are heavy trucks (Hulsey, etc, 1995). During winter months, trucks often use tire chains to cross the bridge. Winter temperatures may drop below -45°C. The structure is an 884.5-m six-span highway bridge (Figures 1,2). The roadway width is 9.2-m, and the deck is supported by two 129.5-cm wide by 414.0-cm deep box girders.

The Yukon River Bridge is on a gravel road between Fairbanks and Prudhoe Bay, Alaska. This bridge is expected to support the Alaska oil pipeline, a future gas line and provide access for heavily loaded trucks hauling freight to the oil field on the North Slope.

Wearing Surface Selection Criteria

A wearing surface over a steel plate should be: a) lightweight, b) sufficiently thick to cover deck irregularities, c) skid resistant, d) abrasion resistant; e) stable and durable over the expected temperature range, f) a corrosion protector; g) bonded to the steel, and h) resistant to rutting and fatigue (Hulsey et al, 1999)

The Yukon River Bridge is on a 6 percent grade. This steep grade combined with winter temperature extremes and heavy truckloads is unusual and different from other orthotropic steel deck bridges. The life of a wearing surface is affected by surface traction, fatigue, permeability, ductility, abrasion, and the shear acting on the fasteners connecting the wearing surface to the steel deck (Troitsky, 1987).

Wearing surfaces over orthotropic steel decks may be thin or thick. Thin is usually less than 25-mm. This thickness will not mask waviness in the steel deck. The life is usually less than 5 years (Hulsey et al, 1999). Thin surfaces consist of a layer of adhesive/cement matrix applied to the steel deck. Traction is achieved by fine grit embedded in the matrix. Installation requires extensive surface preparation and a rigorous adherence to quality control through on-site inspection. Newly developed materials may allow thin surface alternatives to be more successful.

A thick wearing surface might be stiff or flexible and modular or continuous (Hulsey et al, 1995). Thick systems include asphaltic concrete overlays, Portland Cement Concrete (PCC), timber, timber/composite surfaces and propriety systems. Other PCC systems include normal weight or lightweight precast concrete modular elements. Timber decks include solid-sawn, glued-laminated or nail-laminated members. A suitable wearing surface should have (Troitsky 1987):

1. Sufficient ductility so the surface doesn't crack or de-laminate with expansion-contraction;
2. Adequate fatigue strength to prevent flexural cracks caused by deck deflections;
3. Sufficient durability so the surface does not rut, shove or wear;
4. Adequate imperviousness to water, motor fuels and oils; and
5. Sufficient resistance to de-icing chemicals and petroleum distillates.

Alaska Experience (Yukon River Bridge)

About 1976, a two-layer 127-mm temporary timber deck was placed on the Yukon River orthotropic steel deck. Five years later (1981) a middle 6.4-m width of running planks were replaced for the full length of the bridge for $254,250 or $56.46/sq. m. In 1992, the timber deck-wearing surface was removed and the steel deck was repaired and painted. In addition, two experimental wearing surfaces were installed. The length of each was about 15-m. A new two-layer 127-mm untreated timber deck was installed on the remaining steel deck.

Section 1 is 9.1-m wide and 15.2-m long. Cobra X railroad grade crossing modules are fastened to treated timber deck planks. Each module is 457-mm wide and 1,320-mm long. Modules are attached with 22-mm diameter by 102-mm long lag screws mounted to a 76-mm by 305-mm timber plank underlayment.

Section 2 is 9.35-m wide and 15.7-m long. It has treated timber-running planks. Costs for untreated timber deck replacement, steel deck maintenance, and the experimental sections were $1.14 million (see Table 1).

In 1999, test section 1, the Cobra X grade crossing, was left in place. The treated timber running planks in test section 2 was replaced and all untreated timber-running planks were again replaced. Repair work was done on the nose of the pier armor and a new test section was installed (section 3). This test section is a concrete filled grating. Cost of deck repairs was about $680,000 (see Table 1).

The life of a timber deck on the Yukon River Bridge is between 6 and 10 years. The steel deck requires repair and maintenance after 16 years. Costs for a temporary timber deck is about $ 66,000 per year.

Bridge Deck Inspections. In September 1993, the first author conducted an inspection of the wearing surface. Newly replaced running planks experienced some abrasion. The pressure-treated running planks at test section 2 were performing better than the untreated, and the Cobra X grade crossing modules was performing well (Hulsey, et al, 1995).

In 1996, DOT bridge inspection reported that running planks showed some wear and the estimated remaining life was about 5 to 6 years. Except for a broken corner, the Cobra X railroad grade crossing panels were in good condition. However, lag screws used to fasten panels to a timber under-layment were loose and required replacement. The treated running planks in test section 2 were performing somewhat better than the untreated.

A 1998 inspection team found the timber running panels were significantly worn; some had failed and others had lag screws sticking up to cause tire damage. The treated timber running panels were in much better shape than the untreated. The Cobra X railroad crossing test panels was in good shape.

The year 2000 inspection was one year after deck replacement. Findings showed that the Cobra X railroad panels were in reasonably good condition. The experimental concrete grating running planks were in fair condition but showed signs of cracking and appeared to experience excessive deflections.

Experiences by Other DOT's

Findings are based on results from a National Survey (Adhesive Engineering, 1987, Hulsey et al, 1995, 1999, Huval & Associates, Inc., 1992a, 1992b, 1992c, 1992d). A covered bridge in Connecticut has a wearing surface of pressure-treated timber planks over a painted orthotropic steel deck. The bridge is on minimal grade with light traffic. Tire chains are used infrequently.

Two flexible oil modified epoxy-wearing surfaces over orthotropic steel plate decks were studied by Michigan DOT. Traction was poor for both products.

Missouri evaluated wearing surfaces over an orthotropic steel plate deck on the Poplar Street Bridge in St. Louis. Transpo T-48, proprietary epoxy concrete, was recommended and used as the wearing surface replacement.

Louisiana initially studied numerous wearing surfaces for the Luling Bridge. After an initial investigation, four test sections with different wearing surfaces were selected for study. These were 1) Transpo T-48 overlaid with a polymer modified asphalt concrete; 2) asphalt-rubber stress absorbing membrane interlayer (SAMI) (Crumb Rubber); 3) polymer modified concrete with steel fibers overlaid with asphalt rubber; and 4) polymer modified asphaltic concrete. During the study, deck temperatures approached 60°C (140°F) causing researchers to select a stone mastic asphalt (SMA). This was different from the four test cases. Wearing surface performance on the Luling Bridge is still under study.

Orthotropic Steel Deck Flexibility

Static strains for 10 different trucks were reported at 13 locations on the orthotropic steel deck for the Yukon River Bridge (Hulsey et al, 1995, 1999). At the time strains were measured, the deck-wearing surface was overlaid with a 127-mm two-layer temporary timber deck. Results showed that steel deck flexibility was not a design issue.

Maximum strains recorded in the steel deck were 139.1 micro-strain. Minimum strains were -127.7 micro-strain, and the largest range in strain was 186.9 micro-strain. The live load flexural strains were low. This indicated that conventional wearing surfaces might be a viable alternative for the structure.

Wearing Surface Selection Procedure

Below is a rational procedure to assist engineers in selecting a wearing surface for truckloads and weather extremes. This method is based on limiting the tensile strain caused by loading. Strains caused by loaded trucks crossing the bridge are presented in Figure 4. These strains were measured and compared with a finite element model (Hulsey et al, 1995). The extreme temperatures are from Fairbanks data (Hulsey and Powell, 1993).

Only fully bonded wearing surfaces are presented. Figure 4 can be used to select the required thickness. Figure 5, which is based on a finite element analysis for temperature extremes, can be used to determine the magnitude of thermal stresses in the steel deck. Keep in mind, thermal stresses vary with material moduli and thickness of the wearing surface.

Methodology. Determine required thickness for a fully bonded wearing surface.

Step 1. Select design life, the number of load cycles, and design temperature.

Step 2. For the design temperature, select the material modulus; this may be obtained from the supplier or from laboratory tests.

Step 3. Obtain from the supplier or by tests the limiting tensile strain for fatigue.

Step 3. With tensile strain and material modulus, enter Figure 4 and select thickness.

Step4. With material modulus and thickness, enter Figure 5 and determine the magnitude of thermal stresses in the orthotropic steel plate.

Step 5. Calculate the weight of the wearing surface. It should be within the design limit of 1,436 Pa (30 psf).

Rank & select a wearing surface. If different wearing surfaces are to be compared, than:

Step 1. Assign an importance (number between 1 and 10) to each surface condition. An importance number may be needed for: a) the added weight; b) skid resistance; c) abrasion; d) cold weather ductility; e) adhesion to steel, and flexural resistance.

Step 2. Combine the importance factors to give a number that can be used to rank and evaluate this wearing surface with others. Higher numbers are better.

Step 3. Calculate life cycle costs.

Step 4. Find a cost-benefit ratio.

Step 5. Rank the choices and select a wearing surface.

Choose the Wearing Surface Thickness

Two examples are presented for consideration. One is an asphalt AC-5 mix and the other is rubberized asphalt concrete. The Alaska Department of Transportation and Public Facilities (AKDOT&PF) has used these in interior Alaska.

Assume a 0°C design temperature for the Yukon River Bridge Deck and a traffic level of 500,000 load cycles for the life of the wearing surface.

Example 1. An asphalt wearing surface with an AC-5 mix.

1. Enter Table 2 and select a modulus for the 0°C. Modulus is 10,058 mPa.
2. Enter Figure 6 with 500,000 load cycles and select the limiting tensile strain. The strain is 200 micro-strain.
3. Enter Figure 4 with a modulus of 10,058 mPa and a maximum strain of 200 micro-strain and select the required thickness. It is about 76-mm (3-inches).
4. Enter Figure 5 with a modulus of 10,058 mPa and a thickness of 76-mm and find the induced thermal stress on the steel deck. The thermal stress is 32 mPa.

Example 2. A rubber asphalt concrete wearing surface.

1. Enter Table 2 and select a modulus for the 0°C. Modulus is 6,987 mPa.
2. Enter Figure 6 with 500,000 load cycles and select the limiting tensile strain. The strain is 400 micro-strain.
3. Enter Figure 4 with a modulus of 6,987 mPa and a maximum strain of 400 micro-strain and select the required thickness. Use slightly over 51-mm, say 57-mm (2.25-inches).
4. Enter Figure 5 with a modulus of 6,987 mPa and a thickness of 57-mm and find the induced thermal stress on the steel deck. The thermal stress is 42 mPa.

Acknowledgments

The authors wish to thank the AKDOT&PF for their support. They especially wish to thank Mark Miles and Gary Scarborough with the Bridge Design Department at AKDOT&PF.

Summary and Conclusion

Strain in a bonded wearing surface varies with load, truck geometry, thickness and modulus of the wearing surface. Thermal stresses in the steel deck vary with weather, thickness of the wearing surface and the modulus of the wearing surface.

The Yukon River Bridge has a low traffic count, carries heavy trucks, and is subjected to extreme winter temperatures (below -45°C) and a steep grade (about 6 percent).

Other DOT's found the better performing wearing surfaces on orthotropic bridge decks include: Transpo T-48; epoxy asphalt over a coal tar epoxy; stone mastic asphalt (SMA); asphalt rubber stress absorbing membrane interlayer (SAMI); polymer modified concrete with steel fibers overlaid with asphaltic rubber; polymer modified asphaltic concrete; and polyurethane impregnated with stone chips.

A method is presented for selecting a wearing surface to resist flexural fatigue and thermal cracking. Two examples illustrating the method are presented.

References

Adhesive Engineering Company (1987). *Product Data, Technical Bulletin AE 563*, February.

Heins, C.P., and Firmage, D.A. (1979). "Chapter 6. Design of Orthotropic Deck Bridges", *Design of Modern Steel Highway Bridges.* John Wiley and Sons, Inc., pp. 154-226.

Hulsey, J. L. and Powell, D. T. (1993). "Rational Weather Model for Highway Structures", *Transportation Research Record,* No. *1393*, Transportation Research Council, Washington, D.C.

Hulsey, J.L., Yang, L., Curtis, K., and Raad, L. (1995). "Yukon River Bridge, Deck Strains and Surfacing Alternatives", *Transportation Research Center, University of Alaska Fairbanks,* Report No. INE/TRC 94.10 SPR-UAF-92-6, Final Report, Sept.

Hulsey, J.L., Yang, L., and Raad, L. (1999). "Wearing Surfaces for Orthotropic Steel Bridge Decks", *Transportation Research Record*, Paper 991049, National Research Council.

Huval & Associates, Inc. (1992). "Analysis and Engineering Study of Wearing Surfaces, Volume I", Mississippi River Bridge at Luling, Louisiana Department of Transportation and Development, State Project No. 700-29-0060, F.A. Proj. No. BRI-310-4- (134) 223, Dec.

Huval & Associates, Inc. (1992). "Analysis and Engineering Study of Wearing Surfaces, Volume II", Mississippi River Bridge at Luling, Louisiana Department of Transportation and Development, State Project No. 700-29-0060, F.A. Proj. No. BRI-310-4- (134) 223, Dec.

Huval & Associates, Inc. (1992). "Analysis and Engineering Study of Wearing Surfaces, Volume III", Mississippi River Bridge at Luling, Louisiana Department of Transportation and Development, State Project No. 700-29-0060, F.A. Proj. No. BRI-310-4- (134) 223, Dec.

Huval & Associates, Inc. (1992). "Analysis and Engineering Study of Wearing Surfaces, Addendum 1", Mississippi River Bridge at Luling, Louisiana Department of Transportation and Development, State Project No. 700-29-0060, F.A. Proj. No. BRI-310-4- (134) 223, Dec.

Saboundjian S. (1999). "Fatigue Behavior of Conventional and Rubber-Modified Asphalt Mixes", Ph.D. Dissertation, University of Alaska Fairbanks, May.

Troitsky, M.S. (1987). "Orthotropic Bridges, Theory and Design", 2nd Edition, The James F. Lincoln Arc Welding Foundation, Cleveland, Ohio.

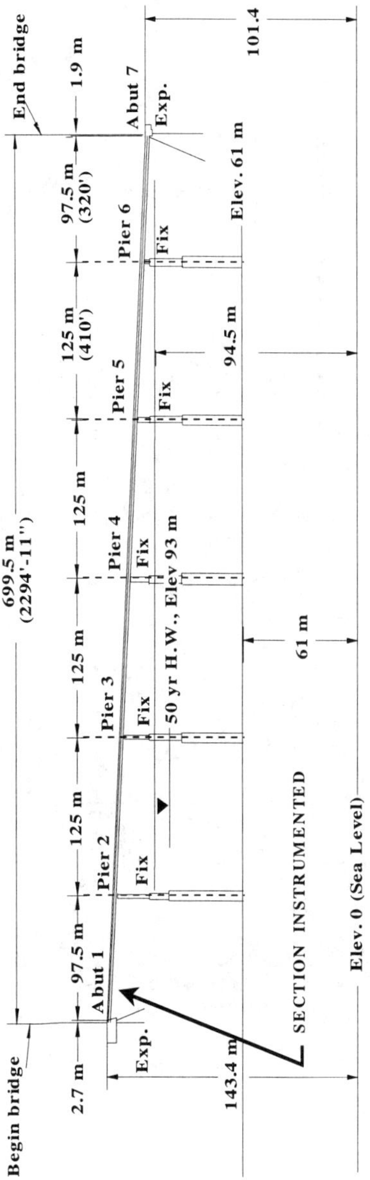

Figure 1. Yukon River Bridge Geometry

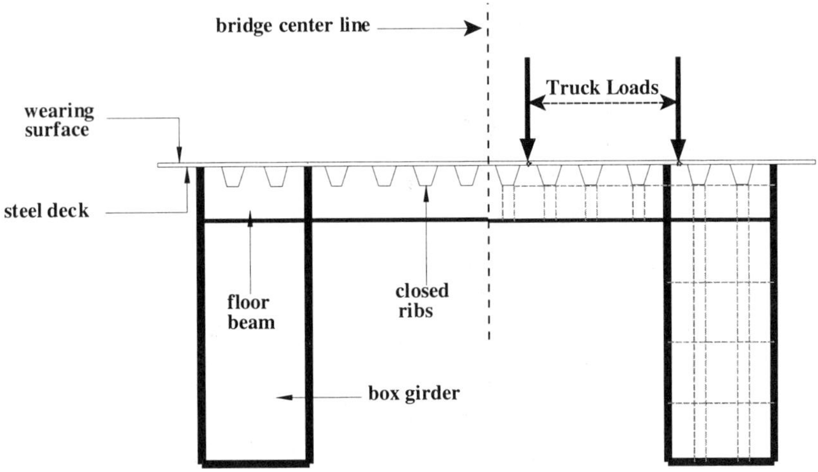

Figure 2. Superstructure Cross-sectional View

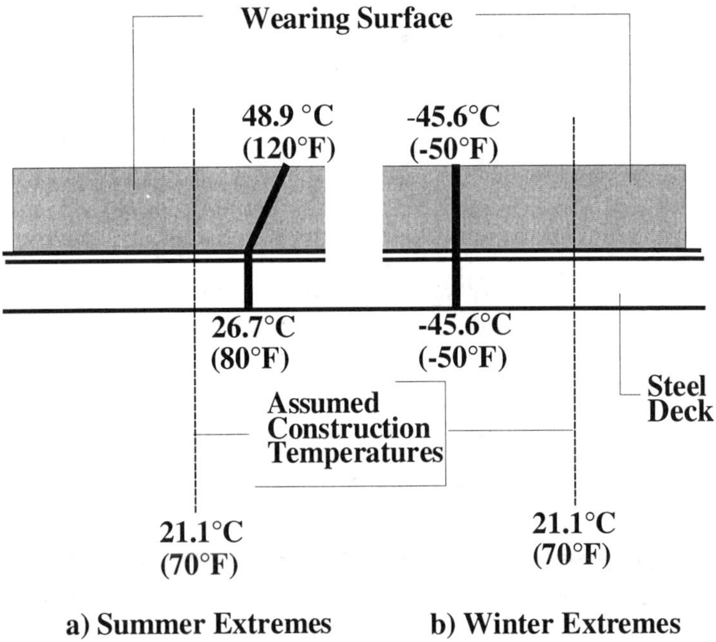

Figure 3. Temperature Exposures for Weather Extremes

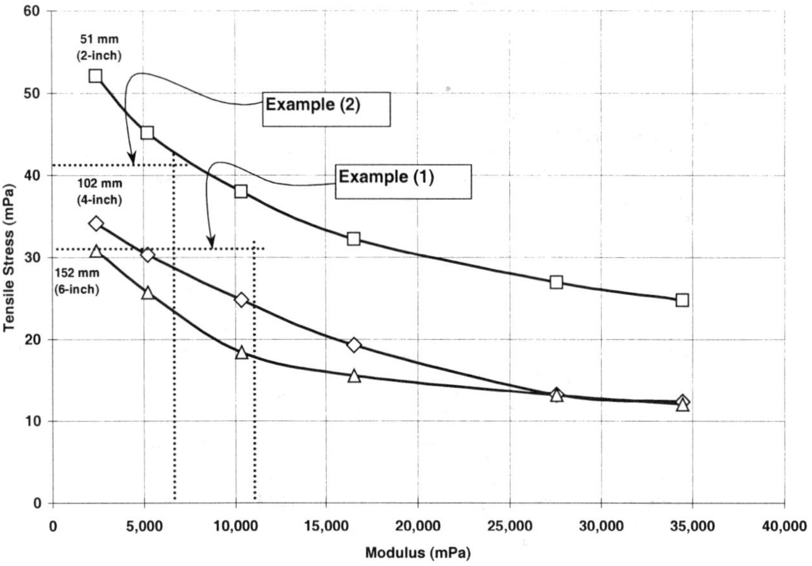

Figure 4. Wearing Surface Live Load Tensile Strain

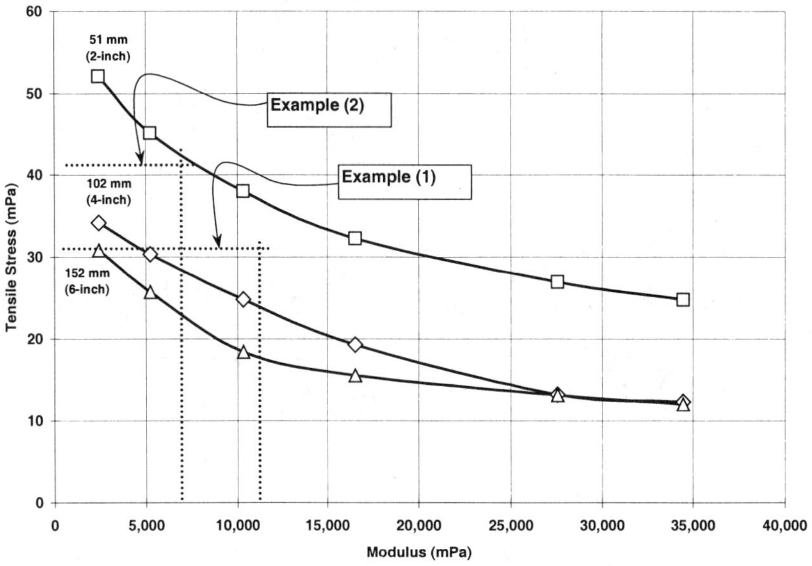

Figure 5. Induced Thermal Stresses in the Steel Deck

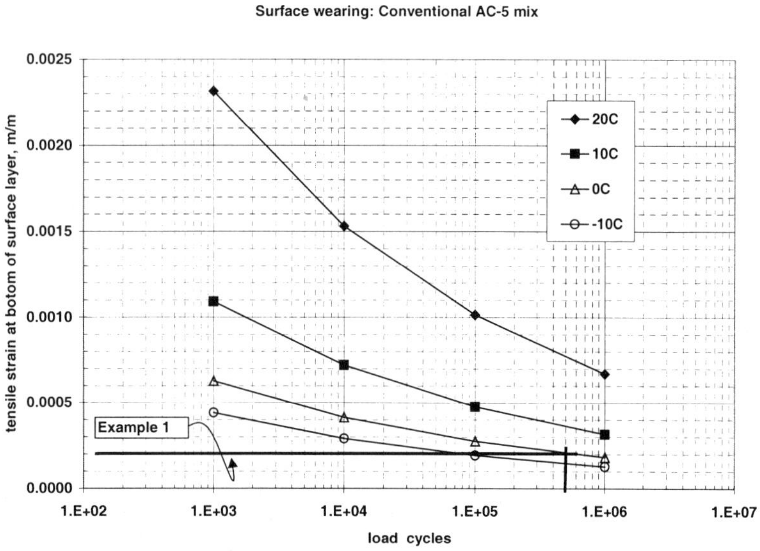

Figure 6. Fatigue Limits for AC-5 Asphalt.
(After S. Saboundjian, 1999)

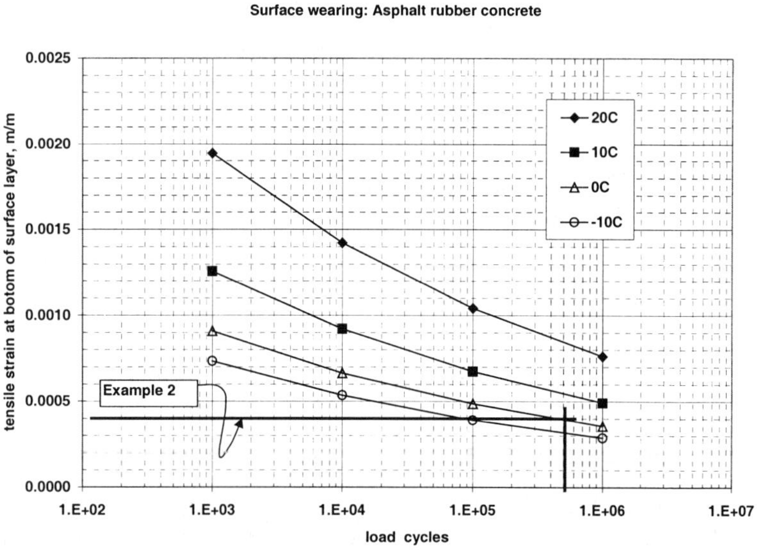

Figure 7. Fatigue Limits for Rubberized Asphalt.
(After S. Saboundjian, 1999)

Table 1. Wearing Surface Repair and Replacement Activity

Dates:	Activity	Quantity	Unit Cost	Cost
Mar-1974 to Jun-1976	New Bridge Construction: • Installed a 2-layer temporary timber deck	6,576 sq. m	--------	-----------
Apr-1981 to Aug. 1982	• Replace a 6.4m width of running planks	4,502 sq. m	$ 55	$ 247,500
May-1986	Nose armor repair at the piers			-----
Aug-1991 to Jun-1992	Bridge Resurfacing (total costs)			$1,140,953
	• Replace timbers, repair & coat steel deck	6,290 sq. m	$ 107	$ 667,821
	• Sec. 1: Cobra X railroad crossing pads	144 sq. m	$ 1,893	$ 272,560
	• Sec. 2: Treated timber running planks	142 sq. m	$ 1,412	$ 200,572
Mar-1999 to Aug. 1999	Bridge Re-Deck (total cost)			$ 681,223
	• Replace running planks	6,179 sq. m	$ 98	$ 608,181
	• Sec 2: Replace treated running planks	142 sq. m	$ 125	$ 17,737
	• Sec 3: New concrete filled grating section	111 sq. m	$ 498	$ 55,305

Table 2. Asphalt Modulus vs. Temperature (After S. Saboundjian, 1999)

	Modulus (mPa)	
Temperature (°C)	AC-5 Mix	Asphalt Rubber Concrete
20	1,215	900
10	4,108	2,909
0	10,058	6,987
-10	17,839	12,466

PERFORMANCE OF KUPARUK RIVER SUBMERSIBLE BRIDGES AND LOW WATER ROADWAY

Alan B. Christopherson[1], M. ASCE; Kenton W. Braun[2], M. ASCE; Dempsey S. Thieman[3], M. ASCE; Jasper H. Hardison III[4]

Abstract

For almost twenty years, road access between the Kuparuk oil field and the infrastructure at Prudhoe Bay/Deadhorse, Alaska was disrupted for an average of 6 weeks during spring flooding of the Kuparuk River. Although a permanent crossing was constructed across the Central Channel in 1981, the East and West Channel roadways were breached annually to allow passage of spring break-up flood waters. In 1999, permanent structures were constructed across the East and West Channels, providing access year-round, except for a maximum closure of one week during spring break-up.

The innovative design of the east and west channel crossings involved a combination of low water roadways and submersible bridge designs. With this design, large spring floods are allowed to pass over low water roadways in all three channels and over the east and west channel bridges during extreme spring flood events.

Since construction, the East and West Channel Crossings have experienced two break-up floods. This paper will discuss the observations and performance of the Kuparuk River Crossings during the 2000 and 2001 spring break-up floods and provide recommendations for future designs of similar crossings.

[1]Senior Vice President, Peratrovich, Nottingham & Drage, Inc., 1506 W. 36th Ave., Anchorage, AK 99503, a_christopherson@pnd-anc.com
[2] Senior Engineer, Peratrovich, Nottingham & Drage, Inc., 1506 W. 36th Ave., Anchorage, AK 99503, k_braun@pnd-anc.com
[3] Senior Engineer, Peratrovich, Nottingham & Drage, Inc., 1506 W. 36th Ave., Anchorage, AK 99503, d_thieman@pnd-anc.com
[4] Hydrologist, Peratrovich, Nottingham & Drage, Inc., 1506 W. 36th Ave., Anchorage, AK 99503, j_hardison@pnd-anc.com

Background

The Kuparuk River, on the North Slope of Alaska, is a braided stream subject to severe environmental conditions including extreme cold, ice floes and high river flows for short durations during spring breakup. In the late 1970's, a spine road was constructed across the Kuparuk River to connect the developing Kuparuk oil field to the existing infrastructure at Prudhoe Bay/Deadhorse. The road crosses the river at a location where three main channels exist in a 3-km (10,000-ft) wide floodplain.

Initial attempts to construct permanent crossings at the Kuparuk River using culverts proved unsuccessful. In 1981, a heavy-duty bridge combined with a low water roadway was constructed across the Central Channel. The road across the east and west channels was deliberately breached each year to allow the spring breakup flood waters to pass. On average, the breaching, flooding and repair of the road, closed access between the Kuparuk oil field and Deadhorse for 6 weeks. Culverts were placed within the East and West Channels at other times of the year to allow summer low flows to pass through the road.

The Central Channel Crossing consists of a 91-m (300-ft) three-span, steel box girder bridge and a 323-m (1060-ft) paved low water roadway. The Central Channel Bridge was designed to support heavy modules and drill rigs, with gross vehicle weights of over 2,000 tonnes (4,400 kips), utilized for development and subsequent support of the Kuparuk field. The bridge, with a deck elevation of 11.6 m (38.0 ft) and a low chord elevation of approximately 9.4 m (30.8 ft) above mean sea level (MSL), was constructed to preserve a minimum of 1.5-m (5-ft) clearance between the 100-year flood water surface elevation and the bridge low chord. By allowing for the low chord clearance below the bridge, river navigability across the Spine Road was maintained. (Christopherson et al., 1991)

The Central Channel low water roadway was constructed to the east of the bridge to provide additional flood capacity during the peak spring breakup events. The low water roadway was paved so that the road surface elevation at both ends was 7.92 m (26.0 ft) with the lowest elevation of the paved roadway at 6.40 m (21.0 ft) above MSL.

Between 1981 and 1997, several concepts were developed to provide permanent crossings at the East and West Channels of the Kuparuk River. In 1997, Peratrovich, Nottingham & Drage, Inc. (PN&D) developed a concept combining paved low water roadway sections with submersible bridges. This concept attempted to match the hydrologic characteristics of the existing Central Channel Crossing to provide a minimum Spine Road closure period and to best match structures already in place in the river floodplain. The main difference between the existing Central Channel Crossing and the new concept was the proposal to allow the new bridges to become submerged during large spring breakup events. Because the road grade could not exceed 3%, submersible bridges were required in order to construct the low water roadways within the narrow channels. This concept was approximately one half the cost of constructing fully elevated bridges across the two channels. Details of this design are described in other publications. (Christopherson et al., 1999 and Braun et al., 2001)

The new East and West Channel Crossings were constructed in 1999. A site plan of the entire Kuparuk River Crossing is shown below in Figure 1. The East Channel Crossing consists of a 64-m (210-ft) bridge with paved low water roadway on each side of the bridge. Total length of the East Channel low water roadways is 427 m (1,400 ft). The West Channel Crossing was constructed with a 46-m (150-ft) bridge and a 210-m (689-ft) paved low water roadway on the east side only. The minimum elevation of low water roadways was set at +6.40 m (21.0 ft) to match the existing roadway in the Central Channel. Bridge lengths were determined to ensure the road overtopping events would not exceed 7 days. A general elevation of the East and West Channel Crossings is depicted below in Figure 2.

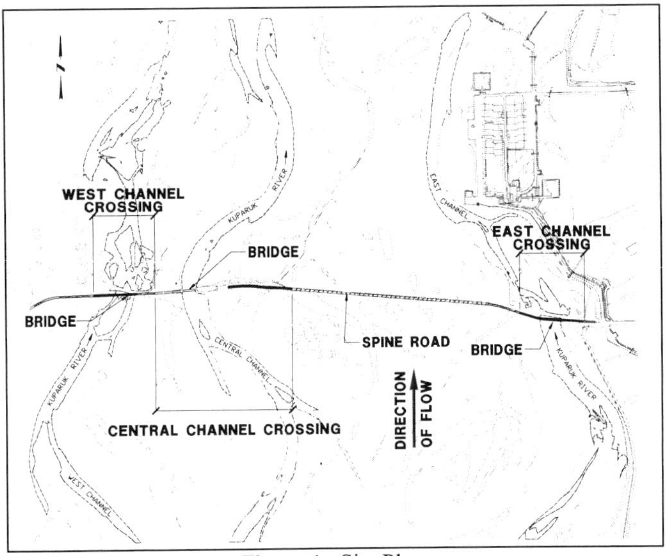

Figure 1. Site Plan.

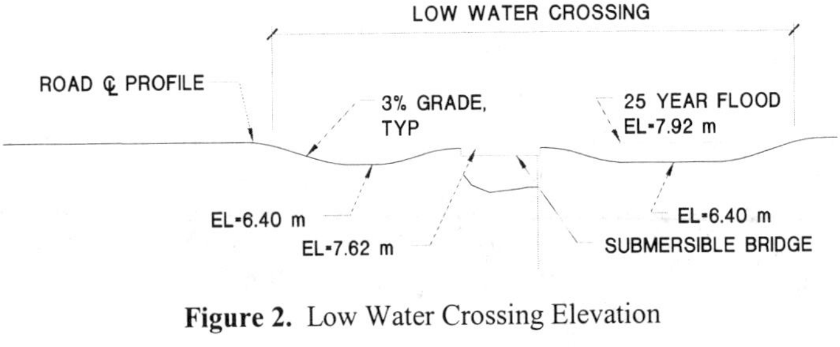

Figure 2. Low Water Crossing Elevation

Hydrologic Characteristics

The Kuparuk River flows north from the foothills of the Brooks Range across the Arctic coastal plain to the Beaufort Sea. A U.S. Geological Survey (USGS) gage 3 km (2 mi) above the Spine Road crossing has been in operation since 1971. The Kuparuk River has a drainage area of 8110 km^2 (3130 mi^2) at the gage station, with 60 percent of this area located on the coastal plain. The flood frequency curve based on 31 years of record is represented by a 2-year flood of 1,300 cubic meters per second (cms) (46,000 cfs), a 50-year flood of 3,400 cms (120,000 cfs), and a 100-year flood of 3,880 cms (137,000 cfs). The peak flood of record is 3,340 cms (118,000 cfs) (Meyer 2001). During the period of record for the USGS Kuparuk River gage all annual peaks have occurred between May 20 and June 15 with an average date of June 8.

Typical flood hydrographs on the Kuparuk River show a very sharp rise, with an increase in daily mean discharge of 420 cms to 570 cms (15,000 to 25,000 cfs) per day for several days after arrival of the breakup front. The recession limb of the curve can take 1 to 2 weeks to fall below 280 cms (10,000 cfs). If snowmelt is interrupted by cold temperatures and cloudy skies the hydrograph can show a double peak.

At the USGS gage site, the Kuparuk River splits between the East and Central Channels. The West Channel splits from the Central Channel shortly downstream of the gage site. During low summer flows, the majority of flow occurs in the Central Channel. During the spring breakup flood, the East Channel has experienced up to 50% of the flow during recent years.

Hydrologic characteristics for the 2000 and 2001 floods are presented below in Table 1.

	2000 Flood	2001 Flood
Peak Discharge	1,800 cms	1,500 cms
Flood Recurrence Interval	Less than 5 yrs	Greater than 2 yrs
Day of Peak Discharge	June 12	June 9
Breakup Front Arrival	June 9	June 6
East Channel Overtop	56 hours	55 hours
Central Channel Overtop	48 hours	17 hours
West Channel Overtop	56 hours	8 hours

Table 1. Summary of 2000 and 2001 flood characteristics. (PN&D 2000 and PN&D 2001)

Performance Of Low Water Crossing - 2000 Breakup

Summary Of Low Water Crossing Performance. The East and West Channel Crossings generally performed as intended during 2000 breakup. Damage did occur to the crossings at one bridge abutment and at some locations on the paved low water roadways. Damage was unfortunate and unexpected, but these structures are prototypes. Fine tuning of the design and construction methods, as well as pre-breakup maintenance procedures for the crossings, is necessary to prevent damage in the future. Damage to the crossings consisted of the following:

- Revetment on the downstream side (spillway) of some paved roadway sections settled and was displaced. (Reference Figure 3)
- Sections of roadway adjacent to some damaged areas of the revetment were undermined. (Reference Figure 3)
- River bed at eastern abutment of the East Channel Bridge scoured to below the toe of the sheet piles, resulting in loss of fill material within the abutment. Upstream sheet piles also settled with receding river bed from loss of support. (Reference Figure 4)

The bridges withstood major ice impacts and performed as intended. Several large ice floes were observed riding up and failing over the ice breaking piers. Some ice jamming did occur when large quantities of ice plugged bridge openings, but the jamming was cleared out by a track hoe on the bridge deck. The East Channel Bridge did experience some overtopping for a brief period of time (approximately 1 to 2 hours). No damage to the bridge structure, support piles or icebreaking piers was discovered.

Factors Contributing to Bridge Abutment Scour. The East and West Channel Bridges experienced extreme amounts of scour during the 2000 Kuparuk River breakup flood. The large amount of bridge scour that was associated with the event was the result of a number of factors. These factors include the following:

- Concentration of flow in subchannels developed during the 1999 breakup flood
- Snow and ice from the previous winter plugged bridge openings, creating pressure flow to the channel bottom
- Large magnitude breakup front
- Ice floe hydraulic effects
- Magnitude and duration of the breakup flood (flow at bridge low chord)

After construction in 1999, the demolished ice workpad at the East Channel Bridge was pushed into a large pile upstream of the bridge. The result was that breakup flows were diverted around the ice/snow pile to the east side of the channel. Flows followed the upstream face of the eastern low water roadway before turning sharply into the bridge opening. This resulted in a subchannel forming that concentrated breakup flows in 2000 along the upstream side of the east abutment of the bridge. Construction of a guide bank would greatly improve performance at this location. However, to date, environmental regulators have been resistant to construction of river training structures.

Frequent winds during the winter of 1999-2000 caused a great deal of snow to be deposited within the openings of the two new bridges. Prior to breakup, both openings were almost completely plugged with ice and snow. When the breakup front arrived at the spine road, flows were channeled under the snow and ice which plugged the openings. The thawed sediments within the scour holes were easily eroded and the scour holes enlarged quickly. These enlarged scour holes under the snow and ice in the bridge openings were evidenced by the persistent whirlpools, and by the disappearance of large ice floes under spans that otherwise appeared plugged.

Figure 3. Damage to downstream revetment at eastern roadway of East Channel Crossing.

Figure 4. East Channel Bridge shortly after 2000 breakup flood peak. Note loss of fill material within eastern abutment. Ice and water on bridge deck indicate recent bridge overtopping.

The enlargement of existing scour holes under the snow and ice plugs within the bridge openings was accelerated by the large magnitude of the breakup front and the swift increase of discharge experienced. The breakup front arrived so quickly and with such force that there was little time for the snow and ice plugging the bridge openings to melt. This was especially evident at the East Channel Crossing where the eastern roadway overtopped within an hour of the arrival of the breakup front.

Observations of the hydraulic interaction of ice floes with the bridge openings suggest that ice floes increase turbulence and scour beyond conditions that would be expected during an ice-free flood. Conditions at the East Channel Bridge were especially dramatic. Large ice floes would hit the ice breaking piers and ride up, temporarily blocking flows into the adjacent bridge span, and creating severe turbulence. Some ice floes, wider than the 9.1-m (30-ft) bridge spans, would become temporarily pinned across two ice breaking piers. The effect of ice floes plugging the bridge spans was to direct flows to the sides and downward towards the channel bottom, creating pulses and jets of turbulent water. Additionally, some large ice floes that were too large to fit through the bridge spans horizontally turned on edge and passed through. It is likely that these ice floes gouged the channel bottom as they passed through the bridge, increasing the scour of the river bed.

The sustained high magnitude flows that were experienced during the 2000 breakup also contributed to the large magnitude of scour. It is expected that maximum scour conditions occur when flows within the bridge openings are at the level of the low chord of the bridge. This was the case for the first five days of breakup at the East Channel. The result was that maximum scour conditions occurred for an extended period of time, maximizing the amount of scour within the bridge openings.

Low Water Roadway Scour. The scour damage at the East and West Channel low water roadways was most severe at the eastern roadway of the East Channel, where the highest flows occurred. From post breakup inspections it was determined that the damage was mainly due to the long linear gaps in the concrete revetment mats and the size of the individual revetment blocks. The gaps in the revetment mats allowed the rushing water to move the gravel fill beneath the geotextile fabric which formed voids. Because the revetment blocks seemed to be too large to settle into the small voids as intended, the small voids were able to grow, undermine adjacent blocks and eventually portions of the roadway.

Because of the large revetment blocks, the mats lacked the flexibility to settle evenly into voids underneath, resulting in severe slope deformation of the downstream low water roadways. The scour damage at the West Channel low water roadway was limited to approximately 10 m (30 ft) measured linearly along the roadway. The scour damage at the west side of the East Channel Crossing was also limited to approximately 10 m (30 ft). The scour damage at the east side of the East Channel extended most of the length of the low water roadway. Additional damage to the east side of the East Channel Crossing was due to large ice floes overtopping the roadway, sliding down the spillway and catching on the large revetment blocks.

Factors Contributing to Low Water Roadway Scour. A number of factors contributed to the scour damage to the low water roadways at the East and West Channels. These factors include:

- Revetment block size and shape.

- Large linear gaps between mats of blocks.
- Large, nearly vertical lip constructed along the edge of paving on the downstream edge of the low water roadway.
- Thaw settlement of gravel fill beneath revetment. Settlement was due to the construction sequence in which revetment was placed onto the newly placed frozen gravel.
- Ice floe impact effects.

Due to schedule requirements, the revetment mats utilized in the construction of the Low Water Crossings project were substituted by the contractor in lieu of the design-specified revetment. All the factors as noted above associated with the revetment mat configuration were the result of this substitution. Further exacerbating the problems with the revetment mats, the vertical constructed "lip" did not follow design requirements and thaw settlement created rough transitions in the slopes. The original design required that the roadway profile remain smooth in areas where overtopping was to occur and specified that gravel fill was to be placed in the summer, thawed condition to alleviate thaw settlement problems. However, schedule constraints drove the owner to construct the road during the winter.

The substituted revetment mats utilized for construction of the East and West Channel Crossings are composed of large square concrete blocks that do not fit tightly together. Large linear gaps are present between the blocks, with the geotextile fabric underlying the blocks being clearly visible. Scour of the river bed at the downstream edge of the revetment pulled gravel down from under the revetment slope along the gaps in the blocks. After breakup, geotextile was observed bulging up between the blocks where gaps had served as conduits for gravel loss under the revetment (reference Figure 5). Because the blocks were so large, they were unable to settle back onto the revetment slope and anchor the remaining gravel.

Ice floes traveling over the low water roadway, especially during the early part of the breakup when there was a large difference in the water surface across the roadway, would slide down the slope of the revetment. As the ice slid down the slope, some would snag on the large gaps between the blocks, in some cases damaging the cable connections between the blocks. Where cable connections were damaged or broken, the blocks would flip up allowing gravel loss from underneath the exposed geotextile fabric. When ice floes became stuck on the downstream slope of the revetment, they also moved the position of the hydraulic jump on the spill apron, sometimes directing it to the downstream edge of the revetment and concentrating scour there.

When the low water roadways were paved, a nearly vertical lip of pavement was formed along the edges of the roadway. This lip was more than 0.3 m (1 ft) high in some places. The effect of this lip was to concentrate the hydraulic jump on the slope of the revetment rather than on the flat spill apron at the bottom of the slope, as intended. A further result of the presence of the lip was to cause ice floes passing over the roadway to drop down onto the slope of the revetment rather than smoothly ride over the transition. This effect caused ice floe impact forces on the cable connections to be even greater, increasing the damage to the revetment mats.

Figure 5. Geotextile bulge along vertical joint in revetment mats.

Lessons Learned From 2000 Breakup Flood.

- Bridge openings need to be cleaned of drifted snow and the ice weakened prior to breakup.
- Ice floes can produce turbulence effects that can increase scour significantly beyond design estimates assuming ice-free conditions.
- Short bridge spans are prone to ice floe jamming and increased turbulence, resulting in increased scour. Slightly thicker superstructures should be considered in order to provide longer spans.
- Size and configuration of revetment mats is critical to performance.
- A smooth transition on the downstream side of the low water roadways is critical to the performance of the spillway.
- Gaps between the revetment mats must be minimized to provide a smooth spillway and reduce local ice impacts to individual blocks.
- Revetment must be placed upon thaw stable (thaw settled) fill.

Performance Of Low Water Crossing - 2001 Breakup

East Channel Crossing Performance. The first 2 days after the breakup front reached the spine road on June 6, flow at the East Channel Crossing was affected by local channel ice and ice floes interacting with the crossing structures. Channel ice in the vicinity of the bridge was in place over most of June 7. The upstream and downstream channel ice was floating and the channel ice under the bridge was still bridge/bottom fast. Water flowed under the upstream section of floating channel ice. This flow surfaced where the ice was scored prior to breakup, just upstream of the ice breaking piers. A portion of the flow traveled under the channel ice beneath the bridge, upwelling downstream of the bridge. On the afternoon of June 7 the channel ice was cleared with the help of a track hoe and rising water levels.

The captured, floating channel ice at the bridge enhanced the formation of ice jams against the bridge. The channel ice upstream of the bridge blocked up to 90% of the channel width, preventing passage of ice floes. The obstructions reduced the velocity of ice floes to the point where the ice breaking piers were ineffective.

An ice jam against the bridge on June 8 caused 0.76 m (2.5 ft) of backwater, overtopping the low water roadways. The ice jam was caused by a large number of ice floes moving down the river in mass that became lodged within the bridge opening. Other observers at the crossing reported that the ice jam extended hundreds of yards upstream and raised the water level above the crossing within 15 minutes. The east roadway had approximately 0.76 m (2.5 ft) of water flowing over it. The west roadway had less than 0.3 m (1 ft) of water, as it was on the downstream side of the ice jam and was not as strongly affected by the backwater. Water flowing over the east low water roadway eroded gravel placed as a temporary repair after the 2000 breakup flood. In addition, the water further undermined the unprotected concrete roadway by as much as 2 m (6 ft) on the downstream side. The west low water roadway performed as intended and sustained no damage.

West Channel Crossing Performance. Flow at the West Channel Crossing structure behaved similar to flow at the East Channel Crossing. Flow during ice conditions was nearly identical, as the channel ice in the vicinity of the bridge impeded ice floes travelling from upstream. Channel ice was in place through the morning of June 8, almost a day longer than at the East Channel due to the slower rise in water surface and timing of work at the bridge by the track hoe. The only significant ice jam of the breakup flood occurred during this time.

Summary of 2001 Pre-Breakup and Breakup Activities. Limited repair work was performed on areas damaged during the 2000 breakup. Rip rap was placed around abutments of the East and West Channel Bridges to protect against further scour. Damaged revetment at the East Channel Crossing was removed in order to reestablish the roadway prism. Where the paved roadway had been undermined during the 2000 flood, pavement was removed and new fill placed to provide a driveable surface. Other than the rip rap scour protection, repairs were temporary and intended only to provide service until long-term repairs could be performed during the summer of 2001. Damage was expected to occur to the East Channel low water roadway if 2001 breakup waters overtopped the road.

Pre-breakup maintenance at the site consisted of removal of snow accumulated under the bridges and scoring of channel ice around the submersible bridges. Snow was excavated from under all three bridges and placed upstream and downstream of the abutments along the low-flow channel margins. Ice scoring was done at the East and West Channel Bridges and consisted of cutting the ice to approximately 70% of its thickness in a series of lines parallel with the channel between the pile bents and perpendicular to the channel just upstream and downstream of the bridges. These maintenance procedures are intended to increase the flow capacity through the bridges during the initial stages of breakup and reduce the occurrence of overtopping the low water roadways due to snow and ice obstructing the bridges.

Maintenance during the breakup flood consisted of using a track hoe to clear ice floes from the East and West Channel Bridges and from the low water roadways at the East Channel. The track hoe was positioned on the bridges to break up ice jams and move individual ice floes through the bridges if they lodged against the upstream piers. The track hoe was also used to keep the channel downstream of the bridges clear so ice floes could move out from under the bridges. At the low water roadways the track hoe was used to remove ice floes that had lodged against the upstream edge of the roadway.

Effectiveness of Pre-breakup Activities. Scoring of the channel ice appeared to help in clearing the bridge of winter ice during breakup. During the first day of flow, as ice upstream and downstream of the bridges floated due to rising water levels, the ice broke along the cuts perpendicular to the channel. The cuts also allowed flow forced under the upstream ice to surface upstream of the bridge.

The track hoe was able to clear the ice from directly in front of the bridge allowing the upstream floe to move to the bridge, be broken on the piers and pass through the bridge spans. The downstream floes could not be reached from the bridge and, therefore, stayed in place until there was sufficient flow to break them up. The downstream floes hindered the clearing of the upstream ice as it blocked ice from moving out under the bridge. Ice floes had to be pushed under the bridge from upstream and then pushed around the large downstream floe. The efforts were effective at increasing flow through the bridge and reduced the length of time that water overtopped the East Channel roadways.

Recommendations for Future Activities. Expanded weakening of the ice upstream and downstream of the bridge may help to clear the local channel ice more quickly. Breaking the upstream channel ice should also reduce the amount of flow forced beneath the ice. Breaking the downstream ice should allow local channel ice to move out sooner and clear the way for ice floes to pass under the bridges unhindered.

Stockpiled snow berms left from clearing beneath the bridge spans affected flow during the first 1 to 2 days of breakup. Snow pile placement, in conjunction with local channel ice, forced flow from the main channel against abutments, contributing to ice jamming. Water was observed flowing under one snow berm and upwelling against the downstream side of nearest abutment. Future snow removal should ensure material is placed clear of bridge openings, channels and abutments.

Track hoe work broke up the local channel ice floating upstream and downstream of the bridges, and helped release channel ice under the upstream edge of the bridge. Removal of ice from the low water roadways does not have as great an effect on the flow. Up to 0.3 m (1 ft) of backwater was observed behind the ice at the upstream edge of the low water roadways, sending more flow through the bridge. Removal of this ice will reduce the local water surface elevation above the roadway, but will not significantly reduce the amount of time that the roadway is overtopped.

Conclusion

Prior to this project, a submersible bridge and low water roadway design had not been attempted in an arctic environment. The original design was an innovative and cost saving approach to providing a long term solution to the crossing of a dynamic river flood plain nearly 3 km (2 mi) wide. Circumstances related to design, owner decisions, construction materials and construction techniques resulted in less than desirable performance in some areas. Information gained through the evaluation and monitoring of the low water crossings performance has been utilized in designing repairs and can be utilized in future projects by engineers, owners and contractors.

Analysis of the damage which occurred to the crossings indicates that following considerations are critical to performance:

- At locations where water is to overtop roadways, road profile must consist of smooth gentle surface with no sharp edges or vertical discontinuities.
- Revetment mats must be placed on thaw stable grade and contain irregular joints.
- Revetment must be capable of settling into voids that form underneath mats.
- Submersible bridges must be designed for far greater scour than that anticipated for conventional, elevated bridges.

References

Braun, K., Christopherson, A. & Thieman, D. (2001). "Kuparuk River Submersible Bridges," *Welding Innovation*, Vol. XVIII, No. 1, 19-21.

Christopherson, A. & Nottingham, D. (1991). "Kuparuk River Module Crossing Case History: A Critical Arctic Oilfield Transportation Link Is Protected by a Refrigerated Foundation," *Society of Petroleum Engineers*, May, 459-469.

Christopherson, A., Braun, K., Thieman, D. & Carn, S. (1999). "Kuparuk River Submersible Bridges and Roadways," *Proceedings of the Tenth International Conference on Cold Regions Engineering*, ASCE, 576-587.

Meyer, D.F., Hess, D.L., Shellekens, M.F., Smith, C.W., Synder, E.F., and Solin, G.L. (2001). *Water Resources Data for Alaska, Water Year 2000: U.S. Geological Survey Water-Data Report AK-00-1.*

Peratrovich, Nottingham & Drage, Inc. (2000). "Kuparuk River Submersible Bridge," Special Award in Special Purpose Category, *Modern Steel Construction*, July, 42-44.

Peratrovich, Nottingham & Drage, Inc. (2000). *Kuparuk River Low Water Crossings Project 2000 Spring Breakup Summary Report.*

Peratrovich, Nottingham & Drage, Inc. (2001). *Kuparuk River Low Water Crossings Project 2001 Spring Breakup Summary Report.*

Orthotropic Steel Deck Bridges Constructed in Cold Regions

Alfred R. Mangus[1]

Abstract

This paper will give an overview of imaginative steel deck bridges currently in operation in cold regions throughout the world as an aid in the "type selection" process.. Imaginative bridges from Canada, Norway, Russia, Sweden, Ukraine and the USA were selected to demonstrate the complete range of steel bridges such as movable bridges; suspension bridges, cable stayed bridges, box girder bridges, railroad bridges, pipeline bridges and floating bridges. The advantages of steel deck bridges in cold regions will be summarized. The names of the owners, designers, fabricators, and contractors have been included to credit their success and contacts to discuss details beyond the scope of this paper. A list of 25 references is provided.

Introduction

Bridges were selected that have innovative approaches to design, construction, fabrication, erection, and aesthetics. Bridges with orthotropic steel decks are very limited, including those in the USA. For example there are about 650,000 bridge structures of all types in the U.S.A. In Europe, there are over 1,000 orthotropic steel deck bridges, but there are less than 70 bridges with orthotropic steel decks in the USA, less than 20 in Canada and less than 10 in Mexico. The selected bridges have a comprehensive range of steel rib types, steel superstructure types and traffic capacity.

The bridges selected demonstrate that steel orthotropic decks have improved the transportation infrastructure in cold regions around the world. Featured bridges from Canada include the Concordia Box-girder Bridge, Pretoria Bascule Bridge, and the A. Murray MacKay suspension bridge. Featured bridges from Norway include the Storda suspension bridge of the Triangular Link project, the Bybrua cable-stayed bridge, and the Nordhordland floating bridge across the Salhus Fjord. Featured bridges from Russia include the Oka River box-girder bridge and the city of Arkhangel, Russia's vertical lift span bridge. Featured bridges from Sweden include the high coast suspension and Tjorn cable-stayed bridge. The South Bridge cable-stayed bridge of the Ukraine is also discussed. Featured bridges from the USA include the city of Valdez, Alaska's floating dock articulating bridge and the Cordova, Alaska Ferry Terminal.

Comparing designs between different countries is more complicated than just translating the languages. (It is also more difficult because engineering slang or jargon varies with each country, so basic language translations may be incorrect). Many

[1] Transportation Engineer Civil, [CALTRANS] California Department of Transportation, 1801 30th Street, Sacramento, CA 95816; phone (916) 445-7325; office Al_Mangus@dot.ca.gov; home mangusalf@aol.com

countries have different vehicle live loading, plus types or grades of steel, charpy criteria varies with temperature ranges at the bridge site, plus numerous types of wearing surface materials which complicates comparing designs. Russian engineers prefer the open rib design and have industrialized this system, while most other engineers prefer the closed rib. Researchers, as well as the owners of orthotropic steel deck bridges, continue to monitor the performance of various rib types (Figure 1).

There is an interesting graphic comparison between code minimum design vehicle loads of Canada, Finland, Sweden, Norway, USA (HS 20), Switzerland, developed by OECD (Figure 2). These complexities make it even more difficult to compare design and maintenance issues. Since orthotropic steel deck bridges are a much smaller category of bridge types, finding similar projects in cold regions can be time consuming. It is well beyond the scope of any single paper to discuss the conversion process from another country's code into an AASHTO code design and details.

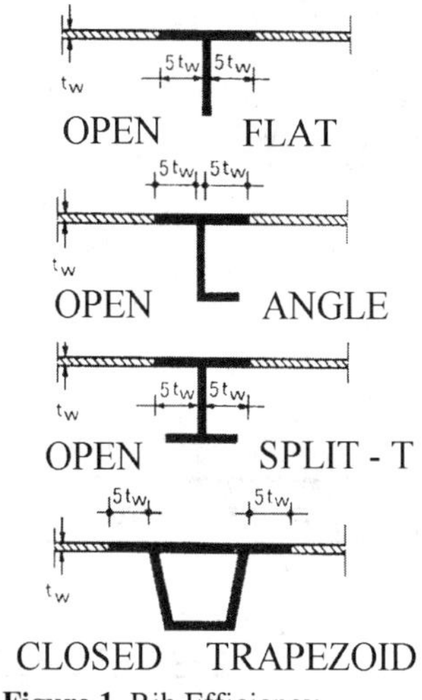

Figure 1. Rib Efficiency

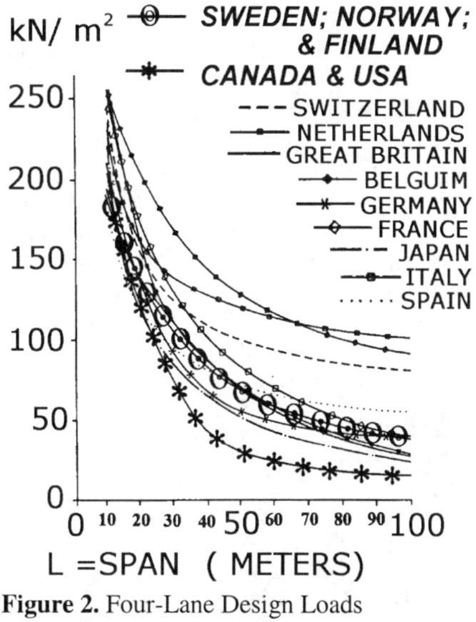

Figure 2. Four-Lane Design Loads

Orthotropic Steel Deck Bridges of Canada

The orthotropic bridges built in Canada include, the Walpole Island swing span bridge (Bowen, and Smith 1970), 1965 Concordia multi-cell box girder bridge, Pretoria Bascule Bridge in Ottawa (Figure 3), 1969 Papineau cable-stayed bridge of Montreal, 1970 A. Murray MacKay suspension bridge (Troitsky 1987), 1964 Port Mann arch bridge Vancouver, Mission Bridge box-girder bridge (Manniche & Ward-Hall 1975), and the Port Carling Bascule Bridge. Redecking of existing bridges with orthotropic steel deck panels include the Lions Gate (Matson 2001), Champlain Bridge of Montreal three-span cantilever truss (Carlin and Mirza 1996), the 1999 Angus L. Macdonald suspension bridge and the 2001 Lions Gate superstructure rebuilding. A Vancouver firm has been promoting steel orthotropic decks for parking garages in cold regions.

Figure 3. Ice-skating below the Pretoria Bascule Bridge Ottawa, Canada

The Concordia Bridge was erected in 1965 over the St. Lawrence River in Montreal and is 690m long, continuous from abutment to abutment. The superstructure consists of three 160m central spans and two 105m end spans. The piers are designed to break the river ice in bending (Figure 4). It has a total width of 28.651m providing two 11.6m roadways, two 1.82m sidewalks and a 1.82m central median. It is a three-cell box-girder bridge with trapezoidal exterior cells with two cantilevered deck sections. The cantilevers extend 4.41m and the three-cell box girder trapezoid is 19.8m, at the top and 17.37m at the bottom. The superstructure depth remains constant at 4.88m for the three central spans and tapers at the end spans from 4.88m at piers to 4.27m at the abutments. The orthotropic steel top-deck plate for the roadway has starts as thickness of 11mm, 16mm over piers one and four and a maximum of 25.4mm over piers two and three. The U-shaped closed ribs are 330mm deep, 8mm thick and spaced 609mm apart. The sidewalk deck plate has a thickness of 5/16 inch and its open ribs are split "T" wide-flange 203mm spaced 457mm apart. The 762mm deep transverse floor beams are spaced 4.57m apart with an 8mm web and a 254mm x 12mm flange. The thickness of the box girder webs and bottom flange, vary respectively from 12mm to 19mm and from 12mm to a 44mm maximum over the piers(Troitsky 1987).

Figure 4. Concordia Bridge, Canada

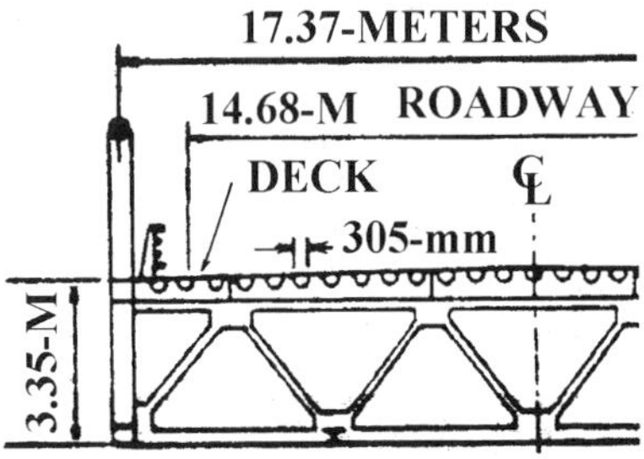

Figure 5. MacKay Bridge, Canada

Orthotropic Steel Deck Bridges of Norway

The first use of orthotropic bridge decks in Norway dates back to 1957 when the Krakeray Bascule Bridge was completed (Troitsky 1987). Norway has a complex aerodynamic wing type orthotropic suspension and cable-stayed bridges. Two unique orthotropic steel deck bridges are the Bergsoysund floating bridge (Mangus, and Sun 1999) and the Nordhordland floating bridge across the fiords of Norway.

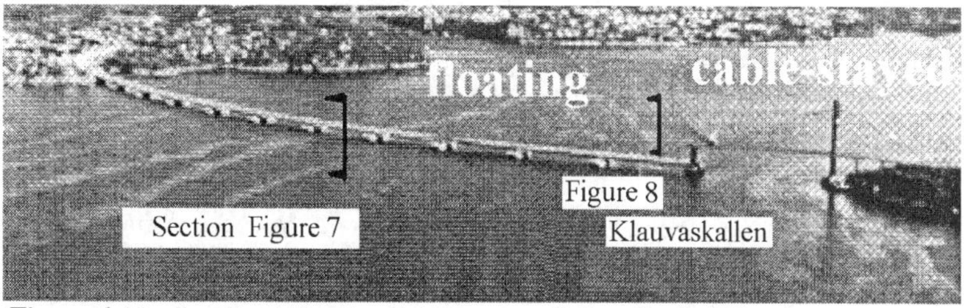

Figure 6. Nordhordland Floating Bridge across Salhas fjord of Norway

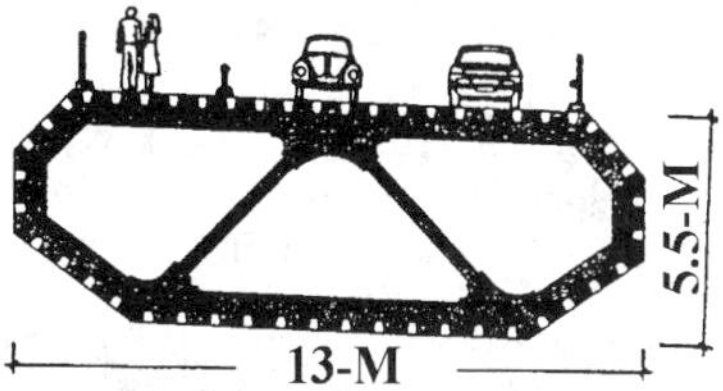

Figure 7. Nordhordland Floating Bridge

Figure 8. Nordhordland Floating Bridge

The Nordhordland Bridge across the Salhus fjord is Norway's second floating bridge and the world's largest floating bridge (Meas, Lande, and Vindoy, 1994). The bridge was opened for traffic in 1994. The total bridge length is 1615m and consists of a high level cable stayed bridge 369m long and a floating bridge 1246m long. The floating bridge consists of a steel box-girder, which is supported on 10 concrete pontoons and connected to abutments with transition elements in forged steel. The main elements are a high-level cable-stayed bridge providing a ship channel and a floating bridge between the underwater rock Klauvaskallen and the other side of the fjord. The cable-stayed bridge provides a clear ship channel. A 350m long ramp is required to transition from the higher bridge deck on the cable-stayed bridge to the bridge deck on the directly on top of the octagonal steel box-girder 11m above the waterline. The steel box-girder of the floating bridge forms a circular arch with a radius of 1,700m in the horizontal plane. The girder is supported on 10 pontoons.

The pontoons are positioned with a center distance of 113.25m and acts as elastic supports for the girder. The girder is designed without internal hinges. The bridge follows the tidal variations by elastic deformations of the girder. The steel box-girder is the main load-carrying element of the bridge (Figure 7). The octagon girder is 5.5m high and 15.9m wide. The free height below the girder down to the waterline is 5.5m allowing for passage of small boats. The plate thickness varies from 14mm to 20mm. The plate stiffeners are traditional trapezoidal shaped and they are spanning in the longitudinal direction of the girder. The stiffeners are supported by cross-frames with center distance of maximum 4.5m. At the supports on the pontoons bulkheads are used instead of cross-frames, because the loads in these section are significantly larger than in the cross-frames. The plate thickness in the bulkheads varies from 8mm to 50mm. The box girder is constructed in straight elements with lengths varying from 35m to 42m. The elements are welded together with a skew angle of 1.2° to 1.3° for accommodation to the arch curvature in the horizontal plane. The cross section dimensions of the octagonal box-girder are constant for the length of the bridge. The stress level varies significantly over the length of the bridge. In the areas with the highest stresses, steel with a yield stress of ReH= 540 MPa is used.

In the remainder bridge, in cross-frames and in bulkheads, normalized steel with yield stress of ReH= 355 MPa is used. The total steel weight of the box girder is 12,500 tons of which approximately 3,000 tons are high strength steel. The elevated ramp is approximately 350m long and has a grade of 5.7 percent (Figure 8). The elevated ramp is constructed with an orthotropic plate-deck 12mm thick and has 8mm and 10mm thick trapezoidal ribs 800mm deep. T-shaped crossbeams support the ribs with a center distance of maximum 4.5m. The main 1,200mm deep box-girders are located one at each edge of the ramp in order to maximize the stiffness about a vertical axis. The lower flanges are 45mm x 900mm and the upper flanges are 50mm x 610mm and the web thickness is 20mm. The minimum yield stress for the structural steel is ReH =355 MPa. The steel weight of the ramp is 1,600 tons.

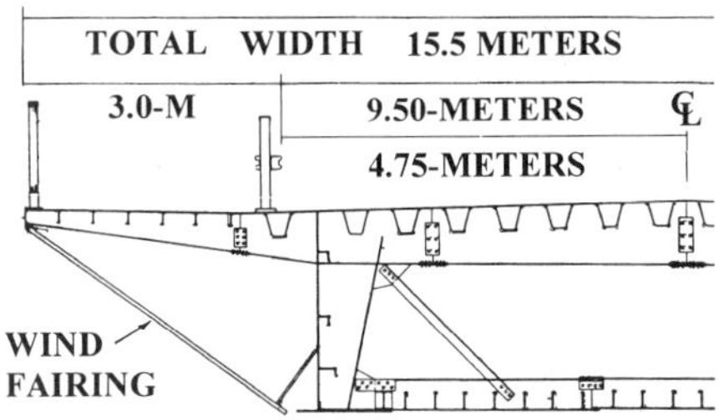

Figure 9. Bybrua Bridge Norway

Figure 10. Storda Bridge Norway

The Bybrua cable-stayed bridge has a main span of 185m (Figure 9). The 15.5m roadway superstructure was fabricated in the shop in 9.0m sections. There is a combined pedestrian plus bicycles area on each side of the three traffic lanes. The cross section of the main span, which is shown in (Figure 9), has a deck-plate of 12mm thickness but this is increased to 16 and 20mm at the cable anchorage. The bottom plate varies between 8mm and 10mm thickness, and the webs between 12 mm and 20 mm. At intervals of 3.0m there are frames supporting the longitudinal stiffening system. In the bridge deck this is made up of standard trapezoidal ribs from the German Steel Company Krupp, and in the bottom flange box section of bulb flats open ribs were utilized (Aune, and Holand 1981). In the longitudinal direction the deck was divided into six fabricated sections, two of these were welded to the web sections. The box bottom was fabricated as three sections. The total steel weight is about 1,100 tons. All elements prefabricated in the shop were welded, as were the field splices in the deck, whereas "Huck" high tensile bolts were used in all other field joints and to some extent ordinary high tensile bolts. All field joints are calculated as friction connection. The whole steel construction is metallized with zinc and painted according to the specifications of the Norwegian Public Roads Administration. The superstructure received the maximum live load stresses during the erection of the bridge. The wearing surface of the bridge deck is the same as developed for the [Little Belt] Lillebelt Bridge of Denmark, by the Danish State Road Laboratories.

The "Triangle Link" project, for the Norwegian National Road Administration, connects three islands off the Norwegian coast south of Belgen to the mainland with three bridges (Larson and Valen 2000). The entire project was completed in April 2001. The two orthotropic steel deck suspension bridges are known as the Storda Bridge and the Bomla Bridge. The Storda Bridge is 1,076m long, has a main span of 677m, with towers 97m high and a vertical clearance of 18m (Figure 10). The Bomla Bridge is 990m long with a main span of 577m and the tower height is 105m. The roadway of both bridges is 9.7m wide. Scanbridge AS of Norway fabricated the Bomla Bridge's steel approach superstructure, which was launched out over the tops of the columns from the shore. The steel components for the main span superstructure of the Storda Bridge were prefabricated in the Netherlands (Figure 10) and the main span superstructure of Bomla Bridge in Italy. The orthotropic ribs for the Storda Bridge were prefabricated in France. The orthotropic sections were transported to site by barge, and were lifted into position by floating crane.

Orthotropic Steel Deck Bridges of Russia

Russian engineers have standardized their orthotropic deck plates using open or flat plate ribs as shown in (Figure 1). They have several launching solutions or standardize methods for pushing the superstructure across a river or gorge. There are a limited number of bridge case histories documented in English, but they provide an overall view of Russian techniques (Blank, Popov, and Seliverstov 1999). The city of Arkhangel, Russia a vertical lift record span bridge of 120.45-m was completed in 1990 (Stepanov 1991). The Berezhkovsky twin parallel bridges are multi-cell box-girder bridges of three spans of 110m + 144.5m +110m. Each bridge has four traffic lanes 3.75m wide. The bridges were the first to be launched with inclined webs (Surovtsev, Pimenov, Seliverstov, and Iourkine 2000).

The four-lane orthotropic twin box-girder bridge crossing the Oka River on the bypass freeway around the city of Gorki, Russia was opened to traffic in 1991 (Figure 11). The 966m long superstructure consists of two spans x 84m + 5 spans x 126m + 2 spans x 84m (Design Institute Giprotransmost 1991). This bridge is a single continuous superstructure with a fixed bearing 420m away from one of the abutments. The total bridge width (29.5m including steel traffic barriers) provides two sidewalks 1.5m wide each, four traffic lanes 3.75m wide each, four 2m wide safety shoulders and a 2m wide center median. The total weight of steel for the super-structure is 10,635 tons, or 373 kg/m2. The orthotropic steel superstructure is comprised five basic elements (Figure 11). There are two main box girders assembled from two L-shaped sections for the bottom face and sides. The intermediate orthotropic plate sections were used for the top flange of the two box girders, as well as the majority of the deck. The end sections of the orthotropic plate were panelized with tapered ends, because only sidewalk loading is required. The transverse diaphragms are steel trusses between the box-girders. The diaphragms required extra steel beams at the bottom flange of the box-girder above the bearings. The main box-girder was shop fabricated in L-shaped sections that are 21m long and 3.6m deep. The intermediate orthotropic welded steel deck plate was shop fabricated in panelized sections of 2.48m wide x 11.5m long.

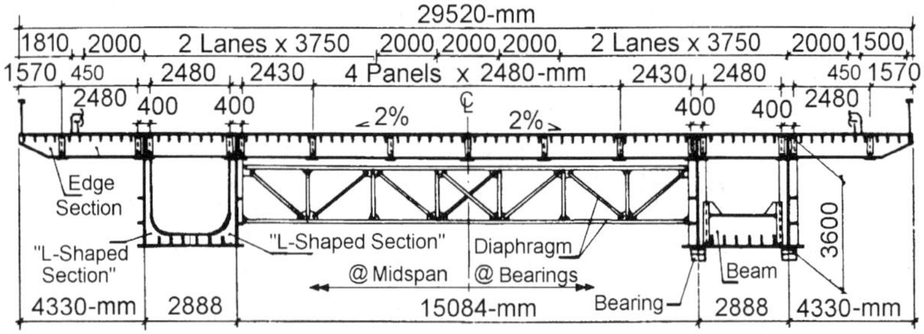

Figure 11 Twin box girder bridge crossing the Oka River, Russia (split – section)

The longitudinal ribs of the orthotropic deck and steel box girders are flat rib plates spaced at 0.35m, and the spacing of transverse ribs is 3.0m for both components. The stiffening ribs of the main girder are located on both sides of every web. The vertical split-T ribs of the box girders were aligned with the transverse ribs of the ribbed plate, thus creating the integral internal diaphragms. The longitudinal stiffening ribs are at a constant spacing along the bridge. Depending on the web thickness, additional vertical stiffening ribs were required between the diaphragms. The superstructure was erected using "continuous launching" from one bank of the Oka River. The shop-fabricated elements were added piece by piece to form a continuous structure at the "erection slip" area on this riverbank. The joints of the horizontal sections of the orthotropic deck and ribbed plates, as well as the joints of the web of the main girder were automatically welded. The joints of the longitudinal ribs of the ribbed plate were manually welded. All the remaining joints used high strength bolts.

The launching was performed without temporary falsework supports in the Oka River. Thus, a temporary strut frame pylon was utilized to support the girder launching. The pylon was 30m high with the baseline of the cable-stays extended 63m to each side. Thus only one temporary strut frame was required. The backstay cables were counterbalance by the weight of the previous span. The pylon technique for the launched box girders is explained in more detail in a textbook(Rosignoli 1998).

Orthotropic Steel Deck Bridges of Sweden

Sweden has complex orthotropic bridges built in the cold regions of their country. The Tjorn cable-stayed bridge across the Askerofjord has a steel main span of 366m (Brodin1982). The deck has a clear width of 15.75m comprising a roadway of three traffic lanes and a combined sidewalk and bicycle area. The orthotropic deck surfaced with asphalt is supported by a welded rectangular box girder of width 8.5m and height 3m and with floor beams at 4m centers. The 12mm thick deck plates of the box and its cantilever extensions, as well as the bottom plate, are stiffened by trapezoidal ribs and the webs by angles.

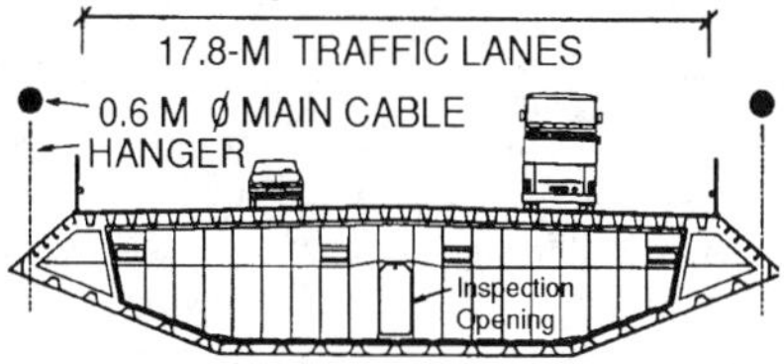

Figure 12 High Coast, Sweden section

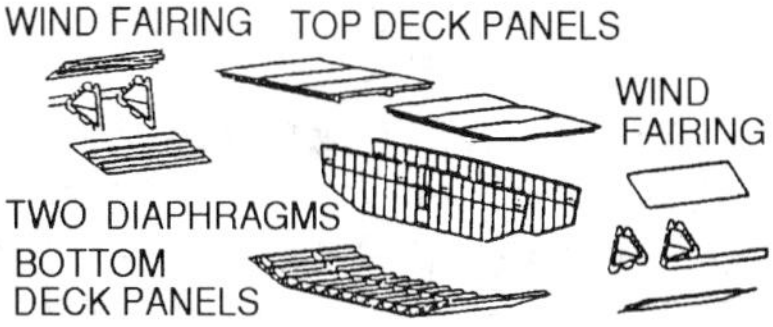

Figure 13 High Coast, Sweden components

The High Coast Suspension Bridge of Sweden is almost the same size of the Golden Gate Bridge (Merging with Nature 1998). The main span is 1,210m long with suspended side spans of 310m and 280m. The width of the roadway is 17.8m allowing for a possible future extension to 4 lanes (Figure 12). The distance between the main cables is 20.8m and there are 20m between the hangers. The girder is continuous through the towers extending 1,800m from abutment to abutment where expansion joints and hydraulic buffers are located. The 48 box girder sections were fabricated at a shipyard in Finland (Pedersen 1997). The standard section is 20m long with two sets of hangers each and weighs 320 tons (Figure 13). The 20m long panels for the deck, sides and bottom were fabricated with a maximum width of 10m. They were fabricated from steel plates, typically 9mm to 14mm thick 10m long and 3 to 3.3m wide. The ribs were 20m long trapezoidal ribs with a plate thickness of 6mm to 8mm.

The plates were placed on a plane and welded in the transverse and longitudinal direction and the trough stiffeners were fitted and welded longitudinally. Connection plates to connect the panels and the diaphragms were welded between ribs. The 20m long edge sections and units for the transverse diaphragm, or bulkhead, were fabricated. The bottom and inclined sides were placed first. Each 4m deep transverse diaphragm or bulkhead was fitted. The edge sections were installed and finally the tow deck panels were placed on top, completing a 20m long subsection. The 31 bridge girder sections for the main span were transported from the fabrication yard in Finland on the three barges in the same way as the sections for the side spans, and erected with a floating sheerleg crane, 130-m boom, starting from mid-span and proceeding towards the towers (Edwards and Westergren 1999).

Orthotropic Steel Deck Bridges of Ukraine

The 1992 signature span of the South Bridge over the Dnipro [Dnepr] River in Kyiv [Kiev], Ukraine is an unsymmetrical cable-stayed bridge with a main span of 271m (Korniyiv, and. Fuks 1994). The main span side of the H-tower is a continuous three-span steel box girder with orthotropic steel deck. The back-span superstructure on the opposite side of the H-tower is a segmental prestressed concrete box section. Concrete construction for the shorter back span of 60-m was used as a counter-weight mass equal to the longer orthotropic main span. The bridge carries a six-lane roadway, two rail tracks and four large-diameter water pipes (Figure 14). The total live load is about 246 kN/m. The three-span (80.5m + 90m + 271m) continuous steel box girders are made of low-alloy steel with a minimum yield strength of 390 MPa.

The bridge was divided into shop-fabricated components that were welded. Field splices were either welded or joined by high-strength bolts. Bolting was used where automatic welding was impractical because of the short length of the weld or difficult access. The cross section of the twin two-cell box-girder bridge consists of six vertical webs, the upper deck plate and the lower box flanges. The narrow cell is for the cable-stayed bridge anchorage. In the central portion of the cross section that carries the two hot water supply pipes, the lower flange was omitted to preclude the undesirable effects of unequal heating inside a closed box. The bearings at the piers permit lateral displacement of the superstructure because of the 41.5m bridge width. The orthotropic decks, bottom flanges of the boxes and the webs have open flat-bar stiffening ribs, a common feature in Ukrainian and Russian bridges. Longitudinal flat or open ribs were placed on the top face of the orthotropic deck plate under the rail tracks, thus avoiding intersections of longitudinal and transverse stiffeners.

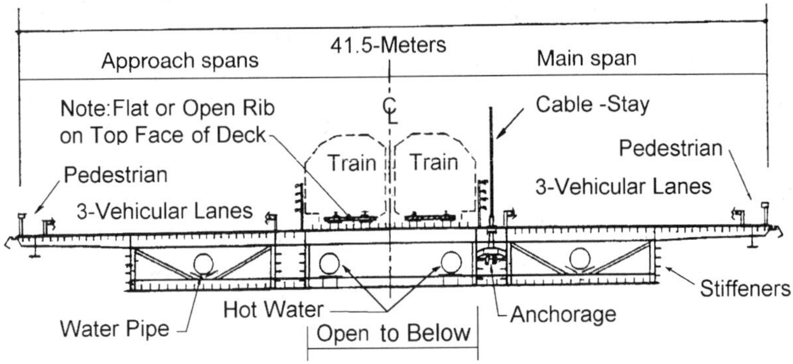

Figure 14 South Bridge Cable-stayed of Ukraine (split–section)

This facilitated fabrication, at the same time precluding stress concentrations at crossing welds that would be susceptible to fatigue under dynamic train loading. Longitudinal ribs under the train tracks have a depth in excess of the design requirements, which permitted longitudinal profile adjustments of the tracks after erection of the bridge superstructure. The steel girders were pre-assembled on the bank of the south and erected by launching. The twin two-cell box-girders were equipped with a launching nose and stiffened with a temporary strut system (Rosignoli1999). Single erection rollers were used at the tops of supports and had a friction factor of less than 0.015. The erection of the 271m main span was accomplished with two false work supports providing three equal spans of about 90m. At the H-tower of the cable-stayed bridge, hinged conditions are provided by supports with limited rotational capability in the vertical and the horizontal planes. The torsional rigidity of the bridge is from the two planes of cable stays, plus the stiffness of the closed box sections. Under one-sided loading of the bridge (three traffic lanes), the deck cross slope of 0.3% was measured in field tests, less than the calculated value of 0.35%. Eccentric hinged connections between the bottom flanges of the stiffening girders and the H-tower were constructed, which considerably reduces the bending moments in the girders.

Orthotropic Steel Deck Bridges United States

The state of Alaska has 13 orthotropic bridges (Mangus and Sun 1999). The orthotropic bridges in Alaska vary in size and type. The largest is the Patton Yukon River Alyeska Trans-Alaska Pipeline Bridge, which was the first built across the Yukon River (Carlson, Platzke and, Dreyer 1976). This bridge is subjected to temperatures of -51 degrees Celsius (Letzring 1997). Concrete must be at or above 4.44 degrees Celsius to properly cure. Although it is physically possible to encapsulate plus heat the concrete construction process, it is usually more expensive (Mangus 1988), (Mangus 1991). The Cordova Ferry Terminal movable orthotropic ramp (Figure 15) (Mangus 2000), city of Valdez Floating Dock Transfer Bridge (Figure 16) (Ozolin, Wilson, and Hutchison 1982), city of Valdez Floating Dock Transfer Orthotropic Bridges (two identical bridges) (Figure 16), the Kake Ferry Terminal, the Metlakatla Ferry Terminal, the Hoonah Ferry Terminal, the Angoon Ferry Terminal, and the Clark Bay Ferry Terminal must resist being near cold seawater. The Rabbit Creek Pedestrian Bridge, the Willow Crest Pedestrian Bridge, the Lathrop Pedestrian Bridge, and the Egan Drive Pedestrian Bridge are located in cold urban areas of Alaska.

Figure 15 Cordova Ferry Terminal USA

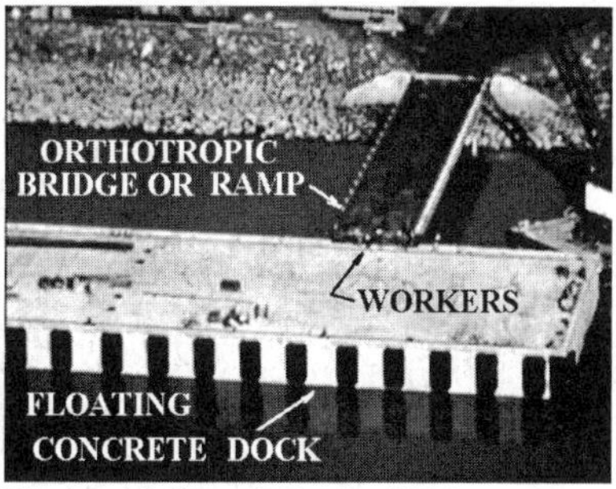

Figure 16 City of Valdez USA

Conclusion – Why use bridges with orthotropic steel decks in cold regions?

Orthotropic steel deck bridges can be economically built in cold regions during any time of the year. Engineers from around the world utilize this practical and economic system for all types of bridges for cold regions. An all-steel bridge can be erected in very cold winter weather. Orthotropic steel deck bridges have proven to be durable in cold regions in cold regions. Most engineers prefer to use a proven durable system. The owners, design engineers, researchers, steel fabricators, contractors, and steel companies have written numerous articles on the advantages of orthotropic bridges. In the 1950's German engineers created the orthotropic system, which was adopted in the 1960's by American engineers. Russia has developed a mass manufactured panelized orthotropic deck system and launching methods for very cold regions. Nordic Road & Transport Research discusses Scandinavian research and is available on line at www.vti.se/nordic.

All of the other reasons used for selecting Orthotropic steel deck bridges are also applicable to cold regions. The Orthotropic steel deck integrates the driving surface as part of the superstructure, and has the lowest total mass of any practicable system. Around the world long span bridges, floating bridges and movable bridges are built with orthotropic steel decks. The lower mass results in lower seismic forces. Hence, large Orthotropic bridges in high seismic zones such as Japan, Alaska, and California. The lower mass requires less energy to move the operable spans of a movable bridge. The mass of every movable span eventually reaches a practical limit due to the size of motor needed to move the bridge. Thus, spans built of heavier reinforced concrete are much more expensive for these bridge types built anywhere. This paper identifies a matrix of rib, superstructure and various bridge types. Desired features can be "mixed and matched" from these case histories for a "solution" in the "type selection" process.

Figure Credits

Figure 1 from Ballio, G., Mazzolani, F. M. 1983, *"Theory and Design of Steel Design Structures,"* Chapman & Hall Ltd. courtesy of Dr. Mazzolani; Figure 2 courtesy of OECD, Organization for Economic Co-Operation and Development Washington D.C. ; Figure 3 courtesy of Sahni, Virendra, "Vinni"City of Ottawa, Canada; Figures 4 & 5 courtesy of The James F. Lincoln Arc Welding Foundation; Figures 6, 7, & 8 courtesy of Dr. Ing. A-Aas–Jakobsen, AS Structural Engineering Consultants, Oslo, Norway; Figure 9 adapted from Aune, Petter, and Holand, Ivar (1981); Figure 10 courtesy of Mr. L. Adelaide of Profilafroid, France; Figures 12 & 13 courtesy of Claus Pedersen of Mondberg & Thorsen AS, Copenhagen, Denmark; Figures 11 & 14 courtesy of IABSE International Association of Bridge Structural Engineers; Figure 15 photo by Bill Gunderson, P.E., Figure 16 courtesy of Wilson, W. of Berger/ABAM Engineers

References:

Aune, P., Holand, I. (1981) *Norwegian Bridge Building – A Volume Honoring Arne Selberg,* Tapir, Norway

Blank, S. A., Popov O. A., Seliverstov, V. A. (1999.) "Chapter 66 Design Practice in Russia," *Bridge Engineering Handbook,* 1st ed., Chen, W.F., and Duan, L. Editors, CRC Press, Boca Raton Florida

Bowen, G. J. and Smith, K. N., (1970)"Walpole Swing Span has Orthotropic Deck," *Heavy Construction News*, Canada

Brodin, S., (1982) "New Tjorn Bridge Across the Askerofjord, Sweden" *Structures*, IABSE, Zurich, Switzerland, C-4, May, 40-41

Carlin, G. P., and Mirza, M. S. (1996) "Replacement of reinforced concrete deck of Champlain Bridge, Montreal, by orthotropic steel deck" *Canadian Journal of Civil Engineering* Canadian Society of Civil Engineers, Montreal, (23), 1341-1349

Carlson, L. A, Platzke, R. and Dreyer, R. C. J., (1976) "First Bridge Across the Yukon River", *Civil Engineering*, ASCE, 47

Design Institute Giprotransmost, (1991) "The Bridge-Crossing over the River Oka on the Bypass Road near Gorki" *Structural Engineering International*, IABSE, Zurich, Switzerland, Vol. 1, Number 1, 14-15

Edwards,Y. & Westergren P., (1999), "Polymer modified waterproofing and pavement system for the High Coast Sweden" *NordicRoad & Transport Research* No. 2, 28

Korniyiv, M. H. and. Fuks, G. B (1994) "The South Bridge, Kyiv, Ukraine" *Structural Engineering International*, IABSE, Zurich, Switzerland, Vol. 4, Number 4, 223-225

Larsen J., and Valen, A., (2000) "Comparison of Aerial Spinning versus Locked-Coil Cables for Two Suspension Bridges(Norway), *Structural Engineering International*, IABSE, 10(3), 128-131

Letzring, M., (1997) "The Trans Alaska Pipeline: 20 Years of Arctic Oil," *KAKM Video PBS Public Television Station,* Anchorage, Alaska, 28 minutes

Manniche, K. & Ward-Hall, G. (1975) "Mission Bridge – Design & Construction of the Steel Box Girder", *Canadian Journal of Civil Engineering*, Canadian Society of Civil Engineers, Montreal, Canada, Vol. 2, N. 2, June 169-192

Mangus, A., (1988) "Air Structure Protection of Cold Weather Concrete," *Concrete International*, American Concrete Institute, Detroit, MI, October 22-27

Mangus, A., (1991) "Construction Activities Inside of Air Structures Protected From the Arctic Environment," *International Arctic Technology Conference*, Society of Petroleum Engineers, Anchorage, AK, May 29-31

Mangus, A, and Shawn, S., (1999) "Chapter 14 Orthotropic Deck Bridges," *Bridge Engineering Handbook,* 1st ed., Chen, W.F., and Duan, L. Editors, CRC Press, Boca Raton Florida

Mangus, A., (2000) "Existing Movable Bridges Utilizing Orthotropic Bridge Decks," *8th Heavy Movable Bridge Symposium Lake Buena Vista Florida*, Heavy Movable Structures, P. O. Box 398, Middletown, NJ

Matson, D. D. (2001), Piece by Piece: Construction of the Lions' Gate Suspension Bridge, Vancouver, B. C., *Civil Engineering*, May, 40-45

Merging with Nature – Hoga Kusten [High Coast] (1998), *Bridge Design & Engineering*, 10, (1), 50 –56

Ozolin, E.; Wilson, W. and Hutchison, B., (1982) "Valdez Floating Dock Mooring System," *The Ocean Structural Dynamic Symposium*, Oregon State University, Corvallis, Oregon, 381-395

Pedersen, C., (1997) *The Hoga Kusten (High Coast) Bridge-suspension bridge with 1210 m main span –construction of the superstructure*, Mondberg & Thorsen A/S Copenhagen, Denmark

Rosignoli, M.,(1998) *Launched Bridges – Prestressed Concrete Bridges built on the ground and launched into their final position*, ASCE Press, Reston, VA

Meas, P., Landet, E., Vindoy, V., (1994) "Design of the Salhus Floating Bridge (Nordhordland Norway)," *Strait Crossings 94,* Balkema, Rotterdam ISBN 90-5410 388-4 3(3), 729-734

Stepanov, G. M., (1991), "Design of Movable Bridges," *Structural Engineering International*, IABSE, Zurich, Switzerland, Volume 1, Number 1, 9-1

Troitsky, M. S., (1987) *Orthotropic Bridges - Theory and Design*, 2nd ed., The James F. Lincoln Arc Welding Foundation, Cleveland, OH

Surovtsev, V., Pimenov, S., Seliverstov, V., & Iourkine, S, (2000) "Development of Structural Forms and Analysis of Steel Box Girders with inclined webs for operation and erection conditions" Giprotransmost J.S.Co, Pavla Kortchagina str. 2, 129278 Moscow, Russian Federation http://ciha.fsv.cz/~eurosteel/158/158.htm

Design, Construction, and Performance of Open Cell Sheet Pile Bridge Abutments

Dennis Nottingham P.E.[1], Dempsey S. Thieman, P.E.[2], Kenton Braun, P.E.[3]

Abstract

Recent innovations in bridge abutment design have resulted in the development of open cell sheet pile bridge abutments. The open cell abutment structure is a cellular flat sheet pile structure driven in a partial circular shape when viewed from above. Open cell structures can easily resist very heavy loads such as extremely heavy drill rig vehicles and module loads, ice attack from arctic river breakup events and high seismic accelerations. Additionally, these abutments uniquely protect the bridge structure from the effects of extreme river scour. These abutment structures are easily constructed without the use of field welding, bolted connections or an independent tieback system, making them less expensive and easier to construct.

The system functions as a membrane, relying solely on the resistance developed between the soil and the sheet pile tail wall. The system does not rely on toe embedment for stability. These structures have significant advantages over typical spill through or reinforced earth abutment structures, especially in arctic conditions.

Over one-hundred open cell sheet pile structures (fifty-nine of which are bridge abutments) have been built throughout Alaska and the rest of the United States, some in very active seismic zones. These structures have performed flawlessly in the twenty years of their history resisting hundreds of earthquakes. Details of seismic performance will be presented.

This paper reviews the general theory, adaptability, construction, and seismic and ice load resistance of the open cell sheet pile abutment design. Examples and history of constructed open cell sheet pile abutments for heavy duty private installations and for public transportation systems will be presented.

[1] Member ASCE, President, Peratrovich, Nottingham & Drage, Inc., 1506 W. 36th Ave., Anchorage, AK 99503; ph. (907) 561-1011

[2] Member ASCE, Senior Engineer, Peratrovich, Nottingham & Drage, Inc., 1506 W. 36th Ave., Anchorage, AK 99503; ph. (907) 561-1011; d_thieman@pnd-anc.com

[3] Member ASCE, Senior Engineer, Peratrovich, Nottingham & Drage, Inc., 1506 W. 36th Ave., Anchorage, AK 99503; ph. (907) 561-1011; k_braun@pnd-anc.com

Introduction

Historically, the most common application for flat sheet piles has been in closed circular bulkhead configurations in various geometrical arrangements. These types of arrangements have proven costly for a number of reasons, such as: close tolerances are required for construction; difficult to hold tolerances when hard driving is encountered; filling a closed structure requires use of conveyors or bucket equipment; and difficult access inside the cell makes compaction efforts expensive.

The open cell sheet pile concept is geometrically similar to the closed cell configuration except that approximately one-fourth of the cell is removed, allowing access to the interior of the cell for fill and compaction operations. The tail walls anchor the face sheets and are buried within the abutment fill, which forms the driving surface of the approach to the bridge. Reference Figure 1.

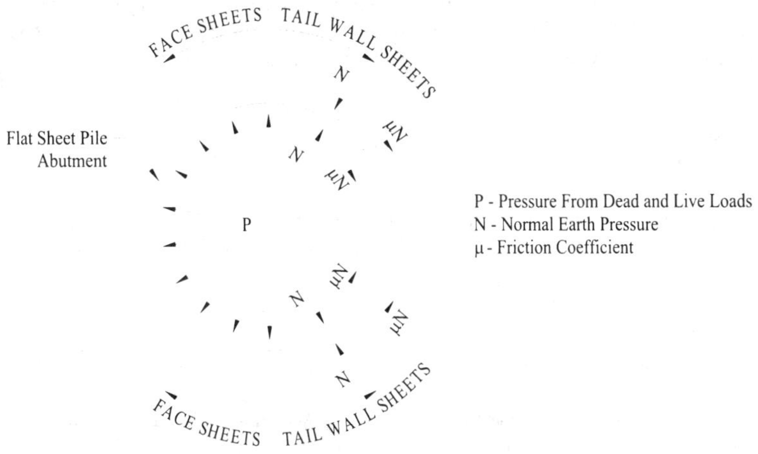

Figure 1. Open Cell Sheet Pile Abutment – Plan View

The flat sheet piles possess important characteristics such as extremely strong interlocks, which allow hard driving and the insurance that they will remain intact. The large structural mass of the interlock provides a substantial corrosion allowance for structures with a long design life. Additionally, the bulbous interlocks act as micro soil anchors providing additional soil / friction resistance in the tail walls. (Reference Figure 2).

In many cases the use of open cell sheet pile abutments is more cost effective than other types of bridge abutment structures. The rounded shape and vertical face of the sheet pile abutments, provide maximum stream flow area for a given bridge length, a significant improvement over spill-through abutments. This fact allows the use of a shorter and less costly bridge superstructure.

Research and Development

Peratrovich, Nottingham and Drage, Inc. (PN&D) began development of the open cell sheet pile abutment concept in 1980 in an effort to produce a low-cost earth retaining structure capable of supporting excessive loads such as oil field modules and large cranes. The idea was to use flat sheet piles as vertical soil friction anchors for a membrane wall system.

An absence of technical information on anchor wall resistance required the development and implementation of a testing and analysis program. A test program was initiated using full size sheet piles in a test box. The apparatus was capable of testing various vertical soil pressures and different soil types. A series of tests were conducted with varying soil types, densities and vertical pressures. Findings confirmed that, due to the micro-anchor interlocks, the soil / sheet pile friction often approaches twice that predicted by classical techniques, depending on compaction (Campbell, 1986) and other soil properties. The classical analysis of tension capacity of a steel plate in granular soil is given by:

$$T = 2FA$$

Where: T = tension force

F = friction capacity = μN

$N = P_C K_O$

P_C = vertical overburden pressure

K_O = coefficient of at rest lateral earth pressure (dependent on compaction, etc)

μ = coefficient of soil/steel friction ($\mu = \tan\delta$, δ=soil/steel friction angle)

A = area of one side of steel plate

The open cell system essentially functions as a horizontally tied membrane that relies entirely on soil resistance against the tail walls to retain soil. The anchor pile installed at the end of the tail wall provides additional resistance and lateral bending stability. Pile embedment of face sheets is not a factor in wall strength or load capacity.

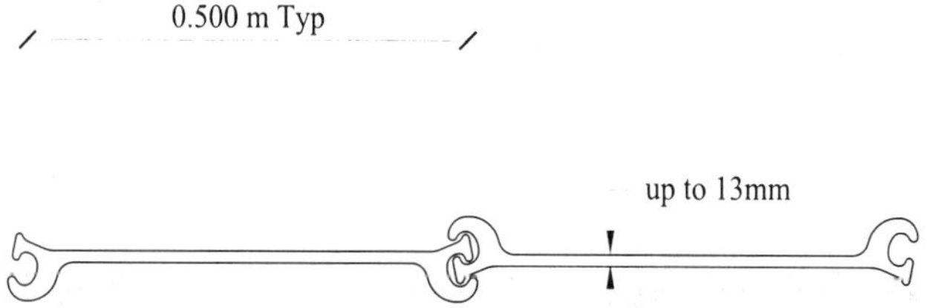

Figure 2. Typical Flat Sheet Pile

In 1981 an opportunity arose to use the new open cell idea for the construction of high-capacity bridge abutments at Prudhoe Bay, Alaska for development of a gravel causeway. The temporary construction bridge would be used to provide access for gravel trucks. During the project, one million cubic yards of gravel were hauled over the bridge in 1.3-MN (130-ton) B-70 trucks, proving the strength of this new abutment design.

Design Considerations

Factors in the design of an open cell sheet pile abutment include conditions such as scour, ice conditions, earthquakes, phreatic water levels, geotechnical information and soil strength, design loads and construction site conditions. By adjusting the length and/or height of the buried tail walls, the open cell abutment system is adaptable to nearly any set of load conditions.

The tail walls rely on the friction at the soil / sheet pile interface to resist the pressures developed on the face of the cell. The amount of friction developed is dependent on factors such soil type, friction angle, compaction, and depth of fill above tail wall.

A brief discussion of major considerations in the design of open cell sheet pile abutment follows:

Scour. Typically, one of the controlling design considerations for bridge abutment design is the predicted depth of scour, a factor that is difficult to estimate accurately. Because the open cell sheet pile design is anchored by the tail walls within the roadway fill, the depth of toe embedment of the face sheets is not a factor in abutment stability. The tip of the face sheets need to remain embedded only to ensure the abutment fill does not escape from within the cell. This is a substantial difference from tied-back or cantilever structures, which depend on toe embedment for structural integrity. For structures located in very active and dynamic river channels, a sheet pile wingwall or flank wall must be designed to prevent scour of the roadway approach fill, protecting the soil which anchors tail wall from scour action.

A spectacular example of the scour resistance of the open cell sheet pile abutment design occurred during the 1999 breakup flood of the Kuparuk River, on the North Slope of Alaska. A large flood event produced river bed scour in excess of 7.5 meters (25 feet), beyond the embedment depth of the face sheet pile. The scouring action of the fast moving water removed most of the gravel fill from within the sheet pile abutment, see Figure 3. The sheet pile settled vertically as the river bed elevation progressed below the tip elevation of the sheets. The hydraulic forces of the moving water and impacts from four-foot-thick ice floes had twisted and somewhat deformed the originally cylindrical abutment shape. Although aesthetically damaged, the structural shell of the sheet pile abutment acted to protect the bridge abutment piles from additional scour and damage from ice floe impacts; the bridge structure and foundation were unharmed by the event. Within a few days after flood waters began to recede, the sheet pile abutment cell and wingwall cell were refilled with gravel, compacted and open to all traffic including heavy drill rigs that weigh over 19.6 MN (4,400 kips). This extreme event shows the resiliency, toughness and stability of the open cell sheet pile abutment design. Conventional tie-back or cantilever walls would have totally failed in this event.

Figure 3. Scour at Kuparuk River Bridge was repaired and fully functional within days of damage.

Seismic Forces. Typically in areas of seismic activity, one of the controlling loads in open cell abutment design is seismic acceleration. Classical wedge or slip circle analysis techniques are used to check the mass stability of the structure. The area of the sheet pile tail wall that extends beyond the failure plane acts to resist movement, providing the safety factor against failure. The length of the tail walls can be adjusted to provide the safety factor required.

Alaska is in an area of high seismic activity, and many open cell structures have experienced significant ground motion acceleration from nearby earthquakes with magnitudes of up to about 7.0 on the Richter scale. The only noticeable change in the affected structures has been some uniform settlement likely due to soil consolidation. Due to the flexibility of this membrane structure, the open cell sheet pile abutments can easily handle differential settlements associated with large contingency earthquakes.

Table 1 presents a summary of seismic rock accelerations experienced by open cell structures to date. The maximum horizontal seismic rock accelerations presented for each site are the average of the values calculated using four different empirical methods, based upon earthquake magnitude, energy attenuation, and distance from the epicenter. This technique was used to reduce the variability between the calculation methods. The table of results present the horizontal rock acceleration for the site. Earth retaining structures will "feel" an acceleration reduced by site effects such as soil type and depth to bedrock. In all cases, the accelerations listed in Table 1 are less than the design acceleration for the projects, however with time, larger design magnitude earthquakes will occur, providing additional information on seismic performance of the open cell structures.

Resistance to Ice Loads. Open cell sheet pile abutments have been installed in numerous arctic locations where ice floes can reach 2 meters (6 feet) in thickness and be driven by fast moving spring breakup flood waters. Due to the compacted fill within the abutment cell, sheet pile damage from ice impact has not been observed under normal conditions. The only exception to this, as discussed above, has been cosmetic ice impact damage occurring when gravel fill within the abutment cell does not exist. Unlike the zigzag shape of a cantilever "Z" sheet pile wall, the rounded shape of the sheet pile cell presents a smooth, resilient surface which does not "catch" the passing ice floes, reducing the possibility of local damage to a sheet pile and deflecting ice forces from the structure.

Settlement and Weak Soils. Due to the flexible nature of this anchored membrane structure, the open cell sheet pile abutment can adjust to extreme global and differential settlements without threatening the stability of the structure. Some projects have been designed for global settlements of over 4.5-meters (15-feet) in an area of consolidating soils. Unlike tied back cantilever abutments where horizontal tie rods can be sheared by the settling soils, the vertical sheet pile tail walls are virtually unaffected by settling soils and actually gain strength with increased soil density (through an increased effective friction angle).

In areas of weak soil, the length of the tail walls can be increased to reduce the unit load to the soil. In addition, the anchor pile at the end of each tail wall provides further resistance, increasing the factor of safety. Vibratory compaction with a modified H-pile can also be used to increase the density and strength of subsurface soils.

Construction

Construction of open cell abutments requires driving sheet piles some distance below the ground surface. The use of a steel template, typically supported by temporary pin piles, ensures the location, alignment and plumbness of the abutment. Standard equipment for abutment installation includes a crane, earth-moving equipment and a vibratory pile driving hammer. A small impact pile driving hammer may also be necessary if difficult driving is encountered. Experience has shown that a typical open cell sheet pile abutment can be installed in less than a week with a crew of about four. After the sheet pile have been driven, standard earth-moving equipment is used to place and compact the fill within the abutment.

Figure 4. Open Cell Sheet Pile Abutment Under Construction

Special thermal modification techniques have been developed for sheet pile construction in arctic permafrost conditions. Typically, a relatively thin and flexible pile could not be driven through frozen soils. To address this problem, a thermal modification technique was developed in which pilot holes are drilled full depth around the perimeter of the abutment cell. The holes are then filled with hot water to thermally modify the frozen subgrade. The sheet pile can then be installed using conventional driving methods, (Christopherson, 1983).

The construction cost of an open cell sheet pile abutment is substantially (up to fifty percent) less than that of other similar types of abutments. The most significant cost difference is due to the fact that here are no walers, tiebacks or other connections to install or inspect in the future. Designs requiring these items are not only more costly, they are failure-prone due to long term corrosion and damage to these critical structural elements that cannot be easily inspected or replaced.

Sample Projects

Heavy Duty Oil Field and Mining Operation Bridge Abutments. Over forty open cell structures have been constructed since 1981 on the North Slope of Alaska. They withstand extreme environmental conditions, operating temperatures that can vary up to 49°C (120°F), massive ice floes up to 2-meters (6-feet) thick, and heavy vehicle crossings of up to 16.9 MN (3,800 kips).

Figure 5. Alaska North Slope Drill Rig Crossing the Kuparuk River Bridge

At a remote location on the Seward Peninsula fourteen open cell bridge abutment structures were constructed in 1985 as part of the Red Dog Mine. These structures withstand heavy traffic from mine dump trucks that haul about 11,700 MN (1.3 M tons) of concentrate per year, and have performed flawlessly since construction.

Public Transportation System Bridge Abutments. Several open cell bridge abutment structures have been designed and constructed for the Alaska Department of Transportation and Public Facilities (ADOT&PF). Although they do not experience the types of heavy loads discussed previously, they are a tribute to open cell design given their acceptance of a "new technology" by a governmental agency.

The Town Creek Bridge in Unalaska, built in 1983 and shown in Figure 6, features the first open cell bridge abutment constructed for Alaska's state highway system. Value engineering of the original design was allowed when it was discovered that, following the failure of a similar project, the original steel crib abutment design was susceptible to scour. Working with an experienced contractor, the new abutment was designed, approved and constructed without incident. It continues to function well today.

Figure 6. Town Creek Bridge in Unalaska, Alaska.
The First ADOT&PF Open Cell Sheet Pile Bridge Abutment, 1983.

Serious damage to the Southcentral Alaska transportation system was caused by a severe rain-induced flood in September of 1995. A bridge along Knik River Road near Palmer, Alaska, at Hunter Creek had been washed out. The ADOT&PF installed a light-duty temporary bridge for emergency use and requested proposals for the design and construction of a more substantial temporary bridge allowing the design process of the permanent structure to proceed at a slower, more cautious pace.

Due to the local availability of materials including flat sheet pile and steel girders, the temporary Hunter Creek Bridge was designed, constructed and open to traffic in just seventeen days. Furthermore, the ADOT&PF turned the "temporary structure" into a permanent structure, negating the need for further design and construction efforts. This example shows the adaptability of the open cell sheet pile concept, resulting in savings of time and money to the public.

Figure 7. C Street Bridge Under Construction - Anchorage, Alaska

The C Street Bridge is a critical connection of urban arterial road to the industrial railroad area near downtown Anchorage, Alaska. The project site has deep deposits of soft clay and an environmentally sensitive salmon stream crossing. The open cell sheet pile abutment was carefully designed to accommodate the sensitive soils structure while providing an "environmentally clean" solution with minimal stream-bank disturbance and with the ability to contain possibly contaminated soils or roadway runoff from seeping into the fish-bearing stream.

Conclusion

The development of the open cell sheet pile abutment structure presents a cost-effective solution that is very adaptable to various site conditions. This bridge abutment structure provides many advantages over other structure types, such as scour resistance, adaptable design, ease of construction, high strength and elimination of buried structural components. These advantages, combined with cost savings and a proven track record, provide a tested, innovative solution to traditional bridge abutment structures.

CONSTRUCTED	NAME	LOCATION	EVENTS	A_{MAX}
1981	Tanana River Bulkhead	Nenana, AK	37	0.08g
1982	West Fork Chulitna Bridge	Colorado Station, AK	158	0.06g
1983	Town Creek Bridge	Dutch Harbor, AK	238	0.09g
1984	North Fork Anchor River Bridge	Anchor Point, AK	299	0.08g
1985	Women's Bay Bulkhead	Kodiak, AK	75	0.07g
1985	Women's Bay Bulkhead Extension	Kodiak, AK	75	0.07g
1986	Valdez Small Boat Harbor Bulkheads	Valdez, AK	96	0.09g
1986	Fish Creek Weir Coastal Trails	Anchorage, AK	245	0.10g
1986	Grouse Creek Bridge	Seward, AK	43	0.06g
1987	SMIC Dock	Seward, AK	38	0.06g
1988	Commercial Fishermans Bulkhead Dock	Valdez, AK	88	0.09g
1988	Portage Glacier Tour Dock Bulkhead	Portage, AK	53	0.08g
1988	Seward Fish Dock Bulkhead	Seward, AK	39	0.07g
1988	UniSea Bulkhead	Dutch Harbor, AK	141	0.09g
1989	Captains Bay Bulkhead	Dutch Harbor, AK	128	0.07g
1989	Unalaska Marine Center	Unalaska, AK	130	0.07g
1990	Ship Creek Point Dock	Anchorage, AK	211	0.10g
1990	C Street Bridge	Anchorage, AK	203	0.10g
1990	King Cove Deep-Water Port	King Cove, AK	58	0.07g
1992	King Cove Breach Bridge Abutments	King Cove, AK	47	0.05g
1992	Salamatof Seafoods Dock	Kenai, AK	126	0.08g
1993	King Cove Inner Harbor Dock	King Cove, AK	37	0.05g
1993	American President Lines Dock	Dutch Harbor, AK	88	0.07g
1993	North Star Barge Dock	Anchorage, AK	141	0.08g
1994	Chevron Bulkhead Wall	Nikiski, AK	88	0.07g
1994	Chevron Bulkhead Wall Extension	Nikiski, AK	88	0.07g
1995	Unocal Bulkhead Wall	Kenai, AK	80	0.08g
1995	Minch Dock	Anchorage, AK	100	0.08g
1995	Homer Spit Dock	Homer, AK	48	0.08g
1995	Hunter Creek Bridge	Palmer, AK	116	0.10g
1995	Solomon Gulch Fish Hatchery Facility	Valdez, AK	47	0.09g
1996	Port Graham Dock Value Engineering	Port Graham, AK	46	0.11g
1997	Cornelius Pass Railroad	Portland, OR	5	0.03g
1998	North Star Dock Extension	Anchorage, AK	44	0.08g
1999	Hawks Beach Erosion Control System	Cook Inlet, AK	32	0.07g
1999	Port MacKenzie	Knik Arm, AK	11	0.05g

EVENTS= the number of times the rock acceleration at the site was computed to be greater than or equal to 0.02g (assumed to be the threshold value for earthquake design) since the structure was built.

A_{MAX}= the average of computed maximum horizontal rock acceleration at the project site during the lifetime of the structure through Aug 15, 2001, using methods presented by Boore et al (1993), Nottingham (1970), and Youngs et al (1988).

Table 1. Brief Seismic History of Existing Open Cell Sheet Pile Structures

References

Boore, D.M., Joyner, W.B., and Fumal, T.E., 1993. *"Estimation of Response Spectra and Peak Accelerations from Western North American Earthquakes: An Interim Report".* USGS Open-File Report 93-509.

Campbell R., Christopherson A., and Nottingham D., 1986. Use of Fins on Piles for Increased Tension Capacity, State of Alaska Department of Transportation and Public Facilities Research Report.

Christopherson A. and Nottingham, D., 1983. *"Pipeline Supported Structures on Driven Piles in Permafrost: State-of-the-Art,"* ASCE Conference on Pipelines in Adverse Environments II, Pipeline Division, San Diego, California,.

Nottingham, D., 1970. Seismic attenuation for Trans-Alaska Pipeline design based on USGS Circular 672, *"Ground Motion Values for Use in the Seismic Design of the Trans-Alaska Pipeline System".*

Youngs et al, 1988, as referenced in Geotechnical Earthquake Engineering by S.L. Kramer (Prentice Hall, 1995).

Preventive Measures Against Snow for High-Speed Train Operation in Japan

Toshishige Fujii[1], Katsuhisa Kawashima[2], Shigehiro Iikura[3], Toru Endo[3] and Ryoji Izunami[4]

Abstract

Since bullet train operation started between Tokyo and Shin-Osaka (the Tokaido Shinkansen line) in 1964, Shinkansen trains have been damaged by snow while running at high speed every winter. Although water has been sprinkled over the track and speed control has been implemented to prevent snow on the track from flying up when trains run, these measures have not been completely. In this study, we reproduced a snow-flying phenomenon by implementing model tests to investigate the relationship between the density and water content of the snow cover. Appropriate speed limits were proposed based on the test results.

It is planned to construct a Shinkansen network throughout Japan. In the present constrained economic situation, it is necessary for Shinkansen lines to be extended to snowy areas to reduce the costs of construction and snow control measures. Aiming at the reduction of running cost for snow control measures, we designed a viaduct for the Hokuriku Shinkansen line that is planned in Hokuriku region, one of the most snowy areas in Japan. In this paper, we describe the concept of the viaduct and demonstrate its effectiveness as a measure against snow through tests with a full-scale mockup and a motorcar rotary.

Introduction

Bullet train (Shinkansen) operation has been developed incrementally in snowy areas. The Tokaido Shinkansen line opened transport service in 1964; the Tohoku Shinkansen line (Tokyo to Morioka) and the Joetsu Shinkansen line (Omiya to Niigata) in 1982; and the Hokuriku Shinkansen line (Takasaki to Nagano) in 1998 as shown in Figure 1.

Before the opening of the transport service of the Tokaido Shinkansen line, measures against snow were not considered. Consequently, trains were subject to snow damage. Snow on the track flies up and accretes to the underfloor equipment when a train runs at high speed. The snow accretion thus built up drops onto the track while the train is running. Because lumps of dropped snow scattered the ballast on the track, and damaged the car body and the environment along the track, water sprinkling has been implemented to wet snow on the track since 1966 (Saito, 1971).

---

[1]D. Eng., Head of Meteorological Disaster Prevention Laboratory, Railway Technical Research Institute, 2-8-38 Hikari-cho, Kokubunji-shi, Tokyo 185-8540, Japan, E-mail:fujii@rtri.or.jp, [2]D. Sc., Senior Researcher, ditto, [3]Senior Researcher, ditto, [4]Manager, West Japan Railway Co., 2-3-8 Matsunouchi, Akashi-shi 673-0016, Japan

Nevertheless, snow is still accreting to the car equipment according to a recent survey carried by Central Japan Railway Company that operates the Tokaido Shinkansen trains. It is pointed out therefore that water sprinkling is not as effective as expected (Fujii et al., 1986). In parallel with water sprinkling, the train speed is limited to prevent the snow on the track from flying up: 70, 120, and 170 km/h. However, the present speed limit is based on the evaluation of various conditions visually made by the train driver or through television cameras on the ground along the track. The conditions to be evaluated include the following: the weather condition (clear, cloudy, sleety, a little snowy, or snowy), the track condition (whether ballast and/or sleeper ladder patterns are visible), the snow depth (whether the snow is higher than the rail surface), and the quantity of accreted snow (how much snow has accreted to the underfloor equipment to the next stop). Since the judgement of an appropriate speed limit is subjective, the train speed limit is not necessarily controlled in a reasonable manner so that trains are often significantly delayed. In this study, we empirically clarified the metamorphic process of surface layer of snow cover due to water sprinkling and the relation between the occurrence of snow-flying phenomenon and the physical properties of snow. After that, we proposed reasonable train speed limits (the critical train speeds that suppress snow flying).

Concerning the Tohoku and the Joetsu Shinkansen line, we implemented measures against snow in consideration of meteorological conditions. Viaducts with spaces for the snow removed by snowplows that are attached to the head of the front car were introduced into the Tohoku Shinkansen line where there is little snowfall along the route. On the other hand, a system of melting snow by water sprinkling was installed on the Joetsu Shinkansen line where the ambient temperature is mild with much snow. These measures have been effective in preventing snow damage.

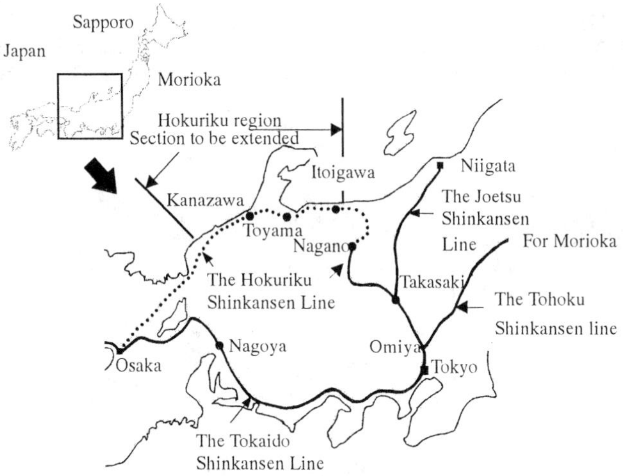

Figure 1. Routes of the Shinkansen lines in snowy areas

Shinkansen lines have established themselves as a snow-proof transport system. However, the snow-melting system is costly because much water and heat are used.

It is planned to extend the Shinkansen system from Morioka to Sapporo in Hokkaido where the temperature is low with much snow (the Tohoku and Hokkaido Shinkansen lines), and from Nagano to Osaka (the Hokuriku Shinkansen line) via the Hokuriku district which is one of the snowiest areas in Japan. Because of the continual extension of Shinkansen lines in snowy areas, more reliable and cost-effective measures are needed against snow for safety and punctual services. Therefore, we developed a new viaduct to mitigate the risk of snow damage for the extended Hokuriku Shinkansen line and demonstrated its usefulness through tests.

Determination of reasonable method for appropriate speed limit in snowy areas under water sprinkling on the Tokaido Shinkansen

Test Equipment and Procedure

We conducted a test at the Shiozawa Snow Disaster Prevention Laboratory in winter. We made test objects first by sprinkling water on a new snow surface and measured its physical properties. After that, we subjected the test objects to a tester to simulate a Shinkansen train running over a snow-deposited track at various speeds.

(1) Test object

We made the specimen with a water permeable block made of plastic (92% voids), with a thickness of 30 mm and an area of 385×490 mm, by letting the natural snow pile up to the height of about 100 mm on the block and sprinkling different quantities of water on the snow. We used a permeable material for this purpose in consideration of the fact that snow on a track lies on the permeable ballast structure.

(2) Water sprinkling

We used two test objects in the testing program. When we prepared the test objects by allowing natural snow to pile up, the air temperature was –3 to 3 °C and the

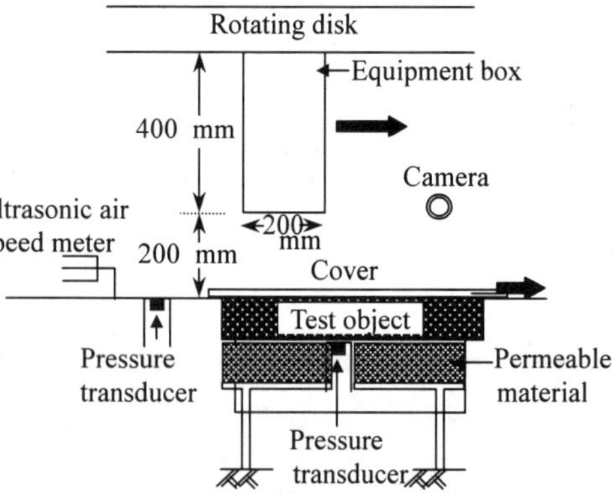

Figure 2. Snow-flying test cross-section

snowfall intensity was up to 5 mm/h. We then placed the objects 8 m distant from a sprinkler and sprinkled water at 5 °C on them at an intensity of about 6 mm/h for 80 minutes. The quantity of water sprinkled was 10 mm. After that, we measured the density and water content of the surface snow layer with one test object and subjected the other to a snow-flying test.

(3) Measuring the physical properties of surface snow layer

As the physical properties of snow should be taken into consideration, we measured the density and gravimetric water content, because they are related closely to the dynamic strength of wet snow and are easy to measure. We measured these variables in the snow to a depth of 20 mm using a cylindrical snow sampler with an inner diameter of 108 mm and a height of 20 mm. To measure the water content, we used a snow-water content meter of the calorimeter type.

(4) Snow-flying tester

Figure 2 is a sketch of the snow-flying tester. It has a rotating disk with a diameter of 5 m on the top and two aluminum boxes (400×200×400 mm) that simulate the underfloor equipment box of Shinkansen car. The disk rotates at the maximum speed of 250 rpm to move the equipment box at the maximum speed of 200 km/h. The equipment box rotates with a test object fixed inside and the upper cover closed. When the speed of the equipment box has reached the specified value, the cover opens instantaneously with a spring force. At the moment when the cover sprang open, we took a picture of the test object surface with a camera and checked it to examine whether the snow flew up. The bottom of the equipment box and the test object surface are 200 mm apart, corresponding to the situation where a snow cover exists approximately up to the rail surface level. We repeated the test at train speeds of 110, 160 and 200 km/h. We will discuss the test results below by taking the rotating speed of the equipment box as the speed of a Shinkansen train.

Test Results

(1) Metamorphosis of surface snow due to water sprinkling

Figure 3 shows the relationship between the quantity of sprinkled water and snow density, and Figure 4 shows the quantity of sprinkled water-snow water content

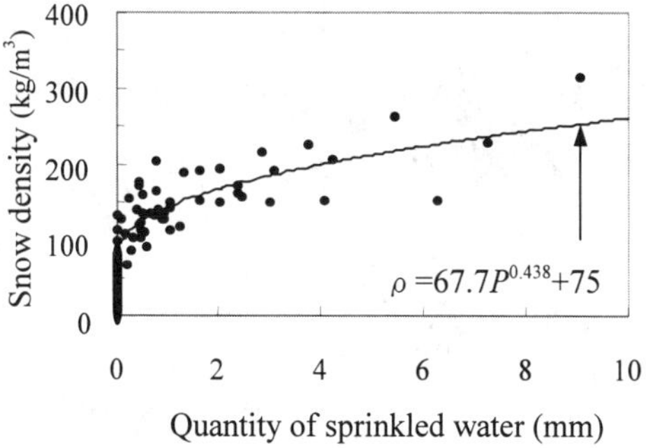

Figure 3. Snow density vs. quantity of sprinkled water

relationship. The density, which was 36 to 132 kg/m^3 before sprinkling of water (initial density), tended to steadily increase as water was sprinkled. Depending on the initial value, however, the values of snow density after water sprinkling vary within a range of 100 kg/m^3. When we denote the density by ρ (kg/m^3), the initial density by ρ_0 (kg/m^3) and the quantity of sprinkled water by P (mm), we can approximate the relationship of the density with the initial density and the quantity of sprinkled water by the following equation.

$$\rho = 67.7P^{0.438} + \rho_0 \tag{1}$$

The water content rapidly increased immediately after water sprinkling started, and reached a peak when the quantity of sprinkled water was 1 to 2 mm. After that, it tended to decrease. When we signify the water content by W (%), we can roughly describe this process by the following three equations:

$$\left.\begin{aligned} W &= 50.2P^{0.626} && (P \leqq 1) \\ W &= 50.2 && (1 < P \leqq 2) \\ W &= 50.2 - 7.20\,(P-2)^{0.280} && (2 < P) \end{aligned}\right\} \tag{2}$$

where the initial snow water content is assumed as 0%. Figures 3 and 4 also show

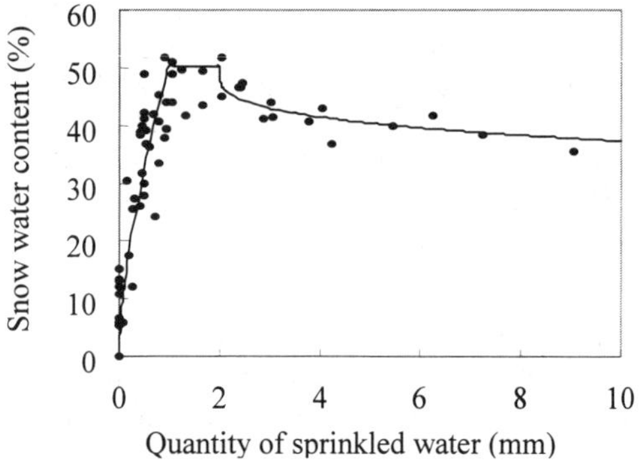

Figure 4. Snow water content vs. quantity of sprinkled water

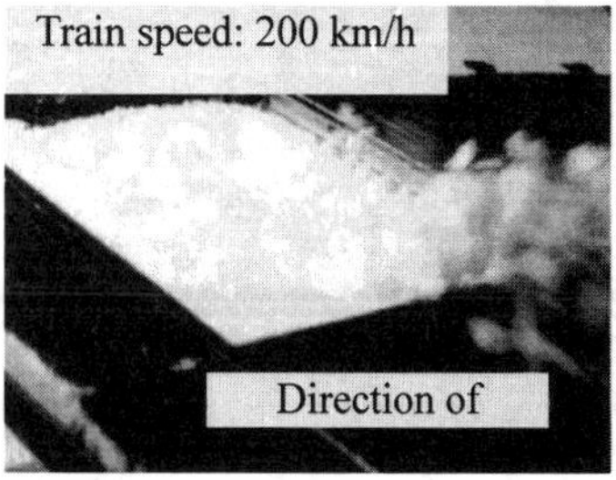

Figure 5. Snow-flying phenomenon

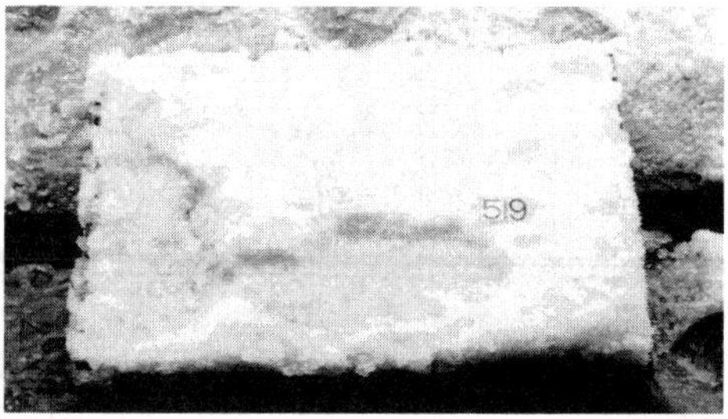

Figure 6. Concave part of surface snow after the test

the curves of equations (1) and (2).

(2) Snow-flying phenomenon

From the pictures and video screens, it was observed that snow-flying mostly took place when the surface snow was metamorphosed by water sprinkling exfoliation. Figure 5 shows a typical scene of flying snow. It also shows that the snow flied up not as particles but as lumps of certain sizes. The concave part of the surface snow on the test object after the test in Figure 6 is the result of partial exfoliation of the surface layer. It has been an accepted view that the snow, metamorphosed by sprinkling of water on the Shinkansen track, flies up just like blowing snow, because the friction stress generated by the train draft moves the snow grains on the surface (Shinojima, 1983). The test results indicate, however, that the snow flying is a phenomenon of exfoliation of the snow cover surface layer. This requires reviewing the conventional theory on the mechanism of the phenomenon, which has been discussed in detail separately (Kawashima et al., 1994).

(3) Relationship of the snow-flying phenomenon with the density and the water content

Figure 7 shows the density and the water content of each test object after water sprinkling and clarifies whether the test object caused snow-flying or not at each test speed. Two lines can section a zone in which the snow-flying phenomenon does not occur, one parallel with the X-axis and the other with the Y-axis. See the zone framed with two thin lines at the upper right corner of Figure 7. These boundary lines move toward higher values of snow density and snow water content as the speed increases.

Determination of Appropriate Speed Limits

The boundary lines to section the snow-flying zone can be specified by the point M or the coordinates of their intersection (W_c, ρ_c) in Figure 7. We call W_c and ρ_c the critical water content and the critical density, respectively. Figure 8 shows their

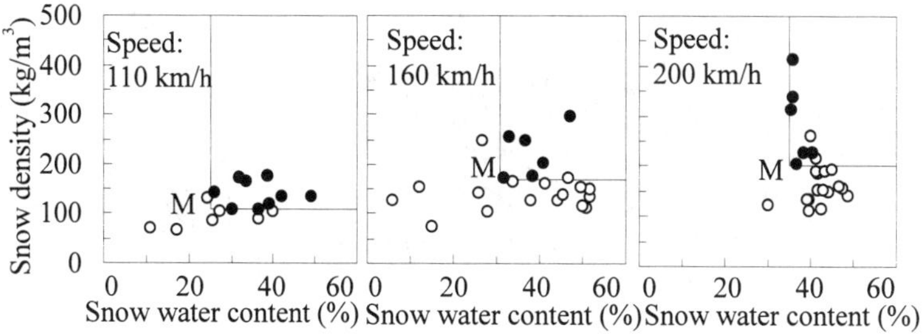

Figure 7. Relationships between snow-flying phenomena, snow density, and snow water content

relationship with the train speed. Although only three data points are available for each of these variables, they give a good linear relation with the train speed V (km/h). These data give the following regression equations.

$$W_c = 0.111V + 12.9 \quad (3)$$
$$\rho_c = 1.01V + 2.05 \quad (4)$$

Equations (3) and (4) enable drawing the boundary lines for any train speed in a water content vs. density chart. Figure 9 shows the boundary lines thus drawn for present speed limits of 70, 120, 170, and the maximum 230 km/h. Figure 9 (speed limit diagram) indicates that the conditions $W \geq 32\%$ and $\rho \geq 174$ kg/m^3 must be satisfied to increase the train speed to 170 km/h, for example. In general terms, this can be described as "snow does not fly up as long as the train runs at a speed lower than V_1 that satisfies the condition $W \geq W_c(V_1)$ and $\rho \geq \rho_c(V_1)$."

Equations (1) and (2), on the other hand, allow estimation of the changes in the density and water content of the surface layer of snow cover. Figure 9 shows the curve of the relationship between the snow density and snow water content when the initial density is 50 kg/m^3 with the quantity of sprinkled water as a parameter.
The combination of this curve and the aforementioned boundary lines provides an estimate of the minimum quantity of sprinkling water that will prevent snow-flying at a given train speed. For example, the minimum quantities of sprinkled water to run without causing flying snow are 0.5, 2 and 4 mm for the present train speed limits of 70, 120 and 170 km/h, respectively. By this method, we can determine the quantity of sprinkled water required for a given train speed and reasonably control the speed limit so that the train can run without being subjected to snow flying.

Since equations (1) and (2) were obtained under limited meteorological conditions, their scope of application is limited at present. It is necessary, therefore, to observe the metamorphic process of the surface layer of snow due to water

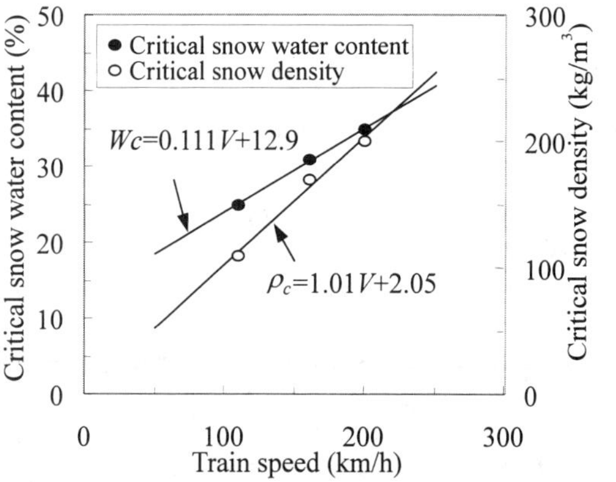

Figure 8. Train speed vs. critical snow density and snow water content

sprinkling under a wider range of meteorological conditions and improve the model's ability describe the changes in the snow density and water content.

The appropriate speed limits can be estimated on the basis of Figure 9 and the fact that the intensity of sprinkled water on the track is 5 mm/h, although we cannot measure the density and water content of the snow on the tracks while Shinkansen trains run. Table 1 shows the appropriate speed limit after water sprinkling started.

Measures for the extended Hokuriku Shinkansen line against snow

<u>**Concepts of the measure against snow**</u>

The Hokuriku Shinkansen line, one of several nationwide high-speed rail lines at the construction planning stage, will start in Tokyo, branch off the existing Joetsu Shinkansen line at Takasaki, and terminate in Osaka. It will pass through Itoigawa, Toyama, and Kanazawa in the Hokuriku region, one of the snowiest areas in Japan (see Figure 1). In winter, the Hokuriku region has comparatively mild temperatures but is hit by heavy snowfall over a cycle of about eighteen years. The monthly average temperature at Itoigawa, Toyama, and Kanazawa in February is about 2-3 °C. The probable maximum snow depth over a twenty-year period is 1.55 to 2.20 m.

On the Hokuriku Shinkansen line, cost-saving measures have to be taken into account. We have carefully investigated various measures for controlling snow on this Shinkansen line by considering the weather conditions described above.

In particular, we have studied the methods of clearing snow on a viaduct that

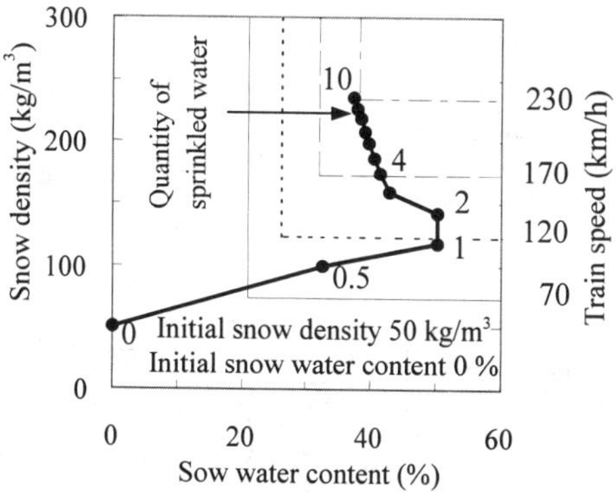

Figure 9. Boundary lines of the snow flying zone for train speed of 70, 120 and 170 km/h, and the curves of snow density and snow water content when ρ_0=50 kg/m^3 (with the quantity of sprinkled water (mm) as a parameter)

Table 1. Appropriate speed after water sprinkling started

Elapsed time *T* (min.)	0≤*T*<18	18≤*T*<48	48≤*T*
Appropriate speed limit (km/h)	70	120	170

was planned to construct along almost all the route excluding tunnel sections.

Proposal of viaduct for snow removal using a motorcar rotary

When compared with methods of controlling snow such as melting with water/electric heaters, constructing snow shelters etc., mechanical snow removal is relatively inexpensive. We decided, therefore, to rely primarily on snow removal using snowplows attached to rail vehicles to push the snow into storage spaces in the viaduct. When these spaces are completely filled with snow, we eject the snow down to the ground using a motorcar rotary, i.e., the snow is thrown over the wall of the viaduct and scattered on the ground. For this reason, a sufficiently wide space to receive the snow is required along the viaduct. In reality, however, there is little space along the ground because so many houses and buildings are adjacent to the route. Accordingly, we also must consider measures against noise.

We finally decided on the method of clearing snow as described above. Next, we had to develop a new kind of viaduct based on the consideration of all the topics above. We designed a viaduct with two parallel pairs of inner and outer walls as shown in Figure 10. The inner wall functions as a noise barrier and as a snow-storage space within the viaduct, and the outer wall serves as a guide for ejected snow when a motorcar rotary removes it. The total width of the upper structure is 15.5 m and the distance between the outer surfaces of the right and left inner walls is 11.7 m. The rail level is 1.205 m above the subgrade concrete. The heights of the outer wall and the inner wall are both 2 m above the rail level. Motorcar rotaries

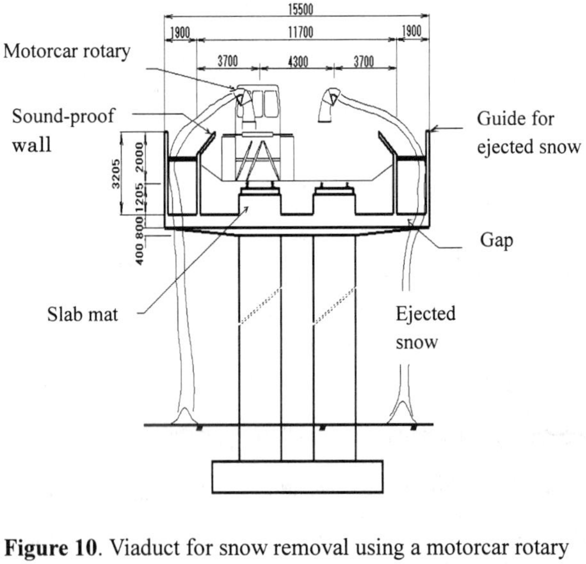

Figure 10. Viaduct for snow removal using a motorcar rotary

eject snow over the inner wall to the gap between the two walls. The type of viaduct described above has never been used in the past. In designing the structure, it is necessary to ascertain the feasibility of snow removal by a motorcar rotary and to clarify the impact load of ejected snow onto the outer side walls before construction. Therefore, we performed snow removal tests using a motorcar rotary.

Test procedure

(1) Test object
We installed a full-scale mockup of a pair of inner and outer side walls beside a sidetrack in a snowy region and excavated a pit for collecting snow outside the outer side wall before winter. The length and height of each wall was 12 m and 2 m, respectively. Four sections including a dummy section separated the length of the outer side wall. The height of the outer side wall was also separated into two sections: upper and lower side walls. Consequently, the size of each outer side wall was 1 m×3 m.

(2) Outer side wall
Four kinds of material were used as the outer side walls at each section along the sidetrack. They included a pre-stressing (PC) plate, expanded metal, wire cloth with 25 mm mesh, fiber reinforced plastic (FRP) grating with 25×100 mm mesh, and FRP grating with 40 mm mesh. We checked on the escape of snow from the crevice of each wall. The impact forces acting on the PC outer wall were measured.

(3) Snow for tests
We piled snow on the sidetrack at depths of 0.2, 0.35 and 0.5 m corresponding to the maximum accumulated snow depth in 2-year, 10-year and 30-year periods in the most heavy snowfall region on the Hokuriku Shinkansen line. The density of the snow for tests was about 500-600 kg/m^3. This value is higher than that the density of snow accumulated on viaducts.

(4) Performance of motorcar rotary
The motorcar rotary was rated at 411 kW. It had four snow blowers, each with a diameter of 1.35 m, and snow blower rotation speeds of 235 or 341 rpm.

(5) Angle of ejected snow to side wall
We plowed the snow and ejected it down through the gap between the inner and outer side walls into the pit with a motorcar rotary. The snow was ejected at an angle of 45 or 90 degrees to the traveling direction.

Figure 11. Testing configuration

(6) Measurement of impact force of ejected snow
We installed pressure gauges on the PC outer side walls to measure impact forces directly. The right-to-left edges of the PC outer side walls were fixed onto right-to-left props through two gauges. Eight gauges were installed on the wall. We also installed strain gauges on the PC outer wall to investigate the relation between stress and strain. We measured the impact forces and strain imposed by the ejected snow at measurement intervals of 2 ms of and videotaped the movement of ejected snow during the tests. Figure 11 shows the testing configuration.

Test results

From the video screens, it was ascertained that snow is easily removed and ejected by a motorcar rotary in both cases at an angle of 45 or 90 degrees to the traveling direction. In addition, it can be estimated that the PC plate causes the least amount of snow to escape from the wall crevice among four kinds of material used as the outer side walls. Analyzing the videotape images, the area hit by the ejected snow was about 1 m^2.

We totaled the impact forces on the wall measured with gauges at the same time. The maximum value of total impact forces recorded was about 20 kN when the ejecting angle was 90 degrees to the traveling direction. Therefore, it can be evaluated that the impact load of ejected snow acting on the PC plate is 20 kPa at the highest. The loads when the ejecting angle was 45 degrees to the traveling direction were $1/\sqrt{2}$ of those at the ejecting angle of 90 degrees. By analyzing the relation between the total impact forces and the maximum displacements of the PC plates, we can obtain the following linear function.

$$Y = 0.1X \qquad (5)$$

where X is the total impact force and Y is the maximum displacement. On the assumption that the PC plate is a simple beam, we checked some models representing the relationship between the total impact forces and the maximum displacements by the pressure gauge span, moment of inertia and Young's modulus of the PC plate. Analyzing the data, we confirmed that the relation could be expressed as a model of uniform concentrated load with a 1-m width.

Conclusions

When a Shinkansen train runs at high speed, the snow on the track flies up. We reproduced this phenomenon first by implementing model tests and studied relationships between the quantity of sprinkled water and snow density, and snow water content. We also evaluated appropriate speed limits. Following results are obtained.

(1) A quantitative relationship was obtained between the quantity of sprinkled water and the changes in the density and the water content of surface snow.

(2) When a train runs at high speeds, the snow on the track flies up not as particles but mainly as lumps exfoliated from the surface layer of the snow cover which has metamorphosed due to sprinkled water.

(3) Whether the snow flies up depends on the density and the water content of the

snow cover surface. We can draw boundary lines to section the snow-flying zone in the water content vs. density chart (speed limit diagram) for a given train speed.
(4) The combination of boundary lines for the snow-flying zone and the curve of the relationship between snow density and water content with the quantity of sprinkled water as a parameter enable reasonable train speed limit control.
In contrast to the existing method, the one proposed here provides a reasonable means to control the train speed limit based on an objective and scientific basis. In the method of estimating the changes in snow density and water content, however, there remain several problems yet to be solved due to the limited range of meteorological conditions to be taken into consideration. Therefore, we will make models for the changes in snow density and water content those cover a wider range of meteorological conditions.

Next, we developed a viaduct, which has two parallel pairs of inner and outer side walls, to facilitate snow removal on the Hokuriku Shinkansen line. To design the viaduct, it is necessary to ascertain the feasibility of snow removal by a motorcar rotary and determine the impact load of ejected snow onto outer side walls that function as a guide for ejected snow when it is removed by a rotary snowplow. We installed a full-scale mockup of a pair of inner and outer side walls and measured the impact forces of snow ejected by a motorcar rotary acting on the outer side wall. By analyzing the data, we obtained the following results.
(5) It was ascertained that we could easily eject snow by a motorcar rotary.
(6) The maximum impact load of ejected snow acting on the PC outer wall was evaluated to be 20 kPa at the highest.
(7) The relationship between total impact forces and maximum displacements of the PC outer wall can be expressed as a model of uniform concentrated load with a 1-m width.
(8) The proposed viaduct was determined to be useful as a measure for viaduct sections against snow.

The proposed viaduct is now under construction on the extended Hokuriku Shinkansen line.

References

Fujii, T., Imai, T. and Endo, T. (1986). "Field investigation on snow accretion to car body on Tokaido Shinkansen." *Railway Technical Research Pre-Report,* A-86-200, 1-35 (in Japanese with English abstract).

Kawashima, K., Endo, T. and Fujii, T. (1994). "Experimental study on prevention of snow-flying phenomena caused by high-speed Shinkansen trains." *Railway Technical Research Report,* 8(3), 7-12 (in Japanese with English abstract).

Saito, M. (1971). "Characteristics of snow damages and their measures on the Tokaido Shinkansen." *J. Japanese society of snow and ice,* 33, 61-78 (in Japanese).

Shinojima, K. (1983). "Study on the prediction of snow accretion to and its dropping from the bottom of Shinkansen train." *Railway Technical Research Report,* 1248, 1-160 (in Japanese).

Modeling the Cutting and Filling Process of Roadbed Construction on Permafrost

Jiankun LIU[1,2], Ya. A. Kronick[3]

Abstract

The permafrost region is sensitive to all engineering activities such as road pavement or railway construction, since any disturb of its original state will cause instability of the thermal and mechanical state of frozen ground. The staged construction process has been proved to have significant influence on the formation of thermal state of roadbed on permafrost. Usually the cuts suffer more from the thermal state change in service time, in this case the thermal regime variation become a dominating influencing factor for roadbed stability. For this reason in this paper the temperature state considering the construction process was studied for the Berrkakit-Tammot-Yakutsk railway, which was constructed on permafrost in east Russia. A typical cut-and-fill section of roadbed on permafrost was selected for numerical simulation to study its thermal regime formation, on which the running stability of roadbed is directly related. The staged construction of roadbed was modeled. A computer program was created for the mentioned problem. The innovative feature is that the construction process including the cutting and filling work were both taken into account. The real construction sequence was adopted as input parameter in the numerical modeling for mesh producing and its subsequent correction. The computed results show realistic temperature distribution.

Introduction

A new railway stretching from Amur to Yakutsk in the far east Russian is located mostly on the permafrost distribution zone (Kondratiev 1997). For some cut-and-fill sections, the construction continued a long period, the thermal state of the roadbed

[1]Associate professor, Ph.D, School of Civil Engineering & Architecture, Northern Jiaotong University, Beijing,100044; [2]Associate professor, Ph.D, State Key Laboratory of Frozen Soil Engineering, Lanzhou ,730000
[3]Professor, Ph.D, Faculty of Soil Mechanics, Base & Foundation, Moscow State University of Civil Engineering, Moscow

and its base needs more time to reach a stable state. It is commonly recognized that the stability of roadbed on permafrost is dominated by its thermal state. The prediction of the stabilizing period is essential for further estimation of the stability. In previous thermal state prediction, the construction sequence was rarely taken into account (An 1988, Feldman 1973, Chistotinov 1964). Kronick (1984) for the first time considered the construction process in thermal state prediction for dam in hydro-technical engineering; the rising layer process in roadbed pavement on permafrost was modelled by Liu (1994). In this paper a typical cut and fill section of the roadbed was modeled, the real construction sequence was simulated in the computer program, the physical domain of computation was allowed to change with time. The computed results are discussed.

Site Description

At 314km of the Berrkakit-Tommot-Yakutsk railway there was a disposed mining site. Construction of the roadbed on the mine fill continued from 1989 to 1994. An icy soil layer of 6.2m high was found in the central part of the designed roadbed. This part was excavated before pavement. The underlying layer is mainly gravel soil with fine inclusion; deeper part is clayey soil. The minimum air temperature in this region is –35.7°C. The maximum air temperature in July reached about 17.3°C. The average air temperature is –8.5°C.
The roadbed at this section was completed in totally 26 stages. In every stage the cut or filled height was 1-2m, the stabilization time for every stage usually took 1-2 months (See Fig 1). After the completion of the roadbed, in wet seasons, water periodically accumulated along the roadside.

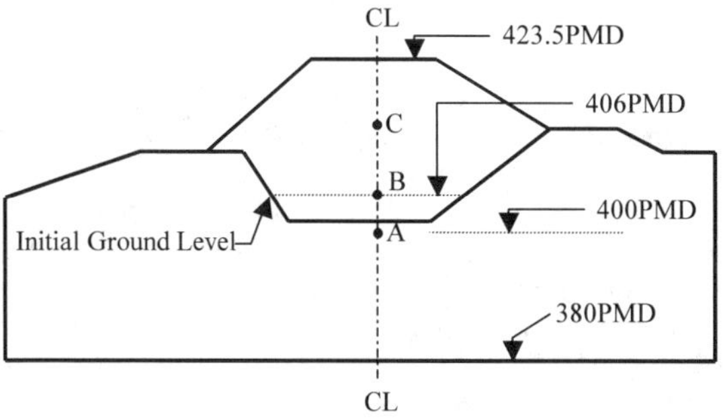

Figure 1. Physical Domain

Mathematical formulation and numerical realization

The governing equation for heat transfer in enthalpy transformation has been proved to be a convenient way for numerical solution (Kronick, 1984):

$$\frac{\partial H}{\partial t} = \text{div}[\ \lambda(\theta)\text{grad}\theta\,] \qquad (1)$$

Where

$H(\theta) = \int C(\theta)d\theta$; θ --temperature, ^{0}C; C(θ)--effective coefficient of heat capacity; J/m^3.^{0}C; λ (θ)--effective coefficient of heat conduction, J/m^{0}C ;

The left and right side of the physical domain was regarded isolated: $\frac{\partial\theta}{\partial x} = 0$, when x=0, 160m.

On the lower boundary of physical domain the temperature was set to a fixed value of -4.0^{0}C according to the field observation (Kondratiev 1997). The initial temperature field was obtained through a long term transient run (10 years) on the given boundary condition for the initial calculation domain.

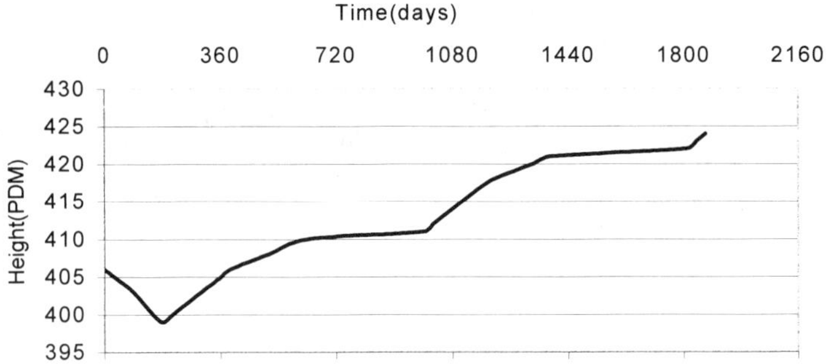

Figure 2 The Roadbed Surface Position according to the Calender Plan of Pavement along Center Line

For the rising layer the key problem is the time-dependency of physical domain, so the main difficulty in numerical analysis is the mesh generation and its subsequent correction according to the planed pavement process. For simplicity, we choose the finite difference method to solve the heat transfer problem for rising soil layer. The initial physical domain was divided into finite rectangles, the new soil layer was also divided into rectangles of the same size, and was then added to the top of the old mesh. In this way the mesh was updated according to construction sequence.

The initial condition for every new layer given by the air temperature at that time:

$$\theta_{new} = \alpha\theta_a \qquad (2)$$

Where θ_{new} is the initial temperature for the new layer, θ_a is the atmospheric temperature, α is boundary layer coefficient. Based on climate data the following expression used for the atmospheric temperature:

$$\theta_a = -8.5 + 26.5\sin(\frac{\pi t}{180} - 2.1156) \quad (3)$$

A program for thermal state prediction in fixed physical domain was modified to deal with the rising layer case in this paper (Liu, 1997).

Soil properties

All needed thermo-physical properties including natural density ρ, effective heat capacity of thawed and frozen soil C_t, C_m, effective coefficient of heat conduction for thawed soil and frozen soil λ_t λ_m, freezing point θ_{hz}; freezing point of pound water θ_{zc}; beginning point of frozen state θ_m are listed in table 1.

Table 1. Thermo-Physical Properties of Soil

Soil type	ρ	θ_{hz}	θ_{zs}	θ_m	C_t	C_m	λ_t	λ_m
	kg/m^3	°C	°C	°C	$10^3 kJ/m^3 \bullet °C$		$W_t/m \bullet °C$	
Gravel with sandy inclusion (filing material)	2000	-0.2	-0.4	-2.0	2.228	2.022	1.51	1.62
Gravel with sandy inclusion	1900	-0.2	-0.4	-3.0	3.110	2.120	2.85	3.02
Icy soil	1250	-0.05	-0.1	-1.0	3.6	2.1	1.57	2.0
Clayey soil	1660	-0.2	-0.5	-5.0	3.020	2.180	1.45	1.57

Results and Discussion

Parts of the computed results are presented in Figures 3-5. Figure 3 illustrates the temperature distribution along the center line of the selected cross-section during the construction period. Because of the relatively long stabilization time in sseasonal layered pavement, on the final stage the temperature distribution along center line has reached a state not far from the long term distribution.

Figure 4 showed the temperature distribution along center line after the construction stage. On the 5-th year the maximum thawing depth reached at about 4 meters, the depth of zero amplitude after the roadbed erection was found at 30 meters below the surface.

Figure 5 showed the temperature variation of point A, B, C (Figure 5) on the center line. At point A, because of the influence of the construction process, the temperature has a substantial change in the first years after completion of pavement, but has stabilizing tendency of with the elapse of time. After 10-15 years, the

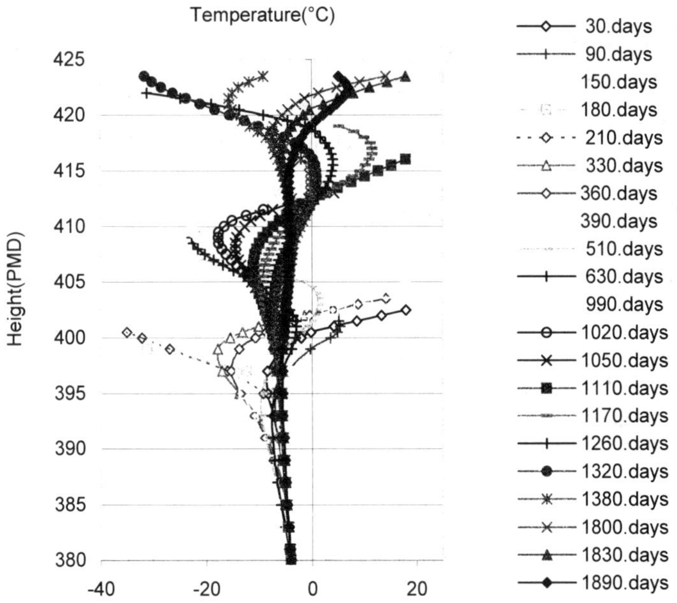

Figure 3. Temperature Distribution along center Line in the Construction Period

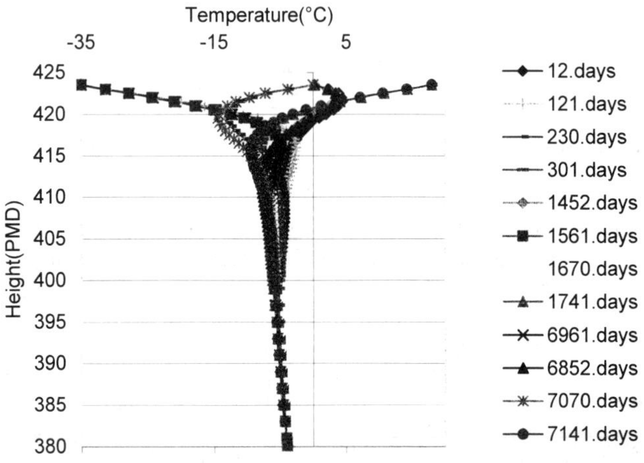

Figure 4. Temperature Distribution along Center Line after Construction Stage

temperature at this point no longer changes, this agreed well with the geocryological engineering experience (Kongdratiev 2000). As for the upper point C then its temperature is affected mainly by the atmospheric temperature.

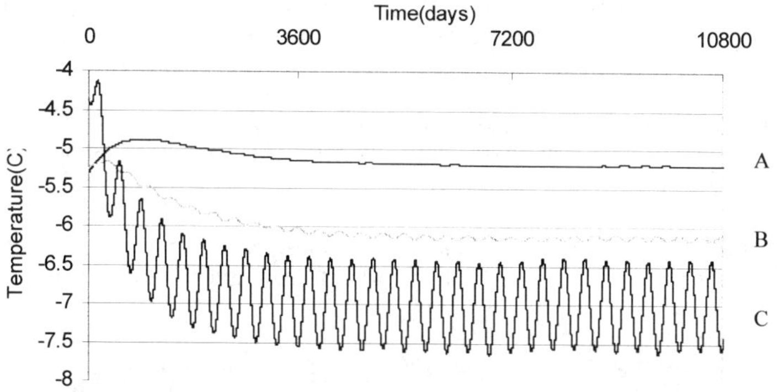

Figure 5. Temperature Variation at Points A, B and C

Conclusion

Based on the numerically simulated results, the following conclusion can be drawn: the stabilization time for the specific section investigated in this paper under the given construction sequence is about 10-15 year, this agreed well with geocryological engineering experience.

Acknowledgement

The research were supported by National Natural Science Foundation of China (No.40171018) and the Foundation for Best Young Teachers of the Ministry of Education.

References

An Weidong (1988), *The interaction of moisture, stress and thermal fields in frozen soils* (in Chinese), Lanzhou University Press.

Chistotinov L. V.(1964) *The thermal process in frozen soils* (in Russian). Publishing house "Science", 151-162

Feldman, G. M. (1973). *Method of thermal state calculation for frozen soil* (in Russian). Publishing house "Science".

Kondratiev V. G. (1997). *Embankment running reliability of Berkakit-Tommot-Yakutsick railway on permafrost*. Transportation construction, No.11, 1997, 3-6.

Kondratiev V. G., Pozin B. A.(2000) *Concepts of geocryological engineering monitoring system for the constructing railway Berkakit – Tommot – Yakutsik.* Zabtransizdat

Kronick Ya.A.(1984). *Thermo-mechanical properties of freezing soil under heavy loading conditions (in Russian)*. Material of conferences in hyrotechnics. Leningrad. Publisher: “ENERGOIZDAT”.

Liu Jiankun.(1994). Numerical studies on thermo-stress-strain state of roadbed considering staged construction. Ph.D thesis, Moscow State University of Civil Engineering.

Liu Jiankun, Ma Wei &Wu Ziwang (1997). *Prediction of temperature regime of roadbed on permafrost considering unsteady seepage of water*. Ground Freezing97. Knutsson(ed.), Balkema, 121-125.

Design and Experience--Controlling Railroad Embankment Deformation on Ice-Rich Permafrost

V.G. Kondratiev[1]

Abstract

This paper presents experimental design and construction data for new methods of strengthening railroad embankments on very icy permafrost soils. Most of these methods are based on adjusting the ratio of cooling and warming factors to reduce the average annual soil temperature and to preserve the original soils' permanently frozen state. One method includes preventive removal of very icy masses from the permanently frozen base, and then refilling the cavities formed with a high consistency soil.

Introduction

A common problem in the cold regions of Russia, USA, Canada, and China is the control of embankment deformation on ice-rich permafrost. Many researchers have studied this problem (Elenevsky & Nizovkin 1936, Hayley 1988, Kondratiev 2001, Peretruhin & Potatueva 1987, Wu Ziwang et al. 1988, Zarling & Braley 1986).

Deformation of embankments on permafrost soils is connected mostly with thermal degradation of the permafrost in the embankment base (Figure 1). The main factors causing permafrost thawing include increased amounts of absorbed solar radiation by the embankment in comparison with the natural surface, infiltration of warm summer precipitation through the body of the embankment, increased thickness of snow cover on the lower part of the embankment and on the adjacent ground surface, the warming effect of the surface, and the presence of groundwater.

The principal measures used to ensure embankment stability on weak bases or those subsiding under thawing usually are the following: Increasing the embankment width and height, placing the embankment on ballast, constructing berms, constructing the embankment with rock, or removing ice-saturated foundation soils and replacement with a free-draining material compacted layer-by-layer (Fyodorov 1977).

---

[1] Dr Geology, Prof. Honorary Transport Constructor, Science head of TransEGEM, 71-31 Bolshaya Academicheskaya, Moscow, 125183, Russia; phone/fax 7-095-482-1871; cp2@mail.ru.

Figure 1. Deformation of Bailkal - Amur railroad, April 1996

These measures are of a passive nature because they are aimed primarily at overcoming the consequences of permafrost degradation, require great expense for track construction and maintenance, and do not solve the problem of ensuring embankment stability on very icy permafrost (Bushin 1992 ; Yakovlev 1992).

Obviously there is a need to develop active methods to ensure embankment stability on very icy permafrost soils aimed at elimination of the basic problem: to prevent permafrost degradation in the embankment base.

We have developed several methods of strengthening embankment bases on very icy permafrost soils. One approach involves adjusting the ratio of cooling and warming factors with the goal of reducing the average annual soil temperature and thereby preserving a permanently frozen state (Kondratiev 2001). Another method consists of preventive removal of very icy masses (ice lenses) from a permanently frozen base and filling the cavities formed with high consistency soils (Kondratiev & Broid 1994). Other proactive alternatives to strengthen the embankment base include:

- snow removal and painting (Kondratiev et al. 1992);
- constructing sheds that protect against sun and precipitation (Kondratiev et al. 1992);
- installation of cooling pipes (Kondratiev et al. 1993);
- installation of a film cover at sections with the terrain cross gradient (Kondratiev 1992);
- installation of cold accumulation devices at sections with the terrain cross gradient (Kondratiev et al.1996);

Methods For Strengthening Railroad Embankment Base On Very Icy Permafrost

Snow cleaning and painting. In zones of permafrost soils subjected to subsidence under thawing, embankments are constructed using techniques that facilitate preservation of base soils in a perennially frozen state: clearing is done in the winter without removing stumps (no grubbing); a moss-vegetation cover is maximally preserved; the embankment first layer is ''end dumped'', thereby not allowing the vehicle traffic on a natural base; and the embankment is compacted layer-by-layer.

In the first half of winter (from the moment of a stable snow cover to the moment of heat flow sign inversion) the snow is regularly removed from the railway, embankment slopes, and the adjacent ground surface, allowing the embankment to cool. (The moment of heat flow sign inversion is established according to data from the nearest weather station when the average daily air temperature becomes consistently higher than that of the soil surface.) The width of a strip near the embankment to be cleared of snow should be, as a rule, more than the width of the embankment side slope. However, it should not exceed the thickness of a layer of soil undergoing significant yearly temperature fluctuations below a cleared surface.

In spring after the snow has melted, the embankment slopes with southern, eastern, and western exposures as well as the top surface are painted white. The painted surface should adequately reflect solar radiation in the spring-summer period. Therefore, if 60% of the protected surface becomes dirty, it should be repainted.

A positive effect is achieved owing to the fact that elimination of the snow insulating influence during the cold period increases heat flow from the embankment body and base into the atmosphere, resulting in significant cooling.

Essential increase of the albedo (a) of the surface coated with new (a = 75%) or old (a=55%) white paint as compared to a natural surface (a= 25%) and, particularly, to the railway surface blackened with coal dust and fuel oil (a = 6%) decreases the quantity of absorbed solar radiation, and reduces summer warming of the embankment and its base. Increased winter cooling and reduced summer warming preserve the embankment base in the frozen state. Prevention of permafrost degradation in the embankment base rules out development of perennial settlement.

The use of this method allows simplification of the embankment design and reduces railway building and maintenance costs; allows use of standard components; avoids the costs of increasing the embankment height and width for greater thermal protection; avoids use of stabilizing berms; avoids the use of costly and not always effective rockfill for embankments; reduces the volume of free-draining material and allows replacement with a less costly ordinary non-frost susceptible soil. This method will maintain the railway capacity, increase the railway service life, and reduce maintenance costs.

Protective shed. Construction of a solar radiation protective shed over an embankment prevents a warming effect from direct solar radiation and summer precipitation as well as drastically increases winter cooling by preventing formation of an insulating snow cover over the embankment. Schematic diagrams of various protective shed types over a railroad embankment are presented in Figure 2.

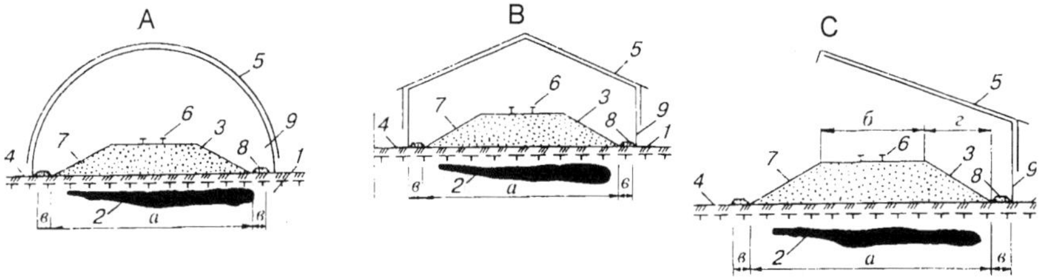

Figure 2. Schematic diagram of a railway track with sun-precipitation protective shed: A - convex arch roof; B - ridge roof; C - simple roof

As shown in Figure 2, at an embankment section with a frozen soil subgrade (1) or with inclusion of underground ice (2), an embankment (3) of standard design is constructed without removal of moss-vegetation cover (4), and a sun-precipitation protective shed (5) is built. The shed design is selected with regard to the geographic latitude and local climatic conditions, slope exposures, and embankment height. The roof profile should be convex of a single-pitch or ridge shape. The shed should protect the railway top structure (6), embankment slopes (7), and the adjacent ground surface (8). The design should include water-retaining banks to protect against rain and direct solar radiation in spring and summer, as well as prevent accumulation of snow in winter. For this purpose the lower part of the shed has air holes (9) facilitating snow removal by airflow beyond the railroad track during passage of trains or during snowstorms. The shed width should be greater than the sum of cross dimensions of the embankment (a) and water-retaining banks (b).

The shed height should be greater than the sum of the obstruction clearance height and the embankment optimum height determined by the results of geothermal analyses for concrete route sections. The above clearance should be included in the shed internal space. Sheds can be built not only over embankment sections but also over zero vertical alignments and in recesses.

A positive effect is achieved owing to the fact that the shed eliminates several principal factors causing permafrost degradation and roadbed subsidence, including the warming influence of snow cover, summer precipitation infiltration, and direct solar radiation.

By maintaining the high strength properties of frozen base soils over the entire railway service life, the need for additional anti-deformation measures no longer arises, the embankment design is simplified, and both the track throughput capacity and the track component service life are increased.

Cross-cooling pipes. Strengthening of the embankment base is achieved by a cooling pipe system installed at the bottom of the embankment that maintains the base soils in a perennially frozen state. The embankment design with the cooling system is illustrated in Figure 3.

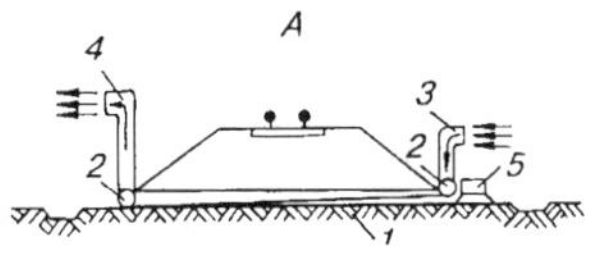

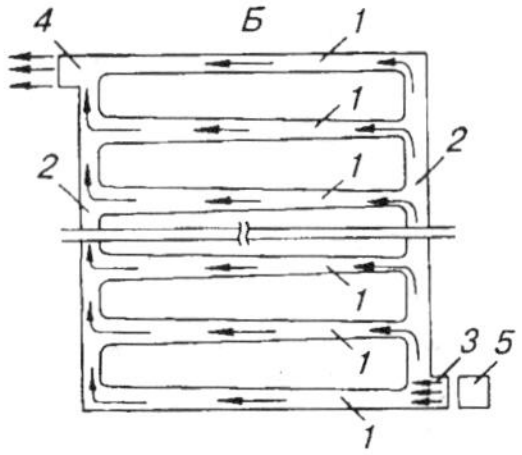

Figure 3. Embankment design with cooling pipes: A - cross section; B - in plan

The cooling system consists of one or several rows of cooling pipes (1) laid across the embankment and used for passing a coolant (cold air, or air-gas mixture) by natural or forced ventilation. The pipe diameter, spacing, and number of rows are determined by geothermal analyses. Ventilation is provided by means of manifold pipes (2), and inlet (3) and outlet (4) connections. To intensify natural ventilation, the holes of inlet and outlet connections are located on different levels. This is facilitated by laying the pipes with a gradient to the embankment base or by the use of tapered pipes. In case of forced ventilation, a fan (5) is also installed. This allows additional cooling in summer with air-gas mixtures, e.g. air-nitrogen, and to ensure embankment stability even in abnormally warm years.

Film cover. To preserve embankment base soils in a perennially frozen state at hill-side sections, it is necessary to prevent a warming effect of not just solar radiation and atmospheric precipitation but also of the surface and above-permafrost waters running downhill to the embankment. The effect of the surface and above-permafrost waters is prevented by constructing an anti-filtration cover (Figure 4) from polymer film (1) on the boundaries of frozen core (2) with draining prism (3) and berm (4) placed from the top side of the embankment. For reliable conservation of the frozen core and very icy base (5) of the embankment the berm is constructed of heat-insulating (6) and nonporous membrane (7) layers, and is cut into the base to a depth exceeding the thickness of a seasonally melting layer (8). The embankment slope from the bottom side is strengthened with rockfill (9) also facilitating conservation of the frozen core and the embankment base. For the case of the slopes with southern exposure, the rock material should have a natural or specially created light-reflecting color. The draining prism base should have a gradient towards thebottom side for surface water run-off.

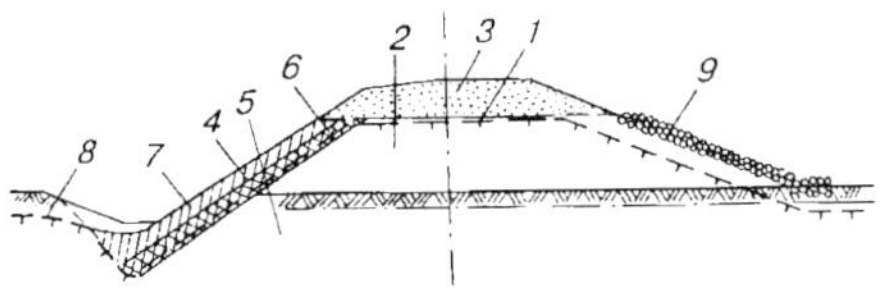

Figure 4. Embankment design with a film cover

Longitudinal cold accumulation devices. The main drawback of the nonporous polymer film cover design is the need for highly labor-intense construction while the railway remains in service. Accordingly, this approach is typically restricted to new construction. The design described below lacks this drawback.

Prevention of the warming effect of surface and above-permafrost waters at hill-side sections is achieved by making a nonporous membrane cover from frozen soils (Fig. 5) using cold accumulation devices (1) like Gapeyev, Long, Makarov, etc. thermopiles placed from the top side of embankment (2).

The devices' air-heat exchanger (3) is installed vertically while a soil heat exchanger (4) extends along the embankment at an angle of 95-100^0 to vertical. For reliable conservation of the frozen and very icy embankment base, an anti-filtration berm (5) is constructed. The berm is cut into the base to a depth exceeding the thickness of the seasonally melting layer (6). The embankment is strengthened with rockfill on the outer slope placed from the bottom side (7). This also facilitates conservation of the embankment frozen core and base. For cases where the rockfill slope has a southern exposure, the rock material should have a natural or specially created light-reflecting color.

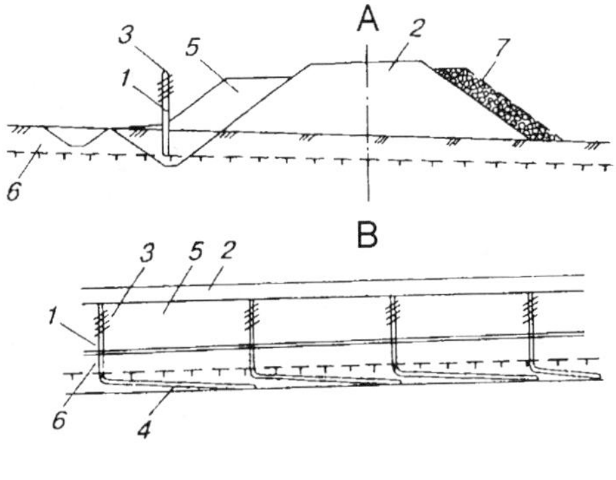

Figure 5. Embankment design with a cold accumulation device:
A - cross section, B - longitudinal section

Substitution of very icy masses for high consistency soils. In some cases, particularly at sections with the soil temperatures close to 0^0C, it may be prudent to take preventive measures by replacing very icy masses with a soil mass not susceptible to thaw subsidence instead of continuously maintaining negative temperatures during railway operation. A method of breaking down and removing very icy permafrost soils or ice masses in the embankment base is offered using a hydrodynamic technique through special wells which are also used to fill the emerging cavities with a thawed soil having required properties, e.g. sand. In this case hazardous settlements of the embankment are ruled out under any thermal conditions or changes in the environment.

Experimental Designs

Experimental design for Berkakit - Tommot - Yakutsk Railway. Experimental embankment sections of the Berkakit - Tommot - Yakutsk Railway are under construction now to test the efficiency of the above technological concepts under operational conditions.

In particular, a sun-precipitation protective shed is anticipated on a 708 km segment of railway between Bottoma and Kirim. The area is situated on the terrace of the Lena River and is underlain by lake-marsh, lake-alluvial, sandy-loam, and loam soils with ice seams and veins ranging in thickness from 10 to 15 m. Suitable embankment soils are not locally available. Suitable soils are available from a materials source 20 km away; suitable rock material is available from a source 48 km away. Thermopiles with rigid insulation board were installed in a section of the test segment to maintain the base soils in a frozen condition over the entire term of usage.

To investigate the effect of the shed, numerical modeling of the temperature regime was conducted (Liu & Kondratiev, 1996) for an embankment and its base consisting of very icy permafrost soils with a temperature near -0.5^0C. The yearly average atmospheric temperature was -8.5^0C; the yearly amplitude was 27.2 ^{0}C. Figures 6 and 7 show the computed temperature fields of the embankment and base on September 1st seven years after construction for both mentioned cases.

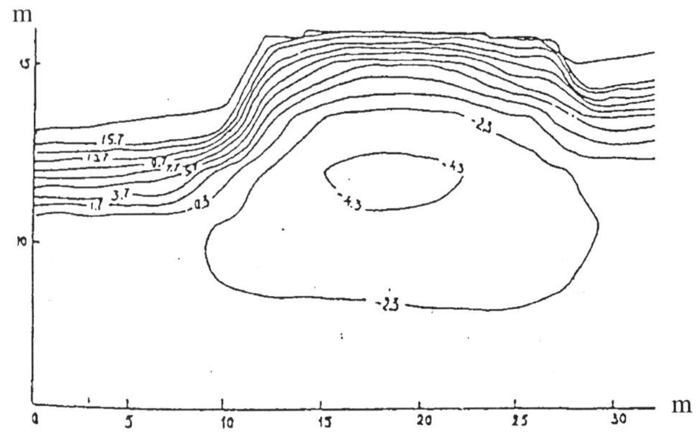

Figure 6. Thermal regime of embankment protected by a shed and its base on September 1st of the 7th year after construction

An obvious low temperature region (-4.3 ^{0}C) formed under the embankment in the protected case (Figure 6). Even at the end of summer, the depth of seasonal thawing was only 1.2-1.5 m for the protected case. But a great thawing zone of 8 m depth has formed in the exposed case (Figure 7; isotherm of -0.3 ^{0}C defines the freezing boundary).

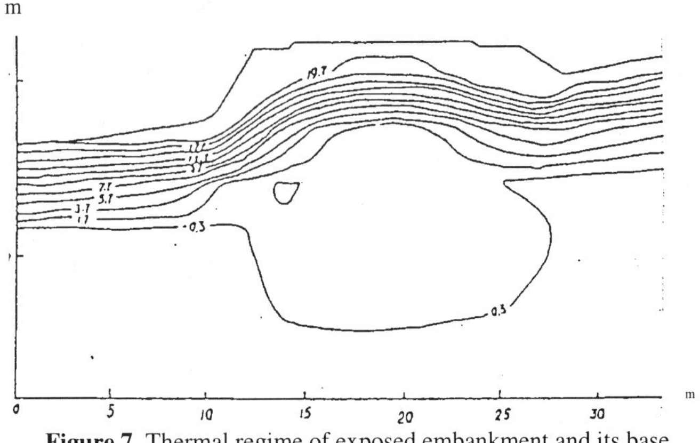

Figure 7. Thermal regime of exposed embankment and its base on September 1st of the 7th year after construction

In this numerical modeling we neglected the precipitation influence; only the solar radiation effect was investigated. Nevertheless, numerical simulations show that a positive effect of the protective shed can be achieved due to the fact that the shed eliminates the warming influence of solar radiation, which results in the embankment base being maintained in a frozen state. By maintaining the high strength properties of frozen base soils over the entire railway service life, the necessity for additional anti-deformation measures no longer arises, the embankment design is simplified, and the track throughput capacity and the component service life are increased.

In the experimental project a protective shed was constructed to maintain the permafrost condition of soils. The shed included a 9.5 m cylindrical covering for a single-track railroad. The foundation consisted of monolithic concrete foundations installed on the embankment.

The cost of the railroad with a shed was determined to be 100 rubles per linear meter lower than with thermopiles (in 1991 prices). The protective shed had the technical advantage of safely ensuring maintenance of the base soils below the entire embankment, versus the thermopiles, which have a radius of influence of only 1.5 m and so do not cover the entire base of the embankment unless spaced very closely. Another advantage of the protective shed is the fact that it is simpler and more reliable over the entire design life.

On a 383 km railway segment in the Tommot region, an additional experimental area will be constructed for checking and comparison of three other technical approaches:

-Use of rockfill embankments with rigid insulation board placed in the base of the embankment, with stone cover on the escarpments;

-Use of long thermo-stabilizers (Kondratiev et al 1996), installed beside the embankment, thus being able to exclude the stone cover on the sides;

-Use of forced aeration by installation of cast-iron pipes in the lower part of the embankment (Kondratiev et al 1993), thereby allowing exclusion of the stone cover on both sides of the embankment and the rigid board insulation from the embankment base.

The Bailkal-Amur railroad has also proceeded with use of new approaches to maintaining the natural permafrost condition of the base soils, with an emphasis on snow clearing.

Experimental design for Transbailkal Railway. On the Transbaikal railroad, 40% of the embankment defects are connected with thawing of the permafrost soils. Defects are present on approximately 85 % of the embankment.

On one such constantly deforming railway section (Figure 8), an engineering-geocryological examination was carried out in 1996 to determine the causes of perennial railroad track deformations and to obtain data for a feasibility study to develop appropriate mitigative measures. This area has gained popularity as "golden" ("a money sink-hole") because of the significant annual expenses incurred in correcting track deformations. Eight repairs were carried out in 1996. The railway incurs constant losses because of the loss of train service during necessary repairs.

Figure 8. Deformations of a Transbailkal railroad track, February 1996.

This segment of the railway was designed for electric locomotives running at a speed of 80 km/hour. The embankment was first constructed in 1902, with additional construction occurring in 1914.

The railway crosses foothill terrain, approximately 30 to 60 m from the toe of the hills, with southern exposures. The railway traverses the high flood plain of the Ingoda River, approximately 500 to 1800 m from the riverbed. The flood plain is relatively level, with a slight slope to the south, and is swampy in some areas. Settlements of the embankment have been noted since 1949. Since 1969, the speed limit for trains has been no more than 40 km/hr, sometimes dropping to as low as 15 km/hour because of the railbed deformations. From 1957 to 1992, numerous examinations of the railbed section were conducted to develop appropriate mitigative measures. However, many of the solutions attempted resulted in more harm than good, often promoting the development of thermokarst processes. The main corrective maintenance action on the line has been lifting and straightening the tracks and adding additional ballast.

The study section is underlain by perennial permafrost, which through long-term technogenic influences during construction and usage of the railway have greatly degraded. Depth to the top of the permafrost soils ranges from 0.5 to 6 m. The frozen soils have a significant ice content, quite often reaching 30 to 50 %. The cryogenic texture of the soils is netlike and flaky. The temperature of soils at the depth of annual zero amplitude ranges from -0.2 to -0.6^{o} C. The maximum thickness of permafrost soils based on geophysical explorations is approximately 35 to 40 m, but is only about 30 m based on soil borings. Frost processes and phenomena reveal themselves in the manner of perennial thawing of the permafrost soils (thermokarst) in the base embankment and on the adjacent ground surface, as well as in the manner of icing and pingo formation in the area.

Regarding engineering-geocryological and hydrological conditions, the nature of the soils and condition of the embankment under the 1600 m length being investigated can be divided into three parts:

- **west**, with permafrost of the least degree of degradation and the greatest thickness, over a total length of 450 m. This part can be divided into two lengths: *potentially dangerous* in connection with possible deformation when permafrost degradation occurs (310 m in length); *dangerous* in connection with the occurrence of permafrost degradation and resulting settlement (140 m in length).
- **central**, without permafrost, approximately 325 m in length. In this part all soils observed are in a thawed condition. Deformations of the embankment are not expected to be a problem.
- **east**, with permafrost that is greatly degraded and contains closed taliks (total length of 825 m). This part can be divided into two lengths: *dangerous*, with perennial deformation, (approximately 275 m in length); *potentially dangerous*, with permafrost soils below the embankment such that thawing could cause deformation of the embankment (approximate extent 550 m).

The earlier measures taken to mitigate track deformations (straightening and lifting the track with new ballast, building an overweighing berm), as well as efforts to deal with track icing (placing sand) promoted a further development of thermokarst processes. The mitigative measures resulted in enlarged bare areas and a dry surface, thereby resulting in increased summer warming compared to the warming that would occur beneath a natural turf-covered and waterlogged surface. Increasing the embankment thickness as a measure of sinking it into the thawed soil and additional ballast placement promoted a growing heat influence below the embankment.

Building a drainage ditch along the upper side of the track has also caused negative consequences. Degradation of permafrost has occurred as a consequence of infiltration of water through the bottom of the ditch adjacent to the embankment.

A water discharge pipe silted up as a result of soils thawing in the base of the embankment. This promoted development of a thermokarst lake adjacent to the upper side of the embankment, rendering an enormous warming influence upon the soils below the lake and below the adjacent embankment, since water from the lake tended to infiltrate under the embankment during the summer.

The general trend of construction and repair efforts, along with the engineering-geocryological conditions of this section, have resulted in expansion and deepening of thermokarst processes. Comparisons made from soil borings taken in 1971 and 1996 show that over 25 years the thickness of the talik has increased by 0.4 to 1.2 m. The depth to the top surface of the permafrost was lowered from 4 to 5.2 m below the natural ground surface. However, the potential for development of thermokarst processes and, accordingly, railroad track deformations, remains high. The thickness of frozen, potentially thaw unstable soils under the embankment ranges from approximately 2.5 to 8 m. Embankment subsidence may also occur due to compression of thawing soils under the embankment weight, weight of railroad track, and train loads.

Without active deformation control actions, the embankment will continue to deform for tens of years, causing continuous maintenance costs and reducing the track carrying capacity.

Analysis of the geocryological conditions in the area indicates that acting upon determined natural and technogenic permafrost-forming factors, and possibly changing the needed direction of the geocryological-hydrological situation, can create optimum conditions for railroad track functioning. It is possible to achieve stability of the embankment by stopping further development of thermokarst processes through cooling the soils or through preventive thawing of icy masses. Moreover, it may be possible to do this using basically the natural thermal resources of cold and heat in Transbaikal, needing only to control those resources.

A feasibility study was designed to evaluate possible deformation prevention actions, including:

· Systematic removal of snow from the embankment surface and the adjacent ground surface in the first half of winter. Also, painting the upper surfaces of the embankment a light-reflecting color in the springtime. The primary positive effect results from removing the insulating snow, thereby sharply intensifying the heat flow from the soils within and below the embankment to the atmosphere, causing their intensive cooling;

· Constructing a protective shed over the railroad track, thereby excluding the thawing influence of snow cover, infiltration of precipitation, and direct solar radiation;

· Installing a cooling pipe system in the embankment base with natural or forced ventilation using cold air or another appropriate cooling agent;

· Placement of an anti-seepage screen made from polymeric film to eliminate the thawing influences of shallow and suprapermafrost water, as well as reduce solar radiation;

· Creation of an anti-seepage screen of frozen soils by means of special devices for cold accumulation;

· Forced preventative thawing of very icy permafrost soils and replacement with competent, thaw-stable soils.

Benchmark analysis of the performance of these various mitigative measures along different lengths of the deforming sections has allowed determination of preferred methods for embankment stabilization, as described below.

On the *west deforming length*, where the thickness of permafrost is near 30 m, a sun-precipitation shed is planned over a length of 144 m. This action will ensure that the soils below the embankment remain frozen over the entire design life of the line.

On *east deforming length*, where the permafrost is currently vastly degraded, 2.5 to 3 m long horizontal thermosyphons will be installed over a length of 275 m. The thermosyphons, in combination with reflective painting and snow removal from the embankment, will prevent further degradation of permafrost in the base of the embankment.

On *potentially dangerous west and east lengths* with ice-rich permafrost soils, reflective painting and snow removal is planned to prevent future thawing.

All recommended methods to prevent deformations should include drainage of surface water and groundwater away from the embankment, as a primary cause of thermokarst development is infiltration of water through the embankment, along with increased absorption of solar radiation. One approach to drainage involves deep well draw down, using available reinforced concrete pipes or installation of new metallic pipe. Improvement of existing shallow drainage systems is also anticipated. Such drainage improvements will reduce icing formation in the area.

The overall estimated capital costs for the stabilization measures described above range from approximately 1770 thousand to 3200 thousand rubles in 1991 prices.

Experimental Design And Construction For Qinghai-Tibet Railway

The Qinghai-Tibet railway in China has been under construction since July 2001. The railway route will pass through an area of permafrost at least 500 km in length. For several decades Chinese specialists have spent considerable time investigating embankment stability in this area. An embankment experiment in Fenghuoshan was conducted by the Lanzhou Branch of the Railway Science Academy.

Since 2000 there has been testing of sun-and-precipitation protective sheds according to the approach described here. A numerical investigation of an experimental shed over the embankment was conducted by Zhang Luxin and Liu Jiankun before 1998. They achieved good results using a computer model. Over a period of 20 years after construction of the shed, the temperature of soils below the embankment at depths of 6 to 9 m reached -8°C (Figure 9), well below the temperature of the soils in natural conditions (-3.5°C). Without the shed for this period, a talik nearly 9 m thick would be formed (Figure 10). In their opinion, the sun-and-precipitation protective shed has proved to be the most efficient measure for stabilization of an embankment on permafrost.

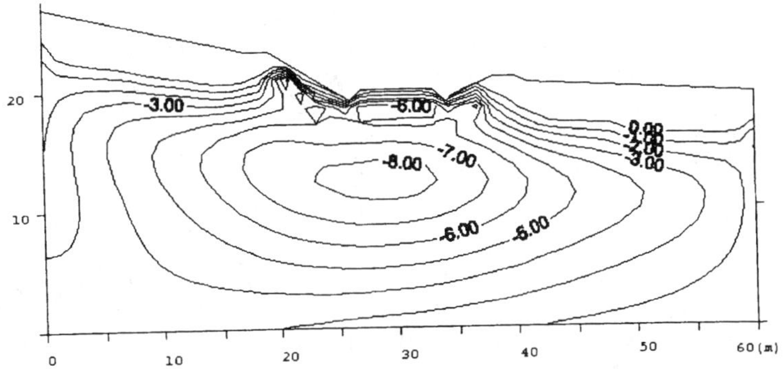

Figure 9. Temperature of soils under embankment with shed over a 20-year period

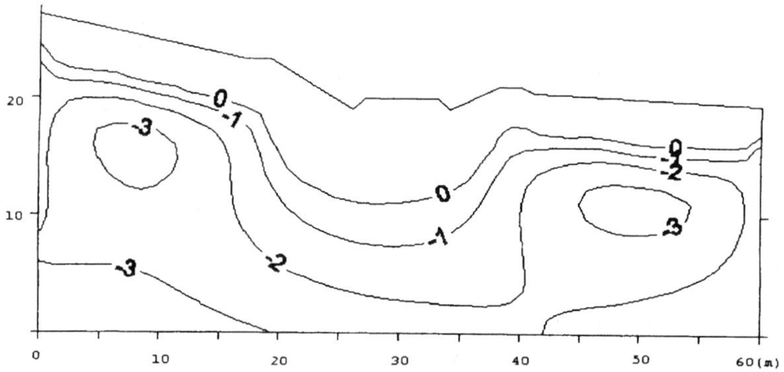

Figure 10. Temperature of soils under embankment without shed over a 20-year period

Since 2001 there has been testing of different strengthening methods for roadbeds on very icy permafrost in Tibet by the First Survey and Design Institute of the Railway Ministry of China. In particular, there has been testing of cross-cooling pipes.

Conclusions

The Russia, Alaska, and China experience provides a favorable outlook for development of strengthening techniques for embankments on permafrost soils based on adjustment of the cooling and warming factor ratios. The primary purpose of this approach is to reduce the average annual temperature of soils and preserve them in a perennially frozen state. For example, according to Zarling & Braley (1986), a protective shed can effectively cool embankments by 3 to 5°C within a few years. The Chinese have obtained good results in Tibet.

References

Bushin, A.V. (1992). "On The Problems Of Ensuring Reliability Of Railroad Bed In Current Conditions." *Railway Transport. Series "Tracks & Track Facilities"* EI/Tsniitei Ministry Of Railways. Issue 5-6, Pp.1-14.

Elenevsky V.V. & Nizovkin G.A. (1936). *Railway Construction On Permafrost.* M.: Transjeldorizdat.

Fyodorov, D. I. (Ed.) (1977). *Surveying And Design Of Baikal - Amur Railway Route.* In Reference-Methodological Manual. -M: Transport.

Hayley, D.W. (1988.). "Maintenance Of A Railway Grade Over Permafrost In Canada": *Proc. 5th Int. Conf. On Permafrost.* Vol. 3, Pp. 43-48.

Kondratiev, V.G. (1992). *Embankment At Sections With Terrain Cross Gradient And Very Icy Permafrost Soils*. Patent No. 1807173.

Kondratiev, V.G. (2001). *Active Methods Of Strengthening The Roadbed Base On Permafrost Soil*. Chita. Zabtrans. 100 Pp.

Kondratiev, V.G., Korolyov, A.A., Karlinsky, M.I., Pozin, V.A. & A.S. Rozanov (1992). *A Method Of Strengthening The Roadbed Base On Very Icy Soil*. Patent No. 1764371.

Kondratiev, V.G., Korolyov, A.A., Karlinsky, M.I., Pozin, V.A. & A.S. Rozanov (1992). *A Railway Track On Very Icy Permafrost Soils*. Patent No. 1740555.

Kondratiev, V.G., Karlinsky, M.I., Savelyev, B.A. & A.V. Sokolov (1993). *Embankment On Very Icy Permafrost Soils*. Patent No. 2010919.

Kondratiev, V.G., Kobzev, V.A. & B.A.Gusakov (1996). *Embankment At Hill-Side Sections With Very Icy Permafrost Soils*. Patent No. 2059752.

Kondratiev, V.G. & I.I. Broid (1994). *A Method Of Strengthening The Roadbed Base On Very Icy Permafrost Soils*. Patent Application No. 94022236.

Kronick, Ya.A. (1982). "Thermal Mechanical Models For Frozen Soils And Cryogenic Processes". In *Soil Reohgology And Engineering Geocryology*. Publishing House Nauka .

Lui Jiankun. (1994). *Thermal And Strain-stress Calculation Of Roadbed Constructed Layer By Layer On Permafrost*. Dissertation Ph.D. Moscow State University Of Civil Engineering.

Liu Jiankun & Kondratiev, V.G. (1996). "The Influence Of Sun-Precipitation Protective Sheds On The Thermal Regime Of Roadbed Base On Permafrost". *Proceedings Of International Symposium Cold Regions Engineering, Harbin, China, 11-14 September 1996*. University Of Harbin Industry Technology Press/

Jia Jianhua, Ran Li, Wang Jiyin (1998). "Choice And Construction Technology Of Foundation Types For Railway Buildings In Permafrost Regions, Gingxai-Xizang Plateau". *Proc. International Symposium On Geocryological Problems Of Construction In Eastern Russia And Northern China*. Vol. 1. Yakutsk. SB RAS Pablishers,

Peretruhin N.A. & T.V. Potatueva (1987). *Interaction Of Embankment And Permafrost*. Tomsk: Tomsk University.

Wu Ziwang Et Al. (1988). *Roadbed Engineering On Permafrost*. Lanzhou University.

Yakovlev, V. Ye. (1992). "Current Maintenance And Major Overhaul Of Roadbed In BAM Railway Conditions". *Railway Transport. Series "Tracks & Track Facilities"*, EI/Tsniitei Ministry Of Railways. Issue 5-6, Pp. 14-28.

Zarling, J.P. & A.W. Braley (1986). *Thaw Stabilization Of Roadway Embankments Constructed Over Permafrost.* Report NO FHWA-AK-RD-81-20, Federal Highway Administration, Washington DC.

Zhang Luxin And Liu Jiankun (1998). "Numerical Investigation Of Experimental Roadbed Thermal Stability On Fenghuoshan Permafrost In Tibet". *Proc. International Symposium On Geocryological Problems Of Construction In Eastern Russia And Northern China.* Yakutsk. Vol. 2. SB RAS Publishers.

Seismic Evaluation And Retrofit Strategies For The University Of Alaska Anchorage Engineering Building

He Liu[1] and William N. Scott, P.E.[2]

Abstract

This paper presents a seismic evaluation of the University of Alaska Anchorage (UAA) Engineering Building (UAAEB). The building was designed and constructed in 1981 to the 1979 Uniform Building Code using moment resisting space frames. The evaluation was performed using current design practice following the 2000 IBC and FEMA 273 requirements. A dynamic time history analysis was performed using a synthesized earthquake record based on the M_w 9.2 "Great Alaska Earthquake of 1964".

Strong shaking exceeded 90 seconds, with felt motion lasting for over 4 minutes (Handler, 1973). Anchorage suffered significant damage due to the event. The peak horizontal ground acceleration on bedrock was estimated at 0.218g. Extensive development has occurred in this area of Anchorage since 1964, including the UAA Campus, and nearby, the Alaska Native Medical Center, and Providence Alaska Medical Center. Geotectnical investigations have identified the UAA Campus as highly susceptible to amplified seismic ground motions on the order of 2.5 or higher (Martiosyan, 2000; Martiosyan, 2001).

A thorough investigation of the building was conducted using the 2000 Edition of International Building Code (ICC 2000); FEMA-273 (FEMA 1997 a), "NEHRP Guidelines for the Seismic Rehabilitation of Buildings"; and FEMA 310 (FEMA 1998), "Handbook for the Seismic Evaluation of Buildings – A Pre-Standard." Pushover analysis studies identified elements of the structure critical to reducing damage. These critical elements were identified for seismic retrofit.

Introduction

The University of Alaska Anchorage (UAA) is located in an area of potentially

[1] Associate Professor, School of Engineering, University of Alaska Anchorage, 3211 Providence Drive, Anchorage, Alaska 99508, Email: afhl@uaa.alaska.edu

[2] Principal Structural Engineer, VECO Corp., 949 E 36th Ave., Suite 500, Anchorage, Alaska 99508. Email: william.scott@veco.com

high seismic risk with a high student population density. Two major critical care hospitals, several schools, and other important public facilities are located in the same seismic area. The Alaska Native Medical Center is the medical service center for the entire Alaska Native population. Anchorage, with a population of more than 260,000 persons, serves as the commercial and industrial hub for Alaska and is one of America's largest airfreight hubs for the Pacific and the Orient. UAA buildings are intended to serve as backup emergency shelters for the Anchorage community. UAA is in a unique position to serve the Anchorage community, especially through cooperative research efforts regarding seismic risk to public facilities. UAA specialists, in collaboration with the Anchorage technical community have begun to assess the vulnerability of UAA campus structures to seismic hazards and recommend strategies to limit property damage, injuries, and loss of life from future strong-motion earthquakes.

Scientists have long recognized Alaska to be one of the most seismically active areas in the world[3]. It is a natural laboratory, where the number of earthquake events exceeds that of the rest of the United States. The earthquake risk is high in much of Alaska, including its largest city, Anchorage. Southcentral Alaska is in a vulnerable location at the junction of the seismic Pacific subduction zone and a strike-slip zone called the Aleutian Trench. During the 1964 Great Alaska Earthquake, the adverse behavior of soils, caused by strong ground shaking, resulted in extensive damage throughout the city and surrounding area. Large earth slides and block movements of earth damaged streets, buildings, utilities, and shore-side structures. A variety of soil-related failures took place, including soil liquefaction, subsidence, mass wasting, consolidation, and formation of cracks and sand spouts. A major earthquake could have an even greater effect on the structures in Anchorage, since the city is more than eight times its size in 1964.

Recent research from a program sponsored by the Alaska Science and Technology Foundation (ASTF), titled, "Seismic Microzonation: Metropolitan Anchorage, Alaska", shows that anomalous ground amplification appears likely in the UAA vicinity. The local soil conditions may amplify seismic ground motions by a factor of 2.5 or higher. These amplification effects would have severe effects on the two to three story buildings, typical of most UAA campus buildings and the two adjacent hospitals. To help protect the UAA facilities, employees, and students, which will further benefit the community at large, a procedure will be developed that can profoundly affect public opinion concerning seismic risk across the state.

---

[3] A 5.1 M_w earthquake occurred under Anchorage while the second author was editing this paper on 2/6/02, followed 45 seconds later by another 4.9 M_w event.

Building Description

The UAA Engineering building is a rectangular three-story structure, 46 m long by 27 m wide by 15 m high. The building frame consists of structural steel moment frames in-filled with steel roof and floor trusses. The floors are composite concrete metal decking. The roof is metal deck. The structure is a regular structure without discontinuities or irregularities. The building is supported on shallow spread footings. The first floor is a slab-on-grade. The design followed the 1979 Uniform Building Code (ICBO 1997) and the 7th Edition of the AISC Manual of Steel Construction (AISC 1975). The design incorporates a moment-resisting space frame (K = 0.67) in both major axes. Figure 1 depicts a typical beam-column joint. The design of the welded moment connections followed the AISC method for "Type 3, Semi-Rigid Framing". The beam web is not attached to the column. Shear is transferred through bearing to stiffened seats. Moments are developed through top and bottom flange plates welded to the columns. Backing plates were not removed.

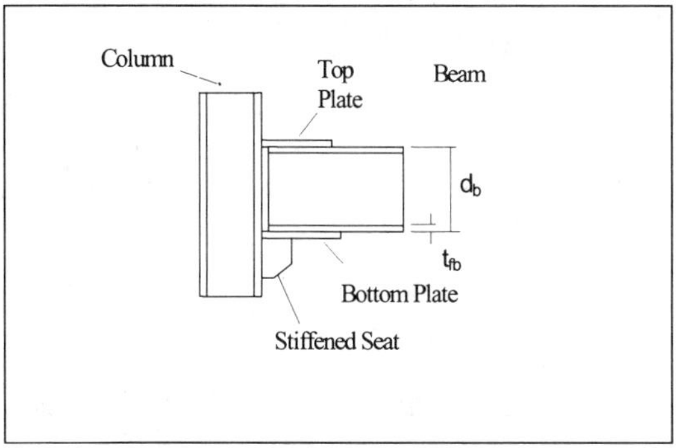

Fig. 1 Typical moment connection

Seismic Evaluation Objective

The seismic evaluation objective is to investigate the building for the Basic Safety Objective (BSO) per FEMA 273. Two levels of seismic events are considered: Life Safety Performance Level (3-C) for BSE-1 (Basic Safety Earthquake 1, 10% probability of exceedance in 50 years, 10%/50 yr) and Collapse Prevention Level

(5-E) for BSE-2 (Basic Safety Earthquake 2, 2% probability of exceedance in 50 years, 2%/50 yr). Seismic accelerations are based on IBC 2000.

Comparison of UBC 1997 and IBC 2000 Seismic Loads

The Engineering Building is the home of the UAA School of Engineering. The building contains offices, laboratories, and classrooms. The seismic importance factor, Is = 1.25, Capacity greater that 300 people.

Table 1. Seismic Gravity Load

Level	Dead[1] (kN)	Snow[2] (kN)	Total (kN)	Elev[3]. (m)
Roof	1,360	1,290	2,650	14.4
Second Floor	2,616		2,616	9.6
First Floor	2,656		2,656	4.8
Total			7,922	

Notes:

1. Floor dead load includes 500 Pa for partitions
2. Snow load is 50% of 2,000 Pa
3. Elevation indicated is above grade

Estimated Building Period. UBC-79 (Eq. 12-3B) and IBC (Eq. 16-40) may be used since the steel moment frames resist all lateral forces and the floor heights are about 4.8m. The estimated natural period is 0.3 second. The natural periods calculated by the SAP2000 (SAP2000 2001) model are shown in Table 2.

Table 2. Calculated building periods, seconds

Mode	Direction	
	North-South	East-West
1	0.309	0.478
2	0.265	0.318
3	0.146	0.299

Seismic loading calculated per UBC 1979 (12-1).

$$V = ZIKCSW = 0.117W = 947 \text{ kN}$$

where:

V = total lateral seismic force
Z = 1 for UBC Zone 4
I = 1.25 (Occupancy > 300)

K = 0.67 for ductile moment-resisting space frame
$C = 1/(15*T^{0.5}) = 0.122$
S = 1.5

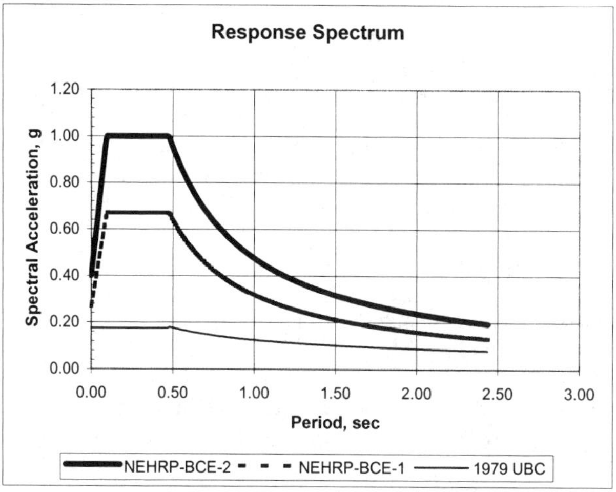

Fig 2. Response Spectrum for UBC 79 and IBC 2000

Seismic loading calculated per IBC 2000. The seismic load specified by the IBC, assuming a Site Class of C, is V = 0.25W. This is based on a calculated structural period of 0.3 second and use of the "Equivalent Lateral-Force Procedure" (IBC Section 1617.4). The site-specific spectra are based on the MCE (Maximum Considered Earthquake, Figure 1615(7)) of S_s = 150%g and S_l = 55%g. Site modification factors were selected from Tables 1615.1.2(1) and Table 1615.1.2(2).

Table 3. Distribution of Horizontal Forces

Level	Weight (kN)	Elev. (m)	W_xH_x	C_{vx}	UBC 1979 (kN)	FEMA 273 (kN)
Roof	2,650	14.4	38,160	0.50	474	990
Second	2,616	9.6	25,114	0.33	313	653
First	2,656	4.8	12,749	0.17	161	337

Using $C_s = 0.31$ and $V = C_s*W = 1980$ kN, the increase of design seismic loading from UBC-79 to IBC 2000 is 253%.

Analysis Description

The building was analyzed by static, dynamic, and push-over methods using SAP2000 (SAP2000 2001). Predictably, the structure is overstressed for seismic load combinations. A dynamic analysis was performed to determine the structural behavior using a synthesized acceleration record for the 1964 Great Alaska Earthquake. The dynamic analysis shows that the connections will experience over 40 stress reversals during a 60-second event. A push-over analysis was then performed to estimate the probable failure sequences of the building elements.

Time History Analysis

Time histories of the North-South horizontal component of the March 27, 1964 Alaska earthquake in the Anchorage area (Papageorgiou 1997) were used in the analysis. This earthquake was significant in its magnitude (9.2), areal extent and long duration, over 5 minutes. The generated time histories are a set of typical strong ground motions for Anchorage that are physically plausible,

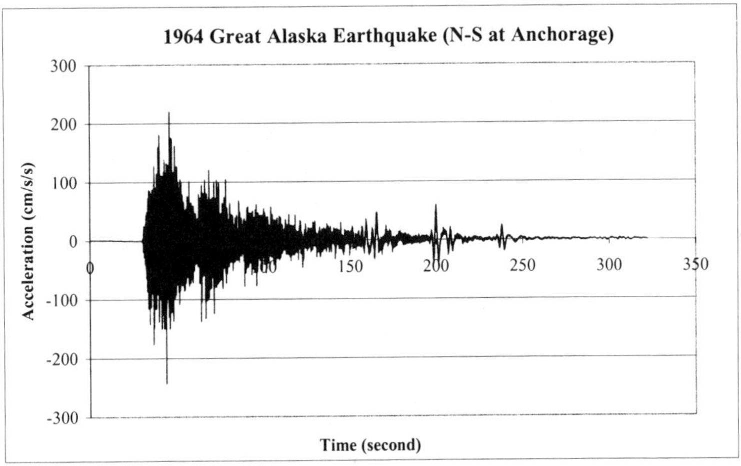

Fig. 3 N-S Ground acceleration at Anchorage, 1964 Great Alaska Earthquake

consistent with the seismological, tectonic and eyewitness accounts, but not necessarily unique. Depending on the frequency range, the discrete wave number representation technique is based on slip models available in the literature, the convolution of Green's functions with far-field pulses and the stochastic modeling approach for generic soil type D, has been used to produce these synthetic ground motions. The North-South ground acceleration time history in the Anchorage area during the synthesized Great Alaska Earthquake is shown in Figure 3.

The analysis used 250 seconds of the record from the time period of 33 seconds to 283 seconds, corresponding to peak values of horizontal acceleration. In the time history analysis, horizontal ground acceleration values were adjusted to 0.4g and applied to both directions of the building.

Evaluation of Beam-Column Connections

Calculations show that the strong beam–weak column connections are stiff and perform in a fully restrained (FR) manner up to the capacity of the connection plates. The demand on the connections is twice the calculated capacity. The estimated moment capacity of the connections is 90% to 95% of the beam capacity, indicating that inelastic yielding will occur in the connections. The resulting behavior is rigid until elastic capacity is reached, then will degrade as the connection plates yield and harden under cyclic loading and yielding of the columns. Further, the the connect has a low fatigue strength (Partridge, 1999). Calculations show the full plastic yielding of the connection occurs at 0.001 radian while developing 90% of the beam moment capacity (Chan and ect. 2000, and Faella et al 2000). The column plastic moment capacity is generally mobilized before the beam-column connections yield.

The building connections do not have sufficient inelastic capacity according to recommendations of the IBC (ICC 2000) and AISC Seismic Provisions (AISC 1997). The connection plates yield at an estimated rotation of 0.001 radian. Inelastic deformation at yield stress will provide additional rotational capacity. The rotational demand by static analysis is 0.01 radian. The maximum rotational demand at a joint, located at the S-E inside corner of the second floor, is 0.0125 radian by linear elastic time history analysis.

The Northridge earthquake showed that moment connection capacity should have an inelastic rotational capacity of 0.01 radian to 0.03 radian for Ordinary, Intermediate, and Special Moment Frames (Page and et al, 1972). Buildings that have insufficient inelastic rotational capacity will experience greater damage and possible instability during seismic events. The moment capacity of the columns was determined to be less than that of the connecting beams. The design of the UAA Engineering Building followed methodology that was recommended prior to Northridge (FEMA 1997 a, FEMA 1997 b, FEMA 1998, Carter et al 1999)

The connection weld on the top plate was held back from the column flange an average of 4 inches, resulting in an unsupported plate. Cracking of the weld or connection plate may occur as a result of the cyclic load reversals. Stability of the building will degrade as connection capacity is lost. The connections do not have sufficient ductility to withstand the estimated 65 cycles of earthquake-induced motion without significant hazard.

Push-Over Analysis

Push-Over analysis is a nonlinear static procedure that can be used for any structure and rehabilitation objective. The analysis is suitable for buildings without significant higher-mode response, like the Engineering Building. Typical ductile behavior was assumed for beam-column connections. It is characterized by an elastic range, followed by a plastic range that includes strain hardening, and a strength-degraded range in which the residual force that can be resisted is significantly less than the peak strength, but still substantial.

In two analysis cases, two push-over force vectors were applied in the N-S and E-W directions separately with the shape proportional to the mass distributions in the corresponding directions. The push-over processes were controlled by joint displacements at two locations in the roof level. The maximum control displacements were about 10 to 14 cm in Case 1. The magnitudes of the maximum control displacements were about 5 to 12 cm in Case 2, pushed in the E-W direction.

In Push-over Case 1, the first plastic hinge occured at the bottom of a column located at the Southeast inside corner of the third floor. In Push-over Case 2, the first three plastic hinges occured almost at the same time: one at the bottom of a column located at the Northeast inside corner at the first floor; one at the end of a beam located at the Northwest inside corner at the second floor; one at the end of a beam located at the Northeast outside corner at the second floor.

To evaluate the possible nonlinear performance of the building in a future major earthquake, it is useful to express the different performance levels in a force – deformation diagram. Figure 4 shows the approximate normalized deformation, in relation to the idealized force versus deformation curve, that is acceptable for Primary (P) and Secondary (S) components for Immediate Occupancy (I.O.), Life Safety (L.S.), and Collapse Prevention (C.P.).

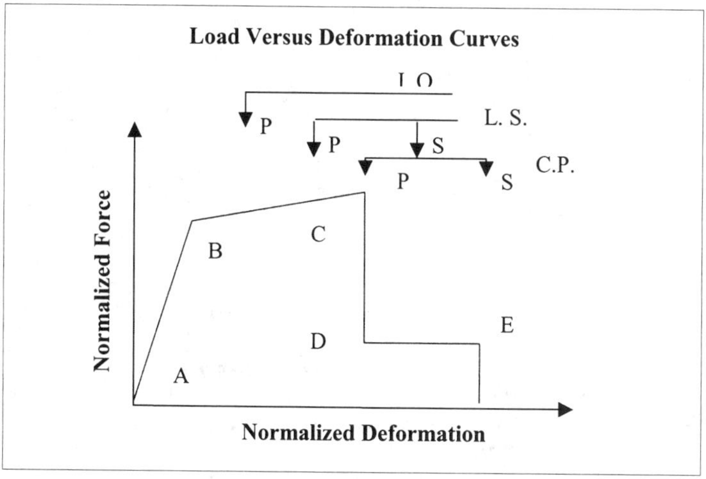

Fig 4. Idealized Component Load versus Deformation Curves for Depicting Component Modeling and Acceptability

The push-over results are shown in Table 4 for Case 1 in the N-S direction and in Table 5 for Case 2 in the E-W direction. For each pushover displacement control-loading step, the base shear force is calculated. The numbers of joints are listed under the different performance levels. The letters A, B, C, D, and E correspond to the typical points in the normalized force versus normalized deformation curve. The letters IO, LS, and CP correspond to different performance levels. Table 4 shows that after the pushover control displacement reaches 10.42 cm and then reverses to –14.03 cm, there would be 12 joints following the B-C portion of the curve reflecting the performance level of Immediate Occupancy, and 9 joints in the performance level between IO and LS Table 5 shows that after the pushover control displacement reaches 12.49 cm and then back to 4.98 cm, there would be 6 joints in the performance level of IO, and 11 joints between IO and LS, one joint between LS and CP, four joints in the level of CP and 10 joints beyond the CP level.

Table 4. Pushover case 1 - N-S direction

Step	Displacement (cm)	Base Force (kN)	Performance Levels Number of Joints			
			A-B	B-IO	IO-LS	LS-CP
0	0.00	0	144	0	0	0
1	5.70	9,362	144	0	0	0
2	8.14	13,300	143	1	0	0
3	10.42	16,828	137	7	2	0
4	-14.03	25,698	123	12	9	0

Table 5. Pushover case 2 - E-W direction

Step	Displacement (cm)	Base Force (kN)	Performance Levels Number of Joints							
			A-B	B-IO	IO-LS	LS-CP	CP-C	C-D	D-E	>E
0	0	0	144	0	0	0	0	0	0	0
1	5.88	20,320	139	5	0	0	0	0	0	0
2	7.11	24,076	132	8	4	0	0	0	0	0
3	12.49	32,995	122	4	12	0	4	2	0	0
4	4.98	16,538	110	6	11	1	2	2	0	10

Proposed Seismic Upgrade Alternative

There are several alternatives available. Strengthening the building would require disruption of building use and removal of architectural finishes. Alternatives include:

1. Strengthen building connections, columns, and foundations.
2. Add bracing to the building framing and upgrade the foundations.
3. Add damping to the building frames to lower the lateral forces to a level that the building framing and connections can resist. Viscous dampers could be installed in the moment frames similar to the bracing of Alternative 2. Alternative 3 requires less strengthening and foundation work than Alternatives 1 and 2.

Discussion and Conclusions

Seismic retrofitting of the UAAEB could improve its predicted seismic performance. Strengthening of connections would not improve the weak column-strong beam condition. The best alternatives for upgrading this structure involve adding bracing or damping.

Bracing could be added to selected frame bays. Bracing will stiffen the structure and increase the IBC 2000 demand. Bracing would also involve modification to the existing foundations to mobilize resisting moment forces.

Twenty percent added damping lowers the structural seismic demand to 65% of the IBC 2000 demand. Added damping is the apparent least cost alternative and has the least impact on the existing structure.

The UAAEB is similar to other structures constructed in Anchorage since 1964, and typical of most urban facilities that have been constructed under various UBC code revisions.

References

AISC (1975), "Manual of Steel Construction, 7th Edition".

AISC (1997), "Seismic Provisions for Structural Steel Buildings".

FEMA (1997 a), "NEHRP Guidelines for the Seismic Rehabilitation of Buildings".

FEMA (1997 b), "NEHRP Commentary to Guidelines for the Seismic Rehabilitation of Buildings".

FEMA (1998), "Handbook for the Seismic Evaluation of Buildings – A Prestandard".

ICC (2000). "International Building Code, 2000 Edition".

ICBO (1979) "Uniform Building Code, 1979"

Carter, Charles J. (1999), "Stiffening of Wide-Flange Columns at Moment Connections: Wind and Seismic Applications", AISC Steel Design Guide Series No. 13, AISC, 1999.

Chan, S.L. and P.P.T. Chui (2000), Non-Linear Static and Cyclic Analysis of Steel Frames with Semi-Rigid Connections, Elsevier.

Faella, C; V. Piluso; G. Rizzano (2000). Structural Steel Semirigid Connections, Theory, Design, and Software, CRC Press.

Handler, Philip (1973). The Great Alaska Earthquake of 1964, Engineering, National Academy of Science.

Martirosyan, A. H., U. Datta, N.N. Biswas, M. Dravinski, A. Papageorgios, and R. Combellick (2000), "Site Amplification Studies for Anchorage, Alaska", Proceedings 6th International Conference on Seismic Zonation, EERI.

Martirosyan, A., U. Dutta, and N. Biswas (2001), "Determination of Site Response in Anchorage, Alaska, on the Basis of Spectral Ratio Methods".

Page, Robert A., David M. Boore, William B. Joyner, and Henry W. Coulter (1972). "Ground Motion Values for Use in the Seismic Design of the Trans-Alaska Pipeline System", USGS Circular 672.

SAP2000 (2001). "Integrated Finite Element Analysis and Design of Structures", Version 7.41, Computers and Structures, Inc. Berkely, California.

Papageorgiou, A.S., Zhang, B., and Biswas, N.N. (1997) *Estimation of Strong Ground Motion from the Great 1964 Prince William Sound Earthquake (Mw=9.2)*, Seismological Research Letters 68 (2), 325.

Partridge, James E., "Stress Reversal within Same Flange of Pre-Northridge Moment Frame Connections", SEAOC 1999 Convention.

Water Storage Tank Response To Earthquake Loads In Alaska – Predictive Modeling By The Finite Element Method

He Liu[1] and Daniel H. Schubert[2]

Abstract

Alaska is one of the most seismically active areas in North America. Large-scale seismic event occurrence is more than ten times greater than in California. Approximately 40-50% of the communities with 60% of the total population are located in seismic zones 3 and 4, which are areas with the potential for high seismic risk. Due to limited access and the isolated locations of many of these communities, a much greater reliance is placed on many of the essential, lifeline facilities. In addition, a much higher per capita investment exists in the facilities, and any recovery effort or needed repairs are expected to be delayed, should wide spread damage occur. Extended periods of snow cover in Alaska increase the likelihood of coincident large snow loads with a seismic event.

A recent trend in earthquake engineering of lifeline facilities is to design structures to a higher and more realistic level of performance. This concept grew out of the extremely large capital expenditures required to restore infrastructure after recent earthquakes in Northridge, California; Kobe, Japan; Turkey; and Greece. Application of advanced dynamic analysis methods has been limited to high cost, large structures. The availability of software for personal computers allows performance based analysis and design of lifeline structures for lower cost projects in small communities. An example of this approach is used to evaluate the performance and response of a ground level steel water storage tank both with and without snow loads to determine whether a significant effect results. A synthetic time history of the horizontal and vertical components of the 1964 Alaska Earthquake is used in the analysis. The computer program ANSYS is used for the finite element analysis (FEA) of the tank and contained fluid. Results are then compared with code values.

[1] Associate Professor, School of Engineering, University of Alaska Anchorage, 3211 Providence Drive, Anchorage, Alaska 99508, Email: afhl@uaa.alaska.edu
[2] Senior Engineer, GV Jones & Associates, Inc., 4141 B St. Suite 309, Anchorage, Alaska 99508, Email: dan_s@gvjones.com

Introduction

Recent earthquakes in the US have been moderate in comparison to other parts of the world. Extensive damage and disruption to water and wastewater service have followed most of the major seismic events outside the U. S. Water service was disrupted to 5 million people following the 1985 Mexico City event. 200,000 people were without water service for one month following the 1990 Philippine earthquake. By contrast, water service was restored to Northridge, California within 4 days (ASCE 1995). Following most major damaging earthquakes, utilities and state governments have responded with renewed interest in methods to mitigate future damage to infrastructure. In areas of large population density and high seismic risk, much emphasis has been placed on evaluation and mitigation measures. In other locations where the risk is not so apparent, preventative measures lag. In areas where very large events are possible, but infrequent, earthquake engineering receives a low priority.

System components are designed in accordance with building codes such as the Uniform Building Code or industry standards such as the American Water Works Association. The design is based on life-safety considerations, not on damage minimization or post-earthquake operations. A major problem with this approach is the high capital cost of post-earthquake repairs. The high costs of repairs and replacement in recent damage in California have cause the public to request engineers to base designs on a "performance level", determined by the owner. In the case of lifelines, higher levels than code-based designs would result.

Two types of problems are related to lifeline earthquake engineering. The first is to determine the expected earthquake ground motion for design, and the second is to determine the expected structural response. The level of ground motion to be used in design can be specified in several ways. Code-based designs use equivalent static formulas, with several parameters incorporating the type of structure, importance, soil conditions, and location relative to seismic zones and faults. Codes make reference to more detailed analyses that use design spectra which incorporate time with peak horizontal ground acceleration or other methods such as time histories of ground motion in a dynamic analysis. These methods permit site-specific factors, and differing degrees of acceptable risk. This usually requires a level of computer modeling, which in the past was limited to large, expensive structures. With the availability of general-purpose structural engineering software, this is no longer the case. The following is an example of applying these methods to the evaluation of water storage tanks.

Water Storage Tanks

Liquid storage tank design in the United States is based on standards prepared by voluntary committees such as the American Water Works Association (AWWA) or the American Petroleum Institute (API). As important components of community lifeline systems, liquid storage tanks should be designed to survive and continue to function under earthquake events. Damage to tanks and other structures in recent earthquakes, and the prevalence of tanks in areas of high seismic risk has continued interest in their behavior under seismic events (Haroun 1991, Lau 1995). Damage to shell walls and roofs, tear-out of anchors, and pipe connection failures can occur. Due to the potential for damage, the value of post-earthquake performance of tanks for water supply and fire suppression is significant. The study of the seismic response of liquid storage tanks has been ongoing for the past 40 years. Housner (Housner 1963) and others (Veletsos 1984, Haroun 1991,) evaluated the hydrodynamic loading assuming the hydrodynamic response to be divided into two components. The impulsive component included a mass of liquid, which moves in unison with the tank structure. The convective component was associated with the liquid surface wave. Tank analysis was modeled as a single degree of freedom oscillator, with an equivalent mass located at different heights above the base. This allowed determination of the overturning moment and base shear, and still serves as the basis for the AWWA design standards.

Finite Element Model Approach

Current knowledge about the behavior of liquid storage tanks is still developing. Recent advances and the availability of faster personal computers using finite element programs allow for more detailed analysis. The fluid-tank system selected for detailed analysis is an 1,893 m^3, at grade, cylindrical steel water storage tank, located in seismic zone 4. The properties of the tank are summarized in Table 1.

In the FEA models of this study, the tank roof system is represented by shell and beam elements, which are placed in the radial and circular directions. The tank wall is modeled by shell elements. The contents are represented as contained fluid elements. The fluid elements at the wall boundary are not attached directly to the shell elements but have separate, coincident nodes that are coupled only in the direction normal to the interface. Relative movements in the tangential and vertical directions are allowed to occur. At the base, fluid element nodes are allowed to move on the surface of the tank bottom plate. Due to symmetry, only one half of tank system is modeled.

Table 1. Tank model parameters

Parameter		Parameter	
Geometric Properties	**Value**	**Material Properties**	**Value**
Height, m	7.37	Fluid Density, kg/m^3	1000
Diameter, m	18.11	Fluid Bulk Modulus	30 x 10^4
Water Level Height, m	6.83	Steel Density, kg/m^3	7.84 x 10^3
Wall thickness, mm	4.76	Steel Poisson Ratio	0.3
Roof thickness, mm	3.42	Steel Yield Strength, kPa	248.2
		Steel Modulus of Elasticity, kPa	206.8 x 10^6
		Steel Tangent Modulus, kPa	20.7x10^6

Large deformations and material yielding in the tank were expected, thus large deformation and material nonlinearities were used in the model. An elasto-plastic stress-strain curve is assumed for the steel wall. Steel exhibits a linear stress-strain relationship up to a stress level of the yield point. Beyond this point, the stress-strain relationship will become nonlinear and plastic. Several options are available for describing plasticity behavior. The simplest Bilinear Kinematic Hardening option was chosen. It assumes the total stress range is equal to twice the yield stress, so that the Bauschinger effect is included. The plastic behavior is nonconservative, path-dependent. The sequence in which loads are applied and in which plastic responses occur affects the final solution results. To achieve an accurate solution in each time step, loads are applied as a series of small incremental sub-load steps, so that the calculation model will follow the load-response path as closely as possible. It is necessary to additionally reduce the load step size after a load step in which a large number of equilibrium iterations was performed or in which a plastic strain increment of greater than 15% was encountered.

Two models were developed and analyzed. One model was without an added snow load, and the second included an 390.6 kg/m^3 snow load, which was represented as an equivalent material mass, with centroid located at the roofline.

Modal Analysis

As an initial step in the evaluation, a modal analysis was performed on the model. The significant natural modes were chosen starting from a comparison of the participation factors, modal coefficients, and mass distribution percentages for each mode extracted. Larger mass distribution percentages usually indicate important modes in the corresponding dynamic response analysis. To ensure the validity of the ANSYS (ANSYS 2001) models, modal analyses for fluid sloshing and a combined fluid-tank system were performed. The results were compared with the available theoretical solutions for flexible tanks of fixed support conditions (Veletsos 1984), results of which are summarized in Table 2. The excellent agreements indicate the validity of the ANSYS models.

Table 2. Modal frequency comparison

	Natural Period (second)		
	Horizontal		Vertical
	1st	2nd	1st
FEA Model	0.1443	0.0811	0.1387
Theory	0.1446	0.0781	0.1430

Effects of Snow Loads on Natural Frequencies

Natural frequencies are proportional to the square root of the stiffness and inverse of mass. No change in structural stiffness would result from the snow load effect, but the added mass would reduce the natural frequencies. Table 3 shows the model results and those determined from theory (Veletsos 1984). As with the tank without snow load, reasonable agreement occurs between the model and the theoretically based calculation.

The differences in time period show a two to three percent increase, which is not significant for design based on a smoothed elastic response spectrum approach. The effect may be significant when using a nonlinear time history analysis due to the change in frequency.

Table 3. Modal frequency comparison- snow load effect

	Natural Period (second)		
	Horizontal		Vertical
	1st	2nd	1st
FEA Model	0.1472	0.0853	0.139
Theory	0.1533	0.0928	0.141

Time History Analysis and Results

Time histories of the North-South horizontal component and the vertical component of the March 27, 1964 Alaska earthquake were used in the analysis. This earthquake was significant in its magnitude (9.2) and areal extent, as well as it's duration of over 5 minutes. The generated time histories are a set of typical strong ground motions for the Municipality of Anchorage that is physically plausible, and consistent with the seismological, tectonic, and eyewitness accounts, but not necessarily unique. Depending on the frequency range, the discrete wave number representation technique based on slip models available in the literature, the convolution of Green's functions with far field pulses, and the stochastic modeling approach for generic soil type D has been used to produce these synthetic ground motions (Papageorgiou 1997).

The analysis used seven seconds of the record from the time period of 39.7 seconds to 46.9 seconds, corresponding to peak values of horizontal acceleration. The complete data record was not used due to computational time constraints. Horizontal ground acceleration values were adjusted to 0.4g. This same factoring value was applied to the vertical time history, with a resulting peak vertical acceleration of 0.3g. Ground acceleration time histories are shown in Figure 1. In the time history analysis, horizontal and vertical ground accelerations were run individually as well as simultaneously combined. Results are summarized in the following sections.

Water Surface Profile. Figure 2 shows the water surface displacement time history at a point on the upper left water surface for the combined case, as well as the individual horizontal and vertical computations. The average surface response frequency is close to the first convective mode, implying that the significant hydrodynamic sloshing motion is dominated by the lower convective modes, which also is consistent with theory. The maximum value of vertical surface displacement is on the order of 312.4 mm, emphasizing the need to provide adequate freeboard to avoid damage to the roof structure. The damage caused by surface waves on roof framing elements and covers has been observed, and supports the usefulness of the model as a predictor of response. There is very limited effect due to the vertical acceleration alone.

Base Shear Time History. Figure 3 summarizes total base shear time history for the combined loading case and the horizontal only acceleration. A maximum value of about 10,044 kN occurs at 5.8 seconds. Results are 2 to 2.5 times the values calculated by the AWWA design code, which is expected. Design standards reduce forces by factors of 3.5 to 4 to obtain the design forces. Tanks, therefore, respond in a nonlinear fashion when subjected to design ground shaking and experience some damage. The time history show that for the first few seconds, the average periods and magnitudes coincide, while the vertical acceleration gradually amplifies the base shear up to 10 percent higher than the horizontal acceleration only. After several seconds, the vertical acceleration significantly amplifies the base shear. After about 4 seconds, the combined horizontal and vertical acceleration base shear is less than the resulting horizontal acceleration results alone, possibly due to a decrease in overall stiffness after yielding under the combined loading condition and phase differences between two components causing canceled out effects.

Large Radial Deformation. The results of the analysis shows the maximum inward horizontal displacement of 164.21 mm at shell wall node 2944, which is located 11-1/4° counterclockwise on the right edge at a distance of 1,706.9 mm from the water surface, at time 4.375 seconds. Photographs showing permanent deformation at the high water level in a water storage tank in California have been documented. The comparison of radial displacement time histories for both the vertical only, horizontal only, and combined for Node 2355 located at a distance of 36 cm from the base is shown in Figure 6.

Plastic Strain Development. The plastic hoop strain contour at time equal to 6.392 second indicate that yield strain are reached or exceed. As noted, the principal strains in a large area of the tank surface are greater than hoop strains. The permanent deformation is evident.

Hoop Stress Distribution. The results of the analysis show the hoop stress contour distribution on the shell wall at times of 4.2 seconds has a maximum hoop stress 2.68 x 10^8 N/m^2. Results show locations where maximum stress occur under the combined horizontal and vertical ground acceleration. Figure 5 shows the hoop stress time histories for the combined and individual horizontal and vertical ground accelerations at selected nodes. Node 2359 is located at 853.4 mm from the tank base. Stresses due to the horizontal component only and the vertical only are both significant, however time phases do not coincide, and in fact, maximum values for each do not coincide. The maximum tensile hoop stress due to combined loading is 249,590 kPa, due to the horizontal only is 280,961 kPa, and due to the vertical only is 229,113 kPa. The maximum compressive hoop stress due to the combined loading is 167,543 kPa, due to the horizontal only is 170,024 kPa, and due to the vertical only is 68,120 kPa. Figure 6 shows the hoop stress time history for node 2358, which is located on the left wall at 1,706.9 mm above the tank base. Maximum values for combined, horizontal only, and vertical only are 285,167 kPa, 276,893 kPa, and 201,327 kPa respectively. As previously noted, yield stress for the model is 248,211 kPa.

Similar analysis using an added mass snow load increases the hoop stress to 40.5 ksi, or a 4.3% increase, occurring at time 7.097 seconds. However, the initial yield occurs as early as at time 0.34 sec.

Axial Stress Distribution Analysis results show axial stress contour distribution on the shell wall at time of 4.491 seconds with the maximum tensile stress of 104,842 kPa on the right side of the tank. The axial stress contour distribution on the shell wall at time of 6.195 seconds with a maximum compressive stress of 114,936 kPa occurs on the right side of the tank.

Water Pressure Distribution. The pressure distribution along the wall height at times of maximum for the combined, horizontal and vertical only ground accelerations were computed. The maximum value for the combined loading occurs at 4.4 seconds, with a value of 238,903 kPa, the time for the horizontal only occurs at 1.06 seconds with a value of 234,422 kPa, and for the vertical only at 6.315 seconds with a value of 153,684 kPa. The occurrence of the maximum pressure for the horizontal acceleration only relatively early in the analysis is noted. The vertical pressure distribution is located on the left side of the tank shows the relative high-pressure at a distance of 1,829 mm above the base of the tank.

Comparison of Results

The effects of snow loads are relatively insignificant in the case under consideration. The following comparison results focus on tanks without snow.

A comparison of the time history of water surface profiles for the combined loading and horizontal loading show very little difference, which is expected due to the natural frequencies of water for the vertical directions. The maximum values of the various results for the combined, the horizontal only and vertical only occur at different times during the analysis. Table 4 summarizes the maximum values and also shows results from the code based calculation for base shear, maximum water surface elevation, and resulting stresses in the vertical and horizontal direction (axial and hoop stresses). For this case, the maximum water surface elevation is less than the code based calculation, but all other parameters are greater, including the base shear, axial stress, and horizontal or hoop stress. Presented results are based on the typical fluid/tank system under the selected segment of synthetic ground motion. The results may vary significantly from case to case. Due to the nonlinear nature of large deformation and path dependence, slight changes in the analysis model may induce unpredictable results.

Conclusions

Results of this study show how the assumptions and material properties selected have a large effect on the modeling outputs. In particular, values of displacement, base shear, and material stresses show substantial variation. The ratio of base shears between nonlinear assumptions and the code-based elastic development is about 3.2 for the combined horizontal and vertical load, and 4.0 for the horizontal load only. This indicates that tanks responding in a nonlinear manner when subjected to scaled design ground motions will probably experience structural damage or permanent deformation. Nonlinear time history analysis under snow load conditions has a small effect on this analysis, but whether this is the case under other geometric tank combinations and different ground excitation inputs will require additional parametric study

Table 4. Comparison of model results

	Combined	**Horizontal Only**	**Vertical Only**	**AWWA D100**
Maximum surface wave height, mm	311.7 (up) 320.6 (down)	329.2 (up) 247.1(down)	3.1 (up) 12.6(down)	800.1
Maximum shell horizontal displacement, mm	164.2 (inward) 60.0 (outward)	113.4 (in) 57.6 (out)	16.8 (in) 6.0 (out)	
Maximum base shear, kN	9,341.3	11,565.4	0	4068.3
Maximum axial stress, kPa	107,827 (T) 127,208 (C)	113,198(T) 120,596(C)	28,151 (T) 27,889 (C)	12,596
Maximum hoop stress, kPa	267,661 (T) 131,393 (C)	277,955 (T) 121,037 (C)	229,244 (T) 49,987 (C)	125,202

Tank response under the combined action of both horizontal and vertical excitations is calculated and compared with responses from each individual excitation. The results show the relative importance of the vertical component of ground acceleration on the overall seismic behavior of the fluid/tank system. The vertical excitation can be transmitted into pulsations of the tank wall resulting in additional stresses and deformation. Most current seismic design codes for ground-based cylindrical tanks neglect the effect of vertical ground acceleration.

Refinements in standard procedures for estimating site-specific earthquake conditions will permit performance-based design of tanks and other lifeline structures. Future research will also permit a greater understanding of the response of structures under various seismic loadings, and permit resources to be directed to ensure that the structures continue to function after large earthquake events.

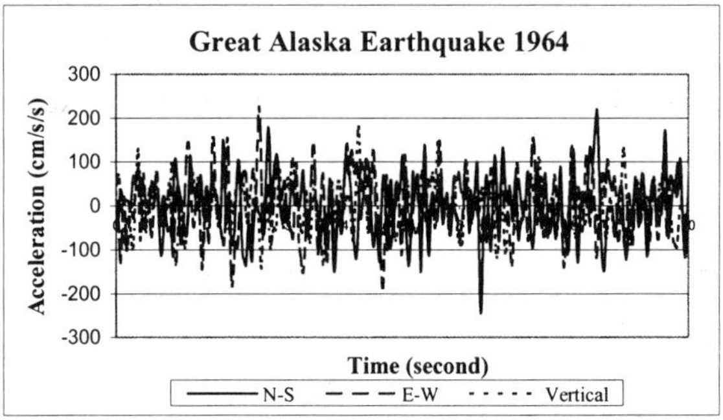

Figure 1. Scaled horizontal and vertical time history, 1964 Alaska Earthquake

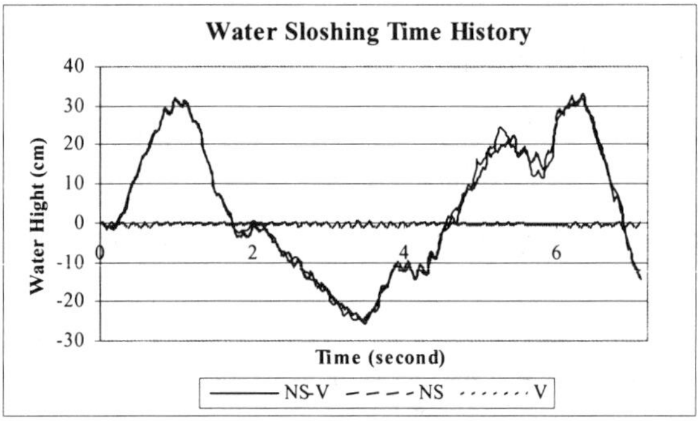

Figure 2. Water, vertical distance UZ at node 123, upper left water surface

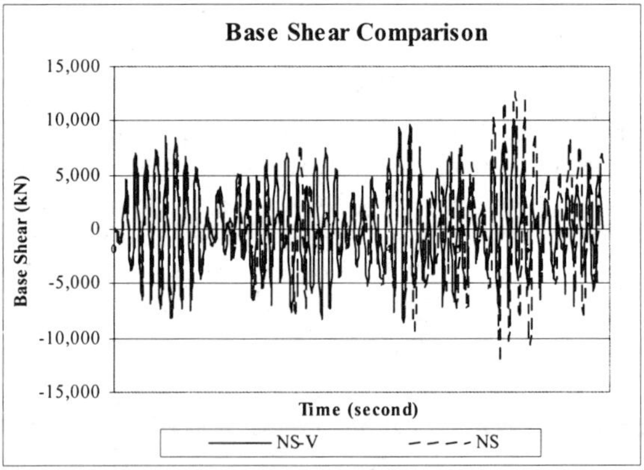

Figure 3. Base shear comparison – combined and horizontal only

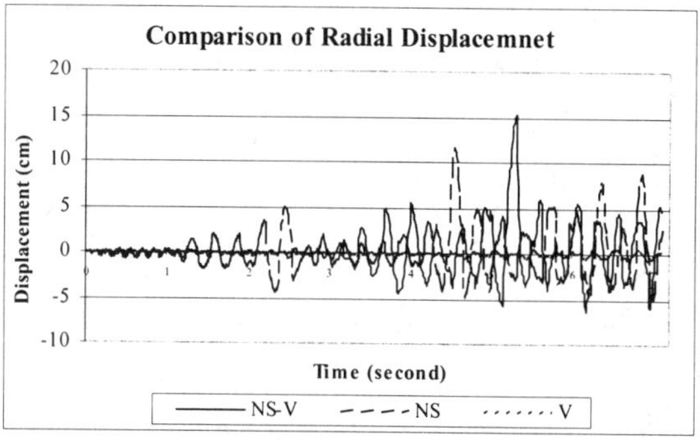

Figure 4. Radial displacement time history at node 2355, 36 cm from base

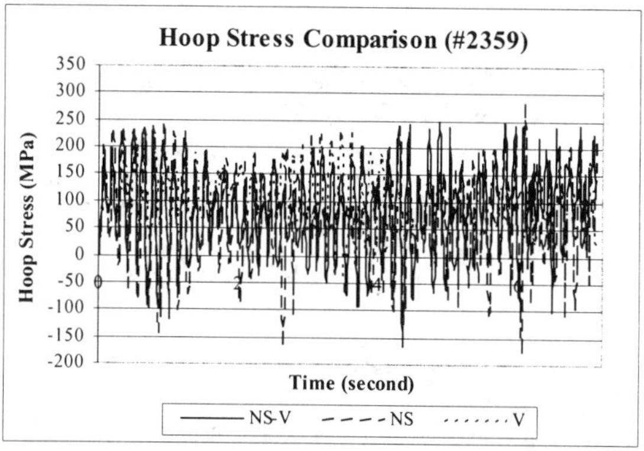

Figure 5. Hoop stress time history at node 2359

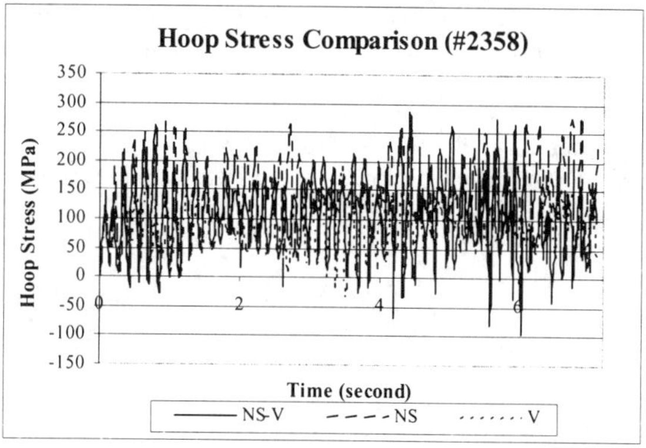

Figure 6. Hoop stress time history at node 2358

References

ANSYS/Multiphysics (2001) is a trademark of SAS IP, Inc., a wholly owned subsidiary of ANSYS, Inc.

ASCE Technical Council on Lifeline Earthquake Engineering, Monograph No. 7, (1995). *Critical Issues and State-of-the-Art in Lifeline Earthquake Engineering*, ASCE.

AWWA Standard for Welded Steel Tanks for Water Storage D100-96, American Water Works Association, Colorado.

Haroun, M. A (1991). *Implications of Observed Earthquake-induced Damage on Seismic Codes and Standards of Tanks*, PVP (223) Fluid-structure Vibration and Sloshing, ASME.

Housner, G.W. (1963). *The Dynamic Behavior of Water Tanks*, Bulletin of Seismological Society of America, 53 (2) 381-387.

Lau, D. T, Tang, A., and Pierre, J., (1995). *Performance of Lifelines During the 1994 Northridge Earthquake*, Can. Jour of Civil Engr 22, 438-451.

Malhotra, P. (2000). *Practical Nonlinear Seismic Analysis of Tanks*, Earthquake Spectra 16 (2), 473-492.

Papageorgiou, A.S., Zhang, B., and Biswas, N.N. (1997) *Estimation of Strong Ground Motion from the Great 1964 Prince William Sound Earthquake (Mw=9.2)*, Seismological Research Letters 68 (2), 325.

Veletsos, A. S. (1984). *Seismic Response and Design of Liquid Storage Tanks*, in Guidelines for the Seismic Design of Oil and Gas Pipeline Systems, ASCE, 255-370, 443-460.

Wozniak, R.S., and Mitchell, W.W., *Basis of Seismic Design Provisions for Welded Steel Oil Storage Tanks*, Proc. Refining Dept, Wash. D.C., API 57, 485-501.

Dynamic Analysis and Design of a Remote Monitoring Tower Subjected To Avalanche Loading

Simon A. Evans[1], He Liu[2], M. ASCE, Michael S. Gaulke[3]

Abstract

Remote Monitoring of environmental conditions has become an increasingly popular means of data collection in part due to the advancements of wireless communication equipment. Recently, remote monitoring stations have been used to collect snowpack and weather data, and to detect avalanches along transportation corridors and other areas of interest. The use of remote monitoring stations in major slide paths have helped transportation maintenance crews predict snow stability, determine when to perform avalanche control activities, and detect when a slide has occurred.

Certain data collection activities require placement of sensors mounted on a monitoring tower, including Doppler radar, directly in or near the avalanche slide path. This will, in some cases, subject the tower to dynamic avalanche forces, snowpack creep forces, as well as other adverse conditions typical of mountainous terrain of cold weather regions. The design of a monitoring tower to resist these forces presents several engineering challenges. This paper reviews some of the advances in avalanche prediction and detection, and examines the dynamic response of a monitoring tower due to avalanche loading conditions using the Finite Element Analysis approach, including time history analyses. Based on the analysis results, suggestions for future work on avalanche loading and the design criteria of remote monitoring towers are presented. A discussion will be given on the dynamic amplification of the tower system as compared to equivalent static loading, and conclusions will be made on the application of monitoring towers in the cold regions.

Introduction

During the second half of the twentieth century, there has been a significant

---

[1]Graduate student, Civil Engineer, Phone: (907) 230-0310, E-mail: simonevansusa@netscape.net
[2]Associate Professor, Civil Engineering Department, University of Alaska Anchorage, Alaska. Phone: (907) 786-1971, Email: afhl@uaa.alaska.edu
[3]Graduate student, Structural Engineer, U.S. Army Corps of Engineers, Alaska District, E-mail: mikegau@hotmail.com

increase in the study of snow avalanche dynamics, snow mechanics, and avalanche diversion, and warning and detection systems. This increase of interest can be attributed to a rise in accidents related to growing populations and development in mountainous areas, and implementation of transportation systems that traverse, mountainous terrain. In Switzerland, an average of 25 people die each year due to avalanches. Since the winter of 1951, when avalanches killed over one hundred people, the Swiss Federal Institute for Snow and Avalanche Research (SLF) has intensified its research of avalanche dynamics (Bartelt & Buser, 2001). Recently, within in the last 20 years, advances in avalanche research, warning, and detection systems have increased dramatically due to technological advances in data collection and processing. The use of remote monitoring stations, which download real time data, provide meteorological data, and can monitor snowpack conditions and detect avalanche activity. Information such as wind speed, air humidity and temperature, ground and snow temperatures, snow height and settlement of the snowpack, as well as avalanche flow speeds can be measured. This information can then be used to predict avalanche occurrence probabilities and project avalanche run out distances (Gubler, 1996).

Technological advances of communications and data collection devices have helped provide engineers with more reliable data to aid in the prediction of avalanches, and the prediction of forces produced by avalanches. For example, Labine (1996) discusses the evolution of remote snow depth monitoring and discusses how the measurement of snow depth has become more reliable under variable conditions. Recent impact pressure studies performed by the Norwegian Geotechnical Institute have used load cells to collect impact pressures as a function of time (Kristensen & Larsen, 1996). This enables a more realistic view of the dynamic behavior of avalanche impact, and allows for the time history analysis of the dynamic response of a structure due to avalanche loading.

There is a great need for further study of dynamic impact of debris flows and avalanches on structures. There is only a limited amount of literature available on these subjects (Armanini, 1997b). Avalanche loading on structures has been treated as a function of release volumes and flow densities (Mears, 1997). Engineering approaches to avalanche dynamic analysis has used the Voellmy model of fluid mechanics, which treats the flowing mass as a fluid, and determines the maximum velocity that an avalanche will reach (Mears, 2001a). Impact pressures can then be estimated based on the computed velocities and estimated fluid densities.

Recent Research and Development in Avalanche Dynamics

The development of numerous analyses for the dynamic behavior of avalanches has greatly affected the understanding of snow avalanche run out and the forces produced by avalanches. Early developments used for determining maximum run out distances and impact pressures of snow avalanches involved predicting jointly the avalanche run out and dynamics, by selecting appropriate friction coefficients for the model. Avalanche models created using this approach involve many unknowns, such as boundary conditions and snow characteristics (McClung & Lied). Currently, avalanche modeling is being performed taking into account that real avalanches are three-dimensional. Different snow types and meteorological conditions are variables that help determine dynamics for a particular slide path. In Norway, a digital mapping system has been developed to show snow avalanche and rock fall hazards based on topographic maps of scale 1:50,000. The use of digital mapping has reduced some of the labor intensity associated with manual mapping (Toppe & Lied 1984). Another model developed in Japan after the Maseguchi avalanche in 1986, models the run out path of a powder-snow avalanche based on the slope topography (Fukushima & Hayakawa). Systems are currently being developed that use remote monitoring stations to alert a command center of avalanche activity. The command center is then able to make road closure recommendations, thus avoiding some of the danger associated with traffic congestion in active avalanche run out zones. The most recent systems are able to determine when a release has occurred, and can then give a warning of avalanche activity as the release occurs. This warning period can be critical to the evacuation of a run out zone.

Avalanche Zoning. Distinguishing different areas of avalanche danger is necessary in populated mountainous areas. Mapping of the hazard is essential in areas of development or potential development because planning locations of permanent facilities requires knowledge of the magnitude and characteristics of an avalanche event (Mears 2001b). For land use planning, Mears rated avalanche zones based on the level of hazard, high, moderate, low and hazard free. The hazard level is based on return period and magnitudes of historic avalanches. Local design codes for structures can then be based upon the level of hazard. Zoning is important in mountainous areas to prevent large-scale disasters and loss of human life.

Avalanche Prediction and Monitoring. Avalanche prediction is difficult for several reasons, but mostly because an avalanche is dependent upon numerous variables, which also carry a level of uncertainty. For example, an unpredicted change in weather such as a sudden increase in air temperature can cause avalanche activity without warning. For long-term predictions related to land use planning, prediction is based on historical records and statistical methods, which give return periods for a slide of a given magnitude (Mears 1992). Avalanche occurrence, has been shown to be related to magnitude of the slide for individual slide paths in the Chugiak Range of Alaska (Hamre & McCarty, 1997).

In the situation that historical records are not available, predictions must be based on observations, and experience. In some areas in Alaska, transportation routes are located along multiple slide paths, where few historical records exist. Avalanche activity during the winter of 1999-2000 caused road closures for several days and left the community of Girdwood without electricity for one week. The use of remote monitoring systems in these areas will allow for better prediction of avalanche occurrence on a real time basis, and will signal the occurrence of an avalanche event. The collection of this data will inform officials of the immediate conditions at a particular slide path. This data can be processed resulting in real time avalanche forecasting without exposing personnel to dangerous conditions, and will assist in making decisions regarding transportation route closures, and avalanche control work.

Remote avalanche warning and detection systems have improved avalanche awareness and safety. Remotely located stations collect data critical to the prediction of avalanche events and make it possible to recognize a release. Remote stations record and transmit data that allows for the direct measurement of snow accumulation, settling, water percolation, and fracture heights within an avalanche release zone (Gubler, 1996). Several new technologies have been used to remotely monitor snowpacks. Snow depth measurements have been performed using an acoustic (ultra sonic) sensor. The sensor determines the distance to a target by sending out ultrasonic sound pulses (Labine, 1997). A sudden change in snow depth can indicate the occurrence of an avalanche. The snow profile radar, which is a Doppler radar system, records snowpack data and automatically recognizes avalanche flow and closes traffic routes (Gubler, 1996).

Avalanche diversion. The use of avalanche diversion structures has been historically implemented in many areas of the world. Diversion of avalanches and debris flow can be divided into two categories, passive and active works (Armanini, 1997a). Active works act to prevent snow loading on a slope and increase slope stability, while passive works deflect the flow or reduce the energy and run out length of an avalanche. Some examples of passive works include avalanche shelters for road crossings, tunnels and mound systems. The Norwegian Geotechnical Institute has been involved in recent studies of how deflecting dams affect run-out distances. The effect of a retaining dam that was located in the slide path was observed for various snow types and avalanche magnitudes (Kristensen & Larsen, 1996). The use of a passive deflection system upslope of a remote monitoring station could possibly help reduce the avalanche loading subjected on the tower.

Impact Studies. Numerous avalanche impact studies have been conducted in the past. Current studies are collecting data that better represent the dynamic impact loading of an avalanche. Sensors are placed in the slide path, and data are recorded as a function of time and location. In the early 1970's, research was conducted at Rogers Pass, British Columbia, where information was recorded concerning the characteristics of avalanches and the loads produced by them on structures (Schaerer, 1973). This study involved the measuring of impact pressures by placing measuring equipment in a well-known avalanche path that receives more than ten avalanche events per year. The topology of the slide path produces avalanches that originate at around 2,100-meters (m) above sea level. The slide path from the release zone to the sensors is 770-m long over steep rocks and a snow filled gully, with an average incline of 33°.

The measuring equipment had two loading cells with surface areas of 645-mm^2 mounted on a 50.8-milimeter (mm) wide steel frame. The study considered the frame to be a minor obstruction, thus not affecting the pressure measurements. The characteristics of dry snow avalanches were described as a mixture of flowing and airborne powder when falling over steep irregular terrain. Visual observations were made of the avalanches during the study noting the flow height and avalanche front velocity. Particles in the avalanches were observed to range from 0.1-mm to 100-mm in diameter.

The recorded impact pressures were limited to the lower half of the dense, flowing portion of the avalanche. The design avalanche is representative of a dry-flowing avalanche with loading occurring over a very short time period, 5 seconds. Because of differences in the particle sizes in the avalanche debris and the steep irregular terrain in the starting zones, snow arrived at the principle gully of the slide path and the location of the sensors at different times, producing peak pressures which occur in several mass waves. Upon initial impact with the loading cells, the peak pressure lasted for approximately 1 second and then reoccurred approximately 2 seconds later. Between the peak pressures, significant pressure drops were recorded, and were likely the result of pressure produced by the light powdery debris typically observed in this type of avalanche.

For the purpose of calculating peak pressures from the observed data, Schaerer assumed that the moving snow was a compressible fluid, and that compaction of the snow and drag resistance with small obstacles such as the load cells and support frame are negligible. Calculation results were in close agreement with the peak pressures measured during the study. The peak pressure was represented by $p_{max} = V^2/2\rho$, which is based on the observed velocity of avalanche front and the density of the avalanche deposits. Further, the study found the average pressure was 30% of the peak pressure or $p_{avg} = 0.3p_{max}$. Schaerer concluded that for practical purposes the impact pressures for a dry snow avalanche could be calculated from the velocity of the avalanche front and the density of the deposited snow.

The avalanches observed during this study had relatively low velocities and were measured on relatively small loading surfaces. Some improvements for future studies were noted to include larger surfaces for recording impact pressures, the measurement of impact pressures from high-speed avalanches, and measurement of the pressure distribution over the full depth of the avalanche. The study concluded that a more uniform pressure with smaller variations of pressure would probably be observed by using larger surface areas for the load cells.

Recent studies have been conducted that investigate the forces on fixed structures from different types of avalanches, and throughout the entire avalanche depth. (Kristensen & Larsen, 1996, Schaer & Issler, 2000). With the new developments in the equipment pertinent to the collection of avalanche data, impact pressures are collected as a function of time and location with readings made up to every hundredth of a second. Measurements are recorded at multiple levels, thus giving a sense of how impact pressures are related to height of flow. These impact pressures can then be applied to a structure, and a dynamic response of a structure can be estimated.

Dynamic Analysis of a Structure Due to Avalanche Loading

Remote monitoring of a slide path for avalanche activity requires the placement of equipment in or near the slide path and if possible in the starting zone of the avalanche. Since some slide paths can occupy an entire mountain slope, in order to monitor for avalanches, the instrumentation must be placed where it could be subjected to avalanche loading during large slide events. For this reason, we have attempted to model the dynamic response of a monitoring tower subjected to avalanche loading, and compare the results to those from static equivalent loading. Our goal in this analysis is to demonstrate the potential of the approach, and is far from exhaustive and will need refinement as better data becomes available. Several simplifying assumptions are made in regards to avalanche loading, and are discussed later. In this paper, we used time history techniques to analyze the response of a structure subject to avalanche loading. The impact data used for this study are based on data collected at Vallee de la Sionne test site in Switzerland, and are merely an interpretation of the actual data collected. At the time of this report, actual data sets were not available (Schaer & Issler, 2000).

Project Description. The study involved analyzing the dynamic effects caused by avalanches on remote monitoring towers placed in avalanche zones. A monitoring equipment package will be mounted at the top of the tower and will consist of

GPS, Doppler radar, transmission antenna, wind gauge, etc. The monitoring tower for this study is modeled after actual towers located in the Chugach Mountains of Alaska, in the Peterson Creek slide path number 2, along the Seward Highway, south of Girdwood. These towers will be accessed remotely to monitor real time snowpack data and the occurrence of avalanches. The monitoring towers were constructed as part of the Alaska Railroad avalanche detection and control program. These towers were not designed to withstand avalanche forces, but were selectively placed to avoid avalanche loading. This study is being conducted for the case where a tower must be placed in the slide path, and thus exposed to avalanche loading.

The slide path has an inclination of 32° in the starting zone at an elevation of approximately 1,160-m. The slide path consists of a broad open slope with limited vegetation. The snowpack depth in the starting zone typically is 6.1-m deep. On a heavy snow year, the slide path typically produces dry-flowing slab avalanches that reach the highway, and is capable of producing large avalanches. For the purpose of our study, we assumed the tower is located near the runout zone, and the snowpack depth around the tower is 2-m.

Model Description. The monitoring tower modeled for this paper was modeled using SAP2000 (SAP2000, 2001), a Finite Element Analysis (FEA) program. The tower has a solid circular cross-section with a diameter of 0.6-m and is 7-m tall. The tower was divided into smaller elements to accurately model the dynamic effects of the avalanche impact. Finally, beam elements were used to model the tower.

The base of the tower was assumed fixed for modeling purposes. Since the model tower will be founded on bedrock and is a cantilever structure, the assumption of fully restrained foundation conditions is accurate. The tower was modeled using reinforced concrete with the following properties: 28-day compressive strength, f'_c, of 27.5 MPa, reinforcing steel yielding strength, f_y, of 415 MPa and a density, ρ_c, of 23.6-kN/m^3.

The loading for the tower consisted of a dynamic component produced by the avalanche load. Snowpack creep loading was assumed to be negligible in comparison to loading produced by an avalanche, and thus was not used in the dynamic analysis. The axial loading of the tower was insignificant compared to the avalanche loading and thus second order effects (P-Delta) analysis of the tower was not required. To compare the usefulness of the dynamic analysis approach, an additional analysis was performed using a static equivalent load distributed uniformly over the monitoring tower. The static equivalent avalanche loading was taken as 30% of the peak pressure. This value was chosen based on the observation that the average pressure of avalanche loading was noted to be about 30% of the peak load during the avalanche time interval due to variations in particle size and specific weight (Schaerer, 1973). The static equivalent avalanche loading was applied uniformly to the tower from the top of the snowpack, assumed to be 2.0-m, to the top of the tower.

The dynamic avalanche loading consisted of applying the avalanche impact pressures as a function of time. The time history of the avalanche pressure was derived by a direct interpretation of a 5-second interval of the peak load and frequency content values from data collected by Schaer and Issler (Schaer, M., and Issler, D. 2000). These values were then reduced for the following 20 seconds to represent the reduction of peak values as the avalanche passes. The derived avalanche time history is shown in Figure 1.

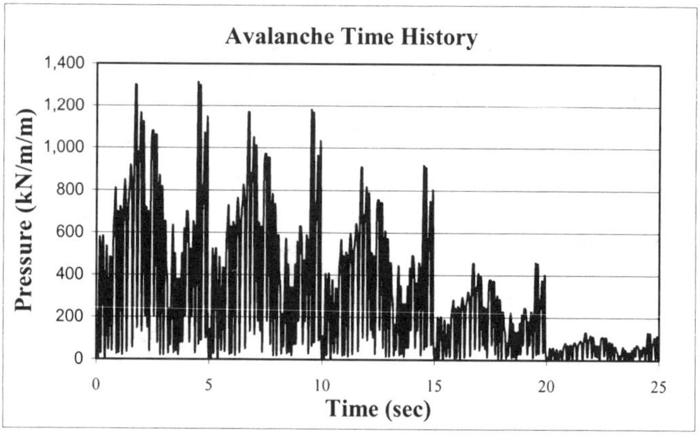

Figure 1. Derived avalanche time history

In the FEA model, the dynamic avalanche loading was represented as a spatially-varying load by a parabolic shape along the height of the avalanche's zone of influence on the tower and a time varying function, with the maximum load occurring at the snow surface. This avalanche loading assumption does not consider effects of friction of the snowpack that would slightly decrease the impact pressures near the snow surface. The frictional effects are ignored for simplicity of calculation, but should be considered in studies of a more comprehensive nature. The assumption of a parabolic loading scheme for the avalanche front is based on the similarity solutions, where the moving and deforming pile of granular material has a parabolic cap shape (Hutter and Nohguchi, 1989). Studies have shown that impact pressures and particle size in an avalanche decrease with height. Dry-snow avalanches can have a three-layered structure, a dense flow on the bottom, a transitional or saltation layer, and a suspension layer on top (Schaer and Issler, 2000). The loading for this study is modeled for the dense flow and transitional layers of the avalanche. Since impact pressures decrease with the height of the avalanche, a parabolic loading is a fair assumption. This assumed loading is for the purpose of comparing a dynamic load with a static equivalent load of the average peak pressures, and is a simplified model of the complicated loading processes of an actual avalanche. This loading function should be refined in future studies when more accurate impact loading data are available, and a better shape function is defined. The avalanche height was assumed to be 8-m, so the parabolic loading on the tower was applied as maximum peak pressure at the top of the snowpack, and as approximately 60% of peak pressure at the top of the tower. The load function is shown in Figure 2.

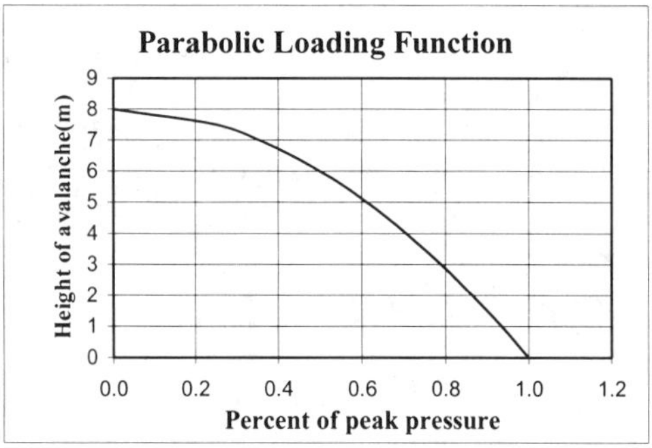

Figure 2. Parabolic loading function, the tower receives 100% of the peak pressure at the top of the snowpack and approximately 60% of the peak pressure at the top of the tower, or at the 5-m height of the avalanche.

Ritz vectors were used to calculate the modal response of the tower. Review of the calculated modes indicated that only the first two modes were significant. The first mode had slightly less than 90% of the mass participating and with the addition of the second mode slightly more than 99% of the mass was participating. Table 1 shows the period, frequency and modal participating mass ratios for the modeled tower.

Table 1. Modal response.

Mode	Period (second)	Frequency (Hz)	Modal Participating Mass Ratios (%)
1	9.53	0.105	89.84
2	1.31	0.764	9.23

In the FEA model, the load component was applied as joint loads. Two primary load combinations were used to determine the effects of the avalanche loading on the tower. The first primary load combination consisted of the self-weight of the tower, the monitoring equipment package dead load, and the static equivalent avalanche loading. The second primary load considered the dynamic avalanche loading only

Analysis Results. Results from this analysis show that the base shear, base bending moment, and the displacement at the top of the tower were increased significantly when subjected to time history dynamic loading, as compared to the static equivalent loading which was based on 30% of peak impact pressure. Table 2 shows a comparison of the two primary load combinations effect on the base shear, base bending moment, and the displacement at the top of the tower. The application of a static equivalent load as a higher percentage of the peak load would obviously decrease the percent difference between the dynamic and static response results. Further studies on avalanche impact loading are necessary to determine if there is a more appropriate static equivalent loading for this type of analysis. The results of this study indicate that a dynamic analysis proves very useful for the design of structures subjected to avalanche loading.

Table 2. Comparison of dynamic avalanche response to static equivalent response.

	Dynamic Avalanche Response	Static + Static Equivalent Avalanche Response	Percent Difference
Base Shear	2,100 KN	980 KN	53%
Base Moment	8,400 KN-m	4,390 KN-m	48%
Top Displacement	0.58 m	0.36 m	38%

Conclusion

The use of remote monitoring stations for the collection of environmental and snowpack information in avalanche zones of mountainous areas of cold regions has many benefits over traditional methods. The largest benefit is that snow safety and avalanche personnel are not exposed as often to dangerous environmental conditions. Technological advances in data collection and communication devices have allowed for the development of real-time analysis of avalanche conditions. Current monitoring systems can detect the occurrence of an avalanche and automatically signal transportation route closures in areas that are exposed to the slide path.

The results of this study indicate that the dynamic avalanche loading does have a significant affect on the design of the monitoring tower. The large base shear, base bending moment, and deflections of the tower could possibly be reduced by optimization of the cross-section, as well as other innovative designs or materials that will help reduce the bending forces and deflections. Also, for future studies of this nature and in the design process for a specific structure, it would be advantageous to analyze avalanche loads typical of all types of avalanches expected for a particular slide path.

An appropriate engineered design of a monitoring tower or other structures exposed to avalanches can be achieved by a combination of analysis of all probable loading schemes, and the possible use of a load reduction or deflection device. The conclusions made by this study indicate that the dynamic loading of an avalanche should be considered in the design of a structure exposed to such loading conditions. The use of remote monitoring stations for avalanche prediction and detection will likely become common within the next decade and is certainly an appropriate means of gathering environmental information in the harsh arctic setting.

References

Armanini, A. (1997a), "Chapter 3 p 163: Control Measures for Debris Flows." *Recent Developments on Debris Flows*, Springer-Verlag Berlin Heidelberg 1997, Germany.

Armanini, A. (1997b), "Chapter 3 p 208: On the Dynamic Impact of Debris Flows." *Recent Developments on Debris Flows*, Springer-Verlag Berlin Heidelberg 1997, Germany.

Bartelt, P., and Buser, O. (2001), "Avalanche Physics Ploughs Ahead." *Physics World p. 25-29, March 2001.*

Fukushima, Y., and Hayakawa, N. (1993), "Analysis of Powder-snow avalanches using three-dimensional topographic data." *International Glaciological Society, Annals of Glaciology 18, 1993.*

Gubler, H. (1996), "Remote Avalanche Warning, Alarm and Control Systems, Fundamentals, Applications and Experience." 1996 ISSW Proceedings, http://www.avalanche.org/~issw/96/art_37_.html.

Hamre, D., and McCarty, D. (1997), "Frequency/Magnitude Relationship of Avalanches in the Chugach Range, Alaska." 1996 ISSW Proceedings, http://www.avalanche.org/~issw/96/art53.html.

Hutter, K., and Nohguchi, Y. (1989), "Similarity Solutions for a Voellmy Model of Snow Avalanches with Finite Mass." *Acta Mechanica, Vol. 82 p. 99 (1990).*

Kristensen, K., and Larsen, J.O. (1996), "The Ryggfonn Avalanche Dynamics Project." http://www.avalanche.org/~issw/96/art54.html.

Labine, C. (1997), "Automatic Monitoring of Snow Depth." 1996 ISSW Proceedings, http://www.avalanche.org/~issw/96/art_39_.html.

McClung, D.M., and Lied, K. (1986), "Statistical and Geometrical Definition of Snow Avalanche Runout." *Cold Regions Science and Technology, Vol. 13 (1987) p. 107-119.*

Mears, A. (1992), "Snow Avalanche Hazard Analysis for Land Use Planning and Engineering." *Colorado Geological Survey Bullitin 49, p55, 1992.*

Mears, A. (1997), "Dry-Slab Density and Thickness During Major Storms." http://www.avalanche.org/~issw/96/art_19_.html

Mears, A. (2001a), "Avalanche Dynamics." http://www.avalanche.org/~moonstone/zoning/avalanche%20dynamics.htm

Mears, A. (2001b), "Avalanche Zoning." http://www.avalanche.org/~moonstone/zoning/AVALANCHE%20ZONING.htm

SAP2000 (2001). "Integrated Finite Element Analysis and Design of Structures", Version 7.41, Computers and Structures, Inc. Berkely, California.

Schaer, M., and Issler, D. (2000), "Particle densities, velocities and size distribution in large avalanches from impact-sensor measurements." *Annals of Glaciology Vol. 32, 2000.*

Schaerer, P. A. (1973), "Observations of Avalanche Impact Pressures." *Advances in North American Avalanche Technology: 1972 Symposium, USDA Forest Service General Technical Report RM-3, September 1973.*

Toppe, R, and Lied K. (1984), "Mapping of Snow Avalanche and Rockfall Hazard by Use of Computer Based Methods" *Norwegian Geotechnical Institute Publication No. 153, 1984.*

Building Pressurization in Extreme Cold Climates

H. Ed Bargar, P.E.[1]
Debendra K. Das, PhD, P.E.[2]

Abstract

Building pressurization, the differential pressure between the interior of a building and its exterior surroundings, is an important design consideration. Pressurization is the driving force in building infiltration/exfiltration. It also affects air flow within building zones. Improper calculation of pressurization can result in undersizing the building's heating and cooling systems, improper operation of air distribution systems, improper operation of elevators, and freezing and failure of water distribution and circulation systems.

Building pressurization is affected by several environmental and design considerations. Primary environmental influences are: wind speed/direction and exterior-to-interior temperature difference. Pressurization due to these two parameters is generally referred to as wind effect and buoyancy or stack effect respectively. In extreme cold climates, the predominant effect is of the buoyancy type. This effect is caused by the temperature, and thus air density, differences between the interior of the building and its exterior surroundings. The larger the temperature difference, the larger the buoyancy effect. In extreme cold climates, large temperature differences often occur at times when wind speed is negligible. Also of critical importance is that while wind effect is transitory, lasting from minutes to hours; extreme temperature differences can be persistent, lasting days, weeks, or longer.

The American Society of Heating, Refrigeration, and Air Conditioning Engineers (ASHRAE) presents an idealized equation to calculate the buoyancy effect. This paper compares differential pressure measurements from an actual building to this idealized model. It also presents several new statistical models based on the collected data. These new models should provide engineers with improved tools to properly account for building pressurization for designs in extreme cold climates.

Introduction

Building pressurization can be viewed as the pressure gradient between the inside and outside of a building or as the pressure gradient between different levels within a structure. These are related phenomenon. The American Society of Heating, Refrigeration, and Air Conditioning Engineers (ASHRAE, 1997) attributes the

---

[1] Instructor, Department of Mechanical Engineering, University of Alaska Fairbanks, Duckering 337, 306 Tanana Loop, Fairbanks, AK 99775-5905; bargarh@asme.org.
[2] Professor, Department of Mechanical Engineering, University of Alaska Fairbanks, Duckering 337, 306 Tanana Loop, Fairbanks, AK 99775-5905; ffdkd@uaf.edu.

pressure gradient to three driving forces: pressure differences across the building envelop created by wind; air density differences between inside and outside air; and operation of mechanical equipment such as combustion equipment and forced ventilation equipment.

As wind flows around a building, external surfaces on the windward side(s) are exposed to positive pressures as the wind stagnates. At the same time, the flow around the building's geometry creates eddies, or lower pressure regions on the leeward side(s). This is known as wind effect pressurization.

Judging the wind speed and direction for use in calculating the wind effect can be difficult. Wind flow in the vicinity of the building is effected by other structures and/or features (trees, outcroppings, etc.) near the site (Lee, et al. 1980). Wind information given in meteorological reports is the speed and direction recorded at a specific height, typically 10m (33 ft). Wind speed typically decreases down to zero at the ground surface and increases up to an elevation of 610m (2,000 ft). Thus it is seldom correct to use the meteorological wind reports directly when dealing with low-rise buildings. However, these wind values may provide acceptable average values when dealing with tall buildings in relatively exposed terrain (Tamura and Wilson, 1968). To try to improve estimation of wind effect pressurization, studies have been conducted for various building configurations and layouts (Swami and Chandra, 1988). From this type of research, equations have been generated to allow estimations given a building's specific geometry and site layout conditions. However, in critical applications, scale model wind tunnel testing may still be required to fully understand wind effect pressurization.

Pressure gradients caused by air density difference across the building envelop are known as the buoyancy or stack effect. Because buildings are not sealed structures, there are air flow paths that allow the exchange of air between the inside and outside of the structure. Buoyancy effect is greatest during the heating season when the temperature difference, and thus the air density difference, between the inside and outside of the building are greatest. Typically, warmer air rises within the building and exits leaving a lower pressure zone at the lower building levels. This causes the cooler, denser outside air to be drawn into the building where it is heated and the cycle continues. The buoyancy effect tends to create a neutral pressure level (NPL) within the building (Tamura and Wilson, 1966). Above the NPL, pressure in the building is positive with respect to (w.r.t) the outside; below the NPL, the building pressure is negative w.r.t the outside.

Flow of air and the location of the NPL is dependent on the flow paths or openings available for air exchange in the building envelop. Building partitions dividing the building horizontally and floor/ceiling assemblies dividing the building vertically also appreciably aid or hinder air flow. A very simplified model (Lee et al., 1985) that had no internal obstructions and only allowed air entry at the lowest level and exit at the highest level showed that fluid mechanics and ideal gas principles could be used to predict the stack effect. An improved model (Lee et al., 1988) showed that the results held even when internal vertical partitioning was introduced.

These simplified models do not realistically represent the air flow paths (doors, expansion joints, penetrations and cracks, etc.) that exist in a 'real world' building. However, ASHRAE presents an equation based on this research to estimate the maximum stack effect pressurization:

$$\Delta P = C_2(\rho_o - \rho_i)g(H - H_{NPL}) = C_2\rho_i g(H - H_{NPL})\left(\frac{(T_i - T_o)}{T_o}\right) \quad (1)$$

where:
ΔP = pressure difference [in. water column (WC)]
ρ = air density [lbm/ft^3]
g = gravitational acceleration, 32.2 ft/sec^2
H = height of observation [ft]
H_{NPL} = height of NPL [ft]
T = average absolute temperature [°R]
C_2 = unit conversion factor, 0.00598
i = subscript denoting indoor
o = subscript denoting outdoor

An improvement to this model accounting for internal partitioning adds a flow coefficient to Equation (1):

$$\Delta P = C_d C_2(\rho_o - \rho_i)g(H - H_{NPL}) = C_d C_2\rho_i g(H - H_{NPL})\left(\frac{(T_i - T_o)}{T_o}\right) \quad (2)$$

This flow coefficient, C_d, is a number between zero and one where one represents no partitioning. The value for a given building is estimated from 'experience'.

The final contributing factor to building pressurization is the operation of mechanical equipment. Appliances drawing their combustion air from within the building and exhaust fans tend to negatively pressurize the building w.r.t the outside. Building Code required outside air ventilation tends to positively pressurize the building. Fans and supply air and return air systems can locally effect pressurization from zone-to-zone within the structure. Care must be taken not to assume that bringing in large quantities of outside ventilation air will cause the entire building to be positively pressurized w.r.t the outside. This ventilation air will change the pressure gradient and NPL within the building but, at extreme cold temperatures, the lower levels will remain negative w.r.t the outside (Bargar and Das, 2001).

The Problem

In extreme cold climates the stack effect tends to dominate building pressurization. During extreme cold temperature periods, wind speed is generally negligible (the mere energy of the blowing wind will raise the ambient temperature). Also the pressure differential attributed to the stack effect is much greater in magnitude than that created by wind effect. And finally, wind effect is transitory while stack effect is more persistent. Extremely cold temperatures may last for days or weeks whereas the wind duration is generally measured in hours at most.

Improperly accounting for stack effect during building design can create costly problems. Infiltration of extremely cold air at levels below the NPL at best reduces productivity by presenting the occupants with uncomfortable conditions when the heating system is unable to keep up with the load. This infiltration can also cause water systems (plumbing, fire sprinkler, hydronic) to freeze and burst rendering the facility unusable until the damage is repaired. A high pressure gradient within the building can cause elevators to malfunction. High air quantities driven by the pressure gradient rush out the elevator doors at upper levels preventing them from closing. Thus the elevator is rendered inoperative.

Due to problems such as those mentioned above, a test building was selected to compare actual differential pressures across the building envelop with those estimated by the models represented in Equations (1) and (2). As noted, the equations are based on some gross simplifying assumptions that do not match actual building conditions. Also, the models were derived under temperate conditions, do they hold under extreme cold temperatures? One could surmise that the laws of fluid dynamics extrapolate to temperatures in the –40°C (-40°F) range. But do the air flow paths presented by the cracks, penetrations, and other building openings extrapolate linearly to the lower temperatures?

Experimental Setup

The Gruening Building on the University of Alaska Fairbanks (Fairbanks, Alaska, USA) campus was chosen as the test building. Gruening is one of the tallest buildings in interior Alaska and is regularly exposed to extremely cold outdoor temperatures for extended periods of time.

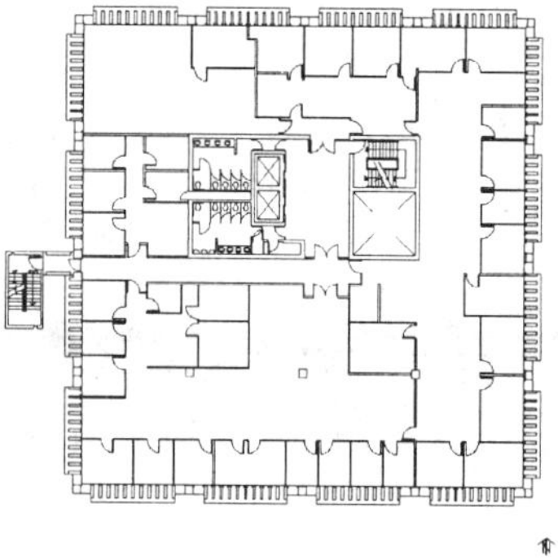

Figure 1. Gruening Building Typical Floor Plan

The Gruening Building is a 9,500 m^2 (102,263 ft^2) classroom/office facility built in 1973. The building's foot print is 36m x 36m (118 ft x 118 ft) oriented as shown in Figure 1. Exterior walls are of heavy reinforced concrete structural members with concrete infill walls. Additional concrete finish panels are hung on the structure for architectural effect. Windows are double-pane permanently fixed units. Double door vestibule entrances are present in each face of the ground level. There is a fire escape door on the west face at each level. Story separation is a 15 cm (6 in) concrete slab with suspended ceiling in the finished areas. Figure 1 shows a typical floor plan. The interior core partitioning at each level is of concrete construction while all other walls are stud and gypsum wallboard construction. Ground level is the third floor; the lower two levels are subterranean. The lower two levels are exposed to the ambient environment to some extent via the fire escape tower on the west face.

Building heat is obtained from campus central plant steam so there is no combustion equipment within the structure. Major mechanical equipment is located on the first level. The all air heating system is comprised of 44,500 L/s (94,300 cfm) supply and return fans which bring in 2,830 L/s (6,000 cfm) of outside ventilation air. Because the data was collected while the building was occupied and during extremely cold temperature it was impossible to shutdown the fan systems and stop the ventilation air system.

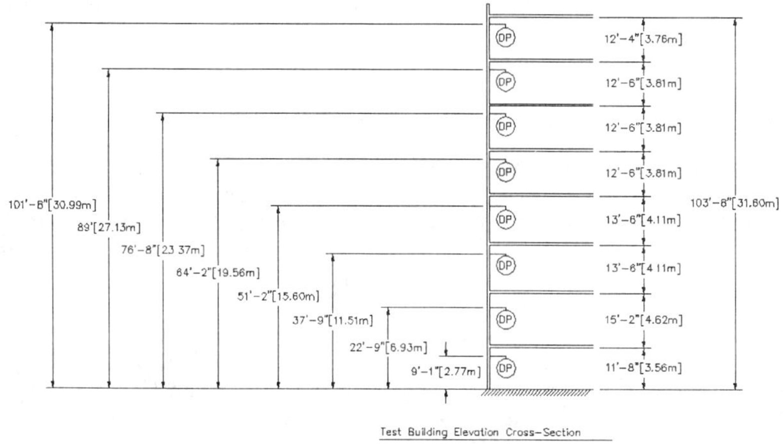

Figure 2. Gruening Differential Pressure (DP) Transmitter Installation

Differential pressure transmitters were installed at each level of the building (Fig. 2) to measure the pressure between the inside and outside of the building to a resolution of 0.6221 Pa (0.0025 in. WC). These transmitters were connected to the building direct digital control (DDC) system which logged the readings every 15 minutes. The DDC system also recorded the outside and inside air temperatures every 15 minutes. A meteorological station nearby on campus maintain by Los Alamos

National Laboratory (LANL) was used as a source of wind data. Date/time stamping of each recorded data point was used to coordinate the data from the disparate sources into synchronous 15 minute time bins. The wind data was used strictly as an elimination factor. Any data recorded while wind pressurization could significantly effect the differential pressure observed was removed from the dataset.

Data Analysis

The building pressurization data and temperature data was logged in the memory of the building's DDC system. This data was downloaded to a laptop computer every one to two days to keep from exceeding the memory capacity of the DDC system. The downloaded data was in raw ASCII text form and also included extraneous information such as command strings to the DDC system to institute the download and standard reporting information from the DDC system not required for data analysis.

The first task in analyzing this data was to import it into a spreadsheet and eliminate all information except for the date/time stamp, temperature reading, and pressure reading at each level. Next the data was reorganized into table format. Each record in the table contained a date/time stamp, 15-minute standardized time bin designation, outside temperature value, and differential pressure values for each of the eight building levels. This form of the data was then saved in a standard comma separated value (.csv) file.

Wind speed and direction data from the LANL instrument package was downloaded from their Web site on a monthly basis in raw ASCII text form. This data was imported to a spreadsheet and extraneous data was removed. For the first eight months of the study year (1999), the wind direction data was reported as north and east wind vectors; the remaining direction data was reported as an angle from north. Spreadsheet manipulation made it easy to convert all direction data to the angle from north form. A spreadsheet routine also converted the LANL instrument time base of Greenwich Mean Time (GMT) to the DDC time base of Alaska Standard Time (AST). The data was then saved to a .csv file with each record containing date/time stamp, 15-minute standardized time bin, wind speed, and wind direction fields.

The remainder of the data analysis was performed using S-Plus (Insightful, Inc.). The two data files (temperature & pressure, wind) were merged and the resulting data file contained only those 15-minute standardized bin records for which both temperature/pressure and wind data existed

The next step was to eliminate any records where wind speed might significantly effect building pressurization. The pressure sensor resolution was 0.6221 Pa (0.0025 in. WC), so any wind speed that would produce a wind effect pressurization equal to or greater than this magnitude resulted in that record being eliminated from the dataset. The ASHRAE Fundamentals Handbook (ASHRAE, 1993) gives the following equation to determine the wind pressure:

$$p_v = C_1 C_p \frac{\rho v^2}{2} \tag{3}$$

where: p_v = wind pressure effect [in. water column (WC)]
ρ = air density [Lbm/Ft^3]
v = wind speed [mph]
C_p = surface pressure coefficient
C_1 = unit conversion factor = 0.0129

The density of air is a function of temperature so a spreadsheet was constructed to calculate the wind pressure for temperature and wind speed ranges of interest to the study. The calculations showed that wind speeds less than 0.9 m/s (2 mph) could not produce a wind pressure large enough to be 'seen' by the pressure transmitters. The dataset was extremely large so a safety factor was incorporated into the elimination criteria. All records with wind speeds exceeding 0.7 m/s (1.5 mph) were removed from the dataset.

Next, the dataset was reconfigured so that each record contained the value of the standardized time stamp, the outside temperature reading, the elevation of the pressure sensor, and the differential pressure reading.

Lastly, the validity of each data point was checked. The Cook's Distance (Cook's-D) method (Jennrich, 1995) was used to evaluate each point. In this method, the data is fit to a statistical model while sequentially removing and then replacing each data point in the dataset. The Cook's-D is a measure of how much each individual point influenced the model fit. Any point with a Cook's-D greater than the test limit should be evaluated to determine if it is an erroneous point

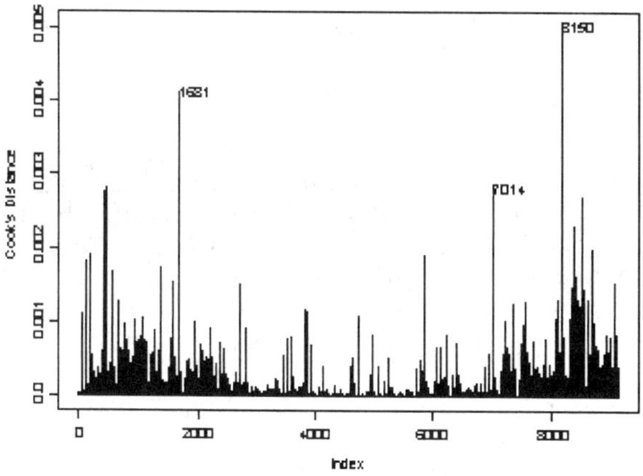

Figure 3. Cook's Distance Evaluation of Data Points

Figure 3 shows that three of the data points (labeled by their point number) were judged to be overly influential based on the Cook's-D test. Each of these points was investigated. It was determined that each appeared to be a bad sensor reading. They were substantially higher than the other pressure readings sharing common elevation and temperature characteristics. This could have been caused by an electrical fault in the pressure transmitter or DDC system or a localized pressure perturbation such as a door opening/closing at the exact time the questionable reading was recorded. These three points were eliminated from the dataset.

Statistical Models

An attempt was made to fit several statistical models to the data collected. The first model was a simple linear regression of the differential pressure as response variable versus the outside air temperature and height from mid-building height as the explanatory variables. The S-Plus linear regression fit for the data set produce the model shown in Equation (4). The t-value and p-value diagnostics showed the degree of fit certainty to exceed 99%.

$$P = 0.0525 + 0.0019h - 0.0006T \tag{4}$$

where: P = differential pressure [in. water column (WC)]
h = elevation from mid-building height [ft]
T = outside temperature [°F]

In creating the linear regression the assumption is made that the data fit a linear relation, the errors (difference between the response variable observed and its value calculated using the model's equation) are normally distributed, that the errors have a random distribution about zero (no error), and the errors have a uniform variance. One of these assumptions failed for the model in Equation (4). A plot of the residuals (errors) versus the fitted values showed a distinct curved relationship (Fig. 4). Therefore, the linear model may not be a good model to use because the assumptions upon which the model is based have been violated.

The model represented by Equation (1) suggests that there might be a nonlinear multiplicative relationship between differential pressure and the explanatory variables (outside temperature and elevation from mid-height). Therefore the data was fit to this type of model resulting in the fit shown in Equation (5). The diagnostics also showed this model to be a good statistical fit.

$$P = \left(4.7337 \times 10^{-2}\right) + \left(3.79001 \times 10^{-6}\right)hT \tag{5}$$

where: P = differential pressure [in. water column (WC)]
h = elevation from mid-building height [ft]
T = outside temperature [°R]

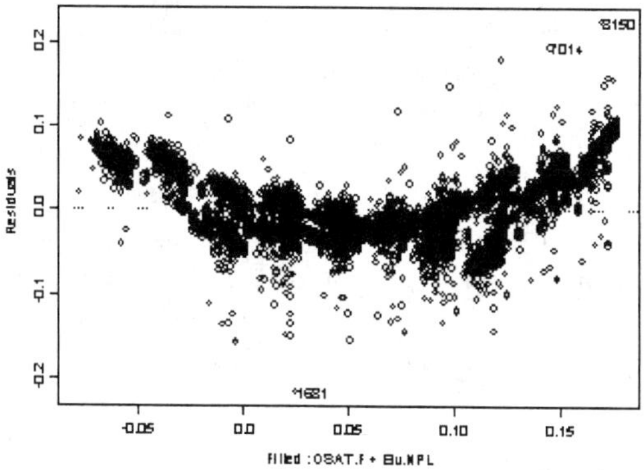

Figure 4. Linear Model Residual vs. Fit Plot

It is important that the absolute temperature scale be used in this equation. Otherwise there is a discontinuity in the model when the temperature transits zero. The model assumptions for this nonlinear model are similar to those for the linear model. The residual vs. fit plot (Fig. 5) shows an even distribution about zero.

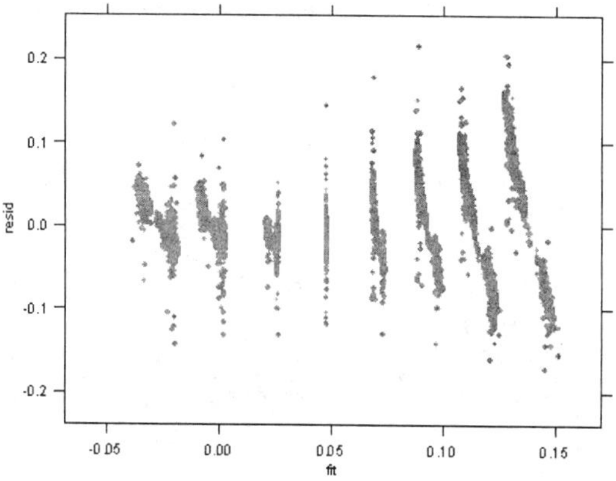

Figure 5. Nonlinear Multiplicative Model Residual vs. Fit Plot

The 'banding' effect is interesting and can be explained. In the dataset, the temperature varies uniformly over its range but the elevation values are at eight discrete values (one for each floor). The eight bands in the plot correspond to these eight discrete elevation values. The variance of the residuals is greater toward the right of the plot. These bands represent the higher levels in the building. Recalling that the two lowest levels are below ground, it is expected that variance at these levels would be less. There are fewer air flow paths and those that exist are not subject to as great a temperature extreme as the above ground air paths. This means variation in air flow and thus differential pressure should be lower at the lower levels as the plot suggests.

Previous research (Lee et al., 1985) suggests that the differential pressure displays an inverse relationship to the absolute temperature. This leads to the last model fit.

$$P = (0.0460443) + (0.935475)\left(\frac{h}{T}\right) \tag{6}$$

where: P = pressure difference [in. water column (WC)]
h = elevation from NPL [ft]
T = outside temperature [°R]

The residual vs. fit plot for this model appears similar to Fig. 5. Thus indicating that the assumptions upon which the model is based have not been violated.

Model and Observed Data Comparison

The last task is to compare the newly created models, Equations (4), (5), & (6), the original models, Equations (1) & (2), to the data observed at the test building.

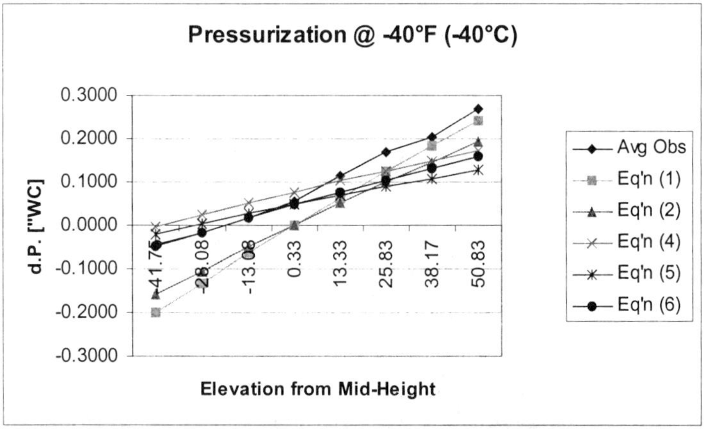

Figure 6. Comparison of Models and Observed Data

The plots shown in Fig. 6 show that the models represented by Equations (1) & (2) do not match the observed data very well. Both over estimate the negative pressure at the lower levels and under estimate the positive pressure at the upper levels. This might be expected since these models are based on assumptions that bear little resemblance to a 'real world' building. Two extenuating circumstances may also help to explain the great differences between values predicted by Equations (1) & (2) and those observed at the test building. The two lowest levels of the test building are below grade. This means that air flow across the envelop at these levels is more restricted than would be expected if they were above ground. This will tend to lower the NPL. Also, excess ventilation was being introduced into the test building while the data was collected. This will tend to positively pressurize the structure, thus lowering the NPL. The lowering of the NPL can also be looked upon as causing the negative pressure w.r.t the outside to be less negative. These two 'problems' with the test building explain only some of the lack of fit between the predicting equations and the observed data.

The statistical models do a better job of fitting the observed data, especially in the lower levels. This is to be expected since they are based on the specific data from the test buildings. However, additional data from additional buildings should result in models applicable to a general building which predict the differential pressure across the envelop better than Equations (1) & (2) since the grossly simplifying assumption have been eliminated.

It is interesting to note that although the linear model represented by Equation (4) is statistically invalid (it violates the assumptions for a linear model), it matches the observed data fairly well.

Conclusions

ASHRAE states that the model represented by Equation (1) results in the maximum pressure differential to be expected. This research supports this fact. This may be viewed as a safety factor when calculating the heating load due to infiltration. However, it may result in greatly oversized equipment.

The fact that the ventilation air could not be shut down while gathering the data confirms the fact that even with large amounts of excess air introduced into the structure, the stack effect will still produce negative pressures w.r.t the outside at the lower levels of a buildings. The designer must include these as zone loads in the heating system design.

The statistical models provide additional tools for the designer in predicting building pressurization. This should help to provide better designs for extremely cold climates.

It is hoped that future research in this area can expand the observed data to many buildings of varying heights and construction types. This should broaden confidence in the statistical models. Also, investigation into the effect of outside air ventilation on total building pressurization and the positioning of the NPL would provide more tools to the designer.

References

American Society of Heating, Refrigeration, and Air Conditioning Engineers. *1993 ASHRAE Handbook – Fundamentals, Inch-Pound Edition*, American Society of Heating, Refrigeration, and Air Conditioning Engineers, Inc., 1791 Tullie Circle, N.E., Atlanta, GA 30329

American Society of Heating, Refrigeration, and Air Conditioning Engineers. *1997 ASHRAE Handbook – Fundamentals, Inch-Pound Edition*, American Society of Heating, Refrigeration, and Air Conditioning Engineers, Inc., 1791 Tullie Circle, N.E., Atlanta, GA 30329

Bargar, H.E. & Das, D.K. (2001). "Stack Effect Infiltration in Extreme Cold Climates". *Proceedings, 52nd Arctic Science Conference*, University of Alaska, Anchorage, AK.

Jennrich, Robert I. *An Introduction to Computational Statistics Regression Analysis*, 1995, Prentice-Hall, Inc., Englewood Cliffs, New Jersey.

Lee, B.E., Hussain M., & Soliman, B. (1980). "Predicting Natural Ventilation Forces upon Low-Rise Buildings" *ASHRAE Transactions* 22(2):35-39.

Lee, K.H., Lee, Y., & Tanaka, H. (1985). "Thermal Effect on Pressure Distribution in Simulated High-Rise Buildings – Experiment and Analysis" *ASHRAE Transactions* 91(2):530-44.

Lee, K.H., Tanka, H. & Lee, Y. (1988). "Thermally Induced Pressure Distribution in Simulated Tall Buildings with Floor Partitions" *ASHRAE Transactions* 94(1):228-42.

Mathsoft, Inc. (Insightful, Inc.). (1997). *S-Plus User's Guide, Version 4.0*, Seattle, WA.

Swami, M.V., & Chandra S. (1988). "Correlations for Pressure Distribution on Buildings and Calculation of Natural-Ventilation Air Flow" *ASHRAE Transactions* 94(1):243-66.

Tamura, G.T. & Wilson, A.G. (1966), "Pressure Differences for a Nine-Story Building as a Result of Chimney Effect and Ventilation System Operation" *ASHRAE Transactions* 72(1):180.

Tamura, G.T. & Wilson, A.G. (1968). "Pressure Differences Caused by Wind on Two Tall Buildings" *ASHRAE Transactions* 74(2):170.

New Developments in Cold-Weather Concreting

Charles Korhonen[1], M.ASCE

Abstract

Cold weather inhibits construction productivity in the northern tier of the United States. Over the years the U.S. Army Engineer Research and Development Center's Cold Regions Research and Engineering Laboratory (CRREL) has developed formulations for cold-weather concrete that facilitate economical placement of concrete at mix temperatures well below freezing. Like many other technological advances, the commercialization of this technology has effectively been blocked by a lack of standards. CRREL is working with several State Transportation Departments to develop a near-term cold-weather concreting capability using off-the-shelf technology. Paralleling this effort, the Civil Engineering Research Foundation is spearheading an initiative to develop an acceptance standard for low-temperature admixtures. The world of cold-weather concreting promises to become a bit friendlier within the next few years and beyond.

Introduction

Cold weather places serious constraints on concrete construction. As temperatures drop, concrete sets more slowly, takes longer to finish, and gains strength less rapidly. If temperatures dip too low, the mix water may freeze and the final product will be irreparably damaged. Because there are times when work must continue despite the weather, the only option available today is to thermally protect the concrete while it cures. This option, though meeting construction's needs, is resource-intensive (labor, time, equipment, and budget). What's needed is a concrete that can be placed and cured in subfreezing weather without thermal protection.

The Cold Regions Research and Engineering Laboratory (CRREL) has made significant progress in developing a technology that will facilitate economical placement of portland cement concrete when its internal temperature is well below the freezing point of water. This is made possible by using chemicals that allow

---

[1] Research Civil Engineer, U.S. Army Cold Regions Research and Engineering Laboratory, 72 Lyme Road, Hanover, New Hampshire 03755-1290.
E-mail: korhonen@crrel.usace.army.mil

concrete to develop strength while it is cold—an impossibility for normal concrete. However, because using chemicals to protect concrete against freezing is not supported by acceptance standards in the United States, little has been accomplished to move this technology into general practice. Recently, however, interest from several northern states' departments of transportation (DoT), the Civil Engineering Research Foundation (CERF), and the U.S. Department of Defense (DoD) has rekindled the effort.

This paper reviews today's cold-weather concreting practice, describes the emerging technology, and discusses the work leading up to current efforts to move this technology into the marketplace.

Current Practice

Current cold-weather concreting practices have remained unchanged since the 1930s. The primary concerns, then and now, have been to maintain an adequate temperature during curing, protect the concrete from freezing, and avoid large thermal shocks to the concrete when it is exposed to the cold. Basically, the concrete must be delivered to the job site warm, any surface that the fresh concrete comes in contact with must be thawed, and the concrete must be kept warm by insulation or heated enclosures while it cures.

The goal in proportioning mixtures for cold weather is to achieve a concrete that cures rapidly. That is, one that sets more quickly and gains strength more rapidly than it ordinarily would. Figure 1 shows that the time of setting of concrete increases as temperature decreases. Likewise, strength develops more slowly at lower temperatures. Though concrete cured in cool weather ultimately becomes stronger than that cured in hot weather, the construction schedule is slowed down. For example, concrete placed and maintained at 5°C, the lowest temperature allowed by the American Concrete Institute (ACI 1988), could take several hours longer to finish and up to a week or more before forms can be safely removed. Delays such as these can result in significant cost penalties.

ACI recommends three ways to shorten the delays caused by cold weather: (1) use more cement, (2) use Type III cement, or (3) use a chemical accelerator. These options are considered equivalent. They cause the concrete to behave as if it were 10°C warmer than it actually is, which can shave a day or two off the schedule. They do not, however, prevent the concrete from freezing. If the concrete freezes after its initial set, but before it has gained much strength, it may lose half its potential strength. Thus, heated enclosures are the best, though the most expensive, option when ambient temperatures drop below the freezing mark. Adding an enclosure can more than double project costs.

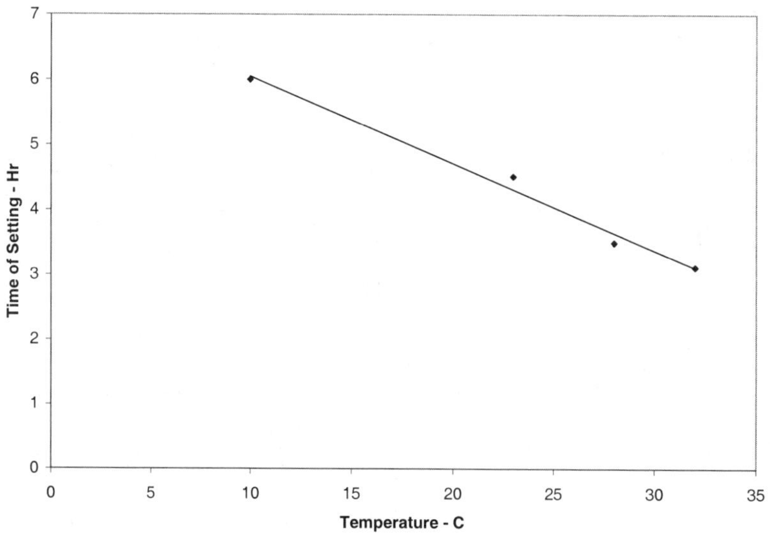

Figure 1. Effect of temperature on the initial time of setting of concrete (Dodson 1994).

Alternate Approach

An alternate approach to cold-weather concreting is to use chemical admixtures that allow concrete to gain strength at subfreezing temperatures. There are numerous chemicals that depress the freezing point of water and many that accelerate the rate of cement hydration—the essential functions of this technology—but the challenge has been to find chemicals that work together.

Figure 2 presents a typical performance of two chemicals in concrete cured at –5°C. The exact chemicals used are not important for this discussion. The intent, rather, is to show the reader what can be expected from low-temperature admixtures, and it is important to know how the concrete was handled. The concrete was prepared at room temperature according to ASTM C192 (1981). After mixing, the concrete was placed into cylindrical molds, vibrated to ensure proper consolidation, and capped with plastic lids to prevent evaporation. Within 40 minutes of the water first contacting the cement, the cylinders were placed into the 20°C and the –5°C curing rooms. The cylinders in the –5°C room reached that temperature well before initial set occurred. Therefore, it is important to recognize that the temperatures shown in Figure 2 represent the internal temperature of the concrete as it gained strength.

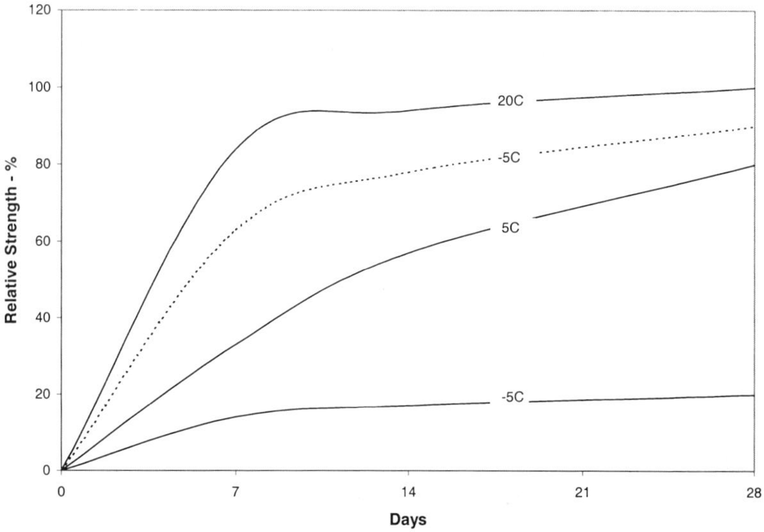

Figure 2. Comparison of normal concrete (solid lines) to that made with admixtures (dotted line).

As Figure 2 shows, the concrete made with the two chemicals and cured at –5°C resisted freezing and gained strength as if it was cured at something well above 5°C. Without benefit of the chemicals, ordinary concrete at this same temperature did not fare too well.

As written elsewhere, a –5°C capability translates into a potential extension of the construction season of up to four months in the United States and a cost savings of one-third compared to conventional cold-weather concreting techniques (Korhonen and Ryan 2000). However, for reasons mentioned earlier, cold-weather admixtures are not commercially available today.

Moving This Technology into the Marketplace

CRREL has spent much of the past decade studying alternate approaches to cold-weather concreting. The first commercial interest in this area occurred in 1992, when CRREL partnered with two major admixture manufacturers in the United States to study low-temperature admixtures. This effort was part of an initiative by the U.S. Army Corps of Engineers to work with private industry on research and development that had potential for advancing the art of construction and for being of value to Corps construction activities.

This cooperative study led to the field demonstration of two prototype low-temperature admixtures in 1994. The first demonstration project was built outdoors at

CRREL, Hanover, New Hampshire, in which a steel-reinforced concrete bin, 3.7 m wide by 4.6 m long with 1.2-m-high walls, 203 mm thick, on a 165-mm-thick slab, was cast in February. In March of that same year, a second field demonstration, conducted in northern Michigan, consisted of replacing several 5.5-m-wide by 6.1-m-long by 150-mm-thick reinforced slabs on grade. In both cases the concrete was easily mixed at low temperature, the admixtures were dosed into the truck during mixing, and the concrete was finished in the usual manner. No special tools or skills were required to work with the concrete and, because external heat was not needed to protect the concrete, a significant amount of non-renewable thermal energy was conserved. The resulting concrete, still in service, is indistinguishable from control concrete.

The prototype admixtures that resulted from the cooperative study were never commercialized, primarily because it was felt that an industry acceptance standard should be in place before this new technology could be opened to general practice.

Though studies continued at CRREL, little happened on the commercial scene until it was discovered that the floors inside the ice condenser rooms of two nuclear power plants had frost-heaved and were close to compromising plant safety. In 1997 CRREL designed a low-temperature concrete mixture made from commercial off-the-shelf admixtures to be placed in the Sequoyah Nuclear Power Plant belonging to the Tennessee Valley Authority. The concrete was pumped more than 100 m horizontally and 10 m vertically and placed, finished, and cured at –8°C. This project showed that it was possible to design and specify low-temperature admixtures from conventional materials without requiring special acceptance standards. This conclusion led to the activity that is discussed below and that continues today.

For the Department of Defense (DoD), it was becoming increasingly clear that there are times when concrete must be placed regardless of the weather, and that insulation or heated enclosures might not be available. This situation presented both a problem and an opportunity. The problem was that any concrete placed unprotected in the cold could be damaged by the cold. The opportunity was that in contingency operations—the battlefield or other emergency—long-term performance may not be a consideration and acceptance standards probably are not important. This allowed a wider scope of chemicals than might otherwise be acceptable for commercial application. In 1999, CRREL evaluated various materials found in a theater of operation for use as expedient concrete admixtures and published a report for the Army for concrete not expected to last more than five years (Korhonen 1999).

More directly related to the success achieved at TVA, low-temperature concrete mixtures were developed at CRREL for the City of New York Department of Design and Construction and Atkinson Construction in Bath, Maine, in 1999 and 2000, respectively. New York wanted the capability to repair its streets and sidewalks later into the fall and earlier in the spring. Atkinson Construction wanted to be able to continue concreting operations during the winter to complete the Naval shipyard it was working on. Neither of these two entities has reported using this technology, as the last two winters have been mild. However, they are ready should the need arise.

Because using chemicals to protect concrete against freezing has been proven technically feasible but is still not practiced in the United States, CRREL embarked on two parallel paths to introduce this technology to a wider audience. The first approach, which began in 2001, is to develop cold-weather admixtures from those that are being used for various purposes in concrete today. This is being done to avoid the necessity of developing a new acceptance standard. It was previously shown that cold-weather admixtures could be formulated by combining existing admixtures, which already comply with industry practice. The goal is to not use more of any single admixture than is recommended by its manufacturer but to use sufficient numbers of admixtures so that the concrete can safely resist freezing down to –5°C. The admixture combination must also force the concrete, when it is cold, to cure as rapidly as control concrete cured at 5°C. Arbitrarily, the initial setting of cold concrete was tentatively established to be not more than twice that of control concrete cured in standard laboratory conditions. This program, started in April 2001, is scheduled to run three years. It is being funded by a consortium of Northern State Departments of Transportation (DoT) and is being monitored by the Federal Highway Association.

The second path has the long term in mind. The Civil Engineering Research Foundation (CERF) is spearheading this one and its intent is to develop a national standard for cold-weather admixtures. As with most emerging technologies, the practice of emplacing fresh concrete in subfreezing weather without the need for heaters has been stymied by the lack of standards. During the fall of 2000, CERF gathered a panel of experts from across the United States to develop a draft of what is hoped will eventually become the standard for cold-weather admixtures. It is recognized that developing new standards, getting them approved, and then accepting them into practice takes time—sometimes, years. Thus, this approach will be a natural follow-on to the DoT effort.

Conclusions

Though not quite there yet, the world of winter concreting promises to become a bit friendlier within the next few years and beyond. Expectations are that the protocol for using existing admixtures to make "true" cold-weather admixtures should become available to the DoT community within three years of this writing. This will permit immediate use of this technology to protect concrete down to –5°C. Following that, a national standard supporting the commercialization of cold-weather admixtures should become available. This standard will specify how concrete should perform when made with the admixtures, and, equally important, that the admixtures may not harm the concrete in the long term. The timing on developing this standard largely depends on how quickly this process can move through committee. Once this is done, it is expected that commercial admixtures, dedicated to protecting concrete down to perhaps –10°C, will enter the market. At that point, cold-weather admixtures will become readily available.

References

ACI (1988) "Cold Weather Concreting," Committee 306. American Concrete Institute, Farmington Hills, Michigan.

ASTM (1981) C192, Standard Method of Making and Curing Concrete Test Specimens in the Laboratory. In *Annual Book of ASTM Standards*, Part 14, Concrete and Mineral Aggregates. American Society for Testing and Materials, Philadelphia, Pennsylvania.

Dodson, V.H. (1994) "Time of Setting," *Significance of Tests and Properties of Concrete and Concrete-Making Materials*, ASTM STP 169C, Chapter 11, American Society for Testing and Materials, West Conshohocken, Pennsylvania.

Korhonen, C. (1999) Expedient low-temperature concrete admixtures for the Army. U.S. Army Cold Regions Research and Engineering Laboratory, Hanover, New Hampshire, Special Report 99-17.

Korhonen, C., and R. Ryan (2000) "New Low-Temperature Admixtures," *Concrete International*, Vol. 22, No. 5, May 2000.

CASE STUDY
Townhome Complex Foundation Failure Due to Frost Action Investigation and Repair

John R. Woodworth, P.E.[1] and Thomas G. Krzewinski, P.E.[2]

Abstract

A large townhome complex north of Duluth, Minnesota experienced frost related problems during construction. The building distortions continued to increase after construction and the damage reached the point that some of the structures had to be removed from service.

Geotechnical and Structural investigations showed highly frost susceptible soils, poor drainage and poor workmanship during construction as well as design deficiencies that affected performance. The investigative work included dismantling some of the more distorted structures. The eventual repairs were massive and included replacement of foundations, replacement of slabs, foundation insulation and installation of drainage systems. The repairs were complicated by most of the building being occupied.

This technical paper will document the project history; the studies carried out prior to remedial designs, the repair designs considered, subsurface exploration, exposure of the existing foundation system, laboratory testing soil samples, thermal analysis of the existing foundation system, structural analysis of the existing foundation system, alternate designs for foundation repairs and the significant reconstruction effort.

---

[1]Principal Structural Engineer, Northland Consulting Engineers, L.L.P., 21 North 21st Avenue West, Duluth, Minnesota 55806

[2]Principal Geotechnical Engineer, American Engineering Testing, Inc., Post Office Box 16008, Duluth, Minnesota 55816

Introduction

The "Meadows of Hermantown" was a HUD financed development on 27 acres of vacant property near the Duluth/Hermantown border (Figure 1). The project was designed in 1992 and constructed in 1993 and 1994. The complex consists of 14 townhome buildings with 12 units in each (166 units total) and a centrally located administration/recreation building. In 1992, a minimal preliminary geotechnical evaluation was completed consisting of 7 borings spread throughout the site. Soil conditions were predominantly silty sand soils with a relatively high water table and significant zones of swamp deposits overlying the silty sands. The site topography was generally bowl-shaped with organic soils present in the bottom of the bowl. The geotechnical report basically said the site was suitable for building, offered some recommendations for allowable bearing capacities and expressed caution about the sensitivity of the on-site silty sands. No infrastructure recommendations and no foundation design details were provided.

The construction proceeded based on the preliminary geotechnical work. The contractor and design professionals were all from southern Minnesota. The city building inspector insisted on a local firm to provide independent testing services during construction. A local geotechnical firm was brought in and, as a first order of business, they reviewed the plans and specifications and discovered many potential performance problems. A letter report was provided to the contractor describing the following issues: highly frost susceptible site soils; the high water table; footings shallower then the anticipated frost depth; unheated garages attached to heated living space with no separation between them; no site drainage provisions; inappropriate on-site material planned as fill; and an inadequate pavement section. The letter report was basically ignored and the contractor proceeded with the comment that the owner was willing to accept the increased maintenance associated with using the on-site soils as fill.

The project was stalled in the spring of 1994 by the inability of construction traffic to access and travel on the site during breakup. The local geotechnical firm was asked to provide a stabilization scheme, which would allow construction to proceed. The roads were subcut several feet; bottoms stabilized with geofabric and non-frost susceptible (NFS) subbase and base compacted into place. The contractor chose only to apply the fix to impassable areas.

The winter of 1995 turned out to be severely cold. Six of the townhome buildings were significantly distorted by frost action. Occupants' complaints included doors jamming, cracks opening, air leaks, heaved/cracked garage slabs, entryways distorted by garage uplifts. The building inspector asked for testing reports from the geotechnical firm and became aware of the development's problems.

The winter of 1996 was also severely cold and more buildings were adversely affected by frost action. The geotechnical firm was asked to investigate the most severely distorted buildings. They found frost down to 1.8 m (6 feet) in March, ice crystals, ice lensing and elevated moisture content in the foundation soils. Recommendations were presented to the owner but no action was taken. In 1997 the City of Hermantown condemned "Building 13" and withdrew the "Certificate of

Occupancy". The owner defaulted on the HUD loan and tenants started vacating the townhomes.

In 1998, HUD auctioned off the complex with the new owner paying \$5.5 million (original construction estimated at \$8 million). The new owner was told by the building officials to hire local professionals to assist them in making the development whole again. The new owner contacted a local structural engineering company as well as the geotechnical engineering company in Duluth to investigate the site and foundation problems that were being experienced at the Meadows. The geotechnical firm hired was very familiar with the site and building problems.

Geotechnical Evaluation

The Duluth/Hermantown area is notorious for cold winters, high water table and a predominance of moderately to highly frost susceptible soils. The mean annual temperature is 3.4° C (38.2° F) and the Design Freezing Index is 1737 degree C-days (3126 degree F-days). These environmental conditions are conducive to the extensive formation of frost. Not surprisingly, the distortions noted at the townhome development were those commonly associated with frost heave.

The geotechnical investigation included 38 borings advanced throughout the site to define the subsurface conditions and to determine where groundwater was located (Figure 2). It also included the dissection of the condemned townhome building #13 to determine what was occurring and what level of repairs to the foundation system were required.

The borings encountered shallow water throughout the site. They also indicated that the on-site silty sand soils were used as the fill material to raise the grade at some of the low spots on site. The fines content in the glacial till soils encountered varied from 30 % to 80 % and the actual USCS classifications varied from sandy silt to silty sand to clayey sand. Some organics were encountered at the fill/native soil interface.

The foundations of building #13 were exposed by a trenching/test pit type operation. The footings exposed were only provided with 0.9 to 1.2m (3 to 4 feet) of cover for frost protection. Water was standing in the excavations at levels approximately 1 foot below the floor slab elevations and the foundation walls were cracked and distorted. The garage slabs had heaved as much as 457.2 mm (18 inches) and were raised 152.4 to 203.2 mm (6 to 8 inches) at the time of the test pitting (summer of 1998). No footing drains were placed during the original construction and a storm drainage system in the streets was not a part of the original civil work. The foundation walls adjacent to garage slabs were displaced outwards in the upper 0.9 to 1.2 m (3 to 4 feet). The displacement experienced by the buildings had left the superstructure visibly distorted with twisted entryways, jammed doors, separations at walls/slabs and around window frames and significant cracks in the finished walls. Some of the floor slabs in the heated structure had moved differentially with respect to adjacent units and garage slabs

Structural Evaluation

The townhome buildings are similar to each other but not identical. Each building has 12 two-story townhomes units. Four large units are located at the corners of the building's footprint with eight smaller units located in the center of the footprint. Each unit has an attached garage. See Figure 8 for a partial, general foundation plan of a typical townhouse building.

During the geotechnical evaluation a local contractor had partially exposed the exterior walls of Building 13 which had been condemned and evacuated. The exterior foundation walls consist of hollow 203.2mm (8 inch) concrete masonry units (cmu) on a concrete footing. Interior foundations consist of a thickened concrete slab beneath bearing walls and columns. The floors of all the garages and the first level of all the units are concrete slab on grade. Foundation insulation and drainage were not provided in the initial construction. Gutters and downspouts were not attached to the eaves and the water from the large roof areas fell directly on the soil adjacent to the building. This soil appeared to slope back toward the building, apparently due to normal consolidation after construction. The individual townhomes are heated year-round. The garages are not heated.

The foundation walls extend seven courses above the footing level and the exterior grade is at or below the top course of masonry. This results in a frost protection depth to the footings of approximately 1.2m (4 feet); a depth less than the minimum 1.5m (5 feet) frost protection depth required by the Minnesota State Building Code; and much less than the 1.8m (6 feet) of cover typically used for frost protection by local professionals. This lack of frost protection, the lack of insulation, the lack of proper drainage and differential heating conditions between the townhomes and garages noted above, and the poor soil conditions detailed in the Geotechnical Evaluation appeared to result in significant differential foundation displacement between the garages and the units themselves. The foundation displacement translated to the timber superstructure framing creating framing distortion in the walls and rooflines. The distortion was accompanied by damage to the finishes including crack formation in the gypsum board walls and ceilings and loss of weather tight-ness at the doors and windows.

Evidence of foundation displacement was severe and substantial. The exterior masonry walls at the garages were permanently displaced in both the horizontal and vertical directions. The displacement often extended into the adjacent foundation walls supporting the heated townhomes. In many cases shattered and cracked masonry accompanied the displaced walls. The concrete garage slabs vertically displaced and cracked to the point that some of the garages could not physically be accessed with an automobile (Figures 3 through 7).

The frost displacement in the garages of the center units was occurring at the exterior foundation wall and the interior foundation separating "back to back" garages located on opposite sides of the building (Figure 8). This foundation was found to be a thickened slab that was attacked by frost from the two, unheated "back-to-back" garages. Evidence of this displacement could be traced upward from the foundation to the roof framing two levels above.

Structural and Geotechnical Repair Recommendations

The structural and geotechnical design professionals recommended a scope of repairs that would reduce or eliminate the frost susceptible soils adjacent to the townhome foundations, remove the damaged masonry foundation walls and garage slabs and reduce the likelihood of future foundation displacement from frost.

The scope of work for the repairs included:

- Removal and replacement of damaged foundation walls in the garage areas as well as the damaged walls in the adjacent heated areas. All undamaged walls and replacement walls were to be core filled with lean grout.
- Replacement of all concrete garage slabs with new concrete slabs on grade. The new slabs were to bear on a 457.2 to 609.6mm (18 to 24 inches) sand cushion and have polyethylene vapor barrier.
- Insulation of all exterior foundation walls with rigid polystyrene insulation. The insulation was to be installed along the face of the upper portions of the foundation walls and out horizontally to a minimum distance of 1.2m (4 feet) from the walls. The horizontal insulation was to be protected with a minimum of 304.8mm (12 inches) of fill.
- Installation of a footing drain system around all exterior footings. Installation of a storm drain system in the streets to collect surface and subsurface drainage and carry it off site. The footing drain system was to connect to the storm drainage system in the streets.
- Installation of gutters and downspouts on all units. The grade adjacent to the units was to be graded to slope away from the buildings.
- Replacement of the backfill along all exterior foundation walls. The new backfill material was to be compacted, non-frost susceptible engineered fill material with a maximum of 7% passing the #200 sieve.
- Installation of baseboard heating systems in all connected interior garages.

The repair scope noted above is shown in Figure 9. The repairs required the lifting and temporary support of each townhome structure. Construction activities would need to be coordinated with townhome occupants, as the buildings would remain occupied during the work.

In addition to the exterior foundation and geotechnical work, repairs were needed to the gypsum board interior finishes, exterior windows and doors and exterior siding. The ventilation and insulation in the buildings was found to be deficient, requiring additional provisions in each structure. The roofs on a number of the buildings were damaged by ice dam formation and were in need of replacement.

Conclusions

The recommended repair work began in the late summer of 1998 and was completed in 2000. The building inspectors have approved the repairs and restored the certificates of occupancy. With the repairs the townhome buildings are performing to the satisfaction of the owner and the units are 100% occupied. The buildings are stable at this time but still somewhat distorted, since the original finishing work occurred as the units were experiencing differential heave.

The cost of the repairs totaled approximately $2.5 Million. The complex has been renamed. The professional time required to make the complex whole again greatly exceeded the amount of professional time that would have been required up front to do the project right in the first place. As it is with most projects that go bad, only the lawyers are happy.

References

Krzewinski, T. G., and Tart, R.G. (1985). "Thermal Design Considerations in Frozen Ground Engineering." ASCE Technical Council on Cold Regions Engineering (TCCRE) Monograph Series.

McFadden, T. T., and Bennett F.L., (1991). *Construction in Cold Regions, A Guide of Planners, Engineers, Contractors, and Managers,* John Wiley & Sons, New York.

National Oceanic and Atmospheric Administration (NOAA), Climatography of the United States No. 81 - Supplement No. 3, "Maps of Annual 1961 - 1990 Normal Temperature, Precipitation, and Degree Days".

Steurer, Peter M. "Methods used to Create an Estimate of the 100-year Return Period of the Air-Freezing Index." U.S. Dept of Commerce, NOAA, National Climate Center, Asheville NC (1989).

Figure 1: Townhome complex. Note frozen pond and cat tails in foreground of top photograph.

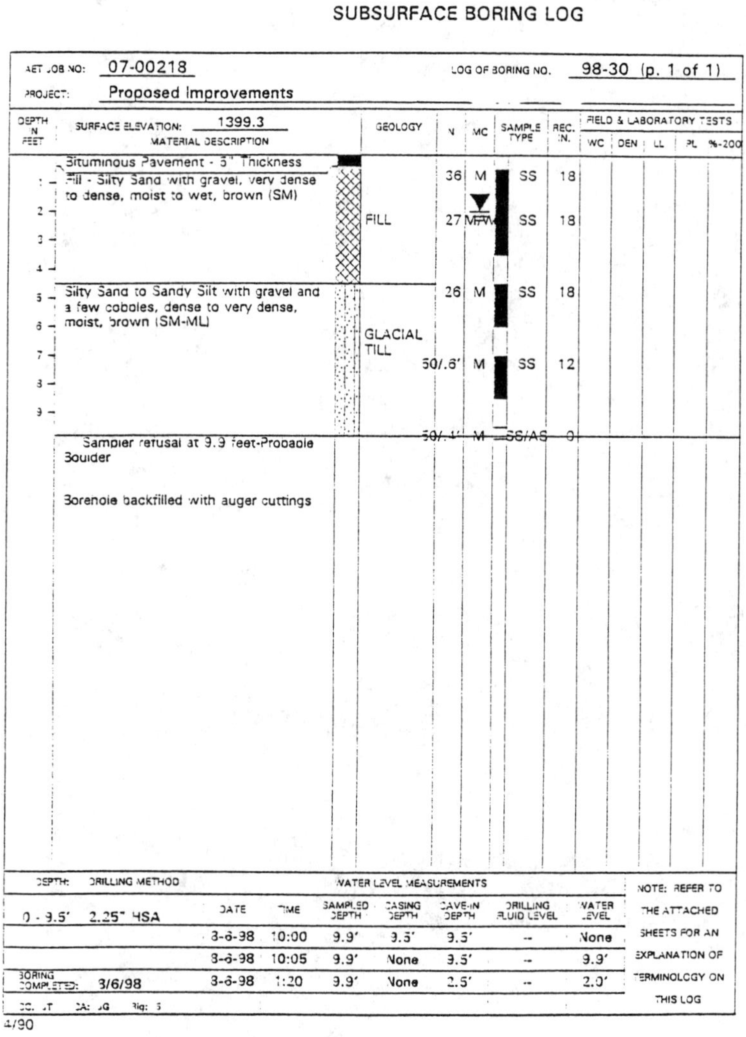

SUBSURFACE BORING LOG

AET JOB NO: 07-00218 LOG OF BORING NO. 98-30 (p. 1 of 1)

PROJECT: Proposed Improvements

DEPTH IN FEET	SURFACE ELEVATION: 1399.3 MATERIAL DESCRIPTION	GEOLOGY	N	MC	SAMPLE TYPE	REC. IN.	WC	DEN	LL	PL	%-200
	Bituminous Pavement - 3" Thickness										
1	Fill - Silty Sand with gravel, very dense to dense, moist to wet, brown (SM)	FILL	36	M	SS	18					
2			27	M/W	SS	18					
5	Silty Sand to Sandy Silt with gravel and a few cobbles, dense to very dense, moist, brown (SM-ML)	GLACIAL TILL	26	M	SS	18					
7			50/.6'	M	SS	12					
			50/.4'	M	SS/AS	0					
	Sampler refusal at 9.9 feet-Probable Boulder										
	Borehole backfilled with auger cuttings										

DEPTH: 0 - 9.5' DRILLING METHOD: 2.25" HSA

WATER LEVEL MEASUREMENTS

DATE	TIME	SAMPLED DEPTH	CASING DEPTH	CAVE-IN DEPTH	DRILLING FLUID LEVEL	WATER LEVEL
3-6-98	10:00	9.9'	9.5'	9.5'	--	None
3-6-98	10:05	9.9'	None	9.5'	--	9.9'
3-6-98	1:20	9.9'	None	2.5'	--	2.0'

BORING COMPLETED: 3/6/98

NOTE: REFER TO THE ATTACHED SHEETS FOR AN EXPLANATION OF TERMINOLOGY ON THIS LOG

4/90

Figure 2: Typical subsurface boring log from Goetechnical Evaluation.

Figure 3: Displaced garage floor slab and threshold.

Figure 4: Separated masonry courses in garage foundation wall.

Figure 5: Separated masonry courses in garage foundation wall corner.

Figure 6: Separated masonry courses and displaced garage slab at garage entrance.

Figure 7: Distorted superstructure framing at interior garage opening.

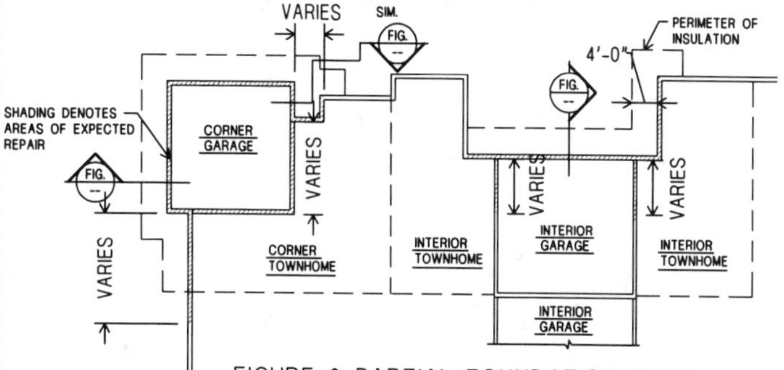

FIGURE 8 PARTIAL FOUNDATION PLAN
NOT TO SCALE

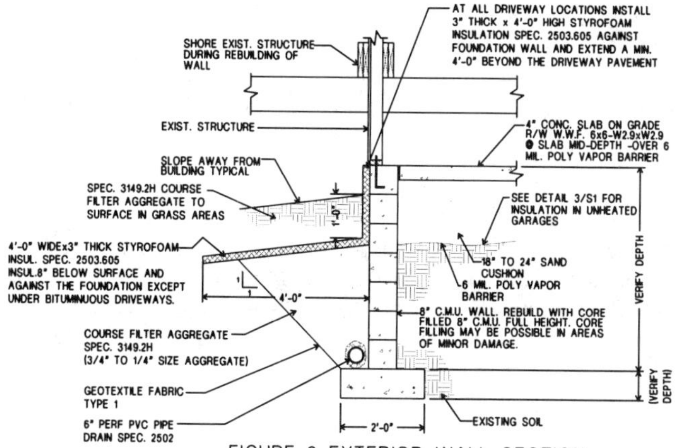

FIGURE 9 EXTERIOR WALL SECTION
NOT TO SCALE

Design of Helical Piers in Frozen Silt

Hannele Zubeck[1], He Liu[2]

Abstract

Helical piers have been used as foundations for board walks, utilidors and fences in rural Alaska in cold and warm permafrost. The piers are becoming more and more popular due to their light-weight, ease of transportation and installation, and resistance against pile jacking. However, engineers have been hesitant to specify them as foundations for buildings due to the lack of design guidelines for frozen ground. To increase the use of helical piers as cost effective foundation alternative, Alaska Science and Technology Foundation funded a study to create these guidelines. This paper presents the development of a finite element model for helical piers in frozen ground and an example of design of helical piers in frozen silt.

Introduction

Helical piers create a light-weight foundation that can be used either in compression or tension. Due to their light weight, ease of installation and little ground disturbance they are becoming very popular in remote building sites such as arctic villages in Alaska (Figures 1 and 2). Conditions in these places often include permanently frozen soils, which changes the dictating design criteria from bearing capacity and consolidation settlement to creep settlement. Current design methods are not suitable for helical piers in frozen ground, which may limit their use. Therefore, the Alaska Science and Technology Foundation (ASTF) is funding research to develop guidelines for design of helical piers in frozen ground. The purpose of this paper is to show how the Finite Element Analysis (FEA) is used to create design curves and aid in design of single helix piers in frozen silt.

Helical Piers

The helical piers come in many shapes and sizes. Typically they consist of a central shaft that is made from square or round sections that can be either solid or hollow. To this shaft is connected from 1 to 4 spiral plates (Figure 3) that are designed to mobilize more soil resistance under normal loading conditions than a conventional pile.

---

[1] Assistant Professor, Civil Engineering Department, University of Alaska Anchorage, Alaska, afhkz@uaa.alaska.edu

[2] Associate Professor, Civil Engineering Department, University of Alaska Anchorage, Alaska, afhl@uaa.alaska.edu

Figure 1. Helical Pier Stockpile in St. Michael, Alaska

Figure 2. Insignificant Ground Disturbance

Applications

The helical piers are used for a variety of applications. In compression, they are used as building foundations for new structures, underpins for existing structures, and as support for boardwalks, utilidors, fence posts, light posts, etc. In tension, the helical piers, often called helical anchors, are used as soil anchors for retaining walls, and to resist wind loads for tall structures.

In Figure 4, the utilidors in Selewik, Alaska, are founded on helical piers, but the boardwalk is founded with conventional methods. The helical piers are well-suited to resist the "pile jacking" that is a common problem with pile foundations in frozen ground. Figure 5 shows helical piers installed for a utilidor in St. Michael, Alaska.

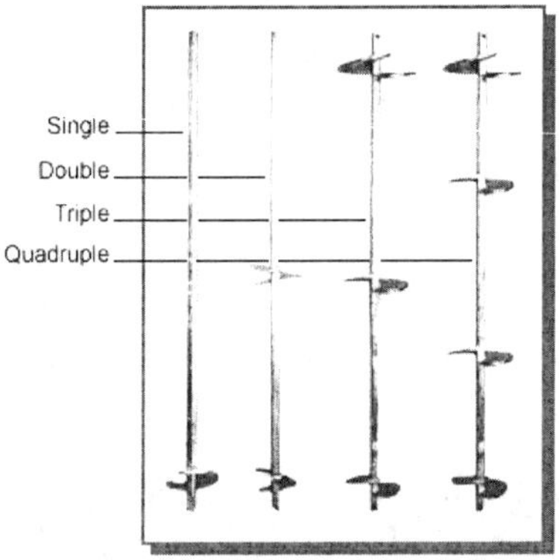

Figure 3. Helical Pier Lead Sections (A.B. Chance Co. 1996)

Figure 4. Typical Uses for Helical Piers: Boardwalks and Utilidors

Installation

Helical piers are installed by applying a torque on the pier shaft. The pier then augers itself into the ground. In frozen ground, additional vertical pressure is often required. First pier installations into very cold permafrost were conducted into 50-mm pilot holes to ensure straight piers. Installers found, however, that piers can be installed correctly even into the cold permafrost without pilot holes.

Figure 5. Utilidor under Construction in St. Michaels, Alaska

Design Methods

The current design method for determining the capacity of helical piers in thawed soils is based on bearing capacity theory. According to Ladanyi and Johnston (1974), it is not appropriate to analyze deep circular pile foundations in frozen soils on the basis of Prandtl-type bearing capacity equations and a separate settlement analysis using Boussinesq's stress-distribution theory and compressibility of soil. In frozen soil, the temperature and undrained creep become predominant in the determination of allowable foundation pressures. Therefore, Ladanyi and Johnson (1974) developed an alternative method for predicting the time and temperature dependent creep settlement and the bearing capacity of frozen soil under deep circular loads. The asymptotic ultimate pullout resistances calculated with the model seem to agree with field results, but it is not clear if the model can be used for long term creep under compressive loads. Therefore, further analysis using FEM is warranted.

The purpose of the project at the UAA is to develop a finite element model that will be used to create design curves with different soil temperatures and soil properties. The analysis developed includes four models that are large model, small model, installation failure model, and creep model (Liu et al. 1999b). The large model analyzes the soil stresses and displacements immediately after the pier is subjected to its design load. These data are critical for the development of more detailed analysis using sub-modeling techniques. The small model is a sub model of the large model. It analyzes the stresses developed within the spiral structure by using results from the large model analysis. The installation failure model is a detailed model of the spiral structure subjected to a torsional load during installation. This model provides insight into the failure mechanism of helical piers during construction. The creep model analyses the long-term displacement and soil stress in frozen ground.

Figures 6 and 7 provide examples of the stresses and deformations in the pier itself and the surrounding soil. The finite element analysis shows that the traditional bearing capacity equations are not adequate design tools for helical piers with multiple helixes (Liu et al, 1999a, 1999b, 2000). The stress distribution in the soil does not agree with the traditional method that assumes the helixes are not connected with each other. Further, the traditional bearing capacity equations do not consider creep in frozen soil. The analysis and results given in the following sections are based on the creep model that utilizes the large model. The creep equation used by Morgenstern et al. (1980) was also used in this analysis (Equation 1):

$$\dot{\varepsilon}_e = \dot{\varepsilon}_c \left(\frac{\sigma_e}{\sigma_{cu\theta}} \right)^n \qquad (1)$$

where: $\dot{\varepsilon}_e$ = strain rate, $\dot{\varepsilon}_c$ = reference strain rate, n = experimental creep parameter, $\sigma_{cu\theta}$ = temperature dependent total deformation modulus, corresponding to the reference strain.

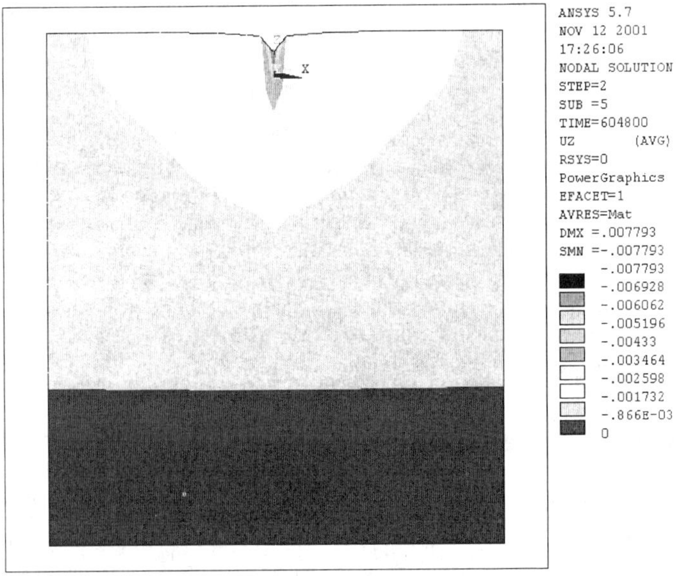

Figure 6. Vertical Displacement Distribution in Soil Volume

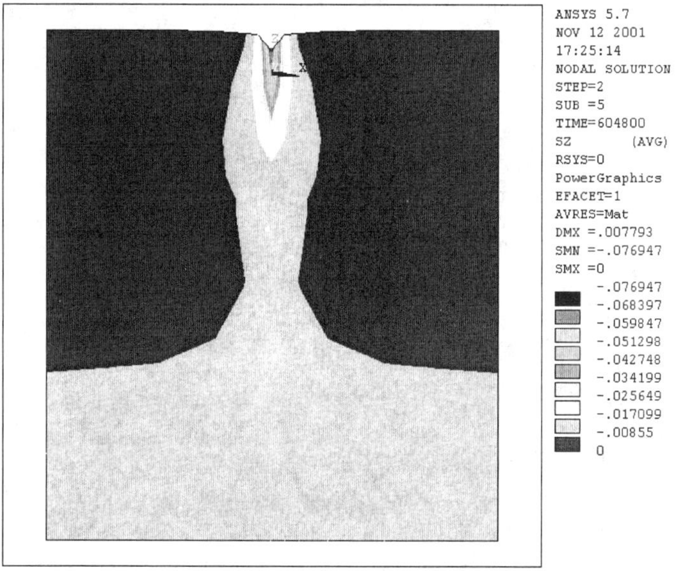

Figure 7. Vertical Stress Distribution in Soil Volume

Materials

The soil properties used for the analysis are given in Table 1. The pier has a pipe shaft with 90-mm outer diameter and 13-mm thick wall, and a 203 mm-diameter, 13-mm thick plate for a single spiral. For the large-scale model, the spiral is modeled as a flat plate in order to limit the number of elements in the model. The element and material properties are provided in Table 2. The distance from the soil surface to the helix is 4.60 m.

Analysis Results

The results of the FEA models are given in Table 3. Creep settlement (displacement) as a function of temperature is plotted in Figures 8 and 9.

Table 1. Soil Properties

	Unit	Temperature	
		-1°C	-5°C
Unit Weight	**kN/m3**	19.07	19.07
	psf	119	119
Cohesion	**kN/m2**	2413	6206
	psi	350	900
Friction Angle	°	25	25
Creep Parameter, n		2.04	2.04
Creep Parameter, $\dot{\varepsilon}_c \left(\frac{1}{\sigma_{cu\theta}} \right)^n$	**1/(h MPa^n)**	$3.81*10^{-7}$	$5.49*10^{-8}$
	1/(s ksi^n)	$5.44*10^{-9}$	$7.84*10^{-10}$
Young's Modulus	**kPa**	1800	7400
	psf	37594	154553
Poisson Ratio		0.2	0.2

Table 2. Pier Properties and FEA Parameters

Material	E (MPa)	ν	ρ (kN/m^3)	Element Type	Nodal Degrees of Freedom
Steel shaft	200,000	0.3	77.0	Beam	6
Steel plate	200,000	0.3	7.88	Shell	6

Table 3. Results of FEA Models

	Displacement (mm)							
Time	**-1°C**				**-5°C**			
(years)	28kN	56kN	83kN	111kN	28kN	56kN	83kN	111kN
0.00	-0.05	-0.11	-0.16	-0.21	-0.01	-0.03	-0.04	-0.06
0.02	-0.05	-0.11	-0.16	-0.22	-0.01	-0.03	-0.04	-0.06
0.08	-0.05	-0.11	-0.18	-0.24	-0.01	-0.03	-0.05	-0.07
0.49	-0.06	-0.14	-0.24	-0.34	-0.02	-0.04	-0.07	-0.10
1.00	-0.07	-0.17	-0.29	-0.42	-0.02	-0.05	-0.09	-0.13
2.00	-0.09	-0.21	-0.37	-0.57	-0.03	-0.06	-0.11	-0.16
3.00	-0.10	-0.25	-0.45	-0.69	-0.03	-0.07	-0.13	-0.19
5.00	-0.11	-0.31	-0.59	-0.92	-0.03	-0.09	-0.15	-0.22
10.00	-0.16	-0.46	-0.87	-1.39	-0.04	-0.11	-0.19	-0.28
15.00	-0.19	-0.58	-1.11	-1.77	-0.05	-0.12	-0.21	-0.31
20.00	-0.23	-0.69	-1.32	-2.10	-0.05	-0.14	-0.24	-0.34
25.00	-0.26	-0.79	-1.51	-2.39	-0.06	-0.15	-0.25	-0.37

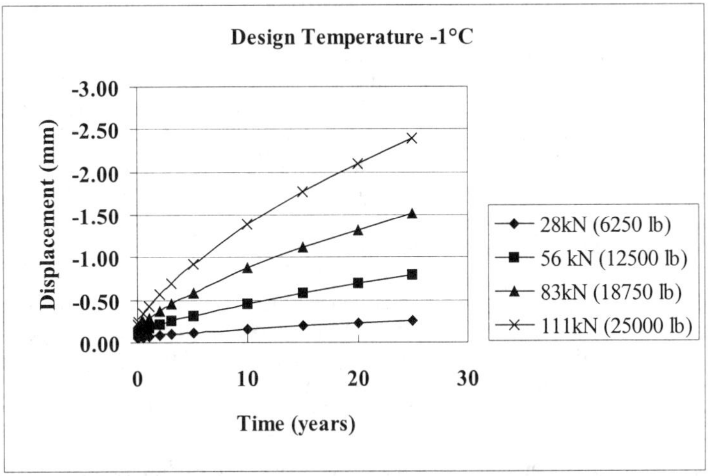

Figure 8. Creep Displacement at Design Temperature of –1°C

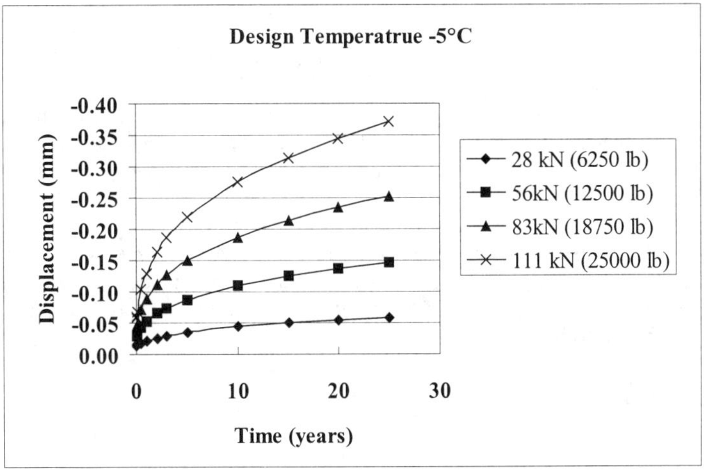

Figure 9. Creep Displacement at Design Temperature of –5°C

Design Example

In order to design single helix piers in the given frozen silt, the following information is needed: design temperature and allowable displacement after 25 years or less. When this information is obtained, the following steps will be performed:

1. Select the proper chart matching the design temperature.
2. Plot the design life and maximum allowable displacement in the chart.
3. Determine the design load that yields a predicted displacement that is smaller than the allowable displacement.

The following example illustrates the design method:

Example: Design a single helical pier foundation for a 100-m long wall with a design load of 20 kN/m. Use a pier with allowable capacity of 111 kN. Design temperature is –2°C and the allowable displacement after 20 years is a) 25mm, b) 2mm (small displacement is selected for illustration).

Solution:

a) Select the chart matching –1°C for conservative results (Figure 8). Note that all loads yield smaller displacement than the allowable of 25mm. For most economical design, select a design load of 111 kN per pier. The required spacing between the piers becomes 111kN/(20kN/m) = 5.55m. If the structural considerations allow this, the total number of the piers is 100m/(5.55m/pier) + 1 pier = 19 piers.

b) Select the chart matching –1°C for conservative results (Figure 8). If the piers were loaded to their capacity of 111 kN, the allowable displacement of 2mm would be exceeded after 18 years. Consequently, smaller pier load is necessary. A load of 83kN per pier would yield 1.5 mm displacement after 20 years, less than the allowable maximum displacement and consequently acceptable. To obtain a load of 83kN per pier, the spacing between the piers becomes 83kN/(20kN/m) = 4.15m and the total number of the piers is 100m/(4.15m/pier) + 1 pier = 25 piers. Again, check if the 4.15-m spacing meets the structural requirements. See Figure 10 for an illustration of this design example. By interpolation, a design load of about 100 kN could also be selected to meet the required allowable settlement requirement.

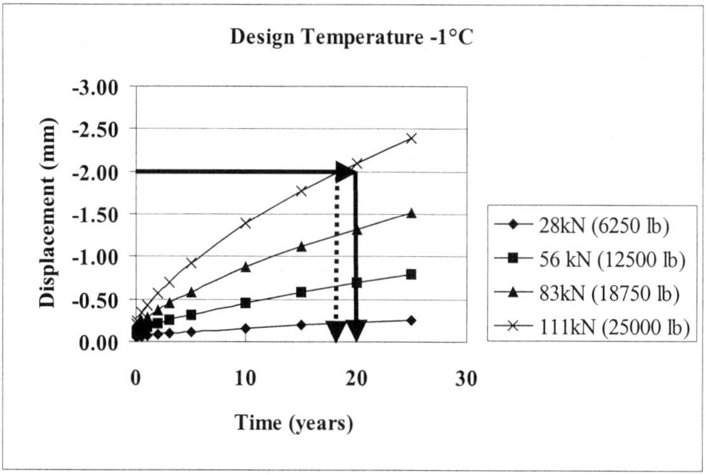

Figure 10. Design Example

Conclusions and Recommendations

Helical piers are widely used in permafrost areas in Alaska. Even if specific design guidelines for frozen ground have not been in existence before, the piers are used successfully as a foundation for boardwalks and utilidors. However, engineers have been hesitant to specify helical piers as building foundations because of the lack of design guidelines. The design method suggested here is very simple and helpful in designing piers for a given frozen soil. The design curves presented do not apply to any other soil than that given in Table 1. Future publications will provide more design curves for other soils. To properly design any foundations in frozen ground, the engineer still needs the creep parameters for the foundation soil. These parameters are not readily available to the engineering community. Therefore, actual soil testing needs to be conducted for various soils to create a library or database of soils encountered in cold regions. These tests would include determination of thawed soil strength parameters, as well as soil creep and strength parameters at several frozen temperatures under various loads.

Acknowledgements

The authors wish to thank a University of Alaska Anchorage student, Paul Kendall, for his prudent work conducting numerous FEA runs.

References

A. B. Chance Co.,1996, *Helical Pier Foundation System Technical Manual*, Bulletin 01-96

Ladanyi, B., and Johnson, G.H., 1974, "Behavior of Circular Footings and Plate Anchors Embedded in Permafrost," Canadian Geotechnical Journal, Vol. 11, pages 531-553.

Liu, H, Zubeck, H. and Baginsky, S, 1999a, "Evaluation of Helical Piers in Frozen Ground," *Proceedings,* Tenth International Conference on Cold Regions Engineering, Lincoln, NH, August 16-19, 1999.

Liu, H, Zubeck, H, Schubert, D, Baginski, S, Hsieh, Y, 1999b, "Behavior of Helical Piers in Frozen Ground, " *Proceedings* NAFEMS World Conference '99 (The International Association for the Engineering Analysis Community), April 1999, Rhode Island.

Liu, H, Zubeck, H, Schubert, D, Baginski, S, 2000, "Application and Analysis of Helical Piers in Frozen Ground", Proceeding of "The 5th Annual Alaska Water and Wastewater Management Association International Research and Development Conference on Rural Sanitation", April 25-26, 2000, Fairbanks, Alaska.

Morgenstern, N, Roggensack, W, and Weaver, J, 1980, "The Behavior of Friction Piles in Ice and Ice-Rich Soils," Canadian Geotechnical Journal, Vol 17, pages 405-415.

Design Considerations for Small Boat Harbors in Cold Regions

Carl Stormer[1], PE, Mike Field[2], PE, and John Daley[3], PE

Abstract

This paper provides a brief discussion of design and construction considerations for small boat harbors in cold regions. Information is offered in general terms and it is left to project engineers to develop appropriate design practices for their respective projects. The paper avoids giving general float design information, which has been exhaustively covered in other reference sources. Rather, the paper focuses on the main design considerations inherent with harbors in cold regions. These main considerations include:

- Types of float systems
- Horizontal and vertical ice loading on structures and icing inside harbors
- Cold climate effects on utility design
- Seasonal construction considerations

Introduction

This paper focuses on harbor design aspects unique to cold regions. For the most part cold region harbor location and construction practices, as well as basic float design are no different than those used in more temperate climates. Because freezing temperatures are the main concern in cold region harbor design, most design practices specific to cold regions are related to attempts to mitigate the affects of freezing within the harbor basin including the possible adverse effects on utilities.

Harbor Users and Needs

Harbor design starts with an examination of the fleet and of the types and needs of the harbor users.

In many cases, the general public will be the main end user for the harbor. These users can be expected to require potable water, electric, and sewer float utility systems. Because educating the entire general public on the use of complicated utility systems is impractical, the system must be simple to operate.

[1] Civil and Marine Engineer, Tryck Nyman Hayes Inc, 911 West 8th Ave. Anchorage Alaska; phone (907) 279-0543; carls@tnh-inc.com.
[2] Civil and Marine Engineer, Tryck Nyman Hayes Inc, 911 West 8th Ave. Anchorage Alaska; phone (907) 279-0543; mikef@tnh-inc.com.
[3] Civil and Marine Engineer, Tryck Nyman Hayes Inc, 911 West 8th Ave. Anchorage Alaska; phone (907) 279-0543; johnd@tnh-inc.com.

Wherever possible, the design should limit involved operational procedures. Public users are more likely to need the utility systems operating during the summer boating season. It may not be necessary to leave all systems 100 percent operational during cold months if usage has dropped off.

Commercial users may require year-round utility service in some harbors. In addition, commercial users may require additional flow and or pressures from water systems and/or they may have additional electric power needs. This necessitates the need to maintain communication with commercial users during utility design phases so that the design can meet their needs.

Because year-round live aboard users will require year-round utility service, utility connections that they use will have to be freeze protected. Water service and sewer standpipe connections can be heat traced and insulated. It may be possible to group live aboards in one area of the harbor. This would allow for the year round services to be maintained on a relatively small section of the float system during the winter.

Some harbors have a full time USCG presence. Due to the year round nature of their mission, the Coast Guard will likely require year-round operation of their utilities. Coast Guard vessels may have special fittings and requirements for water and sewer service. They may also have special electrical and communication requirements. It is important that the Coast Guard be part of the utility planning process so that the final design can meet their needs.

Types of Float Systems

There are several commonly constructed and available float systems types. These include timber, modular concrete, and monolithic concrete float systems.

Timber floats have several advantages for use as a primary float system material including:

- It is naturally flexible and can absorb a lot of wave energy in bending.
- Fabrication can be relatively simple and relies on standard carpentry tools and techniques.
- It is naturally resistant to freeze thaw.
- The dark color of treated timber absorbs solar radiation and will generally dry out fast.
- Repairs can generally be accomplished with standard carpentry tools.
- Properly treated timber will last well over 20 years in the marine environment.

Timber also has some disadvantages including:

- Because timber floats are naturally flexible, they can give the user the feel of being over water on a moving boardwalk.
- Untreated or field cut areas can rot more rapidly.
- Timber fasteners may corrode or become loose over time.

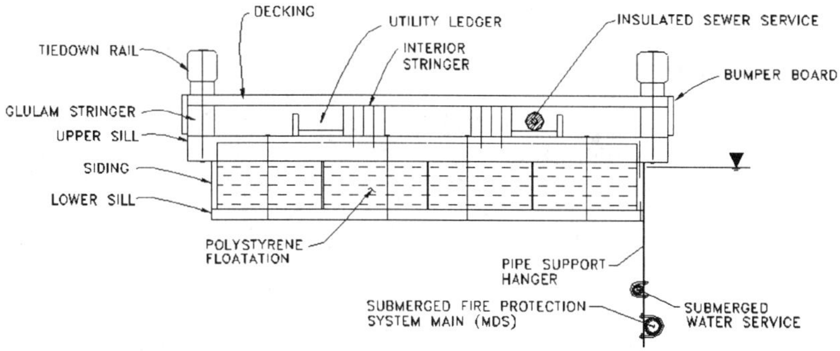

Figure 1. Typical timber float

Concrete is a popular material for float systems. Modular concrete float systems consist of numerous tubs or floats sections typically about 10 feet long. These are connected together with timber whalers and bullrails. Monolithic concrete float are cast in individual structural sections typically 40 to 50 feet long. Concrete has several advantages as a float material including:

- Concrete is very massive and generally rigid. This means that the floats system often gives the user a solid feel similar to walking on a sidewalk on dry land.
- Concrete can be a very durable material requiring very little maintenance.

Of course concrete also has some disadvantages including:

- Repairs to concrete can be difficult, expensive, and require trained personnel.
- The natural white color does not absorb solar radiation and is sometimes slow to dry out.
- Concrete can be susceptible to freeze thaw damage.

Durability is one of the most important requirements for concrete used in the northern coastal environment. Durability is affected by environmental factors such as freeze thaw, wetting and drying, large thermal gradients, and chloride attack. The first requirement for durability is attention to good basic practices. Basics include:

- A low permeability concrete
- A high cement content, low chloride content mix
- Low water to cement ratios
- Good cover over the reinforcing steel
- High quality materials
- Design for good drainage
- Quality control

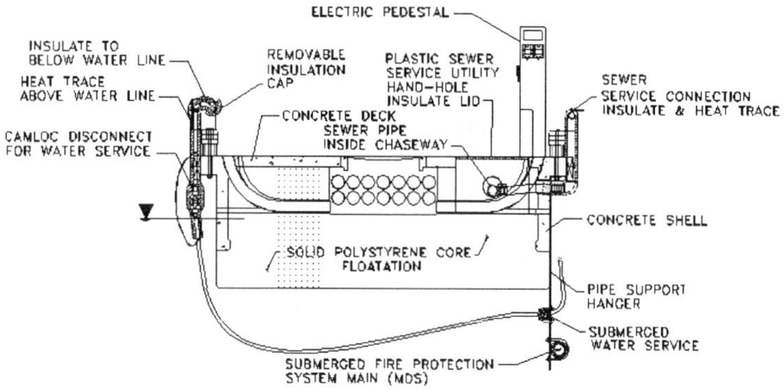

Figure 2. Typical concrete float.

Harbor Icing, Snow and Ice Loads, and Snow Removal

Ice is an important factor in harbor design in cold regions. This is especially true if there is a river or other fresh water source near the area in question, which may be captured within the basin. Sometimes a significant halocline (salinity layer) will form near rivers and other freshwater sources. The resulting fresh water layer is less dense than the underlying salt water and will remain on the surface to freeze without the retarding effects of higher salinity. For this reason, if possible, surface runoff should be diverted to areas well outside the harbor and away from the harbor mouth.

Ice can place significant horizontal loads on harbor structures. The compressive (crushing) strength of ice is of great importance to vertical structures coming into contact with it. Wind or current driven ice that impacts a structure will create a force proportional to it's projected area (width times thickness) and it's crushing strength. Typical design crushing strength values for sea ice range from 1378 to 3447 kilopascals (200 to 500 PSI). Also of consideration is the ice rubble field surrounding a structure. If the ice rubble is grounded it can act to dissipate forces on the structure. If it is free floating, it could act to increase the effective diameter of the structure, thereby increasing the forces on it.

Generally ice is weak in bending and many ice protection structures are designed to take advantage of this. This is generally done by providing some type of ramp that forces the ice to fail in bending under its own weight. A useful rule of thumb is that ice will fail in bending if deflected over 1/10 of its thickness. This is important if the structure is designed to resist ice forces through the use of a ramp. Often a slope value of about 1 in 5 is used for such structures.

Ice can affect structures in other ways as well. It can bond to the structure increasing the dead weight and the effective diameter and hence the wave and tide forces. Ice generally weighs about 640 to 881 kilograms per cubic meter (40 to 55 PCF). Because seawater weighs about 1025 kilograms per cubic meter (64 PCF), ice can produce a significant buoyant uplift force if submerged. This can be significant in some areas with documented cases of ice thickness of many feet encapsulating cold region marine facilities. Bonded ice can also interfere with the interaction of floating and fixed components such as pile restrained floats and gangways.

In addition to the horizontal and vertical loads, ice build up on the surface of the floats can present a slipping hazard to users. In general, snow should be removed as soon as possible after a snowfall from the harbor floats. This will help prevent the snow from compacting and eventually turning into ice. The snow removal methods and equipment should not damage the floats. In some areas snow removal equipment such as "Bobcats" and "four wheelers" can place significant loads on the floats. These loads need to be considered in the design of the floats and deck system.

Deck design should accommodate snow and ice removal by using flush mount hand holes and by avoiding any protrusions or obstacles in main snow clearing areas. Joints between float sections need to be vertically and horizontally well aligned. This will help to prevent damage due to the lip of the snowplow blade or snow thrower from catching on protrusions in the deck. In addition, the floats should be slightly crowned to shed water and, if possible, dark materials should be used for the deck. This will help to absorb solar radiation and dry the deck faster

Float Utility Systems

Float utility systems can include electrical, potable water, fire protection, sewer, and in some cases telephone and cable T.V. Harbor utility systems in cold regions face special design considerations due to the low temperatures in which they must operate. Some harbors have the option of closing and winterizing their freeze-susceptible utility systems making them non-usable during the colder months of the year. If year-round use of utilities is a design requirement, the designing engineer will have to examine each utility system to determine what special features may be required for cold weather operations.

Electrical

One potential problem for electrical systems in cold regions harbor design is icing in and around electrical outlets and fixtures. Placement of electrical fixture is important to avoid water infiltration, and to prevent wind-blown ice and snow from penetrating fixtures and causing corrosion and or electrical short circuits. Electrical components should be constructed of materials that are resistant to corrosion and freeze-thaw cycles, and do not become brittle during extreme cold. It is also important to place electrical outlets in locations that will be protected from snow removing equipment.

As in any harbor facility, the corrosive effects of the marine environment need to be controlled. This may be best accomplished by selecting corrosion resistant materials designed for the marine environment, and by housing electrical components where they are not subject to moisture contact either directly or via condensation. It is important to select fixtures that are sealed from the environment as much as possible. If electrical components are to be housed in heated enclosures, it is important that conduits are sealed and junction boxes are vented to the warm dry atmosphere to avoid condensation buildup and subsequent damage.

Fresh Water Delivery Systems

Piping fresh water in cold weather conditions is not a new problem. There are many piping insulation and heating products available to the designer to help avoid pipe freeze-up. The water piping for a harbor float system may traverse several environments including underground, above ground, under water, and within the float system. Because of tidal and wave movement, the system must include flexible segments. All elements of the water system, including the above water components such as meters, valves, and backflow preventers, must be protected from freezing.

Some piping companies have begun supplying "arctic pipe" (see Figure 3) as a standard "off the shelf" item. Arctic pipe is usually constructed of either a galvanized steel or HDPE inner main pipe with an external channel for holding an electrical heat trace. The inner pipe and heat trace channel are surrounded by 5 to 15.24 centimeters (2 to 6 inches) of foam insulation with some kind of outer jacketing material (often sheet aluminum, PVC, or HDPE plastic). The heat trace system is usually broken up into a number of circuits per total length of run. The number of heat trace circuits and the respective length of run per circuit is dependant on the power requirements of the specified heat trace. Heat trace power requirements usually run between 2 to 8 watts per foot.

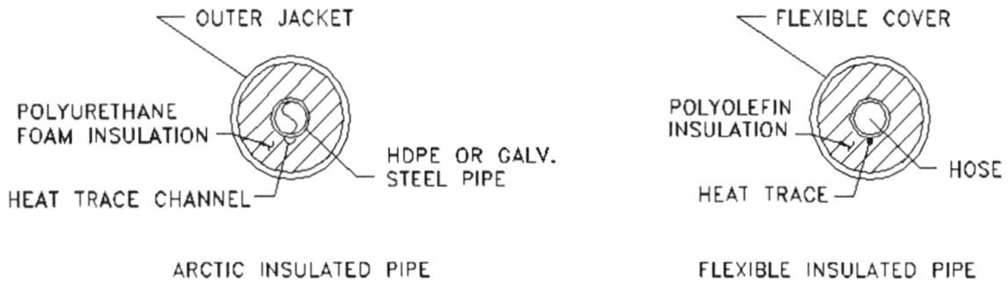

Figure 3. Arctic and flexible insulated pipe.

Water pipes can be routed underground and underwater to avoid insulating and heat tracing. For underground piping, it is important that all un-insulated pipe sections have adequate depth of cover. Depth of cover will vary with location and soil type depending on the anticipated depth of freeze. Because salt water will remain unfrozen in most harbors, submerged piping is a reliable means of freeze protection. It is for this reason that many cold region float systems run the majority of the water piping below the water line.

Various methods have been used successfully to hang submerged waterlines from harbor floats. These vary from tying the pipe off to the floats with rope to fabricating metal hangers that are bolted to the edge of the floats. In all cases, it is important to hang the pipe so that the pipe and/or hangers present no navigational hazards to vessels. The attachment design should consider the possible expansion and contraction of the piping by including expansion loops, and should consider possible buoyant forces if the water is removed from the piping.

One cold region water system design involves keeping the piping completely buried or submerged. Using this system, the piping runs underground into the harbor basin below the water line and then along the bottom of the harbor to the float system. It is important that the piping leave the soil at a depth below the extreme low water elevation and that proper depth of cover is maintained the whole run. Once the piping enters the harbor, it should be weighted to run along the harbor bottom to avoid becoming a hazard to navigation. The float connection should include a flexible link to allow the system to adjust to differing tide levels and to the occasional wave that may occur in the harbor.

Another system design involves running arctic pipe along the gangway to the float system. At the bottom of the access gangway the piping can be connected to the submerged piping via flexible connection. There are at least two interfaces in this system that need special attention, the top and bottom of the gangway. Because of the movement of the gangway with the tides, each of these areas will require a section of insulated flexible hose. Figure 4 shows both a typical underwater system (Alt. 1) and a gangway-mounted system (Alt. 2).

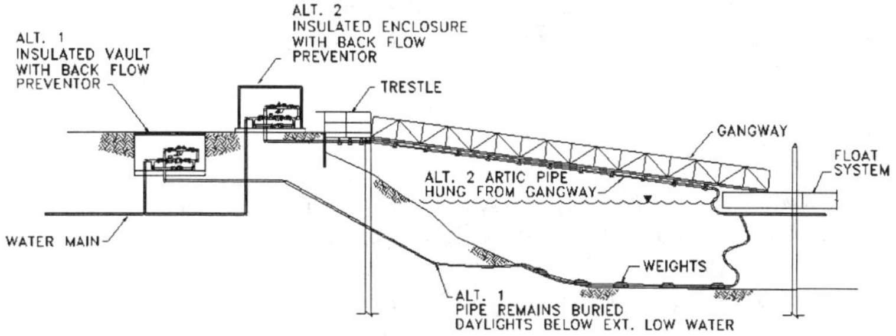

Figure 4. Harbor utility alternatives.

Water system pipes may also be installed inside the floats via cast-in-place conduit or on a utility ledger. Several types of concrete float systems involve a reinforced concrete shell surrounding a foam floatation core. Because the foam floatation can act as an insulating medium, it may be possible to seal the conduit sufficiently so that the piping within the floats is insulated and protected from freezing. In order to insulate the entire pipe, special attention will have to be paid to the joints between float units and to the utility access boxes. These areas will require supplemental insulation. Timber float systems may include a utility ledger on top of the floatation foam. This is typically an above water, below deck, channel designed to accommodate utilities. Arctic insulated piping should be used for these systems.

In both concrete and timber float systems, it is recommended that the entire above water pipe system be heat traced. Figures 1 and 2 show typical cross-sections of floats with internal and external utilities.

End user service standpipes present a problem for winter use as they are exposed to the atmosphere and, therefore, must be freeze protected. All standpipes remaining in use during cold weather must be heat traced and insulated. Because the standpipe usually has a terminal valve or hose-bibb that will experience opening and closing during use, heat trace should be wrapped all the way to the tip of the outfall to insure that residual water in the valve does not freeze. It may be necessary to supply the standpipe with removable insulation "caps" around the valve handle and the outfall end.

One method sometimes employed to help control pipe freezing problems is to simply leave an end service connection open and allow water to remain flowing through the system to prevent freeze-up. This method can work but may create some problems as well. Localized surface ice at the outfall of the running pipe may cause moorage difficulties and freezing fresh water spray can make for a slippery deck surface. There may also be an additional cost for elevated fresh water usage from the local water utility. This method should only be applied in areas not subject to severe or long-duration cold temperatures.

Another method that may be employed is to provide reduced water service for the winter with water available at a few selected areas in the harbor. This would require that special heated and insulated boxes be placed over a few hose-bibbs. The rest of the harbor water system would be shut down and winterized.

Water service standpipes that are shut down during cold periods will have to be easy to drain. Draining can be accomplished by outfitting the system with flexible hose connections at each of the standpipes. A cam-loc type connector works well for this. A shutoff valve is installed on the flex-hose side of the connection. The service standpipes are simply disconnected when shutdown, with the terminal hose-bibbs left open to allow draining. The flex-pipe connection is tethered to the float and allowed to hang below the water line. Restarting shut down water service connections involves, cleaning any accumulated marine growth from the cam-loc flex-pipe connection, disinfecting the cam-loc flex hose connection if the system is used for potable water, and reconnecting via the cam-loc connector to the main water line.

Sanitary Sewer System

More and more harbor facilities are supplying sanitary sewer pump-out stations for vessels. These systems usually consist of a pump that maintains a vacuum in the sewer line to facilitate holding tank pump-out. The pumps can be set to maintain constant line vacuum, or can be turned on by the end user. The end user connects to a standpipe pump-out station using a suction/discharge hose. In cold regions sewer pumps are usually housed inside heated enclosures either on land or on special utility floats. Typically, these pumps are not designed to pump the system completely dry; therefore, the system must be protected from potential freezing.

There are special considerations to keep in mind when designing a sewer system for a harbor. Often mandatory separation distances between the water and sewer lines cannot be met due to space considerations. This will require appropriate agency plan review for obtaining separation distance waivers. In addition, there may be some agency concern for submerged sewer lines, making it impossible to place sewer lines underwater to protect them from freezing.

In the case where the sewer system is routed inside the floats via conduit, design should mirror that of an internally installed water system. As stated previously, floats with foam floatation cores offer some freeze protection to the internally installed piping. In the case of sewer piping, it may be possible to install an internal heat trace directly inside the pipe. Again, all exposed pipe at float joints and inside access boxes must be insulated and heat traced. Figure 3 shows a typical concrete float cross-section with internal and external utility systems.

Fire Protection System

There are wide ranges of technologies available for float system fire protection. This includes regularly spaced hand held fire extinguishers, portable foam units, portable pump systems, and a built in high volume/high pressure water delivery system. Fire protection systems are subjected to the provisions of the National Fire Protection Agency (NFPA) sections 302, 303, 306, 307, and 14 (and other sections) and to local fire regulations.

A popular cold regions fire protection system is the manual-dry standpipe (MDS) system. This type of system consists of a stand-alone water pipe system attached to the floats or dock. The piping is normally kept dry and un-pressurized. It is important that the system is not a part of the potable water system. This is due to the fact that the potable water system is often connected to various harbor users and live aboards via garden hoses etc. Using this type of a system to fight fires can cause difficulties with hoses blowing when the system pressure is increased to fight the fire. Also, because the MDS remains dry except when in use, it remains free of freezing problems. In case of a fire, the fire department hooks up to the piping near the top of a gangway and pressurizes the line. This can be done by a pumper truck or through a connection to a near by fire hydrant. Fire hoses can then be connected to float system hydrants at various locations within the mooring basin.

Design of a MDS system will have to meet agency requirements for spacing, types of equipment, and water flow and pressure. Spacing of hydrants and the location of fire department connections may also be specified by the local Fire Marshall. All connection fittings must be compatible with standardized fittings used by fire departments. A minimum capacity of 1.89 cubic meters per minute (500 GPM) and 689 kilopascals (100 PSI) at the most remote outlet is a standard in many locations.

The MDS system can be installed above or below the water line. The system should be easy to drain after use. Draining can be accomplished via petcocks installed at several low points in the system or the system can be provided with a fitting for blowing out with compressed air.

Because the fire department will require high pressures and flow rates at all points in the float system, it is important to design the system with piping and fittings made to high pressure specifications. As stated previously, a minimum pressure of 689 kilopascals (100 PSI) at the most remote point of the system is usually specified. Since most fire department pumper trucks are capable of supplying much higher pressures then this, it may wise to use design piping pressure values higher than the 689 kilopascals (100 PSI) required for design flow. In general, the piping should be rated at 1103 kilopascals (160 PSI) at a minimum. HDPE - SDR 11.0 (or less) meets this specification. For most installations 15.24 centimeter (6-inch) diameter piping is recommended.

Seasonal Construction Considerations

Construction operations in cold climates are often hampered by the weather conditions. There is a definite construction season in northern areas that must be contended with when planning a harbor project. Winter weather can have an adverse effect on the production rates of crews and on certain types of construction including:

- Cast in place concrete cannot be allowed to freeze until thoroughly cured.
- Coatings, sealants, and glues will not work properly in freezing weather.
- Upland earthwork may be hindered by frozen ground.

There are several activities that may be able to proceed in the winter including:

- Demolition of existing facilities.
- Pile driving in the unfrozen bottom of the harbor area.
- Assembly and installation of prefabricated float units

Scheduling of construction can be problematic when upgrading an existing facility, as the short summer construction season usually coincides with the boating season. When planning any harbor construction project, work schedules will have to accommodate this seasonality. For this reason cold region harbor construction projects are often accelerated operations where crews are working night and day to get the job done in a short time frame. Note that in some areas an added complication is environmental concerns related to seasonal use of the harbor by small fish, which often results in seasonal pile driving restrictions.

Conclusions

In general, basic harbor and harbor facility design in cold regions is similar to harbor design in other parts of the world. However, freezing temperatures make designing harbor utilities challenging and require consideration of user needs, construction scheduling, snow and ice loading, and snow removal. A number of ideas to address these issues have been briefly discussed. As materials and product technologies advance, new means of designing cold region utilities and structures will likely be made available to the design engineer.

References

ASCE (1994). "*Planning and Design Guidelines for Small Craft Harbors.*" American Society of Civil Engineers, New York.

Tobiasson and Kollmeyer, (1991). "*Marinas and Small Craft Harbors.*" Van Nostrand Reinhold, New York.

Kirk Miller, P.E. (1998). "*Design of Small Craft Floating Structures in the Alaskan Environment.*" Anchorage AK.

NFPA (1994). "*Fire Protection Standards.*"

USACE (1984). "*Shore Protection Manual.*" Department of the Army, U.S. Army Corps of Engineers, Washington DC

USACE (1999). EM 1110-2-1612 "*Ice Engineering.*" Department of the Army, U.S. Army Corps of Engineers, Washington DC

State of California, (1984). Department of Boating and Waterways, "*Layout and Design Guidelines for Small Craft Berthing Facilities.*" Department of Boating and Waterways, Boating Facilities Division, California

PIANC, "*Review of Selected Standards for Floating Dock Designs.*" Permanent International Association of Navigation Congress, Brussels, Belgium

RT Allen, (1998). "*Concrete in Coastal Structures.*" Thomas Telford Ltd, London

Daniel E. Smith (1986). "*Cold Climate Utilities Manual.*" Canadian Society for Civil Engineering, Montreal, Quebec.

Robert W. Meyer (1982). "*Structural Uses of Wood in Adverse Environments.*" Van Nostrand Reinhold, New York.

Cold Regions Port Engineering

Orson P. Smith[1], M., ASCE

Abstract

Port engineering brings together many traditional specialties of civil engineering. Difficulties related to freezing temperatures, snow accumulation, and extended darkness compound the challenges of geotechnical, structural, architectural, mechanical, electrical, transportation, and coastal engineering in designing and operating sea and river ports. Previous guidance on cold regions aspects of port engineering has appeared in scattered conference papers and subdivisions of a few general port engineering references. The ASCE Technical Council on Cold Regions Engineering has initiated compilation of a monograph on cold regions port engineering. This paper highlights aspects of cold regions port engineering to be discussed in more detail in the completed monograph. Priorities for future research and development are discussed in light of growing pressures worldwide to develop and export by sea mineral resources of the far north.

Introduction

The world economy in the early 21st century will need the natural resources of Arctic and sub-Arctic regions. Air transport remains unprofitable for mineral payloads and attention to Arctic shipping is growing as a result. Road, rail, and pipeline routes are complicated in the far north by tectonically active glacier-contorted landscapes, low-lying frozen ground, and delicate ecologies. Shorter routes from resource to tidewater minimize terrestrial complications only if a port can be built at the coast. New ice-breaking ship designs are continually improving efficiency of Arctic shipping. Growing evidence of global warming indicates ice-free navigation seasons will be extended and thinner ice will reduce constraints on winter ship transits. More northern port capacity is critical to success of Arctic shipping strategies associated with northern resource development and climate change.

Design guidance for pile-supported offshore structures in ice for oil and gas development is evolving in literature and manuals of practice (*e.g.*, API 1993). Shore protection for artificial islands has received intermittent attention in journals and conferences (*e.g.*, Chen and Liedersdorf 1988). Practical matters of concern to commercial ports in winter have comparatively little documentation scattered about in journals and conference proceedings, such as the International Port and Ocean Engineering under Arctic Conditions series (*e.g.*, Smith and Lee 2001). Consolidated guidance is needed, since many northern ports are now expanding their capabilities.

The following narrative will lead the reader toward various existing sources of practical guidance for planning and design of ports in cold regions. Most of this information will soon be collected in greater detail in a publication of the ASCE

[1] Associate Professor and Arctic Engineering Program Chair, School of Engineering, University of Alaska Anchorage, 3211 Providence Dr., Anchorage, AK 99508; *afops@uaa.alaska.edu*

Technical Council on Cold Regions Engineering, titled the "Cold Regions Port Engineering Monograph" (Smith, in preparation).

Port Site Selection and Layout

Snow, ice, and frozen soils are important considerations for port site selection in cold regions. Road, rail, and air transport modes are affected by freezing conditions, snow, chronic ice fog, and extended winter darkness, causing intermodal logistical constraints to vary with the seasons. The following paragraphs discuss snow, ice, frozen soil, and intermodal logistics for cold regions port site selection and facilities arrangement.

Snow Considerations. Snow loads on port buildings, liquid bulk storage tanks, and cargo containers are critical structural design criteria. Snowfall and snow loads vary over kilometer-scale distances, especially in mountainous areas, as a function of orographic effects on wind speed and direction. Blowing snow reduces visibility and drifts increase structural loading. Snow drifting can be worse in areas where winds are affected by terrain, buildings, parked vehicles, or even chain-link security fences. Attention to regional and local variations in snowfall and wind-driven snow extremes helps to avoid excessive down time, slow time, and snow clearing expense in port areas. Plowed snow cannot be pushed into the sea, since contaminants concentrated on pavement may be forbidden as discharge into rivers or coastal waters. Contained plowed snow disposal areas are necessary for many urban and industrial areas. Practical wisdom regarding layout of facilities to minimize drifting difficulties and accommodate plowing and snow storage is summarized in various cold regions engineering references (*e.g.*, Eranti and Lee 2000).

Floating Ice Considerations. Challenges with ice at ports relate both to icing of upland facilities and to ice sheets formed on the sea or river surface. Dredged approach channels, maneuvering areas, and berths in northern regions should be designed with a view toward the realities of winter service. Policies and practice of pilot associations, shipping companies or vessel insurance underwriters may call for additional keel and bank clearances in navigation channels beyond those allowed in temperate ice-free conditions. The Coast Guard or other regulatory agencies may impose seasonal ice rules. Approach channels, maneuvering areas, and berths should be designed with consideration of the following aspects of ice navigation:

a. Ship maneuverability is retarded by ice,
b. Ice impact forces can divert vessels off course,
c. Low visibility conditions (fog and precipitation) are common,
d. Darkness is more common,
e. Winds can be at their worst,
f. Visual aids to navigation are less effective,
g. Shipboard instruments and communications systems are prone to malfunction,
h. Frazil ice can clog seawater intakes, raising the risk of overheating engines,

i. Repeated passages through icy channels increase presence of brash ice and accelerate new ice growth,
j. Tug assistance is more difficult,
k. Work on deck is less efficient and more dangerous, and
l. Crews are more strained and fatigued in the face of all these challenges.

Channel Design

Ice considerations for channel design primarily involve width and alignment, since depth concerns tend to be mitigated in winter. The entry track to a harbor should not be the same as the departure track. Separate tracks eliminate large turning areas where broken ice can hinder maneuvering vessels. Ice problems in harbors can be reduced considerably if local currents and thermal energy sources are used to advantage. For example, in straits and river mouths the incoming flow may reduce the accelerated growth of brash ice because it will be mixed and contain some thermal energy.

Ship channel widths include sweep path or paths (two-way channels) and clearances for various contingencies, including passing traffic (two-way channels), bank suction, shoaling between dredgings, and inaccuracies of excavation and surveying. A conventional sweep path includes variability of a selected ship's course, as recommended by PIANC (1997).

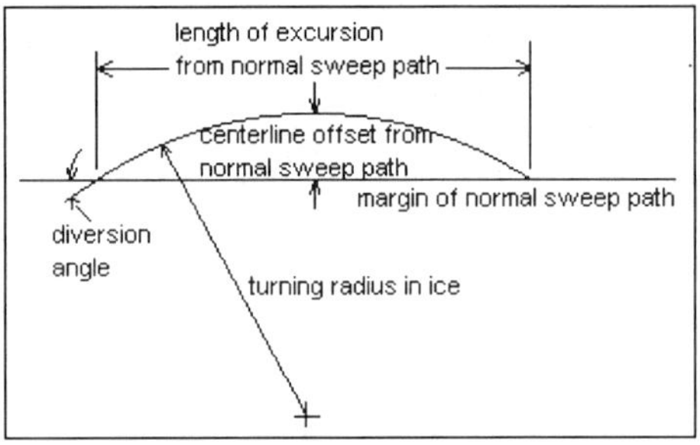

Figure 1. Channel width allowance for ice effects.

Ice effects call for wider sweep path allowances. A ship at the boundary of its normal sweep path that is diverted by ice outside this path will correct in the manner shown in Figure 1. The ship's centerline offset from the normal sweep path, plus half the beam, equates to an extreme sweep path allowance for channels in icy waters (Smith 1999). Cost of excavation for ice allowances may be as much as 40 percent more than corresponding temperate designs, therefore a rigorous investigation is warranted.

Navigation Aids. Automated onshore beacons and lighthouses provide the most reliable navigation aids. Floating aids in approach channels in winter are prone to being sunk or moved out of position by ice. Authorities typically remove most or all floating aids to navigation during the winter ice season. Lights and radar beacons mounted on fixed ice-resistant steel foundations are used where ice would move or destroy floating buoys. These foundations for steel aids-to-navigation are typically round steel piles driven into the seabed to depth sufficient for soil reaction forces to withstand ice and hydraulic forces. The overall size of the steel foundation structure depends on the load exerted by the ice, on the water depth, and on the strength properties of the soil on the seafloor.

Layout of Moorings

Mooring ships in icy waters can present unique challenges to ship, tug, and port operators. Broken ice is a particular problem if a floe becomes jammed between the ship and the dock (Figure 2). This situation can result in excessive forces sufficient to break mooring lines. Sliding the ship forward along the face the dock toward the desired location can avoid this problem. Another method involves bringing the bow in to the desired location at the dock, passing a spring line ashore, and going ahead slow, so propeller wash flushes ice out from between the dock and the ship (Canadian Coast Guard 1992).

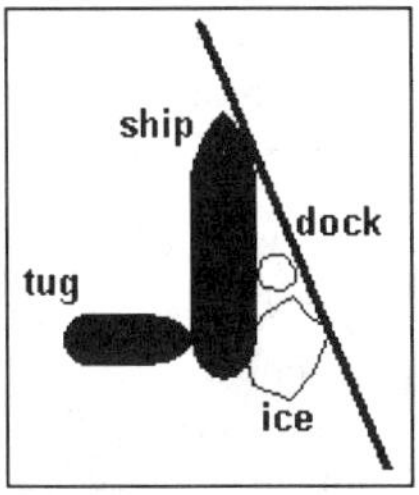

Figure 2. Ice interference with berthing maneuver.

The optimum dock configuration usually is in alignment with prevailing currents, but tidal and wind-induced ice movement can still force ice into ship berths. Natural offsetting currents tend to keep moorings clear of broken ice, but pose their own mooring challenges. Tug propeller wash can be applied ahead of ship arrival to clear the berth of larger ice floes. Tugs with multiple azimuth-drives are especially efficient at this and other port services in icy waters. The extra propeller thrust used for maneuvering ships in ice may also cause severe erosion at the foundations of piers and other structures.

Mooring points (bollards, etc.) should be extraordinarily strong to allow for mooring forces added by ice against the ship. Once moored, it is prudent to keep a continuous watch and have the engine on stand-by for rapid response to ice-related difficulties. Operators of cargo handling apparatus, *e.g.*, container cranes, liquid bulk lines, and roll-on/roll-off gangways, should be ready for quick response to ice-induced ship movement at the dock.

Structural Ice

Water level fluctuations in freezing conditions produce uplift and down-drag forces on docking facilities. Piles and linear structures, such as bulkheads or seawalls, receive uplift loads from intact ice sheets. Downward loads result when water levels fall. Hanging ice may span distances of 5 to 10 m between vertical supports (Figure 3). Other structural members of the docking facility may become thickly coated with ice from breaking waves and tidal action. These accumulations of ice on port structures degrade their utility.

Figure 3. Ice on pilings at Port of Anchorage, Alaska.

Ice Control

Ice control involves means to reduce ice growth in navigational areas and on structures and methods for breaking ice and keeping the broken ice away from navigation channels. These methods can be mechanical, thermal, or hydraulic, such as:

- *Mechanical*: Icebreakers keep the moorings and approach channels open.
- *Mechanical*: Ice booms hold floating ice from entering the mooring area.
- *Thermal*: Ice growth is suppressed by heat brought up from deeper water or by warm water available from industrial plants.
- *Hydraulic*: The river flow pattern is modified so that ice formation in a channel or harbor area is reduced.

Mechanical Ice Control. Broken ice that is detrimental to harbor structures or operations can sometimes be removed by mechanical means, *e.g.,* with a crane with a bucket or rock grapple (cherry picker). Removing ice adhered to structures by manual chipping is slow, labor-intensive, and dangerous for workers. High-density polyethylene panels installed on the face of fendering systems help shed ice, but performance declines as the plastic abrades. Specialized coatings (*e.g.,* Frankenstein and Hanamoto 1983) are also effective, but are costly to apply and require periodic steam cleaning. Steaming (a thermal method) removes ice from a coated surface more easily than from barren concrete or steel.

Icebreaking with ships is accomplished by various classes of icebreakers under a wide range of ice conditions (IMO 2000, Mulherin et al 1994). Icebreakers can be classified according to their working environments, as this often determines the main characteristics of the ships. Harbor icebreakers may often be seen as tugboats with icebreaking capability, designed for the working area of the harbor itself. Harbor icebreaker tasks include towing and assisting ships, breaking ice, and clearing ice away from docks. The existing harbor icebreaker fleet ranges from ice-strengthened tugboats to specially designed icebreakers for operation in harbor areas. These ships have the following characteristics:

- Good maneuverability,
- Good towing capabilities,
- Icebreaking ability suited to local conditions,
- Fire-fighting capabilities, and
- Pollution control (spill response) capabilities.

Shallow-draft icebreakers equipped with multiple azimuth thrusters (Figure 4) are proving extremely efficient in ice control of channels and harbor areas. The ability to direct the propeller wash outward from the ship centerline is an effective means to clear ice from lanes and confined harbor areas. The ships themselves are extremely maneuverable in tight situations and therefore well suited for port service.

High-flow air screens, or "bubble curtains," can be used to keep floating ice away from docking facilities and out of navigation locks. The rising air bubbles accelerate a large volume of water, which must be deflected outwards when it reaches the surface. The resulting divergent surface current is sufficient to prevent ice from being carried across the air screen line. An air bubble curtain can also transport warmer deep water to the surface to retard ice growth. Design parameters for an air curtain system include submerged depth, supply line size, orifice size and spacing, air supply flow rate and pressure, and air screen manifold size and length (Eranti et al 1983).

Figure 4. Shallow-draft icebreaker with multiple azimuth thrusters (Kvaerner Masa-Yards).

Movement of ice in harbors and waterways can be controlled by numerous barrier and diversion systems, such as ice booms, small islands, rubber tire breakwaters, large cells, dolphins, and cluster piles are only a few. The most common of these ice control structures is the ice boom. Ice booms have been used to hold in place ice broken by ship passage and prevent it from blocking channels. Booms are also used to help form stable ice covers to help keep navigation channels free from drifting ice. Booms have been built across an entire channel, with a gap for ship passage, or only on one side of a channel. Many ice booms allow ice to override when the ice load becomes excessive. The cost of building ice booms varies according to their construction and the type of anchors used. The cost of the anchors can amount to over 40 percent of the total. Foltyn and Tuthill (1996) provide a compendium of ice boom design guidance with many practical details.

Thermal Ice Control. Expedient removal of ice adhered to port structures by manual steaming, using a portable boiler, is more effective than manual chipping, but still time consuming. High-pressure hot water washers have the advantages of being portable and more readily available, but are slow if a large area is to be deiced. Water from deeper layers can reduce ice, both through the effect of its higher temperature and, in salt water, by its higher salinity. Warm water outlets from industries, power plants or sewage treatment works are often valuable contributors of heat for deicing purposes (Ashton 1979). Air bubbler systems suppress ice formation and growth when the bubble plume entrains the warmer bottom water and brings it to the surface. Lock gates have been equipped with electric heating systems to reduce ice growth.

Thermal systems have an influence on the environment. The effects of warm water outlets will vary with the local river or ocean conditions. The expected spreading and mixing of outlet water should not create "thermal ponds" or isolated pools of unnaturally warm water. Ineffective mixing of warm water will generate an open, ice-free area near the outlet point. The heat losses from such areas will be large. During cold days, ice fog will be generated, reducing visibility for navigation and causing icing on equipment. A variety of structural ice control measures, both mechanical and thermal, are reviewed by Tuthill (1995).

Erosion Control

Port structures affected by contact with freezing seawater or sea ice include breakwaters, often of cost-effective rubblemound construction. Wave forces tend to dominate design of rubblemound jetties, breakwaters, and revetments, but the potentially erosive effects of ice cannot be neglected. Effects of ride-up and pile-up of sea ice sheets (see Figure 5) on rock "riprap" slopes were studied by Sodhi, Borland, and Stanley (1996) in scale model tests. Recommendations from this work included placing rocks up to twice the ice thickness on slopes of 1V:3H and about 3 times the ice thickness for steeper slopes of 1V:1.5H.

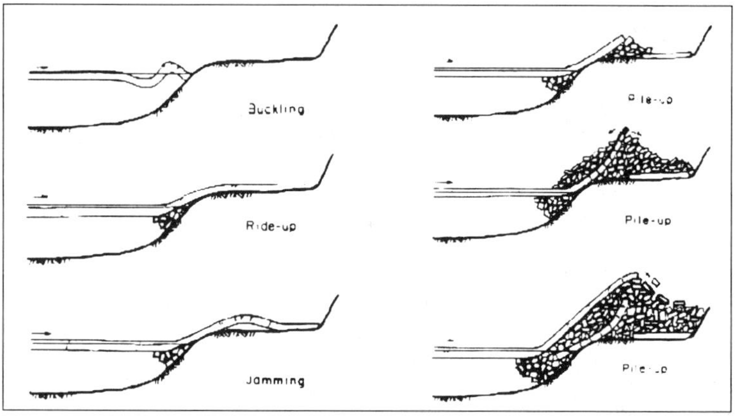

Figure 5. Conceptual sequence of ice ride-up and pile-up (Kovacs and Sodhi 1988).

Most port facilities include artificially steepened slopes or fills that require artificial protection from erosion by waves, currents, and ice. Diverse means have been employed to control erosion on Arctic beaches and slopes of artificial islands for oil and gas developments (Chen and Liedersdorf 1988). Synthetic fabric bags filled with sand or gravel have been reasonably effective in cases where long life is less important, such as islands built for oil exploration. Cranes have been employed to place up to 30 1.5-m^3 bags per hour from a barge-mounted bag-filling apparatus (Sodhi et al 1996). The principal mode of damage is from ice abrasion of the bag cloth, followed by dispersion of sand fill by wave action. Sunlight (ultraviolet radiation) eventually weakens most synthetic sand bag fabric. Damaged bags are

cheaply replaceable, however. Sand bags have proven more effective in shallow water where wave heights are naturally reduced.
Mats of interlocking prefabricated concrete shapes have also been effective. Concrete blocks are more costly to install than sandbags, but they are less vulnerable to abrasion and require less maintenance. They can be connected into mats efficiently placed by cranes on geotechnical fabric-covered plane slopes. Interlocking reduces the unit weight needed to withstand wave forces, resulting in lower weight per unit area requirements. Flexibility of the mats keeps them effective when settlement of the subgrade (Liedersdorf 1988).

Conclusions

This introduction into some specialized aspects of cold regions port design illustrates challenges for engineers who must deal with unique difficulties at the water's edge, as well as with realities of severe winter weather. The availability of a consolidated review in the form of the Cold Regions Port Engineering monograph (Smith, in preparation) will soon provide a starting place for engineers working on their first winter port project and a handy reference for veterans of the northern waterfront.

References

API (1993). "Recommended Practice for Planning, Designing, and Constructing Fixed Offshore Structures in Ice Environments," API Recommended Practice 2N, American Petroleum Institute, Washington, DC

Ashton, G.D. (1979). "Suppression of River Ice by Thermal Effluents," CRREL Report 79-30, Cold Regions Research and Engineering Laboratory, Hanover, New Hampshire

Canadian Coast Guard (1992). "Ice Navigation in Canadian Waters," TP 5063 E, revised edition, Ottawa

Chen, A., and Liedersdorf, C., eds. (1988). "Arctic Coastal Processes and Slope Protection Design," Technical Council on Cold Regions Engineering Monograph, American Society of Civil Engineers, New York

Eranti, E., Lappenen, E., and Penttinen, M. (1983). "Ice control in Finnish harbors," *Proc., International Conference on Port and Ocean Engineering* (POAC 83), Helsinki.

Eranti, E., and Lee, G. (2000). *Cold Region Structural Engineering*, McGraw-Hill, NY

Foltyn, E. P., and Tuthill, A. M. (1996). "Design of Ice Booms," CRREL Technical Digest No. 96-1, Cold Regions Research and Engineering Laboratory, Hanover, New Hampshire

Frankenstein, G. E. and B. Hanamoto (1983). "Design for Ice Forces: Methods of Ice Control," American Society of Civil Engineers, New York

IMO (2000). "Development of Guidelines for Ships Operating in Ice-covered Waters," *Report of the Working Group, Sub-Committee on Ship Design and Equipment*, International Maritime Organization, London

Kovacs, A., and Sodhi, D. (1988). "Onshore Ice Pile-up and Ride-up: Observations and Theoretical Assessment," in Chen, A., and Liedersdorf, C., eds. *Arctic Coastal Processes and Slope Protection Design*, Technical Council on Cold Regions Engineering Monograph, American Society of Civil Engineers, New York

Liedersdorf, C. (1988). "Concrete Mat Slope Protection for Arctic Applications," in Chen, A., and Liedersdorf, C., eds., *Arctic Coastal Processes and Slope Protection Design*, Technical Council on Cold Regions Engineering Monograph, American Society of Civil Engineers, New York

Mulherin, N., Sodhi, D., and Smallidge, E. (1994). "Northern Sea Route and Icebreaking Technology: An Overview of Current Conditions," Cold Regions Research and Engineering Laboratory, Hanover, New Hampshire

PIANC (1997). "Approach Channels: A Guide for Design," International Navigation Association, Brussels

Sandkvist, J. (1983). "Ice in Harbors and Terminals," Proc., Offshore Gouffborg 83, International Conference on Offshore and Marine Technology, Technology for the Arctic"

Sodhi, D., Borland, S., and Stanley, J. (1996). "Ice Action on Riprap: Small-scale Tests," CRREL Report 96-12, Cold Regions Research and Engineering Laboratory, Hanover, New Hampshire

Smith, Orson (1999). "Design of Channels in Icy Waters," *Proc., Port and Ocean Engineering under Arctic Conditions* (POAC '99,) Helsinki

Smith, O., ed. (in preparation). "Cold Regions Port Engineering Monograph," Technical Council on Cold Regions Engineering, American Society of Civil Engineers, New York

Smith, O., and Lee, W. (2001). "Risk Analysis Aspects of the Alaska Sea Ice Atlas," *Proc., Port and Ocean Engineering under Arctic Conditions* (POAC '01), Ottawa

Tuthill, A. M. (1995). "Structural Ice Control: Review of Existing Methods," CRREL Special Report 95-18, Cold Regions Research and Engineering Laboratory, Hanover, New Hampshire

Osprey – A New Platform in Cook Inlet, Alaska

Robert C. Visser, M ASCE[1]

Summary

The paper describes the design, fabrication and installation of a novel platform in Cook Inlet, Alaska. By using engineering creativity and thinking "outside the box" the platform project was completed by Forest Oil Corporation, an independent oil company, at a fraction of the cost of a conventional platform, thereby allowing exploration and subsequent development of a hitherto non-commercial oil field discovery dating back to 1969.

Introduction

The Redoubt Shoal field in Cook Inlet, Alaska, see Figures 1 and 2, was discovered by Amoco (then PanAm) in 1969 but not deemed commercial at the time. Forest Oil Corporation (then Forcenergy) took over the acreage in 1996 and, following a 3D seismic survey, concluded that the field could be commercial. To confirm this, however, required the drilling of at least two and possibly four appraisal wells.

Confirming the commerciality of an offshore prospect is usually done by drilling one or more expendable appraisal wells with a mobile drilling unit. For remote Cook Inlet this method is cost prohibitive because the equipment has to be mobilized from out-of-state and a single appraisal well was estimated to cost some $25 to 30 million. Installing a platform in Cook Inlet the "conventional" way is also cost prohibitive because there is no longer any heavy lifting equipment available on the West Coast and the cost of such a platform was estimated to be $60 to 80 million.

As an alternative Forest decided to investigate the possibility of installing an "exploratory" drilling structure. Requirements for a minimum structure were: (1) capability to support a 20,000-foot [6,100 m] modular drilling rig, (2) installation without heavy lifting equipment, (3) feasibility to be converted into a permanent platform, (4) easy removal in case the prospect did not pan out, and (5) relocatable to another Cook Inlet location.

[1] Belmar Engineering, Redondo Beach, California

A number of alternative platform concepts were developed by Redondo Beach, California consultant Belmar Engineering and further investigated for feasibility by Brown & Root in Houston. The selected concept requires no heavy lifting equipment and uses the tide to install the platform. To keep costs down the platform is located in 45 feet [14 m] MLLW about two miles [3.2 km] from shore rather than in much deeper water, four miles [6.4 km] from shore, where the Redoubt Shoal field is located. The lower platform and future pipeline costs offset the increased extended reach drilling costs.

There were many companies involved in bringing the Osprey platform to reality. These included the design firms of Winzler & Kelly and Ideas in San Francisco for the detailed design, Hyundai Heavy Industries in Korea and Veco in Anchorage for the fabrication, the Dutch firm of Dockwise for transportation and Stolt Offshore, then located in Oxnard, California and Seattle-based Crowley Maritime for installation. Belmar Engineering supervised the engineering and acted as project manager for the design, fabrication and installation. The paper describes the environmental conditions in Cook Inlet and how the platform was designed, fabricated and installed.

Background

Cook Inlet is located in southern Alaska, see Figure 1, and penetrates 170 miles [270 km] inland from the Gulf of Alaska. Table 1 summarizes the area's environmental conditions. Tides are among the highest in the world and range up to 30 feet [9 m]. The resulting currents, up to 13 feet per second [4 m/sec] keep the water turbulent most of the time. Underwater visibility is zero because of suspended sand and glacial flour. Upper Cook Inlet, the area of oil activity, is covered with ice during the winter months. During periods of ice cover, the tides move the ice up and down the Inlet at essentially the same speed as the current. Cook Inlet is in the Pacific earthquake belt and, as evidenced by the widespread damage from the 1964 Good Friday earthquake, platforms in the Inlet are subject to potentially large earthquake loadings.

Including platform Osprey there are now sixteen platforms in operation in Cook Inlet. Fourteen of these platforms were installed in a four year period between 1964 and 1968 when four oil fields and one gas field were developed. A fifteenth platform was installed in 1986 (Visser 1992).

Design

Detail design of the Osprey platform was performed by the design firms of Winzler & Kelly and Ideas of San Francisco. Dimensions of the platform are shown in Figure 3 and the installation procedure is shown in Figure 4. Platform design criteria are shown in Table 2. The tower section has four 14-foot [4.3 m] diameter vertical legs connected four feet [1.2 m] above the bottom with horizontal 10 feet high by 11 feet wide [3 x 3.4 m] box sections. The deck structure is a space truss spanning between the legs. The structure is designed for the 100-year ice condition similar to the design loads on many of the existing Cook Inlet platforms (Visser 1969, 1995, Blenkarn 1970).

Fabrication

The four-column tower, deck and the piles for the platform were fabricated by Hyundai Heavy Industries at their fabrication facility in Ulsan, Korea. Figure 5 shows the completed tower structure and Figure 6 the completed deck. The self-floating tower was equipped with a ballasting and de-ballasting system. The above-water portions of the platform were fabricated from low-temperature steel. The quarters building was fabricated by Veco in Anchorage, Alaska. The Seatrax 6020 pedestal cranes were fabricated in Houston, Texas and shipped to Korea to be incorporated into the deck structure.

Transportation

The fact that the 300-ton [300 t] quarters building was fabricated in Alaska necessitated a convoluted transportation scheme. The quarters building was loaded out in Anchorage on the 240-foot [73 m] long installation "piggyback" barge, see Figure 7. The barge and quarters were then placed on the Dockwise heavy-lift submersible vessel Swan in Kachemak Bay in lower Cook Inlet and taken to Korea, see Figure 8. At Hyundai's yard, the quarters were lifted off the piggyback barge and temporarily stored on the dock. After installing a deck support structure on the piggyback barge the deck structure was placed on top of the barge and the quarters building was placed on top of the deck. After loadout of the tower structure and the foundation piles the Swan returned to Alaska.

Offloading and Mating Operation

Following arrival in sheltered waters in Port Graham Bay in Lower Cook Inlet, see Figure 9, the tower and deck were floated off the Swan. The installation contractor, Stolt Comex Seaway (now Stolt Offshore), took over at this point. The floating tower was ballasted down by flooding until it reached about a 60-foot [18 m] draft. The piggyback barge and deck were then moved in between the legs of the ballasted tower structure, see Figure 10. After the deck was in position, the tower was de-ballasted until the two units were mated. The deck and tower were then welded together. The piggyback barge to deck supports were cut loose so that the mated tower and deck were resting freely on the deck support system. The completed structure, with a draft of 56 feet [17 m], was then towed to location, see Figure 11. Stability of the piggyback barge with the completed platform on top was of concern. To meet Coast Guard stability requirements the platform pedestal crane booms were placed and kept at a specified angle during the three-day transport to the location.

Installation

On-site the barge and platform were held in place at high tide with five tugs. As the tide ran out the tower legs touched bottom and the barge became free, see Figure 12, and was removed from under the platform. Crowley Marine, a subcontractor to Stolt was in charge of what turned out to be a flawless operation.

Pile Foundation

To anchor the platform to the Inlet bottom, seven steel pipe piles; three 36-inch [91 cm] diameter and four 30-inch [76 cm] diameter piles, are located in a 10-foot [3 m] diameter circle in each 14-foot [3.7 m] diameter leg. The piles have a heavier wall section positioned at and below the Inlet bottom to provide the necessary pile moment capacity. The annulus between a 6.5-foot [2 m] diameter inner tube and the outer leg shell was filled with grout to (1) connect the piles to the tower, and (2) to keep the legs from denting during winter ice conditions. The piles subsequently become the first conductor string for the wells. The piles were driven to refusal using the platform cranes and two hammers: an IHC S-90 Hydrohammer and a Delmag D-62 diesel hammer. Both hammers worked satisfactorily. The Hydrohammer, however, was much easier to handle and operate.

Platform Raising

Two 36-inch [91 cm] diameter piles were initially driven to refusal in each leg to provide a base for leveling the structure. During the two-week period between the setting of the platform and the installation of these eight piles the strong currents had washed the gravel bottom material out from underneath the platform legs and it became apparent that the platform was sitting low. With a lack of reference points for comparison, this was actually difficult to determine. Modern DGPS surveying, however, confirmed that the platform was sitting 2.5 feet [76 cm] low. Although wave action would not be a problem, the lower platform elevation could cause a problem in the event of a winter ice pressure ridge hitting the bottom of the deck. Fortunately, the piles had not as yet been grouted to the legs and it was thus possible to raise the platform. This was done by mobilizing eight hydraulic jacks, each with a capacity of 850 ton [7MN] and a stroke of 28-inch [71 cm], from San Francisco. One hydraulic jack was installed on each pile and connected to four 2.5-inch [63 mm] diameter Dywidag rods, which were attached to the platform itself. Figure 13 shows one of the hydraulic jacks in place on top of the pile. The platform, weighing approximately 3,200 ton [3,200 t], was lifted 3.1 feet [95 cm] in a two-stage operation over a 12-hour period. The Dywidag rods were then locked off and the annulus segment around the piles was filled with grout in a three-stage operation.

Pile Installation

In the original concept, it had been planned to install only the three 36-inch [91 cm] diameter piles in each leg. In the event the exploratory drilling was not successful, it would have been easier to remove the platform. The three piles in each leg were sufficient for a "ten-year" ice loading. The piles, however, reached refusal some 30 feet [9 m] higher than anticipated in the very stiff Kenai formation at a penetration of 70 feet [21 m]. As a result, the heavy wall sections of the first two piles in each leg were not at the design elevation. Removal of the soil plug inside one of the piles and then re-driving only gained a few feet [60 cm]. An initial plan to airlift a larger IHC hammer from the Netherlands was abandoned because an analysis by IHC indicated no benefit. The insides of the 36-inch [91 cm] diameter piles already driven were jetted out and filled with grout up to the bottom of the platform legs to provide additional overturning moment capacity. All other piles, including the 30-inch [76 cm] diameter piles were then modified to have the heavy wall thickness at the right place. This provided the platform with the full 100-year ice design capacity.

Scour

Based on experience at other Cook Inlet platforms the design of the platform assumed that the bottom around the legs could be scoured by as much as 5 feet [1.5 m]. Periodic diver surveys taken following completion of the platform installation revealed that there was ongoing scour around the platform legs and that it was approaching the 5 feet [1.5 m] design limit. To correct this potential problem large durabags filled with a sand-cement mix were placed around the legs during the 2002 summer season.

Rig Installation

The drilling rig is modular, meaning that it consists of a number of modules that can be transported to the platform with the supply vessel and placed on the platform using the platform cranes. Figure 14 shows the completed platform with the drilling rig in place. Wells are drilled through the piles, making those piles in effect the first conductor string.

Contingent Relocation Plan

In the event the exploratory wells were not successful (fortunately they were) the structure would have been relocated to another location in the Inlet. The procedure to do this would involve cutting the piles just below the Inlet bottom. The platform would then be lifted off bottom using a barge in a reverse of the installation method. At the new location the grouted 36 and 30-inch [91 and 76 cm] diameter pile segments remaining in the legs would be used as sleeves to drive new 30-inch [76 cm] and 24-inch [61 cm] diameter piles.

Cost

Final cost of the installed platform was $32 million, which includes about $2 million for costs incurred in raising the platform and dealing with the pile-driving problem.

Conclusion

The design and installation of the Osprey platform was innovative for several reasons: (1) using the Cook Inlet tidal action for installation, (2) the self erecting feature, (3) the floating "float-over" mating, and, (4) the capability to be relocated. All of this was accomplished very successfully thanks to a great cooperative spirit and team effort by everyone involved in the project.

References

Blenkarn, K.B. (1970) "Measurement and Analysis of Ice Forces on Cook Inlet Structures." OTC 1261.

Visser, R.C. (1969) "Platform Design and Construction in Cook Inlet, Alaska." SPE 2150, J. Pet. Tech., April 1969, 411-420.

Visser, R.C. (1992) "A Retrospective of Platform Development in Cook Inlet, Alaska." SPE 19160, J. Pet. Tech., February 1992, 146-203.

Visser, R.C. (1995) "Reassessment of Platforms in Cook Inlet, Alaska." OTC 7781.

Table 1. Environmental conditions in Cook Inlet.

Item	US Units	Metric Units
Mean annual temperature	35°F	2°C
Average minimum temperature	25°F	-4°C
Maximum water temperature	55°F	13°C
Minimum water temperature	29°F	-2°C
Extreme tide (Kenai)	32 feet	9.8 m
Maximum Current	13 ft/sec	4 m/s
Maximum wind	80 mph	130 km/h
Maximum wave	28 feet	8.5 m

Table 2. Platform Osprey design criteria.

Item	US Units	Metric Units
Mean Lower Low Water (MLLW)	45 feet	14 m
Extreme high water above MLLW	24.2 feet	7.4 m
Extreme low water below MLLW	6.0 feet	1.8 m
Maximum tide range	30.2 feet	9.2 m
Design current	13 feet/sec	4 m/s
100 year design ice thickness	3.5 ft	1.1 m
Compressive ice strength	300 psi	2.1 Mpa
Minimum ambient temperature	-40°F	-40°C
Design wind	80 mph	130 km/h
Design significant wave height	15.3 ft	4.7 m
Earthquake design criteria	API RP 2A, Seismic Zone 4	
Bottom scour	5 feet	1.5 m

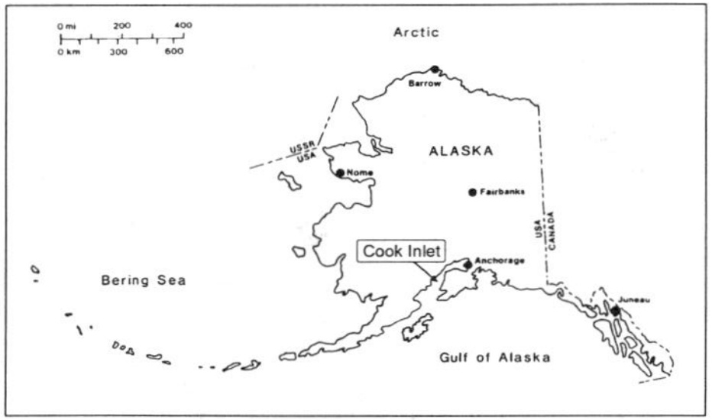

Figure 1. Map of Alaska showing Cook Inlet.

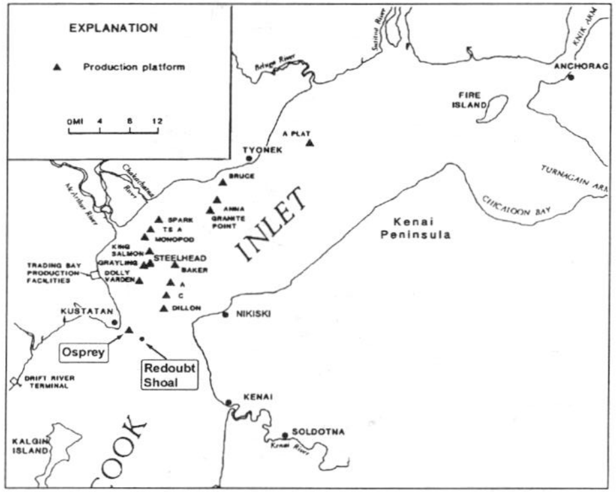

Figure 2. Redoubt Shoal field location in Upper Cook Inlet.

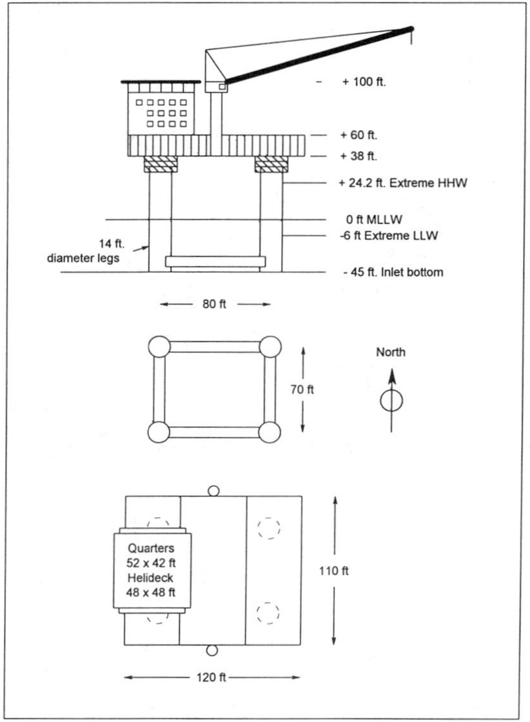

Figure 3. Osprey exploratory drilling platform.

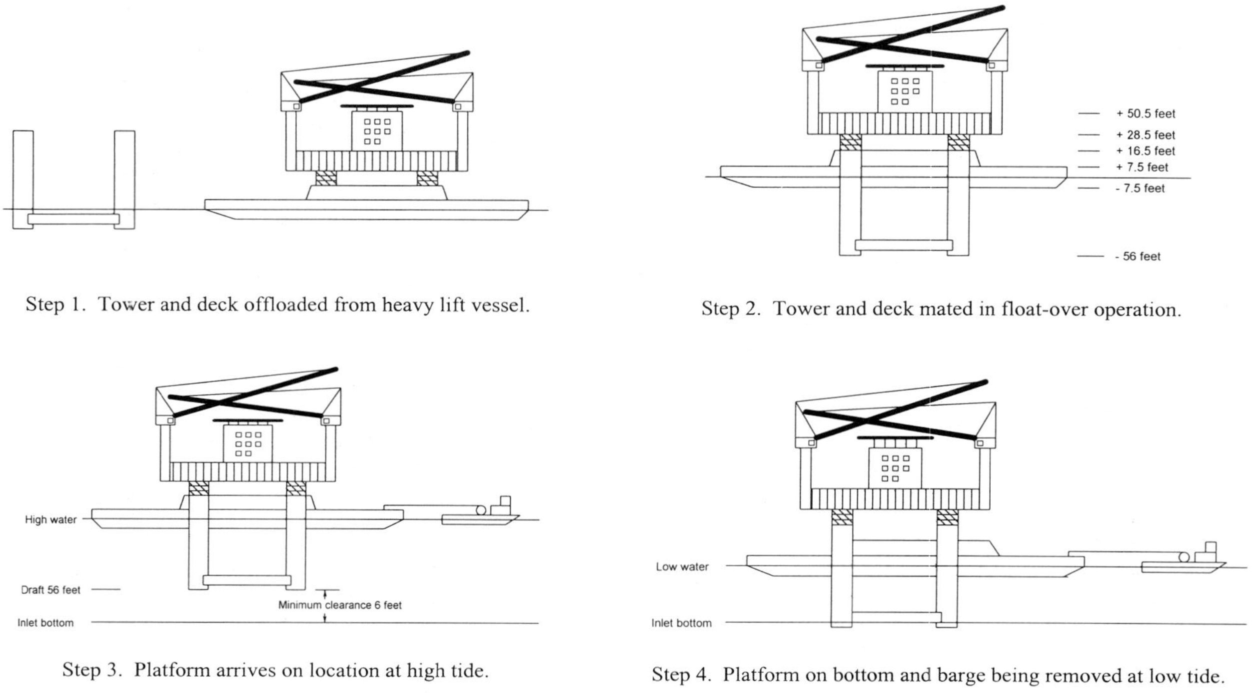

Step 1. Tower and deck offloaded from heavy lift vessel.

Step 2. Tower and deck mated in float-over operation.

Step 3. Platform arrives on location at high tide.

Step 4. Platform on bottom and barge being removed at low tide.

Figure 4. Platform installation procedure.

Figure 5. Completed tower structure at Hyundai yard.

Figure 6. Completed deck structure at Hyundai yard.

Figure 7. Living quarters were fabricated in Anchorage.

Figure 8. Living quarters and barge loaded on Dockwise Swan.

Figure 9. Arrival of the Dockwise Swan in Port Graham Bay.

Figure 10. Mating of tower and deck.

Figure 11. Departure of mated platform from Port Graham.

Figure 12. Platform on bottom and barge being removed.

Figure 13. One of the 850-ton hydraulic jacks used to lift the platform.

Figure 14. Completed platform Osprey.

Pore Structure and Degree of Saturation of Mortar Affected by Seawater and Freeze-Thaw Action

Xin Wang[1] · Koichi Ayuta[2] · Sadamu Ono[3]

Abstract

The combined effect of seawater and freeze-thaw action on the pore structure and degree of saturation of small mortar specimens ($\phi 1 \times 2$ cm and $\phi 0.5 \times 1$ cm) is investigated. The results show that the cement paste becomes porous under the influence of seawater, leading to an increase in the degree of saturation. Seawater immersion and freeze-thaw action leads to the enlargement of pores with radii in the range 10^2 to 10^3 nm, with subsequent occurrence of micro-cracks. As a result, the degree of saturation rises rapidly until the mortar is destroyed.

Introduction

Serious scaling is known to be a common problem afflicting marine concrete structures in cold regions. There are researches[1) 2)] attempts to clarify the mechanism of scaling with a focus on the surface of concrete. These investigations have shown that the pore structure becomes porous and the amount of frozen water increases due to seawater.

[1] Civil Engineer, Doctor of Engineering, C&R Consultant, 2-23-3 Hon-cho, Kokubunji, Tokyo, Japan; phone (81) 42-320-5585

[2] Professor, Doctor of Engineering, Kitami Institute of Technology; 165 Koen-cho, Kitami, Japan; phone (81) 157-26-9474

[3] President, Doctor of Engineering, C&R Consultant, 2-23-3 Hon-cho, Kokubunji, Tokyo, Japan; phone (81) 42-320-5585

However, the process of amount of frozen water rises under the action of seawater and repeated freezing-thawing remains unknown, and it is not known how the degree of saturation, a measure of frost resistance, changes while the pore structure loosened.

In this work, in order to clarify the effects of seawater and freeze-thaw action on scaling, the pore structure, mass, and microstructure of mortar are examined, and the relationship between degree of saturation and frost damage is investigated.

Methodology

Specimens. Specimens were cylindrical castings of mortar measuring $\phi 1 \times 2$ cm and $\phi 0.5 \times 1$ cm. The water-cement ratio of the mixture was 50%. Small specimens were used because they better express the properties of a concrete structure near the exposure surface. According to previous work on chemical combined water[3), the specimens used in this experimental research should express the properties of concrete less than 5 mm from the exposure surface.

Normal Portland cement and standard fine aggregate (as defined in the Japanese Industrial Standards) were used. No AE was added. The flow value of the mortar was 170 ± 5 mm. After placing, the specimens were cured for 24 hours in a room maintained at 20℃ and 90% humidity. They were then immersed in fresh water or sea water at 20℃ up to the age of 28 days. After freeze-thaw tests in seawater or fresh water, the pore structure and mass were measured and the microstructure observed. Three specimens were tested in each experiment.

Experimental Procedure.

Freeze-thaw tests. The temperature was controlled according to the program shown in Figure 1, with variations in the range -20℃ to +10℃. One cycle lasts 4 hours.

Pore structure. The specimens were crushed into 2.5 - 5 mm particles, washed with acetone, and vacuum dried so as to stop hydration. The pore structure was then examined by means of a mercury-intrusion porosity meter.

Microstructure. Prior to freeze-thaw tests and after the tests, smaller specimens ($\phi 0.5 \times 1$ cm) were dried in dry ice (D-dry) to stop hydration. Vacuum evaporation of gold leaf was then carried out to coat the surface. Microstructure observations were implemented using a scanning electronic microscope (SEM) with an acceleration voltage of 20 kV.

Mass. Mass of the specimens was measured in the moist but surface-dried state (w_0) before freeze-thaw testing (age = 28 days), and then every two freeze-thaw cycles

(W_1), as well as after D-dry (W_2). The mass of scaling was also measured after D-dry (W_3). Using Equation (1) below, the amount of scaling (H_n) was calculated. The degree of saturation was calculated using Equation (2).

$$H_n = \frac{W_3}{W_0} \times 100 \qquad (1)$$

Where, H_n : Amount of scaling (%)

$$S_n = \frac{V_w}{V_t} \times 100 \qquad (2)$$

Where, S_n : Degree of saturation (%)
V_t : Total pore volume (mm^3)
$= V \times W_1$
V_w : Volume of water in pores (mm^3)
$= (W_1 - W_2)/\rho$
ρ : Density of water ($\fallingdotseq 1.0 \times 10^{-3}$ g/mm^3)
V: Total pore volume per gram (mm^3/g)

Results and Consideration

Pore Structure. Figures 2 and 3 show the average values of pore structure and total pore volume for three specimens ($\phi 1 \times 2$ cm and $\phi 0.5 \times 1$ cm)immersed in seawater and fresh water up to 28 days of age.

Figure 4 shows the change in pore structure of $\phi 1 \times 2$ cm specimens immersed in seawater with increasing number of frost cycles. Figure 5 shows the total pore volume of the same specimens. Figure 6 shows the pore structure of $\phi 0.5 \times 1$ cm specimens after several cycles of freeze-thaw action in seawater and fresh water. Figure 7 shows the total pore volume of these specimens.

On the basis of the above results, a number of points are clarified.

- The results shown in Figures 2 and 3 demonstrate that specimens immersed in seawater have approximate two times pores in the size range 10^2-10^3 nm than those immersed in fresh water. The total pore volume of these specimens increases also. The total pore volume in the various specimens ranks as follows: $\phi 0.5 \times 1$cm (seawater) > 0.5×1 cm (fresh water) > $\phi 1 \times 2$ cm (seawater) > $\phi 1 \times 2$ cm (fresh water). This clearly shows that the concrete structure becomes more porous nearer to the exposure surface, and this tendency is particularly notable when the concrete is exposed to seawater.

- Figure 4 shows that as the number of frost cycles is increased, pores larger than 10^2 nm increase in number first. This is followed by pores of 10^3 nm radius. These pores are thought break down gradually under the influence of freeze-thaw action. On the other hand, pores of 10 nm radius tend to increase, also.
- As shown in Figure 5, although the number of pores of around 10 nm and several times 10^2-10^3 nm in radius increases with freeze-thaw action in seawater, the total pore volume does not increase with the number of frost cycles.

One possible explanation for this behavior relates to the freezing of pore water in different sized pores. In addition to there is few amount of pore in large radius side, even the water in larger pore freezes at a certain low minimum temperature, the resulting expansion is released into pores with space so no destructive hydraulic pressure is generated. On the other hand, water in smaller pores would not freeze since the temperature does not reach its low freezing point. However, water in medium-size pores generally freezes at normal freeze-thaw temperature, but because the larger pores have already frozen the spacing between unfrozen smaller pores is greater and release of the expansion is difficult. Thus these medium-size pores are easily broken down[4)] and increase in size, leaving fewer pores of the original size. This explains why, in spite of the pore distribution moving in the direction of larger size, the total pore volume remains constant.

- In the case of specimens measuring 0.5 × 1 cm in Figure 6, pores over 10^3 nm in radius enlarge remarkably as the number of frost cycles increases, and the total pore volume clearly rises.

To summarize these results, the pore structure and total pore volume differ with the specimen size, and smaller specimens more easily become porous. It appears that the surface region of a concrete structure, that is the region affected by freeze-thaw action, is very vulnerable to deterioration. In addition, the large amount of pores in several 10^3 nm after freeze-thaw action is supposed to be the cause that the structures affected by both frost action and seawater destroy to pieces.

Picture 1 shows the state of ϕ0.5 × 1 cm specimens after immersion in seawater and fresh water and after 6 cycles of freeze-thaw action. The specimen immersed in fresh water has suffered almost no scaling, while the surface of the seawater specimen is crumbling and shows signs of severe scaling. This demonstrates that the action of seawater and freeze-thaw cycles on concrete causes the concrete surface structure to become porous, making it susceptible to damage.

Microstructure. Picture 2 shows the microstructure of specimens (ϕ 0.5 × 1 cm) immersed in seawater and fresh water for 28 age days and subjected to 6 cycles of freeze-thaw action. In contrast with the specimen immersed in fresh water, whose microstructure is dense, the one immersed in seawater has a comparatively loose structure. Sometimes, when specimens were subjected to freeze-thaw action for 6 cycles in seawater, micro cracks have occurred.

The large numbers of porous pores near the surface of the concrete are destroyed under attack from repeated freeze-thaw action, and as a consequence micro cracks occur. This generation of micro cracks is considered the reason for the scaling.

Degree of Saturation and Scaling. Figure 8 shows the degree of saturation of specimens immersed in seawater and fresh water (ϕ 1 × 2 cm, ϕ 0.5 × 1 cm). Figure 9 shows the degree of saturation with increasing number of frost cycles.

These results lead to the following understandings:

- Compared to specimens immersed in fresh water, the degree of saturation of specimens immersed in seawater is high. This arises because of the exchange of Cl -ions in the seawater with OH+ ions in the cement paste, allowing the seawater penetrate much more[5)] . The degree of saturation of specimens immersed in seawater has reached 95% by 28 days, and the surface of marine concrete structures in this situation is susceptible to destruction.
- The degree of saturation of specimen subjected further to freeze-thaw action rises rapidly.

  Since 12cycles of freeze-thaw action take two days, the cause of the increased degree of saturation is thought to be freeze-thaw action. Repeated freeze-thaw cycles while a specimen is immersed in seawater and subjected to high hydraulic pressure leads to large residual strain[2)] and micro cracking occurs. The seawater penetrates rapidly, and the degree of saturation increases. In the case of freshwater, the specimen hardly changes because its structure is dense and the fresh water is hard to penetrate into the cement.
- In the case of specimens immersed in seawater, the degree of saturation after two freeze-thaw cycles is lower than that after no cycles. This is because the seawater has not yet had time to permeate.
- The degree of saturation of specimens immersed in seawater reaches 100% by the time four cycles have taken place. The penetration of seawater into the measured pores and

micro cracks is the reason for this. It is clear that this generation of micro cracks is the cause of scaling.

The figure 10 shows the relationship between degree of saturation and rate of scaling. When the saturation exceeds 100% (that is, when micro cracks begin to occur) the amount of scaling increases rapidly. It is thought that scaling starts when micro cracks form as a result of freeze-thaw action.

Conclusions

The mechanism of scaling under the influence of both seawater and freeze-thaw action is investigated by looking at the pore structure, degree of saturation, and microstructure of small mortar specimens (ϕ 1×2 cm and ϕ 0.5×1 cm). The following conclusions were arrived at:

- Specimens under the influence of seawater have many pores with radii in the range 10^2-10^3 nm as compared with those exposed to fresh water, and the total pore volume is also greater. This higher pore volume allows seawater to penetrate into the cement paste, leading to a high degree of saturation. A specimen in such a highly saturated state is vulnerable to frost damage.
- If a specimen is further subjected to freeze-thaw action in this state, pores with radii in the range of approximately 10 nm and those in the range of 10^2-10^3 nm grow gradually larger as the number of freeze-thaw cycles rises. The structure of the cement paste loosens, until finally micro cracks occur. This in turn allows much more seawater to penetrate into the cement paste, and the degree of saturation increases rapidly. Scaling then occurs.

References

Wang, X. and Ayuta, K. (1998). "Amount of Frozen Water and Pore structure of Concrete Affected by Seawater", *Cement Science and Concrete Technology*, No.52, pp.684-691, (in Japanese)

Wang, X., Ayuta, K. and Ota, T. (2000). "Amount of Frozen Water and Deformation of Concrete Affected by Seawater", *Cement Science and Concrete Technology*, No.54, pp.404-409, (in Japanese)

Ayuta, K. and Hayashi, M. (1985). "An Experimental Study on Strength of Concrete

Surface by Drying", *Transactions of Japan Society of Civil Engineers*, Vol.15, pp.510-511

Kamada, E. (1981). "Pore Structure of Cement and Frost Damage", *Proceedings of the Japan Concrete Institute*, Vol.19, No.11, pp.36-42, (in Japanese)

Ono, M. and Nagashima, M. (1978). "Some Aspects of the Mechanism of Seawater Attack on the Hardened Cement", *Review of the Thirty-second General Meeting*, the Cement Association of Japan, pp.59-60

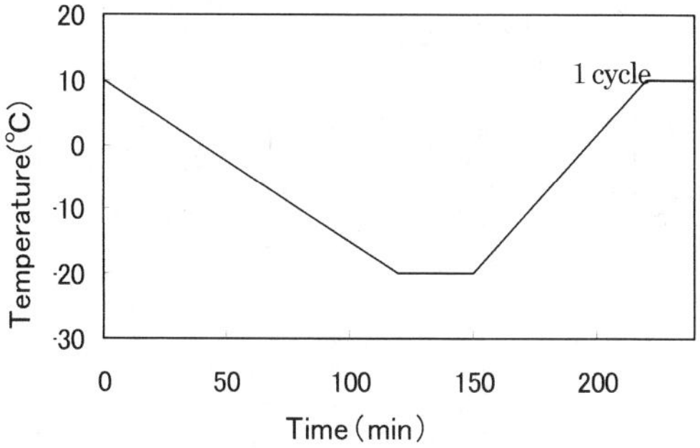

Fig.1 Temperature control program

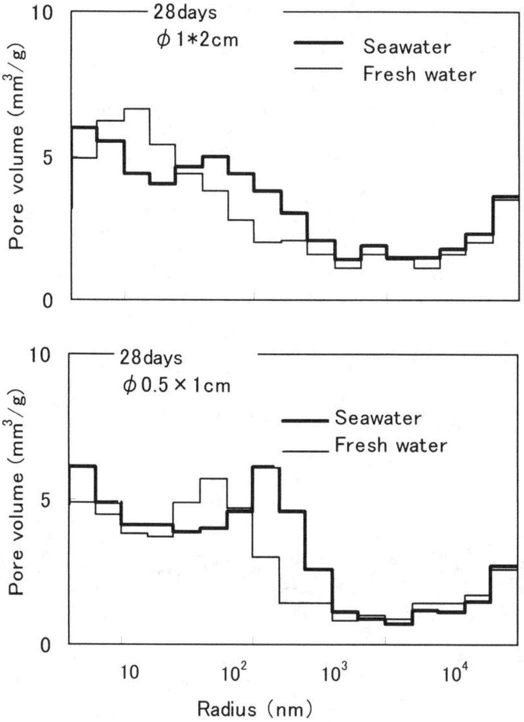

Fig.2 Pore structure of specimen immersed in seawater and fresh water up to 28 days of age

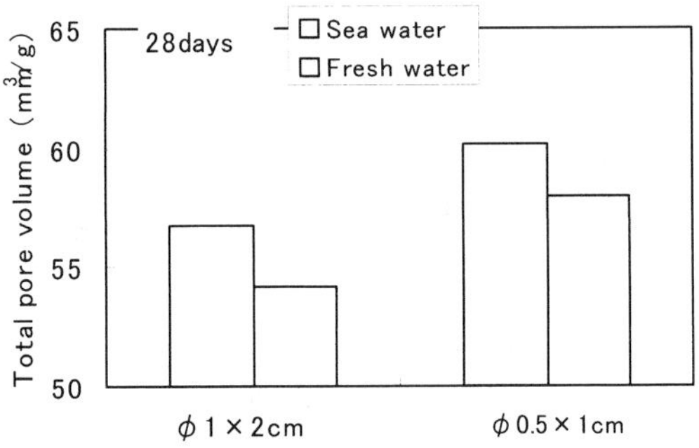

Fig.3 Total pore volume of specimen immersed in seawater and fresh water up to 28 days of age

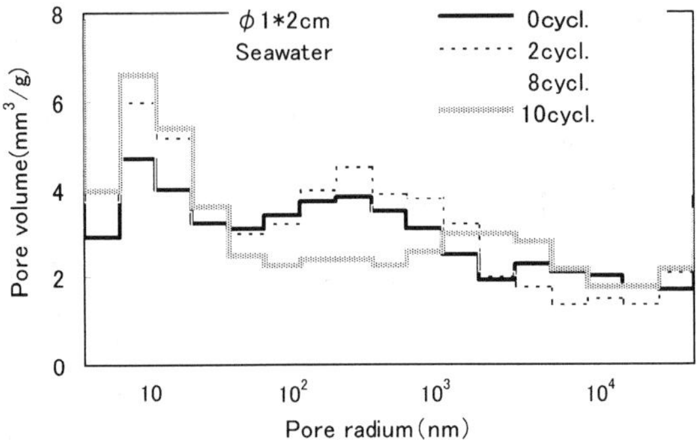

Fig.4 Change in pore structure of specimens immersed in seawater

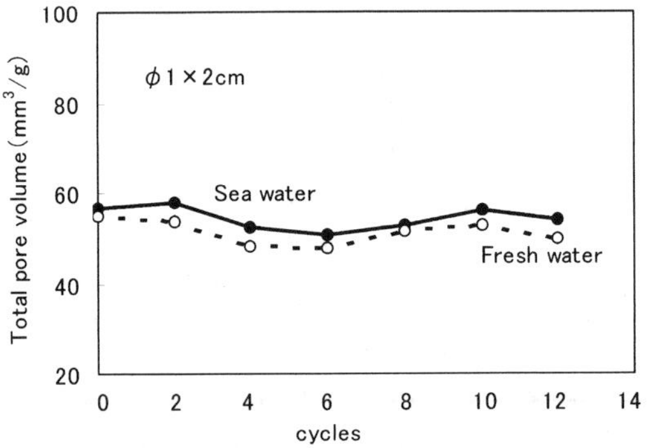

Fig.5 Change in total pore volume of specimens immersed in seawater

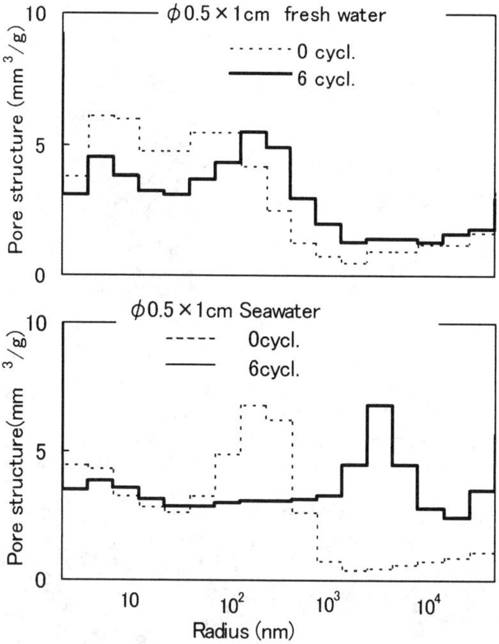

Fig.6 Pore structure affected by freeze-thaw action

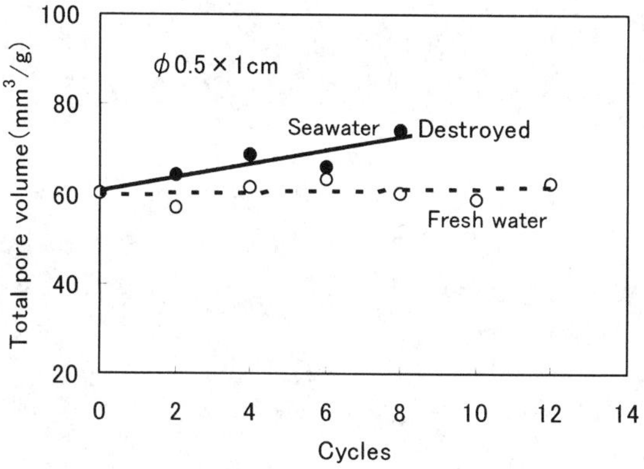

Fig.7 Total pore volume with increase of frost cycles

Picture 1 Deterioration of Specimen (0.5 × 1cm)
(Left, 6 cycles of freeze-thaw action in seawater ;
Right, 6 cycles of freeze-thaw action in fresh water)

Fresh water (28days)

Sea water (28 days)

6 cycles of freeze-thaw action in seawater

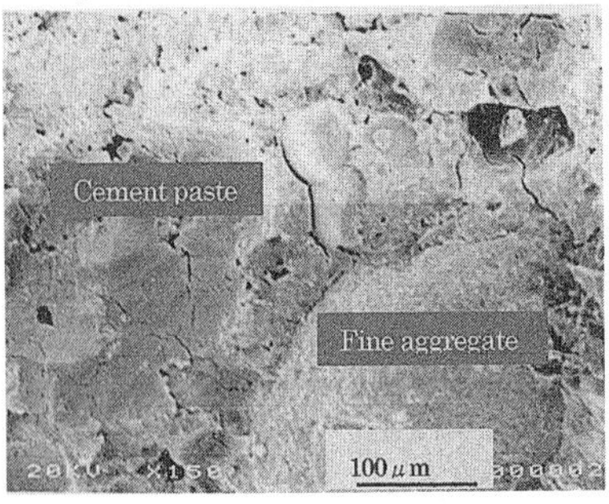

6 cycles of freeze-thaw action in fresh water

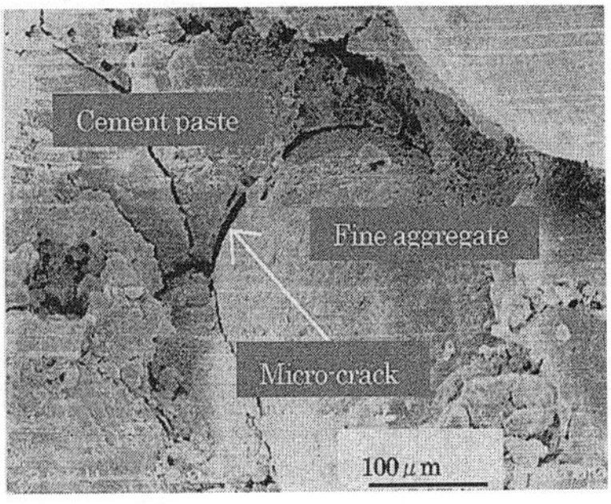

Picture 2 Microstructure

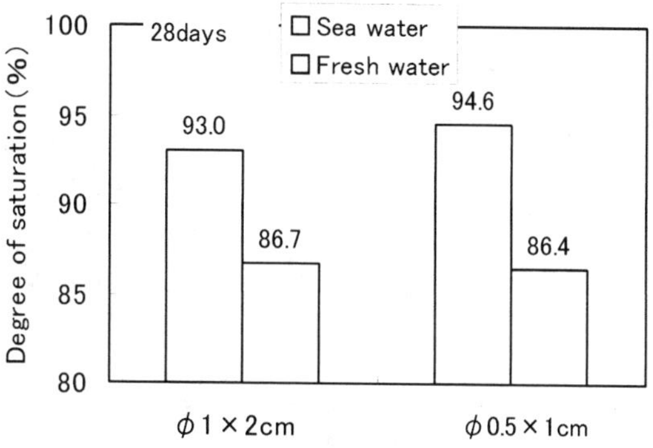

Fig.8 Degree of saturation

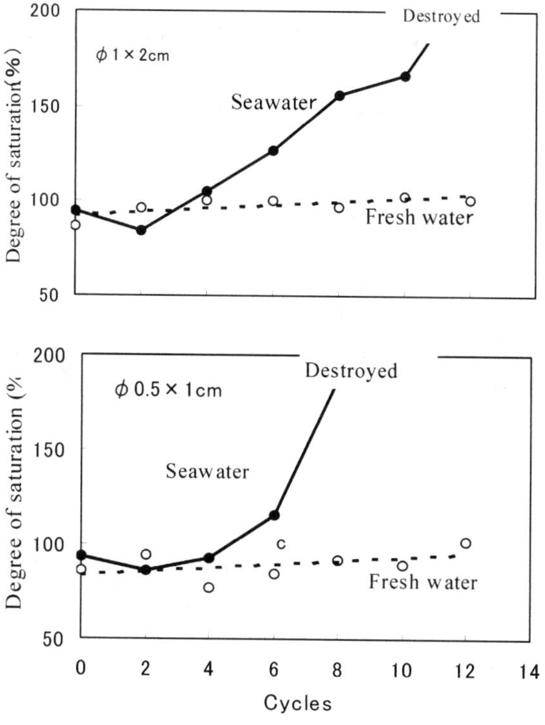

Fig.9 Change of degree of saturation with frost cycles

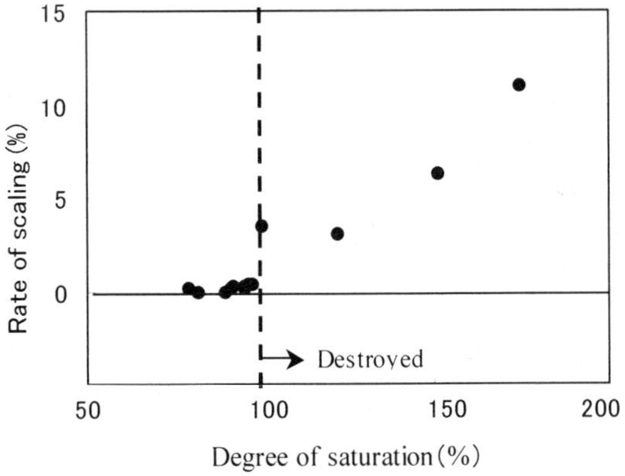

Fig. 10 The relationship between rate of scaling and degree of saturation

Emerging Survey Technologies for Alaska's Coastal Zone

Robert J. Pawlowski[1], Paul D. Brooks[2], & John L. Oswald[3]

Abstract

The coast of Alaska has received major attention for mapping during the last decade. Increased efforts to promote safe navigation through improved nautical charts; map coastal habitats for endangered species and essential fish habitats; land new seabed fiber optic cable systems; and develop port, harbor, and resource extraction infrastructure are all requiring improved cartographic products. Alaska's vastness and remoteness are depending on traditional, new, and emerging technologies to gather the surveying data upon which such projects depend. Horizontal and vertical control has improved through the use of GPS technologies, methodologies, and algorithms. Tidal zoning transfers this control to offshore waters and remote coastal areas, establishing datums for both mean lower low water and mean high water, as required by NOAA and USGS respectively. From this common point, technologies are profiling the land or seabed to comparable degrees of precision and cartographic definition. Photogrammetric technologies, including the use of Side Looking Aperture Radar (SLAR), coupled with interferometry, are defining coastal uplands to new levels of precision, regardless of land cover. Air and space borne Synthetic Aperture Radar is defining coastal lands and land-ocean processes, and river and sea ice parameters, allowing meter resolution that can be tide coordinated regardless of cloud cover. Light Direction and Ranging (Lidar), in an airborne configuration, is bridging the land sea interface, allowing airborne measurement of uplands, shoreline, and nearshore bathymetry at new levels of efficiency. The swath of land and near shore imaging couples with multi-beam echo sounder systems (MBES) producing accurate 3 dimensional cartographic products portraying the coastal bathymetry and extending off the continental shelf. Together, mapping accuracies are being obtained to the submeter level via remote sensing tools.

[1] Alaska Program Manager, Thales GeoSolutions (Pacific), Inc., 911 W. 8th Avenue, Suite 208, Anchorage, AK 99501; phone 907-258-1799; Bob.Pawlowski@thales-geosolutions.com

[2] Governmental Affairs, AeroMap U.S., 2014 Merrill Field Drive, Anchorage, AK 99501; phone 907-272-4495; pbrooks@aeromap.com

[3] Hydrographic Consultant, John Oswald Consulting, 139 E. 51st Avenue, Anchorage, AK 99503; phone 907-273-1815; joswald@lcmf.com

Coastal Surveying and Mapping

Results from Regional forums, including the Marine Transportation System Northwest Regional Dialog, the ASCE TCCRE Marine Transportation and Coastal Infrastructure breakout session, and the UAA Cold Regions Ports and Engineering Conference have identified the need for Alaska coastal baseline data as priorities for coastal morphology, geotechnical characterization, and erosion/accretion rates. Interest has increased recently for use of sophisticated mapping technologies for essential fish habitat, as demonstrated by contract surveys in summer 2001. The need for coastal baseline data dovetails with efforts by federal agencies in improving the mapping of Alaska's coastal areas. Growing interests in state of the art survey imagery for fisheries and coastal zone management, meteorology, oceanography, and international boundaries enables technology development and use. (Monahan, 2001).

NOAA and the Department of Interior agencies employ both public and private assets for state of the art shipborne multi-beam echosounder and airborne lidar survey technology and dovetail with airborne photogrammetry and satellite imagery in the coastal zone. Survey data is collected to IHO Order 1 surveys (IHO, 1988) and to NOAA specifications (NOAA, 2000) or adjusted to DOI National Map Accuracy Standards. Tide stations provide data for a common datum through tidal measurement and zoning. Airborne and spaceborne instruments provide improved determination of shorelines and shoreline features along the southern Alaska peninsula, where synthetic aperture radar enabled tide-coordinated shoreline with cloud cover (Tuell, 1998). Recent compiling of various sources of data by NOAA was able to provide verifiable shoreline for surveys in Aialik Bay, Harris Bay, and the Northwestern Fiord in Kenai Fiords National Park and the southern Alaska Peninsula. MBES surveys, supported with psuedo-sidescan imagery was collected hydrographic survey data at each location. In addition to MBES, Lidar imagery has been collected on the southern Alaska Peninsula, with additional reconnaissance at the Pribilof Islands. These surveys join with photogrammetric surveys to depict coastal features.

Tidal Measurement, Datums, and Zoning

Tidal datums are vertical reference surfaces used for a variety of coastal zone studies. Tide data and the associated reference levels are an integral part of all hydrographic surveying, as well as for determination of the offshore and onshore marine boundaries, and many other GIS related activities in the coastal zone. No hydrographic survey has meaning unless the soundings are corrected to a common datum. The Mean Lower Low Water (MLLW) datum is the reference for sounding data in the United Sates, while other tidal and geodetic datums are generally used for data above MLLW.

In Alaska, the importance of tidal datums reached it zenith in the Dinkum Sands dispute. Dinkum Sands is a low lying gravel feature in the Beaufort Sea, just offshore from Prudhoe Bay. Offshore oil leasing was conducted in this area in 1978 but the federal and state governments disputed ownership of the tracts. Part of the dispute concerned whether the Dinkum Sands was indeed an island and above a specific tidal datum. Because the financial stake of the parties was so great (original lease was $450 million), the case was eventually heard by a Special Master (of the Supreme Court). The U.S. Supreme Court decided the case in 1997 (Reed, 2000), and highlighted the importance of tidal datum data in establishing mean high water and accurate shoreline maps. This case brought to attention the poor distribution of accurate tidal datums and acceptable modern shoreline mapping.

NOAA, National Ocean Service (NOS) has been the primary authority in the United States concerning the tidal datums. This organization maintains a network of 175 operation tide stations in the United States. In Alaska, NOS maintains and operated 16 tide-gauging sites. In addition NOS, has occupied nearly 900 gauging sites throughout Alaska, primarily in support of hydrographic surveying for production of nautical charts. Due to the age and quality of the data, stability of bench marks, and vertical crustal motion NOS only publishes about 110 tidal datums in Alaska, from Prudhoe Bay to Adak to the Tongass Narrow. There are huge linear gaps in these published datums in western and northern Alaska. From Sand Point west and north, only 6 additional published datums are available over more than 3600 miles of coastline, spanning 700 miles between the datum between Dutch Harbor and Nome. NOS maintains data and information at http://co-ops.nos.noaa.gov, which includes tidal benchmark sheets (descriptions and elevations); accepted tidal datums; observed and archived tidal data at operation sites; tidal predictions; and publications.

NOAA and contractors add perhaps a dozen new stations per year in Alaska. Typically, a tide station supporting hydrographic surveying consists of one or more tide gauges mounted on shore attached to sensors that are anchored below the lowest tides. The gauges are electronic and typically interfaced to a satellite telemetry system, with near real time access to the data via the internet. Five bench marks are established to physically preserve the tidal elevations once the data is analyzed. The data is downloaded via the web, at a central processing facility. During hydro operations, the gauges are visited 1-3 times per week to ensure proper operation.

For new sites in Alaska, the tidal datum is computed using 7 to 30 days (or more) of data using the accepted NOAA method of Simultaneous Comparisons. Data from an operation NOAA gauge is used as the "base" or control site, and the new data is mathematically related to this know datum. For re-occupations of existing sites, the surveyor just needs to establish suitable equipment, and recover the bench marks; no comparisons with a control station are then needed.

Tidal zoning is a concept of predicting the time offsets and height differences in a limited project area. Theoretically a tide datum is only adequate in the immediate vicinity of the tide gauge. For practical purposes the surveyor desires to establish as few gauging sites that will be practical and meet project standards. For data points between tide gauging sites a zoning scheme can be made using new and/or historic data. Conventional techniques rely on linear interpolation to define the geometry of the zoning cells. Figure 1 shows a recent project in upper Cook Inlet using a combination of linear interpolation (for the co-range lines) and a wave travel speed formula for the co-phase lines. NOS is also examining more robust methods for zoning using harmonics (TCARI) and GPS, while others are trying to adapt numerical analysis techniques.

Several agencies and countries are beginning to implement GPS for use as a "tide gauge". The GPS is mounted on the hydrographic vessel, and phase data processed in the Real Time Kinematic (RTK) mode or post processed relative to a known elevation. This holds great promise, as the tidal corrector is then observed exactly where needed and not at a shore based tide gauge and then zoned. NOAA recently funded a Small Business Innovative Research grant to adapt GPS in an offshore buoy measuring water levels. One hurdle to the GPS approach is that the fundamental GPS height system (ellipsoidal heights) is different than the tidal datums that are required for most marine work. The relationship between MLLW and ellipsoidal height system is poorly known, in Alaska. As the GPS technology matures more coastal "tide" applications will become commonplace.

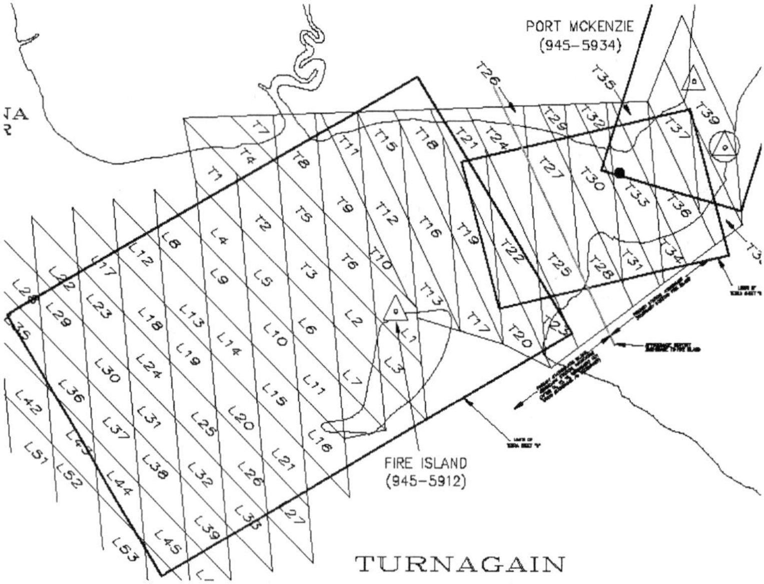

Figure 1. Tidal zoning model for upper Cook Inlet. The N-S lines represent co-range lines while the NW-SE lines represent co-phase lines. Each diamond shaped cell represents a tidal zone with associated time and height offsets to an operating gauge.

Multibeam Echosounder Surveying

MBES is the technological advance from the coupling of single beam echosounder systems and side-scan sonar systems. Controlled by GPS and reduced through heave-motion compensation, MBES enables an accurate swath of the seabed acoustically. However, MBES provides better coverage over the individual systems in that it enables faster data collection rates and overcomes the limits in the area of coverage, as was observed in the data collection in upper Cook Inlet. In previous surveys, tidal currents in excess of 5 knots were routinely encountered and cause difficulties in maintaining data collection speed and side scan sonar towfish depth and position.

MBES enabled a more accurate survey of the area, with high resolution of migratory features like sand waves, ridges, as well as boulder fields and pipelines.

MBES provides a complete survey of the ocean bottom from a motorized vessel versus towed vehicle. Collecting a maximum of 101 beams over 150° swath, an acoustic portrayal of the seafloor is collected. Providing for overlap in accordance with IHO and NOAA standards, a complete acoustic survey of the ocean floor is obtained. Sounding volumes, in excess of 3.2 billlion per 150 square nautical miles are amassed, prior to reduction to a level adequate for accurate digital terrain modeling. This enables resolution of seabed features at the sub-meter level coverage.

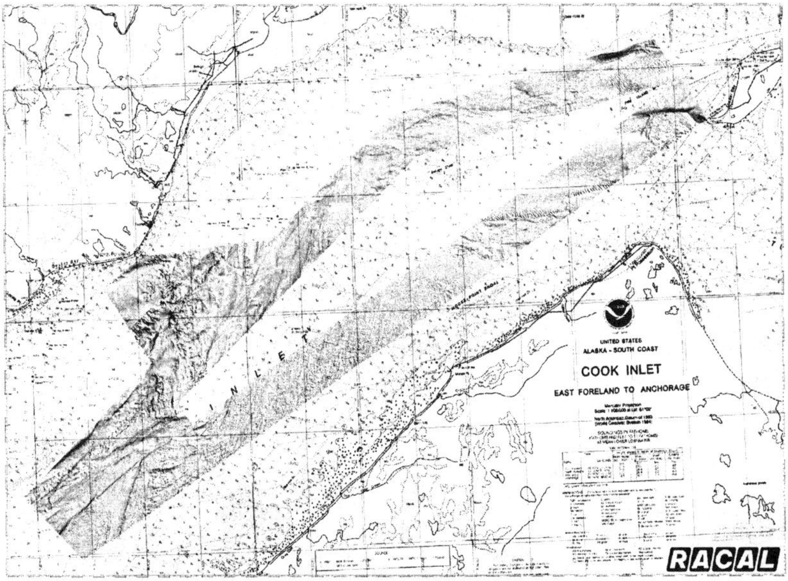

Figure 2. MBES survey of pilot shipping corridors in upper Cook Inlet, as completed for NOAA, 1999.

Figure 2. depicts a survey of Cook Inlet, as produced from high density MBES data collected by the Reson 8111 system, on contract for the NOAA. The survey provides an acoustic portrayal of seabed structures that previously could only be portrayed by side scan sonar imagery. MBES allows the complete understanding of the 3 dimensional features that form the seabed structure. Rocky outcroppings, depositional areas between structures, sand waves, and varying substrate types can be identified. In the case of Cook Inlet, previous unmapped areas of migratory sand waves were identified. Similarly, the extent of boulder fields and glacial erratics could be clearly charted. This presentation also shows the value of combining digital terrain models with the nautical chart in discerning bathymetric detail versus charted soundings, shadings, and contours.

MBES Psuedo-sidescan and Backscatter Imagery

To address fishery habitat questions, a better understanding of sediment types is required. The capability to utilize sound attenuation in determining marine sediment types has been noted for years with the relationship of grain size, porosity, and attenuation defined by Hamilton (1972). Modern multi-beam technology creates psuedo-sidescan imagery through the interpretation of the intensity of sound backscatter. In surveys conducted during 2001, backscatter data was collected with a Reson 8111 aboard R/V Davidson. The Reson 8111 multibeam sonar produces backscatter records along with range and angle packets used for bathymetry. The 8111 can generate backscatter data, which can be collected on a beam-by-beam basis. This backscatter from an individual beam is referred to as a snippet. While a standard sidescan image is produced using one large beam on each side of the sonar, snippets are produced individually from each beam in the multibeam sonar. Snippets can be laced together, end to end, to produce a sidescan type image. The advantage in snippets stems from a large improvement in signal to noise ratio in the image; the result of using a focused beam, rather than a broad beam to sample the backscatter (Lockhart, personal communication). By collecting snippet data as a part of the multibeam survey, greater understanding of the seabed characteristics can be derived.

Figure 3. depicts a combined digital terrain model, derived from the MBES bathymetry data overlaid with the psuedo-sidescan swaths enhanced for backscatter. The image shows the increased distinction of seabed structure derived from the analysis of beam-by-beam backscatter data and provides association with geologic features. This processing enables interpretation for classifying seabed morphology and identifying substrates associated with various habitats.

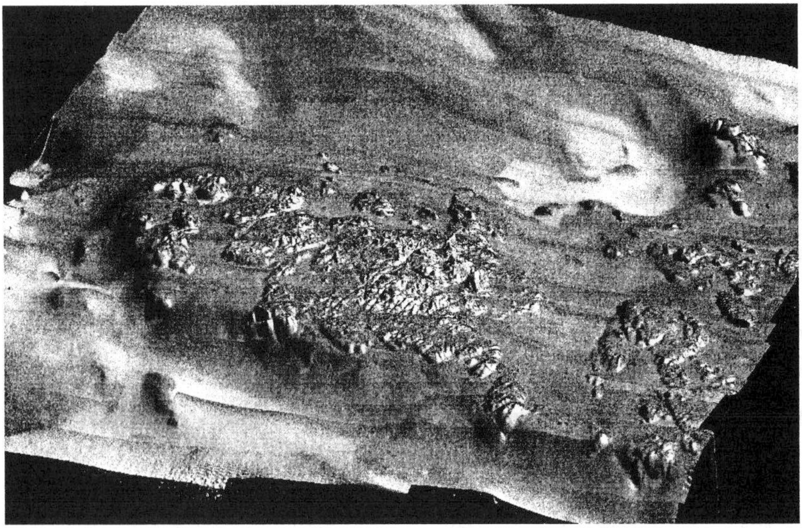

Figure 3. MBES digital terrain model with backscatter drape from Kodiak, Alaska.

Airborne Lidar

Airborne Lidar (Light Detection and Ranging) technology is an emerging technology that is proving itself in addressing coastal issues through Airborne Lidar Hydrography (Barbour, 2001) and the capability for Rapid Environmental Assessment with Airborne Lidar Bathymetry (ALB) (West et. al. 2001). ALB provides the ability to survey in sensitive areas with minimal impact, as demonstrated by the SHOALS system. Comparable applications have been completed during the 2001 field season employing the Tenix Laser Airborne Depth Sounder (LADS) (Sinclair, personal communication). Operating at 1200 feet, surveys were completed over the sensitive nearshore areas of islands located in the Alaska Maritime Wildlife Refuge. The Semidi Islands, along with Chankliutt, Chowiet, and Nakchamik Islands. Surveys were completed from + 20 meters to –23 meters. Depth of survey was influenced by water clarity, turbulence, and lack of surface obstruction, e.g. kelp. Survey data provided cartographic information for nautical chart. However, survey also provided nearshore information on the steepness of beach slope, breadth of intertidal and storm impacted surf zone, and areas of runoff (through turbidity). Masking of signal represented areas of submerged and floating aquatic vegetation and runoff associated with turbidity plumes. Additional reconnaissance imagery provided insight into specific engineering applications. Figure 4. shows the aerial photograph and lidar swath of St Paul harbor, Pribilof Islands, including the entrance channel, breakwater, and harbor basin, as completed by Tenix LADS.

Recognizing the different capabilities based on the technology involved and the platforms flown, the application of Lidar for ALH proves to be an effective tool for working in the nearshore zone, where bathymetric data collection is difficult because of physical factors and environmental conflicts. Utilizing such tools in engineering applications, particularly environmental assessment, may prove to be a

Figure 4. Aerial photograph and LADS lidar imagery from St. Paul harbor, Pribilof Islands during August 2001. Images courtesy of Tenix LADS.

cost effective alternative for large well coordinated projects, as long as the environmental conditions allow adequate penetration into the water column. The

Aleutian Islands and Southeast Alaska provide just such applications. Cook Inlet, and other glacially impacted waterways do not. Season applications may exist for the deltas of Western and Northern Alaska, including Kotzebue Sound and Alaska's north slope, where periods of clear water occur during periods of decreased wind and runoff. The application should be integrated into future decisions, based on magnitude of the area to be surveyed, shallowness of the bathymetry, and costs for mobilization and operations.

Aerial Photography and Airborne Positioning (GPS)

Aerial photography and remote sensing have been employed for assessing completion of integrated airborne and ground-based GPS surveying for many years (Figure 5.). Data collected is developed into photo laboratory reproductions, image analysis and GIS services, cartography, digital image processing and mapping, and software development.

In planning aerial photography missions, remotely sensed satellite imagery is utilized in real time to dispatch flight crews to optimized acquisition of imagery. The efficient dispatch system, coupled with multiple flight crews and aircraft, provide the best possible assurance that projects will be completed on schedule. Flying heights are customized for each flight line, ranging from 1200 feet to 24,000 feet above mean terrain. The versatility of equipment and staff to deal with large, multi-state projects, or small single exposure jobs means critical details such as sun-angle, film type, camera settings, camera exposure stations and environmental conditions, are correct for all projects.

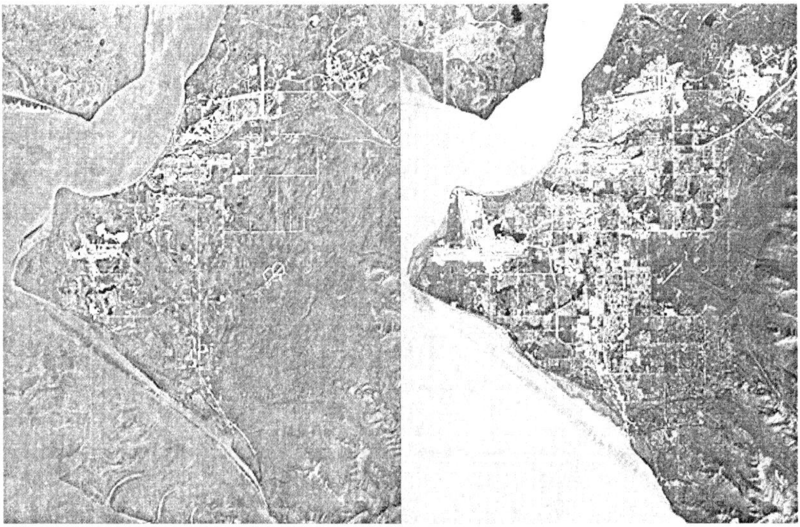

Figure 5. Aerial photograph of Anchorage, Alaska from 1950 and 1999, showing 50 years of growth. Photographs courtesy of AeroMap Inc.

Effective photogrammetry requires the application of precision GPS and attitude determination of airborne platforms. By resolving combined solutions of GPS, airborne mobile and land fixed, integrated with Inertial Measurement instruments and software, project control design is achieved. This provides datum resolution; accurate geodetic ground control; precise GPS positioning of aircraft and mapping sensors; and attitude determination of aerial mapping cameras and other airborne sensors. Present technology allows positioning airborne mapping sensors to accuracies of 5cm, and to determine orientation angles to accuracies of 20 arc seconds. Using the GPS/IMU solutions (Figure 6.), sub-meter accuracy can be carried for distances up to 100 km from the GPS Base Station. Together, advanced positioning systems greatly reduce, or sometimes eliminate, the need for ground survey control for mapping operations.

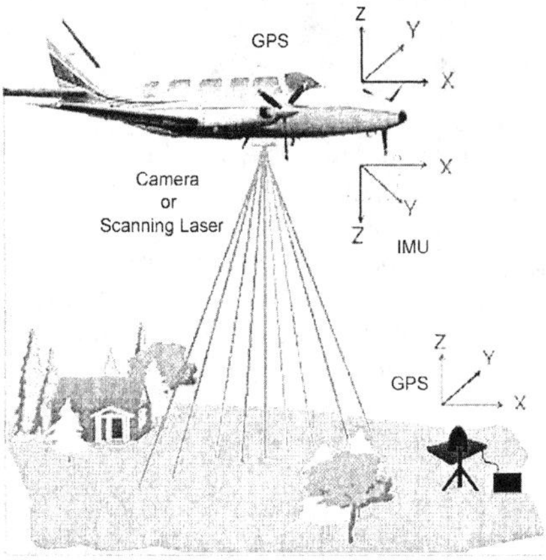

Figure 6. Example of GPS/IMU xyz control solutions required for attitude adjustment during airborne data collection.

In some cases the need for the aero-triangulation step in map production can also be eliminated, leading to greater costs savings and accelerated schedules.
These stated accuracies are in reference to the map product ground control coordinate system and allow the production of the highest accuracy and most cost efficient mapping in the industry. On-going research and development programs continue to improve accuracies and broaden applications with these airborne positioning systems

Remote Sensing and Geographic Information Systems (GIS)

The latest in airborne GPS and inertial systems are used to control both aerial photography and remote sensing acquisitions including digital thermal multispectral, hyperspectral, and LIDAR surveys. Once data for mapping or GIS projects are compiled, setting up the applications, populating databases, selecting an analytical approach, and conducting the analysis is completed. By combining GIS and remote sensing, a wide variety of spatial relationships from simple overlay analysis to complex network and surface models are accomplished. A recent example was completed for Fort Knox Mine in interior Alaska. Employing the Intermap

Technologies, Inc. STAR-3i interferometric synthetic aperture radar products, which include high resolution DEM (0.5 – 3-meter vertical accuracy, 2.5-meter horizontal accuracy and 5-meter posting) and Ortho-rectified Radar Imagery (ORRI), Aeromap conducted image collection for the Fort Knox Mine in interior Alaska (Figure 7).

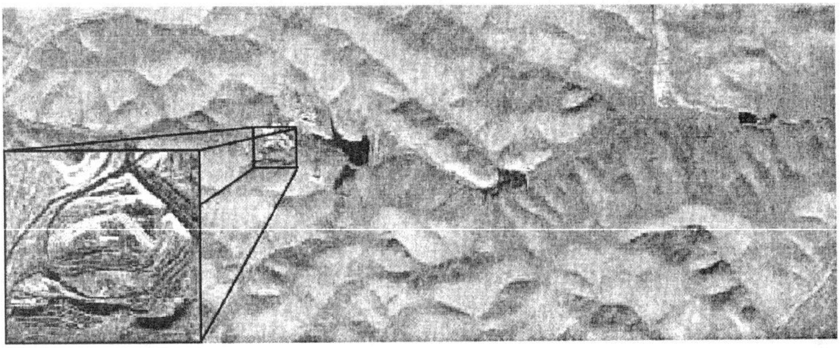

Figure 7. Orthorectified Radar Image of Fort Knox Mine Project. Plot scale was 1:18,000 and enlarged area at 1:5,000-scale, with absolute accuracy of 2.5 meters.

For this project data was acquired, products developed, and samples provided of the products with the original data to Alaskan miners for their evaluation and use as a learning and product-familiarization tool. Below are examples of products that were produced over the Anchorage area to demonstrate the IFSAR technology and the value-added products that can be produced to meet the high accuracy mapping requirements of the State.

Applications of remotely sensed data are not restricted to one sensor, but build upon the myriad of sensors available. Since the 1980's data archives have been privately and publicly available for satellite imagery including LandSAT, SPOT, IRS,

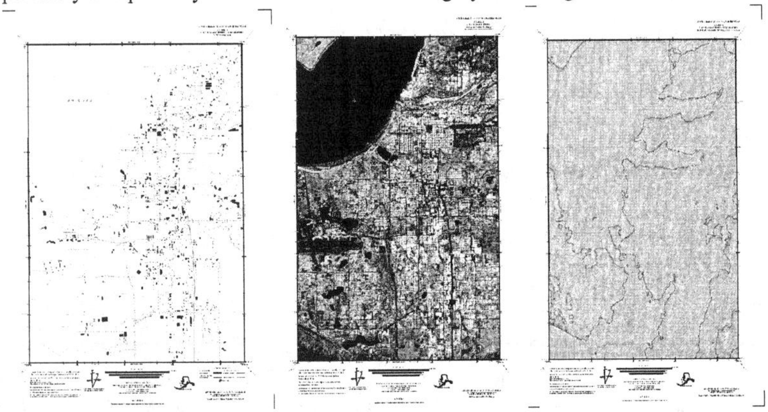

Figure 8. STAR-3i derived products: Topographic map with five-meter contours; orthorectified radar image map; shaded relief with five-meter contours

IKONOS, RADARSAT, and both ERS satellites. The data from airborne sensor systems as well as from satellite platforms, cover the range from the ultraviolet to infrared and thermal spectral bands. This provides the ability to depict the specific feature of interest to the scientists or engineer, as shown in Figure 8. This has enabled low-cost, high-resolution, multispectral solutions using SpecTerra Systems DMSV system, the Daedalus AMS and Positive System's ADAR sensor. Hyperspectral services requiring a selection of more discrete, simultaneously acquired, spectral bands are being applied. This technology is being complimented by the evolution of laser-profiling and LIDAR survey acquisitions to develop precision digital elevation models (DEM) upon which modern GIS systems are dependent.

Mapping, Editing and Image Processing

Editing, Image processing, and GIS applications employ data collection technology with DAT/EM Summit Softcopy and Analytical Stereoplotter systems. Specially modified aero-triangulation software, designed for conventional ground control or airborne GPS control is applied to Alaska mapping projects. With precision image scanning and a full suite of image processing software and equipment, products including digital orthographic imagery, color-matched digital mosaics, thematic maps or high-resolution panchromatic and multispectral satellite imagery are derived. These techniques are applied to image analysis and land classification or environmental impact analysis.

Processing to produce comparable images of the seabed is completed through a comparable specialized suite of software, tailored for resolving acoustically collected bathymetric data. By employing WinFrog, CARIS, and TRITON-ELICS, 3 dimensional DTM's, as well as decimated smooth sheets, contoured for depth are produced. New processing techniques, as applied to backscatter data collected during the psuedo-sidescan mode, is yielding geomorphology and geotechnical data. These processing techniques are expanding the information available from specific acoustic sampling, a limiting factor in surveying beyond the immediate coastal zone.

In the near shore zone, light, in the form of LIDAR, enables bringing specialized image processing to the data. This processing bridges the challenges between high speed data collection and reflectance challenges in the land-water interface. Specialized software enables integrating LIDAR with horizontal and vertical control to insure consistency between flight lines and accuracy in resolving water levels with various stages of the tide.

All coastal zone data sets are controlled for both the horizontal (GPS geo-corrected image) and vertical (tidal datum) control. Tidal zone models are developed to derive correctors that may be applied to MBES, LIDAR or coastal photogrammetry. In either case, rectifying the vertical component is essential in providing an accurate image or DTM for use in cartographic production or GIS interpretation. Once processed, all imagery listed above, can be moved into various GIS applications.

Conclusion

In conclusion, emerging technologies via airborne, spaceborne, and shipborne platforms are providing a visual depiction of Alaska's coastal zone never achieved. The recent application of MBES with 100% bottom coverage yields information on seabed that will benefit not just shippers and engineers, but environmental scientists, fishery biologists, and geologists into the far future. Bridging the shallow coastal zone with Lidar technology will enable cost effective data collection for coastal geomorphology and the angle of repose, upon which decisions on coastal management projects can be made. The juncture of airborne Lidar with aerial photography promises improved continuity of data into the uplands, given use of common elevations and combined technologies. Together, when referenced to accurate horizontal and vertical datums, high resolution cartographic products are available for engineers, scientists, and managers working in Alaska's coastal zone.

References

Barbour, K. E. (2001). "Remote Sensing in Hydrography." *Hydro International.* 5(3), 7-9.

Hamilton, E. L. (1972). "Compressional-wave attenuation in marine sediments." *Geophysics.* 37(4), 620-646.

International Hydrographic Office (IHO). (1988). *IHO Standards for Hydrographic Surveys (S-44).* IHB. Monaco.

NOAA. (2002). "NOS Hydrographic Surveys Specifications and Deliverables." http://chartmaker.ncd.noaa.gov/hsd/specs/specs.htm

NOAA. (2002). "Center for Oceanogrpahic Operations Products and Services" http://co-ops.nos.noaa.gov

Reed, M. W. (2000). *Shore and Sea Boundaries. Volume 3.* U.S. Department of Commerce. Washington, D.C.

West, G.R., Lillycrop, W. J. and Pope, R.W. (2001). "Utilizing Airborne Lidar Bathymetry Technology for REA." *Sea Technology.* 42(6), 10-15.

West, G.R., Lillycrop, W.J. and Pope, R.W. (2001). "Keeping a Low Profile." *Hydro International.* 5(4), 28-31.

Tuell, G. H. (1998). "The Use of High Resolution Airborne Synthetic Aperture Radar (SAR) for Shoreline Mapping." *ISPRS Commission III Symposium.* ISPRS

Monahan, D., Wells, D.E., and Hughes-Clark, J. (2001). "Ocean Mapping: A Slantendicular Look Ahead." *Sea Technology.* 42(6), 44-51.

Numerical Simulation of Ice Interaction with a Wide Cylindrical Pier

Anne Barker[1], Mohamed Sayed[1], and Garry Timco[1]

Abstract

A depth-averaged, hybrid Lagrangian-Eulerian model for ice interaction is described, and used to evaluate ice loads on a 10m-wide, cylindrical pier. Additionally, results previously calculated using a discrete element model, and those from analytical formulas commonly used to estimate ice forces, were compared with the full-scale data. The comparisons are conducted in order to verify the results of these different types of approaches for calculating ice loads. There was good agreement between field observations from similar structures and the hybrid model's calculated values. The results obtained using the discrete element model also compared favourably to full-scale data. In contrast, forces estimated from analytical formulas were much higher than those from the field measurements.

Introduction

The present work concerns recently developed numerical models of ice-structure interaction. These models may save time and expense when used in conjunction with conventional physical models. It is desirable to know how accurate numerical models may be in predicting full-scale features such as ice forces and pile-up heights; a question, often asked concerns the validation of model data. Additionally, it is of interest to compare the results from numerical models with those obtained with analytical solutions in order to evaluate their respective advantages and disadvantages.

Prior to the construction of the Confederation Bridge in Canada, a discrete element model was developed by the Canadian Hydraulics Centre (CHC) to evaluate potential ice loads on the chosen bridge pier design (Sayed, 1995). While discrete element models produce detailed results concerning initial ice failure, these models are computationally demanding, especially for longer simulations. Subsequently, a depth-averaged numerical model of floating ice movements and deformation was developed at the CHC. The model is based on using a Particle-In-Cell (PIC)

[1]Canadian Hydraulics Centre, National Research Council, Ottawa, Ontario, K1A 0R6 Canada, Anne.Barker@nrc.ca Mohamed.Sayed@nrc.ca Garry.Timco@nrc.ca

approach, combined with a viscous plastic ice rheology. It uses discrete particles to model ice advection, while solving the momentum equations over an Eulerian grid. The model, therefore, follows a hybrid Lagrangian-Eulerian formulation. The use of discrete particles reduces numerical diffusion, and improves the accuracy of modelling ice boundary conditions. At the same time, the use of an Eulerian grid makes it possible to utilize an implicit numerical solution scheme for the momentum equations, which substantially increases the computational efficiency.

In the present paper, the depth-averaged, hybrid model is used to estimate ice forces on a cylindrical pier. Comparisons with available field observations are presented. These comparisons are semi-quantitative due to differences in geometry and the limited availability of public data. The results from earlier simulations using a three-dimensional discrete element model and analytical formulas are also compared to the full-scale data.

Particle-In-Cell Numerical Model

Details of the hybrid numerical model were reported by Sayed and Carierres (1999) and Sayed et al. (2000). Only a brief overview of the model is given here. For the Particle-In-Cell (PIC) approach, the ice cover is represented by discrete particles that are individually advected. Particles are used to model advection and keep track of ice thickness and concentration (or area coverage). The momentum equations are solved over a fixed (or Eulerian) grid. Continuum equations are also used to describe the rheology of the ice. In the depth-averaged, two-dimensional PIC formulation, each particle is considered to have an area and a volume. The area of a particle can decrease if the pressure exceeds a certain limit (i.e. ridging pressure). The volume, however, remains constant. Thus, if a particle is subjected to relatively high pressures, its area may decrease, and its thickness would correspondingly increase (constant volume = area x thickness).

Governing Equations

The linear momentum equations are expressed as

$$\rho_{ice} h \left(\frac{d\vec{u}}{dt} + \vec{u}\,\nabla\vec{u} \right) \frac{d\vec{u}}{dt} = \nabla \cdot \sigma + \vec{\tau}_a + \vec{\tau}_w + \vec{\tau}_g \tag{1}$$

where ρ_{ice} is the ice density, h is the ice thickness, $\vec{u}$ is the velocity vector, σ is the stress tensor, $\vec{\tau}_a$ and $\vec{\tau}_w$ are the air and water drag stresses, and $\vec{\tau}_g$ is the grounding stress. The air and water stresses are calculated using common quadratic formulas.

The stress-strain rate relationship is given by

$$\sigma_{ij} = -p\delta_{ij} + 2\eta \dot{\varepsilon}_{ij} + (\zeta - \eta)\dot{\varepsilon}_{kk}\,\delta_{ij} \tag{2}$$

where $\dot{\varepsilon}_{ij}$ is the strain rate, p is the mean normal stress, η is the shear viscosity, and ζ is the bulk viscosity. Note that $\dot{\varepsilon}_{kk}$ is the volumetric strain rate. We assume here that the bulk viscosity, ζ is zero.

The mean normal stress, p, is usually considered to increase with increasing ice compactness, A (area of ice/total area). We use a formula that relates the pressure, p to ice coverage (or aerial concentration) as

$$p = P^* h_{ice}{}^2 \exp(-K(1-A)) \tag{3}$$

where P^* is a reference ice strength, and K is a constant.

The Mohr-Coulomb criterion is introduced by giving the shear viscosity, η, the following value

$$\eta = \frac{(c\cot\phi + p)\sin\phi}{\Delta} \tag{4}$$

where ϕ is the angle of internal friction, and c is the cohesion. The strain rate Δ is given by

$$\Delta = \max\left(\left|\dot{\varepsilon}_1 - \dot{\varepsilon}_2\right|, \dot{\varepsilon}_0\right) \tag{5}$$

where $\dot{\varepsilon}_1$ and $\dot{\varepsilon}_2$ are the principal strain rates and $\dot{\varepsilon}_0$ is a threshold strain rate. For relatively large strain rates, $\Delta > \dot{\varepsilon}_0$, the rheology is plastic and the yield criterion is satisfied. At small rates of deformation, however, the shear viscosity becomes constant, and the corresponding rheology would be viscous. A very small threshold strain rate (typically $\dot{\varepsilon}_0 = 10^{-20}$ s^{-1}) is used in order to maintain a predominantly plastic deformation. A tension cut-off value is further introduced in the model. If a tensile stress develops, the viscosity coefficient, η, is adjusted such that the maximum principal stress is set equal to zero.

Numerical Approach

The preceding set of equations, together with PIC advection, is sufficient to determine the stresses, velocities, and configuration of the ice cover through its interaction with a structure. Starting from an initial configuration, the numerical solution of the above governing equations updates the velocities, pressures, thicknesses, and concentrations at each time step.

The main logic of the solution consists of the following steps: At each time step, the areas of the particles are *mapped* to a fixed grid using linear interpolation; i.e. the areas of the particles are converted to continuum ice concentration values. Similarly, the thicknesses of the particles are converted to continuum ice thickness. Thus, when calculating ice concentration for a node in the fixed grid, particles closer to the node are given higher weight than those farther away from the node. Next, the continuum momentum and rheology equations are solved over that grid. The use of a fixed grid makes it possible to employ *implicit* numerical methods. The solution of the momentum and rheology equations gives velocity values at the nodes of the fixed grid. Those velocities are mapped from the nodes to the particles in a manner similar to that discussed above. An algorithm is used to determine the particles' positions from the velocities - the particles are then advected to their new positions.

Previous Studies

The PIC model has been validated against data from the fixed offshore structure, "Molikpaq" (Kamesaki et al. 2000), the floating conical drilling ship "Kulluk" (Barker et al. 2000a; 2000c), and ice forces on a 2 m wide bridge pier (Barker et al. 2000b). From these and other similar studies, suitable values for the angle of internal friction and cohesion have been determined for various study regions. For this study, an angle of internal friction, ϕ, of 45° and a cohesion value, c, of 50 kPa were used, based upon these previous studies and calibrations.

Particle-In-Cell Model Test Cases

The pier geometry for the test cases was chosen for ease of comparison with the results of a previous discrete element model. An idealized pier represented by a cylinder of 10 m diameter is used. A total of six test cases were chosen for this study. Table 1 shows the details of these cases. The grid had node intervals of 0.5 m in the X and Y directions. The time step was 0.5 s. The driving force was applied by using a constant ice velocity upstream of the study area. The water and wind drag coefficients were 0.005 and 0.002, respectively. As previously mentioned, the angle of internal friction was chosen to be 45°, and the ice had a cohesion value of 50 kPa. The ice cover was represented by a thickness of ice rubble, of either 9 m or 20 m thickness, to represent a thick ridge loading a pier. The results from the study indicated that a small wake formed behind the pier, which remained open for approximately 25 m downstream of the pier, before closing again. This was observed for all of the simulations, with the length of the wake varying slightly. There was a rubble pile-up upstream of the pier for all simulations. For the PIC test cases, one hour of computer run-time was needed in order to simulate 150 s of ice interaction, using a 1GHz personal computer.

Table 1. Test Cases

Run	X Nodes	Y Nodes	Ice Thickness	Ice Velocity
01	400	100	20 m	0.5 m/s
02	500	100	20 m	1.5 m/s
03	400	100	20 m	1.0 m/s
04	400	100	20 m	0.2 m/s
05	400	100	9 m	0.5 m/s
06	500	100	9 m	1.5 m/s

The maximum calculated force ranged from 1.0 MN for a velocity of 0.2 m/s, with a 20 m thickness, to 4.8 MN for the 1.5 m/s velocity, with the same ice thickness. Considering sail height to be 0.1 of the total thickness, the resulting maximum pile-up height varied between 4 m and 10 m for a 20 m initial thickness. A force-time plot of the forces on the structure for the 20 m thickness cases is shown in Figure 1. For the 9 m thickness, the maximum calculated forces were 0.4 MN and 1.2 MN for the 0.5 m/s and the 1.5 m/s cases, respectively. The maximum sail height varied from 4 m to 5 m. The results for the 9 m thickness are shown in Figure 2.

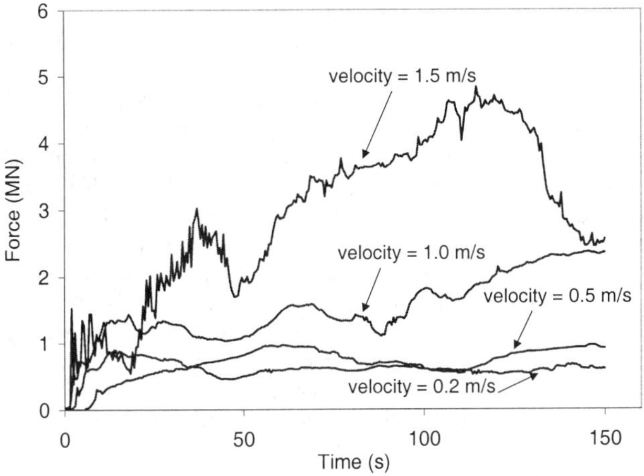

Figure 1. Force versus time plot, 20 m rubble ice thickness

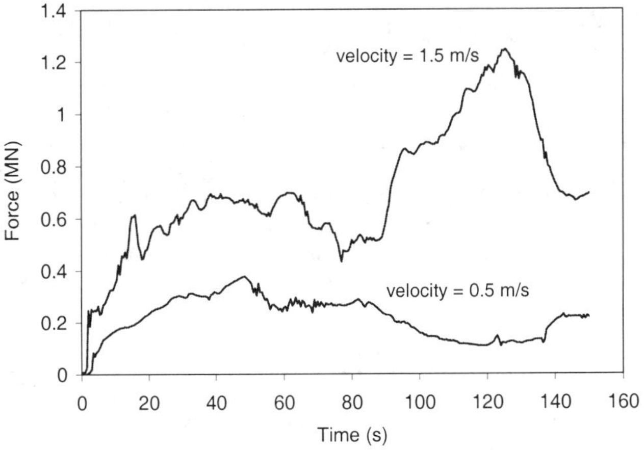

Figure 2. Force versus time plot, 9 m rubble ice thickness

Comparison With Full-Scale Data

Full scale data from a few sources can help in verification of some aspects of model predictions. As usual, such data are limited to specific geometries and conditions, and only parts of the variables of interest are measured. Unfortunately, most data of this nature is proprietary, and therefore of limited availability. For the present problem, data was examined from three sources: the Confederation Bridge in Canada, and the Nordstromsgrund light pier and Kemi-I lighthouse, both in Europe.

The Confederation Bridge, which is 13 km long, has piers with a typical water-line width of 14 m, and an upward-breaking cone at the waterline. There has been a major project for several years to measure the loads on the piers during the winter months. Although most of the data is confidential, Brown (2001) presented results from two winters of ice observation, based upon a full-scale data set of approximately 5800 events. The maximum ice load recorded from these events was less than 5 MN, for keel depths up to 15 m and velocities up to 1.5 m/s. The most frequent loads were found to be between 0.75 and 1.0 MN. The results did not differentiate between different ice features. However, it was observed that for ice depths up to 6 m, the ice force increased with increasing ice depth. At ice depths greater than 6 m, the expected increase in load with increasing keel depth was not observed. Additionally, Brown (2001) showed that there was no obvious correlation between the load and ice speeds up to 1.0 m/s, but there is considerable scatter in the data. Above 1.0 m/s, there was an apparent reduction in ice force. It was also observed that the rubble pile height may decrease with speed and it was commented upon that thicker ice has a tendency to ride-up without the formation of a rubble pile. Kubat et al. (2001) independently measured the ice loads on two of the Confederation Bridge piers and presented peak loads ranging from 0.2 to 0.9 MN, during events recorded on March 5 and 6, 2000. The associated ice velocity was approximately 0.8 m/s, with an ice thickness of between 1 m and 5 m, with some occasions of continuous ice cover.

The Nordstromsgrund light pier, which is in the Gulf of Bothnia, is a concrete cylindrical structure with a 7.2 m diameter at the water line. During the 1970s and 1980s, the light pier was instrumented to measure the structural response from ice loading (e.g., see Engelbrektson 1977), but this data is still not publicly available. More recently, the light pier has been instrumented with seven single load sensors and one 8-segment load sensor along with two underwater sonar units. This work is being done as part of the European initiative to measure and understand low-level ice forces (LOLEIF project – see Schwarz and Jochmann, 2001). Although this data is confidential, they reported on one event with 1.2 m thick hummocked ice moving with a speed of 0.4 m/s. For this ice feature, the measured global load was approximately 1.9 MN.

A test program has been carried out in Finland to investigate the loads on an offshore conical light pier, the Kemi-I lighthouse. The lighthouse was fitted with a 55° conical collar with a cone diameter approximately 10 m at the waterline. Typical annual ice sheet thickness is 0.8 m, and pressure ridges with a keel depth of 12 m are common in this region. Loads were measured throughout the winter months. Määttänen and Hoikkanen (1990) documented maximum loads of 2.5 MN for 0.7 m to 0.8 m level ice and 2.7 MN for 1.1 to 1.3 m thick, rafted ice.

The PIC model data gives results that are in line with these full-scale data observations. For example, the maximum full-scale observed force reported in the paper by Brown (2001) was less than 5 MN, in a data set that included ice depths of up to 15 m and maximum ice velocity of 1.5 m/s. With the simulation producing a maximum force of 4.8 MN for an ice rubble thickness of 20 m with an ice velocity of 1.5 m/s, this indicates that the PIC model's results describe the ice interaction in a realistic manner, despite having a simplified geometry compared with the bridge pier design. The full-scale ice loads from the other two examples of full-scale data were higher than those predicted by the PIC model, given the ice thickness interacting with these structures, although again, there are differences in geometry and the ice velocity in these cases in not known. The model data appears to accurately simulate the movement of ice around the cylindrical structure, including the formation of a narrow wake downstream of the pier, and a rubble pile formation upstream of the pier.

Comparison of Discrete Element and Analytical Models With Full-Scale Data

The Discrete Element Model

Model Overview

Discrete element (DE) models have been used in several studies of ice cover deformation and ridging. Relevant literature is too large to cover here. We only refer to a recent paper by Hopkins (2001), which includes references to earlier work. The following discussion gives a summary of the discrete model formulation. Details of that model and the simulations were given by Sayed (1995).

In the present formulation, spheres are joined in groups to represent irregularly shaped aggregates of ice blocks. The initial configuration consists of equal sized spheres in a tetrahedral arrangement. Each sphere is then given a random displacement up to 20% of the value of its radius along each of the three coordinates. Spheres in contact in that initial configuration are considered to have tensile strength (or "bonded"). At the contact between two spheres that have a tensile bond, the normal force, F_n is calculated using linear springs and viscous dashpots as follows

$$F_n = k\,\delta + b\,v_n \tag{8}$$

where k is the spring constant, δ is the relative normal displacement, v_n is the relative normal velocity, and b is the dashpot constant. Tangential forces were not added at those contacts because bonds form networks, which have normal and shear strength.

If the normal tensile force at a contact exceeds the tensile strength, the bond breaks and no tensile forces are allowed. At the contact between two spheres that are not bonded, the normal and tangential forces are calculated as follows:

$$\begin{aligned}
&compression: && F_n = k\,\delta + b\,v_n \\
&tension: && F_n = 0 \\
&Tangential_forces: && F_t = \mu\,F_n
\end{aligned}$$

where F_t is the tangential force, and μ is the coefficient of friction between ice blocks. The spring constant, k and the tensile strength at each contact were given a random component of up to 20% of the mean values. These random components, together with the randomness of the initial configuration, are introduced in order to account for the irregularity of ice block sizes, shapes, and contact strengths.

Movements of individual spheres are governed by the linear and angular momentum equations. Those equations and solution methods are standard for all discrete element models. The computations are made by keeping track of the positions and velocities of individual spheres. At each time step, the relative displacements between spheres and contact forces are calculated. Contact forces with the pier are then calculated, and body forces added. The body force on each sphere is taken as either the weight and acts downwards, or as the buoyancy and acts upwards, depending on the elevation of the centre of the sphere (above or below water level). Once forces are determined, accelerations are calculated from the resultant force on each sphere. Velocities and positions are next determined by integrating the accelerations. A "neighbours list" for each sphere is kept and updated in order to reduce computer running time.

Program Runs

The numerical simulations carried out by Sayed (1995) were performed on a cylindrical pier, with a 10 m diameter. A trapezoidal, 30 m long ridge with a 20 m keel depth was simulated impacting the pier for approximately 5 s. The program runs varied the ice velocity, which ranged from 0.2 m/s to 1.5 m/s, with a reference case of 0.5 m/s. Additionally, a parametric study was carried out, that changed the keel depth to 9 m while keeping the other parameters the same as the DE reference case. Parameter values used in the DE reference case were:

Sphere diameter	2 m
Velocity	0.5 m/s
Keel spring constant	2.65 MN/m
Consolidated layer spring constant	26.5 MN/m
Keel contact tensile strength	2.3 kN
Consolidated layer contact tensile strength	184.8 kN
Dashpot parameter, b	12.9 kN s/m
Ice-ice friction coefficient	0.2
Ice-pier friction coefficient	0.1

Those parameters were determined by relating contact conditions to continuum ice rubble properties using simulation of uniform bi-axial compression (Sayed, 1995). The value of the spring constant at the contacts was related to the bulk elastic modulus of ice rubble (Sayed, 1995). A range of values for the bulk elastic modulus was determined from available laboratory observations. The time step was 10^{-4} s. This relatively small value was necessary to handle the range of stiffness values.

The DE reference case indicated that the maximum force for ice moving at 1.5 m/s was 9.4 MN, reached after an indentation distance of between 0.5 m and 1 m. However, changing the elastic bulk modulus for the keel rubble to 0.7 MPa from 7.0 MPa resulted in much lower maximum force values. In this case, the maximum force for a velocity of 1.5 m/s dropped to less than 3 MN. When the keel depth was changed to 9 m in the DE model, the maximum force dropped to 2.7 MN for ice moving at 1.5 m/s. The force-time data from the DE model for cases with a velocity of 1.5 m/s are shown in Figure 3, while those for a velocity of 0.5 m/s are shown in Figure 4.

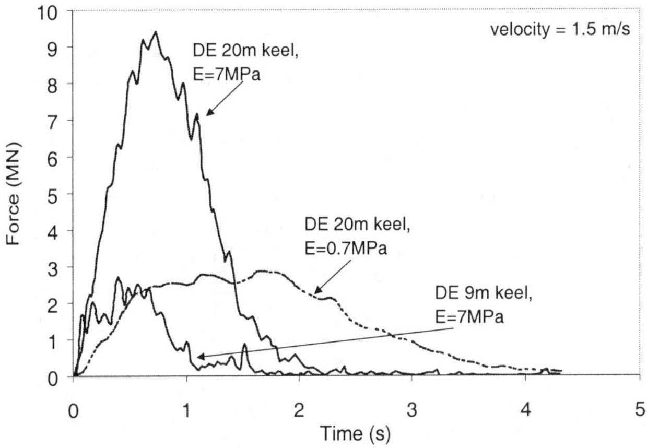

Figure 3. Force versus time, velocity=1.5 m/s, 3D DE simulation

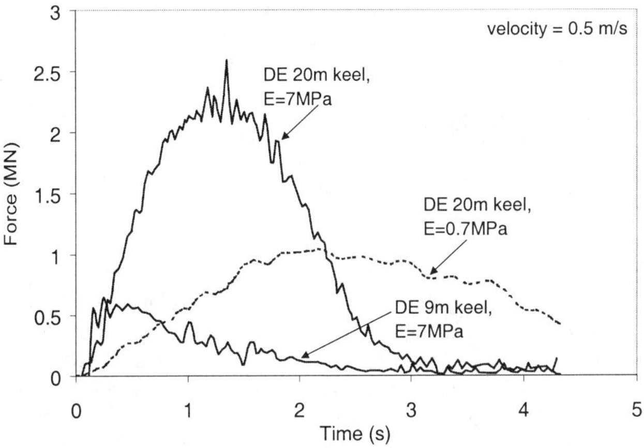

Figure 4. Force versus time, velocity=0.5 m/s, 3D DE simulation

The time step used in the DE model is much smaller than that used for the PIC model ($1x10^{-4}$ s compared with 0.5 s). This results in a difference in the time scales in Figures 1/2 and Figures 3/4. The PIC model is not intended to simulate the initial failure of the ice. Instead, due to its efficiency, it can simulate ice interaction with the pier over a relatively long duration. Alternatively, the DE model can capture details of the initial failure, but cannot be used to follow the interaction past a few seconds, due to the run-time necessary because of the small time step. Comparing the data obtained from the two types of interactions with full-scale data is, therefore, oriented towards examining the magnitude of the maximum predicted forces (rather than patterns of force-time records).

Comparisons between the DE model, the full-scale data and the present PIC model show that the predicted peak forces are in moderate agreement, particularly when the lower range of values for the elastic modulus are used in the DE model (see Table 2). Using the larger elastic modulus results in loads that are almost twice as large as the observed Confederation Bridge loads. This result indicates that the lower values of the elastic modulus may be more representative of continuous failure and the flow of ice against a structure. Considering the differences in model formulations, and geometry of the ice rubble, the predicted forces may be considered in reasonable agreement.

Table 2. Comparison of maximum calculated forces (values in MN)

Model	**Velocity (m/s)**	**0.2**	**0.5**	**1.0**	**1.5**
PIC, 20 m thick		1.0	1.0	2.4	4.8
PIC, 9 m thick		-	0.4	-	1.2
DE, 20 m keel – elastic modulus 7 MPa		0.8	2.5	5.9	9.4
DE, 20 m keel – elastic modulus 0.7 MPa		-	1.0	1.8	2.8
DE, 9 m keel – elastic modulus 7 MPa		-	0.63	-	2.7
Analytical	**Velocity (m/s)**	**?**			
20 m ridge keel		16			
Full-Scale	**Velocity (m/s)**	**?**	**0.4**	**0.8**	**1.5**
Confederation Bridge, max. 15 m thick					5
Confederation Bridge, max. 5 m thick				0.9	
Nordstromsgrund, 1.2 m thick, hummocked			1.9		
Kemi-I, 0.8 m thick, level		2.5			
Kemi-I, 1.3 m thick, rafted		2.7			

Comparison with Analytical Formulas

There have been a number of analytical models developed to predict loads due to the interaction of a ridge with an offshore structure. Estimates of the load are made by summing the loads obtained independently for the consolidated layer, sail and keel portions of the ridge. Loads due to the consolidated layer are usually calculated using the Korzhavin equation. Estimates of the sail and keel loads are usually calculated using plastic limit approaches suggested by Dolgopolov, Prodanovic, Mellor, Croasdale, Hoikkanen or Weaver (see Timco et al. (1999) for a review of this approach). These methods are based on simplistic assumptions concerning ice failure modes and stress distributions.

During the design of the Confederation Bridge, Timco et. al. (1995) conducted a review of the analytical models and predicted a maximum ice load of 16 MN for a 20 m deep ridge interacting with the 14 m wide bridge pier. This load value of 16 MN is considerably higher than the full-scale ice forces (see Table 2), and therefore much larger than load values predicted by the PIC (and DE) model. Timco et al. (1995) noted that the physics represented by the analytical theories, such as failure modes and ice pile-up height, did not match the behaviour observed during physical model studies, nor in field observations of ridge interaction with structures.

Summary

The depth-averaged, Particle-In-Cell numerical model was used to simulate the movement of ice rubble (hummocks and ridges) against a bridge pier. The model produced expected features, such as rubble pile-up formation upstream of the pier and wake formation downstream of the pier. Comparisons were made with available full-scale data, chosen to represent, as much as possible, the cases of interest. The PIC model appears to predict forces in agreement with measured values.

Results previously obtained using a Discrete Element (DE) model were also compared with full-scale data. Predicted peak forces were in general agreement despite differences in model formulation, geometry of ice rubble, and mode of interaction. The DE model provides detailed information regarding the initial contact and failure of ice. The PIC model, however, is oriented towards simulating the continuous failure and flow of ice against a structure. The amount of computer time needed for the PIC model is less than the DE model, as it is less computationally demanding, although there is less detail available about the ice-structure interaction. The full-scale data were also compared to values obtained from analytical formulas that are commonly used to estimate ice forces on structures. These formulas predicted unrealistically high values compared to reported full-scale loads. The purpose of the comparisons was to verify the predictions of the numerical model and determine the range of applicability of each approach.

Future work will include the development of a three-dimensional version of the hybrid PIC model. This would enable accurate modeling of the geometry of the interface between the structure and the ice rubble. The effect of the conical shape of the Confederation Bridge piers, for example, could be evaluated.

Acknowledgements

The support of the Program of Energy Research and Development (PERD) is gratefully acknowledged.

References

Barker, A., Sayed, M. and Timco, G. (2000a). "Numerical Simulations of Broken Ice Interaction with Structures." *PERD/CHC Technical Report HYD-TR-050*, Ottawa, Canada.

Barker, A., Sayed, M. and Timco, G. (2000b). Numerical Simulation of Floating Ice Forces on Bridge Piers. *Proceedings CSCE 2000*, Vol. G, pp. 243-249, London, Canada.

Barker, A., Timco, G., Sayed, M. and Wright, B. (2000b). Numerical Simulation of the "Kulluk" in Pack Ice Conditions. *Proceedings IAHR 2000,* Vol. 1, pp. 165-171, Gdansk, Poland.

Brown, T.G. (2001) Four Years of Ice Force Observations on the Confederation Bridge. *Proceedings POAC'01*, Vol.1, pp. 285-298. Ottawa, Canada.

Engelbrektson, U. 1977. Dynamic Ice Loads on a Lighthouse Structure. *Proceedings POAC'77*, Vol. 2, pp. 654-663, St. John's, Nfld., Canada.

Hopkins, M., Tucker III, W.B. and Mulherin, N.D. (2001) Simulating Ice Drift in Cook Inlet, Alaska. *Proceedings POAC'01*, Vol.3, pp. 1261-1264. Ottawa, Canada.

Kamesaki, K. Shimazaki, K., Kasui, H., Matsumoto H., Frederking, R., Timco, G., Sayed M., and J. Tseng. (2000). Design and Cost Study of Gravity-Based Substructure in the Sea of Okhotsk. *Proceedings ISOPE 2000*, Vol 1. pp. 589-595, Seattle, USA.

Kubat, I. and Frederking, R. (2001) Response of Confederation Bridge to Ice Forces: 2000 Winter Season. *Proceedings POAC'01*, Vol.1, pp. 319-328. Ottawa, Canada.

Määttänen, M. and Hoikkanen, J. 1990. The Effect of Ice Pile-up on the Ice Force of a Conical Structure. *Proceedings IAHR 1990*, pp. 1010-1021, Espoo, Finland.

Sayed, M. (1995) "Numerical Simulation of the Interaction Between Ice Ridges and Bridge Piers." *Controlled Technical Report TR-1995-10*, National Research Council, Ottawa, Canada

Sayed, M., and Carrieres, T. (1999). Overview of a New Operational Ice Forecasting Model, *Proceedings ISOPE '99*. Vol. II, pp. 622-627, Brest, France.

Sayed, M., Frederking, R. and Barker, A. (2000) Numerical Simulation of Pack Ice Forces on Structures: a Parametric Study. *Proceedings ISOPE '2000*, Vol.1, pp. 656-662. Seattle, U.S.A.

Schwarz, J. and Jochmann, P. 2001. Ice Force Measurements within the LOLEIF Project. *Proceedings POAC'01*, Vol.2, pp. 669-680, Ottawa, Canada.

Timco, G.W., Cornett, A.M., Sayed, M., McKenna, R.F. and Frederking, R. (1995) "Ice Ridge Loads on the Northumberland Strait Bridge Piers." *Controlled Technical Report TR-1995-08*, National Research Council, Ottawa, Canada.

Timco, G.W., Frederking, R., Kamesaki, K. and Tada, H. 1999. Comparison of Ice Load Calculation Algorithms for First-Year Ridges. *Proceedings REIFS'99*, pp. 88-102, Mombetsu, Japan.

Convective Cooling in Open Rock Embankments

Douglas J. Goering[1]

Abstract

Preferential natural convection heat transfer in highly-permeable rock embankments and pads is a technique which can be used to generate a passive cooling effect. Such an effect is usually desirable in permafrost areas in order to counteract the warming influence that often accompanies construction activity. Left unchecked, this warming may result in permafrost thaw and failure of the foundation due to of thaw settlement. In non-permafrost areas convective cooling in rock layers and embankments can lead to problems with frost heave due to enhanced winter-time cooling. The current work discusses the characteristics of these rock layers, including techniques for predicting the thermal and fluid mechanical properties. Sample applications are described and experimental and numerical results are discussed.

Introduction

Heat transfer within granular embankments and pads takes on added importance in cold climates due to the possibility of freeze/thaw processes. In permafrost areas, the presence of frozen ground plays a large role in the design, development, and operation of man-made infrastructure. The presence of ground ice, especially massive ice inclusions which are often found in the permafrost layer, makes it imperative that due consideration be given to the thermal effects of any man-made developments. These considerations are particularly important in the more temperate zone of discontinuous permafrost due to the relatively warm temperature of the permafrost layer and the effects that the embankment has on the ground surface energy balance. Removal of native vegetation layers and replacement with paved or gravel surfaces can have a huge impact on energy exchange at the surface and result in warming, thermal instability, and thawing of underlying permafrost. If thawing is allowed to occur, the resulting thaw consolidation typically results in failure of above ground structures. For these reasons, it is normally desirable to employ design and construction methodologies that protect underlying permafrost from thawing.

----

[1]Professor, Dept. of Mechanical Engineering, PO Box 755905, University of Alaska, Fairbanks, AK 99775-5905.

The essential difficulty that must be confronted when planning or designing developments for permafrost areas concerns modification of the pre-existing surface conditions and the associated ground-surface energy balance. The surface energy balance is a complex function of vegetation, surface moisture content, seasonal snow cover, atmospheric radiation, and air temperature, Lunardini (1981). These factors conspire to produce a mean annual surface temperature (MAST) that may differ by several degrees from the mean annual air temperature (MAAT). At undisturbed sites the MAST is typically warmer than the MAAT because of the effect of the insulating snow layer during winter. Disturbance of the native surface conditions due to construction activity often increases the difference between MAAT and MAST, producing warmer conditions that may endanger the thermal stability of the permafrost layer. In cases where the MAST is increased above 0°C, insulation alone will not be effective at limiting permafrost thaw and some sort of active or passive cooling system will be required.

In non-permafrost areas, heat transfer in embankments and pads can also have adverse effects on structures due to the possibility of frost heave during winter months. Many of the Earth's cold regions are underlain by frost susceptible soils that have access to sub-surface moisture. In such cases it is important to eliminate the possibility of frost heave beneath structures by placing foundations beneath the maximum frost level, replacing native material with non-frost susceptible fill, or by providing adequate insulation or embankment heights such that frost will not penetrate into the frost susceptible layers beneath the structure. The third option is often employed in the case of roadway or railway embankments.

Embankments, pads, or layers constructed of granular materials may experience heat transfer via a number of different physical mechanisms. In most geotechnical materials such as soil, sand, and well-graded gravel, heat conduction is the dominant mode of heat transfer. In some cases, a significant amount of heat transfer may be associated with non-conductive heat transfer mechanisms such as pore-fluid convection (e.g. such as that associated with ground-water flow). Structural embankments or pads are usually constructed so as to be elevated above the surrounding surface or ground water elevation and are normally well drained. In this case, the pore space within the granular material will be occupied by air with variable humidity. Under certain circumstances this moist pore air may move through the granular aggregate carrying a significant amount of heat with it. If the pore air velocities are high enough, the convective heat transfer associated with its motion can be much more effective than conduction heat transfer.

In the present work, the fluid-mechanical and thermal characteristics of highly porous granular embankments (sometimes referred to as air convection embankments) are examined. Goering and Kumar, 1996, proposed that embankments such as these could be used to generate a passive cooling effect for application in permafrost areas. These structures are most suitable for roadway, airport or railway embankments or for gravel construction pads, but may also be adopted for use in building foundation designs and hazardous material containment dikes. The technique involves the use of a highly porous embankment material, such as gravel or rock with a low fines content and narrow gradation. During the winter, the embankment is cooled at its upper surface due to low ambient air temperatures. If the cooling is strong enough and the embankment material is of sufficient permeability, natural convection of the pore air will occur during winter months, as illustrated in Fig. 1. During summer the pore-air density gradient is stable and convection does not occur. Thus the embankment acts as a one-way heat transfer device or thermal diode that effectively removes heat from the embankment and underlying foundation material during winter without re-injecting heat during subsequent summers. This passive cooling effect can prevent thaw of underlying permafrost, thus eliminating thaw settlement. In more temperate regions that are not underlain by permafrost, the natural convection can cause enhanced seasonal freezing beneath embankments and exacerbate problems with frost heave in the foundation soils.

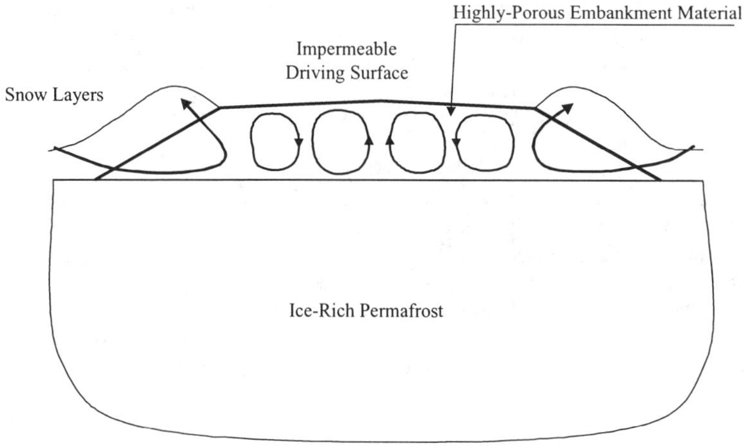

Figure 1. Pattern of winter-time pore air circulation.

Previously Goering and Kumar, 1996, and Goering, 1998, have reported experimental and numerical data that demonstrates the passive cooling capability of air convection embankments. The present paper summarizes the findings of previous studies and attempts to synthesize these into a set of design guidelines. The sections that follow provide background material and a summary of numerical and experimental results. Techniques for predicting material properties and resulting heat transfer characteristics are described and used to formulate approximate design guidelines.

Theoretical Background

Many engineered civil structures employ the use of layered granular aggregate. These include many types of roadway, railway, and airport embankments, as well as construction pads, dikes, material stockpiles, and building foundations. In many respects, the simplest theoretical representation of these structures is a horizontal layer of porous granular material. The porosity of such a material will be a strong function of the particle size gradation, generally increasing as the size gradation becomes narrower. Depending on the site characteristics and elevation, the pore space of the aggregate will often be filled with either moist air, water, or ice. The embankment structures of interest in the present work are typically elevated above local water sources and have very large permeablities that promote free drainage of water. In such cases, the pore space of the aggregate will contain air and the water content will be quite low. Thus, it is possible to think of these structures as horizontal layers of porous material that are saturated with a single pore fluid (air). If the upper surface of the layer is exposed to winter-time atmospheric conditions, the system has characteristics that are similar to a horizontal porous layer heated from below (i.e. cooled from above). As described below, a theoretical analysis indicates that natural convection (buoyantly induced motion of the pore air) will occur in these systems so long as the permeability and vertical temperature gradients are large enough. This convection enhances upward heat transfer through the embankment during winter months. During summer the layer is effectively heated from above and no convection occurs since the resulting density gradients are stable in this case. A number of important heat transfer characteristics can be obtained by examination of the theoretical work which has been done in the past for a horizontal layer of porous material heated from below.

Horizontal porous layer heated from below. Convective heat transfer in porous materials has been the subject of many previous studies. This is a result of the interesting fluid flow and heat transfer characteristics as well as numerous practical applications. As described above, a configuration of particular interest for the present work is a horizontal layer of porous material that is heated from below. It is well known that such a system will experience natural convection of the pore fluid if a critical Rayleigh number is exceeded, Nield and Bejan, 1992. The Rayleigh number is given by:

$$\mathrm{Ra} = \frac{C\beta g K H \Delta T}{\nu k} \tag{1}$$

where C, β, and ν are the volumetric heat capacity, expansion coefficient, and kinematic viscosity of the pore fluid (typically air in this application), g is the acceleration due to gravity, K is the intrinsic permeability of the material, H is the layer height, ΔT is the temperature difference between the top and bottom of the layer (warmer boundary below), and k is the thermal conductivity of the material. A stability analysis shows that a horizontal layer with impermeable boundaries of uniform temperature at its base and upper surface will begin to experience natural convection once the Rayleigh number exceeds $4\pi^2$ ($39.48 \cong 40$). Using this limit, it is also possible to define a critical ΔT:

$$\Delta T_c = \frac{40 \nu k}{C\beta g K H}. \tag{2}$$

Once this critical ΔT is exceeded, warmer pore air from the base of the layer begins rising upward while cooled pore air from above circulates downward. A series of circulation cells arises which increase the heat transfer from the lower to the upper surface (see Fig. 1). Any departure from the ideal horizontal system generally results in reduced stability and convection will begin at smaller Rayleigh numbers. For instance, if the upper or lower surfaces are not at uniform temperature, if the upper or lower boundaries are slightly permeable, or if the system is not perfectly horizontal, convection will set in a lower Rayleigh number. Under actual field conditions it is very likely that these and other factors will conspire to produce convective flow at Rayleigh numbers as much as a half or one quarter of the value listed above.

The most important consideration for the current work is related to the heat transfer enhancement associated with the natural convection flow described above. This enhancement is normally characterized in terms of a Nusselt number which represents a ratio of the heat transfer with convection present to that due to thermal conduction alone. The Nusselt number can be thought of as a measure of heat transfer enhancement due to the presence of convection. For a horizontal porous layer the Nusselt number can be defined as:

$$Nu = \frac{q\,\Delta z}{k\,\Delta T} \qquad (3)$$

where q is the actual upward heat transfer through the layer, Δz is the layer height, and ΔT is the temperature difference across the layer (warmer temperature below).

For summer conditions the layer will be cooler at the base and no convection will occur. For this case the heat transfer occurs by conduction only and Nu = 1 (the absolute lower limit for Nu). For winter conditions, Nu will remain 1 until the temperature difference reaches the critical value given in Eqn. (2) for convection to begin. Once convection begins the Nusselt number will increase. For the ideal case described above Nield and Bejan, 1992, have suggested the following simple Nusselt number correlation which indicates a linear increase in heat transfer rate as Ra increases above the critical value of 40:

$$Nu = \frac{Ra}{40} \qquad \text{for } Ra \geq 40 \qquad (4)$$

Equation (4) correlates reasonably well with experimental data for Rayleigh numbers less than about 200, although there is considerable scatter in the experimental results. For real-world porous layers constructed of rock or gravel, it is most likely that Eqn. (4) will produce Nusselt numbers that are somewhat low. This is a result of the stability considerations described above for field conditions that will likely result in convection for Ra less than 40. The values given by Eqn. (4) can then be thought of as a somewhat conservative (lower) estimate of Nu for real systems.

In field applications it is possible to use Eqns. (1)-(4) to estimate the heat transfer enhancement that would be expected during winter months for a given installation. This can be accomplished by calculating or estimating the quantities in Eqn. (1) in order to obtain an average winter Rayleigh number. This Rayleigh number can then be used in Eqn. (4) in order to estimate the average winter Nusselt number and the associated heat transfer enhancement.

Material Properties

To make the estimates described above for Rayleigh and Nusselt numbers and associated heat transfer characteristics, it is first necessary to accurately determine the thermal and fluid mechanical properties of the embankment material.

Thermal Properties. For the most part, conventional well-known techniques can be utilized to formulate thermal property values for the porous embankment/pad materials and underlying foundation soils. See, for example, Andersland and Ladanyi, 1994. One notable exception concerns the thermal conductivity for dry crushed rock. Johansen,

1975, showed that dry rock materials have higher thermal conductivity values than soils with equivalent porosity and moisture content. He suggested the use of the following correlation for crushed rock:

$$k_{dry} = 0.039\, n^{-2,2} \tag{5}$$

where k_{dry} is dry thermal conductivity and n is the porosity of the material. For freely draining embankments and pads, the moisture content associated with the high-permeability embankment aggregate will be very low, and, thus Eqn. (5) can be used to obtain an approximation of the thermal conductivity of the material as placed in the field. However, even small amounts of moisture can have a large impact on thermal conductivity and it many be necessary, to adjust k_{dry} for a non-zero moisture content, Andersland and Ladanyi, 1994.

It is notable that k_{dry}, as given by Eqn. (5), is a strong function of the porosity of the material. This is also true of the permeability to be discussed below. Accurate knowledge of the porosity is needed in order to predict both thermal properties and permeability. In some cases the porosity may be estimated from gradation distributions, but it is also possible to measure porosity in the lab or to obtain it from knowledge of the solids density and bulk density of the material.

Thermal conductivity, heat capacity, and latent heat are all strong functions of moisture content. As stated above, highly porous rock materials (lacking fines) are very permeable and thus typically have moisture contents of no more than a few percent. However, even small moisture contents can have a significant impact on thermal properties and measurement of moisture content under field conditions may be required to accurately assess these effects.

Permeability. As discussed above, the strength of the convective flow and resulting convective heat transfer is a function of the Rayleigh number as defined in Eqn. (1). Of the various material properties listed in Eqn. (1), many are readily available in tables of thermodynamic properties of air (the pore fluid), and the material thermal conductivity is available from correlations discussed above. One of the most important and most difficult variables to assess is the intrinsic permeablity, K. In general the permeability will increase as mean particle size increases and the size distribution becomes more uniform. There are a variety of semi-empirical relations available in the literature for calculating permeability as a function of particle size, such as the well-known Carmen-Kozney relation, Bear, 1972. However, most of these have been validated for much finer particle sizes than those of interest in the current study and, consequently, must be applied with caution. Fair and Hatch, 1933, proposed a semi-empirical relation that allows the shape of the actual gradation distribution to be included in the calculation of permeability:

$$K = \frac{1}{5}\left[\frac{(1-n)^2}{n^3}\left(\frac{\alpha}{100}\sum\frac{p}{d_m}\right)^2\right]^{-1} \tag{6}$$

where K is the intrinsic permeability, n is the porosity, α is a particle shape parameter, p is the percentage of particles held between adjacent size limits, and d_m is the geometric mean size of those limits. In their paper, Fair and Hatch give guidelines for selecting the particle shape parameter which varies from 6 for spherical particles to 7.7 for angular particles. In practice, the particle size distribution is split into several ranges between 0% passing and 100% passing and the summation shown in Eqn. (6) is carried out over all of the size bins. Due to typically incomplete knowledge of the particle shape parameter as well as inaccuracies in characterizing the gradation distribution of material as placed in the field, calculation of the permeability using Eqn. (6) should be considered approximate.

In some cases it may be possible to directly measure the permeability of the embankment aggregate. This can be accomplished, for instance, by constructing a large flow channel such as a tube, pipe, or square duct, and filling it with the material in question, Goering, 1997. The cross-sectional dimension of the channel should be at least one order of magnitude larger than the maximum particle size and the length must be sufficient to generate a measurable pressure drop when air (or some other pore fluid) flows through the duct at a moderate rate. Care must be taken to ensure that the Darcy flow limit is not exceeded while running the tests. This typically requires very small fluid flow rates and consequently requires highly accurate pressure measuring equipment if useful results are to be obtained. Using the simple Darcy flow law, the intrinsic permeability of the material can then be calculated from the measured pressure drop and flow geometry.

Experimental and Numerical Studies

During recent years a number of investigations have been carried out with the aim of characterizing the thermal behavior of air convection embankments. In 1993 an experimental test embankment was constructed at Brown's Hill Quarry near Fairbanks, Alaska and monitored for a period of two years, Goering, 1998. In addition, the Alaska Department of Transportation constructed a second test embankment through the Experimental Features in Construction Program near Fairbanks, Goering, 1997.

Experimental Results. Experimental data collected from the Brown's Hill test embankment during 1993 through 1995 have been discussed in Goering, 1996, and Goering 1998. Some of the data are summarized here. The embankment was constructed of crushed rock in the size range of 5 to 7.5 cm. It had a height of 2.5 m and a width of 5 m at the top driving surface with 2H:1V side slopes. During the two-year test period, the embankment was operated in two different modes. During the winter of 1993-94 snow was allowed to accumulate naturally on the embankment but no other covering was used on the upper surfaces. In the driving lane the initial snowfall was not removed but packed into an ice sheet to simulate the presence of an impermeable driving surface. The side slope surfaces were left open to the atmosphere other than the natural snow cover. During the winter of 1994-95 conditions were similar with the exception that impermeable plastic sheeting was installed on the embankment side slopes to eliminate the possibility of ambient air intrusion through the snow layers. Figure 1 shows schematic air flow patterns for an embankment without impermeable side-slope coverings. Note that natural snow covers are typically highly permeable, Sturm and Johnson, 1991, and do not, in and of themselves, eliminate the possibility of air circulation in the side slope zones, as shown in Fig. 1. The side-slope circulation shown in Fig. 1 is, however, eliminated if an impermeable barrier is placed directly on top of the side slope.

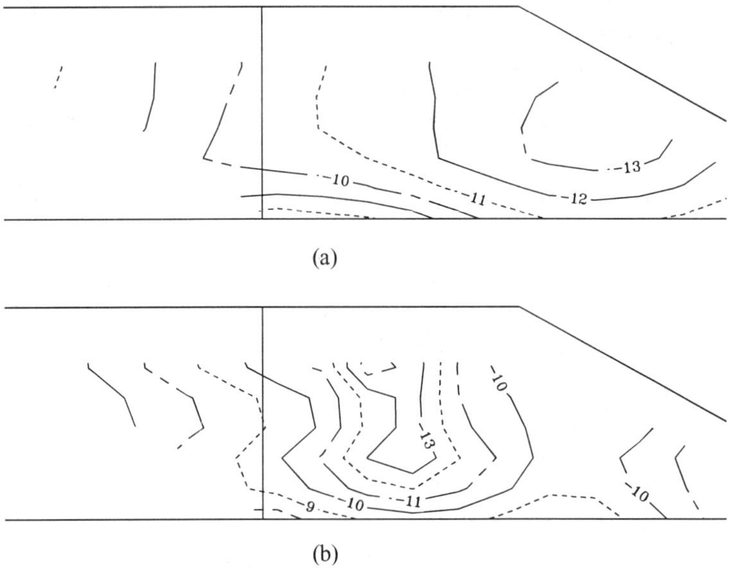

Figure 2. Temperature profiles in the test embankment (°C), a) February 1, 1994, b) February 1, 1995.

Figures 2 a) and b) show temperature profiles obtained from the test embankment in early February of 1994 and 1995, respectively. The temperature profiles shown in these figures are based on a set of 44 thermistor measurement points which were distributed in the embankment over the area shown in the contour plot. In both cases the effect of pore-air convection is apparent, and the presence of plumes or concentrated zones of low temperature are shown in both cases. The main difference between the two data sets is the type of convection pattern which occurs. Figure 2 a) shows the coldest portion of the embankment near the side slope and appears to indicate that ambient air is circulating in the region near the side slope, as illustrated schematically in Fig. 1. Figure 2 b) shows the presence of a more concentrated plume beneath the driving lane and does not indicate much convection near the embankment toe. This is a result of the impermeable plastic cover which was in place on the embankment side slope during the winter of 1994-95. Similar temperature profiles acquired during the summer of both years do not indicate the presence of any convective activity, Goering, 1998.

The effect of winter convection is to increase the heat loss from the entire system during winter months. This, in turn, results in a reduction of the mean annual temperatures in the region at the base of the embankment and in the foundation soil. This is exactly the effect that is desired if there is an interest in maintaining permafrost foundation soils and eliminating the possibility of thaw settlement. The magnitude of the mean temperature depression can be illustrated by plotting mean annual temperatures within the embankment, as shown in Fig. 3 (a) and (b) for the 1993-94 season and the 1994-95 season, respectively. In both cases the figures indicate a gradient in temperature between the base and the upper portion of the embankment. Note that mean temperatures in the upper portion of the embankment are well above 0°C whereas those at the base are below 0°C. Although mean temperatures were not calculated for the upper surface, they are expected to be in the range of +2°C for conditions typical of the Fairbanks, Alaska area, Goering, 2000(a). This is consistent with the data shown in Fig. 3. On the other hand, at the base of the embankment the mean temperatures range from –1 to –3.5 °C for the two cases. During the 1993-94 season, very low mean temperatures were achieved beneath the toe of the embankment (Fig. 3 (b)). This is a result of the side-slope convection that occurred during the 1993-94 season which allowed cold winter air to infiltrate through the snow layer and into the toe of the cross-section, as illustrated in Fig. 1. This convection occurred despite the fact that snow was plowed from the embankment driving surface onto the side slope. This type of embankment geometry should be particularly effective at combating the common problem of shoulder rotation that is caused by the relatively warm temperatures found beneath the side-slope zones of conventional embankments.

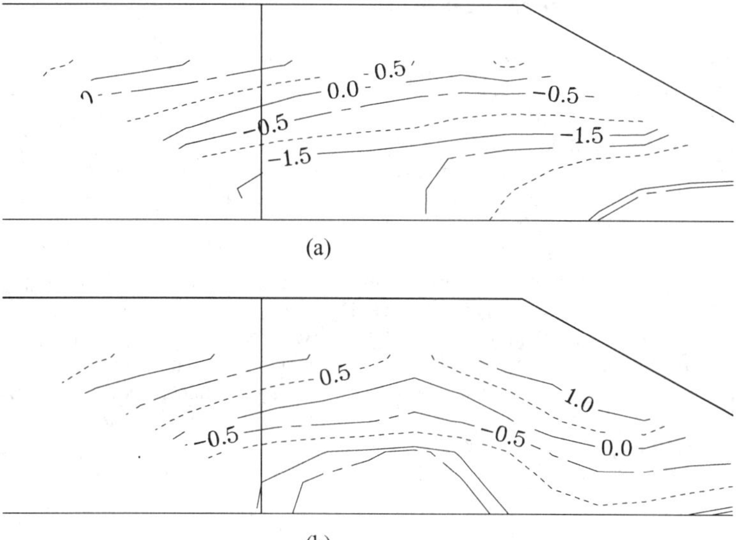

Figure 3. Mean Annual Temperature profiles in the test embankment (°C), a) 1993-94 season, b) 1994-95 season.

One way of characterizing the cooling effect of embankment convection is to examine the thermal offset (reduction in mean annual temperature) that arises in the mean temperature between the upper surface and the base of the embankment structure. Assuming a mean upper surface temperature of +2°C, the experimental results shown in Fig. 3, indicate thermal offsets of approximately 4.5°C beneath the driving surface. Offsets are 5 to 6°C beneath the toe for the 1993-94 season. As shown in Fig. 3, these offsets are effective at reversing the artificial warming that occurs at the upper surface due to alteration of the surface energy balance. If it were not for the offset, mean temperatures at the base of the embankment would be above the freezing level and subsurface permafrost would be at risk of thawing.

Other experimental data are available in Goering, 2001 for the experimental feature that was constructed near Fairbanks in 1996. These reports show that the embankment has been effective at maintaining frozen conditions in the underlying foundation soils.

Numerical Results. In addition to the experimental investigations described above, a number of numerical studies have also been carried out. Goering and Kumar, 1996, completed a numerical investigation of air convection embankments and compared cases with and without convection present. Goering and Kumar, 1999, investigated the impact of embankment material permeability, while Goering, 2000(b), examined the impact of convection in embankment side slopes.

Air Convection Embankment Design

The most important thermal characteristic of embankments, pads, or porous layers that may experience preferential natural convection during winter, is the resulting thermal offset which develops between the base and upper surface of the layer. The ability to predict this offset at the design stage would allow designers to assess the possibility of using such a technique to eliminate problems with thaw settlement. This capability would also be useful for assessing potential problems with frost heave beneath embankments that are not located on frozen ground.

Using climatic data for the design site in combination with available n-factor data (or a more complete surface energy balance analysis), it is usually possible to predict the mean annual surface temperature (MAST) for the surface of the embankment or pad, Goering, 2000(a). For typical conditions, these mean surface temperatures are elevated significantly above pre-construction values and may result in thawing of pre-existing permafrost. An air convection embankment may be able to counteract this warming influence if it can provide a thermal offset of a greater or equal magnitude. Thus, the warming due to alteration of the surface conditions is counteracted by the cooling and associated thermal offset provided by the air convection embankment.

Thermal Offset. Using Eqns. (1)-(6) described above, it is possible to formulate an approximate method for estimating the magnitude of the thermal offset that will arise for a particular embankment under given climatic conditions. This can be accomplished by estimating a mean winter-time Nusselt number and using it to calculate an average vertical temperature difference between the top and base of the embankment layer during winter. To the extent that the mean winter Nusselt number is above one, the temperature difference during winter will be reduced over the summer value. Having a larger temperature difference in summer than during winter months is the mechanism that is responsible for generation of the thermal offset.

In order to formulate an approximate design equation, it is first necessary to make use of an annual energy balance for the embankment layer. In the long term, once the embankment reaches a condition of thermal equilibrium with the foundation soils, the total heat transfer downward into the embankment during summer will approximately equal the total upward heat transfer out of the embankment during winter. Utilizing Eqn. (3), this energy balance can be written in the following form:

$$-q_{winter}\Delta t_w = -\overline{Nu}_w k\frac{\overline{\Delta T}_w}{\Delta z}\Delta t_w = k\frac{\overline{\Delta T}_s}{\Delta z}\Delta t_s = q_{summer}\Delta t_s \qquad (7)$$

where:

$q_{winter,summer}$ = winter or summer heat transfer rate (positive into foundation soils)
$\Delta t_{w,s}$ = length of the winter or summer season
$\overline{Nu}_w$ = mean winter Nusselt number based on $\overline{\Delta T}_w$
k = thermal conductivity of embankment layer
$\overline{\Delta T}_{w,s}$ = mean temperature difference between top and base of the embankment layer during winter or summer
Δz = embankment layer height.

Implicit in the above equation is the assumption that summer heat transfer is governed by thermal conduction, and thus, the mean summer Nusselt number has a value of unity. The signs indicate that heat transfer is out of the system in winter and into the system in summer. After simplifying, Eqn. (7) can be written as:

$$-\overline{Nu}_w\overline{\Delta T}_w\Delta t_w = \overline{\Delta T}_s\Delta t_s \qquad (8)$$

where:

$\overline{\Delta T}_w = \overline{T}_{sw} - \overline{T}_{bw}$
$\overline{\Delta T}_s = \overline{T}_{ss} - \overline{T}_{bs}$
$\overline{T}_{sw}$ = mean temperature at embankment surface during winter
$\overline{T}_{bw}$ = mean temperature at embankent base during winter
$\overline{T}_{ss}$ = mean temperature at embankment surface during summer
$\overline{T}_{bs}$ = mean temperature at embankment base during summer.

In Eqn. (8), Δt_w is the length of the winter season over which the mean Nusselt number and mean winter temperatures are calculated. Probably the most accurate choice for Δt_w is the length of the winter season during which convection actually occurs in the embankment. Then Δt_s would correspond to the remainder of the year. Unfortunately the correct value of Δt_w is difficult to predict accurately at the design stage without carrying out a complex numerical simulation. With some knowledge or estimate of the embankment base temperature it is possible to determine when the critical ΔT will be achieved and, thus, determine the beginning and ending date for the winter convection season. Absent this information, the best choice is probably to estimate these dates based on prior experimental data such as that available in Goering, 1998 or to use the length of the freezing season at the embankment surface. In either case, once Δt_w is determined, the mean temperatures shown in Eqn. (8) must be based on the resulting Δt_w and Δt_s values where $\Delta t_s = 365 - \Delta t_w$ (days).

Using the equations discussed above, it is now possible to formulate a relation for the thermal offset, where this offset is defined as the difference between the mean annual temperature at the upper surface and lower base boundaries of the embankment:

$$\Delta T_{offset} = \frac{\overline{\Delta T_s}\,\Delta t_s}{365}\left(1 - \frac{1}{\overline{Nu_w}}\right). \quad (9)$$

Note that Eqn. (9) produces the expected result in the limits $\overline{Nu_w} \to 1$, and $\overline{Nu_w} \to \infty$. In the first case the thermal offset collapses to zero, as expected, since there is no enhancement of winter heat transfer in this case. In the latter case, the large Nusselt number forces $\overline{\Delta T_w}$ to zero, resulting in a thermal offset that is purely a function of the elevated upper surface temperature during summer.

Using Eqns. (1)-(9) a plausible design scenario is to first use the properties of the embankment in Eqn. (2) to determine ΔT_c required to achieve the critical Rayleigh number of 40. Then, assuming an embankment base temperature near the mean annual air temperature at the site (or near mean foundation soil temperatures) it is possible to estimate when the upper surface temperature will be decreased sufficiently to cause convection to begin. In this way, an estimate for the beginning and end of the winter convection season can be developed and used to calculate Δt_w and Δt_s. Then, using estimates for the embankment base temperature during winter and combining that with the mean upper surface temperature during the convection season (from climatic data), an estimate for the mean winter Rayleigh number can be obtained. The mean Rayleigh number is then used in Eqn. (4) to obtain a value for $\overline{Nu_w}$. Finally, using estimates of the embankment base and surface temperatures during summer, $\overline{\Delta T_s}$ can be determined and used in Eqn. (9) to calculate the thermal offset. Once the offset is known, the success or failure of the design can be judged by subtracting this offset from the mean annual embankment surface temperature. If the resulting value is below the freezing point, the design should be capable of preserving underlying permafrost and avoiding thaw settlement problems.

Perhaps the greatest difficulty to be overcome during the design process suggested above concerns the accurate estimate of average temperatures at the base of the embankment during winter and summer seasons ($\overline{T}_{bw}$ and $\overline{T}_{bs}$). However, it is possible to employ an iterative process to improve the estimates for these average temperatures. For instance, once the suggested procedure has been used to calculate the thermal offset, this offset can simply be subtracted from the mean annual embankment surface temperature to find the mean annual base temperature. The embankment surface temperatures are relatively easy to predict based on known meteorological conditions and surface N-factors, Lunardini, 1981. Note that the assumed base temperatures, $\overline{T}_{bw}$ and $\overline{T}_{bs}$, can also be used to calculate a mean annual base temperature. This value can then be compared with that given by the calculated offset. If a significant disagreement exists

$\overline{T}_{bw}$ and $\overline{T}_{bs}$ can be adjusted and the design calculations repeated until agreement is achieved.

Table 1. Thermal offset calculation for the experiment of Goering, 1998.

Parameter	Symbol	Value	Reference
Permeability	K	6×10^{-7} m^2	Eqn. (6)
Thermal Conductivity	k	0.5 W/m °C	Eqn. (5) Andersland and Ladanyi, 1994
Critical ΔT	ΔT_c	3.8°C	Eqn. (2)
Length of winter convection season and summer season	Δt_w Δt_s	160 days 205 days	
Mean winter base temp. Mean summer base temp.	$\overline{T}_{bw}$ $\overline{T}_{bs}$	-9 °C +3 °C	Eqn. (8) Goering, 1998
Mean winter surface temp. Mean summer surface temp.	$\overline{T}_{sw}$ $\overline{T}_{ss}$	-18 °C +14 °C	
Mean temperature difference during winter	$\overline{\Delta T}_w$	-9°C	Eqn. (8)
Mean temperature difference during summer	$\overline{\Delta T}_s$	+14°C	Eqn. (8)
Mean winter Nusselt number	$\overline{Nu}_w$	2.3	Eqns. (4) & (7)
Thermal offset	ΔT_{offset}	4.5 °C	Eqn. (9)

As an example, it is possible to apply the procedure suggested above to the experimental embankment that was discussed previously in this paper and desribed more fully in Goering, 1998. The calculatoins have been summarized in Table 1. Initially the permeablity is calculated using Eqn. (6) with a uniform gradation curve beginning and ending with particle sizes of 4 and 8 cm (1% fines smaller than 4 cm is assumed), a porosity of 35%, and a particle shape factor of 7.7 (angular particles). The thermal conductivity is calculated using Eqn. (5) for dry thermal conductivity which is then adjusted assuming a 1% moisture content. Using the resulting values of thermal conductivity and permeablity, combined with the physical properties of the pore air, it is possible to calculate a critical ΔT as shown in the table. This indicates that the upper surface of the embankment must be at least 3.8°C colder than the base before convection will begin. Using ΔT_c combined with information regarding the surface and base embankment temperatures, it is possible to establish the beginning and ending of the winter convection season and determine Δt_w and Δt_s. For this case, a harmonic annual temperature variation with a mean value of +1.85°C and amplitude of 27.65°C was utilized. This surface temperature variation was constructed using air temperature data for the test site combined with freeze and thaw N-factors typical of embankment surfaces, Lundardini, 1981. Note that the resulting mean annual surface temperature of +1.85°C appears to be in agreement with the trends shown in Fig. 3. An approximate length for the convection season can then be obtained by assuming that the base of the embankment is near freezing and finding the period of the winter during which the critical ΔT is exceeded. This results in a convection season of about 160 days for this case (beginning on October 23 and ending on March 30). Knowing Δt_w, Δt_s, and the surface temperature variation, it is a relatively easy task to calculate the mean surface temperature during the winter and summer seasons as shown in Table 1. More difficult is estimating the corresponding mean base temperatures during these seasons. In this case the measured data given in Goering, 1998, provides guidance. In a design scenario it may be necessary to simply make educated guesses and then iterate to improve the estimates. Using data provided for winter and summer seasons in Goering, 1998, estimates of -9°C and +3°C were arrived at for $\overline{T}_{bw}$ and $\overline{T}_{bs}$, respectively. $\overline{\Delta T}_w$ and $\overline{\Delta T}_s$ can then be obtained as shown in the table. Finally, applying Eqn. (4) using the magnitude of $\overline{\Delta T}_w$ as a mean temperature difference, a mean winter Nusselt number can be obtained and used in Eqn. (9) to calculate the thermal offset.

Once the calculations are complete, the data in Table 1 can be checked to ensure that it is self-consistent. For instance, the resulting thermal offset can be subtracted from the mean annual surface temperature of +1.85°C to yield a predicted mean annual base temperature of –2.65°C. This value can also be calculated using Δt_w, Δt_s, $\overline{T}_{bw}$, and $\overline{T}_{bs}$. Doing so yields a value of –2.26°C, which is somewhat higher than that given by the calculated offset. At this stage, the estimates of $\overline{T}_{bw}$ and $\overline{T}_{bs}$ can be adjusted and the entire calculation iterated until better agreement is achieved. For instance, the mismatch between the two values of the mean annual base temperature suggests that we may need to lower the estimate of $\overline{T}_{bw}$ (or increase the estimate of $\overline{T}_{bs}$) shown in the Table 1. If $\overline{T}_{bw}$ is adjusted downward from -9°C to –9.5°C, the revised ΔT_{offset} becomes 4.31°C, yielding a new predicted mean annual base temperature of –2.46°C. Recalculating this value using Δt_w, Δt_s, $\overline{T}_{bs}$, and the new value for $\overline{T}_{bw}$, results in a mean annual base temperature of –2.48°C. Thus, much better agreement is achieved.

Comparing the predictions given in Table 1. with actual observations for the experiment, shows reasonable agreement. For instance, the mean annual base temperature of –2.46°C calculated using the final predicted offset, can be compared to the actual data shown in Fig. 3. For both the 1993-94 and 1994-95 season, the data in Fig. 3 shows that the mean annual embankment base temperatures are near –2.5°C, for at least a portion of the base boundary. It is, of course, impossible to predict the details of the variation in base temperatures shown in Fig. 3 using the simple procedure described here. These variations are a strong function of the convective flow patterns that arise in the embankment during winter. The simple design procedure described in the present section is only able to take these flow patterns into account in an average sense. Never-the-less, the calculations shown in Table 1. are probably sufficient from a design point of view in that they approximately characterize the thermal offset that can be expected for this case. The resulting offset of 4.5°C is more than sufficient to ensure sub-zero mean temperatures at the base of the embankment thus providing some insurance against failure due to thaw settlement.

Conclusions

Highly porous rock layers, embankments, or pads can be used to generate a passive cooling effect for use in permafrost areas. The present paper has presented a summary of techniques which can be used to formulate fluid-mechanical and thermal properties for these systems. A simplified design procedure has been formulated based on an appropriate set of parameters and semi-empirical relations. The procedure requires the estimation of a number of operational characteristics. Initial estimates can be based on existing experimental data, some of which has been summarized in the present paper. The design procedure has been applied to the experiment of Goering, 1998, and reasonable agreement with field measurements was obtained. Successful application of the technique provides an estimate of the thermal offset generated at the base of the layer. This offset can then be used in combination with local meteorological data and embankment surface characteristics to estimate the mean annual temperature at the base of the embankment. This estimate can be used to judge whether the resulting design will be likely to have difficulties due to thaw settlement.

References

Andersland, O.B. and Ladanyi, B., 1994, *An Introduction to Frozen Ground Engineering,* Chapman and Hall, New York.

Bear J., 1972, *Dynamics of Fluids in Porous Media,* American Elsevier, New York.

Fair, G.M. and Hatch, L.P., 1933, "Fundamental Factors Governing the Streamline Flow of Water Through Sand", *J. Amer. Water Works Assoc.*, **25**, 1551-1565.

Goering, D.J., 1996, "Air Convection Embankments for Roadway Construction in Permafrost Zones", Proceedings of the Eighth International Conference on Cold Regions Engineering, American Society of Civil Engineers, New York, 1-12.

Goering, D.J. and Kumar, P., 1996, "Winter-time Convection in Open-Graded Embankments," *Cold Regions Science and Technology*, **24**, 57-74.

Goering, D.J., 1997, "Air Convection Embankment Experimental Feature Design Phase I", AKDOT&PF Technical Report #AK-RD-97-02, Fairbanks, AK.

Goering, D.J., 1998, "Experimental Investigation of Air Convection Embankments for Permafrost-Resistant Roadway Design," Proceedings, 7th International Conference on Permafrost, Yellowknife, NWT, 319-326.

Goering, D.J. and Kumar, P., 1999, "Permeability Effects on Winter-Time Natural Convection in Gravel Embankments," *Advances in Cold-Region Thermal Engineering and Sciences*, Springer-Verlag, Berlin.

Goering, D.J., 2000(a), "Passive Cooling for Foundation Stabilization and Contaminant Control in Permafrost Zones", Proceedings of the International Workshop on Permafrost Engineering, Svalbard, Norway, 39-56.

Goering, D.J. 2000(b), "Passive cooling of permafrost foundation soils using porous embankment structures", American Society of Mechanical Engineers, HTD 366-5, 103-111.

Goering, D.J. 2001. Phase II final report - Parks/Chena Ridge air convection embankment experimental feature. Report No. INE/TRC 02.01. pp. 42.

Johansen, O., 1975, "Thermal Conductivity of Soils," U.S. Army Cold Regions Research and Engineering Laboratory, Translation 637.

Lunardini, V.J., 1981, *Heat Transfer in Cold Climates*, Van Nostrand Reinhold, New York.

Nield D.A. and Bejan A., 1992, *Convection in Porous Media,*. Springer-Verlag, New York.

Sturm, M. and Johnson, J.B., 1991, "Natural Convection in the Subarctic Snow Cover," *Journal of Geophysical Research*, **96**, 11,657-11,671.

Thermosyphon Cooling of Chena Hot Springs Road

Anna Forsström[1], Erwin L. Long[2], John P. Zarling[3], Sven Knutsson[4]

Abstract

Thermosyphons have been used to maintain permafrost below structures for over forty years. The Chena Hot Springs Road Test Section became operational during the fall of 1998 and includes three test sections underlain with different types of thermosyphons plus a control section. The test section overlays high-moisture-content permafrost. After two seasons, the thaw depth decreased and the freezing depth increased in all sections, including the control. Pavement edge temperatures showed a decided cooling trend which was slight for the control, more pronounced for the UAF/CRREL units and AFI flat loop evaporator units, and highest for the AFI buried units section. The centerline thermal conditions relate to the condenser size, while the pavement edge or shoulder conditions relate to the type of installation. During the winter season, the AFI buried condenser thermosyphons show surface warming and sublimation of snow closer to that of the control section, while the other two test sections are colder at the surface during the winter season.

Introduction

Construction of buildings and roadways in a permafrost area requires knowledge of cold regions engineering. Slope instability at the shoulders of roadways and settlement of constructions and roadways are common in permafrost areas. To avoid these problems, a thick layer of gravel fill is often used for the construction of roadways and structures. This can be very costly where the construction is built in an ice-rich permafrost area.

To develop and improve thermosyphons with horizontal evaporator sections, a project was funded by the Alaska Science and Technology Foundation, ASTF, Federal Highway Administration, FHWA, Alaska Department of Transportation and

[1] M.Sc., Department of Civil and Mining Engineering, Luleå University of Technology, SE-971 87 Luleå Sweden; phone +46 920 491 000; Anna.Forsstrom@ce.luth.se

[2] P.E., President, Arctic Foundations Inc, Anchorage Alaska

[3] Adjunct Professor, Civil and Environmental Engineering, University of Alaska Fairbanks

[4] Professor, Department of Civil and Mining Engineering, Luleå University of Technology

Public Facilities, AKDOT&PF, the Institute of Northern Engineering of the University of Alaska Fairbanks, UAF, the U.S. Army Corps of Engineers' Cold Regions Research and Engineering Laboratory, CRREL, and Arctic Foundations Inc. Alaska, AFI. An installation for testing various kinds of thermosyphons was completed in November 1998 at Chena Hot Springs Road, northeast of Fairbanks, Alaska. The test section was divided into four equally spaced sections where arrays of three different thermosyphons were installed. Comparisons were made between the three installations and the non-thermosyphon section. The project also addressed the stabilization of roadway embankments constructed over a permafrost area by using thermosyphons and rigid foam plastic insulation. Thermosyphons remove the heat from the subgrade during the winter season. During the summer thaw season, the insulation serves as an element to reduce the heat flowing into the permafrost.

Methods and Materials

Thermosyphons

A thermosyphon is a passive heat transfer device that operates by convection through vaporization and condensation. It consists of a sealed vessel with an upper part working as a condenser and a buried part in the ground functioning as an evaporator. Most condensers have fins to achieve a greater cooling area. A pressurized two-phase working fluid is inside the vessel where both liquid and vapor phases exist. Carbon dioxide (CO_2) is the most commonly used working fluid.

The heat transfer from the ground to the air is driven by a temperature difference across the unit. For the heat transfer to occur, a temperature differential of less than 1°C between the condenser and the evaporator is needed. Below ground, if subjected to temperatures greater than the saturation temperature of the liquid vapor moisture, the working fluid boils and vaporizes. If the vaporized working fluid reaches the condenser and is subjected to cold ambient air, condensation takes place. The condensed fluid gravitates to the evaporator and the cycle is repeated. Whenever the air temperature is greater than the soil temperature the cycle ceases.

Thermosyphons cool the sub-grade, thereby decreasing the ground temperature. The advantages of CO_2 thermosyphons are low maintenance cost, no operating cost, and environmentally friendly. Thermosyphons have been used successfully since the 1960's for stabilizing the ground in permafrost areas. In the past, thermosyphons have mostly been used with a vertical or downward sloped evaporator section (usually on 1 in 10 to 1 in 20 slopes). The flow of fluid was in that way assured along the entire length of the evaporator section. For some installations this requirement resulted in a greater fill thickness to achieve the required slope. In order to use a horizontal evaporator section, the thermosyphon's condensed liquid must be delivered along the entire length of the evaporator section. Failure to accomplish this will result in a phenomenon called dry-out which results in reduced heat transfer.

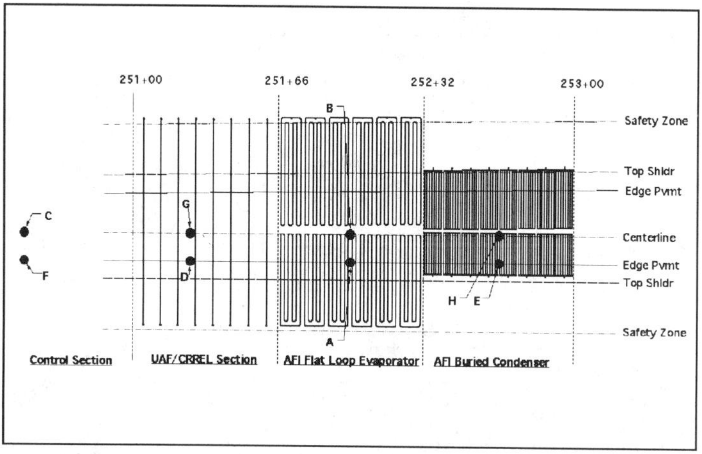

Figure 1. *Layout of Chena Hot Springs Road test section.*

Thermosyphon Installation

The experimental section was constructed just west of Esro Road intersection at mile 4.5, Chena Hot Springs Road, Fairbanks, Alaska. The roadway section is in a valley with wetlands on both sides of the road and has permafrost as the underlying geology. Each individual test section was 20 m (66 ft) in length and completed during the fall of 1998. The plan for the installation is shown in Figure 1.

The first stabilized section, located on the west end of the test section, begins at Station 251+00 and is referred to as the UAF/CRREL thermosyphons. This section was installed with sixteen UAF/CRREL single evaporator units with vertical finned condenser sections. The evaporator pipes consisted of a 65 mm (2.5 in) nominal diameter and a length of either 12.8 m (42 ft) or 14.0 m (46 ft). They were connected to an 80 mm (3 in) nominal diameter, 2.4 m (8 ft) long finned condenser section with an area of 6.5 m^2 (70 ft^2). An internal condensate collector and piping were used to ensure condensate return to the far end of the evaporator section. A "burper" section (a bent pipe) was also incorporated into the design to eliminate vapor locking of the unit. Units had 2.4 m (8 ft) on centers and were the simplest for field installation.

The second section starts at Station 251+66 and is referred to as the AFI flat loop evaporator thermosyphons. This section had twelve Arctic Foundations Inc.'s flat loop evaporator units installed. They consisted of a 2.7 m (9 ft) wide by 12.2 m (40 ft) or 14.2 m (46.5 ft) long horizontal grid of 20 mm (3/4 in) pipes. These were connected to 80 mm (3 in) nominal diameter finned vertical condenser sections of 4.9 m (16 ft) nominal length with an area of 15.8 m^2 (170 ft^2).

The third section starts at Station 252+32 and is referred to as the AFI buried condenser thermosyphons. Sixteen totally buried units were installed at this section with vertical evaporators installed at the edge of the pavement and horizontal condensers installed under the pavement. The condenser grids were 2.3 m (7.4 ft) wide by either 5.5 m (18 ft) or 7.7 m (25.3 ft) long. They consisted of 40 mm (1.5 in)

Table 1. Soil Properties for the roadway material.

Material	ρ_d t/m³ (lb/ft³)	w % (%)	k_f W/m°C (BTU/hr-ft-°F)	k_t W/m°C (BTU/hr-ft-°F)	c_f MJ/m³°C (BTU/ft³-°F)	c_t MJ/ m³°C (BTU/ft³-°F)	L MJ/m³ (BTU/ft³)
Asphalt	2.21 (138)	0 (0)	1.49 (0.86)	1.49 (0.86)	1.88 (28)	1.88 (28)	0 (0)
Crushed aggregate	2.21 (138)	5 (5)	2.58 (1.49)	2.44 (1.41)	1.80 (26.9)	2.04 (30.4)	37 (994)
Type A w=2%	2.21 (138)	2 (2)	1.73 (1.00)	1.97 (1.14)	1.66 (24.8)	1.76 (26.2)	15 (397)
Type A w=5%	2.21 (138)	5 (5)	2.58 (1.49)	2.44 (1.41)	1.80 (26.9)	2.04 (30.4)	37 (994)
Type A w=11.3%	2.21 (138)	11.3 (11.3)	3.95 (2.28)	3.05 (1.76)	2.09 (31.3)	2.62 (39.1)	83 (1627)
Insulation	0.048 (3.0)	0 (0)	0.033 (0.019)	0.033 (0.019)	0.10 (1.5)	0.10 (1.5)	0 (0)
Silt 3-4.5 m depth (10-15' depth)	1.30 (81)	35 (35)	2.58 (1.49)	1.59 (0.92)	1.87 (27.9)	2.83 (42.1)	152 (4082)
Silt 4.5-6 m depth (15-20' depth)	1.17 (73)	42.5 (42.5)	2.53 (1.46)	1.44 (0.83)	1.87 (27.9)	2.91 (43.4)	166 (4468)
Silt 6-10.5 m depth (20-35' depth)	1.01 (63)	55 (55)	2.44 (1.41)	1.06 (0.61)	1.88 (28.0)	3.05 (45.4)	185 (4990)

small diameter pipes that were manifolded and attached to a single evaporator pipe of length 6.4 m (21 ft). The condensers were placed in the non-frost-susceptible (NFS) fill, 0.3 m (1 ft) below the roadway surface and 0.6 m (2 ft) above the insulation. The operating concept for these units is to cool the subgrade and to warm the pavement.

Soil Properties

Soil samples at the site were taken from eight locations, four at the centerline of the roadway (C, G, B and H in Figure 1) and four at the edge of the pavement (F, D, A and E in Figure 1). These are also the locations where thermistors for measuring the ground temperature were placed. The average dry densities (ρ_d) and the moisture contents (w) from the drilling samples are presented in Table 1 along with the thermal conductivities in thawed and frozen condition (k_t and k_f). In Table 1 the volumetric specific heats (c_t and $c_f$respectively) and latent heat of fusions (L) are provided. Data from Kersten (1949) was used as basis for the thermal conductivities of these soils.

A cross section of the centerline of the roadway can be seen in Figure 2. Type A material (Sandy Gravel) was selected as the bedding beneath and filling above the thermosyphons. The evaporators of the UAF/CRREL units and the AFI flat units were installed horizontally on compacted and graded material at a depth of 1.4 m (4.5 ft) below the paved surface. Then 0.4 m (14 in) of type A material was placed above the thermosyphons with 0.1 m (4 in) thick insulation above that. On top of the insulation, 0.8 m (2.5 ft) of type A material was then placed. Crushed aggregate of

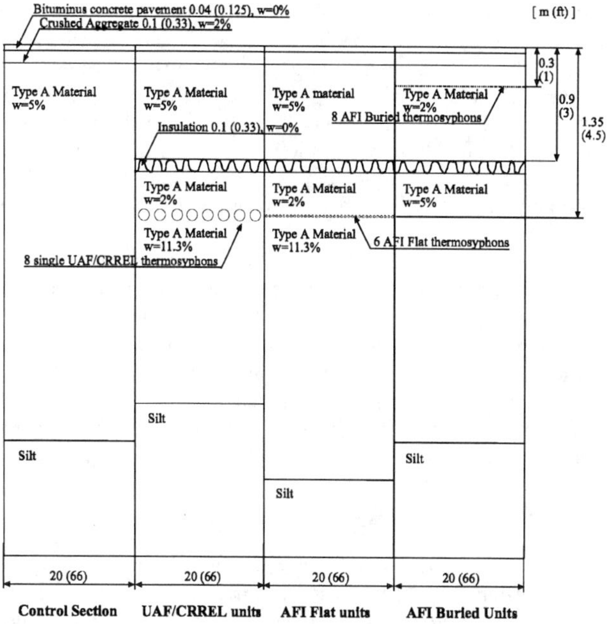

Figure 2. *Vertical cross section of the roadway*

0.1 m (4 in) thickness was added at all section with a 38 mm (1.5 in) thick bituminous concrete pavement on top. The different sections were 20 m (66 ft) wide and the moisture content varied with the different layers of soil as shown in Figure 2.

The dry density of material A was approximately 2.21 t/m^3 (138 lb/ft^3) and moisture contents were 5 % to 11.3 %. A Washington-Dens-O-Meter was used for field density tests and showed compactions ranging from 101% to 97% relative to the 95% compaction specification.

The rigid foam plastic insulation used in the installation was two layers of 50 mm (2 in) thick expanded polystyrene (EPS) board stock. Thermal conductivity listed by the manufacture as 0.029 – 0.031 W/m°C (0.017 - 0.018 BTU/hr-ft-°F). Compressive strength was specified to be at least 276 kPa (40 psi) at 10% deflection.

Temperature readout equipment

The thermistor strings were installed at eight different locations. Figure 1 shows the locations throughout the installation area. Each array of thermosyphons had one string installed at the centerline of the roadway and one string installed at the edge of the pavement at a distance of 7.9 m (26 ft) from the centerline string. The section where thermistor strings C and F were installed was the control section. This section had neither thermosyphons nor insulation installed.

The eight thermistor strings were placed vertically, with the upper most gauge, 1.4 m (4.5 ft) below the pavement. For the UAF/CRREL units and AFI flat units this was just below the plane of evaporators. The thermistor strings were installed in 25 mm (1 in) diameter PVC pipes at 1.4 m, 2.0 m, 3.2 m, 4.7 m, 6.6 m and 8.7 m (4.5 ft, 6.5 ft, 10.5 ft, 15.5 ft, 21.5 ft, and 28.5 ft) below the ground surface. They were capped on the bottom end to prevent inflow of soil or ground water. The annular spaces between the cables and the PVC pipes were filled with dry sifted mortar sand. An additional thermistor, housed in a radiation shield, measured air temperature.

Alpha Thermistor part number 07A001-D3 was used with a nominal 6,000-Ohm resistance at 32°C, and a 0.1 °C tolerance. The data logging system used consisted of a Campbell Scientific CR10X-1M-XR data logger, AM416 multiplexer, SM192-XR-1 storage module, and an ENC-16/18-weather enclosure. Data were initially taken 14 November 1998 and is planned to continue to the fall of 2001. Data were read every minute and hourly averages were recorded. Data are missing for the period of 1999-04-18 to 1999-07-15 and 1999-08-26 to 1999-10-10 due to shorting out of the battery in the data logger.

Test Section Geometry and Thermal Analysis

A two-dimensional finite element code, TDHC for geotechnical thermal analysis, written by Goering and Zarling (1985), was used to arrive at the design of the UAF/CRREL units (see Zarling, 1999). The boundary conditions used in this model were the characteristics for the Chena Hot Springs Road test section. A geothermal heat flux of 0.047 W/m^2 (0.015 $BTU/hr\text{-}ft^2$) was used for the lower boundary of the roadway section based on the nominal geothermal gradient and the thermal conductivity of the sub-grade. Vertical boundaries were specified as zero heat flux boundaries (centerline is a line of symmetry and vertical boundary beyond toe-of-sole is one dimensional heat transfer). For the surface of the pavement, thawing and freezing n-factors 1.8 and 1.0 were used respectively (Andersland B. O. & Ladanyi B. (1994)). The thermosyphon was modeled as a heat extraction node. Parameters changed during the analysis for obtaining the most effective design were the thermosyphon spacing, insulation thickness and thermosyphon conductance.

For designing the installation of AFI Flat units and the buried condenser units a one-dimensional heat transfer analysis was used. The thermal analysis is based on the heat flow rate, q (J/m^2s), equation:

$$q = \frac{\Delta T}{R} \tag{1}$$

where the resistance R ($m^2 {}^\circ C/W$) is in the direction of heat flow and ΔT (°C) is the temperature differential across the resistance (Heuer, Long and Zarling, 1985).

Table 2. *Freezing index, thawing index, and snowfall in Fairbanks for 1998-2000. The average for the years of 1930-2000 are also listed.*

Season	Freezing index °C days (°F days)	Thawing index °C days (°F days)	Snow on ground mm (in)
1997/1998	2776 (4996)		330 (13.0)
1998		1992 (3585)	
1998/1999	3299 (5938)		259 (10.2)
1999		2011 (3620)	
1999/2000	3033 (5460)		417 (16.4)
2000		1789 (3220)	
Average 1930-2000	3006 (5410)	1883 (3390)	404 (15.9)

Results

Weather data

A comparison has been made between the air temperature measured at the site and the air temperature measured by the Weather Service of the National Oceanic and Atmosphere Administration, NOAA, for Fairbanks WSO Airport. For winter 1999/2000 the freezing index in Fairbanks and CHSR was 3033°Cdays (5460°Fdays) and 2992°Cdays (5386°Fdays), respectively. The thawing index for summer 2000 was in Fairbanks 1789°Cdays (3220°Fdays) and 1591°Cdays (2864°Fdays) at CHSR. A difference of 12 percent or less between the two sites justified using the weather data published by NOAA. Table 2 shows the thawing and freezing indexes and snow on ground for Fairbanks for 1997/1998-2000. An average freezing and thawing indices and snow on ground are also listed for the years of 1930-2000. Compared to the average winter during 1930-2000, winter 1997/1998 was warmer than winter 1998/1999 and 1999/2000. Summer 1998 and 1999 was warmer than the average summer during 1930-2000 whilst summer 2000 was colder. The snow on the ground was for winter 1998/1999 and 1999/2000 less than the average but was for winter 1999/2000 greater than the average snow on the ground.

Ground temperature measurements

Ground temperature measurements have been analyzed for the test site from November 1998 to 31 December 2000. In Figure 3 a) – d) and 4 a) – d) the temperatures at depth below the surface of the pavement in meters are plotted versus the time. The temperature measurements are from a depth of 1.4 m to 8.7 m (4.5 ft to 28.5 ft). Isotherms plotted are -16°C, -8°C, -4°C, -2°C, -1°C, -0.5°C, 0°C, 0.5°C, 1°C, 2°C, 4°C, 8°C and 16°C (3.2°F, 17.6°F, 24.8°F, 28.4°F, 30.2°F, 31.1°F, 32°F,

32.9°F, 33.8°F, 35.6°F, 39.2°F, 46.4°F and 60.8°F). Portions of data are missing for most of the summer of 1999.

Center of the roadway

For the first season of the project, 1998/1999, the –0.5°C (31.1°F) isotherm was found at 4.5 m (14.8 ft) on all sections but the AFI buried unit. For 1999/2000 the –0.5°C (31.1°F) isotherm was again at 4.5 m (14.8 ft) for the control section. The –0.5°C (31.1°F) isotherm for UAF/CRREL units and for AFI flat units for winter 1999/2000 was at 6 m (19.7 ft) and 7.5 m (24.6 ft), respectively. The cooling from the AFI flat units were, for the center of the roadway, somewhat greater than the UAF/CRREL units due to a larger fin area. At the buried units, due to extracted heat from the evaporator to the condenser, a shallower depth for the –0.5°C (31.1°F) isotherm occurred at 2.5 m (8.2 ft) for both winter 1998/1999 and 1999/2000.

Thaw depth during summer 1999 was 2.5 m (8.2 ft) for the UAF/CRREL units and for the AFI flat units compared to 4.5 m (14.8 ft) for the control section. For summer 2000, the thaw depth was unchanged for the control section and had decreased to about 2 m (6.6 ft) for the UAF/CRREL units section and the AFI flat units section. For the buried section, the thaw depth was unchanged for summer 1999, but decreased from 4.5 m to 3 m (14.8 ft to 9.8 ft) during summer 2000.

Edge of the pavement

For the first season of the project, winter 1998/1999, the –0.5°C (31.1°F) isotherm was at a depth of 4.5 m (14.8 ft) at the control section, 3 m (9.8 ft) at the UAF/CRREL section, 4.5 m (14.8 ft) at the AFI flat section and 6.5 m (21.3 ft) at the buried section.

The –0.5°C (31.1°F) isotherm for winter 1999/2000 was at a depth of 4.5 m (14.8 ft) for the control section, UAF/CRREL units and for AFI flat units. For the buried units section the –0.5°C (31.1°F) isotherm had moved down to 8 m (26.2 ft). Due to the vertical evaporators located at the edge of the pavement, the buried section cooled to a greater depth than the other sections.

The thaw depth during summer 1998 was for the four sections just below 8 m (26.2 ft). For summer 1999, the thaw depth for the control section was just below 6 m (19.7 ft) and for the other three sections it was about 3 m (9.8 ft). A smaller thaw depth for the control section may depend on a decreased snowfall for 1998/1999, see Table 2. For the other sections, a cooling from the thermosyphons resulted in the shallower thaw depth. The thaw depth of the control section did again increase to 8 m (26.2 ft) in summer of 2000 and decreased slightly compared to 1999 for the other three sections.

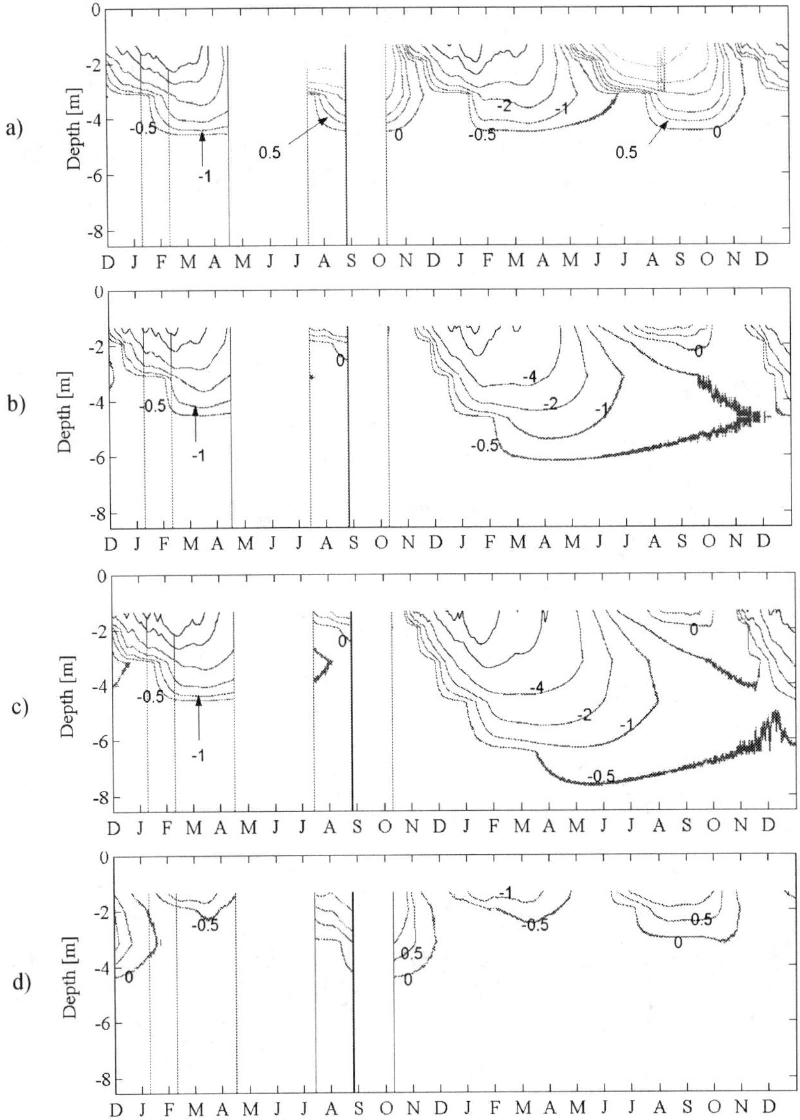

Figure 3 a) – d) *Ground temperature (°C) plotted versus the time Dec 1998 – Dec 2000 for the four sections at the center of the roadway; a) the control section, b) UAF/CRREL units, c) AFI flat units, and d) AFI buried units.*

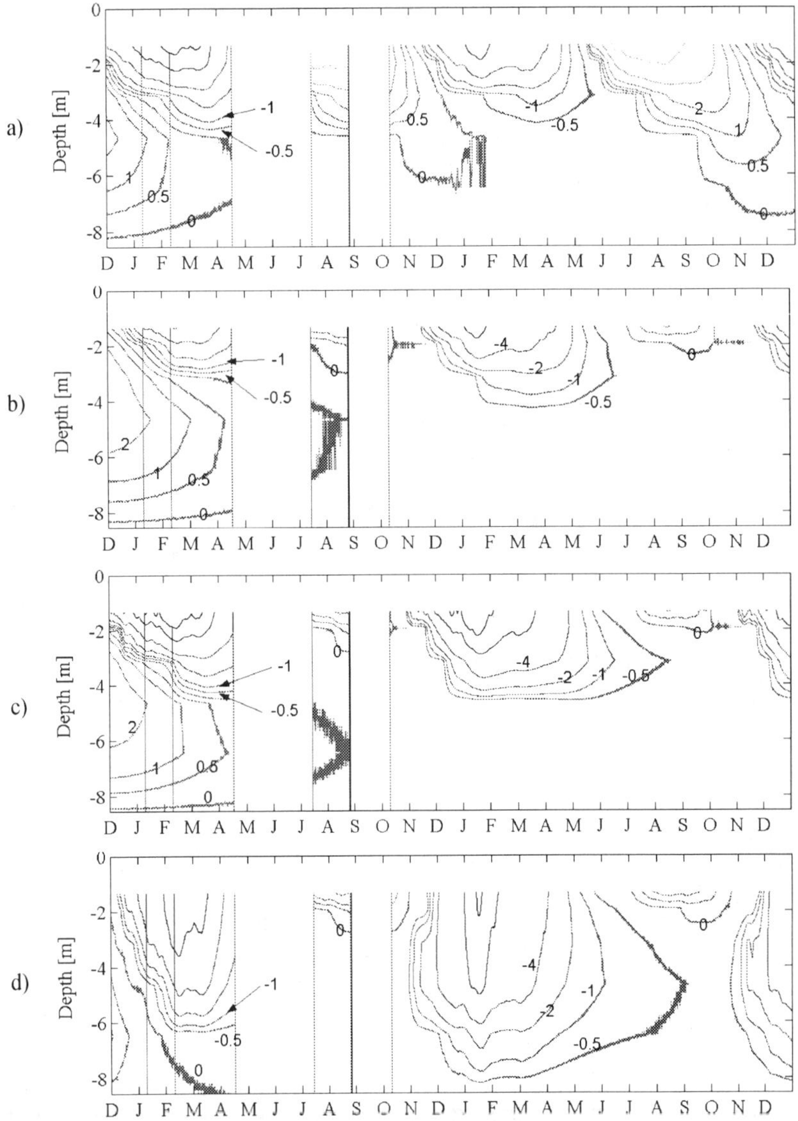

Figure 4 a) – d) *Ground temperature (°C) plotted versus the time Dec 1998 – Dec 2000 for the four sections at the edge of the pavement; a) the control section, b) UAF/CRREL units, c) AFI flat units, and d) AFI buried units.*

Discussion

Thermosyphons are lowering the ground temperature. Data from The Chena Hot Springs Road project shows that thermosyphons are an effective way for stabilizing roadways in permafrost ground.

The UAF/CRREL units and the AFI flat units cool the ground in a very similar manner. The AFI flat units have a larger fin area, which make them more effective than the UAF/CRREL units. Additional charging fluid was added to the UAF/CRREL units in March 1999, which increased their effectiveness (note a lower –0.5 °C isotherm at the edge of the pavement for the UAF/CRREL units compared to the AFI flat units). For winter 1999/2000, the –0.5 °C isotherm was approximately 45 % and 67 % deeper for the UAF/CRREL units and for the flat units respectively compared to the control section at the center of the roadway. Thawing depth was approximately 55 % less for the UAF/CRREL units and the AFI flat units compared to the control section for summer 2000. For the edge of the pavement, the thaw depth was 73 % less for the buried thermosyphons section compared to the control section summer 2000.

The cooling of the buried condenser section thermosyphons can be improved by installing the condensers closer to the surface of the roadway. Advantages of the buried thermosyphons are the non-visible condensers and the sublimation of moisture from the pavement surface.

Acknowledgements

Funding by the Alaska Science and Technology Foundation, Alaska Department of Transportation and Public Facilities, and Arctic Foundations Inc. made this project possible. Luleå University of Technology is acknowledged for financial support during the fall of 2001.

References

Andersland, B. O., and Ladanyi, B. (1994). *An Introduction to Frozen Ground Engineering*, Chapman & Hall, Inc., New York.

Goering, D. J. and Zarling, J. P. (1985), *Geotechnical Thermal Analysis with a Microcomputer*, Civil Engineering in the Arctic Offshore, ASCE.

Heuer C.E., Long E. L., and Zarling J.P. (1985). *Passive techniques for ground temperature control,* Technical Council on Cold Regions Engineering Monogram. Krzewinski, T.G., Tart R. G. Jr (eds). NY ASCE.

Kersten, M. S. (1949). *Thermal Properties of Soils*, Bulletin No. 28, Engineering Experiment Station, University of Minnesota, St. Paul.

Zarling, J. P (1999). "Thermosyphon Installation on Chena Hot Springs Road", Report to Alaska Department of Transportation and Public Facilities.

Recent Developments in Thermosyphon Technology

Edward Yarmak, Jr.[1] and Erwin L. Long[2]

Abstract

Thermosyphons have been used to provide passive subgrade refrigeration for structures founded at-grade on permafrost for approximately 25 years. The majority of these structures utilize thermosyphons installed with conventional sloping evaporators (CSE) beneath subgrade insulation. In the last 5 years, the use of thermosyphons with flat loop-type evaporators (FLE) has been introduced for providing subgrade cooling. FLE units offer versatility for design and their use can simplify construction when compared to installation of CSE thermosyphons. Performance tests show FLE units offer increased heat transfer compared to CSE thermosyphons when installed at windy sites. Advantages and design of subgrade cooling systems using both CSE and FLE thermosyphons are discussed. Critical elements to be considered in the design of either system are listed and a matrix is offered to assist the designer when choosing a thermosyphon system for subgrade cooling.

Hybrid thermosyphons, buried condenser thermosyphons, flat evaporator thermosyphons, and flexible evaporator thermosyphons are also discussed along with their potential applications.

Background

Thermosyphons are passive refrigeration devices that transfer heat against gravity. More particularly, for subgrade cooling applications, thermosyphons remove heat from the ground beneath a structure and release it to the outside ambient air, whenever the air is colder than the ground. The units are typically closed ended tubular vessels charged with a two-phase working fluid. The working fluid fills the entire interior of the vessel with the vapor phase filling the most space and the liquid phase filling the minority of the volume. Typically, a thermosyphon is installed with

[1] Member ASCE; Chief Engineer, Arctic Foundations, Inc., 5621 Arctic Blvd., Anchorage, Alaska, 99518-1667; E-mail, eyarmak@arcticfoundations.com; Telephone (907) 562-2741

[2] Member ASCE; President, Arctic Foundations, Inc., 5621 Arctic Blvd., Anchorage, Alaska, 99518-1667; E-mail, elong@arcticfoundations.com; Telephone (907) 562-2741

a portion of the vessel above ground and the remainder buried. When the above ground portion is cooled below the saturation temperature of the vapor, the vapor condenses on the interior of the cooled portion of the vessel and releases its latent heat of vaporization. This slightly decreases the interior pressure and initiates boiling and vaporization of the liquid whereby heat is absorbed from the ground. The condensate in the upper portion of the vessel flows via gravity to the lower portion of the vessel. Whenever the condensate comes in contact with a warmer portion of the vessel, it evaporates and causes that area of the vessel to cool. The cycle continues as long as the above ground portion is colder than the below ground portion of the thermosyphon. When the above ground portion is warmer than the below ground portion, the cycle ceases and the vapor density gradient within the vessel essentially precludes any heat transfer into the soil from the air except conduction down the vessel walls.

The above grade portion of a thermosyphon is normally termed the condenser and the below grade portion of the thermosyphon is termed the evaporator. For units that are solely used for subgrade refrigeration, the condenser is generally fitted with fins to increase the conductance between it and the air. A section of the vessel known as the riser often connects the evaporator and condenser portions. Generally, the riser is thought to be a nearly adiabatic portion of the thermosyphon that holds the condenser above the level of the snow cover.

Structures founded at-grade on permafrost utilizing thermosyphons for subgrade cooling also have two other very important components of design: subgrade insulation and NFS fill. The typical thermal design methodology for this type foundation economically balances the three major components. The insulation reduces the heat load on the subgrade and limits the thawing of the underlying NFS fill during the summer season when the thermosyphons are not functioning. The NFS fill contains the seasonal thaw and provides a heat sink beneath the structure. The thermosyphons remove heat from beneath the structure primarily during the freezing season and refreeze the NFS fill so that it is completely frozen at the beginning of the thawing season.

CSE Thermosyphons

Conventional sloping evaporator (CSE) thermosyphons are found on the majority of existing passive subgrade cooling system installations. These units utilize a sloped evaporator to insure the condensate will flow from the condenser to the lowest portion of the evaporator. Typically, evaporator slopes are between 10% and 3% with the median being 5%. The thermosyphon manufacturer specifies minimum slopes for different evaporator types.

The first structure constructed utilizing CSE units was the Ross River School in the Yukon Territory, Canada in 1978, (Hayley, 1982). The second CSE system was installed for the OMS Building at the Army National Guard Armory in Bethel, Alaska during the winter of 1978-1979. In the early 1980's, CSE systems were being routinely installed to provide subgrade refrigeration beneath heated at-grade structures throughout the permafrost region in Alaska. Today, there are hundreds of CSE systems in use throughout the permafrost regions of North America.

CSE thermosyphon systems are attractive for a variety of reasons. The systems have a proven track record of over 20 years and conservative property owners like that comfort factor. CSE systems can be installed by virtually any contractor who has experience performing installations of pipe to grade in the ground. CSE units may be fully shop assembled, tested, shipped and installed provided that their dimensions do not exceed shipping limitations. For outsize units, CSE units are readily prefabricated in the shop to the largest piece practical, and then assembled, charged, and tested at the installation site. Typically, the CSE is constructed using welded steel pipe, thereby allowing conventional methods to be used for field assembly. CSE units are generally of sufficient diameter to internally sleeve, should a breech in the evaporator occur, thereby reducing the cost of potential repairs.

There are shortcomings to using CSE units on some projects. When long evaporator sections are used, the drop in vertical as the evaporator slopes under a building can eclipse the capacity of available trenching equipment necessitating either an increase in trencher size or the use of another technique to install the evaporator. Additionally, because it is generally thought that the evaporators should remain in the NFS fill, the volume of the NFS fill can be quite large when long evaporators are used. If the CSE units are installed in a subgrade that may be subject to movement before stabilization occurs, the evaporators may settle or heave out of vertical alignment.

FLE Thermosyphons

Flat loop evaporator (FLE) thermosyphons are a relatively recent development. These units utilize a loop type evaporator that is typically manufactured using 20mm steel pipe. Loop spacing beneath at-grade structures is generally similar to the evaporator spacing used on CSE systems. When looking at FLE systems, there is generally about one half the building footprint area to pipe length ratio compared with a CSE system. The loop type evaporator may be straight from the riser with a U-bend to return or quite circuitous with numerous bends before returning to the riser. A typical plan view of a FLE thermosyphon system is shown in Figure 1.

During the winter of 1993-1994, a prototype FLE thermosyphon was performance tested adjacent to a like sized CSE unit. Both units were installed in unfrozen soil in the prairie just south of Winnipeg, Manitoba with identical condenser sections. After a winter of operation, it was found that the FLE unit froze 1.4 times the volume of soil than the CSE unit. Performance parameters for the FLE units were not upgraded to take advantage of this increase. The first Alaskan deployment of the FLE units was at a 43-unit apartment complex in Kotzebue, Alaska in the summer of 1994. In 1994, there were also 10 deployments of the FLE systems in Canada. (Jardine, 2001) Canadian deployments of the system have outpaced Alaskan deployments, with most passive subgrade cooling systems built for at grade structures since 1994 in Canada being the FLE type. It is estimated that at this time there are approximately 50 to 60 at-grade structures in North America thermally protected with FLE thermosyphons. Just last year, the Ross River School in the Yukon Territory was replaced with a new structure utilizing FLE thermosyphons for subgrade refrigeration.

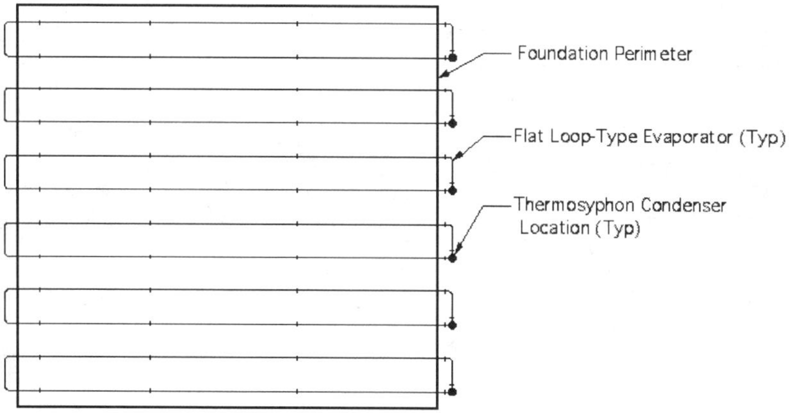

Figure 1. *Plan View of FLE Thermosyphon System*

The FLE thermosyphons utilize proprietary internal components within the riser to force the working fluid to travel in one direction within the evaporator. As the working fluid absorbs heat, it vaporizes and expands. The expansion helps to push the working fluid around the loop. The working fluid moves around the loop in two-phase flow, increasing in velocity as it proceeds around the loop. This allows the loop to be placed on a "relatively" flat plane. The term "relatively" means that there can be some undulations in the vertical profile of the loop. Typical designs allow for as much as 150mm amplitude in the undulations with no detrimental effects on the system. The subgrade can just be graded to a near flat condition within the specifications of the manufacturer and the loops then placed directly on the graded material. Typically the level of the loops is approximately 0.5m to 1.0m below the base of the subgrade insulation. Loop lengths of 150m are not uncommon.

The FLE thermosyphons have many positive attributes: the first being the straightforward evaporator installation. The allowance for the undulations in the evaporator does simplify the installation of the evaporators and reduces the equipment requirements on the job. FLE can be fabricated with a myriad of horizontal bends to concentrate cooling or to miss obstructions near the loop locations. The evaporator plans for the FLE units can be quite circuitous. Designers can position the condensers in locations that are aesthetically appropriate and still where the ambient winter winds can provide maximum cooling. Because the evaporators need not slope like the CSE units, there is no need to increase the NFS fill depth under a structure where FLE units are installed. This is a very important consideration where NFS material is not plentiful. Construction timing is one of the factors that must be taken into account for any subgrade cooling system, and the FLE system offers more versatility than some other system because an allowance for settlement or heave can be built into the design of the system. Additional latitude in scheduling of construction generally gives a lower installed cost. The FLE systems generally weigh less than comparable CSE systems. This reduces the transportation costs to get the subgrade cooling system from the manufacturer to the site.

Table 1. *CSE/FLE Decision Matrix*

Task / Attribute	CSE Thermosyphons	FLE Thermosyphons
Track Record	Since 1978	Since 1994
Shipping Weight	Moderate	Low
Subgrade Preparation	Moderate/High	Low
NFS Fill Quantity	Moderate	Low/Moderate
Fully Shop Assembled	Yes (under 20m)	No
Field Assembly Required	Yes (over 20m)	Yes (always)
Sensitivity to Settlement & Heave	Moderate/High	Low
Prone to Damage by Excavation	Low/Moderate	High
Repair w/o Excavation	Yes	No

FLE units do have some less positive attributes that must be taken into account. The small diameter of the FLE tubing is makes it prone to damage. Longer than standard lengths have a high probability of damage in transit. FLE systems are almost always assembled in the field whereas many CSE systems can be shipped to the site with no assembly required. The assembly crew is another specialized trade at the job-site that must be scheduled into the work. The FLE condensers generally require a bracing system for lateral support because the evaporator tube is not robust enough to provide a firm anchorage. Typically, FLE condensers are braced to the structure or to each other. The tubing size of the FLE is critical to its operation and larger is not always better. Should a tube become damaged, it cannot be readily sleeved like a CSE unit so care must be taken around the installed FLE units. At two projects in Alaska, the FLE tubes were severed during augering that was being performed for the installation of other structural components.

The matrix shown in Table 1 is presented to assist the designer when choosing a FLE or a CSE thermosyphon system.

Hybrid Thermosyphons

Hybrid thermosyphons are units that are designed so that they can be cooled actively as well as passively. By active cooling, it is meant that the some external energy must be expended to promote cooling. Most thermosyphons can be manufactured as hybrid units with the addition of an internal heat exchanger. A typical application for hybrid thermosyphons is to accelerate construction on a project by active freezing. In 1984, hybrid thermosyphons were used to actively freeze a foundation for a communications site in Galena, Alaska. Prior to that time, blowing chilled air across the condensers was the method used for active refrigeration of thermosyphons. Thermosyphon evaporators had also been used as evaporators for active refrigeration, however, the working fluid had to be evacuated for refrigeration to occur and then replaced afterward. Not all hybrid thermosyphons have been actively cooled. In most cases the hybrid units are specified to add additional redundancy to the design of the subgrade cooling system. Often, the heat load on the foundation from the equipment within a structure cannot be accurately estimated, so hybrid thermosyphons are specified to allow retrofitting of an active refrigeration system to the subgrade cooling system.

Notable installations of hybrid units include the Maniilaq Medical Center in Kotzebue, Alaska where all 370 of the structural, i.e. load-bearing, thermosyphons supporting the structure are fitted with internal heat exchangers and the Frozen Soil Barrier at the HRE Pond at the Oak Ridge National Laboratory in Tennessee where 50 hybrid thermosyphons are being actively cooled to contain radioactive contamination. A new deployment of hybrid thermosyphons is ongoing at the Diavik diamond mine at Lac de Gras, Northwest Territories to key a water retention dike into permafrost.

Buried Condenser Thermosyphons

Thermosyphons with buried condensers are designed to release the heat picked up from the subgrade in the near surface soils or embankment. A proposed application for these units is in roadway sections where heat is released just under the pavement. This heat transfer is not designed to thaw ice and snow on the pavement, however, it may cause sublimation that will loosen the bond between the pavement and the ice. An example of this typing of unit is installed at the Chena Hot Springs Road test site, (Forsström, et. al, 2002).

Flat Evaporator Thermosyphons

Thermosyphons have been constructed with flat evaporators. The key to proper operation of a unit with a flat evaporator is to insure uniform distribution of the liquid phase of the working fluid. To accomplish this, either the diameter of the evaporator is sized to be able to accommodate condensate flows from one end or a condensate distribution device is installed within the evaporator. The evaporator grade needs to be flat for these systems to work. The first flat evaporator system deployed in Alaska for subgrade cooling was at the City of Galena's Warm Storage building in 1978. That system relied on the evaporator size to insure condensate distribution along the entire evaporator. More recently, research thermosyphons with flat evaporators and condensate distributors were installed at the Chena Hot Springs Road test site, (Forsström, et. al, 2002).

Flexible Evaporator Thermosyphons

Flexible evaporator thermosyphons have been manufactured since the late 1970's. These units have the capability of being fully charged in the shop and coiled or folded for shipment and then formed into their appropriate configuration for installation at the project site. This minimizes field assembly and charging work by the manufacturer. The materials that have been used for the flexible sections include annealed copper, aluminum extrusions, and coiled steel oilfield tubing. The copper has been the most widely used. These units are only flexible, for a one-time field bend and must be installed appropriately to keep flexure after installation to an absolute minimum. The authors have considered FLE thermosyphons that leave the shop fully charged with flexible evaporators however no units have been manufactured.

New Developments

The latest development in the thermosyphon field is the use of plastics for evaporators or entire units. The authors are currently evaluating a test unit with a plastic evaporator and a conventional steel riser and condenser section. Plastics offer increased corrosion resistance, and perhaps, an increase in flexibility. The plastic evaporator may be used on either CSE or FLE systems, depending upon the out come of the prototype evaluation.

Closure

It is not the scope of this paper to describe all the applications of thermosyphon technology. To date, the authors have been directly involved with subgrade refrigeration on over 700 projects and have attempted to educate the practicing engineer with some of the systems that are available today for passive subgrade refrigeration.

References

Forsström, A., Long, E. L., and Zarling, J. P. (2002). "Thermosyphon Cooling of Chena Hot Springs Road." *Proceedings of the 11th International Conference on Cold Regions Engineering*, May 20-22, 2002, Anchorage, Alaska

Hayley, D. W., (1982). "Application of heat Pipes to design shallow foundations on permafrost." *Proceedings of the 4th Canadian Permafrost Conference*, National Research Council of Canada, Ottawa, 535-544.

Jardine, J. (2001). Personal Communication, Arctic Foundations of Canada, Winnepeg, Manitoba.

Determining Aggregate Frost Susceptibility with the Tube Suction Test

W. Spencer Guthrie[1], E.I.T., Åke Hermansson[2], and Tom Scullion[3], P.E.

Abstract

The Tube Suction Test (TST) was developed in a cooperative effort between the Finnish National Road Administration and the Texas Transportation Institute (TTI) for assessing the moisture susceptibility of aggregate base materials. The moisture susceptibility ranking is based on the mean surface dielectric value of compacted specimens after a 10-day capillary soak, where the dielectric value is most sensitive to the amount of unbound water that exists within the aggregate matrix.

The purpose of this research was to investigate the relationship between the final dielectric value in the TST and the frost susceptibility of aggregates. The laboratory test program included 35 specimens representing 10 aggregate base materials from Indiana, Minnesota, Pennsylvania, Texas, and Virginia. Each sample was subjected to the TST and afterwards to a frost heave test developed at TTI. In the TST, electrical conductivity measurements were also made to evaluate the salinity of the specimens. Sample volumetrics were computed and compared for post-TST and post-freezing conditions.

In this study, aggregates classified in the TST as good performers did not experience significant frost heave, while several materials with higher dielectric values exhibited substantial amounts of heave. These latter specimens were characterized, on average, by greater percentages of fines, higher porosity, and higher moisture contents in the TST. Aggregate samples with high salinity also had high dielectric values in the TST, but they were generally less frost susceptible than other

[1] Graduate Research Assistant, Texas Transportation Institute, Texas A&M University, 3135 TAMU, College Station, TX 77843-3135; phone 979-845-9919, fax 979-845-1701; s-guthrie@ttimail.tamu.edu

[2] Researcher, Swedish National Road and Transport Research Institute, SE-581 95 Linköping, Sweden; phone +4613204072, fax +4613141436; ake.hermansson@vti.se

[3] Research Engineer, Texas Transportation Institute, Texas A&M University, 3135 TAMU, College Station, TX 77843-3135; phone 979-845-9913, fax 979-845-1701; t-scullion@tamu.edu

high-dielectric specimens due to the depression of the freezing point caused by salt ions in the pore water. Samples with high porosity also exhibited reduced frost heave because the additional internal space allowed the formation of ice with less expansion of the aggregate matrix.

Introduction

The Tube Suction Test (TST) was developed in a cooperative effort between the Finnish National Road Administration and the Texas Transportation Institute (TTI) for assessing the moisture susceptibility of aggregate base materials (Scullion and Saarenketo, 1997). The moisture susceptibility ranking is based on the mean surface dielectric value of compacted specimens after a 10-day capillary soak, where the dielectric value is most sensitive to the amount of unbound water that exists within the aggregate matrix. The performance of a material in the TST is largely dependent upon the amount and quality of its fines and upon its salinity, which both affect its affinity for water. While good quality aggregates maintain a strong moisture gradient throughout the test, poor quality materials experience substantial moisture ingress and generally develop uniform moisture profiles during soaking (Guthrie et al, 2001).

If constructed of moisture susceptible materials, pavements subjected to moisture ingress can sustain damage under trafficking when the induced stresses exceed the reduced strength of the material. In Texas, this is a primary distress mechanism and has been the focus of recent TST validation studies (Saarenketo and Scullion, 1995; Guthrie and Scullion, 2000). Moisture damage can be more pronounced in cold climates, where the formation of ice lenses in moisture susceptible layers can cause frost heave of the road structure and supersaturated conditions upon thawing (Taber, 1931). To evaluate the potential of the TST for determining the vulnerability of aggregates to such damage, this research investigated the relationship between the final dielectric value in the TST and the frost heave behavior of aggregates.

This paper first addresses principal factors influencing the moisture and frost susceptibility of aggregate base materials and discusses the implications of dielectric and electrical conductivity measurements used to characterize materials in the TST. Laboratory procedures are then described, followed by an analysis of test results obtained for 35 specimens representing 10 aggregate base materials from Indiana, Minnesota, Pennsylvania, Texas, and Virginia.

Moisture and Frost Susceptibility

The determination of moisture and frost susceptibility classifications for aggregate base materials depends chiefly upon their suction characteristics and hydraulic conductivity as discussed in the following sections.

Suction. Soil suction characterizations are derived from the study of moisture flow through soil media and are composed of both matric and osmotic suction components (Hanks and Ashcroft, 1980). Matric suction is mainly responsible for the capillary phenomenon in aggregate layers, where the radius of curvature of the meniscus in a capillary tube is analogous to the radius of curvature of an air-water interface in an aggregate matrix, and the height of capillary rise to the magnitude of matric suction. The force causing water to rise in a capillary tube depends on the radius of curvature of the meniscus, which is controlled by the radius of the tube and the contact angle with which the water wets the tube, and the surface tension of water (Hanks and Ashcroft, 1980). Matric suction is similarly affected by these factors, where higher surface tension, lower contact angles, and smaller radii give rise to greater suction at the air-water interface.

Because the pore structures and water contents in natural systems directly influence the constitution of these interfaces between soil particles, unique soil-water characteristic curves can be used to relate water content to matric suction for individual soils. For saturated soils, matric suction is essentially zero, but as the water content decreases, suction increases dramatically (Hanks and Ashcroft, 1980). As a consequence of their fine texture promoting smaller pore radii, clay, silt, and till can develop exceedingly high suction values and are typically classified as highly moisture susceptible (Penner, 1962).

Osmotic suction is the suction potential resulting from solutes present in the water portion of a soil or aggregate system (Hanks and Ashcroft, 1980). Water is attracted and diffuses by osmosis into zones of high salt concentrations, driven toward an equilibrium condition of uniform ion concentration within the pore water. Thus, in cold climates where deicing salts are routinely applied to roadways, an increase in moisture content within upper base layers may be observed due to increased osmotic suction as salts leach into the aggregate through time. Along with impacting osmotic suction, the concentration of dissolved solutes affects the surface tension of water, which generally increases with increasing concentrations of inorganic salts such as calcium, magnesium, and sodium chlorides (Weast et al, 1989). Because of its dependence on surface tension as stated earlier, matric suction is intrinsically linked to osmotic suction in this way.

In a freezing system, the salinity of pore solutions enhances the supercooling characteristics of soil water, where increasing solute concentrations generally decrease the freezing point (Anderson, 1967). As a rule, ice forms in a pure state, devoid of any contaminants, and the depression in the freezing temperature is derived from the additional energy required to physically "freeze out" ions during the formation of ice crystals (Panday and Corapcioglu, 1991). When subjected to sufficient suction gradients, water molecules in unfrozen saline films can migrate into freezing soil to create ice lenses (Anderson, 1967).

Hydraulic Conductivity. Given some level of suction and the availability of water, the rate of water ingress is controlled by the hydraulic conductivity of the aggregate layer, which is primarily governed by the mineralogy and compaction of its fine-grained constituents and by the water content (Hanks and Ashcroft, 1980). Higher hydraulic conductivity results from more interconnected void space within the pore structure at higher degrees of saturation that allows water to more easily flow into and through the aggregate matrix. Hydraulic conductivity is an especially important issue in moisture damage mechanisms such as frost heave where water must be able to rapidly respond to changes in suction within the pavement structure (Mitchell, 1976).

Electrical Measurements

Dielectric and electrical conductivity measurements were used in the TST to characterize aggregate base materials in this study. The first is indicative of the amount of unbound water that exists within the aggregate matrix, and the second reflects the salinity of tested specimens. Each property is addressed below.

Dielectric Value. The dielectric value of a material is indicative of its storage capacity in an electrical field (Tareev, 1975). For electrical fields alternating at frequencies between 10^6 and 10^9 Hz, the extent to which a material can store charges, or its polarizability, depends primarily upon the ability of its molecules to be physically reoriented (Gueguen and Palciauskas, 1994). Figure 1 depicts dipole chains formed from orientational polarization of a dielectric material in an electrical field generated by a voltage applied across two condenser plates. At the interfaces, these chains bind, or neutralize, surface charges on the plates so that the number of free charges on the plates is reduced. The bound charges are then said to be stored by the dielectric material. For any material, the dielectric value is equal to the total number of charges, bound and free, on the plates divided by the number of free charges (Von Hippel, 1954).

Materials composed of polar molecules with higher rotational mobility can store greater numbers of charges and will therefore have higher dielectric values than materials whose molecules are more rigidly fixed in position. Because the dielectric value of unbound water is 81, compared to 1 for air, 3 to 4 for tightly bound water adsorbed to mineral surfaces, and 4 to 6 for solid aggregate particles, the dielectric value of a mixture of these constituents is most sensitive to the amount of unbound water in the matrix (Saarenketo, 1998).

Electrical Conductivity. While the dielectric value reflects the ability of a material to store electrical field energy, electrical conductivity is a measure of the field energy

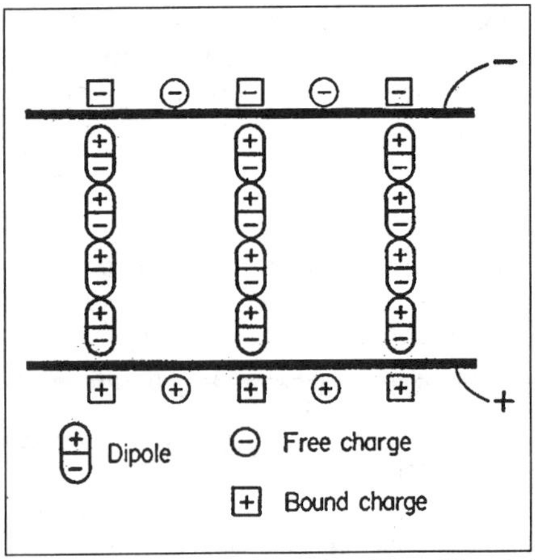

Figure 1. Orientational polarization (Von Hippel, 1954).

lost through induced charge displacements (Powers, 1997). Energy is dissipated as heat through charge transport associated with conduction currents that develop in response to applied electrical fields. These currents appear in materials when free charge carriers are brought into a state of ordered motion by the forces of the applied field (Tareev, 1975). The quantity of electricity passing in a unit time through a unit of a cross-sectional area of a conductive body is the product of the number of charge carriers in a unit volume, the magnitude of charge associated with each carrier, and the average carrier velocity.

Since mineral grains are mainly insulators, electrical conduction in most soils and aggregates is primarily electrolytic (Sharma, 1997). For this reason, in aqueous solutions where salts can dissolve into their constituent ions, the carriers are usually ions originating from ion exchange processes on clay surfaces, for example, or from outside sources such as deicing operations mentioned earlier. Electrical field energy losses result from heat generated by the retarding effect of water on the motion of ions within the fluid. That is, a viscous drag force resists the migration of cations to negative electrodes and anions toward positive electrodes. Some energy is also lost as heat through the slight charge displacements that occur in polarization mechanisms discussed earlier (Powers, 1997).

Laboratory Procedures

The research performed in this study included 35 specimens representing 10 aggregate base materials from Indiana, Minnesota, Pennsylvania, Texas, and Virginia. Each sample was subjected to the TST and afterwards to a frost heave test developed at TTI. Following this testing, specimens were oven dried for mechanical sieve analyses. Specific gravity testing was also performed on each sample, and absorption properties were determined for each of the 10 aggregate types. Procedures for the TST and the frost heave test are presented in the following sections.

Tube Suction Test. TST samples were tested in standard plastic cylinders of 152.4-mm inside diameter and 304.8-mm height. At approximately 6 mm above the outside bottom of the mold, 1.5-mm-diameter holes were drilled around the circumference of the mold at a horizontal spacing of 12.7 mm. One hole was also drilled in each quadrant of the bottom of the mold about 50 mm from the center.

Aggregate samples were then scalped on the 25-mm sieve and compacted at their respective optimum moisture contents in four lifts of 50 blows each with a 4.5-kg hammer dropped from a height of 457 mm to a finished height of about 200 mm inside the pre-drilled mold. After drying at 40 °C for four days, the specimens were placed in a 12.7-mm-deep bath of distilled water at room temperature for a 10-day soak period. Figure 2 depicts the standard TST arrangement. The moisture profiles developed during soaking were considered representative of those likely to exist in the field, given the availability of moisture.

The surface dielectric value and electrical conductivity were monitored daily with the Adek Percometer™ as shown in Figure 3. The Percometer™ operates at 50 MHz, with the electrical field generated between a central node and an outer ring

Figure 2. Typical TST arrangement.

arranged in coaxial fashion. The electrical field extends beyond the face of the probe so that measurements are sensitive to a depth of approximately 25 mm while being completely non-destructive. For materials with high suction and sufficient hydraulic conductivity, substantial amounts of unbound water rise within the aggregate matrix during soaking and lead to higher dielectric values in the TST. Conversely, non-moisture-susceptible materials maintain a strong moisture gradient throughout the test, with little moisture reaching the surface, and have lower dielectric values at the end of the TST (Guthrie et al, 2000). Six surface dielectric readings and six electrical conductivity readings were taken at each measurement time, with five around the perimeter of the sample and the sixth in the center. The highest and lowest readings

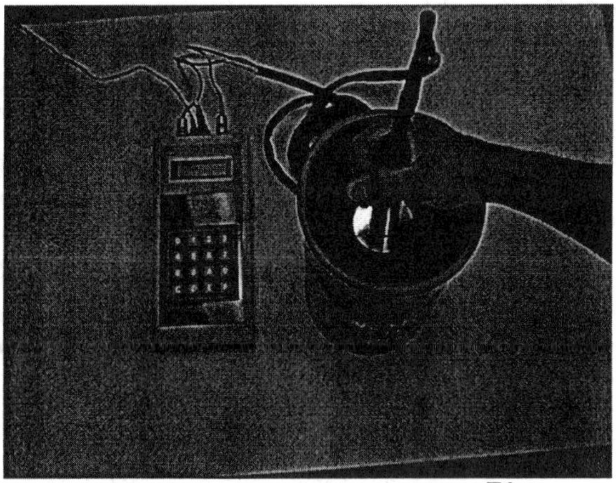

Figure 3. Using the Adek Percometer™.

were discarded as a means of reducing variability caused by surface imperfections, and the remaining four were averaged.

The interpretation of TST results is based on an empirical relationship between the final dielectric value and the expected performance of aggregate base materials (Scullion and Saarenketo, 1997). Aggregates whose final dielectric values in the TST are less than 10 are expected to provide superior performance as base materials, while those with dielectric values above 16 are expected to provide poor performance. Aggregates having final dielectric values between 10 and 16 are expected to be marginally moisture susceptible.

Frost Heave Test. At the conclusion of the TST, specimens were placed in an environmental chamber maintained at –17 °C for continued soaking and frost heave measurements through another 10-day period. As shown in Figure 4, styrofoam insulation was added around the samples to control the direction of frost penetration, and thermostat-controlled heat tape was utilized in the water bath to keep the water from freezing during the test.

A rigid plastic cover with a shallow, 2-mm-diameter hole drilled in the center was placed on the surface of each aggregate sample to facilitate frost heave measurements. In this way, differential frost heaves within a sample were automatically averaged in a single measurement from a rigid metal bar placed in turn across the top of each sample container to the bottom of the center hole in each cover. Measurements were made with a digital caliper to a precision of 0.0025 mm.

The air temperature in the environmental room and the water temperature in the bath were checked regularly, with the latter being consistently around 5 °C for the duration of the testing. With this testing arrangement, surface temperatures just under the sample covers usually did not descend below –4 °C due to constant heat conduction from the bath upwards through the samples. This equates to a maximum temperature gradient of about 0.45 °C/cm in each specimen.

Figure 4. Insulation required for frost heave testing.

Test Results

Because the objective of this study was to evaluate the potential of the TST for determining aggregate frost susceptibility, specimens were categorized by their moisture susceptibility rankings determined from final surface dielectric values measured in the TST. Nine specimens had dielectric values less than 10, nine had dielectric values between 10 and 16, and 17 had dielectric values greater than 16, yielding "good," "marginal," and "poor" ratings, respectively. Typical plots of dielectric value versus time for materials in these categories are given in Figure 5. Then, the degree of saturation and electrical conductivity before freezing, moisture content and porosity before and after freezing, frost heave, and ratio of frost heave to water intake during freezing were calculated for each specimen and averaged in each category as shown in Tables 1 and 2.

The heave-intake ratio was computed by multiplying the measured frost heave by the cross-sectional area of the sample to obtain a volumetric heave, which was then divided by the volumetric water intake. Because water expands nine percent upon freezing, a ratio of 1.09 would indicate that all incoming moisture contributes to frost heave upon freezing. Lesser ratios suggest that sufficient porosity exists in the sample matrix to allow the formation of ice without causing frost heave. Ratios higher than 1.09 designate samples which are nearly saturated or which experience substantial upward redistributions of existing water during the initial freezing process that give rise to measurable heave even before additional water is imbibed from the water bath.

The data presented in Tables 1 and 2 show that, on average, materials with higher dielectric values at the end of the TST are characterized by higher porosity and higher salinity which, in turn, lead to greater moisture ingress and higher degrees of saturation due to increased suction. Upon freezing, samples with higher dielectric values experience more frost heave, imbibe more water, and manifest higher heave-intake ratios than those with lower dielectric values at the end of the TST.

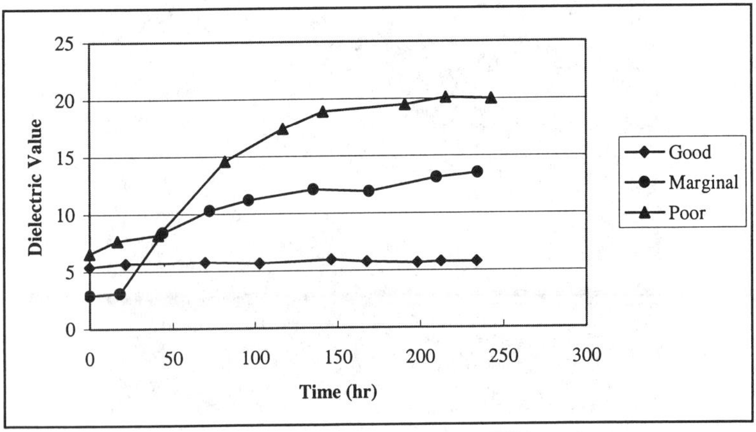

Figure 5. Typical dielectric plots from TST specimens.

Table 1. Post-TST properties.

Dielectric Value	Saturation (%)		Moisture Content (%)		Porosity (%)		Conductivity (μS/cm)	
	Mean	Std. Dev.	Mean	Std. Dev.	Mean	Std. Dev.	Mean	Std. Dev.
ε<10	57.7	10.7	5.1	1.3	19.2	2.2	13	18
10<ε<16	64.1	3.5	6.2	1.7	20.8	4.5	331	267
ε>16	72.3	4.7	8.6	1.0	24.7	3.1	686	307

In order to evaluate the significance of these differences observed between specimens in different moisture susceptibility categories, one-tailed *t*-tests assuming unequal population variances were employed. The null hypothesis in each *t*-test was that no difference existed between the mean values of compared properties, and the alternative hypothesis was that one mean was significantly greater than the other. Unequal population variances were accommodated by using an approximating Student-*t* distribution, where the degrees of freedom in each comparison were adjusted based on sample size and variance (Hogg and Tanis, 1997).

Comparisons of "good" and "poor" performing aggregates yielded significant differences at the 0.05 level of significance for every property presented in the tables. Similar comparisons of "good" and "marginal" materials identified significant differences between values of electrical conductivity, frost heave, and the heave-intake ratio. Significant differences between the "marginal" and "poor" materials exist between values of every property except frost heave and the heave-intake ratio, where, as in other cases in which significant differences could not be detected, relatively high standard deviations and proximate means reduced the ability to discriminate between categories.

The especially high standard deviation for frost heave among "poor" materials can be explained in part by considering the effects of porosity and salt concentrations on frost heave as depicted in Figure 6, where specimens are labeled depending on whether their porosity and electrical conductivity are greater or less than 26 percent or 900 μS/cm, respectively, before freezing. As stated earlier, the importance of higher porosity is that it permits the formation of ice with less expansion of the aggregate matrix. The effects of increased salinity are often manifest in the TST as increased moisture contents caused by higher osmotic suction. As long as the salt concentration is not excessive, such increases in the moisture content can increase the net frost heave experienced by the sample upon freezing. However, for high salt concentrations, where the freezing point of water can be substantially depressed, the pore water is not as readily frozen, and frost heave may be notably reduced. In this study, specimens with porosity greater than 26 percent or electrical conductivity greater than 900 μS/cm did not experience substantial heave for these reasons.

Table 2. Post-freezing properties.

Dielectric Value	Frost Heave (mm)		Moisture Content (%)		Porosity (%)		Heave-Intake Ratio	
	Mean	Std. Dev.	Mean	Std. Dev.	Mean	Std. Dev.	Mean	Std. Dev.
ε<10	4.3	2.0	6.8	1.5	20.9	2.5	0.58	0.23
10<ε<16	7.9	4.6	8.1	2.4	23.7	4.5	0.93	0.30
ε>16	12.4	11.4	11.1	2.1	28.9	4.5	1.09	0.53

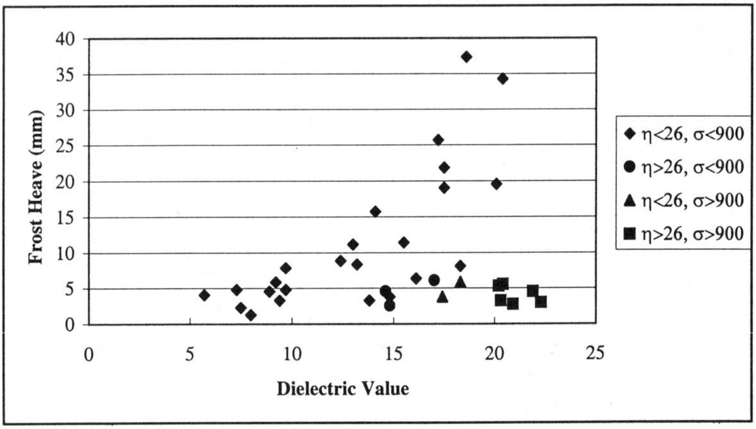

Figure 6. Variation in frost heave.

While wide-ranging frost heave behavior resulting from these effects was observed for specimens characterized in the TST by high dielectric values, the ability to predict the frost heave behavior of low-dielectric specimens was more consistent. For instance, in no cases did a "good" sample experience more than 8 mm of frost heave in laboratory testing. This suggests that aggregate base materials that perform well in the TST may be confidently ranked as neither moisture nor frost susceptible, which is consistent with earlier preliminary findings (Guthrie and Scullion, 2000).

The fines content, or percentage passing the 0.075-mm sieve, followed the same trend as the other listed properties, with average values of 3.9, 7.6, and 11.9 percent for "good," "marginal," and "poor" specimens, respectively. Absorption properties were also determined for each of the 10 aggregate types by measuring the difference in water content between saturated-surface-dry and oven-dry conditions, resulting in average values of 3.6, 4.8, and 6.6 percent for the same categories.

Conclusion

The results of this study provide evidence that the TST is a viable tool for specifying premier aggregate base materials in cold climates. Materials ranked as "good" in the test imbibed significantly less water than higher-dielectric specimens and experienced significantly less frost heave during laboratory testing. Good performers also had lower fines contents, lower porosity, and lower salinity, on average, than higher-dielectric specimens. This study therefore suggests that aggregate base materials that perform well in the TST may be confidently ranked as neither moisture nor frost susceptible.

While all aggregate specimens ranked as "marginal" or "poor" in the TST should be considered moisture susceptible because of their ready moisture transport properties, only part of the samples experienced substantial frost heave during testing. Those with relatively high porosity and salinity were less frost susceptible because of the greater internal space provided for ice formation and the depression of the freezing point caused by salt ions in the pore water. Thus, dielectric values from the TST cannot be used alone to properly determine the frost susceptibility ratings for materials in these categories.

Because of the potential for pavement damage resulting from the use of moisture or frost susceptible materials, this study suggests that such "marginal" or "poor" aggregates should be physically or chemically modified for use in construction of pavements in cold climates. The TST is then recommended for evaluating the efficacy of the selected treatments, where the maximum dielectric value should be less than 10. Because of its success in discriminating among good and poor performing aggregate base materials in Texas, the TST will be implemented statewide by the Texas Department of Transportation in 2001.

Acknowledgments

The authors would like to express appreciation to Stephen Sebesta of TTI for his assistance in preparing many of the samples tested in this study. Lee Gustavus of TTI was instrumental in facilitating laboratory needs.

References

Anderson, D. M. (1967). "Ice nucleation and the substrate-ice interface." *Nature*, 216(5115), 563-566.

Gueguen, Y. and V. Palciauskas. (1994). *Introduction to the physics of rocks*, Princeton University Press, Princeton, N.J.

Guthrie, W. S., Ellis P., and T. Scullion. (2001). "Repeatability and reliability of the tube suction test." *Transportation Research Record*, TRB, National Research Council, Washington, D.C., in press.

Guthrie, W. S., and T. Scullion. (2000). "Assessing aggregate strength and frost susceptibility characteristics with the tube suction test." *Proceedings of the Texas Section, American Society of Civil Engineers, Fall Meeting*, El Paso, Tex., 197-206.

Hanks, R. J. and G. L. Ashcroft. (1980). *Applied soil physics*, Springer-Verlag, New York, N.Y.

Hogg, R. V. and E. A. Tanis. (1997). *Probability and statistical inference, fifth edition*, Prentice Hall, Upper Saddle River, N.J.

Mitchell, J. K. (1976). *Fundamentals of soil behavior*, John Wiley and Sons, Inc., New York, N.Y.

Panday, S. and M. Y. Corapcioglu. (1991). "Solute rejection in freezing soils." *Water Resources Research*, 27(1), 99-108.

Penner, E. (1962). "Ground freezing and frost heaving." *Canadian Building Digest*, CBD-26, National Research Council, Canada, http://www.cisti.nrc.ca/irc/cbd/cbd026e.html. (January 22, 2002).

Powers, M. H. (1997). "Modeling frequency-dependent GPR." *The Leading Edge*, 16, 1657-1662.

Saarenketo, T. (1998). "Electrical properties of water in clay and silty soils." *Journal of Applied Geophysics*, 40, 73-88.

Saarenketo, T. and T. Scullion. (1995). *Using electrical properties to clarify the strength properties of base course aggregates*, Report 1341-2, Texas Transportation Institute, Texas A&M University, College Station, Tex.

Scullion, T. and T. Saarenketo. (1997). "Using suction and dielectric measurements as performance indicators for aggregate base materials." *Transportation Research Record 1577*, TRB, National Research Council, Washington, D.C., 37-44.

Sharma, P. (1997). *Environmental and engineering geophysics.* Cambridge University Press, Cambridge, England.

Taber, S. (1931). "Freezing and thawing of soils as factors in the destruction of road pavements." *Public Roads*, 11, 113-132.

Tareev, B. (1975). *Physics of dielectric materials.* Mir Publishers, Moscow, Russia.

Von Hippel, A. R. (1954). *Dielectrics and waves.* John Wiley and Sons, Inc., New York, N.Y.

Weast, R. C., Lide, D. R., Astle, M. J., and W. H. Beyer, Eds. (1989). *CRC handbook of chemistry and physics.* CRC Press, Inc., Boca Raton, Fl.

Frost Heave Experiment in Open Pit

Jun'ichi Nishikawa[1] and Mitsuru Sakuraba[2]

Abstract

In-situ frost heave test was conducted for the three kinds of soil in open pits. The purpose of this experiment was to observe frost heave of geotechnical materials in natural conditions and to provide fundamental data for future forecasting frost heave at cut slope and road fills.

The experiment of the first season showed the following. (1)The maximum frost penetration depth varied by soil type. The soil which has low thermal conductivity showed small frost penetration. (2)The measured frost heave did not clearly reflect the frost heaving properties in the laboratory frost heave test. Although this might have been due to the difference in saturation, ground water supply and other conditions, further examination will be necessary to clarify this point.
(3)The maximum frost heaving force of 3 to 5 MPa was measured. This was thought to serve as a reference in considering future measures against frost heaving.

Introduction

In cold regions, frost heave occurs in the ground or fill in winter, frequently causing the deformation or failure of the ground, or deformation or destruction of related structures. Although it is possible to

---

[1] Director, Geotechnical division, Civil Engineering Research Institute of Hokkaido, 1-3 Hiragishi, Toyohira-ku, Sapporo, 062-8602 JAPAN; phone +81-11-8415288; nishikaw@ceri.go.jp
[2] Research Engineer, ditto; phone +81-11-8411709; m.sakuraba@ceri.go.jp

evaluate the frost heaving property of soil or other geotechnical materials to some extent by laboratory frost heave tests or other methods, the actual frost heaving quantity in the natural environment is expected to vary depending on the property of geotechnical material, weather, ground water and other conditions. It is therefore difficult to estimate the frost heaving quantity in the natural environment directly from the results of laboratory frost heave tests of geotechnical materials.

An experiment was commenced by preparing open pits to observe frost heave of geotechnical materials in a state as close as possible to the natural environment. In this paper, explanation will be given on the open pits and observation system, and the observation results for the first season will be reported preliminarily.

Specifications of the open pits for frost heave tests

The open pits were set up in the Kashiwabara area on the east of Tomakomai in Central Hokkaido, Japan. The three concrete pits 5 m in length and width and 2 m in depth were set up next to each other. Soil samples are poured into these pits and compacted to the necessary density.

Three tanks for forming and controlling the ground water level, which are installed in the adjoining building, are connected to the respective pits to control the ground water level (Photo 1, Fig. 1). While the tanks supply water automatically when the ground water level becomes lower than the predetermined level, drainage is necessary when the level becomes higher.

Photo 1 An open pit during frost heave test

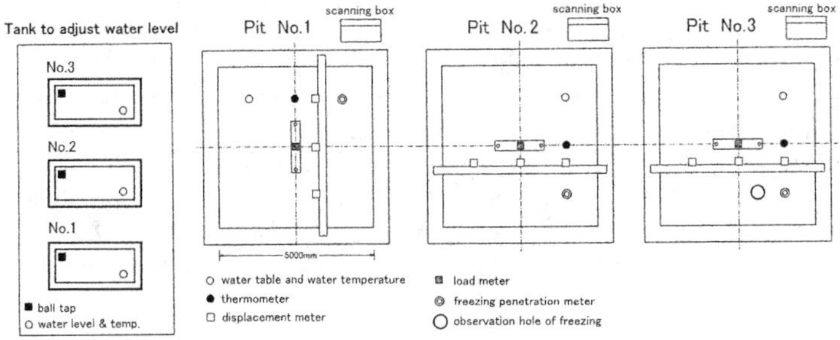

Fig. 1 Plane view of open pits for frost heave test

The factors that can be observed at present include weather conditions such as temperature, wind direction, wind velocity, snow fall and sunshine; underground temperature distribution (a total of 39 points: 8 points at intervals of 2 cm between 1 and 15 cm below the ground; 10 points at intervals of 3 cm down to 45 cm below the ground; 12 points at intervals of 5 cm down to 100 cm below the ground; 9 points at intervals of 10 cm down to 190 cm below the ground); frost heaving quantity; frost heaving pressure and ground water level. These observation data are recorded automatically in the observation building near the open pits and can be stored at the laboratory in Sapporo through the telephone line.

Outline of the test

The subjects of the test were three types of soil (Table 1) and a laboratory frost heave test was conducted in advance. By the test method presented in the Guidelines for Drainage Works (Japan Road Association, 1987), which is generally used at present, the frost heaving rate was 16 to 18% in Pits No. 1 and No. 3, and was as high as 35% in No. 2. By the new proposed test method (Sakuraba and Nishikawa, 2001), the frost heaving speed of Pit No. 2 was not so high.

Table 1 Basic properties of soil for which the test was conducted

sample No	Japanese unified soil classification system	symbol	density of soil particle (g/cm^3)	natural moisture content (%)	optimum moisture content (%)	grain size distribution (%)				maximum dry density (g/cm^3)	liquid limit (%)
						gravel	sand	silt	clay		
No.1	volcanic gravelly sand	SG-V	2.586	74.51	75.00	33.4	58.3	8.3	-	0.792	NP
No.2	clayey gravelly sand	SCsG	2.628	36.25	24.50	15.0	52.6	19.5	12.9	1.450	38.6
No.3	sandy clay	CH-S	2.645	31.90	24.00	1.3	13.7	49.7	35.3	1.514	50.9

Table 2 Results of the laboratory frost heave test

sample No	current method (%)	proposed method (mm/hr)
No.1	17.6	0.12
No.2	34.9	0.14
No.3	15.7	0.11

The soil samples were compacted in open pits. Out of the three types, the moisture content of No. 1 dropped dramatically to 56% at the time of compaction. The moisture content of No. 2 and No. 3 was 30% and 33%, respectively, and close to the natural moisture content. Compaction was conducted to attain a compaction rate of over 90%. After completing the filling, water was supplied from the water-level control tanks to set the ground water level at 1 meter below the ground surface, and measurement was commenced. During the test, snow which accumulated on the surface of soil in the pits was removed as much as possible.

Test results and consideration

The winter of 2000 - 2001 in Hokkaido was colder than usual years. Although the past weather data at the test site were not available, this cold trend was proven by the seasonal freezing index of 673 ℃ · days at the test site, compared with the maximum freezing index between 1968 and 1978 in Sapporo (510 ℃ · days). Figure 2 shows changes in mean temperatures during the season.

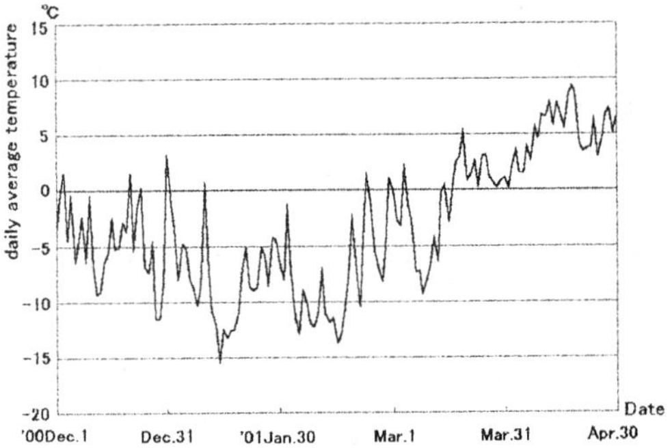

Fig. 2 Changes in mean temperature

In the test pits, freezing began from the ground surface in early December. In Pit No. 1, the frost penetration depth reached its maximum in mid- and late February, thawing began from the surface in early March and the frozen disappeared in late March. In Pit No. 2, the maximum frost penetration depth was observed in late February and early March, followed by the thawing from the surface a little later, and the frozen part disappeared in late April. Pit No. 3 showed a progress similar to that of

No. 2. The maximum frost penetration depth was approximately 60 cm in Pit No. 1, 110 cm in No. 2 and 120 cm in No. 3 (Fig. 3).

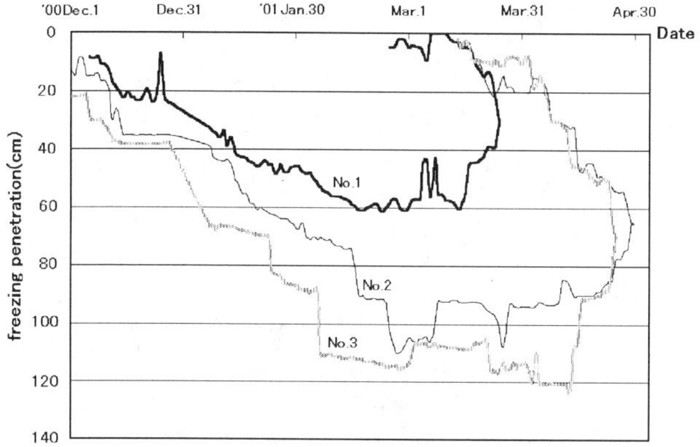

Fig. 3 Changes in frost penetration depth

After the end of winter, thermal conductivity was measured at the ground surface of the pits (Table 3). Changes in the frost penetration depth of each type of soil corresponded to the difference of their thermal conductivity. Out of the three types of soil, the frost penetration depth was small in No.1, which had low thermal conductivity, and large in No. 3, which had high thermal conductivity. Factors of water content considerably lower than the optimum moisture content and large air content at the time of filling No. 1 were thought to be reflected in the low thermal conductivity.

Table 3 Thermal conductivity of soil samples

sample No	thermal conductivity (W/m·K)	moisuture content (%)
No.1	0.35±0.05	39.0
No.2	0.54±0.12	25.2
No.3	0.74±0.21	22.1

For all samples, frost heaving began almost concurrently with the beginning of measurement, peaked in mid-February and had almost finished by the end of March (Fig. 4). At the end of measurement, however, the ground surface level did not return to the level before the beginning of frost heaving. The maximum frost heaving quantity was 70

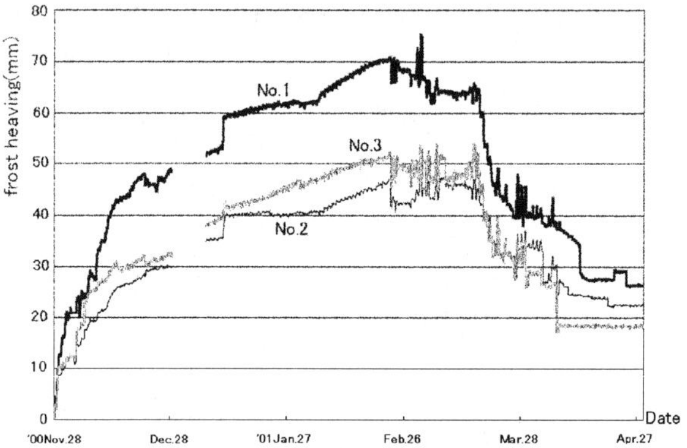

Fig. 4 Changes in frost heaving pressure

mm in No. 1, 45 mm in No. 2 and 50 mm in No. 3. Although the maximum values in No. 2 and No. 3 were almost the same, the value in No. 1 was slightly larger. When the frost heaving quantity almost reached the maximum, core boring was conducted from the surface of the fill in the pits and the section was observed. As a result, formation of an ice lens was observed near the surface of No. 1, although it was not observed in No. 2 and No. 3. The difference between the frost heaving quantities in Pit No. 1 and the other two was presumed to be due to the formation of the ice lens near the surface at an early stage of frost heaving. In the laboratory frost heave test, however, the frost heaving quantity of No.1 was not higher than other two. A great difference with the test using open pits is that, in the laboratory test, the measurements in both methods were conducted in the saturated condition. It can be said that the setting of the ground water level at 1 meter below the ground surface in this test, in particular, was a condition in which water supply was difficult. Although these preliminary test results suggest that the frost heaving quantities in laboratory frost heave test do not necessarily reflect the frost heaving properties under natural conditions, more detailed examination will be necessary in the future.

Two steel posts were fixed at the bottom of open pits and connected at the top. By using this as a reaction force plate, a 10-cm load plate was placed on the surface of the fill to create a condition where frost heaving is prevented, and the frost heaving force was measured by a load cell (Fig. 5). The maximum frost heaving force was 3 MPa in No. 1, 5 MPa in No. 2 and 4 MPa in No. 3. When the same measurement was conducted in a laboratory test using a mold of 15 cm in diameter, frost heaving force ranged from 0.30 to 1.47 MPa (Kaneta, Nishikawa and Tada, 1998). The values measured this time were slightly higher than that. This was due to the difference in soil characteristics, as well as the test scale and conditions, and the difference in value to this extent can be considered insignificant. This tendency is thought to serve as a reference in considering future measures against frost heaving.

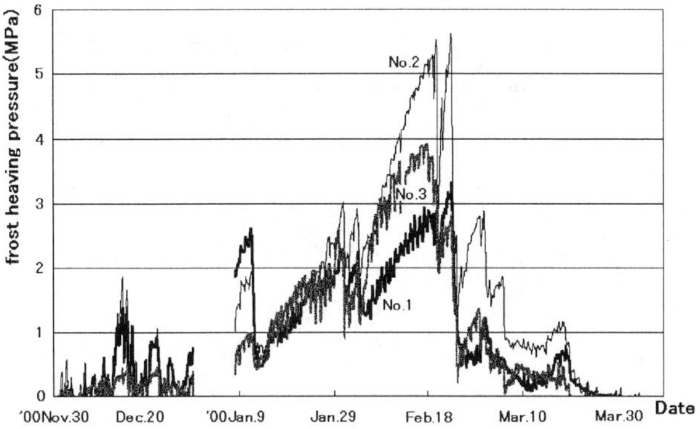

Fig. 5 Changes in frost heaving pressure

Conclusion

Open pits were prepared for outdoor frost heave soil tests, and a test was conducted using three types of soil. The following matters were found from the measurement results in the first winter:

(1) The maximum frost penetration depth varied by soil type. The soil which has low thermal conductivity showed small frost penetration.

(2) The measured frost heaving quantities did not clearly reflect the frost heaving properties in the laboratory frost heave test. Although this might have been due to the difference in saturation, ground water supply and other conditions, further examination will be necessary to clarify this point.

(3) The maximum frost heaving force of 3 to 5 MPa was measured. This was thought to serve as a reference in considering future measures against frost heaving. Frost heaving properties of each soil type may not have been displayed clearly in this measurement in which the ground water level was rather low. Experiments will be continued in the future by reconsidering test conditions.

References

Japan Road Association (1987). Guidelines of drainage works in road earth works, Japan Road Association. (in Japanese)

Sakuraba, M. and Nishikawa, J. (2001). The comparison between the new frost heaving test and the present test, Technical Report of Hokkaido Branch, The Japanese Geotechnical Society, 41, 129-132. (in Japanese)

Kaneta, H., Nishikawa, J. and Tada, S (1998). Frost heave test and design of ground anchor considering frost heave, Monthly Report of Civil Engineering Research Institute, 541, 24-31. (in Japanese)

Behavior of Ice/Rock Mixtures on Slopes

Branko Ladanyi, Fellow ASCE[1]

Abstract.
Behavior of both rock glaciers and conventional glaciers is highly affected by the character and temperature of their bottom deforming layer, usually composed of a mixture of ice and rock elements, the latter ranging from silty sand up to large boulders. Similarly as in the rock mass mechanics, large scale behavior of such mixtures is difficult to evaluate either by direct in-situ measurements or by taking samples. This paper uses some known rock mechanics principles to simulate the large-scale behavior of such mixtures, by considering the role and interaction of each element in the mass.

Introduction

Due to climate warming trends, there has been in recent years an increasing interest in the accelerating creep of rock glaciers and frozen slopes. The term "Rock Glaciers" has been used for various complex landforms of cold mountain areas, from debris-covered glaciers in permafrost-free areas, to steadily creeping perennially frozen and ice-rich debris on non-glaciated mountain slopes. In the field of glaciology, the creep of glaciers has been extensively studied, observed and analyzed for over 100 years. A brief overview of some principal findings in the last 50 years is given in the following section. Many valuable and detailed theoretical models have been proposed through the years for simulating the creep behavior of glaciers, and this paper has no intention to propose another one. Its purpose is only to supply to these models some new geotechnical background, borrowed from the connected fields of frozen ground mechanics, rock mechanics and mechanics of mixtures. It is considered that this "crossbreeding" may be found useful for further development of glacier and rock glacier mechanics.

In particular, this paper attempts to extend some known models of mechanical behavior of unfrozen soil and rock masses to those containing ice, including large-scale creep of ice/rock mixtures and ice/rock interface problems.

---

[1] Professor Emeritus, Dept. CGM, Ecole Polytechnique, CP 6079, Stn. Centre-ville, Montreal, H3C 3A7,Canada, bladanyi@courriel.polymtl.ca

A brief description of basic constituents and the behavior of Rock Glaciers and Glaciers will be given in the following.

Rock glaciers

According to Brown (1973), rock glaciers form under the influence of a periglacial climate in an area lacking the net accumulation of snow required for the formation of a conventional glacier. A permafrost environment must be present (mean annual air temperature lower than 0 °C) to enable the snow and water, which trickles down into the interstices between the rock debris to remain ice. Barsch (1996) defines the term "rock glacier" as bodies of perennially frozen unconsolidated material, supersaturated with interstitial ice and ice lenses, that move down-slope or down-valley by creep. This is thought to be a consequence of the deformation of ice contained in the rock glacier. Because of its relatively high frost temperature, Alpine permafrost is very sensitive to climate changes (Cheng and Dramis, 1992, Haeberli, 1993, Arenson and Springman, 2000). Instabilities of Alpine permafrost are considered to be an issue of national importance in Switzerland (Haeberli et al., 1998). The loss of strength due to melting of ice matrix might combine with an increasing supply of water, leading to stability problems within the rock glacier, which may also become a trigger for debris flow. In addition to Switzerland, rock glaciers are more widely distributed and most intensely studied in Alaska and in the Rocky Mountains bordering United States, Alaska and Canada.

Various definitions of rock glaciers have been proposed in the past, based on their origin and morphological characteristics. However, for the purpose of this paper, the most relevant classification of rock glaciers is that based on their constituent material. From this point of view, a rock glacier may be dominantly: glacier ice, non-glacial (interstitial) ice, ice lenses, permafrost, rubble, or some combination of all these. A schematic classification plot has been proposed by Johnson (1973), Fig. 1, which shows possible combinations of constituents of a rock glacier. In general, this classification indicates that rock glaciers can be considered as mixtures of solids, ice and water, which is one of the subjects of this paper.

In the case of rock glaciers, observations made in Switzerland by Hoelzle et al. (1998) in boreholes drilled to the depth of 20 m, show that the surface movement was mainly due to the creep deformation within the lower third (from –10 to –15 m) of the rock glacier, which was strongly affected by temperature increase of the drift

Two probably the most intensively investigated rock glaciers, Muragl and Murtèl, are situated in the Upper Engadin, Switzerland. Most recent reports on the drilling results and movement observations were published by Haeberli et al., (1998), and by Arenson and Springman, (2000).

The Murtèl rock glacier contains extensive regions of nearly pure ice in the upper 10 to 20 m. The volumetric ice content in this zone is 80-90%, so that there is no contact between solid particles. No such massive ice zone was found in Muragl glacier. A shear layer of ice mixed with fine soil particles was found at the bottom (29-31m) of the Murtèl rock glacier, but predominantly larger particles were common in the Muragl rock glacier, ranging from silty sand up to large boulders of about 2m diameter. The frozen soil extended to bedrock at about 50m in Murtèl.

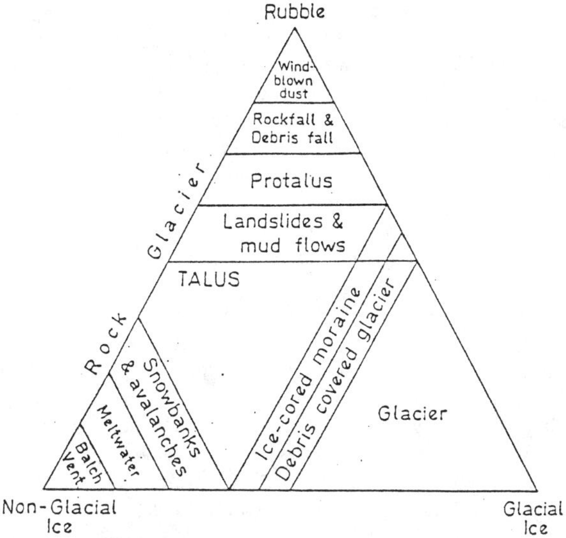

Figure 1 Classification of rock glaciers based on origin of materials (after Johnson, 1973)

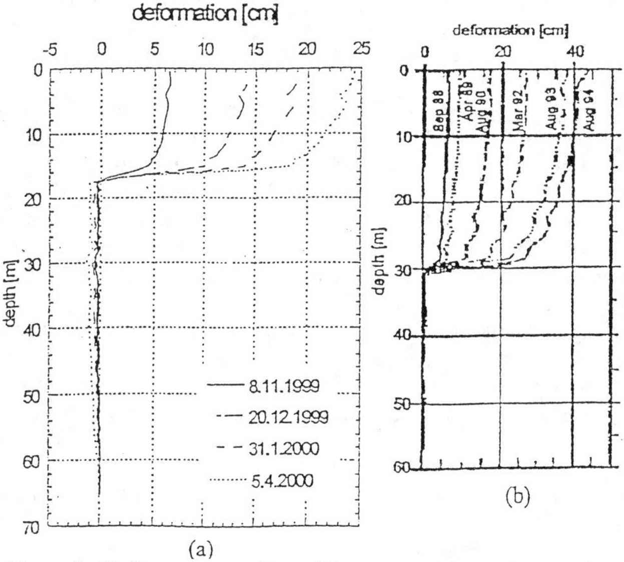

Figure 2. Deformation profiles of the rock glaciers: (a) Muragl rock glacier, BH4; Murtèl rock glacier, BH 1/87 (Haeberli et al, 1998)

Borehole deformations measured with in-slope inclinometers during several years (Wagner, 1992; Haeberli et al., 1998) show for both rock glaciers similar movement profiles: 80-90% of the deformation happens in a bottom shear horizon, with a depth of about two to three meters (Fig. 2). In both cases borehole drilling shows that this deforming layer consists of frozen sediments, mainly of sand (Murtèl) or gravel (Muragl). As for temperatures, at the depth of shear zone, the temperature in Murtèl rock glacier is about –1.6 °C (30m depth), and only about –0.1 °C in the Muragl rock glacier (17m depth) (Arenson and Springman, 2000).

Glaciers

Although glacier sliding down a slope is often theoretically treated as sliding of ice mass down a clean, rigid bedrock surface, in reality more commonly, the base of temperate glaciers consists of a till-like layer containing boulders, pebbles, sand and clay (Engelhardt et al.,1978, 1979; Eichelmeyer and Zhongxiang 1987). Thus, while clean ice-bedrock interface may exist in some cases, it happens more often that glacier bed beneath both temperate and polar glaciers is composed of ice-laden pore water-saturated drift. In the case of temperate glaciers, pore water plays an important role in their deformation, although, at sub-freezing temperatures, this basal drift cannot contain significant quantities of pore water. Clearly, in addition to pore water pressure, the debris content of basal ice plays also an important role in controlling the motion of temperate glaciers by influencing the sliding rate (Hallet, 1981). Basal sliding beneath temperate glaciers occurs by a combination of regelation slip and creep around obstacles, arising from stress concentrations. Under extreme conditions of high subglacial water pressure, catastrophic glacier advances may occur, especially if warming of the basal ice brings it into the temperature range where regelation slip becomes possible (Embleton, & King, 1975).

Consequently, if one applies to the ice-laden drift the theory of viscosity of mixtures, presented in more detail later in the paper, one would expect the drift to be stiffer than pure ice. However, direct observations have shown that at certain conditions, this basal drift deforms more easily than the overlying glacier ice. One possible explanation is that if the drift is composed of frozen silt or clay, which contain relatively large quantities of unfrozen water even down to temperatures of about –3 °C, it may be more deformable than pure ice, at the same temperature, as found by Thompson and Sayles (1972).

As for the basal sliding of glaciers on a relatively clean but rough or undulated rock surface, it is considered to be due to the melting and regelation of ice when moving over rock asperities. (Weertman, 1957; Kamb & LaChapelle, 1964; Kamb, 1970; Nye, 1970, Hallet, 1981). However, due to very large normal pressures under thick glaciers, this regelation mechanism can also be active even at temperatures down to –2 to –5 °C (Eichelmeyer and Zhongxiang, 1987).

In general, the total movement observed at the surface of glaciers is considered to consist of two components – internal deformation or "creep", and basal sliding, the latter being most often preponderant.

Rheology of mixtures

In connection with slope creep and failure, including unfrozen saturated clay slopes, thawing frozen slopes, and already thawed mass sliding on frozen bed or rock, a basic ingredient in the analysis of the stability is the rheological behavior of mixtures rock/clay, rock/ice or rock/water. In the past, pioneering work in the area of rheology of mixtures was done in various fields, including general rheology, soil mechanics, rock-fill mechanics, and frozen soil mechanics. A general conclusion in all these studies was clearly that the behavior of a mixture depends on the character of its solid and liquid constituents and on their relative proportion in the mixture.

In the area of general rheology, one of the earliest work on mixtures was done by Einstein (1906), who proposed a very simple formula, relating the viscosity of a mixture, η_m, to the viscosity of the liquid, η_l

$$\eta_r = \eta_m / \eta_l = (1 + 2.5\, c) \tag{1}$$

where η_r is the relative viscosity, and c is solid grains volume concentration ratio, varying from $c = 0$ (no grains) to $c = 1$ (no liquid). Equation (1) was developed for very dilute suspensions ($c < 10\%$). For denser suspensions, when solid grains start interacting, the linear relation of Eq.(1) has to be modified, and several authors have proposed other empirical formulas, based on laboratory experiments with mixtures, which cover a wide range of solid grain concentrations.

For example, Guth (1945) considered that for larger solids concentrations, Eq.(1) should be replaced by

$$\eta_r = (1 + \alpha_1 c + \alpha_2 c^2 + \alpha_3 c^3 + \dots\dots\dots) \tag{2}$$

and he proposed the following formula for suspensions where grains interact

$$\eta_r = (1 + 2.5\, c + 14.1\, c^2) \tag{3}$$

Through the years, many similar formulas have been proposed by different authors, who were mainly working with suspensions for industrial applications.

In the field of soil mechanics, this work on mixtures has only rarely been mentioned, with the exception of Vallejo (1981), who studied the vane shear strength of sand/clay and gravel/clay mixtures. He compared his results with Einstein's Eq.(1), in which he substituted undrained shear strength s_u for viscosity. It will be seen in Fig. 3 that Eq.(1) is approximately valid only for grain concentrations of less than 56%, and that it does not cover the shear rate effects.

In the area of frozen soil mechanics, pioneering work on sand/ice mixtures was carried out by Goughnour and Andersland (1968), Kaplar (1971), Hooke et al.(1972) and Baker (1979) (Fig. 4), and the findings were summarized by Ting et al. (1983) (Fig. 5). A conclusion from all these studies is that, up to grain volume concentration of about $c = 40\%$, the pore ice governs the behavior, while for $c > 40\%$, the friction intervenes, and, finally, for $c > 60\%$, the dilatancy adds also to the mixture shear strength. On the other hand, as shown in Fig. 5, when a dense sand/ice mixture

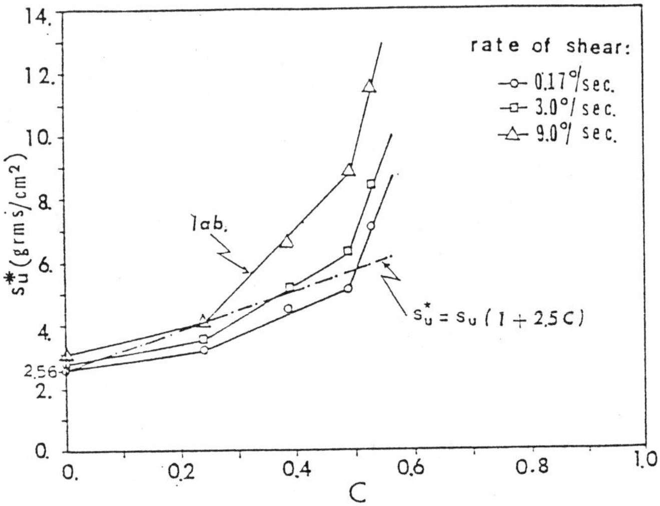

Figure 3. Undrained shear strength of the gravel-clay mixture versus the grain concentration ratio c (after Vallejo, 1981)

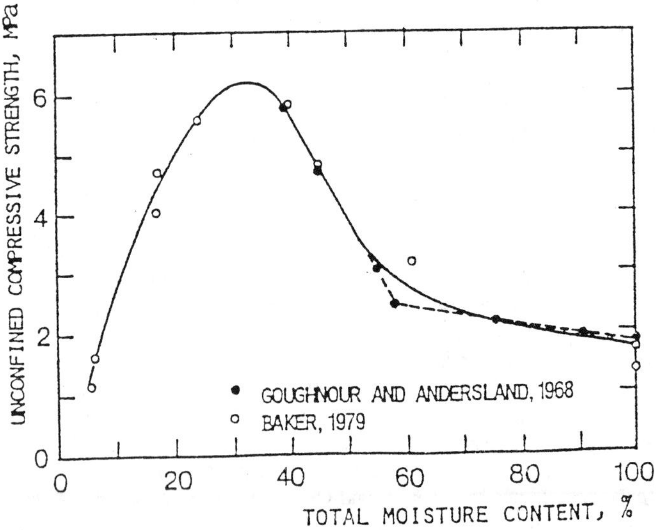

Figure 4. Effect of total moisture content on unconfined compression strength of a frozen fine sand at –12 °C and at a strain rate of 2.2×10^{-6} s^{-1} (after Baker, 1979)

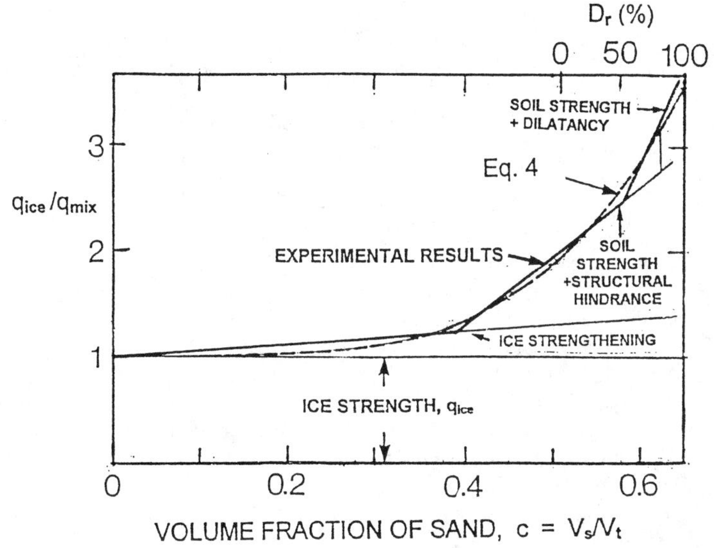

Figure 5. Variation of unconfined compressive strength of frozen Ottawa sand at –7.6 °C and a strain rate of 4.4×10^{-4} s^{-1} (after Ting et al., 1983)

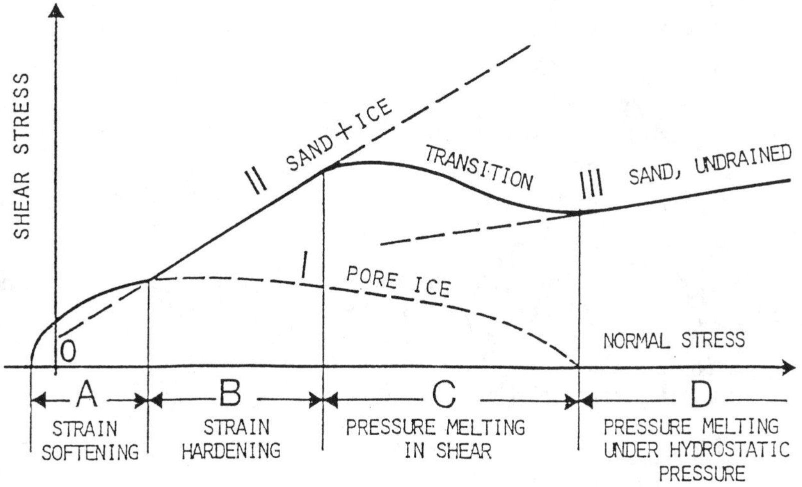

Figure 6. Schematic representation of the whole failure envelope for frozen Ottawa sand (after Ladanyi, 1981)

becomes increasingly unsaturated, the ice cohesion decreases, and eventually the strength of the mixture reverses to that of dry sand. This is also clearly seen in the results of direct shear tests with sand/ice mixtures, published by Nickling and Bennett (1984).

It will be seen in Fig. 5 that the strength of sand/ice mixtures, for $0 < c < 0.65$, can be approximated by the formula

$$Q_r = q_{mix}/q_{ice} \approx (1 + 14.1\,c^4) \tag{4}$$

where q denotes the uniaxial compression strength. This formula implies that rate and temperature effects on the strength of a mixture are the same as those for pure ice.

Clearly, the shear behavior of sand/ice mixtures is also strain rate, temperature and normal pressure dependent, which is shown in the experimental work carried out by Chamberlain et al.(1972). These authors have shown that under a constant temperature, but increasingly high confining pressure, the pore ice melts and eventually the sand/ice mixture starts behaving as water-saturated sand. This general behavior is summarized in Fig. 6 (Ladanyi, 1981), which is a schematic plot of the failure envelope of a frozen sand within a wide range of normal pressures.

Stability of mixtures on slopes

An interesting analysis of cohesionless and cohesive mud- or debris-flows was presented by Vallejo (1980, 1979, 1981, 1989), Takahashi (1978) and Iannachione and Vallejo (2000). This analysis was based on a previous work by Bagnold (1954). According to Vallejo (1989), the shear stress, τ, due to gravity acting on a mixture of large particles and fluid, and measured on the plane of failure (Fig. 7) can be computed from

$$\tau = g\,\rho_{mix}\,d \sin\beta \tag{5}$$

where ρ_{mix} is the density of the mixture, g is the acceleration of gravity, d is the layer thickness, and β the failure surface inclination angle. The value of ρ_{mix} in (5) is given by

$$\rho_{mix} = \rho_f + c(\rho_s - \rho_f) \tag{6}$$

where ρ_f is the density of fluid, ρ_s is the density of solids, and c is the volume concentration of solids.

The shear resistance available at the interface is

$$s = g(\rho_s - \rho_f)c\,d \cos\beta \tan\phi' + (1 - c)\,c_f \tag{7}$$

where ϕ is the angle of shearing resistance between large solid particles, and c_f is the shear strength of the filling fluid. The factor of safety of the flowing mass is then, in terms of total stresses,

$$FS = \frac{\tan\phi}{\tan\beta}\left[\frac{c(\rho_s - \rho_f)}{\rho_{mix}}\right] + \frac{(1-c)c_u}{g\rho_{mix} d\cos\beta} \tag{8}$$

If the interstitial fluid is not water but ice, c_f becomes a function of temperature and strain rate, so FS will be also affected by these parameters. For ice, in the simplest terms,

$$c_u = \frac{1}{2}\sigma_{c\theta}\left(\frac{\dot{\varepsilon}_{1,\min}}{\dot{\varepsilon}_c}\right)^{1/n} \tag{9}$$

where

$$\sigma_{c\theta} = \sigma_{co}(1 + \theta/\theta_o)^w \tag{10}$$

is the reference stress at the reference strain rate $\dot{\varepsilon}_c$, for temperature θ = $-T$ °C, σ_{co} is the reference stress at $\dot{\varepsilon}_c$, when T tends to 0 °C, θ_o = 1 °C, $\dot{\varepsilon}_{1,\min}$ is the minimum strain rate in uniaxial compression, and n and w are creep and temperature exponents, respectively. For example, according to Morgenstern et al. (1980) and Andersland and Ladanyi (1994), for polycrystalline ice, the above parameters would have the values: $n = 3$, $w = 0.37$, and $\sigma_{co} = 0.103$ MPa at $\dot{\varepsilon}_c = 10^{-5}$ $h^{-1.}$

As for the angle ϕ, it is a function of the density of the flowing mass, varying between ϕ_{ice} for $c < 60\%$, and ($\phi_f + \phi_d$), for a dense granular mass sliding over rough bedrock (with sub "f" for friction and sub "d" for dilatancy).

Frozen slope creep due to internal deformation

Similarly to Nye (1970) and McRoberts (1975, 1978), Andersland and Ladanyi (1994, Chap.8) have shown a solution of creep of an infinite frozen slope sliding down a plane, making an angle β to the horizontal. Assuming that the creeping layer is solidly attached to its bottom, and that creep of the mass is governed by the creep power law, valid for an ice-saturated mass (i.e., $c < 40\%$)

$$\dot{\gamma} = \dot{\gamma}_c\left(\frac{\tau}{\tau_{c,\theta}}\right)^n \tag{11}$$

where $\dot{\gamma}$ is the shear strain rate, τ is the shear stress, and $(\dot{\gamma}_c, \tau_c)$ are experimental creep parameters for shear creep. Note that, expressed in terms of uniaxial compression data, and assuming the validity of von Mises flow rule, if $\tau_{c\theta} = \sigma_{c\theta}$, then $\dot{\gamma}_c = 3^{(n+1)/2}\dot{\varepsilon}_c$. The effect of temperature on $\tau_{c\theta}$ can be expressed by Eq. (10).

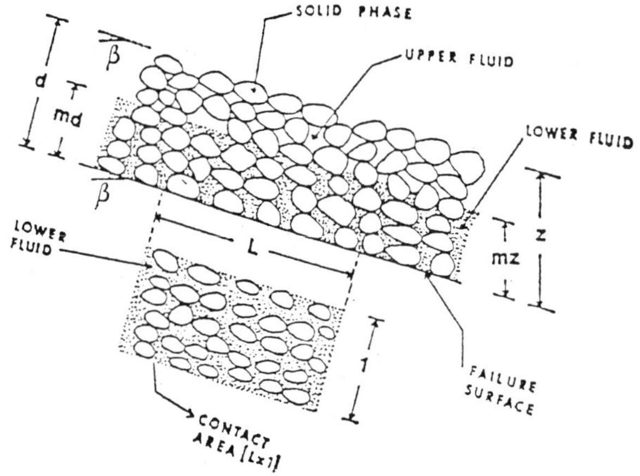

Figure 7. Slope geometry and parameters used in the particulate model of stability analysis (after Vallejo, 1989)

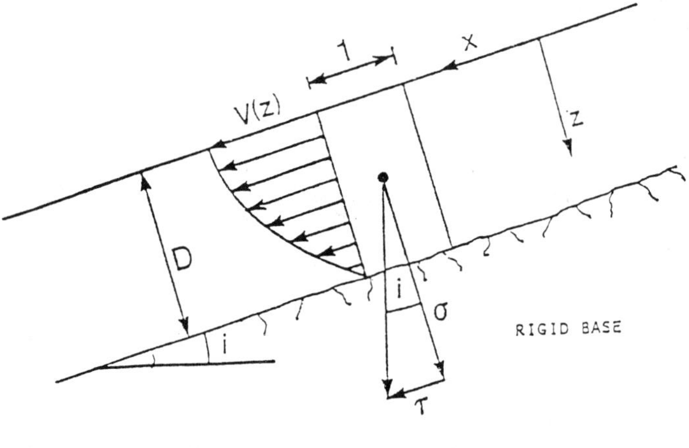

Figure 8. Notation for creep of infinite slope on bedrock (after Andersland and Ladanyi, 1994).

The result, expressed as the creep velocity at any depth $0 < z < D$ in the slope direction, can be written as (Andersland and Ladanyi, 1994)

$$V_x(z) = \frac{\dot{\gamma}_c}{n+1}\left(\frac{g\rho_s \sin\beta}{\tau_{c\theta}}\right)^n \left(D^{n+1} - z^{n+1}\right) \tag{12}$$

This equation defines the creep velocity curve shown schematically in Fig. 8 (Andersland and Ladanyi, 1994) . At $z = 0$, it gives the maximum velocity

$$V_{x,0} = \frac{\dot{\gamma}_c}{n+1}\left(\frac{g\rho_s \sin\beta}{\tau_{c\theta}}\right)^n D^{n+1} \tag{13}$$

It is noted that ρ_s in these equations is, in fact, equal to ρ_{mix} given by Eq.(6).

It was also assumed in Andersland and Ladanyi (1994) that slip will occur at the interface ($z = D$) when the accumulated shear strain attains a critical value, $\gamma_{\max} = \gamma_{crit}$. The time to attain this value is

$$t_{slip} = \left(\frac{\gamma_{crit}}{\dot{\gamma}_c}\right)\left(\frac{\tau_{c\theta}}{g\rho_s D\sin\beta}\right)^n \tag{14}$$

After the slip occurs, there will remain at $z = D$ a non-zero slip velocity, V_{xD}, which must be added to $V_x(z)$ to obtain the total creep velocity down the slope. The interface creep velocity, V_{xD}, can be estimated by the analysis presented in the last section.

Shear strength of ice-filled indented rock joints

Ladanyi and Archambault (1977) have published the results of an experimental study involving direct shear tests with filled, regularly indented, joints, such as in Fig. 9. They used two types of infilling materials (clay and silty sand), three ascending slope angles of irregularities (15°, 30°, 45°), and thickness of filling varying from zero ($t = 0$) to the amplitude of irregularities ($t = a$). The direct shear tests were carried on at 3 normal pressures (σ= 2.90, 8.69 and 14.48 MPa).

The analytical treatment of results involved two cases: The first one in which there is no breakage of irregularities, and the second one involving the breakage irregularities during shear. Since the first case is more relevant to ice-filled joints, it will be presented briefly in the following. A general equation for shear of filled regularly indented joints, developed by Ladanyi and Archambault (1977) joints is

$$\tau = \frac{c_u}{1 - \tan i \tan\phi_u} + \sigma\tan(\phi_u + i) \tag{15}$$

in which i is the ascending slope of irregularities, Fig. 9, and c_u, ϕ_u are the Coulomb's shear strength parameters of the infilling. In Eq.(15) the angle i is not a constant, but varies between i_o and zero, when the relative thickness of infilling (t/a) varies from zero to 100%. This can be expressed by

$$\tan i = M \tan i_o \tag{16}$$

where M is an empirical reduction factor, varying from 0 to 1, which can be computed from (Ladanyi and Archambault,1977)

$$M = \left(1 - \frac{2t}{3a}\right)^2 \tag{17}$$

valid for $i_o \leq 30^{\circ}$, and $\sigma < 0.5\ C_o$, where C_o is the uniaxial compression strength of rock irregularities. For zero infilling thickness, $t/a = 0$, $M = 1$, $i = i_o$, Eq.(15) gives the shear strength of an irregular, clay- or ice-coated joint, which is different from that of a clean, un-coated joint. On the other hand, when t/a attains approximately 1.5, there is no more interference of irregularities ($M = 1$), and Eq. (19) reverses to the strength of the filling alone

$$\tau = c_u + \sigma \tan \phi_u \tag{18}$$

For clay infilling, c_u and ϕ_u are undrained shear parameters in terms of total stresses. For an ice infilling, both c_u and ϕ_u are affected by temperature and strain rate. Figure 10 shows typical experimental results and the proposed approximation for clay-filled indented joints.

For completely irregular joint surfaces, tan i_o in Eq.(20) should, according to Ladanyi and Archambault (1980), be replaced by

$$\tan i_o = \left[1 - \left(\frac{\sigma}{C_o}\right)^{0.25}\right] \tan 33^{o} \tag{19}$$

where C_o is the uniaxial compression strength of the rock asperities.

Equations (15) to (19) take into account only the shear strength of the filling. In order to include also the tensile strength, T_H, of the filling inside the cavities, which is about 0.1 MPa for water and about 1 MPa for ice, Eq. (15) should be written as

$$\tau = \frac{c_u}{1 - \tan i \tan \phi_u} + (\sigma + T_H)\tan(\phi_u + i) \tag{20}$$

which shows that tensile stress generated within the shearing rock cavities ("cavitation") increases the all round confining pressure and produces a dilatation hardening effect in the mass (Ladanyi and Morel, 1990).

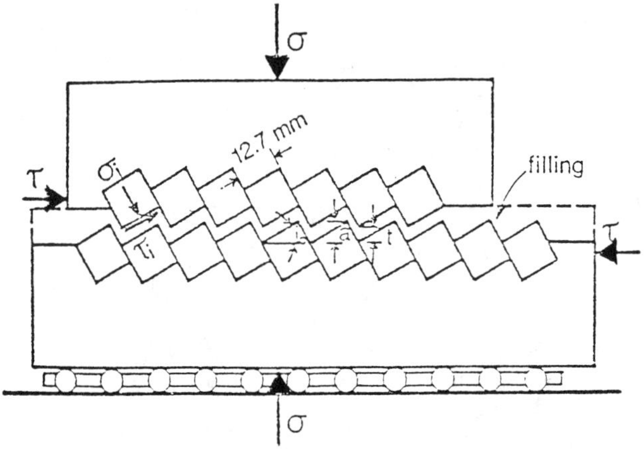

Figure 9. Schematic diagram of the direct shear apparatus for testing filled joints (after Ladanyi and Archambault, 1977).

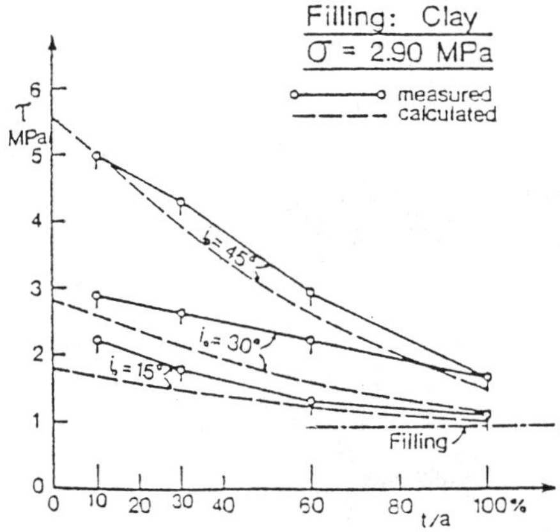

Figure 10. Effect of thickness of clay filling on the strength of joints in direct shear (after Ladanyi and Archambault, 1997).

Creep and strength of ice-filled rock joints

Strain rate and temperature dependent shear strength of ice or ice-rich frozen soil can, from Eq. (11), be expressed by

$$\tau_{mob} = \tau_{c\theta}(\dot{\gamma} / \dot{\gamma}_c)^{1/n} \tag{21}$$

Where τ_{mob} is the mobilized shear strength, where, in terms of uniaxial compression experimental information

$$\tau_{c\theta} = \sigma_{c\theta} = \sigma_{co}(1+\theta/\theta_o)^w \tag{22}$$

With $\sigma_{c\theta}$ and $\dot{\gamma}_c$ defined after Eqs. (10) and (11), respectively.

The shear strength τ_{mob} in Eq. (21) is, in fact, the cohesion portion of the total mobilized shear strength of ice in the filled joint, the second portion, due to friction and work of dilation being expressed by the second term of Eq. (15).

Now, since at the very slow rates of strain as in frozen slopes, the friction in ice may be negligibly small, only the dilation part will remain, and (for $\phi_u = 0$), Eq. 19 will be

$$\tau = \tau_{mob} + \sigma \tan i \tag{23}$$

Equation (23) expresses the total available shear strength of an ice-filled joint. On the other hand, may want to find the shear displacement rate $\dot{s}$ of such a joint under a shear stress τ. Putting Eq. (21) into Eq. (23), and considering that $\dot{s} = \dot{\gamma} t$, one gets

$$\dot{s} = \dot{\gamma}_c t \left(\frac{\tau - \sigma \tan i}{\tau_{c\theta}} \right)^n \tag{24}$$

showing that, for very rough joints and a high normal stress σ, the creep displacements of the joint may not occur at all. An example of calculation of $\dot{s}$, using Eq. (24), is shown in Fig. 11. The following parameters were assumed in the calculation: For the joint: t/a = 0.60, σ = 0.15 MPa. For the filling (ice): n = 3, w = 0.37, σ_{co} = 0.10 MPa, $\dot{\varepsilon}_c$ = 10^{-5} h^{-1}, corresponding to $\dot{\gamma}_c$ = 0.01944 day^{-1}. The computation was made for $0.1 \le \tau \le 1.0$ MPa and θ = 1, 2, and 3 °C (T = -1, -2, and –3 °C)

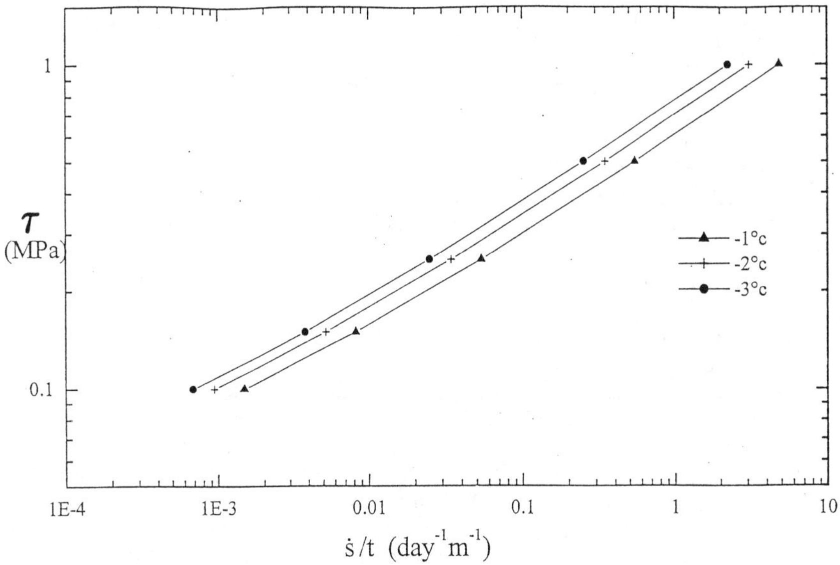

Figure 11. Computed shear creep rate $\dot{s}$ of an irregularly indented ice-filled joint of thickness t under different applied shear stresses τ and three freezing temperatures, after Eq. (24).

Conclusions

The purpose of this paper was to extend to frozen slopes and the frozen debris at the bottom of glaciers and rock glaciers, certain findings and methods borrowed from the fields of mud flow mechanics, rock mechanics and the behavior of mixtures. The analysis shows how the ice component may be included in these methods, leading to the solutions potentially useful for analyzing the stability and motion of frozen slopes under temperature changes.

References

Andersland, O.B. & Ladanyi, B. (1994). *"An Introduction to Frozen Ground Engineering"*. Chapman & Hall, New York, 352 p.

Arenson, L. and Springman, S. (2000). "Slope stability and related problems in Alpine permafrost." *Int. Workshop on Permafrost Engineering*, Svalbard, 185-197.

Bagnold, R.A. (1954). "Experiments on a gravity-free dispersion of large solid spheres in a Newtonian fluid under shear". *Proc. Royal Soc. of London*, Series A, V.225, 49-63.

Baker, T.H.W. (1979). "Strain rate effect on the compressive strength of frozen sand". *First Int. Symp. on Ground Freezing*, Bochum (H.L. Jessberger, ed.) Elsevier, 223-231.

Barsch, D. (1996). *"Rockglaciers"*, Springer, Berlin, 331 p.

Brown, R.J.E. 1973. "Ground ice as an initiator of landforms in permafrost regions." *Research in Polar and Alpine Geomorphology*, 3rd Guelph Symp. on Geomorphology, 25-42.

Chamberlain, E., Groves, C. & Perham, R, (1972). "The mechaical behavior of frozen earth materials under high pressure triaxial test conditions". *Géotechnique* 22 (3), 469-483.

Cheng, G. and Dramis, F. (1992). "Distribution of mountain permafrost and climate." *Permafrost and Periglacial Processes*, 3(2), 83-89

Embleton, C. & King, C.A.M. (1975). *"Glacial Geomorphology"*. Edward Arnold, London.

Eichelmeyer, K. & Zhongxiang, W. (1987). "Direct observation of basal sliding and deformation of basal drift at sub-freezing temperatures". *J. of Glaciology*, 33 (113), 83-98.

Einstein, A. (1906). "Eine neue Bestimmung der Moleküldimensionen". *Annalen der Physik*, Ser. 4, Bd. 19, Ht. 2, 289-306.

Engelhardt, H.F. , Harrison, & Kamb, B. (1978). "Basal sliding and conditions at the glacier bed as revealed by bore-hole photography"" *J. of Glaciology*, 20, 84, 469-508.

Engelhardt, H.F., Kamb, B., Raymond, C.F. & Harrison, W.D. (1979). "Observations of basal sliding of Variegated Glacier, Alaska". *J. of Glaciology*, 23, 89, 406-407.

Goughnour, R.R. & Andersland, O.B. (1968). "Mechanical properties of a sand-ice system". *J. Soil Mech. Found. Div., ASCE* 94 (SM4), 923-950.

Guth, E. (1945). "Theory of filler reinforcement". *J. of Applied Mechanics*, 16, 20-25.

Haeberli, W. (1993). "Research on permafrost and periglacial processes in mountain areas – status and perspectives". *Sixth Int. Conf. on Permafrost*, 2, 111-124.

Haeberli, W., Hoelzle, M. Kääb, A., Vonder Mühll, D. & Wagner, S. (1998). "Ten years after drilling through the permafrost of the active rock glacier Murtèl, Eastern Swiss Alps: Answered questions and new perspectives". *Seventh Int. Conf. on Permafrost*, Yellowknife, 403-409.

Hallet, B. (1981). "Glacial abrasion and sliding: Their dependence on the debris concentration in basal ice". Annals of Glaciology, 2, 23-28.

Hoelzle, M., Wagner, S., Kääb, A. and Vonder Mühll, D. (1998). "Surface movement and internal deformation of ice-rock mixtures within rock glaciers at Pontresina-Schafberg, Upper Engadin, Switzerland." *Seventh Int. Conf. on Permafrost*, Yellowknife, 403-409.

Hooke, R., LeB., Dahlin, B.B. & Kauper, M.T. (1972). "Creep of ice containing dispersed fine sand". *J. of Glaciology*, 11(63), 327-336.

Iannacchione, A.T. & Vallejo, L.E. (2000). "Shear strength evaluation of clay-rock mixtures". GeoDenver, *Slope Stability 2000*, ASCE, 209-223.

Johnson, J.P. (1973). "Some problems in the study of rock glaciers." *Research in Polar and Alpine Geomorphology*, 3rd Guelph Symp. on Geomorphology, 84-91.

Kamb, B. (1970). "Sliding motion of glaciers: theory and observation". *Reviews of Geophysics and Space Physics*, 8(4), 673-728.

Kamb, B. & LaChapelle, E. (1964). "Direct observation of the mechanism of glacier sliding over bedrock". *J. of Glaciology*, 5, 37, 159-172.

Kaplar, C.W. (1971). *"Some Strength Properties of Frozen Soil and Effect of Loading Rate"*. U.S. Army CRREL, Spec. Rep. 159.

Ladanyi, B. (1981). "Mechanical behaviour of frozen soils". *Int. Symp. on Mech. Behav. of Structured Media*, Elsevier, B, 205-245.

Ladanyi, B. & Archambault, G. (1980). "Direct and indirect determination of shear strength of rock mass". *AIME Annual Meeting,* Las Vegas, Preprint No. 80-25.

Ladanyi, B. & Archambault, G. (1977). "Shear strength and deformability of filled indented joints". Int. *Symp. on Geotechnics of Structurally Complex Formations*, Capri, 1, 317-326.

Ladanyi, B. & Morel, J.-F. (1990). "Effect of internal confinement on compression strength of frozen sand". *Canad. Geotech. J., 27,(1),* 8-18.

McRoberts, E.C. (1975). "Some aspects of a simple secondary creep model for deformations in permafrost slopes". *Canad. Geotech. J.*, 12(1), 98-105.

McRoberts, E.C. (1978), "Slope stability in cold regions". Chap. 7 in *Geotechnical Engineering for Cold Regions* (O.B. Andersland & D.M. Anderson, Eds.), McGraw-Hill, New York, 363-404.

Morgenstern, N.R., Roggensack, W.D. & Weaver, J.S. (1980). "The behaviour of friction piles in ice and ice-rich soils". Can. Geotech. J., 17(3), 404-415.

Nickling, W.G. & Bennett, L. (1984). "The shear strength characteristics of frozen coarse granular debris". *J. of Glaciology*, 30,(106), 348-357.

Nye, J.F. (1970). Glacier sliding without cavitation in a linear viscous approximation". *Proc. Royal Soc. London*, Ser. A, Vol.315, No.1552, 381-403.

Takahashi, T. (1978). "Mechanical characteristics of debris flow". *J. of Hydraulics Div., ASCE,* 104 (HY8), 1153-1169.

Thompson, E.G. and Sayles, F.H. (1972). "In situ creep analysis of room in frozen soil." *J. of Soil Mech. & Found. Div., ASCE* 98 (SM9), 899-915.

Ting, J.M., Martin, R.T. & Ladd, C.C. (1983). "Mechanisms of strength for frozen sand". *J. Geotech. Engrg., ASCE* 109(10), 1286-1302.

Vallejo, L.E. (1980). "A new approach to the stability analysis of thawing slopes". *Canad. Geotech. J.*, 17, 607-612.

Vallejo, L.E. (1981)."Determination of the shear strength of granulo-viscous materials using the vane shear apparatus and the two-phase suspension theory". *Int. Symp. on Mech. Behav. of Structured Media,*. Elsevier, B, 373-381.

Vallejo, L.E. (1979). "An explanation for mudflows". *Géotechnique*, 29, 351-354.

Vallejo, L.E. (1989). "An extension of the particulate model of stability analysis for mudflows". *Soils and Found.*, 29(3), 1-13.

Wagner, S. (1992). "Creep of alpine permafrost, investigated at the Murtèl Rock Glacier". *Permafrost and Periglacial Processes*, 3(2), 157-162.

Weertman, J. (1957). "On the sliding of glaciers". *J. of Glaciology*, 3, 21, 33-38.

Alaska Air Temperature Indices - Design 'Warm' Winter

R.L. Scher[1], P.E., M.ASCE

Abstract

Consideration of freezing and thawing air temperatures on foundations, pavements, utilities and structures is integral to civil engineering projects in cold regions. Air-freeze and thaw indices are commonly used to characterize the combined variation, magnitude and duration of annual temperature cycles. Design air temperature indices are typically defined to represent more extreme warm-summer and cold-winter seasons. However, there are situations where a design 'cold-winter' freeze index is of little value; e.g. in permafrost regions when the stability and performance of embankments and foundations dependants upon some magnitude of cooling each winter to re-freeze the seasonal active layer and/or lower the subgrade temperature. For these cases it would be more critical to evaluate the thermal consequences of a 'warm' winter; e.g. reduced bearing capacity and increased creep. Presented herein are design warm-winter freeze indices, as well as updated cold-winter freeze and warm-summer thaw indices for eleven Alaska communities based on daily climate records from 1971 to 2001 including Anchorage, Barrow, Bethel, Bettles, Central, Fairbanks, Gulkana, Kotzebue, Nome, Northway and Tanana; all but the first located in regions underlain by permafrost.

Introduction

The success and long-term performance of foundations, pavements, utilities and structures in cold regions is very dependent upon the consideration given during design to the affects of freezing and thawing air temperatures. Such thermal evaluations commonly utilize air-freeze and thaw indices to represent the combined variation, magnitude and duration of annual temperature cycles.

---

[1]Sr. Geotechnical Engineer, R&M Consultants, Inc. 9101 Vanguard Drive, Anchorage, Alaska 99507 (bscher@rmconsult.com)

Air-Temperature Indices. An air temperature index is determined by summing the degrees that the average daily air temperature, T_{Ave}, is below or above a reference temperature, in this case 0 °C, over a period of time. The air-freeze and thaw index is then defined as the maximum difference in the cumulative air index through the duration of one continuous 'winter' ($T_{Ave} \leq 0$ °C) and summer ($T_{Ave} \geq 0$ °C) period, respectively (U.S. Army, 1966). Maps showing the areal distribution of mean air-freeze and thaw indices, albeit now dated, have been published for North America (Lobacz, 1986), Alaska (Hartman & Johnson, 1984) and Canada (Boyd, 1976; and, Thompson, 1966).

Many thermal evaluations utilize a 'design' air-freeze and thaw index intended to represent the more extreme continuous cold-winter or warm-summer likely to occur during the project life. However, values of these design indices published to-date have not always followed the same definitions.

Most commonly, the design temperature index is defined as the average of the three coldest winters, or warmest summers, recorded over a 30-year period; or when climate records are limited, simply as the most severe winter or summer recorded over the previous ten years (U.S. Army, 1966). Note that Lobacz (1986) recommended this definition specifically for thermal evaluations of "permanent pavements". Maps showing the areal distribution of design air-freeze and thaw indices following this definition have been published for North America (Lobacz, 1986) and Alaska (Hartman & Johnson, 1984). Specific design indices for a number of Alaska communities are also provided in Braley and Connor (1989), and Braley and Zarling (1990). Also for North America, linear graphs correlating the mean and design air-freeze and thaw indices, where the design values represent the extreme index over a 10-year period, are published for arctic and subarctic regions (U.S. Army, 1966) and for Canada (Johnston, 1981).

For permanent structures, Lobacz (1986) recommended that thermal evaluations of foundations be based on design air indices equal to the coldest winter and warmest summer experienced over a 30-year period. Following this definition, a linear graph correlating the mean and design freeze index for the contiguous United States is provided in U.S. Army (1985).

Recently, Steurer (1989) presented a statistical procedure to determine the design air-freeze index corresponding with a 100-year return period for use in detailing shallow insulated building foundations in the contiguous United States. Contour plots of the 100-year design air-freeze index for the United States, including Alaska, are provided in Steurer (1996), and ASCE (2001).

Warm Winter Considerations. In many frozen soils the shear strength, bearing capacity and creep parameters are strongly influenced by temperature. As such, the stability and performance of embankments and foundations on permafrost is often contingent upon achieving some magnitude of cooling each winter to re-freeze the seasonal active layer and/or reduce the ground temperatures. This would be especially true for facilities over warm, ice-rich permafrost that incorporate natural or passive heat transfer schemes (e.g. air-convection embankments, air ducts, thermo-pipes/piles, etc) to maintain some level of minimum ground temperatures. Therefore, for these projects it would be more critical during design to evaluate the consequences of a 'warm' winter; such as reduced refreezing, formation of a talik, or warming permafrost with corresponding reduced bearing capacity and increased creep.

This paper presents design 'warm-winter' air-freeze indices, as well as updated mean and design air-thaw and cold-winter air-freeze indices for eleven Alaska communities, all but one located in regions underlain by permafrost, based on daily climate records from the period 1971 to 2001. Also provided are brief discussions of methods to estimate design air temperature indices at locations with limited climate data, and the apparent climate change in Alaska over the past 30 years as reflected in the calculated mean air temperature indices.

Methods & Results

Climate Data. '*Summary of Day*' records (i.e. minimum and maximum temperature, and precipitation) for the period 1971 to 2001 were obtained from the Alaska State Climate Center for the eleven Alaska communities in Table 1; all but Anchorage being located in regions underlain by continuous or discontinuous permafrost (Ferrians, 1965).

Table 1. Alaska Climate Stations

Station	Latitude	Longitude	Elevation, m	Record
Anchorage	61°10'	150°01'	34	1952-Present
Barrow	71°18'	156°47'	10	1946-Present
Bethel	60°47'	161°48'	38	1946-Present
Bettles	66°55'	151°31'	196	1951-Present
Fairbanks	64°49'	147°52'	133	1946-Present
Gulkana	62°09'	145°27'	479	1949-Present
Kotzebue	66°52'	162°38'	3	1946-Present
McGrath	62°58'	155°37'	105	1946-Present
Nome	64°30'	165°26'	4	1946-Present
Northway	62°57'	141°56'	524	1946-Present
Tanana	65°10'	152°06'	72	1949-Present

Mean Air-Temperature Index. The air-freeze and air-thaw indices were determined for each station in Table 1 following U.S. Army (1966) based on average daily temperatures.

In this method, Equation [1] was used to produce a cumulative air index. The annual air-freeze and air-thaw indices were then calculated for the period August 1970 through July 2001 ($N \geq 11{,}300$ days), as the maximum difference between the cumulative air index through each continuous winter (in Alaska typically September/October through April/May) and 'summer' (in Alaska typically May/June through August/September), respectively. A graphic example of the air-freeze and air-thaw indices are illustrated in Figure 1 based on plots of the cumulative air index determined for Barrow and Fairbanks from 1 August 1994 though 31 December 1995.

Figure 1. Air-Index Plot

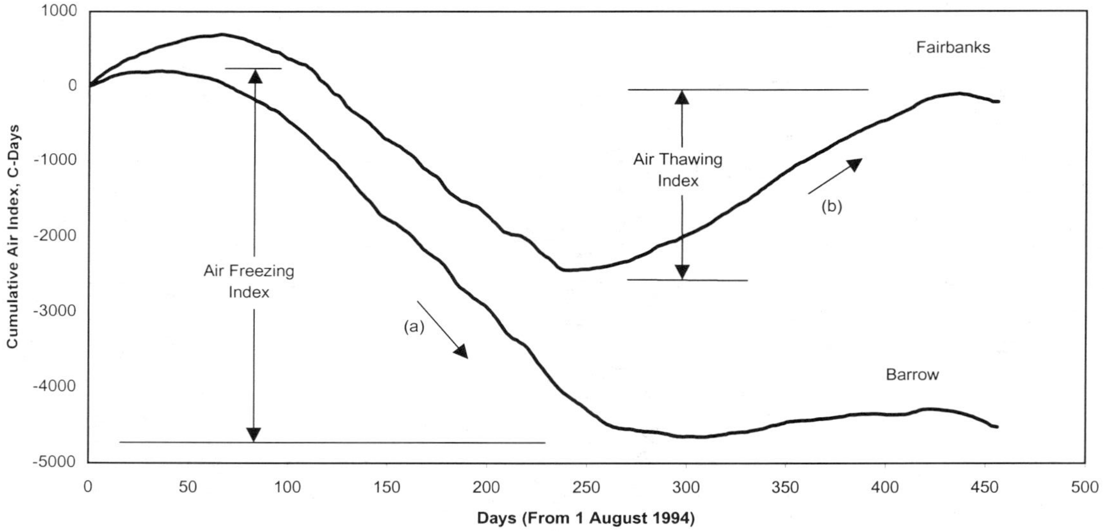

NOTE : A negative slope (a) and positive slope (b) indicate the average daily temperatures are less or greater than 0°C, respectively.

$$AI_{0°C} = \sum_{i=1}^{N} T_{Ave_i} \quad [1]$$

Where : $AI_{0°C}$ is the 'Cumulative Air Index' relative to 0°C; T_{Ave} is the average of the daily maximum and minimum air temperatures (°C) for day i; and, N is the total number of observation (record) days.

Table 2 summarizes the mean air-freeze and air-thaw indices determined for the eleven Alaska communities in Table 1 between 1971 and 2001, including minimum, maximum, mean and standard deviation (count, n = 25 to 30).

Table 2. Annual Air-Temperature Indices Based on Average Daily Temperature (1971 - 2001)

Station	Freeze Index, °C-Days		Thaw Index, °C-Days	
	Min - Max	Mean (*StDev*)	Min - Max	Mean (*StDev*)
Anchorage	345 - 1670	1077 (301)	1621 - 2326	1965 (190)
Barrow	3954 - 5657	4736[(a)] (382)	183 - 631	353 (114)
Bethel	1033 - 2846	1954 (417)	1275 - 1861	1526 (143)
Bettles	2854 - 4542	3536 (417)	1381 - 1902	1683 (139)
Fairbanks	2143 - 3655	2937 (376)	1720 - 2342	2050 (161)
Gulkana	1497 - 3564	2495 (443)	1421 - 1962	1645 (149)
Kotzebue	2412 - 3974	3169 (383)	838 - 1458	1172 (151)
McGrath	2029 - 3747	2894 (394)	1544 - 2137	1766 (155)
Nome	1466 - 3208	2246 (412)	908 - 1481	1215 (154)
Northway	2923 - 4503	3556 (380)	1240 - 2007	1674 (162)
Tanana	2559 - 4063	3169 (363)	1448 - 2070	1802 (169)

(a) 4905 °C-Days (*StDev* =366 °C-Days) in Steurer & Crandell (1995) for SOD records from 1951 to 1980.

Design Air-Temperature Indices. The design air temperature indices were determined for each of the communities in Table 1 following U.S. Army (1966) by averaging the indices calculated from the average daily temperature data for the three warmest winters, coldest winters, and warmest summers between 1971 and 2001.

Table 3 summarizes the recommended design warm-winter air-freeze indices, as well as the updated design cold-winter and warm-summer freeze and thaw indices, respectively, for the communities in Table 1. For comparison, Table 3 also includes the respective 'cold-winter' design freeze and 'warm' thaw indices published by others for these communities.

Table 3. Design Air-Temperature Indices (1971 - 2001)

Station	Freeze Index, °C-Days			Thaw Index, °C-Days	
	'Warm'	'Cold'	BERG2[a]	'Warm'	BERG2[a]
Anchorage	558	1544	1778	2271	2222
Barrow	4109	5508[b]	5500	586	500
Bethel	1271	2743	2778	1800	1944
Bettles	2891	4371	≈4450[c]	1883	≈1550[c]
Fairbanks	2222	3573	3833	2299	2222
Gulkana	1709	3164	3389	1939	1833
Kotzebue	2589	3868	4278	1412	1333
McGrath	2270	3616	≈3300[c]	2041	≈2000[c]
Nome	1580	3112	≈3000[c]	1433	≈1400[c]
Northway	3008	4243	4444	1894	1778
Tanana	2593	3874	≈3900[c]	2042	≈1860[c]

(a) Braley & Connor (1989)
(b) '100-Year Return AFI' = 5655 °C-Days in Steurer & Crandell (1995) for SOD records from 1951 to 1980.
(c) Estimated from Hartman & Johnson (1984)

Discussion

Estimating Mean Freeze & Thaw Indices from Average Monthly Temperatures. In Alaska, relatively few communities have daily climate records spanning 30 years. However, there is often sufficient data for many more communities to at least characterize the mean freeze and thaw indices based on historic average monthly temperatures. In this approach, the air index is determined for each month by multiplying the average monthly temperature (°C) by the number of days. The mean air freeze and thaw indices are then determined by summing the air index for the months with average temperatures either above or below 0°C, respectively.

Numerous authors have pointed out the inaccuracies of determining air indices based on average monthly temperatures; principally due to the 'change-over' month(s) during which the average daily temperatures may rise or fall below freezing several times (e.g. Boyd, 1976; Haas, 1984; and, Steurer & Crandell, 1995). Data in Haas (1984) and Steurer & Crandell (1995) indicate that the magnitude of this error, as a percent of the annual air freeze or thaw index, is more or less inversely proportional to the length of the particular season. That is the error in air-index determined from average monthly temperatures will increase as the length of the respective season decreases. Graphic and empirical methods have been proposed to reduce the air-index error during the change-over months (U.S. Army, 1966 and, Boyd, 1976, respectively).

To quantify the magnitude of error in arctic and subarctic Alaska, the mean freeze and thaw indices were estimated as described above for each of the eleven Alaska communities in Table 1, using the historic average monthly air temperatures for the period 1971 to 2001, without any correction for the change-over months. These values were then compared with the respective mean indices in Table 2 determined from the average daily temperatures. In general, the value of the freeze indices based on mean monthly temperatures were slightly less (range 1.7 to 9.2 %, average 3.1 %) than those in Table 2. Similarly, the thaw indices based on mean monthly temperatures were slightly less (range 2.5 to 7.8 %, average 4.6 %; excluding Barrow which was about 16 % lower) than those in Table 2.

Estimating Design Air Freeze and Thaw Indices. There are at least two approaches for estimating design air freeze and thaw indices characteristic of a project location lacking 30-years of climate data. First, Lobacz (1986) suggested that the design temperature indices for pavement evaluation could be based on the most extreme winter and summer over the past ten years; and for structural foundations using a ratio of *design* to *mean* index determined at the closest stations which do have at least 30-years of temperature data. Second, a number of publications have illustrated a strong linear correlation between the *design* and *mean* air temperature indices (e.g. Johnston, 1981; U.S. Army, 1966; and, U.S. Army, 1985).

Following the second approach, Figure 2 illustrates plots of the mean (Table 2) and design (Table 3) air temperature indices determined for each of the stations in Table 1. Equation [2] represents the general form of a straight line fit through these points using a regression analysis.

$$TI_D = m \times TI_M + b \qquad [2]$$

Where : The *design* and *mean* air temperature indices (TI_D and TI_M) are in °C-Days; and, m and b are linear coefficients, summarized in Table 4.

Regional Climate Change. Linear regression analysis of the annual air-freeze and thaw indices determined for the eleven communities in Table 1 indicate a general, but not surprisingly variable (i.e. $r^2 = 0.003$ to 0.297) warming trend across much of Alaska over the past 30 years. For example, at Barrow (Figure 3), the annual air-freeze index has been decreasing since 1971 about 43 °C-Days per year, and the annual air-thaw index has been increasing about 10 °C-Days per year. However, at Fairbanks the air-temperature indices suggest a much 'weaker' warming trend (Figure 3), where since 1971 the annual air-freeze index has been decreasing only about 14 °C-Days per year, and the annual air-thawing index has been increasing less than two °C-Days per year.

Figure 2. Mean vs Design Air Temperature Indices

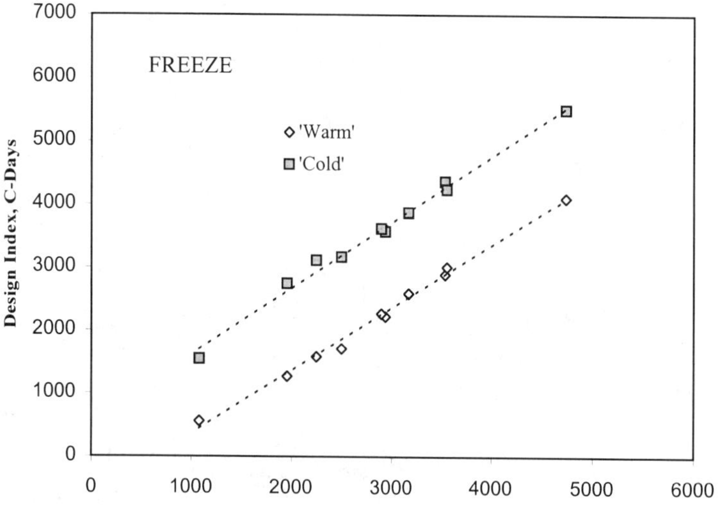

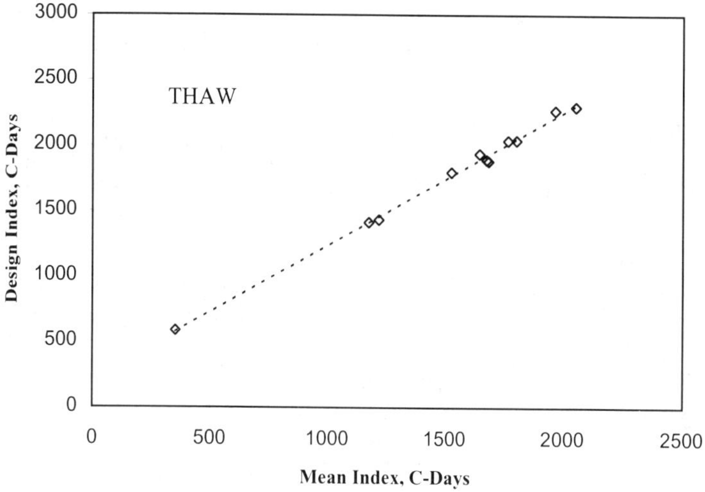

Figure 3. Air Temperature Index Trends (1971 - 2001)

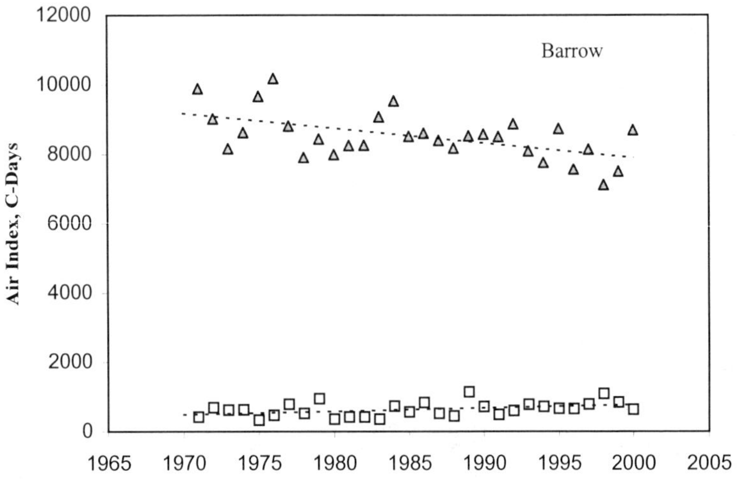

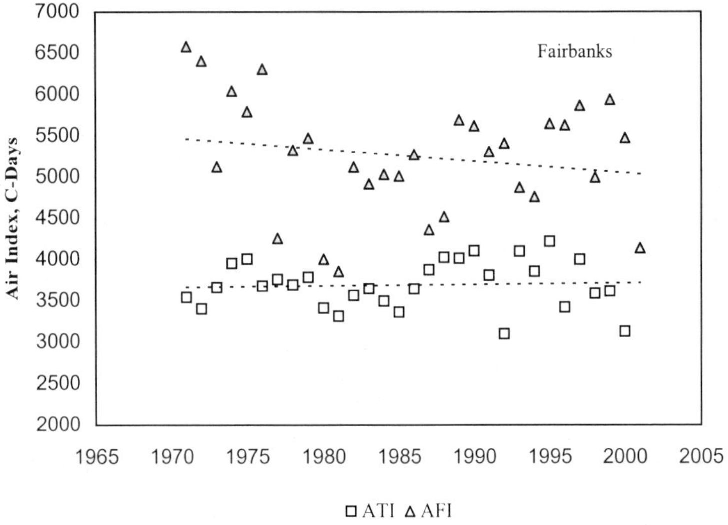

Table 4. Design vs Mean Air Index Coefficients
(r^2 = *Coefficient of Determination*)

Design Index	Equation [2] Coefficients		
	m	b (°C-Days)	r^2
Average 3 Extreme Seasons in a 30-Year Period[(a)]			
Freeze - Cold	1.05	570	0.991
Warm	1.00	-636	0.993
Thaw	1.02	214	0.996
Extreme Season in a 30-Year Period[(b)]			
Freeze - Cold	1.06	703	0.986
Warm	1.04	-885	0.988
Thaw	1.03	258	0.992

(a) Design index for "permanent" pavements (Lobacz, 1986)
(b) Design index for foundations of "average permanent structures" (Lobacz, 1986)

Summary

A new design air-freeze index is proposed to represent an extreme 'warm' winter. This new index could be used to evaluate the consequences of a warm winter on the stability and performance of structural foundations and embankments over permafrost, such as reduced shear strength and bearing capacity, and increased settlements, especially in warm ice-rich subgrades. The updated mean and design air indices for 11 Alaska communities, all but one in regions underlain by permafrost, are presented in Tables 2 and 3, based on average daily temperatures from 1971 though 2001.

For locations in arctic and subarctic Alaska lacking 30-years of climate data, simple methods are discussed to develop reasonable estimates of the mean air indices based on historic average monthly temperatures, and design indices based on the mean indices. Despite the inaccuracies of these methods, they should be adequate to at least characterize the mean and design freeze and thaw indices for many projects.

Finally, the annual air freeze and thaw indices determined for the eleven communities in Table 1 all indicated a general, but not surprisingly variable warming trend across Alaska over the past 30 years.

Acknowledgments

The author is in debt to Mr. Dwight D. Pollard, Ph.D., State Climatologist with the Alaska State Climate Center for guidance in using the *Summary of Day* climate data.

References

American Society of Civil Engineers (ASCE; 2001). "*Design and Construction of Frost-Protected Shallow Foundations (FPSF)*". ASCE Standard.

Boyd, D.W. (1976). "Normal freezing and thawing degree-days from normal monthly temperatures. *Can. Geotech. J.*, 13:176-180.

Braley, W.A., and B. Connor (1989). "BERG2, Micro-Computer Estimation of Freeze and Thaw Depths and Thaw Consolidation". *Report AK-RD-90-01*. Alaska Department of Transportation and Public Facilities, Statewide Research.

Braley, W.A., and J.P. Zarling (1990). "MUT1D, Multilayer User-Friendly Thermal Model in 1 Dimension". *Report AK-RD-90-02*. Alaska Department of Transportation and Public Facilities, Statewide Research.

Ferrians, O.J. (1965). "Permafrost map of Alaska". *U.S.G.S. Miscellaneous Geologic Investigations Map* 1-445.

Haas, W.M. (1984). "Errors in freezing index values calculated from monthly average temperatures". *Proc. Cold Regions Engr. Spec. Conf., Montreal, Quebec*, CSCE, pp 1075-1089.

Hartman, C.W. and P.R. Johnson. (1984). *Environmental Atlas of Alaska*. University of Alaska.

Johnston, G.H. (1981). *Permafrost Engineering Design and Construction*. John Wiley & Sons.

Lobacz, E.F. (1986). "Arctic and Subarctic Construction - General Provisions". *U.S. Army Cold Regions Research & Engineering Laboratory, Special Report 86-17.*

Steurer, P.M. (1996). "Probability distributions used in 100-year return period of air-freezing index". *J. Cold Regions Engineering*, 10(1):25-35.

Steurer, P.M. (1989). "Methods used to create an estimate of the 100-year return period of the air-freezing index." *Frost-Protected Shallow Foundation Development Program - Phase II Final Report, Appendix A*. Prepared for the Society of Plastics Industry, NAHB Research Center, Upper Marlboro, MD.

Steurer, P.M., and J.H. Crandell (1995). "Comparison of methods used to create estimate of air-freezing index". *J. Cold Regions Engineering*, 9(2):64-74.

Thompson, H.A. (1963). "Air temperatures in Northern Canada with emphasis on freezing and thawing indexes". *Proc 1st Int'l Permafrost Conf., Lafayette, Indiana, NAS-NRC Publ.* 1287, pp 272-280.

U.S. Army (1985). "Pavement Design for Seasonal Frost Conditions". *Department of Army Technical Manual TM 5-818-2.*

U.S. Army (1966). "Calculation Methods for Determination of Depths of Freeze and Thaw in Soils". *Department of Army Technical Manual TM 5-852-6.*

Probabilistic Analysis of Thaw Penetration in Fairbanks, Alaska

Ted S. Vinson[1] and Yoon-Shin Bae[2]

Abstract

A probabilistic approach may be adopted to predict freeze and thaw depths to account for the variability of (1) material properties, and (2) contemporary and future surface energy input parameters (e.g. air temperatures, cloud cover, snow cover) predicted with global climate models. To illustrate the probabilistic approach, an example of the prediction of thaw depths in Fairbanks, Alaska, is considered. More specifically, the Stefan equation is used together with the Monte Carlo simulation technique to make a probabilistic prediction of thaw penetration. The simulation results indicate that the variability in material properties, surface energy input parameters, and temperature data can lead to significant uncertainty in predicting thaw penetration.

Introduction

A problem frequently encountered in cold regions engineering is to estimate the depth of freeze or thaw penetration beneath the ground surface. Our ability to assess the actual freeze or thaw penetration depth depends on the reliability/variability of the material property and surface energy parameters input to the governing predictive equations. The material property and surface energy input parameters are often assumed to be fixed values. In reality these values are not fixed values but can exhibit substantial variability. A probabilistic analysis may be performed using a deterministic model (i.e., predictive equation) and the probability density function for the input parameters. A probabilistic analysis supports a statement of the likelihood of occurrence of a given outcome for a given set of variable input parameters.

Recently concern for global climate change in the arctic and sub-arctic has received attention relative to the performance of existing structures and as an additional design criterion for new projects. While a consideration of global climate change may be appropriate (Serreze et al, 2000) it is not appropriate to consider it in a deterministic manner (e.g., a rate of temperature increase over a specific future time interval at a given location). The reason to treat global climate change probabilistically is obvious when it is recognized that no fewer than thirty-one credible global climate models (GCMs) are available to predict future global climate change. Consequently, the prediction of the rate of temperature increase for a specified future time interval, for an assumed greenhouse gas concentration increase, at a given location, may have thirty-one different values.

----

1 Professor of Civil, Construction and Environmental Engineering, Oregon State University, Corvallis, Oregon 97331

2 Formerly Graduate Student, Oregon State University, Corvallis, Oregon

A probabilistic approach provides a means of evaluating the combined effects of uncertainties, and a means of distinguishing between conditions where uncertainties are particularly high or low. Therefore, a probabilistic approach may be adopted to predict freeze and thaw depths to account for the variability of (1) material properties, and (2) contemporary/future surface energy input parameters and air temperatures predicted with GCMs.

The objectives of the study presented herein are to:

- Investigate the variability of air temperatures predicted with GCMs in Fairbanks, Alaska.
- Develop a probabilistic approach that supports the determination of thaw penetration in Fairbanks, Alaska.

More specifically, a probabilistic analysis of thaw penetration using the Monte Carlo simulation technique is presented. The probabilistic analysis was performed using commercially available software (Palisade Corporation, 2000).

Global Climate Models and Climate Predictions for Fairbanks, Alaska

A GCM is used to simulate the many processes that produce climate. As noted by McGuffie and Henderson-Sellers (1997), the aim of a GCM is the calculation of the full three-dimensional character of the climate for the earth. By 1995, there exist 31 global climate modeling groups (Gates et al., 1998). The fundamental equations employed in GCMs are conservation of energy, conservation of momentum, conservation of mass, and the ideal gas law (McGuffie and Henderson-Sellers, 1997). The first step in obtaining a GCM solution is to divide the earth's surface and atmosphere into a series of columns and specify grid points at the column corners. Next, conditions are specified at each grid point for the surface of the earth and several layers in the atmosphere. Finally, the fundamental governing equations are solved at each grid point using numerical techniques (McGuffie and Henderson-Sellers, 1997). Gates (1992a) noted that the horizontal resolution of the columns ranges from 2.5° latitude × 3.75° longitude to 8° latitude × 10° longitude. Because the column resolutions are different for each model, all GCMs are not on the same grid. The accuracy of the GCM depends, in part, on the spatial resolution of the grid points. To predict temperature for a specific location in a grid cell, each model uses a timestep approach and an interpolation scheme (McGuffie and Henderson-Sellers, 1997).

GCMs are used for atmospheric research and long range forecasting through simulation of general circulation. It is possible to predict the climate with idealized conditions (e.g. doubling the concentration of CO_2 over a specific time interval CO_2, planet covered all over with ocean, etc.) as well as simulation of real atmosphere.

The validity of GCM predictions can only be tested against the past or present climate. Usually when a GCM is developed an initial objective is to ascertain how well it compares with the present climate. Thereafter it may be used to simulate past climate. Assuming a reasonable good comparison, the model can be used to predict possible future climates under various scenarios.

McGuffie and Henderson-Sellers (1997) noted that even within one particular GCM there are many features and evolutionary stages of development. Because

GCMs share a common purpose, it is possible to apply different climate model types to the same prediction task. This necessary verification work has motivated the climate modeling community to initiate a series of model intercomparisons and evaluations of performance.

The Atmospheric Model Intercomparison Project (AMIP) undertook the systematic validation, diagnosis and comparison of the performance of atmospheric GCMs (Gates, 1992b). The decade 1979-1988 was selected as the AMIP test period. "The choice was a compromise between the desire to use as long a simulation period as possible in order to obtain representative results, and the increasing difficulties with observed global data sets as one proceeds to earlier years" (Gates, 1992b). A common list of standard monthly-averaged GCM output consists of the monthly mean global geographical distribution of selected surface and vertically integrated variables (sea-level pressure, ground and surface air temperature, cloudiness, precipitation, soil moisture and snow mass).

By 1995 there were 31 atmospheric GCMs participating in AMIP (Gates et al., 1998). Figures 1 and 2 show the predicted monthly air temperatures at 2 meters above the surface from 27 GCMs for Fairbanks, Alaska (four GCMs were excluded because of unreasonable results/outliers).

McGuffie and Henderson-Sellers (1997) state "no one model performs well in all the evaluations employed; no one test evaluates all aspects of the participating models; and the model group mean (after excluding unreasonable results/outliers) outperforms any one individual model". This fact is illustrated in the results shown in Figures 1 and 2. However, there is no basis to select the results of one model over another. The results illustrate the variability of GCM predictions that can be incorporated in a probabilistic approach to thaw prediction. AMIP results which would suggest the variability in results for the prediction of future climate under a specific scenario (e.g., doubling CO_2 during 100 yrs) are not available at this time, but it is logical and prudent to assume that the variability would be greater than the variability shown in Figures 1 and 2 for the prediction of past climate.

Statistical Characterization of Input Parameters

To perform thaw penetration, information is needed on the following geotechnical parameters: dry unit weight(s) of soil(s) (γ_d) and water content(s) (ω). These parameters are assumed to be statistically independent; i.e. there is no correlation relationship between the parameters. With knowledge of γ_d and ω, one can calculate thermal conductivity, K, and volumetric latent heat capacity, L.

The statistical variability in an input parameter is reflected by its probability density function. It is useful to plot a histogram to select a probability density function. However, the method is very subjective, and for small samples (e.g., less than 20 observations) it is not accurate to construct a histogram for the purpose of identifying the distributions because the shape of the distribution will depend on the number of class intervals (Picard, 1988). The distributions in common use are the normal distribution, the lognormal distribution, and the extreme value distribution.

Phoon and Kulhawy (1999) identify three primary sources of geotechnical uncertainties: inherent variability, measurement error, and transformation uncertainty.

Statistical information on measurement error is rather limited and rigorous statistics on transformation uncertainty generally are not available. Consequently, only inherent variability was considered for the dry unit weight and water content of soils. It is reasonable to assume these variables are normally distributed. The mean values and coefficient of variation (COV) for γ_d and ω are given in Table 1.

The surface energy input parameters required for a thaw penetration calculation include the air temperature (and related thawing index) and the ratio of surface index to air index, termed the "n-factor". Direct determination of the n-factor for a specific location requires concurrent observations of air and surface temperatures throughout a number of complete freezing and thawing seasons (Air force manual No. 88-19, 1966). The average n-factor for a given surface and location can change significantly from year to year (Andersland and Ladanyi, 1994). Considering this information it is reasonable to assume the probability density function, mean value and coefficient of variation (COV) for the n-factor given in Table 2 (the range of n factor was truncated at 1.1 and 2.3).

In the absence of thirty years of continuous recorded air temperatures, the design air thawing (or freezing) index is defined as the air-thawing (or freezing) index for the warmest summer (or coldest winter) in the latest 10-year period (Air force manual No. 88-19, 1966). The design air thawing indexes based on the air temperatures predicted by twenty-seven GCMs for the period (1979-1988) are given in Table 3.

The three candidates for probability density function of design air thawing index are shown in Figure 3. The probability distributions were ranked by the Excel add-in @RISK (Palisade Corporation, 2000) and the results are given in Table 4. Using a Chi-square test (χ^2), the extreme type-I distribution was selected as the most plausible probability density function.

Development of a Probabilistic Model for Thaw Penetration

The use of probabilistic methods in civil engineering has been presented to the profession for nearly three decades (Ang and Tang, 1975; Whitman, 1984) and has recently been reaffirmed as a very useful tool for geotechnical engineering practitioners (Duncan, 2000). The use of probabilistic methods to assess thaw penetration is new. The distribution of thaw penetration may be determined by combining the probability density functions of the basic random variables. One method to accomplish this is the Monte Carlo simulation technique. The random numbers must have the same probability density function as the variable. Each input to the deterministic thaw penetration model is a sample within the probability density function created by the random values. The results of repeated combinations of variables (Monte Carlo passes) constitute the population for the possible outcomes of thaw depth.

Figure 4 presents a flowchart of a probabilistic analysis for thaw penetration based on the Stefan equation (Lunardini, 1981) as the deterministic model. The dry unit weight of the soil, water content, n-factor and maximum air thawing index are randomly selected from their assumed probability density functions (Frame 1). The thermal conductivity of the unfrozen soil and latent heat are calculated using the randomly selected properties (Frame 2). Next, using the calculated and randomly selected variables (thermal conductivity of unfrozen soil, latent heat, maximum air thawing index and n-factor) thaw penetration is calculated (Frame 3). The procedure is repeated 10,000 times and the probability distribution of thaw penetration is estimated (Frame 4).

To identify the specific type of probability distribution both the Chi-squared test (χ^2) and K-S test were employed. Based on these tests, the lognormal distribution was selected as the most plausible probability distribution. The probability density function for thaw penetration based on the Stefan equation is given in Figure 5 and probability distributions ranked by Excel add-in @RISK (Palisade Corporation, 2000) are given in Table 5.

As noted by Rosowsky (1997), the Monte Carlo simulation provides an exact solution to the probability of "failure". The concept is to generate sets of outcomes of the random variables in the deterministic model and to record the number of times the outcomes are less than the failure value. The estimate of the probability of failure is simply the number of failures divided by the total number of outcomes.

The graphical illustration of probability of failure using the concept of factor of safety (FS) is given in Figure 6. With failure defined as a factor of safety of one, the probability of failure is the population of values with a factor of safety of one or less divided by the total population of factor of safety values.

Figure 7 presents an example based on the Stefan equation. If the allowable thaw depth is 2.44m, the probability of failure (i.e., the likelihood that the thaw depth will be equal to or greater than 2.44m) is 20 %. (The probability of failure is equal to the area under the probability density function for $X \geq 2.44$m divided by the total area under probability density function.) Conversely, the probability of successful performance (i.e., the likelihood that the thaw depth will be equal to or less than 2.44m) is 80 %, or 100% minus the probability of failure.

Summary and Conclusions

The results presented herein illustrate a probabilistic analysis applied to a conventional cold region's engineering problem with relatively small additional effort compared to a conventional deterministic analysis. An example of the prediction of thaw depths in Fairbanks, Alaska, was considered. Input parameters were generated from the assumed probability density functions estimated from extensive published data, or a probability density function determined from observational data with statistical tests. The Stefan equation was used together with the Monte Carlo simulation technique to make a probabilistic prediction of thaw penetration. The simulation results indicate that the randomness/variability in material properties, surface energy input parameters, and temperature data lead to significant uncertainty in predicting thaw penetration. It is clear that further work is needed to reduce the errors of GCM predictions, as well as other input parameters, if the uncertainty in predicted thaw penetration is to be reduced.

References

Andersland, O. B. and Ladanyi, B. (1994), An Introduction to Frozen Ground Engineering, Chapman & Hall.

Ang, A.H-S and Tang, W.H. (1975), Probability Concepts in Engineering Planning and Design, Volume I, John Wiley & Sons Inc.

Departments of the Army and Air Force (1966), "Arctic and Subarctic Construction, Calculation Methods for Determination of Freeze and Thaw in Soils", U.S. Army Technical Manual TM-5-8526m, U.S. Air Force manual AFM 88-19, Chapter 6.

Duncan, J. M. (2000), "Factor of Safety and Reliability in Geotechnical Engineering", Journal of Geotechnical Engineering, ASCE, Vol. 126

Gates, W. L. (1992a), "The Validation of Atmospheric Models" PCMDI Report No. 1, Lawrence Livermore National Laboratory.

Gates, W. L. (1992b), "AMIP: The Atmospheric Model Intercomparison Project", *Bulletin of the American Meteorological Society*, Vol. 73.

Gates, W. L., Boyle, J. S., Covey, C. C, Dease., C. G., Doutriaux, C. M., Drach, R. S., Fiorino, M., Gleckler, P. J., Hnilo, J. J., Marlais, S. M., Phillips, T. J., Potter, G. L., Santer, B. D., Sperber, K. R., Taylor, K. E., and Williams, D. N. (1998), "An Overview of the Results of the Atmospheric Model Intercomparison Project (AMIP)" , PCMDI Report No. 45, Lawrence Livermore National Laboratory, April.

Kersten, M. S. (1949) "Thermal Properties of Soils", Engineering Experiment Station, University of Minnesota, St. Paul, MN Bulletin 28.

Lunardini, V. J (1981), Heat Transfer in Cold Climates, Van Nostrand Reinhold Company.

McGuffie, K. and Henderson-Sellers, A. (1997), A Climate Modeling Primer, John Wiley & Sons Inc.

Palisade Corporation (2000) "Guide to Using @RISK – Risk Analysis and Simulation Add-In for Microsoft Excel", Version 4, March.

Phoon, K.-K., and Kulhawy, F. H., (1999), "Characterization of Geotechnical Variability", *Canadian Geotechnical Journal*, Volume 36.

Picard, H. (1988), "A probabilistic Model for Coastal Landslides", Master of Science Report, Dept. of Civil Engineering, Oregon State University, Corvallis, Oregon.

Rosowsky, D.V. (1997), "Structural Reliability," Chapter 26, Structural Engineering Handbook, edited by Chen, W.F., CRC Press.

Serreze, M. C., Walsh, J. E., Chapin, F. S., Osterkamp, T., Dyurgerov, M., Romanovsky, V., Oechel, W., Morison, J., Zhang, T., Barry, R. G. (2000) "Observational Evidence of Recent Change in the Northern High-Latitude Environment", *Climate Change*, Volume 46, Kluwer Academic Publishers

Whitman, R. V. (1984), "Evaluating Calculated Risk in Geotechnical Engineering", *Journal of Geotechnical Engineering*, ASCE, Vol. 110 (2)

Table 1. Variability of dry unit weight and water content of soils (source: Phoon and Kulhawy, 1999)

variables	probability distribution	mean	std. dev.	COV (%)
γ_d	Normal distribution	1600 (kg/m^3)	112 (kg/m^3)	7
ω	Normal distribution	29 (%)	5.22 (%)	18

Table 2. Variability of n-factor

variables	probability distribution	mean	std. dev.	COV (%)
n	Normal distribution	1.7	0.3	17.6

Table 4. Probability distributions of maximum air thawing index ranked by Excel add-in @RISK

	First rank	Second rank	Third Rank
χ^2-test	Extreme Type - I Distribution	Lognormal Distribution	Normal Distribution

Table 5. Probability distributions of thaw penetration ranked by Excel add-in @RISK

	First rank	Second rank	Third Rank
χ^2-test	Extreme Type - I Distribution	Lognormal Distribution	Normal Distribution
K-S test	Extreme Type - I Distribution	Lognormal Distribution	Normal Distribution

Table 3. Design air thawing indexes based on the air temperatures predicted by twenty seven GCMs for the period (1979-1988) (unit: °C*hours)

bmr	ccc	cnr	csi	csu	der	dnm	ecm	gfd
15793	30014	19402	38959	9895	18617	21474	17213	26077
gis	gla	gsf	jma	lmd	mgo	mpi	mri	nca
44294	44521	33766	25728	23980	34590	18171	17958	46322
nmc	nrl	rpn	sng	sun	ucl	uiu	ukm	yon
30212	36153	15362	34474	14478	19967	17304	34331	32903

Legend:

bmr: Bureau of Meteorology Research Centre, Melbourne, Australia
ccc: Center for Climate Modeling and Analysis, British Columbia
cnr: Centre Natl Recherches Meteorologiques, France
csi: Commonwealth Scientific and Industrial Research Organization, Melbourne, Australia
csu: Colorado State University
der: Dynamic Extended Range Forecasting, GFD Lab, Princeton, NJ
dnm: Dept of Numerical Mathematics, Moscow, Russia
ecm: European Center for Medium Range Weather Forecasts, United Kingdom
gfd: Geophysical Fluid Dynamics Lab, Princeton, NJ
gis: Goddard Institute for Space Sciences, NY, NY
gla: Goddard Laboratory for Atmospheres, Greenbelt, MD
gsf: Goddard Space Flight Center, Greenbelt, MD
jma: Japan Meteorological Agency, Tokyo, Japan
lmd: Laboratoire de Meteorologie Dynamique, Paris, France
mgo: Main Geophysical Observatory, St. Petersburg, Russia
mpi: Max Planck Inst for Meteorology, Hamburg, Germany
mri: Meteorological Research Institute, Tsukuba, Japan
nca: Natl Center Atmospheric Research, Boulder, CO
nmc: Natl Centers for Environ Prediction, Washington, DC
nrl: Naval Research Laboratory, Monterey, CA
rpn: Recherche en Prevision Numerique, Quebec, Canada
sng: SUNY-Albany; NCAR, Boulder
sun: State Univ. of NY (SUNY) - Albany
ucl: Univ. of Calif – Los Angeles
uiu: Univ. of Illinois - Urbana
ukm: United Kingdom Meteorological Office
yon: Yonsei University, Seoul, South Korea

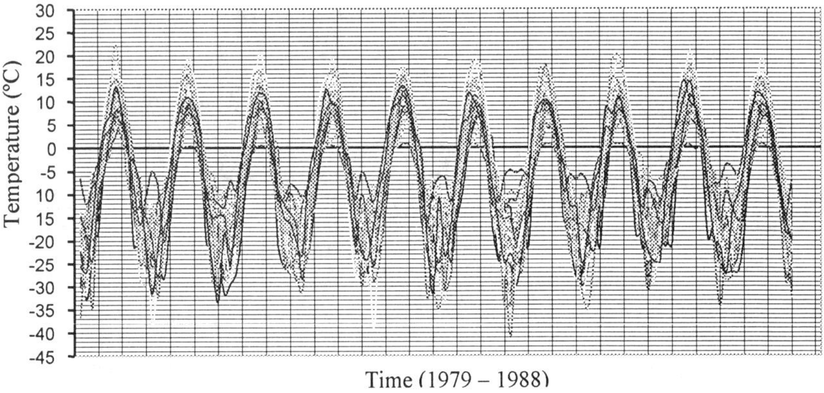

Figure 1. Predicted air temperatures at 2m above the surface from 27 GCMs for Fairbanks, Alaska

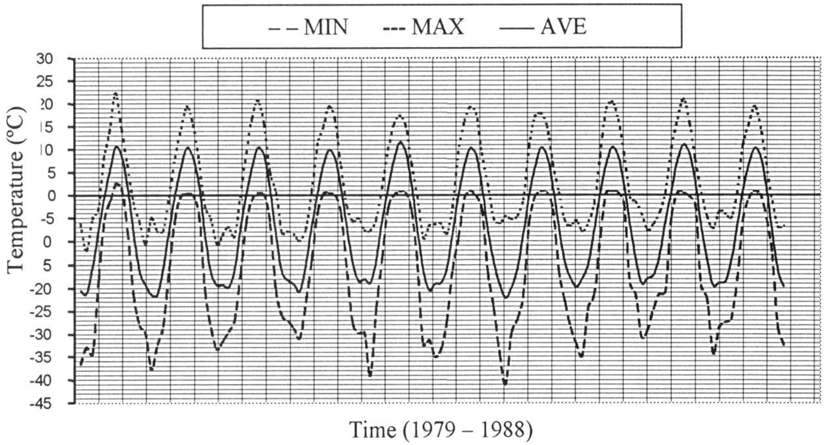

Figure 2. Average, minimum, and maximum air temperatures predicted from 27 GCMs for Fairbanks, Alaska

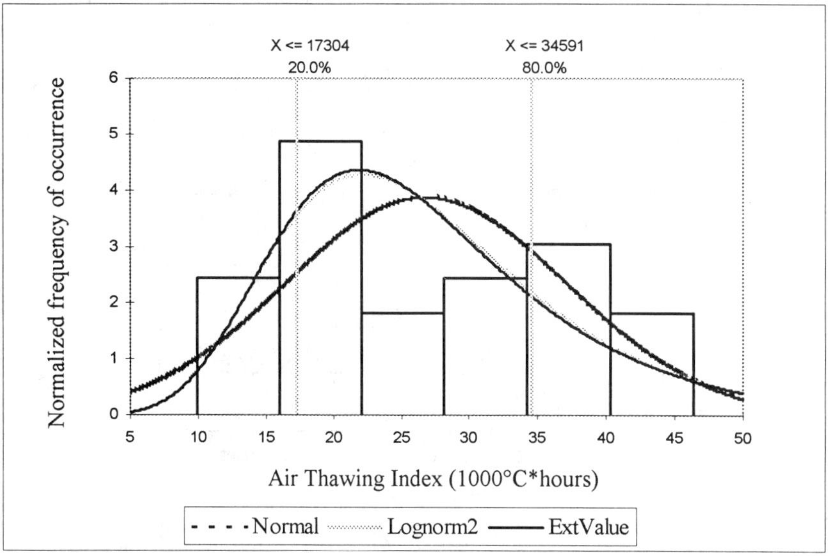

Figure 3. Candidate probability density functions for maximum air thawing index

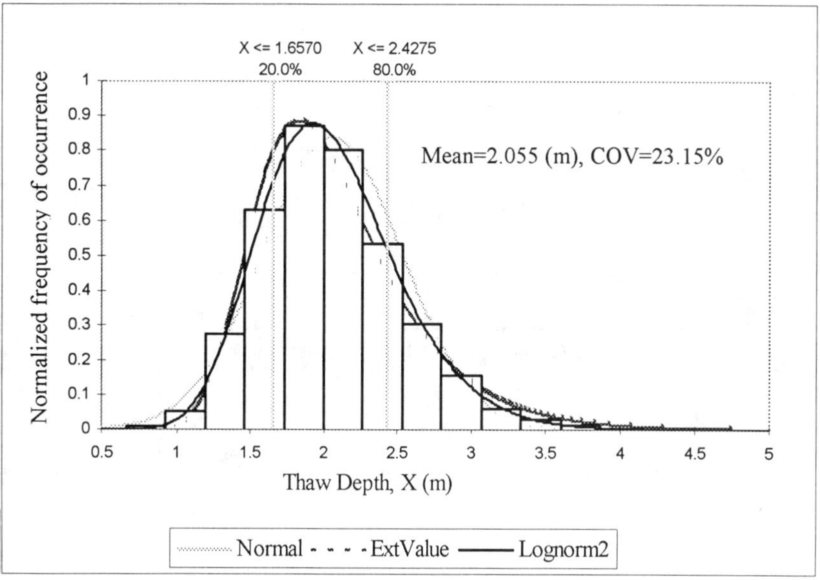

Figure 5. Probability density function for thaw penetration based on Stefan equation

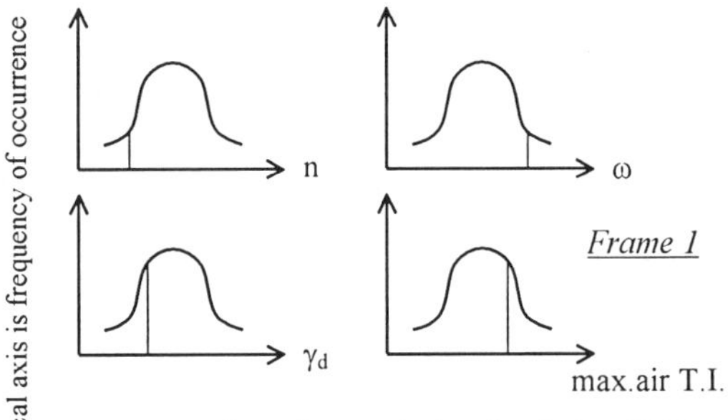

Frame 1

n, ω, γ_d, and maximum air thawing index (T.I.) are randomly selected 10,000 times

The 10,000 values of K_u and L are calculated from the following equation (Kersten, 1949)

(note: equations for γ_d in pcf) *Frame 2*

$$\left(K_u = \frac{(0.7\log\,\omega + 0.4)10^{0.01\,\gamma_d}}{12} \right) \qquad (L = 1.44\gamma_d\omega)$$

The 10,000 values of thaw penetration, X, are calculated from the Stefan equation (Lunardini, 1981):

$$\left(X = \sqrt{\frac{2 \cdot K_u \cdot n \cdot (\text{max. air T.I.})}{L}} \right)$$

Frame 3

Generated from 10,000 values

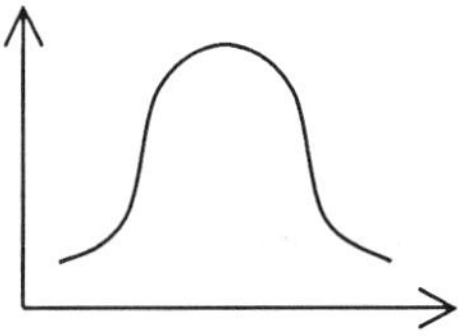

Frame 4

Thaw penetration

Figure 4. Flow chart of a probabilistic analysis for thaw depth based on Stefan equation

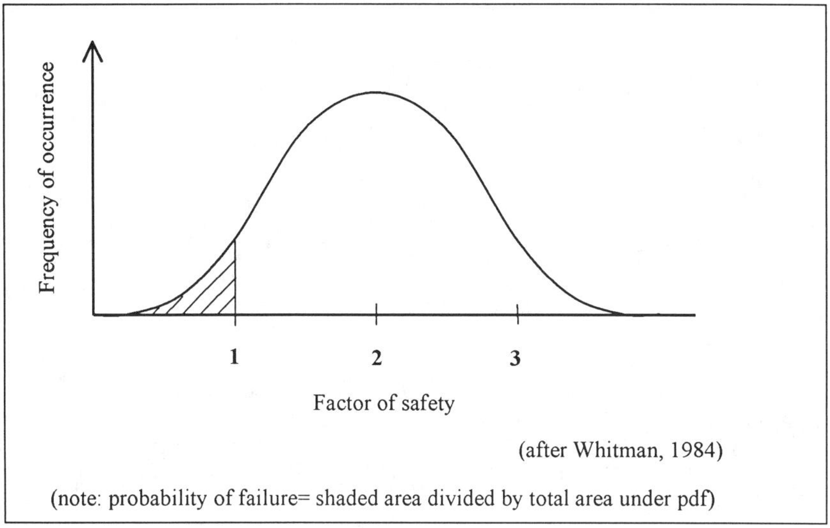

Figure 6. Graphical illustration of probability of failure with factor of safety

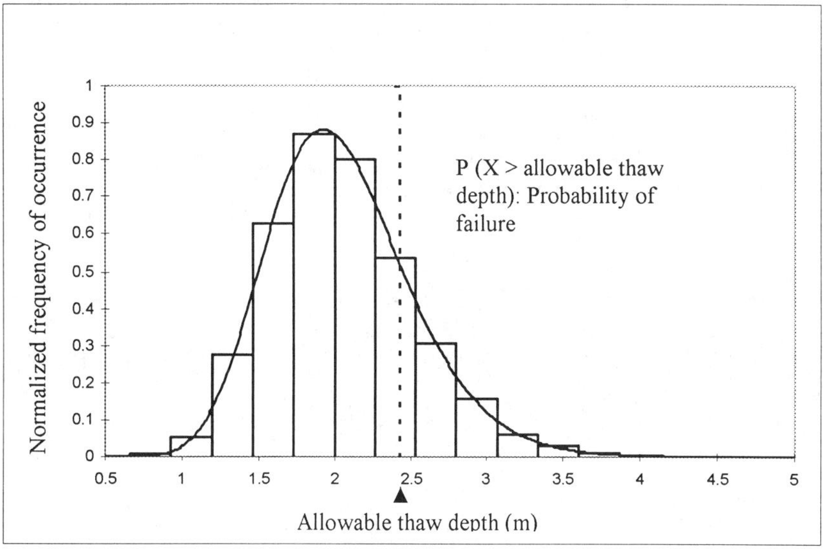

Figure 7. Graphical illustration of probability of failure based on Stefan equation

Ground Temperatures Across the Old and New Roads at Mile 130, Richardson Highway During 1954-62

Huijun Jin[1, 2], Max C. Brewer[3], and Robert A. Perkins[2]

ABSTRACT

Year-round studies of the geothermal impacts of road construction in a "warm" (-1 to -2°C) permafrost area were undertaken during 1954-1962 at six road sections across the Richardson and Glenn Highways, in the vicinity of Glennallen, Alaska. The most extensive studies were at Richardson Highway Mile 130 where nine 6.1-m vertical thermistor cables were installed beneath the newly constructed road, beneath a 40-year-old roadway, and in undisturbed areas. The cables, installed beneath the centerlines, shoulders, and the adjacent ditches of the highways, as well as beneath undisturbed areas, provided weekly temperature measurements at 13 depths. A tenth cable, with thermistors at 0.3-m spacing, was installed horizontally at a depth of 1.2-m across the new road.

The mean annual ground temperatures in the ice-rich lacustrine silty clays in the Glennallen area appeared to approximate –1.2°C, about 1.8°C warmer than the mean annual air temperatures. The seasonal amplitudes of temperature changes were greatest beneath the centerline of the new road; increasing the depth of seasonal thaw from 1.87 m, versus about 0.7 m in the undisturbed areas, in 1954 to 3.0 m in 1957. The old road centerline and ditch then had similar depths of thaw, about 4.9 m, after almost 40 years of light usage. This suggests that differential settlement in the new roadway would continue until at least an additional 1.8-m-depth of thaw had been completed. While the permafrost at the 6.1-m depth beneath the centerline of the new road was warming, the permafrost at the same depth beneath the old, abandoned road was cooling slightly during the study period.

INTRODUCTION

The highways in the Glennallen area, in southeastern interior Alaska, have experienced difficulty in construction and from costly maintenance for the past half century due to frost heave and differential thaw settlement (Jin, 2001). To assist the highway engineers, the Alaska Road Commission (predecessor to the Alaska Department of Highways) and the U. S. Geological Survey instrumented six road

[1]Cold and Arid Regions Environmental and Engineering Research Institute, CAS, 260 West Donggang Road, Lanzhou 730000, China.

[2]Department of Civil and Environmental Engineering, University of Alaska Fairbanks, PO Box 755900, Fairbanks, AK 99775 USA.

[3]Former Geophysicist with U. S. Geological Survey, 4200 University Drive, Anchorage, AK 99508 USA.

sections during 1954-62 in order to obtain the temperatures, and changes in temperatures, in the permafrost beneath and adjacent to the highway sections.

STUDY AREA

The Glennallen study area lies within a generally flat portion of the Copper River Basin at an elevation of approximately 460 m. The Copper River Basin is generally underlain by permafrost, from near the natural surface to depths of 30 to 40 m, with a maximum determined thickness of 183 m, and areas of predominantly fine-grained deposits (Ferrians, 1965). The mean annual ground temperatures (MAGTs) in the area range from –0.5 to –2.0°C at depths of 3-9 m (Greene *et al.*, 1960; Nichols and Yehle, 1961). The depth to the top of permafrost (permafrost table) in undisturbed areas ranges from 0.3 to 1.8 m depending on the type of vegetative cover, type of mineral soil encountered, surface drainage, frequency of summer precipitation, and the thickness and duration of snow cover. The vegetative cover includes black and white spruce, poplar, quaking aspen, alder and willow. The dominant species depend on whether the vegetation is growing in an area of moist sphagnum moss, 10- to 15-cm thick, or in a muskeg environment often with standing water.

The primary and most extensive weather records for the Glennallen area are those from the Gulkana Airport (62°09'N, 145°28'W, 658 m) at approximately Mile 119 on the Richardson Highway (about three miles north of the junction of the Glenn Highway Mile 189 and Richardson Highway). The average mean annual air temperature (MAAT) reported during 1954-62 was –3.1 °C (Table 1), -3.3°C if the unusually warm year of 1957 were factored out. This, when compared with the Gulkana MAAT of –3.4°C for the years 1950-75, suggests that, during 1954-62, variations in the air temperatures had little warming effect on the permafrost beneath the roads. However, a significant warming trend, began in 1976 with the MAATs averaging –1.95°C between 1976 and 1993, an average increase of 0.08°C/year. This increase occurred because of warmer winters, rather than warmer summers (Table 1).

Table 1. Average MAAT (°C), Gulkana Airport, Alaska (US Weather Bureau, 1950-99)

Study Period	1954-62	1950-75	1976-93	1950-93
Mean Avg. Ann. Max. Air Temp.	+3.3	+3.1	+4.0	+3.4
Mean Avg. Ann. Min. Air Temp.	-10.9	-10.8	-7.5	-9.4
Mean Avg.Ann. Air Temp.	-3.1	-3.4	-2.0	-2.8

The mean annual precipitation during 1950-89 was 27.8 cm of which an average of 7.2 cm water equivalent, or about 25%, fell as snow from November through March (U. S. Weather Bureau, 1950-99). The data are deficient in detailing summer precipitation when frequent drizzles, as differentiated from infrequent substantial rains, and the resulting evaporative cooling can have the same impact on ground temperatures as an unusually cool summer.

Six roadway sections, Mile 99, 128 and 130 on the Richardson Highway, Mile 181.7 and 187.5 on the Glenn Highway, and Mile 0.5 on the Tok Cutoff (now a continuation of the Glenn Highway), were selected for intensive study of the geothermal regimes of the underlying permafrost to depths of 6.1 m. The roadway sections, most about 6.1 m in width, were in areas of new road construction, reconstruction, relocation or where differential settlement from deterioration of the underlying permafrost, and/or frost heaving had caused very heavy maintenance problems. The northernmost roadway section at Richardson Highway Mile 130 was chosen for discussion in this report because it was the most extensively instrumented and also had a nearby section of old road for comparison.

STUDY SITES

The vegetation in the muskeg on the eastern side of Mile 130 Richardson highway was dominated by dwarfed, scraggly black spruce and grasses. The 5-m-wide ditch along that side of the highway had standing or running water during the warm seasons. On the western side, in addition to well-developed white and black spruce, quaking aspen, alder, willow and poplar were common. Moist to dry sphagnum moss, about 15 cm thick, was prevalent, providing excellent thermal insulation for the underlying soils. The drainage on the western side was much better because of the westward sloping topography. In July 1954, five thermistor cables, each 6.1 m in length, were installed vertically in the new road section beneath the centerline, west shoulder, east ditch, undisturbed ground to the east, and undisturbed ground to the west. A sixth 6.1-m vertical cable was installed beneath the centerline of the old road. The two parallel roadways were separated by about 25 m of vegetated, undisturbed ground. A seventh cable, 8.2 m in length, was installed horizontally at a depth of 1.2 m (at about the base of gravel fill—top of the original surface) beneath the new road.

Construction of the new road at Mile 130 began over undisturbed ground in 1951. The roadway was hand cleared, and the trees were laid parallel to the centerline below the road fill on the undisturbed vegetation mat of sphagnum moss and low brush. In 1952, about 0.9 m of relatively clean, coarse gravel and sand was spread over the section with minimum disturbance. However, by July 1954, when the six thermistor cables were installed, differential settlement had necessitated placing another 0.5 m of gravel over the road in order to provide a relatively smooth surface. Drilling for installation of the cables revealed a subsurface of icy, pebbly, clayey silt, overlain by a thin layer of sand. This section was paved in 1956 and has been patched and re-paved several times since. A good riding surface seldom lasted more than a few years. The permafrost beneath the road, originally at a depth of about 0.7 m, degraded from about 1.9 m when the thermistor cables were installed, to 3.1 m in November 1957, largely as a result of the abnormally warm summer in 1957.

The nearby old road was built during the pre-WWII period, beginning in 1916 as little more than a passable trail for vehicles and widened as needed. When originally built, the roadway was cleared and stripped to the mineral soil. Additional gravel was placed on the surface due to the increased WWII traffic. In 1954, when the centerline thermistor cable was installed, the road surface was about level with the surrounding terrain, and the ditches were only slight depressions. The permafrost soil moisture contents for these holes ranged from 10.9% to 71.1% by dry weight, averaging 31%; about 21% where the soils had thawed. In 1954, the permafrost table beneath the centerline was at about the 4.9-m depth and seasonal freezing of the ground extended to a depth of about 2.3 m. By 1959-60, the depth to the permafrost table appeared to be about 5.1 m.

MEASUREMENT AND DATA STRUCTURE

The thermistor cables did not penetrate to the same depth below the pre-existing undisturbed ground surface, nor did they penetrate to equal depths in the geological formation at this highway section. The cables, 1.8 cm in diameter, were black neoprene-jacketed, 20-conductor, 22-gauge-stranded copper, and had 13 thermistors inserted, each at a particular depth. The observational depths generally were 0.0, 0.15, 0.3, 0.6, 0.9, 1.2, 1.5, 2.3, 3.0, 3.8, 4.6, 5.3, and 6.1 m.

The sensitivity of the thermistors was about 160 ohms change per 1°C change in temperature at 0°C. Therefore, measurements, using a skilled operator and clean contacts in the switch box and the bridge resistor contacts and poles, could be made to 0.006°C. The thermistor cables generally were read on a weekly basis, although there were gaps in the data collections. Destruction of some cables occurred during the following road maintenance/reconstruction projects. Snow depths were recorded since they influenced the underlying temperatures.

RESULTS AND DISCUSSIONS

Vertical Temperature Profiles Beneath New and Old Road Centerlines

The end-of-winter and end-of-summer temperatures-depth profiles beneath the centerline (CL) of the new (NR) and old (OR) roads, and beneath undisturbed ground (UG) between those roads, are provided in Figure 1. Of immediate importance from Figure 1 is the fact that the permafrost temperatures deeper than 2 m beneath the centerline of the new road had decreased, three years after placement of the first fill in 1952, and were colder than those beneath the old road or the undisturbed area. This resulted primarily from three factors: 1) the clearing of snow from the road during the winter and the loss of the insulating value from that snow; 2) the loss of the insulating value of the original 10-15 cm thick layer of sphagnum moss, because of its compression to a thickness of only 1-2 cm, and 3) the lower moisture content of the added fill, allowing a more rapid penetration of the winter cold wave and hence a longer period of time that heat could escape from the subgrade, thus greater cooling of the permafrost. While those factors provide the basis for construction of all-season gravel roads and protection of the "cold" (-6 to -

10°C) permafrost (Brewer, 1974), they are not practical for employment in "warm" (-0.5 to –2°C) permafrost, such as Glennallen, where excessive thicknesses of fill would be required.

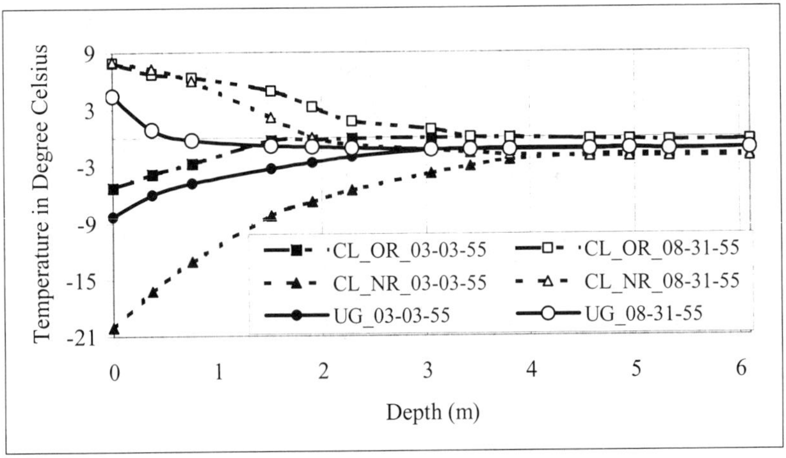

Figure 1. Early March and late August 1955 temperature-depth profiles beneath centerlines (CL) of the new (NR) and old (OR) roads, and an undisturbed ground (UG), Mile 130 Richardson Highway.

The amplitudes of the temperature changes were consistently greatest, warmer during the summer and colder during the winter, beneath the centerline of the new road (Figure 1). The warmer summer temperatures, especially those during an unusually warm summer such as 1957, gradually deepened the active layer thawed zone reaching a depth of 3.0 m by November 1957. Based on the depths of annual freeze-back beneath the old road, this appeared to approximate the average ultimate depth of the active layer beneath a 6.1-m wide roadway in the Glennallen area. In succeeding years, during a warmer winter when the freeze back did not reach that depth, some of the heat in the still-thawed zone was dissipated to the underlying permafrost causing a marginal amount of additional thawing, as well as some warming of the underlying permafrost. By 1960, that type of mechanism had begun to warm the 3-6 m temperature profile beneath the centerline of the new road moving the temperature profile towards that beneath the old road.

The drilling encountered permafrost at 4.9 m when the thermistor cable was installed beneath the centerline of the old road in July 1954. The temperature at that depth, upon return to equilibrium, was –0.20°C, suggesting that was the freezing temperature of the soil at that depth. The temperature of the permafrost was –0.27°C at 6.1 m. By May 1962, the temperature at 6.1 m beneath this abandoned road had gradually decreased to –0.64°C. With vegetation gradually invading the road, the temperature at 6.1 m appeared to be moving towards the temperature at a similar depth beneath the undisturbed vegetated area. That temperature would appear to be about –1.25°C under then existing climatic conditions (Figure 1).

Temperatures at the 1.2 m Depth Under the New Road

An additional 0.5-m thickness of gravel was placed and the new road graded just prior to the July 1954 installation of the horizontal thermistor cable at a depth 1.2 m beneath the road surface. The length of the cable was completely within the thawed fill, including shoulders, which was thicker on the western side of the road near the zero thermistor position because of the terrain slope. As the thickness of the fill decreased eastward from the centerline, the material at depth was less well drained and thus less thermally sensitive as reflected each year in the lower July temperatures (Figure 2). The temperatures beneath the western lane of the road were consistently several degrees warmer during July and correlated with the air temperatures.

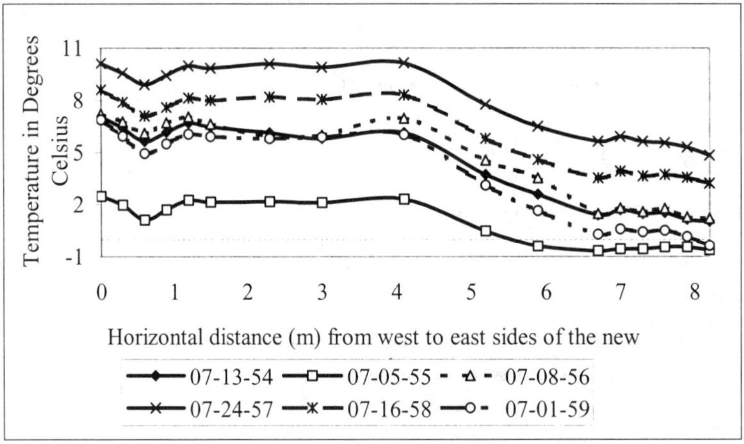

Figure 2. July temperatures at the 1.2-m depth under the new road

The more contrasting January mid-winter temperatures, reaching minimums of about –11°C beneath the centerline and adjacent traffic lanes in 1955, 1956 and 1959 (Figure 3), showed the result of keeping the road cleared of snow thus removing its insulating effects. The winter temperatures beneath the western shoulder were consistently colder than beneath the eastern shoulder because of the increased thickness and lower moisture content, thus greater thermal sensitivity of the fill. The temperatures beneath both shoulders also showed the insulating effects of the snow that had drifted against the edges of the road or had been bladed there.

The road was paved with asphalt in August 1956, two years after the temperature measurements began, and there was little evidence that the paving *per se* had an appreciable thermal effect at a depth of 1.2 m.

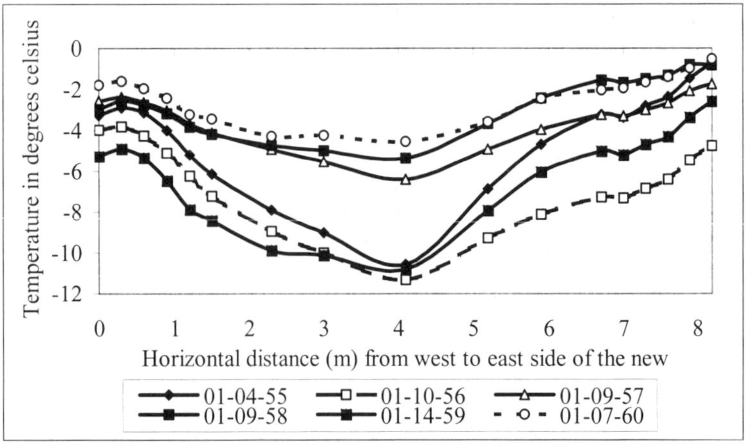

Figure 3. January temperatures at the 1.2-m depth under the new road

Ground Temperatures at a Depth of 3.0 m

The permafrost temperatures, beneath the undisturbed area between the new and the old roads, seasonally cycled between –1 and -2°C, without establishing any trend, during the 1954-62 study period (Figure 4). There was a lag time of about 3.5 to 4.2 months before the reversal of the temperature trends reached that depth, i.e., the cooling resulting from freeze-up began about the beginning of the following February. The time-temperatures also evidenced a very pronounced asymmetry in the cooling curves, since the cooling of the permafrost could not begin until the active layer, about 0.7 m in thickness, had completely frozen, allowing the escape of heat from the underlying permafrost to the ground surface (zero curtain effect).
Figure 4 indicates that the seasonal thaw penetrated to the depth of 3.0 m beneath the centerline of the new road for the first time in November 1957. That was 5.5 years after the first 0.9 m of gravel and sand fill had been placed on the timber-cut, but otherwise undisturbed surface. The 3-m cable depth was about 2.3 m into the original mineral soil, since the original vegetation mat would have been compressed to about 2-3 cm in thickness. The compressed organics lose their insulating value and, usually being saturated, become good heat conductors, especially when frozen. Thus, it is questionable whether the extra expense was warranted in attempting to protect the organic mat within the narrow confines of the roadway in an area of warm (-1 or –2 °C) permafrost. However, such protection may be warranted at the edges of the road since the organic mat offers some protection from erosion.

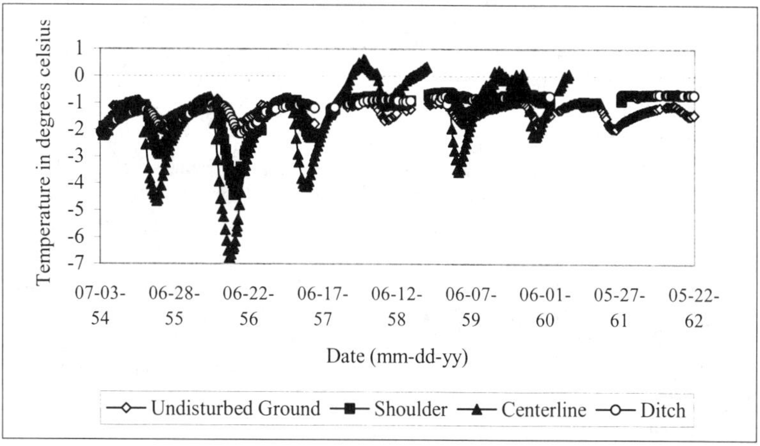

Figure 4. Time-temperatures at the depth of 3.0 m beneath the new road

Ground Temperatures at a Depth of 6.1 m

The initial temperatures from all six vertical cables installed in 1954 showed the effects of the heating generated by the drilling and backfilling, ranging from about 0.05°C beneath the centerline of the old road, to 0.13°C at the centerline of the new road, to as much as 0.30°C at the undisturbed site where a 0.46-m diameter hole was drilled (Figure 5). The time for recovery to the original equilibrium temperatures, after drilling and installation of cables, ranged from six weeks at the centerlines to almost three months at the undisturbed site.

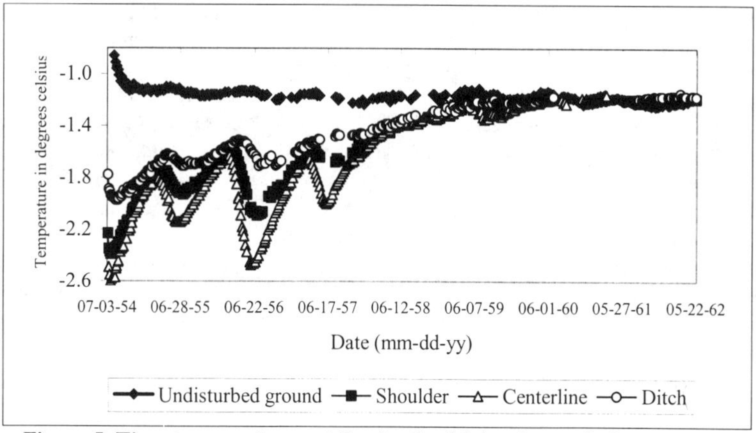

Figure 5. Time-temperatures at the depth of 6.1 m beneath the new road

The ground temperatures at the 6.1-m depths beneath the new road centerline, shoulder and ditch showed marked seasonal changes between 1954 and mid-March 1958, with the greatest amplitude, about 0.8°C, occurring beneath the centerline. Distinct warming trends were noted during the same period. After 1958, the seasonal changes and the warming trends were muted at all three sites. The temperatures at the undisturbed site between the roads showed a gradual cooling of about 0.15°C (to –1.25°C), with only minor seasonal variations, during the 7.5 years after reaching the original equilibrium temperature. During the same period, the ditch temperatures adjacent to the east side of the road displayed a warming of about 0.78°C (to –1.18°C), the shoulder a warming of about 1.24°C (to –1.14°C) and the centerline a warming of about 1.41°C (to –1.16°C). The temperature, not shown in Figure 5, beneath the upslope undisturbed site also exhibited a warming, of about 0.30°C (to –1.30°C). While not proven, there is a suggestion that minor changes in surface drainage might have been responsible for a warming beneath the undisturbed area on the upslope side of the road and a cooling beneath the undisturbed area on the downslope side.

The time lag for the effects of fall freezing at the surface to reach the 6.1-m depth at the undisturbed upslope site was about 7.5 to 8.5 months. In contrast, it took only 4.2 to 6.2 months for the effects of the initiation of thaw at the surface to reach the same depth. The difference in lag times in warming and cooling was caused by the delay in the cold wave penetrating to the same depth in the fall due to the zero curtain effect.

Ground Temperatures at Three Old Road Sites

There were four thermistor installations for the old road section: undisturbed ground west, centerline, ditch and edge of stripping-berm. The centerline cable was installed in 1954, the other three cables in 1956. The stripping berm cable was installed for study of the active layer and natural re-vegetation, and the results are not immediately applicable for the present study. The zero curtain temperatures (the steady temperatures reached before the effect of fall freezing in the active layer, or cooling in the permafrost) began at the depth of 1.5 m beneath the old road centerline and the ditches, ranged between –0.3°C and about –0.7°C (Figure 6). That suggests that the sediments contained minor amounts of salts of unknown chemical compositions, since the actual freezing temperatures of most soils is in the range of –0.2°C and the soils at these sites were thawed well below the 1.5-m depth. The coldest temperature (-4.5°C) recorded at the 1.5-m depth occurred beneath the undisturbed area. That resulted because, once freezing had penetrated to the permafrost table, thermal conduction allowed for the rapid escape of heat to the surface. The asymmetry in the temperature curves for the undisturbed site are very pronounced with the 1.5-m warmest temperatures reaching –0.7°C, the assumed zero curtain temperature, year after year.

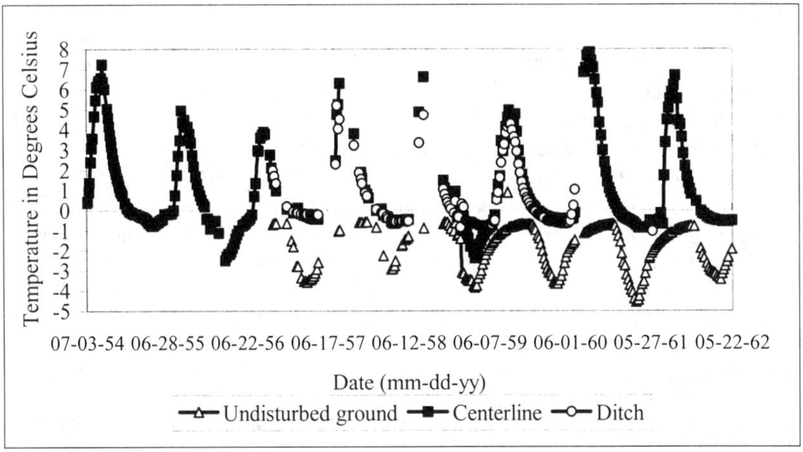

Figure 6. Time-temperatures at the depth of 1.5 m beneath the old road

The winter minimum temperatures at the 1.5-m depth beneath the centerline of the old road ranged from 5 to 8°C warmer than at the same depth beneath the new road, largely because the former had a snow cover, whereas the latter generally was kept snow-free. More significantly, however, the summer maximum centerline temperatures at that depth beneath the old road also tended to be about 4°C warmer than beneath the new road during the period of study. This condition prevailed both during the period when the new road surface was gravel and after it was paved. This was unexpected as the new roadbed was elevated and had better drainage, and thus the fill should have been more thermally sensitive. While not documented, it is theorized that the effects of evaporative cooling at the surface, coupled with air turbulence in this more-open roadway area, may have been responsible for the cooler temperatures at this depth.

The temperatures at 6.1 m beneath the centerline were –0.24 to –0.26 °C after temperature equilibrium had been re-established in the fall of 1954 (Figure 6). At that time, the permafrost table was at a depth of about 4.9 m, similar to that found beneath the ditch when the cable was installed two years later. Thus, the almost four decades of light usage of this road had resulted in about 4 m of thaw of the permafrost. There also was evidence of a gradual cooling at the 6.1-m depth beneath the centerline. That cooling amounted to about 0.35°C (from –0.3°C before 1956 to about –0.55°C, then dropping 0.09°C suddenly in November 1960 before stabilizing at about –0.65°C) during the following years. At the same time there was a gradual warming at a depth of 5 m. This suggests that while the upper depths were receiving heat from the surface, at 6.1 m there was a small amount of heat being lost to the deeper depths as the vertical temperature profile adjusted to abandonment of the usage of the road.

The Mean Annual Ground Temperatures (MAGTs) at the 6.1-m Depth

The MAGTs at the 6.1-m depths during 1954-1962 are summarized in Table 2. The results show that the old road ditch had the highest MAGT (-0.37°C). The centerline of the old road had the second highest MAGT (-0.45 °C). The coldest MAGTs were found beneath the shoulder (-1.62°C) and centerline (-1.61°C) of the new road. There were differences in the MAGTs beneath the undisturbed sites, which varied from –1.17°C at the undisturbed ground between the two roads, -1.23°C west of the old road, and –1.40°C east of the new road. The MAGTs west of the old road were stable; the MAGT west of the new road appeared to be cooling while MAGTs east of the new road were definitely warming. The linear distance separating the three undisturbed sites was about 72 m. The MAGTs beneath the new road centerline, shoulder and ditch were all warming. The MAGTs beneath the old road centerline were definitely cooling, while those beneath the old road ditch remained unchanged.

Table 2. MAGTs at a depth of 6.1 m at Mile 130, Richardson Highway, 1954-62

MAGT (°C)	UG_OR	CL_OR	DI_OR	UG_NR	SH_NR	CL_NR	DI_NR	UG_NR
1954		-0.27		-1.08	-2.16	-2.12	-1.90	-1.58
1955		-0.29		-1.14	-1.82	-1.90	-1.68	-1.52
1956	-1.24	-0.31		-1.14	-1.83	-2.03	-1.60	-1.47
1957	-1.24	-0.36	-0.36	-1.18	-1.62	-1.76	-1.52	-1.47
1958	-1.18	-0.45	-0.37	-1.18	-1.40	-1.40	-1.36	-1.37
1959	-1.21	-0.50	-0.37	-1.17	-1.27	-1.29	-1.24	-1.32
1960	-1.23	-0.55	-0.38	-1.19	-1.22	-1.18	-1.21	-1.30
1961	-1.26	-0.64		-1.21		-1.16	-1.19	-1.31
1962		-0.65		-1.22			-1.17	-1.30
Mean	-1.23	-0.45	-0.37	-1.17	-1.62	-1.61	-1.43	-1.40

CONCLUSIONS

- Thawing of the permafrost beneath a 6.1-m wide roadbed in the Glennallen area proceeded in a non-linear manner, dependent on summer temperatures, length of summer and the soil moisture contents, until an active layer thickness of about 3.0 m was achieved. This required about 5.5 years after the first fill was placed for the new road at Richardson, Mile 130.

- Unusually warm summers, late freeze-ups, or both allowed deeper thaw and the development of a "talik" to a depth approaching 5 m after 40 years (Richardson, Mile 130 old road), in area having relatively good surface drainage. The temperature of the upper permafrost below the "talik" gradually cooled (Table 2) after the road was abandoned.

- While thawing proceeded more rapidly beneath the centerline of the new road, the ultimate depths of thaw beneath the centerline, shoulder and ditch were similar.

- No significant, or even measurable, differences in summer temperatures were detected beneath gravel and paved surfaces.

- Hand-clearing and the protection of the surface vegetation prior to placement of fill did not appear to affect permafrost temperatures, to be cost-effective, nor useful except as it might slow erosion along the shoulders or ditches.

ACKNOWLEDGEMENTS

Analyses were supported by an EPSCoR Fellowship, University of Alaska Fairbanks. The authors wish to acknowledge the numerous USGS and ARC employees who contributed to the data acquisitions; particularly to Dr. Donald R. Nichols (USGS), for his dedication to data acquisition, to the late John R. Watson (ARC), whose engineering vision made the data acquisition possible, and to Professor John P. Zarling for his insightful advice.

REFERENCES

Brewer, M. C. (1974). "Permafrost, its impact on development, in Man's Impact on Arctic and Subarctic environments." *Arctic Institute of North America*, 19-27.

Ferrians, O. J. (1965). *Permafrost Map of Alaska (1/2,500,000).* US Geological Survey.

Greene, G. W., Lachenbruch, A. H., and Brewer, M. C. (1960). Some thermal effects of a roadway on permafrost, in *Short papers in the geological sciences: U. S. Geological Survey Professional Paper* 400-B, 141-144.

Jin, H. (2001). "Ground temperatures of permafrost along the highways in the Glennallen area, Alaska and their engineering applications". *A Master of Civil Engineering Degree Project*, Department of Civil and Environmental Engineering, University of Alaska, Fairbanks, 10 August 2001, 1-96.

Nichols, D. R., and Yehle, L. A. (1961). "Highway construction and maintenance problems in permafrost regions," in *Symposium on Geology as Applied to Highway Engineering, 12th Ann., 1961: Tennessee Univ. Eng. Expt. Sta. Bull.* 24, 19-29.

U. S. Weather Bureau (1950-99). *Climatological Data, Alaska.*

Shore Erosion in Russian Arctic

Yuri Shur[1], Alexander Vasiliev[2], Michael Kanevsky[2], Vitaly Maximov[2], Sergei Pokrovsky[3], and Viacheslav Zaikanov[4]

Abstract

The authors have been studying shore erosion in Russian Arctic since 1973. Numerous monitoring sites have been established at sea, river, and lake shores in Russian Arctic. Different methods of evaluating the rate of erosion have been tested. Two of these methods, monitoring along stationary transects and comparison of aerial photographs of reasonable scales taken at different times, were found to be most productive. Data of shore erosion monitoring from a few to more than 30 years are presented.

Study Sites

Since 1973 we have established more than 70 monitoring sites in West Siberia and North Yakutia to measure erosion at the Laptev Sea and the Kara Sea shores, shores of thaw lakes, and shores of Yana and Indigirka rivers in North Yakutia (Shur and others, 1978, 1984a, Pavlov and others, 1995, Vasiliev and others, 2001). Almost half of the sites have not been revisited yet. Some sites have been monitored from a few to dozens of years. One site, located at the Marre-Sale Arctic Meteorological Station on the west shore of the Yamal Peninsula, has been monitored regularly since 1978.

Methods

The rate of erosion can be evaluated if the shore retreat for a specific time is known. The shore retreat is measured as the distance between some stable features on shore away from the shoreline and points on the shoreline at different moments spaced in time. To define points on the shoreline is not always easy because the shoreline can be a wide area between the top and the foot of a cliff. In evaluations of the rate of erosion, the top of a cliff is typically considered as a shoreline. All data presented here are referred to the top of a cliff.

The following methods were used to evaluate the rate of erosion:

1. **Comparison of measurements of the distance from a stable ground feature on shore (a lighthouse, an electrical pole, a building, etc) to the edge of a cliff with measurements previously made by others**. We found it difficult to use the method when we tried to repeat measurements made by others because the shoreline is uneven and the direction of previous measurement is usually unknown.

---

[1] Associate Professor Depart. of Civil and Environmental Engineering Univ. of Alaska Fairbanks
[2] Earth Cryosphere Institute. Siberian Branch of Russian Academy of Sciences, Tumen
[3] Graduate Student Depart. of Geology and Geophysics Univ. of Alaska Fairbanks
[4] Institute of Geoecology of Russian Academy of Sciences, Moscow

2. **Comparison of maps and survey plans created at different time.** We found it practically impossible to compare a shoreline presented on old maps (prior to 1970) that were developed from a survey with a shoreline on new maps which are based on aerial photographs. Maps of scale 1:100,000 can be used to evaluate extremely great rates of erosion for a long period of time. Maps of smaller scale are essentially useless. Similar difficulties were experienced by E. Leffingwell – the father of American Permafrost science in the early 1900s. "The maps of the early explorers are on such a small scale that it necessary to depend to a great extent upon the narratives in discussing changes of the coast in historic time" (Lefingwell, 1919). He mapped Flaxman Island with the expectation that following generations can use his map for evaluation of the rate of shore erosion. The value of the map for such purpose is yet to be checked.
3. **Study of sequential aerial photographs.** To evaluate the rate of shore erosion, aerial photographs at scales from 1:10,000 up to 1:25,000 spaced at time intervals of 20 and more years have been compared. Measurements were made with a stereo comparator. The obvious advantage of the method is its independence from fieldwork and ability to evaluate erosion along any part of a shoreline. The precision of the method depends on the scale of aerial photographs and so a small rate of erosion can be overestimated as well as underestimated. The history of erosion between surveys is not known and the rate of erosion is averaged throughout the period between aerial photography. As a result, the rate of current erosion can be underestimated. Data on shore erosion published during the last 5 to 10 years are mainly obtained by this method.
4. **Sequential surveys.** We surveyed the shoreline in an area that was surveyed 30 years before by others for land use purpose. We found that even a survey at a large scale does not provide a precise description of the cliff position at the time of survey if the shoreline is not the main subject of survey. Evaluation of erosion rate could be greatly improved if stable ground features were included in the previous survey so that measurement can be made along straight lines connecting such features and crossing the shoreline at an angle more than 45°. We also repeatedly surveyed the shoreline of two thaw lakes over several years and found it time consuming.
5. **Measurements along permanently established profiles**. This has been the main method used in our studies to date. The method is precise and easy to use. Initial installation of profiles can be time consuming but monitoring is easy and does not require survey equipment. Data obtained with this method are specific to the study area but can be extrapolated to other sites with a terrain analysis.
6. **Interviews of local people**. Local experience is very valuable and can be a good source of information. It can also be a source of confusion because people often better remember extraordinary events, for example great storms. The greatest rate of shore erosion in Russian Arctic, 150-200 m/year over a 10-year period, was included in the scientific literature (Kal'anov, 1934)

based on an interview with a manager of the local fishery and can not be considered as reliable. The most productive approach is to involve local people in studies. The famous Russian Arctic explorer A. Middendorf obtained the first long-term monitoring data on permafrost temperature in the middle of 19th century in Yakutsk with the assistance of a local salesman. We received great help from a local bus driver, Viktor Orekhov, who has measured the shore retreat at our experimental site close to the Chokurdakh Village in the Low Indigirka River during several years when we did not have an opportunity to visit this remote site.

7. **Studies of resistance of frozen and thawed soil to erosion.** Although such study does not characterize the rate of erosion which also depends on other factors, it helps to understand the physics of erosion.

Results

Frozen soil resistance to erosion. To compare resistance of frozen and thawed soils to erosion, we used techniques proposed by G. Bastrakov (1977). A portable water cannon was used. Resistance to erosion was computed from Bastrakov's equation based on the size of hole created by a spurt, time of erosion, water pressure, distance from the cannon nozzle to eroding soil, and area of the cannon nozzle. Each experiment was repeated from 5 to 10 times. Water temperature during field experiments was about 3°C. Field experiments were done at a cliff at the experimental site Marre-Sale on the Kara Sea.

Soil data and resistance to erosion are shown in Table 1. We consider data in the table as relative resistance of soils to erosion. As can be seen from Table 1, resistance of frozen soil to erosion is significantly greater than resistance of thawed soil. Very large ratios of R_{fr}/R_{th} show that permafrost thawing is the limiting factor.

Table 1
Resistance to erosion of frozen and thawed soils

Soil	Unit Weight kg/m^3	Plasticity Index	Water content %	Resistance To Erosion Of Frozen (Rfr) And Thawed (Rth) Soils (kN)		R_{fr}/R_{th}
				R_{fr}	R_{th}	
Sand, fine	1600	-	29	$1.3*10^5$	$5*10^3$	$5*10^3$
Silt	1830	6	33	$4*10^5$	$3.3*10^4$	$3.3*10^4$
Silty clay	1900	10	30	$4.5*10^5$	$6*10^4$	$6*10^4$
Silty clay		11		$>10^6$		

Sand is more susceptible to erosion than silt and clay in an unfrozen state as well as in a frozen one. Resistance of frozen soil to erosion depends on water temperature; roughly it is inversely proportional to water temperature. Resistance to erosion in Table1 is correlated with soil composition. The other very important factor is the cryogenic structure of frozen soil that defines patterns of ice distribution in the soil. The cryogenic structure is an important characteristic of frozen soil and defines the properties of frozen soil as well as the properties of thawing soil. For example, vertical elements of the net of ice inclusions in the reticulate cryogenic structures define the lines of zero cohesion in thawing soil.

Our numerous observations of the impact of cryogenic structures on resistance of frozen soil to erosion were mainly quantitative. Depending on cryogenic structure, we distinguished eight groups of frozen soil resistance to erosion (Table 2). Cryogenic structures in the Table 2 were described by Shur and Jorgenson (1998). Resistance to erosion increases with an increase of the group number.

Table 2
Cryogenic structures of frozen soil and its resistance to erosion

Soil	Soil Cryogenic Structure	Resistance To Erosion
Sand	Massive (pore ice)	II
Silty clay (silt with similar cryogenic structure has resistance to erosion one number lower and clay one number greater than silty clay)	Massive (pore ice)	VI
	Layered	V
	Reticulate	III
	Incomplete reticulate	IV
	Ataxitic (soil inclusions in ice matrix)	II
Ice		I (if water temperature well above the melting point of ice in soil); VIII (if water temperature at or below the melting point)
Peat	Massive (pore ice)	VII

Frozen soil is strong in a mechanical sense but susceptible to thermal weathering. It loses strength with a phase change from ice to water. Erosion of frozen soil is always started with thawing of soil and removal of thawed soil. One of the well-known forms of shore erosion in permafrost region is undercutting of big blocks of frozen soil and their collapse usually along ice-wedges. Some blocks are 10 m wide and tens of meters tall. Even this type of erosion is possible because of thawing of soil at the contacts with waves, running, or still water and removal of thawing soil. Removal of thawing soil greatly increases the rate of thawing and as a result the rate of erosion. The limiting situation is the permanent contact of frozen soil with atmosphere and water. This situation becomes possible only in a case of ablation in which thawed material is instantly and constantly removed from the surface of the contact. In such case, erosion is defined by the rate of thawing with ablation, which is enormous and rarely occurs (Shur, 1977).

River Erosion

We have studied erosion of shores of the Indigirka and Yana Rivers in North Yakutia since 1974 when we reestablished the monitoring site near Chokurdakh Village of in the Low Indigirka River 200 km from the East-Siberian Sea. This shore segment had originally been studied by A. Tolstov (1962) in the 1950s. The site is located on a high floodplain between two small tributaries of the Indigirka River. Soil on a site is syngenetecly frozen ice-rich silt with wide ice wedges. The monitoring site is more than 4 km long and includes 17 transects. According to Tolstov (1962), the average rate of shore retreat was 7 m/year in the 1950s. The site had been visited several times between 1974 and 1990. In some years when we were absent, Mr. Victor Orekhov made measurements of the shore retreat. In total, the site has been monitored for more than 30 years. The average rate of retreat during this period has been about 6.5 m/year, with a range of 3 to 10 m/year. The greatest rate of retreat takes place at the mouth of a tributary, where up to almost 30 m/year has been measured in some years. The lowest rate of retreat of 2 m/year occurs in the central part of the site. The data show that a monitoring site has to cover a significant part of the shore to be representative. Data from measurements along any one transect can be misleading if they are extrapolated over a large segment of the shore.

Three smaller monitoring sites were established in the Indigirka River Delta close to the Village Poliarnoe, about 20 km from the West Siberian Sea. Each site consists of several transects. Two sites are located on the high floodplain and the third on the first terrace of the Indigirka River next to the village Polarnoe. The village is the oldest Russian settlement in North Yakutia. It has been severely affected by erosion and relocated many times during its 250 year history. Monitoring data are presented in Table 3. The average rate of erosion at these sites ranges from 1.6 to 3.2 m at the high floodplain and is about 5 m/year at the first terrace in the village vicinity.

We have also studied erosion of shores in the Low Yana River. One of the sites is located 190 km from the Laptev Sea close to the village Kazachie on the third terrace of the Yana River. This site is well-known in Russian permafrost literature because of an exposure of massive ice of the Pleistocene age. This exposure had been intensively studied by the Russian Permafrost Institute in the early 1950s when the river eroded the high vertical cliff. We visited the site for the first time in 1975 and installed four transects to measure the retreat of the top of the cliff (Shur and others, 1984a). The height of the cliff was about 100 m then. The height of its vertical part with an exposure of massive ice was an average 15 m. We estimated then that since the early 1950s, the top of the cliff has retreated with an average erosion rate of 6 to 6.5 m/year (Shur and others, 1984b). The site was revisited in 1980 and practically every year until 1990. Measurements along transects showed that the average rate of retreat was about 6.5 m/year, which correlates well with the preliminary conclusions of 1975. The rate of retreat varied from 3 to 9 m/year, depending directly on the height of the ice exposure and the climate conditions of a specific summer. Retreat of the top of the cliff at this site has continued for almost 40 years after the river stopped eroding the foot of the cliff.

More than 10 monitoring sites were established along the Yana River at the intensively eroded floodplain and the first terrace. Some of the sites have not been revisited yet. The others have been monitored during several years. The monitoring data are presented in Table 3. The rate of erosion at the sites varies from 0.3 to 3.5 m/year with average for all sites of about 2 m/year, which is 3 times smaller than the rate of retreat of the top of the cliff at the Mus-Khaya site. There, the rate of erosion is not influenced by the river erosion but depends only on interaction of the extremely ice-rich soil and massive ice with the atmosphere. This type of erosion is known in Russian Permafrost literature as thermal denudation (Are, 1980).

Table 3
Rate of shore erosion on Yana and Indigirka Rivers (North Yakutia)

Monitoring Site	Terrain Unit	Observation Period, years	Average Rate Of Erosion, m/year
Yana River 190 km from the sea	Middle flood plain	5	0.3
	Lower level of high flood plain	5	1.8
	Upper level of high flood plain	5	2.7
	III terrace above flood plain	36	6.5
Yana River 136 km from the sea	Lower level of high flood plain	5	2.6
	Upper level of high flood plain	5	2.1
	I terrace above flood plain	5	3.5
Yana River Delta, 36 km from the sea	Middle flood plain	5	0.7
	Low flood plain	3	0.6
Yana River Delta, 7 km from the sea	Middle flood plain	8	3
Indigirka River Delta, 20 km from the sea	High flood plain	3	1.5
Indigirka River Delta, 20 km from the sea	I terrace	6	5
	High flood plain	6	3.2
Indigirka River, 200 km from the sea	High flood plain	32	6.5

Thermal Denudation

We have studied thermal denudation of steep slopes at numerous gold pits in the Kular Gold Bearing District in North Yakutia over a 10 year period. The overburden soil there is syngenetic ice-rich permafrost with ice-wedges penetrating the entire overburden. Volumetric ice content of soil including ice-wedges varies from 70 to 90%. Volumetric ice content of soil without ice-wedges is usually more than 50%. The genesis and the age of the overburden are similar to Fairbanks silt, but ice content of the soil is greater than in Fairbanks silt. Slopes of pits with a height ranging from 10 to more than 20 m were almost vertical (70° and steeper) in the beginning. The toe of slope remained stable while the upper part has retreated due to slope processes. The retreat process is similar to the retreat of the cliff top if the rate of its retreat is much greater than the retreat of the cliff toe eroded by the sea or by the river. The rate of the retreat was measured by different techniques including regularly repeated measurements along more than 200 permanent transects, by survey, and by photo survey.

According to long-term monitoring, the rate of the retreat of the top of slopes varied from 6-8 to 30-35 m/year. The slope orientation is one of the factors defining the difference, but is not the most important one. The south facing slopes retreat 1.2 to 1.5 times faster than the north facing slopes. The other important factors are height of a slope, soil type, ice content and size of massive ice, surface water, and human impact on the surface behind the edge of the slope. The last factor has the greatest impact on the rate, the distance of the retreat, and the final angle of the stabilized slope. At sites where the soil surface has been affected by heavy equipment, the rate of erosion greatly increases (see Figure 1).

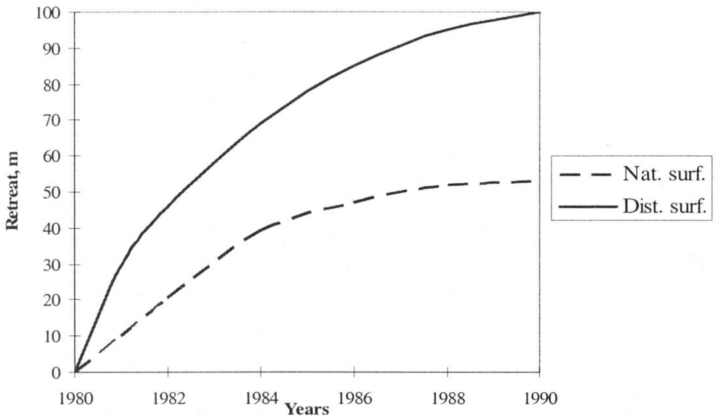

Figure 1. Retreat of edge of open pit with natural and disturbed surface

To study the reason for such an increase in control conditions, we purposely removed vegetation in a wide area behind the edge of the slope at one of our experimental sites. The great increase of the active layer in the area with removed vegetation leads to thawing of the massive ice. Cohesion of the saturated unconsolidated thawing soil with melting ice is close to zero. Chunks of thawing soil slide into a pit even at a small slope angle, and the top of the massive ice becomes exposed to the atmosphere. Melting of ice-rich soil and ice occur from two surfaces (from the top and from the side) and the rate of the retreat of the edge increases significantly. If the natural surface is damaged in the areas of surface water drainage, or if the slope retreat reaches thaw lakes, the rate of erosion would greatly increase (Figure 1). At such sites, it reaches 35 m/year and more. Similar increase of thermal denudation takes place on sea and river shores where soil surface behind a cliff is damaged by vehicles. One of the measures that can reduce the rate of shore erosion in areas with high cliffs and ice-rich permafrost is a protective zone along a shore to preserve the natural surface behind a cliff.

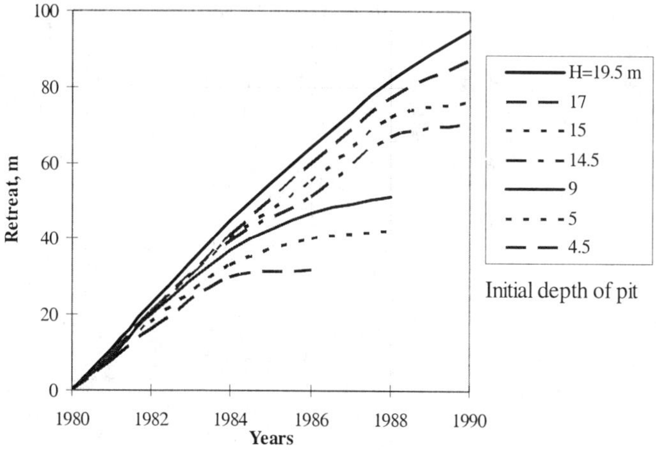

Figure 2. Retreat of the edge of open pit

The retreating slope in the areas with massive ice has a sloping lower part where massive ice and frozen soil are covered with thawed soil and a vertical cliff of ice or ice rich soil. This cliff is also called the thermal cliff. The rate of retreat depends on the height of the thermal cliff and decreases considerably with decrease of the thermal cliff height below 4.5 m (Figure 2). The angle of stabilized slopes with similar soil conditions varies from 5° to 60° depending on the initial height of the slope and the surface conditions behind the edge of the slope.

Thaw Lakes And Reservoirs

Shore erosion of two small reservoirs in an area of ice-rich permafrost with massive ice was studied in North Yakutia. Reservoirs with diameter of a few hundreds of meters were in use during one summer for gold mining operations. In some parts of the shore, vertically exposed massive ice was in direct contact with water. In other areas, a thick layer of previously thawed soil covered the shore. Waves are not developed in such small reservoirs and erosion is a result of the thermal interaction of still water and thawing ice-rich permafrost and massive ice. The rate of erosion of the reservoir shores varied from zero to a few centimeters per summer in areas without exposures of massive ice. At parts of the shore with exposed massive ice, the rate of erosion reaches 25 m and more in one summer. Erosion occurred with formation of deep (several meters) niches in massive ice and ice-rich permafrost at and below the water surface, resulting in large blocks of ice and frozen soil breaking off. One of the blocks was 8 m wide. Break off always occurs along ice wedges. Erosion of thaw lake shores in areas without massive ice in West Siberia was found to be on the order of 0.1 m/year or less. In areas with thick horizontal massive ice, the rate of retreat of the top of the cliff depends on the rate of thermal denudation and reaches 10 m/year.

Sea Erosion

In 1978, we began monitoring erosion of the Kara Sea shore in the vicinities of the Arctic meteorological station Marre-Sale, on the west shore of the Yamal Peninsula. The station is one of the oldest in Russian Arctic and has provided continuous data for almost 100 years. The station, which has numerous facilities, was originally built at the beginning of the 20th century and has been severely damaged by sea erosion and rebuilt from scratch at least twice. The geologic, geomorphic, and permafrost conditions of the area were studied by the authors and others. The experimental site is located on the second and third marine terraces at elevations ranging from 10 to 30 m in the continuous permafrost zone. Soil types include sand, silt, silty clay, and clay of a questionable genesis. The content of visible ice in the soil is generally less than 30%. It occasionally reaches 60% and more in the upper layers of permafrost. Sands have massive cryogenic structure (invisible pore ice), while the cryogenic structure of silt, silty clay and clay varies from massive to reticulate and layered. Ice-wedges occur irregularly. Their thickness varies from tens of centimeters to 2 meters and they reach depth of 3 to 5 m. Several inclusions of horizontal massive ice with thickness from 0.5 to 3 m occur in sand and clay mainly at elevations from 2.5 to 4 m.

The 4500 m long monitoring grid includes 60 profiles spaced from 20 to 100 m. Evaluation of the erosion rate is based on annual measurements of the retreat of the top of the cliff. The average rate of erosion for the period 1949 - 1978 was found by comparing our 1978 survey and the previous one of 1949. The average rate of erosion for that period was estimated as 1.4 meter per year (Shur and others, 1984b) for the part of the site next to the meteorological station. The retreat of the top of the cliff has sufficient spatial and temporal variations. The greatest retreat has reached 65 m for 22 years, the smallest only 11 m. As was mentioned above, even greater difference in the rate of erosion was found at the experimental site at the Chokurdakh Village on the Indigirka River.

The average rate of retreat during 1978 to 2000 has been 1.7 m/year for the entire site. The average rate of retreat for the site in each year is shown in Figure 3. In the warm summer of 1990 the average rate of shore retreat for the site was 3.3 m/year, which is about two times greater than average for the period of measurements. The curve in Figure 3 can be interpreted as half of the periodical. Final conclusions about the periodicity of erosion can be made in the future after data for the entire cycle are accumulated.

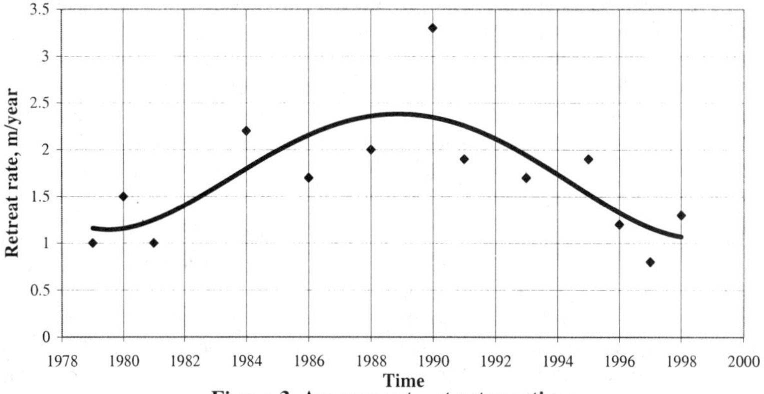

Figure 3. Average retreat rate vs. time Marre Sale (Kara Sea)

Variations of the rate of shore erosion were compared with variations of the thawing index. A strong relationship between them was found. It is expected that the average summer air temperature will increase in the monitoring area about 1.0 to 1.5°C due to climate change. According to the monitoring data, the average rate of shore erosion will increase 1.15 to 1.25 times and will reach about 2.5 m per year. This prediction does not take into account the increase of sea level as a result of the climate change. The change of sea level will have greater impact than change of the thawing index. We expect that the average rate of shore erosion will reach 4 to 5 m/year (Vasiliev and others, 2001), if the sea level increases by 1 meter, as is predicted.

In 1980, we established 15 monitoring sites at different areas of the Laptev Sea shore. These sites have not been revisited yet, but observations during site installations provide a good understanding of the possible rate of shore erosion. The Laptev Sea shore in a geological sense consists primarily of the remnants of the Pleistocene plain, which was formed in the periglacial conditions during last glaciation. Plain deposits include syngenetic ice-rich permafrost with thick and tall ice wedges. In the end of the Pleistocene and during the Holocene, the plain was partly destroyed by thaw lakes. The terrain currently consists of hills with ice-rich permafrost, so-called edoma, and wide low areas of old drained lakes (alasses) with soil, which has much lower ice content than edoma. In our more than 150 km field trips along the shores of the Laptev Sea, we have not found any correlation between locations of currently eroded parts of the shore and soil ice content. More than that, the foot of the cliffs at alasses and edoma are eroded of the same rate. But the rate of retreat of the tops of cliffs differs greatly. The top of the cliff at an alass site retreats at practically the same rate as the foot of the cliff. The top of the cliff at edoma site retreats much faster than the foot of the cliff. Erosion of the foot of the cliff at the edoma triggers thermal denudation which continues independently for a long time. In some areas the horizontal distance between the top and the foot of the cliff in edoma reaches more than 100 m. This shows that retreat of the top of the cliff proceeds much faster than retreat of the foot. The retreat of the top of the cliff at edoma sites does not depend on sea erosion. Rather, it is influenced primarily by thermal denudation. The rate of retreat of the top of the cliff at the Laptev Sea shore is about 4 m/year (Are, 1980). This means that the rate of the erosion at the foot of the cliff is much slower than 4 m/year

Conclusions

Monitoring data show great spatial and temporal variations in the rate of shore erosion. Data also show that the rate of river erosion is greater than the rate of sea erosion. Erosion of the foot of a cliff is always a trigger for the retreat of a cliff but does not always control the rate of the retreat of the top of the cliff. In areas of extremely ice-rich permafrost with massive ice, thermal denudation is the leading factor influencing cliff retreat. Retreat of the top of the cliff depends on surface conditions behind the cliff. Any impact on this surface such a vegetation removal can greatly increase the rate of erosion. A protected zone along the shoreline can greatly reduce the rate of erosion of high cliffs in areas with ice-rich permafrost. In a short period of time (a few decades) retreat of the top of the cliff can be independent of the retreat of the foot of the cliff. In a greater scale of time (a hundred years and more) the rate of erosion is equal to the rate of retreat of the foot of a cliff.

References

Are, F, E., (1980) Thermal abrasion of seashores. Nauka Press, Moscow, 157 p. (In Russian)

Bastrakov, G.V., (1977). Soil resistance to erosion. *Geomorphology, 2*, pp. 52-55. (in Russian)

Kal'ianov, V.P., (1934). Geomorphic and hydrogeological observations from the boat "Al'batros" in summer 1932. *Zemlevedenie*, vol. 36, issue 3. (in Russian)

Leffingwell, E., (1919). The Canning River region northern Alaska, United States. Geological survey. Professional paper 109, Washington, Govt. print. off., 251 p.

Pavlov, A., Vasiliev, A., and Shur, Y. (1995). Monitoring of permafrost conditions of the West part of the Yamal Peninsula. In: The 25th Arctic Workshop. Centre d'Etudes Nordiques. University Laval Quebec, Canada. March 16-18, p.144-147

Shur, Y.L., (1977) Erosion of reservoirs shores in permafrost region. In: Studies of shores of reservoirs in Siberia. Nauka Press, Novosibirsk, p.117-121. (In Russian)

Shur, Y. and Jorgenson, M.T. (1998). Cryostructure development on the floodplain of Colville River Delta, Northern Alaska. In; Permafrost, Proceedings 7th International Conference, Yellowknife, Canada, pp. 993-1000.

Shur, Y.L., Petrukhin, N.P., and Slavin-Borovskiy, V.B. (1978). Shore erosion in permafrost region. In: Cryogenic processes. Nauka Press, Moscow, p. 57-73 (In Russian)

Shur, Y.L., Vasiliev, A.A., Veisman, L.I., Zaikanov, V.G., Maksimov, V.V., and Petrukhin, N.P., (1984a). Methods of study of rate of shore erosion. In: Shore processes in permafrost region. Nauka Press, Novosibirsk, p. 5-12. (In Russian)

Shur, Y.L., Vasiliev, A.A., Veisman, L.I., Zaikanov, V.G., Maksimov, V.V., and Petrukhin, N.P., (1984b). New data of monitoring of shore erosion in permafrost region. In: Shore processes in permafrost region. Nauka Press, Novosibirsk, p.12-19. (In Russian)

Tolstov, A.N. (1962). Some data on shore erosion in the Low Indigirka River. *Geography of Yakutia*, issue 2. (In Russian)

Vasiliev A.A, Pokrovsky S.I., and Shur Y.L. (2001) "Dynamics of the coastal thermoerosion of West Yamal". *Earth Cryosphere. V. 5, N 1,* pp 44-52 (in Russian).

Advances in Icing Control at Navigation and Hydroelectric Structures

Robert B. Haehnel[1]
Nathan D. Mulherin[1]
Andrew M. Tuthill[1]

Abstract

Recent advances in de-icing and anti-icing technologies have been evaluated in the laboratory and field to assess their applicability for use at hydraulic (navigational and hydroelectric) structures operated by the U.S. Army Corps of Engineers. In addition to innovative heat panel designs, the U.S. Army Cold Regions Research and Engineering Laboratory (CRREL) has evaluated low-adhesion paints and coatings and electrolytic and electro-expulsive shedding (EES) technologies.

Comparisons of engineered low-adhesion coatings to standard paints used by the Corps show that little improvement is gained for some coatings while for others gains of nearly 50% can be realized. Plastic materials (used as cladding) offer a slight advantage over coatings in terms of their ability to reduce ice adhesion strength. The energy requirements for electrolytic and electro-expulsive shedding are as much as 50 times less than heat tracing. However, corrosion of the electrodes using electrolytic methods can be substantial. Electro-expulsive shedding technology is being used on aircraft for in-flight deicing and shows promise for use on hydraulic structures as well.

Introduction

Adhesion of ice to surfaces creates problems for many industries, including hydropower and navigation. At present, ice removal techniques are costly, hazardous, and time consuming. Andersson and Andersson (1992) reported that at one hydropower station in Sweden, ice-related costs averaged $0.2 million/year over a 10-year period. Annual maintenance costs at Corps of Engineer projects due to ice problems were estimated as $33 million in 1992 (Haynes et al. 1993). Recent advances in de-icing and anti-icing technologies have been evaluated in the laboratory and the field to assess their applicability for use at hydraulic (hydropower and navigation) structures operated by the U.S Army Corps of Engineers. At Corps projects, considerable resources are expended annually to keep ice off steel and concrete structures to maintain operations through the winter months. Comparisons of the performance of these new technologies with current practice are given.

Heat panels

Heat has been used for years at hydraulic structures to control icing. Typical methods include steam lances, heat tracing embedded in concrete walls, and mineral insulated (MI) heaters installed under steel side-seal rub plates. Though these have proven effective, they all have demonstrated limitations. Steam lances pose the same

[1] U.S. Army Cold Regions Research and Engineering Laboratory, 72 Lyme Road, Hanover, NH 03755-1290, rhaehnel@crrel.usace.army.mil

problems as pike poles and other manual methods in that they are very labor intensive and slow, and they frequently place personnel in hazardous locations on the lock or dam. Placing heat where it is needed using heat tracing is desirable, yet the practice of embedding heat tracing in concrete typically provides only a short-term benefit, since the heaters burn out after relatively short service and cannot easily be replaced. Placing side-seal rub plate heaters that have replaceable electric elements on dam gates was beneficial, but the limited area that was heated did not entirely prevent the gates from freezing shut.

For example, Figure 1 shows ice accumulated on the downstream side of a tainter gate at Gavins Point in Yankton, SD. These gates have heaters to prevent the side seal from freezing to the steel rub plate. However, because of the low heat conduction of the surrounding concrete, the heat from the 10-kW elements remains confined to the vicinity of the rub plate. Water leaking by the side seal freezes and then bridges between the gate and pier wall, freezing the gate solidly in place. To prevent the gate from freezing, the heater needs to extend over an area large enough to melt most, if not all, of the ice off of the wall and gate. The enclosure for the heater must have high conductivity (e.g. that of aluminum or steel) so that the heat is distributed uniformly over the area. Furthermore, provision for easily replacing the heater elements when they fail must be designed into the enclosure. The heaters are turned on in the fall and are thermostatically controlled throughout the winter.

Figure 1. Icing on a tainter gate at Gavins Point project, Yankton, SD, prior to heater panel installation (left) and after installation on the pier wall (right), showing the panel's outer covering removed (parallel white lines are troughs for the heater tracing).

The first proof-of-concept heater panel was installed in 1993 on a miter gate recess on Starved Rock Lock and Dam at Ottawa, IL. This 0.9- m (3-ft) × 2.4-m (8-ft) aluminum panel proved to be robust and was successful in keeping a protected portion of the wall ice free throughout the winter (Haynes et al. 1997). Figure 1 shows an aluminum heater panel installed in 1997 at the Gavins Point Project. A similar heater panel was installed in 2001 on the spillway gates of the Turner Falls Project, Montague, MA (Haehnel 2001). For both the Gavins Point and Turner Falls projects, ice formation on the pier wall and gate was a persistent problem because of water leaking past the tainter gate side seal, making the gate inoperable. At the Heywood Generating Station, St. Catherines, Ontario, Canada, their submerged tainter gates freeze in place because of frazil ice accumulation in the upstream bypass channel. In this case, heaters installed on the gate face and sides were needed to keep the gate from freezing in place. In this design the heaters were protected by steel enclosures that were an integral part of the gate. Table 1 gives a summary of the details of these three installations.

Table 1. Summary of heater panel installations.

Location	Year installed	Heater type	Total power per gate (kW)	Approx. power density (kW/m^2)
Gavins Point, Yankton, SD	1997	MI	16	0.7
Heywood Station, St. Catharines, Ontario	1998	MI immersion type	40	4.4
Turner Falls, Montague, MA	2001	MI	16.8	0.57

Low-adhesion materials and coatings

Numerous materials, coatings, and paints are commercially available that are advertised to have low-friction or non-stick properties. Some of these coatings are also marketed as ice-phobic (i.e. significantly lowering the adhesion strength of ice). We have measured the ice adhesion strength for many of these coatings and materials in the laboratory to rank their relative performance (e.g. Haehnel and Mulherin 1998). Our most recent study focused on the suitability of these materials and coatings for controlling icing at hydraulic structures. We measured the ice adhesion strength of common paints used by the Corps of Engineers to protect steel members on hydraulic structures and compared their performance to the low-adhesion coatings. Both vinyl-based paints (used at freshwater projects) and epoxy paints (used mainly for salt/brackish water applications) were evaluated. Since the paints used by the Corps have been primarily developed for their high durability, it was considered unlikely that the low-adhesion coatings would replace them but instead would be applied over the Corps paints to reduce ice adhesion to the surface. Consequently our laboratory tests were designed to simulate this condition, and ice-phobic coatings were layered over samples already having the Corps paints applied.

An alternate means of protection might be to clad an area with a low-adhesion material. Consequently several candidate plastic cladding materials, such as Teflon™, acetal, and polyurethane, were evaluated in this study as well. Table 2 lists the paints, low-adhesion coatings, and materials evaluated in this study. Also listed in Table 1 are coatings and materials evaluated by CRREL in the past.

We evaluated the materials and coatings in the laboratory at CRREL using a shear test apparatus (Fig. 2). The apparatus configuration is a cone test, which is typically used in evaluating the performance of adhesive joints (Anderson et al. 1977). In this configuration an adhesive is used to bond concentric cones of variable angle, and then an axial load is applied so that the cones are pulled apart. By varying the cone angle the relative amounts of shear and tension being applied to the adhesive joint can be controlled. We test using a cone angle of 0° (two concentric cylinders), which predominately loads the adhesive in shear. In our tests ice is used as the adhesive. The inner cylinder (or pile) is either made of the material to be evaluated or coated with a candidate ice-phobic material. Once the sample is frozen, it is placed in the test apparatus (Fig. 2) and then loaded at a constant rate of displacement until the ice–pile bond fails. The measured load at the time of bond failure is used to compute the shear strength of the joint (the maximum force divided by pile–ice contact area). This is our indicator of the adhesive strength of the ice bonded to the material of interest. Details of this test procedure are given in Haehnel and Mulherin (1998). These tests were all conducted at –10°C.

Figure 3 shows the relative adhesive strengths of the ice bonded to the coatings and materials listed in Table 2. Test results from this study are indicated in light gray. In Figure 3 we grouped our results according to the test pile's base material: plastic, stainless steel, carbon steel, or aluminum. The height of each bar indicates the average failure stress measured for each material, while the vertical lines indicate the range in measured values. In these tests, all of the Corps paints except the V102-e exhibited relatively scattered results, making it difficult to know whether there was a statistically significant benefit afforded by the ice-phobics. Teflon™ and polyethylene had the lowest average adhesion strengths, as did the commercial coatings WC-1-ICE and Kiss-Cote, which were each applied to aluminum piles. On carbon steel piles, we found that the V-103c vinyl paint performed as well as PSX700. Furthermore, using the Wilcoxon Ranked Sum test (a non-parametric analysis of variance test), we found that there was no statistical difference, at $\alpha = 0.05$ (95% confidence level), between the performance of the V-103c black and any of the coatings applied to it. However, there was a statistical difference between C-200a alone and C-200a with TroyGuard applied over it.

Although the V-103c paint appears to have very low ice adhesion, this is not generally true for the other paints typically used by the Corps. The white and gray vinyl paint V-766e and aluminum vinyl paint V-102e had average adhesion strengths that were only 20–25 percent less than bare carbon steel, and the epoxies C-200a and MIL-P-24441C Type III had adhesion strengths about the same as bare carbon steel. The engineered coatings appeared to reduce the adhesion strength for the C-200a coal tar epoxy, with both PSX700 and TroyGuard reducing the adhesion strength by about 40 percent.

Table 2. Materials and coatings evaluated at CRREL using a 0° cone test to measure the adhesive shear strength of ice.

Material	Composition
Paints and coatings	
Kiss-Cote	Kiss-Cote 1083 (polydimethyl siloxane) used on aluminum samples and Kiss-Cote MegaGuard (polydimethyl siloxane) used on steel samples
Polyurethane paint	BMS (Boeing Material Spec) 10-60 polyurethane over BMS 10-11 epoxy primer
Rod-Coil-A	Developmental Foster-Miller rod-coil-A
Rod-Coil	Developmental Foster-Miller rod-coil coating
Wearlon	Water-based, methyl silicone copolymer epoxy
PSX-700	Siloxane and polyurethane epoxy
TroyGuard acrylic urethane	Fluoropolymer suspension and mineral spirits in clear urethane
TroyGuard/ polyurethane	Fluoropolymer suspension and mineral spirits in BMS 10-60 polyurethane
Inertia 160	Trimethyl hexamethylenediamine epoxy
Envelon	Resin-based ethylene acrylic acid copolymer thermoplastic
Slip Plate #1	Natural graphite coating in mineral spirits
WC-1-ICE	Saturated polyester resins in fluoropolyol with PTFE and organofunctional silicone fluid additives, modified with a fluorotelomer intermediate, and activated with a trimer of HDI
SA-RIP-4004	Saturated polyester resins modified with fluorotelomer intermediates activated with a biuret of HDI
Corps Paints	
V-103c	Vinyl resin, type 3 (20), carbon black (1.5), diisodecyl phthalate (3.4), methyl isobutyl ketone (36.0), toluene (39.1% by weight)
V-766e	Vinyl resin, type 3 (5.6) and type 4(11.6), titanium dioxide and carbon black (13.0), diisodecyl phthalate (2.9), methyl isobutyl ketone (32.0), toluene (34.7) ortho phosphoric acid (0.2 % by weight)
V-102e	Vinyl resin, type 3 (18.2), aluminum powder (8.3) diisodecyl phthalate (3.1) methyl isobutyl ketone (33.8), toluene (36.6 % by weight)
C-200a	Coal tar epoxy
MIL-P-24441C Type III	Polyamide epoxy
Materials	
Teflon™	Polytetrafluoroethylene (PTFE) thermoplastic
Polyethylene	Ultra-high-molecular-weight polyethylene thermoplastic
Acetal	Acetal copolymer thermoplastic
Dupont Delrin™	Polyoxymethylene homopolymer thermoplastic
Carbon Steel	Cold-rolled 1018
Stainless Steel	Type 410
Aluminum	Type 7075

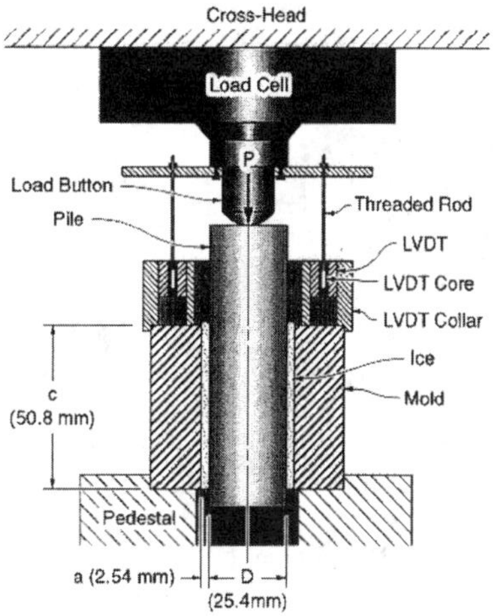

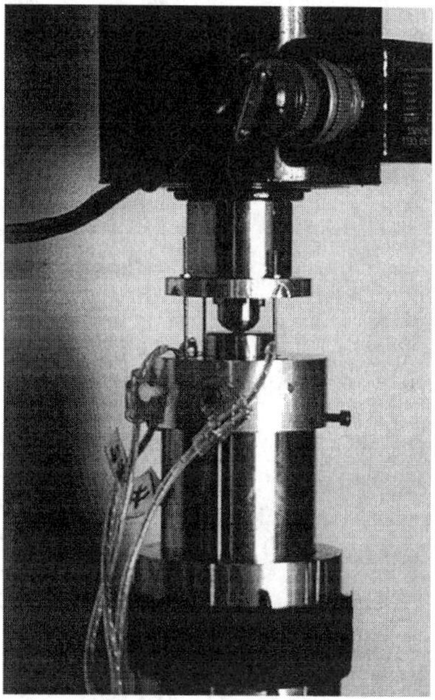

Figure 2. 0° cone test apparatus.

Most of our tests were conducted on pristine samples. However, a few of the plastics had been previously mounted outside on a tower during the winter at the top of Mt. Washington, NH, and were exposed to atmospheric and ultraviolet weathering. The number of days of weathering is indicated in the parentheses following the plastic's name (Fig. 3). Samples that were not weathered (pristine) have no parenthetical numbering associated with them. These initial results showed that the effect of weathering could significantly increase ice adhesion strength. Most notably the adhesion strength for polyethylene increased by over a factor of two after being exposed to sunlight and atmospheric conditions for only 28 days. The same trend was evident for the weathered acetal and Delrin™ samples, though the differences were not as great. The differences were statistically significant ($\alpha = 0.05$) for polyethylene but not for acetal and Delrin™. To further study the effect of weathering on ice adhesion, the pristine plastic and coated carbon steel samples were mounted in a navigation lock chamber on the Mississippi River near St. Louis, Mo. (Lock and Dam 25) and exposed to field conditions, including cyclical wetting and drying and abrasion from moving ice, sediment, and debris, for the duration of the 2001-2002 winter and spring seasons. In late summer 2002 they will be returned to CRREL for retesting so that their adhesion strengths can be compared to the results in Figure 3. Concurrently 3-m × 2.4-m test patches of PSX-700, WC-1-ICE, and TroyGuard EX527 will be applied to the lock chamber wall at Lock and Dam 25; their durability and performance will be monitored over the upcoming years.

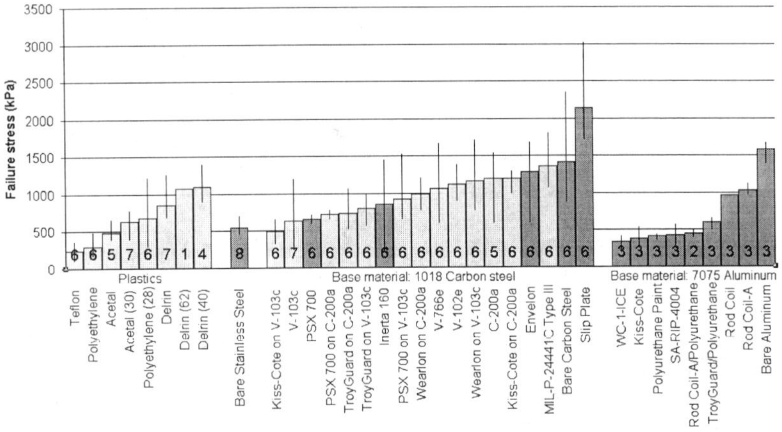

Figure 3. Shear adhesion strength of ice bonded to various materials and coatings. Light gray bars indicate new tests done under this study. The numbers in the bars indicate the sample size. The numbers in parentheses indicate the number of days the samples were weathered atop Mt. Washington, NH.

All of the tests conducted to date have shown that the adhesive shear strength of ice bonded to a variety of materials and coatings varies by less than an order of magnitude. For the pristine plastic samples the variation in adhesion strength between Teflon™ (lowest bond strength) to Delrin™ (highest bond strength) is less than a factor of four. Similarly the bond strength of ice to carbon steel painted with V-103c is approximately three times lower than that of bare carbon steel, and the adhesion to bare aluminum is about five times higher than to WC-1-ICE over aluminum. While these reductions in ice strength are significant, they are not enough to eliminate the need for additional methods of ice removal. To have the ice shed off a vertical surface under its own weight, the adhesion strength must be less than the shear stress that the ice exerts on the wall. For example, the adhesion strength of the ice to the substrate would have to be less than 5 kPa for a 0.6-m-wide ice collar to fall off a lock wall under its own weight. This requires a factor-of-50 reduction in ice adhesion strength in comparison to the best available coatings and materials. Consequently low-energy materials should be considered system enhancements for other methods, such as heat or EES panels, steam lances and pike poles.

Electrolytic shedding

Recent work by Petrenko et al. (e.g. Petrenko and Qi 1999, Petrenko and Courville 2000, Haehnel et al., in review) shows that low DC voltages applied through ice can cause it to release by electrolytic decomposition of the ice. Laboratory experiments show that as little as 6 V is needed to cause the ice to separate into atomic oxygen and hydrogen. These gases coalesce into bubbles at the electrode, and over time these bubbles cover a significant portion of the ice–substrate contact area. Given sufficient time, all of the ice in contact with the substrate is completely converted to gas and the bond is eliminated. Petrenko and Qi (1999) and Haehnel et al. (in review) found that with a voltage application (20–40 VDC) of only one minute, the ice bond strength to stainless steel could be reduced by 60–80% or more. Figure 4 shows an electrode configuration used to remove ice from a flat surface (Petrenko and Courville 2000).

This technology relies on the conductivity of the ice to complete the circuit between the two electrodes. For saline ice the conductivity is quite high, and the electrode spacing can be quite large (1 cm or more). As the purity of the ice increases, its conductivity declines, and the electrode spacing must be reduced, or the applied voltage increased, for the electrolyzing circuit to be completed. Haehnel et al. (in review) found that, for ice formed from Mississippi River water, the electrode spacing needed to be as small as 0.51 mm (0.020 in). Using an electrode spacing of 100 μm, Petrenko and Courville (2000) were able to produce the effect in pure ice; in these experiments micro-grids had to be transferred onto an insulating substrate using printed circuit board techniques. Table 3 provides a comparison of the power requirements to produce electrolysis in these experiments.

Haehnel et al. (in review) observed corrosion of their stainless steel electrodes as a result of the electrolytic process, which was more pronounced when saline ice was tested. Petrenko and Courville (2000) were able to eliminate corrosion by plating their electrodes with gold. Future work requires developing robust micro-grid systems that can withstand the rigors of abrasion and impact in a field application.

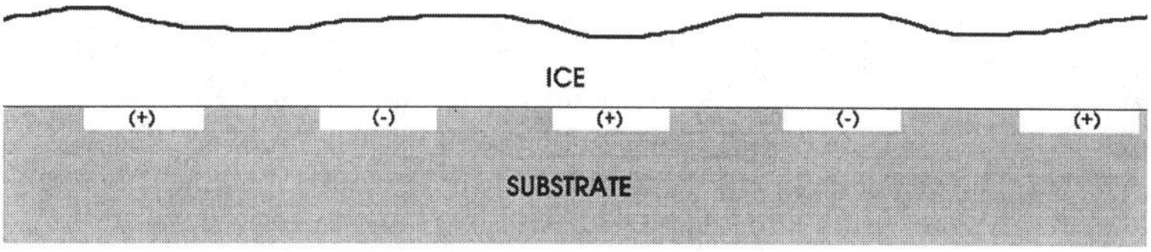

Figure 4. Schematic of the electrode configuration used for removing ice from a flat surface by electrolysis.

Table 3. Power requirements for electrolytic shedding with various electrode spacings and water types.

	Water type	Applied voltage (VDC)	Electrode spacing (cm)	Power density (kW/m^2)
Petrenko and Qi (1999)	Saline, 0.5%	21	1	0.7
Haehnel et al. (in review)	Saline, 0.1%	20–30	0.25	3
Haehnel et al. (in review)	Mississippi River water	20–60	0.051	3–6
Petrenko and Courville (2000)	Pure	10–30	0.01	0.2–0.8

Electro-expulsive shedding

Another method for shedding ice—electro-expulsive separation, or EES—applies a short burst of current (approximately 3 ms in duration) in opposite field directions through adjacent conductors (Haslim and Embry 1989, Embry and Friedman 1989). This produces opposing electromagnetic fields in the conductors, which rapidly forces them apart. The conductor configuration can be enclosed within low profile, flexible panels and mounted on a surface needing protection from ice buildup. Ice that accumulates on the panel surface is fractured and released when the panel skin flexes. The burst of current is provided from a bank of capacitors, which are charged to approximately 500V and then rapidly discharged into the copper foil conductors, expanding the blanket and shattering the ice. This cycle can be repeated until all of the ice is completely removed from the EES blanket. It has been used for deicing ship hatch covers and removing ice from the leading edges of airplane wings. Power consumption for the EES system is on the order of 700 W/m^2 (during the 15 s it takes to charge the capacitors) and has successfully dislodged ice accumulations up to 4–5 cm thick (Embry and Friedman 1989).

Laboratory tests conducted at CRREL evaluated the use of an EES system to shed ice grown on a vertical lock wall. In these tests a 1.22-m × 1.04-m flat EES panel was mounted to a concrete wall on one side of a 2.48-m long × 0.89-m wide × 1.22-m-deep chamber located in a coldroom (air temperature maintained at –10°C). Water was pumped into and out of the chamber to simulate the cyclical wetting and freezing on a lock wall. These tests showed that the EES panel was effective at removing a 2.5-cm-thick ice layer and the associated ice collar from the wall. Four cycles of the EES system were required to completely remove all of the ice from the panel. Figure 4 shows the initial ice accumulation on the panel and the panel clear of ice after cycling the system the fourth time.

Also as part of these laboratory tests we tried to remove the ice from the wall with the EES panel submerged under water. In these tests we found that the ice did not separate from the panel, and the water rendered the ESS completely ineffective. Thus, this system would only work when the ice is in air. (e.g. EES panels would not work on ice frozen to the upstream skin plate of a dam gate).

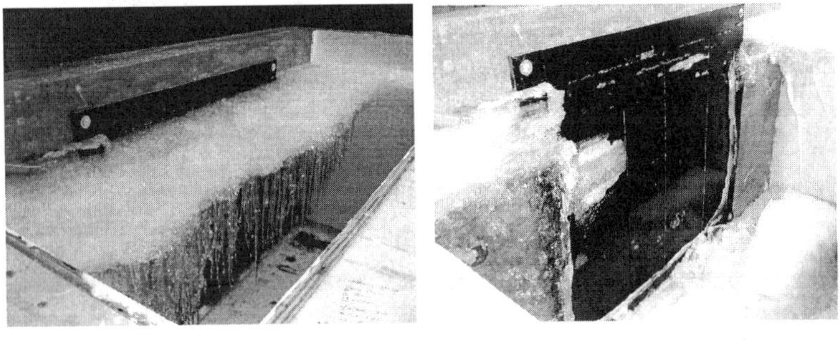

Figure 4. Shedding ice using electro-expulsive separation (EES). The initial ice collar on the EES panel (black) is shown at left, and the panel clear of ice after four cycles is shown at right. The panel measures 1.2 m wide and 1.0 m tall.

The ESS panel has been deployed at Lock and Dam 25 on the Mississippi River for field trials during the winter of 2001-2002, where the performance and durability of this system will be evaluated.

Efficiency of electrical methods

Since the low-adhesion coatings that are commercially available do not reduce the bond strength sufficiently for the ice to shed under its own weight, some type of active system (e.g. pike poles, electric heating, etc.) still needs to be employed to actually remove the ice from the surface. Table 4 gives a comparison of the active systems discussed in this report, namely heaters, electrolytic shedding, and electro-expulsive separation. Though heated trash racks were not discussed in this report, for comparison we included in Table 4 the performance of trash rack heaters (obtained from the literature). The power requirements for each of these systems has been given elsewhere in this report but are repeated here for convenience.

Table 4 shows that, in general, the power requirement is the same for all of these systems, with the systems operating in air requiring 0.2–0.8 kW/m^2. Systems that operate under water (immersed) have power requirements about an order of

Table 4. Power and energy requirements for complete removal of ice using various methods applied to different applications.

Method	Power density (kW/m^2)	Energy density (kJ/m^2)	Approximate response time (min)
Conventional Heaters			
Heater panels on riverine structures (in air)[1]	0.6–0.7	540–630	15
Heater panels on riverine structures (immersed)[2]	4.5	3900	15
Trash rack heaters (immersed)[3]	2–6.7	-	Used in continuous anti-icing mode
Electrolysis			
River water[4]	3–6	300–3000	3–8
Saline ice[4,5]	0.7–3	40–540	1–3
Pure ice using micro-grid electrodes[6]	0.2–0.8	12–48	1
Electro-expulsive Shedding			
Lock wall panel (present work)	0.7	75	18

[1]Haynes et al. 1997, Bockerman and Wagner 1998
[2]Haehnel and Clark 1998
[3]Billfalk 1987, Reid 1928, Samsioe 1924, Ruths 1924, Logan 1974, Daly et al. 1992
[4]Haehnel et al., in review
[5] Petrenko and Qi 1999
[6] Petrenko and Courville 2000

magnitude higher. Of greater concern to operators, because it affects annual operational costs, is the energy consumed to remove the ice. In this case we see a much broader separation in performance, with EES and optimized electrolytic systems consuming a fraction of the energy needed to operate conventional heaters. Another factor of great concern to operators is how quickly the ice can be removed to restore their facilities to full service (response time). De-icing is typically an acceptable mode of operation for the electrolytic and EES systems, with their very rapid response times.

Conclusions

There are a number of new ice-phobic coatings commercially available that provide a significant reduction in ice bond strength Some of the coatings and paints that we have tested to date have one-half to one-third the ice adhesion strength of bare metal. Plastic cladding materials perform slightly better. Yet it appears that the adhesion strength needs to be reduced by a factor of 50 or more below that of the best available ice-phobic materials and coatings for the ice to shed from a vertical surface under its own weight. The long-term performance or durability of both plastics and coatings has still to be evaluated. Durability tests are in progress.

Electrolytic and electro-expulsive (EES) de-icing methods show great promise in removing ice quickly while reducing energy consumption by one to two orders of magnitude in comparison to conventional electric heaters. Future work will be needed to engineer these technologies for potential application at navigation and hydroelectric structures. They must be made to fit the areas and components that need protecting, and they must be hardened to withstand rigorous field conditions. CRREL is conducting field tests of an EES prototype panel during the 2001–2002 winter season.

References

Anderson, G.P, Bennet, S.J., and DeVries, K.L. (1977) *Analysis and Testing of Adhesive Bonds*, Academic Press, New York.

Andersson, A., and Andersson, L. (1992) Frazil ice formation and adhesion on trash racks. In *Proceedings, 11th International Symposium on Ice, IAHR,* 15-19 June, Banff, Canada, vol. 2, p. 671–682.

Billfalk, L. (1987) Ice problems in hydro power plants. Report U(L) 1987/15, Vattenfall Alvkarlebylboratoriet, Alvarkarleby, Sweden.

Bockerman, R.W., and Wagner, P.A. (1998) Solutions to Gavins Point tainter gate icing problems. In *Proceedings, Engineering Technology Transfer and Training Conference*, Kansas City, MO, June 2–4.

Daly, S.F., Haynes, F.D., Garfield, D.E., and Clark, C.H. (1992) Field tests of a surface-heated trash rack to prevent frazil ice blockage. *Proceedings, 11th International Symposium on Ice, IAHR,* June 15–19, Banff, Alberta, Canada, vol. 1, p. 71–78.

Embry, G.D., and Friedman, R. (1989) Electro-expulsive ice removal application to vertical launch system hatch and to personnel access hatch. In *Proceedings, 1989 U.S. Navy Symposium on Arctic/Cold Weather Operations of Surface Ships,* 29–30 November, 1 December, vol. 1, p. 307–314.

Haehnel, R.B. (2001) Solutions for icing on gates at Turner Falls Dam, Montague, MA. ERDC/CRREL LR-01-25, U.S. Army Cold Regions Research and Engineering Laboratory, Hanover, NH.

Haehnel, R.B., and Clark, C.H. (1998)Maintaining operation of bypass tainter gates during frazil ice conditions at the Heywood Generating Station. Contract Report prepared for St. Catharines Hydroelectric Commission, St. Catharines, Ontario, U.S. Army Cold Regions Research and Engineering Laboratory, Hanover, NH.

Haehnel, R.B., and Mulherin, N.D. (1998) The bond strength of an ice-solid interface loaded in shear. *Ice in Surface Waters, Proceedings of the 14th International Symposium on Ice,* Potsdam, NY, 27–31 July, 1998, Balkema Press, Rotterdam, p. 597–604.

Haehnel, R.B., Ryerson, C.C, Mulherin, N.D., Clark, C.H., and Petrenko, V.F. (in review) Application of an interfacial DC charge to reduce ice adhesion strength. *Cold Regions Science and Technology.*

Haslim, L., and Embry, G.D. (1989) Electro-expulsive deicing system. In *Proceedings, Arctic Technology Workshop,* (ed. J. Richter-Menge, W.B. Tucker III, and M.M. Kleinerman), 20–23 June, Hanover, NH. Special Report, No. 89-39, U.S. Army Cold Regions Research and Engineering Laboratory, p. 292–302.

Haynes, D., Haehnel, R., and Zabilansky, L. (1993) Icing problems at Corps projects. REMR-HY-10, Waterways Experiment Station, Vicksburg, MS.

Haynes, D., Haehnel, R., Clark, C., and Zabilansky, L. (1997) Ice control techniques for Corps projects. REMR-HY-14, Waterways Experiment Station, Vicksburg, MS.

Logan, T.H. (1974) Prevention of frazil ice clogging of water intakes by application of heat. Report REC-ERC-74-15, U.S. Bureau of Reclamation, Denver.

Petrenko, V., and Courville, Z. (2000) Active de-icing coating for aerofoils. *Proceedings, 38th Aerospace Science Meeting and Exhibit*, AIAA 2000-0632, 10–12 Jan., Reno, NV.

Petrenko, V., and Qi, S. (1999) Reduction of ice adhesion to stainless steel by ice electrolysis. *Journal of Applied Physics*, vol. 86, no. 10, p. 5450–5454.

Reid, C.R. (1928) Electric heating of rack-bars in hydroelectric plants. *The Engineering Journal*, April, p. 263–265.

Ruths, A. (1924) Ice troubles in Norwegian power plants. *Trans., First World Power Conference*, London, p. 771–787.

Samsioe, A.F. (1924) Measures taken in Sweden against ice troubles at water power plants. *Trans., First World Power Conference*, London, p. 806–811.

Statistical Description of the Accretion Phase of Atmospheric Icing Events

C. Portaix[1] and J. Druez[2]

In cold climates, atmospheric icing is an important factor for the design and reliability of overhead transmission lines. For strength and reliability assessments, a data base is necessary to characterize the statistics of mechanical and meteorological conditions during accretion. Since the early 1980's, a natural icing test site located on Mt.Valin (Quebec), Canada, has been used to record such icing data. This paper presents results, with respect to the accretion phase, obtained from our analysis of this extensive database. The four types of ice defined were rime, glaze, wet snow and mixtures of these types. A total of one hundred and three icing events were selected on each of the four cables located at the test site. Forty-eight events produced rime, thirty-six produced glaze, six produced wet snow and thirteen produced mixtures of these types of ice. We then analyzed the distributions of the icing rate, the air temperature and the wind speed. By comparing the main test line and the perpendicular second test line (which is oriented along the prevailing wind direction) icing rates were found to be significantly lower in the latter case. This result indicates that for all types of ice accretion the cable orientation is an important factor on the accretion properties. The study of the icing rate distributions suggest that the normal probability law does not estimate high icing rates well. Other probability models that should be further investigated are the Weibull, gamma, beta and exponential distributions.

---

[1] Present Affiliation: Rowan Williams Davies & Irwin Inc., Guelph, Ontario, N1K 1B8
Phone: 519-823-1311, Email: cgp@rwdi.com
[2] Département des Sciences Appliquées, Université du Québec à Chicoutimi, Québec, G7H 2B1 Phone: 418-545-5011, Email: jdruez@uqac.uquebec.ca

Introduction

In cold climates, atmospheric icing causes severe damage to human activities every winter. For power lines and communication networks, which are very vulnerable to icing, this is an important criterion for their design and reliability.

Modeling can be used to better understand atmospheric icing and predict extreme and also more frequent situations. The two approaches used to model atmospheric icing are: physical modeling, based on a mathematical description of the processes, and empirical modeling, based on statistical relationships between measured parameters. Empirical modeling relies on natural icing data and is, therefore, specific to the testing site and sensitive to local meteorological conditions. As for physical modeling, several models have been developed, tested and implemented throughout the icing community (Makkonen, 1996).

Due to the stochastic nature of atmospheric icing, empirical modeling has become a greater component of structure icing analysis (McComber et al., 1998, Druez et al., 1999). In order to develop empirical models, databases are necessary to describe the mechanical and meteorological parameters related to atmospheric icing. We have developed such a database at a well exposed test site located on Mt. Valin, Quebec, Canada, for several years. This type of database is rare because the instrumentation needs to be highly reliable during the harshest months of the year and functional for several seasons to obtain statistically significant data (Govoni and Ackley., 1983, Druez et al., 1999, Takebe et al., 1998).

This paper consists of an analysis of the statistical properties of this extensive data base over the 1991-98 icing period, with respect to the accretion phase of one hundred and three icing events observed on each of the four conductors of the test lines. Typical distribution values of icing rate on cables, average air temperature and average wind speed are presented and compared to determine the effects of cable diameter, cable orientation, and type of ice.

Description of the Test Site

The test site is located on Mont Valin, 45 km north-east of Chicoutimi, Quebec, Canada, at an altitude of 902 m. Due to its northern latitude and the influence of the Saguenay Fjord at its south, this site experiences icing events from October to April.

The station includes a main test line and a perpendicular second test line (Figure 1). The main test line spans 96.5 m and is composed of one stranded 12.5 mm steel cable (used on high voltage transmission lines as a ground wire), one stranded ACSR 35 mm conductor, and a bundle of four 35 mm ACSR conductors. This line is nearly perpendicular to the prevailing East-West wind direction. The second test line, composed of one stranded 7.93 mm conductor, extends 31.4 m in span almost perpendicular to the main line.

On both test lines, each conductor is equipped with a load cell that measures the mechanical tension at one extremity, induced by the wind and ice loads. From this mechanical tension, the ice mass is computed using the catenary and equilibrium equations, assuming that the ice distribution is uniform along the cable. This assumption is correct for such short test spans located at the top of a mountain.

The meteorological instruments are a temperature sensor (shielded from wind and sun), a heated anemometer, a heated snow and rain gauge, and an icing rate meter. The anemometer is composed of a rotor and a direction vane that measure respectively the wind velocity and the wind direction. The wind velocity component normal to the main test line is calculated from the wind direction, the wind speed and the line orientation. The snow and rain gauge measures the equivalent water precipitation by melting the snow or ice it captures. Finally, the icing rate meter determines the occurrence of icing events by measuring the shift in resonant frequency of a vibrating cylindrical probe.

The data acquisition and transmission system is located in a building close to the lines. The data are collected at 30 minutes intervals and transmitted via telephone to a computer at the University of Quebec at Chicoutimi daily. These data are checked for validity and can be re-transmitted accordingly if the connection failed.

Measurements differ with the type of data. The ice load data are computed by averaging the signals from 180 samples (one every 10 s) on the each load cell. The temperature, the equivalent water precipitation and the alarms from the icing rate meter are measured at the time of acquisition. Finally, the wind direction and wind velocity are averaged from 200 samples recorded during the 10 minutes prior to acquisition time. Also, a frame by frame video recorder films the main test line at 4 s intervals.

Types of Atmospheric Icing Events

Several types of atmospheric icing events can create large ice accumulations on structures, such as glaze caused by freezing rain or freezing drizzle, rime due to icing fog or in-cloud icing, wet snow, and mixtures of these types of ice. Due to large differences between these types of icing events, it is necessary to distinguish each one before analyzing them in a data base. This is done by comparison of data from the precipitation gauge, the icing rate meter and the temperature sensor (Druez et al., 1995). The three main types of ice are described below.

Glaze is a very compact ice of transparent appearance. Its surface is smooth and its density is about 0.8 to 0.9 g/cm^3. It is produced by supercooled rain droplets, at air temperatures around the freezing point (Makkonen, 1996). This type of ice is frequent throughout North America.

Rime is formed by icing fog or in-cloud icing. Hence, the size of the drops are smaller than for glaze. Rime can be classified into two types: hard rime and soft rime. Hard rime is compact ice of opaque or transparent appearance, depending on the amount of air it contains, and its surface can be smooth or irregular. The density of hard rime depends on air temperature and wind speed, and varies between 0.5 and 0.8 g/cm^3. Air temperatures during hard rime events generally lie around -3°C to -15°C. The second type of rime (soft rime) forms at temperatures lower than hard rime. Soft rime is rough and has a density smaller than 0.5 g/cm^3. It is encountered in cold, dry and generally mountainous regions, whereas hard rime is observed in more humid areas.

Lastly, wet snow is slushy and sticky ice that bonds to structures. Its density is high (from 0.5 to 0.8 g/cm^3) and air temperatures have to be above freezing for this type of ice to form.

At the Mt. Valin icing test site, all these types of icing events, and mixtures of these occur. To distinguish between these types of ice, the meteorological and mechanical parameters measured and the video camera are used. For the 1991-98 period, we observed one hundred and three significant icing events (over 170 g/m). Forty eight events produced rime, thirty six produced glaze, six produced wet snow and thirteen produced mixtures of these types of ice.

Design of the Accretion Phase Database

Test site measurements have shown that atmospheric icing is history dependent (Druez, et al., 1995). Consequently, all three possible phases of an icing event (accretion, persistence and shedding) have to be considered when studying icing on a cable.

The first phase of an icing event is the accretion phase, starting with an ice-free conductor. The meteorological conditions and the thermo-mechanical properties of the cable during this initial phase have an important impact on the physical characteristics of the accreted ice, and will also affect the evolution of the ice on the conductor. The important meteorological parameters during this phase are air temperature, wind velocity and direction, relative humidity and droplet size (Minsk, 1980). Depending on these meteorological conditions, the accretion phase may end with shedding or it may continue with the persistence of the ice.

To identify the end of the accretion phase of an icing event, we have developed a criterion based on our database. The criterion is formulated as follows. During the accretion phase, if the mass of ice shed and/or the duration of ice persistence is greater than a determined threshold, then the accretion phase is finished. We have determined these thresholds based on the analysis of all the icing events from the 1991-98 icing period. The criterion was then used to select the accretion phases and build the database.

The database includes the three types of icing events defined above, and mixtures of these, and is composed of the following fields recorded or calculated during the accretion phase: average air temperature, average wind speed perpendicular to the main test line, phase duration, ice mass at the end of the accretion phase, number of alarms from the icing rate meter, equivalent water precipitation and icing rate.

Presentation of Results and Discussion

The first part of this discussion concentrates on icing rate statistics, while the second part discusses meteorological parameters during accretion.

Icing Rate

The icing rate of an event on a structure is the variation of the ice mass that accumulates on this structure per unit time. Parameters that can influence it are the meteorological conditions during the event (i.e., air temperature, wind speed and direction, droplet size, relative humidity, collection efficiency) and the structure's geometry and exposure.

Results for the average and the maximum values of the icing rate during accretion are presented in Table 1. To better understand the accretion process, this table has been partitioned by type of ice and by cable.

The results in this table indicate that, for all types of ice, the icing rates are generally at least two times smaller on the second test line (i.e., the 7.93 mm diameter cable) than on the main test line. The differences do not seem to depend on the type of icing event (e.g., glaze, rime, wet-snow or a mixture of these).

This table also shows that the icing rate is larger for glaze and wet snow events, as compared to rime and mixtures. In fact, the maximum icing rate observed was 140 g/m.h during a 5 hour freezing rain event on the 35 mm diameter cable in October 1993. During this event, the average temperature was slightly below freezing (-0.6°C) and the average perpendicular wind speed was the highest of all icing events recorded (44 km/h).

Comparisons between the cables demonstrate that for rime, the average icing rate is 35% greater on the 12.5 mm cable (14.3 g/m.h) than on the 35 mm cable (10.6 g/m.h). This difference was also observed previously (Druez et al., 1995) and can be explained by the smaller torsional flexibility of the smaller ground cable. On the contrary, for glaze, the average icing rate is slightly larger on the 35 mm cable (39 g/m.h) than on the 12.5 mm cable (37.3 g/m.h). This is also true for wet snow events. However, due to the smaller number of occurrences of precipitation icing events, these results should be considered with caution. Therefore, we can not determine the effect of cable diameter for these precipitation icing events. Lastly, for mixtures, the two cables have identical average icing rates. More data is also necessary for this type of event to better understand the relation between the conductor's diameter and the icing rate.

The remainder of our analysis focuses on the characteristics of the icing rate distributions. These distributions are presented in Figures 2 to 5 for all four cables and for rime, glaze, mixtures, and wet snow. In the case of rime (Figure 2), several features can be highlighted. Firstly, at least 80 % of icing rates recorded lie between 2 g/m.h and 20 g/m.h for all cables. Secondly, a comparison between conductors demonstrates that the icing rate range is greater on the 12.5 mm cable than on the more concentrated distributions on the other conductors. Finally, when considering the normal law distribution approximation, low occurrences of the icing rate are generally over estimated, whereas higher occurrences are under estimated. This feature is not surprising within two or three standard deviations of the average where additional data would surely reduce this difference.

Figure 3 presents the icing rate distributions for glaze. A comparison between all four conductors shows certain similarities and differences. A common feature to all is that close to half of the icing rate observations are located within one class of the average, which coincides with the maximum frequency class. On the other hand, a notable difference between the conductors is the fact that the range of the icing rate is greater on the 35 mm conductors, both single and in a bundle, as compared to the smallest cables; this result is opposite to the one observed above in the case of rime. In contrast, one characteristic similar to that of rime is related to the normal law distribution approximation. Just as in the case of rime, low occurrences of the glaze icing rate are over predicted by the normal law approximation, except at the higher values, where they are under predicted.

The icing rate distributions for mixtures are presented in Figure 4. Our first observation is that, unlike the other types of ice discussed above, the average icing rate is characterized by a small frequency of occurrence. This is explained, at least on the main test line, by the fact that the distributions are composed of two sub-groups of data, each seemingly concentrated about its own average icing rate. The statistical properties of each group do not correspond to any particular type of ice, even though they both seem to be closer to rime.

Lastly, Figure 5 presents the icing rate distribution for wet snow on the 35 mm diameter cable. The normal law has been overlaid on the distribution for reference only. It is interesting to note that the icing rates for this type of event are of the same order of magnitude as for rime. As the small number of events shows, wet snow rarely occurs at the test site where rime is the dominant type of ice observed.

In this discussion, we have used the normal law distribution as a first approximation to compare the data to probability models. This analysis showed that asymmetry and occurrences at high icing rates are important characteristics of the distributions. Therefore, models such as the Weibull, gamma, beta and exponential laws should be investigated to catch these particular features.

Temperature and Wind Speed

In this section, we discuss the statistical characteristics of temperature, and then wind speed, during the accretion phase of the icing events recorded.

For this, Table 2 presents the average, standard deviation, maximum and minimum of temperature observed during accretion for all types of ice. In decreasing order, the average temperature was recorded as follows : wet snow (0.2°C), glaze (-1.9°C), mixtures (-4.5°C) and rime (-6.2°C). The highest average temperature during accretion was 0.8°C for a wet snow event in April 1995. Temperatures during this event varied between 0°C and 2.5°C and perpendicular wind speeds reached 40 km/h. The icing rate was typical (26 g/m.h) on the 35 mm diameter cable. As for the lowest temperature observed during the accretion phase, the equipment recorded an air temperature of –25.6°C during a rime event in January 1994. These low temperatures are frequent at the test site between December and March, but they are rarely associated with ice accretion. The icing rate for this event was low (5 g/m.h) on the 35 mm diameter cable and wind speed averaged 17 km/h.

Temperature distributions are presented in Figure 6 for the four types of ice observed. From the temperature distribution for rime, it is worthwhile to note that temperatures quite frequently reach -2°C (at least 30% of the occurrences) and that 90% of observations lie between 0°C and -8°C. For glaze, the temperature distribution shows that occurrences do not vary as much as for rime and that the range of temperatures is smaller (from 0°C to -4.5°C). Furthermore, for this type of ice, 80% of temperature observations lie between 0°C and -2°C. In the case of mixtures, the temperature distribution shows two distinct distributions that can be explained by the two major types of ice observed at the test site: rime and glaze. Finally, the temperature distributions for wet snow events show that all temperatures recorded for this type of ice are positive.

Statistics of the wind speed perpendicular to the main line are also computed for analysis. These results are presented at Table 3 for the four types of ice. This table shows that perpendicular wind speeds for the types of ice observed are, in decreasing order, 25 km/h for glaze, 21.8 km/h for wet snow, 18 km/h for mixtures and 17.1 km/h for rime. Events resulting from in-cloud icing (e.g., rime and mixtures) have lower average wind speeds than the ones from precipitation (e.g., glaze and wet snow). The highest average wind speed observed was 44 km/h in the case of glaze, corresponding to the maximum icing rate event discussed in the previous section.

Wind speed distributions are presented in Figure 7 for the four types of ice. This figure shows that the range of wind speeds is large for both rime and glaze.

Conclusions

1. Comparisons between the two test lines show that the line perpendicular to the main wind direction has a overall greater icing rate for all types of icing events. For rime, the average icing rate is higher on the 12.5 mm diameter cable as compared to the 35 mm cable. This has not been verified for glaze.

2. The maximum icing rate (140 g/m.h) was observed during a 5 hour freezing rain event on the 35 mm diameter cable. Also, extreme average temperatures observed are -25.6°C during an in-cloud icing event and 0.8°C during a wet snow event. Lastly, the highest wind speed observed was 44 km/h in the case of glaze, corresponding to the maximum icing rate event.

3. In this analysis, we have used the normal law distribution as a first approximation to compare the data to probability models. This has demonstrated that asymmetry and occurrences at high icing rates are important characteristics of the icing rate distributions. Therefore, models such as the Weibull, gamma, beta and exponential laws should be investigated to catch these particular features.

Ackowledgments

This work was funded by the National Sciences and Engineering Research Council of Canada, the FCAR (Québec), Hydro-Québec and the Université du Québec à Chicoutimi Foundation. The authors would like to thank Marco Boudreault and Serge Gauthier for their technical contributions and Sophie Gouzy for her assistance with the database.

References

Druez, J., McComber, P., Laflamme, J., (1995), "A comparison of glaze and in-cloud icing events on a test line". Proceedings of the 14th International Conference on Offshore Mechanics and Arctic Engineering, Copenhagen, Denmark, pp. 93-100.

Druez, J., McComber, P., Farzaneh, M., (1999), "Analysis of atmospheric icing events observed at the Mount Valin test site during the 1995-96 season", Proceedings of the 9th International Offshore and Polar Engineering Conference, Brest, France, pp. 574-580.

Govoni, J., Ackley, S., (1983), "Field measurements of combined icing and wind loads on a wire". Proceedings of the 1th International Workshop on Atmospheric Icing of Structures, Hanover, USA, pp.205-215.

Makkonen, L., (1996), "Modelling power line icing in freezing precipitation". Proceedings of the 7th International Workshop on Atmospheric Icing of Structures, Chicoutimi, Canada, pp.195-200.

McComber, P., Delafontaine, J., Laflamme, J., Druez, J., Paradis, A., (1998), "A neural network system to estimate transmission line icing". Proceedings of the 8th International Workshop on Atmospheric Icing of Structures, Reykjavik, Iceland, pp.101-106.

Minsk, L.D., (1980), "Icing on structures", CRREL Report 80-31, 18 pages.

Takebe, T., Fukazawa, M., Kiushi, S., Kojima, Y., Oka, T., (1998), "Observation of icing on Mt. Takaishi test line". Proceedings of the 8th International Workshop on Atmospheric Icing of Structures, Reykjavik, Iceland, pp.259-265.

Table 1: Icing rate statistics (g/m.h)

	Conductor			
	perp. to prevailing wind			**prevailing wind**
	12.5 mm cable	**35 mm cable**	**one cable of 35 mm bundle**	**7.93 mm cable**
			Rime	
Average	14.3	10.6	8.8	5.6
Std deviation	9.2	6.4	5.4	4.3
Maximum	38.1	26.4	24.2	16
			Glaze	
Average	37.3	39	37	14.2
Std deviation	24.8	30.9	30.4	10.3
Maximum	120.6	140	131.8	47.4
			Wet snow	
Average	21	24.4	22.7	5.6
Std deviation	14.7	10.6	9	1.7
Maximum	46	42.9	39.4	6.8
			Mixtures	
Average	12.5	12.3	12	5.6
Std deviation	6.5	4	8.8	3.2
Maximum	28.2	28.2	31.4	12.5

Table 2: Air temperature (°C) statistics

	Type of ice			
	Rime	**Glaze**	**Wet snow**	**Mixtures**
Average	-6.2	-1.9	0.2	-4.5
Std deviation	3.7	1.3	0.2	2.9
Maximum	-1.1	-0.1	0.8	-1.2
Minimum	-25.6	-4.5	0.1	-10.3

Table 3: Perpendicular wind speed (km/h) statistics

	Type of ice			
	Rime	**Glaze**	**Wet snow**	**Mixtures**
Average	17.1	25	21.8	18
Std deviation	7.7	10.5	9.8	5.9
Maximum	37.2	44.1	31.9	31.4

Figure 1: Plan view of the experimental site main and second test lines

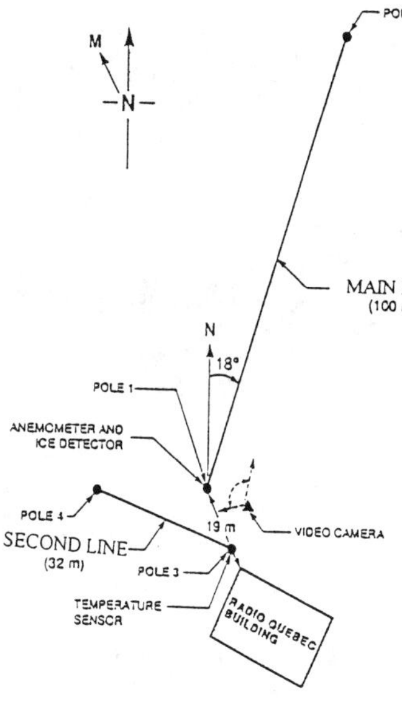

Figure 2: Icing rate distributions for rime compared to a normal law distribution (solid line)

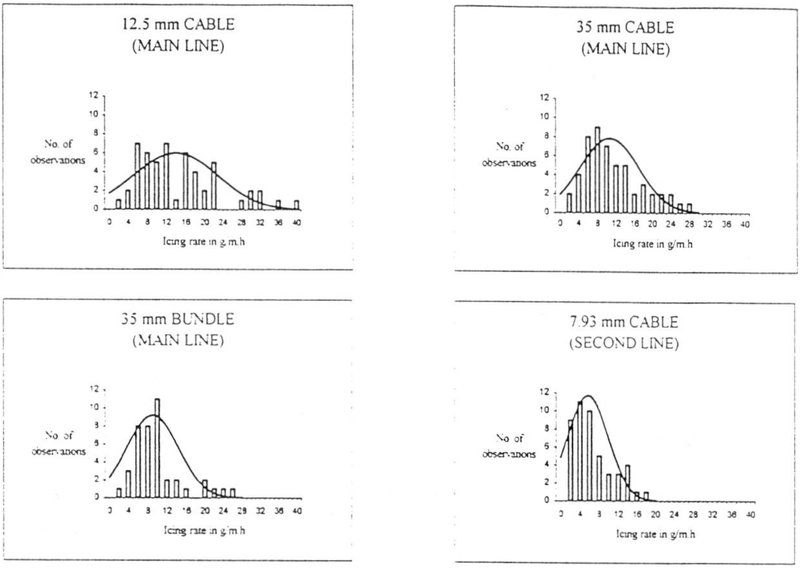

Figure 3: Icing rate distributions for glaze compared to a normal law distribution (solid line)

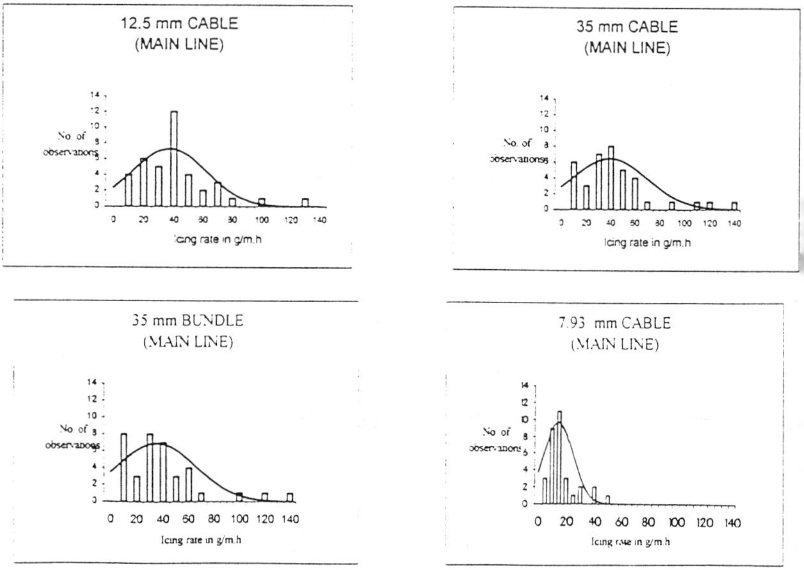

Figure 4: Icing rate distributions for mixtures compared to a normal law distribution (solid line)

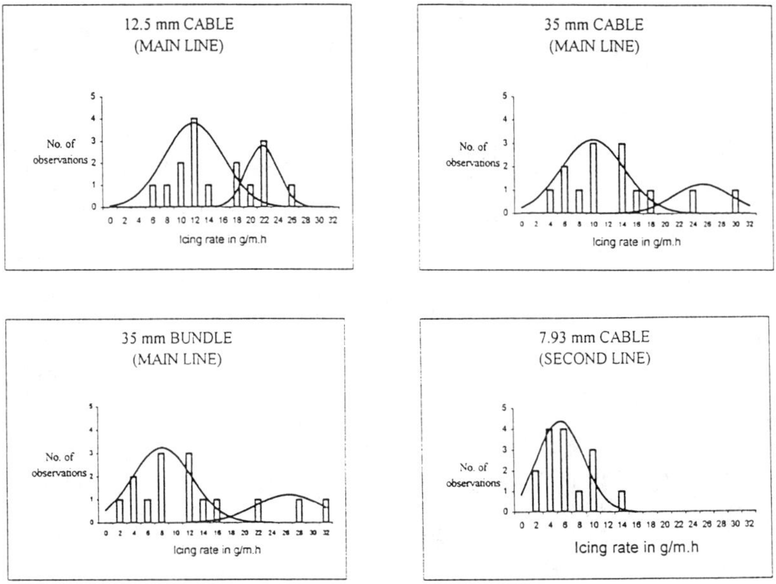

Figure 5: Icing rate distributions for wet snow compared to a normal law distribution (solid line)

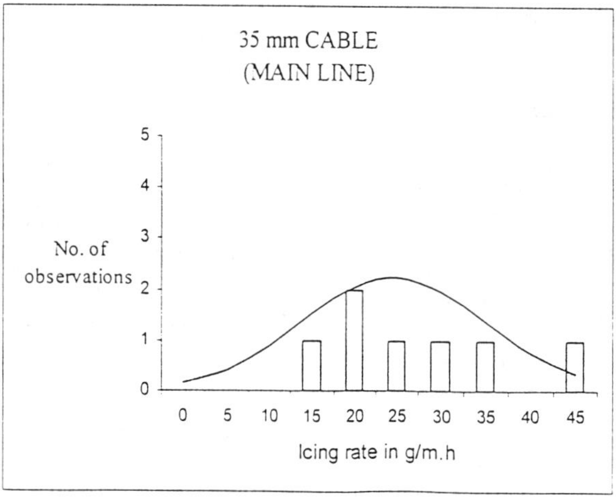

Figure 6: Temperature histograms

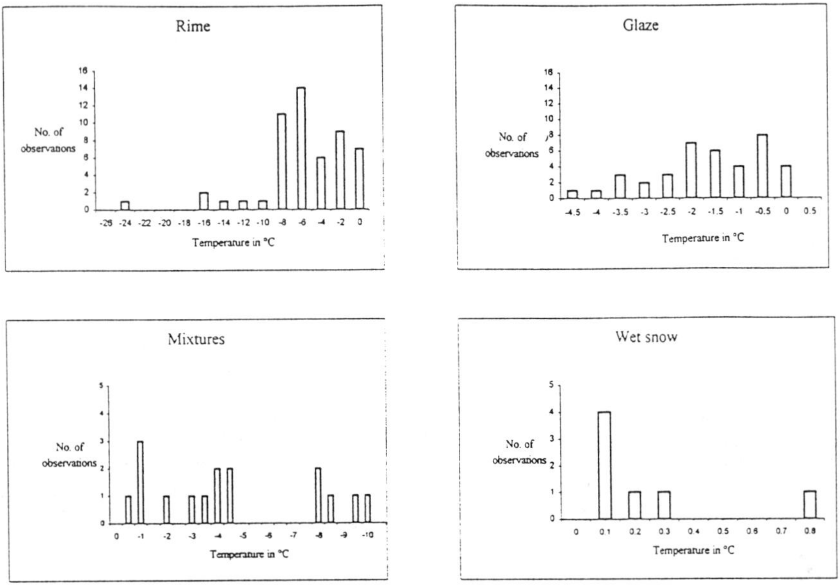

Figure 7: Wind speed histograms

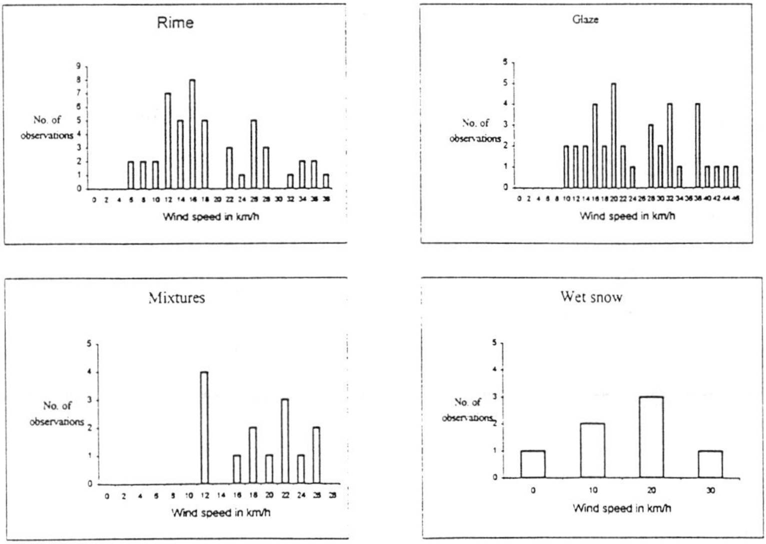

Ten Years of Ice Force Measurements on the St. Regis River, Hogansburg, NY: 1989–1999

Robert B. Haehnel[1]
Kathleen D. White[1]

Abstract

Ice loads were measured for ten years on the east pier of the NY Route 37 bridge over the St. Regis River in Hogansburg, NY. This record comprises the longest documented history of ice loads on any bridge in the world. Statistics on the loads for the 10-year period of record are presented along with statistics on ice thickness and strength. The measured ice strength at this site agrees well with other data measured in Canada and Europe.

The largest measured ice force on the pier was 356 kN, which is only 15% of the design ice load. Statistical analysis of the data indicates that the return interval on the design load appears to be greater than 1000 years, suggesting the design may be overly conservative. Probabilistic approaches to determining the appropriate design load may help prevent overdesign of future bridges, thereby reducing cost while still maintaining a reliable transportation infrastructure. Field measurements of ice loads on a wide range of river sizes extending over many years will be required to use probabilistic approaches with reasonable levels of uncertainty.

Introduction

The existing American Association of State Highway Transportation Officials (AASHTO) bridge design code for ice loads is largely based on data obtained on two bridges located on the Athabasca and Pembina Rivers in northern Canada (AASHTO 1996, 1998, Lipsett and Gerard 1980). These bridges are located at latitudes 54 and 55° N, where the ice thickness is typically almost twice that seen in the continental United States. Furthermore, ice break-up at these sites in northern Canada typically occurs in early to mid-April, when the ice can still be very strong. Because guidance based on these data may be conservative, it may impose undue additional cost for bridges constructed in the continental United States. To verify the reasonableness of the existing code for U.S. bridges, field data of ice forces on bridges located in the lower 48 states should be obtained. Furthermore, the new statistically based Load-Resistance Factor Design (LRFD) methods recommended by AASHTO (1998) for bridge design requires information on load probability. To date, there is no information on ice load probability, and few records exist from which such information can be derived.

[1] U.S. Army Cold Regions Research and Engineering Laboratory, 72 Lyme Road, Hanover, NH 03755-1290

To fill these gaps in knowledge, a load panel to measure ice forces was installed by the U.S Army Cold Regions Research and Engineering Laboratory (CRREL) in December 1989 on the east pier of the NY Route 37 bridge over the St. Regis River in Hogansburg, NY (Fig. 1).

The St. Regis River is located in northern New York State and flows north into the St. Lawrence Seaway. The Route 37 Bridge is located about 7 km upstream of the confluence with the Seaway. The St. Regis River is about 90 m wide at the bridge, and the 50-year discharge for this river is about 600 m^3/s. This river can be characterized as representative of a typical medium-sized river found in the northern continental United States. This location is ideal for measuring ice forces because ice forms every winter on this river and some winters experience more than one break-up event. Ice forces were measured at this site during ten winters, spanning December 1989 through March 1999. This report summarizes the ice force measurements, compares them to current design code, and recommends future revisions based on the data.

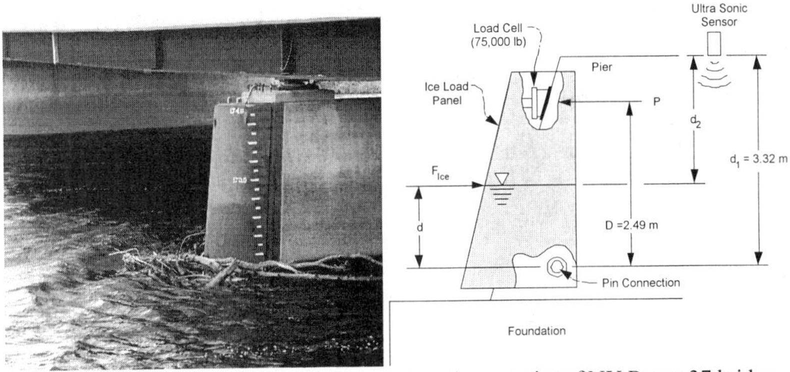

Figure 1. Instrumented load panel located on the east pier of NY Route 37 bridge.

Measurement equipment

Ice force measurements. The ice forces were measured using a load panel mounted on the upstream nose of the pier (Fig. 1). The panel is a simply supported beam pinned at the base with a reaction support at the top. The base of the panel is mounted on bearings and attached to a shaft at the base of the pier, forming the pin connection. A 330-kN load cell is mounted at the reaction support at the top of the panel. The force panel was installed in December 1989. A detailed description of the panel is given in Haynes et al. (1991).

The ice force on the pier, F_{ice}, can be calculated by summing the moments about the pin connection (Fig. 1) such that

$$F_{ice} = P\frac{D}{d} \tag{1}$$

where P is the load measured by the load cell at the reaction support, D is the distance from the pin to the load cell, and d is the variable distance from the pin to the ice. An ultrasonic sensor was mounted on the underside of the bridge at a fixed height d_1 above the pin connection. This device measures the distance to the water level/ice surface, d_2, by bouncing sound waves off the surface, so that $d = d_1 - d_2$.

A data acquisition system was set up in the basement of a building on the east bank of the river. The load and water level data were recorded on a magnetic tape storage unit. The system was unmanned and recorded data automatically when load exceeded 2.2 kN on the panel. A time and date stamp was recorded on a datalogger at the start of each triggered event. During the last three years of the project, a camera was mounted on the east bank of the river and was trained on the load panel. This camera was triggered by the datalogger to take two pictures at the start of each load event.

Local weather conditions and ice cover. Air temperature data from Massena, NY (located about ten miles from the bridge site) was used to estimate the ice thickness throughout the winter. Ice growth resulting from heat transfer processes is commonly estimated using the modified Stefan equation (USACE 1999):

$$t = a\sqrt{S_f} \tag{2}$$

where t is ice thickness, S_f is the accumulated freezing degree-days, and a is an empirical constant. S_f is a summation of the average daily degrees below freezing and can be calculated from reported air temperature data. This is analogous to heating degree-days used for estimating heating requirements. The coefficient a can be determined from historical measurements. The bridge sits in the backwater to the Niagara Mohawk Power Corporation's Hogansburg Power Station, where sheet ice typically occurs rather than a shoved ice cover. Thus, this method is reasonable in estimating ice thickness at this location. Through least-squares regression we found that using a =1.36 (for S_f in °C-day and t in cm) gives a calculated ice thickness that differs from the observed ice thickness by 20% or less. This value is within the range suggested for a sheltered small river given by USACE (1999).

Measured forces and pressures

Four failure modes of the ice were observed in the force records and from photographs: bending, crushing, impact, and splitting. Two of these failure modes can be seen in Figure 2: crushing and splitting.

The peak forces measured each year are given in Table 1. The highest load (356 kN) was measured in the first year of the study. Observed loads have been higher than 250 kN in five out of the ten years. Also given in Table 1 is the calculated effective ice strength for each peak load. The range in the effective ice strength on the St. Regis River was 0.36 to 1.1 MPa. This is about the same as the effective strength measured on the Athabasca River in northern Canada (0.45 to 1.4 MPa). On the Pembina River (also located in northern Canada) the measured effective strength was higher, ranging from 0.9 to 2.3 MPa (Lipsett and Gerard 1980). This suggests that the measured ice strengths in Canada during spring break-up are generally comparable to what is seen in the northern continental United States.

Figure 2. Ice failing by crushing (left) and splitting (right) after impacting the pier. Both events occurred on 29 March 1998.

Table 1. Annual measured peak ice force, effective strength, and ice thickness on the St. Regis River at Hogansburg, NY.

Year	Date of peak force	Peak ice force, F (kN)	Effective ice strength, p (MPa)	Ice thickness, t (cm)
1990	17 Mar	356	1.0	25–30
1991	07 Mar	285	0.71	30–36*
1992	11 Mar	172	0.37	35–41*
1993	02 Apr	187	0.38	41*
1994	02 Apr	235	0.36	53
1995	16 Jan	269	1.1	20*
1996	19 Jan	303	0.75	33*
1997	–	<2.2	–	30–36*
1998	28–30 Mar	310	0.95	25–30
1999	–	<2.2	–	30*

* Estimated using Eq. 2.

The ice force record indicates that the ice cover breaks up and runs in this river reach during January and March or early April. There were two winters (1997 and 1999) in which the ice forces on the panel never exceeded the trigger level (2.2 kN). Both of these years had very mild springs, and the ice essentially rotted in place. Additionally, during 1997 the ice did not form on the river until late January, which significantly reduced the growth time for the ice.

The load data and photographs (e.g. Fig. 2) indicate that the ice floes can be quite large. Photos obtained during the spring ice-out in 1998 showed floes ranging in size from 3 m across to over 60 m (or more than 2/3 of the river width); this range of sizes is well reflected in the force record. Impacts that last for 30 seconds to 1 minute or more are typical for large floes interacting with the pier and result in very high loads. Small floes tend to result in short-duration loads that are typically much smaller. In addition, estimates from the photo record suggest that it is not unusual to have ice floe velocities of 0.6 to 1.5 m/s, which is comparable to the floe velocities measured by Lipsett and Gerard (1980) at the two Canadian bridges.

Statistical analysis of force measurements

The load data obtained between 1990 and 1998 were analyzed to determine the probability distribution of the measured loads. The goodness-of-fit between the load distributions and a number of standard distributions was determined using the Anderson–Darling test (Palisade Corporation 1997). This test, which places emphasis on the tails of the distribution, was thought to provide the best estimate of fit for the force data. Appropriateness was also checked by evaluating the difference between the input and calculated distributions as well as P–P and Q–Q plots. The analyses indicate that the best fit is the Inverse Gaussian distribution with parameters μ = 79.091 kN and λ = 130.335 kN. This distribution is defined as follows (Evans et al. 1993):

$$f(x \mid \mu, \lambda) = \sqrt{\frac{\lambda}{2\pi x^3}} e^{\left(-\frac{\lambda(x-\mu)^2}{2\mu^2 x}\right)}, \; x > 0, \; \mu > 0, \; \lambda > 0. \quad (3)$$

The goodness-of-fit between the Inverse Gaussian distribution and the observed data is evident in Figure 3.

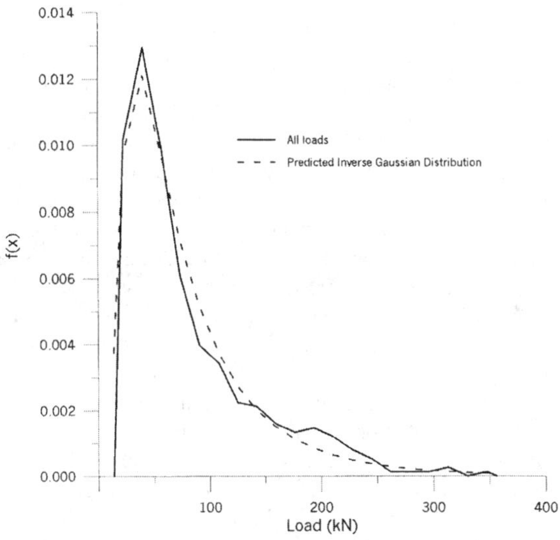

Figure 3. Comparison between probability density functions for observed data and Inverse Gaussian distribution.

While providing important insight into the probability distribution of seasonal ice loads, the distribution of peak annual loads cannot be inferred from this distribution of all loads. Furthermore, the number of peak annual loads that have been measured during the observation period at the St Regis River Bridge (eight) is too small to develop a distribution of annual peaks with any confidence. This is evident from Figure 4, which presents the probability plot of peak annual loads given in Table 1. Although it appears that the peak loads are unlikely to reach the design load of 2400 kN during standard design periods of 50 or 100 years (percent chance exceedance of 2 to 1, respectively), there are insufficient peak annual load data available at this time to predict how unlikely this event would be (i.e, it is not prudent to postulate the 100 or 500 year events based on extrapolation from only 10 years of data).

Discussion of results

In the AASHTO Standard Specifications for Highway Bridges (AASHTO 1996) the horizontal ice force is calculated using

$$F = C_n p t w \tag{4}$$

where C_n is a coefficient for nose inclination, p is an effective ice strength, t is the ice thickness, and w is the pier width. The effective ice strength, as defined by AASHTO, is the pressure at which the ice fails by crushing or splitting on the pier. The specification allows for a wide range in this ice strength, from 690 kPa (100 psi) to 2800 kPa (400 psi). The ice thickness in Eq. 4 is chosen based on knowledge of historical ice records, while the effective strength is often based on experience.

Unfortunately AASHTO's Standard Specifications for Highway Bridges (1996) has no capability to compute the probability of occurrence of the computed force. This basic issue led to the development of the LRFD Specification (AASHTO 1998). Using LRFD, the bridge designer can take into account the probabilities of load, P(F), and resistance, P(R), to design a reliable and robust bridge system. There is no guidance provided in the LRFD specification for computing the correct statistics for ice loads. In addition, the LRFD specification assumes that loads are normally distributed, which does not appear to be applicable for all ice loads, especially ice loads as indicated by Figure 3. Factors that affect the design ice load and that may contribute to the over design of current bridges with respect to ice loading should be addressed by LRFD in the future.

The independent variables in Eq. 4 are the ice thickness and effective strength. Using the force data and estimated ice thickness, the effective ice strength can be calculated from Eq. 4. Probability plots of measured data for ice thickness and effective strength (based on peak loads, Table 1) are given in Figures 5 and 6, respectively. For comparison, the values of effective ice strength and thickness used for the bridge design are shown on Figures 5 and 6. It is evident that the design values for these variables are reasonable given the design life of the structure. For the ice thickness and effective strength the design value could occur as frequently as 1 in 50 years to 1 in 200 years; these recurrence intervals are well within the design life of the bridge. This suggests that the ice thickness and effective strength design values for the Route 37 Bridge are not overly conservative. Yet, choosing these values to determine the design ice force computed in Eq. 4 assumes that both the design ice strength *and* thickness will occur at the same time. Assuming that ice thickness and effective ice strength are independent random variables, the probability of the design values occurring at the same time is

$$P(t = t_d, p = p_d) = P(t_d)P(p_d) \tag{4}$$

where the subscript d indicates design values of t and p. Based on Figures 5 and 6, rough estimates of $P(t_d)$ and $P(p_d)$ could be 0.005 and 0.02, respectively. Thus, the probability of both occurring at the same time would be on the order of 0.0001 (a recurrence interval of 10,000 years).

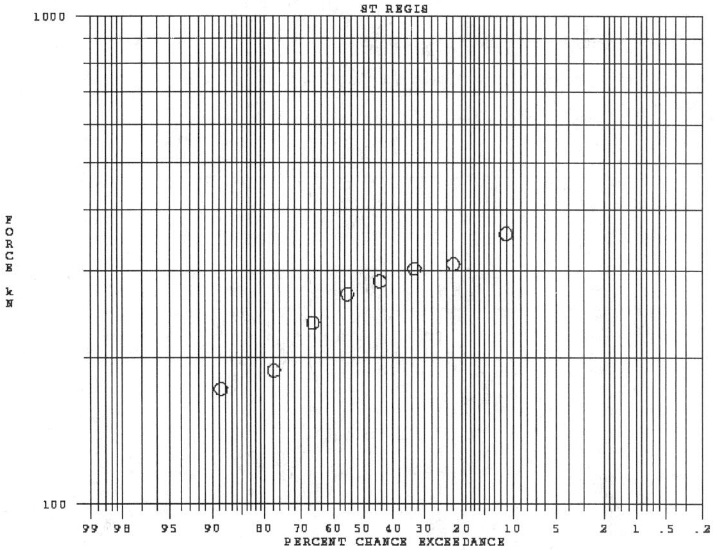

Figure 4. Probability of occurrence of annual peak loads.

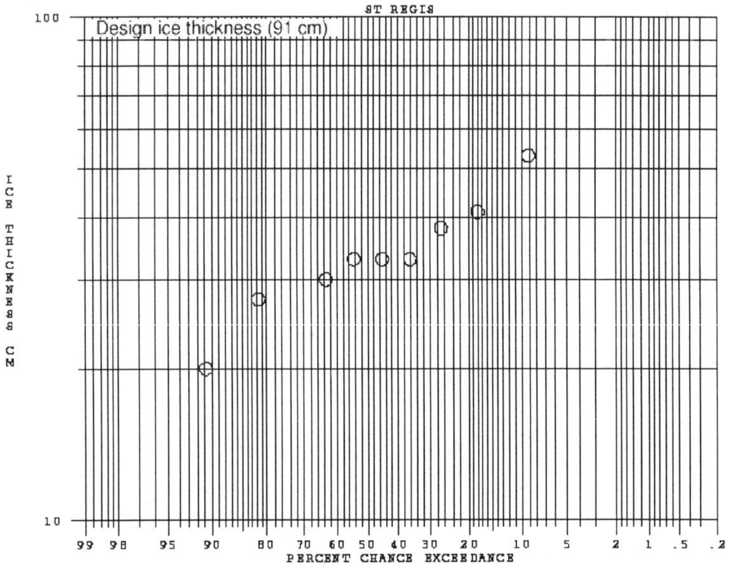

Figure 5. Probability plot of ice thickness at the NY Route 37 Bridge, Hogansburg, NY.

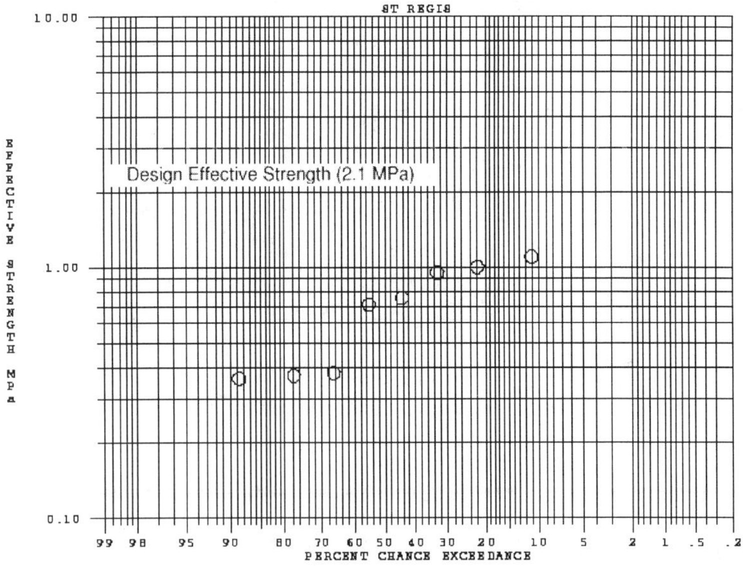

Figure 6. Probability plot of peak effective ice strength measured at the NY Route 37 Bridge, Hogansburg, NY.

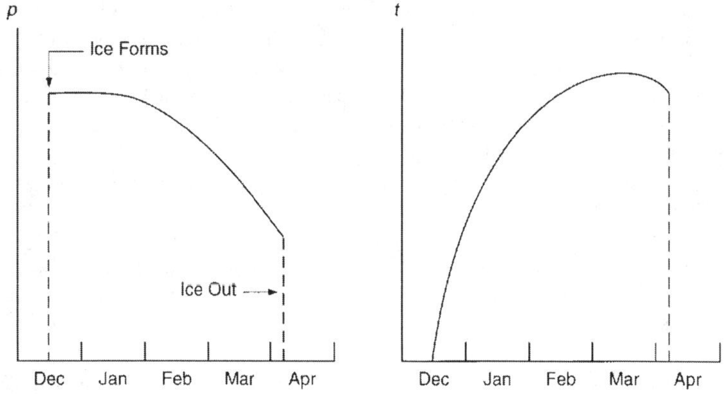

Figure 7. General trends of ice strength, p, and thickness, t, throughout a typical winter season.

Factors that make it unlikely that the maximum ice strength and thickness would occur at the same time have to do with the ice growth and decay process. During initial ice formation early in the winter season, air temperatures are generally quite low and the resulting ice is very strong yet thin. With continued low temperatures the ice grows rapidly and can reach 30 cm or more in the course of a few weeks. Once this ice layer is formed, it inhibits heat transfer from the atmosphere to the water, slowing further growth of the ice. The addition of snow cover on the ice also acts as an insulator, further retarding ice growth. Thus, thickening to 50 cm or more takes extreme cold over a long duration. Though ice is very strong at the outset, its strength can decline quite rapidly and is closely tied to the ice surface temperature, with lower air temperatures yielding stronger ice. In a general way these effects are depicted in Figure 7.

The ice cover at the Route 37 Bridge often breaks up and goes out during a mid-January thaw, in addition to the typical ice run in March or early April. During a January event the ice is strong but not very thick. During this study, 30 cm of ice has been measured at this site in mid-January (a typical ice thickness for that time of year on the northern tier of the lower 48 states). As the winter progresses, the ice thickens, yet the ice strength begins to decline as the air temperature moderates, the sun angle increases, and the hours of sunlight increase. As spring approaches, air temperatures rise above freezing and the ice strength declines rapidly (especially when a heavy rainfall rapidly warms the ice surface to freezing). Haehnel et al. (1999) showed that in just the three-week period leading up to spring break-up, the ice flexural strength on four rivers in North Dakota and Minnesota dropped by a factor of 2 or more (from 1 to 0.4 MPa). Ashton (1985) and Prowse et al. (1990) report similar rapid reductions in ice strength (from 1 to 0.2 MPa) prior to spring break-up. So during a spring ice event the ice can be thick but not very strong. Thus, it is rare to have coincidence of maximum ice thickness and ice strength. The likelihood of strong, thick ice being present during a late-season ice run is further diminished by the occurrence of January ice runs at this site, which force the ice formation process to re-start in mid-winter. This was the case in 1990, when the peak annual ice force at the site occurred. The ice thickness during the March event was only 25–30 cm, yet the ice was very competent in comparison to other years (Table 1).

Historical data may be useful for determining the appropriate probabilities for ice thickness. Using Eq. 2 and available air temperature records, one can estimate the maximum expected ice thickness over the period of record. It is not unusual to have air temperature records that extend back 20–40 years, which will give a reasonable estimate of the probability of the extreme event. Based on data from Wadhams (1981), it is likely that this probability distribution will be found to be non-normal as well.

It is more difficult to obtain such information for estimating the probability of the effective ice strength. There are limited ice force data available from which effective ice strength can be derived (e.g. Schwartz 1970, Neill 1976, Lipsett and Gerard 1980, Haynes and Sodhi 1983, Sodhi et al. 1983, Haynes et al. 1991, Zabilansky 1996, Brown et al. 1998, Johnston et al. 1999). These data sets span a wide range of river sizes and climates, but most have only 1–3 years of data. The longest records are ten years on the St. Regis River, seven years on the Athabasca River, six years on the White River, and six years on the Rideau River. From these records, preliminary effective ice strength probabilities can be developed. However, much longer periods of record are needed to develop a robust database upon which designers can draw. Of the sites where ice forces have been measured, only two riverine sites are still active and continuing to collect data: the St. Regis River in Hogansburg, NY and the White River in White River Junction, VT. One site on the Northumberland Straight in Canada may provide additional information on seasonal changes in ice impacts, albeit not for strictly freshwater ice. The small data set underscores the need to keep existing sites active, as well as adding new sites to expand the available data. Presently the St. Regis site is funded to continue data collection through 2004, which will give a total of 15 years of data at this site.

Conclusions and recommendations

Ice forces have been measured for ten years on the NY Route 37 Bridge over the St. Regis River in New York. The maximum ice force measured during this time was 356 kN (15% of the design ice load). From the ten-year record we estimate that the design load may have a recurrence interval greater than 1000 years at this site. This suggests that application of the current guidance can lead to over design, even though the design values for ice thickness and effective ice strength are reasonable for this location. However, the sample size provided by ten years of data at one site is too small to fully develop guidance for applying statistically based ice load design criteria. A probabilistic approach to determining the appropriate design ice load could prevent needless over design of a bridge while still providing a reliable structure throughout its design life.

These measurements also show that the assumption of a normal load distribution may be inappropriate for ice loads, and adoption of methods to handle non-normal load distributions need to be considered for the LRFD to be a robust design specification.

Future studies should also seek to expand the available ice load data with the objective of lengthening the period of record and expanding the database in terms of river size, morphology, and climate. Also, provisional effective ice strength probabilities should be computed from the currently available force data and made available for the engineering community for developing realistic design ice loads.

Acknowledgments

This work was funded by the states of New York, Illinois, Michigan, Minnesota, and Ohio and the Federal Highway Administration.

References

Ashton, G.D. (1985) Deterioration of floating ice covers. *Journal of Energy Resources Technology*, Vol. 107, June, p. 177–182.

American Association of State Highway and Transportation Officials (AASHTO) (1996) *Standard Specifications for Highway Bridges.* 16th Edition, Washington, D.C.

American Association of State Highway and Transportation Officials (AASHTO) (1998) *LRFD Bridge Design Specifications.* 2nd Edition, Washington, D.C.

Brown, T., Croasdale, K., Bruce, K., and Azarnejad, A. (1998) Observations from the Confederation Bridge Ice Monitoring Program. *Proc., 8th International Offshore and Polar Engineering Conference (ISOPE'98)*, Montreal, Canada, p. 438–442.

Evans, M., Hastings, N., and Peacock, B. (1993) *Statistical Distributions.* 2nd edition, John Wiley and Sons, Inc., New York.

Haehnel, R.B., Zabilansky, L.J., and Clark, C.H. (1999) Evaluation of the use of dusting for the prevention of ice jams in Minnesota. Contract Report CON-128, U.S. Army Cold Regions Research and Engineering Laboratory, Hanover, NH.

Haynes, F.D., Sodhi, D.S., Zabilansky, L.J., and Clark, C.H. (1991) Ice force measurements on a bridge pier in the St. Regis River, New York. Special Report 91-14, U.S. Army Cold Regions Research and Engineering Laboratory, Hanover, NH.

Haynes, F.D., and Sodhi, D.S. (1983) Ice forces on a Yukon River bridge pier. Internal Report 868, U.S. Army Cold Regions Research and Engineering Laboratory, Hanover, NH.

Johnston, M.E., Timco, G.W., and Frederking, R. (1999) An overview of ice load measurements on bridge piers. *River Ice Management with a Changing Climate: Dealing with Extreme Events, Proc., 10th Workshop on River Ice,* Winnipeg, Manitoba, June 8–11.

Lipsett, A.W., and Gerard, R. (1980) Field measurements of ice forces on bridge piers, 1973–1979. Report No. SWE 80-3, Alberta Research Council, Edmonton, Alberta.

Neill, C.R. (1976) Dynamic ice forces on piers and piles. An assessment of design guidelines in the light of recent research. *Canadian Journal of Civil Engineering*, No. 3, p. 305–341.

Palisade Corporation (1997) *BestFit™ Probability Distribution Fitting for Windows. User's Guide.* Palisade Corporation, Newfield, NY.

Prowse, T.D, Demuth, M.N, and Chew, H.A.M. (1990) Changes in the flexural strength of ice under radiation decay. *Nordic Hydrology*, Vol. 21, p. 341–354.

Schwartz, J. (1970) The pressure of floating ice-field on piles. *Proc., IAHR Symposium, Ice and its Action on Hydrualic Structures,* paper 6.3.1-12, Reykjavik, Iceland.

Sodhi, D.S., Kato, K., and Haynes, F.D. (1983) Ice force measurements on a bridge pier in the Ottaquechee River, Vermont. CRREL Report 83-32, U.S. Army Cold Regions Research and Engineering Laboratory, Hanover, NH.

USACE (1999) *Engineer and Design Manual: Ice Engineering.* Engineer Manual 1110-2-1612, U.S Army Corps of Engineers, Washington, D.C..

Wadhams, P. (1981) Sea ice topography of the Arctic Ocean in the region 70°W to 25°E. *Phil. Trans. Roy. Soc. London,* A302, p. 1464–1504.

Zabilansky, L.J. (1996) Ice force and scour measurements for the White River, Vermont. CRREL Report 96-6, U.S. Army Cold Regions Research and Engineering Laboratory, Hanover, NH.

Evaluation of Dusting for Preventing Ice Jams

Leonard J. Zabilansky[1], Robert B. Haehnel[2], and Charles H. Clark[3]

Abstract

A demonstration of the effectiveness of dusting for advancing the deterioration of river ice was carried out on the Red River of the North Basin shared by Minnesota and North Dakota. A sand spreader suspended beneath a helicopter was used to apply sand on the ice. The ice strength and thickness was monitored in dusted and control areas at five sites over a three-week period from the time of dusting to ice breakup. These tests showed no increased deterioration in terms of reduced ice strength or thickness in the dusted ice in comparison to the undusted ice. The failure of the dusting to produce any effect on ice deterioration is attributed to the sustained subfreezing air temperatures experienced throughout the test period and the light color of the sand used for dusting.

Introduction

Ice jam flooding occurs frequently on many rivers within the Red River of the North Basin in Minnesota and North Dakota, causing extensive damage to businesses, residences, and farmlands. During the spring of 1996 there were nine major ice jams on the Red River of the North Basin alone. Nonstructural ice control techniques that have been employed to mitigate the ice-jam-related flood damage include mechanically removing or weakening the ice, perforating the sheets, or enhancing solar melting by applying a dark material on the ice surface. Several states, including Alaska and Nebraska, have used dusting as a means to reduce ice jam severity by accelerating the decay of river ice in problem reaches. Dusting is the application of a dark material (e.g., sand or leaves) to an ice or snow surface to increase solar absorption. Agencies within Minnesota and North Dakota, in cooperation with the re-

[1] Research Civil Engineer, Cold Regions Research and Engineering Laboratory (CRREL), 72 Lyme Rd, Hanover, NH 03755
[2] Research Mechanical Engineer, CRREL
[3] Electronics Technician, CRREL

spective Army National Guards, planned field trials of dusting in the spring of 1999 to evaluate its effectiveness of dusting for reducing ice strength and thickness and subsequently reducing jam severity. This report is an overview of the investigation; details are presented in Haehnel et al. (1999a, 1999b)

Background

Air temperature, water temperature, solar radiation, snow cover, and precipitation affect the springtime deterioration of river ice. The quantity and timing of these weather conditions play a large role in the melting rate of the ice cover, and they determine whether or not there will be sufficient ice to form a jam when the spring freshet occurs. Dusting is intended to accelerate the melting of the snow cover, which protects the ice from warm air temperatures and solar radiation. By applying a dark material to the white snow and snow-ice on the river, more solar energy is absorbed, and this protective layer melts faster. Depending on the dusting material used, the albedo (the ratio of reflected solar radiation to incoming solar radiation) of the white surface can be reduced from 0.5–0.9 (for snow-ice and freshly fallen snow) to 0.1–0.2, thereby significantly increasing the absorbed solar energy. Clear or black ice is optically quite clear and readily absorbs much of the solar radiation striking it. It typically has an albedo of 0.1–0.2. Once the overlying protective white layer of snow is removed, dusting does little to increase the deterioration of black ice. Consequently the objective of most dusting operations is to accelerate the removal of the snow layer so that sunlight can start deteriorating the black ice sooner.

One factor affecting ice jam formation is the size of ice floes present at breakup. As an ice cover deteriorates, it loses thickness and strength. The ice thickness and the flexural strength of the ice control the size of ice floes during breakup. In general, thin, weak ice forms small ice floes, while thick, strong ice results in large ice floes; generally the larger the ice floes, the more likely that a jam will occur. Dusting of the ice not only melts the snow cover, but observations made by Arnold (1961), Spetsov (1965), and Williams (1967) indicate that the dust particles penetrate into the surface of the ice and may weaken the ice structure. By comparing the flexural strength between the control and test sites, the actual reduction of the flexural ice strength due to dusting can be determined, and a positive correlation between dusting and reduction in ice jams can be inferred. Though numerous studies have documented the effects of dusting on snowmelt and particle penetration into an ice cover, there have been no efforts to study the loss of ice strength caused by dusting, the objective of this investigation

Site Selection

To evaluate the benefit of dusting, control and test sites were selected at the planned dusting locations. A control site was used to monitor the deterioration of the ice without the influence of dusting. The test site was an area within the dusted reach where deterioration was monitored. Comparison of the test and control give an indication of any acceleration in deterioration of the dusted sections over natural conditions. To maintain the dusting as the primary variable between control and test sites, we selected sites in dusted and undusted areas that were close together and had similar hydraulic conditions, river morphology, sun exposure, and ice thickness. Areas with abnormalities that would affect the local rate of melting or ice weakening (water inlets, contaminated or altered ice surface, etc.) were not used as test or control sites.

The general locations selected for the dusting were based on historic ice jam occurrences within the basin (Fig. 1). The two rivers in Minnesota that most frequently experience ice jams are the Red Lake River and the Thief River (a tributary to the Red Lake River). From 1910 to 1997, no fewer than 89 ice jams occurred on these two rivers alone; these jams occurred most frequently in the cities of Crookston and Thief River Falls. The metropolitan districts of both cities lie along these flat-sloped rivers, and ice jams occur as the ice tries to pass through the river meanders. Three sites with a history of ice jams were selected, including upstream of Thief River Falls, 6.7 km upstream on Crookston, and just downstream of Crookston. The two sites in North Dakota were on the Park River in the city of Grafton and on the Goose River in the city of Hillsboro. The sites are characteristic of tributaries for the Red River of the North in that they are relatively narrow meandering channels that commonly jam with ice in the spring.

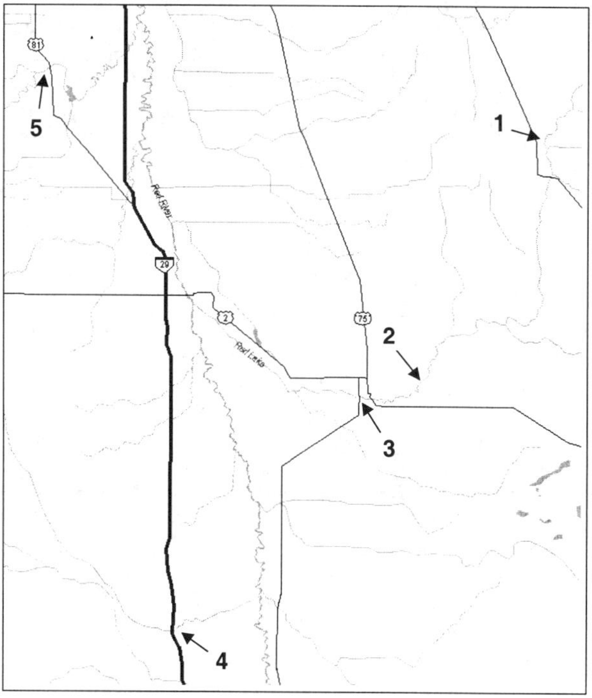

Figure 1. Locations of dusting sites. 1–Thief River Falls, MN; 2–Upstream of Crookston, MN; 3–Downstream of Crookston, MN; 4–Hillsboro, ND; 5–Grafton, ND.

Experimental Approach

Dusting Procedure

A variety of materials have been used for river ice dusting, including sand, fly ash, dyes, and leaves. In general, the darker the material, the better. Also, the material should have a density greater than water to keep from being readily washed away, and it needs to be environmentally benign. Choosing a material that is indigenous to the area generally satisfies this last requirement. The dust is normally applied using aerial crop-dusting equipment. Such equipment is usually inexpensive, and application is quick; on wide, straight rivers these aircraft work very well. On narrow, winding rivers, such as the Red River of the North, it is difficult to dust with these aircraft. The alternative used for this application was a conventional 1-ton sand spreader suspended below a helicopter. The density of the sand coverage was a function of the settings of the mechanical stops (used to regulate the volume of sand delivered to the rotating spreading disc) and the ground speed of the helicopter.

Meteorological and River Conditions

The effectiveness of the dusting is a function of the weather conditions from the time of dust application to ice out. The important variables that affect dusting are incoming solar radiation, air temperature, and snowfall. Of additional interest is the water temperature, which provides an indicator of the melting taking place at the water–ice surface. Meteorological stations were set up in Thief River Falls and at the Hillsboro sites to continuously monitor air temperature, water temperature, and solar radiation. The water and air temperatures were measured with thermistors, and the incident solar radiation was measured with a pyrometer. To quantify the increased solar energy absorbed by the dusted ice, the albedos of the control and test sites were measured using a pair of portable pyrometers. Snowfall data were obtained from the Grand Forks Airport in North Dakota.

Ice Strength

To measure the bending strength of the ice, we used a cantilever beam test technique, which is briefly described here. A full description of the test apparatus and procedure appears in Haehnel et al. (1999a, 1999b). For a cantilever beam tests, three sides of a rectangular beam are cut free from the parent ice sheet, leaving a tongue of ice with the fourth side or root end of the beam remaining attached to the parent sheet. The free end of this cantilevered beam is loaded until the ice beam breaks at the root. Because dusting would only affect the top surface, the beams were tested by pushing down on the free end. This loading configuration puts the top surface in tension and the bottom in compression. A minimum of three beams was tested for each set of cantilever beam tests. Beam width (b) and length (l) were 1.5 and 6–7 times the ice thickness (h), respectively. The beams were prepared by first drilling a 20-cm hole at the root end of the beam with an ice auger; the radius at the root of the beam reduces the stress concentration at the root. The three beams were cut adjacent to each other. The outside beams were separated from the parent ice sheet with a single vertical chainsaw cut. The center beam was isolated from the adjacent beams by a 5- to 15-cm slot between the beams. A 15-cm slot was also cut at the free end to ensure that the beam did not bind with the parent ice sheet as the tip of the beam deflected under the load.

The ice beams were loaded using a simply supported beam apparatus with a hydraulic ram reacting on the parent ice sheet. The hydraulic pressure was monitored using a pressure gage in series with the ram. The force acting on the free end of the beam, P_{ice}, was calculated by using simply supported beam equations with the reaction distances and load from the hydraulic ram. The flexural strength σ of the ice is then calculated using cantilever beam equations (eq. 1) using l as the distance from the failure crack to the centerline of the load blocks, P_{ice} as the computed failure load, and width b and thickness h as measured at the failure face:

$$\sigma = \frac{6P_{ice}l}{bh^2} \quad (1)$$

Results and Discussion

Dusting Application

The sand was applied 2 March 1999 on the Thief River. A problem with that application was that the sand was applied cold and frequently froze in the hopper. As a result, frozen clumps of sand plugged the feed to the spinner disc before the entire load could be dispersed on the ice. Also, because some frozen clumps landed intact on the ice, the sand on the Thief River had a mixture of discrete sand particles and sand clumps distributed over the ice surface. Though this was an unintended result, we observed that after a couple of days the clumps penetrated deeper into the snow-ice than individual particles. In many areas on the Thief River, the rotor wash blew the sand into the tracks left in the snow by the snowmobiles, concentrating the sand in those areas rather than evenly distributing the sand over the surface. However, the bulk of the ice was covered with an evenly distributed sand layer.

The test site in Crookston was dusted on the morning of 3 March. To eliminate the clumping problem experienced in Thief River Falls, the Minnesota Department of Transportation kept the sand-loaded trucks inside a heated garage overnight. This warm sand did not freeze into clumps during application and was evenly distributed on the ice surface. The rotor wash from the helicopter picked up the loose snow on the surface, mixed it with the sand, and blew the loose snow and sand off the flat ice; it collected in snowmobile tracks in the snow. After a couple of hours of sun, the sand that collected in the snowmobile tracks showed evidence of melting the snow.

The reach of the river upstream of Crookston was dusted on the sunny afternoon of 3 March. The ice surface was swept clean by the rotor wash, and the snow and sand collected in pockets, crevices, and imperfections on the ice surface. The sand from one load had sufficient time to melt into the surface to provide a rougher surface, which then captured more sand when the operation resumed.

Snow fell a couple of days after completing the dusting; the snow depth varied from a few centimeters in Crookston to 20 cm in Thief River Falls. In areas where the sand had collected (snowmobile tracks, imperfections in the ice surface, and frozen clumps), the snow melted quickly. Sand was also brought back to the surface in areas with high snowmobile traffic.

Table 1. Summary of sand applications at each of the evaluation sites.

Site	*Distance dusted (km)*	*Volume of sand applied (m^3)*	*Application density (m^3/km)*
Thief River Falls, MN	4.2	5.7	1.4
Upstream of Crookston, MN	2.5	5.0	2.0
Crookston, MN	2.2	3.6	1.6
Grafton, ND	0.120	0.35	1.7
Hillsboro, ND	0.220	0.5	2.3

The sand was applied 11 March 1999 at both Grafton and Hillsboro sites, immediately following a snowstorm. During the dust application the rotor wash picked up the newly fallen snow and mixed it with the sand as it was being applied.

Meteorological and Ice Conditions

The meteorological data measured at Thief River Falls showed that the air temperature did not rise above freezing for any significant length of time until after 7 April. The incoming solar radiation was high throughout the test period, and on average the incident radiation was over 600 W/m^2 at Thief River Falls and around 500 W/m^2 at the Hillsboro site.

The dusted reaches had lower albedos than in the control section, even after the snowfall covered both the test and control areas (on 5 March). However, in some cases the differences were only slight. The reason that the albedo was not much lower in the dusted reaches is likely a result of the light color sand used for dusting. The albedo of the sand itself was in the range of 0.2–0.3; ideally materials used for dusting should have an albedo around 0.1.

Breakup occurred at both sites in Crookston on the night of 30 March. The ice broke up in Thief River Falls around 2–3 April. The ice sheet in Hillsboro broke up into floes prior to 28 March, but they remained in the river reach until early April. Ice-out on the Park River in Grafton occurred on 31 March.

It is interesting to note that some of the ice was cloudy because of sediment that had mixed with the ice as it formed. In a case such as this, once the overlying snow is removed, the mud entrained in the ice would be far more effective at accelerating the deterioration of the ice than artificial dusting. In other areas natural dusting occurred as soil from adjacent farmlands was blown onto the ice. Where possible, such natural processes should be taken into consideration when planning ice mitigation operations.

The loss in ice thickness measured at all of the sites through the test period is summarized in Table 2. This loss was computed by subtracting these measured ice thicknesses from the ice thickness measured during the reconnaissance We find that there is no strong trend suggesting that the ice thickness preferentially melted faster in the test sections than in the control at any of the five sites evaluated. Of particular note is that the ice in the test section at Grafton appeared to melt slower than in the control. This result may be because of the natural variability of melting along the river or because the dust blocked the solar radiation. If the dust layer is too thick, the sand actually acts as an insulator and shields the ice from solar radiation.

Table 2. Summary of ice thickness loss and albedo of the ice in the test and control areas.

Site	*Average loss in ice thickness (cm)*		*Average albedo*	
	Control	*Test*	*Control*	*Test*
	Thief River Falls, MN			
4 March			0.79	0.70
10 March	–8			
30 March	1	13	0.39	0.36
1 April	13		0.55	0.45
	Crookston 1, MN			
3 March			0.53	0.42
6 March		11		
8 March (after snowfall)	–2		0.91	–
26 March	13	18	0.36	0.16
	Crookston 2, MN			
3 March			0.64	0.52
5 March (after snowfall)	–6	–17	0.81	0.79
27 March	23	11	0.21	0.20
	Grafton, ND			
9 March	17.4			
29 March	30	13.5	0.19	0.16
	Hillsboro, ND			
9 March	1		0.87	
28 March			0.23	

Areas selected for beam tests were chosen to be representative of the majority of the surrounding ice in the area (with respect to snow depth, snowmobile tracks, etc.), and existing cracks in the ice were avoided. Subsequent tests in the same section were upstream of the previous strength measurement sites to avoid any possibility of the ice being affected by a change in hydraulics or solar heating caused by the previous test hole.

The flexural strength of the ice in both the test and control areas for all five sites is plotted in Figure 2. The filled symbols are measurements taken in the control areas, and the open symbols are measurements taken in the test sections. The lines indicate the average strength measured for a particular site (solid lines are the control areas and the dashed lines are test sites). Several trends are apparent in this plot. First, the scatter in the data decreases greatly with time. The scatter in the data measured early in March is characteristic of cold ice. Though this cold ice can be very strong, it is also brittle and is therefore sensitive to small imperfections in the ice that can create stress concentrations and premature failure. As the ice warms, it becomes more ductile and less sensitive to small imperfections in the ice. Thus, the later tests gave more repeatable results. We also note that the ice strength from site to site was very similar throughout the test period.

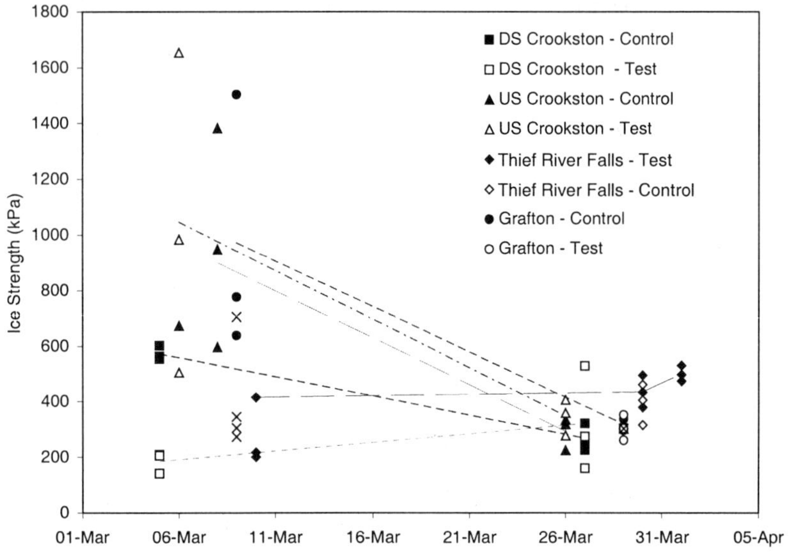

Figure 2. Ice strength in the control and test areas as a function of time.

During midwinter and early spring, the ice is cold, with the surface of the ice exposed to the air nearly at equilibrium with the subfreezing ambient air temperature. At the bottom of the ice cover, the ice is at 0°C and in thermal equilibrium with the water. During this period, ice is at its maximum strength; this cold, strong, freshwater ice can have flexural strengths as high as 1500–2000 kPa; however, values closer to 500–1000 kPa are more typical (Ashton 1986). The ice strength declines from these midwinter values as the ice warms to a uniform temperature of 0°C throughout its thickness. Icestrength tests conducted by Prowse et al. (1990) on a pond near Saskatoon, Saskatchewan, Canada, showed the flexural strength during the early decay period, as the ice warms to 0°C, to be in the range of 300–700 kPa. Once the ice becomes isothermal throughout its thickness at the freezing temperature, the ice strength drops to around 200–300 kPa. Prowse et al. (1990) and Ashton (1985) showed that the ice strength decays further as the ice becomes porous as a result of solar radiation causing internal melting at the grain boundaries.

Our results follow these trends quite well. In early March, with nighttime air temperatures well below freezing, the flexural strength of the ice was as high as 1400–1600 kPa. Yet by late March the ice strength was typically 200–500 kPa. Comparing the latter strength values with those of Prowse et al. (1990), we conclude that the ice at our test sites was nearing the point where it was isothermal at freezing but likely had not yet started to become porous.

Data from Thief River Falls show that during the study period the average air temperature was subfreezing until about 27 March, only about one week before ice-out there. Bonin and Teichmann (1949) and Williams (1967) showed that, until the average daily air temperature is above –2 to 0°C, dusting has virtually no effect on increasing the surface melting of the snow or ice and accelerating the decay of the ice cover. These observations agree with Ashton (1985) and Prowse et al. (1990), who did not observe increased porosity of the ice until the cover was isothermal at the freezing temperature. These independent works indicate that until the average daily air temperature is above freezing, the solar energy absorption into the ice is small in comparison to the loss of energy from the ice cover to the water and air through convection, advection, evaporation, and radiation.

Comparing the ice strength (Fig. 2) and thickness (Table 2) in the test and control sections, we see that there is no difference in the flexural strength or ice thickness in dusted or undusted sections of the river. Two factors were likely the cause of the lack of difference. First, the measured albedos in the control and test areas were not much different because of the light-colored sand used in this study. Second, the air temperature in the basin remained subfreezing until only about a week before the spring freshet, and there was not enough time for the solar energy to cause any significant decay in the ice cover before the ice broke up.

The questions then arise: How does the spring of 1999 compare with the average year for this basin, and can anything that we have learned be applied generally? According to the historical record, the river ice in the Grand Forks, ND. area of the basin breaks up around the second week in April. In a typical year the average daily air temperature at Grand Forks, ND, rises above freezing around the last day of March, about two weeks before the historical ice-out date. During the spring of 1999 the ice broke up in the Grand Forks area about two weeks earlier than would be expected in a typical year, and the air temperature rose above freezing only about a week before breakup. Thus, the breakup was early during this study period, and the time the air temperature was above freezing before the ice went out was about half what might be expected in a normal year. For dusting to be successful, typically about one month with an air temperature consistently at or above freezing before ice-out is required. Hence, even in a normal year, two weeks may not be enough time for dusting to have a large effect. We conclude that trying to advance the deterioration of an ice cover by enhancing the absorption of solar energy may not be a viable option in the Red River of the North Basin.

Conclusions

The ice thickness and flexural strength were monitored in control (undusted) and test (dusted) areas at five sites during a three-week period to determine the effectiveness of dusting. The differences between the dusted and control sections were indistinguishable when comparing both ice thickness losses and differences in the flexural strength of the ice. North of Fargo, ND, on the Red River of the North Basin—where the average daily air temperature remains well below freezing until just before ice break-up—dusting and other methods that rely on solar energy to weaken the ice will likely have little or no effect on advancing the deterioration of an ice cover.

Acknowledgments

The Minnesota Departments of Public Safety, Natural Resources, Military Affairs, and Transportation, the Federal Emergency Management Agency, and the North Dakota Army National Guard funded this work.

References

Arnold, K.C. (1961) "An investigation into methods of accelerating the melting of ice and snow by artificial dusting." *Geology of the Arctic: Proceedings of the first international symposium on arctic geology, 11–13 January 1960, Calgary, Alberta, Canada,* Vol. II, University of Toronto Press, Toronto, 989–1013.

Ashton, G.D. (1985). "Deterioration of floating ice covers." *Journal of Energy Resources Technology,* 107 (June), 177–182.

Ashton, G.D., Ed. (1986). *River and lake ice engineering,* Water Resources Publications, Littleton, Colorado.

Bonin, J.H., and O.E.Teichmann (1949). "Investigation of solar energy for ice melting." Armour Research Foundation, Technology Center, Final Report, Project No. 90-688 D.

Haehnel, R.B, L.J. Zabilansky, and C.H. Clark (1999a). "Evaluation of the use of dusting for prevention of ice jams in Minnesota." Contract Report for the Minnesota Department of Public Safety, U.S. Army Cold Regions Research and Engineering Laboratory, Hanover, New Hampshire.

Haehnel, R.B, L.J. Zabilansky, and C.H. Clark (1999b). "Evaluation of the use of dusting for prevention of ice jams in North Dakota." Contract Report for the North Dakota Army National Guard, U.S. Army Cold Regions Research and Engineering Laboratory, Hanover, New Hampshire.

Prowse, T.D, M.N. Demuth, and H.A.M. Chew (1990). "Changes in the flexural strength of ice under radiation decay." *Nordic Hydrology,* 21, 341–354.

Spetsov, F.A. (1965). Weakening of ice in spring season by darkening its surface. In *Trudy Koordinatsionnykh Soveshchaniy po Gidrotekhnike (Transactions of the Coordinating Conference on Hydraulic Engineering)*, 17, 212–218.

Williams, G.P. (1967). "Ice dusting experiments to increase the rate of melting of ice." Technical Paper No. 239, National Research Council, Ottawa, Canada, January.

Ice Cover Effects on Bed Scour: Case Studies

Leonard J. Zabilansky[1]

Abstract

The effect of an ice cover on the rate of erosion and deposition of riverbed sediments has been the subject of much speculation. At the center of the debate is the lack of field measurements correlating the hydraulic and ice conditions with changes in bed elevation. With the recent development of robust scour probes using time-domain reflectometry (TDR) technology, it is now possible to obtain field measurements that are independent of surface conditions. Arrays of TDR probes were installed at three sites to monitor the bed elevation from freeze-up through break-up. The sites were in the Missouri River in eastern Montana, in a navigation pool on the Mississippi River, and upstream of a bridge pier in the White River in Vermont. The probes were connected to a data acquisition system to monitor the scour–fill cycle in real time during the winter, along with temperatures and water pressure. Measurements from the three sites imply that the rate of erosion or deposition of bed material depends on the variation in flow from the discharge at the time the ice cover formed.

Introduction

Sediment movement is driven by the stream hydrograph; erosion or scour usually occurs on the rising limb, whereas redeposition typically occurs on the falling limb. During one scour cycle the bed may be excavated, or a scour hole created, and then refilled. In cold regions the presence of ice adds another dimension to the already complex scour process. An ice cover approximately doubles the wetted perimeter of the river, thereby adding flow resistance. Conveying a similar open-water discharge under ice requires an increase in stage, mean velocity, or both. Typically the stage has to increase two to four times the ice thickness before the ice sheet breaks up (Donchenko 1975). As long as the stage is below this threshold, the ice cover defines the water surface elevation. Therefore, any increases above the freeze-up discharge but below the break-up threshold are reflected in increases in the mean water velocity.

Currently numerical models used to forecast sediment movement are based on extensive laboratory tests with very little field data for validation. Numerical and physical models seldom consider the effects of an ice cover or ice and debris accu-

[1] Research Civil Engineer, U.S. Army Cold Regions Research and Engineering Laboratory, 72 Lyme Road, Hanover, NH 03755-1290; phone 603-646-4319; Leonard.J.Zabilansky@erdc.usace.army.mil

mulations on the sediment transport process. This is due, in part, to the lack of instrumentation to monitor the erosion and deposition of sediments in real time under a variety of conditions, such as flooding, debris accumulation, and ice covers.

Scour Instrumentation

The U.S. Army Cold Regions Research and Engineering Laboratory (CRREL) utilized time-domain reflectometry (TDR) technology to develop and patent an instrument capable of real-time monitoring of a full scour cycle that is independent of water velocity and water temperature. Once installed in the riverbed, these robust but non-intrusive cable-based probes can tolerate impact from ice and debris. Because the probes are independent of a structure, they can be deployed to monitor bed elevations around buried utilities, constriction and abutment scour at bridges, and the stability of contaminated sediments.

The development and calibration of the TDR probes is presented in Yankielun and Zabilansky (1998) but briefly reviewed here. TDR operates by propagating an electromagnetic pulse along a transmission line. The pulse travels down the transmission line at a velocity that is a function of the electrical and physical characteristics of the transmission line and the surrounding medium. At any boundary condition along the transmission line (e.g., air/water and water/sediment), a dielectric discontinuity exists. This discontinuity is also an electrical impedance boundary. As a pulse traveling down the transmission line encounters a boundary, a portion of the pulse energy is reflected back to the source; the rest of the energy continues through the boundary until another boundary condition, or the end of the transmission line, causes part or all of that remaining energy to be reflected back towards the source. The time that the pulse takes to propagate down and back a length of the transmission line is called the "round-trip travel time." The travel time is related to the dielectric constants of the surrounding materials.

CRREL's TDR probe has two parts: the sensor and the anchoring section. The sensor or transmission section is fabricated from two parallel 3.17-cm- (1.25-in.-) diameter pipes spaced 7.6 cm (3 in.) on center. A plastic block on the top of the probe serves as a junction box for connecting the coax instrumentation cable to the sensor and keeps the pipes parallel but electrically isolated. The length of the sensing section is a compromise between the anticipated scour and the required resolution controlled by the instrumentation; we have used sensing lengths up to 1.5 m with a resolution of 1 cm. The bottoms of the parallel pipes are welded to the lower anchoring section, which provides the lateral support for the sensing section as the bed material is eroded, exposing the pipes to the current. The length of the anchoring section is a function of the lateral resistance of the bed material; it is typically on the order of 2 m. Probes are installed into the river bottom such that the top of the probe just protrudes above the bed (when measuring scour) or at the maximum level of anticipated deposition.

To avoid variations in the propagation velocity associated with the sediment layers surrounding the sensing section, the pulse is introduced at the top of the probe. Calculating the bed elevation with respect to the top of the probe is then a function of propagation velocity in water, which is nearly constant. Basing the travel time on the velocity in water, rather than the soil surrounding the probe, allows the probes to be deployed in conductive soils where the first boundary is the water/soil interface. The conductivity of the soil surrounding the buried portion of the TDR significantly reduces the level of the returned signal, making it difficult to identify subsequent boundaries.

Field Applications

The TDR probes have been deployed at three locations to assist in characterizing the effects of an ice cover on river morphology and scour around bridges. The studies were conducted on the Missouri River in eastern Montana, in a navigation pool on the Mississippi River, and upstream of a bridge pier in White River Junction, Vermont. The sites, conditions, and scour measurements are summarized below.

Missouri River, Culbertson, Montana. The objective of this study was to identify the variables that trigger sediment transport and result in significant changes in the river system between fall and spring. Five sites along the Fort Peck Reach of the Missouri River in eastern Montana were monitored during the winter of 1998-99 (Zabilansky et al., in prep.). To correlate the changes with various ice cover processes (i.e. formation, growth, and break-up), periodic bathymetric and hydraulic surveys were conducted to bracket each phase: during open water in October, following ice formation in January, prior to break-up in February, and during open water in April. To correlate changes in bed elevation with the hydrograph and ice conditions, the bed elevations at seven locations were monitored continuously at the Culbertson, Montana, site using TDRs. This reach of the river can be characterized as a meander-

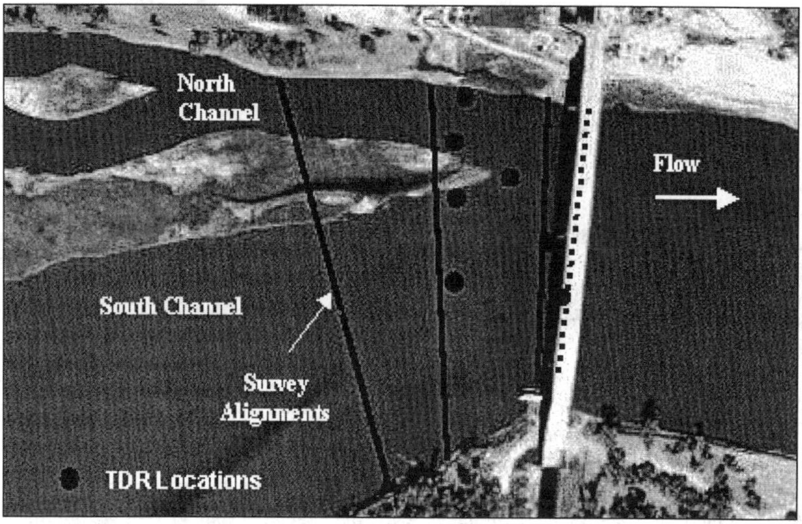

Figure 1. TDR locations and survey alignments, Missouri River.

ing channel with two sub-channels, with nominal channel depths of about 3 m in open water. The channel bed in this reach of the river is primarily sands and silts. Figure 1 shows the survey alignments used for periodic bathymetric measurements and the locations of the TDRs. Prior to freeze-up in December the average daily flow was approximately 280 m^3/s, which increased to 340 m^3/s for January and February following freeze-up and tapered to 250 m^3/s by early March. Superimposed on the average discharge are the hydroelectric releases. Prior to break-up in March the nominal ice thickness was 50 cm.

Ice cover formations occurred by juxtaposition of drifting frazil-ice slush and pans and pieces of skim ice. The ice formation process triggered a shift in the primary channel or thalweg from the south channel to the north channel. Initially the south channel was the thalweg and conveyed the majority of the flow and ice pieces. At one point the ice arched across the entrance to the north channel at the upstream end of the island, forcing the ice to pack and thicken in the south channel. As the ice pieces continued to accumulate in the south channel, a thermally grown ice cover formed downstream of the arch in the north channel. The underside of the ice accumulation in the south channel was hydraulically rougher than the smooth, thermally grown cover in the north channel. These differences in channel roughness biased the flow towards the north channel, which led to enlarging of the northern channel section by erosion of the bed and banks. Bed lowering was also measured by the TDRs along the north channel. The combined process of smoothing of the bottom of the ice, which reduced the resistance to flow, and enlarging of the north channel, which reduced the mean velocity and the rate of bed erosion, re-established equilibrium between discharge and flow area. The TDR readings of the bed elevation in the north channel remained stable until break-up, which occurred in March. As the thalweg shifted to the north channel, the velocity decreased in the south channel and the TDRs in the area measured a slight increase in bed elevation. The bed elevations at all the monitoring locations suddenly decreased in the early stage of ice break-up, even though the ice cover predominantly melted in place.

As the channel melted out upstream, ice blocks broke free from shore and floated down to the upstream edge of the ice cover. The turbulence of transitioning from open-water to ice-covered flow combined with the submergence of ice blocks floating downstream disrupted the velocity profile in the transition zone. Shifting the velocity profile increased the near-bed velocities, with a subsequent increase in erosion of the bed. Once the ice sheet fractured, the stage increased, the water velocity decreased, and the bed elevation started to increase. Although the TDR measurements were localized, the trends correlated with the global periodic bathymetry surveys taken at all the sites.

Mississippi River, Andalusia, IL. The objective of the study was to document the changes in bed elevation during the winter to calibrate a numerical sediment model. The instrumented site is at the confluence of a dredged channel with a side channel in a navigation pool on the Mississippi River downstream of Rock Island, Illinois. The instrumented site is shown in Figure 2, with the dredged channel (which provides water to a wildlife reserve) on the lower right, the side channel in the middle of the figure, and the navigation channel on the opposite side of the island along the top

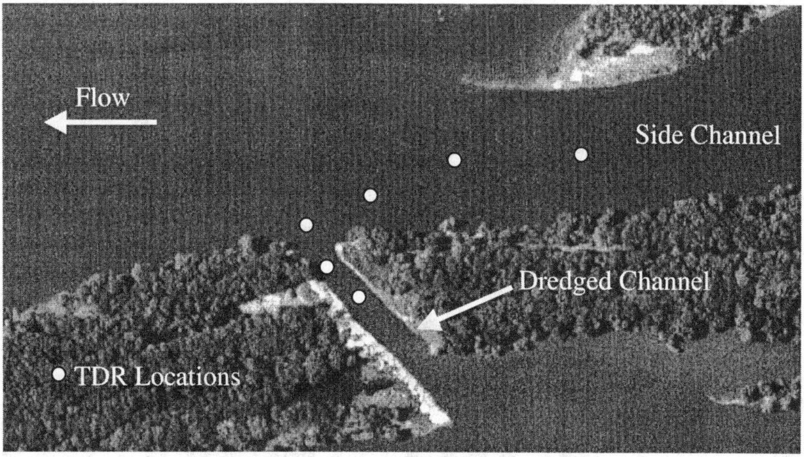

Figure 2. TDR locations, Mississippi River.

edge of the figure. Bed elevations were monitored at six locations using the TDR probes during the winters of 1999-2000 and 2000-01 (Zabilansky, in prep.). Probes were installed in the side channel and the dredged channel for the reserve; channel depths were 3 and 1.5 m, respectively. Water velocities within the pool were relatively low, and a thermally grown ice cover formed in place with a relatively smooth bottom surface. Being a regulated pool, the stage was not allowed to increase as the ice cover formed. The maximum ice thickness in the confluence area was on the order of 22 cm, and ice completely covered the width of the river. There was no appreciable change in bed elevation in the side channel during the winter, but material was deposited in the dredged channel. Ice melted in place starting in the navigation channel. As the ice melted in the navigation channel, material started to deposit in the side channel and continued to deposit in the dredged channel. Deposition continued until the side channel was ice-free.

White River, White River Junction, Vermont. To document the cumulative effect of scour around bridge piers, three TDR probes were installed in the White River immediately upstream of a bridge pier. The river is an unregulated shallow watercourse with a gravel bed. The probes were installed immediately upstream of one of the piers of the Route 5 bridge pier in White River Junction, Vermont, in September 1996 and have been monitored since (Zabilansky and Yankielun, 2000) The initial ice cover is composed primarily of frazil accumulations confined by shore ice growing laterally from the riverbanks. Typical midwinter flows are about 30 m^3/s with a water depth of 0.4 m, but discharge can jump to 110 m^3/s when melting occurs in the basin. Break-up in the White River is triggered by an increase in stage due to rainfall or snowmelt in the basin. As stage rises, the bed recedes as the system tries to maintain equilibrium between the flow area and discharge, with the ice cover fixing the water

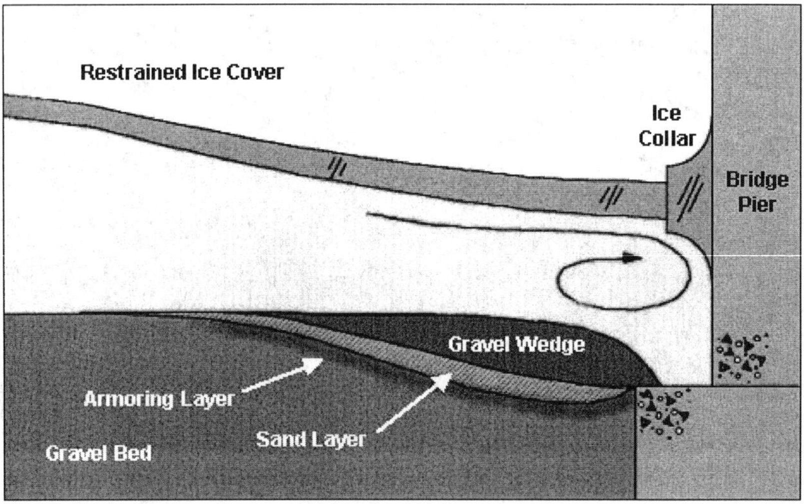

Figure 3. Restrained ice cover and sediment layers.

surface elevation. Scour continues until the ice cover breaks up, at which time deposition starts.

During the summers, scour chains installed in conjunction with the probes were recovered and the maximum depth of scour documented before resetting the chains in preparation for the next winter. Repeatedly, the upper portions of the chains were found in a 4- to 15-cm sand layer covering a very dense layer of armoring rock. The sand layer was capped with a gravel wedge, which is thickest at the pier and tapers upstream. It can be surmised that the once the ice cover breaks up, the stage increases, decreasing the water velocity, and the suspended sediment can no longer be carried through the low-velocity zones associated with the scour holes. As the open-water phase of the spring hydrograph continues, the gravel being transported as a bed load deposits on top of the sand layer, restoring the gravel wedge upstream of the pier by early summer (Fig. 3).

Discussion

Near-bed water velocity and water depth dictate the fate of suspended sediment. For a particular reach of river in open-water conditions, the maximum water velocity (V_{ow}) occurs at the surface, and the water depth (d_{ow}) is dictated by discharge (Q_{ow}) but depends on the bed roughness and channel slope as well. For a given reach, once a stationary ice cover forms, the velocity at the ice/water interface becomes zero, which shifts the velocity profile and the maximum velocity (V_{ic}) downward. The location of the maximum velocity is controlled by the friction at the ice/water and water/bed interfaces. To convey an equivalent open-water discharge, the water depth (d_{ic}) must increase and the open-water stage–discharge relationship is no longer valid. The increase in stage at freeze-up does decrease the water surface slope and the ability to carry sediment. As the ice cover thickens over the winter, it becomes more rigid and strongly attached to the shoreline and midchannel structures, which impedes the free response of the ice cover to changes in water level. To some extent this controls the flow area, as long as discharge does not increase the stage enough to cause break-up. Any fluctuation in the discharge below this threshold level has to be accommodated by an increase in the mean velocity. If the mean velocity (V) is greater than the velocity at freeze-up (V_{ic}), the increase in the near-bed velocity can trigger bed erosion.

At the Mississippi site, when the entire river was ice covered, the velocity profiles across the river were similar, with comparable near-bed velocities. As the navigation channel melted out, it was able to accommodate a larger portion of the flow, reducing the discharge and water velocity in the side channel. The lower velocity cannot carry the suspended sediment, and it precipitated out, detected by the increase in bed elevation at the monitoring points in the side channel. This observation is in agreement with Al-Abed (1989), who reported that a free-floating ice cover reduces the bed and suspended load capacity to 34% and 50% of the sediment capacity of an open-water condition, respectively. When the ice cover allows the stage to increase in response to an increase in discharge, the ice cover is classified as "free-floating."

At the Culbertson site, discharge at the time of freeze-up was 280 m^3/s, and the ice cover that formed at the corresponding stage elevation (datum) defined the cross-sectional flow area for the remainder of the winter. The ice cover remained stationary, and only shoreline cracks showed evidence of some flexing as the system attempted to accommodate an incremental increase in discharge above the freeze-up datum. With the ice cover defining the water surface elevation, the available flow area was also defined at freeze-up (A_{ic}). Ignoring creep, which would relieve the bending stress in the ice cover, any increase in discharge above the freeze-up discharge has to be accommodated by an increase in mean velocity. Increases in the velocity naturally lead to lowering of the bed elevation if the material is erodable. When the ice cover prevents the stage from rising in response to an increase in discharge, the ice cover is classified as "confining."

The confining effect of the ice cover was also detected by the probes upstream of the bridge pier in White River as a discharge increase prior to break-up. Characterizing sediment movement under an ice cover is difficult, but bridge piers increase the complexity. Olsson (2000) conducted a laboratory study using floating Styrofoam ice sheets to assess the influence of an ice cover on the depth of scour around bridge piers. He reported that the floating cover increases the scour depths up to 35% compared with open-water conditions. In the field the ice is not free floating because the piers provided intermediate anchor points for the ice cover.

The ice cover is thicker adjacent to the bridge pier because of the thermal conductivity of the pier and natural water-level fluctuations. The vertical face of this ice collar is jagged because the channel skim ice repeatedly breaks and refreezes; the upper and lower surfaces of the ice collar gradually curve into the vertical pier (Fig. 3). This curvature redirects the current towards the bed with little loss of momentum.

In shallow rivers this localized jetting action may dramatically increase the resuspension of particles, especially the non-compacted material deposited in the scour hole following a previous event. The cyclic flooding–freezing process around the pier also artificially thickens the ice surrounding the pier. In the near vicinity of the pier, the ice sheet that forms between piers and abutments is typically frozen to or has sufficient shear interlock with the ice collar to resist uplifting water pressure. As the inter-pier ice thickens and becomes more rigid, it spans between the bridge piers or abutments as a structural beam, preventing the water from rising in the vicinity of the bridge. The flexibility of the ice cover can be quantified by its characteristic length (l), which can be approximated by $16h^{3/4}$, where h is the ice thickness in meters. The radius of the area influenced by deflecting or restraining the ice cover is four times the characteristic length. With the bridge pier spacing ranging between 60 and 100 m, it is conceivable that the areas of influence for the piers overlap. For example, a 0.5-m-thick ice cover has a characteristic length of 9.5 m, or a radius of influence of 38 m. Given a 60-m span between piers, the radii of influence would overlap, and the interaction between the piers would prevent the ice sheet between the piers from freely responding to the increased discharge, thereby "confining" the flow area between the piers. Upstream of the bridge the ice cover is only fixed at the shore, and the mid-channel ice can flex as the water level increases. Thus, the ice sheet upstream of the bridge serves as a funnel, capturing the flow and confining it under the ice sheet at the piers. If the ice sheet is continuous around the bridge, any increases in discharge are reflected in the velocity and accelerated scour.

Conclusions

An ice cover can have a significant influence on the sediment transport process, particularly in the vicinity of bridge piers. Sediment transport and scour analysis needs to consider the deviation in the discharge from the time of freeze-up; if the discharge is greater than at freeze-up, the ice cover may to some extent confine the flow. The mean velocity then increases to accommodate the additional discharge, with a subsequent increase of the near-bed velocities and possible acceleration of the erosion of the riverbed.

Acknowledgments

The Omaha District of the U.S. Army Corps of Engineers (COE), the Rock Island District of the COE, and the Vermont Agency of Transportation funded studies on the Missouri River, Mississippi River, and White River, respectively.

References

Al-Abed, N.M.A. (1989). "Sediment transport under ice covered channels." Thesis submitted to University of Kashmir.

Donchenko, R.V. (1975). "Conditions for ice jam formation in tailwaters." U.S. Army Cold Regions Research and Engineering Laboratory, Draft Translation 669.

Olsson, P. (2000). "Influence of ice-cover on local scour at circular bridge piers." Water Resources Engineering, Lulea University of Technology, Lulea, Sweden.

Yankielun, N.E., and Zabilansky, L.J. (1998). "Innovative instrumentation techniques for detecting and measuring effects of sediment scour under ice." *ASCE Water,* 98.

Zabilansky, L.J. (In prep.). "Effects of ice cover on sediment movement." Contract report for Rock Island District, Corps of Engineers.

Zabilansky, L.J., and Yankielun, N.E. (2000). "Scour measurements under ice in real time." *Proceedings of the Transportation Research Board Meeting 2000*, TRB ID 00-420, Washington, DC, 12-14 Jan 2000.

Zabilansky, L.J., Ettema, R., Yankielun, N.E., and Wuebben, J.L. (In prep.). "Survey of river-ice influences on channel stability along the Fort Peck Reach of the Missouri River, Montana, Winter 1998–1999." U.S. Army Cold Regions Research and Engineering Laboratory, Technical Report.

Reservoir Operations and Their Effect on Ice Regime

Jon E. Zufelt, PE, PhD, M.ASCE[1] and Steven F. Daly, PE, PhD, M.ASCE[2]

Abstract

The Baltimore District of the U.S. Army Corps of Engineers recently undertook a study to determine whether a change in the operation of one of its flood control reservoirs would affect the degree of flood protection downstream. The intended operational change entails eliminating the current practice of providing an additional winter drawdown, which provides increased flood control storage for winter and spring runoff. The study indicated that the proposed change would not affect the current level of flood control downstream of the reservoir. Questions remained as to whether the ice regime would be changed and flooding increased near the upstream end of the reservoir. The paper describes the analysis used to determine the effects of the reservoir operational changes on the ice regime.

Introduction

Whitney Point Lake is a Corps of Engineers project located on the Otselic River in Broome County, New York, 32 km (20 miles) north of Binghamton, New York. The primary purpose of the project is to provide flood control for the Tioughnioga River downstream of Whitney Point Village, the lower reach of the Chenango River, and the Susquehanna River downstream of Binghamton. A secondary purpose of the project is recreation. A recent study conducted by the Corps of Engineers investigated the feasibility of enhancing in-stream aquatic habitat downstream of Whitney Point Lake and within the lake itself (U.S. Army Corps of Engineers 2000b). The primary change proposed to the operation of the Whitney Point Lake project is to eliminate the current 2.13-m (7-foot) winter drawdown. This drawdown provides additional flood control storage in the project for winter and spring runoff. The U.S. Army Corps of Engineers reported that, despite the reduction of winter storage as a result of the proposed change, there would be no change in the relationship among flood frequencies, flood flows, and flood stages downstream of Whitney Point Lake for

[1] Research Hydraulic Engineer, CRREL-AK, P.O. Box 5646, Ft. Richardson, AK 99505-0646; Jon.E.Zufelt@erdc.usace.army.mil

[2] Research Hydraulic Engineer, CRREL, 72 Lyme Road, Hanover, NH 03755-1290; Steven.F.Daly@erdc.usace.army.mil

flows up to and including the 100-year exceedence frequency flow event. Reducing the winter storage will have an effect on the stages of Whitney Point Lake and the Otselic River immediately upstream of the project.

These changes upstream of the project were not addressed in the Corps of Engineers report because they do not have an economic impact. The land that will be inundated was acquired when the project was built, or flow easements have been obtained. However, the Upper Lisle Bridge, located 6.9 km (4.3 miles) upstream of the Whitney Point Lake dam, is a traffic artery of Broome County, New York, that will be affected by this proposed change. Historically, the Upper Lisle Bridge has been periodically inundated by runoff entering Whitney Point Lake. However, concern was raised that eliminating the Whitney Point Lake winter drawdown would make the Upper Lisle Bridge more susceptible to damage by ice in the Otselic River and the lake during winter and spring runoff events. The Baltimore District of the U.S. Army Corps of Engineers enlisted the help of the U.S. Army Engineer and Development Center, Cold Regions Research and Engineering Laboratory (CRREL), to analyze the changes in the reservoir operation and their effects on the ice regime and jamming at the upstream end of Whitney Point Lake.

Description of Reservoir and Operations

The primary purpose of the Whitney Point Lake project is to minimize downstream flooding. When not being operated for flood control, the project's releases are adjusted to maintain a stable lake level for recreation or to maintain a minimum downstream flow, or both (U.S. Army Corps of Engineers 1996). To maintain a stable pool level, releases from the dam are adjusted to approximately equal inflow. During the winter, when the lake level is held at elevation 294.4 m (966 ft) NGVD (National Geodetic Vertical Datum), the stable lake level can be maintained up to an inflow rate of approximately 62.3 cubic meters per second (2200 cubic feet per second) using the three gates of the Whitney Point Lake outlet works. When the inflow increases beyond approximately 62.3 m^3/s (2200 ft^3/s), the level of Whitney Point Lake begins to rise but the outflow through the three fully open gates also increases because of the increased head on the outlet works. During high inflow conditions, project releases are limited, reduced, or completely eliminated, based on downstream river stages. Inflow into the lake in excess of the outlet flows is then stored for release at a later time when the downstream stages have receded. During these periods, the lake elevation rises as a result of the stored inflow. The amount of elevation change is determined by the storage capacity of the lake as a function of elevation, and the amount of excess inflow. The outflow of the lake is determined by the downstream conditions, and therefore, a relationship between the outlet flow and the lake elevation cannot be described by a rating curve during these times.

The current operation scheme is to maintain the Whitney Point Lake level at the winter pool elevation 294.4 m (966 ft) NGVD from 10 December to 1 May to provide full flood control capabilities. From 1 to 15 May, the lake is filled an additional 2.13 m (7 ft) to a summer pool elevation of 296.6 m (973 ft) NGVD to increase recreational opportunities during the primary recreation season. From 1 to 10 December, the lake level is lowered to its winter elevation. Of course, the actual lake levels may vary, especially during events when the inflow rises above 62.3 m^3/s (2200 ft^3/s) or when the outflow is curtailed.

The primary change proposed to the operation of the Whitney Point Lake project is to eliminate the current 2.13 m (7 ft) winter drawdown. This will both enhance the habitat within the lake itself and provide storage for augmenting flow during low flow periods to enhance habitat downstream of the lake. Eliminating the winter drawdown, however, will reduce the flood control storage available in Whitney Point Lake, causing the lake to rise faster and to have higher peak stage elevations during high flow events.

Flow and meteorological time series data are available in electronic format from the USGS and the Corps of Engineers. The USGS reports the average daily flow in the Otselic River at Cincinnatus, which is located several miles upstream of Whitney Point Lake. The Corps of Engineers measured the hourly outflow and the hourly water surface elevation of Whitney Point Lake at the dam. They could then estimate the inflow into the lake based on this information and the elevation-storage curve of Whitney Point Lake (U.S. Army Corps of Engineers, Whitney Point Lake Area/Capacity Table, Revised in June 1998 based on 1996–1998 sedimentation surveys, Carolyn Brumfield, personal communication). The Whitney Point Lake inflows and stages were available in electronic form from 1986 to the present. Graphical data are available from 1970 when the dam was put into operation.

Daily meteorological measurements are made by the National Weather Service and are available on CD (National Weather Service 1998). These data include daily maximum and minimum air temperature, and the total daily precipitation. Data from the First Order Station at Binghamton, New York, and the coop station at Norwich, New York, were used in the analysis.

Reservoir Flooding Hydrology

A simple hydrologic routing model of the lake was developed in order to estimate the water surface elevations of Whitney Point Lake that would have occurred for the historical period of 1986 to the present if the 2.13-m (7-foot) winter drawdown were eliminated. The actual measured flows would be used as input to the model with the only change in reservoir operation being the change in minimum pool level. The model output provides the number of times that the bridge would be submerged, at what depth, and for how long under the proposed plan.

The model is based on the continuity equation, which states that changes in the elevation of Whitney Point Lake result from differences between the flow into and out of the lake.

$$\frac{\partial S(h)}{\partial t} = I - O \qquad (1)$$

where S = the storage of the Whitney Point Lake; t = time; I = the instantaneous inflow; and O = the instantaneous outflow. The storage of Whitney Point Lake is a function of the depth of the lake, h, and can be determined from a storage-elevation curve. Hourly measurements are available for 1986 to the present.

The proposed project conditions were estimated by starting the simulation for each winter at 296.6 m (973 ft) rather than the observed 294.4 m (966 ft). Sample results for an event in January 1966 are shown in Figure 1.

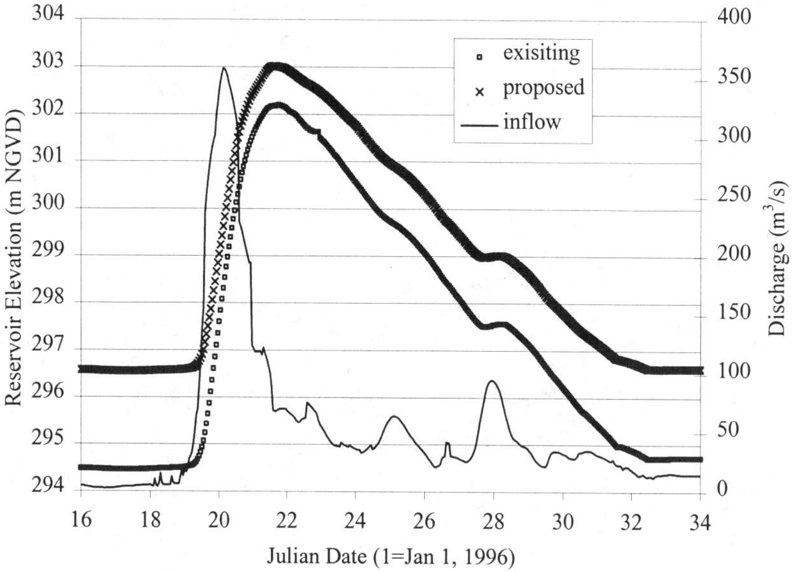

Figure 1. January 1996 event showing the hourly values of Whitney Point Lake elevation, the estimated Whitney Point Lake elevation if the 2.13 m (7 ft) winter drawdown were eliminated, and the estimated inflow.

The results can be seen in Table 1, which shows the periods of time when the bridge was submerged, the maximum depth at the bridge and the differences between the observed, and the proposed project conditions. The period of submergence occurs when the elevation of Whitney Point Lake is above the low steel elevation of the Upper Lisle Bridge, 298.1 m (978 ft) NGVD. As a result of the elimination of the 2.13 m (7 ft) winter drawdown, the number of times that the Upper Lisle Bridge would be impacted for the period 1986 to the present would increase from 8 to 12. The average time of submergence would increase approximately 53 hours for each event and the maximum depth of submergence would increase by 1.13 m (3.7 ft).

Analysis of Ice Jam Hydraulics

The effects of changes in the reservoir operation on the hydraulics and ice regime of the reservoir and river were investigated using the Hydrologic Engineering Center River Analysis System (HEC-RAS). HEC-RAS is a computer application for simulating steady and unsteady flow in river systems, including the effects of bridges, channel modifications, flooding encroachments, and ice covers and jams (U.S. Army Corps of Engineers 1998).

Table 1. Estimated changes in submergence time and submergence depth for the Upper Lisle Bridge for current and proposed operation.

Year	Total Time Submerged (Hours)			Maximum Submergence (meters)		
	Observed	Project	Difference	Observed	Project	Difference
1987	0	60	60	297.92	298.57	0.65
1988	55	126	71	298.78	300.05	1.28
1989	54	118	64	298.72	299.44	0.73
1990						
1991	0	74	74	297.66	299.10	1.44
1992	0	3	3	296.53	298.11	1.58
1993	696	730	34	305.90	306.30	0.40
1994	372	491	119	302.53	303.61	1.09
1995						
1996	165	234	69	302.18	303.02	0.83
1997						
1998	149	200	51	301.55	302.74	1.19
1999	45	106	61	298.79	300.13	1.34
2000	137	140	3	300.53	301.86	1.33
2001?	0	24	24	296.88	298.49	1.61
Average	139.42	192.17	52.75	299.83	300.95	1.12

The geometry for the HEC-RAS model was developed from a GIS (geographic information systems) database that included topographic data at 3-m (10-foot) contours and hydrographic data at 0.6-m (2-foot) contours. Cross sections were abstracted from this information and entered into HEC-RAS using the GEO-RAS extension of ArcView (U.S. Army Corps of Engineers 2000a). All cross sections were extended to a contour elevation of at least 305 m (1000 ft) NGVD, if possible. The information on the Upper Lisle Bridge was taken from the maintenance drawings developed by the Broome County, New York, highway department. The channel roughness was estimated on the basis of accepted ranges for rivers with this plan form and appearance and from field observations.

Estimated Ice Thickness. The estimated ice thickness on Whitney Point Lake and the Otselic River upstream of the reservoir was obtained on the basis of a heat transfer analysis using the Stefan equation (U.S. Army Corps of Engineers 1999). This approach assumes that the ice growth rate is controlled by the heat transfer rate through the ice cover. It assumes a steady state temperature profile through the ice with the bottom of the ice at the ice/water interface temperature of 0°C (32°F) and the surface of the ice at the daily average air temperature, as long as the daily average air temperature is below 0°C (32°F). The maximum ice thickness, η_{max}, for a winter can be estimated as

$$\eta_{\max} = \alpha\sqrt{AFDD_{\max}} \tag{2}$$

where $AFDD_{max}$ = maximum Accumulated Freezing Degree-Days recorded from the onset of freezeup for each winter. The number of Freezing Degree-Days (FDDs) for any day is found by subtracting the average air temperature from 0°C (32°F). The coefficient α varies with the units (°C or °F) and takes account of the insulating properties of snow on a growing ice cover. Based on the meteorological record of snowfall, a value of $\alpha = 0.02$ m(°C-day)$^{-1/2}$ was used in this study.

The maximum ice thickness was estimated for each year from 1978 through 1997. These estimates were based on the air temperatures measured by the NWS at the Norwich, New York, coop station. The annual maximum ice thickness ranged from 0.30 to 0.57 m (12 to 23 in.) for this period, with an average of 0.42 m (16.88 in.). In all subsequent analyses, we used a thickness of 0.46 m (18 in.) which, while it is a slightly conservative estimate of ice thickness for this case, is typical of average ice conditions in the northeast United States.

Estimating Breakup Discharge. The process of river ice cover breakup begins with an increase in discharge, usually caused by a spring rain or significant snowmelt in the drainage basin. As the flow rises, the cover is lifted vertically and the stage increases. Longitudinal cracks along the shoreline flex and additional longitudinal cracks may form along the channel centerline or third-points. Continued stage rise causes the ice cover to be lifted free from its connection to the shore-fast ice. The combination of loss of connection and increased top width attributable to the stage rise results in the cover being able to float freely in the river and move downstream. Transverse cracks form and the cover breaks into increasingly smaller pieces as it travels downstream.

If an area is reached where ice transport capacity is less than the supply of ice moving downstream, such as a channel constriction, tight bend, or intact ice cover, a jam may form. Jams will thicken such that their internal strength and shear resistance at the banks is great enough to withstand the downstream-acting forces of gravity and shear of the water flow on the jam underside. Once a jam forms, water levels rise quickly upstream of it. With continued increases in discharge, the jam will continue to thicken to resist the downstream-acting forces. Water levels upstream of an ice jam often result in overbank flooding, which reduces the discharge traveling beneath the jam and thus reduces shear on the jam underside. Although water passes onto the floodplain, the ice jam is typically confined to the main channel as trees along the banks retain ice in the channel.

A jam will fail when the downstream-acting forces increase beyond its ability to resist them. This could be caused by a continued increase in discharge, loss of shear resistance at the banks, or loss of resistance at the jam toe (downstream end). Increases in discharge result in increased shear stress on the jam underside and can also result in considerable melting of the ice in the jam. Loss of shear resistance at the banks can be the result of melting or water level increases significantly above the top of bank, such that ice is able to move into the floodplain. Toe support can be reduced by melting or significant increases in stage such that the resistance to downstream movement is lost. Once a jam fails, the ice travels downstream until it melts or reaches another area where ice transport capacity is reduced.

For most of the winter, discharge in the Otselic River is fairly low and a competent ice cover forms quickly. As stated above, the temperature records for Norwich indicated that the ice cover on the river would reach a thickness of 0.46 m (18 in.) in an average winter. The Otselic River discharge increases quickly as air temperatures rise and snow melts in the basin. Discharges able to induce breakup and movement of the ice cover can occur in any month of the winter (December through April). If a breakup occurs early enough in the season (mid-December to mid-January), it is possible that another ice cover can form on the Otselic River such that a second breakup could occur later in the spring.

Following an inspection of the discharge records, it was determined that the normal winter freezeup flow is approximately 7.1 m^3/s (250 ft^3/s) at the inlet to Whitney Point Lake. This discharge was used as the base flow rate. Figure 2 shows a profile plot of the water levels and 0.46-m (18-in.) ice cover on Whitney Point Lake at the current winter reservoir level of 294.4 m (966 ft) NGVD and the planned level of 973 ft NGVD. The figure shows that the backwater reaches only to 1500 m (5000 ft) below the bridge at the 294.4-m (966-ft) NGVD level but about 1200 m (4000 ft) upstream of the bridge at the 296.6-m (973-ft) NGVD level.

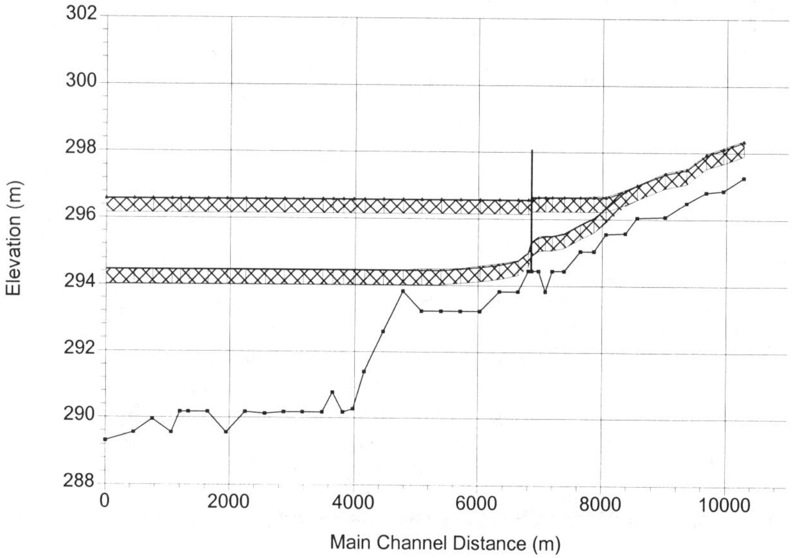

Figure 2. Whitney Point Lake profile showing the typical freezeup levels at a discharge of 7.1 m^3/s (250 ft^3/s) for Existing 294.4-m (966-ft) NGVD and Proposed 296.6-m (973-ft) NGVD reservoir elevations.

HEC-RAS was used to model the Otselic River–Whitney Point Lake system at increasing discharge levels to determine at what discharge the 0.46-m (18-in.) ice cover would break up and where the cover was most likely to jam. A downstream rating curve was required to simulate these various discharges in the Otselic River. To construct a representative rating curve, the actual stages in the Whitney Point Lake during the initial rising limb of high-flow events were plotted against the observed inflows. This data set was analyzed to develop a representative rating curve for both the existing and proposed operating conditions.

Analysis of field observations of several similar-sized rivers in the northeastern United States (Zufelt et al. 2001) indicated that the ice cover would begin to break up and move downstream when a combination of increased stage and top width allowed the cover to float freely from the support of the shore ice. The criteria for ice cover breakup for this river–reservoir system were found to be

$$\Delta WSL \geq 2t_i \tag{3}$$

$$B/B_{pb} \geq 1.20 \tag{4}$$

where ΔWSL = the change in water surface level from the base discharge of 7.1 m^3/s (250 ft^3/s); t_i = the ice cover thickness, B = the top width; and B_{pb} = the pre-breakup top width at 7.1 m^3/s (250 ft^3/s).

The modeling indicated that the target water level increase required to break up the sheet ice cover is reached between 28 and 57 m^3/s (1000 and 2000 ft^3/s) for the reach of the Otselic River upstream of the backwater section. Examining the changes in top width and the velocity at each cross section indicated that the breakup flow for the initial reservoir elevation of 294.4 m (966 ft) NGVD was 51 m^3/s (1800 ft^3/s). At this inflow discharge, the ice cover would break up and move downstream from the upstream end to approximately 5800 m (19,000 ft) above the dam. A similar analysis was conducted for a beginning reservoir elevation of 296.6 m (973 ft) NGVD. This showed that the upstream reaches would exhibit a similar behavior, with the ice cover breaking up at approximately 51 m^3/s (1800 ft^3/s). Owing to the higher initial backwater, however, the cover would only break up and move downstream to 7900 m (26,000 ft) upstream of the dam.

Ice Jamming. Ice jamming occurs when the downstream movement of ice is slowed or stopped by obstructions or abrupt changes in river plan form. In the case of the Otselic River and Whitney Point Lake, the ice sheet in the backwater of the reservoir impedes the downstream movement of ice. River plan form also is important in determining where the moving ice stops and begins to jam. At a winter reservoir level of 294.4 m (966 ft) NGVD, the ice will continue to move downstream into the reservoir backwater. At this point, the reservoir significantly widens and deepens, resulting in an associated drop in energy slope and thus the ability to transport ice.

At a winter reservoir level of 296.6 m (973 ft) NGVD, the backwater extends a considerable distance upstream. The moving ice reaches the backwater of the reservoir at a location coincident with a 180-degree meander and local scour hole. This combination of restrictions to ice movement will result in the ice jamming at this location. Figure 3 shows the plan form of the upper portion of Whitney Point Lake and the Otselic River and points out the locations where moving ice would jam at winter reservoir elevations of 294.4 and 296.6 m (966 and 973 ft) NGVD.

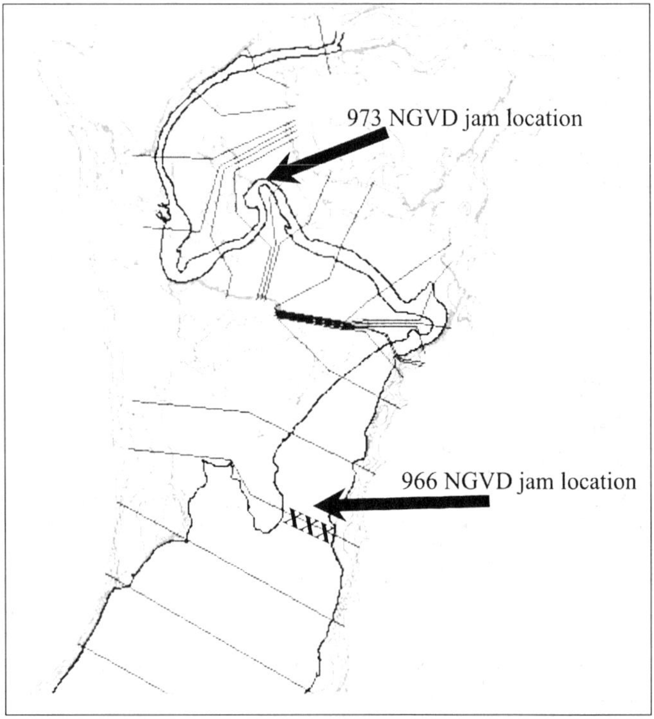

Figure 3. Upper portions of Whitney Point Lake and the Otselic River denoting locations where ice jamming occurs at winter reservoir elevations of 294.4 and 296.6 m (966 and 973 ft) NGVD.

Figure 4 compares the water surface elevation and ice thickness profiles when the jam first occurs at 51 m^3/s (1800 ft^3/s) for the two reservoir elevation conditions. At the 294.4-m (966-ft) NGVD reservoir elevation, the jam forms below and builds through the bridge opening, causing a significant rise in water elevation. The velocities in the narrow bridge opening are quite high and may prevent the formation of a thick accumulation directly in the opening. While water levels are high, ice does not appear to contact the low steel of the bridge. For the 296.6 m (973 ft) NGVD reservoir elevation, however, the sheet ice remains in place through the bridge opening and upstream. The jam forms about 1200 m (4000 ft) upstream of the bridge and quickly assumes the same thickness profile and water levels as the 294.4-m (966-ft) NGVD jam at these upstream locations.

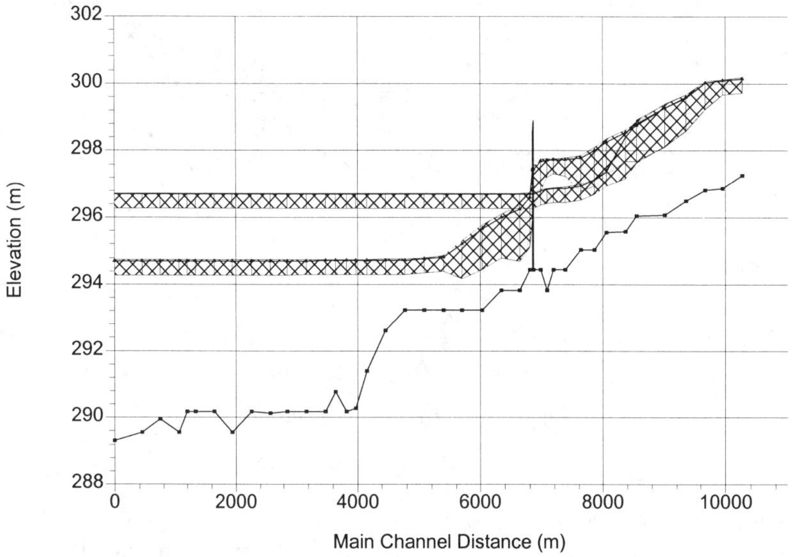

Figure 4. Water surface and ice thickness profiles when the jam first occurs at 51 m^3/s (1800 ft^3/s) for Existing 294.4-m (966-ft) NGVD and Proposed 296.6-m (973-ft) NGVD initial reservoir elevations.

The river is generally tree-lined upstream of the bridge, although there are a few reaches where wide, open floodplains exist. The ice jams were modeled as being confined to the main channel section, with water allowed to flow into the floodplains and around the jams. Water levels and jam thickness will increase with increasing discharge. At some point, water levels will be high enough to cause overbank flooding of low-lying areas. With ever-increasing discharge, the jams will eventually fail as they lose bank support or the confining ice cover as the toe erodes.

Figure 5 shows the water surface and ice thickness profiles at a discharge of 113 m^3/s (4000 ft^3/s). At this point, the ice thickness and water levels below the bridge have increased significantly for the 294.4-m (966-ft) NGVD reservoir elevation. The level of the ice has increased significantly but is still below the low steel at the bridge opening. This increased constriction results in water flowing around the bridge and over the Upper Lisle Road. Figure 5 shows that the reservoir level has increased by over 0.9 m (3 ft) at a flow of 113 m^3/s (4000 ft^3/s), and, thus, the jam is likely to fail due to loss of support at the toe. Most of the ice will pass through the bridge opening upon failure, clear of the low steel. Some ice may pass over the Upper Lisle Road or become stranded on the floodplain.

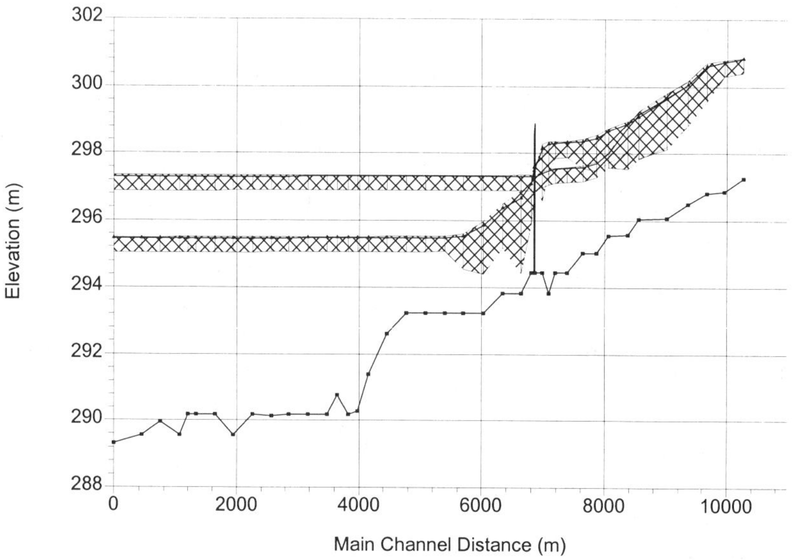

Figure 5. Water surface and ice thickness profiles at 113 m^3/s (4000 ft^3/s) for Existing 294.4-m (966-ft) NGVD and Proposed 296.6-m (973-ft) NGVD initial reservoir elevations.

For the initial reservoir elevation of 296.6 m (973 ft) NGVD, water levels upstream of the jam continue to rise and significant overbank flooding occurs. Some floodwaters will short-circuit the jam, cutting off the meander where the jam formed. Although there is significant overbank flooding upstream of the bridge, the water levels are not high enough to induce flooding over the Upper Lisle Road. At this discharge, the water levels from the bridge to the toe of the jam have increased from 0.9 to 1.2 m (3 to 4 ft), resulting in a loss of support of the toe of the jam. Increases in discharge above 113 m^3/s (4000 ft^3/s) will induce jam failure, with the ice passing through the bridge opening and into the reservoir. There appears to be no danger of ice hitting the low steel of the bridge in this scenario.

Conclusions

In an average year, 0.46 m (18 in.) of ice will form on the Otselic River and Whitney Point Lake. This cover will break up and travel downstream at an inflow discharge of approximately 51 m^3/s (1800 ft^3/s). The ice will jam at the reservoir backwater and cause water levels upstream to increase. Jams will remain in place and thicken up to a discharge of about 113 m^3/s (4000 ft^3/s). At this point the water level and top width near the toe will have increased to a point that there will no longer be support of the toe of the jam and failure will occur. With continued increase in the discharge and reservoir levels, ice will flow freely into the reservoir.

The main difference in the ice jamming process between the two initial reservoir levels of 294.4 and 296.6 m (966 and 973 ft) NGVD is the location of the jam. At an initial level of 294.4 m (966 ft) NGVD, the jam will form at the backwater about 1500 m (5000 ft) downstream of the Upper Lisle Bridge. At an initial level of 296.6 m (973 ft) NGVD, the jam will form approximately 1200 m (4000 ft) upstream of the Upper Lisle Bridge. In both cases, under normal breakup conditions, the jams will fail prior to reaching water or ice elevations that would impact the low steel of the bridge. In both cases, overbank flooding will occur with ice deposited on the floodplains. This analysis shows that water and ice will pass over the Upper Lisle Road prior to jam failure at an inflow discharge of 113 m^3/s (4000 ft^3/s) when the initial reservoir level is 294.4 m (966 ft) NGVD. Water and ice are less likely to pass over the road prior to jam failure for the initial reservoir level of 296.6 m (973 ft) NGVD, as the water levels immediately upstream of the bridge will be less.

The change in the operation of the reservoir will lead to increased periods of submergence of the Upper Lisle Bridge at the upstream end of Whitney Point Lake. The total flood control capacity of the reservoir will be reduced because of the increase in reservoir elevation during the winter months. The proposed changes, however, do not appear to increase the likelihood of ice impacts on the Upper Lisle Bridge.

References

National Weather Service (1998). *Cooperative Summary of the Day*, U.S. Department of Commerce, NOAA, National Climatic Data Center, Asheville, NC.

U.S. Army Corps of Engineers (2000a). *HEC-GeoRAS: An extension for support of HEC-RAS using ArcView, Users Manual,* Hydrologic Engineering Center, Davis, CA April 2000.

U.S. Army Corps of Engineers (2000b). *Section 1135 Whitney Point Lake Project Modification, Draft Feasibility Report and Environmental Assessment,* U.S. Army Corps of Engineers Baltimore District, Baltimore, MD, October 2000.

U.S. Army Corps of Engineers (1999). *Ice Engineering Manual,* EM 1110-2-1612, U.S. Army Corps of Engineers, Washington, DC 20314-1000, April 1999.

U.S. Army Corps of Engineers (1998). *HEC-RAS River Analysis System, User's Manual, Version 3.0*, Hydrologic Engineering Center, Davis, CA.

U.S. Army Corps of Engineers (1996). *Master Manual for Reservoir Regulation, Susquehanna River Basin - Upper Basin, Appendix A Whitney Point Lake, New York,* U.S. Army Corps of Engineers Baltimore District, Baltimore, MD, October 2000.

Zufelt, J.E., Reed, C.J., and Weed, J.L. (2001). "Modeling Snowmelt Runoff and the Effects of Climate Change in Central New York State." *Proceedings, 11th Workshop on River Ice*, Ottawa, Ontario, Canada, May 2001.

Ice Passage at Gated Navigation Projects in the Upper Ohio River Basin

Andrew M. Tuthill[1] and Steven F. Daly[1]

Abstract

River ice periodically impedes winter tow traffic on the major inland waterways in the U.S. During ice periods, moving vessels continually re-break the ice cover on the navigation channel creating an accumulation of broken ice pieces known as brash ice. In the pools above navigation dams where water velocity is typically low, brash ice congestion may become severe enough to bring tow traffic to a standstill. This nearly occurred in January of 2001 on the upper Ohio and Monongahela Rivers where the Pittsburgh District of the US Army Corps of Engineers operates a series of 15 locks and dams. In response, the towing industry asked the Corps about the possibility of releasing extra water at dams to clear brash ice downstream. At the district's request, the Cold Regions Research and Engineering Laboratory (CRREL) studied the problem in an effort to determine what flow levels and durations would be needed and to clear brash ice from the pools under a variety of scenarios. Based on observation and theory, CRREL developed relationships between discharge, time and brash ice volume cleared from the navigation channels above each gated dam in the Pittsburgh District.

Introduction

On rivers with winter barge traffic, brash ice can seriously hinder tow movement on the pools upstream of navigation dams. Under average mid-winter discharges, water velocities are typically low enough to allow sheet ice to form on the navigation pools. The tow traffic operates within a broken-out channel that is usually several hundred feet wide and bordered on its sides by the sheet ice. With repeated tow passages, the broken floes, or brash ice can congest the navigation channel to the point where the tows can no longer maneuver.

Although winters have been relatively mild in the Upper Ohio Basin for the last 15 years, the winter of 2001 was closer to the long-term average in terms of coldness. In 2001, the maximum accumulated freezing degree (Centigrade) day value of 213° was slightly below the 50-year average of 221. As a result of the cold temperatures and below-average river discharge, some of the navigation channels on the Upper Ohio and

[1] U.S. Army Cold Regions Research and Engineering Laboratory, 72 Lyme Road, Hanover, NH 03755-1290, atuthill@crrel.usace.army.mil, sfdaly@crrel.usace.army.mil

the Monongahela in particular became ice congested to the point where tows had trouble maneuvering.

Due to the difficulties navigating the pools on the Monongahela, the towboat industry asked the Pittsburgh District if it would be feasible to clear out brash ice by passing it at navigation dams. The district's initial response was that, at the low river flows, clearing significant amounts of ice would not be possible without loss of upstream pool depth. The concept of storing extra water in tributary flood control reservoirs then using it to flush ice was also discussed. This practice carries the risk of reducing reservoir storage capacity should a mid-winter runoff event occur however. As a result of the January 2001 ice conditions and the questions about passing ice, Pittsburgh asked CRREL to study the issue and develop relationships between discharge and ice passage capability at its gated navigation projects.

Fig. 1 shows the locations of navigation projects in the Pittsburgh District. Gated projects on the Ohio are Hannibal, Pike Island, New Cumberland, Montgomery and Emsworth Locks and Dams. Gated dams on the Monongahela are Mon 4, Maxwell, Point Marion, Morgantown, Hildebrand and Opekiska. The remainder of the Pittsburgh District navigation dams on the Ohio, Monongahela and Allegheny, have overflow weirs with no gates and therefore do not control the passage of brash ice. Table 1 provides data on the navigation pools upstream of the dams. The following observations on ice passage are based on interviews with lock and dam operators in the Pittsburgh District of the Corps of Engineers.

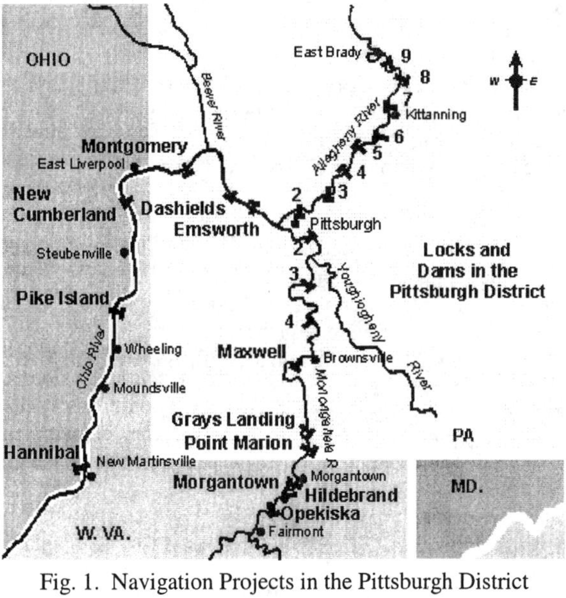

Fig. 1. Navigation Projects in the Pittsburgh District

Table 1. Pittsburgh District Navigation Pool Data

River	*Project*	*River Km**	*Pool Length (km)*	*Normal Upper Pool Elevation (m MSL)*	*Pool Storage by Pittsburgh H&H (km^3)*
Ohio					
	Hannibal	204	68	189.9	160.3
	Pike Island	136	48	196.3	110.1
	New Cumberland	88	37	202.5	91.3
	Montgomery	51	30	207.9	70.9
	Dashields (weir)	21	11	210.9	
	Emsworth	10	28	216.4	52.7
Monongahela	L/D 2 (weir)	18	20	219.0	
	L/D 3 (weir)	38	29	221.5	
	L/D 4	67	32	226.6	31.0
	Maxwell	99	33	232.6	32.9
	Gray's Landing (weir)	132	14	237.1	
	Point Marion	146	18	242.9	14.2
	Morgantown	164	10	248.1	7.6
	Hildebrand	174	12	254.5	9.4
	Opekiska	186	21	261.2	17.8

*Measured from Pittsburgh in the downstream direction for the Ohio, and in the upstream direction from Pittsburgh for the Monongahela.

A common solution to brash ice congestion of navigation channels is to clear the ice by passing it through dam gates. Necessary conditions are: 1) Sufficient river discharge to draw the ice downstream through the pool to the gates on the dam, and 2) Warm enough air temperatures to prevent brash ice pieces from freezing together. Under marginal ice passage conditions, vessel traffic is needed to break ice arches in the navigation channel upstream of the dams and keep the ice moving. Free towboats or "light boats" are critical in these ice passage operations.

At gated dams on the Monongahela and upper Ohio Rivers, ice is passed beneath tainter gates in the raised mode or passed over submergible tainter gates. In the underflow mode, these tainter gates must be raised a minimum of about 1.2 to 1.5 m to draw ice pieces beneath. A submergible gate opening must be greater than the thickness of the ice pieces in order to pass them (usually 0.6 to 0.9 m is sufficient). A third of a meter of tainter gate opening in underflow is equivalent in discharge to about one meter depth of flow over a submergible gate. Also, at most dams in the Pittsburgh District, it is not possible to open one or two tainter gates beyond about 1.5 m under low tailwater conditions without risking damage to the downstream bed protection. Due to these factors, significantly more discharge is needed to pass ice in underflow mode than in overflow.

Often at projects equipped with submergible gates, more flow is required to draw ice towards the dam than is needed to actually pass the ice over the gates. The opposite may be true if the dam is equipped with only underflow gates. In this case, the opening required to pass the ice and provide adequate tailwater depth to avoid scour may be well in excess of the flow needed to draw the ice towards the dam.

Passing ice at dams has a number of associated difficulties. First, extreme cold and heavy ice often coincide with periods of low flow and the discharge needed to pass the ice is not available. Second, if one navigation pool is seriously ice-congested, it is likely that the same conditions exist in the pools downstream, especially in rivers that run to the north such as the Monongahela. One possibility is to use the hydrograph from either a natural runoff event or a reservoir release to flush ice from pools in a downstream sequence. The difficulty with this practice is that the water wave typically travels faster than the moving ice and, as the wave attenuates, the ice stalls at downstream locations.

Average winter discharges on the Monongahela and Upper Ohio are typically low and variable On the Monongahela at Charleroi, PA the average winter discharge is 386 m^3/s, with a standard deviation of 187 m^3/s. On the Ohio at Sewickley, PA the winter average flow is 1267 m^3/s, with a standard deviation of 571 m^3/s. Not only are the variations in flow great, but also the changes can occur rapidly. For example, in a period of only a few days in late January of 2001, the rivers went from a low flow, ice-congested condition to a high discharge situation where all the ice moved downstream system-wide. During these higher flow events, the dam gate openings required to pass the water discharge are typically great enough to draw the ice beneath the gates and pass it as well.

Approach

Based on field interviews with project operators, two important ice passage scenarios were identified, going from low to high flow. In all cases, we assumed that the ice is well broken and loose, because, according to the operators, if the brash ice is frozen in a solid mass, near-flood level discharges are needed to get it moving. We first estimated a minimum flow needed to pass the loose brash ice through the navigation dam. Second we estimated the minimum river flow needed to continuously move the loose brash ice in the navigation channel downstream to the dam. When possible, these estimates were based on discussions with project personnel. Where field information was lacking, estimates were based on comparisons to nearby projects and theoretical criteria such as threshold water velocities and ice accumulation force balance calculations. Table 2 lists the estimated minimum and continuous ice passage discharges. Although upstream water velocity is higher, we assumed that the water velocity near the dam actually controls the rate at which ice passes the dam gates. Also, because the upstream water velocities are higher one would expect the drifting ice accumulation to compress somewhat in length as it moved downstream.

In most cases during the winter period, loose brash ice fills the broken-out navigation channel while sheet ice covers the channel sides. Based on this, and interviews with operators, we assumed a typical condition of a 0.3-m-thick accumulation of loose brash ice filling a centrally aligned 76-m-wide navigation channel for the Monongahela and a 90-m-wide channel for the Ohio, bordered on its sides by shore-fast sheet ice. It was assumed that the ice moves at the average velocity of the water in the channel. For each gated navigation project, under the two flow scenarios, we developed relationships between the length of pool cleared of ice and the time required for the ice to move past the dam.

The HEC-RAS model (US Army 1998) was used to calculate water velocities in the pools, based on river geometry provided by Pittsburgh Hydraulics and Hydrology (H & H) Section. HEC-RAS is a one-dimensional gradually varied flow model that calculates water surface profiles on rivers, with and without an ice cover. We modeled an open channel case and a centrally aligned open channel bordered on the sides by 15-cm-thick sheet ice (n = 0.015). According to District personnel the 15-cm sheet ice thickness represents relatively severe winter conditions for the region and the Mannings roughness of 0.015 is reasonable for a sheet ice cover (US Army 1999). The simulations used the minimum and continuous ice passage discharges listed in Table 2.

Table 2. Discharges and HEC-RAS-Calculated Average Channel Water Velocities for Ice Passage

		Minimum for Ice Transport			Continuous Ice Transport		
River	*Project*	*Total Discharge* *(m^3/s)*	*Average Water Velocity Above Dam* *(m/s)*	*Average Water Velocity @ Head of Pool* *(m/s)*	*Total Discharge* *(m^3/s)*	*Average Water Velocity Above Dam* *(m/s)*	*Average Water Velocity @ Head of Pool* *(m/s)*
Ohio	Hannibal	790	0.24	0.52	1020	0.34	0.67
	Pike Island	765	0.27	0.49	1020*	0.49	0.85
	New Cumberland	565	0.15	0.40	1020*	0.24	0.61
	Montgomery	540	0.18	0.43	1105*	0.37	0.79
	Emsworth	565*	0.46	0.67	1020	0.82	0.98
	Average Ohio		0.26	0.50		0.45	0.78
Monongahela	L/D 4	480	0.40	0.67	850	0.67	0.94
	Maxwell	480	0.37	0.82	710*	0.58	0.98
	Point Marion	310	0.37	0.76	710*	1.22	1.74
	Morgantown	340	0.43	0.70	710	0.94	1.34
	Hildebrand	310	0.37	0.70	710	0.82	1.28
	Opekiska	340	0.40	0.85	680	0.82	1.31
	Average Monongahela		0.39	0.75		0.84	1.26

* Inferred

Fig. 2 shows HEC-RAS calculated velocity profiles for the pool above Montgomery Dam. The average water velocities in central navigation channel are slightly higher than those calculated in the completely open channel case because the ice covers decrease the conveyance of channel sides relative to navigation channel. Table 2 lists the calculated velocities at the downstream and upstream ends of the pools for the ice-bordered navigation channel case.

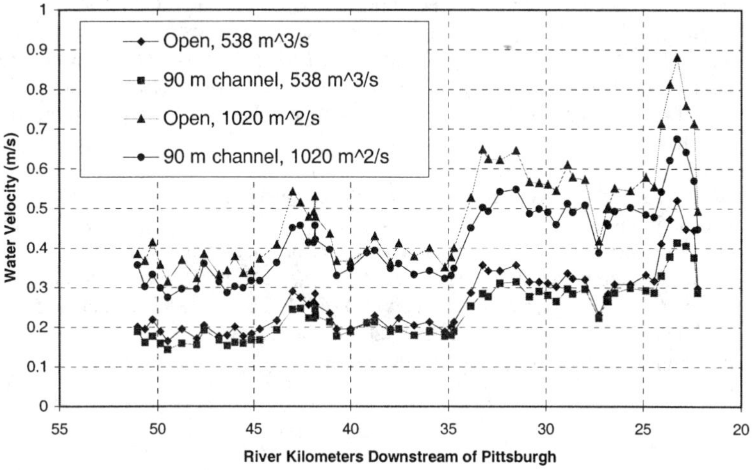

Fig. 2. HEC-RAS calculated water velocities for Montgomery Pool

The minimum water velocity needed to move an ice accumulation can be estimated based on the wide river ice jam ice jam force balance developed by Pariset et al. (1966) and presented in U. S. Army (1999) as:

$$\frac{d\left(\overline{\sigma}_x t_i\right)}{dx}B + 2\tau_b t_i = (1-e)\gamma_{ice} t_i S_f B + \tau_w B$$

where

t_i = ice accumulation thickness
τ_b = shear resistance along the banks
B = ice accumulation width
τ_w = water shear on ice underside
S_f = slope of water surface
γ_{ice} = specific weight of ice
e = ice accumulation porosity
$\overline{\sigma}_x$ = longitudinal stress within the ice accumulation in the streamwise direction.

The force balance theory treat the ice accumulation as granular material with internal stresses in the vertical, streamwise and lateral directions. Neglecting the first term, the previous equation simplifies to:

$$F_g + F_w = F_b$$

to say that, at the point of incipient motion, the downstream-acting forces of a brash ice accumulation due to gravity F_g and water shear F_w are resisted by shear along the channel sides F_b. The rationale for ignoring the first term is that, in the situation of a relatively flat pool above a dam, the water shear and water slope are low and internal stresses and ice accumulation thickness are relatively constant in the streamwise direction.

As a test of the theory, a force balance calculation was made for typical brash ice conditions at the minimum ice passage flow in the Pool 4 on the Monongahela. Based on interviews with lock and dam operators, accumulations of loose floating brash in this area are typically no more than 0.3 m thick. Assuming an accumulation thickness of 0.3 m, an ice accumulation porosity (e) of 0.5 (Tuthill, 1998) and an angle of internal friction of 45 degrees (Liu and Shen, 2000), the resisting force due to ice friction along the channel sides F_b is 65 N/m by the following calculations:

Vertical stress; $\overline{\sigma_Z} = \frac{1}{2}\rho' g(1-s)(1-e)t_i = 56.6$ Pa; where s = specific gravity of ice (0.916) and ρ' = density of ice (916 kg/m^3)

Longitudinal stress; $\overline{\sigma}_x = k_x \overline{\sigma}_z = 328$ P; where $k_x = \tan^2\left(45°+\frac{\phi}{2}\right) = 5.8$

Lateral stress; $\overline{\sigma_y} = k_l \overline{\sigma}_x = 109$ Pa; where k_l = coefficient relating longitudinal to lateral stress (0.33 used)

Shear stress along the banks $\tau_b = k_0 \overline{\sigma_y} = 109$ Pa; where k_0 = coefficient of friction = *tan* $\phi = 1.0$

Resisting force due to bank shear $F_b = 2\tau_b t_i = 65$ N/m.

The average river depth upstream of Lock and Dam 4 is 4.9 m, and, at the minimum observed ice passage flow of 480 m^3/s, the HEC-RAS-calculated average water velocity U is 0.4 m/s (Table 2). A representative brash ice channel width B of 76 m was used. A Mannings roughness n_i of 0.04 was used for the 0.3-m-thick brash ice cover (Nezhikovskiy, 1964) and a riverbed roughness n_b of 0.02 used to represent fine sediment material. The hydraulic radius R was approximated as the average channel depth of 4.9 m. The portion of the hydraulic radius contributing to shear on the ice underside R_{ice} was estimated from:

$$R_{ice} = \left[\frac{1}{1+\left(\frac{n_i}{n_b}\right)^{-\frac{3}{2}}} \right] R = 3.6 \text{ m}$$

The downstream force due to water shear on the ice underside F_w was estimated from ice roughness and water velocity:

$$F_w = \tau_w B = \frac{f}{8} \rho U^2 B = 70 \text{ N/m}$$

where $f = \frac{8gn_i^2}{R_{ice}^{\frac{1}{3}}} = 0.082$ and $\rho =$ density of water = 1000 kg/m^3

The downstream force on the ice accumulation due to gravity was calculated from:

$$F_g = (1-e)\gamma_{ice} t_i S_f B = 2 \text{ N}$$

Total downstream forces $F_w + F_g$ were 72 N/m, slightly greater that the calculated resisting force due to ice friction along the channel sides of 65 N/m. This shows the observed minimum ice passage hydraulic conditions at Dam 4 to be similar to the theoretical lower threshold.

Fig. 3 shows calculated minimum average water velocities needed to move brash ice accumulations in the 15 to 60 cm thickness range, for channel widths ranging from 30 to 90 m The curves illustrate that the potential for arching increases as the channel narrows, but there is little difference between the threshold ice passage velocities for the 76 and 90 m navigation channels.

Results

Based on discharges and calculated average water velocities near the dam, distance of brash ice cleared against time were plotted, for both the minimum ice passage and the continuous ice passage scenarios. Fig. 4 shows the plot for Monongahela Lock and Dam 4. We assumed that the calculated average water velocities for the pools near the dams controlled the rate at which ice could be cleared from the pools. Table 3 summarizes the results, listing total time (t) needed to clear the pools (including ungated upstream pools) under the two discharge (Q) scenarios. These times are conservative in that they do not account for compression of the ice accumulation due to the faster currents at the upper ends of the pools. Also listed are the water storage volumes (V) required to move the ice calculated as V = Qt.

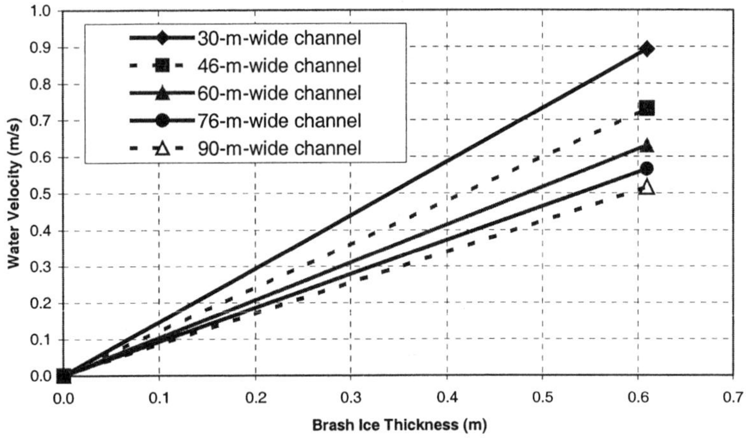

Fig. 3. Calculated water velocities thresholds for passage of brash ice for a range ice accumulation thicknesses and channel widths

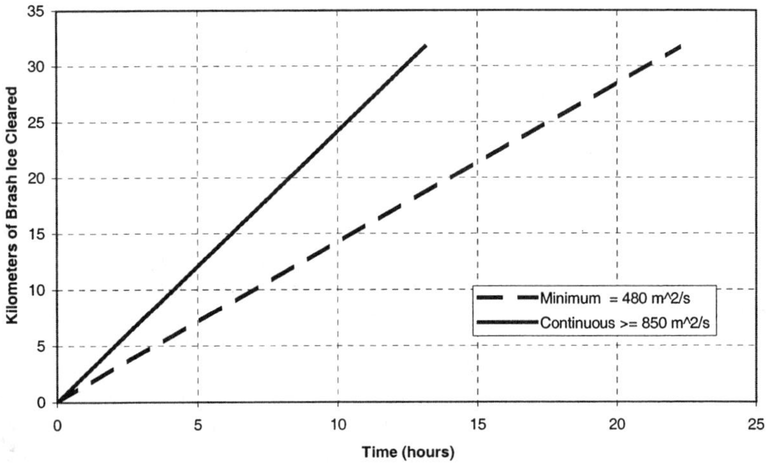

Fig. 4. Kilometers of brash ice cleared verses time for the Monongahela L&D 4 Pool, at the minimum and continuous ice passage discharges.

Times range from a maximum of 77 hours to clear the Hannibal Pool under the minimum ice passage discharge, to a minimum of 3 hours to clear the Morgantown Pool under the continuous ice passage discharge. An interesting result is that about the same storage volume is needed to clear the ice under both the minimum and continuous discharge scenarios. This suggests that, if the goal is to clear all the ice from the pool, the higher discharge is preferable because it requires less time. Also, the ice is less likely to arch at the higher flow so the need for assistance from navigation will be less.

Operators interviewed at Monongahela Lock and Dam 4 recalled that in 1993, they were able to clear about 9.6 km of broken loose ice in 4 to 6 hours. This data point falls between the minimum and continuous curves in Fig. 4.

The calculated water volumes needed to clear ice from the pools is somewhat in excess of the pool storages listed in Table 1. This results from the assumption that the ice is cleared at the average water velocity near the dams rather than the average water velocity for the entire pools, which is in most cases higher.

Times required to clear ice from pools generally increase in the downstream direction. This suggests that, unless one is willing to pass a flow volume in excess of the total of all the pool storages listed in Table 3, the ice cleared from upstream pools will tend to accumulate in downstream ones.

Table 3. Times and Storages Required to Clear Brash Ice from Navigation Pools

River	*Project*	*River Km*	*Pool Length (kilometers)*	*Minimum for Ice Transport*			*Continuous Ice Transport*		
				Flow (m^3/s)	*Time to Clear Pool of Ice (hours)*	*Storage Required (million m^3) met*	*Flow (m^3/s)*	*Time to Clear Pool of Ice (hours)*	*Storage Required (million m^3)*
Ohio	Hannibal	203.5	67.9	793	77	221	1019	56	206
	Pike Island	135.6	48.0	764	49	133	1019	27	100
	New Cumberland	87.6	36.5	566	67	129	1019	42	153
	Montgomery*	51.0	29.6	538	62	121*	1019	23*	86
	Emsworth**	10.0	28.0	566	47**	95**	1019	26**	95
Monon-gahela	L/D 4	66.8	31.7	481	22	38	849	13	41
	Maxwell***	98.5	33.5	481	36***	63***	708	23***	58
	Point Marion	146.2	18.0	311	14	15	708	4	10
	Morgantown	164.2	10.0	340	7	8	708	3	8
	Hildebrand	174.2	11.6	311	9	10	708	4	10
	Opekiska	185.8	21.4	340	15	17	680	7	17

* Including Dashields Pool, ** Including Mon 2 and Mon 3 Pools, *** Including Gray's landing Pool

Travel times from the head of the Opekiska Pool to Hannibal L&D are 16 and 10 days respectively for the minimum and continuous ice passage discharges listed in Table 2. This implies that, if flows were at the thresholds for continuous ice passage, then it would take about ten days to clear the entire system of ice. It is assumed that no melting occurs en route and the ice accumulations do not compress due to the higher velocities at the upper ends of the pools. A less conservative approach would be to use the average pool velocities to calculate the travel times. These faster velocities would account to some extent for melting en route and compression of the ice cover. Under the average velocity scenario, it would still take about 8 days to clear the system of ice if flows were at the continuous ice passage discharges.

Conclusions

In planning a flow release to clear brash ice, the first step would be to determine the amount of ice to be cleared. This would depend on the amount of flow available. In a low discharge situation, the goal might be to simply clear the upper lock approach and a few miles of the upstream navigation channel to facilitate lockages and improve vessel maneuverability in the approach. With higher discharges, clearing significant portions of the pool becomes an option and it is important to determine if the ice clearing operation is to be limited to specific project or to be a system wide effort. It is also critical to assess what the downstream effects the ice releases will be. If discharge is forecast to be at or above the estimated level for continuous ice passage for a period approaching eight days, then clearing the entire system of ice is feasible.

At the minimum ice passage discharge, we assumed that the goal would be to clear brash ice from the upper approach and a limited distance of the navigation channel upstream. Gates on the lock side of the dam are therefore favored because they are closest to the broken-out navigation channel, and may help clear ice from the upper approach.

It is important to restate that the flows needed to clear an entire pool of ice, let alone the entire system, are usually not available during the winter months without possible augmentation from reservoir storage. Reservoir releases for ice clearing are unlikely however due to the other operational constraints such as flood control. In order to impound water in for the purpose of ice flushing, the reservoirs would have to forfeit some of their flood storage capacity.

Acknowledgements

Many contributed to this study, which was funded by the Pittsburgh District. Francis Doyle and Frank Campbell participated in the field visits to the projects. Dave Turscanyi and Dave Sneberger shared their experience and knowledge of the rivers and projects, as did the lock operators interviewed. Mark Zaitsoff, Walt Leput and others in Pittsburgh H&H reviewed the report, and John Gagnon of CRREL collected and analyzed the meteorological data.

References

Liu, L. and Shen, H. T. (2000) Numerical Simulation of River Ice Control with Booms U. S. Army Cold Regions Research and Engineering Laboratory ERDC/CRREL Technical Report 00-10.

Pariset, E., Hausser, R, and Gagnon, A. (1966) Formation of ice Covers and Ice Jams in Rivers, ASCE Journal of the Hydraulics Division, Vol. 92, No. 6, pp 1-24.

Nezhikovskiy, R. A. (1964) Coefficient of roughness of bottom surfaces of slush ice covers on rivers. Soviet Hydrology Selected Papers, No. 2, p. 127-150.

Tuthill, A. M., Wuebben, J. L. and Gagnon, J. J. (1998) ICETHK Users Manual U. S. Army Cold Regions Research and Engineering Laboratory CRREL Special report 98-11. 72 Lyme Rd. Hanover, NH.

U. S Army (1998) HEC-RAS, River Analysis System, Version 2.2, U S Army Corps of Engineers Hydrologic Engineering Center, Davis, CA.

U. S Army (1999) Engineer Manual 1110-2-1612, Ice Engineering, U. S. Army Corps of Engineers, Washington, DC.

Inflatable Crest Gates at Mississippi River Lock and Dam 1

George Waldow[1] and Brad Bystrom[2]

Abstract

Frigid winters, severe ice conditions, and dramatic spring floods exact a yearly toll on dams and hydraulic infrastructure. Inflatable gates or "rubber dams" are a non-rigid water control alternative that can provide significant benefits in cold regions. This relatively recent technology can be successfully employed at both new and existing water retaining structures. Inflatable crest gates were installed to replace traditional flashboards at a Mississippi River power and navigation dam. This paper discusses the motivations for improved crest gate technology, economic justification for inflatable gates, installation procedures, and system performance to date.

Background

The St. Paul District of the U.S. Army Corps of Engineers completed Mississippi River Lock and Dam 1 in 1917. The facility was constructed between the Twin Cities of Minneapolis and St. Paul, Minnesota to extend commercial navigation to the historic St. Anthony Falls milling district located 9 km upstream. Navigation facilities including two 17 m-wide x 122 m-long lock chambers, operations buildings, and a visitor interpretive center form the right dam abutment. A 16 MW hydroelectric power station, owned and operated by Ford Motor Company, forms the left dam abutment. A 175 m-long x 10 m-high concrete overflow spillway spans the river channel between the lock structures and the power station. Figure 1 is an overall view of the spillway, power station, and Ford truck assembly plant from the limestone bluff above the navigation locks.

----

[1]M. ASCE, Vice President and Sr. Project Manager, HDR Engineering, 6190 Golden Hills Dr., Minneapolis, MN 55416

[2]Engineering Supervisor, Twin Cities Assembly Plant, Ford Motor Company, 966 South Mississippi River Blvd., St. Paul, MN 55116

Figure 1. View of dam and power station.

Winters can be severe in the Twin Cities with temperatures occasionally falling below –35 C. Ice conditions on the river typically curtail navigation activity for eight to twelve weeks each winter. Ice expansion pressures, freeze/thaw cycles, fluctuating pool levels, and impacts from large ice floes combine to inflict damage on project structures and adjacent shorelines.

Flashboard Issues

When Ford built its hydroelectric station in 1924, a system of hinged steel flashboards was installed across the overflow spillway. This system was designed to maintain established navigation pool levels while diverting up to 170 m^3/sec of flow to the four hydroelectric turbines. Hydraulic head available for energy generation was approximately 11 m under normal flow conditions. The 4.9 m-long x 0.6 m-high flashboard panels were held erect with tension rods that were designed to break when overtopping depth approached 0.5 meters.

Winter ice pressure or spring high water caused the rigid flashboard system to collapse nearly every year. Following a collapse, the hinged panels typically could not be raised for one to three months. When river conditions permitted, the pool level was drawn down slightly below the spillway crest. Then Ford maintenance personnel worked from small boats to raise the 16 flashboard panels and secure them to embedded eye-bolts with new tension rods. Sometimes the panel raising process had to be repeated several times in a single year. Occasionally flow conditions were such that the panels could not be raised for an entire year, seriously reducing energy production at the low head generating station.

Shortcomings of the original flashboard system included the following:

- The system leaked, reducing water available for energy production.
- Ice and floating debris caused flashboard panels to collapse.
- Significant energy production was lost because flashboard panels could not be raised for extended periods following a collapse.
- Raising collapsed panels was hazardous for personnel working in small boats at the spillway crest.
- Lowering the pool below the spillway crest for flashboard maintenance interfered with commercial navigation.
- Flashboard panels could not be remotely collapsed or raised.
- Controlled sluicing of debris or ice floes over the spillway was not possible.
- Flashboard maintenance was labor intensive and increasingly costly.

Alternative Systems Analysis

During a comprehensive hydropower improvement program initiated in the early 1990s, Ford wanted to determine whether a controllable gate system could economically resolve some or all of the problems associated with the flashboard system. Of particular interest was the potential increase in annual hydroelectric generation. This incremental energy would reduce the cost of energy purchased from the local utility and justify capital investment in a new gate system.

A comparative generation analysis was performed, assuming the presence of a hypothetical two-position crest gate system that could be raised and lowered in response to specific river flows and pool elevations. The analysis also assumed removal of old weir remnants creating head loss in the power station tailrace.

Run-of-river hydroelectric energy potential is basically a function of available turbine flow and the difference in water surface elevations across a dam. An energy simulation model was developed that accounted for actual power station generating characteristics and river management parameters. Utilizing historical records of daily water levels, stream flow, and energy generation, the model calculated the incremental changes in daily energy production that would have been possible with the hypothetical crest gate system in place.

Modeling results showed that potential increases in annual energy production ranged from 2,539 to 5,950 MWH during the fifty-year analysis period. The average annual increase in energy production with controllable crest gates and weir removal was calculated to be 4,454 MWH.

Based on the economic value of increased energy generation, four controllable crest gate alternatives were selected for further evaluation. These included hydraulically actuated steel crest gates, two types of inflatable rubber crest gates, and an inflatable/steel hybrid gate system. Vendors were asked to recommend an appropriate configuration for their product and to provide budgetary cost information.

Each alternative system appeared to be at least marginally feasible when subjected to Ford's very demanding cost-justification criteria. However, based on comparative analysis of installed cost estimates, operational characteristics, maintenance issues, warranty terms, system complexity, and construction considerations, only the two inflatable rubber gate alternatives were selected for more detailed design and competitive proposals.

The existing spillway was discovered to be well suited for inflatable gate installation for several reasons:

- The spillway required no more than one or two inflatable gate segments.
- Required bladder length, height, and overtopping depth were well within the range of industry experience.
- The spillway crest concrete was smooth and in good condition.
- An existing recess in the spillway crest for collapsed flashboards could be readily adapted for mounting inflatable gates.
- The hollow Ambursen-type spillway dam would simplify installation of air and drainage piping.
- The adjacent powerhouse was an available source of electric power and an all-weather shelter for air blowers and system controls.
- Pool level monitoring and recording equipment was already in place.

Inflatable Rubber Gates

The inflatable gate or "rubber dam" concept originated in the United States tire industry but was made commercially successful by the Japanese during the 1980s. Inflatable dams have been installed for various purposes throughout the world. Cold regions applications include Norway, Japan, Canada, and the Northern United States.

Inflatable rubber gates are simply linear tubes formed by folding a long strip of nylon-reinforced rubber fabric over on itself and clamping the two edges to a concrete foundation or dam crest. When the flat tube is inflated with air, it rises into a bulb-like cross-section, forming a "rubber dam". Each inflatable gate system is custom designed for its specific application. Special geometries and clamping connections are required at abutments and intermediate piers.

Required inflation pressures are normally very low (0.08 kg/cm^2), allowing the gate systems to be filled with air by low-pressure blowers rather than air compressors. A simple arrangement of piping and valves is required to inflate and deflate the bladders. Drains are also necessary to release any accumulated condensate or leakage water from inside the bladders. Depending on the application, system controls can be as basic as a single float switch or as complex as a networked PLC.

Most bladders are designed for only a small amount of overtopping flow depth before they must be deflated to prevent damage. Unsteady pressure conditions tend to develop at the downstream base of an inflated bladder during overtopping. Unsteady pressures induce oscillations and vibrations that can stress the fabric and abrade the bladder surface where it moves against concrete. Proper ventilation of the downstream gate face can significantly moderate pressure fluctuations against the bladder and thus allow greater overtopping depth. One manufacturer offers an integral slotted fin on the bladder for this purpose. An overtopping depth of approximately one half the bladder height is permissible with the optional fin.

Authorization and Permits

Permission from several regulatory agencies was needed before the flashboards could be removed and new crest gates installed. The Federal Energy Regulatory Commission, which licenses the hydroelectric facility, was agreeable to the change, provided that the Corps of Engineers did not object.

The Corps of Engineers required complete design information and a hydraulic analysis to demonstrate that the proposed crest gate installation would not impact dam stability or measurably increase upstream flood elevations. St. Paul District engineering staff performed a thorough review of the project design to be certain that it satisfied all relevant Corps standards before authorizing construction to proceed on the federal spillway. A state construction permit for modification of a dam had to be obtained from the Minnesota Department of Natural Resources.

System Installation

In the summer of 1995, bids to furnish and install an inflatable crest gate system and remove the old weir remnants from the power station tailrace were requested from pre-qualified general contractors. Bidders were required to submit two complete sets of pricing, based on pre-approved designs of two inflatable dam manufacturers. Based upon best functional utility and lowest evaluated cost, a contract was awarded to Maertens-Brenny Construction Company to install a two-gate system supplied by the Bridgestone Engineered Products Company. A strategic decision was made to install the inflatable gates during warm weather and low flow conditions. Detailed design, fabrication and delivery of the inflatable bladders and mounting components took approximately five months. Actual construction at the project site did not begin until the summer of 1996. Figure 2 shows a rubber bladder on its shipping spool.

Since the navigation pool had to be maintained at or above crest elevation during construction, a cofferdam was required. Vertical holes were drilled on 1.2 m centers along the spillway crest. Steel pipes were then inserted to support lumber-reinforced plywood panels. This method allowed safe and convenient access from boats moored against the spillway. Moreover, it could easily be repeated for future maintenance or bladder replacement. To minimize risks associated with high water, only half the spillway was coffered at a time.

Following flashboard removal, the first concrete abutment and the center pier were constructed. These features sloped 1V on ½ H along the dam axis and provided a properly angled surface to secure the ends of the inflatable bladders. Air supply and exhaust pipes were extended through drilled holes in the crest from the gallery in the hollow dam and cast into the pier and abutment. Drainpipes were installed to direct any condensate or leakage water down through the crest and into existing drainage outlets inside the dam.

Figure 2. Rubber bladder on shipping spool.

Holes were drilled approximately 0.2 m on center along the axis for mounting the bladder. Stainless steel bolts secured with epoxy grout were used to anchor steel base plates to the spillway, center pier and abutments. A barge-mounted crane was used to unroll the 4,900 kg rubber bladders onto the spillway crest. Figure 3 shows the pre-drilled bladder material being unrolled directly onto the base plate and embedded anchor bolts. After the bladder was properly positioned, ductile iron clamping plates were placed over the anchor bolts and secured with nuts and spring washers. Figure 4 shows the first bladder completely installed (but not yet inflated). The center pier and cofferdam are ready for installation of the second bladder.

Figure 5 is a view of the electrical and mechanical equipment and local controls for the inflatable gates. This equipment is located inside the power station just behind the concrete wall visible below the spillway in Figure 4. Each bladder has its own 1.9 kW blower, dedicated piping system, and controls. Manifolds and valves allow for dual or crossover operation of the gates during maintenance or emergencies. Automated system monitoring and control functions are accomplished from the power station control room.

The only concession to cold weather, other than locating electrical and mechanical equipment in the power station, was to heat trace critical segments of the external drainage piping. Figure 6 is a view of the completed inflatable gate project showing the aerating effect of the bladder fins during overtopping.

Figure 3. Placement of rubber bladder on spillway.

Figure 4. Bladder installed but not inflated.

Figure 5. System electrical and mechanical equipment.

Figure 6. Finished project showing aeration effect of fins.

Performance and Reliability

Ford has operated the inflatable crest gate system at Lock and Dam 1 since it was completed in 1996. System performance has been virtually problem free through six Minnesota winters and two significant Mississippi River flood events. The flexible gates expand and contract in direct response to pressures from sheet ice. In spite of passing large trees, ice floes, and sharp debris such as dock wreckage, the bladders have experienced no punctures or leaks. In fact, they show no noticeable damage. Exterior surfaces of the rubber bladders exhibit no obvious cracking, discoloration, cuts, or abrasions. With the exception of algal growth and stains on the wetted surfaces, the exposed bladder material generally appears in "like new" condition. Figure 7 shows the inflated crest gates experiencing no problems supporting a derelict tree deposited on the dam and coping with river ice during moderate winter conditions.

Performance of electrical and mechanical components has also been very satisfactory. The gate system has never experienced an unscheduled deflation event. Blowers, valves, and system controls have performed without fail. System maintenance has been limited to periodic testing, inspections, and blower lubrication. Repairs to date involved adding a couple of drip legs and fixing a minor airline leak caused by the flood of 2001 when nearly three meters of overtopping was experienced at the site.

Figure 7. Inflated crest gates in moderate winter conditions.

Conclusion

Ford Motor Company's experience with inflatable crest gates on the Mississippi River demonstrates that sometimes there is a better way to manage water levels and deal with ice and debris in harsh winter conditions. A non-rigid inflatable gate system has proven to be functionally superior, safer and more economical than the outdated flashboard system it replaced. Performance has been virtually flawless and the system exhibits no noticeable deterioration after six winters of operation.

Annual operation and maintenance costs have been far less than those experienced with the original flashboard system. The additional hydroelectric energy generation made possible by the inflatable gates quickly offset their capital cost. More importantly, the increased energy production will help stabilize the St. Paul Assembly Plant's purchased power costs for years to come.

References

Graves, S., et al, (1995) *Submission of Rubber Dam Equipment Supply Information to Ford Motor,* Bridgestone Engineered Products Company, Nashville.

Graves, S., et al, (1995) *Causes and Countermeasures for Oscillation & Vibration in Rubber Dams,* Bridgestone Engineered Products Company, Nashville.

Johnson, J.C. and Waldow, G. M., (1993) *Hydraulic Head Enhancement Study – Ford Motor Company*, Stanley Consultants, Minneapolis.

Rohlf, J. H. and Waldow, G. M., (1997) "Cost-Justifying Project Improvements in the Industrial Sector." *Proc. Waterpower '97*, ASCE, New York, 1650-1658.

Waldow, G. M. and Bystrom, B. J., (2001) *Application for New License for the Ford Hydroelectric Project, FERC Project No. 362*, HDR Engineering, Minneapolis.

Waldow, G. M., Bystrom, B. J., and Rohlf, J.H., (2001) "More Power from the Mississippi River: Upgrading the Ford Hydroelectric Station." *Hydropower in the New Millennium,* Swets & Zeitlinger, Lisse, Netherlands, 517-525.

Analysis of Rainfall Frequency Data for Fairbanks Alaska

Rachel Armstrong[1] and Robert F. Carlson, PE, Ph.D.[2]

Abstract

Rainfall frequency information is the essential input for many water resources evaluation projects such as stormwater management, transportation drainage and urban hydrology. The last general study of rainfall intensity-duration-frequency (IDF) information for Alaska was completed in 1963. The last published IDF curve set for Fairbanks was in 1955. Since rainfall storm characteristics may vary over time, an updated rainfall IDF model will be a useful input tool for water resources studies in interior Alaska.

As part of a stormwater management study with the City of Fairbanks to map and model the stormwater drainage system in the downtown area, we performed a rainfall frequency study using 37 years of data from 1950 to 1999. (There are several missing years in this series.) We assembled a partial duration series of the 45 highest independent events with 1 to 24 hour durations and 1.5 to 50 year return periods. The return period was estimated using the Blom Equation. The 5, 10, 15, and 30-minute durations were estimated by calculations from the 1-hour value. Following the standard conversion from depth to intensity values, the IDF relationships are shown in both table and graphical form and may be easily used in standard stormwater runoff calculation programs.

In comparing the intensity values from this study, we found a decrease to about 60% of that from the 1963 value for the 50-year return period and 5-hour duration storm. In other combinations we found similar decreases. Since this type of analysis can be accomplished with data that is readily available and easily downloaded from Internet sites, we recommend that similar calculations be performed for other Alaskan cities and used in place of the older 1963 study.

---

[1] Research Assistant, Department of Civil and Environmental Engineering, University of Alaska Fairbanks, Fairbanks, Alaska 99775

[2] Professor and Chair, Department of Civil and Environmental Engineering, University of Alaska Fairbanks, Fairbanks, Alaska 99775

Introduction

The modeling of rainfall is important in determining design constraints. For Fairbanks, Alaska, the most recently developed IDF curves are those that can be extrapolated from the statewide rainfall IDF maps presented in the 1963 Technical Paper of the U.S. Department of Commerce and Weather Bureau (Miller). The U.S. Department of Commerce and Weather Bureau also published a plot of the rainfall IDF curves for Fairbanks in a 1955 publication (U.S. Weather Bureau).

The University of Alaska Fairbanks is participating in a project with the city of Fairbanks to map and model the stormwater drainage system of downtown Fairbanks. Since rainfall trends change over the years, updated rainfall IDF curves for Fairbanks will provide a more accurate model and prove valuable to other professionals as well. To achieve this end, readily available one-hour Internet rainfall data from the Fairbanks International Airport was compiled and Microsoft Excel's spreadsheet capabilities employed for analysis.

Objectives

This study provides IDF rainfall values as well as a normal and logarithmic presentation of the IDF curves for the Fairbanks area, including 5, 10, 15, 30-minute and 1, 2, 3, 6, 12, 18, and 24-hour durations and return periods of 1.5, 2, 5, 10, 25, and 50 years.

Methods

From the National Climatic Data Center's Online Store, 37 years of clock hour precipitation data from the Fairbanks International Airport gauging station was obtained (2000). This data spanned the years from 1950-51, 1963-70, and 1973-99. The missing years in this sequence contained either wholly or partially incomplete records and were thus discarded. Partially missing years were not included because this would decrease the accuracy of the probability analysis. However, the record from 1991 did not include the month of April and was included in this analysis because rainfall values large enough to be significant did not occur in April in all other years of the study. Although the years of record were not consecutive, the advantage of a greater length of record outweighed the disadvantages of a broken record. Rainfall data was available for two other gauging stations in Fairbanks but only for the years of 1976-78, which did not prove useful for this study.

Using consecutive intervals, rainfall data was compiled for durations greater than one hour. The method for this can be demonstrated by considering the two-hour duration data. To find the first two-hour period in a day the first and second one-hour periods are added together. Then the second two-hour period is found by adding the second and third one-hour periods together. This is continued until the 24th hour of the day is added to the first hour of the next day. Using this method, the one-hour periods overlap, especially in the longer durations.

After compiling rainfall durations, the largest rainfall events for each duration needed to be chosen using either the annual maximum or partial duration series. In the annual maximum series, the largest rainfall event from each year is chosen, while for the partial duration series the largest overall rainfall events are chosen, which can result in more than one event being selected from each year. Both methods yield similar results. The partial duration series was used for this analysis, which meant that a method was needed to find independent (or non-overlapping) storms. In this study, storms were determined to be independent by insuring that each significant storm event was separated from another significant storm by at least the length of its duration. For example, for two-hour duration storms at least a two-hour period separating one storm event from the next validates that the two storms can be considered independent of each other. This method of determining independence was also used in developing IDF curves for the Michigan Department of Transportation (Trefry et al 2000). The top periods of rainfall were tested for independence to provide a compilation of the 45 greatest independent rainfall events for each duration. In addition to being tested for independence, precipitation events were also discarded if they occurred in the winter months when precipitation in Fairbanks is more likely to occur in the form of snow than rain; the months of October through March were considered winter months and excluded from the compilation of largest rainfall events.

When examining plots of intensity (or depth) versus duration, it was apparent that the relationship between the two variables was a log-normal distribution. A formal examination to confirm this, however, was not performed. According to David Chin, "[i]n hydraulic applications, the log-normal distribution has been found to reasonably describe such variables as daily precipitation depth" (2000). The Blom formula is an accepted probability formula for hydrologic data with normal and log-normal distributions (Ponce 1989). From the 45 top storm events, the probability of their recurrence was calculated by the Blom formula: $P = (m-0.375)/(n+0.25)$, where P is the probability of reoccurrence, m is the rank of descending values, and n is the number of years in the series.

The return period of each storm was calculated by taking the inverse of the probability, and the intensity was calculated by dividing the rainfall depth by the duration. Having found the return period and intensity of each storm event, next intensities for the specific return periods chosen for our study were necessary. Depth versus return period was plotted with both a scatter plot and with a logarithmic scale. Two different plots were used in order to ensure greater accuracy. Then two depth values were obtained for each return period of 1.5, 2, 5, 10, 25, and 50 years through local extrapolation on each plot. The average of the two values was used to find the depth value for each specific return period and duration. Next, graphs were created which display intensity versus duration with a specific frequency: one for each return period and one plot containing all the return periods combined.

As may be expected intuitively, the trend of the graphs is that for a given duration the larger the return period (or smaller the frequency) the greater the intensity of the rainfall. However, these original plots had a connect-the-dots-type appearance. Plotting the IDF relationship logarithmically and fitting linear equations to each return period smoothed the final plots. Table 1 displays the equations of the

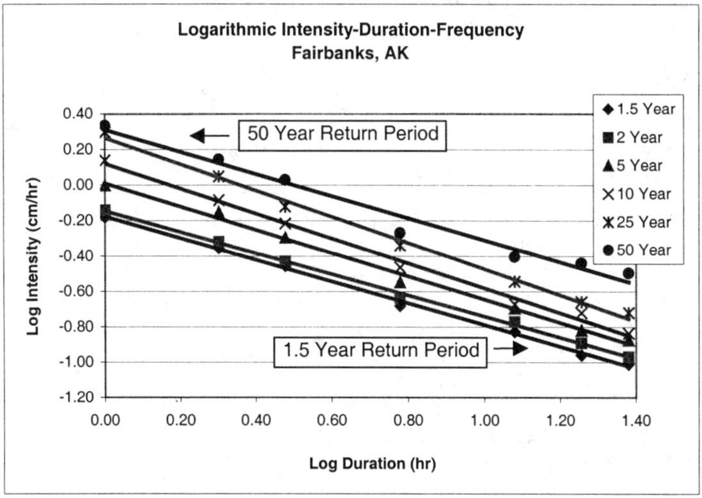

fitted lines along with their R^2 values and Figure 1 shows the fitted lines along with a scatter plot of the data points. The R^2 value for the 50 year return period is the

Figure 1: Lines fitted to the logarithmic scatter of data points.

Table 1: Equations and R^2 values of lines fitted to logarithmic scatter plot in Figure 1.

Return Period	R^2	Equation
1.5	0.998	y = -0.6118x - 0.1766
2	0.998	y = -0.5947x - 0.1452
5	0.994	y = -0.6549x + 0.0102
10	0.994	y = -0.7017x - 0.1213
25	0.995	y = -0.7371x + 0.2649
50	0.975	y = -0.6210x + 0.3113

smallest because this includes the 1967 flood for the 24-hour duration, which is likely a rainfall event with a much greater return period than could be determined with our 37 years of data.

One-hour was the smallest duration of rainfall data for Fairbanks available from thc National Climatic Data Center. In order to create a more complete rainfall model, rainfall intensities for durations smaller than one-hour were calculated with equations from the U.S. Weather Bureau (Miller 1963). The factors supplied were 0.79 to obtain 30-minute data from one-hour data, and 0.72, 0.57, and 0.37 to obtain 15, 10, and 5-minute data respectively from 30-minute data. Since these intensities were calculated, they were not used in the probability analysis, but were plotted in the final IDF curve.

Daily, as well as hourly, rainfall is available from the Fairbanks International Airport gauging station. In his 1963 publication on Alaska, John Miller states that

1.13 is the average factor for converting observational day to true 1440-minute maximum rainfalls (1963). In a study performed in Oklahoma in 1999, the factor

Table 2: Comparison of 24-Hour and daily rainfall intensities.

Return Period (yr):	1.5	2	5	10	25	50
24 Hour Intensity (cm/hr):	0.087	0.093	0.114	0.134	0.160	0.298
Daily Intensity (cm/hr):	0.098	0.107	0.134	0.146	0.191	0.320
Factor:	1.12	1.16	1.17	1.09	1.19	1.07
Average factor of difference: 1.13						

used for this purpose was 1.12 (Tortorelli et al 1999). In this study, the probability analysis was performed on both the daily data and 24-hour data. 24-hour data consists of the largest consecutive 24-hour period of rainfall starting at any time in a particular day and including the appropriate portion of the next day. The rainfall intensity for daily and 24-hour rainfall was compared for each return period and the average factor of difference found to be 1.13, which agrees with the 1963 finding. Table 2 shows the factor of intensity difference for each return period.

Conclusions

The results of this study include a final presentation of the intensity-duration-frequency plot developed for Fairbanks, Alaska, along with a table of plotted values (See Figures A-1, 2 and Table A-1). These charts are created from the set of equations found in Table 1 fitted to the logarithmic scatter plots of the data. Also included in the appendix for comparison is Table A-2, which shows the intensity-duration-frequency data developed before the equations were applied to the data.

Upon comparing the results of this study with the 1963 rainfall map from the US Weather Bureau (Miller 1963), it is clear that rainfall intensity in Fairbanks has decreased by an average of 20% (See Figure B-1 and Tables B-1, 2 for the IDF curve from 1963 and a comparison of the two IDF curves). This is a significant decrease in value. The difference may be caused by the limited data available in the 50's and 60's when previous studies were conducted. The intensity-duration-frequency plots developed in this study may be used as an updated rainfall model for the city of Fairbanks to assist in design and water resources evaluation projects.

Recommendations

The authors recommend that IDF rainfall models be updated across the state. Other methods can be used, such as probability paper or moment analysis; however, the method laid out in this paper is more useful because of its simplicity and effectiveness. Use of this method could swiftly provide many Alaskan cities with updated rainfall intensity-duration-frequency models.

References

Chin, David A. (2000). *Water-resources engineering*, Upper Saddle River, NJ.

Miller, J.F. (1963). "Probable maximum precipitation and rainfall-frequency data for Alaska for areas 400 square miles, durations to 24 hours, and return periods from 1 to 100 years," U.S. Weather Bureau Technical Paper No. 47, Washington, D.C.

National Climatic Data Center, http://www.ncdc.noaa.gov> (June 5-7, 2000).

Ponce, V.M. (1989). *Engineering hydrology: Principles and practices*, Englewood Cliffs, NJ.

Tortorelli, R., Rea, A., and Asquith, W. (1999). "Depth-duration frequency of precipitation for Oklahoma," Water-Resources Investigations Report 99-4232, Oklahoma City, OK.

Trefry, C.M., Watkins, D.W., and Johnson, D.L. (2000). "Development of regional rainfall intensity-duration-frequency estimates for the state of Michigan." *Proc. ASCE Joint Conference on Water Resour. Engrg. and Plng. and Mgmt.*, Minneapolis, MN.

U.S. Weather Bureau. (1955). "Rainfall intensity-duration-frequency curves for selected stations in the United States, Alaska, Hawaiian Islands, and Puerto Rico," Technical Paper No. 25, Washington, D.C.

Stormwater Management Model Development for Fairbanks, Alaska

Stephanie Gould, EIT[1], David Barnes, PE, Ph.D.[2], and Robert F. Carlson, PE, Ph.D.[3]

Abstract

This joint government-university stormwater management development project will eventually aid in compliance with Phase II National Pollution Discharge Elimination System (NPDES) requirements. The project is enabling the inventory and initial modeling and investigation of management techniques for the stormwater drainage system of the interior Alaskan city of Fairbanks.

The City of Fairbanks is the lead agency and end user of this project and is primarily coordinating, mapping and inventorying their stormwater system. The Civil and Environmental Engineering Department at the University of Alaska Fairbanks (UAF) is using the information gathered by the City to model the system. The hydraulic model created will be a comprehensive tool required for understanding the system and is a necessary foundation for continued development of a stormwater management program. Data is also being collected for eventual support of a water quality component to the model. Ancillary to this data collection, jet truck cleaning, the primary means of system maintenance, is being investigated for its potential as a water quality management technique.

Introduction

Fairbanks, Alaska is a city of over 30,000 people in interior Alaska (FNSB, 2001). Temperatures are extreme, ranging from –52 to 36 °C (–62 to 96 °F) with an average annual precipitation of 276 mm (10.87 in) (ACRC, 2001). The City is anticipating the need to apply for a Phase II NPDES permit for its stormwater, which flows untreated into the Chena River.

In 1999, the City of Fairbanks and UAF initiated a study to develop and evaluate stormwater management practices with the requirements of an NPDES permit in mind. The City, in conjunction with UAF, is developing an accurate map, inventory, and model of the City's stormwater drainage system. This comprehensive tool is required for a detailed understanding of the system and is a necessary foundation for continued development of a stormwater management program.

[1] Research Assistant, Department of Civil and Environmental Engineering, University of Alaska Fairbanks, Fairbanks, Alaska 99775

[2] Assistant Professor, Department of Civil and Environmental Engineering, University of Alaska Fairbanks, Fairbanks, Alaska 99775

[3] Professor and Chair, Department of Civil and Environmental Engineering, University of Alaska Fairbanks, Fairbanks, Alaska 99775

Using information gathered by the City, UAF is creating a storm drainage model. StormCAD (Haestad 2000), a stormwater flow calculation program developed by Haestad Methods, is being used to model three drainages: Noble, Wickersham and Lacey, in the downtown core area. These drainages each discharge into the Chena River from the south and include both residential and commercial zones.

Since all of the City's stormwater flows untreated into the Chena River, water quality is also being addressed. No thorough evaluation of the effluent water quality or of the effect of specific stormwater management practices on water quality for this region has been found. This project is the beginning of the necessary evaluation.

Background

The StormCAD portion of the project involves the input of system inventory and location data, as well as rainfall information and general landcover parameters (rational coefficients, drainage areas, etc.). The StormCAD model then generates hydraulic information (flows, velocities, etc.) based on selected modeling schemes and design parameters. The rainfall study done for the model is covered in a separate paper (Armstrong and Carlson, 2002).

Since the StormCAD model is being developed for three drainages, water quality sampling is being restricted to the same area. The drainages are in the City's core area bounded by the Chena River to the North, Airport Way to the south and the Steese Highway to the east. The Noble drainage runs along the east edge of town and includes both a small residential area and the business district. Lacey is the smallest drainage, with few branches, and runs between the Noble and Wickersham systems, down Lacey Street in the core business area. The Wickersham system is the largest of the three, with many branches wandering through both residential and business areas.

In FY01, grab samples of water and sediment were taken throughout the model area to give an indication of the nature of runoff water and sediment quality. Two samples were drawn initially just prior to freeze up (October 10, 2000) - one in the business area on Lacey Street at 3rd Avenue and one in a residential area of the Wickersham drainage at Kellum & 6th. To increase the reconnaissance effort, seven additional sites were then added after spring thaw so that a far end, mid-stream and the outfall manholes of the three model drainages were sampled at twice (May and June 2000).

Table 1. Summary of Reconnaissance Results (FY01)

Parameter	Test Method	Results	Statutory Limit	Method Reporting Limit (MRL)
Arsenic	EPA 200.9	<0.004 to 0.013 mg/L	0.006 mg/L	0.004 mg/L
Cadmium	EPA 200.9	<0.0006 to 0.0008 mg/L	0.00088 mg/L	0.0006 mg/L
Chromium	EPA 200.9	<0.002 to 0.012 mg/L	0.1 mg/L	0.002 mg/L
Lead	EPA 200.9	<0.004 to 0.11 mg/L	0.001 mg/L	0.004 mg/L
Zinc	EPA 200.7	<0.04 to 4.61 mg/L	0.048 mg/L	0.04 mg/L
Iron	EPA 200.7	0.78 to 10.1 mg/L	0.3 mg/L	0.04 mg/L
Copper	EPA 200.7	<0.0009 to 1.14 mg/L	0.0036 mg/L	0.009 mg/L
Oil & Grease	EPA 1664	<2.22 to 37.7 mg/L	Visible sheen or 15 μg/L	2.22 mg/L (varies with volume)

All samples were drawn using manholes for access, as apposed to inlets, to represent the composite of the system water in the area. These twenty-one samples (a quality assurance duplicate is included with those listed previously) were tested for arsenic, cadmium, chromium, lead, zinc, iron, copper, and oil/grease. The results from this sampling activity, summarized in Table 1, have been used to develop the current sampling plan.

Present Effort

With only twenty-one data points available, additional water quality information is needed to evaluate programs and plans. To better understand how to proceed with best management practices (BMP) schemes as a part of stormwater management issues, the current sampling effort was focused with the specific objectives of:

1. Continuing to build database of water quality information for use in water quality modeling once the flow calculation module in StormCAD is available, as well as, to determine trends over time.
2. Determining if a correlation exists between turbidity (a test that can be conducted in the field) and metal concentrations (specifically lead, zinc and/or copper).
3. Beginning to examine the effectiveness of the current jet truck cleaning operations being done to the stormwater system.

The jet truck cleaning operation was chosen as the focus for this effort because it is currently the primary maintenance being done to the system. The process involves allowing a high-pressure hose to work its way through the pipes between inlets. As the water is shot back towards the operator, the water and debris from the pipes are suctioned out and dumped back at the maintenance facility. This procedure is performed around the City on an as-needed basis, with need dictated by flooding problems, particularly in the spring. This is done as maintenance, but since it removes sediment and any adsorbed contaminants, the question has been raised --could this procedure also be used to improve water quality?

Method

The following sections outline the sampling program decisions. The basic goal of the program is to create baseline data for a series of sites, then clean the middle site, and look for effects in water quality in the manhole cleaned and downstream. This will be compared to any changes in the upstream site, which will not be cleaned.

With this basic plan in mind, parameters for testing and sites were selected, the quality assurance project plan (Gould 2001) required by ADEC written and sampling begun.

Parameter Selection

Previous efforts in FY01 gathered twenty-one samples, which were tested for arsenic, cadmium, chromium, lead, zinc, iron, copper, and oil/grease. Since the goal of the program is to prepare for eventual compliance with an NPDES permits, a set of regulatory limits was selected and compared with the test results and reporting limits possible with the test protocols being used (see table 2).

Table 2. FY02 Method Reporting and Statutory Limits

Parameter	Test Method	Statutory Limit	Method Detection Limit (MDL)
Arsenic	EPA 200.9	0.006 mg/L	0.004 mg/L
Cadmium	EPA 200.9	0.00088 mg/L	0.0006 mg/L
Chromium	EPA 200.9	0.1 mg/L	0.002 mg/L
Lead	EPA 200.8	0.001 mg/L	0.2 μg/L
Zinc	EPA 200.8	0.048 mg/L	1.4 μg/L
Iron	EPA 200.7	0.3 mg/L	0.3 μg/L
Copper	EPA 200.8	0.0036 mg/L	0.4 μg/L
Oil & Grease	EPA 1664	May not cause visible sheen or exceed 15 μg/L	2.22 mg/L varies with volume

Three metals; lead, copper and zinc, were chosen for the detailed study. Cadmium and chromium were immediately ruled out because the levels in previous samples were not high enough to be of concern. Oil and grease, an acknowledged issue, were also eliminated in favor of focusing on sediment removal and metals. Of the remaining parameters: arsenic, lead, zinc, iron, and copper; iron was eliminated because it is not a priority pollutant and arsenic was also dropped because it had few incidences of regulatory concern. This left lead, copper and zinc as the focus for testing.

Site Selection

Six sampling sites were originally selected to create two lines, representing an upstream, downstream and site to be cleaned. In choosing the sites the following criteria were considered:

1. Is it a simple line without untested branches? This criterion was to control some of the variables and allow a psuedo-mass balance approach to be applied (psuedo because manholes are being tested not inlets).
2. Can the line include a site that was previously sampled? This is to capitalize on existing data if possible. This was difficult, as the previously sampled sites were not selected with this testing scheme (criteria 1) in mind.
3. Will the sites be able to follow the sampling plan? Many sites were removed from consideration because they had recently been cleaned or cleaning was required for other reasons.

The specific sites chosen included one previously sampled (Kellum and 6th) and are all at an end of the Wickersham drainage, in the residential area west of the downtown, business core. Primarily because of criteria three, these sites were the only ones available.

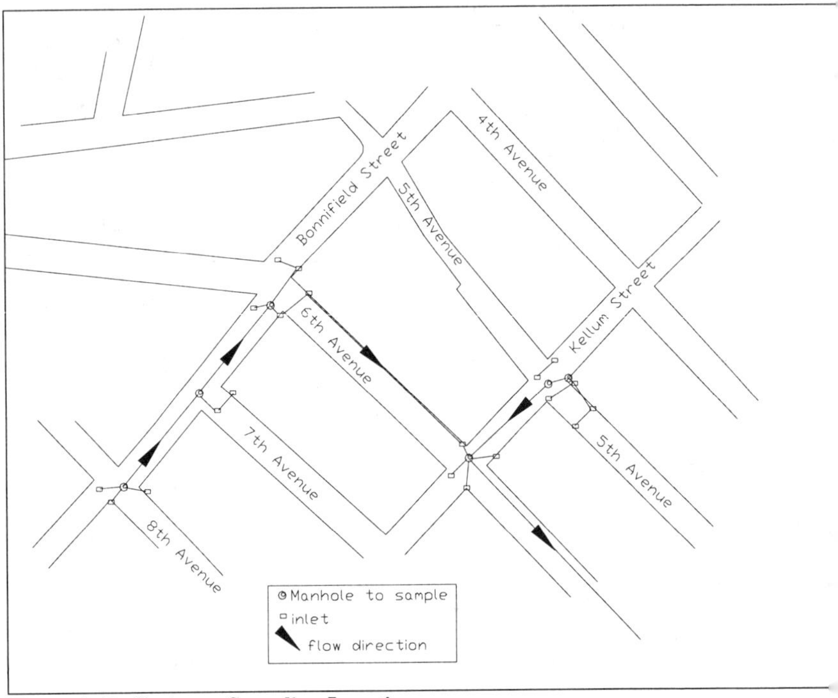

Figure 1. Preliminary Sampling Locations

The background data collected previously at Kellum and 6th Avenue was then examined to determine if the parameters selected for testing were still appropriate. There was some concern that the levels of lead and copper might be too low for effective analysis, since FY01 results at Kellum and 6th Avenue fell at or below laboratory reporting limits. Switching methods, and thereby the limits, addressed this problem. Since previous reports indicated only that the value was below laboratory reporting limits leaving no way to determine how close to the limit the value might be, the laboratory also agreed to report an actually value as low as the practical quantitation limit and flag those that fell below the method reporting limit which is established based on statistical quality measures. Using the sample results available (Table 3) the plan was approved with the selected parameters.

Table 3. FY01 Results at Kellum & 6th Avenue

Parameter	Matrix	Analytical Method	FY02 Method Detection Limit	Previous Values at Kellum & 6th
Zinc	Water	EPA 200.8	1.4 μg/L	1.01, 2.90 and 1.24 mg/L
Lead	Water	EPA 200.8	0.2 μg/L	4 μg/L, <4 μg/L, <9μg/L
Copper	Water	EPA 200.8	0.4 μg/L	10, <9 and <9 μg/L

When the first site visit was made August 16th, it was discovered that the manhole at Kellum and 5th Avenue (north) was not a typical manhole but was instead an earth-lined hole at the end of the line. This showed a need to consider adjusting the parameters and procedures for the City's inventory. Since this was the upstream site, the second cleaning site was removed from the study. Kellum and 6th (the downstream site in that line) was left in the study because it had been sampled previously and is also directly in-line with the others (see Figure 1). The names of the sites actually sampled are given in table 4.

Table 4. Current Sampling Sites

Site	Role in Study
Kellum Street and 6th Avenue	Sampled in FY01 (3 times) and immediately downstream of Bonnifield test line
Bonnifield Street and 6th Avenue	Downstream of Bonnifield and 7th
Bonnifield Street and 7th Avenue	Middle site to be jet truck cleaned
Bonnifield Street and 8th Avenue	Upstream of Bonnifield and 7th

Water Testing Plan

From each of the four remaining sites, water samples were drawn each Thursday until freeze up. UAF staff took temperature, pH, dissolved oxygen (DO), conductivity, settleable solids and turbidity measurements on three samples from each site, while CT&E Environmental Services, Inc. was contracted to test for lead, copper and zinc. The parameters being collected and the methods used are summarized in Table 5.

Table 5. Measurements, methods and frequency

Parameter	Matrix	Method/Equipment	Schedule
pH	Water	Hach SensIon 2 meter	Weekly, until freeze
Conductivity	Water	Hach SensIon 5 meter	Weekly, until freeze
DO	Water	Hach SensIon 6 meter	Weekly, until freeze
Turbidity	Water	Hach 2100P Turbidimeter	Weekly, until freeze
Settleable Solids	Water	Imhoff Cones	Weekly, until freeze
Zinc	Water	EPA 200.8	Weekly, until freeze
Lead	Water	EPA 200.8	Weekly, until freeze
Copper	Water	EPA 200.8	Weekly, until freeze
Zinc	Sediment	SW6020 ICP-MS[3]	Before cleaning
Lead	Sediment	SW6020 ICP-MS	Before cleaning
Copper	Sediment	SW6020 ICP-MS	Before cleaning

Sediment Sampling

The focus of this project is on water quality, however since sediment samples had been drawn previously, one sediment sample was drawn from each site the Thursday prior to cleaning. This gives an indication of what is being removed from the system by the cleaning method. These samples were tested for copper, lead and zinc, and the results are given in Table 6.

Table 6: Soil Results from September 6, 2001

	Copper (mg/kg)	Lead (mg/kg)	Zinc (mg/kg)	Total Solids (%)
Bonnifield & 8th	24.7	29.0	1640	60.1%
Bonnifield & 7th	24.2	29.1	1170	31.4%
Bonnifield & 6th	51.8	79.2	4490	30.1%
Kellum & 6th	48.1	46.5	1590	27.5%

Current Status and Preliminary Results

This is an ongoing research project that will require additional data before final conclusions can be drawn. Seven sample sets were created. Two weeks (September 27 and October 4) after the manhole at Bonnifield and 7th Avenue was cleaned, sampling wasn't done because the manhole remained dry during the September 13 and 20 sampling visits. It was noted at those visits that the water levels in the sumps were dropping. Rather than continue to sample the stagnant water it was determined that further sampling would wait until there was additional runoff. This finally occurred October 11 with melt runoff from a light snowfall. The precipitation (and temperature) information for the Fairbanks area is shown in Figure 2. October 11 became the last sampling visit as the snow signaled the end of the field season. Table 7 shows a summary of the data collected.

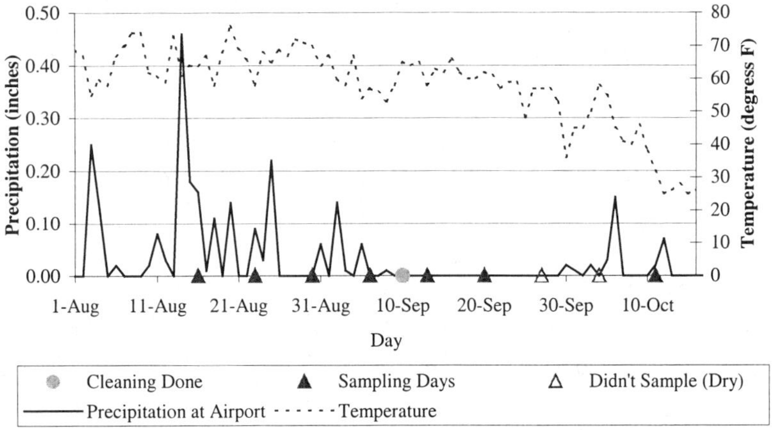

Figure 2. Fairbanks Weather during Sampling Season (data from ACRC-2, 2001)

Table 7. Ranges of Preliminary Results

Parameter	Bonnifield & 8th	Bonnifield & 7th	Bonnifield & 6th	Kellum & 6th
Copper [mg/L]	1.59 - 10.3	6.54 - 31.20	3.76 - 15.3	3.72 - 53.3
Lead [mg/L]	0.400U - 7.71	7.48 - 49.2	4.5 - 15.4	3.53 - 17.0
Zinc [mg/L]	154 - 951	589 - 3000	734 - 2320	757 - 2240
DO [mg/L]	5.00 - 9.00	5.69 - 9.55	1.13 - 9.13	1.97 - 9.59
Conductivity [ms/cm]	25.90 - 92.4	22.2 - 282	33.60 – 259	28.60 - 246
pH	7.29 - 8.37	7.02 - 7.44	6.80 - 7.90	6.76 - 7.38
Turbidity [NTU]	3.26 - 14.6	10.7 - 26.1	10.8 - 21.9	9.35 - 21.8

U - Indicates the analyte was analyzed for but not detected

The results were examined first for a correlation with turbidity. Using the regression analysis features in MS Excel polynomial, power, logarithmic and exponential trends were examined. The highest correlation obtained was R^2=0.5326 which is not significant. This relationship was found for a power series relationship between zinc and turbidity as shown in Figure 3.

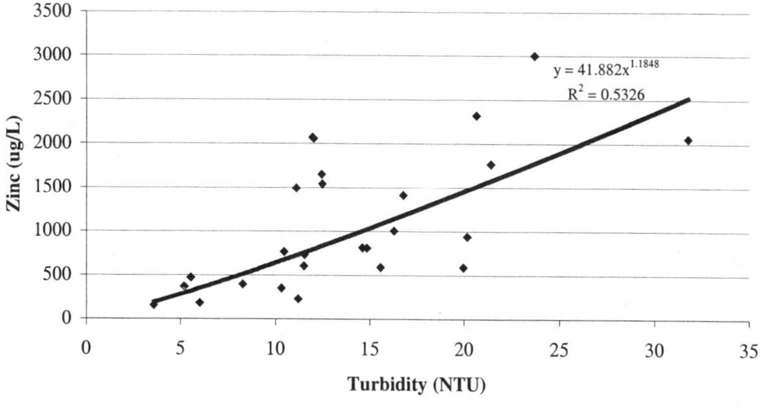

Figure 3. Zinc versus Turbidity

These preliminary indications show no correlation between turbidity and metal concentrations but this result cannot be made conclusively because of the small dataset.

The data was also examined for trends resulting from the jet truck cleaning. Here the problems with the short timeframe of the project become most evident. The dry spell in the middle of data collection and immediately after cleaning (see Figure 2) makes it likely that first flush effects overwhelm any cleaning effects. The data has been plotted by week in Figures 4-6. Note that cleaning was done after week 4.

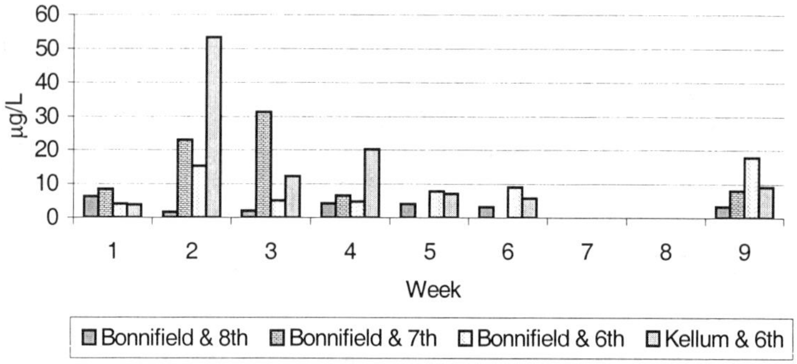

Figure 4. Copper Results by Week

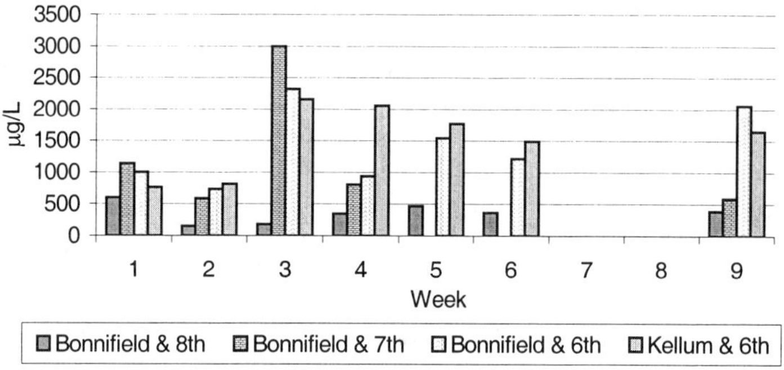

Figure 5. Zinc Results by Week

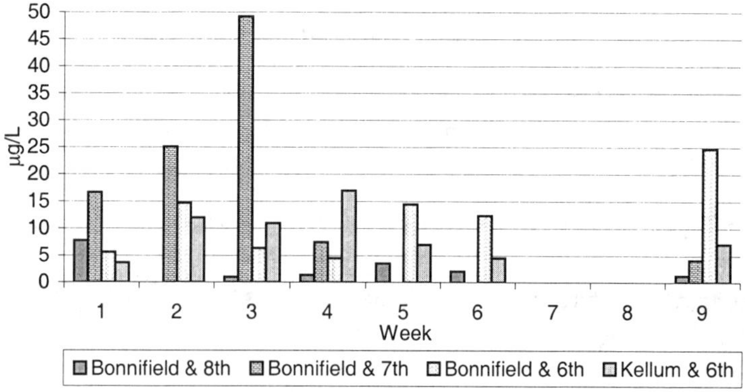

Figure 6. Lead Results by Week

Future Plans

Although there is always the hope that it will, the small dataset used in this study was not expected to provide clear results. Additional sampling is planned for the spring of 2002, with particular attention to the break up season. It is hoped that funding will continue, allowing the effort to process a full field season's worth of data. These preliminary results, and the knowledge accumulated by collecting the data, are being used to refine the study for such a continuation.

Acknowledgements

The authors would like to acknowledge the City of Fairbanks for their participation in and support of this project. Larry Crouder, Gerald Colp, Chris Haigh, Dave McNary, Henry Bratcher, Dave Jacoby, John Lemberger, Dave Weaver, Tim Springer and Barry Williamson have all worked to make this project possible. Several students at the University of Alaska Fairbanks have also been instrumental in accomplishing the necessary field and laboratory work for this project: Chad Hosier, Serena Markey, and Rachel Armstrong. We also thank Amy Ash and Joyce Beelman at the Alaska Department of Environmental Conservation for their time and support.

Funding for this project has been provided by a Community Water Quality Grant, held by the City of Fairbanks under the Alaska Department of Environmental Conservation's Nonpoint Source Pollution Program.

References

Armstrong, Rachel and Carlson, Robert (2002). "Analysis of Rainfall Frequency Data for Fairbanks, Alaska", Conference Proceedings from the 11th International Conference on Cold Regions Engineering.

Alaska Climate Research Center (ACRC, 2001), "Alaska Monthly Temperature Record" and "Normal Monthly Precipitation, Inches" http://climate.gi.alaska.edu/climatology/index.html#monthly.

Alaska Climate Research Center (ACRC-2, 2001), "2001 Daily Temperatures, Precipitation, and Snowfall for Fairbanks, Alaska" http://climate.gi.alaska.edu/FaiData/Fai0801.html.

Fairbanks North Star Borough Community Research Center (FNSB, 2001), *Community Research Quarterly*, Vol. XXIV, No. 2.

Gould, Stephanie (2001). *Fairbanks Stormwater Management Development Quality Assurance Project Plan*, Revision 1, August 13, 2001.

Haestad Methods, Inc. (1995-2000). *StormCAD for AutoCAD 2000, v. 4.1*, Waterbury, CT.

US Environmental Protection Agency (EPA 2000). *Methods and Guidance for the Analysis of Water*, Version 2, CD-ROM.

US Environmental Protection Agency, Office of Solid Waste, SW-846, http://www.epa.gov/epaoswer/hazwaste/test/main.htm.

Icing Regulation of the Groundwater Flow in the South of East Siberia

Boris M. Shenkman[1]

Abstract

By estimating the groundwater flow drained by rivers, it was established that most hydrographs of the cold period are essentially deformed. The reason for the deviation of the actual hydrograph from a true one is that liquid flow is partially or entirely due to the formation of river icing. Such withdrawal of flow distorts significantly the seasonal and long-term dynamics of natural resources of subsurface water, and this especially applies for water of high frequency (for example 95%), thereby impairing the reliability of hydrogeological forecasts. With the purpose of compensating for the influence of icing regulation, a method of reconstructing the hydrograph of a cold period is suggested.

Introduction

As to hydrogeology, the southern part of East Siberia is represented largely by a major (500000 km^2) Angara-Lena artesian basin of first order. The way of structure, it has a rather complicated composition and consists of six relatively autonomous basins. Within the area of draining influence of the water course systems of the rivers Angara and Lena, predominant occurrence corresponds to terrigenous, terrigenous-carbonate and carbonate rocks ranging in age from Jurassic to low-Cambrian. Rocks of terrigenous type are characterized by pore-fissure stratum and middle water. Terrigenous-carbonate sediments predominantly include karst-fissure and mid-depth water. The aquifers of such a kind locate in the internal field of the artesian basin. In the peripheral region of the hydrogeological structure, the artesian slope is dominated by water of karst massifs (Maximovich 1972; Pinneker and Schenkman 1991) (Figure 1). Depending on lithofacies composition, the water content of rocks in plan undergo substantial changes, which is expressed by the variation of the value of

[1]Doctor, Senior Scientific Researcher, Institute of the Earth's Crust SB RAS, 128, Lermontov str., Irkutsk, 664033, Russia; phone 3952-467-777; bshenk@crust.irk.ru

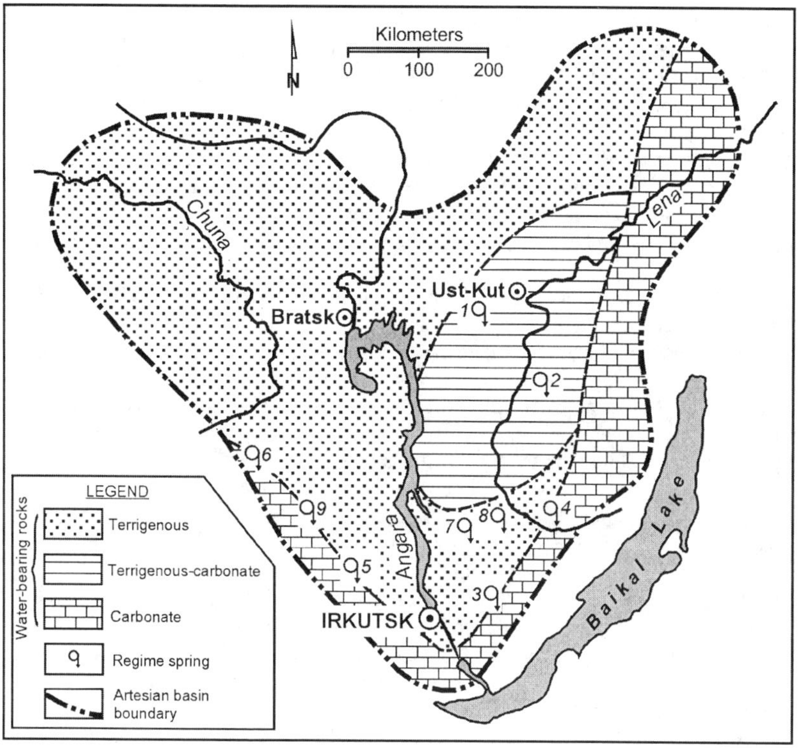

Figure 1. The scheme of dominating distribution of water-bearing rocks.

groundwater flow modulus from 0.3 to 20 l/s·km^2. Such a powerful influence of water-bearing rocks almost entirely masks the zonality indications of groundwater flow and imparts it the features of irregular discreteness.

The water content of rocks drops rapidly with depth, irrespective of composition. Zonal filtration surveys showed that the decrease in capacity to boreholes is approximated by the hyperbolic equation:

$$\frac{q_i}{q_0} = \frac{23.1}{H} - 0.06$$

where q_i is the capacity of the ith interval; q_0 is the capacity of the interval first from the surface; and H is the middle of the testing interval (50 m).

It is easy to see that the character of the decrease in rock permeability with depth is independent of starting water availability. A discrete, layer-by-layer decrease in water availability is taking place quite rapidly. Thus, in the depth range

50-150 m specific water inputs decrease by a factor of five with respect to the starting interval, and by a factor of 15 within a further 100 m. Finally, at depths of 250-300 m the filtration parameters decrease 40 times.

The concentration of the main bulk of ground water in the upper part of the hydrogeological section led to the fact that the subsurface flow came to be almost entirely under control of the riverbed system of the basins of the Angara and Lena rivers. The drainage region, however, shows a significant vertical nonuniformity of the groundwater flow. The variability is due mainly to two factors: the increased rugged topography, and the presence of one or few regionally widespread impervious beds.

Five erosion surfaces are usually traceable in the basins of the Angara and Lena rivers. The first and second of them are not tapping of the groundwater reservoir. The third erosion surface comes to intersect the groundwater table, which ensures constant groundwater feed of the rivers. At the third erosion surface there is taking place only partial drainage of the zone of free water exchange. Therefore this area exhibits relatively low values of the modulus of groundwater flow, which increases rather rapidly with the increasing catchment area.

Discharge of the basic volume of groundwater flow occurs on the erosion surface of fourth order. Especially full discharge of groundwater corresponds to catchment areas where impervious beds are opened by the erosion network. In this case the value of the modulus of groundwater flow remains stable with the increasing area of the river basin. Thus the river course network of the largest rivers receives groundwater predominantly in the form of a transient surface flow. Consequently, groundwater feeding of the rivers is exhausted mainly on the fourth (less frequently - third) erosion surface, which is represented by basins of minor rivers whose areas do not exceed 5000 km^2 (Kudelin and others 1966).

A study of the genetically uniform flow of these rivers that exists in a pure form only in the cold (October-March) season showed that actual hydrographs of winter low water are often strongly deformed (Figure 2). The curve of hydrograph

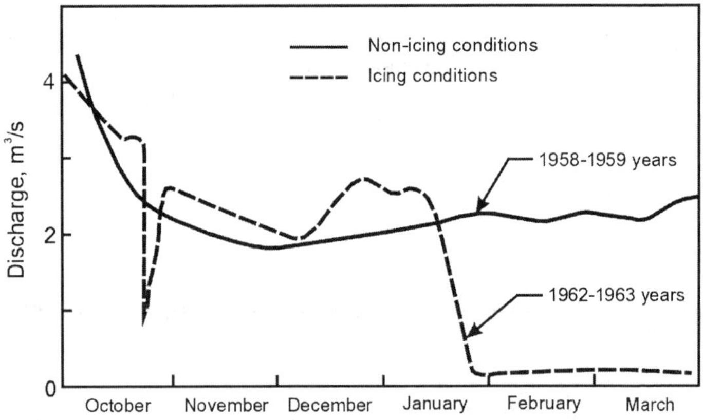

Figure 2. Hydrographs of winter low water Tangui-Udinski.

decline repeatedly undergoes abrupt gaps up to a total cessation of the flow. An intercomparison of hydrographs of winter low water and a comparison of them with depletion curves of flow rates of springs for the period of independent feeding regime gives evidence of an icing regulation of the groundwater flow. An important feature of the withdrawal of liquid flow is its strict irregularity. Not only does the formation of an icing affect the reliability of assessing the natural resources of fresh groundwater, but it distorts substantially the seasonal and long-term dynamics of groundwater flow, of high frequency in particular. However, it is this category of natural resources of groundwater that is the most valuable since it determines the lower limit of reliable operation of subsurface water intakes.

Methodology

It is necessary to appreciate the volume and dynamics of withdrawal of the liquid flow for the distorting influence of the icing to eliminate. During a hydrometric survey in the period of winter low water, large losses of water were estimated (Table 1). So far as the groundwater outcrop from the stream beds is excluded, the negative

Table 1. River flow losses into the formation of an icing (tributaries of the Angara river)

River	Period of observations	River gauge	Drainage area, km^2	Flow intensity, m^3/s		Increment, m^3/s
				on artesian basin boundary	on outlet	
Belaya	1971/72	11	2300	45.5	23.3	-22.2
Belaya	1973/74	10	2300	40.5	27.1	-13.4
Iya	1970-1975	5	4100	24.5	20.2	-4.3
Oka	1971/72	10	13400	32.8	32.0	+0.8
Kuda	1973/74	7	5220	2.10	0/60	-1.5
Kuda	1960-1965	6	4860	3.40	0.70	-2.7

regime is totally caused by the formation of icings. In fact, the formation of icings is higher than the differences as the increments of flow rates (positive and negative) since the groundwater influent between river gauges to ignore. Thus, a hydrometric survey confirms the very fact of the formation of icings and their approximate scales. But this method of accounting does not give any idea of the dynamics of the process, in spite of the significant labor input and cost.

It is quite obvious that the problem would have been solved in full measure if it were possible to reproduce the true outlook of the hydrograph winter low water. In the general case the discharge regime in the period of no feeding is satisfactorily described by the Maillet-Boussinesq equation:

$$Q = Q_0 e^{-\alpha(t_2 - t_1)}$$

where Q_0 is the stream discharge at the date t_1 (when stream flow is entirely due to groundwater discharge); Q is the stream discharge at the date t_2 (beginning of ice failure); and α is the coefficient that characterizes the depletion rate of water reserves over the catchment area.

For the conditions of the southern part of East Siberia, the starting river discharge (Q_0) is determined simply and enough exactly. This is river discharge when air temperature passes over meaning "-5°C". A more complicated situation is with the groundwater depletion coefficient whose value is variable and depends primarily on the composition of water-bearing materials, the type of collector, the regulated volume of the aquifer, and on the amount of groundwater reserves over the catchment area in a current year and for several recent years. Hence we are justified in anticipating that quite a definite meaning of α is inherent in a concrete catchment area or in a group of such areas with similar formation conditions of the groundwater flow. But an idea of the true meaning of α can be provided only by observations at water sites with genetically uniform feed over a yearly cycle. Such a requirement is met, above all, the phreatic discharge toward this end observations for nine large springs were run, the output dynamics is in best agreement with the groundwater feeding hydrograph of the river.

To confirm this thesis, it would be useful to detect - in long-term records of total river discharge - hydrographs of winter low water with a minimum liquid flow losses (or its total absence) for the formation of an icing. Hydrographs of such a kind were expected as the only temperature resource itself, the air freezing index Σ(-T°C), is insufficient for a regular formation of an icing. Moreover, the existing irregularity of withdrawal of liquid flow in significant volumes within a separate low-water period or over a series of years is, to a significant extent, determined by accompanying conditions: water reserves of a current and recent year, the accumulation dynamics Σ(-T°C), the date of establishment of snow cover, and by its rate of rise. If the accompanying elements of a current hydrometeorological situation intensified the effect of the main factor, Σ(-T°C), then there was a substantial or total withdrawal of liquid flow for an icing formed (see Figure 2). When the events show a reversed evolution, the winter low-water hydrograph remains virtually undistorted and reflects a nearly true dynamics of depletion of water reserves over the drainage basin. Thus the irregularity of formation of an icing is in good agreement with the region's location in the intermediate zone between areas of sporadic permafrost (southern limit situate just northern the considered territory) and permafrost islands. The irregularity of an icing formed, by virtue of poor forecasting, serves not only for correct estimation groundwater flow. In an extreme situation, the formation of an icing can lead to catastrophic consequences, as took place in the Lena basin in the spring of 2001. Very hard winter 2000/2001 years contributed to intense icing. Ice thickness in river bed reached 3-4 m. Ice jam was formed in beginning of spring. Water level rised on 20 m. In this case the Lensk town and a lot of small village was wash away.

To detect undistorted winter low-water hydrographs in the time series, it was necessary to construct actual hydrographs of total river flow for all rivers and for the entire period of observation summary length about two thousand years. Such is the methodological basis for estimating the formation of an icing, and hence of the groundwater flow of the cold season.

Discussion

Observations of the output dynamics of large springs made it possible to establish several immutable facts. Primarily, in the range of mean annual output values of 50-11000 l/s the groundwater depletion coefficient of the period of independent feed of groundwater (mean values) varies from 98×10^{-5} to 287×10^{-5} (Table 2). In other

Table 2. Groundwater depletion coefficients (α) of regime springs

Springs number on Fig.1	Spring name	Dominating composition of water-bearing rocks	Time of observations, year	Spring output, l/s	α, $n\cdot10^{-5}$		
					max	min	mean
1	Ustkutskii	terrigenous-carbonate	3	250	165	138	156
2	Korshunovskii		3	50	141	65	98
3	Reshetnikov	carbonate	6	400	180	110	153
4	Ilikta		5	1440	230	160	190
5	Talniki		3	3650	190	120	150
6	Tankasi		6	3500	120	100	110
7	Bolshaya Iret		17	11000	191	45	116
8	Hromovskii	terrigenous	7	240	330	260	287
9	Butakovo		6	210	230	190	210

words, α varies from 0.001 to 0.003 (Shenkman 1978). Secondly, the composition of water-bearing rocks and the type of collector do not affect very strongly the value of α. And thirdly, the ratio $\alpha_{max}/\alpha_{min}$ fluctuates mainly between 1.20 and 1.60, which indicates moderate variability of the dynamic behavior parameter under consideration.

It is obvious that the groundwater depletion coefficients for the groundwater flow that is drained by rivers, must closely approach, in their values, to the α of springs. And this coincidence will be the largest for hydrographs from those years when the formation of icings was negligibly small or totally lacking. About 237 such hydrographs were detected in the basins of the Lena and Angara rivers (within the confines of the artesian basin) or about 12% of the total number. Only a part of the sample is presented in Table 3. Only those rivers were taken into account, in the time series of which undistorted hydrographs of a cold season occur at least three times. A reduction of the amount of sampling by a factor of four did not lead to any appreciable impairment of reliability: the data presented in Table 3 encompass 176 "true" hydrographs, or 74% of their total number. By comparing Table 2 and 3, it becomes clear that the mean values of the groundwater depletion coefficients of spring output and discharge flow are similar: 0.001-0.003 and 0.002-0.003,

Table 3. Groundwater depletion coefficient (α) of groundwater feeding of rivers in the absence of losses of liquid flow for the formation of an icing (small sample)

River	Drainage area, km^2	Dominating rocks composition on drainage area	Number of winter hydrographs: total	Number of winter hydrographs: undeformed	Average long term subsurface feeding, m^3/s	α, $n \cdot 10^{-5}$: max	α, $n \cdot 10^{-5}$: min	α, $n \cdot 10^{-5}$: mean
Kuta	11200	terrigenous	38	12	21.1	350	200	277
Mura	9320		23	3	3.30	320	277	300
Ilga	7960		43	11	17.1	388	294	346
Chuksha	2130		29	10	2.60	290	208	215
Igirma	4480		34	6	4.20	300	156	263
Vihorevka	3840		20	12	3.90	325	236	280
Ilir	2660		12	10	3.60	308	173	246
Tangui	2610		12	7	3.20	303	214	267
Toporok	1190		39	13	2.70	350	161	264
Kadui	1140		21	9	1.10	417	198	285
Tunak	812		19	6	1.80	328	210	279
Ilim	7380	terrigenous-carbonate	12	9	28.4	300	256	285
Kochenga	3830		12	3	15.3	356	230	287
Kupa	2220		23	5	8.10	231	186	210
Korshuniha	197		8	4	0.83	219	91	155
Polovinnaya	175		26	14	0.45	358	200	270
Lena	4690	carbonate	24	12	17.6	372	208	269
Manzurka	3280		25	5	4.50	326	175	258
Tagna	1770		25	10	9.40	215	82	161
Harat	288		20	4	1.20	346	210	246
Zikur	26		14	11	0.10	390	170	260

respectively. The calculated values of α are sufficiently reliable as they were calculated for the time period from 120 to 180 days. As in the case of the spring output, the groundwater depletion coefficient of groundwater feeding of rivers during a cold season depends on the composition of water-bearing rocks. But a weak tendency toward an increase of α still exists in the series of terrigenous-carbonate sediment. Yet, on the average, they are not beyond the limits of 0.002-0.003. The amplitude of the fluctuations is somewhat smaller when compared with the springs, which seems to be associated with a certain asynchronous character of the flow processes in catchment areas of the rivers. About the same influence on the value of α is exerted by the size of the catchment area, and by the discharge and modulus of groundwater flow. Thus it was succeeded to prove that the values of α obtained reflect a nearly true character of depletion of groundwater natural resources in the catchment area. And, secondly, it was established that the mean value of the groundwater depletion coefficient does not leave the limits of 0.001-0.003. Moreover, the amplitude of the fluctuation even of maximum values lies in the range 0.002-0.004. The observed stability of α, nevertheless, is quite acceptable. Calculations indicate that when reconstructing the deformed hydrograph of a cold season, the deviation of the value of α by ±0.001 gives an error in estimating the reconstructed low-water flow rate not exceeding ±5%. The source of sizable errors in estimating the icing flow, and hence of the winter low-water flow can, most often, be due to the incorrectly selected start point of calculation. Usually, such difficulties arise at the moment the river passes to groundwater feeding, the water reserves over the catchment area reach maximum amounts. Such a situation in the catchment area occurs at the period of abundant autumnal feeding when the most fissure rocks become water-bearing. In these cases the depletion of water reserves proceeds at an accelerated pace, with the groundwater depletion coefficient ranging from 0.004 to 0.007. And only upon reaching the basic level by the water reserves, a deceleration of water loss sets in, and the quantity α acquires a statistically stable value. In other words, the depletion curve is functionally uniform, but in some specific cases it can consist of two intercepts with different parameters of the Maillet-Boussinesq equation. A gross error is excluded in the case of a correct determination of the conjugacy point of the two branches of the depletion curve.

Upon subjecting the winter low-water hydrographs to the reconstruction procedure, it was possible to assess the icing regulation of subsurface flow. The formation of icings within the Angara-Lena artesian basin was generally significant and highly nonuniform. This nonuniformity is clearly conspicuous in terms of the icing flow coefficient (ratio of the mean long-term formation of icings to the measured winter low-water flow). The proportion of the icing component in the low-water flow varies over a wide range and, not infrequently, approaches the measured liquid flow. Under especially favorable (for the formation of an icing) conditions the coefficient increases to 110%. In some years, during especially cold winters when the liquid flow in January-March almost entirely goes into an icing, the formation of an icing can exceed the low-water flow by a factor of 2-3. Irregular, extremely variable withdrawal of the liquid flow led to a disturbance of the natural regulation of the basic component of the overall river flow. This process can be quite well illustrated. It is known that the basic flow (groundwater feeding of rivers) possesses a higher regulation compared with the total river flow. Consequently, the coefficient of variation of the winter low-water flow ($C_{V_2} = \frac{\bar{x}}{\sigma}$) must not exceed that of a total river flow (C_{V_1}). In actual fact, for most time series C_{V_2} exceeds C_{V_1} by a factor of 2-3 (Figure 3). Such a behavior of high-frequency flow unambiguously indicates the influence of a powerful factor that has nothing to do with the behavior of a normal depletion of water reserves over the catchment area. Furthermore, the excessive value of C_{V_2} indicates not so much the scale of formation of an icing as its irregularity. Indeed, if the withdrawal of liquid flow were regular as takes place sometimes in the area of continuous permafrost, the value of the variability index of

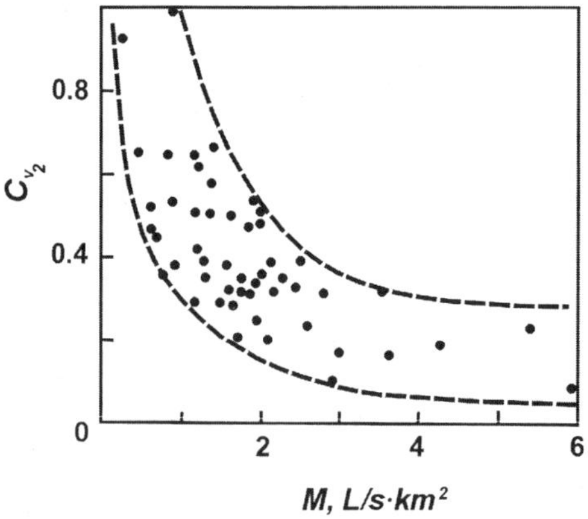

Figure 3. A ratio of the variability indexes of mean annual total discharge (C_{V_1}) and winter mean low water discharge (C_{V_2}).

winter low water would not exceed the values of C_{V_1}. Additional evidence for the instability of the icing process is the weak relationship between the flow discharges of adjacent months of the cold season. Because of the pulsating character of the icing process, the correlation coefficient (*R*) of discharge for October and January-March, for example, does not exceed 0.60 for most time series (Figure 4a). As would be expected, the decrease of *R* is caused by an increase of the icing flow coefficient. The disturbance of the order subsurface runoff is also confirmed by the character of the

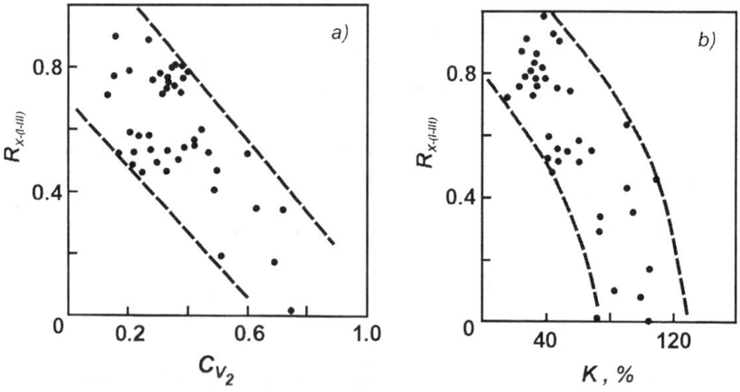

Figure 4. Change of icing flow coefficient (K) and variability index of an actual measured winter low water as the case on value of a correlation coefficient of the discharge of October and January - March.

relation between $R_{X\text{-}(I\text{-}III)}$ and C_{V_2} : the cold period flow increases in variability with a decrease of the intraseries connection between the flows of the preceding and subsequent periods (Figure 4b).

The space-time nonuniformity of the icing formed in the Angara-Lena artesian basin manifests itself in a rather contrasting manner. Figure 5 is a schematic representation of the distribution of values of the icing flow coefficients. The icing process is the least intense over the drainage area composed by terrigenous-carbonate (Angara-Lena plateau) and especially carbonate (Fore-Sayan and Fore-Baikal) rocks. Here the value of the icing flow coefficient only rarely reaches 20%. The highest variability of this index is characteristic for the region of predominant occurrence of terrigenous formations (from 10 to 110%). Such a large range of fluctuations is connected with the filtration nonuniformity of water-bearing formations: from pore and fissure sandstones and tufas to argillite-aleurite strata. If a zonality of the formation of an icing does exist, then this occurs in a masked form only. Even the territory of permafrost islands does not show any enhancement of the formation of icings (see Figure 5). On the contrary, this territory exhibits an extensive region with a minimum (10-20%) value of the coefficient of formation of an icing. Thus, within

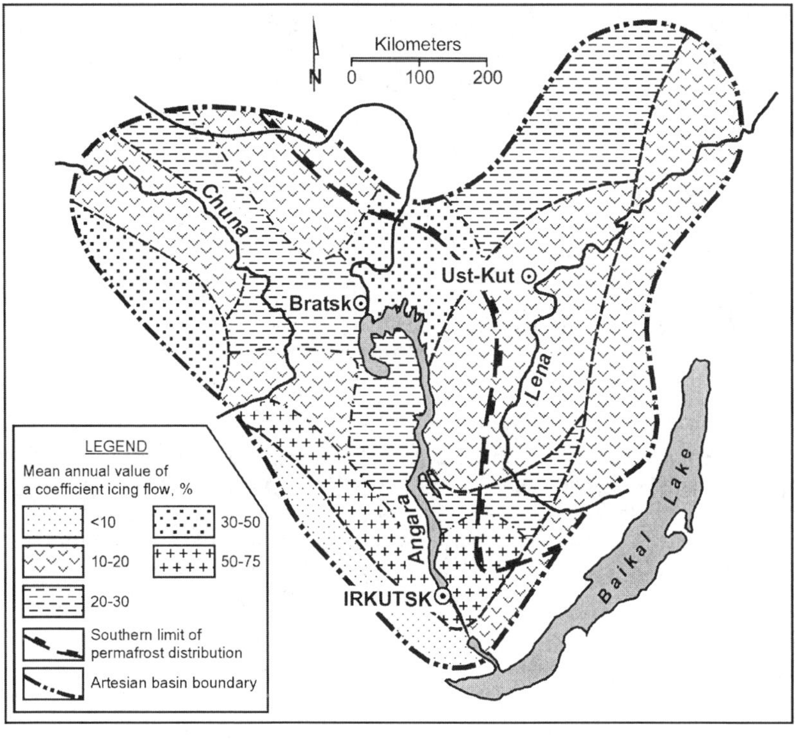

Figure 5. The schematic map of mean annual value of a coefficient icing flow.

the artesian basin, the manifestation of the main zonal factor (air freezing index) is quenched by the influence of local conditions. The greatest influence on the intensity of icing formed is exerted by the lithofacies composition of rocks in the aeration and saturation subzones, and by humidification of the territory. The combination of these two main flow-forming factors largely determines the actual distribution of resource over drainage basin. Consequently, there must exit a sufficiently close link between groundwater feeding and the formation of an icing. Indeed, it is traced most fully when comparing the variability index of the actual measured winter low-water flow with the mean annual modulus of subsurface flow (Figure 6). The field of binary

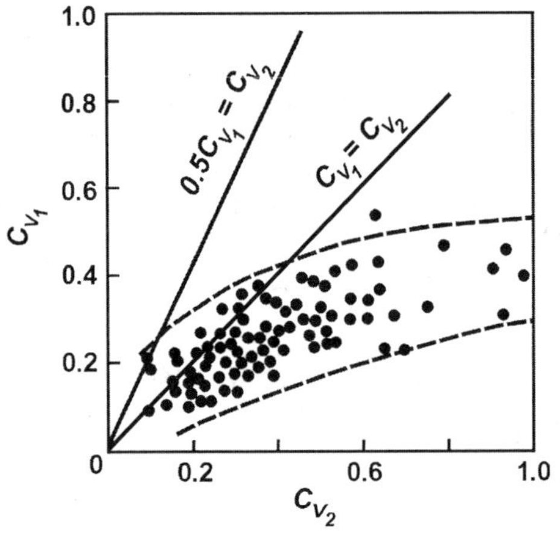

Figure 6. Character of link between a variability index of an actual measured winter low water and mean annual module of a groundwater runoff.

relations is limited by hyperbolic envelopes, which allow us to consider two groups of differing drainage areas. The first group includes basins with a flow modulus *(M)* as large as 1.5 $l/s{\cdot}km^2$ and $C_{V_2} > 0.4$. The second aggregate involves catchment areas with M>3.0 $l/s{\cdot}km^2$ and $C_{V_2} < 0.4$.

Estimations of the formation of an icing from the icing flow coefficient provides only a relative idea of the withdrawal of liquid flow. In the case of high modules of groundwater flow, losses of 10-20% are not affect its estimation in the cold season. However, with small modulus characteristics, errors of determination can be as high as 200-300%. In extreme cases, one may well gain a wrong idea that a total depletion of subsurface water reserves over the catchment area occurs in the cold season.

Conclusion

Through the study reported here, it was possible to determine the place of the formation of an icing when estimating the subsurface flow, especially of the part of it, which falls within the cold season. Reliable data on the winter low-water flow are exceptionally valuable as they make it possible to calculate the resources of groundwater of high frequency. This is particularly true for minimum 30-day stream flow of 95%- frequency, the value of which can serve as a guarantor of a reliable operation of water intakes based on subsurface water. Above all, the findings of this study can be used in areas with a low groundwater flow modulus because it is here where the greatest economic activity is taking place.

Acknowledgment

This study was financially supported by a grant from the Russian Foundation of Basic Research (Project 01-05-64012). The author sincerely thanks two anonymous reviewers for their constructive comments.

References

Kudelin, B.I., and others (1966). *Underground flow on USSA territory,* Moscow State University, Moscow, 289 p. (in Russian).

Maximovich, G.A. (1972). "Subterranean carbonate karst waters in the artesian and hydrogeological folded regions of the USSR//Karst in carbonate rocks." *Transactions of the Moscow Society of Naturalists,* Moscow State University, Moscow, 29-35 (in Russian).

Pinneker, E.V., and Schenkman, B.M. (1991). "Die karstwässer der Irkutsker amphitheaters." *Sonderdruck aus Steirische Beiträge zur Hydrogeologie.* Bd 42, Graz, Austria, 103-129.

Shenkman, B.M. (1978). "Groundwater runoff mapping on the basis of winter hydrometric investigations." *Regional Hydrogeology and Engineering Geology of Eastern Siberia,* Nauka, Novosibirsk, 135-150 (in Russian).

Impacts of Urban Waste on Irkutsk Groundwater

Irina B. Shenkman[1]

Abstract

The initial state of Irkutsk groundwater has been studied in connection with the reconstruction and extension of the municipal wastewater treatment plant, in the vicinity of which there are dumps of solid domestic waste and snow. Pollution sources are located on the surface of the right-bank high flood plain of the Angara River valley. It was ascertained that excessive leakages and infiltration feed under the municipal wastewater treatment plant and dumps leds to the formation of a hydrogeodynamic domes of brackish water with anomalously high content of NH_4, Mn, Fe, Cl and other components of chemical composition. Pollutants are carried with the groundwater flow to the Angara River bed, but in significantly lower concentrations as a consequence of intense dilution with pure groundwater. An increase in the capacity of the municipal wastewater treatment plant, with the dumps persisting, will inevitably lead to an enhancement of groundwater contamination. To minimize the potential environmental damage, a package of mitigating measures is proposed.

Introduction

In the study area, the first from the surface is the aquifer of recent alluvial Quaternary deposits. Subconfined and confined water of Jurassic coal-bearing rocks open ubiquitously under them. Groundwater is directly adjacent to water-bearing formations of the floodplain, which occur on valley slopes of the Angara River. This water is hydraulically associated with the alluvial aquifer, and the latter is associated with the Angara River.

The aquifer of flood plain deposits occurs in gravel-shingle sediments with sandy filler bedding on sandstones and aleurolites of the bedrock. From the surface,

---

[1]Senior Scientific Researcher, Institute of the Earth's Crust RAS, 128, Lermontov str., Irkutsk, 664033, Russia; phone 3952-467-777; bshenk@gpg.crust.irk.ru

the aquifer is overlaid by the flood plain alluvium, although in areas of occurrence of a low flood plain these sediments are also water-bearing. The thickness of the aquifer varies within 3-6 m. Filtration parameters of gravel-shingle sediments, determined from the pumping-test data (Filuk and others 1957), are generally high (maximum values of hydraulic conductivity, characteristic for the near-fluvial band, are 200-400 m/day), but decrease rapidly toward the valley slope (Table 1). Immediately in the

Table 1. Hydrogeological Parameters of Quaternary Alluvial Aquifer (average)

Zones-stripes	Static water level, m	Aquifer thickness, m	Specific yield, l/sec	Hydraulic conductivity, m/day	Transmissivity, m^2/day
Riverside	1.5	2.7	25	990	2670
Middle	2.5	4.4	10	290	1280
Pre-slope	3.9	5.5	2	40	220

area where the treatment plant and the dump of solid domestic waste are located (Figure 1), the hydraulic conductivity and transmissivity do not exceed 40 m/day and

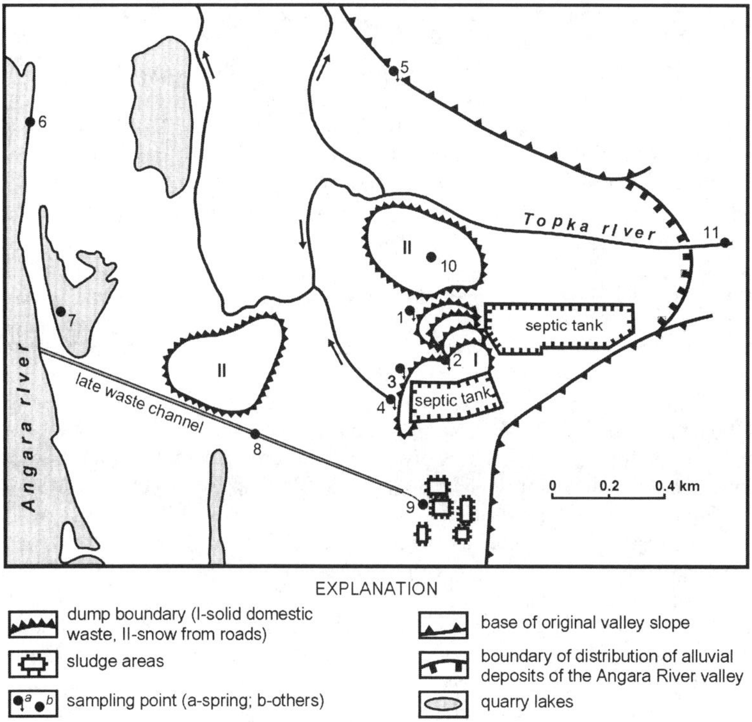

Figure 1. Location of the municipal wastewater treatment plant and dumps.

200-300 m^2/day, respectively. The bedding depth of the aquifer is 1.5-2.5 m. However, in disposal areas of solid domestic waste and solid sludge residuals, because of the large thickness of made deposits, water table is a 12.0 m below the surface. In these cases the aquifer can have a local hydraulic head. The groundwater table represents a smooth surface with a gradient of 0.006 toward the Angara River. In this situation, a hydrogeodynamic dome with amplitude of up to 3.0 m is clearly traceable under the municipal wastewater treatment plant. The existence of groundwater high was determined by means of drilling (Archive of JSC "The Eastern Siberian Board of Engineer and Constructive Researches" 1992). A formation of such a kind exists also under the closed dump of solid domestic waste. The presence of the dome is quite well indicated by a chain of springs around the periphery of the dump. An important feature of the aquifer is the disturbance of its natural bedding. The near-fluvial part is opened by numerous gravel-shingle quarries (the largest of them are shown in Figure 1) nearly to the bedrock. In the near-slope band, the aquifer is overlaid by technogenic deposits, on which the sludge areas of the municipal wastewater treatment plant are located.

In a natural state the aquifer of flood-plain deposits is represented by hydrocarbonate magnesium-calcium water with a mineralization of 0.25-0.3 g/l. Groundwater of the valley slopes of the Angara River approaches background values as regards the level of mineralization (up to 0.5 g/l) and chemical composition. At the same time they are contaminated with pollutants inherent in the private sector: with nitrites, phosphates, and especially with nitrates (up to 70 mg/l).

The groundwater temperature around the periphery of the dump of solid domestic waste during the cold period is 0.3-0.4° C, which is one order of magnitude lower than the zonal value. This gives unambiguous evidence for the existence of permafrost in the core of the dump. Its appearance is in quite good agreement with the annual total negative air temperatures (2850°C).

Methodology

In February-March 1999, eleven water samples were taken from boreholes, springs, quarry lakes and shallow streams that are fed uniquely by groundwater.

Samples from boreholes were taken with a steel siphon-type sampler from a fixed depth. Water from the sampler was expelled with nitrogen into plastic bottles of different volumes. This made it possible to perform a hermetic transfer of the sample into a container corresponding to the analysis. Samples from the springs were collected from the spring discharge.

The bottles were rinsed (2×) with the groundwater (surface water) before the sample was collected, filled with groundwater (surface water), and stored in cleaned plastic containers in the dark at 4°C.

All plastic and glassware were cleaned and soaked in 2 M HNO_3, for at least 12 h, then rinsed with distilled water and dried at 60°C in a convection oven.

Unstable components of chemical composition and state were analyzed 2-4 hours after the sampling. Redox potential and pH were measured in the field using a pH Meter (Hanna Instruments HI 9025). Water samples for determining heavy metals were filtered through the "Blue ribbon" paper filter 2-4 hours after the sampling. A full chemical and gas analysis of water were carried out within several days by applying methods of International Organization for Standardization (Fomin and Cheskis 1992) at the geochemical laboratories of Institute of the of Earth's Crust.

Heavy metals in solution were determined by graphite furnace atomic absorption spectrophotometry (Karl Zeiss Iena, AAS-1N) after solvent extraction by 1.0% Na diethyldithiocarbamate thihydrate in 4-methylpentan-2-one. All samples were acidified to pH<1.5 (conc HNO_3) before solvent extraction. Na, K, Rb were determined by flame atomic absorption spectrophotometry (Karl Zeiss Iena, AAS-3). Gas composition of groundwater was determined by chromatograph LHM-72 using a detector of thermal conductivity.

Discussion

Groundwater at the site of the municipal wastewater treatment plant was likely contaminated prior to their construction due to the discharge of untreated domestic sewage during 1934-52 years (Municipal archive of Irkutsk 1953). Once the municipal wastewater treatment plants were put into operation, the areal extent of groundwater contamination likely increased, but concentrations decreased due to dilution. According to the survey data from the end of the 60s, water salinity did not exceed 0.83 g/l (Archive of JSC "The Eastern Siberian Board of Engineer and Constructive Researches" 1968-71). At that time a down gradient plume of pollutants was observed (Authors' data).

With the formation of dumps of solid domestic waste and snow, there was an expansion of the contamination area. Groundwater mineralization was gradually increasing, and chemical composition underwent appreciable changes. In the beginning of the 90s, water salinity in the influence area of the dumps and clearing facilities increased to 1.3 g/l (Archive of JSC "The Eastern Siberian Board of Engineer and Constructive Researches" 1981, 1982). Furthermore, hydrocarbonate ammonium, chloride-hydrocarbonate and sulphate-hydrocarbonate water formed due to excessive leakages from the municipal wastewater treatment plant and dumps (Figure 2). The pre-existing low-mineralization hydrocarbonate water almost disappeared. Where the chemical type persisted, mineralization increased to 1-2 g/l. In the direct influence area of the municipal wastewater treatment plant and of the dump of solid domestic waste, it was significantly higher. Maximum salinity of solutions at the site of the municipal wastewater treatment plant reached 6.5 g/l; in the body of the dump of solid domestic waste and in the fragment of the aquifer bedding underneath, it was 3-4 g/l. In spite of the surface location of pollution sources, the entire water-bearing thickness of alluvial sediments, as deep as the bedrock, was contaminated. However, the penetration of pollutants into groundwater of Jurassic deposits is hampered by a ubiquitously occurring eluvium represented by a layer of high-plastic clays 2-3 m thick (do not show on the cross section).

By the year 2000, a relatively stable hydrogeochemical regime had formed in the area of the municipal wastewater treatment plant and dumps (Shenkman and Shenkman 1999). Groundwater in the area of direct influence of pollution sources represent to some extent diluted leachates. They are characterized by high concentrations of HCO_3, Fe^{2+}, Mn^{2+}, CO_2 and NH_4^+ (Table 2).

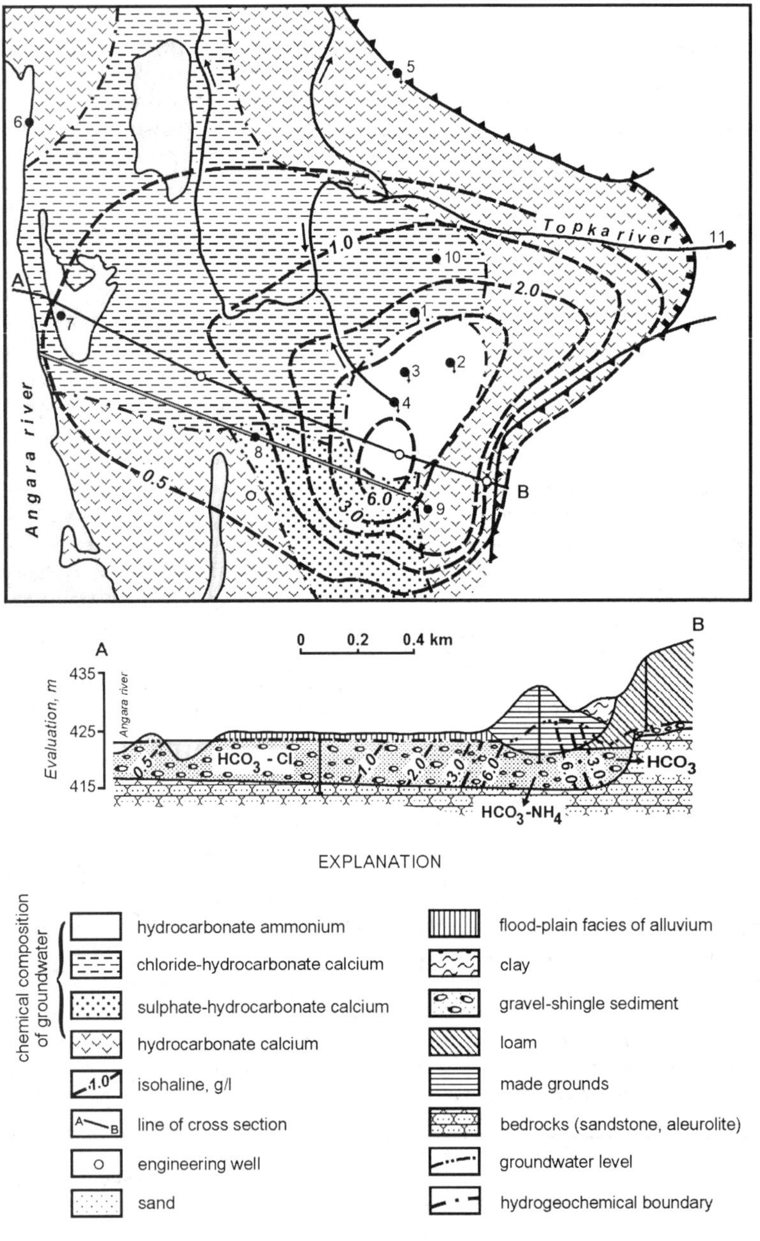

Figure 2. Hydrogeochemical map and cross section.

Table 2. Groundwater Mineralization and Chemical Composition near Main Sources of Contamination (average, mg/l), 1999

Object	Water salinity, g/l	O_2 / CO_2	Fe^{2+} / NH_4	Na / K	Ca / Mg	Cl / SO_4	HCO_3 / HPO_4	Oxid-ability, mgO/l
Treatment plant	6.5	- / -	3 / 560	550 / -	40 / 413	923 / 4	3965 / -	-
Dump of solid domestic wastes (closed)	3.3	0 / 500	17 / 300	80 / 73	194 / 125	65 / 19	2930 / 0.99	31.4
Dump of snow (open)	1.3	- / -	0.3 / 30	220 / -	59 / 42	195 / 65	650 / -	-
Background area	0.22	- / -	2.4[a] / 0.2	8[b]	30 / 11	3.5 / 6.6	159 / -	-

[a]$\Sigma(Fe^{2+}+Fe^{3+})$
[b]$\Sigma(Na+K)$

Oxygen that goes into oxidation processes and is consumed by microorganisms is extracted from water, nitrates, sulphates, Fe and Mn oxides, and from other oxygen-rich compounds. As a consequence of this, the thickness consisting of waste develops a reduction zone with high content of nitric compounds (NH_4, organic nitrogen), which creates an especially strong and long-lasting pollutant potential. Mineralization of organic substances in an oxygen-free medium is accompanied by an intense formation and release of CH_4 (methane phase of transformation of solid domestic waste (Dementiev 2000)). Coming up into upper layers of the dump and entering the aeration zone, methane partially oxidizes to transform into CO_2 (Table 3).

Table 3. Gas composition of groundwater under dump of solid domestic wastes

Springs number on map	Date of sampling	Content	Components of composition				
			O_2	N_2	CO_2	CH_4	Σ
1	11.02.99	%%	0.96	10.45	25.21	63.38	100
		m/l	0.10	11.24	213.80	68.17	293
3	11.02.99	%%	0.00	45.12	37.26	17.62	100
		ml/l	0.00	41.30	376.7	18.95	437
Background area	11.02.99	%%	20.50	75.27	4.23	0.00	100
		m/l	8.50	24.20	12.40	0.00	45

Note: H_2, He, Ar, CO, H_2S are absent

Although the solid domestic waste dump has been officially closed for a long time, geochemical processes are still occurring in this area with sufficient intensity. This is due in part to the continued, un authorized disposal of process sludges in this area. It is also due to a lowered decay rate of solid domestic waste because of the permafrost within the body of the dump. For that reason, the influence of this conventionally closed dump will continue for a very long time.

As to heavy metals, groundwater samples show high values only for manganese and strontium (Table 4).

Table 4. Heavy Metals Concentrations of Natural Waters near the Municipal Wastewater Treatment Plant

Sampling point and its number on map	Content, mcg/l								
	Co	Cu	Zn	Ni	Cr	Pb	Cd	Mn	Sr
Spring 1	-[a]	-	14	-	-	-	-	5980	2400
Spring 3	-	-	63	-	-	-	-	1840	2100
Spring 5	-	-	4	-	-	-	-	-	660
Angara River, below sewage discharge, 6	-	-	9	-	-	-	-	-	103
Quarry lake, 7	-	-	8	-	-	-	-	1140	660
Late waste channel, 8	-	-	4	-	-	-	-	1200	770
Topka River, 11	-	-	5	-	-	-	-	82	118
Detection limit	5	3	2	5	3	10	2	3	10

[a]Below detection limit

In the influence area of the pollution sources, four fragments were identified, with the predominant occurrence of some or other chemical type of groundwater (see Figure 2). ***Sulphate-hydrocarbonate calcium water*** forms directly under the municipal wastewater treatment plant. They have a limited occurrence area because of the instability of sulphates in the reduction condition. ***Hydrocarbonate ammonium water*** that has the highest mineralization, occurs under the solid domestic waste dump. The high initial concentration and relative geochemical stability of the ammonium ion led to remote migration of the pollutant: at a distance of nearly 2 km from the emission site, the content of NH_4 in groundwater not below 1 mg/l (Figure 3). The largest areal occurrence was received by ***chloride-hydrocarbonate calcium water*** that forms mainly in snow disposal areas. This is associated with pollution of snow by rock salt used to spray on roads in the winter season (98 tons in 2000). The preservation of chlorides in the filtration flow is determined by the conservative character of the chloride ion, and it is due to this that it was detected almost throughout the survey area downstream the groundwater flow from the pollution source.

Conclusion

Under cold climate conditions, the decay process of solid domestic waste slows down; therefore, closed dumps can serve as a contamination source for groundwater for a long time. At present a total polluted area from 0.5 g/l isohaline is estimated at about 0.3 km^2. The area of the pollution sources inside of this contour hardly reaches 0.06 km^2. Such a correlation of the areas plus the large increment of the space between 0.5...1.0 g/l isohalines suggest that there is taking place a rapid decrease of the influence of the municipal wastewater treatment plant and dumps on groundwater

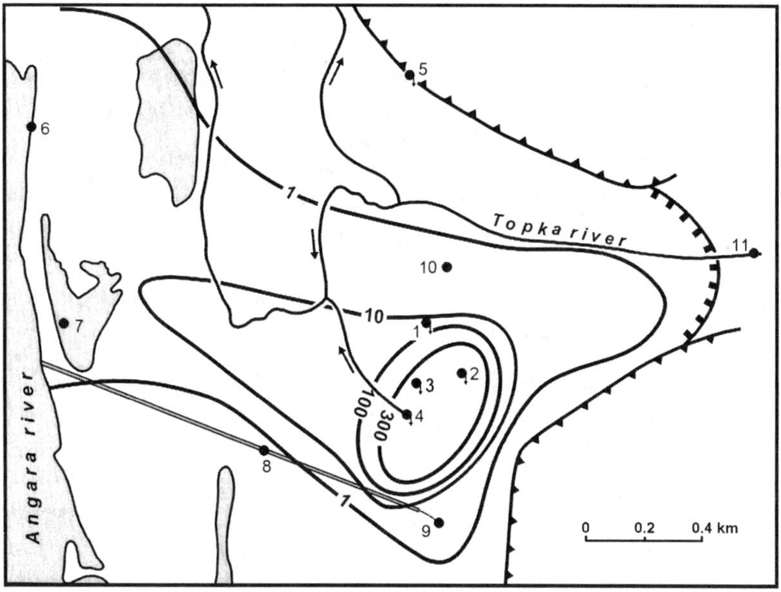

Figure 3. Ammonium contours map (mg/l).

and hence only minor contamination of the Angara River bed. This seems to be caused by two factors. Firstly, the amount of leachates that form from pollution sources is relatively small. Secondly, at the sites of the municipal wastewater treatment plants and dumps the filtration properties of water-bearing rocks are rather low, and the gradient of the groundwater flow is small. On the contrary, outside the dumps and municipal wastewater treatment plant the aquifer has high filtration properties and hence increased dynamic properties. Therefore, the dilution of mineralized solutions is occurring quite intensively. However, with the extension of the municipal wastewater treatment plants themselves, the increasing volume of snow disposal, and with the continuing disposal of process sludge on the closed dump of solid domestic waste, a total effect of contamination may increase. To avoid the deterioration of environmental conditions, it is expedient to take the following nature conservation measures:

1. screening indiscriminately all "wet" objects of municipal wastewater treatment plant;
2. removal of the dump of solid domestic waste to dry hydroisolated sand quarries, and its recultivation;
3. transfer of the dump of polluted snow beyond the water protection area of the Angara River;
4. transfer of areas that have been recultivated from dumps, to citizens for gardening purposes.

Acknowledgments

This study was financially supported by a grant from the Russian Foundation of Basic Research (Project 01-05-64012). The author sincerely thanks two anonymous reviewers for their constructive comments; and Mr. V. Kachura (Director, JSC "The Eastern Siberian Board of Engineer and Constructive Researches"), and Mr. V. Yakubovsky (Mayor of Irkutsk) for possibility of using the archive data. The author is grateful to Mrs. L. Durban for her help in carrying out the water analyses.

References

Dementiev, S.Yu. (2000). "Transformation of conversion processes of solid domestic waste." *Geoekologiya*, No.1 (in Russian).

Filuk, A.G., and others (1957). Unpublished report "Filtration investigations in territory of Irkutsk and suburbs for 1954-1956 years." Mosgidep, Angarsk dispatch office, Irkutsk (in Russian).

Fomin, G.S., and Cheskis, A.B. (1992). "Water. The control of chemical, bacterial and radiation safety under the international standards." *Handbook,* Gelikon, Moscow, 392 (in Russian).

Shenkman, I.B., and Shenkman, B.M. (1999). "A qualitative condition of groundwater in region of Irkutsk right-bank municipal wastewater treatment plant." *State paper on the environmental situation of Irkutsk region,* JSC SPI "Oblmashinform", Irkutsk, 145-147 (in Russian).

Hydrology of Northern Russia: A Case Study with Application for a Pipeline Project

Vladimir A. Shutov [1]

Abstract

Landscape structure and numerical cluster analyses were applied to hydrological regionalization of the environment of a constructed gas pipeline in the North of European Russia. The studies are to improve runoff calculation on the rivers along the pipeline route to mitigate the damage caused by spring floods. As a result the basins are determined to be analogues for the numerous ungauged rivers by flood flow calculation. Calculations were performed using adopted manuals with detail evaluation of input variables such as snow water content in those basins located on hilly uplands. Regional correlation between runoff and conditioning factors were obtained and taken into account for the area. A great attention was paid to the development a model for how to take the wetland-related impact on runoff regime of the region.

Introduction

There are large gas pipelines laid (and are to be laid) in northern Russia. They are made of steel pipes about 1.6 m in diameter. The pipeline usually lies under ground where it is subject to many natural factors which are hazardous for it and, incidentally for the environment. When the pipeline passes over numerous small rivers and brooks, it can act like a dam due to inaccurate prediction of maximal flood water stage. Such cases in wide flat-bottom valleys lead to durable inundation of surrounding area, particularly by spring flood. Hence, the urgency of exact calculation of the spring floods is clear, and is the main objective of the applied hydrology projects.

By examining the environmental situations on a number of pipelines in Russia we arrive at a conclusion that the hydrology-related factors can provoke ruptures of the pipeline and accidental releases of the gas (often explosive) in the atmosphere. Our investigations in applied hydrology were initiated about six years ago (Shutov and Kapotov, 1998, 2000) and firstly were only to support decision making in project design. Later on they were developed mainly in following studies:

- hydrological regionalization using combined landscape and cluster analyses;
- searching for regional relationships between spring flood runoff, climatic variables and watershed characteristics;

----

[1] Senior scientist, Valday Branch of State Hydrological Institute.
175400, Valday, Pobeda Street 2, Russia; hydroshpere@mail.ru

- development of the methods for calculation of the spring flood discharges for the small rivers crossed by pipeline route.

This paper presents the methods of regional hydrology used to assist those project designers with only fundamental knowledge in hydrology. We exemplify this by using the results obtained mainly for the *Arkhangelsk* region and Republic of Karelia where the author has been an investigator at the Wetland Research Station. The main goal of this paper is to provide hydrological support of the pipeline engineering, but the method developed below may further advance flood warning in general. It must also be noted that these examples may pertain to broader engineering applications.

Hydrological Regionalization

By landscape analysis five regions have been distinguished (Figure 1) along the examined pipeline route in northern Russia (Shutov and Kapotov, 1998). According to previous summaries (Isachenko and Shlyapnikov, 1989) and also to our personal experience they may be described as follows: (1) moraine plains of middle-Quaternary age with alluvial lake-glacial lowland plains (intrazonal areas), (2) upper-Quaternary hills so-called *Njandoma* upland, (3) limestone karst plateau, (4) the White Sea littoral lowland and terraces, and (5) lake abundant region.

The first aim of the project was to provide general outlook at the route prior to dealing with sophisticated models for runoff calculations. We determined two sets of attributes for the drainage basins along the route. The first consists of physiographic characteristics: total drainage area, mean elevation, partial area occupied by lakes (*FL*), wetlands (*FW*) and, that appears to be very important, sandy soils (*FS*) obtained by a digital overlay of the soil maps. The second data set consists of runoff characteristics with particular attention to spring floods. These are: fraction of spring runoff in annual runoff volume, mean spring runoff depth (h_0, mm), the «sharpness factor» of flood which is the ratio of peak specific discharge to h_0-value, the «shape factor» described the asymmetry through the ratio of monthly averaged discharges for April and June and, the variation factor (ratio of the square root deviation to average runoff depth). About the same attributes were proffered in framework of the European research program FRIEND (Gustard *et al.*, 1990). We also tried to observe those factors and methodologies earlier accepted in Finland since it is adjacent to the region (Kuusisto, 1985, Seuna, 1983).

We developed so-called belt-diagrams (Figure 2) to display a route. This is a convenient way of tracing a long-sized profile by using both reference data and special investigations data. Besides, the regional maps were drawn for spring runoff depth, precipitation amount and snow water equivalent. Of particular interest is that wetland and sandy soil percentages as well as spring runoff depth increase along the route from south-east to north-west. As was also found, there are basins with excessive fraction of pervious sandy soils that dumping runoff. Coupled effect of wetlands and sands increases northwards that regulates runoff effectively. Forest percentage does not affect flow, since up to 90% of the area is forested evenly.

Cluster Analysis

The cluster analysis is proficiently used to choose the representative basins and homogeneous regions (Burn, 1989, Nathan and McMahon, 1990). We also implied it, with some improvements, to the project (Shutov, 2000a). The Euclidean metric in multidimensional space of the basin attributes was adopted as the measure of similarity. Table 1 represents an example of a matrix of similarity for the rivers in one of the defined regions. As the Euclidean metric of compared watersheds decreases, the similarity of them increases. The regional analogue-basins were selected through this procedure as those watersheds which are the most similar with others within the regions. The Euclidean metric was defined as follows:

$$L_{ij}^2 = \Sigma (Y_{ik} - Y_{jk})^2 \tag{1}$$

where the index k corresponds to the basin attributes, indices i and j associated with classified basins.

We assumed as input variables Y_k the values of the attributes divided by their averages. We used that method of cluster analysis (Hosking and Wallis, 1997) when those objects are at-first segregated in a «nucleus» (see such a nucleus featured with bold font in **Table 1**) within initially compiled correlation matrix whose L-measures are the lowest, these constitute a primary class. Then, the boundary objects, which have higher L-values are replaced by turns into one or another class. We should do so until the mean L-values within each class can reach to their minimum value and, at the same time, the differences between the classes become maximal. These conditions can be given as below:

$$L_A = min, \quad L_B = min, \quad \{Y_A - Y_B\} = max \tag{2}$$

where L_A and L_B are the mean Euclidean metrics of two classes examined, $\boldsymbol{Y}_A$ and $\boldsymbol{Y}_B$ are respective vectors of mean values of the attributes, i.e. $\boldsymbol{Y} = \{Y_1, Y_2, \ldots Y_k\}$.

Table 1. Matrix of the Euclidean metrics for the II-nd hydrological region

##	River names	8	9	10	11	12	13	14	15
8	*Vel'*	0	0.92	1.31	0.85	1.07	1.17	1.14	1.21
9	*Podjuga*		0	1.13	0.97	1.03	0.89	1.18	1.57
10	*Voloshka*			0	1.11	1.14	1.00	1.32	1.52
11	*Sulanda*				0	**0.55**	**0.50**	**0.52**	1.23
12	*Lepsha*					0	**0.59**	**0.56**	1.26
13	*Lim*						0	**0.59**	1.51
14	*Iksa*							0	1.60
15	*Mosha*								0

Thus, the general idea is how to classify river basins by combining *a-priory* geographic information with numerical cluster analysis based upon Euclidean metric. The cluster and the previous landscape analyses have provided quite comparable results represented schematically in Figure 1.

The III-rd and V-th regions have relatively low spring runoff. This results from the karst in the first case, and from the lakustrine influence in the second case. The IV-th region of littoral lowland abounds in wetlands and it has the greatest runoff. The similarity index L_{ij} is reduced by the impact of regional phenomena. It is lowest of all in the V-th region with large lakes. There is, for instance, the *Suma* river basin which consisted of a cascade of five lakes that attenuate floods most effectively. The same is to be said about the *Vyg* basin whose runoff is regulated by man-made reservoirs to provide navigable water stages all the year round.

Another example of this kind is the basin of the *Mosha* river where there are tributaries flowing across the pipeline. It may be annexed by L_{ij} either to the II-nd or to the III-rd region, but by its landscape structure it is evidently upper-Quaternary area, i.e. the II-nd region. The III-rd region watersheds are composed of limestone aquifers and well pervious soil cover, that greatly affects annual runoff and, particularly, the flood hydrograph making it flatter.

Two principal outcomes were received: some nuclear structure of the regions and non-clear boundaries among them. Obviously, if the classifying basins could be less and numerous, then the regions could be defined more exactly. But, there is not a continuous network of small gauged basins, therefore we cannot hope to identify the regions absolutely. To find an elusive borders among them, we may propose to distinguish the regions only by natural borders (river watersheds) in spite of the fact that anomalous basins within the regions can be segregated (see Figure 1).

Spring Runoff Calculations

Regional flood calculations are performed (Sokolov *et al.*, 1978) by using the following equation:

$$Q_p = h_0 K_o \delta_L \delta_S \delta_W F/(F+1)^n \qquad (3)$$

where Q_p is maximal river discharge (m^3/s) of given exceedance probability p %, h_0 is runoff depth (mm), K_o is an empirical factor which indicates sharpness of the flood flow hydrograph, δ_L, δ_S and δ_W are reductive factors of lakes, sandy soils and wetland partial areas respectively, F is area of the drainage basin, n is an empirical regional exponent (for given region $n = 0.17$).

This simple procedure based on reference data is accomplished with a number of previously developed regional relationships between runoff characteristics and physiographic factors. The key factor, K_0 is determined by the data of the analogue basins. The h_0 values are evaluated using regional maps which have been drawn by observation data. It can be improved by implication of the relationships between runoff depth and watershed characteristics such as elevation or sandy soil partial area (Figure 3). We proffered the following regional relationships between spring runoff depth h_0 and mean elevation Z of the drainage basin and sandy soil percentage FS:

$$h_0 = 0.58\,Z + 65 \quad \text{and} \quad h_0 = -1.91\,FS + 183 \tag{4}$$

as well as the below empirical equations received for the wetland-related reduction.

We should take into account the climate conditioned trends in runoff characteristics. Time series showed that the spring runoff depth is slightly increasing for most of the examined rivers except those regulated with reservoirs. This quite corresponds to the results obtained for the Baltic Sea basin (Bergström and Carlsson, 1994). Fraction of the spring runoff in annual runoff volume is firmly decreasing, e.g. for the *Mosha* basin from 59% in 1950s to 51% by 1980s. At the same time, the warming climate observed during the last two decades provides here less intensive peak flows that contradicts with Fennoscandia (Krasovskaja and Gottschalk, 1993). Probably, this fact is due to a less intensive radiant snowmelt by cloudy weather compensated for some higher temperatures. Hence, we recommended to choose some partial time series for temporally averaged data.

Wetland-Related Impact On Flood Runoff

A critical issue is how to reveal the reductive influence of wetlands on spring runoff. We have used for this aim a re-evaluation of the factor δ_w by Eq. (3) for those rivers with gauged sites. The reduction of the peak flow under wetland influence is usually described as follows:

$$\delta_w = 1 - B\,Log(0.1F_w + 1) \tag{5}$$

where F_w is partial wetland area, B is empirical factor which is depending on the types of wetlands and on where they are situated within watershed area.

The results received by using this methodology showed that four types of river basins can be distinguished based on their response to the wetland conditioning flood reduction. As was clearly found, the so-called *aapa* wetlands in Karelia affect flood flow in an unusual way. They do not attenuate peak flow, but in opposite, amplify it: the maximum specific runoff increases. Thus, the reductive factor δ_w will be more than one in this case. This arises due to the lowlands, in which the *aapa* are situated, are generally stretched valleys acting as natural macro-scale drains within the river basins. Melt waters flow over deeply frozen ground or above ice which covers the *aapa*. Thereby, the spring runoff rapidly concentrates and the flood hydrograph becomes sharp. The empirical Eq. (5) must be substituted with the following:

$$\delta_w = 1 + 0.0033\,F_w \tag{6}$$

These equations can be convenient to use as an addition to the accepted manuals (Sokolov *et al.*,1978). The errors by calculating the peak flow without this corrective factor can achieve 30 % in general case. The peak runoff can be underestimated that eventually leads to the expected river stage for the pipeline passage being too low and the pipe laid across that stream being flooded.

Ground Water Flow In Wetlands

We examined the ground water flow conditions for an area of wetlands and described them by the so-called Local Drainage Factor (LDF) which was determined here as the ratio between water yield and infiltration intensity for the area where the pipe can strongly impact the ground water flow. We outlined the segment by using the flow path network drawn on aerial imagery of the wetlands as was recommended (Saveljeva, 1987). The LDF value in hours is the time that a soil column of several depths (0.2 m as in given case, Table 2) can be drained of existing water. For compacted fill in the pipeline trench, the LDS values can be much more than those given here for a native peat with a permeable upper layer (turf of *shpagnum* moss).

Hence, the pipe laid under ground can become an obstacle for ground water flow. We used a method to calculate the possible ground water level under the influence of a pipeline. The method is based upon the water balance equation and so-called hydro-morphology theory of wetland water flow (Ivanov, 1975) and determining the snow water equivalent and snowmelt intensity (Shutov, 1993, 2000b) as input and evaporation and drainage flow as output. The example in Table 2 shows that the total value of LDF for a 1m layer is about 11 hrs., with considerable non-homogeneity in its vertical profile: the LDF for upper friable layer is three times less than that for the compressed peat layer below. Observations showed that the ground water level for native wetlands of the region vary from 10 – 20 cm beneath the surface just after snowmelt to only 50 - 70 cm by late summer. There can be a leakage from saturated peat layers to lower-layered aquifers and a lateral flow to the adjacent forest productive areas on mineral grounds. The pipe with its embankment make this water transfer unfeasible that leads to a local inundation and, gradually year-by-year, to enlargement of the wetland areas.

Table 2. An example of calculation the local drainage factor (LDF)

z, cm	*SM*	*HM*	μ	*K*, cm/s	LDF, hours
20	0.65	0.11	0.43	0.145	0.82
40	0.42	0.10	0.32	0.029	3.06
60	0.31	0.09	0.22	0.022	2.78
80	0.23	0.09	0.14	0.017	2.29
100	0.17	0.09	0.08	0.011	2.02
				Total:	10.97

SM and *HM* are saturated and hygroscopic soil moistures respectively, μ is water yield coefficient, *K* is the saturation hydraulic conductivity

Conclusion

Proceeding from the above analysis of watershed physiographic factors for an area near gas pipeline routes, sections were identified with combined environmental hazardous phenomena. Before the pipeline construction three sections were recommended for special field investigations: (1) the route segment that passed through karst regions in the *Dvina* and *Onega* river basins; (2) the wetland landscapes with numerous lakes near the White Sea; (3) rivers such as *Vel'* and *Mosha* with their wide, flat and forested valleys.

It was discovered, that the environmental hazard zones concentrated at these sections of the pipeline are particularly subject to flooding due to enlargement of wetland landscapes. Another hazardous phenomena is the ability of water saturated and frozen ground to subside and to break under load of any construction. Pertinent factor actual to the north region is that the forest ecosystems suffer from impact of construction due to disturbance in ground water flow.

Based upon the above methodology principles, we suggest a program for investigation of the environmental hazardous factor *in-situ*, in those watersheds cross-secting by an arterial pipeline, preferably in the Valday Hills area as having a complex relief, various soils, along with a high degree of knowledge that has yet been obtained on the hydrological regimes by long-term observation at Valday supposed as valid for a wide range of conditions.

References

Bergström, S., and B. Carlsson. 1994. River runoff to the Baltic Sea. – Ambio, 23, 280-287

Burn, D. 1989. Cluster analysis as applied to regional flood frequency. – Water Resour. Planning and Management, 11, No. 5, 567-582

Gustard, A., L. Roald, S. Demuth, H. Lumadjeng, and R. Gross. 1989. Flow regimes investigations from experimental and network data (FREND). 2 Vols., Inst. of Hydrology, Wallingford, UK

Hosking, J.R.M. and J.R.Wallis, 1997. Regional frequency analysis. - Cambridge Univ. Press

Krasovskaja, I., and L. Gottschalk. 1993. Frequency of extremes and its relation to climate fluctuations. – Nordic Hydrology, 24, No.1, 1-24

Kuusisto, E. 1985. Estimation of mean maximum spring runoff in Finland. – Aqua Fennica, 15, 47-51

Isachenko, A.G. and A.A. Shliapnikov, 1989. Landscapes. - Mysl', Moskow (in Russian)

Ivanov, K.E., 1975. Water transfer in wetlands. - Leningrad, Gidrometeoizdat. (in Russian)

Nathan, R.J., and T.A. McMahon. 1990. Identification of homogeneous regions for the purposes of regionalization. – J. Hydrol., 111, No. 1, 217-238

Saveljeva, T.S., 1987. Comparison of structures of wetland drainage basins by aerial photographs. - Trans. of State Hydrol. Institute, 329 issue, 71-85 (in Russian)

Seuna, P. 1983. Influence of physiographic factors on maximum runoff. - Publ. of the Water Research Inst., No. 50, Helsinki, Finland, 5-28

Shutov, V.A. 1993. Snowmelt calculations. - Meteorol. and Hydrol., No. 4, 20-27 (in Russian)

Shutov, V.A., 2000a. Cluster analysis and numerical taxonomy of river basins. - Meteorol. and Hydrol., No. 1, 90-100 (in Russian)

Shutov, V.A. 2000b. Snow cover in various relief and landscape conditions - In: D.L.Kane (Ed.) Water Resources in Extreme Environment. - Proc. of AWRA Spring Specialty Conference, Anchorage, Alaska, USA, May 1-2, 2000, 19-24

Shutov, V.A., and A.A. Kapotov, 2000. Applied hydrology for gas pipeline engineering. - Ibidem, 275-280

Shutov, V.A. and A.A. Kapotov, 1998. Application of regional hydrology for the design of a pipeline route in north Russia. - In: H. Wheater and C. Kirby (Eds.) Hydrology in a Changing Environment. - John Wiley and Sons, Chichester. Vol. III, 247-252

Sokolov, A.A., S.E. Rantz, and M. Roche (Eds.). 1976. Flood flow computation. Methods compiled from the world-wide experience. - Gidrometeoizdat, Leningrad (in Russian).

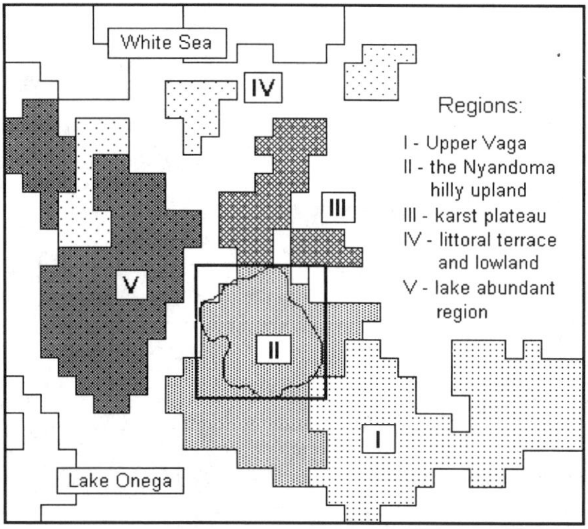

Figure 1. Scheme of the hydrological regions (I to V) in Northern Russia

The *Mosha* river basin is outlined as an example of that area where runoff depends on elevation

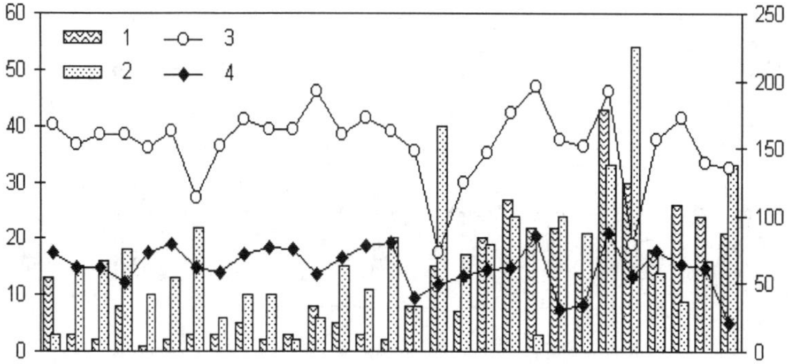

Figure 2. Belt-diagram of several physiographic and runoff characteristics along the pipeline

Partial area (%) of wetlands (1) and sandy soils (2), spring runoff depth (3), in mm (at the right scale) and the ratio between peak (maximal) and mean specific runoff (4)

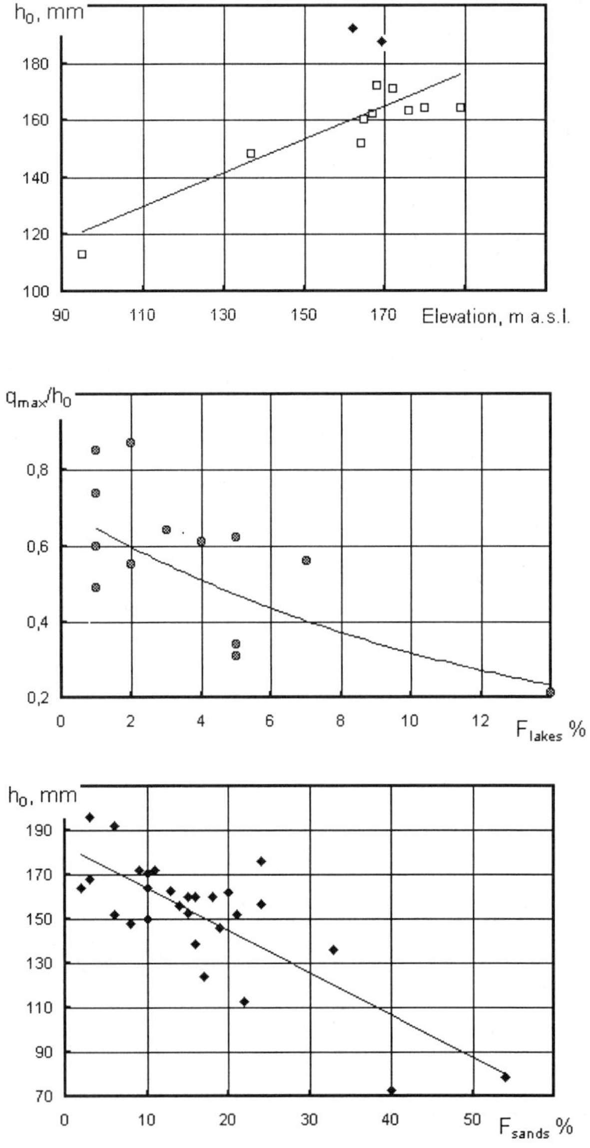

Figure 3. Empirical relationships between runoff and some physiographic features

Applications of Study Results in Snow Hydrology

Vladimir A. Shutov [1]

Abstract

Snow cover spatial distribution and snow melting have been studied at Valday Hills and in many other regions of Russia. An expert appraisal is proposed of how the snow cover obstructs industrial areas. To improve flood flow forecasts, a correlation has been established between snow cover and terrain elevation. Landscape conditions, forest at first, have been investigated as affecting snow cover distribution. An original method is proposed for quantifying the local scale distribution by the data on relief, land use and forest species. Snow cover distribution is examined as a factor influencing construction engineering. Snow melting processes were explored as conditioned by energy flux. Surface albedo of intact and contaminated snow was determined and appointed to be a key factor in energy exchange. Data of long-term observations were gathered to analyze the time series of snow water content and net water yield from a melting snow pack. Extreme values of snow water content were used as an input for hydrological calculations of drainage flow. Proposals were made of how to renew and improve the observation of snow cover as one of the main components of global climate and environment.

Introduction

Seasonal snow cover prevails over the winter in the landscapes of Russia, Canada and many other countries. It has a powerful effect on environmental processes and on human activity. Results in snow hydrology identify many implications to industrial activity besides their importance to pure geographic and climatic studies. These are: spring runoff forecasting, special support of agriculture and water projects, construction engineering. Fighting against snow drifting over roads and protection of fertile soils from erosion are not possible without information on the snow cover.

Land-based and aerial snow surveys, as well as satellite data allow the estimation of the snow cover extent at a large scale. To investigate the local spatial distribution of snow, we have to undertake experimental surveys either landscape-based, or grid-based in the area of interest. From long-term observations in watersheds at the Valday Hills in north-western Russia, regularities in snow accumulation and snowmelt by various climate and landscape conditions have been

----

[1] Senior scientist, Valday Branch of State Hydrological Institute.
175400, Valday, Pobeda Street 2, Russia; hydroshpere@mail.ru

discovered (Shutov, 1994, 1995, 1996, 1997, 2000a). This paper investigates the problem of how to provide a systematic approach for their use in construction engineering.

To expound upon the importance of snow cover investigations we used the method of expert evaluations (Wright and Green, 1987). It provides an appraisal of how much effect any natural phenomenon has on an industrial environment. The experts determine the degree of such an impact based on their intuition and experience. Each of the individual evaluations are presumed as based on knowledge, which has already been accepted (Barry, 1981, Gray and Male, 1981, Kuusisto, 1984).

We examined here human activity including hydro-meteorology, and take into account the following attributes of the snow cover: snow coverage SC, snow depth SD, bulk snow density BD, snow water equivalent SWE, snow evaporation SE, snow melt rate SMR, mechanical firmness MF, water capacity WC, water permeability WP, snow surface albedo SA.

We appraised the degree of impact on the industrial areas as presented in Appendix 1. By compiling this matrix of conformity snow – industrial areas we can come to a conclusion that SC, SWE, and SMR are the major features of snow and that runoff forecasting, agriculture (including soil reclamation) and climate assessment are the most sensitive to the limitations related to the snow cover features.

Certainly, this evaluation of several areas is based upon personal experience and can probably be questioned. However, it is well considered that agriculture is responsive to almost all components of the water cycle including the snow cover. High rank position of winter recreation (skiing) is due to its evident sensitivity to the snow cover features, for construction such as pipeline engineering which is subject to soil freezing, and for road traffic which can be interrupted due to heavy snow drifting. Fisheries can suffer indirectly, only by a scarce solar light going into the lake waters through an opaque snow-ice cover.

The author has experience in many of the above applications. So, by using experimental data, researches was conducted to determine the environmental impact assessment for a gas pipeline in northern Russia (Shutov and Kapotov, 2000), and for agricultural snow management in Russia, Kazakhstan and Mongolia (Shutov, 1990, Kaljuzhny and Shutov, 1998, Shutov and Nazagdorj, 1988).

Snow Cover Data Available For Flood Prediction

Snow cover is initially distributed throughout the plains under climatic conditions, and then is re-distributed due to aeolian drifting in relation to the local relief, land use and vegetation. The same factors play over hills and mountainous terrain, where other factors such as elevation and slope aspect define snow distribution. Such regions in Russia and some neighboring countries (Ukraine, Poland, Sweden and Germany) are fertile for crops, heavily settled, and industrially developed. They are subjected to dangerous spring floods. Therefore, we have to make an effort for the study of snow extent and snow melting to provide reliable prediction of flood flow.

A significant influence of the Valday Hills on snow distribution has been found by an analysis of data observed in the *Polomet'* river basin (Shutov, 1996, 2000a), which is 2200 km^2 in area with elevations from 40 to 300 m a.s.l. These data give a detailed snow cover distribution on a watershed scale, contrary to the simple interpolation of the standard survey data provided by surrounding weather stations. By using these data, we obtained detailed patterns of the SWEs in this drainage basin (Figure 1). The spatially averaged SWEs based on the detailed patterns are found to be much greater (170 mm vs. 110 mm for those interpolated by standard network data). The latter value is much more than its regional averages (110 mm) adopted in forecasting, that is to be implicated to spatially distributed hydrological models.

The closed elevation - snow relationships have been found by observations in the *Moskwa* river basin (nearby the city of Moscow) and in the *Urals* Mts. Respective correlation coefficients are in the order of 0.85-0.95 and, lapse rates vary from 30 to 45 mm per each 100 m height. Hence, snow map development has been proposed as a step-wise procedure. Firstly, the observed values are converted (according to the lapse rate) to a basic elevation level specified previously. These zero-level values have been interpolated into the grid cells by using geostatistical (e.g. kriging) procedures. The lapse rate was then taken into account and, as a result, detailed spatial patterns of snow has been drawn as a map (Shutov, 1996, 1998).

Snow And A Forest Canopy

Snow accumulation under forests depends on snow retention (interception) on tree crowns. Thereby, the SWEs vary significantly even in small drainage basins. The SWE-values under a forest canopy can be estimated by a corrective factor *KF* which is determined as the forest - open land ratio of the respective SWEs. Based upon observations in European Russia, we found that *KF* values increase southwards. This depends on climate conditions, but most of all on the intensity of episodic snow melt during the winter. As is obvious, the latter is different for a field and a forest canopy due to differences in incoming solar radiation. Thus, the forest - open land ratios depend on both structure of the tree canopy, solar radiation, and air temperature. We obtained an empirical equation which is based on such a correlation:

$$KF = A\,[\,\Sigma(T_{>0}) / H_0\,] + B \qquad (1)$$

where $\Sigma(T_{>0})$ is the sum of positive air temperatures during a winter, H_0 is the SWE for an open area, A is so-called degree day melt factor, as it was found locally A = 4.1 mm/degree a day (Shutov, 1995, 1997), B is the factor describing the net snow storage under a tree canopy (see Table 1).

Table 1. The net snow storage under trees (interception is defined as I = 1 – B)

Region	Birch	Pine	Spruce	Alder	Mixed
Valday Hills	0.96	0.89	0.80	0.93	0.85
Moscow	0.96	0.90	0.73	0.95	0.81

This factor is changing with forest age that is the most inherent to a fir (spruce) stand. There also is a correlation between interception rate I and forest density D (Kaljuzhny and Shutov, 1998) described as follows:

$$I = 0.402\ D^2 + 0.088\ D \tag{2}$$

Such relationships are found similar for Russian and Canadian forests (Ffoliott *et al*., 1989). This is presented as useful to evaluate the water yield when planning for a silviculture and forestry.

Informative-Mapping Analysis (INFOMAP)

The author has developed the so-called «informative-mapping analysis» (INFOMAP) (Shutov, 1994, 1995, 2000a). This method is based upon the data of detailed snow surveys over two small (some 0.5 km^2 in area) watersheds: *Usadjevsky Log* with with prevalent open grassland area and *Tajezhny Log* covered with coniferous and mixed forests. The data showed that the snow distribution patterns are persistent and repeated year after year. Snow is concentrated mainly within hollows, under bushes, at the forest edge with the greatest SWEs accumulated on east-faced (leeward) slopes. Conversely, the SWEs near the watershed boundaries, on hilltops and at west-faced (wind-exposed) slopes decrease. As to a forest canopy, the SWEs vary in respect for tree species (see above) being also slightly variable along with local relief.

It is hardly possible to evaluate an individual effect of each snow accumulation factor, since they are affecting together. It seems to be more reasonable to arrange different topographic and landscape elements according to the local snow accumulation coefficient (LSAC) which is the ratio of the at-cell value of SWE and its average evaluated through a representative survey. Then each specified cell of the surface is described by an index which attributes the snow accumulation capacity. These indices (Appendix 2) are collected in four groups: relief elements, slope aspects, plant cover (or land use) and forest species. Inside these groups, each index ranks as a digit M from 5 to 1 in respect to snow retarding capacity of the cell.

Mean rank of indices reflects the probability of maximum snow retention (maximal LSAC value) for a cell. This can be rendered by the classical Shannon's definition of the entropy as the measure of information, which is proportional to a binary logarithm of the mean rank value, and can be expressed empirically based upon snow survey data with the following equations:

$$\mathrm{LSAC} = 0.375\ \mathrm{Log}_2\,(M) + 0.50 \tag{3a}$$

$$\text{LSAC} = 0.235\ \text{Log}_2\,(M) + 0.65 \tag{3b}$$

where the first equation belongs to a field (open) area, the second one to a forested watershed. The correlation coefficient is equal to 0.82-0.84.

Significant biases detected in respective empirical scatter plots are explained by the remaining uncertainties. To diminish them, we must assume in addition the dynamic attributes such as wind direction and wind velocity affecting snow drifting. By this, the model would become dynamic. Although, it must be noted that the snow transport models suggested as dynamic (Liston and Sturm, 1997) comprise such a static element as empirical wind-topography indices assumed spatially constant.

The INFOMAP allows us to reconstruct gridded spatial distributions of SWE (see Appendix 2) by allocating those SWE-values observed at a representative site with topographic and landscape maps. As to forested areas, we must quantify both the tree species and the surface relief (Shutov, 1994). Such an analysis given an opportunity for calculation of the SWEs and snow depth that is to be applied to the problems of snow management and environmental engineering, for instance, when the local snow distribution has to be retrieved nearby any large construction (railway, pipeline, dike, etc.).

Snow And Soil Frost Depth

Let us look at why pipeline construction is mentioned in Appendix 1. In the normal design of a pipeline there is a demand to specify the depth of the pipe as limited by soil freezing depth inherent to a region and soil type. The latter depends on snow depth which insulates the underlying soil from low temperature influence. The effect of a snow blanket is to decrease the temperature amplitudes both seasonal and diurnal that can generally be expressed in exponential form:

$$A = A_0\, exp\,(-\pi Z\, C\, \rho / \lambda\, t) \tag{4}$$

where A and A_0 are the temperature amplitudes at the snow pack bottom and on the surface respectively, Z is snow depth, C its heat capacity, ρ is snow density, λ is thermal conductivity of snow, t is the period of undulation, π is the pi-number.

The Eq. (4) together with the values of λ specified empirically are simplistic fundamentals in applied snow physics. Combining this with the spatial distribution of the snow depth received from the above INFOMAP analysis, we render the temperature distribution at the soil surface under snow. It should be assigned as the boundary conditions for simulation of the soil freezing depths. Up-to-date models with graphical output use a snow distribution data to provide imaging of soil thermal conditions and soil frost (Hinzman *et al.*, 1998).

Another, quite simple approach has been proposed by the author (Shutov and Belolutskaja, 1990) where it was found by observations that the diurnal temperature undulations cannot penetrate a deep snow pack and, are usually equal to 5 –

10 days. Given the temperatures above and under the snow pack observed at a weather station, we may compute the temperature under snow pack of any depth by using only simple equations like Eq. (4) and the thermal conductivity λ of the snow. The latter (in W/m·degree) has been approximated in the above paper as following a function of the snow density ρ (in g/cm^3):

$$\lambda = 0.115 \exp(1.9\rho) \tag{5}$$

Such a convenient way has allowed us to convert the spatial distribution of snow depth over a small experimental watershed to a distribution of soil temperature and to the soil freezing depth.

Snow Water Contents As The Input For Design Drainage

Designing tile drainage in agriculture or culverts in road construction near large buried pipelines, for instance, requires data on the frequency of water inflow to the drains by rainstorms and snow melt from long-term observations. From this, as well as from soil water properties, we can determine the inter-drain spacing and drainage depth. However, two main issues remain. First, the evaluation of the frequency of the water inflow rates, and second, detailed investigation of the soil water properties over the area in relation to soil types and soil texture.

Resulting from a summary of long-term observation data, we found that snow water content plus rainfall during the snow melt period are the major input variables for determining the net inflow rate. Snow water contents affecting the spring flood vary spatially particularly due to non-homogeneous micro-relief and forest (Shutov, 1994). The local peculiarities, such as deep snow drift at leeward side of the pipeline embankment can create artificial ecotones within a landscape.

Frequency of the snow water equivalent H is used as an input for calculation of possible water inflow to the drainage by snow melt. The Gumbel distribution function (Figure 2) has been used to estimate the extreme values as widely proved long ago (Viessman *et al.*, 1977). After this approach we extrapolated the tail of the distribution to extreme values by the following empirical equation:

$$H = 27.7\,G + 83.5 \tag{6}$$

where $G = -Ln\,\{Ln\,[1/(1 - P)]\}$ is normalized Gumbel variate, P is exceedance probability (%) and respective return period.

Hence, for instance, the snow water content for $P = 1\%$ ($G = 4.6$), i.e. for the return period of 100 years, is approximately equal to 210 mm. The same approach was employed to evaluate (by observations and using indirect method based on energy budget) the extreme snowmelt rate that was found to be 67 mm/day.

Snow Management Practices

Snow management practices were developed mainly for agricultural applications in order to increase the water content of the soil. Artificial snow retention promotes water sufficiency over open plains due to decreasing snow drifting and provides a soil water supplement in arid lands (Shutov, 1990, Kaljuzny and Shutov, 1998). It is also used to improve the criptovegetation of winter crops, to keep soil water, to delay overland flow and to prevent contamination of waters and soil erosion. Another way is to control the water regime through regulation of the snowmelt by strewing dust or ashes. This method has been put into practice in Russia. Terms and conditions for this are predictable.

Schematically the runoff creating surface on a slope is visualized as altered tracts with different infiltration capacity. Melt water percolates into thawing soils under the snow ridges built by special snow plough or below the snow drift caught behind their lee side. At the same time, the areas between may by impervious by cold winter. Field studies showed, that water supplement takes place (Table 3) particularly by severe winters. It is possible to predict this after a snow and soil moisture survey. By mild winter frost such a snow retention can not be efficient. Moreover, one can expect an additional water loss due to increasing evaporation from rugged and dirty snow particularly from steep south-faced slopes.

Table 3. Water budget components found in five year observations

Water budget components	Snow retention	Intact snow cover
Snow + Precipitation	210.2	182.8
Snow evaporation	26.2	17.2
Water supplement	156.8	119.8

Evaporation and solar radiation measurements were carried out, as well as snow surface albedo. We have mapped (Shutov, 1990) the evaporation losses and the soil water supplements as corresponding to the indices of drought for the corn production areas in Central Russia, Ukraine, southern Siberia and northern Kazakhstan.

Snow melt rate has been detected indirectly by energy fluxes in the snow – atmosphere system. To detect all the components of the energy budget, we have implemented the profile method of observation by automatic devices. It was found, that the surface albedo is of great importance to regulate snow melt, since it is doubtless that dirty snow pack melts more intensively. We have studied this on special plots with intact and contaminated snow (Shutov, 1997). The resulting relationship between albedo α and soil particles and slack concentrations k is as follows:

$$\alpha = \alpha_0 [1 - 0.076 (Log\, k)^{2.44}] \tag{7}$$

Snow pack initially detains some portion of liquid water, then it releases it as water yield depending on snow grain size and, respectively, on snow permeability. We have studied these events using impermeable plots and special devices such as shallow «pan» lysimeters. Based upon investigations, we proposed a simple approach to evaluate snow melting empirically through so-called «radiation degree day factor» (Shutov, 1997) and we found them varied, significantly different from that degree day factor used in hydrological simulation models.

Conclusions and Proposals

Construction and environmental engineering require reliable data on the soil properties since the soil is an element of the constructions of roads, channels, underground pipelines, etc. These properties are spatially distributed conformable to the landscape and vegetation conditions. The same is to be said regarding the snow cover as unevenly distributed over a landscape. Since most of the issues pertaining to the snow cover and the soils are affected by both natural (climate, hydrology) and man-made factors (construction, in particular) we must examine the soil and snow together. Many problems remain for future investigations and development of snow hydrology. They can be reduced to the following proposed study topics:

1. Measurement and data acquisition technique. The main problem is how to retrieve a virtual picture of snow cover distribution both from land based surveys and satellite imagery. By lacking coupled satellite and land-based data, we cannot derive true SWEs using only remotely sensed brightness temperature of the snow. We consider two directions: the first is to carry out the investigations in snow physics, the second is to examine the snow cover distributions within some key pixels of the satellite imagery. The INFOMAP is one of the possible ways to retrieve a true spatial picture of snow cover.

2. Simulation and forecasting of dangerous snowmelt-induced floods, including those endangering constructions. Satellite data must be collated with radar technique and land surface observations. In Russia and other countries the radar-based precipitation measurements are used to detect snowfall intensity and SWE (Shutov, 2000b). Based on our experience, the problem seems to be clarified when spatial patterns of SWE may be obtained comparable with radar or satellite data output.

3. Applied research and snow management. There is a complex interaction between the crystal structure in the snow pack and the chemical compounds such as sulfur dioxide or heavy metals delivered mainly from the atmosphere. They migrate into the soil and further to stream water during snowmelt period. The problem is apparently to identify the sources and sinks of industrial contaminants which can spread off the funnel and accumulate within the snow pack. Besides, during the last two decades in Russia and abroad answers to environmental problems are requested. Thereafter, it could be of great importance to undertake investigations to evaluate a possible effect of snow management for protection of stream waters and lakes near construction.

References

Barry, R.G. (1981). Mountain weather and climate. - Methuen, London and New York.

Ffoliott, P., G. Gottfried and M. Backer (1989). Water yield from forest snow pack management. - Water. Resour. Res., 25, No. 9, 1999-2007.

Gray, D.M. and D.H. Male (1981). Handbook of Snow. Principles, processes, management and use. - Pergamon Press, London.

Hinzman, L.D., D.J. Goering, and D.L. Kane (1998). A distributed thermal model for calculating temperature profiles and depth of thaw in permafrost regions, - J. of Geophys. Res., 103, 28975-28991

Kaljuzny, I.L. and V.A. Shutov (1998). Modern status and problems of field investigation of snow cover. - Water Resour., 25, No. 1, 34-42 (in Russian).

Kuusisto, E. (1984). Snow cover and snowmelt in Finland. - Publ. of Water Research Inst., 55, Helsinki.

Liston, G.E., and M. Sturm (1997). A blowing and drifting snow model applied to Arctic environments. – In: Proc. of the Northern Research Basins 11th Int'l Symposium, Pruhoe Bay/Faibanks, Alaska, USA, August 18-22, 1997, vol. 1, 165-177

Shutov, V.A. (1990) Methods for estimating hydrological role of snow retention and snow management. - Water Resour., 17, No. 1, 44 -51 (in Russian)

Shutov, V.A. (1994). Snow water storage distribution on watersheds of forest zone. - Meteorol. and Hydrol., No. 9, 85-92 (in Russian)

Shutov, V.A. (1995). An experience of snow cover field investigation and modelling studies. - Water Resources., 22, No. 6, 645-653 (in Russian)

Shutov, V.A. (1996). Interpolation and mapping of snow storage on river basins with allowance for altitudinal gradient. - Meteorol. and Hydrol., No. 10, 67-73 (in Russian).

Shutov, V.A. (1998). Multy-scale spatial analysis of snow cover. - Bulletin of Russian Academy of Science, Geographical series, No 1, 122-132 (in Russian)

Shutov, V.A. (2000a). Snow cover in various relief and landscape conditions. - In: Douglas L. Kane (Ed.) Water Resources in Extreme Environment. Proc. AWRA's 2000 Spring Speciality Conf., Anchorage, Alaska, USA, 19-24

Shutov, V.A. (2000b). Processing the radar precipitation data for engineering hydrology – Ibidem, 275-280, pp. 315-321

Shutov, V.A., and A.A. Kapotov. (2000). Applied hydrology for gas pipeline engineering. - Ibidem, 275-280

Shutov, V.A. and H. Natzagdorj (1988). Studying the snow cover and estimation of its usage for agriculture in Mongolia. - Trans. of the Mongolian Hydrometeorol. Inst., 13, 102-110 (in Russian).

Viessman, W., J.W. Knapp, G.L. Lewis and T.E. Harbaugh (1977). Introduction to Hydrology. 2nd Ed., Harper & Row, N.Y., 704 pp.

Wright, D.S. and G.D. Green (1987). An environmental impact assessment methodology for major resource development. - J. Env. Management, 24, No. 1, 1-16.

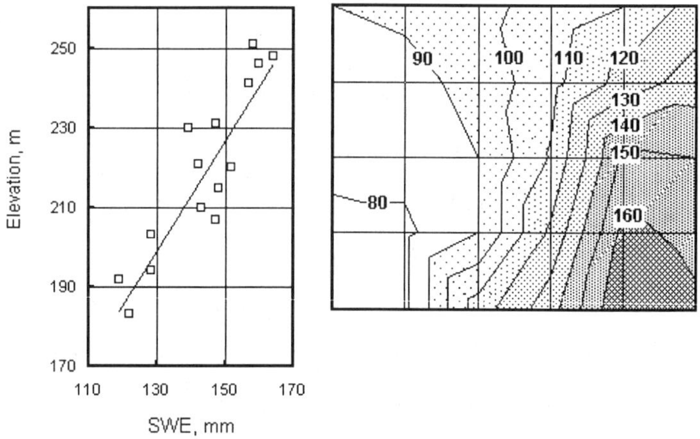

Figure 1. Spatial distribution of the snow water equivalent on the Valday Hills
Left: elevation – snow water equivalent (SWE) relationship
Right: schematic map of SWE (mm) for the small river basin

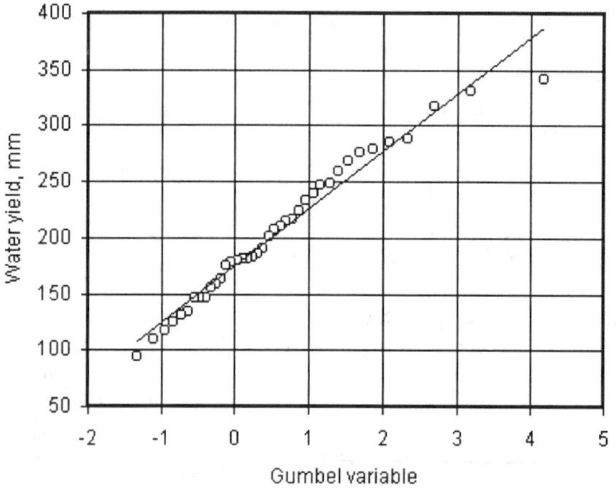

Figure 2. Frequency curve for the snowmelt water yield observed at Valday Hills

Appendix 1. Expert evaluation of the influence of snow cover on industrial areas
(1, 2, 3 are respectively feeble, significant and strong influence, 0 - no impact)

	Industrial area	SC	SD	BD	SWE	SE	SMR	MF	WC	WP	SA	
1	Runoff forecasting	3	2	2	3	1	3	0	2	2	1	19
2	Climate assessment	3	2	1	3	2	3	0	0	0	3	17
3	Agriculture	3	3	2	3	1	2	1	1	1	0	17
4	Recreation (skiing)	3	2	2	0	0	2	3	2	0	2	16
5	Road traffic	3	3	1	2	0	2	2	1	1	0	15
6	Timber industry	3	3	1	1	0	2	2	1	1	0	14
7	Pipeline construction	2	3	2	2	0	1	2	0	1	0	13
8	Turf mining	3	2	1	2	1	2	1	0	0	0	12
9	Soil protection	2	1	1	2	0	3	0	0	1	0	10
10	Railway transport	2	3	1	2	0	1	0	0	0	0	9
11	Fishery	2	2	1	1	0	0	2	0	0	1	9
		42	38	22	35	9	32	17	8	8	8	

Appendix 2. Explanation of the INFOMAP procedure

A. Ranking local snow accumulation factors

→ Relief element	M	Slope exposure	M	Land use	M	Forest species	M
Hollow	5	East	5	Bushes	5	Bushes	5
Convex slope	4	North	4	Stubble	4	Deciduous	4
Plain surface	3	Plain surface	3	Forest	3	Pine	3
Concave slope	2	West	2	Meadow	2	Mixed	2
Hilltop	1	South	1	Fallow	1	Spruce (fir)	1

B. An example of INFOMAP analysis for a snow covered area

Relief

2	3	2
4	4	4
5	2	3

Exposure

1	3	1
2	4	4
5	4	3

Land use

2	2	3
3	5	3
3	3	1

Forest types

		2
2	5	1
4	1	

Mean sign value →

1,67	2,67	2,00
2,75	4,50	3,00
4,25	2,50	2,33

$Log_2 M_{mean}$ →

0,74	1,42	1,00
1,46	2,17	1,58
2,09	1,32	1,22

LSAC →

0,78	1,03	0,88
1,05	1,31	1,09
1,28	1,00	0,96

SWE, mm

109	145	123
148	185	154
181	140	135

Snow water equivalent at homogenous field snow survey course = 141

Evaluating Capping Alternatives for an Arctic Landfill

Cory Hinds, P.E. [1], Jennifer LeMay, P.E. [2], Kim Stricklan, P.E. [3]

Abstract

This paper presents an evaluation of capping alternatives for an 18.2 hectare landfill located in Barrow, Alaska. When closure planning for the Barrow Landfill began, excavation and removal or treatment of wastes was included in the initial evaluation of potential closure options. However, due to Barrow's remote location and the high costs of waste excavation, transportation, and/or incineration, capping was identified as the most cost-effective closure method. Three general types of caps were evaluated in a comparative analysis, including a freezeback cap, silt cap, and a composite soil/geomembrane cap. A numerical selection process introduced by Peggs and Thiel was used for rating and selecting a specific geomembrane. Cold weather shear testing results for Linear Low Density Polyethylene (LLDPE) and Polypropylene (PP) are presented. Based on the evaluation process and the cold weather testing of geomembrane samples, the recommended capping alternative was a buried PP geomembrane.

Introduction

Barrow is the northernmost community in the United States (see Figure 1). The severe climate, remote location, and scarcity of quality material sources make earthwork a challenge.

Barrow's climate is an arctic desert and consists of a combination of extremely low temperatures, minimal precipitation, and persistent winds. The mean annual temperature is –12.5°C. Thermal monitoring at the Barrow Landfill indicates that thaw penetrates 1.5 to 2.1 meters into unvegetated landfill soils. The ground surface temperature is usually below freezing by late October and remains frozen until late June. Permafrost in this region reaches depths exceeding 305 meters. Precipitation averages 12.7 centimeters annually, including approximately 71 centimeters of snow. Although precipitation is light, soils are frequently saturated with water in the summer because of low evaporation rates and no loss of surface water to groundwater due to the permafrost barrier.

Local material sources (soils) are limited to 1 meter of ice-rich silt over poorly graded sands and limited gravel with silt.

[1] Project Manager, URS Corporation, 2700 Gambell Street, Anchorage, AK 99503, Phone (907) 562-3366; cory_hinds@urscorp.com
[2] Project Engineer, URS Corporation, 2700 Gambell Street, Anchorage, AK 99503, Phone (907) 562-3366; jennifer_lemay@urscorp.com
[3] Associate VP, Environmental Dept, Arctic Slope Consulting Group, 301 Arctic Slope Ave., Suite 300, Anchorage, AK 99518, Phone (907) 349-5148; kstricklan@ascg.com

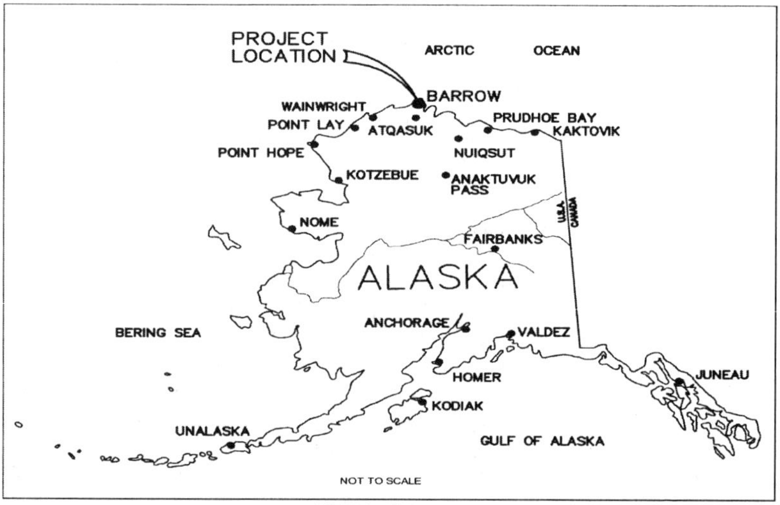

FIGURE 1: VICINITY MAP.

Evaluation and Selection of Capping System

The Barrow Landfill property will be designated as open space after closure is complete, except for the area around a solid waste incineration building. No heavy equipment movement or operations will be conducted on the landfill surface after closure.

Excavation and removal or treatment of wastes was included in the initial evaluation of potential closure options. However, due to the size of the landfill, Barrow's remote location, and the high costs of waste excavation, transportation, and/or incineration, on-site containment via capping was identified as the most cost-effective closure method. Three general types of containment were evaluated: a freezeback cap, a silt cap, and a composite soil/geomembrane cap. Schematics of each cap are shown on Figure 2.

Containment Alternative 1: Freezeback Capping

Under this alternative, a soil cap would be placed to raise the permafrost level, causing freezeback of the landfill wastes in place. The existing cover of the Barrow Landfill is approximately 1 meter thick. A soil layer ranging from 1.2 to 2.1 meters thick would be placed over the landfill area to be capped. Approximately 0.15 meter of topsoil would be placed over the soil layer and vegetated. Soil cover would also be placed in a lagoon located within the active landfill area, reducing the area of the thaw bulb and keeping the waste frozen.

Because the landfill waste/soil is frozen, it is not likely that methane would be produced. On-site gas sampling confirms this. Therefore, a vent system is not needed with a freezeback cap.

Containment Alternative 2: Standard Low - Permeability Soil (Silt) Cap

This alternative consists of a multi-layer soil cap. From bottom to top, the conceptual cap section would consist of: 0.3 meter of gravel fill, a minimum of

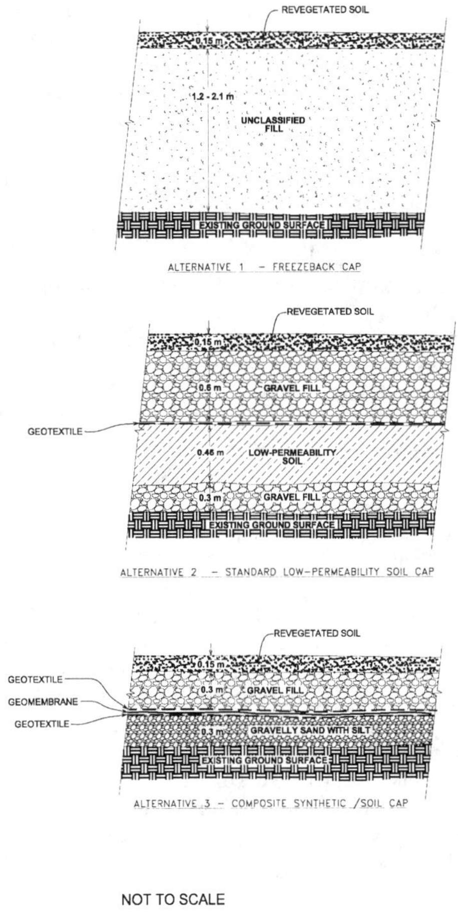

FIGURE 2: CAPPING SCHEMATIC FOR ALTERNATIVES 1, 2, AND 3.

0.46 meter of low permeability (1 x 10^{-5} cm/s) soil, a permeable filter fabric, 0.6 meter of gravel fill, and a revegetated 0.15 meter topsoil layer.

A network of collection piping would be installed below the cap to passively vent excess landfill gas generated by the anaerobically decomposing organic wastes. As with the freezeback alternative, the majority of the wastes would freeze in place. However, wastes immediately adjacent to the lagoon thaw bulb (i.e., below water level at shallow depths) would remain thawed.

The purpose of the bottom 0.3 meter layer is for surface leveling and is standard for all capping alternatives. The 0.46 meter layer of low-permeability fill directly satisfies Alaska Department of Environmental Conservation (ADEC) requirements for capping a Class II landfill. The 0.6 meter gravel cover on top of the low-permeability silt is necessary for construction. Because the silt is an ice rich material and has high moisture content, it is extremely difficult, if not impossible to place, spread, and compact without the additional gravel cover. The additional gravel cover provides a workable surface for compaction requirements over the very wet, soupy, low-permeability soil layer.

Containment Alternative 3: Synthetic Liner System

This alternative consists of a multi-layer, soil/geomembrane capping system. From bottom to top, the conceptual cap section would consist of: 0.3 meter of gravelly sand with silt; a 12 ounce non-woven geotextile cushioning layer; a 1.016 mm (40 mil) unreinforced PP geomembrane; a 12 ounce non-woven geotextile cushioning layer; 0.3 meter of gravel fill; and a revegetated 0.15 meter topsoil layer. The 0.3 meter of gravelly sand with silt would have the lowest permeability available from workable local material sources. When installing the cap, the liner edges would be secured in anchor trenches. As with the freezeback alternative, a majority of the wastes would freeze in place. However, wastes immediately adjacent to the lagoon thaw bulb (below water level at shallow depths) would remain thawed.

All three capping alternatives would allow for closure of the Barrow Landfill in accordance with the ADEC regulations and employ conventional technologies. Material for implementing Alternatives 1 and 2 is located in the City's gravel pit and overburden stockpile (called Moolin Mountain). Due to the extremely high water content of the Moolin Mountain silts (up to 78% due to ice), there was significant uncertainty whether the silt could be successfully dewatered and compacted to meet the required hydraulic conductivity. The level of effort required to work this material was also uncertain, which increased the risk of construction problems and delays. Additionally, quality control for the placement of the silt would be limited. These concerns were confirmed during the 2001 field season as the material liquefied, and even low ground pressure machinery was unable to maneuver on it.

Alternative 3 uses synthetic materials that would require specialized technicians to install and seam the synthetic liner together. However, a synthetic cap is easier to construct because there is no material to dewater, and better quality control during installation yields superior long-term performance. In addition, not only is Alternative 3 the least expensive of the three capping alternatives, it also appears to be the most effective.

Geosynthetic Liner Selection Process

Preliminary research identified advantages and disadvantages of numerous types of geosynthetic liners that may be suitable for use at Barrow and are discussed below.

Geosynthetic Clay Liner (GCL)

- Reported good performance under freeze/thaw conditions in laboratory testing (EPA, 1997).
- Under normal conditions, self-healing and easy to install (EPA, 1997).
- Suitability as the primary liner may be questioned by regulatory authorities (ADEC, 2001).
- Significantly heavier than other products, resulting in increased transportation costs (EPA, 1997).
- Careful consideration should be exercised before using as sole hydraulic barrier under freeze-thaw conditions (ADEC, 2001 and Kraus, 1997).
- Relatively expensive material cost (see Table 1).

High Density Polyethylene (HDPE)

- Highly crystalline molecular structure (>52% crystalline). Higher crystallinity means the membrane is harder (better puncture and chemical resistance) but also more brittle and harder to work with. HDPE has less flexibility than LLDPE and significantly less flexibility than PP (DPS, 2001).
- Brittle temperature reported at between –70°C and –90°C (GSE, 2001 and Poly-Flex, 2001).
- Elongation at a yield of approximately 12% (GSE, 2001 and Poly-Flex, 2001).
- Able to be welded at low temperatures, but reduced "welding window" could increase difficulties during construction (GSE, 2001 and Poly-Flex, 2001).
- Multiple examples of successful cold climate installations. Most often installed as liners to exposed ponds (Polar Supply, 2001).
- Cheaper material costs than PP, comparable material costs to LLDPE (see Table 1).

Linear Low Density Polyethylene (LLDPE)

- Semi-crystalline molecular structure (33% to 40% crystalline). This material is less crystalline than HDPE, but more crystalline than PP. LLDPE is more flexible than HDPE, but less flexible than PP (DPS, 2001).
- Brittle temperature reported at between –70°C and –90°C (GSE, 2001 and Poly-Flex, 2001).

- Elongation at a yield of approximately 300 to 400% (GSE, 2001 and Poly-Flex, 2001).
- Able to be welded at low temperatures, but reduced "welding window" could increase difficulties during construction (GSE, 2001 and Poly-Flex, 2001).
- Multiple examples of successful cold climate installations. Most often installed as part of capping systems due to increased flexibility over HDPE (Polar Supply, 2001).
- Cheaper material costs than PP, comparable costs to HDPE (see Table 1).
- Easier to handle than HDPE at low temperatures (GSE, 2001 and Poly-Flex, 2001).

<u>Polypropylene (PP)</u>

- Low-crystalline molecular structure (<10% crystalline). PP is more flexible than both HDPE and LLDPE (DPS, 2001).
- Brittle temperature reported at approximately –40°C. [However, there appears to be industry concurrence that PP simply has not yet been tested at lower temperatures] (GSE, 2001 and Poly-Flex, 2001).
- Elongation at a yield of approximately 700% (GSE, 2001 and Poly-Flex, 2001).
- Able to be welded at low temperatures. Increased "welding window" could reduce difficulties during construction. Preferred over HDPE and LLDPE (GSE, 2001 and Poly-Flex, 2001).
- Multiple examples of successful cold climate installations (Polar Supply, 2001, and Rollin, 2001).
- More expensive material costs than both HDPE and LLDPE (see Table 1).
- Able to be partly prefabricated in the factory reducing field seam requirements and installation time (GSE, 2001 and Poly-Flex, 2001).
- Less expansion and contraction with extremes of temperature than HDPE or LLDPE (GSE, 2001 and Poly-Flex, 2001).
- Preferred by the majority of installation contractors on the basis of its easier workability (GSE, 2001 and Poly-Flex, 2001).

Approximate material and shipping costs for these geosynthetic liners are presented in Table 1.

All of the suppliers who have any reported experience with regard to the installation of liners in cold climates recommend either LLDPE or PP (Polar Supply, 2001, GSE, 2001, and Poly-Flex, 2001). The GCL was eliminated from consideration due to high material and shipping cost. LLDPE and PP were evaluated further. For PP, there are two choices: reinforced or unreinforced. Scrim reinforcement increases tear resistance, puncture resistance, and further decreases the coefficient of linear thermal expansion. However, both cost and weight increase with added reinforcement. For this application, the increased strength was not required, so unreinforced PP was evaluated.

TABLE 1
APPROXIMATE MATERIAL AND SHIPPING COSTS OF GEOSYNTHETIC LINERS

Liner Type	Material Cost ($\$/m^2$)	Shipping Cost ($\$/m^2$)	Total Cost ($\$/m^2$)
GCL	4.84	2.91	7.75
1.524 mm (60 mil) HDPE or LLDPE	3.34	0.75	4.09
1.016 mm (40 mil) HDPE or LLDPE	2.37	0.64	3.01
1.016 mm (40 mil) UPP	4.20	0.64	4.84

Notes:

1. Material costs for HDPE and LLDPE are essentially the same.
2. GCL material cost is based on CETCO Bentomat ST.
3. Material costs obtained from Polar Supply, AK, and NW Linings, WA. Weights were obtained and verified from CETCO, Gundle, and Poly-Flex.
4. Shipping involves trucking to Prudhoe Bay, then barge to Barrow.
5. m^2 = square meters.

There are no geomembrane landfill covers in the Arctic, and no applications duplicating the proposed installation. However, the following are several known successful applications of each liner in northern regions:

<u>1.016 to 1.524 mm (40 to 60 mil) LLDPE</u>

- Alaska Regional Native Heritage Center (pond liner) 18,600 m^2
- Anchorage Cold Storage Facility (vapor transmission liner) 9,300 m^2
- Nixon Fork Mine (dam face) 46,500 m^2
- Wildwood Correctional Facility (soil liner) 37,000 m^2

<u>1.016 mm (40 mil) PP</u>

- Yellow Knife Mine (tailing pad liner – in progress) 26,700 m^2
- Pogo Mine (tailing pad liner) 26,000 m^2

There have been several geomembrane failures on the North Slope. The failures have included brittle failure of exposed liners under impact (such as vehicle traffic at low temperatures) or excessive stress (liners in a dam face under tension) (Polar Supply, 2001).

Table 2 presents a comparison of geomembrane properties. One of the most important factors for membrane applications in areas of large temperature fluctuation is weld strength. Weld strength is important because membranes have the potential to shrink after installation and generate large stresses at the welds. Heat application during field welding may change the molecular structure of the material at the welds and make it more crystalline (e.g., more brittle and more susceptible to failure). It is standard to compare 1.016 mm (40 mil) PP to the 1.524 mm (60 mil) LLDPE because there is less possibility of crystallizing the weld in the PP. The chances of crystallizing the weld on the thinner but less crystalline PP is comparable to the chances of crystallizing the weld on the thicker but more crystalline LLDPE (Kilius, 1993). In addition, 1.524 mm (60 mil) PP is rarely used (GSE, 2001 and Poly-Flex, 2001 – Serrot does not manufacture 1.524 mm PP). An explanation of each of the properties in Table 2 follows.

TABLE 2
IMPORTANT GEOMEMBRANE PROPERTIES

Property	ASTM Test Method	1.524 mm (60 mil) smooth LLDPE	1.016 mm (40 mil) PP
Brittle Temperature	D746-B	- 70 °C	-40 °C*
Thermal Expansion Coefficient	D696	1.0×10^{-4} cm/cm·°C	2.0×10^{-5} cm/cm·°C
Low Temperature Flexibility	D2136	-40 °C	-40 °C
Break Elongation	D638	800%	600%
Tear Resistance	D1004	15 kg	5 kg
Puncture Resistance	D4833	38 kg	16 kg
Workability	N/A	reasonable	preferred
Warranty	N/A	10 years	10 years

Note: There appears to be industry consensus that PP has not yet been tested at lower temperatures. Actual brittle temperature of PP may be lower than listed.
Source: GSE, 2001 and Poly-Flex, 2001.

- Brittle Temperature per ASTM Test Method D746-B. In this test, material samples withstand a specified impact at the listed temperature. Serrot International (geomembrane manufacturer) and local material representatives believe that the brittle temperature for PP is lower than –40°C. Apparently PP has not yet been tested at lower temperatures. Since the proposed application includes a geotextile cushioning layer, maintenance of soil cover, and institutional controls to prevent traffic, the liner should not receive impacts. Therefore, this test is of limited applicability.

- Thermal Coefficient of Expansion per ASTM Test Method D696. This is a measure of expansion/contraction of the liner due to changes in temperature. The data show that PP will expand/contract approximately five times less than LLDPE. A lower thermal coefficient of expansion is preferred because less movement means less stress in the welds. However, with the proposed 0.45meter soil cover constraining the liner, the expansion/contraction should be minimal.
- Low Temperature Flexibility per ASTM D2136. Both geomembranes will remain flexible at –40°C, the minimum expected temperature at the liner burial depth.
- Break Elongation per ASTM D638. This test shows percent increase in length before failure. Geomembrane elongation may be necessary if heaving/settling or excessive expansion/contraction occurs after installation. Results show that both materials are extremely flexible.
- Puncture and Tear Resistance per ASTM Test Methods D4833 and D1004. Due to its higher crystallinity, LLDPE is harder and more durable than PP. However, with the geotextile cushion and maintained soil cover, the liner should be protected from puncture or tear.
- Workability. There is no ASTM test for workability. Both LLDPE and PP have been successfully seamed (welded) at low temperatures. However, PP has an increased welding window and is preferred due to its increased flexibility which leads to easier workability.
- Warranty. Membrane manufacturers will offer a standard 10 year warranty (against manufacturer's defects when the liner is properly installed for use as designed) for both LLDPE and PP in the proposed application. Longer warranties may be obtained; however, any failure due to defect would be expected to happen within the first year of installation. Therefore, a 10 year warranty is considered adequate.

Both geomembranes were evaluated based on the previously discussed criteria. First, the relative significance of each of the factors was identified on a Scale of 1 to 5 with 5 being "most significant" and 1 being "of little (but still some) significance." Next, the two liners were rated for each factor on a scale of 1 (poor) to 10 (excellent). Finally, the relative significance of the rating was multiplied by the rating. The most technically appropriate material is the liner with the highest final score. Table 3 presents the rating evaluation.

The highest total score identifies the most technically appropriate material for the proposed application. Although the material scores were very close, PP scored the highest and is therefore the most technically appropriate material. Membrane manufacturers agree that either of these liners would be successful in the proposed application. Serrot International tested these two materials to evaluate their performance under stress at low temperatures. Welded seams, being the weakest part of the material, were tested under tension at temperatures of –70°C. Testing results indicated that PP would perform more consistently at low temperatures (see standard deviation for test results in Tables 4 and 5). Based on the evaluation, 1.016 mm (40 mil) unreinforced PP was recommended, and the cap was designed with this geomembrane material.

TABLE 3
RATING EVALUATION

Parameter	ASTM Test Method	Significance on Scale of 1 to 5	1.524 mm (60 mil) smooth LLDPE		1.016 mm (40 mil) unreinforced PP	
			Rating	Adj	Rating	Adj
Brittle Temperature	D746-B	3	10	30	9	27
Thermal Expansion Coefficient	D696	5	7	35	10	50
Break Elongation	D638	5	10	50	9	45
Tear Resistance	D1004	2	10	20	5	10
Puncture Resistance	D4833	2	10	20	5	10
Workability	N/A	5	7	35	10	50
Warranty	N/A	3	10	30	10	30
	Total			220		222

Adj =Performance is adjusted depending on significance of parameter
(This rating scheme is adapted from Peggs and Thiel, 2000)

TABLE 4
MATERIAL PROPERTIES TESTING RESULTS
1.016 MM (40 Mil) UNREINFORCED PP SEAM
PROJECT: SEAM TEST AT– 70°C

BONDED SEAM STRENGTH				SEAM PEEL ADHESION		
SHEER STRENGTH (kg/cm)		Elongation @ Break (%)	BREAK CODE	PEEL STRENGTH (kg/cm)	Elongation @ Break (%)	BREAK CODE
15.9		1134	SE1	11.1	1080	SE1
15.9		681	SE1	12.5	880	SE1
16.8		1538	SE1	14.9	1320	SE1
16.1		1737	SE1	13.6	920	SE1
16.3		1241	SE1	13.6	900	SE1
Avg.	16.3	1266		13.1	1020	
S.D.	2	405		8	185	

SE1 Break at outer edge of seam.

TABLE 5
MATERIAL PROPERTIES TESTING RESULTS
1.524 MM (60 Mil) LLDPE SEAM
PROJECT: SEAM TEST AT– 70°C

BONDED SEAM STRENGTH				SEAM PEEL ADHESION		
SHEER STRENGTH (kg/cm)		Elongation @ Break (%)	BREAK CODE	PEEL STRENGTH (kg/cm)	Elongation @ Break (%)	BREAK CODE
34.5		1655	SE1	19.1	40	SE1
35.1		2491	SE1	8.6	40	SE1
36.7		2541	SE1	37.8	80	SE1
34.7		2604	SE1	33.1	120	SE1
39.7		1893	SE1	5.2	40	SE1
Avg.	36.1	2237		20.8	64	
S.D.	12	433		81	36	

SE1 Break at outer edge of seam.

Conclusions

As part of planning the closure of the Barrow Landfill, different capping alternatives were evaluated, including a freezeback cap, silt cap, and a composite soil/geomembrane cap. Both the freezeback cap and silt cap were considered unfeasible for construction due to the lack of acceptable cover material available in Barrow. For evaluation of the composite soil/geomembrane cap, several liner materials were evaluated, including GCL, HDPE, LLDPE, and PP. Evaluation criteria included cost, brittle temperature, thermal coefficient of expansion, puncture and tear resistance, workability, and warranty. Test results indicate that while LLDPE and PP would both perform well in an arctic climate, PP would perform more consistently at lower temperatures. Therefore, a composite soil/ geomembrane cap, utilizing a PP liner, was chosen as the selected capping alternative for the Barrow Landfill.

References

Alaska Department of Environmental Conservation (ADEC) (2001), personal communication.

Bombardier L., D. Jette, M. Piche, and A. Rollin (1999), "Unprotected PP Liner for Storage of Paper Mill Black Liquor in Cold Region," Geosynthetics Conference paper presentation.

Comer A.I., M.L. Sculli, and Y.G. Hsuan (1995), "Effects of Freeze-Thaw Cycling on Geomembrane Sheets and Their Seams," proceedings of GRI, pp 853-866.

Department of Polymer Science, University of Southern Mississippi (DPS) (2001), http://www.psrc.usm.edu/macrog/crystal.htm. (December, 2001).

Environmental Protection, Inc. (EPI) (2001), www.geomembrane.com (December 2001).

Gundle/SLT Environmental, Inc., (GSE) (2001), www.gseworld.com (December 2001) and personal communications (December 2001 and January 2002).

Kilius, D.E. (1993), "Cold Temperature Performance of Polypropylene Geomembrane and Seams," proceedings of GRI, pp 191-201.

Kraus, Jason, Craig Benson, Allan Erickson, and Edwin Chamberlain (1997), "Freeze-Thaw Cycling and Hydraulic Conductivity of Bentonitic Barriers", Journal of Geotechnical and Geoenvironmental Engineering, pg 229 (March 1997).

Peggs, Ian and Rick Thiel (2000), "Selecting a Geomembrane Material", www.geosynthetica.net (December 2001).

Polar Supply (2001), personal communication.

Poly-Flex, Inc. (2001), www.poly-flex.com (December, 2001), and personal communications (December 2001 and January 2002).

Rollin, A. (2001), personal communication with Brett Flint, URS 2/26/01.

Shah, A., and R.K. Frobel (1995), "Polypropylene Geomembranes – the Alternative Containment Solution," proceeding of Geosynthetics, pp1081-1092.

Serrot International, Inc. (Serrot) (2001), www.serrot.com.

U.S. Environmental Protection Agency (EPA) (1993), "Presumptive Remedy for CERCLA Municipal Landfill Sites".

U.S. Environmental Protection Agency (EPA) (1997), "Geosynthetic Clay Liners Used in Municipal Solid Waste Landfills".

Rethinking Landfill Development and Operation in Permafrost Regions

Gregory L. Magee, P.E.,[1] and William J. Rice, P.E.[2]

Abstract

Freezeback landfills, a special regulatory category in Alaska for landfills located on permafrost, are discussed. Since most lands in the Arctic are generally wet, poorly drained, and underlain by ice-saturated permafrost, aboveground landfills are preferred over other types. Design goals for an aboveground freezeback landfill are encapsulation and freezeback. Thermal modeling guides the design for new landfills underlain by permafrost. Fluid management, fill sequence, and corrective action plans are critical to successfully operate a landfill located on permafrost. Because most landfills in the Arctic are small and often expensive to build and operate, achieving complete freezeback by placing a thicker final cover may not be practical or affordable. Instead, designing for partial freezeback may be more appropriate.

Introduction

Landfilling municipal solid waste in Alaska's permafrost regions presents unique challenges. For many years, landfills in the Arctic were built and operated much like conventional landfills in the continental United States. However, little consideration was given to the effects of the arctic environment on landfills, particularly when permafrost was disturbed.

Today, many of these landfills are considered dumps, and have become environmental liabilities and hazards to public health and safety. During the short summer months of the far north, these dumps become quagmires of garbage and water caused by melting permafrost and ground settlement. Because of these problems, there is greater awareness among owners, engineers, and regulators about constructing and operating new landfills in permafrost regions. Future landfills located on permafrost need to be designed, constructed, and operated to prevent permafrost from thawing, and to use the frozen nature of permafrost and the arctic environment to contain waste far into the future.

---

[1] Principal Engineer, MWH, 4100 Spenard Road, Anchorage, AK, 99517, USA. Member ASCE. Phone (907) 248-8883; gregory.l.magee@mwhglobal.com.

[2] Project Engineer, MWH, 4100, Spenard Road, Anchorage, AK, 99517, USA. Member AWRA. Phone (907) 248-8883; william.j.rice@mwhglobal.com.

This paper reassesses development and operation of landfills located in permafrost regions. Freezeback landfills, a special regulatory category for landfills in Alaska, are discussed. Design goals for aboveground freezeback landfills are proposed, the use of thermal modeling as a design tool is discussed, and the importance of operational plans to maintain thermal stability is emphasized. The paper ends with a discussion of closure and post-closure requirements for freezeback landfills.

Permafrost and the Fragile Active Layer

Permafrost is permanently or perennially frozen ground. Defined by temperature only, it is a combination of soil, rock, water, and other buried materials that have been frozen naturally at 0 degrees Celsius (32 degrees Fahrenheit) or less for 2 consecutive years or more (ADEC, 1999). Permafrost may include various amounts of frozen soil and ice, ranging from relatively dry frozen gravel to virtually pure ice or brackish water that remains liquid at temperatures less than freezing (Selby, 1990).

Permafrost is continuous in Northern Alaska and discontinuous in most of the central and southern parts of Alaska (Figure 1). For permafrost to form and persist, average annual air and ground surface temperatures must be below 0 degrees Celsius (32 degrees Fahrenheit). It can extend to depths of 2,000 feet below ground surface in the continuous permafrost zone across Alaska's Arctic Coastal Plain (USGS, 1999). The average annual temperature of the ground surface, thermal properties of the subsurface materials, and geothermal gradient of the earth primarily determine the thickness of permafrost.

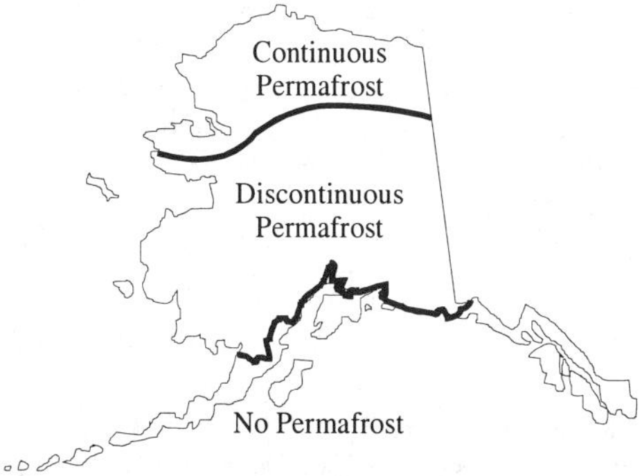

Figure 1 – Permafrost Map of Alaska
(Source: Modified from USGS, 1965)

Other factors that affect the occurrence and thickness of permafrost include the vegetation layer, snow cover, slope and aspect of the surface, surface water, and groundwater. Removing vegetation and its insulating effect usually causes the surface temperature to rise and underlying permafrost to thaw during summer months. The insulating effect of deep snow tends to prevent the formation of permafrost.

Southward-facing slopes receive more solar radiation and are less likely to be underlain by permafrost than are northward-facing slopes.

Warming effects of streams, rivers, lakes, and oceans can cause permafrost beneath these water bodies to be thin or absent (USGS, 1999). Permafrost is impervious to infiltration of water and affects the movement and discharge of surface water and groundwater. Water on top of the permafrost table can perch in the active layer (suprapermafrost aquifer) and can surface above ground. Discharge of groundwater beneath the permafrost (subpermafrost aquifer) is only possible through unfrozen zones. Sand and gravel deposits may contain flowing groundwater that conducts sufficient heat to melt the permafrost or keep permafrost from forming.

Soil type and water content affects the presence of permafrost, as different types of soils and rocks conduct heat at different rates. In discontinuous permafrost areas, silt in alluvial and glacial deposits is more likely to contain permafrost than sand and gravel embedded in silt.

The top of the permafrost is called the permafrost table. Above the permafrost table is the active layer, layers of vegetation, and soils that freeze in winter and thaw during summer. Thickness of the active layer depends on soil type and presence of vegetation on top. Depth of thaw varies greatly with latitude, climate, and soil type.

The active layer is very fragile. It is highly susceptible to damage if disturbed during summer. When the active layer is removed or disturbed, permafrost is no longer insulated from the summer heat. Within the first 2 years of disturbance, for example, settlement from thawing permafrost can be 10 percent to 25 percent of the original active layer depth, depending on actual water content and soil type (NSB, 2000a; USAF, 2000).

Disturbing the active layer also disrupts the thermal balance, which may take years to restabilize. During that time, ponds may develop from melting ice wedges and ice-rich permafrost. This can alter terrain and change vegetation cover over decades of time. Results of this process are similar to thermokarsting, which is a natural phenomenon in permafrost regions. Thermokarsting is caused by melting and freezing cycles that gradually cause ground to subside, especially in ice-saturated soils. This natural process is extremely slow but can result in widening depressions that fill with water (Selby, 1990).

Current and Proposed Regulations

The State of Alaska recognizes three classes of municipal solid waste (MSW) landfills: Class I, Class II, and Class III. These classes are grouped by the mass of MSW received on an annual daily average and the remoteness of a site with respect to other disposal sites. Landfills that accept more than 22.1 metric tons (20 tons) daily are classified as Class I. A Class II landfill accepts between 5.5 metric tons (5 tons) and 22.1 metric tons (20 tons) daily and a Class III facility accepts less than 5.5 metric tons (5 tons) daily. The Oxbow Landfill near Prudhoe Bay is the only Class I landfill on the North Slope of Alaska. Landfills in Barrow and Kotzebue are Class II and other landfills above the Arctic Circle are Class III.

According to current Alaska Department of Environmental Conservation (ADEC) solid waste regulations, 18 Alaska Administrative Code (AAC) 60, a new

Class I or II landfill located on continuous or discontinuous permafrost can be constructed without a liner and a leachate collection system, and operated without groundwater and methane gas monitoring. These special regulatory exemptions are granted under 18 AAC 60.325, *Permafrost Landfills*, if the owner or operator can demonstrate that the landfill would be developed and operated to prevent underlying permafrost from thawing, to allow freezing of unfrozen waste, and to keep waste frozen after closure.

New regulations have been recently drafted by ADEC to clarify landfill development on permafrost, and will revise 18 AAC 60.325, *Permafrost Landfills*. The draft regulations require that all landfills located on permafrost are designed and operated to minimize thawing and accumulation of free liquid and to control subsidence. For a Class I or II landfill located on permafrost to be constructed without a liner and monitoring for groundwater and methane, the current requirements of 18 AAC 60.325 will still apply, but will be found under a different name, "*Freezeback Landfills*." This title change makes clear that a Class I and II landfill located on permafrost can be constructed and operated either with or without a liner and monitoring requirements, but to be exempt from liner and monitoring requirements, freezeback needs to be demonstrated (Sonafrank, 2002). For this paper, "freezeback landfill" is used instead of permafrost landfill to make this distinction.

To demonstrate freezeback, a thermal analysis is required, which determines the thickness of insulating soil cover necessary to keep waste frozen. Also, monitoring landfill temperature is required to detect thawing of waste and underlying permafrost during the active life of the landfill and post-closure periods (ADEC, 1999).

Although a liner and a leachate collection system and groundwater and methane gas monitoring are required for Class I or II landfills in Alaska (unless permitted as a freezeback landfill), these requirements do not apply to Class III landfills unless ADEC determines that they are necessary to protect public health or the environment. If these requirements do apply, and the Class III landfill is located on continuous and discontinuous permafrost, special exemptions under 18 AAC 60.325 may be granted and the landfill can still be designed and operated to meet conditions for freezeback landfills (ADEC, 1999).

Closure standards for Class I or Class II landfills are also different from Class III in terms of the composition of the minimum 0.6-meter (2-foot) final cover. Class I or Class II landfills require an infiltration layer of at least 0.45 meters (18 inches), with a permeability less than or equal to the permeability of the bottom liner or natural soil, and no greater than 1×10^{-5} centimeters per second (3.9×10^{-6} inches per second). Permeability of permafrost meets and exceeds this requirement as the bottom liner. Class III landfills have no permeability requirements for the 0.6-meter (2-foot) final cover. Also, if a landfill is permitted as a freezeback landfill, the final cover needs to be thick enough that the active layer does not reach buried waste. Finally, post-closure care requirements are similar, with the exception of post-closure durations; 30 years are required for Class I and Class II landfills and 5 years for Class III landfills (ADEC, 1999).

Design Goals

Design goals of an aboveground freezeback landfill are encapsulation and freezeback (NSB, 2001b). Encapsulation contains waste from:

- beneath the landfill, by minimizing disturbance to the underlying active layer and permafrost.
- the sides of the landfill, with soil berms wide enough to freeze and stay frozen at the core year-round.
- above the landfill, with a final cover of sufficient thickness and vegetation to insulate and prevent thawing of buried waste.

This configuration can produce conditions necessary to induce freezeback. Freezeback is a process where *in-situ* permafrost and freezing ambient air temperatures are used to permanently freeze waste. Over time, buried waste eventually becomes permafrost. Freezeback effectively retards decomposition of wastes, thus reducing formation of leachate and methane gas, and eliminating the possibility of contaminating the surrounding environment. Figure 2 illustrates conceptual encapsulation and freezeback for an aboveground freezeback landfill.

Although freezeback of MSW has not been documented, freezeback of one-time disposal of drilling waste in permafrost has been achieved. Although it seems possible to freeze back thawed waste and keep it frozen, collecting and analyzing more thermal monitoring data and performing more research are necessary to validate the freezeback process.

Landfill Type

There are three types of landfills: aboveground, excavation, and depression. Aboveground landfills are generally used where terrain is unsuitable for excavation or where high groundwater conditions exist and adequate supplies of construction and cover materials are locally available. An excavated landfill is preferable in areas where construction and cover materials are in short supply and the water table is not near the surface. Similar to excavated landfills, land depressions, such as canyons, ravines, dry borrow pits, and quarries, are used for landfills.

In the Arctic, aboveground landfills are proposed over excavation or depression landfills. Most lands in the Arctic are generally wet, poorly drained, and underlain by ice-saturated permafrost. An aboveground landfill can be developed and operated to prevent disturbing the permafrost and provide sufficient insulation to minimize thawing of permafrost underneath the landfill. In contrast, an excavated or depression landfill may require extensive water drainage controls to prevent runoff from mixing with waste, which is strictly prohibited by ADEC. ADEC currently supports an aboveground design for landfills located on permafrost that keeps the insulating mat in place by constructing a gravel pad and soil berms over existing ground (ADEC, 2001).

An aboveground landfill requires constructing a working pad or surface and berms. The tops of berms should accommodate equipment access and fencing.

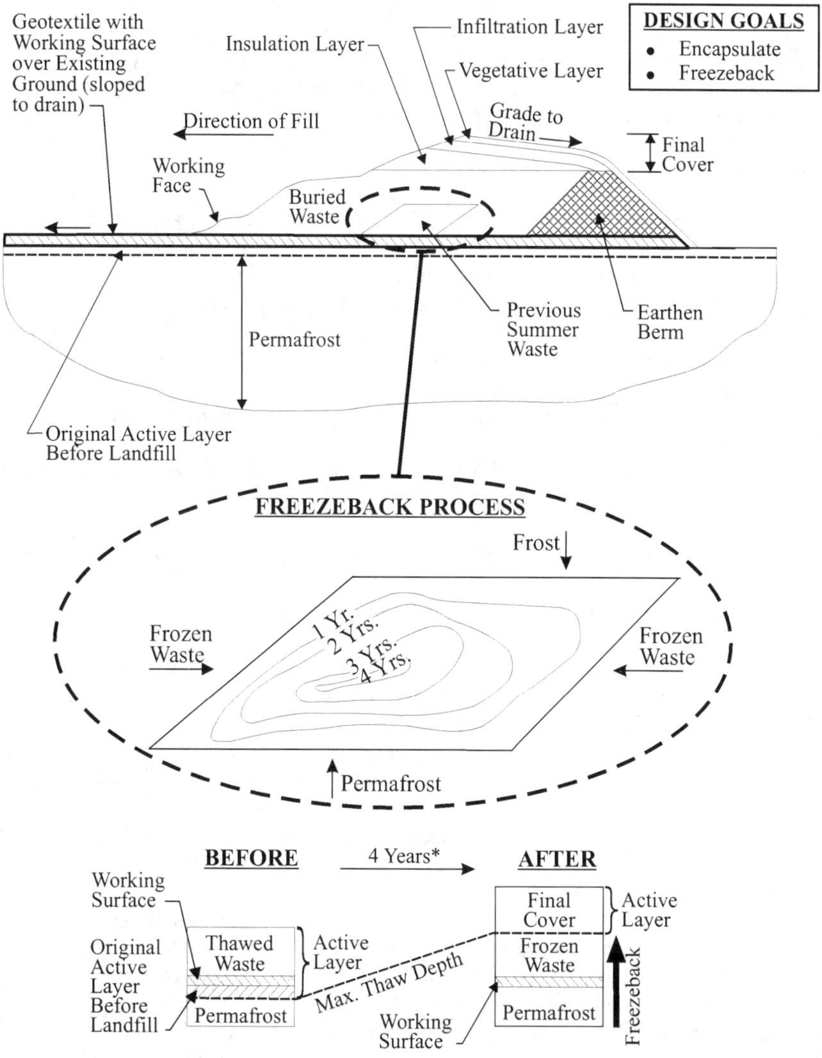

*In this example, it takes 4 years to "freeze back" thawed waste completely. Freezeback times vary widely throughout a landfill. Depending on site conditions, it may take several years to completely refreeze a landfill cell after it is closed.

Figure 2. Freezeback Concept of an Aboveground Landfill Located on Permafrost

Thermistors are installed in the berms, working surface, and in each closed cell to monitor temperatures for freezeback and thawing. Thermistors should be placed through the active layer and into the permafrost to record freezeback. Figure 3 gives conceptual plan and section views of an aboveground freezeback landfill.

Thermal Modeling

A thermal analysis should be conducted for every new landfill in permafrost regions to safeguard against potential problems or failure. The thermal model guides the landfill design. Thermal analysis can determine a working surface thickness that minimizes thawing of underlying permafrost or keeps summer thaw depths of open working surface areas within the original active layer. It can also establish soil thickness for the final cover necessary to maintain buried waste in a permanently frozen state. For an excavated landfill, the analysis would show the extent of permafrost thaw that can be expected from surrounding embankments and final cover thickness.

For freezeback landfill status under ADEC regulations, thermal modeling is required to demonstrate the following:

- new landfill development will not result in long-term thawing of the permafrost.
- buried waste will freeze back after closure.

Many thermal models have been developed, ranging from one-dimensional analysis using the modified Bergen equation to two-dimensional analysis using finite difference and finite element analysis (AKDOT, 1990; GSI, 2001). Three-dimensional models are also used. The type of model used depends on the scope, budget, and time for landfill development. Currently, ADEC has no guidelines for thermal modeling.

The following three steps are used to create a model to determine the time it takes for waste to freeze back:

1. An analysis of placing a working surface on existing ground is performed to determine maximum seasonal thaw depth.
2. A second analysis determines the thickness of the closure cover necessary to prevent waste from thawing seasonally.
3. A third analysis determines freezeback of a closed cell, which combines results from the working surface and final cover analyses with thermal properties of the waste.

Typically, landfill cells are filled and closed in a systematic manner. Thermal modeling considers the worst-case scenario within a landfill cell. The worst-case scenario would occur during summer when the active layer is deepest and deposited waste has highest internal latent heat. After summer waste is encapsulated, it begins the process of freezeback. Freezeback of summer waste takes more time

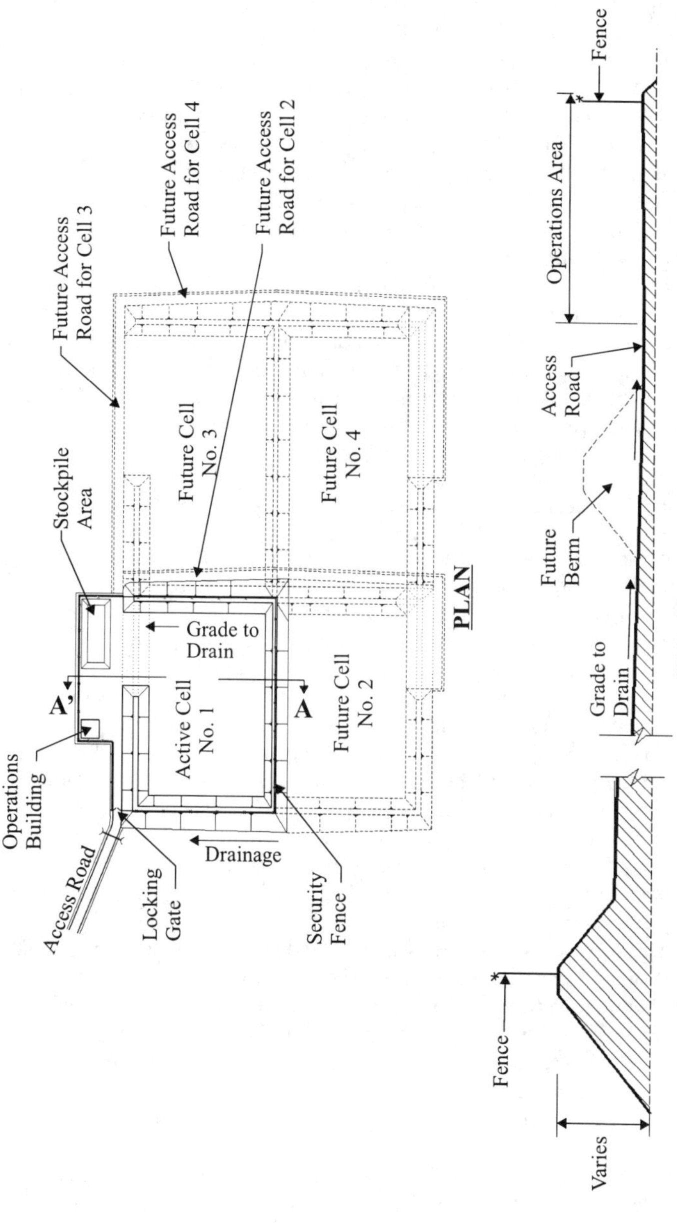

Figure 3. Conceptual Aboveground Freezeback Landfill (NSB, 2001)

than freezeback of winter waste, which is generally frozen before being covered. The number of worst-case cells would vary in a landfill. With a fill sequence plan, it is possible to surround summer-placed cells with winter-placed cells.

During winter, if the surface is not insulated with snow, freezeback occurs at a faster rate than with snow. During summer, a dark surface collects more heat, causing more thaw. Moisture content of soil factors heavily in the freezeback rate because of the amount of energy needed to freeze water. Moisture content of operational cover soils determines, to a large degree, freezeback time of the buried waste. For instance, a silty soil cover would decrease freezeback time due to relatively higher moisture content. Conversely, a silty soil cover prevents summer heat from penetrating as deep into a landfill as a more gravelly soil cover.

Although difficult to quantify, thermal properties of composite waste are an important factor in the model. One method is to take the geometric mean of the thermal properties of all materials in the waste. However, this method does not take in account how voids influence latent heat transfer; more research is needed to improve this method. Another factor is the heat-generating capacity of MSW; again, more research is needed to determine the degree that MSW decomposes in arctic climates, particularly in a freezeback landfill design.

Operational Considerations

Operational plans are an integral part of a successful landfill in permafrost regions. Normal landfill operations should be modified to help the arctic climate freeze waste and keep it frozen. Every landfill in permafrost regions should have fluid management, fill sequence, and corrective action plans to help promote freezeback.

A fluid management plan is critical to thermal stability. The plan should state standard operating procedures that minimize potential for water and waste to come into contact. For example, the plan should preclude water ponding at the working surface and require that the landfill be free-draining, away from the working face and berms. Also, the plan should indicate that snow be removed from the entire landfill area before spring breakup. During winter operations, snow should be removed near the working face prior to waste placement. Bulk liquids and oily wastes are not compatible with freezeback and should not be placed in the landfill.

Another important operational control is a fill sequence plan. The direction of fill in each cell should be from the higher elevation to the lower elevation of the landfill surface to promote drainage away from the working surface. Prior to spring breakup, an intermediate or final cover should be placed over waste that has accumulated during the winter. This cover helps insulate winter waste from summer thaw, maximizing the amount of waste that is frozen, and minimizing infiltration during summer months. Cells should be sequenced so that summer-placed waste is surrounded by winter-placed waste to maximize freezeback.

Lastly, to correct problems resulting from thawing and settlement over time, a corrective action plan is necessary. Corrective measures are required if visual and temperature monitoring detect evidence of thermal instability and subsidence. In general, corrective measures may include placing additional waste or fill for insulation, and using foam board insulation or thermosyphons for localized areas.

Closure and Post-Closure Care

Landfills in permafrost regions can be designed for either partial or complete freezeback, depending on the thickness of the final cover. Given that most landfills in the Arctic are small and often expensive to build and operate, achieving complete freezeback by placing thicker final cover may not be practical or affordable. Instead, partial freezeback may be more appropriate.

For a freezeback landfill, final cover should be thick enough that the active layer does not reach and thaw buried waste during summer. Depending on the area and slope of cover, required thickness can vary from 1.2 to 1.8 meters (4 to 6 feet). The standard 0.6-meter (2-foot) final cover would only provide for partial freezeback, with the topmost layer of waste thawing each summer. Figure 2 illustrates final closure of an aboveground freezeback landfill.

For freezeback landfills, the thermal model determines thickness of the final cover. Types of soil and moisture content are critical to minimizing thickness. To obtain this objective, a thermal balance is necessary between materials high in moisture content and low in silt content to achieve freezeback and keep buried waste frozen. Vegetation is also a factor.

Depending on the class of landfill, freezeback landfills in Alaska are monitored for 5 to 30 years after final closure. During this post-closure care period, a landfill is monitored for temperatures as it refreezes and stabilizes. Any erosion, water ponding, and settlement of final cover should be corrected for continued thermal stability.

Conclusions

In the past, little consideration was given to effects of the arctic environment on landfills, particularly when permafrost was disturbed. Rethinking development and operation of landfills located in permafrost regions is needed to avoid these problems. Because of the high potential for water and settlement problems, it is necessary to design, construct, and operate landfills located on permafrost that minimize disturbances to the active layer and underlying permafrost.

Since most lands in the Arctic are generally wet, poorly drained, and underlain by ice-saturated permafrost, aboveground landfills are preferred over other types. An aboveground landfill consists of a working pad and berms above the existing ground.

Design goals of an aboveground freezeback landfill are encapsulation and freezeback. These conditions induce freezeback where *in-situ* permafrost and freezing ambient air temperatures are used to permanently freeze waste. Over time, the buried waste eventually becomes permafrost. A thermal model is used to meet these goals and should guide design for every new landfill in permafrost regions, whether or not it is designed as a freezeback landfill. More research is needed to determine the thermal properties of MSW.

Operational plans are an integral part of a successful landfill in permafrost regions. These plans are necessary to promote freezeback and to ensure thermal stability over time. Fluid management, fill sequence, and corrective action plans are all critical to successfully operating a landfill in arctic regions.

Landfills located on permafrost can be designed for either partial or complete freezeback, depending on thickness of final cover. Freezeback landfills require complete freezeback. Given that most landfills in the Arctic are small and generally expensive to build and operate, achieving complete freezeback by placing a thicker final cover greater may not be practical or affordable. Instead, partial freezeback may be more appropriate.

During the post-closure care period, the landfill is monitored for temperature as it refreezes and stabilizes. Any erosion, water ponding, and settlement of the final cover should be corrected for continued thermal stability.

Finally, freezeback of MSW has not been documented. Although it seems possible to freeze back thawed waste and keep it frozen, collecting and analyzing more thermal monitoring data and performing additional research is necessary to validate the freezeback process.

References

Alaska Department of Environmental Conservation (ADEC). 1999. Solid Waste Management Regulations, 18 AAC 60, as amended through July 11, 1999.

Alaska Department of Environmental Conservation (ADEC). 2001. Landfills in Permafrost. Solid Waste Management Program. February. 2 pp.

Alaska Department of Transportation and Public Facilities (AKDOT). 1990. MUT1D – Multilayer User-Friendly Thermal Model in 1 Dimension. December. 56 pp.

Geo-Slope International (GSI). 2001. TEMP/W finite element geothermal analysis model, version 5. Calgary, Alberta Canada. http://www.geo-slope.com/ 403-269-2002. October.

North Slope Borough (NSB). 2000a. Class III Municipal Solid Waste Application for the Anaktuvuk Pass Landfill, Permit No. 9931-BA014. Prepared by MWH, 4 pp. plus attachments.

North Slope Borough (NSB). 2000b. New Barrow Landfill Site Selection. Department of Capital Improvement Program Management. Prepared by MWH, 45 pp. plus attachments.

North Slope Borough (NSB). 2001. Project Analysis Report, New Barrow Landfill at Site E. CIP No. 13-206, Contract No. 2001-181. Department of Capital Improvement Program Management, Prepared by MWH, 43 pp. plus attachments.

Selby, M.J. 1990. Earth's Changing Surface. Oxford University Press, New York. 607 pp.

Sonafrank, Nancy. 2002. Alaska Department of Environmental Conservation. Personal communication with Gregory L. Magee. January18.

U.S. Geological Survey (USGS). 1999. Ground Water Atlas of the United States, Alaska, Hawaii, Puerto Rico and the U.S. Virgin Islands, HA 730-N. http://capp.water.usgs.gov/gwa/chn (November 25, 2001).

U.S. Geological Survey (USGS). 1965. Permafrost Map of Alaska. Miscellaneous Geological Investigation, Map I-445. Scale 1:2,500,000 feet.

United States Air Force (USAF). 2000. Solid Waste Application – Cape Lisburne LRRS, Alaska, Coverage Under Renewal Permit No. 9631-BA009. Performed by MWH. 6 pp. plus attachments.

Design Considerations for GCLs Used in Alaska Landfills

Kelly Merrill, P.E.,[1] Henry Friedman,[2] and Greg Goodale[3]

Abstract

This paper describes design considerations associated with the use of a geosynthetic clay liner (GCL) in both the base lining system of a planned new cell and in the final cover of an existing unlined cell at the Central Landfill in Palmer, Alaska. The Cell 2B base lining system will include a composite geosynthetic liner consisting of an 80 mil (2 mm) textured HDPE liner overlying a GCL on both the sideslopes and base of the cell. A similar base lining system has been used successfully at the nearby Anchorage Regional Landfill. The use of this lining system in a Seismic Zone 4 area on relatively steep side slopes requires careful evaluation of both internal shear strength of the GCL and interface friction between the GCL and textured HDPE. Summarized results of shear strength testing performed on several types of GCL at both high and low normal stresses are presented here. Design considerations for use of a GCL in the final cover over unlined Cell 2A at the Central Landfill are also discussed. Requirements for GCL conformance testing and construction quality assurance are briefly described. Examples of GCL use in other Alaska landfills are also referenced.

Introduction

The Central Landfill, located in the Matanuska-Susitna Borough in Southcentral Alaska, is the third largest landfill in the state. It is located off the Palmer-Wasilla highway within a 250-hectare designated landfill reserve. The first unlined landfill cell (Cell 1) was closed in 1988. The currently active cell is unlined Cell 2A. The waste in Cell 2A will reach final design elevation within the next 5 years and will be capped with a final cover when the cell is full. The next phase of the Central Landfill development plan will be Cell 2B, a lined cell immediately to the south of Cell 2A. Cell 2B has been designed and is planned for construction in 2003. The location of various landfill facilities can be seen in Figure 1.

The designs for both the Cell 2A cover and the Cell 2B base lining system include a geosynthetic clay liner (GCL). The GCL planned for use on these projects consists of a layer of bentonite clay sandwiched between two nonwoven geotextiles.

---

[1] Geotechnical Engineer, CH2M HILL, 301 West Northern Lights Blvd., Suite 601, Anchorage, Alaska 99503

[2] Solid Waste Specialist, CH2M HILL, 777 108th Avenue NE, Bellevue, Washington 98004

[3] Solid Waste Project Manager, Matanuska-Susitna Borough, 350 E. Dahlia Avenue, Palmer, Alaska 99645

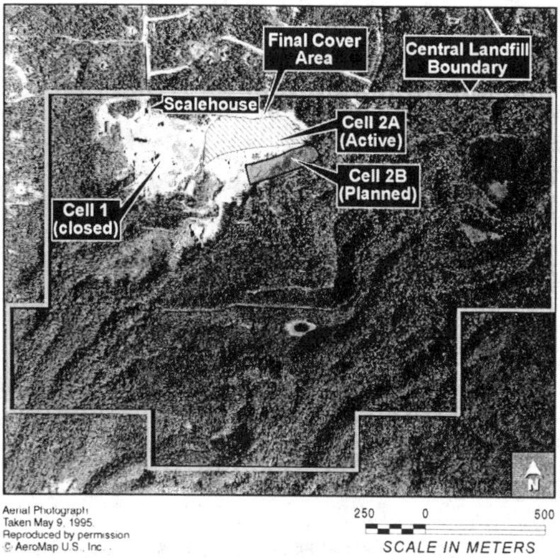

Figure 1. Central Landfill

The GCL is reinforced by needlepunching through the geotextiles to provide enhanced shear strength. Other types of reinforced GCLs are also available, with reinforcing provided by stitch-bonding or adhesives. The double nonwoven needle-punched GCL was selected for the Central Landfill project to provide maximum internal and interface shear strength for stability reasons, primarily because of the high seismicity of Southcentral Alaska.

The Central Landfill is in the Transitional climatic zone of Alaska, with mean annual precipitation of approximately 460 mm and a mean annual temperature of 0° C. Near-surface geologic units underlying the Central Landfill site include both glacial drift and glacial outwash deposits. Soils within these units consist primarily of relatively clean sand and gravel with frequent cobbles and boulders.

Other Uses of GCLs in Alaska

GCLs are particularly well-suited to use in Alaska and other cold regions because of their relative insensitivity to freeze-thaw cycles, wet-dry cycles, and ease of construction during the short, wet summers typical of many cold regions. GCLs have been successfully used in many other applications in Alaska; a few examples follow:

- **Anchorage Regional Landfill (ARL) Cells:** Composite geomembrane/GCL base lining systems have been used at the ARL, the largest landfill in Alaska, in the last four cell expansions dating back to 1990. GCLs were initially used only on the bottom of each cell because of concerns about stability associated with use of GCLs on the side slopes. However, improvements in GCL reinforcement and resulting increased shear strength has allowed the successful use of GCLs in both the bottom and side-slope lining systems for the last two cells at ARL.

- **ARL Leachate Lagoon:** Two leachate lagoons were constructed at ARL to store and pre-treat leachate before delivery to the municipal wastewater treatment plant. The lagoons were designed with a double geomembrane lining system with an underlying leak detection system. One of the lagoons leaked excessively when tested with water during construction. The contractor solved the problem by adding a composite geomembrane/GCL liner above the existing lining system. Since the composite liner was added, there has been no perceptible leakage through the lining system.
- **Fairbanks North Star Borough (FNSB) Landfill Cell:** The first lined cell at the FNSB landfill, Cell 1, was constructed in 1998 and incorporated a composite geomembrane/ GCL liner on both the bottom and side slopes of the cell.
- **St. George Island Landfill:** A GCL alone (with no overlying geomembrane) was used as the resistive barrier in a final cover over an unlined landfill on St. George Island.
- **Kodiak Island Landfill:** A GCL alone was used in a final cover for an unlined landfill on Kodiak Island.
- **Adak Island:** A GCL alone was used in a cover over a contaminated soil stockpile.

GCL Use in Base Lining Systems

The design liner system for the bottom and side-slopes of Cell 2B at the Central Landfill is shown graphically in Figure 2. Key issues that were considered during design of the base lining system are described in the following paragraphs.

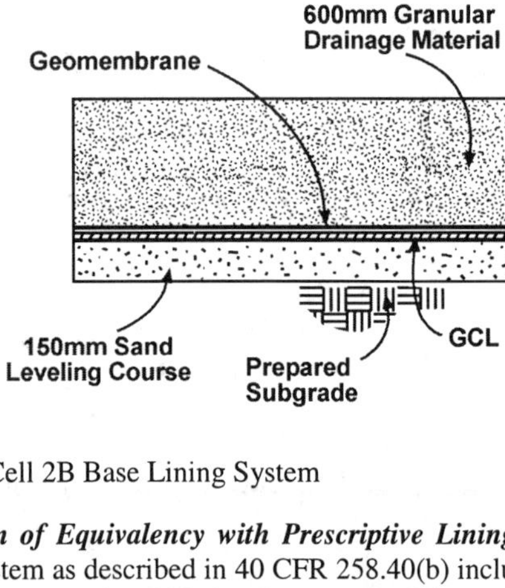

Figure 2. Cell 2B Base Lining System

Demonstration of Equivalency with Prescriptive Lining System. The prescriptive base lining system as described in 40 CFR 258.40(b) includes a compacted clay liner (CCL). Construction of a CCL is often not feasible in cold regions because of the lack of local suitable clay materials, the difficulty of placing a CCL during the short, wet construction season common to cold regions, and the difficulty inherent in quality control/quality assurance for construction of the liner system. A GCL-based system has been demonstrated to be an equivalent or better lining system in several cases in Alaska and the Lower 48 (Koerner and Daniel, 1994). The Alaska Department of Environmental Conservation (ADEC) requires proof of equivalency before approval of an alternative lining system is granted.

Composite Action with Geomembrane. The GCL is designed to interact with the geomembrane to form a resistive barrier with a very low effective permeability. The GCL alone has a permeability of approximately $1x10^{-8}$ to $1x10^{-10}$ cm/sec depending on the level of confining stress. If a leak occurs in the geomembrane, the bentonite in the GCL will locally swell at that leak location and effectively plug the leak. The GCL also serves as a cushioning layer to protect the geomembrane from damage resulting from subgrade protrusions.

Resistance to Freeze-Thaw Cycling. Base lining systems in cold regions may undergo several freeze-thaw cycles before waste is placed to a depth such that no further freezing would occur, particularly on side slopes where waste might not be placed for a number of years. Tests have shown that GCLs are very resistant to freeze-thaw cycles (Shan and Daniel, 1991; Hewitt and Daniel, 1997; Kraus, et al, 1997). Test results by numerous researchers indicate that GCLs do not experience an increase in hydraulic conductivity after undergoing numerous freeze-thaw cycles because of the self-healing characteristics of bentonite. Laboratory specimens of GCL have undergone up to 20 freeze-thaw cycles with no increase in hydraulic conductivity observed; field GCL specimens have been exposed to one or two winters of freeze-thaw cycles with no adverse impact on hydraulic conductivity. (Kraus, et al, 1997). Compacted clay liners (CCLs), in contrast, have shown significant increases in hydraulic conductivity due to the presence of small cracks throughout the soil structure in both laboratory and field tests (Daniel, 1993; Benson, et al, 1995).

Chemical Compatibility with Adjacent Materials. The hydraulic conductivity of GCLs has been shown to significantly increase when exposed to concentrations of certain divalent cations, such as calcium chloride (Vasko, et al, 2001). Chemical compatibility of GCLs and adjacent materials must be considered for each application. Ion exchange with adjacent soil or liquid can increase hydraulic conductivity of the GCL, particularly in calcium-rich environments. Studies have shown minimal impacts on GCL hydraulic conductivity when the GCL is exposed to MSW landfill leachate (Ruhl and Daniel, 1997).

Puncture Resistance. Puncture resistance is critical for a GCL lining system because of the thin geosynthetic layers. Accordingly, a minimum 300-mm thick overlying granular drainage layer over the geomembrane/GCL liner is recommended for protection during construction and waste placement. A 600-mm drainage layer is used at the Central Landfill and ARL specifically to increase the puncture protection for the liner.

Shear Strength. Both internal and interface shear strength must be carefully considered during design of a GCL lining system. Reinforced needle-punched GCLs with nonwoven geotextiles on both sides have been developed to increase internal shear strength of the GCL and interface strength between the GCL and the adjacent geomembrane. GCLs have often been installed with a woven geotextile against the geomembrane to promote intimate contact between the two geosynthetics and thus enhance leakage resistance. However, the friction angle along the woven geotextile/ geomembrane interface is lower than the friction angle along the nonwoven geotextile/geomembrane interface. One reason for this difference is the potential for bentonite extrusion through the relatively porous woven geotextile. Use of a nonwoven geotextile reduces the bentonite extrusion, thus increasing the interface friction without significantly compromising the leakage resistance of the system (Wilson-Fahmy and Koerner, 1995).

The EPA recently sponsored a full-scale field testing program near Cincinnati that included geomembrane/GCL composite lining systems (Daniel, et al, 1998). The program involved construction and monitoring of fourteen full-scale test plots using five different commercially available GCLs. The program was focused on assessing the internal shear strength of the GCLs under low normal stresses (900 mm of cover soil was placed above the geomembrane for all test plots). Under static loading conditions (dead load of the cover soil only, no seepage or equipment loads) the test results indicated that a composite geomembrane/reinforced GCL liner has a static factor of safety of at least 1.0 on a 2:1 (H:V) slope and at least 1.5 on a 3:1 slope, both with respect to the GCL internal shear strength.

These results assume that the internal strength of the reinforced GCL is the lowest strength in the lining system. If a weaker interface strength or internal strength from another material exists in the lining system, lower factors of safety would result. In fact, the EPA test plot program experienced two unexpected slide failures resulting from weak interface strengths on the 2:1 test plots. In both cases, the failure surface was the interface between the woven geotextile side of the GCL and the overlying textured geomembrane. The lower than expected interface friction angles were attributed to bentonite extruding through the porous woven geotextile. This unexpected occurrence emphasizes the need for material-specific testing on all projects. A GCL with nonwoven geotextiles on both sides did not experience a failure on a 2:1 test plot slope.

A similar composite geomembrane/double-nonwoven GCL lining system was successfully installed on the 3:1 side slopes and bottom of Cell 6 at the ARL during the summer of 1999. No problems have been encountered with that lining system during or after construction.

Material and project-specific shear strength testing is strongly recommended for design of any GCL lining system. Testing should include internal and interface strength tests of all lining system components to identify the weakest zone for design purposes. Direct shear tests should be run with test conditions modeling field conditions as closely as possible. For example, sample hydration, consolidation, and shearing phases should be conducted at rates and under normal pressures representative of field conditions. A summary of both interface and internal shear strength results from direct shear tests conducted on double-nonwoven GCLs for

ARL Cell 6 is included in Table 1. Results are available for both peak and large deformation strength values, under both high and low normal stresses. These values can be used for preliminary design, but project-specific testing is recommended for final design.

For ARL Cell 6, shear strength test results indicated that the Bentomat DN GCL met the specified strengths at both low and high normal stresses, so it was approved for use on both the side-slopes and bottom of the cell. However, the Bentofix NW GCL met the specified shear strength at low normal stresses, but did not meet the specified strength at high normal stresses. Accordingly, Bentofix NW was approved for use on the cell side-slopes but not on the cell bottom where high normal stresses under the waste mass are anticipated. The Cell 6 contractor elected to use Bentomat DN on the cell bottom and Bentofix NW on the cell side-slopes.

Slope Stability. Both static and seismic stability of the base lining system must be carefully evaluated using project-specific material properties as discussed in the preceding section.

Granular Drainage Material on Side Slopes: The critical design case for stability of the cell side-slopes is the temporary period during or after construction and prior to waste placement with the granular drainage layer in place. The side-slope stability analyses should be conducted for the following cases:

- Self-weight of lining system and granular drainage material (dead loads)
- Dead loads plus loads from seepage forces in the granular drainage layer during or after heavy rainfall events.
- Dead loads plus equipment live loads
- Dead loads plus equipment live loads plus seepage forces

It is critical that the granular drainage layer meet specifications regarding fines content and permeability to avoid increasing the height of seepage and reducing the factor of safety to possibly unacceptable levels.

Design for seismic stability of the cell side-slopes is not normally required because the slopes are typically only exposed without waste buttressing for a short period of time.

Waste Static Stability: The static stability of waste placed in a cell must also be evaluated to aid determination of allowable waste height and waste configuration. The strength of the base lining system is a critical part of this analysis because of the potential for low strengths within the lining system relative to native soil and waste strengths. Slope stability analyses should be run on representative waste cross-sections considering two separate strength cases, as follows:

- Case 1: Residual strength (e.g., material or interface strength at large deformations)
- Case 2: Peak strength (e.g., material or interface strength at small deformations)

The material strength used in the analysis for a particular case should be the most critical (lowest) of the various interface and internal strengths in the lining system.

Table 1. Summary of GCL Shear Strengths[1]

	Interface Friction				GCL Internal Strength							
	GCL/Textured HDPE				Reinforced GCL							
	Peak		Large Deformation		Peak		Large Deformation		Normal Stress Range	Displacement Rate		
GCL Material Type[2]	c, psf[4]	Delta, Deg.	c, psf	Delta, Deg.	c, psf	Phi, Deg.	c, psf	Phi, Deg.	psf	In[4]/Min	Test Lab	Date
High Normal Stress												
Bentomat DN	1005	26	915	14	1640	28	620	8	3600-14400	0.04	Geosyntec	Aug-1999
Bentofix NW	483	22	425	8	1206	20	445	5	3600-14400	0.04	Precision Geo	Apr-1999
Bentofix NW	915	17	805	4	2100	22	320	8	3600-14400	0.04	Geosyntec	Jun-1999
Bentomat DN	545	26	520	10	790	26	280	9	3600-14400	0.04	Geosyntec	Nov-1997
Low Normal Stress												
Bentofix NW	128	26	71	16	394	51	494	17	144-576	0.04	Precision Geo	Apr-1999
Bentofix NW	95	28	60	22	460	39	NA[3]	NA	144-576	0.04	Geosyntec	Jun-1999
Bentomat DN	99	32	35	25	345	45	NA	NA	150-600	0.04	Geosyntec	Apr-1998
Bentomat DN					600	30	NA	NA	150-600	0.04	Cetco	Nov-1997

Notes: 1) Results obtained from testing completed for ARL Cell 6 project during design and construction.
2) Bentomat DN and Bentofix NW--double-nonwoven GCLs.
3) NA reported for GCL--large deformation internal strength indicates results not available because test apparatus was unable to fail GCL through midplane of the bentonite at low normal stresses. Accordingly, interface strength controls at low normal stresses for these reinforced GCLs.
4) 1 psf = 47.88 kpa; 1 in = 25.4 mm

Stark and Poeppel (March 1994) advocated use of residual strengths on the side-slopes and peak shear strengths on the cell bottom due to low induced deformations in the base compared to the side-slopes. The side-slopes are subjected to downdrag shear stresses associated with waste settlement; such stress increases the potential for development of residual strength. The cell bottom, in contrast, is subject to much less induced shear stress. The decision on whether to accept a particular factor of safety is highly dependent on whether residual or peak strengths are used for design.

Waste Seismic Stability: The first step in assessing the seismic stability of waste in a cell is to select appropriate earthquake design criteria. EPA Landfill Regulations, Subtitle D, specify that all landfill containment structures be designed to resist the maximum expected horizontal acceleration with a 90 percent or greater probability of not being exceeded over a design period of 250 years. This criterion applies to the final landfill configuration.

The waste sections analyzed for a given cell typically represent a temporary construction case which will cease to be a concern when adjacent cells are filled due to buttressing effects. Accordingly, a much lower design period can be associated with this temporary construction case compared to the 250 year design period associated with the final landfill configuration. Design periods can be approximated as the time interval between completion of filling in each cell. For example, Cell 2C at the Central Landfill is planned to be filled 5 years after filling in Cell 2B is completed, resulting in a design period of 5 years for Cell 2B.

To design for the same level of risk as specified in Subtitle D, peak horizontal ground accelerations with a 90 percent probability of not being exceeded over the selected design periods need to be determined. Ground motions can be determined from a site-specific seismic risk assessment such as one carried out for the ARL site (Earth Mechanics, April 1994). Current seismic hazard maps developed by the USGS (USGS 1998) or similar sources may also be referenced to obtain appropriate design ground motions.

Once appropriate design ground motions have been determined, the critical sections determined from the static stability analyses are evaluated for seismic stability. Average pseudostatic seismic coefficients equal to one-half the peak horizontal acceleration values were used in the Cell 2B analysis (Richardson and Kavazanjian, 1995) to determine seismic factors of safety.

Seismic factors of safety of 1.0 or greater using the methodology described here indicate that permanent deformations are expected to be limited to less than 300 mm under the design seismic loading (Richardson and Kavazanjian, 1995). Maximum allowable deformations of 150 to 300 mm have typically been used in practice for design of geosynthetic liner systems (Richardson and Kavazanjian, 1995). If seismic factors of safety are less than 1.0, deformation analyses can be carried out using an approach suggested in the literature (Seed and Makdisi, August 1977; and Newmark, 1965).

Constructability. A GCL-based lining system is easier to construct than a CCL-type liner system, primarily because of reduced requirements for field installation and quality control/quality assurance (QC/QA). The GCL can be installed quickly in a controlled and consistent manner, while still maintaining a high level of quality control and quality assurance. A CCL takes longer to install than a GCL and the soil gradation and moisture content must be controlled to achieve the desired low permeability. Changes in the weather can require drying or wetting of the CCL soil to achieve the optimum moisture content for compaction.

GCLs cannot typically be installed during periods of significant precipitation to avoid premature hydration under zero normal load, which would lead to swelling and possible loss of GCL strength. Under the geomembrane, the GCL will slowly hydrate as it draws moisture from the subgrade soils. Hydration typically occurs within several days to several weeks, depending on the moisture content of the subgrade. It is important to ensure that the GCL is not allowed to hydrate without an overlying normal pressure applied. Accordingly, the granular drainage material over the geosynthetics should be placed as soon after liner placement as possible.

GCL Use in Cover Systems

Equivalency to Prescriptive Cover. The prescriptive final cover as defined in 18 AAC 60.395(a) for an unlined municipal landfill consists of an infiltration layer at least 450 mm thick with a permeability no greater than 1×10^{-5} cm/sec, and an erosion layer at least 150 mm thick of earthen material capable of sustaining native plant growth. Regulation 18 AAC 60.395(b) allows ADEC to approve an alternative final cover design that will protect public health and the environment if it includes the following:

1) An infiltration layer that achieves an equivalent reduction in infiltration as the prescriptive infiltration layer, and
2) An erosion layer that provides equivalent protection from wind and water erosion as the prescriptive erosion layer.

An equivalent GCL-based cover was proposed to ADEC for Cell 2A at the Central Landfill because of the difficulty of finding local materials that would meet the prescriptive cover permeability requirements. The proposed alternative cover is shown in Figure 3. This type of final cover has been approved and successfully implemented for both unlined and lined landfill cells in Alaska and in the Lower 48 (Powell and Griffin, 1996; EPA, 1997; Miller and Kaul, 2000; Brown, et al, 2001).

Most of the design considerations described previously for GCLs used in base lining systems are also applicable for GCLs used in final covers. Points to emphasize for cover applications are noted below.

Reduction in Infiltration. The permeability of GCLs is typically around $1x10^{-8}$ to $1x10^{-10}$ cm/sec. This permeability is much lower than the prescriptive requirement of $1x10^{-5}$ cm/sec for the infiltration layer over

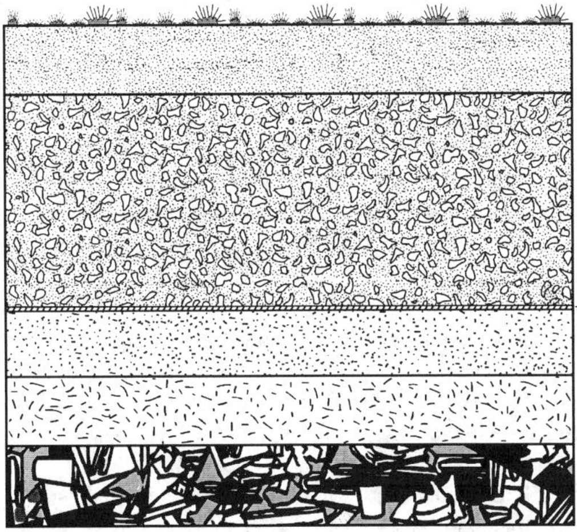

Figure 3. Cell 2A Final Cover Cross-Section

unlined cells. Although the GCL is thinner than the 450 mm prescriptive layer, the reduction is infiltration will be much greater than in an 450 mm layer at $1x10^{-5}$ cm/sec.

Resistance to Freeze-Thaw Cycling. The excellent resistance of GCLs to repeated freeze-thaw cycles is even more important in a cover system, as the near surface GCL will likely be subjected to numerous freeze-thaw cycles each winter over the entire lifetime of the landfill cover.

Resistance to Wet-Dry Cycles. GCLs have also shown good resistance to permeability increases resulting from successive wetting and drying cycles when permeated with de-ionized water (Lin and Benson, 2000). Desiccation cracks appear upon drying of the GCL, but upon rehydration and swelling, no increase in hydraulic conductivity has been found even after numerous wet-dry cycles.

Slope Stability. The GCL final cover system must also be evaluated for static and seismic stability as described above for base lining systems. Analyses related to a veneer failure of the cover system should be conducted for the following cases:

- Self-weight of cover system (dead loads)
- Dead loads plus loads from seepage forces in the drainage layer above the GCL during or after heavy rainfall events
- Dead loads plus equipment live loads
- Dead loads plus equipment live loads plus seepage forces
- Dead loads plus seismic loads (for the post-construction case).

Lower factors of safety than are typically considered suitable may be acceptable for veneer-type potential failure surfaces, particularly under the very high seismic design loads common to many parts of Alaska. Unlike the base lining system, the cover system is at the surface and can be repaired relatively easily if excessive movements were to occur under a very large earthquake event, for example.

Global stability of the overall waste mass, including the cover, should also be evaluated for both static and seismic stability as described previously.

Drainage. Provisions for drainage must be provided at the toe of the slope to allow water to drain from the granular layer over the GCL in the final cover. It is important to provide positive drainage away from the landfill slopes to avoid water backing up into the liner system and waste mass, potentially becoming a driving force for a veneer failure on top of the GCL surface.

Differential Settlement. GCLs can accommodate large deformations as typically occur in a municipal solid waste landfill (La Gatta, et al, 1997). Adjacent GCL panels are typically overlapped 150 to 300 mm or more to accommodate the anticipated magnitude of settlement.

Landfill Gas Venting. Landfill gas venting is necessary because the low-permeability GCL will not allow gas to pass through freely after it has hydrated. The system proposed for Cell 2A consists of passive vent pipes protruding through the GCL near the midpoint of the side-slope. Bentonite clay granules will be used to fill the seam around the pipe penetrations. The vent will have a shepherd's hook or cap to prevent water from entering. The vents will be connected by perforated HDPE pipes in the operational cover material beneath the GCL. The vents will be spaced at approximately 60 m on center to provide adequate venting.

Construction QA/QC

The effectiveness of any base or cover lining system will depend on how well it is constructed; therefore, thorough quality assurance/quality control (QA/QC) procedures related to the GCL must be included in the construction specifications. An effective QA/QC program should address the following areas:

- GCL material properties
- GCL subgrade preparation
- GCL storage, handling, and placement
- Quality of GCL seam construction
- Cover material quality tests
- Depth and grading of cover layers
- Proper installation of gas venting structures

Well-written specifications detailing a thorough QA/QC program and conformance testing prior to and during construction are necessary to ensure the adequacy of the installed GCL, as follows:

1) **Prior to GCL Manufacturing**--Information should be requested that describes the manufacturer's product and documents that the proposed GCL material conforms to the project specifications. Typically, standard manufacturer's literature comprises much of this submittal for initial material approval. Additional testing may be required at this stage if test data are unavailable for project-specific test parameters. The manufacturer's quality assurance/quality control plan and material samples should also be requested at this time.

2) **During GCL Manufacturing**--Factory test results on samples from representative rolls of the material as it is manufactured for the project should be required. Proposed test parameters for strength and permeability testing should be reviewed and preapproved by the project engineer to ensure that field conditions are accurately modeled. Factory material samples should be submitted to the engineer for independent testing to confirm and/or supplement the contractor's data.

3) **During Liner Construction**--Conformance tests should be run on samples of the material delivered to the site. Tests can be run on any properties that may be of particular importance on a given project.

Detailed guidelines for quality assurance and conformance testing of GCLs are available from several sources (EPA 1993; ASTM 1994; ASTM D5889, 1998).

Results of typical construction conformance QA/QC peel strength tests are included in Table 2 for the double nonwoven GCLs used for ARL Cell 6. Peel strength tests, run in accordance with the modified ASTM D 4632 standard, provide an indirect indicator of the internal strength of the GCL. Direct correlations between internal shear strength values as presented in Table 1 and peel strength values as presented in Table 2 have not been developed, but the relatively inexpensive peel strength test can be used as a strength index test for samples where direct shear tests are not run. For cases where high internal GCL strengths are required, a minimum peel strength of 30 lbs should be specified. The GCL manufacturers may need to provide additional reinforcement of their GCL product through increased needle-punching to consistently meet a peel strength of 30 lbs. Review of the shear strength results from the ARL Cell 6 GCLs in Table 1 and the peel strength results from those same GCLs in Table 2 allows a comparison to be made of the general relationship of these test results.

Table 2. GCL Conformance Index Tests
Peel Strength Results--ARL Cell 6

			Peel Strength(1) (ASTM D 4632) (LBS)(2)	
GCL Type	**Lot #**	**Roll #**	**MD(3)**	**TD(4)**
Bentomat DN	199920020	682	54	61
Bentomat DN	199920020	701	33	46
Bentomat DN	199920020	711	48	59
Bentomat DN	199921020	748	48	51
Bentomat DN	199921020	784	70	81
Bentomat DN	199921020	805	39	51
Bentomat DN	199921020	829	81	73
Bentofix NW	R9042610	6975	40	34
Bentofix NW	R9042610	7031	36	33
Bentofix NW	R9050310	7039	34	32
Bentofix NW	R9050310	7080	34	34
Bentofix NW	R9060210	7590	44	40
Bentofix NW	R9060210	7609	39	45
Bentofix NW	R9060710	7651	34	34

Notes: (1) Average Value From 5 Tested Specimens.
(2) 1 Lb = 4.448 N
(3) MD = Machine Direction
(4) TD = Transverse Direction

Summary and Conclusions

In conclusion, following are the key points discussed in this paper:

- GCLs are well-suited to use in Alaska and other cold regions because of their relative insensitivity to repeated freeze-thaw and wet-dry cycles.
- GCLs are also advantageous in cold regions with short construction seasons because of the relative ease and speed of construction.
- GCLs have been used successfully in numerous applications in Alaska and the Lower 48, including landfill cell base lining systems, landfill covers, lagoon liners, wetlands, fish ponds, and secondary containment applications.
- GCLs can be used in combination with geomembranes or alone, depending on specific application requirements.
- Project-specific shear strength testing should be carried out on GCLs proposed for any given application. Testing of GCL internal shear strength as well as interface friction of anticipated low-strength interfaces should be carried out. Test parameters should model expected field conditions, including degree and rate of hydration, degree and rate of

consolidation, and applied normal stresses under which hydration or consolidation may occur.

- Interface and internal shear strength tests should be carried out under the range of stresses expected in the field, since failure envelopes are typically nonlinear. Additionally, different interfaces or materials may control design (provide the lowest shear strength) at different stress conditions.
- Static and seismic stability analyses should be carried out to address all load cases pertinent to the base lining system (bottom and side slope) and cover lining system. Consideration should be given to each of the various stages of landfill development, including construction, waste filling, precipitation events, seismic events, and throughout the design life of the landfill.
- Acceptable factors of safety for slope stability analyses should be selected with due consideration given to whether peak or residual (large deformation) strengths are used in the analyses. Lower factors of safety may be appropriate if residual strengths are used.
- To design for the same level of risk specified in EPA Subtitle D for the final landfill configuration, seismic stability analyses of interim landfill configurations can be conducted using a reduced design earthquake acceleration corresponding to the interim period under consideration.
- If the factor of safety for landfill seismic stability is less than 1.0, seismic deformation analyses can be carried out to estimate the magnitude of movement and assess the potential impact on landfill facilities.
- Quality assurance and conformance testing of GCLs is important and should cover all project phases, including before and during GCL fabrication, and during construction as the material is brought to the site.
- Peel strength of GCLs is an important conformance index test and provides a quick, relatively inexpensive method to frequently screen GCLs delivered to a project for adequate internal shear strength.

References

American Society for Testing and Materials. *ASTM Standards and Other Specifications and Test Methods on the Quality Assurance of Landfill Liner Systems.* ASTM 1994.

American Society for Testing and Materials. "ASTM D5889, Standard Practice For Quality Control Of Geosynthetic Clay Liners." ASTM 1998.

Benson, C.H.; Abichou, T.H.; Olson, M.A.; and Bosscher, P.J. *Winter Effects on Hydraulic Conductivity of Compacted Clay.* ASCE, Journal of Geotechnical Engineering, Vol. 121, No. 1, January 1995.

Brown, Damon; Brummer, Christopher J.; and Delmanowski, Michael. *Cost Effective Alternative To An Unreinforced GCL For Landfill Final Cover Systems,* Geosynthetics Conference 2001.

Daniel, D.E., *Geosynthetic Clay Liners (GCLs) in Landfill Covers.* 31st Annual Solid Waste Exposition, Solid Waste Association of North America. San Jose, CA. August 1993.

Daniel, D.E., Koerner, R.M., Bonaparte, R., Landreth, R.E., Carson, D.A., and Scranton, H.B., *Slope Stability of Geosynthetic Clay Liner Test Plots*, ASCE, Journal of Geotechnical and Geoenvironmental Engineering, Vol. 124, No. 7 pp. 228-637, 1998.

Earth Mechanics, Inc. *Seismic Hazard Evaluation, Anchorage Regional Landfill.* April 30, 1994.

Hewitt, R.D. and Daniel, D.E. *Hydraulic Conductivity of Geosynthetic Clay Liners after Freeze-Thaw.* ASCE, Journal of Geotechnical and Geoenvironmental Engineering, Vol. 123, No. 4, April 1997.

Koerner, R.M. and Daniel, D.E. *Technical Equivalency Assessment of GCLs to CCLs.* Proceedings GRI-7 Conference on Geosynthetic Liner Systems, Industrial Fabrics Association International, St. Paul, MN, pp. 255-275. 1994.

Kraus, J.F.; Benson, C.H.; Erickson, A.E.; and Chamberlain, E.J. *Freeze-Thaw Cycling and Hydraulic Conductivity of Bentonitic Barriers.* ASCE, Journal of Geotechnical and Geoenvironmental Engineering, Vol. 123, No. 3, March 1997.

LaGatta, M.D., Boardman, B.T., Cooley, B.H., and Daniel, D.E., *Geosynthetic Clay Liners Subject to Differential Settlement*, ASCE, Journal of Geotechnical and Geoenvironmental Engineering, Vol. 124, No. 7 pp. 228-637, 1997.

Lin, Ling-Chu, and Benson, Craig H. *Effect of Wet-Dry Cycling on Swelling and Hydraulic Conductivity of GCLs.* ASCE, Journal of Geotechnical and Geoenvironmental Engineering, pp. 40-49, January 2000.

Miller, Brad and Kaul, Joe. *GCL Closure Cover Case Studies in the Arid West.* Third SWANA Arid Climate Landfill Symposium in Albuquerque, NM. April, 2000.

Newmark, N.M. *Effects of Earthquakes on Dams and Embankments.* Geotechnique, Vol. 15, No. 2, pp. 139-160, 1965.

Powell, Lee A. and Griffin, James L. *Experience with Geosynthetic Clay Liners for Landfill Closure at the Tomoka Farms Road Landfill, Daytona Beach, Florida*. September 1996 SWANA Conference.

Richardson, Gregory and Kavazanjian, Edward. *RCRA Subtitle D (258) Seismic Design Guidance for Municipal Solid Waste Landfill Facilities.* U.S. Department of Commerce. April 1995.

Ruhl, Janice L. and Daniel, D.E., *Geosynthetic Clay Liners Permeated with Chemical Solutions and Leachates*, ASCE, Journal of Geotechnical and Geoenvironmental Engineering, Vol. 123, No. 4, pp. 369-381, April 1997.

Seed, H. Bolton, and Makdisi. *A Simplified Procedure for Estimating Earthquake Induced Deformation in Dams and Embankments.* UCB/EERC 77/19, August 1977.

Shan, H.Y. and Daniel, D.E. *Results of Laboratory Tests on a Geotextile/Bentonite Liner Material.* Geosynthetics '91, Vol. 2, Industrial Fabrics Association International, St. Paul, MN, pp. 517-535. 1991.

Stark, T.D. and Poeppel, A.R. *Landfill Liner Interface Strengths from Torsional Ring-Shear Tests.* ASCE, Journal of Geotechnical Engineering, Vol. 120, No. 3, March 1994.

U.S. Environmental Protection Agency. *Geosynthetic Clay Liners Used in Municipal Solid Waste Landfills*, EPA530-F-97-002, July 1997.

U.S. Environmental Protection Agency. *Technical Guidance Document: Quality Assurance and Quality Control for Waste Containment Facilities.* EPA, September 1993.

U.S. Geological Survey. *National Seismic Hazard Mapping Project.* 2001. http://geohazards.cr.usgs.gov/eq/

Vasko, Scott M.; Jo, Ho Young; Benson, Craig H.; Edil, Tuncer B.; and Katsumi, Takeshi. *Hydraulic Conductivity of Partially Prehydrated Geosynthetic Clay Liners Permeated with Aqueous Calcium Chloride Solutions.* Geosynthetics Conference 2001.

Wilson-Fahmy, R.F. and Koerner, R. M. *Leakage Rates Through a Hole in a Geomembrane Overlying a Geosynthetic Clay Liner.* Proceedings of Geosynthetics '95, Industrial Fabrics Association International, St. Paul, MN, pp. 655-668. 1995.

Effect of Freeze-Thaw Cycling on the Hydraulic Conductivity of Soil-Bentonite Mixtures

Gary J. Foose[1] and Robert A. Thomson, Jr.[2]

Abstract

Soil-bentonite mixtures are often used for liners and covers for waste containment systems in cold regions. In these applications, soil-bentonite mixtures can be exposed to environmental stresses such cycles of freezing and thawing, which can increase the hydraulic conductivity of soil-bentonite mixtures. In this study, hydraulic conductivity tests were conducted to assess how freeze-thaw cycling affects the hydraulic conductivity of a soil-bentonite mixture. The permeating fluids were tap water and fluids containing divalent cations at concentrations representative of naturally occurring field conditions. Results indicate that the hydraulic conductivity of soil-bentonite mixtures can increase by several orders of magnitude as a result of exposure to cycles of freezing and thawing, in the presence of divalent cations. The hydraulic conductivity increase is due to the inability of the bentonite to swell and fill the void spaces in soil-bentonite mixtures. These findings suggest that soil-bentonite mixtures may not be as resistant to damage from freeze-thaw cycling as previously reported by other investigators. Thus, designers and regulators must carefully consider the hydraulic performance of soil-bentonite mixtures in cold regions where they may be exposed to cycles of freezing and thawing.

Introduction

Engineers are often confronted with designing earthen hydraulic barriers that perform adequately in cold regions. Many laboratory and field investigations have shown that the hydraulic conductivity of compacted clay increases when exposed cycles of freezing and thawing (e.g. Zimmie and La Plante 1990, Chamberlain et al. 1990, Benson and Othman 1993). Mixtures of sand and bentonite are an alternative to compacted clay liners and have been used to construct earthen hydraulic barriers

---

[1] Assistant Professor, Department of Civil and Environmental Engineering, University of Cincinnati, Cincinnati, OH 45221, gary.foose@uc.edu, ph:513-556-3781, fx: 513-556-2599

[2] Graduate Assistant, Department of Civil and Environmental Engineering, University of Cincinnati, Cincinnati, OH 45221

(e.g. Garlanger et al. 1987, Chapuis et al. 1992, O'Sadnick et al. 1995). Wong and Haug (1991) and Kraus et al. (1997) found that compacted sand-bentonite liners are resistant to cycles of freezing and thawing.

The objective of this study is to quantify the effects of freezing and thawing cycles on the hydraulic conductivity of sand-bentonite liners permeated with solutions containing low concentrations of divalent cations. The solutions used in the study included, tap water, 0.01-M $CaCl_2$, and 0.05-M $CaCl_2$. These solutions are representative of natural soil pore waters that occur in the field (Davis and DeWiest 1966).

Background

Field and laboratory investigations have shown that exposure of compacted clay liners to cycles of freezing and thawing results in an increase in hydraulic conductivity. Based on data compiled by Othman et al. (1994) for compacted clays having a hydraulic conductivity $<1x10^{-7}$ cm/s, the hydraulic conductivity of compacted clay subjected to cycles of freezing and thawing can increase by 1 to 3 orders of magnitude. The increase in hydraulic conductivity is thought to occur as a result of cracks forming as water migrates to the freezing front and ice lenses form. Upon thawing, these cracks do not completely heal and they become preferential pathways for flow.

Several testing programs have shown that sand-bentonite liners are resistant to damage caused by exposure to cycles of freezing and thawing. Wong and Haug (1991) found that for a mixture of Ottawa sand and Na-bentonite the hydraulic conductivity actually decreases when exposed to cycles of freezing and thawing. Wong and Haug (1991) attributed the ability of the sand-bentonite mixture they tested to resist damage from cyclic freezing and thawing to the swelling capacity of the Na-bentonite which allowed the bentonite to fill any voids in the sand matrix.

Similar findings were reported by Kraus et al. (1997) based on laboratory and field tests on a sand-bentonite liner exposed to cycles of freezing and thawing. Kraus et al. (1997) also performed a morphological investigation on a sand-bentonite liner constructed in the field and subjected to two cycles of freezing and thawing. They did not find any visual evidence of cracks in the sand-bentonite mixture when the mixture was thawed. Kraus et al. (1997) also visually examined the macrostructure of the sand-bentonite liner while the liner was still frozen. Samples retrieved from the field test, using a core barrel, and sections cut from laboratory specimens did not reveal any evidence of the formation of ice-lenses. Based on these observations, Kraus et al. (1997) concluded that sand-bentonite liners are resistant to cyclic freezing and thawing because ice-lenses do not form during freezing in these types of hydraulic barriers.

The individual sand grains of a sand-bentonite liner are relatively impervious compared to the Na-bentonite. Therefore, the hydraulic conductivity of a sand-bentonite liner can be attributed to the low hydraulic conductivity of the bentonite fraction of the mixture. The bentonite typically used to construct sand-bentonite liners is Na-bentonite, which is primarily composed of montmorillonite having sodium as the predominant adsorbed cation. Na-bentonite is characterized by a high cation exchange capacity (80-150 meq/100 g) (Grim 1953, Mitchell 1993). When hydrated, Na-bentonite adsorbs a large number of water molecules and cations which form a thick diffuse double layer. This property is responsible for the large swelling capacity of Na-bentonite.

The volume of bound water for Na-bentonite is correlated with the hydraulic conductivity, with lower hydraulic conductivities associated with thicker layers of bound water. An indirect measurement of the volume of bound water is the swelling capacity. Therefore, the hydraulic conductivity of bentonite is an indirect function of the swelling capacity, with lower hydraulic conductivities associated with increased swelling capacity. Polyvalent cations (e.g. Mg^{2+}, Ca^{2+}, Al^{3+}) are preferentially adsorbed by bentonite. Adsorption of polyvalent cations in place of Na^{+} results in a decrease in the thickness of the bound water layer, which reduces the swelling capacity of the bentonite, and results in an increase in the hydraulic conductivity.

Materials and Methods

The soils used in this study consisted of bentonite and sand. The physical and chemical properties of the soils are summarized in Table 1. The bentonite component of the mixture was PureGold Gel, a powdered sodium bentonite with no organic or polymer additives, marketed by CETCO. The sand component of the mixture consisted of clean fine quartz sand.

The sand-bentonite mixture was prepared in the laboratory by manually mixing proportions of the two soils using a trowel. Based on a battery of hydraulic conductivity tests, the mass fraction of bentonite was selected to be 10% by mass. This mass fraction of bentonite was sufficient for the hydraulic conductivity of the sand-bentonite mixture to tap water to be $<1\times10^{-7}$ cm/s, the common design specification for liners used in waste containment. A compaction curve for the mixture using modified Proctor effort (ASTM D 1557) is shown in Figure 1.

Specimens used for hydraulic conductivity testing were compacted using modified Proctor effort and compacted at a water content of 20%. Cincinnati tap water was used to hydrate the sand-bentonite mixture by incrementally spraying the soil with a mist of water and repeatedly mixing the soil with a trowel. Following compaction, the specimens were extruded from the compaction molds, using a hydraulic jack, and then assembled into flexible wall permeameters (ASTM D 5084). Back-pressuring was not used because field cover systems are not subjected to back-pressuring when wetted and to permit easy sampling of effluent pore water from the tests. Flow was driven using the falling head-constant tail water method and the average hydraulic gradient was 11.5. The average effective stress was 18.8 kPa, which is representative of the light overburden that exists in landfill cover systems.

Table 1. Physical and Chemical Properties of Sand and Bentonite Soils.

Property	Value
Na-Bentonite	
Liquid Limit (ASTM D 4318)	436
Plastic Limit (ASTM D 4318)	36
Plasticity Index (ASTM D 4318)	400
Fines (<0.075 mm) (%) (ASTM D 421)	98
USCS Classification (ASTM D 2487)	CH
Cation Exchange Capacity (meq/100 g) [a]	45.9
pH (Saturated Paste) [a]	8.9
Soluble Salts (mmhos/cm) [a]	5.27
Sand	
Liquid Limit (ASTM D 4318)	-
Plastic Limit (ASTM D 4318)	NP
Plasticity Index (ASTM D 4318)	NP
G_s (ASTM 854)	2.65
Sand (0.074 - 4.75 mm) (%) (ASTM D 421)	98.5
Fines (<0.075 mm) (%) (ASTM D 421)	1.5
USCS Classification (ASTM D 2487)	SP

[a] Analyses performed by Agavise Laboratories, Northwood, ND.

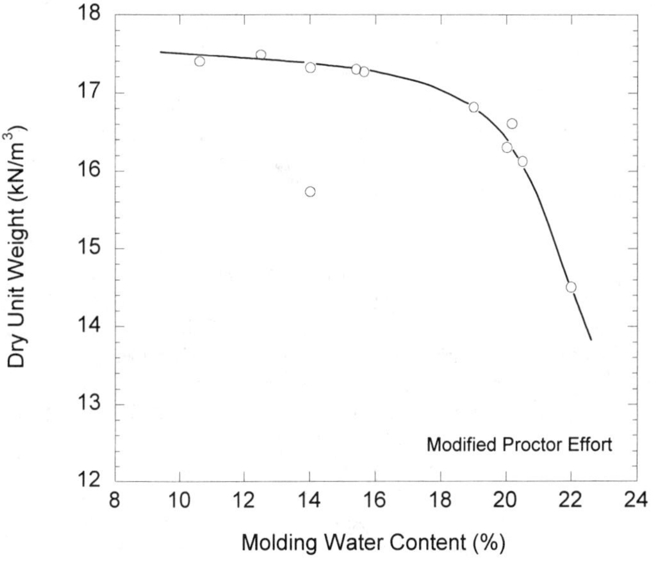

Figure 1. Compaction Curve for Sand-Bentonite Mixture.

After compacting the specimens, the hydraulic conductivity of the sand-bentonite mixtures to tap water was measured following procedures described by ASTM D 5084. Permeation with tap water continued until the hydraulic conductivity remained steady for four consecutive readings varying by less than ±50% of the mean of the last four measurements, no trend in hydraulic conductivity was evident.

After the initial hydraulic conductivity to tap water was measured, the specimens were carefully removed from the permeameters and frozen using the free standing closed system procedure described by Othman et al. (1994). Specimens were sealed in a plastic bag to prevent desiccation and placed in a laboratory freezer for cooling to -15 °C (ASTM D 6035). The closed system is representative of field conditions in which there is no free access to water below the frost line. After being in the freezer for at least 24 hrs, the specimens were removed and placed back into the permeameters, permeated with the chemical solution for two weeks during which the hydraulic conductivity was measured, and then frozen again. Two weeks of permeation at a hydraulic gradient of 11.5 corresponds to approximately 3.8 months of flow at a gradient of 1.5, which would occur in covers having a gravel drainage layer that becomes saturated. A control specimen was continuously permeated with 0.05-M $CaCl_2$ after the initial hydraulic conductivity of the control specimen to tap water was measured. This control specimen was not subjected to cycles of freeze/thaw.

Three different permeant liquids were used in this study: Cincinnati tap water, 0.01-M $CaCl_2$, and 0.05-M $CaCl_2$. Lin and Benson (2000) report that the ionic strength of a 0.0125-M $CaCl_2$ solution is representative of the ionic strength of pore fluid from surface soils found in Wisconsin and pore waters having this strength or greater naturally occur in desert environments. Solutions of calcium chloride were prepared by dissolving calcium chloride in deionized water. Properties of Cincinnati tap water are listed in Table 2.

Results

The hydraulic conductivities of the sand-bentonite mixtures subjected to cycles of freezing and thawing and permeated with low concentration solutions of $CaCl_2$ are shown in Figure 2. The initial hydraulic conductivities of all three of the specimens permeated with tap water are essentially the same (Number of Cycles of Freeze/Thaw = 0). Within the first five cycles of freeze/thaw, the hydraulic conductivity of all three specimens remains low and no general relationship between hydraulic conductivity and cycles of freeze/thaw is clear. These results contrast those reported by Wong and Haug (1991) who found that freeze/thaw cycles slightly decrease the hydraulic conductivity of sand-bentonite liners. These low hydraulic conductivities can be attributed to the large swelling capacity of the bentonite, which allows the bentonite to swell and fill the voids in the sand-bentonite mixture created by the expansion of the pore-water during freezing.

Table 2. Chemical Properties of Cincinnati Tap Water.[a]

Property	Concentration (mg/L)
Na	29
Ca	127
Mg	38
Al	0.146
Cl	30
Fl	0.97

[a] Data supplied by Greater Cincinnati Water Works.

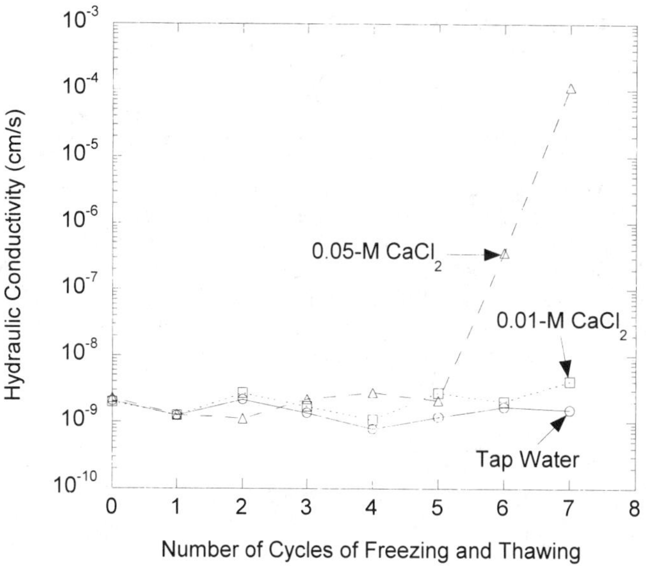

Figure 2. Hydraulic Conductivity of Sand-Bentonite Liner Exposed to Freezing and Thawing.

After the sixth cycle of freeze/thaw, the hydraulic conductivity of the specimen permeated with 0.05-M $CaCl_2$ increased by nearly two orders of magnitude to 3.6×10^{-7} cm/s. After the seventh cycle of freeze/thaw, the hydraulic conductivity of the same specimen increased to 1.1×10^{-4} cm/s. Prior to freezing for the next cycle, the specimen was carefully examined for evidence of sidewall leakage, but no evidence was found. It is hypothesized that the voids created during freezing did not entirely heal upon thawing, resulting in high hydraulic conductivity. In contrast, the specimens permeated with tap water and 0.01-M $CaCl_2$ and subjected to cycles of freeze/thaw maintained low hydraulic conductivity after six cycles of freezing and thawing.

The duration of permeation corresponding to the hydraulic conductivity measurement reported for the specimen permeated with 0.05-M $CaCl_2$ after 7 cycles of freeze/thaw is 14 weeks. The hydraulic conductivity of the control specimen continuously permeated with 0.05-M $CaCl_2$ and not subjected to cycles freeze/thaw after 14 weeks of permeation was 3.2×10^{-9} cm/s. Thus, the increase in hydraulic conductivity of the specimen subjected to cycles of freeze/thaw is primarily caused by damage due to freeze/thaw cycling.

The exact mechanism that caused the increase in the hydraulic conductivity of the specimen permeated with 0.05-M $CaCl_2$ and subjected to cycles of freeze/thaw is not entirely understood. One possible explanation is that cation exchange of calcium for sodium on the bentonite reduced swelling capacity. A graph of thawed volume versus cycles of freezing and thawing is shown in Figure 3. The unfrozen volume of the specimen permeated with 0.05-M $CaCl_2$ decreased by >7% after 7 cycles of freezing and thawing. This decrease in volume is due to shrinkage of the bound water layer. The reduced swelling capacity perhaps prevents the bentonite from fully filling the void space within sand matrix which is manifested as an increase in the hydraulic conductivity. The electrical conductivity of the effluent solution from this test was essentially the same as that for the influent solution, which indicates that the solute had entirely permeated the specimen after 12 weeks of permeation. Solute transport in the specimen occurring via diffusion would explain how it is possible for the solute to have completely permeated the specimen, thus making the entire specimen vulnerable to damage due to freeze/thaw cycling. This hypothesis is being further evaluated by additional testing.

The specimen permeated with tap water did not increase in hydraulic conductivity after being subjected to cycles of freezing and thawing. This finding is consistent with results reported by Kraus et al. (1997) and Wong and Haug (1991). In contrast to the results reported by Wong and Haug (1991) and similar to the results shown by Kraus et al. (1997), cycles of freezing and thawing did not result in a decrease in hydraulic conductivity. One possible reason that cycles of freezing and thawing do not result in an increase in the hydraulic conductivity of sand-bentonite mixtures permeated with tap water is that tap water contains a variety of monovalent and divalent cations. Therefore, exchange of calcium from solution for adsorbed sodium on the bentonite does not occur to a degree such that the swelling capacity of the bentonite is significantly reduced. As shown in Figure 3, the thawed volume of the specimen permeated with tap water decreased by <5%.

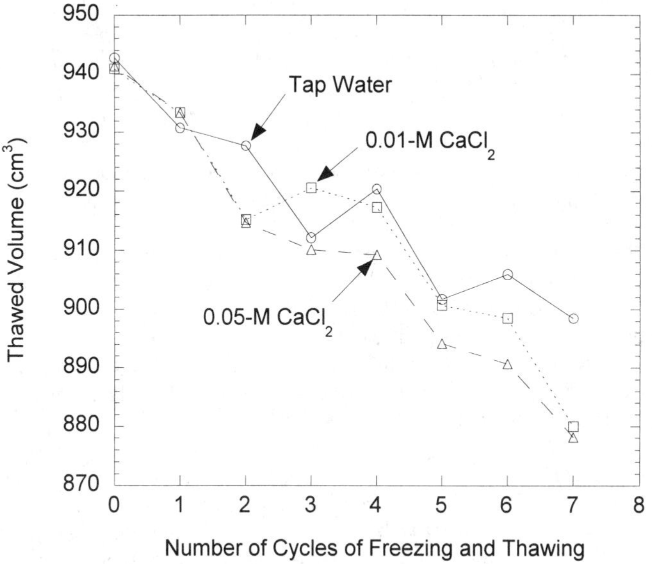

Figure 3. Thawed Volume of Sand-Bentonite Mixture Subjected to Cycles of Freezing and Thawing.

The hydraulic conductivity of the specimen permeated with 0.01-M $CaCl_2$ remained low after being subjected to 7 cycles of freezing and thawing. The thawed volume of the specimen has decreased by 6%. Further testing on this specimen is presently being performed. It is anticipated that exchange of calcium from solution for adsorbed sodium on the bentonite will continue to occur and the swelling capacity of the bentonite will decrease further. This is likely to result in an increase in the hydraulic conductivity with additional cycles of freezing and thawing.

Conclusions

Results from the laboratory tests show that sand-bentonite mixtures can be adversely affected by freezing and thawing in a closed system and permeated with solutions containing divalent cations. All of the specimens had low hydraulic conductivity ($\sim 2\times10^{-9}$ cm/s) during the first five cycles of freezing and thawing. During the seventh cycle of freezing and thawing, the hydraulic conductivity of the specimen permeated with or 0.05-M $CaCl_2$ increased from 2.3×10^{-9} cm/s to 3.6×10^{-7} cm/s and increased further to 1.1×10^{-4} cm/s after the seventh cycle. For the specimen continuously permeated with 0.05-M $CaCl_2$ for the same duration, the hydraulic conductivity did not increase. Therefore, the increase in hydraulic conductivity of the specimen permeated with 0.05-M $CaCl_2$ and subjected to cycles of freezing and thawing is the result of damage to the specimen incurred during the freezing and thawing processes. The large increase in hydraulic conductivity is attributed to cation exchange of calcium in solution for sodium adsorbed by the bentonite fraction of the mixture. Exchange of divalent cations for monovalent cations results in a decrease in the swelling capacity of the bentonite fraction of the mixture which manifested itself as an increase in hydraulic conductivity. Measurements of the thawed volume of the specimen are consistent with this hypothesis.

The hydraulic conductivity of specimen permeated with either tap water or 0.01-M $CaCl_2$ did not increase after 7 cycles of freezing and thawing. For the specimen permeated with tap water, it is believed that the variety of monovalent and divalent cations in the permeant limit cation exchange. Additional testing is being performed on the specimen permeated with 0.01-M $CaCl_2$ to determine if the hydraulic conductivity of this specimen will also increase as calcium in solution exchanges for adsorbed sodium.

Results from this study suggest that the hydraulic conductivity of sand-bentonite liners exposed to cycles of freezing and thawing with pore waters that contain divalent cations may increase due to the reduction in the swelling capacity of the bentonite. Therefore, the durability of sand-bentonite liners subjected to cycles of freezing and thawing may not be as great as previously reported. There are a variety of conditions, including natural ones, in which sand-bentonite liners may be permeated with liquids containing divalent cations and few monovalent cations. Therefore, use of a sand-bentonite liner in situations where they may undergo cycles of freezing and thawing must be considered carefully.

References

Benson, C. H. and Othman M. A. (1993), "Hydraulic conductivity of compacted clay frozen and thawed in situ," *Journal of Geotechnical Engineering*, ASCE, 119(2), 276-294.

Chamberlain, E. J., Iskander, I., and Hunsiker, S. E. (1990), "Effect of freeze-thaw on the permeability and macrostructure of soils," *Proceedings of the International Symposium on Frozen Soil Impacts on Agricultural, Range, and Forest Lands, Spokane*, WA, 145-155.

Chapuis, R. P., Lavoie, J., and Girard, D. (1992), "Design, construction, performance, and repair of the soil-bentonite liners of two lagoons," *Canadian Geotechnical Journal*, 29(4), 638-649.

Davis, S. N., and DeWiest, R. J. M. (1966), *Hydrogeology*, John Wiley & Sons, Inc. New York, NY.

Garlanger, J. E., Cheung, F. K., and Bishar, S. T. (1987), "Quality control testing for a sand-bentonite liner," *Geotechnical Practice for Waste Disposal '87*, R. D. Woods, ed., ASCE, Reston, VA, 488-499.

Grim, R. E. (1953), *Clay Mineralogy*, McGraw-Hill, New York, NY.

Kraus, J. F., Benson, C. H., Erickson, A. E., and Chamberlain, E. J. (1997), "Freeze-thaw cycling and hydraulic conductivity of bentonitic barriers," *Journal of Geotechnical and Geoenvironmental Engineering*, ASCE, 126(1), 229-238.

Lin, L. and Benson, C. H. (2000), "Effect of wet-dry cycling on swelling and hydraulic conductivity of GCLs," *Journal of Geotechnical and Geoenvironmental Engineering*, ASCE, 126(1), 40-49.

Mitchell, J. K. (1993), *Fundamentals of Soil Behavior*, 2nd Ed., John Wiley and Sons, Inc., New York, NY.

O'Sadnick, D. L., Simpson, B. E., Kasel, G. K. (1995), "Evaluation and performance of a sand/bentonite liner," *Proceedings of Geoenvironment 2000*, Y. B. Acar and D. E. Daniel, eds., ASCE, Reston, VA, 688-701.

Othman, M. A., Benson, C. H., Chamberlain, E. J., and Zimmie, T. F. (1994), "Laboratory testing to evaluate changes in hydraulic conductivity of compacted clays caused by freeze-thaw: State of the art," *Hydraulic conductivity and waste contaminant transport in soils, ASTM STP 1142*, D. Daniel and S. Trautwein, Eds., American Society for Testing and Materials, Philadelphia, PA, 227-254.

Wong, L. C., and Haug, M. D. (1991), "Cyclical closed-system freeze-thaw permeability testing of soil liner and cover materials," *Canadian Geotechnical Journal*, 28(6), 784-793.

Zimmie, T. F., and LaPlante, C. M. (1990), "The effect of freeze-thaw cycles on the permeability of a fine-grained soil," Proceedings of the 22nd Mid-Atlantic Industrial Waste Conference, Technion, Inc., Lancaster, PA, 580-593.

The Peculiarities of Utilization of the Technogenic Waste into Permafrost

Sergey Alexeev[1], Ludmila Alexeeva[2]

Abstract

Hydrogeochemical peculiarities of the processes of the ecologically dangerous waste dumping into the perennially frozen rocks in the Yakutian diamond-bearing province (Western Yakutia) are considered. The expediency of this utilization way applied in the cryolite zone was settled. The results of the simulation of the physical-chemical processes of the brine-enclosing rocks and brine-ground ice interaction are represented. The prospects of high-mineralized solutions burial into the permafrost without negative ecological influence are demonstrated.

Introduction

During the last years the problem concerning the way of the ecologically dangerous waste dumping in the perennially frozen rocks was widely discussed. The opinion prevails that the long-term and reliable burial of such waste in frozen rocks is not only possible, but is also more preferable than into unfrozen rocks in many parameters. According to the modern viewpoints permafrost is not a continuous fluid-impervious screen. Ice-rich zones are formed in the upper part of permafrost. Beneath this layer rock cavities are filled with ice or gas entirely or partly. Thickness of the free-ice collector zone can reach tens of meters. The permafrost permeability is confirmed by many real facts: the presence of free ice cavities in the drill cores, the intensive drill solutions absorption at the well sinking, the gas shows, the reactions to the air pressure changing (air emission or suction). Permafrost does not undoubtly confine bed for gases and moreover is accessible for the filtration of the mineralized solutions not freezing at the natural rock temperature. Also ice-rich rock, permeable in thawed condition, is not a barrier for strong brines. Reserve capacity of a frozen rock massif is defined not

[1] Doctor, head of Laboratory of Hydrogeology, Institute of the Earth's Crust SB RAS, Lermontov street 128, Irkutsk, 664033, RUSSIA; FAX: (7-3952) 462900, phone (7-3952) 460637, salex@crust.irk.ru

[2] Doctor, senior scientific researcher, Institute of the Earth's Crust SB RAS, Lermontov street 128, Irkutsk, 664033, RUSSIA; FAX: (7-3952) 462900, phone (7-3952) 467777, lalex@crust.irk.ru

only by connected free ice cavities, but also by the creation of H_2O-transition zones in the liquid phase (about 9% ice volume in the rocks). Moreover, the air porosity of nature ice, which is 3% on average, has real importance (Saveliev, 1971).

Background

The ice-brine interaction proceeds according to the mode: 1) equilibrium; 2) crystallization of H_2O from solution into ice at progressive cooling; 3) ice melting if environment temperature is lower than 0°C but higher than cryoeutectic solution. The most widespread one under the contemporary climate condition is the third system state especially under the cryo-hydrogeological conditions of the Siberian platform and the Arctic Zone. In these regions contact of permafrost with negative-temperature chloride brines (cryopegs) is traced everywhere. Cryopegs are intra- and subpermafrost solutions with mineralization 100-300 g/L and more. They don't freeze at the normal temperatures in subsurface (as usual -5...-2 °C, minimum -14...-10 °C), because deeper cooling (-21.4 °C for NaCl, -55 °C for $CaCl_2$) is required for the diluted salt crystallization. Such solutions are not in equilibrium with ice and can transmit it in the liquid phase, eliminate fluid-impermeability of the permafrost and migrate through cryogenic rocks.

The essence of the processes taking place at the waste utilization in the perennially frozen rocks, is: 1) ice melting occurs more extensively in blocks containing highly mineralized waters; 2) as waste is dumped, the permeability of the ground increases as joints are formed by thaw of ice in rock pores and fractures; 3) the main part of summary permafrost capacity is "additional capacity", formed as a result of the ice melting in the porous blocks (Alexeev, 2000).

To estimate the potential burial sites, and the application of this method in certain regions, the presence of optimum cryo-hydrogeological, structural-tectonic and hydrogeochemical conditions is necessary. First of all the continuous permafrost has to be widely spread, its considerable thickness (more than 100-200 m) and the bed position tens of meters lower than the nearest valleys bottoms are necessary. Permafrost temperature has to be low (colder than -2 °C), and the age of permafrost must be at least 100 years. These features are inherent especially in the permafrost of the northern type. Epigenetic sedimentary lithified depositions, intensively fractured crystalline massif and coarsely grained (clastic) rocks with isotropic permeability increase the useful capacity. The cryogenic rock masses in the rugged relief areas drained during the freezing period have the best characteristics to be used as a burial reservoir. For burying it's worthwhile to choose the tectonic crushing zones.

Thus disposition and location depth of underground storages must be determined by type and kind of waste, waste influence on thermal and mechanical stability of frozen rocks and on their screening conditions. Evaluating natural geological-geographical (including geocryological) storage environments and analyzing its development during the required period of isolation is also important.

The similar method of drainage brine burial has been applied in the development of the diamond depositions in Western Yakutia. The exploitation of the Udachnaya kimberlite pipe was accompanied by the drainage waste utilization into the near-bottom permafrost part during 15 years. By the end of the year 2000 about 10^6 m^3 of brines have been dumped. Nowadays, a new burial area creation is being planned.

Presentation and discussion of results

To estimate the scale and the consequences of dumping of high-mineralized technogenic solutions into permafrost the physical-chemical simulation of the dilution-precipitation processes and the different water types mixing during the drainage waste burial was carried out. The simulation was conducted using software HydrGeo designed at the Tomsk Department of Institute of Geology of Petroleum and Gas SB RAS (Bukaty, 1997, 1999).

In the first stage, the interaction between drainage brines and enclosing sedimentary rocks was simulated: the high-mineralized water drained by the Udachnaya pipe quarry was brought into contact with the primary rock-forming minerals – calcite, dolomite and gypsum with average open porosity of 10 %. The second modeling stage included mixing of the drainage brines and saline ground ice.

The results of the analysis of the brine samples and ground ice melts represented in Table 1 are used as initial data.

Table 1: Initial solution composition

			Drainage brines	Ground ice	
				Sample 1	Sample 2
pH			5.85	7.2	6.8
Density, kg/m^3			1229	1000	1000
Components	g/L	K^+	13.09	0.0067	0.16
		Na^+	20.1	0.007	0.40
		Mg^{2+}	12.56	0.106	0.38
		Ca^{2+}	67.22	0.251	0.59
		Cl^-	199.59	0.443	2.48
		Br^-	3.96	0.006	0.04
		SO_4^{2-}	0.17	0.677	0.53
		HCO_3^-	0.36	0.098	0.09
	mg/L	Li^+	173.87	0.44	2.3
		Rb^+	17.36	0.049	0.17
		Sr^{2+}	1186.3	5	10.42
TDS, g/L			317.05	1.72	4.67

Before simulation, the initial data have been preliminarily recalculated to correspond to the principle of "electroneutralness" (to keep sum balance of cations and anions taking into account the measured pH and the equilibrium in carbon system - HCO_3^-, CO_3^{2-} и CO_{2free}) for T=22 °C and P=0.1 MPa (lab condition).

If there was supersaturation of initial solution relative to the above referenced minerals the procedure of its elimination was executed. For that purpose, a simulation was made to evaluate the interaction between solution and "rock-glass" in conditional rock containing 0.0 solid phases and having 100%

open porosity. In this case, nothing is dissolved however if the solution is supersaturated the sediment can precipitate from it.

After that another recalculation of the chemical composition of solution was made for the real PT conditions under which dilution-precipitation processes will occur (the layer conditions: T= -3 °C and P=2.5 MPa). Each solution was checked for the presence of supersaturation relative to the referenced minerals. Information about the real solution composition after the preliminary recalculation and the changing of PT conditions is represented in Table 2.

Interaction of drainage brines and enclosing sedimentary rocks. Solution of the problem of dumping drainage water and enclosing rocks equilibrium is based on the simulation of liquid and solid phase interaction. The drainage quarry water was used as liquid phase (see Table 1). The calculation results showed that the initial drainage water is unsaturated relative to the minerals – calcite, dolomite and gypsum. When waste water is pumped into the permafrost the interaction with enclosing rocks causes the waster water to become saturated with some minerals, however precipitation of some salts from liquid phase is not observed. It provides a reason for discussing the absence of chemical colmatage processes. Moreover under real conditions during the constant technogenic water dumping into the permafrost massif mainly consisted of carbonaceous rocks, leaching of calcite and dolomite and their transmission into the liquid phase would be observed in the near-bore space. At the same time the environment porosity is possible to increase and the collecting properties of enclosed rocks might improve.

Table 2: The real solution composition brought in equilibrium at the layer conditions (T= -3.0 °C, P =2.5 MPa)

Components	Drainage water, g/L	Ground ice, mg/L	
		Sample 1	Sample 2
H^+	2.5E-5	0.000835093	0.000216581
Na^+	21.03	153.51	412.6
K^+	13.09	6.67	156
Mg^{2+}	12.56	106.4	377
Ca^{2+}	67.22	250.5	591.2
Li^+	0.17	0.44	2.29
Rb^+	0.0174	0.049	0.171
Sr^{2+}	1.186	5	10.42
OH^-	2.3E-6	0.00306384	0.000125697
Cl^-	199.59	443.2	2482
Br^-	3.96	6.41	39.1
$(SO_4)^{2-}$	0.17	677.3	530
$(HCO_3)^-$	0.36	97.6	91.5
pH	5.85	7.2	6.8
Salinity, g/L	319.33	1.747	4.692
Density, kg/m^3	1229.7	1000	1000.3
Ionic strength, mol/L	0.659	0.0037	0.1195
CO_2, mol/L	0.0059	0.0016	0.0015

The mixing of drainage brines and saline ground ice. This simulation included the mixing of the drainage quarry water with the saline ground ice melts in the following proportion: 90:10, 80:20, 70:30, 60:40, 50:50, 40:60, 30:70, 20:80 and 10:90 %. The results of modeling are represented in Tables 3 and 4.
It's been established that all obtained mixtures have different levels of saturation of the examined minerals; moreover the unsaturation degree increases with growth of mixed water mineralization (the growth of the drainage water amount in the mixture). Under the layer conditions the interaction of dumping brines and saline ground ice will result in ice melting and solution dilution, but the modification of water chemical composition won't occur. The causes of precipitation and decreases in filtration properties of rock massif are also absent.

Conclusions

1. To estimate the potential burial projects and the application of this method of storage in permafrost regions the presence of optimum cryo-hydrogeological, structural-tectonic and hydrogeochemical conditions is necessary. The underground storages must be disposed in the continuous permafrost region, with considerable thickness of frozen rocks (more than 100-200 m) and with the bed position tens of meters lower than the nearest valleys bottoms. Cryogenic rock masses located on rugged relief areas that were drained during the freezing period have the best characteristics to be used as a burial reservoir. Tectonic crushing zones or an intensively fractured crystalline massif and coarsely grained (clastic) rocks with sufficient value of the useful capacity are good for storage of waste water.

2. The simulation results show that during the waste dumping into the permafrost interaction of technogenic water and enclosing rocks would lead to the physical-chemical equilibrium in the water-rock system and to the elimination of the solution unsaturation relative to the minerals of the enclosing rocks. Only the quantitative relations of the components in the solution will change. The leaching of calcite and dolomite and their transmission into the liquid phase would be observed in the near-bore space. Both pure drainage brines and their mixture with low mineralized ground ice melts (in any proportion) are unsaturated relatively to dolomite and calcite.

3. The interaction of the drainage brines dumping into permafrost and saline ground structure-forming ice under the layer conditions would occur only with ice melting and brine dilution without decreasing of filtration-capacity properties of the surrounding rock massif.

Acknowledgements

The study has been carried out with the financial support of Russian Fund for Basic Research (grant's number 01-05-64012). The authors appreciate all anonymous reviewers whose helpful work improved this paper.

References

Alexeev, S.V. (2000). *The cryogenesis of groundwaters and rocks (on an example of the Daldyn-Alakit region of Western Yakutia).* Publ. House of SB RAS, Novosibirsk (in Russian).
Bukaty, M.B. (1997). "The working out of the software in the petroleum-gas geology." *J. Razvedka i ohrana nedr.* N 2, 37-39 (in Russian).

Bukaty, M.B. (1999). *The advertise-technical description of the software HydroGeo.* Russian Scientific-Technical Informative Center, № 50980000051 PC, Moscow (in Russian).

Saveliev, B.A. (1971). *Physics, chemistry and structure of the ice cover of the sea and fresh basins.* Moscow State University Publ. House, Moscow (in Russian).

Table 3: Composition (g/L) of the solution, obtained at the interaction drainage quarry water (A) and melted ground ice (sample 1) (B)

Components	Proportion A:B								
	90:10	80:20	70:30	60:40	50:50	40:60	30:70	20:80	10:90
H^+	2.3E-6	3.2E-6	3.4E-6	3.2E-6	2.9E-6	2.4E-6	2E-6	1.7E-6	1.2E-6
Na^+	18.9	16.8	14.7	12.7	10.6	8.5	6.4	4.3	2.2
K^+	11.8	10.5	9.2	7.9	6.5	5.2	3.9	2.6	1.3
Mg^{2+}	11.3	10.1	8.8	7.6	6.3	5.1	3.8	2.6	1.4
Ca^{2+}	60.5	53.8	47.1	40.4	33.7	27.0	20.3	13.6	6.9
Li^+	0.157	0.139	0.122	0.105	0.087	0.07	0.053	0.035	0.018
Rb^+	0.0157	0.0139	0.0122	0.0105	0.0087	0.007	0.0053	0.0035	0.0018
Sr^{2+}	1.068	0.95	0.83	0.71	0.60	0.48	0.4	0.24	0.12
OH^-	1.3E-7	1.97E-7	2.9E-7	4.3E-7	6.3E-7	9.1E-7	1.3E-6	1.8E-6	2.5E-6
Cl^-	179.7	159.8	139.8	119.9	100.0	80.1	60.2	40.3	20.4
Br^-	3.56	3.17	2.77	2.38	1.98	1.59	1.19	0.8	0.4
$(SO_4)^{2-}$	0.22	0.27	0.32	0.37	0.42	0.47	0.53	0.58	0.63
$(HCO_3)^-$	0.33	0.31	0.28	0.26	0.23	0.20	0.177	0.15	0.12
H_2O	2048.23	3186.79	4325.34	5463.92	6602.46	7741.02	8879.6	10018.1	11156.7
pH	5.99	6.1	6.3	6.4	6.5	6.7	6.8	6.9	7.1
Salinity, g/L	287.6	255.8	224.1	192.3	160.5	128.8	97.0	65.3	33.5
Density, kg/m^3	1207.5	1185.2	1161.3	1138.0	1112.6	1090.1	1065.8	1042.5	1018.6
Ionic strength, mol/L	3.47266	1.98558	1.28138	0.870662	0.601597	0.411679	0.270465	0.161349	0.074504

Table 4: Composition (g/L) of the solution, obtained at the interaction drainage quarry water (A) and melted ground ice (sample 2) (B)

Components	Proportion A:B								
	90:10	80:20	70:30	60:40	50:50	40:60	30:70	20:80	10:90
H^+	5.0E-7	4.8E-7	4.6E-7	4.5E-7	4.3E-7	4.0E-7	3.8E-7	3.4E-7	2.9E-7
Na^+	18.9	16.9	14.8	12.77	10.7	8.6	6.6	4.5	2.5
K^+	11.8	10.5	9.2	7.9	6.6	5.3	4.0	2.7	1.4
Mg^{2+}	11.3	10.1	8.9	7.7	6.5	5.3	4.0	2.8	1.6
Ca^{2+}	60.6	53.9	47.2	40.6	33.9	27.2	20.6	13.9	7.3
Li^+	0.16	0.14	0.12	0.11	0.09	0.07	0.05	0.04	0.02
Rb^+	0.016	0.014	0.012	0.011	0.009	0.007	0.005	0.004	0.002
Sr^{2+}	1.068	0.951	0.833	0.716	0.598	0.481	0.363	0.246	0.128
OH^-	1.6E-7	2.2E-7	2.9E-7	3.9E-7	5.2E-7	6.8E-7	8.7E-7	1.1E-6	1.2E-6
Cl^-	179.9	160.2	140.5	120.7	101.0	81.3	61.6	41.9	22.2
Br^-	3.69	3.17	2.78	2.39	2.0	1.61	1.22	0.82	0.43
$(SO_4)^{2-}$	0.21	0.24	0.28	0.31	0.35	0.39	0.42	0.46	0.49
$(HCO_3)^-$	0.33	0.31	0.28	0.25	0.23	0.2	0.17	0.15	0.12
H_2O	918.2	926.9	935.5	944.0	952.7	961.3	969.8	978.4	987.0
pH	5.9	6.0	6.1	6.2	6.3	6.4	6.5	6.61	6.7
Salinity, g/L	287.9	256.4	224.9	193.5	162.0	130.5	99.1	67.6	36.2
Density, kg/m^3	1206.1	1183.3	1160.4	1137.5	1114.7	1091.8	1068.9	1046.0	1023.2
Ionic strength, mol/L	7.75412	6.84238	5.94819	5.07019	4.20754	3.36078	2.52899	1.71176	0.90861

Operation of Soil Vapor Extraction In Cold Climates

David L. Barnes[1], Elizabeth Cosden[2], Bryan Johnson[2], Karol Johnson[2], Stina Stjärnström[3], Karin Johansson[3], Dennis Filler[4]

Abstract

Soil Vapor Extraction has proven to be a viable method for reducing the mass of volatile compounds from soils that have been impacted by contamination. The theory, design and use of SVE systems have been fairly well documented, however the use of SVE systems in cold environments such as Alaska has not been adequately investigated to date. This paper examines the use of SVE systems in cold environments. The affect of low soil temperatures on volatile compound removal by SVE is discussed. Methods used to overcome the impact low temperatures have on the operation and effectiveness of SVE systems are examined in detail. Possible low cost strategies such as barometric pumping during winter months are presented as solutions to reducing SVE operating costs in cold regions.

Introduction

Over the past several decades significant effort and resources have been directed towards cleaning up soil and groundwater contaminated by releases of petroleum products and organic solvents. The volatile organic compounds (VOCs) associated with these releases include chlorinated solvents such as trichloroethylene (TCE) and tetrachloroethylene (TCA), and petroleum compounds such as benzene, ethylbenzene, toluene, and xylene. Once released to soil, VOCs can provide a continuing source of soil contamination by volatilization and diffusion, and of ground water contamination by infiltration and dissolution,

----

[1]Assistant Professor, University of Alaska Fairbanks, PO Box 755900, Fairbanks, AK 99775; phone 907-474-6126; ffdlb@uaf.edu [2] Fairbanks Environmental Services, [3] Luleå University of Technology, [4] Shannon and Wilson, Inc.

until the mass of the compounds in the soil has been depleted. Soil contaminated with VOCs can become a health and explosive hazard if the vapors migrate to buildings containing basements, and drinking water from an aquifer impacted by these contaminants poses a potential risk to human health and the environment.

In recent years soil vapor extraction (SVE) has become accepted technology for removing VOCs from unsaturated soils. SVE is an effective and cost efficient method for remediating *in situ* soil using blowers or vacuum pumps attached to extraction wells screened in unsaturated soil. The system may contain multiple wells, either horizontal or vertical, with each well connected in parallel to the vacuum/blower system. The SVE system extracts soil-gas contained in the contaminated soil, which is subsequently treated above ground to remove the VOCs before being discharged to the atmosphere. SVE systems also enhance desorption of contaminants from the soil particle surface, increase volatilization of nonaqueous phase liquids (NAPL), and volatilize dissolved groundwater contaminants. While the operation of SVE treatment systems in temperate climates is well understood, the operation in cold climates presents circumstances that require further investigation.

This paper explores the affects cold air and soil temperatures have on the operation of SVE systems, and presents possible solutions for optimizing system operations. The influence of temperature variation on mass removal rates created by SVE systems is examined. Next, the experience gained from operating SVE systems in cold climates is discussed. Finally, possible strategies for optimizing SVE system operation in cold regions are presented.

Influence of Temperature on VOC Mass Removal Rate

The physical and chemical properties of contaminants influence the fate and transport of VOCs in the subsurface by influencing the distribution of VOCs among the four phases they exist in soil (vapor phase, dissolved phase, adsorbed phase, and liquid). Liquid phase volatile compounds are considered immiscible compounds or NAPL even though they are slightly soluble in water. Their density in relationship to water further categorizes NAPL into light nonaqueous phase liquids, often referred to as LNAPLs, (immiscible compounds with densities less than water) and dense nonaqueous phase liquids, or DNAPLs (immiscible compounds with densities greater than water). Contaminants released into the subsurface can migrate through the unsaturated soil and collect in the vadose zone as residual NAPL, retained by capillary forces between solid particles. A fraction of the NAPL will partition into existing soil-water, adsorb onto the surface of soil particles, and volatilize into pore space. Light nonaqueous phase liquids that reach the ground water table will accumulate in the nearly water saturated zone above the water table.

Conversely, DNAPLs will sink below the water table and collect in bedrock depressions or other impermeable surfaces.

Removal of VOCs from subsurface unsaturated soils by SVE is accomplished by advection and diffusion. The rate of removal by these transport mechanisms is controlled by the physical-chemical properties of the compound, mainly vapor pressure, solubility, air-water partitioning coefficient, and the water-solid partitioning coefficient. Temperature variations have the greatest influence on vapor pressures and air-water partitioning coefficients (represented by Henry's Law constant). Montgomery (2000) presents vapor pressures and Henry's Law constants for common VOCs.

A logical question one might ask is: what affect does low soil temperatures have on the ability to remove VOCs from cold soil versus more temperate soils? Calculating the time required to reach different removal fractions as a function of soil temperature will provide insight into answering this question.

To make these calculations, an analytical model is used to calculate the time required to reach these different removal percentages (McWhorter, 1993). The purpose of presenting this model is to illustrate the influence soil temperature will have on VOC mass removal rate. The model uses the method of images (McWhorter and Sunada, 1977) to derive a function for a family of curves that describe the flow patterns to the extraction well created by a pressure induced at the well. The curves are called streamlines, and the function is called a stream function. The equation for the stream function (ψ) corresponding to a pressure induced by a horizontal SVE well is as follows:

$$\psi(x,y) = \frac{Q_w}{2\pi}\left[\tan^{-1}\left(\frac{x}{a-y}\right) - \tan^{-1}\left(\frac{x}{a+y}\right)\right] \quad (3)$$

where Q_w is the flow rate at the well, a is the depth from ground surface to the horizontal well, and x and y are spatial coordinates with the origin at the ground surface (horizontal axis) and the centerline of the well (vertical axis). For this analysis, the depth to ground water is sufficiently deep to have negligible impact on soil-gas flow.

The time required for VOCs to travel along the flow paths defined by the stream function to the extraction well is simply the volume between two adjacent streamlines (called a streamtube) divided by the flow rate through the streamtube. Basic geometry is used to determine the streamtube volume and the flow rate is the same through each streamtube (total flow rate divided by the number of streamtubes). The number of streamtubes is arbitrarily selected.

A mass balance analysis is used to determine the concentration of VOCs in the soil-gas at the well.

To evaluate the impact soil temperature has on the removal of VOCs consider soil contaminated with benzene. For this example, benzene is dissolved in the soil-water, sorbed to the soil particles, and volatilized into the soil-gas; an assumption is made that no liquid phase benzene is contained in the soil pores. In reality, most soils contaminated by an immiscible fluid will contain the liquid phase contaminant (NAPL) in the pore space at some saturation greater than zero. However, the goal of this example is to simply illustrate the effect of soil temperature on mass removal rate. Thus, the assumption is adequate for purposes of this example.

A horizontal well is installed and screened at 5 m below the ground surface. A 10-m^3/min flow rate is established at the well using a vacuum blower. Soil porosity is 0.35, and the water-solid partitioning coefficient is 0.83 L/kg. The soil is homogeneous and isotropic. A mass balance is used to calculate concentration in the evacuated soil-gas as a function of time.

The removal of benzene was evaluated at two different soil temperatures (10 °C and 5 °C) using the method described above. Figure 1 shows the resulting log scale curve for concentration (C/C_o, where C_o is the initial concentration) versus time for the two soil temperatures. The curves shown in Figure 2 illustrate the typical trend of decreasing VOC removal rate with time. At a soil temperature of 10 °C, the time required to remove 52% of the benzene is approximately 7 days. Reducing the soil temperature to 5°C results in an increase in the time to remove 52% by approximately 4 days. To remove 92% of the benzene from the contaminated soil at a soil temperature of 10 °C requires approximately 551 days, compared to 832 days when the soil temperature is 5 °C.

This example illustrates the influence soil temperature will have on mass removal rate created by SVE systems. In the early stages of VOC removal by SVE the greatest influence on mass removal rates will be the flow rate created by the reduction of pressure at the extraction well. In the later stages of removal when the mass removal rate decreases (as shown in Figure 1 for both temperatures), the flow rate will have less of an influence and the impact of soil temperature will be greater in comparison to early in the life of a SVE system.

Consider the effect seasonal soil temperature variations have on the rate of contaminant removal from impacted soils. Subsurface soil temperature lags above ground air temperatures by several weeks in Interior Alaska. As the freezing front moves down through the vadose zone, heat is liberated as soil water freezes. Through winter, subsurface cooling continues to a depth at which the geothermal gradient governs temperatures. Consequently, the largest impact of temperature on contaminant removal by SVE will be in near surface soils, where soil temperature has the greatest response to air temperature changes. Conversely, removal of VOCs from soil located at depths influenced little by air temperature fluctuations will be less variable. Furthermore, active SVE in winter has the potential to channel heat from the ground to the surface, thereby accelerating subsurface freezing.

In concert with potential reductions in contaminant removal rates due to fluctuating soil temperatures is the operational challenge of keeping near surface and above ground SVE piping free of ice during winter operation. Experience has been gained from operating SVE systems for the last decade in Interior Alaska. The practical side of operating these systems in cold regions is discussed in the following section.

Cold Regions Operation and Maintenance

Many system modifications are required to continually operate a SVE system in extremely cold climates. Additionally, air sparging (AS) systems that are often installed together with SVE systems to enhance contaminant removal can also experience difficulties in cold climates. AS systems inject clean air into the saturated subsurface (below the ground water table) to treat contaminated soil and groundwater by enhancing volatilization and microbial degradation. The volatilized compounds are then removed from the unsaturated soil above the ground water table by soil vapor extraction. Components of both the AS and the SVE system must be protected from freezing during extremely cold temperatures. The following sections present several methods for continuing treatment system operation under freezing conditions.

Above Surface Material

Above ground system components that must be protected from the elements include blowers, knockout drums, electrical equipment, associated above ground piping, and remote monitoring enclosures (RME). Typically blowers and knockout drums are housed inside an enclosure, a connex for example, which must be insulated and equipped with a heater to maintain a minimum constant temperature. Electrical equipment should also be housed inside the enclosure when space is available.

Remote monitoring enclosures are above ground enclosures used to control and direct airflow from the individual SVE probes to the main lines. During the winter months, the heat from the air sparge manifold will provide warmth to the RME; however, additional modifications must be made for the RME to withstand extremely cold temperatures. An effective method of protecting the RME is to install insulation prior to the onset of winter. Insulation is placed inside the enclosure walls and ceiling. The cost of this process is primarily associated with labor hours, typically two days per year, and the insulation. While effective, insulation alone may not completely protect the SVE lines inside the RME from freezing. Above ground piping as it enters the connex may be heat traced and surrounded with sprayed urethane foam insulation or, if an AS system is operating with the SVE system, then backflushing can be conducted to dislodge ice from the lines. Backflushing with air sparge is discussed below.

Below Surface Material

Below ground system components requiring protection from the elements during year-round operation include SVE main air distribution lines running from the vapor extraction blower to the RME, SVE air distribution lines running from the RME to the individual SVE probes, and the SVE probes. Each of these components is susceptible to freezing problems. Water occasionally enters the air distribution lines via condensation and/or is drawn into the SVE probes when groundwater levels are unnaturally high. The water subsequently freezes causing the air distribution line to become fully or partially blocked, restricting airflow. Not only does this cause system operational problems due to increased pressure on the blowers, it also inhibits effective remediation. Two technologies are available to mitigate this problem: (1) backflushing the system as necessary throughout the winter and (2) heat tracing and insulating the piping.

Backflushing is a technique used to remove ice blockage from the SVE lines and probes. Backflushing can only be used on remediation systems that combine both AS and SVE. During the backflushing process, the SVE system is temporarily shut down, cross connections between AS lines and SVE lines are opened (Figure 2, photographs A and B), and warm air from the AS blower is forced into the SVE lines. This process can be conducted as often as necessary, whenever SVE flow is reduced and blower pressure is increased, typically once every one to two weeks. Backflushing may be conducted for anywhere from a couple of hours to an entire day, depending on the severity of the air stream blockage. The costs associated with this process are primarily the labor hours required to manually put the system into backflush mode, approximately four hours per event.

The second method used to facilitate year-round operation of SVE involves the application of heat trace and sprayed urethane foam insulation to above ground and below ground air distribution lines (Figure 2, photograph C and drawing D). Heat trace consists of an electrical wire set that is applied to the outside of the air distribution line during system installation. When supplied with an electrical current, the wire set generates the heat that is used to increase thermal conduction. The sprayed foam insulation increases the efficiency of the heat trace by reducing the amount of electricity necessary to generate the heat. Backflushing, using AS/SVE cross connections, may also be incorporated into this type of system to allow a fast economical method for clearing the horizontal lines of obstructions such as ice and or water.

This method has a higher installation and operation cost; but theoretically, has a lower maintenance cost. The largest portion of the cost is purchasing the heat trace and sprayed foam insulation. Typical costs for heat trace and sprayed foam insulation are approximately $12.00 per linear foot (this includes 3-inch thick insulation around the circumference of the pipe).

Labor costs are approximately 40 hours for 500 linear feet of air distribution line, which includes taping the heat trace to the air distribution lines, preparation time to properly bed the SVE line and electrical connections, and testing the line. If remote power is needed, installation of conduit, junction boxes and electrical wire is also required. Operation costs for heat trace is approximately 5 watts of electrical energy per linear foot of heat trace. Every 200 feet of heat trace utilizes one kilowatt of electrical energy at an approximate cost of $0.11 per kilowatt-hour.

In comparison to the previous method of backflushing, this method is most cost effective in situations when the contractor performing operation and maintenance of the remediation system does not reside in the city where the remediation system is located. The reduced need for human intervention to perform maintenance decreases labor hours. By not relying on backflushing, treatment system operation doesn't experience the time lost due to SVE system shutdown for applying the cross connection process.

System Optimization

It has been shown that 1) low soil temperatures reduce the rate of contaminant removal because of decreasing contaminant vapor pressure and air-water partitioning coefficient and 2) additional expenses, above typical operation costs, are typically incurred when operating SVE systems in cold climates. Therefore, can SVE systems be optimized to operate more efficiently and cost effectively in cold regions? Past experience illustrates that complications of operating SVE systems in areas that experience cold climates and freezing soil temperatures are mainly due to freezing condensation in air distribution lines, above ground equipment failures due to freezing, and possible reduced removal capacity due to low soil temperature. Optimizing SVE systems to increase operation efficiency and reduce operation costs relies on a basic understanding of SVE system operation in cold climate and must take into consideration the degree of remediation that has been completed.

Figure 3 shows a typical relationship between air temperature (above ground surface) and soil temperature (below ground surface) for a location in

Interior Alaska. These curves illustrate that there is a lag time between the point in time when the lowest (and highest) air temperature is reached and when the lowest (and highest) soil temperature is reached and that this lag time is a function of depth. Two time periods in Figure 3 are notable: (1) June when the average temperature of the soil has reached a minimum, but the average air temperature has increased from the seasonal low, and (2) January when the average soil temperature is still relatively warm, but the air temperature has reached an average annual low.

Concentrating first on the later condition (relatively high soil temperatures and low annual air temperatures), throughout this period one will experience problems with the equipment near and above the ground surface such as, previous discussed, freezing pipes and high costs of keeping the system in operation. During this same period of time, the rate of VOC removal should still be relatively favorable owing to the relatively warm soil temperatures. As operation of the SVE system progresses, the average air temperature increases, however the corresponding average soil temperature is still low (period 1 above). The reader should note that in the previous simplified example a 5°C decrease in soil temperature resulted in a 51% increase in the time required to remove 92% of the VOC from the soil by SVE. Thus, instead of having problems in this period of operation with the near surface and above ground surface equipment one might now get lower VOC removal rates.

According to the above analysis one can draw the conclusion that the VOC removal rates will still be favorable during periods of low air temperatures, but because of higher operational cost it might be expensive to remove the contaminants. During periods of increasing air temperatures the conditions have changed, but the affect on the cost of system operation may be similar. Due to low temperatures in the soil the VOC removal rate decrease but warmer air temperatures reduces operational expenses. Under these conditions, a high cost of VOC removal relative to the costs during periods when both the soil temperature and air temperature are relatively high (typically mid to late summer) will still be encountered.

To optimize the operation of SVE systems with respect to cost, the stage of VOC removal and the annual temperature changes should be taken into consideration. In the early stage of operation (refer to Figure 1), when removal rates are relatively high, it is most likely more efficient to let the system run regardless of season and soil temperature. As the rate of mass removal decreases (portion of the curves shown in Figure 1 corresponding to large time), operation of SVE systems throughout the year may become inefficient. As discussed previously, during periods of annually cold air temperatures the soil temperatures are still relatively favorable for VOC removal, however an analysis of the cost of operating a SVE system in comparison to the return in VOC removed may show that it is more economical to shut the system off during this period and reestablish operation when the air temperatures conditions are more favorable. Considering the lag time between annually low air temperature and soil temperature, questions can still be raised as to the appropriate time to reestablish SVE operation. From the analysis above, the rate of VOC removal may still be hindered due to low soil temperatures when the air temperatures are favorable for SVE operation. Solutions to the reduction in VOC removal rates due to relatively cold soil temperatures may be as simple as placing a layer of insulating material over the ground surface to slow the progression of the cold front through the subsurface.

Shutting down a SVE system during periods of cold air temperatures does not necessarily mean that the movement of the VOCs through the pore space has stopped. In heterogeneous unsaturated soil, the dominant transport mechanism of VOCs from areas of low soil permeability to areas of high soil permeability is diffusion. This makes intuitive sense since the airflow created by a SVE system will flow through areas of relatively high permeability leaving diffusion as the dominant transport mechanism in the relatively low permeability regions.

The rate of VOC diffusion from areas of relatively low permeability to areas of relatively high permeability is controlled by the magnitude of the difference in VOC concentration between the two areas. During periods of SVE operation, this concentration gradient is kept high resulting from the constant removal of VOC from the pore space in high permeability regions by the induced airflow. Once the SVE system is shut off halting the airflow, diffusion of VOC from low permeability regions to high permeability regions will still occur – albeit at a steadily decreasing rate due to the ever decreasing concentration gradient. Once the operation of the SVE system is reestablished during periods of favorable air temperature, the contaminant that has diffused to the relatively high permeable soils will be subsequently removed by the flow of air created by the SVE system. For this result to be effective, operation of the SVE system has to have progressed to the point where a majority of the VOC has been removed from the regions with relatively high soil permeability to allow for diffusive transport. In a heterogeneous soil, this point is most likely achieved when there is a minimal change over time in the VOC concentration measured in the off-gas collected at the extraction well. When conditions in the soil favor initial high rates of diffusion, shutting off the SVE system during periods of low air temperature seems to be an effective option to reduce the operational costs while still making some progress towards removing the contaminant from the subsurface.

Another option during periods of unfavorable air temperatures is to switch from actively evacuating air by mechanically placing a vacuum on the extraction wells to allowing atmospheric barometric pressure changes create the pressure differential between the atmosphere and the subsurface soil. This passive method of soil vapor extraction is known as barometric pumping. Given the reduction in airflow rate under a barometric pumping scheme in relationship to a traditional active SVE system, the rate of condensed water accumulation and subsequent ice production in exposed piping during periods of cold air temperatures will be reduced. Under this operation scenario, removal of VOC from subsurface soil will still progress, but the operational costs will be reduced.

Summary

This manuscript has presented issues related to the operation of SVE systems in cold regions. Operation of these systems are challenged by maintenance issues such as protecting system components from the effects of freezing and from issues associated with reduced VOC removal rates due to annually low soil temperatures. Possible low cost strategies have been presented as solutions to reducing operating costs for SVE system in cold regions. By understanding the fundamentals of SVE systems and the associated costs of operating these systems, operational adjustments can be made to maximize the rate of VOC removal while minimizing the cost.

Acknowledgements

This study was funded by the Alaska Science and Technology Foundation (ASTF) and we would like to thank ASTF Grants Administrator James Palin and Technology Administrator Dr. Robert Chaney for their continuing support.

References

Alberty, R.A. and R.J. Silbey. (1992) *Physical Chemistry,* John Wiley & Sons, New York.

McWhorter, D.B. 1993. Personal Communication.

McWhorter, D.B. and D.K. Sunada. (1977) *Ground-Water Hydrology and Hydraulics,* Water Resources Publications, Fort Collins, Co.

Montgomery, J.H. (2000) *Groundwater Chemicals Desk Reference,* Lewis Publishers, Boca Raton.

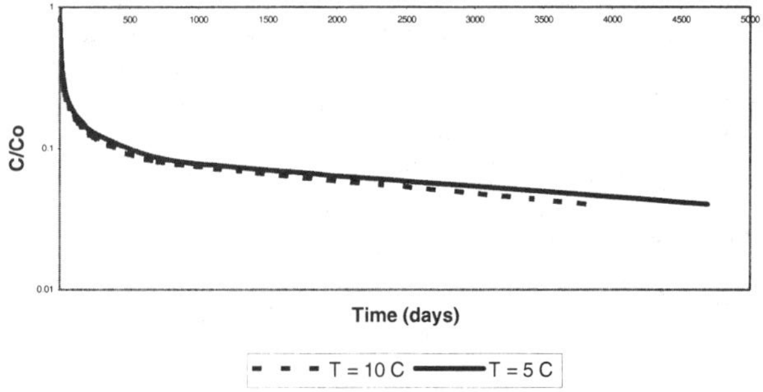

Figures 1. Benzene concentration versus time calculated at the extraction well for two similar SVE systems, one operating at a soil temperature of 10°C and the other operating at a soil temperature of 5°C.

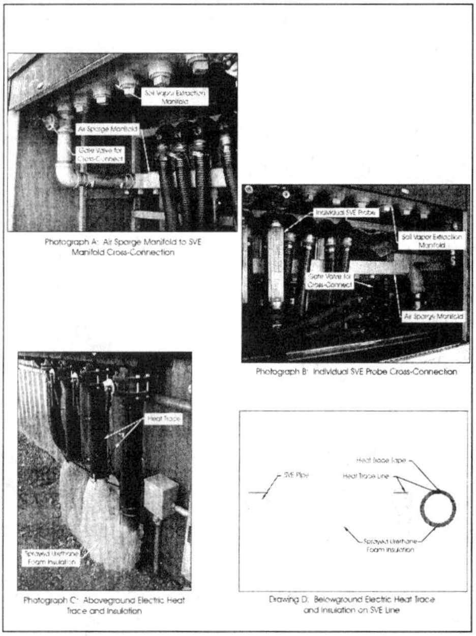

Figure 2. Examples of typical cross-connections and heat trace for SVE systems operating in cold climates.

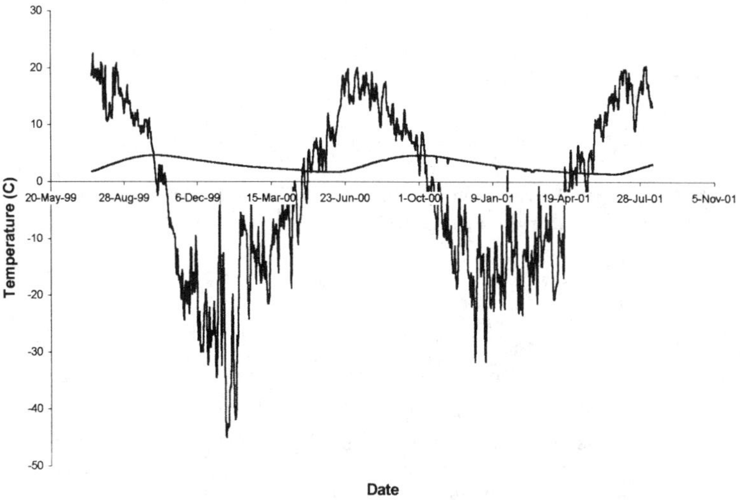

Figure 3. Air temperature (above ground surface) and soil temperature (below ground surface).

Anaerobic Biodegradation Of Benzene Under Conditions Favorable To Iron-Reduction Under Environmental Conditions Of Sub-Arctic Groundwaters

Nada I. Raad[1]

Abstract

This research was conducted to detect and understand anaerobic degradation of benzene in a sub-arctic aquifer environment. The major point in this paper is that indigenous microorganisms from anaerobic groundwater wells contaminated with BTEX (benzene, toluene, ethylbenzenes, xylenes) can use benzene as a carbon or energy source under conditions favoring iron reduction. Microcosms were prepared and incubated under anaerobic conditions, with hematite (Fe_2O_3) added to some to permit growth of iron reducing bacteria. Several concentrations of benzene were used as a substrate, and the microcosms were incubated at temperatures of 21 °C and 4 °C.

Results indicated that the aromatic hydrocarbon, benzene, was anaerobically transformed under both psychrophilic (4 °C) and mesophilic (21 °C) conditions. The temperature and substrate concentration played a role in determining the rate at which benzene was biodegraded under conditions favorable to iron-reducing bacteria. For example, under conditions favoring iron-reducing bacteria, benzene biodegraded faster at 21 °C than at 4 °C in most microcosms during the first two months. Benzene degradation slowed when the concentration of benzene became less than 7 ppb under all conditions. Phenol and propionic acid were detected as intermediates for this transformation.

Problem and Background

The soil and groundwater in Operable Unit 5 (OU5) at Fort Wainwright, Alaska, have become contaminated with petroleum hydrocarbons as a result of leakage from underground tanks and accidental spills. This site is located in the continental climate zone of interior Alaska and has an average annual ground temperature of 4 °C and, hence, offered the opportunity to collect bacteria living in this cold anaerobic environment and study BTEX degradation *in vitro*. For the

[1] Nada I. Raad, University of Alaska Fairbanks, Civil and Environmental Department, 306 Tanana Drive, Fairbanks, Alaska 99775; phone 907-474-1933; ftnir@uaf.edu

purpose of this research, aquifer materials were obtained from the former area, well FWM 6894.

Indigenous microorganisms in BTEX-contaminated groundwater can use the contaminants as a source of carbon and energy. When aerobic bacteria use the contaminants in this way, they consume dissolved oxygen as a terminal electron acceptor, which often leads to anaerobic conditions. Anaerobic bacteria require alternative electron acceptors such as sulfate (SO_4^{-2}) and iron (Fe^{+3}). Study of the degradation of benzene under anaerobic and low temperature conditions is needed to better evaluate the potential contribution of anaerobic bioremediation for contaminated site restoration.

Zobell and Agosti (1972) and Kerry (1990) found that psychrophilic microorganisms have the ability to degrade a variety of contaminants aerobically. The research of Edwards and Gribic`-Galic` (1992, 1994), Evans et al. (1991), and Haag et al. (1991) indicate that bioremediation of groundwater contaminated with aromatic hydrocarbons, including benzene, is possible using mesophilic anaerobic bacteria. Alternate electron acceptors, such as nitrate (Kuhn et al., 1985), ferric iron (Lovely and Lonergan, 1990), and sulfate (Edward et al., 1992; Reinhard et al., 1997) support microbial degradation of aromatic hydrocarbons. However, anaerobic biodegradation of aromatic hydrocarbons at temperatures below 10 ^{o}C, in the presence of electron acceptors such as Fe^{+3}, has not been studied in detail.

Anaerobic microcosm experiments conducted under field conditions at several temperatures greater than 9 °C indicated the potential degradation of benzene (Baedecker et al., 1993). Low temperature does not necessarily prevent the anaerobic degradation of aromatic hydrocarbons in general, but anaerobic benzene biodegradation at temperatures less than 9 °C has not been observed. In addition, evaluation of anaerobic degradation of benzene at a variety of different sites is essential to understand variations in the activities of bacteria because of differing environmental conditions, soil types and climate. Previous studies did not associate the disappearance of benzene with factors other than biodegradation, such as adsorption, diffusion, evaporation, and chemical reactions.

Studies done in the past did not systematically investigate a variety of electron acceptors and the degradation of benzene at both high and low temperature under the same conditions. Among the environmental factors that could affect benzene biodegradation under anaerobic conditions are temperature, nutrient concentrations, and substrate concentration. Better understanding of the range of environmental conditions conducive to the biodegradation of benzene *in vitro* could help to predict the fate of hydrocarbon contamination in soil and groundwater in cold regions.

Objectives

This research was conducted *in vitro* to detect and understand the anaerobic degradation of benzene in a sub-arctic aquifer environment. Benzene was used as a carbon and energy source. Experiments were conducted under conditions favoring iron reduction with incubation at temperatures of 21 °C and 4 °C. Understanding the transformation of benzene and other compounds could help engineers predict the fate of contaminants in soil and groundwater in cold regions.

Research Approach and Methods

The concentration of benzene, reduction of terminal electron acceptors, and production of intermediates are among the factors that were measured *in vitro* to detect the transformation and mineralization of benzene. A sterile, disposable Teflon bailer and autoclaved serum bottles were used to collect groundwater samples. The serum bottles were kept in a bag-type glove chamber (Thomas Scientific, Swedesboro, New Jersey) that was purged continuously with nitrogen. The serum bottles were closed tightly using butyl rubber stoppers crimped with aluminum seals (Bellco Glass Inc., Vineland, New Jersey). The samples were immediately transported in a cooler to the Water and Environmental Research Center and stored at 4 °C until the following day.

Laboratory studies were conducted in the dark under anoxic conditions favoring iron- and sulfate reduction and methanogensis conditions. For the purpose of this paper, benzene degradation under iron-reducing bacteria will be elaborated. Microcosms were prepared as described in microcosms and experimental design section and in Figure 1.

Modified EPA method 602- purge and trap capillary gas chromatography (5890 Hewlett- Packard Company, Palo Alto, California; Federal Register, 1984) was used to analyze for BTEX in all the samples. Analysis was done using another HP 6890 gas chromatograph interfaced with a HP 5973 mass spectrometer. A modified standard method-Fe phenanthroline assasy was used to measure ferrous and total iron concentration.

Since proper preservation of culture was important, the culture was maintained by regular subculture on nutrient media and by drying and deep freezing, following the methods described by Gerhardt et al. (1994) and Edwards and Gribic`-Galic`(1994).

Microcosms and Experimental Design

Enrichments of microorganisms were prepared the day following sample collection were prepared. The medium (Lovely 1986a) was used to support the growth of iron-reducing bacteria. All serum bottles and rubber stoppers were autoclaved before they were used. Aquifer samples were transferred to the set of 12, 160-ml serum bottles containing medium. Each experimental serum bottle received a 50-ml aquifer sample and 50 ml of the prepared medium. Four control serum bottles were aseptically prepared, and 50 ml of the medium alone was added to each. All bottles were closed tightly using butyl rubber stoppers and crimped with aluminum seals. A set of six experimental and two control serum bottles were incubated at room temperature (21 °C) and another set of six experimental bottles and two control serum bottles were incubated at 4 °C upside down in the dark. The active microcosms were fed with approximately 2 µl of benzene (99.9% pure) every month for 10 months as recommended by Edwards and Gribic`-Galic` (1994). After periods of 7 months, 8 months, and 9 months, quantitative measurements were done to confirm the degradation of benzene and to check the production ferrous ions in each set.

After an acclimation period of one year, modified medium was used to support the growth of iron-reducing bacteria in separate incubators (Raad, 2001). A summary of the experimental setup is given in Figure 1. The initial concentrations in the experimental samples were 10, 50, and 200 ppb and in the controls, 50 ppb. The bottles were shaken twice daily. Benzene, ferrous and ferric iron, and protein concentrations were analyzed using triplicate samples from every group, over a period of six months (Hutchins et al., 1991, Lovely and Phillips, 1994; Lovely et al., 1994, and Anderson et al., 1998). The triplicate samples were sacrificed at each sample point.

All preparations were done in a bag-type portable glove chamber that had been purged overnight with N_2/CO_2 (80%/20%) in a cold room at 4 °C.

Results

In the experiment, iron was the electron acceptor and benzene served as the substrate. The degradation in the concentration of benzene under iron-reducing conditions over a period of six months is summarized in Figure 2 and Table 1.

The concentration of benzene in the headspace of each bottle was negligible as calculated by using Henry's law (Mackay and Shiu, 1981: Bamford et al., 1998). At the end of the six month period, there was no significant loss of benzene to volatization or other non-biological processes (Raad, 2001).

Benzene loss was 100 % within 6 months at 10 and 50 ppb benzene, and approximately 98 % at 200 ppb, at both 21 °C and 4 °C. During the first two months, the concentration of benzene dropped rapidly, but the decrease in concentration slowed after the second month, when the remaining benzene concentration was less than 7 ppb. At 4 °C, ferrous iron was detected during the second month in the bottles that had been amended with 200 ppb benzene. Also, ferrous iron was detected during the third and fourth months in the bottles that had been amended with 10 and 50 ppb of benzene at 4 °C.

Initially, the rate appeared faster at 21 °C than 4 °C, in some of the bottles with initial benzene concentrations of 50 and 200 ppb (Figures 2). After two months (when the concentration of benzene was low) the loss of benzene was slower at 21 °C than 4 °C, for initial benzene concentrations of 50 ppb and 200 ppb (Figure 2), but temperature had no effect at the other 10 ppb initial concentration. Dissolved ferrous iron was detected in the experimental active bottles but not in the controls. Neither the chemically killed nor the autoclaved controls showed any degradation of benzene (Raad, 2001).

The sulfide added to the medium apparently precipitated with ferrous ions as ferrous sulfide. The black precipitate was greater at higher initial benzene. The occurrence of the black precipitate seemed to occur earlier at 21 °C than at 4 °C. The black precipitate was not observed in the control bottles.

The GC/MS was used to identify intermediates in active serum bottles. Under iron-reducing conditions, both phenol and propionic acid were found in samples that contained detectable ferrous iron (Raad, 2001).

Benzene Biodegradation over 30 Days

In most of the experiments, the extent of benzene biodegradation was not measured until after one month. In Figure 3, biodegradation of 25 ppb of benzene over a period of 35 days is shown, for iron-reducing conditions at 4 °C. Benzene was completely degraded over a period of 35 days, with a lag of 15 days (see Figure 3).

Discussion

In this work, it was shown that benzene could be degraded at 21 °C and at 4 °C under strictly anaerobic conditions in the presence of ferric iron. Highly crystalline hematite (Fe_2O_3) added to the medium for iron-reducing bacteria had low solubility. The rate of biological ferric iron reduction tends to decrease with an increasing degree of crystallinity. This indicates that the organisms must come into direct contact with Fe (III) in order to reduce it to Fe (II) (Munch and Ottow, 1983). Measuring the dissolved Fe (II) was not sufficient to measure the rate of Fe (III) reduction, since substantial Fe (II) precipitated in solid form as FeS.

Substantial benzene removal by non-biological processes can be ruled out. At the end of the six month period, there was only about 5% loss of benzene in the autoclaved or poisoned control bottles.

The transformation pathway of benzene apparently includes ring oxidation. In this research, GC/MS was used to detect, qualitatively, transient intermediates. A trace amount of phenol and propionic acid was detected in the biologically active benzene microcosms that was not detected in the controls.

The temperature and substrate concentration played a role in determining if benzene was biodegraded under conditions favorable to iron-reducing bacteria (Figure 2). Ferrous iron was detected in the first month at 21 °C. At 4 °C, ferrous iron was detected during the second month in the bottles that had been amended with 200 ppb benzene. Also, ferrous iron was detected during the third and fourth months in the bottles that had been amended with 10 and 50 ppb of benzene at 4 °C. Since most of the benzene degradation occurred in the first two months, the ferrous iron detected during the first 2 months was the result of benzene degradation and that measured during the remaining 4 months was mainly due to degradation of the other intermediate compounds. In the case of iron-reducing bacteria, benzene degraded completely under most conditions.

Summary and Conclusions

The anaerobic microbial populations utilized in this study degraded benzene under conditions favorable to iron reduction at both low and moderate temperatures. Benzene degradation was nearly complete in the first month; this is very rapid compared to earlier observations, probably because of the lengthy acclimation period. Phenol was detected as an intermediate.

The *in vitro* research conducted on the anaerobic biodegradation of benzene provided new insight into the anaerobic degradation processes that occur at cold temperatures associated with sub-arctic environments. The results of the laboratory experiments reported here are specific preparing microcosms from well FMW 6894, Fort Wainwright, Alaska. These findings, however, will help to explain the biodegradation of benzene under iron- reducing conditions in similar field situations. Additional research is needed to calculate the rate of degradation of benzene by itself and in the presence of other contaminants.

Acknowledgment

The author would like to thank Water and Environmental research Center at University of Alaska Fairbanks for providing the equipment to do the research.

References

Anderson, R.T., Rooney-Varga, J.N., Gaw, C. V., and Lovely, D. R. (1998). "Anaerobic benzene oxidation in the Fe(III) reduction zone of petroleum-contaminated aquifers." *Envir. Sci. Technol.*, 32, 1222-1229.

Baedecker, M. J., Cozzarelli, I.M., and Eganhouse, R.P. (1993). "Crude oil in a shallow sand and gravel aquifer-III. Biogeochemical reactions and mass balance modeling in anoxic groundwater." *Appl. Geochem.*, 8, 569-586.

Bamford, H.A., Baker, J.E., and Poster, D.L. (1998). "Review of methods and measurements of selected hydrophopic organic contaminant aqueous solubilities, vapor pressure, and air-water partition coefficients." *United states Department of Commerce Technology Administration; National Institute of Standards and Technology. NIST Special publication*, 928, 19-29.

Edwards, E.A. and Gribic`-Galic`, D. (1994). "Anerobic degradation of toluene and o-xylene by a methanogenic consortium." *Appl. Environ. Microbiol.*, 60, 313-322.

Edwards, E.A., and Gribic`-Galic`, D. (1992). "Complete mineralization of benzene by aquifer microorganisms under strictly anaerobic conditions." *Appl. Environ. Microbiol.*, 58, 2663-2666.

Edwards, E. A., Wills, L. E., Reinhard, M., and Gribic`-Galic`, D. (1992). "Anaerobic degradation of toluene and xylene by aquifer microorganisms under sulfate-reducing conditions." *Appl. Environ. Microbiol.*, 58, 794-800.

Evans, P. J., Mang, D. T, and Young, L.Y. (1991). "Degradation of toluene and m-xylene and transformation of o-xylene by denitrifying enrichment culture." *Appl. Environ. Microbiol.*, 57, 450-454.

Federal Register (1984). *Rules and Regulations*, U.S. Government Printing Office, Washington, D.C., 49 (209), 43272-43280.

Gerhardt, P., Murray, R. G. E., Wood, W. A., and Krieg, N. R. (eds.). (1994). "Methods for general and molecular bacteriology." *American Society for Microbiology*, Washington, D.C., 21-281.

Haag, F., Reinhard, M., and MaCarty, P.L. (1991). "Degradation of toluene and p-xylene in anaerobic microcosms: evidence for sulfate as a terminal electron acceptor." *Environ. Toxi. Chem.*,10,1379-1389.

Hutchins, S.R. Sewell, G.W., Kovacs, D.A, and Smith, G.A. (1991). "Biodegradation of aromatics hydrocarbons by aquifer microorganisms under denitrifying conditions." *Environ. Sci. Technol.*, 25, 68-76.

Kerry, E. (1990). "Microorganisms colonizing plants and soil subjected to different degrees of human activity, including petroleum contamination in the Vesfold Hills and MacRobertson Land, Antarctica." *Polar Biology*, 10, 423-430.

Kuhn, E. P., Collberg, P. J. , Schnoor, J. L., Wanner, O. , Zehnder, J.B. and Schwarzenbach., R. P. (1985). "Microbial transformation of substituted benzene during infiltration of river water to groundwater: Laboratory column studies." *Environ. Sci. Tech.*, 19, 961-968.

Lovely, D.D., and Philipis, E. J.P. (1994). "Novel Processes for anaerobic sulfate production from elemental sulfur by sulfate-reducing bacteria." *Appl. Environ. Microbiol.,* 60, 2394-2399.

Lovely, D.D., Woodward, J.C., and Chapelle, F.H. (1994). "Stimulated anoxic biodegradation of aromatic hydrocarbons using Fe(III) ligands." *Nature*, 370, 128-131.

Lovely, D. R., and Lonergan, D.J. (1990). "Anaerobic oxidation of toluene, phenol, and p-cresol by dissimilatory iron-reducing organism, GS-15." *Appl. Environ. Microbiol.*, 56, 1858-1864.

Lovely, D.R. (1986a). "Organic matter mineralization with the reduction of ferric iron in anaerobic sediments." *Appl. Environ. Microbiol.*, 51, 683-689.

MacKay, D., and Shiu, W.Y. (1981. "A critical review of Henry's Law constants for chemicals of environmental interest." *J. Phys. Chem. Ref. Data*, 10, 1175-1199.

Munch, J.C., and Ottow, J.C.G. (1983). "Reductive transformation mechanism of ferric oxides in hydromorphic soils." *Ecol. Bull (Stockholm)*, 35, 383-394.

Raad, N. I. (2001). "Anaerobic biodegradation of benzene under environmental conditions of sub-arctic groundwater." *Thesis*.

Reinhard, M., S. Shang, P.K. Kitanidis, E. Orwin, G. Hopkins, D. and Lebron, C.A. (1997). "In situ BTEX biotransformation under enhanced nitrate- and sulfate-reducing conditions." *Environ. Sci. Technol.*, 31, 28-36.

Zobell, C. E., and Agosti, J. (1972). "Bacterial oxidation of mineral oil at sub-zero Celsius." *Abstracts, 72nd Annual Meeting of the American Society for Microbiology.*

Table 1: Cumulative degradation (%) over a period of six months under conditions favorable to iron-reducing bacteria. The concentrations of benzene are the mean of the three replicates measured every month.

Iron-Reducing Bacteria			Iron-Reducing Bacteria		
Benzene 10 ppb at 21 °C			Benzene 10 ppb at 4 °C		
Month	Benzene (ppb)	Σ Degradation %	Month	Benzene (ppb)	Σ Degradation %
0	10	0	0	10	0
1	6.8	32	1	7.1	29
2	3.2	68	2	5.2	46
3	3.5	65	3	3.7	63
4	3.5	65	4	3.7	63
5	3.9	61	5	5.5	45
6	0	100	6	0	100
Iron-Reducing Bacteria			Iron-Reducing Bacteria		
Benzene 50 ppb at 21 °C			Benzene 50 ppb at 4 °C		
Month	Benzene (ppb)	Σ Degradation %	Month	Benzene (ppb)	Σ Degradation %
0	50	0	0	50	0
1	3.9	92	1	33.2	33
2	6.1	88	2	4.4	91
3	5.4	89	3	3.2	93
4	6.0	88	4	4.1	92
5	4.3	91	5	0	100
6	0	100	6	0	100
Iron-Reducing Bacteria			Iron-Reducing Bacteria		
Benzene 200 ppb at 21 °C			Benzene 200 ppb at 4 °C		
Month	Benzene (ppb)	Σ Degradation %	Month	Benzene (ppb)	Σ Degradation %
0	200	0	0	200	0
1	3.5	98	1	26	87
2	18.4	91	2	6.3	97
3	4.3	98	3	9.0	95
4	12.6	94	4	4.1	98
5	6.5	97	5	8.3	96
6	3.9	98	6	5.4	98

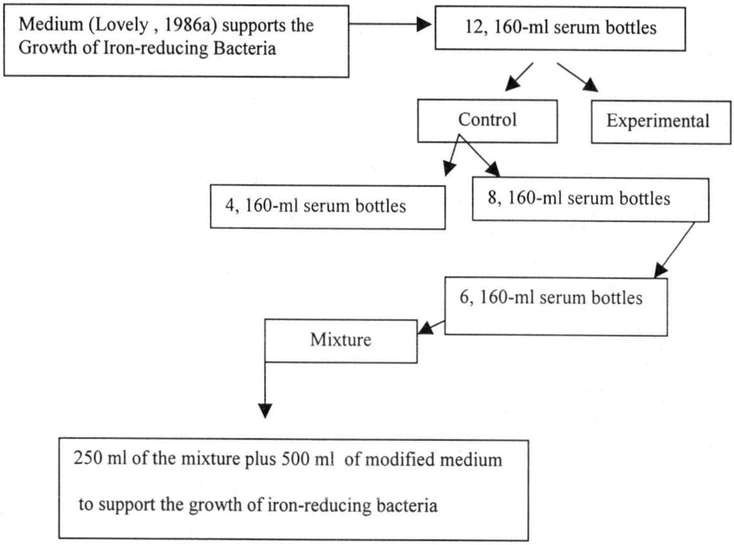

A set of 180, 30-ml serum bottles were prepared and incubated as follows:

- 18, 30-ml serum bottles; 30 ml fresh medium/each+ 0.5mg/250 ml mercuric chloride/each bottle + 50 ppb benzene/each incubated at 21 °C
- 18, 30-ml serum bottles; 30 ml fresh medium/each + 0.5mg/250 ml mercuric chloride/each bottle + 50 ppb benzene/each incubated at 4 °C
- 18, 30-ml serum bottles; 30 ml fresh medium/each + autoclaved/each + 50 ppb benzene/each incubated at 21 °C
- 18, 30-ml serum bottles; 30 ml fresh medium/each + autoclaved/each + 50 ppb benzene/each incubated at 4 °C
- 18, 30-ml serum bottles; 30 ml fresh medium/each + 10 ppb benzene/each incubated at 21 °C
- 18, 30-ml serum bottles; 30 ml fresh medium/each + 10 ppb benzene/each incubated at 4 °C
- 18, 30-ml serum bottles; 30 ml fresh medium/each + 50 ppb benzene/each incubated at 21 °C
- 18, 30-ml serum bottles; 30 ml fresh medium/each + 50 ppb benzene/each incubated at 4 °C
- 18, 30-ml serum bottles; 30 ml fresh medium/each + 200 ppb benzene/each incubated at 21 °C
- 18, 30-ml serum bottles; 30 ml fresh medium/each + 200 ppb benzene/each incubated at 4 °C

Figure 1: Experiment setup.

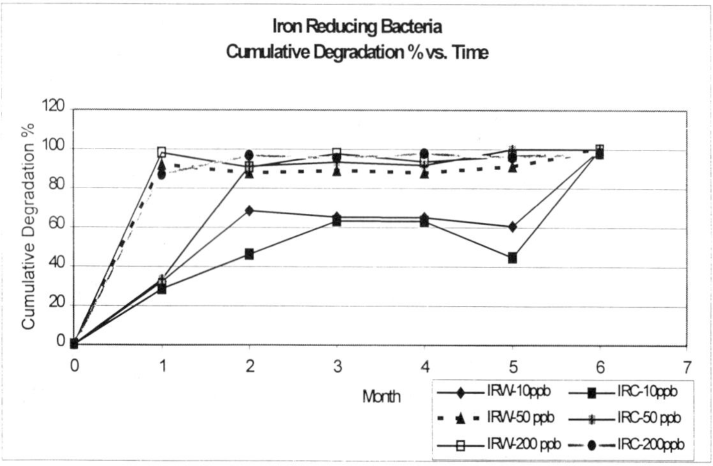

Figure 2: Cumulative degradation (%) over a period of six months under conditions favorable to iron-reducing bacteria.

The percentages of benzene are the mean of the three replicates measured every month.

IRW-10 ppb = Iron-reducing conditions, amended with 10 ppb benzene at 21 °C
IRC-10 ppb = Iron-reducing conditions, amended with 10 ppb benzene at 4 °C
IRW-50 ppb = Iron-reducing conditions, amended with 50 ppb benzene at 21 °C
IRC-50 ppb = Iron-reducing conditions, amended with 50 ppb benzene at 4 °C
IRW-200 ppb = Iron-reducing conditions, amended with 200 ppb benzene at 21 °C
IRC-200 ppb = Iron-reducing conditions, amended with 200 ppb benzene at 4 °C

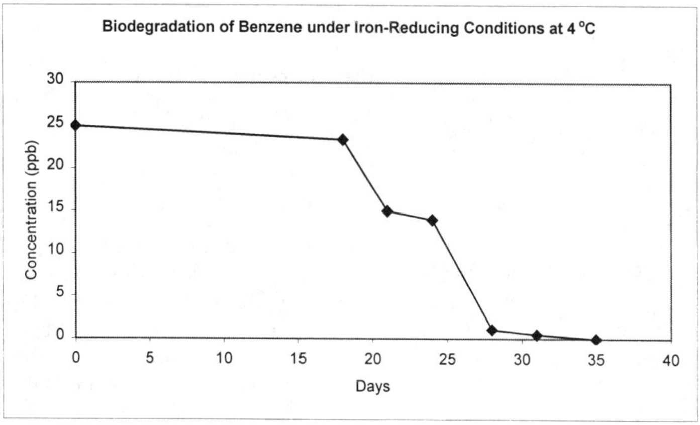

Figure 3: Biodegradation of 25 ppb of benzene at 4 °C in a period of 35 days.

Dynamics of a Fuel Release in Frozen Alluvium

Dennis M. Filler[1], David L. Barnes[2]

Abstract

A release incident introduced 2,380 L (630 gal) of diesel fuel into frozen alluvial soils within the floodplain of the Tanana River, in interior Alaska. The winter release occurred in a residential community, with potential to impact drinking water wells. Mitigation of the problem was confounded by snow cover, proximity to a site structure, and shallow groundwater. Understanding spill dynamics in the subsurface is important toward spill response, assessing impact, and minimizing property devaluation.

Two questions needed answering for this investigation:

1) Was 2,380 L (630 gal) of diesel fuel really released?
2) What is the magnitude of groundwater impact?

The spill zone is documented, with focus on the partially frozen active layer. Two approaches to answering the investigative questions are discussed, with emphasis on a mass-balance evaluation of the unsaturated zone. Soil parameters are evaluated for impact analysis of the vadose zone. Soil microstructure is briefly discussed relative to translocation and alteration by hydrocarbon contamination. Soil hydraulic properties are briefly discussed relative to gauging groundwater impact. Knowledge of contaminants in freezing ground in furthered.

Introduction

The focus of responders to fuel spills on land is generally cleanup and corrective action. However, haphazard spill response often occurs; a protocol for terrestrial spills in Alaska was implemented only this year. Depending on location, proximity to water (surface water bodies and groundwater), and ecological sensitivity, spill response techniques used have included dike and boom containment, soil washing/flushing, excavation and removal, and both controlled and uncontrolled burns (Alaska Clean Seas, 1999). There is difficulty in applying an all-encompassing spill-response manual because of Alaska's diverse geography and climate. Tundra and other northern terrain sometimes require innovative spill response and cleanup techniques.

[1] Project Manager/Sr. Engineer, Shannon & Wilson, Inc., 2055 Hill Road, Fairbanks, Alaska 99709; Affiliate Faculty, Dept. of Civil & Environmental Engineering, University of Alaska Fairbanks, P.O. Box 755900, Fairbanks, Alaska 99775-5900; (907) 479-0600; dmf@shanwil.com.

[2] Assistant Professor, Dept. of Civil & Environmental Engineering, University of Alaska Fairbanks, P.O. Box 755900, Fairbanks, Alaska 99775-5900; (907) 474-6126; ffdlb@uaf.edu.

One shortcoming of cold-regions spill response is understanding, and perhaps acknowledgement, of the importance of spill dynamics. This paper considers fundamentals of subsurface contaminant migration and mass-balance analysis in evaluating diesel impact to the groundwater table at a site in interior Alaska. The evaluation is used to answer the relevant questions and address residual contamination in the subsurface.

Background

The residential spill site is located approximately 50 miles southeast of Fairbanks (Figure 1), within the floodplain of the Tanana River, in Salcha, Alaska. A remnant slough lies approximately 46m (150ft) east of the residence spill. Arctic grade heating oil (#2 diesel) from two aboveground storage tanks was discharged to the ground over two days in early December 2000, as apparent acts of vandalism. Based on fill records and the volume of fuel removed from the tanks before decommissioning, approximately 2,380 L (630 gal) was released. Tank 1 was emptied through the in-line filter petcock and Tank 2 discharged via a tap located on the fuel supply line. The ground below the tanks was snow covered; spill patterns were not visibly discernable at ground level. At Tank 1, a dog hole 20cm (8in) is diameter was observed in the ground, approximately 0.3m (1ft) southwest of the tank outfall. The ground surface by Tank 1 sloped away from the house. The ground surface under Tank 2 sloped towards the house. Groundwater was approximately 2.5m (8.2 ft) below the ground surface. Cleanup began within 48 hours of the releases, and was completed eighteen days later. Sub-freezing air temperatures prevailed throughout the incident.

Site Characterization

A backhoe equipped with a frost-bucket was used to break frozen ground and excavate contaminated soil. Field screening with a photoionization detector (PID) was used throughout the soil cleanup effort to segregate soil and guide sampling and analysis.

Two excavations were dug relative to the individual tank releases. The excavations were separated by a thin bench of uncontaminated soil approximately 1m (3ft) across (Figure 2). At Tank 1, removal of contaminated soil resulted in an excavation with maximum dimensions 9.5m (31ft) long, 4.6m (15ft) wide, and 3.4m (11ft) deep. The Tank 2 excavation measured approximately 4.9m (16ft) long, 2.75m (9 ft) wide, and varied between 0.3 and 2.9m (1 and 9.5 ft) deep. Excavation depths varied depending upon levels of soil contamination, as depicted in Figure 2. Maximum excavation depths reflect soil removal to accommodate recovery well installations.

A crescent-shaped excavation resulted from the Tank 1 spill, wrapping around a northern expanse of the residence (Figure 2). The Tank 2 excavation undercut the northeast corner of the house; two hydraulic jacks were installed to support foundation timbers in place of the corner post. The corner post was a 15cm (6in) square freestanding railroad tie, embedded 1.2m (4 ft) deep.

The backhoe was used to place a thirty centimeter (12in) diameter, corrugated metal recovery well south of Tank 1; another was installed southwest of Tank 2 (RW1 and RW2, respectively). Each 3.6m (12ft) long recovery well was torch-slotted around the bottom half, and embedded approximately 1m (3.2 ft) below the groundwater surface to accommodate product recovery and monitoring. A 5cm (2in) diameter, PVC monitoring well (MW1 in Figure 1) was installed to 4.6m (15ft) depth, downgradient of the excavations for monitoring purpose.

Soil (confirmation samples inclusive) and groundwater samples were collected from the excavations for analyses of diesel range organics (DRO) and benzene, toluene, ethylbenzene, and xylenes (BTEX) according to Alaska Method AK 102 and EPA Method 8021B, respectively. The analytical results and applicable cleanup levels are summarized in Tables 1 and 2; select results are shown in Figure 2. Bulk soil samples were collected from strata to classify soils and determine soil parameters.

Table 1. Summary of Soil Analytical Results (in mg/Kg soil)

Sample No.	Depth m (ft)	DRO	Benzene	Toluene	Ethylbenzene	Total Xylenes
Tank1 Excavation						
1	1.4(4.5)	23000	-	-	-	-
2	1.1(3.5)	1470	0.135	2.79	7.16	32.7
3	1.5(5)	82.5	<0.02	0.305	1.06	4.88
4	1.8(5.8)	36.6	<0.02	<0.07	<0.07	<0.07
Tank2 Excavation						
5	0.8(2.6)	33100	0.276	7.97	16.5	70.6
6	Stockpile	7280	0.229	7.01	14.8	63.7
Cleanup Level		200	0.02	5.4	5.5	78

Table 2. Summary of Groundwater Analytical Results[1] (in mg/L water)

Location	DRO	Benzene	Toluene	Ethylbenzene	Total Xylenes
Tank 1 Excav.	15.5	0.018	0.265	0.272	1.61
Tank 2 Excav.	2310	0.425	2.29	0.894	4.73
Cleanup Level	1.5	0.005	1.0	0.7	10

[1] Initial assessment results included here to substantiate that groundwater had been impacted.

Vadose zone soils were consistent for both releases; a composite of silts, over poorly-graded sand, above a mixture of gravel sands (Figure 3) to the water table. The composite silt layer comprised 0.3m (1ft) of sandy-silt, above an organic lense underlain by a thin (<2.5cm or 1in), broken silty-clay lense, with uniform silt below. Spill volumes and soil parameters for the three layers are summarized in Table 3.

Table 3. Soil Parameters and Release Quantities

Parameters		Values
Mass of contaminated Soil (approximate)	Tank 1 Excavation	150 tons
	Tank 2 Excavation	70 tons
Volume of fuel released (estimated)	Tank 1	1,135 L (300 gals)
	Tank 2	1,250 L (330 gals)
Soil Layer 1 (frozen)	Thickness	0.76m (2.5ft)
	Classification	SM and ML
	Dry unit weight	10.2 kN/m^3 (65 pcf)
	Specific gravity (solids)	2.65
	Moisture content	21.0%
	Hydraulic conductivity	10^{-5} to 10^{-3} cm/s* ($3.3x10^{-6}$ to $3.3x10^{-4}$ ft/s)
Soil Layer 2	Thickness	0.55m (1.8ft)
	Classification	SP-SM and SP
	Dry unit weight	12.2 kN/m^3 (77.5 pcf)
	Specific gravity (solids)	2.68
	Moisture content	4.6 – 5.3%
	Hydraulic conductivity	10^{-4} to 10^{-2} cm/s* ($3.3x10^{-5}$ to $3.3x10^{-3}$ ft/s)
Soil Layer 3 (to water table)	Thickness	1.2m (3.9ft)
	Classification	GP
	Dry unit weight	18.7 kN/m^3 (119 pcf)
	Specific gravity (solids)	2.72
	Moisture content	1.2 – 1.3%
	Hydraulic conductivity	10^{-1} to 10^{2} cm/s* ($3.3x10^{-2}$ to 3.3 ft/s)

*From Bouwer (1978). SM is sand-silt mixtures; ML is inorganic silt; SP-SM is poorly-graded sand with silt; SP is poorly-graded sand; GP is poorly-graded gravel with sand.

In Table 3, soil classifications were made using the Unified Soil Classification System (American Society for Testing and Materials (ASTM) method D-2488 & 2487-90). Soil moisture content and specific gravity (and soil density) are defined by ASTM test methods D-2216 and D-854, respectively. Grain-size distributions (by ASTM D-421) were also performed on the individual layers to substantiate soil classifications in applying hydraulic conductivity ranges.

Spill Evaluations and Conclusions

A spill event can be examined through a hydrodynamic evaluation or mass-balance analysis of the unsaturated zone in answering the investigative questions posed. A thorough hydrodynamic evaluation requires determination or soil hydraulic properties (frozen and unfrozen moisture contents, hydraulic conductivities, and soil-water potentials), examination of soil macro- and microstructure, and fundamentals of flow of immiscible fluids through porous media. A basic hydrologic evaluation can be performed through measuring key soil parameters, *in situ* hydraulic conductivity, and/or *in situ* soil-water potentials. However, funding for hydrodynamic or hydrologic evaluation is not often attached to spill cleanup and recovery.

With little additional cost to spill response, a mass-balance analysis can be performed with knowledge of specific soil properties from the spill zone. For the Salcha incident, spill response and cleanup were completed without undergoing freeze-thaw cycling. The short-term event was thought to produce minimal alteration of soil structure, precluding microscopic evaluation.

Considering the Tank 1 release, observations around the excavation walls suggested that fuel was largely retained within the upper frozen silt stratum. Fuel infiltrated the uppermost sandy-silt portion, and the intermittent, thin organic lense, before residing as interbedded lenses of fuel-laden silt at the bottom of the frozen layer (Figure 3). A mass-balance analysis is applied to the frozen silt layer to answer the first investigative question.

In performing a volumetric calculation based on soil dry density (γ_d) and Table 3 soil parameters for Layer 1, soil porosity (n_1) is calculated as

$$n_1 = 1 - \frac{\gamma_d}{G_s \gamma_w} = 0.607 \qquad [4]$$

where G_s is specific gravity of soil solids and γ_ω the unit weight of water. The weighted-average dry density ($\gamma_{d(avg)}$) for the Tank 1 excavation is based on the individual layer dry densities and their proportional thickness with respect to the average excavation depth. The average excavation depth, taken directly from measurements shown in Figure 2, was 2.04m (6.7ft). $\gamma_{d(avg)}$ is calculated as 17.9 kN/m^3 (114.3 pcf). The total volume of soil excavated (V_T), based on this weighted-average dry density and the total mass of contaminated soil removed (150 tons) is 74.3 m^3 (97.2 cy). The volume of soil excavated from Layer 1 (V_1) is proportional to the total excavated volume as the ratio of Layer 1 thickness to that of the average excavation depth, or

$$V_1 = \frac{0.76}{2.04} V_T = 27.7m^3 (36.2cy) \qquad [5]$$

The total volume of voids in Layer 1 is

$$V_{v1} = V_1 n_1 = 16.8m^3 (22cy) \qquad [6]$$

of which 21 percent was water-filled. Therefore, the available pore space for diesel fuel in Layer 1 was 13.3 m^3 (17.4 cy). This does not imply that the entire available pore volume will be completely filled with fuel after release; the volume is merely *available* for filling. Since less than 1.6 m^3 (2 cy) of space is required to hold 1,135 L

(300 gal) of fuel, the available pore space in Layer 1 was more than sufficient to contain the fuel released.

Similarly for the Tank 2 excavation, the total excavated soil volume was 34.6 m^3 (45.3 cy), of which 10.5 m^3 (13.8 cy) was from the frozen silt layer. At 21 percent moisture content, the volume of voids available for fuel was 5 m^3 (6.6 cy), again sufficient to hold the estimated 1,250 L (330 gal) released.

The previous analyses consider the total soil moisture content as unfrozen. However, much of the soil-water was likely frozen at the time of the releases. Nevertheless, when one considers the degree of soil saturation (36 percent for Layer 1) and allowing for volume expansion, the silt layer had the capacity to hold the volumes of fuel released.

Fuel deposition within the frozen silt layer as depicted in Figure 3 is a curious phenomenon. Liquid flow through soil will not be uniform, tending to concentrate along preferential flow paths. These paths (e.g., cracks, fissures, and macro-channels - paths between aggregate soil) are often found in fine-grained soil above coarser materials (Bowman et al., 1987 and Jaynes et al., 1988). As water-diesel solution traveled down through the silt layer, micro-spatial variability and variability in moisture content and consolidation may have been responsible for the horizontal "lensing effect" realized. In winter, as the freeze-front moves downward soil moisture is wicked toward the progressing front. Preferential pressure(s) as a function of degree of soil saturation likely contributed to development of the phenomenon. The degree of saturation (and consolidation) in the lower realm of the frozen silt layer in December 2000 may have been less than that of upper regions. Interestingly, this lensing effect occurred below soil mounding and drainage slope that restricted overland flow of fuel at this location. Surface conditions can be deceptive with regard to subsurface transport of contaminants.

In considering groundwater impact (the second investigative question), Table 2 analytical results clearly indicate that fuel reached the water table from both releases, but at significantly different magnitudes. The difference between the levels of contamination reaching the groundwater table from each tank is explained by the different pathways fuel traveled.

At Tank 1, topography and the degree of water saturation in the top silt layer allowed for broad surface spreading of diesel; rapid infiltration likely occurred only at the ground hole beneath the tank. The analytical data collected in the immediate vicinity of this tank (Table 1, samples 1 and 2 for soil, and Table 2 groundwater sample) support this assertion. Once in the subsurface, fuel adsorbed to soil particles, dissolved into soil water, and volatilized into soil gas. Owing to the heterogeneous nature of relative permeability of the medium to free-phase product, the migration rate of product will be highly spatially variable. At the time of the release, high moisture content and percent frozen moisture in the silt layer was likely significant enough to favor lateral spread over vertical migration. Figure 3 illustrates the retention capacity of the silt layer at different levels.

In certain areas of the impacted soil horizon, dissolved phase hydrocarbons likely traveled at or greater than pore velocity (channeling) early on. In general, as water (or water-hydrocarbon) content increases in soil, soil-water potential decreases and unsaturated hydraulic conductivity increases (Mualem, 1976; Case et al., 1983, and Panian, 1987). This helps explain why fuel was not found concentrated at the top of the silt layer. Cold temperatures act to reduce soil hydraulic conductivity in moist soils. Constantz (1982) demonstrated that a temperature drop from 25°C to 2°C (77°F to 36°F) can cause unsaturated hydraulic conductivity in fine-grained soils to decrease by one order of magnitude. By the time fuel reached the lower realm of the silt layer, the driving hydraulic head had been significantly reduced. Previously we hypothesized why fuel resided in soil lenses at the bottom of the frozen silt layer.

The majority of fuel released from Tank 2 pooled around the foundation post at the northeast corner of the house. Installation of this post created a pathway for vertical migration along the post. Without recompaction, reduced consolidation in soil can be expected a short distance away from the post in all directions. Hydraulic conductivity would be higher in this porous zone, relative to that of the normally consolidated silt surrounding it. The resulting cone of influence formed by infiltrating fuel is depicted in Figure 4. Because the post was embedded to the base of the silt layer, there was little soil structure to impede rapid infiltration into the underlying drier, granular soils. The magnitude of DRO in groundwater below the foundation post was nearly 150 times the concentration detected in groundwater below Tank 1.

Acknowledgements

Shannon & Wilson is grateful to the Alaska Department of Environmental Conservation for their support and recognition of the scientific value in investigating the impact of contaminants in freezing ground.

References

Alaska Clean Seas. 1999. "Alaska Clean Seas Technical Manual, Volume 1: Tactics Descriptions." Developed for Alyeska Pipeline Service Company, Exxon Company, BP Exploration, and ARCO Alaska, Inc.

Bouwer, H. 1978. *Groundwater Hydrology.* McGraw-Hill Book Co., New York, New York.

Bowman, R.S., H. Bouwer, and R.C. Rice. 1987. The role of preferential flow phenomenona in unsaturated transport, *Proceedings of Specialty Conference on Environmental Engineering,* ASCE New York: 477-482.

Case, C., M. Kautsky, P. Pearl, R. Goldfarb, S. Leatham, and L. Metcalf. 1983. Unsaturated flow properties data catalog, Publication 45033, Water Resources

Center, Desert Research Institute, University of Nevada.

Constantz, J. 1982. Temperature dependence of unsaturated hydraulic conductivity of two soils, *Soil Science Society of America Journal* 46, no.3: 466-470.

Jaynes, D.B., R.S. Bowman, and R.C. Rice. 1988. Transport of a conservative tracer in the field under continuous flood irrigation, *Soil Science Society of America Journal* 52:618-624.

Mualem, Y. 1976. Catalogue of hydraulic properties of unsaturated soils, Technion, Haifa, Israel.

Panian, T.F. 1987. Unsaturated flow properties data catalog, Volume 2, Water Resources Center, Desert Research Institute, University of Nevada, Publication 45061, NTIS, U.S. Dept. of Commerce, Springfield, VA.

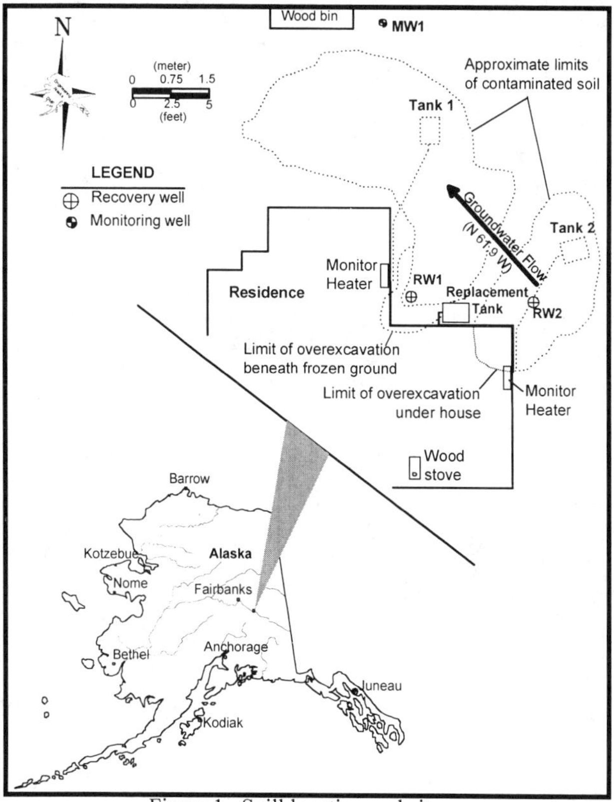

Figure 1. Spill location and site map.

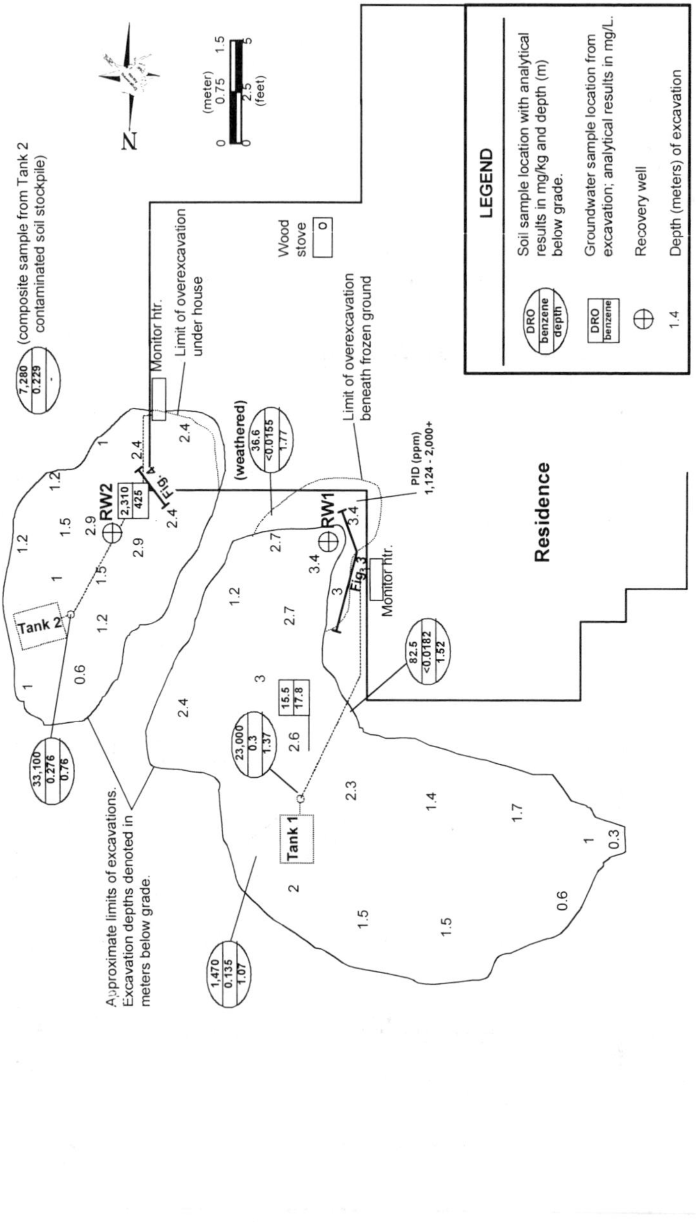

Figure 2. Excavations and soil and groundwater contamination levels.

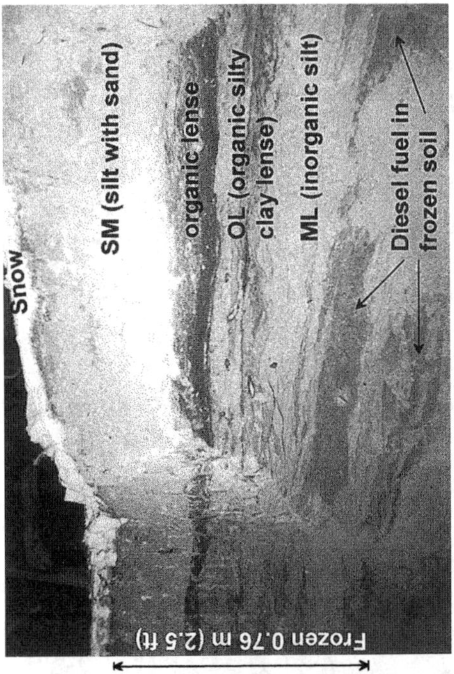

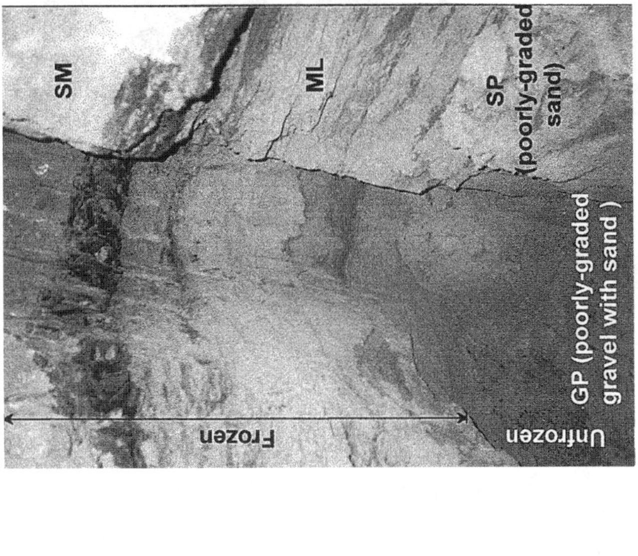

Figure 3. Tank 1 release: frozen silt laden with "fuel lenses" (top) adjacent diesel-contaminated soil column to ground-water (left).

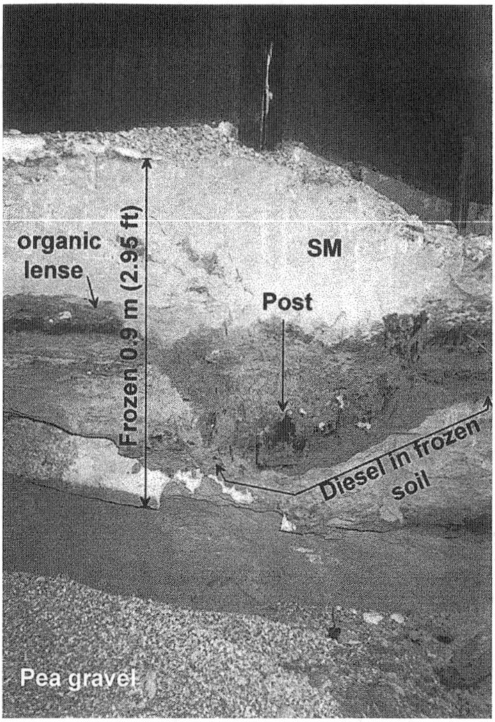

Figure 4. Diesel fuel infiltration at foundation post (pea gravel at bottom is backfill material).

Investigations into disinfection by-product formation and control for small drinking water systems in Northern Regions

Daniel M. White[1], D. Sarah Garland[2], Craig R. Woolard[3], Jasprit Narr[2]

Abstract

Disinfection by-products (DBP) are formed by the reaction between dissolved organic carbon (DOC) in water and a disinfectant, usually chlorine. Raw water samples from 17 small water systems in Alaska were used to study disinfection by-product formation. UV absorbance at 254 nanometers (UV 254) served as an excellent predictor of DBP formation potential (DBPFP) for surface waters. UV-254 was combined with other standard methods such as color and DOC to determine the nature of organic matter contributing to DBPs. No other methods showed the strong correlations between UV-254 and DBPFP. DOC, on the other hand, showed only a fair correlation to DBPFP. For each of the 17 Alaska waters studied, the total trihalomethane formation potential (TTHMFP) and haloacetic acid formation potential (HAA5FP) ranged from 37-1200 ug/L. For each source, the TTHMFP and HAA5FP were essentially equal.

Introduction

Due to the technical and geographical constraints of many of Alaska's small drinking water systems, meeting EPA rules under the Safe Drinking Water Act is challenging. The Alaska Small Systems Training and Technical Assistance Center (ATTAC) was established by the EPA in 1998 to address some of the challenges facing Alaska's small systems. ATTAC was charged with providing training and technical assistance to village operators as well as conducting technology demonstration and research to address the unique problems associated with small remote drinking water systems. In an effort to address problems salient to these small systems, a study was initiated to investigate the

---

[1] Corresponding author. Department of Civil and Environmental Engineering, University of Alaska Fairbanks, Fairbanks, AK 99775-5900, Phone: (907) 474-6222, Fax: (907) 474-6087, Email: ffdmw@uaf.edu

[2] Water and Environmental Research Center, University of Alaska Fairbanks, Fairbanks, AK 99775-5861 (At time work was conducted)

[3] School of Engineering, University of Alaska Anchorage, Anchorage, AK 99503.

organic matter ubiquitous in northern surface waters and its relation to disinfection by-product (DBP) formation.

Materials and Methods

Sample collection and processing

A geographically diverse group of community water systems with highly organic drinking water supplies were invited to participate in the study. Several other systems were included on their own request. A total of 15 systems using surface waters and 2 systems using groundwater agreed to participate. Figure 1 shows the approximate locations of the 17 systems. Table 1 lists the raw water source, seasonal availability of raw water, and the population served. The category "river" includes streams of all sizes; "lake" includes all surface impoundments.

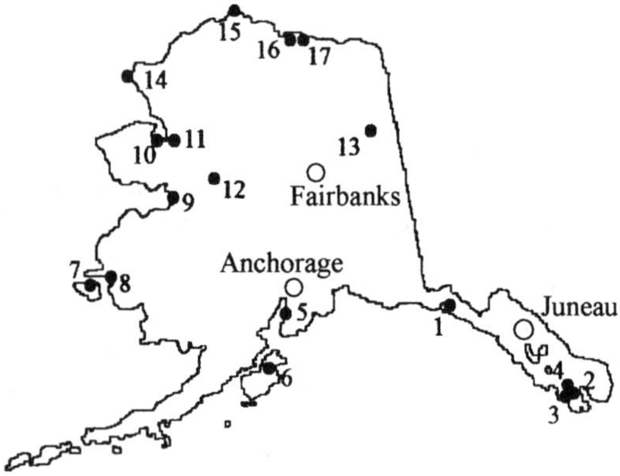

Figure 1. Location of water systems participating in study (figure modified from Narr et al. 2001).

Water sample collection was described previously in Narr et al., (2001). In summary, samples were collected in 1-liter amber glass bottles by the water treatment plant operators, packed in coolers with icepacks and shipped by air to the University of Alaska Fairbanks. All samples were stored in amber bottles at 4°C until further processing.

Sample Analysis

Analytical procedures for color, pH, DOC, UV-254, and disinfection by-product formation potential (DBPFP) were described in Narr et al., (2001) and are summarized below. Apparent and true color were determined for unfiltered and filtered aliquots using an Orbeco-Hellige 975MP Water Analysis System according to (Clesceri et al. 1998). A VWR Scientific model 8015 pH meter was used to measure pH. All color and pH tests were performed on samples at 22 to 25 °C. Sample pH ranged from 6.8 to 8.5 and was not adjusted before testing for color.

Table 1. Description of water systems in Fig. 1 (modified from Narr et al., 2001).

#	Water source		Season	Pop. served
1	Surface	River	all year	summer 1500, winter 150
2	Surface	River	all year	50
3	Surface	River	all year	150
4	Surface	Lake	all year	600
5	Groundwater	280 ft.	all year	5000-7000
6	Surface	Lake	all year	250
7	Surface	River	July--Oct.	150
8	Surface	Lake	June--Sept.	300
9	Surface	River	all year	1000
10	Surface	River	July--Oct.	150
11	Surface	River	all year	500
12	Groundwater	262 ft.	all year	580
13	Surface	River	all year	90
14	Surface	Lake	June--Sept.	750
15	Surface	Lake	all year	3500
16	Surface	Lake	July--Sept	500-600
17	Surface	River	all year	1000

Dissolved organic carbon (DOC) was determined using a Shimadzu 5000 TOC autosampler by the combustion-infrared method, as described in the *Standard Methods 3510 B*. DOC was only performed on 11 of the 15 surface waters. In all cases where DOC results are reported, data were derived from the 11 sample set. The ultra-violet (UV) absorbance at 254 nanometers (nm) was measured using a high-precision Beckman spectrophotometer (DU 520). A quartz cell with a path length of 1 cm was used to hold the sample.

DBPFP testing requires chlorination of raw water samples according to specific conditions called the uniform formation conditions (UFC) specified in the Information Collection Rule (USEPA 1996) Analysis of TTHMs and HAA5s were conducted by a certified commercial laboratory according to EPA methods 524 and 522, respectively.

Results and Discussion

Table 2 provides a set of summary statistics for the raw water samples. Most of the waters were visibly colored both before (i.e., apparent color) and after filtering (i.e., true color). The pH reported in Table 2 was measured when each sample arrived in the lab, and was not necessarily the pH of the water at the source. In a recent study of 50 groundwater samples, the pH measured in the lab exceeded the pH measured in the field for every sample by an average of 0.6 pH units (Miller et al. 1990). Although color and UV 254 of natural waters are both known to vary with pH, all but two of the water samples had pH between 7.0 and 7.9 and half of the samples had pH between 7.0 and 7.3. The DOC of the sample set ranged between 1.3 and 18 mg/L with an average of 8.5

mg/L. TTHM formation potential (TTHMFP) ranged from 37 μg/L to 1050 μg/L and HAA5 formation potential (HAA5FP) from 37 μg/L to 1200 μg/L. Both the median and the mean TTHMFP and HAA5FP values would exceed levels set by the Stage 1 D/DBP Rule (80 and 60 μg/L, respectively).

Table 2: Summary statistics for water quality parameters (modified from Narr et al., 2001)

	Apparent Color (CU)	True Color (CU)	pH	DOC (mg/L)	UV-254 (1/cm)	TTHMFP (μg/L)	HAA5FP (μg/L)
Mean	96	74	7.4	8.5	0.295	327	337
Max	299	257	8.5	18	0.826	1050	1200
Min	6.3	0.2	6.8	1.3	0.044	37	37

Listed in Table 3 are summary statistics and correlations for specific ratios of common water quality parameters in the ATTAC samples. The second to the last line in Table 3 lists the correlation coefficient between the numerator and denominator of the ratio in each column. The last line shows the correlation between the numerator and denominator when the two groundwater samples were excluded from the correlation calculation. Although the ratio of true to apparent color ranged from 0.04 to 0.97, high apparent color was without exception associated with high true color and the relationship was linear. The high correlation between apparent and true color was not affected by the exclusion of the groundwaters. On average, 70% of color remained after filtering. Color, measured at 420 nm, was also very highly correlated with UV absorbance at 254 nm (r = 0.97 for all samples).

Table 3: Summary statistics for ratios of basic water parameters.

	True/app color	true color/ UV-254	SUVA ($m^{-1}l/mg$)	true color/ DOC
Mean	0.70	230	3.3	8.0
Max	0.97	439	5.4	18.7
Min	0.04	5	1.8	0.2
Correlation r	0.97	0.97	0.87	0.75
r (no GW)	0.97	0.97	0.89	0.80

UV-254 is commonly used as a surrogate for both bulk DOC and DBP precursors (Edzwald et al. 1985). Although UV-254 generally increases with DOC, the exact relationship varies between water sources. The parameter, UV-254 v. DOC is commonly referred to as specific UV absorbance (SUVA), expressed as UV 254/DOC. SUVA is generally thought to be a good predictor for DBPFP (Krasner, 1999). UV-254 and DOC were moderately well correlated for the set of waters (r = 0.87) and slightly better correlated for the set of surface waters alone (r = 0.89).

As shown in Figure 2, HAA5FP was nearly equal in value to TTHMFP and thus was very strongly correlated to TTHMFP (r = 0.99). This result was interesting since in some other studies many more TTHMs were found compared to HAA5s.

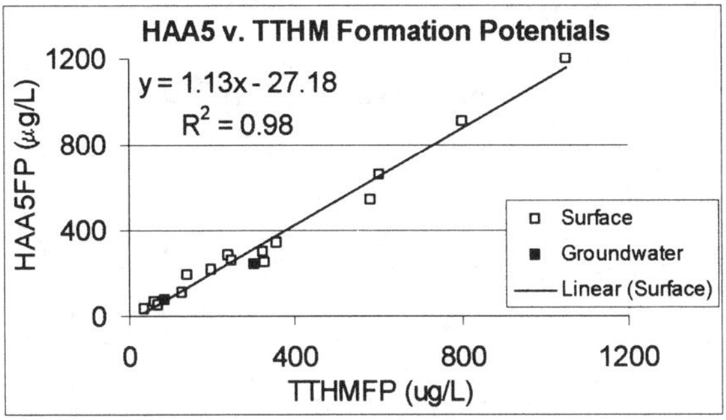

Figure 2. HAA5FP vs. TTHMFP in 17 community water systems

As shown in Figure 3, the DOC of a water was only moderately well correlated to DBPFP, particularly at the high and low ranges of DOC. In particular, a DOC below approximately 5 mg/L did not necessarily have a nil DBPFP as the trendline would indicate. The two groundwaters were within the variability in Figure 3 suggesting that DOC was equally good at predicting DBPFP in surface water as groundwater.

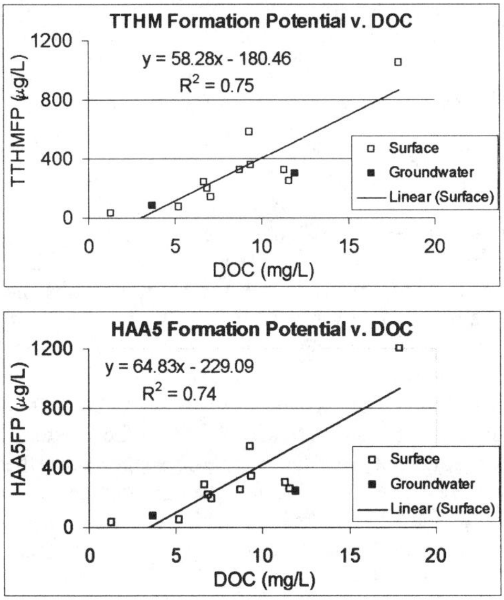

Figure 3. Comparison of TTHMFP and HAA5FP vs. DOC

In Figure 4, the strong correlations between DBPFP and UV-254 confirm previous studies that showed UV-254 to be a good indicator of DBPFP. For both TTHMFP and HAA5FP, however, the two groundwater samples did not fall on the same line as the surface waters. It may be possible with more groundwater samples to show a different relationship, however, we can say that if a relationship exists, it is potentially different than that for surface waters.

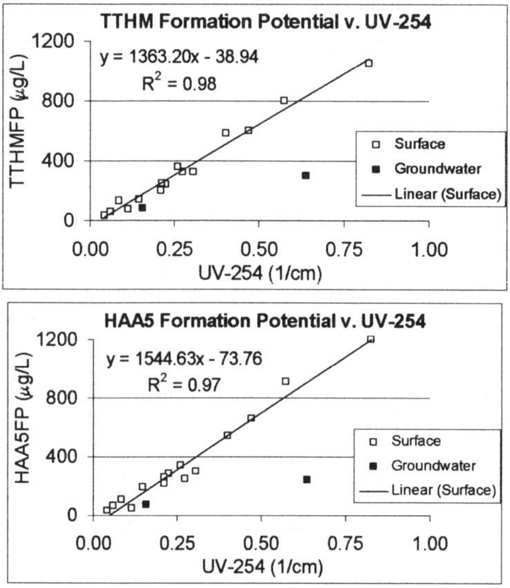

Figure 4. TTHMFP and HAA5FP vs. UV-254 in 17 community water systems.

As shown in Figure 5, normalizing both DBPFP and UV-254 by DOC reduced the range of the data by an order of magnitude, causing the data points to have a greater relative scatter and therefore a lower correlation coefficient "r". UV-254 divided by DOC is referred to as SUVA. SUVA does not provide any information about the concentration of DBPFP one might find in a source water, rather it suggests whether or not the organic matter in water is enriched in DBP forming substances (on a per gram basis). In Figure 5, the two groundwaters were clearly separated from the surface water trend lines. This result suggests that high SUVA values tell less about the enrichment of DOC in DBP precursors for groundwaters than surface waters. The same conclusion could be drawn from Figure 4. For a given TTHMFP or HAA5FP, a much greater UV-254 value was required for groundwaters than for surface waters. Since UV-254 is normally a measure of aromaticity of DOC, the result suggests that the aromaticity in the groundwaters was less reactive with chlorine than similar chemical structures in the surface waters. This effect would be observed if the aromatic structures in groundwater were more highly polymerized. No studies were conducted to test the polymerization of DOC in groundwaters vs. surface waters.

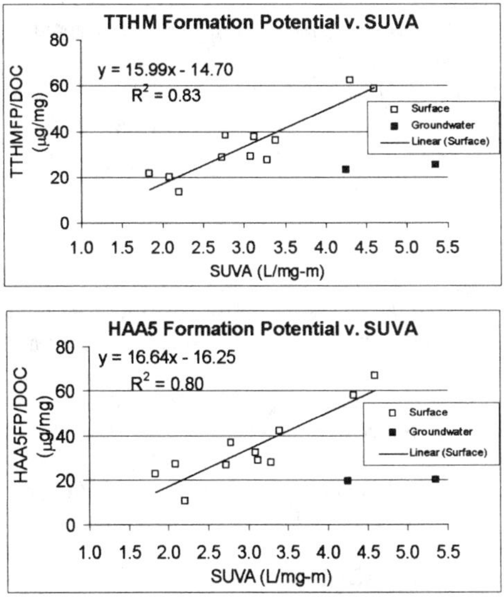

Figure 5. TTHMFP and HAA5FP vs. SUVA

Conclusions

UV absorbance was found to be the best indicator of TTHM formation potential for raw Alaskan drinking waters of surface origin, in spite of differences in geographic origin, soil type, and vegetation. The two groundwaters in the study did not fit the pattern evident in the surface waters based on UV-254. A larger number of groundwaters would need to be tested to determine whether the relationship between UV absorbance and TTHMFP follows a pattern for high-DOC Alaskan groundwaters. The results suggest that UV-254 monitoring of raw water may be useful for many applications in water treatment including:

- Monitoring raw water to determine if it contains DBP precursors
- Monitoring raw and treated water to determine if a treatment process effectively removes DBP precursors
- Determining if DBP precursors persist or are being formed in storage and distribution systems.

UV-254 monitoring is a simple and inexpensive procedure that could be conducted on or off-line in the same manner as chlorine residual and turbidity. A study of the effects of different treatment processes and storage practices on DBPFP in Northern water systems is currently being conducted as part of ATTAC.

Acknowledgements

The authors gratefully acknowledge support from the EPA funded Alaska Technical and Training Assistance Center and the US Geological Survey (through the Water and Environmental Research Center at the University of Alaska Fairbanks). The research could not have been conducted without the generous support of water treatment plant operators throughout the state.

References

Clesceri, L. S., Greenberg, A. E., and Deaton, A. D., eds., (1998), *Standard Methods for the Examination of Water and Wastewater, 20th ed.* American Public Health Association, American Water Works Association, Water Environment Federation, Washington, D. C.

Collins, M.R., Amy, G.L., and King, P.H., (1985). Removal of organic matter in water treatment, *Journal of Environmental Engineering*, Vol. 111(6), 850-864

Edzwald, J. K., Becker, W. C., Wattier, K, L., (1985). Surrogate Parameters for Monitoring Organic Matter and THM Precursors, *AWWA Journal,* Vol 77, no.4, 122-132.

Krasner, S. W. Chemistry of Disinfection By-Product Formation, *Formation and Control of Disinfection Byproducts*; AWWA, 1999, *27-52.*

Menough, personal communication.

Narr, J., S. Garland, D. White and C. Woolard, (2001), Experience with DBPs in Alaska's small drinking water sources", In. Proceedings of the American Water Works Association Annual Conference, Washington DC.

Shorney, H.L., S. Radtke, P. Hargette, W. Knocke, A. Dietrich, R. Hoehn and B. Long, (1999), Removal of DBP precursors by enhanced coagulation and lime softening, AWWA Research Foundation, Denver, CO.

USEPA (1996). *ICR Manual for Bench- and Pilot-Scale Treatment Studies*, EPA Document # EPA 814-B-96-003, April 1996, page 2-37.

Investigation of the Fate of Wastewater Phosphorus within the Processed Kimberlite Containment Area at BHP's EKATI™ Diamond Mine

Karen Graham[1], Daniel W. Smith[2], Selma E. Guigard[3]

Abstract

At the Ekati™ diamond mine, located 300 km northeast of Yellowknife, wastewater consisiting of sewage from the housing facility, flow from the vehicle washing bay and wastewater from neighbouring worksites was previously collected, treated and discharged into a local lake. The discharge was observed to have negative impacts on the lake which included an increased algae growth and lowered dissolved oxygen levels in the lower portion of the lake as early as 1997. Part of the cause of this impact was believed to be increased phosphorus concentrations in the lake originating from the treated wastewater discharge. There was concern that the increased phosphorus levels may affect downstream lakes, in the form of eutrophication and decreased winter dissolved oxygen concentrations. To minimize the impact, BHP Diamonds Inc. relocated the discharge of treated wastewater to its processed kimberlite containment system.

The purpose of this study was to understand the fate of phosphorus in the treated wastewater when added to the processed kimberlite containment area. The study involved conducting laboratory investigations to develop adsorption isotherms for phosphorus on the processed kimberlite. The phosphorus adsorption isotherms can be used to evaluate the fate of phosphate in the processed kimberlite containment area. The effects of parameters such as temperature, coagulant, and flocculent addition were determined.

The quantity of phosphorus adsorbed per gram of kimberlite was found to be low. At a pH of 7.0, 0.5 kg of kimberlite is needed to reduce the concentration of phosphorus from 10 mg/L to 0.1 mg/L in a volume of 1L of wastewater (i.e. 0.2 mg phosphorus removed/g of dried kimberlite). Even though the kimberlite adsorption of phosphorus is low, there is more than adequate kimberlite available in the tailings to remove the phosphorus loading when the pH is 7.0.

---

[1] MSc Student, University of Alberta; Currently working; Research Analyst, DPRA, 338 Wall Street, Princeton, NJ 08540; phone 609-279-2123; karengraham@dpra.com

[2] Professor and Director, Environmental Engineering Program, 304 Environmental Engineering Building, Department of Civil & Environmental Engineering, University of Alberta, Edmonton, Alberta, Canada T6G 2M8; phone 780-492-4138; dwsmith@civil.ualberta.ca

[3] Assistant Professor, Environmental Engineering Program, 304 Environmental Engineering Building, Department of Civil & Environmental Engineering, University of Alberta, Edmonton, Alberta, Canada T6G 2M8; phone 780-492-8585; selma.guigard@ualberta.ca

Introduction

BHP's Ekati™ Diamond mine is located 300 km northeast of Yellowknife. It is the first operating diamond mine in Canada. The population of the permanent camp is approximately 400 people and it may reach higher numbers during the summer when construction is occurring. All of the wastewater from the camp and neighbouring worksites is collected and treated. In the past, the treated wastewater effluent was discharged into a nearby lake. As early as 1997, the nutrients in the effluent were observed to increase algae growth and decrease dissolved oxygen levels in the lake (Rescan, 2000). There was concern that the increased nutrient levels, particularly phosphorus, may affect downstream lakes, in the form of eutrophication and decreased winter dissolved oxygen concentrations (Rescan, 2000). The discharge of the treated effluent was therefore relocated to the processed kimberlite containment system to minimize the impact.

Very little is known about the interaction between the clays and minerals that make up the processed kimberlite mixture and wastewater effluent components, such as phosphorus. A preliminary bench scale study conducted by BHP found that total suspended solids (TSS), phosphate (PO_4^{3-} and total phosphorus) and turbidity were reduced, beyond a straight mass calculated value, in the supernatant from a mixture of 1.9 parts wastewater to 100 parts processed kimberlite washwater (BHP, 1999).

Diamond Mine Information. The majority of natural diamonds have been derived from deposits of kimberlite (BHP and DIA MET, 1995). For mineral processing involving diamonds there are three main operation units that are used in the reduction and diamond recovery process to separate the diamond from the surrounding kimberlite. The kimberlite ore is crushed and scrubbed, concentrated by heavy medium separation and concentrated by X-ray sorting (BHP and DIA MET, 1995). The concentration of diamonds in the kimberlite ore is only one to two parts per 10,000,000 parts at the Northwest Territories deposit (BHP and DIA MET, 1995). This results in large amounts of material that must be processed to recover the diamonds.

Crushing, scrubbing and screening are done to reduce the size of the rock to be processed and to remove soft fines and clay minerals from the crushed ore. The specific gravity of diamond is 3.52 while the waste rock is only 2.7 and the kimberlite is 2.2 to 2.4 (BHP and DIA MET, 1995). This allows the use of density separation processes. The heavy medium separation (HMS) step separates high-density particles (including the diamonds) from low-density particles through gravity separation. In this step, ground ferrosilicon and water is added to the ore to form slurry in the HMS unit. The ferrosilicon is a suspension medium of high density that enhances the separation of high and low-density particles (BHP and DIA MET, 1995). The majority of the ferrosilicon is recovered and reused (BHP and DIA MET, 1995).

The recovery process involves further concentration by X-ray sorting. Diamonds luminesce under X-rays, which allows separation of luminescent material (mostly diamonds) from non-luminescent material (no diamonds).

Ekati*^{TM} *Mine Tailings. Tailings are produced in each of the concentration steps. Fine tailings and coarse tailings are treated separately at the BHP site (BHP and DIA MET, 1995).

The fine tailings are thickened to allow process water to be recycled and to reduce the volume of tailings to be disposed. A coagulant is added to the thickener to encourage flocculation and settling. The thickener overflow is recycled and used as process water. The underflow is sent to a storage tank prior to disposal. The treated sewage and other effluents are added to the holding tank prior to pumping to the containment area. The fine mine tailings consist of kimberlite rock, coagulants and minor amounts of acid generating materials and heavy metals (BHP and DIA MET, 1995). Approximately 60 percent of the total ore that is processed becomes fines.

The tailings must be treated and/or disposed. The most common method used by mines is the use of tailing ponds (also called containment areas or impoundment ponds) to store the processed waste rock. If the tailings do not contain enough water, water must be added so the tailings can be transported as slurry through pipes from the process plant to the tailings pond. Currently, BHP uses Long Lake as its fine mine tailings disposal area.

Ekati*^{TM} *Wastewater Treatment System. An extended aeration system is used to treat the wastewater generated at the Ekati^{TM} site. This system was chosen due to its simple design and operation. There are three basic parts to the system: an aeration tank, a clarifier and a sludge holding tank. In the aeration tank, microorganisms use components of the wastewater as a food source. The solids formed tend to clump together to form sludge. Oxygen is added to this tank to support microbial respiration. The sludge is separated by gravity in the clarifier and a portion of the sludge is recycled back into the aeration tank to speed up the treatment process. The excess sludge is held in an aerated sludge holding tank until land disposal.

Objectives

The purpose of this study was to understand the impacts of the addition of treated wastewater to the processed kimberlite containment area by conducting controlled laboratory investigations. These investigations were performed under two different temperature conditions to represent the temperature extremes in the lake throughout the year. In particular, the study focused on the fate of phosphorus in the processed kimberlite containment area.

Equilibrium time experiments were completed at two different temperatures to determine the contact time required for the phosphorus concentration to reach equilibrium. Adsorption isotherms were also developed to examine the adsorption of phosphorus by the processed kimberlite. The phosphorus adsorption isotherms can then be used to evaluate the fate of phosphate in the processed kimberlite containment area.

Materials and Methods

Batch studies in the present investigation comprised of equilibrium time studies and adsorption isotherm studies.

General Procedures. All chemicals used were of analytical reagent grade. All glassware was soaked in a solution of 1% Nitric acid for three hours and rinsed twice with de-ionized water prior to use to ensure no contamination by phosphorus. All solutions were prepared with de-ionized water. A stock solution of phosphate was prepared using potassium phosphate in de-ionized water. Orthophosphate has been used since it is anticipated that the phosphorus in the treated wastewater effluent at Ekati™ is, for the most part, in the form of orthophosphate. This assumption is based on the fact that literature suggests that other forms of phosphate such as polyphosphates and metaphosphates are converted to orthophosphate when they come into contact with microorganisms that are found in raw wastewater (Metcalf and Eddy, 1991). This breakdown of other forms of phosphate to orthophosphate by the microorganisms occurs in a relatively short period of time (Metcalf and Eddy, 1991) and is therefore assumed to occur in the extended aeration tank at the Ekati™ site.

Kimberlite Preparation. Processed kimberlite tailings were obtained from BHP's Ekati™ Diamond mine. The processed kimberlite mine tailings were obtained prior to any flocculant or coagulant addition. The tailings were in slurry form that settled during shipping. Before use, the tailings were mixed with a hand rotor to form a uniform slurry. A subsample of the tailings slurry was centrifuged at 6,500 rpm for 20 minutes to separate the kimberlite from the water. The supernatant was decanted, and de-ionized water was added to rinse the solids prior to a second centrifuge at 6,500rpm for another 20 minutes. The supernatant was decanted again and the remaining solids were dried at a temperature of 65°C. The solids were placed in a dessicator to cool to room temperature. The solids were finally ground with a mortar and pestle to return them to a more uniform size prior to being used in the adsorption experiments. The composition of the prepared kimberlite tailings as determined by ICP spectroscopy is shown in Table 1.

Equilibrium Time Experiments. The equilibrium time experiments were carried out using the procedure outlined below:

1. 190 mL of de-ionized water was placed in a 250 mL bottle
2. Three different samples served as controls:
   (i)De-ionized water
   (ii) De-ionized water with kimberlite
   (iii) De-ionized water with phosphorus
3. For most samples prepared kimberlite was added to the bottle For the control samples (i) and (iii) no kimberlite was added.
4. The bottle was capped and sealed with Teflon tape
5. The bottles were loaded into one of two custom-built rotary tumblers (used to keep the samples well mixed)

Table 1. Chemical Composition of Thickener Feed (from ICP Spectroscopy)

Analyte		Result	Unit
Silicon	SiO_2	47.8	% dry
Magnesium	MgO	23.8	% dry
Iron	Fe_2O_3	6.32	% dry
Aluminum	$Al2O_3$	5.38	% dry
Calcium	CaO	2.61	% dry
Potassium	K_2O	1.23	% dry
Titanium	TiO_2	0.353	% dry
Sodium	Na_2O	0.228	% dry
Phosphorus	P_2O_5	0.19	% dry
Barium	BaO	0.156	% dry
Manganese	MnO	0.0902	% dry
Strontium	SrO	0.04	% dry
Zirconium	ZrO_2	0.0038	% dry
Total Oxides		101	% dry
Loss on Ignition		12.9	% dry
Chromium		602	μg/g dry
Zinc		94.3	μg/g dry
Vanadium		77	μg/g dry
Cobalt		46	μg/g dry
Molybdenum		13	μg/g dry
Berylium		1.6	μg/g dry

6. The tumblers were rotated at 10 revolutions per minute for 24 hours
7. 10 mL of 200 mg/L phosphate stock solution was placed in each 250 mL bottle (bringing the initial phosphorus concentration of the entire solution to 10 mg/L)
8. The bottles were reloaded into the tumblers and rotated at 10 revolutions per minute
9. Three bottles were removed at each time interval. The bottles were removed after contact times ranging from 10 minutes to 14 days.
10. The solutions from the bottles were filtered through a 2 mm glass fiber pre-filter and then filtered through a 0.45 μm membrane filter to remove the kimberlite. Filters were pre-soaked prior to use following the procedure outlined in *Standard Methods* (APHA, 1992). Pre-soaking involved soaking the filters for one hour in de-ionized water, changing the water and soaking for another three hours.
11. The filtrate was collected and analyzed in duplicate for phosphate using the Vanadomolbydophosphoric Acid Method described in *Standard Methods* (APHA, 1992). The Vanadomolybdophosphoric Acid Colourimetric method (APHA, 1992) was used to determine the phosphorus concentrations. The

Vanadomolybdophosphoric Acid Colourimetric method was chosen due to its simplicity and its applicable concentration range. To ensure that the total phosphorus concentration was being measured the Persulfate Digestion Method (APHA, 1992) was used to release all the phosphorus as orthophosphate, which is the form of phosphorus that is measured by the Vanadomolybdophosphoric Acid Colourimetric method. A UV-visible spectrophotometer was used for the measurement of absorbance.

Adsorption Isotherm Experiments. The adsorption isotherm experiments were carried out using a procedure similar to that outlined in the equilibrium time experiments. Steps 1 through 7 were completed. After a set contact time, all the samples were removed from the tumblers and steps 9 through 11 were completed.

Results and Discussion

Equilibrium Time Studies. The equilibrium time studies were completed to determine the contact time required for the adsorption of phosphorus by the kimberlite to be complete. The kinetic studies were conducted at two different temperatures, 18°C and 5°C, to represent the two extremes of lake temperatures near the mine site. The tumblers were placed in temperature controlled rooms, set at 18 ± 0.5°C for the warmer temperature boundary and 5 ± 0.5°C for the lower temperature. The equilibrium time determined from these experiments was used as the contact time for the adsorption isotherm studies.

At the start of the experiments the pH was measured and found to be 7 ± 0.5 and the dosage of kimberlite was 2.5 g/L. This set of conditions was chosen for simplicity. The initial phosphate concentration was 10 mg P/L of orthophosphate.

The phosphorus concentration at both temperatures decreased rapidly during the first six hours of contact time with the kimberlite (Figure 1). The data presented in Figure 1 represents the average of two separate runs for each temperature. After the first six hours, the rate of adsorption decreased until the phosphorus concentration remained constant after approximately one full day of contact. It was assumed that at this point equilibrium had been reached. This assumption was reinforced when the concentration of samples removed after a contact time of 14 days were found to have an average concentration of 8.8 mg/L. As can be observed from Figure 1, the time to reach equilibrium and the equilibrium concentration was found to be the same for both temperatures. The equilibrium concentration was found to be 8.8 mg P/L for both temperature conditions. This translates to a removal of 0.5 mg of phosphorus per gram of kimberlite.

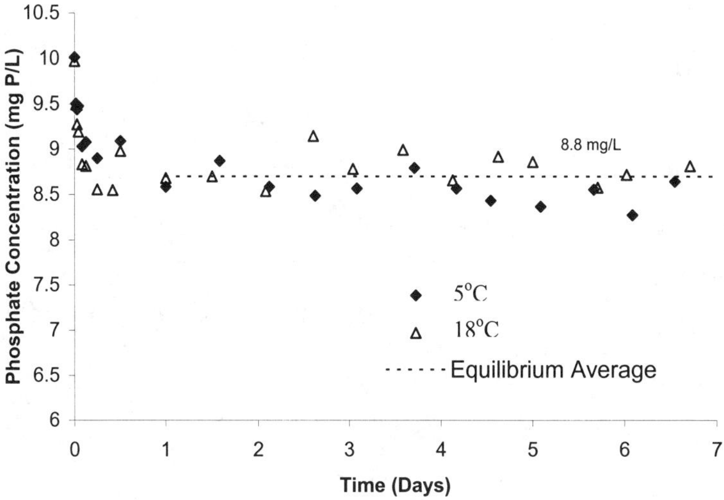

Figure 1. Equilibrium Study Results

Adsorption Isotherms. To ensure that equilibrium has been reached, the contact time for the adsorption isotherm studies was set at 48 hours.

The adsorption capacity of kimberlite tailings was modeled using the linearized form of the Freundlich isotherm as given in equation 1.

$$\log\frac{x}{m} = \frac{1}{n}\log C_e + \log K_f \qquad (1)$$

where C_e is the concentration of phosphorus remaining in solution at equilibrium (mg P/L); x is the mass of phosphorus adsorbed (mg P); m is the mass of kimberlite (g dried kimberlite); n and K_f are constants. The experimental equilibrium data for both temperatures was fitted with the Freundlich isotherm equation and is presented in Figure 2. The Freundlich isotherm constants for temperatures of 18°C and 5°C are summarized in Table 2.

Table 2. Freundlich Isotherm Constants and Experimental Conditions

Temperature Condition	1/n	K_f	Number of Data Points	r^2
18°C	0.68	0.10	31	0.92
5°C	0.58	0.10	13	0.85

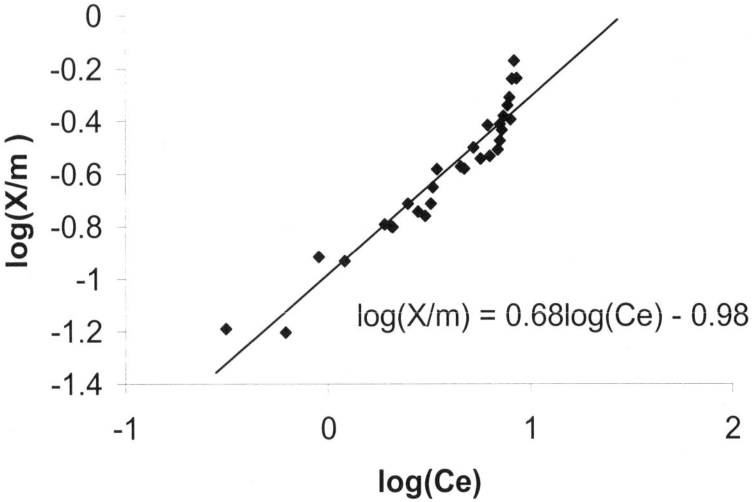

Figure 2a. 18°C Adsorption Isotherm

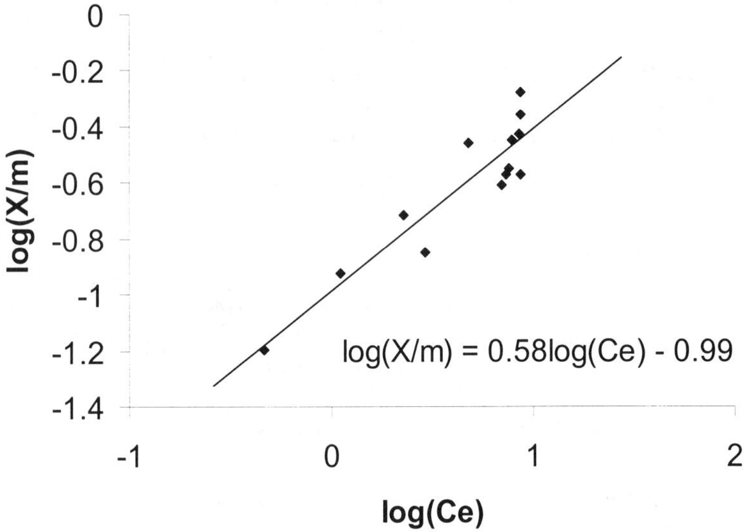

Figure 2b. 5°C Adsorption Isotherm

Significance of Results. The Ekati^{TM} mine site currently has no regulations on the amount or concentration of phosphorus that can be released into the environment. From the isotherms, however, a rough calculation can be used to determine how much dried kimberlite would be needed to adsorb a certain amount of phosphorus. According to the results presented here, to reduce the concentration of phosphorus from 10 mg/L to 0.1 mg/L approximately 0.5 kg of dried kimberlite per litre of wastewater would be needed at a temperature of 18°C and approximately 0.4 kg per litre of wastewater at a temperature of 5°C.

An estimated maximum of 320,000 L of wastewater could be produced in a day at the mine site based on a camp population of 800 with a per capita production rate of 400 L of wastewater per person per day. If the treated wastewater has a phosphorus concentration of 10 mg/L and at a temperature of 18°C, it would require approximately 160 tonnes of dried kimberlite per day to reduce the phosphorus concentration to 0.1 mg/L.

Just after start-up, the mine site was processing 9,000 tonnes of ore per day and is currently in the process of increasing the amount to 18,000 tonnes per day. When the ore is processed approximately 60 percent of the ore becomes fines. This translates to the current disposal of fines being roughly 5,400 to 10,800 tonnes of dried kimberlite per day. This amount is well above the 160 tonnes needed to reduce the phosphorus concentration of 320,000 L of wastewater from 10 mg/L to 0.1 mg/L. Even though the kimberlite adsorption of phosphorus is low according to the results presented here, with proper mixing of the wastewater effluent with the fine tailings, there is more than an adequate amount of kimberlite available to remove the phosphorus when the pH is 7.0 and the temperature is 18°C.

All of the conclusions discussed in this section are based on the preliminary results of the adsorption isotherm studies. These studies have not included the effects of the addition of coagulant and flocculant or the possibility of desorption. These factors must be studied before a final conclusion on the fate of phosphorus by kimberlite mine tailings can be reached.

Future Experiments. The kimberlite loading in the mine tailings has been found to be more than enough to remove enough phosphorus from the wastewater to reduce the phosphorus concentration to 0.1 mg/L. However, there are other conditions that must be studied to determine if the adsorption is affected by coagulant and flocculant addition as well as the affect of changes in pH. Other parameters such as the presence of chemical agents and heavy metals may also have an affect but will not be included in this study.

More experiments are currently being completed to determine the affect that coagulant and flocculant have on the adsorption of phosphorus by kimberlite. It will be determined if the coagulant and flocculant can remove any phosphorus. It will also be determined if the coagulant and flocculant affect the adsorption of phosphorus by the kimberlite.

The possibility of desorption will be studied. The main causes of desorption are changes in temperature, pH or liquid concentration. The results should help determine the potential for phosphorus desorption from the kimberlite.

Conclusions and Recommendations

From the preliminary results presented in this paper, it appears that adding the wastewater effluent to the kimberlite mine tailings may be an effective method to dispose of the treated wastewater effluent.

The adsorption of phosphorus by kimberlite at a pH of 7.0 reaches equilibrium after approximately one day of contact time at temperatures of 18°C and 5°C. The equilibrium concentration was 8.8 mg P/L when the initial concentration was 10 mg/L and 2.5 g dried kimberlite/ L was added. This translates to a removal of 0.5 mg of phosphourus per gram of kimberlite.

To reduce the concentration of phosphorus from 10 mg/L to 0.1 mg/L, approximately 0.5 kg of dried kimberlite per litre of wastewater is needed at a temperature of 18°C and approximately 0.4 kg per litre of wastewater at a temperature of 5°C.

Based on the preliminary results presented in this paper, even though the adsorption of phosphorus per gram of dried kimberlite is low, there is more than an adequate amount of kimberlite available to remove the phosphorus loading from the addition of treated wastewater to the mine tailings when the pH is 7.0 and the temperature is 18°C.

Coagulant and flocculant are added to the mine tailings prior to the addition of the treated wastewater. The coagulant and flocculant may have an affect on the adsorption of phosphorus by the kimberlite. This affect will have to be determined before a final conclusion on the ability of the mine tailings to adsorb the phosphorus loading from the addition of treated wastewater.

There is also the possibility of the phosphorus desorbing when the environmental conditions around the kimberlite are changed. The main causes of desorption are usually due to changes in temperature or pH. The affect of changes in pH and temperature will be studied to determine the potential of the phosphorus to desorb from the kimberlite.

References

American Public Health Association (APHA) (1992). *Standard Methods for the Examination of Water and Wastewater*, 18th ed., American Public Health Association, Washington, D.C.

BHP Diamonds Inc. (1999). *Bench Scale Study: Treated Sewage Effluent Mixed with Tailings,* BHP Diamonds Inc.,Vancouver.

Metcalf & Eddy (1991). *Wastewater engineering : treatment, disposal, and reuse*, 3rd ed., McGraw-Hill, New York.

Rescan Environmental Services Ltd. (2000). *1999 Kodiak Lake Sewage Effects Study Technical Repo*rt, Rescan Environmental Services Ltd., Yellowknife, NWT.

Laboratory and Field Investigations of Low Temperature Microbial Activity in Rivers

K.D. White[1], C.M. Reynolds[2], D. Ringelberg[3], K.L. Foley[4], and L.B. Perry[4]

Abstract

Adequate levels of dissolved oxygen (DO) in rivers are essential to sustain aquatic life and maintain acceptable water quality. The most critical period for dissolved oxygen deficits is typically assumed to be summer because low flows often coincide with high water temperatures. However, previous research has documented the occurrence of wintertime oxygen deficits in ice-covered rivers. Despite these observations, water quality models neglect the effects of reaeration and deoxygenation at the low temperatures occurring during winter, resulting in uncertain model results for the winter period. This study provided high-temporal-resolution water quality data for an intermittently ice-covered, relatively nutrient-rich river in Vermont. Field data and laboratory studies were performed to evaluate reaeration and deoxygenation coefficients, and microbial diversity was examined. Our field data showed that DO patterns in a small Vermont river are different during ice-covered periods than during open-water periods and that DO levels can be at or above saturation levels for the ice-covered period. In addition, diurnal cycling of DO persists during the ice-covered period and appears to be related to photosynthetically active radiation (PAR) in a manner that suggests that photosynthetic activity may be contributing to reaeration. Microbial communities were found by two independent measures to be different under an ice cover than in open water.

Introduction

Dissolved oxygen (DO) deficits have been known to coincide with ice cover formation (Schreier et al. 1980, Whitfield and McNaughton 1986), and DO deficits can increase until ice cover breakup (Shallock and Lotspeich 1974). Rapid decreases in winter DO have been observed at river–reservoir confluences (James et al. 1992, McBean et al 1979, Reitner at al. 1997), with large oxygen deficits reported in deep

[1] Research Hydraulic Engineer, Member, ASCE; [2] Research Physical Scientist; [3] Microbiologist; [4] Physical Science Technician, U.S. Army Engineer Research and Development Center, Cold Regions Research and Engineering Laboratory, 72 Lyme Rd., Hanover, NH 03755-1290, phone 603-646-4187; Kathleen.D.White@erdc.usace.army.mil.

stratified reservoirs (Gamble 1971). Degraded water quality manifested as reduced DO can also result from shock loadings that occur during periods when the stability of a river's microbial community is low because of the lack of diversity. Shock loadings can include rainfall or snowmelt runoff containing high biochemical oxygen demand (BOD) highway and airfield de-icers, as well as unusually large outflows from combined sewer systems. These effects can be magnified in rivers that experience low flow during winter. Seasonal microbial diversity and community structure changes that might influence the dominant metabolic processes occurring in a river have not been well characterized.

DO processes in ice-covered rivers have received little attention because the solubility of oxygen in water is high at the lower water temperatures experienced during winter. Thus, the same magnitude of oxygen depletion due to biochemical oxidation has less effect during the winter than during the summer. Despite previous research that has documented the occurrence of wintertime oxygen deficits in ice-covered rivers, water quality modelers often neglect the reaeration and deoxygenation impacts caused by microbiological processes at low temperatures. Understanding the relationships among seasonal river changes and river microbial community structure will lead to improved predictive capabilities for winter water quality. This study provided high-temporal-resolution water quality data for an intermittently ice-covered, relatively nutrient-rich river in Vermont. Field data and laboratory studies were performed to evaluate reaeration and deoxygenation processes, and microbial diversity was examined. The study is described in detail in White et al. (2001).

Methods

Field Measurements. Field data were collected from 22 December 1999 to 31 March 2000 on the LaPlatte River at Hinesburg, VT. A YSI 600X probe [with sensors measuring DO (mg/L and % saturation), water temperature (°C), conductivity (mS/cm), and pH] and a LICOR spherical quantum sensor (LI-193SA) measuring photosynthetically active radiation (PAR) ($\mu mol/m^2s$) were placed in the river at the same location described in White et al. (2001). PAR measures solar radiation in the form of visible light for wavelengths between approximately 400 and 700 nm, which is used by photosynthetic organisms located in the water column. Both the probe and the sensor were connected via cable to a Campbell CR10X data acquisition system located on the riverbank. The data were collected at 10-minute intervals for storage in the data acquisition system. Stored data were accessed and downloaded by computer via a modem and cellular phone at intervals ranging from daily to every seven days. A web camera was added to monitor site conditions on a near-real-time basis. A solar cell and battery provided power for the system. Air temperature and battery voltage were also monitored. Ice conditions (including the depth of snow on the ice, ice morphology, and bottom surface structure) were monitored periodically.

Water samples were obtained at intervals, stored in sterile gallon bottles, and transported to CRREL for further testing. Microbial samples were also collected during open-water and early ice-covered periods by suspending a glass-bead-packed tube just below the surface of the water or ice (White et al. 2001). This method allowed sampling of the biofilm on the glass beads for microbial characterization, in addition to the open-water samples. The microbial samplers were autoclaved prior to sampling. Microbial samples were obtained by aseptically rinsing microorganisms from the glass-bead columns with a known volume of sterile water into a sterile vessel.

Laboratory Measurements. The laboratory portion of the study used CRREL's temperature-controlled respirometer, which has been modified to allow for simultaneous monitoring of samples incubating at different temperature regimes. Each temperature regime can be cycled in a defined amplitude and frequency schedule or held constant, and samples are incubated under low-light conditions. Respiration measurements using evolved carbon dioxide (CO_2) are an estimate of general, non-photosynthetic microbial activity. Respiration tests were conducted on water samples collected for intermittent and permanent ice cover conditions (6 January and 15 January 2000) from the LaPlatte River to examine whether river microbial community structure varied during the winter. CO_2 evolution at 1° and 25°C were continuously monitored. Appropriate sterile controls were included so fluctuations caused by temperature-dependent CO_2 solubility in water could be subtracted.

In addition to general heterotrophic activity estimated from the CO_2 evolution measurements, microbial diversity was characterized and compared by a modified fatty acid methyl ester (FAME) procedure using whole cells. A modification of the procedure was selected in which the entire lawn of microorganisms was treated as a sample, rather than individual isolates (Sasser and Wichman 1991). This approach provides an indication of bacterial community structure changes that could be used as an estimate of changes in the bacterial diversity of the river system. The community structure in the LaPlatte River was compared at three sampling dates using culturable microorganisms recovered from the glass-bead sampling tubes. These sample times represent well-established ice cover (25 February), immediately following ice breakup (February 29), and after ice out (24 April). Principal component analysis (PCA) techniques were used to compare fatty acid profiles among sample times.

Results and Discussion

Water quality data. Water quality data were collected continuously at the Hinesburg site between 22 December 1999 and 31 March 2000. Intermittent ice covers occurred between 24 and 26 December and on 28 December, 29 December, 31 December, and 6 January. A permanent seasonal ice cover began on about 12 January 2000. By 15 January the ice cover was 10 cm thick, and substantial amounts of frazil ice were deposited beneath the solid ice. Algae were visible both within the frazil ice and in accumulations at the surface in the open-water area. This ice cover remained in place until between 27 and 28 February 2000, when the ice cover broke up following a snowmelt and rainfall event. Open water was observed on 29 February.

Statistical analyses of the water quality data were performed using SAS® statistical analysis software (SAS® Institute 1996). Descriptive statistics for the water quality variables are provided in Table 1. The sample populations between the ice-covered and open-water periods are statistically significantly different ($p \leq 0.05$) for all of the measured variables as determined using the Wilcoxon rank sums test. Water temperature, DO, pH, and conductivity conditions observed during the ice-covered period are relatively stable compared to open-water conditions and are similar to those reported by White and Melloh (1999). The data demonstrate a stabilizing effect of ice cover on water quality parameters that may provide some biological benefits to lotic communities as suggested by Hynes (1970).

Correlation tests were also performed, and during the open-water periods, water temperature was moderately correlated with DO percent saturation (r^2= 0.53), pH (r^2= 0.69), and air temperature (r^2= 0.66). Percent saturation DO was moderately correlated with pH (r^2= 0.77). Conductivity and pH were also correlated with probe depth (r^2= –0.57 and –0.70, respectively). During the ice-covered period, poor correlations were found among water temperature, PAR, DO, pH, and conductivity, except in three cases: probe depth was found to be moderately correlated with DO concentration (r^2 = 0.50), DO percent saturation (r^2 = 0.50), and pH (r^2 = –0.56).

The PAR measurements during open-water periods indicate that adequate incident sunlight is reaching the sensor level (approximately 1 ft deep) to allow for photosynthetic activity. Open-water mean integrated irradiance was 7.6 mol/m^{2}day, with a daytime average integrated irradiance of 16 mol/m^{2}day. The maximum observed PAR was 1731 μmol/m^{2}s on 18 March. These levels are comparable to open-water wintertime values reported by Guasch et al. (1998) as well as to summertime measurements for shaded rivers.

During the ice-covered period, the mean integrated irradiance was 0.9 mol/m^{2}day, with a maximum observed PAR of 348 μmol/m^{2}s on 24 December. After the formation of the permanent seasonal ice cover on 12 January, the mean integrated irradiance was 0.7 mol/m^{2}day. The daytime average integrated irradiance was 1.4 mol/m^{2}day, with a mean observed PAR of 16.2 μmol/m^{2}s. The maximum observed PAR during the permanent ice-covered period was 269 μmol/m^{2}s on 13 January, just

Table 1. Descriptive statistics for water quality data.

Variable	N	Mean	Std Dev	Minimum	Maximum
Open-Water Period					
Water temperature (°C)	6747	1.93	1.83	–0.02	8.03
Conductivity (mS)	6747	0.15	0.03	0.06	0.26
DO (percent saturation)	6747	113. 2	5.96	89.60	127.90
DO (mg/L)	6747	15.68	0.78	11.16	17.57
Depth (m)	6747	0.23	0.15	–0.04	0.93
pH	6747	8.10	0.09	7.84	8.38
PAR (μmol/m^{2}s)	6747	87.91	168.05	0.60	1731.00
Air temperature (°C)	6747	1.67	6.23	–18.54	19.37
Ice-Covered Period					
Water temperature (°C)	7733	0.01	0.03	–0.06	0.20
Conductivity (mS)	7734	0.21	0.04	0.08	0.30
DO (percent saturation)	7734	107.54	2.73	100.60	114.60
DO (mg/L)	7734	15.70	0.40	14.66	16.76
Depth (m)	7734	0.39	0.19	0.21	1.39
pH	7734	7.97	0.06	7.81	8.10
PAR (μmol/m^{2}s)	7734	10.36	21.97	–2.39	348.10*
Air temperature (°C)	7734	–8.68	8.30	–32.54	9.66

*occurred during period of intermittent ice cover before permanent ice cover formation

after ice cover formation. These values are somewhat lower than those reported by White and Melloh (1999), possibly due to frazil ice deposited beneath the ice cover. They are comparable to measurements by Vincent et al. (1998), who reported PAR values of about 80 μmol/m^{2}s at 10 m (32.8 ft) depth in Lake Vanda and about 8 μmol/m^{2}s at 5 m (16.4 ft) depth in Lake Bonney, both freshwater Antarctic lakes exhibiting primary productivity. Thus, despite the relatively low PAR levels observed in the LaPlatte River during the ice-covered period, photosynthetic activity could be expected to continue. Numerous researchers have reported that aquatic plants and other biota adapt to photosynthesis under low light levels (e.g., Marchand 1996).

The observed DO levels were supersaturated during the ice-covered period, averaging 15.7 mg/L and 107.5% of saturation. These values compare favorably to data reported by White and Melloh (1999), which averaged 16.4 mg/L and 113.7% of saturation during the late-winter ice-covered period. The generally high DO levels may reflect highly efficient reaeration at low temperatures, photosynthetic input, or both. Diurnal cycling of DO was observed throughout the ice-covered period.

The low-amplitude DO fluctuations observed are comparable to those reported by Simonsen and Harremoes (1978) and White and Melloh (1999). Fluctuations in water temperature do not fully explain the observed cycling, indicating a biological component. Typical diurnal DO cycles during the ice-covered period resemble that shown in Figure 1 for 20 January 2000. DO levels were lowest during the nighttime hours, began to rise through the morning hours, and peaked in early to mid-afternoon. As the ice-covered period progressed, the DO cycle became less well defined (Fig. 2), but the trend remained the same. The December open-water curves resembled the typical curve shown in Figure 1, as do those for April and May.

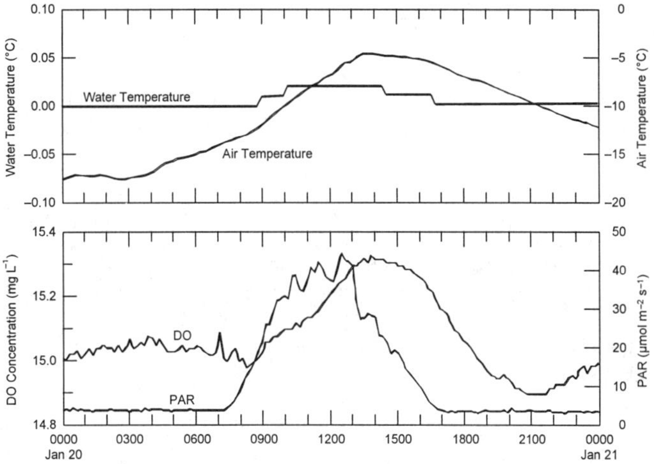

Figure 1. Typical observed DO, PAR, and water temperature during the ice-covered period (20 January 2000).

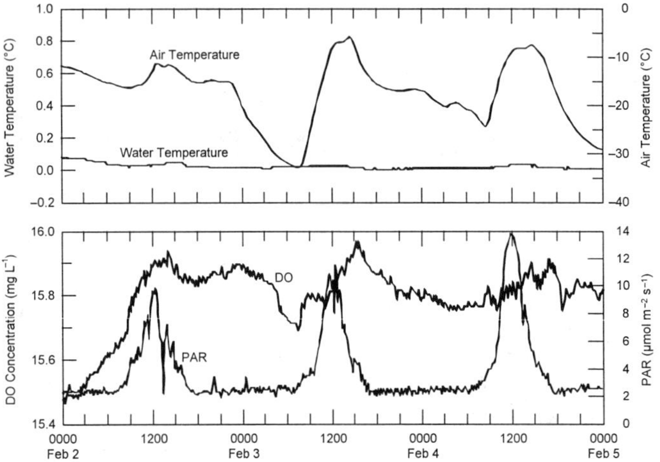

Figure 2. Less-well-defined DO cycling during later ice-covered periods, as illustrated for the period 2–4 February 2000.

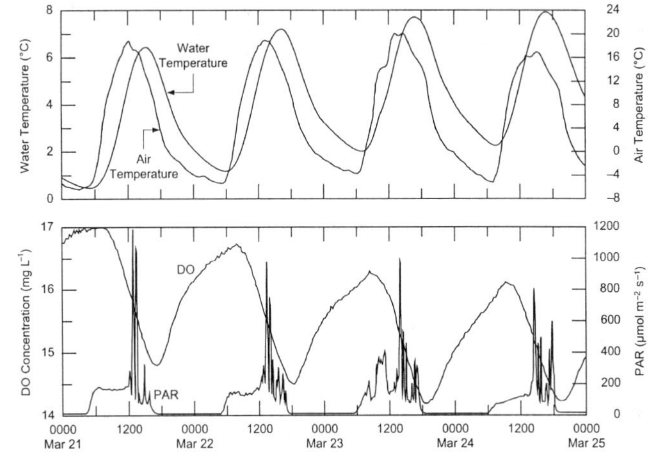

Figure 3. Typical DO, PAR, and temperature data following ice cover breakup (21–24 March 2000).

However, DO cycling during the period following ice cover breakup (after 28 February) did not follow this trend. Instead, DO reached its lowest concentration at approximately sunset and increased to a peak in the morning (Fig. 3). DO concentration and water temperature were also more strongly correlated during this period ($r^2 = -0.82$) than during the ice-covered period ($r^2 = -0.08$), since the barrier to heat transfer provided by the ice was absent. Following ice cover breakup, DO peaked earlier in the day than during the ice-covered period. This is an indication that ice cover breakup could have induced changes in the system that significantly changed the microbial dynamics. Such changes could include the loss of microbial communities living within the frazil ice deposit and the scouring or erosion of microbial communities living on bed or bank surfaces. The DO cycling observed immediately following breakup was similar to that observed by White and Melloh (1999) in the relatively pristine Sleeper's River, VT, where biological activity was thought to be minimal.

Respirometer Data. Cumulative evolved CO_2 respiration data obtained from the respirometer study are shown in Figure 4. Water samples collected on 6 and 15 January were evaluated at 1° and 25°C. On 6 January the river had only intermittent ice cover, but by 15 January the ice cover was approximately 10 cm thick at the sampling site. Both samples had similar respiration patterns at 1°C. This was shown by a lag period of approximately 100 hr, followed by an increase in respiration from 100 hr to 270 hr to a higher but relatively constant rate. At 1°C, samples from both 6 and 15 January behaved similarly with respect to CO_2 evolution, each evolving a cumulative total of approximately 50 μL CO_2/mL during the 270-hr incubation and each having a lag time of approximately 100 hr.

Additional aliquots of the water samples collected on 6 and 15 January were also incubated concurrently at 25°C and 1°C using constant-temperature baths (Fig. 4). Data for both sample collection times again showed a lag period of approximately 100 hr. However, from 100 to 270 hr, there was greater CO_2 evolution from the water sample collected on 15 January under the ice cover than from the sample collected on 6 January in open water. Cumulative CO_2 evolution values were approximately 330 μL CO_2/mL and 170 μL CO_2/mL for the 15 January and 6 January samples, respectively. The significantly different CO_2 evolution response to the warmer incubation temperature between samples collected on 6 and 15 January strongly suggests that

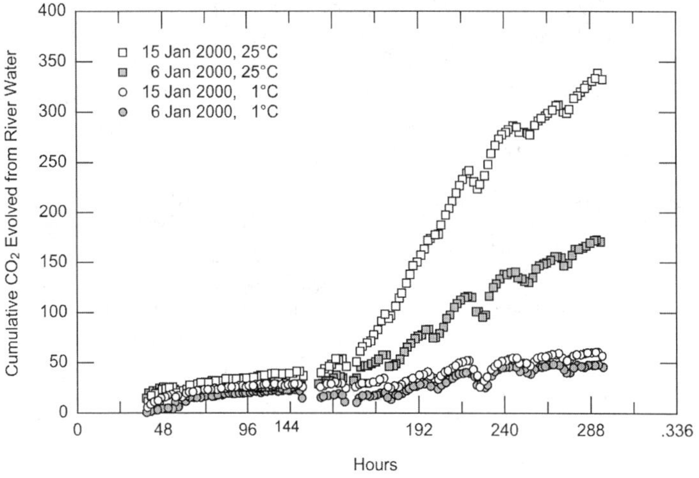

Figure 4. Cumulative respiration data for samples obtained 6 and 15 January 2000.

there were different microbial communities in the river during open-water and ice-covered periods.

One explanation for greater CO_2 evolution from the early ice-covered sample than from the open-water sample is that a frazil ice matrix had formed during the early ice-covered period. This frazil ice matrix provided a structure for biofilm attachment and growth, as confirmed by visual observations on 15 January that revealed algae both within the frazil ice deposited beneath the ice and in frazil accumulations at the surface in open-water areas. Algae or bacteria present in such a biofilm would supply carbon to the system. Because the respirometer has insufficient light to support algal photosynthetic processes in the incubation vessels, catabolic bacterial processes would dominate during respiration studies. Hence, greater release of CO_2 from the sample taken on 15 January would support the idea of greater biomass carbon, whether algal, bacterial, or both, in the ice-covered sample relative to the open-water sample.

Microbial Data. In addition to the respiration tests, we characterized microbial communities by using a modified fatty acid-based procedure, FAME. This comparison is based on culturable microorganisms that were recovered from the glass-bead samplers and integrates the fatty acid profiles of all culturable microorganisms from a sample into a single profile. In effect, this approach treats an entire sample as an "organism" and then compares the similarity or dissimilarity among samples. Because this approach requires growth on 0.1X TSA, the initial data we have gathered favors bacterial changes over fungal or algal changes. These samples we collected represented periods with a well-established ice cover (25 February), immediately following ice breakup (29 February), and over six weeks after ice out (24 April).

The PCA results for the fatty acid data obtained from culturable microorganisms recovered using the glass-bead samplers are shown in Figure 5. Ellipses have been added to better show groupings. For our purposes the importance of these data is that the culturable microbial communities from the river clearly separate into three distinct groups among the three sampling times based on whether they are indicative of gram-positive or gram-negative bacteria. These data corroborate the respirometer data and support the concept that there are seasonal, ice-cover-related microbial community structure differences in the river.

Conclusions and Recommendations

The purpose of this study was to investigate riverine dissolved oxygen concentrations during winter, including the ice-covered period, with emphasis on microbial processes contributing to reaeration and deoxygenation. Our data showed that DO patterns in a small Vermont river are different during the ice-covered period than during open-water periods. In addition, diurnal cycling of DO persists during the ice-covered period and appears to be related to PAR in a manner that suggests that photosynthetic activity may be contributing to reaeration. Visual observations at the time of ice cover

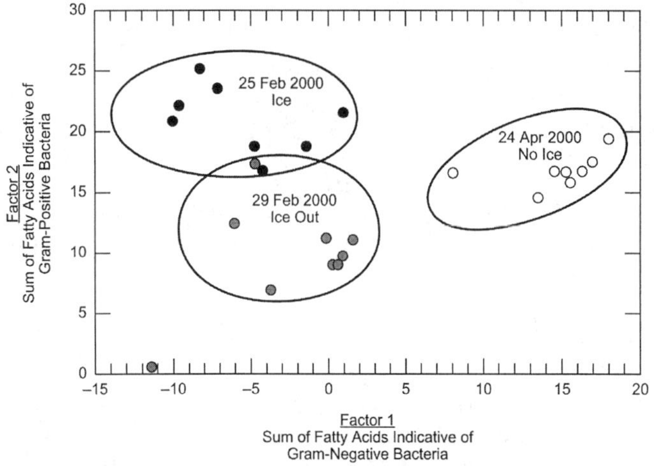

Figure 5. Results of the principal component analysis of fatty acid data obtained from culturable microorganisms collected by the glass bead samplers during the ice-covered period, immediately after ice cover breakup, and well after ice cover breakup.

formation showed algae present in the ice cover and within the frazil matrix trapped beneath the ice cover. It is possible that the porous frazil structure provides a surface for biofilm formation that supports microbial communities in a different manner than the typical lotic environment. Microbial communities were found by two independent measures to be different under an ice cover than in open water.

The results of this study indicated that photosynthetic activity could be important in winter water quality as a source of reaeration. The presence of an ice cover did not prevent photosynthetic activity. These data support the concept that improved understanding of the interactions between frazil ice, ice-covered water, and the microbiology of ice and water may suggest ways to improve winter water quality. The use of phospholipid ester-linked fatty acid analysis of samples extracted directly from water may avoid problems with culturing. Water samples within and below the frazil accumulation should be obtained in order to more clearly identify potential community differences. Additional field data and modeling are required to investigate other reaeration processes, including the effects of air temperature, and to develop a range of reaeration and deoxygenation coefficients useful to modelers. With additional field investigations, this data could lead to improved winter predictions of the riverine effects of winter discharges of high-BOD effluents such as deicing fluids.

Acknowledgments

The authors gratefully acknowledge Mr. Rocky Martin, Public Works Director, Town of Hinesburg, VT, for his assistance in locating and installing the field sampling site, and Ms. Susan E. Hardy, Environmental Science Branch, CRREL for her input and suggestions to this report. Funding was provided in part by the Water Operations Technical Support program of the U.S. Army Corps of Engineers.

References

Gamble, R.W. (1971) *Water Quality on the St. John River Under Winter Conditions.* New Brunswick Department of Fisheries and Environment: Fredericton, NB.

Guasch, H., J. Armengal, E. Marti, and S. Sabater (1998) "Diurnal variation in dissolved oxygen and carbon dioxide in two low-order streams." *Water Research,* Vol. 32, No. 4, p. 1067–1074.

Hynes, H.B.N. (1970) *The Ecology of Running Waters*. University of Toronto Press: Toronto.

James, W.F., H.L. Eakin, D. Gunnison, and J.W. Barko (1992) "Sediment-overlying water relationships affecting wintertime dissolved oxygen conditions in the Big Eau Plaine Reservoir, Wisconsin." Water Quality Research Program Technical Report W-92-2, U.S. Army Engineer Waterways Experiment Station: Vicksburg, MS.

Marchand, P.J. (1996) *Life in the Cold: An Introduction to Winter Ecology*. University Press of New England: Hanover, NH.

McBean, E., G. Farquhar, N. Kouwen, and O. Dubek (1979) "Prediction of ice-cover development in streams and its effect on dissolved oxygen modeling." *Canadian Journal of Civil Engineering*, Vol. 6, p. 197–207.

Reitner, B., A. Herzig, and G.J. Herndl (1997) "Microbial activity under the ice cover of the shallow Neusiedler See (Austria, Central Europe)." *Hydrobiologia,* No. 357, p. 173–184.

SAS Institute, Inc. (1996) "SAS® Software Version 6.12." Cary, NC.

Sasser, M., and M.D. Wichman (1991) "Identification of microorganisms through use of gas chromatography and high-performance liquid chromatography." In *Manual of Clinical Microbiology*, 5th edition, eds. W.J. Hausler, Jr., K.L. Herrmann, H.D. Isenberg, and H.J. Shadomy, American Society of Microbiology: Washington, DC, p. 111–118.

Schreier, H., W. Erlebach, and L. Albright (1980) "Variations in water quality during winter in two Yukon rivers with emphasis on dissolved oxygen concentration." *Water Research*, Vol. 14, p. 1345–1351.

Shallock, E.W., and F.B. Lotspeich (1974) "Low winter dissolved oxygen in some Alaskan rivers." US Environmental Protection Agency Arctic Environmental Research Laboratory: College, AK.

Simonsen, J.F., and P. Harremoes (1978) "Oxygen and pH fluctuation in rivers." *Water Research*, Vol. 12, p. 477–489.

Vincent, W.F., R. Rae, I. Laurion, C. Howard-Williams, and J.C. Priscu (1998) "Transparency of Antarctic ice-covered lakes to solar UV radiation." *Limnology and Oceanography*, Vol. 43, No. 4, p. 618–624.

White, K.D., and R.A. Melloh (1999) "Diurnal variation in dissolved oxygen measurements during late winter ice-covered period, Sleepers River, VT." *Proceedings, 10th International Conference on Cold Regions Engineering* [August 16-19, Lincoln, NH], p. 550–560.

White, K.D., C.M. Reynolds, J.P. Laible, K.L. Foley, and L.B. Perry (2001) "Low-temperature microbial activity in river systems," *Water Quality Technical Notes Collection* (WQTN AM-11), U.S. Army Engineer Research and Development Center, Vicksburg, MS. *www.wes.army.mil/elpubs/wqtncont.html*

Whitfield, P.H., and B. McNaughton (1986) "Dissolved-oxygen depressions under ice cover in two Yukon rivers." *Water Resources Research*, Vol. 22, No. 12, p. 1675–1679.

Application of Vacuum Sewer in the North Slope Borough

Tim Bergin, P.E.[1], Richard Reich, P.E.[2], and James Crum, P.E.[3]

Abstract

Alaska's North Slope Borough elected to go with vacuum sewer systems for four of the seven communities in which it is installing piped water and sewer. One of these community systems (Point Hope) has been in operation for over two years. These systems are unique in that they are the first systems in Alaska that use total conventional plumbing (instead of vacuum toilets and gray water valves). This paper presents the planning process which led to the decision to install vacuum sewers, generalized design aspects of the Point Hope system, and operating experience accumulated over two years operation at Point Hope.

Introduction

Vacuum sewer technology has been used to transport sewage in an extreme arctic environment on many Alaskan sanitation projects during the past 30 years. Some of these have been constructed for industrial development (oil field and mineral development) but the major use has been to provide waste collection systems in smaller Native villages in remote areas of the State. The vacuum systems under construction serving the North Slope Borough (NSB) communities of Point Hope, Atqasuk, Nuiqsut and Kaktovik are unique from other systems of this type being built by the state's Village Safe Water or federal Public Health Service programs. The unique features of the NSB projects have developed in response to the desires of the communities, character of the planning process, availability of funding from municipal bond sales, and involvement of the regional government in initial funding and continued operation of the utilities. These are discussed as they apply to the vacuum sewer system component of the 4 NSB village systems with a particular emphasis to the Point Hope system. This system serving a population of 700 is completely operational with most of the house services completed. The other three systems are in various phases of construction, so are not yet complete. Some operational characteristics are provided.

[1] Director-Environmental Services, Florida Keys Aqueduct Authority, Key West, Florida.

[2] Business Analyst, ASCG, Anchorage, Alaska.

[3] Senior Review Engineer, ASCG, Anchorage, Alaska.

The objective of this paper is to provide the planning background of the Point Hope utility system leading to the decision to install the vacuum sewer system and generalized aspects of this project's vacuum design.

Vacuum System Operational Characteristics

A typical vacuum sewer collection system consists of four components: the vacuum station, the main lines, service lines / house connections, and vacuum interface valve. The vacuum station is located in a central facility with vacuum pumps and a receiver tank. The pump automatically maintains negative pressure in the entire system. The waste rapidly flows into the system through the service lines and mains when the interface valve is activated and opens. Sewage flow speeds in excess of 48 Km/hr (30 mph) are normal in a system. Sewage flows in the system through "saw tooth" pockets created by piping grade changes that aid in the vacuum transmission process. The pipe grading to these "saw tooth" steps allows the liquid to flow forward in the system between each vacuum "flush" that rapidly propels the liquid toward the vacuum receiving tank. When sewage in the vacuum-receiving tank at the central facility reaches a predetermined level, a sewage pump is automatically activated to transport the waste to the treatment plant by pressure pipeline.

The interface valve is a critical component of the system. It separates the gravity flow house plumbing from the vacuum transmission system. It can be two separate valves with one located within a special toilet and a second gray water valve serving a small sump located within the heated space of the house. The household gray water plumbing drains to the sump where the interface valve automatically opens when the sump reaches capacity, drawing waste into the vacuum system. The toilet valve is manually activated to draw waste into the system. Some consider a system of this type (with vacuum toilet) a "true vacuum system " in that it uses a minimal amount of water (one quart per toilet flush). Small arctic communities with this type of system use less than 114 lpcpd (30 gpcd). Another approach is to use conventional household plumbing fixtures draining to an external sump where a single vacuum interface valve provides the transition between gravity and vacuum parts of the system. (Costello, 2001). Small arctic communities with these types of systems generally use more water to provide a similar service - over 170 lpcpd (45 gpcd) is normally used. Point Hope uses this type of arrangement.

Arctic Site Conditions Ideal for Vacuum Sewers

There are several site conditions and vacuum system attributes that make it an ideal choice for serving a "typical" village located in an arctic environment. These include:

- The village setting is usually in a flat local topography near a water body that has historically provided a means of transportation and is integral to subsistence activities. The flat topography, combined with high ground water in areas with unfrozen soils, results in difficulty in constructing deep gravity sewers. In areas where permafrost is present, excavation for deep gravity sewer lines can be cost prohibitive.

- Many of the village locations have poor geotechnical conditions for construction of buried pipelines. The soils are fine grained with ice-rich permafrost conditions and often are "warm" permafrost that can be easily thawed. These soils provide poor foundations for buried utilities.
- The vacuum sewer system is not grade critical. The system can transport through uneven grades and provides a limited degree of lift within the system. If the piping shifts due to intermittent freeze thaw cycles, the system can continue to operate.

Background

The NSB is the largest single municipal service area in the United States, larger in terms of landmass than several of the contiguous states. The extent of the NSB is shown on Figure 1. A vast majority of its residents are Inupiaq Eskimos, who have learned to survive in the harshest of environments known to man. In total, the region includes approximately 7,700 residents that live in eight villages. Barrow, which is the economic, transportation and administrative center for the NSB, is the largest village with approximately 4,700 residents.

The NSB was incorporated as a First Class Borough on July 2, 1972 and the residents subsequently adopted a Home Rule Charter on April 30, 1974. Incorporation as a Borough enabled the new municipality to overcome years of neglect and isolation, and to address overwhelming public needs. The NSB began to levy property taxes on the oil and gas facilities in the Prudhoe Bay Area, marking the beginning of a new era of economic growth and development of an infrastructure based on local control. This new tax base has generated essential funding for an array of services that include water and sewer systems, power and heating systems, housing and school facilities, public health facilities, transportation infrastructure, police and fire protection. Virtually none of these services were available prior to 1972, where conditions had been comparable to those of third world countries.

The quality of life for the people of the North Slope has been greatly improved since the incorporation of the NSB. The investments made in capital improvement programs are evident in the quality and quantity of programs and services available to the residents.

Basis for Design

Planning on Borough sanitation improvements started during the 1970's. In Point Barrow, the NSB decided early on that it wanted a buried water and sewer system that was based on existing technology. The Borough first decided to construct buried utilities in Barrow, primarily because of local resources and a higher population base. The remaining seven villages would be upgraded to a truck haul system, with subsequent improvements implemented at a later time. Across the Borough, reliable water sources were developed and solid waste facilities, including sewage lagoons for disposal of honey buckets, were constructed.

Construction on these initial sanitation goals proceeded during the 1980's providing substantial improvements within the region. However, the construction effort was not without its problems. The Barrow utility design resulted in an underground utilidor project that was a challenge to the political, engineering, and construction entities involved. The wooden underground walk-through utilidor, constructed in the ice rich permafrost soils, became a multi-year construction challenge that eventually had to be reevaluated due to its extreme construction costs. Eventually, this resulted in the decision to proceed with an alternative design for extensions serving the outlying subdivisions. The design was modified with improvements to include the direct burial of water and sewer piping rather than construct buried utilidors.

The sanitation construction effort in the smaller communities proceeded at this same time to provide truck haul water/sewer systems along with waste disposal sites. Although this was not an ideal service, these projects provided substantial improvements to the sanitation conditions. There were complaints about honey bucket service, availability of water, and conditions created by disposal of gray water near the homes. It was noted that the haul system was easier to build and could be expanded at a reasonable cost to serve new houses in a growing village.

Revised Design Approach

The construction of the Barrow underground system proceeded incrementally each year and continued to provide a significant boost to the Barrow economy. The smaller villages desired the same benefits that the residents of Barrow enjoyed by having a piped system. The local economic boost that Barrow experienced was also desired in the small villages. The use of honey buckets for human waste disposal collected by truck in the smaller villages was considered a reduced form of service. It was identified that the disposal of gray water in the immediate vicinity of the dwellings was a health hazard and an undesirable feature of the original haul system. Some of the villages were converting their systems to add an individual wastewater storage tank and a flush toilet that was periodically pumped and transported by vacuum truck.

Village Piped Water and Sewer Project

The expressed needs of the smaller villages resulted in political decisions to proceed with planning for improved services in the smaller villages. A preliminary analysis indicated "residents preferred and expected a piped water and sewer system which would provide a level of service equivalent to Barrow, Fairbanks, or Anchorage"(NSB,1993). The Borough approved a $313 million bond sale to provide financing for this goal, which would be repaid by future property tax revenues.

The initial phase of preliminary engineering to serve the 7 villages was completed in 1993. This effort presented a conceptual scope of work and cost for each village to upgrade the existing sanitation facilities to support a piped system along with distribution and collection piping. It's important to note that the report recommendation specifically stated that a buried vacuum sewer collection system was not as reliable as a pressure sewer system and provided no specific advantages to a gravity sewer system. In all villages, the vacuum sewer system was rejected as an alternative for further consideration. More importantly, the communities made it known that they wanted a system exactly like Barrow and that they were not interested in any experimental or unproven system. The engineering analysis supported these views.

At this time, the vacuum systems in Noorvik and Emmonak were operational and PHS was planning other similar systems. The operational experiences of these systems may not have been sufficient to persuade inclusion of the vacuum system in the preliminary engineering work analysis. In addition, it was acknowledged that over 20 percent of the NSB village homes had been converted to water and sewer holding tanks – truck haul system with conventional plumbing within the home so connecting these facilities to a gravity sewer system would not require extensive interior plumbing modification. The cost of household plumbing changes was another factor confronting planners in that the municipal bond funding would not allow any work to be performed in the interior envelope of a private structure. Any interior plumbing upgrades would need to be funded by the homeowners and they were reluctant to accept increased costs for vacuum plumbing fixtures and related operational responsibilities. In addition, there was concern by the engineers whether gray water sumps located within the interior envelope met the plumbing code requirements for venting.

With the sale of municipal bonds for the Village Water and Sewer Project approved, the NSB's Department of Capital Improvement Program Management (CIPM) was tasked with the design and construction of the Project and selected the engineering firm of LCMF Incorporated (formerly BTS/LCMF) through a competitive process as the design manager. ASCG Incorporated and CH2M Hill were sub consultants to LCMF and Shiltec Alaska participated as a technical consultant to CIPM. An aggressive schedule for design and construction was developed based on the scope prepared under the preliminary engineering report. This work included the design and construction of water intake structures, raw water pipelines, water treatment plants, water storage reservoirs, buried water distribution and wastewater collection pipe networks, wastewater treatment plants and effluent outfall piping and discharge facilities. One of the main goals for the Project was to standardize the water and sewer systems across the Borough and it was envisioned that this project would take seven years to complete, from initiation of design to completion of construction activities.

As the detailed engineering started on the first villages, it became apparent that more of the project resources would be required to provide water supply and waste treatment improvements compared to the preliminary planning. Some villages expressed environmental and land use concerns over needs to expand sewage lagoons, and most of the water sources were found not adequate for supplying a piped system. This resulted in a change in approach where several of the villages would now be provided with a completely new water treatment facility along with a packaged secondary sewage plant contained in a suitable arctic environment structure. Through use of standardized designs for several villages, some design cost savings resulted, but, overall, the project funding contingencies were being consumed.

The existing water supply systems and waste disposal systems were incorporated into the new standard designs where possible. For example, the existing Point Hope water source was converted to supply greater quantities of water to the new treatment plant and the existing sewage disposal lagoon was converted into a large subsurface disposal field for treated effluent.

The desired deep gravity sewer collection system recommended in the preliminary engineering report (NSB, 1993) for Point Hope was revisited by the design team in 1995 when it appeared that a vacuum sewer system would be best for technical and economic reasons. At that time another evaluation report was completed that again concluded that the deep gravity system should be used, but the report did not consider the economics of construction or the potential damage during excavation to the village roads, buildings, and utilities located adjacent to the trenches. The loose gravel "ball bearing" soils in the village would create unstable slopes and wide trenches during excavation. During the review of the report findings in village meetings, a deep concern was expressed about proceeding with this approach. The design team was required to evaluate the Point Hope project again to considering three collection methods: Deep gravity, Shallow gravity (<2m (<6 feet) bury), and Vacuum sewers (NSB, 1996 & 1997). Also several design issues were evaluated to provide information on the proposed Point Hope design. This data was presented to the CIPM and the village and a decision made in 1996 to proceed with the design of a vacuum system, but the system would have specific features that had not been extensively used in a severe arctic environment.

These included:

- The system would use standard plumbing fixtures in the home, which would drain by gravity to the vacuum interface pit located remote from the house.

- The vacuum interface valve (between the house plumbing and vacuum system) would serve up to 4 homes. It would require a separate power supply for heating and control. Operation of the valve pit would be a utility responsibility.
- The vacuum system would be buried. Local desires required that all systems be buried.

The revised project direction to abandon the deep gravity sewer system and proceed with a shallow bury vacuum sewer resulted in a substantial amount of project savings with the Point Hope construction budget dropping from $77 million dollars to under $50 million. With the project now completed, the expended construction funding totals about $55 million.

Vacuum Test Facility

The revised project direction did create some technical challenges to the design team. A major concern developed over the capability of a vacuum system to operate in the severe arctic environment when large quantities of super cold air are drawn into the system during each interface valve flush cycle. The vacuum interface valve requires drawing in-rush air (approximately 25-40 l (7-10 gallon) volume per flush) from the atmosphere above the valve pit which could be –45°C (–50 °F) at some times of the year. This had been considered a potential problem with past vacuum system designs in Alaska that could freeze up the sewer. Previous designs utilized vacuum toilets and separate gray water tanks. The gray water-flushing tank was vented to the interior of the home and drew in-rush flushing air from the heated dwelling space. The vacuum toilet did not require venting - its in-rush air is drawn from the warm house at the time of flushing.

The proposed NSB vacuum systems would have the interface valve outside the house with cold in-rush air. This initiated a testing program to determine if the vacuum interface valves would handle the cold air in-rush. Also, the test facility evaluated several mock-up buried valve pit / sewer service line configurations to determine design parameters. The testing program would also allow the evaluation of three different vacuum system valves that are on the market (Iseki, Roediger, and Airvac).

A test facility was built in Anchorage, Alaska. The facility consisted of a modified refrigerated shipping container (conex). Temperatures in the conex were held at –18°C (0°F) in order to simulate soil temperatures that were expected, air intake temperatures were set at –45°C (–50°F) and the water supply temperature was set at 7°C (45°F). The conex contained mock-up facilities for the sewer service lines, buried interface valve pit, and an alternative above ground interface valve box that would serve a single house. The facility incorporated vacuum pumping equipment and a computerized data gathering system. Several weeks of tests were conducted during 1997. The conclusions included:

- The vacuum interface valves could operate without freeze-up coming from a short intake vent on the valve pit with inrush air of –45°C (–50°F).
- The inrush air did not create a freezing problem in the service line piping and intermittent conditions of low temperatures could be controlled with a heat tape
- An above ground interface valve box mounted on a pedestal foundation proved to be a more acceptable design. It was incorporated into the Point Hope design and would also be used in Atqasuk, Nuiqsut, and Kaktovik. The concept of a buried interface valve pit was generally eliminated (NSB,1998).

Final System Design

The water systems serving the NSB villages are fill and draw systems, i.e., the water supply for the year is filled up in the summer and then drawn off for the rest of the year. This design constraint was apparent to the designers throughout the development of the project. As the community grows in future years, the water supply will become more critical. The current use of a vacuum system for waste transport results in minimal water use for this purpose and future water consumption can be reduced even more. Retrofitting the system to install vacuum toilets to handle black water can reduce Point Hope's water consumption further and utilizing the existing pedestal mounted vacuum sumps to handle gray water.

The system built in Point Hope generally follows established practices for arctic utility construction. The major features are presented and shown on Figure 2. The pipe used for water and sewer mainlines is HDPE with urethane insulation and HDPE outer carrier pipe. The Point Hope vacuum system design consists of two 200mm (8-inch) vacuum mains with three laterals coming off each main. Five of the laterals are 100mm (4 inch) diameter and one serving a future growth area was 150mm (6 inch).

An access vault is installed where each lateral intersects the main line. This vault contains the diversion valve for each lateral along with an access point for the heating system controls. Each sewer lateral consists of an arctic insulated pipe with a glycol supply and return line located in the lower portion of the insulated carrier pipe. Heat is supplied from a heating facility located in the wastewater treatment building. Glycol heated to 15°C (60°F) in Point Hope and 7°C (45°F) in other villages is pumped by a 1 KW (1-1/3 HP) centrifugal pump into a 75mm (3 inch) heating line installed along the 200mm (8 inch) mainlines. At each access vault, a temperature control valve varies the flow in each lateral, and therefore the heat admitted to that service area. The heating system is only used between November and May and is only used at the lowest flow ranges. The heating system is monitored to keep the system thawed, but also to not allow thawing of the adjacent permafrost (this is not critical in Point Hope, but is in the other villages with ice rich permafrost). The sewer service lines are 100mm (4")HDPE with 100mm (4") of insulation and an outer PE jacket. Service line heat tapes are supplied from the electric panel at the service box adjacent to each house (Hansen, 2000). The schematic and photograph of the service box are shown on Figures 3 and 4. Access to the vacuum sewer main line is from clean-outs located in the access vaults.

The Point Hope sewage treatment plant was already under design when the decision to proceed with a vacuum system was made. This resulted in the sewage plant being sited near the original sewage treatment ponds that were on a "hill" about 2m (7 feet) higher than the rest of the community. The vacuum plant was located within the wastewater treatment plant building. As a result, the vacuum mainlines require a stair step of piping going into the plant. In order to alleviate potential waterloging problems that may be caused by this configuration, air admission valves were installed at access vaults one and two. A waterlogged system occurs when an excessive amount of liquid collects in the piping system normally filled substantially with air. When this occurs, the operation becomes sluggish, the vacuum is reduced and activated valves are prevented from operating. The special air admission valves sense a low vacuum upstream and intermittently introduce additional flushing air to clear out water logging. The system design generally uses Airvac design standards.

The Vacuum pump installation was a skid mounted pumping plant with stainless steel piping and tank and 18.7KW (25 HP) duplex Busch vacuum pumps and 2.2KW (3 HP) duplex Cornell non-clog sewage pumps.

Construction Highlights

Construction on the project was phased to coordinate with design document completion and capacity of the contractor forces. The first phase of construction was completed in 1997-98 and consisted of improvements to the existing facilities to support a new piped system along with several major components of the new system. These consisted of improvements to the water intake structure, rehabilitation of the 9.6 Km (6 mile) raw water supply pipeline, construction of a new water treatment plant building and treatment plant, construction of a new 15,000 Cum (4 million gallon) treated water reservoir, construction of a water distribution-pump facility within the old water treatment plant, and construction of a new wastewater treatment building complete with treatment plant, vacuum skid, and heating plant.

The second phase generally consisted of all below ground structures and piping, including all water distribution piping, all vacuum mains and laterals, access vaults, service lines, and service box pedestals at each house. This work was completed during 1999. The third phase generally consisted of connection of dwellings to the service box pedestals, installation of electrical controls and power. This work was completed between 1999 and 2001.

All the major construction on this project was preformed by a joint venture between Tigara/Conam. Tigara is the local village corporation.

Operation and Maintenance

The vacuum system was placed in operation in August 1999 and has operated very well during the last two years. Periodic maintenance reviews indicate it requires less maintenance than a comparable gravity system. There have been no major trouble calls on the sewage system. The glycol heating system works well and it has been noted that a minimal amount of heat has been needed.

The system is operated and managed by the village corporation working cooperatively with the NSB Department of Municipal Services. The charges for service are based on a flat rate system and are generally $70 per month for residential water and sewer service.

The Point Hope system has proven a success. It has proven that buried vacuum systems that utilize standard plumbing fixtures can be built in an extreme arctic environment. It has established some new parameters for design in that the interface vacuum valves are now installed much closer to in-rush intake air vents than previously thought possible. The use of the pedestal mounted service box to contain the vacuum interface valve for each home is a unique feature of this system. Its use allowed the vacuum system to be constructed and still utilize standard plumbing fixtures in the homes. If the communities wish to retrofit the system at some future date to install vacuum toilets, it can be done at a nominal expense. This would result in additional water savings that are critical in these arctic communities.

References

Costello, P., Schubert, D., and Lore, E., (2001), "Design, Construction, and Operation of Vacuum Sewer Systems in Rural Alaska", *Sixth Annual Research and Development Conference on Rural Sanitation*, Anchorage, Alaska.

Hansen, K.E., and Martin, R. W., (2000), "Vacuum Sewer / Circulating Water Service Connections For the North Slope Borough's Village Water & Sewer Project", *Fifth Annual Research and Development Conference on Rural Sanitation*, Fairbanks, Alaska.

North Slope Borough, Capital Improvement Project Management Department, (1993), "Water and Sewer Project – Preliminary Engineering Report – Phase I", Barrow, Alaska.

North Slope Borough, Capital Improvement Project Management Department, (1996), "Sewage Collection Alternative Analysis Report – Point Hope", Barrow Alaska.

North Slope Borough, Capital Improvement Project Management Department, (1997), "Vacuum Sewer Service Alternative Analysis – Point Hope", Barrow Alaska.

North Slope Borough, Capital Improvement Project Management Department, (1998), "Final Report- Vacuum Test Facility", Anchorage, Alaska.

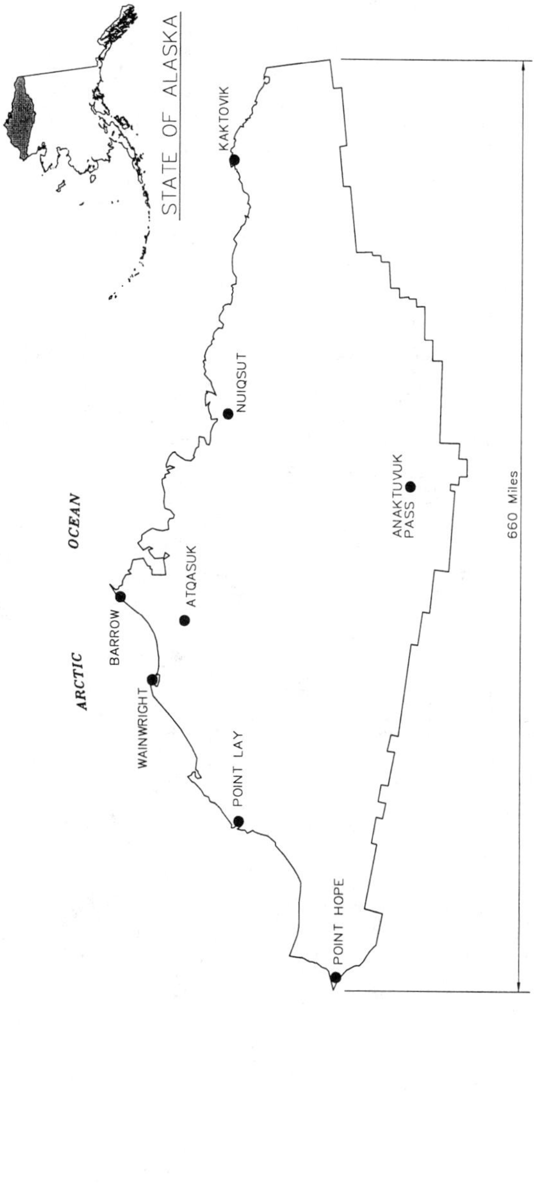

Figure 1 – Map of North Slope Area and 8 Villages

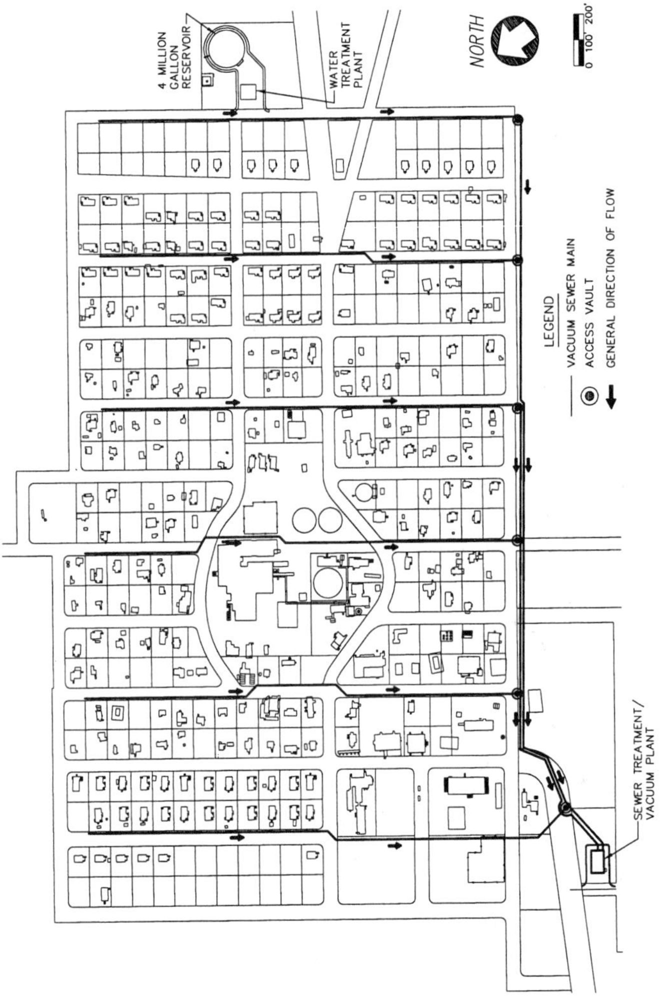

Figure 2 – Plan View of Point Hope Vacuum System

Figure 3 – Typical Home with Vacuum Service Pedestal

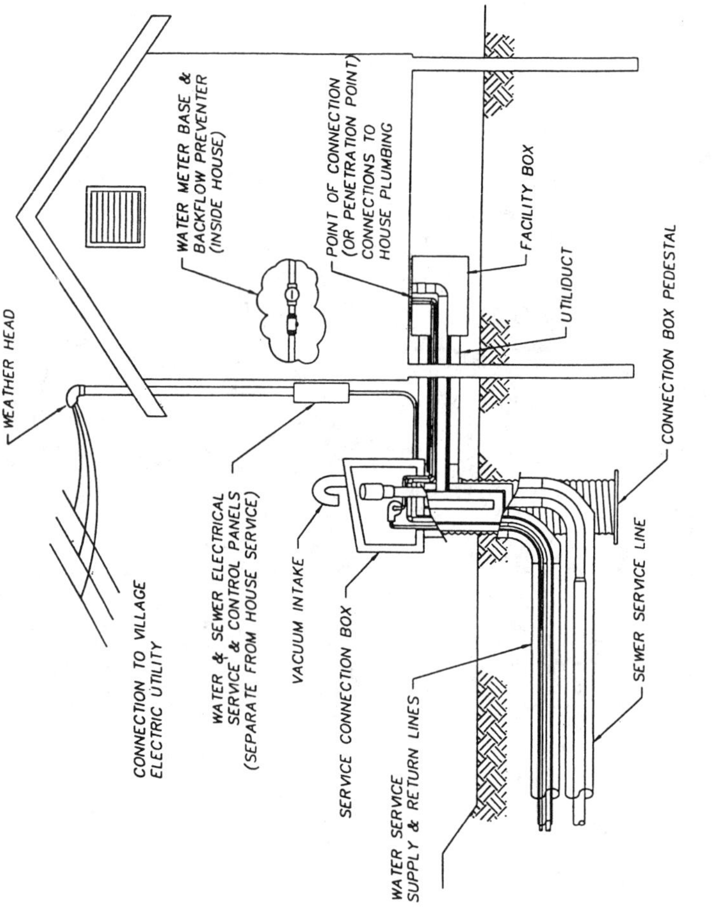

Figure 4 – Drawing of Vacuuum Service Pedestal and Service Line

Buried Utilities for Alaska: Dewater or Winter Work?

J. L. Hulsey[1], Nada Raad[2], Steve Saboundjian[3], Randy Brand[4]

Abstract

This paper describes a construction project in Fairbanks, Alaska where new sewer lines and a lift station are to be installed below the water table. Typically, a contractor would dewater the site, excavate, and install the underground facilities, then backfill and compact. However, the discovery of contamination near the project site required a restrictive dewatering permit by the Alaska Department of Environmental Conservation (ADEC). The first attempt to dewater the site was unsuccessful due to abundant summer rain producing a rapid increase in water table levels.

A second ADEC permit stipulated that the dewatering discharge be regularly monitored for contamination levels. The paper describes the quality control plan developed for this purpose. It includes hourly effluent sampling, GC testing and water table depth measurements.

By the time the contractor excavated and installed a new sewer line, the effluent contamination level rose above acceptable limits and dewatering was terminated. The rest of the project was completed during winter.

Introduction

Improvements to 1.4 km of Old Steese Highway in Fairbanks, Alaska were started in the summer of 2000. The work included street widening, curb and gutter, traffic signals, traffic lights, a railroad crossing and underground utilities. New underground utilities included water, storm sewer, sanitary sewer, a lift station, telephone, and power. The lift station and most of the first 340 m of sanitary sewer to be installed was below water table. Traditional site dewatering and pipe installation was planned. However, benzene contamination near the site precluded conventional dewatering.

It is the intent of this paper to show how a project can be successful when restrictive permits near contaminated sites are in-place. The paper illustrates how excavations for a lift station and sections of the sewer below water table were dewatered and installed.

[1]Associate Professor of Civil Engineering, Univ. of Alaska Fairbanks, ffjlh@uaf.edu,
[2]Environmental Engineer, Great Northwest, Inc.,
[3]Research Engineer, Alaska Department of Transportation and Public Facilities,
[4]Vice President, Great Northwest, Inc., General Contractors, Fairbanks, Alaska

Overview

Hydrocarbon contamination near the construction site impacted excavation dewatering for a proposed concrete lift station and a new sanitary sewer. The dewatering effort for the sewer was between the lift station and the Wendell Street Bridge (Figure 1). The contractor, Great Northwest Inc. (GNI), installed a new storm sewer between Noyes Slough and College Road and constructed a discharge outlet at the Noyes Slough and Minnie Street (Figure 1). These features were used to transfer water pumped from dewatering wells to the Noyes Slough.

GNI submitted a permit to use five wells to dewater the lift station excavation. During permit review, it was disclosed that a significant amount of contamination existed near the construction site. Recorded contamination was on private property outside of the state right-of-way. Eventually restricted permits were granted by ADEC. Within hours of installing the precast concrete lift station heavy rains came and caused the Chena River to flood. Subsequently, dewatering was terminated.

A second permit was submitted to dewater the excavation for the sanitary sewer between the lift station and the Wendell Street Bridge. A restrictive permit, approved by ADEC, required regular monitoring of the groundwater discharge at Minnie Street. Sampling, testing and water table height measurements were done at one monitoring well at each of the four sites (Figure 1).

Hourly sampling combined with testing for contamination using a portable GC enabled GNI to excavate and install some of the new sewer line. However, the dewater effluent contamination level rose above acceptable levels and pumping was terminated.

New dewatering permits were rejected. The rest of the work was completed during November when the ground was frozen. Materials excavated from the site were placed back in the ditch.

Project Details

Lift Station. A concrete lift station was installed at 4th Street (Figures 1 and 2). Two contamination sources (at 3rd Street and Old Steese Hwy) were about 122 m from the wells. Figure 3 shows a cross section of the site. A dewatering scheme aimed at lowering the groundwater level to an elevation of 125-m [i.e. 1.2 m below bottom slab of the proposed lift station]. Five wells were installed in a circular pattern (Figure 2) around the lift station [radius = 9 m].

The combined effect of pumping from five wells was determined using Theiss' solution and the superposition principle (Fetter, 1994). For a given pumping time (t, hrs) and pumping rate (Q, m^3/hr), the drawdown (d, m) at the lift station was found as:

$$d = \sum_{i=1}^{5} \frac{Q}{4\pi\ K b} W(u_i) \quad ; \ W(u_i) = \text{well function} \quad ; \quad u_i = \frac{r_i^{\ 2}\ S}{4 K b t} \qquad (1)$$

Where: K = hydraulic conductivity of aquifer = 0.32 m/hr

b = thickness of aquifer = 27 m
S = storage coefficient = 0.005
r_i = distance from well i to point in question [for lift station, r = 9 m]

Values of K and S reflect the type of soil at the site (Fetter, 1994).

Expected drawdown due to pumping at the proposed lift station is presented in Figure 4. The graph shows how the groundwater elevation is expected to vary with pumping rate (Q) and time (t) for five wells. Table 1 shows three alternatives.

Effect of pumping on Noyes Slough water level was a concern to the ADEC. ADEC required that the slough water elevation be maintained at about 129.5 m (Figure 3). Distances (r_i) from the wells to the slough were: 167.6 m for well 1, 178.3 m for wells 2 and 5, and 190.5 m for wells 3 and 4. Again, using Equation 1, drawdown at the slough due to pumping from the five wells is determined. The elevation of the water level in the slough is shown in Table 1. This water level was maintained at about 129.5 m.

An ADEC permit allowed GNI to proceed with site dewatering. It was implemented successfully and the lift station was installed. No fluctuations in the water level at the Noyes Slough were detected during the dewatering process. After the section of the lift station above the water table was in-place, the crew began installing the pipe to connect the lift station. It was during the pipe installation when rains came and the Chena River flooded causing the ground water to rise faster than the dewatering effort. Consequently, the pumps were turned off.

Sanitary sewer lines. Five equally spaced wells (27.4 m center-to-center) were installed to dewater the site where a sewer pipe was to be laid 3.7 m below the road surface, i.e. elevation 129 m. At that time, the groundwater elevation was 130-m. The wells were longitudinally aligned along the shoulder of the roadway. Figures 5 and 6 show the location of the wells (and manholes) in the initial dewatering phase.

Once a section of the sewer pipe was laid between two manholes (say between manholes 1 and 2), the plan called for shutting the well closest to manhole 1 and installing another at the farthest end of the well line (next to manhole 3, 27.4 m from well 5). This procedure was to be repeated until the entire pipe was laid.

Equation 1 was used to develop a relationship between the pumping rate (Q) and the time (t) required to reach a drawdown of 1.52 m at different manhole locations between wells (Figure 7). For example, if the pumping rate is 23 m^3/hr (70 gpm) from each of the 5 wells, it will take 45 min, 1 hour and 2 hours to get a drawdown of 1.52 m at manholes 2, 1 and 3, respectively.

An ADEC permit was obtained for this dewatering plan. The permit stipulated that if the effluent contained contaminants above acceptable levels, dewatering should be stopped. Although 5 wells in a line were proposed for use during dewatering, the construction crew would install one well and dewater. During construction, excavation and well installation occurred simultaneously. Work continued until well 4 was installed and test results showed contamination in the effluent. Consequently, the pumps were turned off and further dewatering of the site was called off.

Permits, Testing and monitoring

The first permit. GNI complied with the water discharge permit requirements issued by the ADEC, ADOT&PF, Department of Natural Resources (DNR), and Department of Fish and Game (F&G).

During the process of applying for permits to dewater the construction site, GNI found that the site was contaminated with benzene, toluene, ethylbenzene and xylene (BTEX) and possibly other contaminants. This was due to leakage of underground storage tanks (UST) and spills that migrated to the groundwater level. An existing air-sparging system is at Site 1 (Figure 1). This system was in-place and approved by ADEC as part of a previous clean-up plan. Table 2 shows ADEC reported contamination levels for the two sites affecting the work. The data were based on analytical BTEX results from groundwater samples at Site 1 and groundwater monitoring and analytical data at Site 2 (Figures 2 and 5).

ADEC issued restrictive water discharge permits based on levels of groundwater contamination that existed at the time of construction. GNI planned to dewater the excavation around the proposed lift station at 4th Street and the Old Steese Highway. Therefore, GNI initially applied for a General Wastewater Disposal Permit to discharge dewatering effluent into the Noyes Slough.

As part of the permitting process and prior to dewatering, ADEC requested that GNI install a test well at the lift station. Groundwater samples showed that no contamination existed. However, turbidity (740 NTU) and settleable solids (4 ml/l) test results were high. This was because the well was newly installed. Results indicated that the level of total iron was 62.9 mg/l. This was compared to the background samples taken from the Noyes Slough below the Illinois Street Bridge. Background sample tests showed that turbidity was 100 NTU, settleable solids were 10 ml/l and total iron was 55.5 mg/l.

ADEC was concerned that dewatering might spread contamination and impact water quality. Therefore, ADEC issued a restrictive permit to dewater. Restrictions included:

- Ground water elevation was to be monitored at sites 1, 2 and 3 (Figure 1).
- If the ground water elevation drops below the lowest recorded natural elevation levels of 128.9 m, dewatering would cease.
- The water discharged into Noyes Slough shall not be allowed to flow upstream and discharge into the Chena River. So, GNI installed a temporary dam to restrict the backflow.
- GNI was asked to monitor all general permit parameters on a daily basis and for three days following the dewatering effort. Limitations of the permit included monitoring the discharge at the Noyes Slough at the Minnie Street Bridge. The permit was based on the following criteria (ADEC, Fairbanks Public records):
  - total aqueous hydrocarbons (15 μg/l)
  - Total aromatic hydrocarbons (10 μg/l)
  - pH (between 6.5-8.5)

  - flow (500 gpm for 7 days, except for one 48-hour period where the flow can be up to 5,000 gpm; not to exceed 4,320,000 gpd)
- Besides those restrictions, GNI was required to monitor total iron, turbidity, and settleable solids and meet the following limits:
  - Total iron (2 μg/l), limit was increased. High levels of iron exist naturally.
  - Turbidity (5NTU's above background)
  - Settleable solids (0.2 ml/l)
- In addition to monitoring the well water effluent at the Minnie Street discharge pipe, ADEC also requested that the Noyes Slough be monitored at the Illinois Street Bridge and the inlet and outlet to the Chena River. Dewatering activities were to be stopped if ADEC received complaints from the public regarding the quality and quantity of the discharge water.

Five wells were used for dewatering (Figure 2). During dewatering, discharge was monitored for hydrocarbons, turbidity and settleable solids, total iron, and pH. The discharge was sampled at the discharge pipe at the Noyes Slough (Figure 1) and at several Noyes Slough downstream locations. Monitoring was implemented to verify water quality was maintained during the dewatering effort. Samples were laboratory tested in accordance with the requirements stipulated by the permit.

ADEC suggested using the Noyes Slough as a settling basin for any iron (Fe III) that might be present in the discharged effluent. Permission was granted by DNR to use the slough for this purpose.

Iron CHEMetrics Test kits (Iron), pH strips, and a Portable TDS Meter (turbidity), were used to measure field variables. The initial discharge occurred on August 11, 2000, with an expected duration of seven days. The method of dewatering was through dewatering wells installed with gravel around slotted casings covered with screen wire. Water was discharged into the city storm-drain then directed towards the Noyes Slough through the newly installed pipe (culvert) protruding at the bank of the channel covered with riprap. GNI requested permission to measure groundwater levels at sites 1, 2 and 3 (Figure 1). Electrical tape with a sensor (E-tape) was used to measure water level in the monitoring wells. The tape was decontaminated after each measurement and before it was used again.

Dewatering began on August 11, 2000 with an intention to install a lift station and sanitary sewer pipes (Figure 2). On August 14, 2000, the Chena River began to flood causing ground water to rise faster than the site could be dewatered. Thus, the wells were turned off on August 16, 2000. Under this permit, discharge water test results for total aqueous and aromatic hydrocarbons were negative.

The portion of the lift station below water table was installed. However, sanitary sewer and sanitary manholes between the lift station and the Wendell Street Bridge were not installed.

The second permit. After the wells were shut down, the ground water elevation was monitored periodically. For example, the water level on August 22, 2000 was around 130-m above sea level. Future dewatering was postponed because the ground water elevation was higher than previously recorded.

Later, GNI applied for a permit to dewater the area between the lift station and the Wendell Street Bridge (Figure 5). This permit was more restrictive than the first. Dewatering began around September 10, 2000, with an expected duration of two weeks. The maximum rate of discharge was about four million gallons per day. GNI proposed to dewater with 5 wells operating at a time. These wells would be installed in a line with gravel around slotted casings covered with screen wire (Figure 5). The location of these wells was to change during the processes of installing the sanitary sewer pipes. There were 4-16,047 cal (25 HP) and 2-9,628 cal (15 HP) pumps used for water withdrawal. Water was to be discharged into the city storm-drain then directed towards the Noyes Slough through a recently installed pipe (Figure 1); the method of discharge was the same as the previous permit.

Since the proposed dewatering was closer to the contaminated sites, ADEC was more strict in their second permit requirement and stated that it was "highly likely that the department will take enforcement action" if dewatering causes the water to exceed the State of Alaska Water Quality Standards. In order to prevent such a conflict, GNI sampled and analyzed the water hourly before and after being discharged into the Noyes Slough. The water was analyzed for contaminates using two portable GC's. In addition, other variables were tested frequently at various locations along the Slough.

Dewatering activities were carried on until September 14, 2000, when discharge showed the presence of BTEX. All dewatering activities were shut down. Immediately samples were collected from different locations in the Noyes Slough and analyzed for BTEX. The contamination was diluted and the effect to human health and environment was determined to be negligible in the Noyes Slough.

Following this incident, ADEC rejected permit requests to dewater. Work was completed in November. At that time, the water table dropped the bottom of the excavation. Frozen excavated material removed from the roadway was stockpiled and placed in the ditch after the sewer pipe was installed.

Conclusions

This paper illustrates an effective way of dewatering near contamination plumes. It also shows how sampling and monitoring may be an effective method for moving groundwater from point-to-point. More research is needed to study the most efficient and economical procedure to follow and prevent the spread of contamination during dewatering in a groundwater contaminated area. When contaminates are near dewatering pumps other construction methods may be needed.

References

Alaska Department of Environmental Conservation, *Fairbanks Public Records*.

Bellandi R. (1988). *Hazardous Waste Site Remediation*. O'Brien & Gere Engineers. Van Nostrand Reinhold, New York.

Fetter C.W. (1994). *Applied Hydrogeology*, 3rd. Ed., Prentice-Hill.

Manahan S. E. (1994). *Environmental Chemistry*. Lewis Publishers, Ann Arbor.

National Research Council (NRC) (1993). *In Situ Bioremediation*. National Academy Press, Washington, D.C.

Treatment Technologies Screening Matrix, http://www.frtr.gov/matrix.

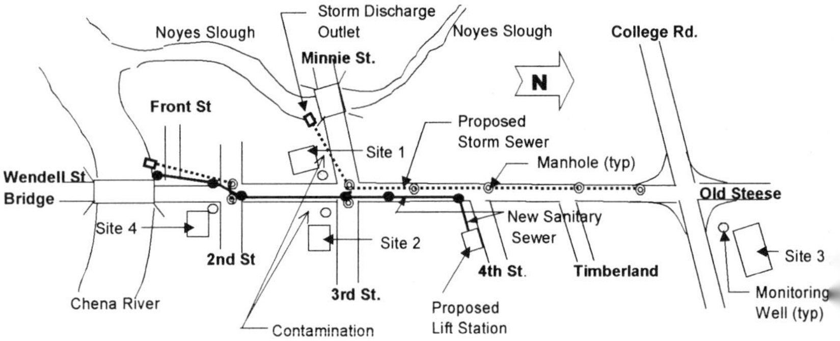

Figure 1. Project site conditions

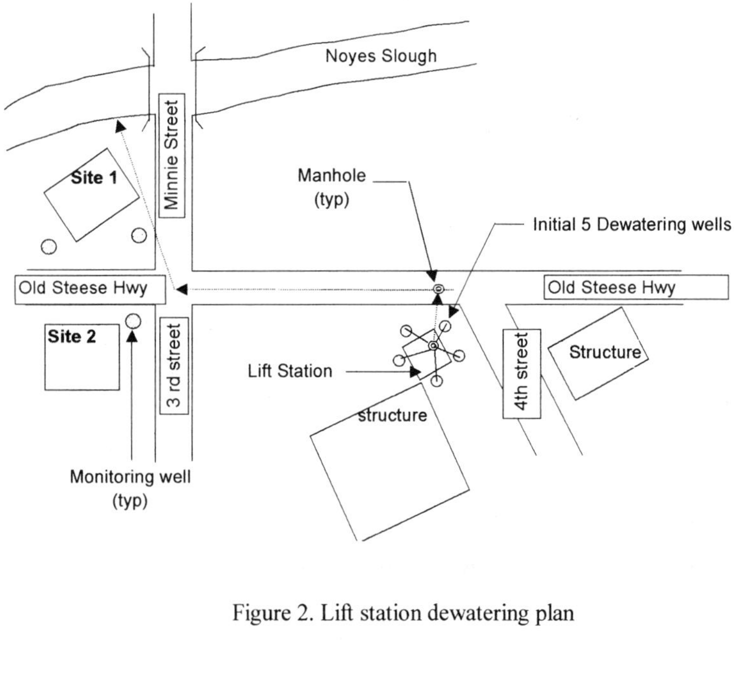

Figure 2. Lift station dewatering plan

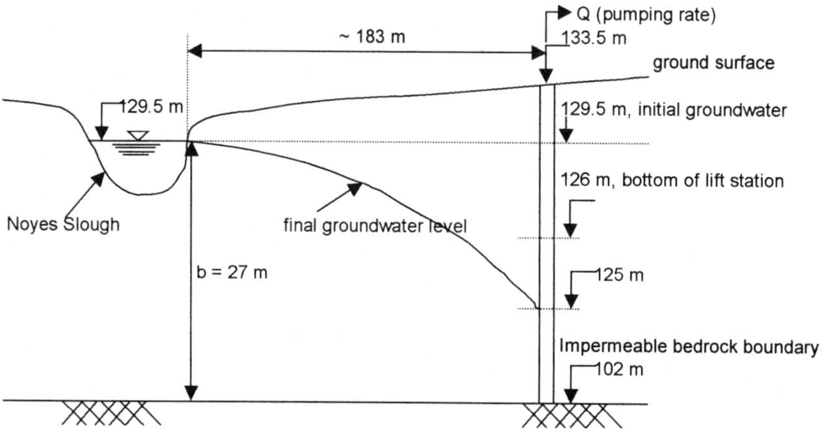

Figure 3. Water level at the Noyes Slough (not to scale)

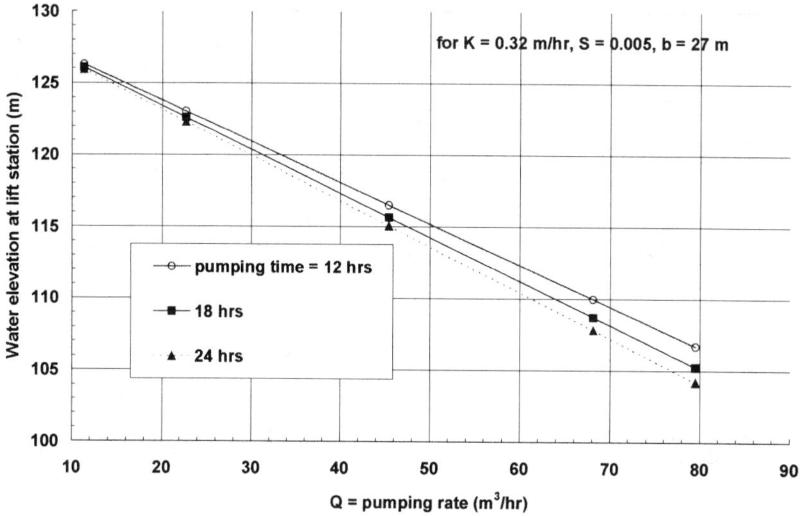

Figure 4. Water elevation at lift station due to pumping

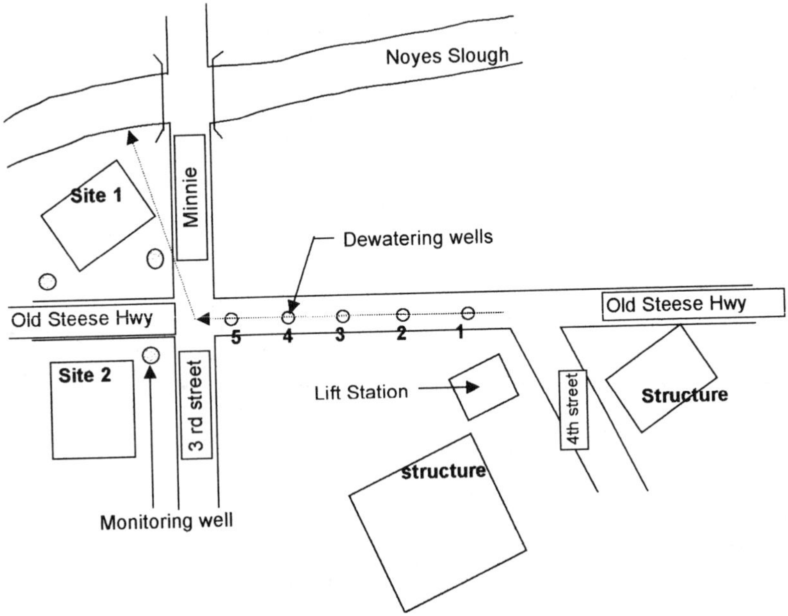

Figure 5. Well locations for sewer pipe installation

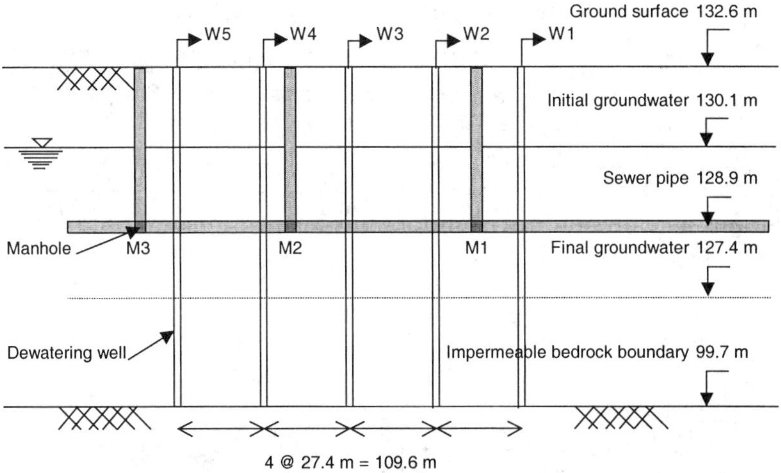

Figure 6. Dewatering for the sewer pipe installation (not to scale)

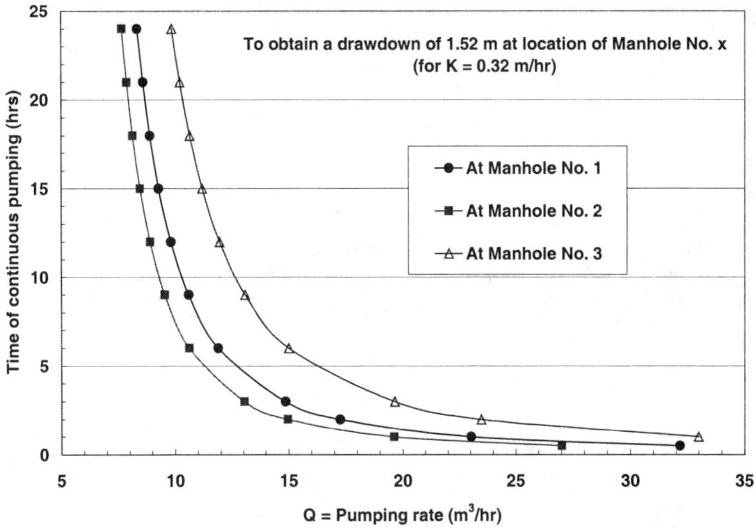

Figure 7. Pumping rates vs. time for a 1.5-m drawdown at a given manhole

Table 1. Pumping rate vs. elevation at the Slough

Alternative	Pumping rate, Q m^3/hr (gal/min)	Pumping time, t (hrs)	Water elevation at the Noyes Slough, m (ft)
1	14 (62)	24	128.6 (422)
2	15 (65)	18	128.9 (423)
3	16 (70)	12	128.9 (423)

Table 2. Contamination levels for sites 1 and 2 (Year 2000)

Date	Contaminant		
	Type	Regulatory Limit (μg/L)	Concentration at Monitoring well (μg/L)
Site 1:			
Jan 7, 2000	Benzene	5	27.4 – 190
Jan 7, 2000	Toluene	1,000	<DL (0.5) - 1
Jan 7, 2000	Ethyl benzene	700	<DL (0.5) ≤ 1
Jan 7, 2000	Xylenes	10,000	<DL (1.5) – 2.84
Site 2:			
Feb, 2000	Benzene	5	2.47 – 3,760
Feb, 2000	Toluene	1,000	< 1.0 - 498
Feb, 2000	Ethyl benzene	700	< 4.89 – 326
Feb, 2000	Xylenes	10,000	8.81 – 1,290

DL = Detection limit

Teaching Arctic Engineering Online

Orson P. Smith[1], M. ASCE

Abstract

The University of Alaska Anchorage (UAA), School of Engineering began web-enhanced graduate instruction of arctic engineering and related topics in 1998 and plans to provide a series of Arctic Engineering graduate courses completely online by 2004. The program begins with the popular keystone course CE A603 "Arctic Engineering," which is associated with engineering registration with the State of Alaska. Many of the evolving principles of online education are taken to their limits in the preparation and delivery of such specialized computation-oriented topics. Engineering graduate students are typically motivated learners with a practiced toolkit of study skills that lends itself to asynchronous online learning. The UAA experience has revealed important lessons regarding web delivery of graduate-level technical education that have led to continuing improvements in modular learning materials and interactions between instructors and students.

Introduction

The UAA School of Engineering began web-enhanced instruction of engineering courses in 1998, encouraged by a campus-wide effort to create course web pages and to incorporate Internet use in all levels of university instruction. The University office responsible for conventional distance learning was renamed Academic Technology Services and provided with resources to guide faculty in web applications for instruction and learning assessment. The School of Engineering first participated in this evolution with expansion of its web site to include syllabi and detailed course materials for viewing or download, reducing the need for tedious reproduction of handout papers. A part-time Webmaster was hired to organize web pages for information on School programs with links to course web sites. This mode of web use for course work was constrained by the requirement that faculty take time to train themselves in use of web editing software to create and maintain their own course web pages. By the end of the academic year, some UAA engineering courses had elaborate web sites, while others had no more than a simple announcement created by the Webmaster.

UAA Academic Technology Services in 1999 first provided campus-wide use of and faculty training in the commercial Blackboard software platform (Copyright © 2001 Blackboard, Inc.) for online course management. A number of competitive software systems are in use at other universities. The availability of the Blackboard system at UAA has made web-enhanced or 100 percent online teaching much easier for faculty with minimal training in web page creation and maintenance. This system,

[1] Associate Professor and Arctic Engineering Program Chair, School of Engineering, University of Alaska Anchorage, 3211 Providence Dr., Anchorage, AK 99508; *afops@uaa.alaska.edu*

like similar competitive commercial software, provides ready-made course web sites with point-and-click course management options (Figure 1) including:

- Password access only to enrolled students,
- Course building and authoring tools,
- Assessment tools including a question pool and grade book, and
- Synchronous and asynchronous collaboration tools.

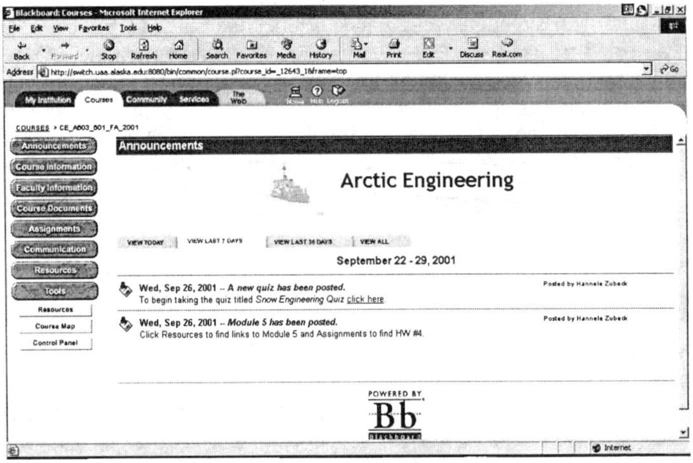

Figure 1. Home Blackboard web page for CE A603 online, Fall 2001.

The UAA School of Engineering has offered the graduate course CE A603 "Arctic Engineering" in conventional lecture format every semester for over 25 years. The class satisfies a requirement for professional registration in Alaska, which has assured steady enrollment on the order of 40 to 50 students per semester. Enrollments during the late 1970's and 1980's at times exceeded 100 per semester, corresponding to design and construction of the Trans-Alaska Pipeline System, North Slope oil and gas exploration and development, and Alaska infrastructure build-up made possible with oil revenues. Enrollments declined during the early 1990's, but have increased again in recent years. This resurgence seems to relate to growing attention worldwide toward development of mineral resources in the far north and transport of these resources southward to market. International companies recognize growth of northern mineral developments, transportation systems, communities, military outposts, and communications systems as a strategic market for the future.

The UAA Provost targeted teaching CE A603 online in 1999 as an opportunity to experiment with this mode of distance delivery. Already using extensive web resources via the School of Engineering web server, this course appeared ideal for an experiment in online course delivery. The opportunity gained the attention of the University of Alaska President in 2000. The President's office allocated special resources for development and initial delivery of the course 100 percent online. Course materials were developed for implementation on Blackboard during the summer of 2000 and tested during the fall 2000 semester, which also had class meetings and conventional lectures. Feedback from students and other lessons learned were applied to enhance course materials for the first 100 percent online delivery of the course in the Spring 2001 semester.

The Alaska regulation governing registration and licensing by the State Board of Registration for Architects, Engineers, and Land Surveyors (12 AAC 36.110) reads "...An applicant for registration as an architect, engineer, or landscape architect must have successfully completed a board-approved university level course in arctic engineering or its equivalent...." The Board previously approved CE A603 Arctic Engineering for this purpose and the Board's Licensing Examiner has been routinely notified by the UAA School of Engineering of students who pass the course. The Board endorsed the proposal for online delivery of CE A603 as meeting the statutory requirement, following a presentation by UAA faculty at its November 2000 meeting.

Synchronous versus Asynchronous Interaction

Preparations for teaching Arctic Engineering online included readings and short courses in online teaching (*e.g.*, LERN 2001 http://www.lern.org/). The UAA School of Engineering faculty was thus introduced to tenets of the new specialty of "electronic pedagogy." This area of academic endeavor is changing as fast as the computer hardware and software used for educational purposes. The effectiveness of online education remains a subject of debate.

Online learning is not the impersonal end to thousands of years of face-to-face education. Given the anonymity of many students in the large classes that prevail at most modern universities, online communications actually increase interaction through email exchanges. Communications by email are now familiar to everyone. The Blackboard system also offers an email-like "Discussion Board" feature, which can be organized into topical forums for questions and answers shared by the class, with optional anonymous participation. A roster of email addresses encourages inter-student messages and formation of work groups.

Concerns about interaction have split web-based learning into two major categories, *synchronous* and *asynchronous*, each with significantly different hardware and software requirements. Synchronous learning involves simultaneous participation of the class in the form of a teleconference or conference telephone call. Web-based systems with visual and audio, or just audio, participation are available. The prototype Internet 2 broadband interconnection among select universities (http://www.internet2.edu/ 2001) accommodates simultaneous audio and video interaction among widespread, properly equipped users. A number of commercial vendors provide software designed to use more commonly available hardware in synchronous teleconferencing mode, including simultaneous text, graphics, and audio exchanges (*e.g.*, http://www.centra.com/ 2001).

The UAA School of Engineering chose an exclusively asynchronous mode of delivery for CE A603 for the following reasons:

- Students in all time zones of the world might potentially enroll, making scheduling of regular synchronous meetings impractical,

- Engineering professionals tend to work long hours and travel a great deal. Web-based asynchronous teaching accommodates delivery anytime to anyplace with Internet access.

- Students in asynchronous mode can choose their own optimum learning time to review course materials and interact with the class. This advantage has proven useful for all students at UAA, in combination with face-to-face class meetings.

Structure of CE A603 Online

The web-based CE A603 Arctic Engineering course follows the conventional 16-week semester schedule at UAA. New material and assignments are posted weekly. Otherwise, the course is asynchronous with no scheduled class meetings, physical or virtual. Homeworks, papers, and exams are transmitted via email to the instructor or via the email-like Blackboard "Digital Dropbox."

A well-chosen textbook gains value in online teaching by providing common reading and tested organization of course topics. A good text for a graduate engineering topic typically has the added value of a reference for esoteric formulae and associated physical properties presented in tables and graphs. Text readings reduce, but don't eliminate need for online materials to explain course topics. With a good text, online materials can be focused to add context, regional perspective, and explanation of practical applications. The textbook currently in use for CE A603 online is by Eranti and Lee (2000) and may be ordered online from the UAA Bookstore. Readings in this text are supplemented with readings from websites, some with downloadable publications, such as reports of the US Army Cold Regions Research and Engineering Center (CRREL 2001 http://www.crrel.usace.army.mil/).

Learning Modules

The semester pace of CE A603 online lends itself to development of modules equivalent to weekly classroom lessons. Specialized digital learning module files can be applied with little or no modification in other courses where the same topic is relevant, *e.g.*, the CE A603 module about "global climate." The modular approach also accommodates presentations by a number of instructors in the same course. The main instructor becomes a conductor of diverse material prepared by specialized colleagues. Since digital materials can be prepared essentially anywhere, collaborations with co-instructors off-campus, out-of-state, or overseas are equally feasible. A library of valuable specialized learning modules is rapidly accumulated.

The primary mode of module presentation in CE A603 is by PowerPoint slide shows, incorporating text notes and audio narration. Slow telephone modem connections do not readily accommodate the most sophisticated PowerPoint presentations with notes and audio. PowerPoint files are translated to html and presented as collections of individual slides, each appearing in the manner of separate web pages. Though users need only download one slide at a time this way, audio files of less than one-minute narration are sizeable enough to bog down a slow Internet connection. The convention adapted for CE A603 to accommodate users with slow connections is to make available PowerPoint presentations with and without narration. Narration scripts for each module are presented as separate text files. Reading of a script assures efficient use of narration time. Notes equivalent to the narration script can also be incorporated in the PowerPoint slide show, at the discretion of the presenter. Participants in the first CE A603 course online soon requested readily printable copies of PowerPoint slides, which were provided as separate "pdf" files of printable multi-slide handout sheets.

Questions of intellectual property rights abound in online teaching. Materials prepared for online delivery in CE A603 include PowerPoint slides, syllabi, schedules, notes, and example computations presented as text files and annotated links to pertinent web sites. No methods exist to completely prevent plagiarism of materials presented online. Blackboard's password-protected access to the website prevents access to non-students. Another step to discourage theft of this valuable intellectual property is to name the university, course, module author, and date in the footer of every PowerPoint slide, page of text, and graphic. Converting text and graphics to "pdf" format for viewing by Acrobat Reader freeware (Copyright ©2001 Adobe Systems, Inc.) also provides modest discouragement. The "pdf" format also assures a common appearance to files that otherwise might not match the software brand or version onboard a particular computer.

Electronic Homework

The computational assignments associated with engineering education, essential to learn practical skills of the profession, are a particular challenge with online education. Online students must submit "ehomework" as a digital file that can be attached to an email message. An undesirable substitute is to fax hand computations. For digital computations, both as examples presented to students and as ehomework submittals, software with the chalkboard emulation of MathCad (Copyright © 2001 MathSoft Engineering & Education, Inc.) is very useful. Spreadsheets are useful for graphs, but don't show formulae well. Equation editors of word processors can be used for notes in the format of a technical paper, but do not actually perform computations in live-document mode, as does MathCad and its closest competitors. Solutions to homework assignments (Figure 2) and exams have more lasting value as live documents that can be slightly modified to illustrate parameter sensitivity or to create a new problem altogether.

Students in the UAA online Arctic Engineering class are encouraged to use software suited to each particular task, as long as the ehomework submittal is complete within a single file. Ehomework files are opened with compatible software by the grader, annotated in red, and returned to the student, following the age-old traditions of paper homework submittals. Classes with 40 or more students are most practically conducted with an instructor's assistant to help with grading homework and an extraordinary volume of email communications associated with homework exchanges and other course details.

Compute the R value of insulation that must be added to the surface of the water at Lake Spenard in Anchorage to limit the total ice thickness to two feet. Assume a freezing index of 2000 deg F-days and assume the ice is just beginning to form when the insulation is added. Neglect the thermal resistance of the air film and assume no snow buildup on the insulation.

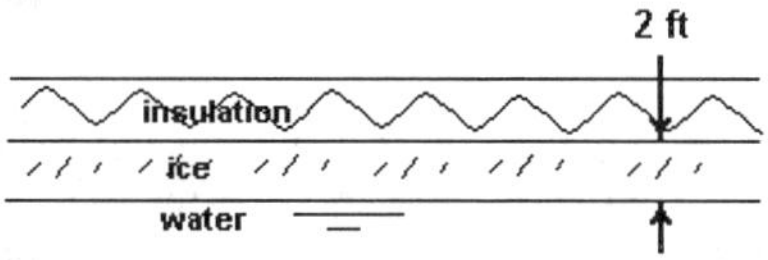

Assume: $FI := 2000\,degF \cdot day$ $\quad k_{ice} := 1.3 \cdot \frac{BTU \cdot ft}{hr \cdot ft^2 \cdot degF}$ $\quad L_{ice} := 144 \cdot \frac{BTU}{lb}$ $\quad \gamma_{ice} := 57 \cdot \frac{lb}{ft^3}$

$$\frac{FI}{\gamma_{ice} \cdot L_{ice}} = R_{ins} \cdot (X_2 - X_1) + \frac{\left(X_2^2 - X_1^2\right)}{2 \cdot k_{ice}} \qquad X_1 := 0 \cdot ft \qquad X_2 := 2 \cdot ft$$

$$R_{ins} := \frac{\frac{FI}{\gamma_{ice} \cdot L_{ice}} - \frac{\left(X_2^2 - X_1^2\right)}{2 \cdot k_{ice}}}{X_2 - X_1} \qquad R_{ins} = 2.155\,hr \cdot ft^2 \cdot \frac{degF}{BTU}$$

Figure 2. Example MathCad solution to an Arctic Engineering problem, including an embedded digital sketch.

Assessment

Ehomework is both a learning tool and, by scoring submittals, a means of assessing learning progress. Other means of assessment should be applied, in the form of quizzes, exams, project reports, papers, or surveys. The prevailing wisdom regarding student plagiarism and cheating on online exams is to use as many means as possible throughout the course of assessing individual learning. Proctoring exams is tedious and costly in time and effort of everyone involved. Only the most dedicated cheaters can enlist help with 2 exams, several quizzes, 8 ehomeworks, and a technical paper, assigned in CE A603 online. Experience with this and other online classes, as well as the advice of electronic pedagogy experts, is that multiple means of assessment will usually reveal the true level of an individual's learning. We expect, of course, that cheating is a rare problem among engineers involved in graduate education.

Online exams for CE A603 are equivalent to traditional take-home exams. A set of problem statements is presented online with a deadline for submittal of solutions several days later. Blackboard provides easy-to-use means for presenting exams with multiple choice, multiple answer, true-false, or fill-in-the-blank answers. Good exam questions using multiple choice or true-false format should anticipate common errors of understanding in design of alternate answers. Quizzes and exams should include enough problems so an inadvertent mistake in submittal or minor ambiguity in the problem statement does not result in a drastic reduction of overall score. Ambiguities in exam problem statements rapidly develop into email exchanges rivaling Supreme Court cases. Tested problems are best for important exams. New problems or minor changes to old ones should be tested with an assistant or colleague before transmittal to students online.

Blackboard provides an online point-and-click answer sheet for submittal of quiz, exam, and survey responses. Students complete all the answers, and then click a "submit" button. The Blackboard system instantly transmits the overall score and a review of questions missed, including correct answers and comments the instructor may have added for those missing a particular question. A practical difficulty with this arrangement occurs when a student is partway through the answer sheet and loses connection to the Blackboard web site. CE A603 students are warned several times to have all their answers ready before opening the answer sheet and to move carefully, but expeditiously through the sheet before clicking "submit." The system can be adjusted to allow multiple attempts to complete the answer sheet, but this option precludes feedback of scores, correct answers, and comments. The take-home mode and single access to the answer sheet appears to be most suitable for graduate engineering courses, such as CE A603.

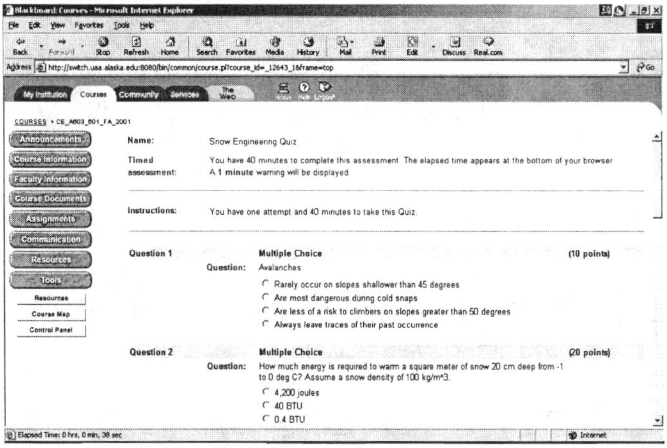

Figure 3. Example excerpt from a timed quiz answer sheet.

Arctic Engineering Graduate Program Online

The UAA School of Engineering is preparing to offer an online graduate program of Arctic Engineering leading to a Master of Science degree. Interest in the program of conventional graduate studies has increased in recent years. Specialized Arctic Engineering subjects to be available online by 2004 include:

- Arctic Engineering
- Frozen Ground Engineering

- Ice Engineering
- Snow Engineering
- Arctic Hydrology and Hydraulic Engineering
- Arctic Utility Distribution
- Arctic Heat and Mass Transfer
- Arctic Materials Engineering

Conclusions

Teaching specialized graduate engineering online appears to be an acceptable approach for a university with limited resources to make its most unique educational opportunities available to a worldwide market. Modern computer hardware and software provide affordable means for distance delivery with minimal investment in instructor use of these tools. Engineers are extraordinarily prepared through prior education and professional practice to take advantage of this opportunity for personal fulfillment and career enhancement through graduate education online.

References

Blackboard, Inc. (2001). http://www.blackboard.com/ (Sept. 30, 2001).

Internet 2 (2001). University Corporation for Advanced Internet Development, http://www.internet2.edu/ (Sept. 30, 2001).

Centra Software, Inc. (2001). http://www.centra.com/ (Sept. 30, 2001).

LERN (2001). Learning Resources Network, http://www.lern.org/ (Sept. 30, 2001).

Eranti, E. and Lee, G. (2000). Cold Regions Structural Engineering, McGraw-Hill

CRREL (2001). US Army Cold Regions Research and Engineering Center, http://www.crrel.usace.army.mil/ (Sept. 30, 2001).

Successful Government Funded Projects In Rural Alaska: The 3 C's: Challenges for Community Involvement

By Kerry E. Walker[1]

Abstract

Successful rural Alaska projects are dependent upon the relationship developed during project development between the government funding agency and the rural community. Most rural Alaska communities are Alaska Native villages. This paper identifies the challenges to effective community involvement during project development. These challenges are reviewed from three viewpoints, "The 3 C's":

1) Complexity of the issues that are being dealt with;
2) Cultural factors impacting the relationships; and
3) Communication challenges the relationship must work through.

Introduction

Government agency people are sometimes surprised when they fly out to help an Alaska Native village and rather than being received with open arms are instead met by apathy or outright hostility. The natural inclination is to take this rejection personally, either withdrawing or becoming defensive, or, more actively, aggressive.

Upon hearing several stories from colleagues about such encounters and being indirectly involved in several others, the author determined to take a look at relations between Alaska Native villages and government agencies to see if some light could be shed on the causes of both the relationships that didn't seem to be working and those that seemed to be working well.

This paper identifies the challenges to development of effective community involvement during project development. These challenges are reviewed from three viewpoints, "The 3 C's":

1) Complexity of the issues that are being dealt with;
2) Cultural factors impacting the relationships; and
3) Communication challenges the relationship must work through.

[1] Value Engineering Officer, U.S. Army Corps of Engineers, Alaska District, CEPOA-SA-VE, P.O. Box 898, Anchorage, AK 99506-0898; phone 907-753-5725; kerry.e.walker@poa02.usace.army.mil

Examples from which to frame these viewpoints are drawn from experience the author (a current employee of a government agency based in Anchorage) gained working on projects in rural Alaska.

Framing Village/Agency Relations: The 3 C's

1) *Complexity of the Issues.* The issues that take government agency personnel into Alaska Native villages are, by definition, complex. The easy problems are handled intra-village or with minimal governmental interaction. The complex problems require a longer, more intense, more enduring relationship.

To illustrate the type of complexity involved in such issues, consider the environmental restoration process taking place near many villages. The Department of Defense is committed to cleanup of hundreds of military installations used during WWII, the Cold War, and communications systems build-up in Alaska

The process used is similar to the Superfund process. It is called the Defense Environmental Restoration Program (DERP). It is very methodical and can take from 3-7 years to implement depending on funding ups and downs and the nationwide priority of the site.

Throughout the 1990's federal and state agencies have been receiving significant funding for clean-up projects in rural Alaska. However, to date, only a handful of the hundreds of environmental restoration sites have received 'site closure' from the state, and agreement that the site has been cleaned up sufficiently.

To resolve any of these issues requires input from interested and affected parties (in particular, the affected community).

Identifying the Problem. Lack of agreement and clarity in identifying the problem at a clean-up site leads to lack of clarity in communicating about it. This highlights the need for agencies and community residents to achieve consensus about what the problem is prior to determining what to do about it.

Site Closure-What is it? The first difficulty we have concerning reaching site closure is the fact that there is no consensus on what site closure is. There are numerous definitions of 'site' and 'closure' (e.g. Webster's definition of each of these terms; Environmental Protection Agency's (EPA's) Comprehensive Environmental Response Compensation and Liability Act (CERCLA) definition of 'site' as 'fence to fence; Alaska Department of Environmental Conservation's (ADEC's) definition of underground storage thank (UST) closure; ADEC's definition of 'adequate' cleanup as cleaning up 'to the discretion of the Department').

However, regarding 'site closure', there is one thing on which we can all agree. That is that there is not a universal definition of 'site closure' covering all sites and all projects. It is apparent that to come to site closure at a site will first require a site-specific consensus of what site closure means on that specific site. The definition of 'site closure' must be determined for each village for each federal environmental restoration program addressing that village.

However, just because there is not a 'universal' definition for site closure does not mean there is not a universal 'theme' for site closure. Based on review of the federal environmental legislation establishing environmental restoration programs we identify a recurrent theme. This theme rings out in the often repeated eight words that define the objective of environmental restoration: "to ensure protection of human health and the environment". The State of Alaska expands on this definition by inserting the word "welfare". Their covenant is 'to ensure protection of human health, welfare, and the environment."

To "ensure protection" is to keep "safe".

Risk – What is it? In discussing the concept of a 'safe' site we find that all our discussions center around the opposite of safe. They center around a term we use to measure the opposite of 'safe'. We use the term 'risk'. But now what is risk? And what constitutes acceptable risk? Many believe that this is a 'universal' 'technical' question/problem and that by doing more 'science' and becoming smarter we eventually will 'know', beyond a reasonable doubt, what levels of pollution humans and the environment can tolerate without adverse impact. However, the fact of the matter is, there is no universal agreement on what constitutes 'acceptable risk' (Fischhoff, 1981).

Indeed, if we look to where the term 'risk' came from when discussing pollution, we can understand why a strictly 'objective' solution to the question of 'How clean is Clean Enough?' will not be sufficient. Indeed, the answer to the question will require both scientific 'objective' input as well as site-specific 'subjective' input (Sandman, 1991).

Mary Douglas (1966, 1982, 1992) has done extensive research on cultural views of risk and pollution. She discusses the changing use of the term 'risk': "Originally the idea of 'risk' was neutral, it took account of the probability of losses and gains, and has been used extensively in the context of gambling. However, currently, in the context of clean-up projects, risk now means 'danger'. High risk means lots of danger. However, the word 'danger' does not have the aura of science or the pretension of a possible precise calculation" (Douglas, 1992).

'Danger' always has a context. There must be someone to feel 'danger' for it to occur. However, we all have a different perception of danger and, additionally, each culture has a different perception of danger.

Facts Versus Values. The 'values' we hold impact how we see the 'facts' in a situation. The 'values' of a culture impact significantly their view of the 'facts'. It is theorized that each culture has a dominant value orientation but that variant value orientations within a culture are both required and permitted (Kluckhohn, 1961).

The 17 Inupiat values as presented in the September 1996 insert to the *Arctic Sounder* follow: Respect for Elders, Hard Work, Hunter's Success, Humor, Respect for Nature, Family Roles, Portrait of People, Knowledge of Family Tree, Knowledge of Language, Love for Children, Respect for Others, Avoid Conflict, Cooperation, Sharing, Responsibility to Tribe, Humility, Spirituality. These can be contrasted to the values of the Army as identified on performance appraisal forms as: Commitment, Competence, Candor, Courage, Loyalty, Duty, Selfless Service, and Integrity.

Differentiating "normative" (value-laden) questions from "technical" questions is often very difficult particularly for those trained in Western science.

The perception of the world (worldview) a villager has may differ markedly from the perception of a town-based scientist.

Subsistence foods, which may be impacted by pollution from agency sites, are significant to Alaska Natives for reasons far beyond its nutritional value. They regard subsistence food as an essential element of health and well being which cannot be substituted by imported food. Harvesting, processing, and consuming subsistence food are defining social and cultural activities, and are key elements of Alaskan Native identity (adapted from Usher, 1995).

Administrative Complexities. Administration of the Defense Environmental Restoration Program (DERP) in Alaska is complicated by other federal programs that impact, overlap and/or contradict the DERP program. Examples follow:

President Clinton (Clinton, 1994) recently issued the Environmental Justice Executive Order. DOD strategy for implementing this order calls for closer attention to sites with minority populations. EPA Region X completed a report in August 1996 (USEPA, 1996) titled "The Alaska Military Sites Projects" recommending increased emphasis on Restoration Advisory Boards (a federally initiated board to "assist" (but not make) decisions on cleanup projects). The goals of the environmental justice program do not fit smoothly into the DOD goal of "risk reduction" in the DERP program.

Litigation. What is the motive for the federal agencies 'cleaning – up' Alaska? Is it indeed the altruistic motive identified in the recurrent eight words in all of the environmental laws to 'ensure protection of human health and the environment'? Indeed, to 'ensure protection' may not require any 'clean-up' at all.

Or perhaps, is the motive for this activity to preclude litigation? Does the federal government think that if any of our pollution hurts someone then we could get sued? If so, this perception leads us to an entirely different approach to the problem. Rather than cleaning up, so that there is no danger to the community, the attitude becomes one of cleaning up only to the point where litigation would not be able to prove the federal government's at fault for any mishap. Now, instead of addressing 'danger to the community', we would be addressing litigation mitigation (danger to the federal government).

Economic Issue. It is difficult for villagers to observe all the federal monies spent on the 'clean-up' process and realize that less than 10% of that reaches the village (Colt, 1993).

2) *Cultural Factors.* "Culture is that complex whole which unites a group of people through shared patterns of behavior. Culture includes language, technology, economic organization, political organization, worldview and religion, art and play, social organization and the enculturation of its members" (Davis, 1976).

A culture can be defined as a group of people with a 'learned' set of behaviors. With this in mind, federal agencies, state regulatory agencies, as well as Alaska Native villages can be viewed as cultural groups.

Agency Issues. Federal agencies providing cleanup in Alaska are relatively young cultures. As an example we'll look at the Air Force. The Air Force, although it has only been in existence for less that half a century has developed a strong rich culture. Based on the fact they have their own culture and have developed their own 'language' (many acronyms and terms specific to their missions and programs), they can be described as being their own 'tribal unit'. The purpose of life (mission) of the Air Force 'tribal unit' is "to keep 'em flying". In recent years, a strong sensitivity to the environment has, through legislation, filtered down and been assimilated into the Air Force culture. To ensure they will be able to continue to fulfill their 'mission', they have taken on sub-missions of becoming stewards of the environment and minimizing environmental impacts.

Rules. There is a very distinct, definitive social and political structure within the Air Force. Decision-making authority to allocate funds to implement decisions does not lie with the 'tribal unit' residing in Alaska. This unit can make recommendations to 'higher headquarters', but not commit funds.

Each action the Air Force makes is governed by numerous rules. There are 'sharing' rules regarding contracting that have been mandated by Congress and are implemented through the Federal Acquisition Regulations (FARs). They mandate things such as competitive bidding, small business set-aside, and non-discriminatory hiring practices for contractors.

In addition, there are 'environmental regulator' rules guiding their behavior. It is not considered sufficient to follow the 'spirit' of the environmental laws. The 'spirit' is identified in almost every piece of legislation as 'ensuring protection of human health and the environment'. One can get so caught up in the multiplicities and contradictions of so many detailed regulations (some 13 federal environmental compliance programs, and, in Alaska, some six environmental compliance programs) that it is easy to lose sight of the forest for the trees. I have seen many a manager with a case of what I diagnose to be 'regulatory paralysis' (Walker, 1997). There are so many do's and don'ts that one fears for personal liability unless they can find a specific detailed regulation giving them the green light to do what, ethically and technically they know is right. This fear has been multiplied with the recent passage of the Federal Facilities Compliance Act of October 1992. Now, your agency can be 'fined' by state or federal regulators. There is no longer sovereign immunity for certain aspects of environmental legislation. If you inadvertently error or interpret a regulation differently than a regulator might, your agency may get fined and may end up firing you. It's surprising that any action gets taken at all.

Short Term Funding. Many of the decisions for the clean-up program need to be made quickly to be able to obligate funds before they are 'lost'. The concept of 'losing' funds is part of the federal bureaucratic culture. Certain funds are provided by Congress with deadlines by which it must be spent. If it is not spent by that date it must be returned to higher headquarters and/or Congress for other use (but not to pay the federal deficit).

When the Air Force receives funding to be used on Alaska projects there often is a high urgency to make decisions on how to spend it to ensure it will not be lost. The inclination is to minimize the number of people involved in the decision to ensure a decision is, indeed, made. Having to coordinate with additional people may result in 'lost' funding due to lack of 'timely' decisions.

Obligation rate is one of the significant performance factors on every federal employees annual appraisal.

Personnel Turnover. Part of the military culture, both uniformed and civilian, is that the members of any unit are only around for a short time. To outsiders it appears as though there is no long term commitment to the project or the peoples involved. Putting forth effort to establish a trusting relationship often seems futile when one knows that at any time that person may be reassigned to other projects or leave the position.

Village Issues.

Intra-Village Demands on Time. Within each community there are many demands on the time of the villagers. Besides jobs and subsistence activities there are significant activities related to kinship obligations and tribal obligations. There are a diversity of boards and other formal organizations to attend to. Some villages have a village council, a tribal council, school boards, and subsistence boards. There are activities and demands associated with the regional corporation and with the village corporations established under the Alaska Native Claims Settlement Act (ANCSA). Then there is always a continuous stream of agencies, tribal ones like the Tanana Chiefs Council (TCC), State ones like those with roads and safe water, and federal ones of all types from fish and wildlife to housing to clean-up.

Intra-Village Politics. As with any village there are intra-village factions and diverse opinion on matters. Balancing kinship responsibilities with responsibilities to various boards and corporations is challenging to say the least. In some cases, various boards, councils, etc., within the village have been able to define which problems/issues/areas of the village will be handled by one organization versus another.

Village Leadership. Davis (1969) notes: "in one village the chief was reportedly beaten up after an agency man left, the complaint being he didn't speak up enough. This introduces the topic of leadership qualifications and the special demands put on leaders whenever any major project, program or other innovation is introduced into a community. On the one hand, the characteristics of a good chief for the "outside" world include reading and writing English, efficiency and skill at diffusing information, eliciting and reporting "decisions," the ability to get the council together on short notice, and behaving with the appropriate deference in the presence of visiting official. On the other hand, key qualifications from the village standpoint include the ability to speak the Native language, high moral conduct, good hunting or fishing record the ability to speak up to the white man plus a willingness to take the blame for everything that goes wrong."

Village/Agency Interactions.

Cultural Diversity (Big Picture). Alaska is a vast subcontinent with diverse environments that have been home to a very different people and cultures. Each of Alaska's Native peoples have their own history and development in Alaska prior to following contact with Euroasian peoples (Langdon, 1993; 1996).

Alaska was settled during the last ice age, some 15,000 to 30,000 years ago. During the settlement process, a number of different cultures evolved. Each of these cultures adapted to the unique environment in which they lived and developed a separate language and culture (Krauss, 1975). At the time of contact with Europeans, (contact date is generally accepted as 1741 when Vitus Bering arrived from Russia), 20 unique cultures existed in "The Great Land".

There is some similarity between some of the cultures. As the Romance language group of Europe includes Spanish, French, Italian, and Portuguese, there is a language group in Alaska called Athabaskan that includes 13 of the 20 indigenous cultures. It includes Koyukon, Tanaina, Tanana, Kutchin, Ahtna, Tanacross, Upper Tanana, Ingalik, Holikachuk, Tlingit, Eyak, Upper Koskokwim, and Han. The difference between these cultures can be likened to the difference between the nations of Europe. Also, within each of these nations several tribes can be found each with its own dialect. Presently, although you can often find people within each community that speak English, the 10,000 year old culture is still quite intact, albeit assimilating some aspects of the newcomer's culture.

Another language group within Alaska is the Eskimo-Aleut which includes Aleut, Central Yupik, Alutiiq, Siberian Yupik, and Inupiaq.

The remaining 2 cultures are unique; Haida and Tsimshian.

As an example of the 'diversity' of cultures that one federal program may work with, consider the Air Force group that handles the clean up for rural active Air Force sites in Alaska.

The 41 Air Force remote cleanup sites listed in their "Environmental Smart Book" lie within 11 different culture groups as identified above. The number of clean up sites for each culture group is as follows: Aleut (4), Alutiiq (3), Yupik (7), Inupiat (14), Koyukon (5), Tanaina (1), Tanana (3), Kutchin (1), Ahtna (1), Upper Tanana (1), Upper Kuskokwim (1).

Historic Village/Agency Interactions. Although the original/linguistic/cultural affiliation of the people in the vicinity of the installation may be important, perhaps we will find the proximity of the sites to villages and the history of relationships with the villages/towns will be even more important. Also, recent history of relationship with Native non-profit and profit corporations, borough, state and federal agencies figure significantly in contemporary issues. Ongoing changes within the Air Force (and the U.S. Army Corps of Engineers), must also be part of the larger understanding.

The following excerpt is from the paper "A Village View of Agencies" by Nancy Yaw Davis (1969). Dr. Davis wrote the paper after making observations in villages impacted by the 1964 Great Alaska Earthquake and the aftermath of government 'assistance' to villages.

"We are going to declare a three month
vacation and none of you white folks can come".

> "This statement was made by a newly elected chief in a village which had just survived a year during which many different agencies were intensively involved in village affairs. He was expressing a condition I have labeled "agency saturation". The point of agency saturation may be revealed in a number of ways. Villagers may make recurrent references to confusion over agency purposes, express anger about unfulfilled promises and nominal consideration of village preferences, engage in community-wide amusement over inner-agency conflicts, and display increased intra-village political difficulties directly related to new problems inadvertently instigated by outside intervention. A key indication of agency saturation is marked concern over apparent loss of village autonomy and an expressed desire for its return."

Davis says there are several symptoms of "agency saturation": 1) confusion over the real purpose of the agency; 2) lack of 'real' consideration of village preference; and 3) increased intra-village political conflict.

Villages are perceptive of 'interagency conflict'. "Interagency conflict appears most obvious when more that one agency is working simultaneously in the same village.

Then the fact one agency has a machine the other needs but cannot use becomes both puzzling and amusing. In traditional village way of life sharing of goods and equipment is expected. Further, it is reported that if more than one agency is in the village and problems develop, the agencies will blame one another."(Davis, 1969)

The reaction when villages are saturated by agencies and change is being pushed upon them varies by village. "At any one time the villages of Alaska represent a wide range of positions concerning change, and desired direction of change" (Davis, 1976).

Some villages are resisting by refusing to respond to agency requests for information and participation in project planning and execution.

Western Science and Traditional Knowledge. Patricia Cochran (1996), Executive Director of the Alaska Native Science Commission explains that "for years, a common complaint of Natives has been that science, the government and other western institutions don't give proper credit and respect to traditional indigenous knowledge. The observations made by the Native communities are different, often anecdotal". Cochran explained, "They {villagers} are the ones that say our Laborador tea is tasting different, or our clams look different." The information is valuable but doesn't fit the Western mold of scientific study.

"The burning question is how do we incorporate that local and traditional knowledge into the Western mode so that both the scientific and Native Communities benefit," she said. "Everybody agrees it is time to do that".

It has been emphasized that Alaska Natives have other ways of knowing and learning about things other than through the western scientific method (Kawagley, 1995; Fienup-Riordan, 1994).

It is the elders in a community who are turned to for advice on whether it's safe to eat the fish and berries. A scientific risk assessment must merge traditional Native knowledge for the results to be acceptable to the community.

3) *Communication Challenges.*

Logistics of Face-to-Face Meeting. The Air Force management staff of the clean up at remote Air Force bases is located in Anchorage, Alaska. Face-to-face communication with the villagers is hampered by distance and logistics challenges. If Alaska were overlaid the contiguous 48 states, Barrow would lie in Minnesota, Attu (mid-way out the Aleutian Chain) would lie in Stockton, California, and the panhandle would stretch to Savannah, Georgia. (Looking at this Alaska map overlay, one might note, as a recent Bush Cabinet member did on a trip to Alaska, that 'there are no roads west of Minnesota!' His shock was apparent and he wanted to know if anyone else in Washington DC 'knew' that.) One Air Force project manager may manage site cleanup projects a thousand miles apart. More than likely none of their sites will be road accessible.

Logistics to gain access to a site are challenging in that most of the sites do not have road access nor regular air service. It becomes necessary to charter aircraft or boats. The charter process is inhibited greatly by the federal regulations regarding permitting/licensing of charter companies.

For the villagers to come to Anchorage face-to-face meetings is prohibitively expensive and funding is not authorized under the DERP program (Beauchamp, 1997).

One of the most expensive U.S. cities to fly to and from Anchorage is Cold Bay, Alaska on the Aleutian Chain.

Program/Scientific Languages Barrier, Communication is an essential ingredient for successful cleanup. There is difficulty in finding "common ground" from which to communicate. Translating the language of science, "scientese'; the language of the regulations, "EPAese" even into "layman's English", let alone Yupik, Inupiat, Athabaskan, Aleut or any other aboriginal language, is a formidable task.

Interethnic Communications. All cultures have a timing/rhythm to their speech patterns (Oleksa; 1994). Many Alaska Native cultures speak with great articulation and, what a 'Type A' personality person would consider, substantial pauses. Many English speakers consider a half second pause to allow for the other guy to jump in to the conversation sufficient. This is not true in many other cultures.

Ron and Suzie Scollon (1980) have studied cross-cultural linguistics extensively. In one of their books "Interethnic Communication" published by the Alaska Native Language Center at University of Alaska, Fairbanks in 1980 they compiled a table comparing how an Athabaskan views an English speaker versus how an English speaker view an Athabaskan.

Decision-Making Authority. What is the role of patron/client versus partners?

"Under the best of circumstances, communicators face challenging ethical questions in balancing the community's right to know, the need to protect public health, the cost of unnecessarily alarming people, and the uncertainty of much of the scientific information. When communication about contaminants involves a marked power imbalance between public authorities, who are part of a multimillion dollar research industry, and people at the community level who are poor by comparison, ethical concerns should be paramount". Cases have been identified (Usher, 1995) "in which information was released to the media before communicated to the local village under study, apparently for professional and personal advancement".

This concept of not speaking for the others in the village is emphasized in a recent consensus document produced by the village of Kaktovik titled Kaktovik People (1996). In the living document they identify a stance on issues for which 'consensus' has been reached. They also identify areas in which 'consensus' has not yet been reached. They emphasize the document will change periodically as the issues evolve.

Communication. Ecological issues can pose communication difficulties among stakeholders, risk managers, and risk assessors. These individuals do not share common language systems and may not share common value systems. These differences are often not recognized and this can lead to problems throughout the assessment process. A better understanding of these differences is needed in order to learn how the groups can communicate more effectively.

To have any chance at all of succeeding, people can't rely on what we commonly refer to as communication as the dictionary defines communication: "transmitting information". Just exchanging information won't work because there has to be a useful context to put it in. These people have to use words, symbols, images, models whatever they can lay their hands and minds on to construct relevant meanings from both the available information and their individual expertise.

"Organizations that attempt to substitute increased communication for increased collaboration will learn the hard way that there is a tremendous difference. Flooding someone with more information doesn't necessarily make him a better thinker. Creating a shared understanding is simply a different task than exchanging information. It's the difference between being deeply involved in a conversation and lecturing to a group. The words are different, the tone is different, the attitude is different, and the tools are different. Collaboration takes communication back to its roots. The act of collaboration is an act of shared creation and/or shared discovery." (Schrage, 1990)

Conclusion

There are significant challenges to meet to ensure success in rural Alaska projects. Whereas there may be a tendency to focus on 'technical challenges', which can be considerable in themselves (e.g. building on permafrost, accessing gravel resources, logistics of getting heavy equipment and materials to the site, accessing water), there are other formidable challenges permeating each and every project.

Beyond the 'technical' challenges, there is another level of challenges. The solutions for these challenges will not be found 'at the boards' or in an engineer's manual. The solutions will come during the process of developing a strong partnership with the community based on trust and respect.

These challenges to effective community involvement were reviewed from three viewpoints "The 3 C's":

1) Complexity of the issues that are being dealt with;
2) Cultural factors impacting the relationships; and
3) Communication challenges the relationship must work through.

Building stronger partnerships based on trust and respect is a prudent and essential course of action.

In conclusion, the main policy implications of this paper are, briefly:

1. There is a need for careful assessment of current situations in each village to be directly or indirectly affected by a project.

2. There is a need for development of constant and thorough involvement of villagers on project issues and alternatives, including discussion of consequences.

3. There is a need for increased awareness by all Alaskans of profound regional,village and agency differences.

4. There is a need for increased awareness that working through our differences we can come to shared understandings based on fundamental values of trust and respect and grounded in our shared humanity.

Recommendations

Working with communities is essential to effective environmental restoration in rural Alaska. Effective communicators seek to understand another's views before seeking to be understood (Covey, 1995). An essential step to closing the communication gap with others is to learn more about others and how they view the world. We recommend cultural diversity training and communications training for all involved on these projects. In addition, we recommend implementation of a motivation system for proactively initiating and maintaining communications. This motivation system would recognize the challenges facing staff outlined in this paper.

References

Beauchamp, S. M., Walker, K., Tuttell, M., and Thomas, B. (1997). "A Case Study in Community Involvement in Rural Alaska." *ISCORD '97, Proceedings of the Fifth International Symposium on Cold Regions Development*, Anchorage, Alaska, May 1997.

Clinton, W. (1994). *Government-to-Government Relations with Native American Tribal Governments.* https://www.denix.osd.mil/denix/Public/Native/Outreach/Memos/execmemo.html

Cochran, P. (1996). *Presentation to Arctic Division, American Association for the Advancement of Science.* Girdwood Alaska, September 1996.

Colt, S. (1993). *ANCSA and Rural Alaska An Economic Reality Check.* Institute of Social and Economic Research, University of Alaska, Anchorage.

Covey, S. R. (1995). *The Seven Habits of Highly Effective People.* Covey Leadership Center, Inc.

Davis, N. Y. (1969). *A Village View of Agencies.* Presented at the Alaska Science Conference, AAAS Meetings. Fairbanks.

Davis, N. Y. (1976). *Steps Toward Understanding Rapid Culture Change in Native Rural Alaska, Commission Study #16.* Federal-State Land Use Planning Commission for Alaska. Anchorage.

Douglas, M. (1966). *Purity and Danger: An Analysis of the Concepts of Pollution and Taboo.* Routledge, London.

Douglas, M. (1982). *Risk and Culture: An Essay on the Selection of Technological and Environmental Dangers.* University of California Press, Berkeley.

Douglas, M. (1992). *Risk and Blame-Essays in Cultural Theory.* Routledge, London.

Fienup-Riordan, A. (1994). *Boundaries and Passages: Rule and Ritual in Yup'ik Eskimo Oral Tradition.* University of Oklahoma Press, Norman, OK.

Fischhoff, B., Lichtenstein, S., Slovik, P., Derby, S.L., and Keeney, R.L. (1981) *Acceptable Risk.* Cambridge University Press, New York.

Kaktovik People (1996). *In this Place: A Guide for Those Who Would Work in the Country of the Kaktovikmiut.*

Kawagley, A. O. (1995). *A Yupiaq Worldview: A Pathway to Ecology and Spirit.* Waveland Press, Inc, Prospect Heights, Illinois.

Kluckhohn, F., and Strodtbeck, F. (1961). *Variations in Values Orientation*. Ron, Peterson, and Company, Evanston, Illinois.

Krauss, M. (1975). *Alaska Native Languages: Past, Present, and Future*. University of Alaska Press, Fairbanks.

Langdon, S. J. (1993). *The Native People of Alaska*. Greatland Graphics, Anchorage.

Langdon, S. J. (1996). *Course notes as interpreted by Kerry Walker, Anthropology of Alaska Natives*. University of Alaska Anchorage, Fall 1996.

Oleksa, Father M. (1994). *Communication Across Cultures (Video Series)*. KTOO-TV, Juneau.

Sandman, P. (1991). *Risk = Hazard + Outrage: A Formula for Effective Risk Communication*. Video Series, American Industrial Hygiene Association.

Schrage, M. (1990). *No More Teams! Mastering the Dynamics of Creative Collaboration*. Doubleday, New York.

Scollon, R., and Scollon, S. B. K. (1980) *Interethnic Communication*. Alaska Native Language Center, University of Alaska, Fairbanks.

State of Alaska (1997). *Community Profile*. Department of Community and Economic Development. http://www.dced.state.ak.us/mra/CF_COMDB.htm.

Stern, P. C., and Fineberg, H.V., eds. (1996). *Understanding Risk; Informing Decisions in a Democratic Society*. National Academy Press, Washington D.C.

USEPA (1996). *The Alaska Military Sites Project*. United States Environmental Protection Agency. EPA910-R96-004.

Usher, P. J., Baikie, M., Demmer, M., Nakashima, D., Stevenson, M.G., and Stiles, M. (1995). *Communicating about Contaminants in Country Food: The Experience in Aboriginal Communities*. Inuit Tapirisat, Canada.

Walker, K. E. (1997). "Beyond Regulatory Paralysis (Coping with the Environmental Compliance Maze)." *ISCORD '97, Proceedings of the Fifth International Symposium on Cold Regions Development*. Anchorage, Alaska, May, 1997.

Subject Index

Page number refers to the first page of paper

Author Index

Page number refers to the first page of paper